T0186298

TUBULAR STRUCTURES XV

PROCEEDINGS OF THE 15TH INTERNATIONAL SYMPOSIUM ON TUBULAR STRUCTURES, RIO DE JANEIRO, BRAZIL, 27–29 MAY 2015

Tubular Structures XV

Editors

Eduardo Batista
Civil Engineering Program, COPPE, Federal University of Rio de Janeiro, Brazil

Pedro Vellasco & Luciano Lima
Structural Engineering Department, State University of Rio de Janeiro, Brazil

CRC Press
Taylor & Francis Group
Boca Raton London New York Leiden

CRC Press is an imprint of the
Taylor & Francis Group, an **informa** business

A BALKEMA BOOK

COVER PHOTOGRAPHS:

Front: *Carioca Wave*© – Casa Shopping, Barra da Tijuca, Rio de Janeiro, Brazil
Photographed by Luciano Lima

Back: Experimental tests
Photographed by Eduardo Batista & Luciano Lima

CRC Press/Balkema is an imprint of the Taylor & Francis Group, an informa business

© 2015 Taylor & Francis Group, London, UK

Typeset by MPS Limited, Chennai, India
Printed and bound in Great Britain by CPI Group (UK) Ltd, Croydon, CR0 4YY

All rights reserved. No part of this publication or the information contained herein may be reproduced, stored in a retrieval system, or transmitted in any form or by any means, electronic, mechanical, by photocopying, recording or otherwise, without written prior permission from the publishers.

Although all care is taken to ensure integrity and the quality of this publication and the information herein, no responsibility is assumed by the publishers nor the author for any damage to the property or persons as a result of operation or use of this publication and/or the information contained herein.

Published by: CRC Press/Balkema
                P.O. Box 11320, 2301 EH Leiden, The Netherlands
                e-mail: Pub.NL@taylorandfrancis.com
                www.crcpress.com – www.taylorandfrancis.com

ISBN: 978-1-138-02837-1 (Hbk + CD ROM)
ISBN: 978-1-315-67549-7 (eBook PDF)

Table of contents

Structural behaviour of cross-sections and members

Tubular Structures XV – Batista, Vellasco & Lima (eds)
© *2015 Taylor & Francis Group, London, ISBN 978-1-138-02837-1*

Preface

This book contains the papers presented at the 15th International Symposium on Tubular Structures (ISTS15) held in Rio de Janeiro, Brazil, from May 27th to 29th, 2015. The Symposium, now regarded as the key international forum for the presentation and discussion of research, developments and applications in the field of tubular structures, was organised by Federal University of Rio de Janeiro and State University of Rio de Janeiro in collaboration with the International Institute of Welding Sub-commission XV-E. The fourteen previous symposia, held between 1984 and 2012, are described in the "Publications of the previous symposia on tubular structures" section of this book. Throughout its 31-year history the frequency, location and technical content of all the symposia has been determined by the IIW Sub-commission XV-E on Tubular Structures.

The Symposium was sponsored by Vallourec, ABCEM, FAPERJ & UFRJ.

A total of 85 technical papers, each of which has been reviewed international experts in the field, are included in the proceedings. One of these papers relates to the invited 'Kurobane Lecture', given, at this Symposium, by Prof. Yoo Sang Choo from the National University of Singapore, Singapore. Prof. Choo was selected by the IIW Sub-commission XV-E. The Kurobane Lecture is the International Symposium on Tubular Structures Keynote Address which was inaugurated at the ISTS8 in 1998.

The editors would like to express their sincere gratitude to the reviewers of the papers for their hard work and expert opinions. The editors also wish to thank the international programme committee and the local organizing committee. Particular thanks are owed to Vallourec, ABCEM, & CrEAct.eve for their much appreciated support and efforts.

The information provided in this publication is the sole responsibility of the individual authors. It does not reflect the opinion of the editors, supporting associations, organizations or sponsors, and they are not responsible for any use that might be made of information appearing in this publication. Anyone making use of the contents of this book assumes all liability arising from such use.

The editors hope that the contemporary applications, case studies, concepts, insights, overviews, research summaries, analyses and product developments described in this book provide some inspiration to architects, developers, contractors, engineers and fabricators to build ever more innovative and competitive tubular structures.

This archival volume of the current "state of the art" will also serve as excellent reference material to academics, researchers, trade associations and manufacturers of hollow sections in the future.

Editors
Eduardo Batista
Federal University of Rio de Janeiro

Pedro Vellasco
Luciano Lima
State University of Rio de Janeiro
2015

Tubular Structures XV – Batista, Vellasco & Lima (eds)
© 2015 Taylor & Francis Group, London, ISBN 978-1-138-02837-1

Publications of previous international symposia on tubular structures

L. Gardner (Ed.) 2012. Tubular Structures XIV, 14th International Symposium on Tubular Structures, London, United Kingdom, 2012. Boca Raton/London/New York/Leiden: CRC Press/Balkema.

B. Young (Ed.) 2010. Tubular Structures XIII, 13th International Symposium on Tubular Structures, Hong Kong, China, 2010. Boca Raton/London/New York/Leiden: CRC Press/Balkema.

Z.Y. Shen, Y.Y. Chen & X.Z. Zhao (Eds.) 2009. Tubular Structures XII, 12th International Symposium on Tubular Structures, Shanghai, China, 2008. Boca Raton/London/New York/Leiden: CRC Press/Balkema.

J.A. Packer & S. Willibald (Eds.) 2006. Tubular Structures XI, 11th International Symposium and IIW International Conference on Tubular Structures, Québec, Canada, 2006. London/Leiden/New York: Taylor & Francis (including A.A. Balkema Publishers).

M.A. Jaurrieta, A. Alonso & J.A. Chica (Eds.) 2003, Tubular Structures X, 10th International Symposium on Tubular Structures, Madrid, Spain, 2003. Rotterdam: A.A. Balkema Publishers.

R. Puthli & S. Herion (Eds.) 2001. Tubular Structures IX, 9th International Symposium on Tubular Structures, Düsseldorf, Germany, 2001. Rotterdam: A.A. Balkema Publishers.

Y.S. Choo & G.J. van der Vegte (Eds.) 1998. Tubular Structures VIII, 8th International Symposium on Tubular Structures, Singapore, 1998. Rotterdam: A.A. Balkema Publishers.

J. Farkas & K. Jármai (Eds.) 1996. Tubular Structures VII, 7th International Symposium on Tubular Structures, Miskolc, Hungary, 1996. Rotterdam: A.A. Balkema Publishers.

P. Grundy, A. Holgate & B. Wong (Eds.) 1994. Tubular Structures VI, 6th International Symposium on Tubular Structures, Melbourne, Australia, 1994. Rotterdam: A.A. Balkema Publishers.

M.G. Coutie & G. Davies (Eds.) 1993. Tubular Structures V. 5th International Symposium on Tubular Structures, Nottingham, United Kingdom, 1993. London/Glasgow/New York/Tokyo/Melbourne/Madras: E & FN Spon.

J. Wardenier & E.P. Shahi (Eds.) 1991. Tubular Structures, 4th International Symposium on Tubular Structures, Delft, The Netherlands, 1991. Delft: Delft University Press.

E. Niemi & P. Mäkeläinen (Eds.) 1990. Tubular Structures, 3rd International Symposium on Tubular Structures, Lappeenranta, Finland, 1989. Essex: Elsevier Science Publishers Ltd.

Y. Kurobane & Y. Makino (Eds.) 1987. Safety Criteria in Design of Tubular Structures, 2nd International Symposium on Tubular Structures, Tokyo, Japan, 1986. Tokyo: Architectural Institute of Japan, IIW.

International Institute of Welding 1984. Welding of Tubular Structures/Soudage des Structures Tubulaires, 1st International Symposium on Tubular Structures, Boston, USA, 1984. Oxford/New York/Toronto/Sydney/Paris/Frankfurt: Pergamon Press.

Editors
Eduardo Batista
Federal University of Rio de Janeiro

Pedro Vellasco
Luciano Lima
State University of Rio de Janeiro
2015

Tubular Structures XV – Batista, Vellasco & Lima (eds)
© 2015 Taylor & Francis Group, London, ISBN 978-1-138-02837-1

Organization

This volume contains the Proceedings of the 15th International Symposium on Tubular Structures – ISTS15 held in Rio de Janeiro, Brazil, from 27th to 29th May 2015. ISTS15 has been organised by Federal University of Rio de Janeiro, State University of Rio de Janeiro & the International Institute of Welding (IIW) Sub-commission XV-E.

INTERNATIONAL PROGRAMME COMMITTEE

Prof. J.A. Packer (Chair), University of Toronto, Canada
Prof. E. Batista, Federal University of Rio de Janeiro, Brazil
Prof. D. Beg, University of Ljubljana, Slovenia (in memorian)
Prof. M.A. Bradford, University of New SouthWales, Sydney, Australia
Prof. D. Camotim, IST, Portugal
Prof.Y.Y. Chen, Tongji University, Shanghai, China
Prof. S.P. Chiew, Nanyang Technological University, Singapore
Prof.Y.S. Choo, National University of Singapore, Singapore
Prof. D. Dubina, The Polytehnica University of Timissoara, Romania
Prof. L. Dunai, Budapest University of Technology and Economics, Hungary
Prof. A. Elghazouli, Imperial College London, UK
Prof. L. Gardner, Imperial College London, UK
Dr S. Herion, Karlsruhe Institute of Technology, Germany
Mr G. Iglesias, Instituto para la Construccion Tubular, Spain
Prof. J.P. Jaspart, University of Liège, Belgium
Prof. U. Kuhlmann, University of Stuttgart, Germany
Mr M. Lefranc, Force Technology Norway, Sandvika, Norway
Prof. P.W. Marshall, MHP Systems Engineering, Houston, USA and Singapore
Prof. A.C. Nussbaumer, Ecole Polytechnique Fédérale de Lausanne, Switzerland
Prof. R.S. Puthli, Karlsruhe Institute of Technology, Germany
Prof. J.A. Requena, State University of Campinas, Brazil
Prof. A. Sarmanho, Federal University of Ouro Preto, Brazil
Prof. M. Serrano, University of Oviedo, Spain
Prof. B. Schafer, Johns Hopkins University, USA
Mr T. Schlafly, American Institute of Steel Construction, USA
Prof. L.S. da Silva, University of Coimbra, Portugal
Prof. T. Ummenhofer, Karlsruhe Institute for Technology, Germany
Prof. B. Uy, University ofWestern Sydney, Australia
Dr G.J. van der Vegte, Delft University of Technology, The Netherlands
Prof. P. Vellasco, State University of Rio de Janeiro, Brazil
Prof.Y. Wang, University of Manchester, UK
Prof. J. Wardenier, Delft University of Technology, The Netherlands and National University of
    Singapore, Singapore
Prof. B. Young, The University of Hong Kong, Hong Kong, China
Prof. X.L. Zhao, Monash University, Melbourne, Australia

LOCAL ORGANIZING COMMITTEE

Eng. A. de Araújo, V&M, Brazil
Prof. E. Batista, Federal University of Rio de Janeiro, Brazil
Prof. C. Baságlia, University of Campinas, Brazil
Prof. R. Fakury, Federal University of Minas Gerais, Brazil
Prof. A. Landesmann, Federal University of Rio de Janeiro, Brazil

Prof. L. Lima, State University of Rio de Janeiro, Brazil
Prof. M. Malite, University of São Paulo, Brazil
Prof. I. Morsch, Federal University of Rio Grande do Sul, Brazil
Eng. R. Pimenta, CODEME Engenharia, Brazil
Prof. J. Requena, State University of Campinas, Brazil
Prof. A. Sarmanho, Federal University of Ouro Preto, Brazil
Prof. P. Vellasco, State University of Rio de Janeiro, Brazil

Tubular Structures XV – Batista, Vellasco & Lima (eds)
© 2015 Taylor & Francis Group, London, ISBN 978-1-138-02837-1

Acknowledgements

The Organising Committee wish to express their sincere gratitude for the financial assistance from the following organisations: Vallourec, ABCEM, FAPERJ & UFRJ.

The technical assistance of the IIW Sub-commission XV-E is gratefully acknowledged. We are also thankful to the International Programme Committee as well as the members of the Local Organising Committee. Finally, the editors want to acknowledge the following ISTS15 reviewers:

M. Eekhout
G. Iglesias
M. Bradford
R.H. Fakury
X.L. Zhao
A.Y. Elghazouli
Y. Wang
A. Heidarpour
M. Lefranc
L. Gardner
T. Ummenhofer
L. Borges
O. Fleisher
M.A. Serrano
S.P. Chiew
J.A. Packer
A. Nussbaumer
N. Boissonnade
D. Camotim
L.S. da Silva
A.T. da Silva

R. Keays
R. Stroetmann
Y.Y. Chen
D. Lam
M.L. Romero
J. McCormick
A. Landesmann
A. Santiago
Y.S. Choo
B. Young
T.M. Chan
J.A.V. Requena
P. Vellasco
L. Lima
T. Björk
P. Ritakallio
L.W. Tong
N. Silvestre
E. Batista
J. Becque

Editors
Eduardo Batista
Federal University of Rio de Janeiro

Pedro Vellasco
Luciano Lima
State University of Rio de Janeiro

ISTS Kurobane Lecture

Tubular Structures XV – Batista, Vellasco & Lima (eds)
© *2015 Taylor & Francis Group, London, ISBN 978-1-138-02837-1*

Tubular joints – NUS research and applications

Y.S. Choo

Centre for Offshore Research & Engineering and Department of Civil & Environmental Engineering,
National University of Singapore, Singapore

ABSTRACT: This paper presents background details and selected highlights of the research and development (R&D) on tubular joints conducted at the National University of Singapore (NUS). The research covers plate reinforced joints, thick-walled joints, fabricated trunnions, elliptical hollow section joints, enhanced partial joint penetration welded joints, grouted joints and nonlinear responses of tubular frames. The approach adopted in the projects involves particular experiments on test specimens to provide physical evidence and reference results for numerical and theoretical investigations. Some of the R&D results have provided insights and bases for code development, and have been applied to offshore and onshore structural design and construction. The integrative efforts, with significant input from our international collaborators, motivate the team to further contribute towards more cost-effective and safe solutions for the industry and society.

1 INTRODUCTION

The author is honoured to deliver the 8th "Kurobane Lecture" at the 15th International Symposium on Tubular Structures. He dedicates this paper to the late Professor Michael Horne, his mentor and research supervisor at the University of Manchester. He salutes the distinguished experts who have presented the previous Kurobane Lectures, with Prof. Yoshiaki Kurobane inaugurating the Lecture series.

1.1 *Background*

This paper presents background details and highlights some of the research and development projects on tubular joints conducted at the National University of Singapore (NUS), which has provided insights and bases for code development, with some applied to offshore and onshore structural design and construction.

An introduction on offshore structures as a system is first presented and the motivations for research into the behaviour of various joint types and sub-systems is documented. The author is privileged to have worked closely with some of the international experts who delivered the previous Kurobane Lectures. Dr. Nick Zettlemoyer initiated the research on plate-reinforced joints. Prof. Jaap Wardenier contributed appreciably to the research on thick-walled joints, elliptical hollow section joints and grouted joints. Prof. Jeff Packer provided extensive contributions on elliptical hollow section joints. Prof. Peter Marshall has been instrumental in initiating research on enhanced Partial Joint Penetration (PJP+) joints and provided important contributions on grouted joints.

The approach adopted in our tubular joint research and development projects involves particular experiments on test specimens to provide physical evidence and reference results for numerical and theoretical investigations. The integrative efforts, with significant input from our international collaborators, motivate us to further contribute towards more cost-effective and safe solutions for the industry.

1.2 *Offshore structures*

The first steel template type platform was installed in 6 m of water offshore of Louisiana in the Gulf of Mexico in 1947 to support drilling for oil (Lee 1968). Since then, thousands of offshore platforms have been installed worldwide for oil & gas production. The development of specialized mobile drilling platforms, generally known as jack-up rigs, has enabled cost-effective exploratory and production drilling in water depth up to 120 m. Figure 1 shows a jack-up rig with its drilling derrick positioned above the wellhead platform, drilling a production well.

It can be noted from Figure 1 that each of the three legs of the jack-up rig is constructed from an assemblage of members, with the profiled ends of the horizontal and diagonal tubular braces welded to the vertical chords, which consists of two half pipes pre-welded to each of the vertical geared-teeth rack plates that accommodate the jacking system that elevate the hull. In general, the braces provide the shear capacity of the leg, while the chords provide the axial and flexural stiffness. The requirement to minimize the environmental forces from waves and currents on the tubular braces results in smaller diameters (d) and associated larger wall thicknesses (t) (i.e. thick-walled tubes, with $2\gamma = d_0/t_0 \leq 10$).

For the wellhead platform adjacent to the jack-up rig in Figure 1, larger diameter tubular members are adopted, with corresponding higher 2γ ratios. The recommended practice and standards for design, construction and installation of fixed offshore platforms include API RP2A-WSD (2000) and ISO 19902 (2007). Wardenier (1982) and Marshall (1992) provide extensive background on the research and development of welded tubular joints.

An offshore platform consists of a substructure, generally termed a jacket, with piles installed offshore, supporting the topsides. The jacket and topsides are constructed in a fabrication yard and then transported to the offshore location for installation. Figure 2 shows a close-up view of a jacket lifted by a crane barge and being lowered to the supports, seen at two positions of the near side of the transport barge. It can be noted from Figure 2 that the construction team is monitoring the placement of the jacket leg onto a doubler-plate cradle support, which will subsequently be connected through fillet welds to the jacket leg. For this connection, the loads acting on the two braces above the jacket leg will be transferred through the leg (as chord) to the support braces on the barge.

2 PLATE-REINFORCED JOINTS

The research at NUS on doubler plate-reinforced tubular joints was initiated by Dr. Zettlemoyer of Exxon-Mobil in 1996 due to lack of test evidence in the published literature. A series of experiments and associated numerical studies on plate-reinforced T-joints subjected to brace axial loads was thus started. Choo et al. (2005) presented details of the experimental investigations and Vegte et al. (2005) described the numerical study and comparisons with test results for T-joints.

Figure 3 illustrates the schematic arrangement and geometric parameters for an X-joint reinforced with doubler or collar plates. For a doubler-plate reinforced joint, as shown in Figure 3a, the brace is welded directly to the plate through a penetration weld, whereas the doubler plate is fillet welded to the chord. This is the scheme adopted for the jacket leg in Figure 2. An alternative scheme, as shown in Figure 3b, is the collar-plate reinforcement. This is an effective external reinforcement scheme for an under-designed joint, as described by Choo et al. (2004c). In this case, the brace is first welded directly to the chord. The collar plate parts are subsequently fitted and welded to the chord and brace.

In Figure 3, the chord and brace diameters are respectively d_0 and d_1, with the corresponding wall thicknesses t_0 and t_1. The length of chord and brace are respectively l_0 and l_1. The reinforcement plate is assumed to be square, with length l_d and thickness t_d (for doubler) and l_c and t_c (for collar). The geometric parameters are: brace-to-chord diameter ratio $\beta = d_1/d_0$, chord diameter-to-thickness ratio

Figure 2. Placement of jacket onto prepared supports with doubler-plate arrangement.

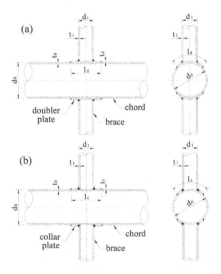

Figure 3. Schematic arrangement and geometric parameters for (a) doubler-plate and (b) collar-plate reinforced X-joints.

Figure 1. Jack-up platform with drilling rig above wellhead platform (photograph courtesy of Keppel Offshore & Marine Ltd).

$2\gamma = d_0/t_0$, brace-to-chord thickness ratio $\tau = t_1/t_0$, reinforced plate-to-chord thickness ratio $\tau_d = t_d/t_0$ (for doubler) and $\tau_c = t_c/t_0$ (for collar), plate length-to-brace diameter ratio l_d/d_1 (for doubler) and l_c/d_1 (for collar).

2.1 Test programme for plate reinforced T-joints

Choo et al. (2005) presented details of the test programme for twelve specimens sub-divided into six pairs, whereby each pair considers either brace compression or tension. Figure 4 shows the test set-up for each of the T-joints using the 2,000 kN test rig in NUS Structural Engineering Laboratory. It can be noted that large pins resting in elongated holes provide roller supports at the chord ends, and the actuator imposes brace axial load incrementally. The chord length between the pin supports is $l_0 = 2840$ mm, the brace length is $l_1 = 1100$ mm and the reinforcement plate length is $l_d = l_c = 305$ mm (with $l_d/d_1 = l_c/d_1 = 1.37$). Details of three selected pairs of specimens, with measured dimensions and material properties, are tabulated in Table 1 to highlight the effect of strength enhancement of collar-plate or doubler-plate reinforcement.

Figure 4. General test arrangement for plate reinforced T-joint under brace axial loading.

2.2 Strength enhancement for plate reinforced T-joints subjected to brace axial loads

Figure 5 shows the load-ovalisation curves for the two selected pairs of T-joints (EX-03 & EX-04 with collar-plate, and EX-07 & EX-08 with doubler-plate). The ovalisation is based on the measured change in chord diameter under brace axial loading, with ovalisations up to 80 mm (or $0.2d_0$). It can be noted that brace

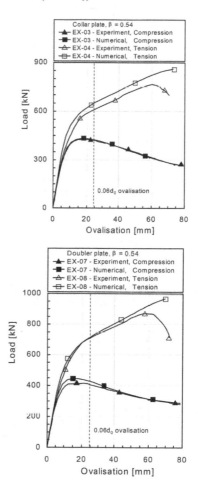

Figure 5. Experimental and numerical load-ovalisation curves for pairs of T-joint specimens under brace compression or tension (EX-03, EX-04 and EX-07, EX-08).

Table 1. Selected specimen details for doubler- and collar-plate reinforced T-joints (based on measured dimensions and material properties).

Specimen	Type	d_0 mm mm	d_1 mm	t_0 mm	t_1 mm	t_d mm	$\beta = d_1/d_0$	$2\gamma = d_0/t_0$	$\tau = t_1/t_0$	$\tau_d = t_d/t_0$	f_{y0} (MPa)	Brace loading	$F_{u,test}$ $F_{u,test}$ (kN)	$R_{reinforced}/R_{ref}$
EX-01	Un-reinforced	409.5	221.9	8.1	6.8	–	0.54	50.6	0.84	–	285	C	305.1	–
EX-02	Un-reinforced	409.5	221.9	8.1	6.8	–	0.54	50.6	0.84	–	285	T	543.2	–
EX-03	Collar	409.5	221.9	8.1	6.8	6.4	0.54	50.6	0.84	0.79	285	C	425.6	1.39
EX-04	Collar	409.5	221.9	8.5	6.8	6.4	0.54	48.2	0.8	0.75	276	T	609.2	1.05
EX-07	Doubler	409.5	221.9	8.5	6.8	6.6	0.54	48.2	0.8	0.78	276	C	415.8	1.28
EX-08	Doubler	409.5	221.9	8.2	6.5	6.4	0.54	49.9	0.79	0.78	312	T	708.0	1.16

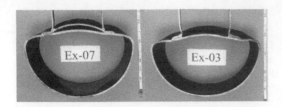

Figure 6. Section views of tested reinforced T-joint specimens EX-07 (doubler) and EX-03 (collar).

compression on EX-03 or EX-07 leads to a peak in the load-ovalisation curve, and is selected as $F_{u,test}$ in Table 1. For brace tension on EX-04 or EX-08, the chosen $F_{u,test}$ is based on the deformation limit of $0.06d_0$, as recommended by Lu et al. (1994) and subsequently adopted by IIW XV-E.

In Table 1, the reference strength $F_{u,test}$ for the unreinforced specimens EX-01 and EX-02 is based on the same criteria as that of EX-03 to EX-08. It can be noted from the $R_{reinforced}/R_{ref}$ ratio that strength enhancement ranges from 1.28 to 1.39 (for brace compression), and 1.05 to 1.16 (for brace tension). The larger ratio for EX-08 can be attributed to its higher yield stress.

2.3 Observed behaviour of doubler and collar plate reinforced T-joints

Figure 6 shows section views of two reinforced T-joints after tests, for doubler-plate joint EX-07, and collar-plate joint EX-03. For each joint, the location with significant deformations (yield hinges) in the chord occurs just outside the welds between the brace and the chord, or reinforcing plate (depending on the joint type) and at two locations on the upper half of the chord cross section.

An interesting deformation pattern is noted for doubler plate joint EX-07, where the portion of the doubler plate within the brace footprint hardly deforms, whereas the outer surface of the chord wall is pushed down just under the welds between the brace and the doubler plate, resulting in separation of the doubler plate and the chord. It is observed in Figure 6 and Table 1 that T-joint EX-03, with the collar plate welded and integrated externally to the brace and chord, is more effective in load transfer.

2.4 Finite element modelling and analysis for tubular joints

This sub-section summarises the numerical modelling and analysis strategy that has been adopted by our research team at NUS for tubular joints utilizing the finite element software ABAQUS (2001). More specific details to particular joint types are reported in our papers, such as Vegte et al. (2005) and Choo et al. (2004a, 2004b) for plate-reinforced joints and Choo et al. (2003a, 2003b) for thick-walled joints. The importance of verification and benchmarking with referenced experimental, numerical and/or theoretical results is emphasized.

For a particular joint subjected to associated loading, the appropriate symmetry in geometry, loading and boundary conditions has to be considered to determine the FE model for the analysis. For example, a uniplanar T-joint with simply supported ends subjected to brace axial load may be idealized by a quarter finite element (FE) model.

For each FE model, the element size is varied in such a way that relatively smaller elements are used where the stress gradient is more critical. Therefore, the mesh density decreases from the vicinity of the brace-chord intersection to the end of the brace or chord. Mesh convergence studies are essential to ensure the analysis produces accurate results. For the joints investigated, two layers of quadratic, solid elements through the thickness of brace or chord are adequate for members with $2\gamma > 30$.

The use of three-dimensional solid elements enables the simulation of weld geometries with high accuracy.

When a plate-reinforced joint is loaded, contact may occur between the bottom surface of the doubler or collar plate and the chord outer surface. Contact interaction plays an important role in the load transferring mechanism of plate-reinforced joints and thus non-linear contact analysis is required. Since both the reinforcing plate and the chord wall are deformable bodies, a deformable-deformable contact interaction is defined using a "master-slave" algorithm in ABAQUS. No friction between the contact surfaces is assumed.

In line with the plate-reinforced T-joint experiments, axial loads are applied using the displacement control method by prescribing the vertical displacement of the nodes of the brace tip. Nodes in a plane of symmetry are restrained against displacements perpendicular to their respective plane of symmetry. The axial brace load is reacted by restraining the vertical displacements of the nodes coinciding with the centre of the pin, thus enabling rotations and horizontal sliding of the chord ends.

Since the incorporation of material non-linearity in ABAQUS requires the use of a true stress-true strain relationship for each material, the measured engineering stress-strain curve is represented by a multi-linear relationship and subsequently converted into a true stress–true strain relationship. The welds are assumed to have the same material properties as the base metal.

2.5 Strength enhancement for plate reinforced X-joints subjected to in-plane bending

Choo et al. (2004a, 2004b) further investigated the behaviour of doubler- and collar-plate reinforced X-joints subjected to in-plane bending for a wide range of geometric parameters. For an X-joint subjected to in-plane bending, a quarter FE model with appropriate boundary conditions at symmetry planes and load specifications is appropriate.

Figure 7 shows comparative deformed shapes for the doubler- and collar-plate reinforced X-joint (for parameters $\beta = 0.64$, $2\gamma = 31.8$, $\tau = 1.0$, $\tau_d = 1.0$ and $\tau_c = 1.0$), with plate extent of $l_d/d_1 = l_c/d_1 = 1.25$ and

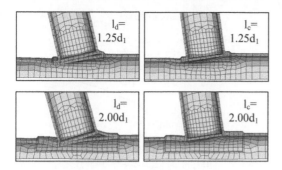

Figure 7. Comparison of deformed shapes between doubler and collar plate reinforced X-joints under in-plane moment (for $\beta = 0.64$, $2\gamma = 31.8$, $\tau = 1.0$, $\tau_d = 1.0$ and $\tau_c = 1.0$).

2.0. The results support the observations in Figure 6 and Table 1.

It can be observed, for the cases of $l_d/d_1 = 1.25$ and 2.0, that the doubler plate separates from the chord outer surface on the tensile side of the brace. Because each collar plate part is welded to the chord surface around its edges, the collar plate deforms similarly to the chord wall on both compression and tension sides. The collar plate reinforced joint is thus stiffer than its doubler plate counterpart, and is expected to provide higher strength enhancement. It has been observed that the collar-plate reinforced X-joints are capable of sustaining significantly higher in-plane bending moments than the corresponding doubler-plate reinforced X-joint. Choo et al. (2004a, 2004b, 2004c) presented results on joint strength enhancement for doubler- and collar-plate reinforcement joints for a wide range of geometric parameters and loading conditions.

The improved deformation capacity and associated enhanced strength of collar-plate reinforced T-joints has been confirmed by cyclic loading tests reported by Shao et al. (2011) which demonstrated larger ductility and dissipation of energy than corresponding un-reinforced joints.

3 THICK-WALLED JOINTS

The three-dimensional framing of each of the three legs of a jack-up platform, shown in Figure 1, consists of thick-walled tubular members to minimize the effects of environmental forces while providing requisite strength. For these thick-walled tubes, a lower diameter-to-thickness ratio of $2\gamma < 10$ is utilised. The research was initiated because design codes and recommendations, for example API (2000) and ISO (2007), limit the applicability of their tubular joint equations to $2\gamma > 20$.

Choo et al. (2003a, 2003b) presented research results for thick-walled X-joints under brace axial loads. In this study, four layers of quadratic 20-noded solid elements, termed C3D20R in ABAQUS, over the wall thickness were adopted after a mesh convergence study. Figure 8 shows the equivalent plastic strain contours plotted on the deformed geometry for the same

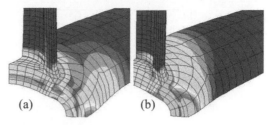

Figure 8. Equivalent plastic strain contours plotted on deformed geometry for thick-walled X-joint (with $2\gamma = 7$, $\beta = 0.7$) subjected to chord stress ratio (a) $n = -0.9$ (compressive) and (b) $n = 0.9$ (tensile).

X-joint, with $2\gamma = 7$ and $\beta = 0.7$, subjected to high compressive or tensile chord stress. Extensive yielding around the brace-chord intersection, but with no distinct yield hinge, can be observed from Figure 8. This behaviour is different from joints with $2\gamma > 20$ that can be represented by the ring model, which Wardenier and Choo (2006) elaborated.

Choo et al. (2003a) adopted the plastic limit load approach by Gerdeen (1980) to define the reference strength for thick-walled joints in view of the different load-displacement responses from joints with larger 2γ ratios. Figure 9 compares the deformation limit proposed by Lu et al. (1994) to that of the plastic limit approach. The lower reference deformation level for the limit load approach is consistent with the conservative lower bound assumption adopted.

Figure 10a shows the normalized strength of thick-walled joints for $2\gamma < 14.5$ obtained by Choo et al. (2003a) with the results for joints with $2\gamma > 14.5$, as reported by Vegte (1995). The trend lines for joints with three β ratios are found to be consistent. Figure 10b shows large variation in joint strength for thin-walled joints (with $2\gamma > 20$) for higher β ratios, while the strength for thick-walled joints (with lower 2γ ratios) is seen to be converging. This is consistent with the explanation by Wardenier and Choo (2006) that the load transfer mechanism is dominated by membrane action for thick-walled joints with higher β ratios.

Further research results for other thick-walled joint types and loading have been published. Choo et al. (2004d) presented results on X-joints with different included angles and chord stress levels. For X-joints with a low included angle θ in combination with a large β ratio, chord shear can reduce the joint strength. Choo et al. (2006) reported the effects of boundary conditions and chord stresses on the strength of K-joints. Qian et al. (2007) presented results of X-joints subjected to brace moment loadings.

Qian et al. (2009a) evaluated the application of IIW (2012) strength equations for thick-walled CHS X-, T- and K-joints and found that the calculated strength provides close prediction of the joint strength obtained using the limit load approach as documented by Choo et al. (2003a) for brace axial loads. For thick-walled joints loaded by brace in-plane bending or brace out-of-plane bending, the finite element results are 4–7% lower than the calculated strength. Vegte et al.

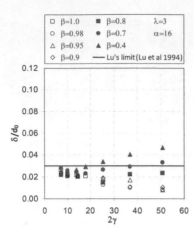

Figure 9. Comparison of deformation limit proposed by Lu et al. (1994) to the deformation level corresponding to joint strength defined by limit load approach by Choo et al. (2003a).

(2010) conducted further evaluation of the moment capacity of CHS joints.

4 FABRICATED PIPE TRUNNIONS FOR HEAVY LIFT

The author received funding and worked closely with industry partners in Singapore to research various aspects of heavy lifting through field measurement of sling forces, laboratory testing of fabricated trunnions, numerical and theoretical analyses, and software development. The research team recognized that the lift capacity of the crane vessel, rigging configuration, sling selection, and lift point design are all important factors in heavy lift evaluation and design for ensuring safety during lifting operations. The research results have been documented in various papers, including: a design support system for heavy lift (Choo et al., 2000), modelling of frictional slip (Lee et al., 2003) and strength of fabricated pipe and plate trunnions (Choo et al. 2001, 2002a & 2002b).

The author provided technical advice towards the installation and integration of three large module structures, each weighing more than 1000 tonnes that formed the deck of a semi-submersible drilling vessel. Figure 11a shows the overall rigging arrangement adopted for lifting one of three deck modules using the crane barge. The fabricated pipe trunnion concept was proposed and implemented, facilitating the lift installation. Figure 11b shows the lower portion of the fabricated trunnion integrated into the deck structure, and the upper portion transferring the double sling loads via shear forces on both sides of the trunnion.

A research programme to investigate the performance of fabricated pipe trunnions was set up at NUS in view of the design recommendations which are empirical and based on engineering experience and judgment (see Brown & Root, 1991; Shell UK, 1991).

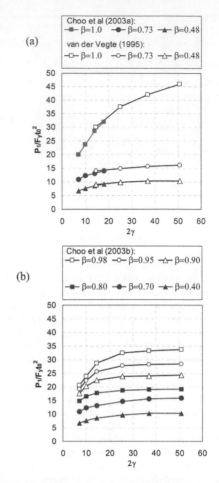

Figure 10. X-joint strength variation with respect to 2γ for different β ratios ($\alpha = 16$, $d_0 = 4406$ mm) (a) comparison with van der Vegte (1995), and (b) strength variation with 2γ and β.

Figure 11. (a) Overall rigging arrangement for lifting of module structure, and (b) detail of fabricated trunnion and slings.

These two documents recommend the total sling load to be transferred by the provided shear plate alone to the chord, and disregard the contribution of the trunnion brace.

Figure 12 shows the set up of a Type A fabricated trunnion specimen, with side braces, in a 10,000 kN capacity test rig. The actuator applies a compressive vertical load P to the chord, and the two saddle supports provide reaction forces, each of P/2 that are transferred as predominantly shear force and associated in-plane moments to the braces. An alternative scheme, with the chord pre-cut with two holes for placing a through brace before welding at the brace-chord intersection, is designated as a Type B specimen. Choo et al. (2001) presented details of the experiments, and selected results are tabulated in Table 2 to highlight the different behavioural characteristics and relative performance of trunnions with side braces and a through brace.

In Table 2, the actual dimensions and material properties, and associated geometric ratios of specimens CT3, CT5 and CT7 are tabulated with the failure mode and maximum applied load, $P_{u,test}$. The shear yield load of the side or through brace is designated as $P_{y,p}$. It is found that the ratio of $P_{u,test}/P_{y,p}$ provides a consistent indicator of the trunnion failure mode, with $P_{u,test}/P_{y,p} < 1$ indicating chord plastification. Figure 13 presents the comparisons of experimental and numerical results of specimens CT3, CT5 and CT7. Specimen CT5 failed by chord plastification, while CT3 and CT7 failed by brace yielding, with $P_{u,test}/P_{y,p}$ ratios of 1.27 and 1.21 respectively. The significant strength of Type A specimens CT3 and CT5 (with side braces) shows that the recommendations by Brown & Root (1991) and Shell UK (1991), which neglect the contribution of side braces, are overly conservative, if consistent material and welding are adopted to ensure no lamellar tearing of the chord wall.

It may be noted in Table 2 and Figure 13 that CT7 with a through brace scheme (with $2\gamma = 40.6$) achieves very similar ultimate load $P_{u,test}$ as CT3 (with a significantly thicker chord with $2\gamma = 24.8$). In CT7 the through brace provides a direct load path while a thicker chord is required for CT3 to resist the in-plane bending moment. This confirms the approach by Wardenier (2001) to follow the forces, and identify potential failure modes and locations, while accounting for compatible deformations and material behaviour. The approach is also applicable to the trunnion attachment to the structure.

5 ELLIPTICAL HOLLOW SECTION JOINTS

In recent years, architects have been increasingly attracted to explore various uses of Elliptical Hollow Sections (EHS) due to the unique shape, with major-to-minor outside dimensions of 2:1, and the ability to achieve different images from different perspectives. While the performance of EHS members has been researched by Gardner and Chan (2007), Chan et al. (2010) and Haque et al. (2012), information on joint design was lacking. The Comite International pour le Developpement et l'Etude de la Construction Tubulaire (CIDECT) supported a collaborative

Figure 12. Test set up of fabricated trunnion in 10,000 kN rig.

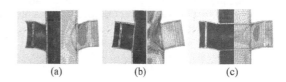

Figure 13. Comparisons of experimental and numerical results with sectioned views and failure modes for fabricated trunnion specimens (a) CT3 (with side braces), (b) CT5 (with side braces) and (c) CT7 (with through brace – Type B).

Table 2. Selected pipe trunnion specimens with dimensions and properties, and comparison between test and reference brace strength.

Trunnion specimen		Chord			Brace			Geometric ratios				Test and comparison			
Label	Type	d_0 (mm)	t_0 (mm)	f_{y0} (MPa)	d_1 (mm)	t_1 (mm)	f_{y1} (MPa)	β	2γ	$2\gamma_1$	τ	Failure Mode	$P_{u,test}$ (kN)	$P_{y,p}$ (kN)	$P_{u,test}/P_{y,p}$
CT3	A	508	20.5	417	406	12.5	376	0.80	24.8	32.5	0.6	brace	5420	4280	1.27
CT5	A	508	15.2	420	406	17.0	360	0.80	32.0	23.9	1.1	chord	4580	5500	0.83
CT7	B	508	12.5	350	406	12.5	376	0.80	40.6	32.5	1.0	brace	5160	4280	1.21

Table 3. Dimensions, material properties, geometric parameters and experimental results of EHS X-joints as reported by Shen et al. (2013 and 2014a).

Test	Chord [mm]				Brace [mm]				Non-dimensional parameters					Expt. load [kN] (Governing load load in **bold**)		Failure mode*
	b_0	h_0	t_0	f_{y0}	b_1	h_1	t_1	f_{y1}	θ_1	b_1/b_0	2γ	η	τ	Ultimate	3%b_0	
X90-1T-UT	220	110	5.94	402	110	220	5.94	402	90	0.50	37.0	1.00	1.00	339.1	**187.9**	CP, CT
X90-1C-UT	220	110	5.94	402	110	220	5.94	402	90	0.50	37.0	1.00	1.00	202.4	**150.5**	CP
X90-2T-UT	220	110	5.94	402	220	110	5.94	402	90	1.00	37.0	0.50	1.00	596.8	**574.5**	BF, SW
X90-2C-UT	220	110	5.94	402	220	110	5.94	402	90	1.00	37.0	0.50	1.00	**539.7**	537.8	SW, GB
X90-3T-UT	110	220	5.94	402	110	220	5.94	402	90	1.00	18.5	2.00	1.00	1557.0	**1188.8**	SW, BF
X90-3C-UT	110	220	5.94	402	110	220	5.94	402	90	1.00	18.5	2.00	1.00	**555.1**	459.8	SW

*Failure modes: BF – brace failure; CP – chord plastification; CT – chord tear-out; GB – global buckling (overall); SW – chord side-wall failure.

research programme between the National University of Singapore, the University of Toronto and Delft University of Technology to perform research on EHS joints.

Figure 14 shows four possible types of EHS joints, designated by Choo et al. (2003c), and EHS X-joint dimensions and parameters. Haque and Packer (2012) reported test results from an extensive set of EHS T- and X-joints that provided key reference for calibration of the nonlinear finite element (FE) analyses. Six of the X-joint test results are referenced in Table 3 to highlight the different behavioural characteristics, failure modes and associated reference strength for the four joint types.

Figure 15 compares the load-displacement curves obtained from experiments by Haque and Packer (2012) and nonlinear FE analyses for Type 1 and Type 3 X-joints subjected to brace axial compression and tension, as reported by Shen et al. (2013, 2014a). As a result of extensive studies, EHS joint resistance equations were then proposed by Shen et al. (2013, 2014b). It was found that the EHS equations for Types 1 and 2 joints (with the braces welded to the wide sides of chord) are consistent in formulation to the resistance expressions for RHS joints. The EHS equations for Types 3 and 4 joints (with the braces welded to the narrow sides of chord) are found to be consistent in formulation to the resistance expressions for CHS joints.

Wardenier et al. (2014) have further investigated and reported a package of design recommendations for the four types of axially loaded elliptical hollow section X and T joints.

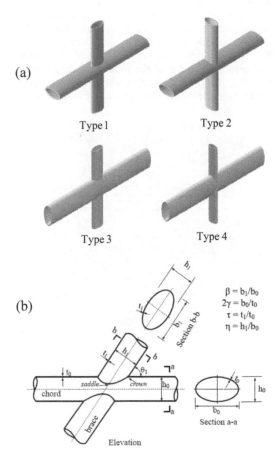

Figure 14. Designations and definitions: (a) types of EHS joints according to Choo et al. (2003); (b) nomenclature for a typical EHS X joint.

6 JOINTS WITH PJP+ WELDS

Marshall initiated NUS research into new welding details, termed enhanced partial joint penetration (PJP+), that ensure simple but high quality control welds, while satisfying safety requirements on fatigue life for tubular joints (Qian et al., 2009b). The PJP+ welds reduce the requirement on workmanship, compared to Complete Joint Penetration (CJP) welds

adopted in offshore jacket structures. Figure 16 shows selected welding details A, B and D, with the brace in direct contact with the chord outer surface. The concept is to provide larger groove profile angles and weld volumes to ensure adequate static and fatigue performance.

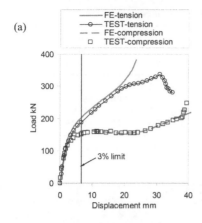

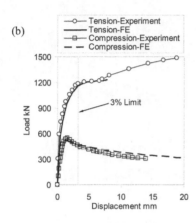

Figure 15. Comparisons between load-displacement curves for brace tension and compression for specimens: (a) X90-1T-UT and X90-1C-UT and (b) X90-3C-UT and X90-3T-UT.

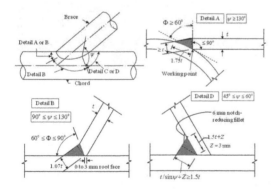

Figure 16. Enhanced Partial Joint Penetration (PJP+) weld detail proposed by Qian et al. (2009).

Figure 17 shows the geometric configuration and arrangement for fatigue tests of tubular X-joints subjected to in-plane bending for the PJP+ project. The brace-chord included angle of $\theta = 60°$ is selected, together with compatible dimensions of the chord and a removable loading fixture, so that each specimen can

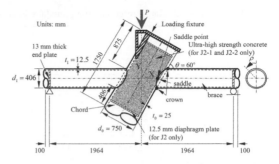

Figure 17. Geometric configuration and test arrangement for tubular X-joint under in-plane bending for PJP+ project.

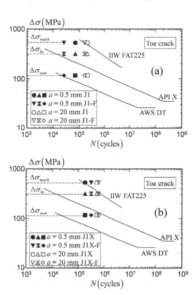

Figure 18. Comparison of the experimental data with the S–N curves for: (a) toe cracks in the chord for J1 and J1-F; and (b) toe cracks in the chord for J1X and J1X-F.

be tested twice by flipping the specimen after the first fatigue test. Figure 17 shows the chord with ultra-high strength concrete provided for specimens J2-1 and J2-2 only.

Qian et al. (2013b) presented results of the fatigue performance of the X-joints with PJP+ welds. The first specimen J1, with its flipped configuration J1F was tested without surface treatment on the weld surface or weld toe. The second specimen J1X had a ground weld surface performed after fabrication. After the first fatigue test on specimen J1X, the weld toes of both the brace and chord were treated with burr grinding with a radius of 3 mm before the second cyclic test was applied to the flipped specimen J1X-F.

Figure 18 summarises the excellent fatigue performance of the two specimens J1 and J1X (and results of the flipped positions J1F and J1X-F) as compared to the AWS (2008) DT curve for nominal stress range, $\Delta\sigma_{nom}$ and the API (2000) X curve for hot spot stress range $\Delta\sigma_{hs}$. In the figure, the solid symbols represent the number of cycles to crack initiation with a crack

Table 4. Specimen details for reference and grouted X-joints subjected to in-plane bending.

Specimen	Type	d_0 (mm)	t_0 (mm)	f_{y0} (MPa)	d_1 (mm)	t_1 (mm)	f_{y1} (MPa)	f_{grout} (MPa)	E_{grout} (GPa)	β	2γ	τ	$M_{CI,g}$ or $M_{CI,ref}$ (kN.m)*	$M_{max,g}$ or $M_{Max,ref}$ (kN.m)*	$\frac{M_{CI,ref}}{M_{CI,ref}}$	$\frac{M_{Max,ref}}{M_{Max,ref}}$
X1	Reference	508	14.4	295	408	21.7	277	–	–	0.8	35.3	1.5	323	355	–	–
X1G	Grouted	508	15.1	334	407	22.9	309	184	62	0.8	33.6	1.5	**808**	**914**	2.5	2.6
X2	Reference	407	20.7	334	406	21.0	333	–	–	1.0	19.7	1.0	769	769	–	–
X2G	Grouted	406	21.8	316	406	22.3	333	203	66	1.0	18.6	1.0	**1161**	**1293**	1.5	1.7

*Note: $M_{CI,g}$ and $M_{CI,ref}$: Moment for observed Crack Initiation (CI) for grouted and reference joint
$M_{Max,g}$ and $M_{Max,ref}$: Maximum (Max) moment observed in test for grouted and reference joint

depth of 0.5 mm, while the hollow symbols refer to the number of cycles causing a deep crack penetrating to about 80% (20 mm) of the chord wall thickness. The interactions of built-in residual tensile stresses and compressive stresses caused by in-plane bending at locations around the brace-chord intersection do not have an observable effect on the fatigue lives.

Qian et al. (2014) further investigated the fatigue performance and residual strength of specimen J2-1 (and flipped configuration J2-2) with the chord filled with ultra-high strength concrete. Qian et al. (2013c) conducted residual strength tests of the PJP+ welded specimens and reported on brittle failure caused by lamellar splitting in one specimen, and Qian et al. (2013d) reported a ductile tearing assessment.

7 GROUTED JOINTS

The research on grouted joints at NUS was initiated due to minimal guidance in codes (e.g. API, 2000; ISO, 2007) to account for potential strength enhancement through infilling ultra-high strength grout (with $f_{grout} > 180$ MPa) to the chord. Choo et al. (2007) reported on tests of two pairs of X-joints under in-plane moment using the 10,000 kN test rig at NUS. The specimen details for the reference (as-welded) and grouted X-joints are tabulated in Table 4. The length for each of the braces is 1.6 m. Figure 19 shows the grouted X2G joint specimen after in-plane bending test.

In Table 4, M_{CI} corresponds to the moment for observed crack initiation and M_{max} for the maximum moment attained, with subscripts g (for grouted) and ref (for reference). The factor $M_{CI,g}/M_{CI,ref}$ shows the strength enhancement due to grouting. It is observed that the infilled grout has enhanced the crack initiation load of joints X1 and X2 by a factor of 2.5 and 1.5 respectively. Figure 20 provides a comparison of joints X1 and X1G, showing deformed shapes of the computed and test sections in pairs. It can be observed in Figure 20 that the strength enhancement of specimen X1 results from the infilled grout minimizing ovalisation and plastification of the chord.

In 2009, the NUS research team together with industry partners embarked on a Joint Industry Project to investigate the static strength of tubular X-joints reinforced with ultra-high performance grout subjected to brace axial compression or tension. A total of

Figure 19. Grouted X-joint X2G specimen after in-plane bending test.

(a)   (b)

Figure 20. Section views showing deformed shapes of un-reinforced and grouted specimens, comparing numerical and experimental results for (a) X1 and (b) X1G.

21 specimens, with $d_0 = 324$ mm or 457 mm, and combinations of geometric ratios of $2\gamma = 25.9$, 40.5, 57.1 and β = 0.4, 0.7, 1.0 were tested for three configurations: as-welded (for reference strength), fully grouted chord and double-skin grouted chord. The team conducted an extensive parametric study after careful comparison and calibration with the experimental results. The research findings are reported in the Ph.D. theses submitted by Shen (2011), Chen (2012) and Wah (2013). Shen and Choo (2012) reported on the fatigue assessment of a tubular T-joint with a grouted chord.

The team continues to investigate the fatigue performance of X-joints under brace bending, and the static strength of K-joints, for different configurations. These research results are being consolidated and progressively submitted to journals for reviews as the confidentiality period for Phase 1 of the Joint Industry Project closes.

8 NONLINEAR RESPONSES OF TUBULAR FRAMES

Bolt et al. (1994) and Bolt & Billington (1995) reported the extensive large scale ultimate strength tests on two-dimensional (2D) jacket frame structures to investigate the effect of joint behaviour and framing redundancy on X-braced frames, and the effect of local joint behaviour (including fracture failure) on K-braced frames. Bolt & Billington (2000) presented results from ultimate load tests on three-dimensional (3D) jacket type structures to investigate the reserve and residual strength of the 3D frame subjected to different loading positions and directions. The 2D and 3D frame tests reveal the capacity of steel tubular frames to redistribute and sustain loads beyond the first component (joint or member) failure.

Figures 21a and 21b show the schematic views of the 3D frame supported by a self-reacting frame for Load Case II with a large actuator imposing incrementally a vertical load in the plane of Panel E. The author was privileged to participate as a benchmark analyst for the 3D frame tests, and witnessed the test sequence for Load Case II. It is noted that Panel E of the 3D frame structure has very similar dimensions, framing pattern and tube sizes as the 2D Frame III reported by Bolt et al. (1994), but with the critical X-joint designed as a weak joint to fail before brace buckling of the diagonal member. Figure 21b highlights the two braces in Panel E that buckled after the X-joint failure. In order to simulate the failure sequence of the critical components (joint and member) within the frame, it is therefore important to represent the nonlinear joint characteristics and member behaviour consistently.

One of the research efforts led to the paper by Qian et al. (2013a) that developed a nonlinear load-deformation formulation for tubular X- and K-joints in push-over analyses. The nonlinear joint characteristics are specified in the USFOS (Soreide et al., 1992) program to compute the global load-displacement response for the 3D Frame Load Case II. Figure 22 shows that the computed response for the proposed joint representation matches closely that from the test. As the critical joint is flexible, a rigid joint assumption results in a stiffer response before a first peak of about 870 kN due to brace buckling and then a lower frame strength as a result of a different sequence of component failures. The MSL joint representation is seen to provide a close match to the test response but the solution terminates at about half of the global frame displacement.

The investigations on tubular joints (including strength and load-deformation response of as-welded,

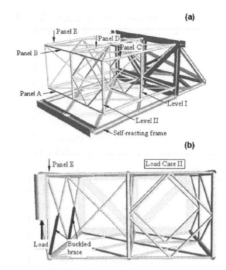

Figure 21. Schematic views of BOMEL 3D Frame Test with (a) self-reacting frame and labelled 2D panels, and (b) applied load in the plane of Panel E for Load Case II.

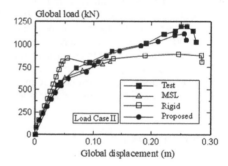

Figure 22. Comparisons of global load-displacement response for Load Case II of BOMEL 3D frame test with proposed joint formulation by Qian et al. (2013a) and other joint representations.

reinforced and grouted joints) are part of the on-going NUS research programme on structural integrity management of offshore structures. Asset integrity is becoming increasingly important in view of the need to ensure safe and continued operations of aging platforms, as elaborated by Zettlemoyer (2010). An important facet is to enhance understanding of the environmental load effects, predominantly imposed by waves and currents. The collaborative research with Prof. Paul Taylor at Oxford University on current blockage, as reported by Taylor et al. (2013) and Santo et al. (2014a, 2014b), is providing improved knowledge to the scheme proposed by Taylor (1991), and incorporated in API (2000) and ISO (2007).

9 CONCLUDING REMARKS

The author has been privileged to conduct research and development (R&D) on tubular structures and joints with some of the pioneers and industry. The R&D efforts at NUS have benefited significantly from their extensive knowledge, unique perspectives and expert input.

In this paper, background considerations and selected highlights of the R&D on plate-reinforced joints, thick-walled joints, fabricated pipe trunnions, elliptical hollow section joints, enhanced partial joint penetration welded joints, grouted joints and nonlinear responses of tubular frames were presented. Wardenier (2001) recommended that designers should follow the forces, and identify potential failure modes and locations through considerations of relative stiffnesses, compatible deformations and material behaviour. It can be seen from the presented examples that Wardenier's recommendation provides a rational approach to identify a potential weak link or location that requires strengthening or modification. The proper strengthening through direct load transfer will then lead to a more robust and cost-effective design.

The author recognizes that systematic investigations of selected test specimens or frames with comprehensive instrumentation will provide an essential reference for calibration of numerical and theoretical models. Large scale tests, such as the BOMEL 2D and 3D frame tests (as reported by Bolt & Billington (1995 and 2000)), are costly and require support from funding agencies and industry. International collaboration, with appropriate support from institutions, public and private sectors, is thus essential to address key "missing gaps" that will benefit the society at large.

ACKNOWLEDGEMENTS

I am honoured to have been selected by the International Institute of Welding, Subcommission XV-E, to present the 8th "Kurobane Lecture" at the 15th International Symposium on Tubular Structures (ISTS). Professor Yoshiaki Kurobane has had an exceptional career, and he is a pioneer and role model to all younger researchers in the field of tubular steel structures. Dr. Addie van der Vegte and I were privileged to host the 8th ISTS in Singapore when Prof. Kurobane delivered the Inaugural "Kurobane Lecture". I express my heartfelt thanks to Prof. Jaap Wardenier, Prof. Peter Marshall, Prof. Jeffrey Packer, Prof. Ram Puthli, Prof. Xiao-Ling Zhao, Dr Nick Zettlemoyer, Mr. Nigel Nichols and Mr. Trevor Mills for being my mentors, collaborators and good friends.

I am honoured to serve as Lloyd's Register Foundation Chair Professor in National University of Singapore. I would like to recognize the significant contributions of my colleagues and former research students, including Prof. Richard Liew, Prof. Paul Taylor (Oxford University), Prof. Xudong Qian, Dr. Feng Ju, Dr. Matthew Quah, Mr. Juxiang Liang, Mr. Xuekun Qian, Dr. Wei Shen, Dr. Zhuo Chen, Dr. Yi Feng Wah, Dr. Yang Zhang and Dr. Harrif Santo.

I am also appreciative of the strong support of our industry partners and Singapore government agencies, including (in alphabetical order): Agency for Science, Technology and Research (A*STAR), ABS, BP, CIDECT, ClassNK, Continental Steel, Densit, Exxon-Mobil, Keppel Offshore & Marine, Lloyd's Register, Maritime & Port Authority, McDermott International, Nautic Group, Petronas Carigali, Singapore Maritime Institute and Tata Steel. The Professorship and research funding provided by Lloyd's Register Foundation and scholarships from National University of Singapore are gratefully acknowledged. Lloyd's Register Foundation is a charity with a mission to protect life and property and to advance transport and engineering education and research.

REFERENCES

ABAQUS. 2001. *ABAQUS Standard User's Manual*. Rhode Island, USA. Version 6.2.

API RP2A-WSD. 2000. *Recommended practice for planning, designing and constructing fixed offshore platforms – WSD*, 21st Ed., American Petroleum Institute.

American Welding Society (AWS). 2010. *Structural welding code – steel. AWS D1.1/D1.1M:2010*, 22nd Ed.

Bolt, H.M., Billington, C.J. and Ward, J.K. 1994. Results from large scale ultimate load tests on tubular jacket frame structures. *25th Offshore Technology Conference*, BOMEL Ltd.

Bolt, H.M. and Billington, C.J. 1995. Result from large scale ultimate strength tests of K-braced jacket frame structures. *26th Offshore Technology Conference*, BOMEL Ltd.

Bolt, H.M. and Billington, C.J. 2000. Results from ultimate load tests on 3D jacket type structures. *31st Offshore Technology Conference*. Houston

Brown & Root. 1991. *Joint industry project: Heavy lift criteria*. Brown & Root Vickers Technology Ltd, U.K.

Chan, T.M., Gardner, L. and Law, K.H. 2010. Structural design of elliptical hollow sections: A review. *Structures and Buildings, Institution of Civil Engineers*, Vol. 163: 391–402.

Chen, Z. 2012. *Static strength of tubular X-joint with chord fully infilled with high strength grout*. Ph.D. thesis. National University of Singapore.

Choo, Y.S., Ju. F., Li, L. and Li, M. 2000. A design support system for heavy lift. *Trans. Institute of Marine Engineers*. Vol. 112, Part 2: 43–51.

Choo, Y.S., Quah, C.K., Shanmugam, N.E. and Liew, J.Y.R. 2001. Fabricated trunnions for heavy lift. *Trans. Institute of Marine Engineers*. Vol. 113, Part 2: 64–76.

Choo, Y.S., Quah, C.K., Shanmugam, N.E. and Liew, J.Y.R. 2002a. Static strength of plate trunnions subjected to shear loads – part I. Experimental study. *J. Constructional Steel Research*. Vol. 58: 301–318.

Choo, Y.S., Quah, C.K., Shanmugam, N.E. and Liew, J.Y.R. 2002b. Static strength of plate trunnions subjected to shear loads – part II. Computational study and design considerations. *J. Constructional Steel Research*. Vol. 58: 319–332.

Choo, Y.S., Qian. X.D., Liew, J.Y.R. and Wardenier, J. 2003a. Static strength of thick-walled CHS X-joints – Part I. New approach in strength definition. *J. Constructional Steel Research*. Vol. 59: 1201–1228.

Choo, Y.S., Qian. X.D., Liew, J.Y.R. and Wardenier, J. 2003b. Static strength of thick-walled CHS X-joints – Part II. Effect of chord stresses. *J. Constructional Steel Research*. Vol. 59: 1229–1250.

Choo, Y.S., Liang, J.X. and Lim, L.V. 2003c. Static strength of elliptical hollow section X-joint under brace compression. *Tubular Structures X, Proc.10th Intern. Symp. on Tubular Structures*, Madrid, Spain, 253–258.

Choo, Y.S., Liang, J.X., Vegte, G.J. van der and Liew, J.Y.R. 2004a. Static strength of collar plate reinforced CHS X-joints loaded by in-plane bending. *J. Constructional Steel Research*. Vol. 60: 1745–1760.

Choo, Y.S., Liang, J.X., Vegte, G.J. van der and Liew, J.Y.R. 2004b. Static strength of doubler plate reinforced CHS X-joints loaded by in-plane bending. *J. Constructional Steel Research*. Vol. 60: 1725–1744.

Choo, Y.S., Liang, J.X. and Vegte, G.J. van der. 2004c. An effective external reinforcement scheme for circular hollow section joints. In: *Connections in Steel Structures* V. Amsterdam: AISC & ECCS, Bouwen met Staal.

Choo, Y.S., Qian. X.D. and Foo, K.S. 2004d. Static strength variation of thick-walled CHS X-joints with different included angles and chord stress levels. *Marine Structures*. Vol. 17: 311–324.

Choo, Y.S., Vegte, G.J. van der, Zettlemoyer, M., Li, B.H. and Liew, J.Y.R. 2005. Static Strength of T-Joints Reinforced with Doubler or Collar Plates. I: Experimental Investigations. *ASCE J. Structural Engineering*. Vol. 131: 119–128.

Choo, Y.S., Qian, X.D. and Wardenier, J. 2006. Effects of boundary conditions and chord stresses on static strength of thick-walled CHS K-joints. *J. Constructional Steel Research*. Vol. 62: 316-328.

Choo, Y.S., Chen, Z., Wardenier, J. and Gronbech, J. 2007. Static strength of simple and grouted tubular X-joints subjected to in-plane bending. *5th Int. Conf. Advances in Steel Structures, Special Symposium on Tubular Structures*, Singapore.

Gardner, L. and Chan, T.M. 2007. Cross-section classification of elliptical hollow sections. *Steel Compos. Struct.*, Vol. 7 Issue 3: 185–200.

Gerdeen, J.C. 1980. A critical evaluation of plastic behavior data and a united definition of plastic loads for pressure components. *Welding Research Council Bulletin*. New York.

Haque, T. and Packer, J.A. 2012. Elliptical hollow section T and X connections. *Canadian J. Civil Engineering*, Vol. 39(8): 925–936.

Haque, T., Packer, J.A. and Zhao, X.L. 2012. Equivalent RHS approach for the design of EHS in axial compression or bending. *Adv. Struct. Eng.*, Vol. 15 Issue 1: 107–120.

ISO 19902. 2007. *Petroleum and natural gas industries – fixed steel offshore structures*. BS EN 19902.

International Institute of Welding (IIW). 2012. *Static design procedure for welded hollow section joints – recommendations*. 3rd Ed., IIW Doc. XV-1402-12, Paris.

Lee, G.C. 1968. Twenty years of platform development. *Offshore* (June).

Lee, K.H., Choo, Y.S. and Ju, F. 2003. Finite element modeling of frictional slip in heavy lift sling systems. *Computers and Structures*. Vol. 81: 2673–2690.

Lu, L.H., de Winkel, G.D., Yu, Y. and Wardenier J. 1994. Deformation limit for the ultimate strength of hollow section joints. *Tubular Structures VI, Proc. 6th Intern.Symp. on Tubular Structures*, Melbourne, Australia, 341–348.

Marshall, P.W. 1992. *Design of welded tubular connections – basis and use of AWS code provisions*. Elsevier Science Publishers B.V.

Qian. X.D., Choo, Y.S., Liew, J.Y.R. and Wardenier, J. 2007. Static strength of thick-walled CHS X-Joints subjected to brace moment loadings. *ASCE J. Structural Engineering*. Vol. 133: 1278–1287.

Qian. X., Choo, Y.S., Vegte, G.J. van der. and Wardenier, J. 2009a. Evaluation of the new IIW CHS strength formulae for thick-walled joints. *Tubular Structures XII, Proc. 12th Intern. Symp. on Tubular Structures*, Shanghai, China, 271–279: Taylor & Francis Group, London.

Qian, X., Marshall, P.W., Cheong, W.K.D., Petchdemaneegam, Y. and Chen Z. 2009b. Partial joint penetration plus welds for tubular joints: fabrication and SCFs. *Proc. IIW Int. Conf. Advances Welding and Allied Technology*. 16–17 July 2009.

Qian, X., Zhang, Y. and Choo, Y.S. 2013a. A load-deformation formulation for CHS X- and K-joints in push-over analyses. *J. Constructional Steel Research*. Vol. 90: 108–119.

Qian, X., Petchdemaneengam, Y., Swaddiwudhipong, S., Marshall, P., Ou, Z. and Nguyen, C.T., 2013b. Fatigue performance of tubular X-joints with PJP+ welds: I – Experimental study. *J. Constructional Steel Research*. Vol. 90: 49–59.

Qian, X.D., Ou, Z.Y., Swaddiwudhipong, S. and Marshall, P.W. 2013c. Brittle failure caused by lamellar splitting in a large-scale tubular joint with fatigue cracks. *Marine Structures*. Vol. 34: 185–204.

Qian, X.D., Li, Y. and Ou, Z.Y. 2013d. Ductile tearing assessment of high-strength steel X-joints under in-plane bending. *Engineering Failure Analysis*. Vol. 28: 176–191.

Qian, X., Jitpairod, K., Marshall, P., Swaddiwudhipong, S., Ou, Z., Zhang, Y. and Pradana, M.R. 2014. Fatigue and residual strength of concrete-filled tubular X-joints with full capacity welds. *J. Constructional Steel Research*. Vol. 100: 21–35.

Santo, H., Taylor, P.H., Williamson, C.H.K. and Choo, Y.S. 2014a. Current blockage experiments: force time histories on obstacle arrays in combined steady and oscillatory motion. *J. Fluid Mech*. Vol. 739: 143–178.

Santo, H., Taylor, P.H., Bai, W. and Choo, Y.S. 2014b. Blockage effects in wave and current: 2D planar simulations of combined regular oscillations and steady flow through porous blocks. *Ocean Engineering*. Vol. 88: 174–186.

Shao, Y.B., Lie, S.T., Chiew, S.P. and Cai Y.Q. 2011. Hysteretic performance of circular hollow section tubular joints with collar-plate reinforcement. *J. Constructional Steel Research*. Vol. 67: 1936–1947.

Shell UK Ltd. 1991. *Guidelines for lifting points and heavy lift criteria*. Engineering Reference Document No. EM/039.

Shen, W. 2011. *Behaviour of grout infilled steel tubular members and joints*. Ph.D. thesis, National University of Singapore.

Shen, W. and Choo, Y.S. 2012. Stress intensity factors for a tubular T-joint with grouted chord. *Engineering Structures*. Vol. 35: 37–47.

Shen, W., Choo, Y.S., Wardenier, J., Packer, J.A. and Vegte, G.J. van der. 2013. Static strength of axially loaded EHS X-joints with braces welded to the narrow sides of the chord. *J. Constructional Steel Research*. Vol. 88: 181–190.

Shen, W., Choo, Y.S., Wardenier, J., Packer, J.A. and Vegte, G.J. van der. 2014a. Static strength of axially loaded elliptical hollow section X joints with braces welded to wide sides of chord. I: Numerical investigations based on experimental tests. *ASCE J. Structural Engineering*. DOI: 10.1061/(ASCE)ST.1943-541X.0000791

Shen, W., Choo, Y.S., Wardenier, J., Packer, J.A. and Vegte, G.J. van der. 2014b. Static strength of axially loaded elliptical hollow section X joints with braces welded to wide sides of chord. II: Parametric study and strength equations. *ASCE J. Structural Engineering*. DOI: 10.1061/(ASCE)ST.1943-541X.0000792

Soreide, T.H., Amdahl, J.A., Eberg, E., Holmas, T. and Hellan, O. 1992. USFOS – a computer program for progressive collapse analysis of steel offshore structures. Theory manual. SINTEF STF71 F88038, Trondheim, Norway.

Taylor, P.H. 1991. Current blockage: reduced forces on offshore space-frame structures. In *Offshore Technology Conference, OTC 6519*. Houston, U.S.A.

Taylor, P.H., Santo, H. and Choo, Y.S. 2013. Current blockage: reduced Morison forces on space frame structures with high hydrodynamic area, and in regular waves and current. Ocean Engineering. Vol. 57: 11–24.

Vegte, G.J. van der. 1995. The static strength of uniplanar and multiplanar tubular T- and X-joints. *Doctoral Dissertation*. Delft University of Technology, The Netherlands.

Vegte, G.J. van der, Choo, Y.S., Liang, J.X., Zettlemoyer, N. and Liew, J.Y.R. 2005. Static Strength of T-Joints reinforced with doubler or collar plates. II: numerical simulations. *ASCE J. Structural Engineering*. Vol. 131: 129–138.

Vegte, G.J. van der., Wardenier, J., Qian. X. and Choo, Y.S. 2010. Re-evaluation of the moment capacity of CHS joints. *Proc. Institution of Civil Engineers – Structures and Buildings*. Vol. 163, Issue 586: 439–449.

Wah, Y.F. 2013. *Static strength of double-skin grouted X-joint infilled with high strength grout*. Ph.D. thesis. National University of Singapore.

Wardenier, J. 1982. *Hollow section joints*. Delft: Delft University Press.

Wardenier, J. 2001. From a tubular morning mist to the tubular morning glow. 2nd ISTS Kurobane Lecture. *Tubular Structures IX. Proc 9th Int. Symp. Tubular Structures*. Dusseldorf, Germany.

Wardenier, J. and Choo, Y.S. 2006. Recent developments in welded hollow section joint recommendations. Advanced Steel Construction. Vol. 2: 109–127.

Wardenier, J., Choo, Y.S., Packer, J.A., Vegte, G.J. van der and Shen, W. 2014. Design recommendations for axially loaded elliptical hollow section X and T joints. *Steel Construction*. Vol. 7, No. 2: 89–96.

Zettlemoyer, N. 2010. Life extension of fixed platforms. 6th ISTS Kurobane Lecture. *Tubular Structures XIII. Proc 13th Int. Symp. Tubular Structures*. Hong Kong, China.

Architecture, applications and case studies

Tubular Structures XV – Batista, Vellasco & Lima (eds)
© 2015 Taylor & Francis Group, London, ISBN 978-1-138-02837-1

Tensegrity chandeliers for a shopping street in the Hague NL

M. Eekhout
TU Delft, Octatube Delft, The Netherlands

ABSTRACT: Tensegrity structures are exciting structures to design, but the most complicated structures to analyze structurally and to build. They have become popular as works of art by Kenneth Snelson. They have been used extensively as stabilizations of frameless glazing facades and roofs. One project design in The Hague NL, contained three tensegrity structures, each with more than 250 tubes and 1500 cables to be connected as chandeliers above a crowded shopping street. It appeared to be an impossible dream. But the reasoning behind the impossibility is interesting enough to be taken up for future design & build projects.

1 INTRODUCTION

In 2013 the engineering, production and building of a number of 'tensegrity' chandeliers was tendered by the city of The Hague in The Netherlands. The architect was Lana du Croq, (architects office LEV), who had won a design competition some 5 years before. The idea behind the urban redesign of the 'Grote Marktstraat', one of the busiest shopping streets in town, was to change this street into a giant family living room, with stone tapestry, steel couches, 3 large suspended chandeliers and a series of slender suspended chandeliers at the shop fronts.

The architect worked on the design for more than 4 years. She hired the Dutch office of Arup in Amsterdam to do a sample engineering. she had the staff members and students at TU Eindhoven made a model and prepared her tendering package. The client, the The Hague municipality went for an official European tendering, regarding the tensegrity chandeliers and the lanterns as complex lightening devices. A number of the tendering parties were indeed lighting producers. In May 2013 the tendering process started after a pre-qualification process, which selected 6 tendering parties (see fig. 1, 2).

Shortly after the start of the tendering process the Municipal Aesthetics & Monuments Commission judged the design as undesirable and aborted the design after 5 years of design and engineering. It was a blessing in disguise as Octatube, as one of the most experienced tensegrity designers and builders in Europe, discovered quite fast that the tensegrity structure was not buildable at all in the situation of a busy shopping street where life had to go on. Yet the challenge how to realize this type of complicated tensegrity structures remains and is interesting enough to be revealed in this article and to be discussed in an international forum.

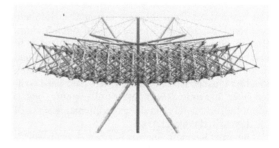

Figure 1. Octatube design proposal of one of the chandeliers.

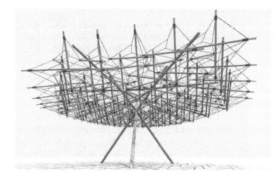

Figure 2. Tensegrity model of a chandelier made by Octatube.

2 TENSEGRITY STRUCTURES AS ARTWORK

A tensegrity structure is a structure which is composed of linear compression elements and linear tension elements, where all compression elements are only connected by means of cables. Not one compression element (usually a tube) touches another one directly.

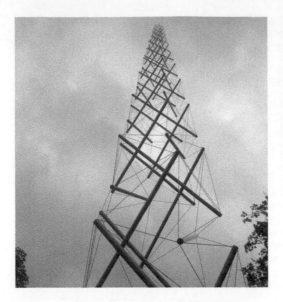

Figure 3. The Snelson tensegrity tower in Hoenderloo.

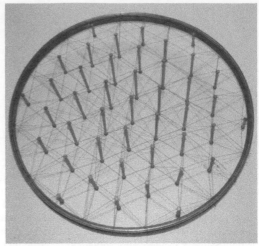

Figure 4. The bicycle wheel tensegrity structure made in 1972.

The name dates from the time that student Kenneth Snelson (1927) worked together with tutor Richard Buckminster Fuller (1895–1981) in the late 40s to make new types of structures. Kenneth claims he had thought out the concept, though 'Bucky' was so quick to adopt it as his finding. Long letters have been written since that time. Buckminster Fuller surely tried to adopt these structures in the experimental designs of his domes and cupolas.

The artist Kenneth Snelson has been known worldwide because of his sculptures in stainless steel and aluminum and stainless steel cables which he built on numerous places and exhibitions. His structures are quite elegant and abstract in its detailing.

The Snelson Tower (fig. 3) is a 30 m high tensegrity tower made by Kenneth Snelson and his studio in the Kröller-Müller park in Hoenderloo, NL. It has one flaw, however. The pre-stress in the cables is not high enough to keep the structure upright through the stormy autumn season. The flaw is in the way the pre-stress is applied and partly is also reduced during final assembly. The tensegrity tower of Snelson has fallen several times in autumn during storms and is stabilized by guying cables in winter or demounted and laid flat on the ground in winter. I will return to that point later in this contribution.

3 SNELSON TENSEGRITIES

The Snelson structures were popular amongst architects when I was a student in the late 60s and early 70s. In fact they were regarded as the highest class of structural designs at the TU Delft faculty of architecture. I was a student-assistant of the late professor Dick Dicke (1924–2003) at the Chair of Mechanical Engineering, who introduced out-of-the-box thinking

on new materials and new structures. He gave lectures how to make a reliable structure with unfit material, even with candy, or with 'Zappi', unknown materials with unrealistic properties that forced the students to think clever and not on the automat. I organized two seminars with tensegrity structures in 1972 of 2 weeks with 20 students. One of these students, now professor Dirk Sijmons, even remembered this seminar in his inaugural speech as a landscape professor 40 years later. We made tensegrity structures in the rim of a bicycle wheel (fig. 4). Indeed it was quite difficult to solder these models evenly pre-stressed. It was for most of us the last design exercise before starting with our final studies. We kept to designing and building models, soldering mostly. But tensegrities were difficult.

4 SANTANDER TENSIGRITY ROOF IN MADRID

In my early years of Octatube, in the 90s, I would develop a whole generation of tensegrity structures for stabilization of glass facades and glass roofs. The bicycle wheel tensegrity structures from my student's time were realized around 35 years later as the 30 m circular roof for the Santander Bank in Madrid, as designed by American architect Kevin Roche.

We designed a circular ring of RHS 350 × 350 mm as a bicycle wheel structure of pre-stressed upper cables and lower cables, distanced by vertical steel posts at regular distances. This collection of vertical studs kept the upper and lower cables stable. The cables were pre-stressed stainless steel rods. The vertical compression studs were made in stainless steel tubes, but were prepared as glass tubes as an alternative 'Plan B' during the tender stage. The bicycle system is a prototype of a 'closed' system: a closed entity of compression and tension components. There

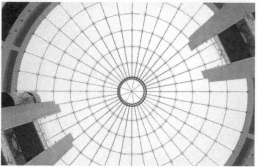

Figure 5 and 6. Two interior views (perspective and straight up) of the tensegrity roof structure for the Santander Bank in Madrid.

Figure 7. Outer view of the 52 m high tensegrity façade for the OZ building.

Figure 8. Internal view of the same.

are other systems that are called 'open systems'. In those cases the compression elements and tension elements are connected, but the edges are fixed on outer structures, like the concrete structure of a building. The edges of open systems are always anchored on outdoor building structures.

5 OPEN SYSTEM FOR TENSEGRITY FAÇADES

For the OZ Building in Tel Aviv, designed by Avram Yaski, a vertical façade was designed and built in 1995 with a width of 16 m and a height of 52 m. This glass surface was stabilized by horizontal tensegrity trusses of 16 m span on each floor height at 3.6 m distance (See fig. 8). Each horizontal truss is composed of two pre-stressed cables kept at proper distance by compression tubes, perpendicular on the glass surface. The inward curved cable takes care of wind compression on the façade, the outward curve takes care of wind suction. In fact these trusses are open trusses (reaction forces to be generated by the building are high), leading to an extremely efficient tensile tensegrity structure.

6 SQUARE TENSEGRITY GRIDS FOR ROOFS

In a series of structures in The Netherlands, square roof modules of 8×8 m were realized by 16 glass panels

of 2×2 m, which were connected by Quattro nodes on compression tubes each 2×2 m. The tubes are stabilized in two directions by a system of upward and downward directed cables, leading to the square CHS grid of 8×8 m. All of these structures are closed structures, as they are surrounded by a tubular compression grid at 8 m distance. So Octatube has ample experience with all kind of tensegrity structures, (see fig. 9,10).

7 THE HAGUE TENDER FOR TENSEGRITY CHANDELIERS

However, the invitation for tendering of the tensegrity structures in The Hague opened up a 'box of Pandora'. Tensegrities are the most difficult structures

Figure 9. Tensegrity roof structures for the Droogbak, Amsterdam(arch. Joop van Stigt).

Figure 10. Roof for the Natural Museum, Leeuwarden, (arch. Jelle de Jong).

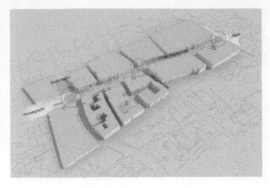

Figure 11. Town scale design pedestrian street with 3 tensegrities.

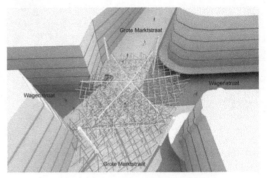

Figure 12. Detail design development by Octatube.

known. They had been studied thoroughly according to the architect, otherwise she would not go for tender. Yet these structures were extremely complicated in its composition. We had never seen such a large number of masts and cables in tensegrities before. At least this large entity would be a very experimental structure. The client just went for European tender as if it were a standard building project. How could we realize an experimental tensegrity concept that had been advised by the best advisors around as a reliable structure? (See fig. 11, 12).

8   DUTCH TENSEGRITY ARTWORKS

In the past I designed and built several tensegrity structures. The first of the experiments were artworks designed, developed and realized with the artist Loes van der Horst (1919–2012) and Krijn Giezen (1939–2011) as tensegrity structures, combined with membranes which made them more heavily loaded as structures. The first one was made in 1975, when I just left TU Delft with a diploma as Dutch building 'ingenieur', having worked for a while at Frei Otto's Institute

for Lightweight Structures in Stuttgart. Loes van der Horst was an artist in weaving. Together we designed and developed the 'Bijlmobiel', a structure of 2 masts and 6 cable nets spanned between the masts, situated in a new town near Amsterdam, the Bijlmermeer. This artwork had been redesigned as 'Yellow Wings' and is under construction in 2015. A later artwork (see fig. 13) was a real tensegrity artwork, won in a design competition. It was made of 6 masts, 4 booms and 3 compression studs. These were all steel compression tubes stabilized by 105 cables. The roof planes were made by 3 membranes with the associated wind and snow loads on the tensegrity structure. The structural calculations were rather difficult. It was 1978. The Ices Strudl program still was operating with punch cards. The 1 to 20 scaled model proved to be more helpful than the more than 50 computer runs with large deformations with which the computer program could not cope. The sculpture was built in 3 weeks time, although the planning was only 3 days. It appeared extremely complicated to establish the exact geometry with the analyzed pre-stresses.

In those times I had a discussion with Peter Rice in 1991 at the Amsterdam Academy, whether he would know exactly how high the pre-stresses in the cable structure in the Louvre Pyramid were. I did not get a satisfactory answer. Most probably the builders

Figure 13. Model of the artwork.

Figure 14. The actually realized Hemweg tensegrity structure on the terrace with membranes.

Figure 15. Outside of the Louvre Pyramid.

Figure 16. Inside of the tensegrity pyramid, Paris, designed by I.M.Pei and Peter Rice.

Figure 17. Sculpture at the Maritime Museum, Amsterdam.

stopped pre-stressing when the glass panels were visually seen as flat. (See fig. 15, 16). This all added to the mystery of tensegrity structures.

A few years later I made a sculpture in the inner court yard of the Amsterdam Maritime Museum with krijn Giezen. It existed of 4 twisted masts stabilized by 12 cables, fixed on 4 foot points with two large twisted sails (fig. 17). People were afraid of the masts falling down, they walked around the sculpture instead of walking through it, as was the logical way. Which was quite visible in the winter snow. The sculpture was removed 5 years ago to enable the inner courtyard to be covered by a glass roof.

So, from our experiences we learned that complicated tensegrity structures were highly indeterminate as structures. Once one single cable is stressed, all other cables will be influenced in their stresses. The tensegrity structure at Hemweg had 13 tubes and 105 cables. It took 5 times as long as anticipated to build and yet we found it was not 100% reliable (Fig. 14).

9 THE HAGUE TENSEGRITY CHANDELIERS

Let us go back to the challenge. In the tensegrity clouds of The Hague around 250 tubes of 5 m length were designed, connected by 1500 cables. One can understand that an entity of fixed cables and compression tubes can be analyzed. The real question was how to reach this stage with properly pre-stressed cables? The

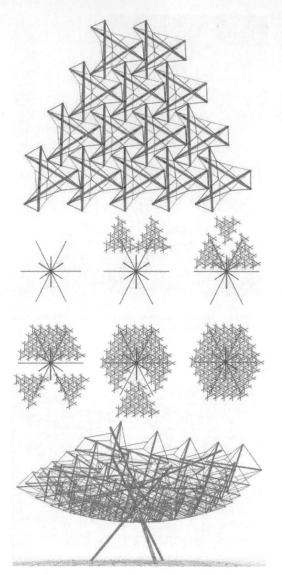

This challenge on the desk in 2013 seemed improbable and impossible. Yet the item of tensegrity structures challenged Octatube with its daring structural engineers and architects. So we studied the possibilities seriously.

Making tensegrity compression elements is not a challenge. Circular tubes, even in stainless steel, can be made as regular components. Cable elements with two eye ends and turnbuckles for post-stressing can be easily produced. Normally we would look for prefabrication off-site. But prefabrication in larger units was impossible as the compression components measure around 5 m long and overhead tramway cable lines at 4.5 m height obstruct the traffic of these units. So prefabrication was not an option.

The structure had to be built on an enormous scaffold platform on the street, with free height for pedestrian and cyclist traffic and for the fire brigade. (fig: 11). As per specs the building time was restrained to 3 months only. There was not a single modus, not even in my wildest dreams, I could think of pre-stressing basic tensegrity units and then combine and post-stress them together on the site.

10 BUILDABILITY

So the basic problem of the tensegrity structure was that the buildability of such a complex structure with so many compression tubes and an overwhelming amount of tension cables on top of a busy shopping street in a restricted time, under pressure of a European tender as if it was a normal tender, was impossible. And this claim will stay for another generation, unless more than regular money would be available to take the proper time for experimentation, ample time and computer assisted tension gauges to keep track of the pre-stress process of the individual stresses in the elements and the geometry.

I leave this for future experimentations, by whom?

Figure 18–20. Unitized and prefabricated tensegrity structures as drawn on the computer.

biggest problem was how to erect, how to install and how to pre-stress to the phase of a fully reliable structure, suspended in 25 tons of deadweight above the shopping street.

REFERENCES

Horst, L. van der, 2009. *Voor de horizon*. Thieme.
Snelson, K., 2009. *Forces made visible*. Hudson Hills.
Buckminster Fuller, R., 2008. *Operating Manual for Spaceship Earth*. Lars Mueller.

Tubular Structures XV – Batista, Vellasco & Lima (eds)
© 2015 Taylor & Francis Group, London, ISBN 978-1-138-02837-1

New velodrome in Medellín (Colombia)

X. Aguiló
BAC Engineering Consultancy Group, Madrid, Spain

J. Gomà
BAC Engineering Consultancy Group, Barcelona, Spain

ABSTRACT: The proposed structural roofing system is based on the connection between seventeen circular umbrellas of different diameter and height. They follow the principles of a bicycle wheel system tied up horizontally. This includes a central prop, a tensioning double mesh of two inverse geometries generated from faceted conoids, and a compression ring. All compression rings are at the same level, connecting them at tangency points. In order to obtain a space free of columns, vertical supports are only provided along the perimeter umbrellas. As a consequence, the two main central umbrellas are suspended from the perimeter ones. Pre-stressing tension forces are introduced in rods so to ensure that the resultants are always tension forces. In order to guarantee the most efficient solution, and comply with the aesthetics requirements imposed by the architect's concept design, a solution based on large diameter lattice columns made of small diameter hollow sections was chosen.

1 INTRODUCTION

1.1 Bio

Location: Medellín, Colombia
   Architect: Giancarlo Mazzanti (El Equipo de Mazzanti) and Fake Industries
   Membrane consultant: BAT Spain
   Structural consultant: BAC Engineering Consultancy Group
   Developer: Medellín Municipality (INDER)
   Project: 2013
   Total area: 13.200 m²

1.2 Location

The project site is located within the boundaries of Medellin Airport aircraft approaching zone. Therefore, the height of the roof is limited in several locations depending on the area in question. The project site is in the order of 165 m × 100 m, with the lightweight roof extending to an approximate area of 13.200 m².

1.3 General description of the architectural project

The Medellin Velodrome project was the winning proposal of an international competition hosted by Medellin Municipality (INDER) which drew the attention of a respectable number of well recognized architectural firms.

The roof is comprised of 17 umbrellas in the shape of cycling wheels, conventionally circular in plan, with varying diameters ranging from 8 m to 64 m, as well as varying heights between 10 m and 16 m, covering the court as well as the pavilions. These umbrellas are typically formed by a central column and a perimeter compression ring which are connected to each other by means of a double radial mesh of preloaded tension rods. The perimeter ring is always placed at the same level (+8.00), to easily allow connection between adjacent umbrellas at their tangency points.

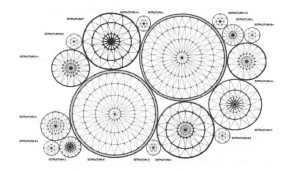

The cycling court as well as the pavilions has to be free of columns. Further, the combined width of the court and stands is larger than the diameter of the largest central umbrellas. As a result, these umbrellas are void of central columns (they only have a prop inside the envelope) and are suspended from the surrounding umbrellas. Consequently, the total clear span consists of the diameter of the supported umbrella in addition to half the diameter of the supporting umbrellas. Hence, there are two families of umbrellas: 2 supported and 15 supporting.

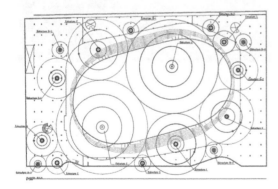

Additionally, there is another important factor to consider when shaping the varying umbrellas. In the case of a cycle wheel, it is clear that the "thicker" it becomes the stiffer it will be. Nevertheless, there are two geometrical limits to the design of the roof. On one hand, there is a clear height under the roof that needs to be respected; whilst on the other hand, as mentioned before, the total height of a specific umbrella depends on its relative position.

2   DESCRIPTION OF THE STRUCTURE

2.1   *Preliminary considerations on the structural response*

The project raises an oneiric association between the function of the structure (cycle racing) and the primary component of the roof structure (cycle wheel). But the fact is that there are significant differences between the demands and responses of both elements.

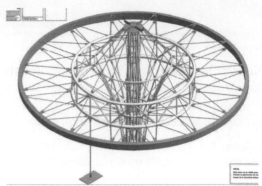

– Regarding the umbrellas, self-weight is clearly the prevalent load (wind and seismic loads will be discussed later), whilst in the case of the cycling wheel the main forces are those caused by the rider's (rider's self-weight and dynamic breaking effects) at the contact point between the ground and the wheel itself.
– Due to the variable nature of these dynamic loads (in quantity as much as direction), all cycle wheel spokes need to be heavily pre-stressed; otherwise, under certain conditions, some of them will be forced to work under compression, which is not compatible with their slenderness ratio. This scenario does not fully apply to the umbrellas structure; once installed at their final locations, only minor changes in sign (from tension to compression) are expected in about 50% of their "spokes" due to the effect of live loads; this can be easily controlled by means of a slight preloading. Major changes are not generally expected (only locally and under seismic circumstances).
– Unlike the cycle wheels (whose envelop is composed of two flat, symmetrical conoids aligned base to base), in the case of the umbrellas this is only true in the case of the smallest ones; in the majority of the cases, each one of these conoids is really formed up by one or two conic trunks (for medium and larger sizes respectively) topped by single cone. Obviously this shape is much more flexible than the cycle wheel, since the remaining geometric figure in a transverse section is not a couple of perfect triangles, but each one of them is "decomposed" in the addition of this perfect triangle plus one or to adjoined irregular quadrilateral. This is especially true for the case of anisotropic load cases and, for local heavy actions.
– Regarding the fabrication process of a cycle wheel, nowadays common industrial techniques are employed which allow us to preload all the wheel spokes at a single time. This is not possible using similar techniques in the case of the umbrellas structure which are tens of meters in diameter.
– Further pertaining to the umbrellas, whilst taking into consideration that, as mentioned before, self-weight is the main load case. Therefore, it's quite clear that the final distribution of forces in the

framework very much depends on the construction process.

Conversely, and taking into account the aerodynamic shape of any individual umbrella or the set in general, the application of prescriptive codes based on shape factors will not lead us to a reasonable scenario, since wind forces obtained will be no doubt overstated.

Therefore, a performance based analysis was developed in order to carefully evaluate these wind forces by means of a computational wind tunnel in which different wind directions were considered while accounting for varying air circulation conditions within the umbrellas themselves.

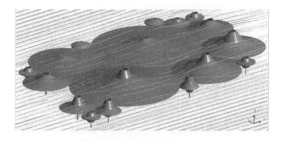

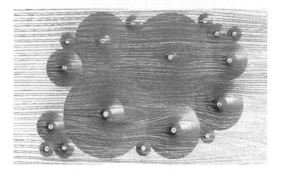

In the free portions of the perimeter ring, the circumstances are equivalent to the entire perimeter of the supporting umbrellas: tension forces in the upper face and compression forces in the lower face. On the contrary, in the support areas the condition is just the opposite: tension-down, compression-up.

In order to solve these problems, preloading all spokes will guarantee that they keep functioning under tension, regardless of the construction process. However, where the medium and large sized umbrellas are concerned, this option has to be promptly disregarded due to its extremely high cost. Therefore, the best solution to the problem consists of taking advantage of the intrinsic weight of the structure components (and especially the perimeter rings) to reduce, at a minimum, the number and scope of preloading operations. This project is a result of that approach.

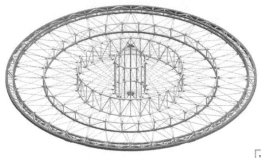

2.2 Design and construction strategy

As was previously stated, there are to families of umbrellas: the supporting units (15 in total) and the supported units (2 in total). As a result, the differing conditions pertain solely to the support scheme.

In relational to the common umbrellas, the perimeter rings are suspended from the central column; in such a situation, the "spokes" of the upper face tend to work under tension, whilst the ones of the lower face tend to work under compression; it then follows that, the latest is not possible due to the extreme slenderness ratio of the rods; as a result they will need to be preloaded. In such a situation, when a central umbrella is connected to the perimeter ring of the former, tension forces increase in the upper face rods next to the connection point, whilst compression forces would also tend to increase in the lower face rods (in fact, if they have been preloaded, tension forces tend to decrease).

The large, central umbrellas present further complexity; two different conditions coexist within them.

In this scenario, the design process cannot be based solely on the global model, but further, the simulation of the construction process. To that effect, sensitive variables of each individual umbrella first need to be identified. Otherwise, it is almost impossible to control the performance of the whole set, by modifying mechanical characteristics of specific components, once it has been assembled.

Additional topics taken into account throughout the design process include:

– As before mentioned, the upper/lower conoid of a medium/large size umbrella is formed up by the assembly of one or two quite flat conic trunks in addition to a cone (the lower cone does not exist in the case of the suspended main umbrellas). In order to control the flexibility of the whole framework, circumferential trusses need to be provided at each slope change of the skin membrane surface, in such a way that the set (upper ring, lower ring and trussing members) perform as a rigid cylinder.

- The anisotropic nature of the loads (shear forces between couples of umbrellas) makes it advisable to use different sizes for equivalent rods, depending on their specific location with respect to the connection points.
- The umbrella's perimeter beams not only have to support huge compression forces, but also relevant values of shear forces (both vertical and horizontal), bending moments (also in both vertical and horizontal planes) and torsional moments. Therefore, it was designed as a hollow boxed section whose sloped sides extend to the interior in order to facilitate easier connection of the rods. This box varies in the plate thickness depending on the size of the umbrella, and also in the width of its sloped sides. But the vertical external plate has always the same width in order to easily accommodate/connect two adjacent umbrellas.

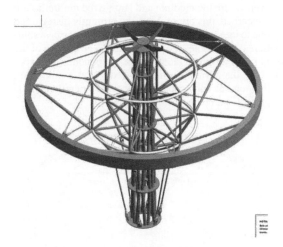

- The external medium size umbrellas show a very flexible response due to the fact that their loads are extremely eccentric; in order to control these movements, retention rods are provided to counterbalance the reactions of the central main umbrellas.

In closing, we have been systematically referring to rods instead of cables. The reason of that choice is that cables with end fittings do not develop a pure elastic response unless they have been tested (preloaded) prior to their installation. The problem is that they can experience permanent inelastic elongations during this calibrating process. In a case like that, where tolerances are very stringent, this performance is completely unacceptable.

3   THE CONSTRUCTION PROCESS

The construction process is based on the principle of maximizing the inherent benefits presented by the laws of physics (gravity) by preloading the rods (passive preloading), and thus limiting the active preloading operations. In the case of the common umbrellas (15 units), the upper tension ring will be temporarily placed at an appropriate height relative to the ground (by means of temporary props). Subsequent rings are then suspended using the permanent rods from top down. Once the lowest tension ring has been installed, the trussed circumferential inner cylinders will be completed.

Thereafter, rods forming the upper cone are connected to both the upper plate and tension ring. The set as a whole is then heavy lifted using the permanent column and so driven to its final position. In the next step of the process, rods forming the lower cone are connected to both the lower tension ring and plate. The set of lower rods (lower cone) just as the upper rods (upper cone) are preloaded by tightening the lower plate against the base plate.

Once all the supporting umbrellas have been installed and leveled, assembly of the two supported umbrellas will begin. First, the perimeter ring is placed at its final location, supported by temporary props.

From this moment on, the construction process is developed by repeating the same operation from the outside inwards. The first trussed vertical cylinder (closest to the perimeter ring) is placed at its final position using temporary propping while the upper and lower rod meshes are installed. Thereafter, the props are removed and the next cylinder is assembled in the same way. Finally, the umbrella's central permanent prop is installed thus generating the upper and lower cones that top the upper and lower conoids.

To guarantee tautness of rods under varying load cases (live loads, wind, seismic), an exact preloading process will need to be carried out, only affecting

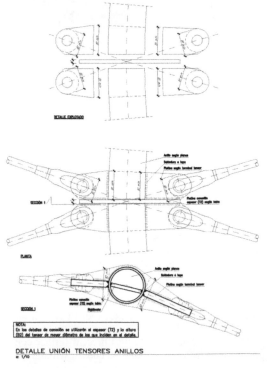

DETALLE UNIÓN TENSORES ANILLOS
e 1/10

certain rods. Finally, the suspended umbrellas are connected to the supporting units and the support props are removed.

The image above is the connection of the rod bracing the steel tubes.

4 CONCLUSIONS

The use of structural hollow sections was the result of a design approach focused on guaranteeing the most efficient solution from the capacity and construction point of view.

As it is well known, the structural hollow sections concentrate most of its mass around their perimeter, therefore resulting on a larger radius of gyration compared to UB/UC sections with the same mass. As a consequence, the slenderness of the hollow section is lower, reducing the impact of the buckling effects on its design. This makes them ideal for slender elements under compression forces.

It should be noted that the CHS sections have the same properties along all axis. This brings great benefits to the design, allowing a very efficient design of slender elements under compression forces with the same buckling length in all directions. It used has allowed the designer to optimize the design of some of the key elements of the structure, such as the main columns and some of the perimeter rings.

The use of CHS also rationalizes the detailing and construction of the most complex connections resulting of the building intricate geometry. (i.e. the connection between the tension bars against the inner rings, with each one at a different angle).

On the other hand, the design has also included fabricated steel box hollow sections along the main perimeter ring. These sections have been designed deeper than wider as the bracing is less effective along the vertical direction. The detailing is simplified by the fact that all the tension members connected to the main ring are attached at the same angle.

As mentioned early on, the use of hollow sections allowed us not only to ensure that the most efficient structure was provided, but also to simplify and rationalize the detailing and construction of this complex structure.

REFERENCES

It seems interesting to conclude this paper with some references to earlier bicycle wheel structures:

- Court yard floor in La Pedrera, Barcelona (Spain), 1907 Antonio Gaudí.
- Utica Auditorium, New York (EEUU), 1955, Lev Zetlin.
- New York State Pavillion, New York (EEUU) 1962, Lev Zetlin & Associates.
- Madison Square Garden, New York (EEUU), 1968, Charles Luckman & Fred Severud.
- Olympic Games Gym Pavilion, Seoul (South Korea), 1984 David H. Geiger.
- Sony Center, Berlin (Germany), 1996, Helmut Jahn.

Tubular Structures XV – Batista, Vellasco & Lima (eds)
© *2015 Taylor & Francis Group, London, ISBN 978-1-138-02837-1*

Reconstruction of a school building in Wolfsburg, Germany

H. Pasternak & M. Moradi Eshkafti
Brandenburg University of Technology, Cottbus, Germany

T. Krausche
ipp, Braunschweig, Germany

ABSTRACT: Heinrich-Nordhoff School in Wolfsburg, Germany, was built in 1971/72 as a concrete building. In 2013, the roof of the existing building was opened and over 2 floors a skylight roof was built with a supporting tree structure from circular hollow sections. The building is characterized by a high light penetration into the interior of the building. The reconstruction work required an intervention into the existing load transfer. The significant incision into the building had to be done by additional supports (steel frames) as well as a glass roof supported by a tree structure. This reconstruction may be and attractive example of how to insure a high penetration into the interior of the building. A special focus, within the welded tree structure was put on the formation of the nodes during the static design. A nonlinear Finite Element Analysis of the nodes, including the welding simulation, was carried out in order to study the behavior of complex joints under loading in presence of residual stresses.

1 INTRODUCTION

The Heinrich-Nordhoff School is located in the urban area of Wolfsburg. In the fall of 2008, renovation measures of parts AC of the building began (Figure 1).

A great level of attention has been paid to the conversion of part D, in which a steel tree structure with a glass roof in the center of the building is used. A workshop was drafts drawn in school several times. In September 2010, the concrete plans by the City of Wolfsburg were presented at school. In February 2011, the reconstruction of part D started. This conversion also included the construction of the cafeteria that could be ready by the start of school in 2013. The constructions work has been done separately where all the parts involved in the projects could be finished by the mentioned date.

Figure 2 shows one of the models shown in the workshop in February 2010. This model belongs to part where the tree structure was supposed to be built.

2 CONSTITUENT BUILDING – HOUSE D

The existing school building part D, with an area of 38.50 m × 33.80 m and height of 12.35 m, built in solid composite construction in the year 1971/1972. The massive ribbed slabs with round steel carrier KT 610 and a height of 28.5 cm span girders and walls of reinforced concrete and masonry structures. The reinforced concrete girders run – perpendicular to the ceiling clamping direction – on reinforced concrete columns. The floor slabs have no adequate

Figure 1. Aerial view IGS Wolfsburg (Heinrich Nordhoff Gesamtschule 2013).

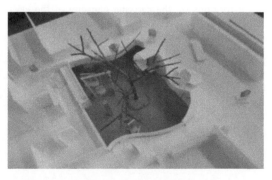

Figure 2. Design workshop tree, Team 2 (Heinrich Nordhoff Gesamtschule 2013).

Figure 3. Opening of the decks.

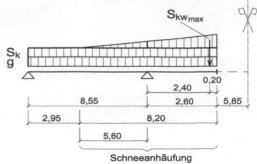

Figure 4. Static system after opening the roof deck.

concrete cover and were partially corroded. In order to guarantee the sustainability of the existing structure, an investigation of the compressive strength of concrete by core drilling was necessary. The compressive strength of concrete was determined in the MPA Braunschweig according to DIN EN 12504-1 and DIN EN 12390-3. As an average compressive strength of 34.7 N/mm^2 was tested and was thus within the estimated area. To prepare the concrete cover again, shotcrete was applied over a large area.

3 CONVERSION WORK

The majority of the renovation work consisted bringing over two floors and an opening in the ceiling plane. The ceiling of the 1st floor (roof level) and the ground floor (Figure 3) were opened. This opening of 11.90 m × 13.35 m required a modified load transfer. Different complementary measures were performed within individual areas of the building.

Individual ceiling sections were intercepted by support structures made of steel (steel frame) and directed into the wall plate below it. In order to prevent uncontrolled cracking in the support region, the top of the ceiling was cut 5 cm deep. Furthermore, the modified structural system of the ceiling of a 4-field continuous beam to a single-span beam with cantilever (Figure 4) was investigated. The analysis of the average force profiles and the existing level of reinforcement showed no restrictions on the carrying capacity of the ceiling. The same applies to those affected by the opening main girder. Meanwhile, existing static system (2-span beam with cantilever) was reduced to a one-span beam with cantilever. For the recording of the roof loads from the dome light a 100 cm heavy reinforced concrete coating was located at the southern edge of the roof. This is not supported on the relatively soft ribbed ceiling, instead. Upon the existing reinforced concrete bars.

4 ROOF – SKYLIGHT

In order to fill the interior of the building the upper stage with natural light, a light roof, consisting of a

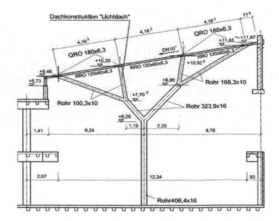

Figure 5. Tree structure.

desk-shaped roof structure and a supporting tree on the model is provided. It represents the core of the construction project. Over the ceiling opening is at an angle of 10°, a desk-like roof construction curious that is covered with glass elements. The roof is made up of 4 main bars with hollow sections. These are supported by a tree-like structure made of pipes. The tree structure in main and secondary branches and a trunk are presented in Figure 5.

The cross sections of the tubes are as in nature to a tree. From top to bottom bigger and adapt according to the force progression. The tube connection is rigidly welded to 5 nodes. The tribe consists of a tube of diameter 406.4 mm and a thickness of 16 mm. the four outgoing tubes from the root, the next larger branches have a diameter of 323.9 mm and 16 mm thickness. The upper branches to support the pitched roof. To connect the ground floor to the tree structure, a bridge structure made of steel profiles has been introduced to the tribe. Figure 6 shows the screw-bridge structure.

5 CALCULATIONS OF THE TREE STRUCTURE

The light roof was calculated as a lattice framework with rigid connections (Figure 7). The tree trunk

Figure 6.   Bridge construction with trunk.

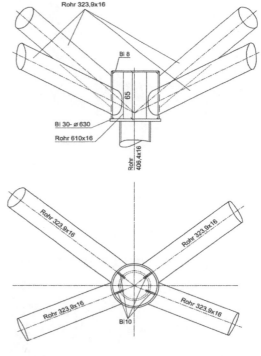

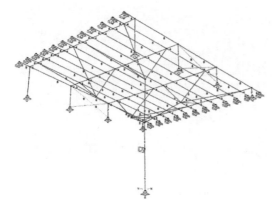

Figure 7.   Static system – Lattice framework.

Figure 8a.   Node upper branches.

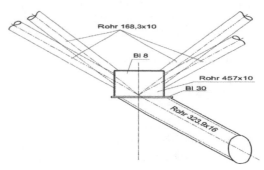

Figure 8b.   Node at trunk.

stored centrally on an existing reinforced concrete column, which is located directly below the basement ceiling. The roof load was assumed according to DIN 1055 (current DIN EN 1991:2010-12), with the extraordinary load combination of the northern German lowlands was considered for the snow load case.

For the calculation of the nodes, a FE analysis was necessary because this type of node is not regulated in the standards (EN 1993-1-8:2010). Both EUROCODE and CIDECT cover standard geometries by giving some failure modes which could not be used for studying the behavior of such a complex joint even when simplified. The calculation was performed with a geometrically and materially nonlinear finite element analysis of the nodes based on a stress-strain relationship of the bilinear material. There were 3 different authoritative nodes examined. The lowest on the trunk and two upper nodes, those are not identical due to the geometric conditions. The nodes were formed from tubular sections as "pot" from S355. The bottom pot has a diameter of 610 mm and is 25 mm thick (Fig. 8a). The upper nodes are with a 457×20 mm pipe (Figure 8b) executed.

Furthermore, the pots closed with a lid plate and additionally reinforced with internal stiffeners. Figure 9 shows the stress distribution of the lowest

node on the trunk where the maximum value was 290 MPa. The biggest value for compressive stress was in the lower part of the pot as shown below.

6   EXECUTION OF THE TREE STRUCTURE

The executed structure in S355, partially in S235, is shown in Figure 10. The assembly of the steel profiles was limited within the building, due to the cutout space, for a larger pre-fabrication in the workshop. Therefore, a significant part of the local structure was assembled by assembly welds. Both the steel construction tolerances according to DIN 18800-7:2008-11 (current DIN EN 1090:2012-02) as well as interfaces

Figure 9.   Stresses at the lowest node.

Figure 10a.   The executed structure.

Figure 10b.   One of the nodes in the tree structure.

Figure 10c.   The tree structure.

Figure 11a.   Final view – First floor.

Figure 11b.   Final view – Ground floor.

to inventory building had to be respected. The on con-nection were welded by the method 135 metal active gas welding (MAG) without additional preheating. The entire tube structure is sealed and has received an external corrosion protection with a white topcoat.

7   CONCLUDING

The concept of light-filled atrium in school has paid off (Figure 11a and b). The desired effect of open-ing the building and the associated incidence of light,

ranging in the basement, creates a special atmosphere in the IGS and is used as a multifunctional space with cafeteria.

8   WELDING SIMULATION

In addition a welding simulation (Radaj, 2002) has been carried out where the number of braces (branches) has been reduced to only one which is shown in Figure 12. The welding simulation was done to show the great importance of welding imperfections when dealing with load capacity analysis.

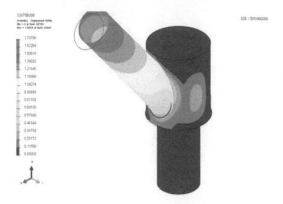

Figure 12a. Deformed shape and the initial shape.

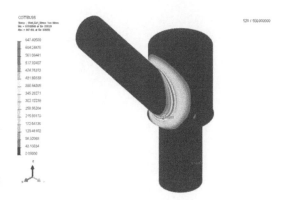

Figure 12b. Von Mises stress distribution.

8.1 Theory

The welding simulation has recently developed strongly. Enabling understanding complicated operations during the welding, and thereby a targeted optimization of the design. A key assumption is the weak coupling between thermo-physical and thermo-mechanical part model. The thermal analysis is performed by solving the heat conduction equation. The welding simulation will give you directly distortion and residual stresses which occur during the production process. For the simulation there are solid elements as well as shell elements available. Due to the highly time consuming calculation with the solid elements, an analysis with shell elements will shorten the expense. Welding simulation was done by using SYSWELD, where the geometries where imported from ABAQUS and meshed before setting the parameters and running the simulation. Arc welding was chosen for joining the parts. Welding speed was chosen to be 20 mm/s, Current 290A and Voltage 29V as the parts are quite thick so the energy input had to be high in order to make sure about the quality of the welded zones and also to make sure that that the parts are joint properly. There is a need of a parametric study to evaluate the appropriate parameters or use of an optimization program.

Figure 12a shows the distortion with scale in 1:50 for the system after cooling time which was 500 seconds. The initial and deformed shapes are both shown. The maximum deformation has been observed on the edge of the brace. The maximum distortion which was captured on the brace was about 1.73 mm.

Figure 12b presents the Von Mises stress distributions from the welding simulation after 500 seconds (cooling period) which could be later used for the load levels in ABAQUS.

8.2 Results & comparison

In Figures 12a and 12b, distortion and stress are shown after welding of one pipe to the knots. This calculation was already very time consuming but shows clearly the possibility to predict the distortion and stress precisely, assuming the knowledge of the welding parameters. The simulation also shows that the imperfection given by a eigenvalue analysis are not the same as the distortion from the actual weld for more complex geometries such as from the tree structure. To take all aspects into account there is a need of further investigations. All pipes and welds within the order of the production process need to be considered. In order to take into account all the relevant factors, the calculation time will increase. In meaningful engineering those assumptions are desirable or the interface between welding and load capacity analysis should be improved.

ACKNOWLEDGEMENTS – MAIN PROJECT PARTNERS

Client: Stadt Wolfsburg, Wolfsburg
Architect: Wolfsburg Consult GmbH, Wolfsburg
Design: M. Niemann, NIEMANN INGENIEURE, Magdeburg
Construction steel: Stahlbau Behrens GmbH & Co. KG, Vahldorf
Verification: Prof. H. Pasternak, T. Krausche, Braunschweig

REFERENCES

Anon., 2013. *Heinrich Nordhoff Gesamtschule.* [Online] Available at: www.hng-wob.de [Accessed 18 12 2013].
M. Niemann, H. Pasternak, T. Krausche 2014. Umbau eines Schulgebäudes der Heinrich-Nordhoff-Gesamtsschule in Wolfsburg. *Bauingenieur,* Volume 89 (2014), pp. 182–185.
Radaj, D., 2002. *Eigenspannungen und Verzug beim Schweißen.* ISBN: 978-3-87155-194-9
EN 1993-1-8:2010-12, EC3: Design of steel structures – Part 1–8 Design of joints.

Tubular Structures XV – Batista, Vellasco & Lima (eds)
© 2015 Taylor & Francis Group, London, ISBN 978-1-138-02837-1

Structural design of the roof structure for the SwissTech convention center in Lausanne

C. Pirazzi, M. Bosso & G. Guscetti
INGENI SA, Lausanne and Geneva, Switzerland

O. Fleischer & S. Herion
CCTH, Karlsruhe, Germany

ABSTRACT: This article treats of the structural concept and the technical challenges encountered during the realisation of the wide span spatial steel roof for the new convention centre at the Swiss Federal Institute of Technology in Lausanne (EPFL). With its main audience hall, designed to accommodate up to 3000 people, this convention centre is the largest of its kind in Switzerland and one of the largest in Europe. Its principle function is to receive scientific and academic events. The construction in the northern part of the EPFL campus began in 2011 and was finished in early 2014.

1 INTRODUCTION

The new convention centre at the EPFL is part of a building complex, including student housing for 500 students, a commercial centre, a hotel, a restaurant and administrative facilities. An international architectural competition was arranged in late 2006 and won by the total contractor HRS with the Lausanne architectural office Richter – Dahl Rocha & Associés in association with the structural engineering office INGENI.

The architectural project, an interpretation of an edge chamfered, multi-faceted stone, progressively detaching itself from the ground, is convincing by its strong expression and clear design. The heart of the convention centre is incontestably the main audience hall with 3000 seats. Its ground level can automatically be changed from an audience configuration with fixed, terraced seating into a flat levelled floor for a dinner gala or other events. This sophisticated system can be changed one into the other within 20 minutes.

A huge foyer is destined as an exhibition area. At the inferior level, multifunctional and modular conference rooms can hold up to 1400 people.

2 STRUCTURAL CONCEPT

2.1 Bearing structure

The structural highlight of the project is the steel roof with its trapezoidal geometry, covering the building over a total length of 116 m and a width of 67 m and cantilevering the foyer by a span of 40 m.

The chosen bearing structure follows the trapezoidal roof geometry and assimilates the inner spaces.

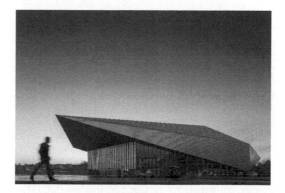

Figure 1. View from the square.

It uses a minimum number of support points on existing elements like the concrete cores. The resulting multi-faceted bearing structure exploits at best its structural potential thanks to its generous static height.

Its primary bearing elements are formed by two huge spatial steel girders, the so-called mega-girders. They are disposed laterally and supported by a total of six supports on the four concrete cores, which are situated in the north and in the centre of the building, forming a static system of a two-span beam with cantilever. The mega-girders are composed of ten facets, consisting of steel frameworks with chords, struts and ties.

In order to reduce dead load deformation, the mega-girders are conceived with a vertical pre-camber of up to 130 mm at their cantilevered extremity.

Circular hollow sections (CHS) in S355 steel quality have been chosen for the mega-girders. In comparison with common I-shaped or box-shaped cross

Figure 2. Main audience hall.

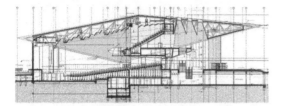

Figure 3. Longitudinal cross section.

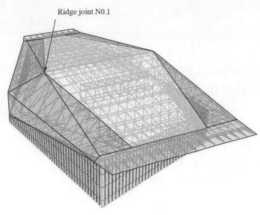

Figure 4. Structural model with lateral mega-girders, transversal girders and façade.

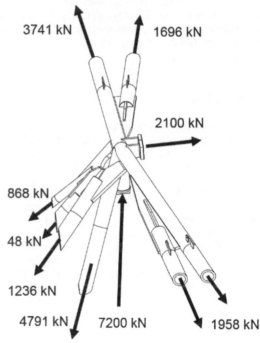

Figure 5. Ridge-joint No. 1 with axial design forces.

sections, CHS highly simplifies joint geometry. Additional material costs can thus be justified. The chords sections have diameters of 298.5 mm, the struts and ties sections vary between 168.3 mm and 219.1 mm except for some heavy loaded main struts which show the same diameter as the chords. The wall thickness varies between 7.1 mm and 40 mm.

Due to the complex trapezoidal roof geometry, the design of some of the joints proved to be especially difficult. At the supported ridge joint No. 1 for example, the arrival of ten structural bars with design forces of up to 7200 kN has to be managed as one single joint (Figs. 4–5). All arriving elements, except the vertical support bar, are submitted to tension forces. The design of this joint has been done in close collaboration with an external expert for tubular construction (CCTH). The solution found considers geometry and load level, giving priority to higher loaded members. The four most loaded members, which are the two upper plane chords, the main diagonal and the support bar, are cut to fit by CNC and butt welded. All other members, less loaded, are connected to this joint by means of thick plates ($t = 80$ mm).

Even though the fatigue does not govern the structural design, special attention was paid to weld quality and weld control. However, because of the large transverse dimensions of the mega-girders, only few tubes could be welded together in shop. Most tubes were assembled and welded on site.

The erection of the mega-girders was done by means of temporary supports under each node of the lower chords and under some nodes for the upper chords. The temporary supports are up to 25 m high, their stability is ensured by struts and ties.

In the transversal direction, the mega-girders are linked by means of conventional truss girders, spanning up to 40 m with a static height of 3.2 m and a

centre-distance of 5.5 m. The truss girders are prefabricated and brought on site. Some truss girders have a static height up to 5.5 m. These trusses are prefabricated in two parts and bolted on site prior to erection. The girder system provides an essential transversal stiffness, which blocks the mega-girders horizontally. Without this stiffness, the mega-girders, especially in the cantilevered part of the roof, would be subjected to large horizontal deformation. For this reason, all transversal girders have to be mounted before the removal of the temporary supports of the mega-girder.

In-plane girders at the roof's extremities provide horizontal stiffness due to frame effect. These

Figure 6. Prefabricated ridge-joint No. 1 in workshop, turned upside-down.

Figure 7. Slide bearing as horizontal connection between façade and roof.

structural elements and the transversal truss girders have an essential function for global roof stability.

2.2 Façade

The building façade with a height of up to 17 m is conceived as a self-supporting structure. No vertical roof loads are transmitted to the façade. However, horizontal wind loads have to be transmitted from the façade to the roof.

For this purpose, a sophisticated linking system has been elaborated in close collaboration with the steelwork company. This system is based on a piston with free vertical deformation capability, but able to transmit horizontal loads (Fig. 7). Overall forty slide bearings are disposed along the façade, linking both, façade and roof, allowing the roof to deform freely in vertical direction (Fig. 8).

The façade structure is composed of slender trapezoidal shaped columns, spaced 1.5 m and linked together horizontally by a system of tie-bars.

Depending on their position relative to the roof and the cantilever, the slide bearings are subjected to different vertical deformation under life loads. In the

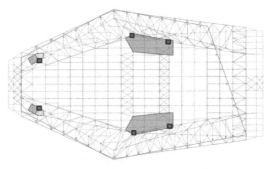

Figure 8. Distribution of slide bearings along the façade (little dots) and main structural bearings (rectangles).

cantilevered part of the roof, vertical deformation is much larger than in nearby abutments.

According to the numerical model, the slide bearing situated on the outermost position under the cantilevered part of the roof, for example, is subjected to a downward deformation of 50 mm due to snow load and 40 mm upward deformation due to wind suction. A maximum characteristic deformation amplitude of 90 mm could therefore theoretically occur. In order to reduce this deformation and to improve dynamic behaviour, the cantilevered part of the roof is subjected to a vertical downward pre-deformation, which also can be interpreted as a pre-tensioning of the roof.

2.3 Pre-deformation of the roof

The pre-deformation is realised by means of two cables with a diameter of 38 mm. They are hidden in the hollow façade columns in the south façade, fixed on the lower chord of the cantilevered part of the mega-girders and anchored in the concrete beams of the floor slab. The application of vertical tensioning forces pulls the roof down, but it also changes its static system, adding two elastic bearings. When snow load occurs, the tensile forces in the cables decrease, whereas they increase under wind suction load (Fig. 9). The pre-tensioning is designed in order to reduce vertical deflection at the decisive slide bearing of about 25 mm due to snow load and to 30 mm due to wind load, respectively. Furthermore, the dynamic behaviour of the roof is improved, as the cables act as damping devices.

The measurement of roof deflections and cable forces during the building service will enable to verify major hypotheses assumed for the design.

3 NUMERICAL INVENSTIGATIONS OF THE MAIN JOINTS

3.1 General

Due to their complexity, the design resistances of hollow section joints located in the mega-girders cannot

A – Dead load and camber

B – Pre-tensioning

C – Wind load

D – Snow load

Figure 9. Principles of vertical pre-deformation of the cantilevered roof, A) dead load and pre-cambering, B) pre-deformation of the roof, C) system under wind load, D) system under snow load.

Figure 10. Structural steelwork during construction.

has not been clear if the connected members offer rigid or pinned joint behaviour, nodal displacements and rotations of both have been used in the numerical investigations.

The loading of the joints results from applied nodal displacements and rotations to the supports of the neighbouring joints.

3.3 *Discretisation of the joints*

Except joint No. 1, the joints are directly welded without eccentricity. The members are modelled with nominal dimensions provided by INGENI. Joint No. 1 consists of nine CHS and a member connecting the mega-girder with the secondary framework. Very small angles between members and the complex inter-sections of a directly welded joint would hamper the cutting and welding or even make it impossible. Therefore, joint No. 1 has been realised using gusset plates (Fig. 11).

Since the wall-thickness of the sections have always been bigger than 8 mm, only full-penetration butt welds were considered in the numerical work. A linear elastic material ($E = 210$ GPa, Poisson's ratio $\nu = 0.3$) and linear geometric behaviour neglecting large deformations was used.

Due to the complex geometry of the joints, free mesh technique with fully integrated 4-node linear tetrahedron elements (C3D4) has been used for meshing. The lengths of the CHS have always been taken to $l_i = 5 \cdot d_i$, the connection to the neighbouring joints has been meshed with linear line elements (B31) in order to

be determined based on EN 1993-1-8 (EC3) or recommendations, for example the CIDECT Design Guide 1 (Wardenier et al. 2008). Therefore, the stress distributions and joint deformations of the most critical joints have been determined by CCTH (KoRoH GmbH) in order to ensure sufficient joint resistance. These joints have been identified by INGENI during the design process of the roof structure. Additionally, the optimisation of the joints with regard to fabrication aspects has been a part of the target specifications of CCTH.

3.2 *Loading*

The nodal displacements and rotations of the mega-girders have been determined by INGENI based on framework analyses for various load cases. Since it

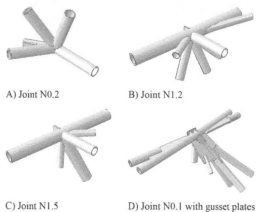

A) Joint N0.2   B) Joint N1.2

C) Joint N1.5   D) Joint N0.1 with gusset plates

Figure 11.   Numerically investigated joints.

reduce the number of degrees of freedom. Multipoint constraints (MPC) implemented in ABAQUS are used for coupling the end surfaces of the members with the line elements.

3.4   *Results*

The assembly sequences of the sections as well as the optimum design of the gusset plates have been defined with the help of the numerical investigations in a close collaboration of INGENI, CCTH and Zwahlen & Mayr, which has been responsible for the fabrication of the steelwork.

With the numerical investigations of CCTH, it has been shown that the joints offer sufficient resistances. Marginally increased stresses, mostly limited to a single node, are strictly localised to the intersections (welds) and result from numerical singularities (Fig. 12).

Considering more realistic weld geometries as combinations of full-penetration butt welds and fillet welds would minimize the singularities and further reduce stresses in these regions.

In general, complex joints like the ridge joint No. 1 are well suited to be made out of cast steel instead of be welded. However, this option was not pursued because of the unique geometry of each joint and the associated costs applying this technique in case of lack of repetition.

4   CONCLUSION

The roof structure of the SwissTech Convention Center on the Campus of the EPF in Lausanne is conceived as a spatial steel structure with a cantilever of 40 m.

Figure 12.   Von Mises stresses of joint No. 1.

Its principle bearing structure is composed of two spatial steel girders. The structural system follows the multi-faced shape of the architectural project and thus uses its form's potential in terms of rigidity (master of deformation) and efficiency (safe material).

Circular hollow sections have been used in order to simplify joint geometry. Most of the joints are welded on site, with welding thicknesses of up to 40 mm. Special attention was paid to weld quality and weld control.

In order to reduce vertical deflections under life loads, the cantilevered part of the roof is subjected to vertical pre-deformation by means of two cables. Their tensile force decreases when snow load occurs and increases under wind load. Vertical deflection and dynamic behaviour of the roof can thus be improved.

The total amount of steel used for the roof and the façade structure is about 1000 tons.

Principal participants:

General investor and building owner: Credit Suisse Real Estate Fund LivingPlus (CS REF LivingPlus) and Credit Suisse Real Estate Fund Hospitality (CS REF Hospitality)

Total contractor: HRS Real Estate SA

Architect: Richter-Dahl Rocha & Associésarchitectes SA

Structural engineer convention centre: INGENI SA

Steelwork Company: Consortium of Zwahlen & Mayr SA and Hevron SA

Principle tenant: EPFL Swiss Federal Institute of Technology, Lausanne – SQNE

Copyright of photos: Fernando Guerra, Yves André

Tubular Structures XV – Batista, Vellasco & Lima (eds)
© *2015 Taylor & Francis Group, London, ISBN 978-1-138-02837-1*

MyZeil Frankfurt – Design and execution of the architectural building envelope

R. Stroetmann
Technische Universität Dresden, Institute for Steel and Timber Construction, Germany
Krebs und Kiefer Beratende Ingenieure für das Bauwesen GmbH Dresden, Germany

ABSTRACT: In the city centre of Frankfurt the PalaisQuartier was built – a complex of buildings for manifold use. An essential part of it is the 'MyZeil' with approximately 78000 m^2 of gross floor area that sets standards by its vanguard architecture. In view of the urban architecture competition and the overall project the present article describes the planning and execution of the large free formed roof having a size of about 13500 m^2 as well as the special façades. Particular attention is paid to the specific features of the structural design and calculations, building execution and quality assurance [1].

1 INTRODUCTION

In the city center of Frankfurt/Germany the project PalaisQuartier with a gross floor area of 226,000 m^2 was erected, a building complex with a multipurpose use. Substantial part is the shopping mall MyZeil with a floor area of approximately 78,000 m^2, which provides due to the avant-garde architecture accent lighting. To attract attention and interest to visit the shopping mall already due to the architecture, in the description of the competition the request for an expressive design was formulated.

In the competition design of the architecture office M Fuksas Arch (Rome, Frankfurt) for the roofing the idea of a canyon was created, who crosses the building regions north, west and east like a river bed. Due to the positioning of a steel-glass-funnel in the center of the free formed roof landscape natural light can reach the ground floor of the 43 m height building and the view to the different storeys for the visitor is possible. In the region of the penetrations the openings of the floors were designed with organic shapes, which are fitting to the funnel-geometry. Beginning from the fifth floor, the edges of the floors are following the shape of the canyon.

The design of the roof geometry and the bearing structure was an interactive process, which was continuously adapted to the development of the functional, technical and artistically requirements. It should be provided a filigree, homogeneous bearing structure as a triangle steel grid with rectangular steel hollow profiles. Anchoring with ropes in the flat regions of the canyon as well as the design of stronger bearing elements should be avoided. These boundary conditions required an adaptation of the geometry of the canyon and the transition zones to the funnel.

Figure 1.   PalaisQuartier Frankfurt – Urban design concept of KSP Engel und Zimmermann Architects, Frankfurt.

In the context of the design competition and the integration of the shopping mall MyZeil in the whole building complex PalaisQuartier the following paper describes the planning and execution of the approximately 13,500 m^2 free formed roof and the special facades. Thereby characteristics of the structural design and static calculation as well as the construction work and quality insurance is emphasized.

2 DESIGN COMPETITION AND OVERVIEW OF THE ENTIRE PROJECT

In the course of an architectural peer review process, seven out of more than three hundred internationally renowned architecture offices were selected and commissioned to develop a building concept for the former

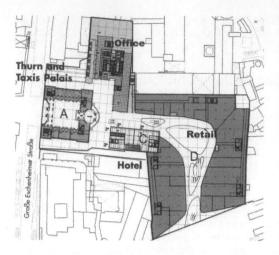

Figure 2. PalaisQuartier Frankfurt – plan view with names of buildings.

Figure 3. Animation of PalaisQuartier Frankfurt – view from Große Eschenheimer Straße.

premises of Deutsche Telekom in Frankfurt am Main. The property is located in a prominent downtown location and adjoins the Zeil, which is one of the three busiest shopping streets in Germany. As part of the compulsory architectural design proposal, the urban connection to the neighboring buildings of the Kaufhof and Zeilgalerie had to be checked, the heritage building remains of the Palais Thurn und Taxis had to be integrated, and a plan for the premises of the Frankfurter Rundschau had to be developed. The optional design proposal included the planning of an authentic reconstruction of the palace and its integration into the overall concept.

In the peer review process, KSP Engel und Zimmermann – Frankfurt prevailed ultimately with its design of the building ensemble over its competitors Mario Bellini – Milan, Coop Himmelb(l)au – Vienna, Christoph Langhof – Berlin, Christian de Portzamparc – Paris, Massimiliano Fuksas – Rome, and Richard Rogers – London (Fig. 1). Since Massimiliano Fuksas could convince with his avant-garde design of the retail building, his office was commissioned to plan and design the architecture of MyZeil.

Following the architectural competition, the other specialized planners were selected. For the planning of the structural design, several engineering offices with broad experience and high capability were invited. The consortium of Krebs und Kiefer and Weischede, Herrmann und Partner could compete successfully in a conceptual design competition. The consortium was entrusted with the structural design of the entire project.

The PalaisQuartier, which is characterized by two high-rise buildings, three public squares, a multifunctional building, and the reconstruction of the historical Palais Thurn und Taxis, was built on the former premises of Deutsche Telekom between the Zeil and the Großen Eschenheimer Straße until the year 2009. On the $17,400\,\mathrm{m}^2$ site, a total of $226,000\,\mathrm{m}^2$ of gross floor area were provided for retail and office space,

restaurant and hotel space, sport and entertainment as well as parking and service areas. The complex consists of four separate buildings A to D, which share a common basement (Figures 2 and 3). In four of the six underground levels there is an underground car park with 1,390 parking spaces.

3 DESIGN OF THE ROOF AND FACADES

To attract attention and arouse interest in visiting of MyZeil by a unique architecture, the client wished a highly expressive design of the building, which was already expressed in the invitation to tender. The building should stand out from the surrounding buildings by special accents and act as a magnet for visitors to the Frankfurt city center.

The idea for the building roof developed by M Fuksas Arch is inspired by a canyon, which, similar to a riverbed, runs through the three building areas North, West, and East (Fig. 4). A funnel in the center of the free form roof landscape provides natural light down to the ground floor of the building and allows visual links to the top. In the area of penetration, the building floor slabs are fitted with organically shaped openings and adapted to the geometry of the funnel. Above the fourth level, the contours of the floor slabs follow the course of the canyon.

The development of the roof geometry and support structure was an interactive process, which had to be continuously adjusted to the updated design, functional and technical requirements. The initial topology of the free form surface was created by heating and plastic deformation of a Plexiglas sheet on the building model, in which indentations were formed subsequently. The geometry was then digitally developed further with the 3D graphics program "Rhino". In

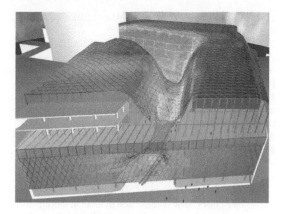

Figure 4. Building model of Zeilforum – perspective view from south.

Figure 5. Further developed roof geometry – view of roof and Zeil façade.

this design/modeling tool, so-called NURBS (Non-Uniform Rational B-Splines) are used for the free form surface. The amount of data required to describe the geometric shape is comparatively small, and the data exchange with other programs can take different forms.

For the design, a filigree, homogeneous as possible support structure consisting of steel truss members arranged in a triangular mesh was intended. Trussed beams as well as heavy supporting members should be avoided in the planar areas of the canyon. These boundary conditions required an adjustment of the geometry of the canyon as well as the transitions to the later remaining funnel.

The form of the structure of the canyon is based on catenaries, thus the structure is mainly subjected to tensile forces. The transition between the canyon and the funnel was made continuous so that the funnel acts as a large column that transfers the loads from the surrounding roof areas to the concrete structure by compressive forces (Figures 4 and 5).

The supporting structure of the roof over the building parts East, West, and North acts as a grillage. There, hinged columns are placed, which are connected to the system nodes of the steel lattice structure. Because of the generally not congruent position with the subjacent concrete columns, the uppermost floor slab transfers the column loads to the building axes. To accommodate the horizontal loads from the hanging canyon areas as well as the wind and stabilizing forces, cross braces are used, which are placed between two hinged columns and aligned to the canyon.

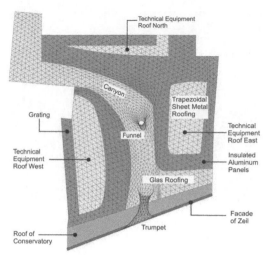

Figure 6. Top view of the roof with declaration of regions and allocation of cladding materials.

At a later stage, a connection between the free form roof and the diamond-shaped facade of the Zeil was planned by including a so-called trumpet, which forms a continuous transition from the canyon and flows like a funnel into the facade of the Zeil. The trumpet thus provides a visual links from the Zeil to the free form roof (Fig. 4 and 5).

In addition to the integration of the design elements canyon, funnel, and trumpet, the roof topology had to be adapted to the concrete structures and the areas for building services and had to be provided with a sufficient slope for drainage.

The definition of the mesh of the steel lattice structure was made in consideration of harmonious lines and a continuous sequence of triangles, which had to be adjusted in their size and the angles between the axes of the members to the topology of the roof. In this way, mesh sizes of less than one square meter were generated in the areas of high curvature of the funnel and trumpet. However, also mesh sizes of more than three square meters were generated in the flat roof areas. The member lengths vary from one to three meters and are on average about 2.15 m. In designing the lattice structure, different conditions had to be observed (Fig 6). These include

– the arrangement of nodes above the posts of the vertical facades,
– the connection of the trumpet to the nodes of the diamond facade at the Zeil,
– the arrangement of a set of members along the bottom of the canyon so that there the drainage works and harmonious lines result,
– the integration of the lines of the guide rail for the roof cleaning device and the snow guards at the junctions of the roofs for building services and the canyon,
– the formation of triangles also at the roof edges,
– the design of the contours in accordance with the perspective of building users.

Figure 7. Diamond-shaped façade at the Zeil – view of supporting structure.

In a first step, a triangle mesh was generated automatically according to defined specifications using the mesh generator of the program package ANSYS. This was used for the pre-dimensioning of the support structure and the determination of the expected profile dimensions. Here, the effects of the geometric specifications became visible in the component dimensions and the demand for steel.

After an optimization of the topology of the roof, the mesh was developed according to the design specifications of the architect using a surface-oriented CAD program in the next step. Since the mesh refers to the surface of the roof and thus maps the joints between the roof panels, the geometric structure of the supporting members had to be derived from the mesh for the calculations. The cross-sectional axes of the members are aligned with the bisector of the adjacent triangular surfaces. The position of the member axes is given by their distance from the surface.

At a later stage, the mesh geometry as shown in Figure 4 to 6 was further optimized by taking into account planning updates compared to the planning permission application and approved static calculations.

The surface of the roof was covered with different materials (Fig. 6). Insulating glass (solar and heat protection glazing) is used above the mall in the areas of the canyon and funnel. The composition is, from outside to inside, toughened glass (thickness typically 8 mm), a 18 mm wide space, and laminated safety glass consisting of partially pre-tensioned glass with a PVB interlayer (thickness typically 2×8 mm). If there were larger angular differences between adjacent panes, stepped edge glazing was used, which give sufficient glass fitting on the steel profiles or limit the joint width between the outer toughened glass panes.

The transition areas between the mall and the building parts North, West, and East were given a covering of triangular thermally insulated aluminum panels. Above the so-called technical equipment roofs of the building parts, trapezoidal sheet metal roofing was used because of the lower architectural requirements. Grating was used for the cover on the roof edges, which

allows the air exchange with the air conditioning of the building.

The diamond-shaped steel and glass facade on the Zeil extends from levels 1 to 3 and has a height of 15 m. The structure consists of welded rectangular profiles with cross-sectional dimensions of about 160×80 mm. The facade is suspended from the edge of the floor slab above the level 3 and transfers the vertical loads there. It can move freely in the horizontal direction and is supported in the floors below at the nodes of the floor slab edges (Fig. 7). Above the main entrance, the facade is connected to the roof structure via the trumpet without a joint gap (Figs. 4 to 6).

Also on the south side of the Zeilforum, there is the conservatory, which is located in level 4. Above the diamond facade, poles are connected that support the front edge of the roof. The recessed facade at the rear edge of the roof of the westerly part of the building is penetrated above the conservatory by the concrete structure of the fitness area (Fig. 12). Above this, it supports the edge of the free form roof. At a later stage, various changes were made in relation to the planning permission application:

The roof of the conservatory was fitted with a steel lattice structure with triangular meshes to match the main roof. In addition, the main roof was pulled down seamlessly from the MusicHall of the eastern part of the building to the conservatory. The facade directed towards the Palais Thurn und Taxis has, in the levels 1 to 3, a design identical to that of the diamond-shaped facade of the Zeil. Above this, similar to the east side of the building, post and beam facades are provided.

4 STATIC CALCULATION

4.1 Assumption of loads

Due to the complex geometry of the free form roof, expert reports were ordered to determine the snow and wind loads. Frankfurt is located in zone 1 of the German wind zone map ($v_{ref} = 22.5$ m/s, $q_{ref} = 0.32$ kN/m^2). For the reference height of approximately 40 m, a peak velocity pressure of 0.94 kN/m^2 is used. The distribution of the wind pressure was determined in the boundary layer wind tunnel of I.F.I. Institute for Industrial Aerodynamics GmbH, Aachen. For this, a model in scale 1: 500 was created that included the entire PalaisQuartier and the adjacent buildings. To study different wind directions, the model was mounted on a turntable (Fig. 8).

Because of the interference effect with the two high-rise buildings for office and hotel use, the investigation of the Zeilforum was carried out in the building ensemble. Because of wind build-up effects in front of the high-rise buildings and the nozzle effect resulting from the wind flow between these buildings, greater pressure and suction loads were expected in comparison to a free flow around the Zeilforum. This was also reflected in the results of the pressure measurements.

To capture the wind pressure distribution, the roof was equipped with 230 pressure taps. The pressure

Figure 8. Wind tunnel investigation – model of PalaisQuartier with adjacent building area and perspective view.

distribution was measured in 30° increments for 12 wind directions. The evaluation led to wind load maps for the global wind effect at different flow directions as well as to "envelopes" of pressure and suction distributions for the local design of the roofing. Corresponding investigations were also carried out for the facades of the building. In addition and in superposition with the wind loads on the building exterior, positive and negative pressures were considered inside the building.

In determining the design snow loads, in particular, the possibility of snow accumulations in the areas of the canyon and funnel had to be considered. Due to drifts of snow and snow sliding down inclined surfaces, load concentrations and dynamic effects occur, which are relevant for the design of the supporting structure and the glass panes. In the area of the canyon, characteristic values of up to $2.6 \, \text{kN/m}^2$ are to be taken into account for the snow loads.

As constructive measures, snow guards are installed at the junction of the building parts North, West, East and the canyon as well as at the transition between the canyon and the funnel. The latter prevent sliding of snow to the main entrance at the Zeil. In addition, heating devices were placed at the funnel to reduce or avoid snow accumulation by continuous melting.

Because of deformations and possible constraining forces, the jointless supporting structure of the roof had to be investigated for temperatures deviating from the temperature during construction of the structure. The dimensions of the roof projected

onto the horizontal plane are approximately 134 m in north-south direction, 120 m in east-west direction, and 155 m in the direction of the northwest-southeast diagonal.

When determining the temperature differences for the summer and winter case scenarios, it was to distinguish between the external and internal support structure and between transparent and opaque roofing. In the calculation, the maximum differences +40 K (summer case scenario, indoor, glazed roof area) and −39 K (winter case scenario, outdoor) were used.

In addition to the dead loads of the supporting structure and the cladding, the actions from the deformation of the supporting concrete structures, live loads of the cleaning equipment and suspended advertising panels, water accumulation in the drainage trough at the foot of the funnel as well as seismic loads were considered.

4.2 Calculation and design of the structure

For the calculation of the structure, a spatial truss model was developed that represents the special facades and their interaction with the roof structure. The model is based on the plans of the building application and consists of approximately 8250 members (Fig. 9). The joints of the members with the nodes of the roof structure were classified as rigid and as full-strength according to EN 1993-1-8, Section 5.2 [2]. In this way, the calculation of state variables could be done without considering the flexibility of the nodes, and the members could be designed without considering reductions attributed to classifications as partial-strength joints.

The steel structure is supported in the area of the technical equipment roof by hinged columns and bracing connected to the concrete structure, at the foot of the funnel by joints to the steel truss ring connected to the mezzanine (between ground floor and first basement floor), and at the facades by the connection to the floor slabs. The deformations at the support points depend only partially on the size of the roof loads. They depend to a greater extent on the time (creep and shrinkage of the reinforced concrete construction, building settlement) and the load on the concrete structures. For this reason, rigid support conditions were assumed in the directions of transferred forces with the exception of the foot of the funnel. The deformations of the support points calculated for the concrete structure were taken into account by imposed deformations.

The calculation of the structure for the relevant combinations of actions was performed according to second order theory using equivalent geometric imperfections. The shapes of the imperfections were specified by scaling the mode shapes of a modal analysis. For large, highly statically indeterminate structures of the present form, it has to be considered that often the scaled mode shapes have an influence on the dimensioning of only parts of the structure. Therefore, for a structure with different member dimensions,

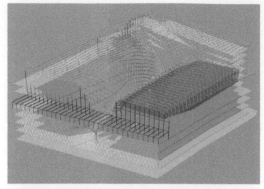

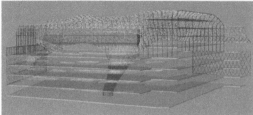

Figure 9. Building model with supporting structure of roof and special façades.

all mode shapes with an associated critical load factor α_{cr} smaller than 10 have to be taken into account, which corresponds to the "10% rule" according to EN 1993-1-1.

For the roof and facade constructions, mainly RHS members of steel grade S355 were used. Circular hollow sections were used for the columns on the technical equipment roofs. The highly loaded and later not visible truss ring under the funnel consists of welded hollow sections, I-beams for the web members, and flats for the nodes and connection elements (Fig. 10).

The limitation of the deformations of the structure results from static and functional needs. Deformations due to permanent loads are compensated by cambers. Because of the resulting constraining forces in the glass, the deflection of the members below the glass panes had to be limited to 1/200 of the member length. A sufficient slope for the roof drainage is to be observed also under the presence of the deformations. For variable actions, the maximum values of the deformation at the roof edges are in the range of 60 to 70 mm in horizontal as well as in the vertical direction.

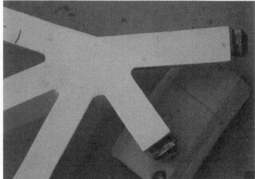

Figure 10. Construction of the nodes – a) star shaped node of the roof, b) milled node, c) node of the façade.

5 MANUFACTURING AND ASSEMBLING

The structure of the roof and the diamond-shaped facade was prefabricated in the shop as single members or in segments with straight bars. Continuous welds were used for the welded rectangular sections: bevel-groove welds below the glazing and fillet welds towards the building interior. Profile and node geometry were chosen such that the required supply lines (sprinkler and energy supply) could be integrated and covered. To get a homogeneous design of the roof the width of 60 mm was chosen for all cross-sections. The height and the wall thicknesses of the individual members were elected according to the static requirements.

The members are connected to the nodes by single-bevel butt welds, double-bevel butt welds and bevel-groove welds in the form of half Y-groove or two half Y-groove butt welds, partially in combination with fillet welds. For the joints welded on the site, the geometric adjustments and weld preparations of member ends and nodes were performed in the shop. For the joints, milled solid steel nodes, star-shaped elements made of solid material, and splice plates were used (Fig. 10).

Figure 12.   View to the roof with glass and aluminum panels.

Figure 11.   Erection face of steelwork – a) adjusting of the ladders, b) view southwards through the canyon.

The welds had partially to be grinded flat and smooth. This was especially true for the nodes that can be seen from the shopping mall. When the weld quality had to be proved, the criteria of quality level B according to ISO 5817 [3] had to be observed.

As preparatory work for the assembly of the steel construction, the anchorage elements for the columns and bracings as well as the facade connections had to be integrated into the concrete slabs. In addition, the steel truss ring, which was shipped to the construction site in three parts, was mounted for the connection of the funnel.

Since the basement floors of the PalaisQuartier were constructed in lot 1 and the mounting of the roof started with the foot of the funnel, delivery and installation of the truss ring and the associated embedded steel mounting parts lay on the critical path. The three ring segments were first welded together and then connected to the steel mounting parts.

For assembling the roof support structure, birdcage scaffoldings were used, which had to be continuously adapted to the assembly progress. The bases for the system nodes were measured taking into account the camber and the deformations due to permanent actions. In this way, it could be expected that, after removing the scaffoldings, the geometry of the roof structure complied with the target geometry.

The lifting assembly was performed with the existing on-site tower cranes. For access of the assembly personnel, boom lifts, scissor lifts and mobile scaffolds were used. The assembling of the roof was performed according to the following scheme:

– placing and aligning of the elements,
– installation of the additional members between the elements, tacking and welding, mending, applying, and completing the corrosion protection coating,
– assembling of the insulating glass panels, roof panels, trapezoidal sheets and gratings,
– releasing the scaffolding by lowering the support points and subsequent disassembly of the scaffolding, further corrosion protection work, particularly on the columns, which were possibly damaged during subsequent work in the construction progress.

For corrosion protection, mainly coating systems according to ISO 12944 [4] were used. With respect to the corrosive attack, it had to be differentiated between components indoor (heated buildings with neutral atmosphere: corrosivity category C1 "very low") and outdoor (urban and industrial atmospheres, moderate pollution by sulfur dioxide: corrosivity category C5). Accordingly, two and three layer coating systems with dry film thicknesses of $160 \,\mu m$ or $240 \,\mu m$ were used. The color tint papyrus white (RAL 9018) was chosen.

6   CONCLUSIONS

The execution works on the free form roof and the special facades of the Zeilforum were taken forward under high effort. The completion of these works took place in the spring of 2009. For the roof, about 600 tons of steel, $6{,}000 \, m^2$ glazing, $4{,}100 \, m^2$ opaque surfaces of aluminum panels, and $3{,}400 \, m^2$ of trapezoidal sheet metal and grating catwalks were used. With the completion of the construction project, the city of Frankfurt has been enriched by a building that sets accents in its design and functionality.

7   PROJECT PARTICIPANTS

Client:
PalaisQuartier GmbH & Co. KG, represented by

Bouwfonds MAB PalaisQuartier GmbH, Frankfurt

Architects:
KSP Engel und Zimmermann GmbH, Frankfurt

Massimiliano Fuksas Architekten, Frankfurt/Rom (Zeilforum)

Structural engineers:
Arbeitsgemeinschaft Krebs und Kiefer & Weischede, Herrmann und Partner;
Knippers und Helbig Beratende Ingenieure, Stuttgart

Project management:
Drees & Sommer GmbH, Frankfurt
Expert for wind and snow loads:

I.F.I. Institut für Industrieaerodynamik GmbH, Institut an der Fachhochschule Aachen

Construction company for the roof and steel-glass-facades:
Waagner Biro Stahlbau AG, Wien/Österreich

8 PHOTOGRAPHS AND FIGURES

Figures 1, 2, 3 – Bouwfonds MAB Development GmbH

Figures 4 to 7 and 9 to 12 – KREBS+KIEFER Ingenieure, Germany (www.kuk.de)

Figure 8 – I.F.I. Institut für Industrieaerodynamik GmbH, Institut an der Fachhochschule Aachen

REFERENCES

[1] Stroetmann, R., Istel, R.; Hanek, D.: PalaisQuartier Frankfurt: Zeilforum – Planung und Ausführung der architektonischen Gebäudehülle. Ernst & Sohn, Stahlbau 77 (2008), Heft 10, S. 696–707
[2] EN 1993: Design of steel structures – Part 1-8: Design of jonts (Issue June 2006)
[3] ISO 5817: Welding – Fusion-welded joints in steel, nickel, titanium and their alloys (beam welding excluded) – Quality levels for imperfections (ISO 5817:2003 + Cor. 1:2006)
[4] ISO 12944: Paints and varnishes – Corrosion protection of steel structures by protective paint systems (ISO 12944-1998)

Where tubular structures fail – examples from one engineer's experience

R.H. Keays
Keays Engineering, Melbourne, Australia

ABSTRACT: The efficiency of tubes for transmission of axial loads makes them a natural choice for truss structures and for bracing framed structures. Their streamlined aerodynamic shape means they are also the natural choice for cantilever towers. The wide range of sizes readily available in higher strength steels leads to efficient minimum-weight structures. But there are challenges to this efficiency. Design of connections is not easy, and higher working stresses may lead to problems with vibrations. Through over 40 years' experience in the design, fabrication, and erection of steel structures, and latterly in the forensic analysis of collapses, the author has seen many cases where structures failed. This paper documents a number of these where tubes were the primary structural element, attempting to draw conclusions on the root cause of each failure.

1 INTRODUCTION

The efficiency of tubes for transmission of axial loads makes them a natural choice for truss structures and for bracing framed structures. Their streamlined aerodynamic shape means they are also the natural choice for cantilever towers. The wide range of sizes readily available in higher strength steels leads to efficient minimum-weight structures. Wardenier (2010) gives many examples of the successful use of tubes in structures.

But there are challenges to this efficiency. Design of connections requires careful consideration of local three-dimensional effects, joint eccentricities, and access for welding. The higher working stresses result in higher deflections, lower natural frequencies and limited damping, which may combine to give problems with vibrations.

Occasionally, the structural design does not rise to these challenges, and a failure results. Sir Arthur Pugsley (1966) lists eighteen lessons that can be drawn from structural failures, one of which is "Endeavour to ensure that all accident investigations lead not only to remedies for the kind of structure concerned, but also benefit other related types of structure". This paper is intended to address that lesson.

The definition of failure used here is broad – "an unacceptable difference between expected and observed performance" (Leonards, 1982).

All the samples here are from the author's direct experience as the erector's engineer or an observer/reviewer post-event.

The observed failures are grouped into five categories:

– Wind-induced vibration
– Bolted field splices
– Butt welds
– Tube/tube connections
– Dynamic overload

Each instance will be summarized, with a discussion of the group and potential remedies to reduce the chance of future occurrence.

2 WIND-INDUCED VIBRATION

2.1 *General comments*

Vortex shedding from wind flow across tubulars is a well-known phenomenon; it leads to resonant vibration when the shedding frequency matches a natural frequency of the structure. Cantilevers are more susceptible than members supported at both ends, but the author has seen examples of both.

2.2 *Cantilevers*

The 57-storey office building at 101 Collins Street features a pair of spires (600 mm diameter steel tubes) rising 60 m off the roof to make it the (then) tallest building in Melbourne. The top 20 m was unbraced. An electrician installing flood-lights observed a dark line on one spire, and noticed it opening and closing with the sway of the spire. The dark line seen by the electrician was a fatigue crack, starting from the weld of a step iron.

The vibration problem was relieved by the addition of vertical chains within the vertical tubes. The chains were selected to act as a tuned mass, with damping from chain impact on the tube wall.

The 51-storey Bourke Place office tower featured a micro-wave antenna mast as part of the capital decoration bringing its overall height to just 6 m short of 101. The author was tasked with creating the erection

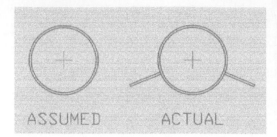

Figure 1. Section on antenna mast.

Figure 3. B of the bang sculpture, manchester.

Figure 2. Sticks on western approach to City Link.

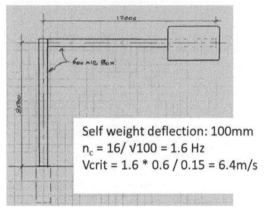

Self weight deflection: 100mm
$n_c = 16/\sqrt{100} = 1.6$ Hz
Vcrit = 1.6 * 0.6 / 0.15 = 6.4m/s

Figure 4. Freeway overhead sign gantry.

scheme, and checked the structural engineer's drawings for vortex shedding. Unbeknown to him at the time was the addition of two fins for attachment of the antennas.

When the electrician came to install the antennas, he found cracking at the corners of the cabinet welded into the truss at the base of the mono-pole. It transpires that the pole had been vibrating when the wind direction aligned with the fins. The solution in this case was simple – the addition of antennas was sufficient to "confuse" the vortex shedding.

The 39 red sticks in Figure 2 form part of an architectural gateway on the freeway from Melbourne Airport to the city. The designer was well aware of the possibility of vibration, and included a tuned mass damper in the head of each stick. Problems with supply of the damper led the constructor to devise a temporary framework of plastic pipe which acted to minimize vibration in the wind. Its effectiveness was proved by its omission on one stick. That afternoon's breeze was sufficient to excite vibration.

Ten years on, one stick was observed to vibrate. Inspection of the damper showed that a spring holding the damper ball had broken.

Spikes fell off the massive iconic sculpture, "B of the Bang" in Manchester before it could be officially opened. The author visited the site 6 months later, and observed vibration in moderate winds (<5 m/s). The spikes were tapered, which typically is sufficient to reduce the magnitude of vortex induced forces, but clearly not in the case.

A Freeway Overhead Sign Gantry had been in service for about 5 years when locals reported "the sign is vibrating more than it had before". Inspection showed that one foundation bolt had broken. The author's analysis suggested it had been suffering from wind-induced vibration.

Given the length of time it had been in service, it appeared likely that vibration occurred on only a narrow range of wind directions, possibly associated with flow past an adjacent building. Accelerometers and an anemometer were fitted. It took two months' recording to find the right combination of speed and direction. The issue was resolved by trimming 3 m off the length of the cantilever.

Figure 5. Collins square sculpture.

2.3 *Members supported at both ends*

Melbourne Exhibition Centre features a row of slender SHS columns (125 × 6 SHS × 8.5 m). These were observed during construction to vibrate when the wind blew along the line of the columns. The problem went away when the paving was completed; its effect was to build in the column base, and provide some damping.

The moving roof on the Docklands Stadium is supported by four double-arch trusses, each spanning 165 m. Chords are 508 × 24 to 508 × 12 CHS in Grade 450. Webs are 323 × 13 down to 168 × 5 CHS and up to 16 m long. During construction it was noted that the web members vibrated noticeably in the wind.

A damper was designed, but not installed due to an oversight. Some nine months later, an inspector checking paintwork noticed a crack in the weld joining the diagonal to the bottom chord, "large enough to fit a $20 note through it". With the crack transverse to the tension in the bottom chord there was potential for a fatigue failure in the not too distant future.

The crack was repaired, and Stockbridge Dampers were then added to this and a number of similar members.

A similar problem occurred during erection of the next such moving roof truss at Hisense Arena, but the vibration stopped once the structure was complete.

The roof on the MCG Northern Grandstand cantilever forward some 40 m from the rear support, with cables to masts and backstays carrying its weight. The end bays require bracing between the end and adjacent masts. This was done with 273 × 6.4 CHS, about 16 m long. It started to vibrate the afternoon after erection. Stockbridge Dampers were subsequently added to this and a number of similar members.

Dion Horstman's *Supersonic* sculpture connects the forecourts of two city buildings. It was fabricated from 273 CHS, typically 4.8 mm wall but thicker at places, with the longest straight length at 31 m. The author was engaged by the erector to determine the erection sequence. Calculated self-weight deflections exceeded 300 mm at one point, and over 100 mm at several others.

It started vibrating in modest breezes during construction. It has been subsequently stiffened, and the vibration appears to have been ameliorated.

This structure had other defects, which are discussed later.

2.4 *Analysis*

The fundamental natural frequency of a simply supported beam is given by:

$$f_n = \frac{\pi}{2}\sqrt{\frac{EI}{wL^4}} \qquad (1)$$

where E = Youngs Modulus, I = Moment of Inertia, w = Mass/Unit Length, and L = span.

The critical velocity for vortex shedding (when "lock-in" occurs) is given by:

$$V_{crit} = f_n b / S_r \qquad (2)$$

where b = breadth, and S_r = Strouhal Number.

Substituting (1) into (2), and noting that the radius of gyration, r, is about $b/\sqrt{8}$ for circular tubes, gives:

$$V_{crit} = \frac{\pi}{2}\sqrt{\frac{8E}{\rho}}\left(\frac{L}{r}\right)^{-2} \qquad (3)$$

For circular tubes, the Strouhal Number is normally taken as 0.2. Using this, V_{crit} is about 5 m/s when the Slenderness Ratio (L/r) is 150, and 10 m/s at 100. In the local context, Melbourne experiences average winds between 5 and 10 m/s when the sea breeze comes in on a sunny summer afternoon.

For square or rectangular tubes, the Strouhal Number is lower (between 0.11 and 0.15), resulting in a higher V_{crit} for the same Slenderness Ratio.

Axial load in the tube has an effect on the natural frequency, with equation (1) modified by the expression $(1 + P/P_{cr})^{1/2}$ (reaching zero at the buckling load). This serves as a possible explanation of the cases where the vibration was reduced once the structure was complete.

2.5 *Suggested remedial action*

Given the extent of this problem, a paragraph in Wardenier (2008) or Wardenier (2010) explaining the issue should suffice. A simple warning about tubes with an (L/r) greater than 100 might suffice. Reference might be made to Robinson & Hamilton (1992), which provides extensive guidance.

3 INADEQUATE BOLTED FIELD SPLICES

3.1 *General comments*

In Australia, responsibility for design of connections generally lies with the steelwork designer, not the fabricator, so there are few specialist connection designers. Also, contractors and erectors are reluctant to use field-welded joints, making efficient design of bolted joints critical to production.

Figure 6.   Deformed knee-brace joint.

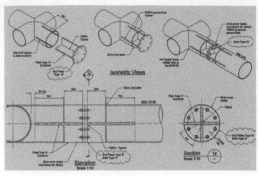

Figure 7.   Connection detail drawing – AAMI Stadium.

3.2   *Lapped splice in knee-brace*

The large warehouse (600 m × 110 m) structure has 219 × 6.4 CHS knee-brace struts to the three internal columns to reduce bending moments in the rafters. The joint shown below deformed under gravity loads at the time cool-room insulation was being installed in the suspended ceiling. Fortunately for the 250 workers fitting out the building at the time, the building did not collapse.

This failed at the confluence of two defects. The primary defect was that the designer's model failed to account for eccentricities associated with the lap joint. The second was that the plate supplied had a yield strength of 285 MPa, when the specification called for 350 MPa for all plate in the job. By the author's assessment, the "blame" for the failure rested 80% with the designer and 20% with the fabricator.

Following this incident, the author was advised of two similar cases. Warnings were issued by relevant authorities.

3.3   *Grade 12.9 bolts in tension*

The AAMI Stadium features a shell roof, with tubular ribs (273CHS) and support beams (508CHS). Although there is considerable shell action, the support beams carry significant bending moments, and the field splices were a challenge.

The contractor chose to use cruciform joints with bolts in tension, as illustrated below. The support beam joints used eight M42-12.9 "Unbrako" bolts, tensioned turn-of-nut. This was decided despite the designer being provided with a copy of Keays (2006).

Some bolts failed within months of installation. The author suspects excessive initial tensioning followed by the addition of further gravity loads, and possibly cyclic wind loads precipitated the failures. Remedial work required the addition of perimeter welds round the flange plates.

Grade 12.9 bolts are known to have a susceptibility to stress-corrosion cracking, and are not normally used in structural applications. Similar bolts were used in the runway beam splices on the Docklands Roof (noted earlier); one of these had failed after 6 years in service

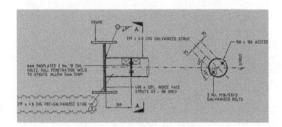

Figure 8.   Connection detail – sound tube.

from the combined effect of cyclic loading from roof movement and corrosion from ingress of moisture in an ill-fitting joint.

3.4   *Flush joint splice detail*

The Sound Tube on the freeway from Melbourne Airport to the city serves to isolate adjacent highrise apartments from traffic noise, as well as an architectural feature of the elevated roadway.

To isolate the arches from bridge deflections, they are coupled by struts with sliding connections at one end, allowing limited movement of one arch relative to its neighbors. Just 3 braced bays are provided for overall stability.

The design drawings showed this detail for the strut joints – 6 mm end plate with just 2 M16 bolts. This same detail was inadvertently used for the braced bays. The author drew attention to its inadequacy for the braced bays, and the fabrication drawing adjusted to use a more appropriate connection.

As fabrication had already started, the erection procedure was changed to start at an unbraced bay, with chain-block guys to hold the structure upright. One rigger working on another aspect of construction "borrowed" one guy. That afternoon the structure collapsed in 10–15 m/s wind.

3.5   *Jack-up barge leg joint*

The "flush" joint connection in the two cases above was determined from aesthetic considerations. This example is one where function dictated the form.

Workers check the steel ribs that toppled over on the western part of City Link. Picture: PENNY STEPHENS

Figure 9.   Sound Tube collapsed- from *The Age*.

Figure 10.   Sideson 2 in operation.

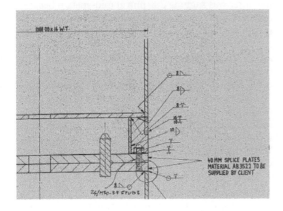

Figure 11.   Leg splice detail drawing.

"Sideson 2" is a jack-up drilling barge which can be dismantled for truck transport. The legs are 1 m diameter fabricated tubes, in three 11 m lengths, with flush joints to pass through collars on the barge. Jacks press against tabs on the tube faces to lift the barge clear of waves while on station.

The operators failed to notice they had not locked off one jack before leaving the barge for the weekend. The barge slid down the leg, producing a bending moment in the leg beyond the capacity of the splice. The splice then failed by tearing of the plates above the pocket for the bolts.

The design was subsequently modified to increase the thickness of the plates round the pocket, sufficient

Figure 12.   Typical splice – collins square sculpture.

for that to be stronger than the bolts at the flanged connection. This was done in appreciation that in such circumstances something had to break, and it was better to break something easily replaced – the bolts.

3.6   *Collins square sculpture*

The artist for this sculpture desired "seamless" connections, which the structural engineer provided by adopting a "flush" joint connection detail. This had cruciform plates with a Tee-bar connection to the tube.

The cap plates were only 12 mm thick, which led to a stress concentration at the plate/tube junction such that the plastic capacity of the joint in bending was only 20% of the 273×4.8CHS's capacity. Worse, yielding to form the plastic hinge could start at about 12% of the tube capacity. With long spans it would be quite easy for the riggers to induce such moments in the joints, just by forcing the far end to fit. The difficulties in design of the erection sequence were resolved by (a) designing a "Jim Crow" to straighten a deformed splice, (b) making the permanent cover plates structural members with full penetration butt welds, and (c) ensuring that the client was aware of misgivings about the adequacy of the design.

3.7   *Suggested remedial action*

Eccentric lapped joints are easy to erect, but their use should be limited to single angle braces in tension. For Hollow Section compression struts, double cover plates are simple and easy to design, and have no problems with eccentricities. Erection efficiency is achieved by using common sizes for cover plates, and laying out the joint so that the cover plates can be attached to the strut with a single bolt during erection. Their use should be promoted as the preferred method for connecting such members.

Flush joints for CHS members are frequently specified for aesthetic reasons, but it is clear from these examples designers need reminding that connections should be robust. Perhaps design issues with these could be addressed by expanding Keays (2006), with clear recommendations on what is acceptable.

Figure 13. Fractured CHS butt joint.

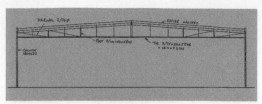

Figure 14. Section on roof truss.

Figure 15. Fractured RHS butt joints.

The local design code was recently changed to make it clear that connection strength should be commensurate with the member's capacity (>30% in simple construction and >50% in rigid construction), not simply adequate for the design loads. This should help resolve an on-going problem.

4 DEFECTIVE WELDED JOINTS

4.1 CHS truss tension chord butt weld

This was a simple butt weld joining two stock lengths in 6.4 mm wall CHS, the bottom chord of a 25 m roof truss. After 10 years in service, the joint fractured as a crane placed material for a new air-conditioning system on the roof. Fortunately, the load remained on the crane hook, and the lapped Z-purlins had sufficient capacity to hold the roof weight without the benefit of the truss bottom chord. Inspection showed the truss chord had fractured round the full perimeter.

This detail shows there was only 2 to 3 mm of penetration in the butt weld, and that there was no bevel preparation of the pipe ends necessary to achieve full penetration.

The welds had been subjected to in-process inspection by an experienced weld inspector. However, he was not present full-time and it seems unlikely he would have viewed the preparation prior to welding.

The architect had called for the welds to be ground flush, but this aesthetic requirement was not repeated on the structural drawings. Based on the authors experience in similar situations, it is most likely the welder was not aware that the weld was to be ground flush, and he might have compensated for not preparing the pipe ends by excess weld reinforcement.

4.2 RHS truss tension chord butt weld

This was a case where one mistake led to another. It was a building within a building – a sound-proof studio for production of a TV series. The original concept for the roof structure was 460UB Rafters and 150UC Columns with a Tie between the Rafter/Column junctions. Space required for air-conditioning and other services led to the Tie being offset 800 mm below the junction.

After installing the insulation, the builder noticed the rafter was sagging 200 mm at mid-span, it became apparent that there was a design defect – the Tie was in the wrong place, and not nearly strong enough for its task. The solution to this problem was to add $2/75 \times 50 \times 5$RHS either side of the original tie, with diagonal braces to gain some truss action with the rafter and relieve bending moments in the Column.

The erector chose to install the 24 m long RHS members in three 8 m lengths, making the butt welds in the final position. The builder then proceeded to install the air ducts, fire-sprinklers, lights and cables. When this work was almost complete, one RHS butt weld fractured, and the building collapsed.

Examination of the joint showed limited penetration, and weld omitted where access was limited.

This would not have happened if the welder had made the RHS/RHS joints before lifting the 24 m length into position. But then it would not have happened if the structural engineer had designed the building correctly in the first place.

4.3 Suggested remedial actions

In both cases, the tradesmen involved made an unforgivable error in not preparing the joints for a full-penetration weld, and their supervisors/inspectors deserve a rap on the knuckles for missing the bad workmanship.

The only suggestion that comes to mind is for designers to insist that the preparation of butt-welded joints should be examined by the supervisor and/or inspector prior to welding the first pass.

Figure 16. Southern star observation wheel, 2009.

Figure 17. Fractured joint at Inner Ring (from Welding-Web.com).

5 TUBE/TUBE CONNECTIONS

5.1 *Fin plate to CHS*

This connection was on a large Ferris Wheel where circumferential members balanced the loads from inclined spokes. These were subjected to stress reversal with each turn of the wheel. The structure had a significant degree of redundancy, as is desirable for one subject to fatigue, and checks had been made to ensure the structure would survive with any one member completely fractured.

Cracks developed in the fin-plate connections to the radial members after only two months in service. The plates were 28 mm, and the tubes 323 × 6.4 and 9.5 CHS. The plates were simply butt-welded to the faces of the CHS, with no consideration given to hoop bending of the CHS wall. A simple alternative would have been to slot a single plate through the CHS wall connecting one intersecting member directly with its counterpart on the far side.

Welds were inspected during fabrication, but there was no inspection of the structure during the commissioning phase. It was noted that the secondary rod bracing was not sustaining its initial tension, but no advice was sought from the designers.

Figure 18. Weak mitre joints at collins place.

The structure has been redesigned and the wheel totally rebuilt with the aim of achieving a 50 year life without visible cracking.

5.2 *Collins square sculpture*

To add to the concerns with wind-induced vibrations and low-strength splice joints, mitre joints and intersections were detailed without stiffening plates. Wardenier (2008) includes charts giving the moment capacity of mitre joints in 273 × 4.8 CHS at about 20% of the tube capacity.

5.3 *Other cases*

The author recalls a simple case of his own design.

A truss joint on the nose of a launching girder suffered excessive deformations at initial loading causing work to stop. Member sizes were 250x9SHS chord and 150SHS web. It subsequently became obvious the loads imposed were 60% higher than they should have been (through using a 100t jack where a 60t jack was nominated).

5.4 *Suggested remedial actions*

The CIDECT Guides, Wardenier (2008) and Packer (2009) comprehensively address the design issues, but are daunting to the uninitiated. Perhaps there is a need for a simpler approach, such as:

- Start a truss design with web members at 75% of the chord size.
- Always include a diagonal plate in a mitre joint.

6 DYNAMIC OVERLOAD

This was a tragic case that led to the death of a basketball player. He was practicing on an open-air court with fabricated steel ring support. He slam-dunked the ball, then grabbed hold of the ring and bounced it. The post failed at its base, and the ring hit his throat.

The post was an 89 × 5 mild steel pipe. By the author's calculation, deflection at the outboard end of the ring under the weight of the backing board and a 75 kg person would have been about 50 mm. It was clearly very flexible. A video of players bouncing the ring was played at the coroner's inquest.

There was evidence of corrosion at the post base; this, along with the significant number of cycles of

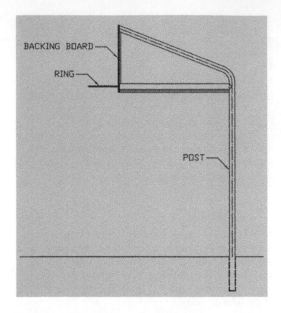

Figure 19.   Basketball hoop support post.

bending stresses at the base of the post, was considered sufficient to explain the failure.

The base of the post was contained by a socket cast into a concrete footing. Such arrangements are common with davits. The author has always added a flange to the davit post immediately above the socket to eliminate combined bending and bearing as a failure mode, but is not aware of any code prescribing such reinforcement, except for saddles on pressure vessels. Perhaps anticlastic bending contributed to this incident.

Playground equipment in Australia is covered by a Standard which limits deflections of a simply supported beam to Span/240. If the post had been sized to comply with this standard, it would have been more than strong enough for the applied loads. Its increased stiffness would have made the "bounce" much less, which would in turn discourage the players from bouncing.

7   CONCLUSIONS

This paper has summarized 20 different structural failures in tubular structures. From the examples seen, it is reasonable to conclude there is not a major problem. A few refinements to the guidance literature should suffice.

The common theme with all the failures discussed here is that every case could have been predicted by an engineer experienced in that aspect of structural engineering. The importance of independent reviews cannot be over-emphasized.

It is pleasing to note that tradesmen and the general public have an ability to note and report defects in structures, but it is concerning that they frequently have an unwarranted trust in the abilities of engineers.

REFERENCES

Keays, R.H. (2006). Field joints for tubulars – Some practical considerations. In Packer & Willibald (eds), *Tubular Structures XI:* 309–316, Taylor & Francis, London.

Keays, R.H. (2008). Structural failures – Lessons for designers from one engineer's experience. *Australian Structural Engineering Conference,* 2008. Melbourne.

Leonards, G.A. (1982). Investigations of Failures. *Journal of the Geotechnical Engineering Division*, ASCE, 108 (GT2):187–246.

Packer, J.A., Wardenier, J., Zhao, X.-L., van der Vegte, G.J., & Kurobane, Y. (2009) Design Guide for Rectangular Hollow Section (RHS) Joints under Predominantly Static Loading. Geneva: CIDECT, 2nd Edition.

Pugsley, A. (1966). The Safety of Structures. Edward Arnold, London.

Robinson, R.W., & Hamilton, J. (1992). A Criterion for Assessing Wind Induced Crossflow Vortex Vibrations in Wind Sensitive Structures. *Health & Safety Executive,* Report OTH 92 379, London.

Wardenier J., Kurobane Y., Packer J.A., van der Vegte G.J., & Zhao X.L (2008), Design guide for circular hollow section (CHS) joints under predominantly static loading. Geneva: CIDECT, 2nd Edition.

Wardenier, J., Packer, J.A., Zhao, X.L. & van der Vegte, G.J. (2010). Hollow sections in structural applications, Delft: Bouwen met Staal.

Tubular Structures XV – Batista, Vellasco & Lima (eds)
© *2015 Taylor & Francis Group, London, ISBN 978-1-138-02837-1*

PREON box – The speedy tool for industrial hall constructions

N. Genge & C. Remde
Vallourec Deutschland GmbH, Düsseldorf, Germany

K. Weynand & J. Kuck
Feldmann + Weynand GmbH, Aachen, Germany

ABSTRACT: The worldwide demand for wide-span industrial buildings is constantly growing, especially in terms of logistics centers and hangars, just to name a few. To meet this trend PREON box, a modular construction system, has been developed. It is an in-house development including a patented steel roof frame system that enables the economic realization of large spans up to 100 meters. PREON box allows combining standardized manufacturing with high flexibility to respond to customer's needs. To make even the planning process more efficient, a software tool has been developed specifically for the design of this system. The software is based on a specification defining the entire load bearing structure (parameterized components, girder types, preferred sections, standardized joints, requirements/limitation given by Eurocode). After entering the system data, which is mainly the dimension of the hall and a few boundary conditions for the loads, the structure is generated automatically. As a result of the subsequent iterative calculation, a complete design note and the bill of material including cost estimation is received. By using the software it is possible to save material and cut process time. The further development of the software is planned until 2016 and includes the automated preparation of workshop drawings, as well as various options for individual adaption.

1 INTRODUCTION

1.1 General

Emerging markets and cross-border trade are becoming faster, and in accordance investments have to be carried out faster, more flexible and more economical. Changing customer requirements presuppose innovative approaches in planning and design. These new methods of approach can benefit from the modular construction system PREON box. It offers economical solutions especially for structures with large spans and high load bearing capacities. This is the case when buildings such as logistic facilities, production halls or aircraft hangars are planned. Beside economic aspects and standardization, especially flexibility, speed and quality will be considered here.

1.2 Motivation

To meet the customer's needs the PREON system was created. From 2005 on the PREON system has been realized in several projects (Josat 2010). One of them was the construction of a new steel mill in combination with a subsequent rolling mill in Jeceaba/Brazil. The facility consists of 7 bays, a span of 35 and 50 meters respectively, and a total floor space of 224,000 square meters. The length of the building is 700 meters. The main attention on this project was the timeline. Some 15,000 tons of hot rolled seamless tubes were transported in eleven ship loads to Rio de Janeiro. A number of local steel construction companies welded the steel sections together on the basis of the construction drawings elaborated in Brazil to create the PREON modules, which were then transported to the construction site as needed in accordance with the coordinated erection planning. Although the static calculation was carried out properly some of the used hollow sections were chosen according to their availability, due to the time pressure in the course of the project. In the end the project was completed successfully. However, a certain potential of optimization was seen in the design phase, which led to the idea to develop a particular design tool for the PREON system.

1.3 Objectives

The present paper describes the scope and the actual status of the development of this design tool called "PREON designer". Based on a detailed system specification, see section 3, the software tool PREON designer consists of a generation module which builds a complete detailed structural system based on few parameters like number of bays, bay width, length and height of a structure. For this generated structure, all cross section are evaluated in an automatic design loop as described in section 5.2, and the cost of the structure are immediately estimated, see section 6.

The aim is to provide a software tool which simplifies the design of hollow section structures and which gives economic solutions in overall height and weight especially for large span structures with high load bearing capacities, where plate girders are no more economic.

2 PREON BOX STRUCTURES

What is a PREON box structure?

A PREON box structure is a complete hall, or a substructure of it, (mainly) composed of MSH sections (hot finished structural hollow sections according to EN 10210). A substructure may be a complete frame or just a girder. The main structural elements are lattice girder structures – used for main and secondary girders or, if required, for columns. Figure 1 shows a typical PREON box structure.

The patented PREON roof girder (Josat 2010) is a combination of two different types of girders. In the low shear loaded middle area of the girder a Vierendeel part with variable bracing distance is used. At the edges of the girder K-trusses are provided. One advantage of the Vierendeel part in the middle is the economic orthogonal cut of the braces. Furthermore, the Vierendeel part serves as the "compensation part", so that the K-trusses are more flexible in their length and hence, can be optimized in view of geometrical parameters such as brace angles. This yields in a structure which

Figure 1.   A typical PREON box structure.

is cost-effective thanks to its adaptation to the distribution of forces and thanks to its large range of spans. The majority of the components are hot finished square and rectangular hollow sections. With this, the system represents a cost-effective and versatile solution for industrial building construction.

3 SYSTEM SPECIFICATION

In order to design an industrial hall in an automatic way by means of a computer program, it is essential that a detailed technical description of all components is available. Therefore, in a first step, a so-called system specification has been developed. The system specification is a kind of knowledge base and it describes all technical details of the PREON box system like geometry, generation and design algorithms as explained hereafter more in detail.

PREON box is a modular construction system. In contrast to other typical construction systems, all structural elements, such as a complete girder or a connection detail, are parameterized components. This means that a component, e.g. a plate, is not specified for example by a particular fixed length, width and thickness, rather than by the parameters L, B and T, examples are seen in Figure 2 for a girder and a girder support and in Figure 3 for a typical column for high rise halls with crane. The advantage of this approach is the fact that, on one side, all components are standardized and can therefore be implemented as predefined types in the design software and, on the other side, the actual sizes or dimensions are very flexible to be adapted to the needs of the client.

Because the content of the specification should be implemented in a design software, it is important that also all requirements and limitations for the use of each element is specified, e.g. minimum and maximum girder spans, maximum crane loads, etc. In other words: the system specification says exactly what is possible and what is not possible.

For each element such as girders, columns or joints, the system specification provides one or more solutions, i.e. different types or details of an element. For a particular construction, the selected type or detail

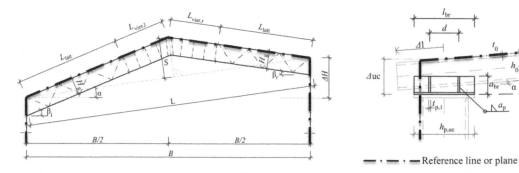

Figure 2.   Examples of parameterized components (PREON girder and girder support).

depends either on a pre-design procedure, see section 5.1, or on the choice of the user. But the system specification includes not only all possible types of components with parameterized dimensions. It contains also algorithms for the automatic generation of the structural system. This is explained more in detail in section 4.

The present version of the system specification contains all structural components for the primary steel structure including girders and columns, purlins, horizontal and vertical bracings and all connection details. Crane runway girder and its supporting structure are in development and will be available by end of 2015.

4 DESIGN TOOL

The computer program developed for the design of PREON box structures is called PREON designer. With the help of this design tool, the whole process from the definition of the global layout of a steel structure till the cost estimation of the steel structure can be done with a few mouse clicks. The user will perform 3 steps:

Firstly, some information related to the location of the site like its height above sea level, wind and snow zone must be given. These data will be used to generate climatic actions like wind loads and snow loads. Imposed loads and self-weight of the roofing and cladding must be provided by the user. Self-weight of the steel structure is considered automatically. In the

section database, individual sections are marked as so-called preferred sections. When the user selected the option to use "preferred sections only", this could lead to a slight increase of some section sizes during the design process, but due the better availability or lower prices, the total costs of final structure could decline. Also, the user may limit the list of available section in order to avoid the use of specific sections.

In a second step, the user specifies the global layout of the steel structure by means of number of bays, bay height and width, distance of the frames, inclination of the roof, etc. Through those data, the structure is defined by so-called reference lines or planes respectively, see Figure 2 and Figure 3. As an example, a hall with two bays is shown in Figure 4.

After the data input, the user will run the generation and design process.

This last step will start with the automatic generation of the structure. Based on the initial choice of elements evaluated in the pre-design, see section 5.1, all individual members of the primary steel structure are generated. To determine the correct position, so-called reference lines or planes respectively, see Figure 2 and Figure 3, are used. Requirements for connections details are directly taken into account. For example, for the generation of all lattice girders, the positions of braces are chosen in such a way that all K joints will be, if possible, gap joints. The requirements for the size of the gaps specified in EN 1993-1-8 are directly considered. Also eccentricities in the joints, if required, are taken into account when modeling the structure. Using gap joints will allow easy and cost effective fabrication.

The generated system for the example shown in Figure 4 can be seen in Figure 5.

Then the automatic design will be executed as described in detail in section 5.2. In an iterative procedure, optimized sections will be evaluated. Verification of the steel structure is made according to the Eurocodes.

When the design procedure is successfully terminated, the software will generate a full design note and a material list. Based on this bill of material, cost estimation is made. More details about the cost estimation are given in section 6.

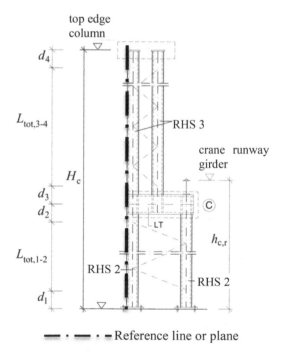

Figure 3. Examples of parameterized components (column with crane support.

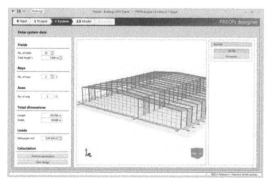

Figure 4. PREON box structure defined by reference lines.

61

Figure 5.   Generated PREON box system.

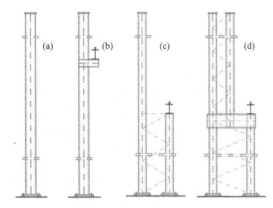

Figure 6.   Example of column types.

5   DESIGN OF THE STRUCTURE

5.1   *Pre-design*

Due to the automatic and efficient design procedure described in the next section, the pre-design of the structure in the classical sense in not really that important. However, a realistic guess of the section sizes will reduce the number of iteration in the design procedure and hence, reduces calculation time.

A more important aspect is a reasonable choice of the type of components, for example the column type (Figure 6) being only a single section type (a) or (b) or a lattice girder type (c) or (d). The type to be used depends on several parameters like height of the column and roof and crane loads, etc.

This is important because the type of component will normally not change during the design procedure. This must be done manually in case the design procedure cannot find a solution.

To determine realistic assumption for the selection of the types of components, parameter studies are undertaken to classify typical structures. Basis for the classification are the user defined parameters which define the global layout of the steel structure, such as bay height and width, distance of the frames, crane loads, etc. As a result, a frame can be classified for example as light, medium of heavy. Based on this classification, appropriate types of column can be selected in the pre-design. This selected type is used to generate the model of the steel structure in the next step.

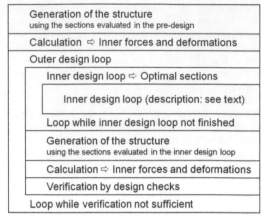

| Generation of the structure |
| using the sections evaluated in the pre-design |
| Calculation ⇨ Inner forces and deformations |
| Outer design loop |
| Inner design loop ⇨ Optimal sections |
| Inner design loop (description: see text) |
| Loop while inner design loop not finished |
| Generation of the structure |
| using the sections evaluated in the inner design loop |
| Calculation ⇨ Inner forces and deformations |
| Verification by design checks |
| Loop while verification not sufficient |

Figure 7.   Structogram of the general design procedure.

5.2   *Design*

To design manually, in an economical way, a lattice girder structure consisting of hollow sections could be a time consuming work. Why? Because the individual members (braces and chords) cannot be design independently due to the fact that the resistance of the joints depend mainly on the cross section properties of the connected members. In fact, this is the reason why in practice, where the design of the members is often made by the structural engineer while the design of the joints is made by the steel constructor, it is not easy or even not possible to find an economic solution for the global layout of a hollow section structure. To enable the implementation of an efficient design process in a computer program Feldmann + Weynand developed a general procedure that delivers optimized solutions for different types of structures, i.e. not only for the patented PREON girder but also for other typical structures (e.g. different types of lattice girders used as beam or columns, frame structures).

The flow chart in Figure 7 shows the general design procedure.

The first steps of the design procedure are the automatic generation of the structure and the calculation of internal forces and moments. The generation of the structure is described in detail in section 4. The calculation of the internal forces and moment is performed by a classical global frame analysis respecting the provisions given in the Eurocodes. The provisions concern for example loads, load case combinations or imperfections. As a result of these two initial steps, beside the design values of forces and moment acting on the members and joints, deformations of the pre-designed structure are evaluated.

After this, the so-called outer design loop, see Figure 7, can be started. The outer design loop will immediately start the so-called inner design loop. This inner design loop is the most relevant step of the procedure with regard to the challenge to design automatically structural systems like lattice girders.

In the inner design loop, the members of the structure have first to be classified in different types. This

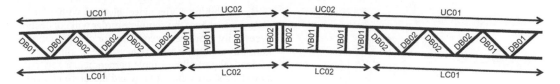

Figure 8.   Different member types in a structure (UC = upper chord, DB = diagonal brace, VB = vertical brace, LC = lower chord).

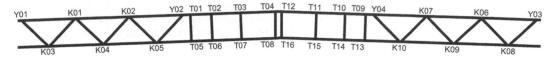

Figure 9.   Joints in a Structure (Y = Y joint, K = K joint, T = T joint).

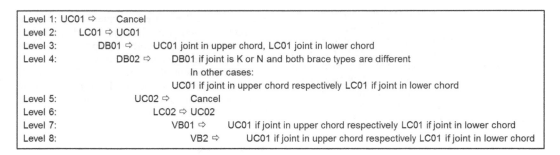

Figure 10.   Table of dependencies for a structure.

step avoids that for example a chord, which could be fabricated as one continuous member, will be separated into different parts with different cross sections. Figure 8 shows this classification using the example of a PREON girder. For each of these member types an appropriate cross section will be evaluated at the end of the inner design loop. Appropriate cross section means here that, from a list of all available cross section, a set of cross sections is selected which of course fulfil the cross section checks and members checks respectively for all member types but also fits with the restriction related to the design of the joints, for example the brace to chord with ratio β.

In other words: In case of a lattice girder made of hollow sections, special joint checks have to be done in addition to the member checks (cross section checks and stability checks). In these joint checks, the interaction between the members depending on the particular joint configuration (e.g. K joint, Y joint, T joint) is considered. Figure 9 shows the joints which have to be checked for the example.

As said before, the joints have not only to resist the applied inner forces and moments. In addition special requirements concerning geometry have to be fulfilled. Hence there are a lot of possible combinations of cross sections for the member. To select the most efficient cross sections a table of dependencies has to be elaborated. Figure 10 gives an example for this.

Each level represents a member type (e.g. Level 1 represents UC01). In a recursive procedure the checks have to be done level by level. In each level the best possible cross section for the related member type has to be found. Therefore the possible sections for the considered member are sorted, from the best economic section to the worst. Then member checks are performed for the different cross sections. If the verifications of all member types connected to a particular joint are valid, the joint itself has to be checked. This happens in Level 3, 4, 7 and 8 in this example. For the verification of the joints the software COP (Weynand et al. 2010) is used internally.

If the checks in one level are not valid for any cross section, the algorithm jumps back to that level which is related to the member type indicated by "⇒" in Figure 10. "Cancel" means: no solution found for this particular cross section or combination of cross sections. In the case that all checks from Level 1 up to the last level are valid, an optimal design has been found for the generated structure and the inner design loop is finished.

In order to decide about an economic solution for a certain set of cross sections, a material list is created for each case. Based on this material list the weight of the structure or the cost of the structure (see section 6) can directly be estimated. Typically, the lowest total weight of the steel structure will be used to find the most economical solution. However, the steel price must not be

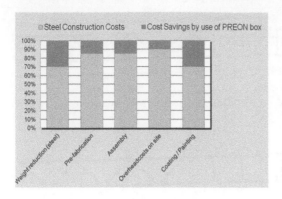

Figure 11. Cost savings with PREON box.

By combining a Vierendeel element with a lattice truss the weight of the complete girder will not decline necessarily. However, accounting for cheaper fabrication, the total costs may reduce. The T joints of the Vierendeel part allow simple perpendicular saw-cuts, as well as short welding lines and therefore less filler material and welding time. The position of the linking head plates of the separate modules is chosen skillfully, in order to avoid welding on the construction site. This shortens fabrication time and simplifies erection. Due to the smaller surface of hollow sections compared to open sections, costs for coating can be reduced. Figure 11 shows the savings reached in one of the executed projects.

direct proportional to the weight. Also, if other factors like fabrication and erection costs are taken into account, the most economical solution may not necessary the lowest weight. Those parameters may be taken into account by cost estimation tool, see section 6.

With the sections evaluated in the inner design loop a new generation of the structure, a new calculation of the internal forces and moments and a verification of the structure will be done. If the verification is sufficient, the design of the structure is finished.

6 COST ESTIMATION

During and after the design, see section 5.2, a material list is created including weight and surface of the cross sections. A simple software tool called PREON costor (Cost estimator) is developed which combines the material list and a price list.

A price list is a set of parameter where the user specifies for each individual cross section (or as default values for all cross section) a weight dependent price and optionally a surface dependent price. The user can create one or more price lists. Dependent on the situation, the prices can include material costs, fabrication and erections costs, surface treatment like corrosion protection, or just account for the buying of the sections.

As long as realistic prices are available, the cost estimation for a structure is available as soon as the design, see section 5.2, is done.

7 ECONOMIC ASPECTS

From 2005 until 2013 almost over 20 different projects have been realized with PREON. During this time the system has proved its competitiveness with regard to efficiency, especially in spans larger than 30 meters.

8 SUMMARY AND CONCLUSION

This paper describes the enhancement of an existing construction solution by developing dedicated software. This design tool enables to respond to the need of the quickly changing industry sector.

The paper presents the technical benefits of PRON box as well as some studies made on the economic aspects by comparing traditional design solution with the new PREON box system.

REFERENCES

EN 1993-1-1. 2005. *Eurocode 3: Design of Steel Structures – Part 1–1: General rules and rules for buildings.* Brussels: CEN.

EN 1993-1-8. 2005. *Eurocode 3: Design of Steel Structures – Part 1–8: Design of joints.* Brussels: CEN.

Weynand, K.; Kuck, J.; Herion, S. 2014. *Systemspezifikation PREON box – Vallourec Spezifikation für Hallentragwerke aus MSH Profilen.* Internal documentation, Vallourec Deutschland GmbH.

Josat, O. 2010. *PREON – the flexible standard in hall construction.* Tubular Structures XIII – Young (ed), Proceeding s of ISTS 13, University of Hong Kong.

Weynand, K.; Kuck, J.; Oerder, R.; Herion, S.; Fleischer, O.;Josat, O.; Schneider, M. 2010. *Design tools for hollow section joints.* Tubular Structures XIII – Young (ed), Proceeding s of ISTS 13, University of Hong Kong.

Composite tubular structures

Tubular Structures XV – Batista, Vellasco & Lima (eds)
© 2015 Taylor & Francis Group, London, ISBN 978-1-138-02837-1

Study on the cracking behavior of Concrete Filled Steel Tube (CFST) for tall bridge piers subjected to horizontal cyclic loading

M. Zhou, X.G. Liu, J.S. Fan & J.G. Nie

Key Laboratory of Civil Engineering Safety and Durability of China Education Ministry,
Department of Civil Engineering, Tsinghua University, Beijing, China

ABSTRACT: Composite tubular structures are now widely used in bridge engineering field in China. Among them, the Concrete Filled Steel Tube (CFST) bridge pier for large-span tall-pier rigid frame bridges are popular, especially in mountain regions. This paper reports the concrete cracking behavior of the CFST for tall bridge piers under horizontal cyclic loading. A scale model test was carried out. The cracking behavior of the CFST was observed, involving the concrete crack spacing, concrete crack width and the steel strain distribution along a length of the crack spacing. A theoretical formula for calculation of the concrete crack spacing of CFST was proposed. It was found that the concrete crack spacing is dependent on the size of the core concrete, but not the steel ratio.

Keywords: Concrete filled steel tubes (CFST); tension; crack; crack spacing; crack width

1 INTRODUCTION

The concrete filled steel tube (CFST) is widely used in composite structures, including buildings and bridges. The in-filled concrete cracks when the CFST is subjected to tension. A few of experimental studies are reported in the literature (Han et al. 2011). Rare theoretical study has been carried out in the previous researches.

In this study, the cracking behavior of the CFST in tension is mainly studied. A theoretical formula was proposed for the prediction of the crack spacing. A scale model test of the CFST pier was carried out. The cracking behavior of its CFST column was investigated, including the crack spacing, residual crack width and steel strain distribution. The test result of the crack spacing was compared with the prediction of the proposed formula.

2 FORMULAS

Figure 1 illustrates the stress sharing of the CFST in tension. Lm is the average crack spacing. Considering the equilibrium relationship along a length of the crack spacing gives

$$\bar{\tau} \cdot P_c \cdot \frac{L_{max}}{2} = f_t \cdot A_c \qquad (1)$$

where $\bar{\tau}$ is the average bond stress along a length of the crack spacing, f_t is the concrete tensile strength, S_c is the perimeter of the concrete, A_c is the area of the

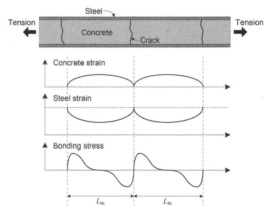

Figure 1. Stress sharing of the CFST in tension.

concrete, L_{max} is the maximum crack spacing. Then, Equation (1) leads to

$$L_{max} = 2 \frac{f_t}{\bar{\tau}} \cdot \frac{A_c}{P_c} \qquad (2)$$

For a simple solution, it is assumed that the distribution of the bond stress along the length of the crack spacing is linear. This gives

$$\bar{\tau} = \frac{\tau_{max}}{2} \qquad (3)$$

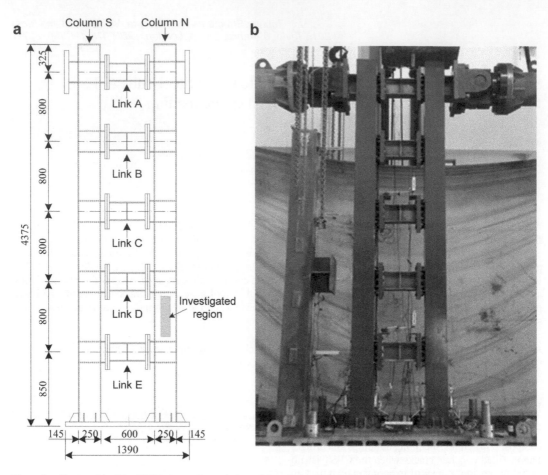

Figure 2. Test model of the CFST pier. a, General view of the specimen (unit: mm). b, Picture of the model.

where τ_{max} is the maximum bond stress, taken as $\tau_{max} = f_t$. Substituting Equation (3) into Equation (2) leads to

$$L_{max} = 4 \cdot \frac{A_c}{P_c} \tag{4}$$

For CFST with rectangular cross section, Equation (4) gives

$$L_{max} = 2 \cdot \frac{ab}{a+b} \tag{5}$$

where a and b is the length and width of the rectangular cross section, respectively. For CFST with square cross section, it gives

$$L_{max} = a \tag{6}$$

where a is the side length of the square cross section. Noting that $L_{min} = L_{max}/2$, it gives

$$\frac{f_t}{\tau} \cdot \frac{A_c}{S_c} \leq L_m \leq 2\frac{f_t}{\tau} \cdot \frac{A_c}{S_c} \tag{7}$$

where L_{min} is the minimum crack spacing. Substituting Equation (3)(5)(6) into Equation (7) leads to

$$\begin{cases} \dfrac{ab}{a+b} \leq L_m \leq 2 \cdot \dfrac{ab}{a+b}; \text{Rectangular cross section} \\ \dfrac{a}{2} \leq L_m \leq a; \text{Square cross section} \end{cases} \tag{8}$$

Equation (4) and (8) imply that the crack spacing of the CFST in tension is independent with the steel ratio or steel strength. It is only related to the size of the in-filled concrete. This characteristic is very different from that of the reinforced concrete (RC) structures, whose crack spacing is highly influenced by the reinforcement ratio.

3 SCALE MODEL TEST

A scale model test of the CFST pier for tall bridges under horizontal cyclic loading was carried out. The cracking behavior of its CFST column was investigated.

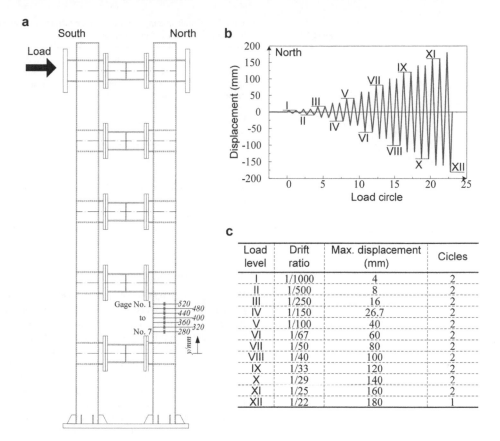

Figure 3. Measuring devices and loading procedure. a, Arrangement of measuring points. b, Loading procedure. c, Loading detail.

3.1 Test model

Figure 2 shows the overall layout of the test model. The scale ratio is 1:6. The model consists of three parts: CFST columns, steel links and the bottom plate. The two CFST columns, column S and column N, are coupled by five steel links, link A to E. Columns are fixed to the bottom plate, and the bottom plate is fixed to the test ground. The pier model is 4375 mm in height and 1390 mm in width. The bottom plate is 50 mm in thickness, and the spacing of the steel links is 800 mm. The cross section of the CFST column is 250 mm × 250 mm with wall thickness of 6 mm. The horizontal load is applied at the height of link A, as shown in Figures 2a, b. During the cyclic loading, two CFST columns will be subjected to tension alternately. Cracking of the in-filled concrete will take place when the tension strain is large enough, especially near the foot of the column. With this concept in mind, the region of column N between link D and E is selected as the investigated region, as shown in Figure 2a.

3.2 Measuring arrangement

Since the crack width or crack development can hardly be directly observed during the test, the steel strain is measured to reflect the cracking behavior of the in-filled concrete. Figure 3a gives the arrangement of the measuring points. The points are centralized in the investigated region. Seven strain measuring points are placed in a line along the central axis of column N. The points are numbered from 1 to 7 from top down. Point 4 is located at the center of the investigated region, that is, 400 mm offset from the horizontal central axis of link D and E. The spacing of the measuring points is 40 mm. Thus, the measuring region is 240 mm in length. Since the crack spacing of the column is predicted as 119–238 mm based on Equation (8), the measuring region covers a length of crack spacing.

3.3 Loading procedure

The horizontal load is applied at the height of link A (Fig. 3a). Two actuators are used for the loading (Fig. 2b). Twelve loading level are designed according to the drift ratio (max. displacement divided by loading height), as illustrated in Figures 3b, c. Each loading level contains two loading circles except for the last one (Fig. 3c). North is set to be the positive direction of the loading.

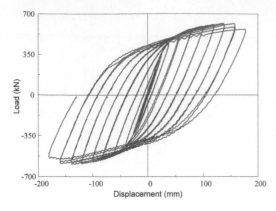

Figure 4. Horizontal load-displacement curve of the pier.

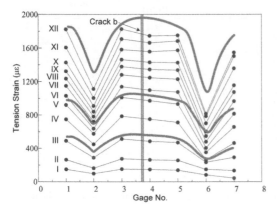

Figure 5. Longitudinal strain distribution in the investigated region.

4 RESULTS AND DISCUSSIONS

Figure 4 gives the horizontal load-displacement relationship of the test model. It is shown that the pier remains elastic for the first four loading levels. Obvious plastic deformation appears since the fifth loading level. The maximum load increases continuously from the first to the eleventh loading level. It descends at the last loading level. In the selected investigated region, the cracking behavior of CFST column will be studied through following aspects.

The strain distribution, crack spacing and residual crack width.

4.1 Strain distribution

Figure 5 gives the tested tension strain distribution of different loading levels in the investigated region. Three bold lines illustrate the tendency of the distribution. The strain tendency shows a wave shape characteristic. It is found that the crest appears between Gage 3 and 5. This implies that a crack of in-filled concrete may emerge in the region between Gage 3 and 5. Furthermore, two troughs exist in the distribution

Table 1. Crack spacing gained from the test, strain distribution, and formula predicted results.

Crack spacing name	Test result mm	Strain distribution result mm	Predicted result using Equation (8) mm
Spacing a-b	216.3	160	119–238
Spacing b-c	281.6	160	119–238

and their distance is four times of the gage spacing, that is, 4×40 mm. This implies a crack spacing of 160 mm, which is between the minimum and maximum crack spacing predicted by Equation (8), 119 mm and 238 mm.

4.2 Crack spacing

From Figure 4, it can be told that the residual deformation of the test model is over 100 mm to the south after the test. This ensures the observation of cracks in the investigated region, which is located on the northern column. So after the test, the eastern side steel plate was cut off to expose the in-filled concrete, as shown in Figure 6. Three obvious cracks are found, named Crack a, b and c from top to bottom. Crack a and c locate at the top and the bottom of the investigated region, respectively. Crack b locates at the middle of the region. The crack spacing is measured along the central axis of the column, as illustrated in Table 1 and Figure 6a. Spacing of Crack a and b is 216.3 mm, larger than the value implied by the strain distribution result (160 mm), but within the range predicted by Equation (8), that is, 119–238 mm. Spacing of Crack b and c is 281.6 mm, exceeding the predicted range. This may be caused by the influence of the joint area of Link E and the column, since Crack c lies very close to the top plate of the joint area. Besides, Crack b cross the central axis at the point of $y = 406.6$ mm based on the coordinate system determined in Figure 3a. This tells that the crossover point of Crack b and central axis locates near Gage 4 on the side of Gage 3, as shown in Figure 5. This observation agrees with the strain distribution result, which predicts a crack existing between Gage 3 and 5 as discussed before.

4.3 Residual crack width

On the basis of the automated assessment of cracks on concrete surfaces using adaptive digital image processing (Liu et al. 2014), Crack b is further assessed. Figure 7 illustrates the details of Crack b. Figure 8 gives the result of the residual crack width. It is shown that the widest part of Crack b reaches 0.24 mm, and most part of Crack b exceeds the width of 0.1 mm. This is a preliminary application of the digital-image-processing crack assessment method. It is found that this method is effective in improving the measuring accuracy and reducing the measuring labor.

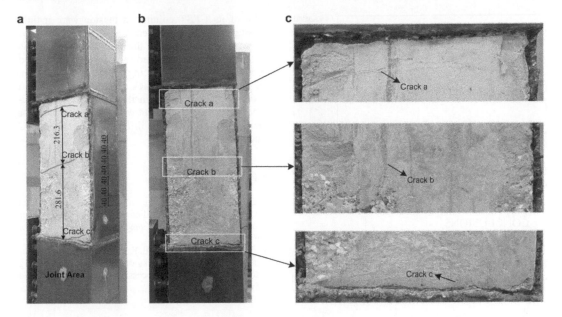

Figure 6.   Concrete cracking result in the investigated region. a, Crack distribution (unit: mm). b, Elevation of the crack distribution within the investigated region. c, zoom-in picture of the cracks.

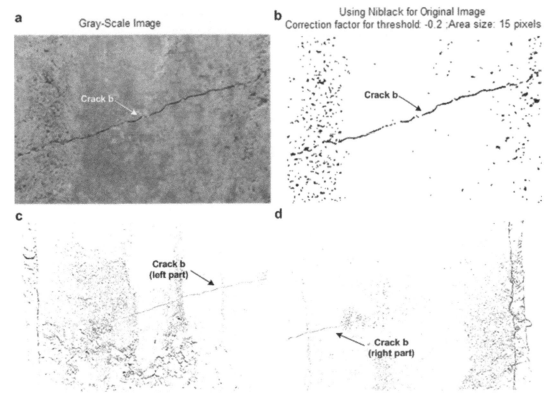

Figure 7.   Details of Crack b using digital-image-processing crack assessment method. a, Grey scale picture of Crack b. b, Crack b. c, Left part of Crack b. d, Right part of Crack b.

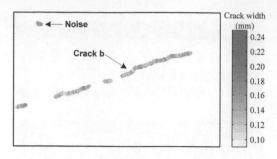

Figure 8. Crack width distribution of Crack b using digital-image-processing crack assessment method.

5 CONCLUSIONS

This paper proposed a theoretical formula for predicting the crack spacing of CFST in tension. A scale model test of CFST pier was carried out. The steel strain distribution, crack spacing and residual crack width of its CFST column in tension were mainly investigated. The measured crack spacing matches well with the prediction of the proposed formula. It is concluded that the proposed formula can give a good prediction of the crack spacing of the CFST in tension. It is also found that the crack spacing of the CFST in tension is mainly determined by the size of the in-filled concrete, independent with the steel strength or steel ratio, which is very different from that of RC structures.

REFERENCES

Han, L. H., He, S. H. & Liao, F. Y. 2011. Performance and calculations of concrete filled steel tubes (CFST) under axial tension. *Journal of Constructional Steel Research*, 67: 1699–1709.

Liu, Y. F., Cho, S. J., Spencer, B. F. Jr. & Fan, J. S. 2014. Automated assessment of cracks on concrete surfaces using adaptive digital image processing. *Smart Structures and Systems*, 14(4): 719–741.

Tubular Structures XV – Batista, Vellasco & Lima (eds)
© 2015 Taylor & Francis Group, London, ISBN 978-1-138-02837-1

Circular double-tube concrete-filled tubular columns with ultra-high strength concrete

M.L. Romero, A. Espinós & A. Hospitaler
Universitat Politècnica de València, ICITECH, Valencia, Spain

J.M. Portolés & C. Ibañez
Universitat Jaume I, Department of Mechanical Engineering and Construction, Castellón, Spain

ABSTRACT: This paper presents the results of an experimental campaign where the buckling resistance of twelve double-tube concrete filled steel tubular (CFST) columns is obtained. While there are some papers in the literature which have investigated concentrically loaded stub columns of such typology, no investigations on slender columns have been found up to date. The tests presented in this paper are the preliminary results of an extensive experimental campaign (28 tests) where the effects of two parameters are analyzed: strength of concrete (normal strength and ultra-high strength concrete) and the ratio between the thicknesses of the outer and inner steel tubes. The buckling load at room temperature of the specimens is analyzed in terms of the strength of concrete and the appropriate distribution of the steel in the composite column. By maintaining the same total area of steel, two combinations are initially studied: 'thick outer tube-thin inner tube' or 'thin outer tube-thick inner tube'.

1 INTRODUCTION

Concrete-filled steel tubular (CFST) members are being increasingly used worldwide as composite columns in new building developments. In addition, the use of high strength concrete (HSC) is becoming popular thanks to the reduction of its technological costs, in such a way that even ultra-high strength concrete (UHSC) has been introduced recently. The use of this material presents great advantages, mainly in members subjected to considerably high compressive axial forces, as it occurs in columns of high-rise buildings and bridge piers. Nevertheless, the usage of high strength materials in columns reduces their cross-section while increasing the member slenderness, with the consequent buckling problems and detrimental effect to the fire resistance. In addition Eurocode 4 (CEN 1994) is still limited to concrete grades up to 50 MPa.

For this reason, new innovative solutions are needed in order to guarantee that this type of columns produce the appropriate structural response.

In this paper, the initial results to characterize a novel type of cross-section (double-tube) are presented, which solves some of the previously referred problems and shortcomings, being possible to employ different concrete grades in the inner core and outer concrete ring. This would help to prevent spalling problems associated to UHSC by being subjected to lower temperatures in those parts of the column where it is found more useful. At the same time, the presence of an internal steel tube with a lower temperature would help to resist the second order effects in slender columns.

Up to now the main work on this topic has been performed by the groups of Prof L.H. Han, Prof. X.L. Zhao and co-workers, where several papers have been published at room temperature and fire about the so-called "double-skin" CFT columns (Zhan and Han 2006, Elchalakani et al 2002, Lu et al 2010, Zhao et al 2010, Lu et al 2010, Huang at al 2010, Lu et al 2010b, Tao et al 2004, Tao and Han 2006), where the inner CHS is empty.

However, other authors have proposed the solution to embed massive steel sections inside the core (Neuenschwander et al 2010, Schaumann and Kleibömer 2014) maintaining a high axial capacity for multi-story buildings.

In turn, Liew et al (2011) have recently tested a new typology of cross sections called "double-tube" where the inner tube is filled also with concrete.

This paper presents the results of an experimental program where both double-skin and double-tube slender concrete-filled steel tubular columns have been tested and compared. Given the reduced number of experimental results found in the literature, the main objective of this paper is to compare the behaviour of such innovative cross-sections subjected to axial compression.

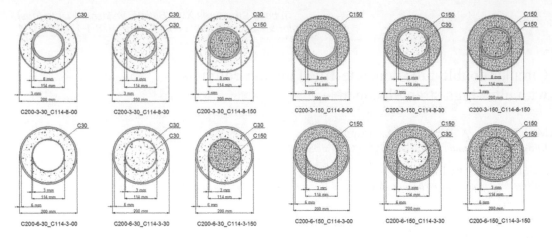

Figure 1. Cross-sections series 1.

Figure 2. Cross-sections series 2.

2 EXPERIMENTAL CAMPAIGN

2.1 General

The authors have performed several experimental campaigns to study the buckling resistance at room temperature of slender CFST columns with circular, square, rectangular or elliptical cross-sections (Portolés et al. 2011, 2013, Hernández-Figueirido et al 2012. However, CFDST sections have not been tested until now.

The tests presented in this paper are the initial results of an extensive experimental campaign (28 tests) where the effects of two parameters are analyzed: strength of concrete (normal strength and ultra-high strength concrete) and the ratio between the thicknesses of the steel tubes. However, up to date only 12 experiments have been tested at room temperature.

The sections were selected so as to maintain the total steel area in all the tests (±4%), being equal to that of a CFST column previously tested by the authors. Eight of the column specimens were filled with concrete in the inner core (normal or ultra-high strength concrete, i.e. double-tube), while the other four columns were only filled in the outer concrete ring (i.e. double-skin). The dimensions of the typical cross-sections can be seen in Figure 1 and Figure 2. Nominal plain C30 and C150 grade concretes and steel S355 were used, although the real strengths obtained from the material tests are summarized in Table 1.

It is worth noting that the first three sections from each series had a thick inner tube and a thin outer tube, while in turn the other three sections had a thin inner tube and a thick outer tube.

This variation was decided given the interest to investigate what is more valuable from the practical point of view: to use a thin or thick outer tube. In a fire event it can be expected that the first group of sections (i.e. thick inner tube) perform better, while the second group of sections (i.e. thick outer tube) should be able to sustain higher loads at room temperature. This opposed behaviour needs to be confirmed, as an equilibrium in the design process should be reached.

Table 1. Detail of specimens.

Name	D_{ext} mm	t_{ext} mm	$f_{y,ext}$ MPa	$f_{c,ext}$ MPa	D_{int} mm	t_{int} mm	$f_{y,int}$ MPa	$f_{c,int}$ MPa	N_{exp} kN	N_{EC4} kN
Series 1										
NR1	200	3	300	36	114	8	377	00	1418	1674
NR2	200	3	332	45	114	8	403	42	1627	1990
NR3	200	3	272	43	114	8	414	134	1774	2213
NR4	200	6	407	35	114	3	343	00	1644	1912
NR5	200	6	377	44	114	3	329	40	1964	2156
NR6	200	6	386	43	114	3	343	123	2076	2543
Series 2										
NR7	200	3	300	138	114	8	377	00	2571	2567
NR8	200	3	332	139	114	8	403	44	2862	2728
NR9	200	3	272	139	114	8	414	141	3077	2893
NR10	200	6	407	137	114	3	343	00	2612	2833
NR11	200	6	377	139	114	3	329	45	2793	2952
NR12	200	6	386	140	114	3	343	140	3093	3305

2.2 Column specimen and set-up

All the specimens were manufactured at Universitat Politècnica de València (Spain) and tested later at Universitat Jaume I in Castellón (Spain). The buckling length of the columns was 3315 mm in all tests being tested under pinned-pinned (P-P) end conditions where a $300 \times 300 \times 15$ mm steel plate was welded to both ends of the columns. All the specimens were tested in a 5000 kN testing frame in a horizontal position, Figure 3. More details of the test setup can be found in Portolés et al. 2011. Linear variable displacement transducers (LVDTs) were used to measure the deflection at five points along the column (0.25L, 0.375L, 0.5L, 0.625L and 0.75L). Once the specimen was put in place, displacement control tests were carried out in order to measure the post-peak behaviour.

2.3 Material properties

2.3.1 Steel tubes

Circular steel hollow sections were used in the experimental program for both the inner and outer steel tubes.

Figure 3. Test set-up.

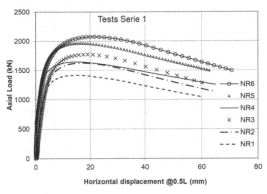

Figure 4. Load versus mid-span displacement in Series 1.

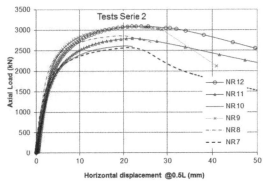

Figure 5. Load versus mid-span displacement in Series 2.

The real yield strength of the hollow steel tubes (f_y) was obtained for each column specimen by performing the corresponding coupon test. The modulus of elasticity of steel was set following the European standards with a value of 210 GPa.

2.3.2 Concrete

In this initial experimental program, two types of concrete were used, with nominal compressive strengths of 30 MPa and 150 MPa. The concrete batches were prepared in a planetary mixer. In order to obtain the real compressive strength of concrete (f_c), sets of concrete cylinders were prepared and cured in standard conditions during 28 days. All samples were tested on the same day as the column was tested, as shown in Table 1.

2.4 Experimental results

The maximum axial load of all specimens (N_u) is listed Table 1 and the axial force versus mid-span displacement response for the tests are presented in figure 4 and figure 5.

2.4.1 Behaviour of tests from Series 1

The general trend of the curves results as expected: when the thicker tube is located in the outer part of the section, the maximum load increases although the load is concentric and the total steel area is approximately constant. This is due to the increase in the moment of inertia of the total section, which reduces the influence of the second order effects. Besides, if the

inner CHS is filled also with concrete, the maximum load also increases. However, it must be highlighted that similar maximum loads are obtained for the tests C200-3-30_C114-8-30 and C200-6-30_C114-3-00, where only small differences are found due to the diverse values of the steel yield strength.

Consequently, it can be stated that a double-tube column with the thicker tube in the inner position can reach almost the same buckling load (1627 kN), Table 1, as a double-skin CFT column with the thicker steel tube in the outer position and without inner concrete (1644 kN).

Moreover, it is worth noting the reduced effect of using UHSC in the inner core in comparison with NSC (i.e. tests C200-3-30_C114-8-150 and C200-3-30_C114-8-30). In fact, an increase of only a 9 % in the load bearing capacity of the column is found despite the elevated cost of UHSC, which is approximately 5 times the cost of NSC.

2.4.2 Behaviour of tests from Series 2

Again, the general trend is observed and as better concrete is poured in the inner core as higher is the axial load. However, the behavior of tests with the external ring of UHSC is somewhere surprising, as null effect is observed if the thicknesses of the steel tubes are varied from outer to inner. Tests NR7 is equal to NR10, while NR8 = NR11 and NR9 = NR12 respectively.

From these results it is inferred that the stiffness of the composite column is governed by the UHSC outer ring while the steel tubes has lower effect.

3 DISCUSSION ON EUROCODE 4

In this section, the simplified method of design in Clause 6.7.3 of EN 1994-1-1 (CEN 2004) will be assessed against the results of the room temperature tests carried out in this experimental program. For members in axial compression, the design axial buckling load at room temperature can be obtained through the corresponding buckling curves according to Clause 6.7.3.5(2) from Eurocode 3. It defines that the maximum axial capacity of a CFST column ($N_{b,Rd}$) should be computed as:

$$N_{b,Rd} = \chi \cdot N_{pl,Rd} \qquad (1)$$

where χ is the reduction factor given in EN 1993-1-1 in terms of the relative slenderness $\bar{\lambda}$ from the corresponding buckling curve, and $N_{pl,Rd}$ is the plastic resistance of the composite section according to 6.7.3.2(1).

The relative slenderness $\bar{\lambda}$ is defined in Eurocode 4 as:

$$\bar{\lambda} = \sqrt{\frac{N_{pl}}{N_{cr}}} = \sqrt{\frac{A_c f_c + A_s f_y}{\frac{\pi^2 EI}{L^2}}} \qquad (2)$$

where $EI = E_s I_s + 0.6 E_{cm} \cdot I_c$, and I_s and I_c are the second moment of inertia of the steel tube and the concrete core respectively; E_s is the modulus of elasticity of steel; and E_{cm} is the secant modulus of elasticity of concrete.

As this code does not allow the possibility to use double-tube configuration, the inner steel tube has been considered as an inner reinforcement, selecting accordingly the curves "a" or "b" from Table 5.5 in EN 1994-1-1 (CEN 2004) in terms of the geometrical steel ratio. The results are summarized in Table 1 and Figure 6.

As it can be seen, the method in EC4 provides unsafe results on average for all the columns with an error around 15% for cases from Series 1 but more accurate results are observed for cases of Series 2. No special trend is observed in terms of the steel distribution or strength of concrete in the inner core.

With all the exposed above, it can be concluded that the method in EC4 for members in axial compression does not provide reliable results for evaluating the buckling resistance of double steel tube concrete-filled tubular columns if the outer ring is composed by NSC but surprisingly is quite accurate if the outer ring is filled up with UHSC. Further tests, both numerical and experimental, would be needed for evaluating the accuracy of EC4 method.

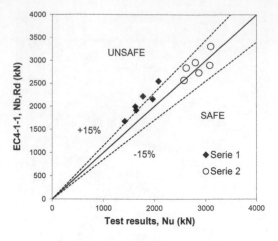

Figure 6. Comparison of Eurocode 4 and test results.

4 CONCLUSIONS

The results of an experimental program on slender dual steel tubular columns filled with normal and ultra-high concrete under room and elevated temperatures have been presented in this paper. Given the reduced number of experiments found in the literature, this work provides novel results to the research community. The test parameters covered in this experimental program were the influence of the strength of concrete in the inner and outer ring and the variation of thicknesses of the outer and inner steel tubes, being the objective to study the load bearing capacity.

From the test results, it was found that a double-tube column with the thicker tube in the inner position can reach almost the same buckling load than a double-skin CFT column with the thicker steel tube in the outer position and without inner concrete.

It results also noteworthy the reduced effect presented by the UHSC in the inner core in comparison with NSC.

For those tests with UHSC in the outer ring, it was found that the variation of the thicknesses in the outer and inner steel tube did not have the expected result.

Using the tests results, the design rules in Eurocode 4 have been assessed both for room temperature. Within the limited cases of this study, it can be initially inferred that the method in EC4 Part 1.1 for members in axial compression provide unreliable results for evaluating the buckling resistance of CFDST columns, although further tests are needed for evaluating the method in a more accurate manner.

However, a reduced number of tests have been performed up to date, and more experiments and extensive numerical modelling should be performed to achieve reliable results.

ACKNOWLEDGEMENTS

The authors would like to express their sincere gratitude to the Spanish Ministry of Economy and

Competitivity through the project BIA2012-33144 and to the European Community for the FEDER funds.

REFERENCES

CEN: Comité Européen de Normalisation 2004. EN 1994-1-1, Eurocode 4 2004: Design of composite steel and concrete structures. Part 1-1: General rules and rules for buildings. Brussels, Belgium.

Elchalakani, M., Zhao, X.L. & Grzebieta, R. 2002. Tests on concrete filled double-skin (CHS outer and SHS inner) composite short columns under axial compression. Thin-walled structures 40(5):415–441.

Hernández-Figueirido, D., Romero, M.L., Bonet, J.L. & Montalvá, J.M. 2012. Ultimate capacity of rectangular concrete-filled steel tubular columns under unequal load eccentricities, Journal of Constructional Steel Research 68:107–117.

Hernández-Figueirido, D., Romero, M.L., Bonet, J.L. & Montalvá, J.M. 2012. Influence of Slenderness on High-Strength Rectangular Concrete-Filled Tubular Columns with Axial Load and Nonconstant Bending Moment, J. Struct. Eng. 138(12): 1436–1445.

Huang, H., Han, L.H., Tao, Z. & Zhao, X.L. 2010. Analytical behaviour of concrete-filled double skin steel tubular (CFDST) stub columns, Journal of Constructional Steel Research 66(4): 542–555.

Liew, R. & Xiong, D.X. 2011, Experimental investigation on tubular columns infilled with ultra-high strength concrete. In: Tubular Structures XIII. Boca Raton: Crc Press-Taylor & Francis Group; 637–645.

Lu, H., Han, L. & Zhao, X. 2010. Fire performance of self-consolidating concrete filled double skin steel tubular columns: Experiments. Fire safety journal 45(2):106–115.

Lu, H., Zhao, X.L. & Han, L.H. 2010b. Testing of self-consolidating concrete-filled double skin tubular stub columns exposed to fire, Journal of Constructional Steel Research 66(8–9): 1069–1080.

Neuenschwander, M., Knobloch, M. & Fontana, M. 2010. Fire behaviour of concrete filled circular hollow section columns with massive steel core, Proceedings of the International Colloquium Stability and Ductility of Steel Structures SDSS, September 8–10, 2010, Rio de Janeiro, Brazil.

Portoles, J.M., Romero, M.L,, Bonet, J.L. & Filippou, F.C. 2011. Experimental study of high strength concrete-filled circular tubular columns under eccentric loading. Journal of constructional steel research 67(4):623–633.

Portoles, J.M., Serra, E. & Romero, M.L. 2013. Influence of ultra-high strength infill in slender concrete-filled steel tubular columns. Journal of constructional steel research 86:107–114.

Schaumann, P. & Kleibömer, I. 2014, Experimentelle Untersuchungen zum Trag- und Erwärmungsverhalten von Verbundstützen mit massivem Einstellprofil im Brandfall"/"Experimental investigations on the structural and thermal behaviour of composite columns with embedded massive steel core" (In German), 19. DASt-Kolloqium, October 2014, Hannover, Germany (In press).

Tao, Z., Han, L.H. & Zhao, X.L. 2004. Behaviour of concrete-filled double skin (CHS inner and CHS outer) steel tubular stub columns and beam-columns, Journal of Constructional Steel Research 60(8): 1129–1158.

Tao, Z. & Han, L.H. 2006. Behaviour of concrete-filled double skin rectangular steel tubular beam–columns, Journal of Constructional Steel Research 62(7): 631–646.

Zhao, X. & Han, L. 2006. Double skin composite construction. Progress in structural engineering and materials 8(3): 93–102.

Zhao, X.K., Tong, L.W. & Wang, X.Y. 2010. CFDST stub columns subjected to large deformation axial loading, Engineering Structures 32(3): 692–703.

Tubular Structures XV – Batista, Vellasco & Lima (eds)
© 2015 Taylor & Francis Group, London, ISBN 978-1-138-02837-1

Experimental study of beam to concrete filled elliptical steel column connections

T. Sheehan, J. Yang, X.H. Dai & D. Lam
University of Bradford, Bradford, UK

ABSTRACT: Concrete-filled steel tubular columns are becoming increasingly popular in tall building structures owing to their excellent strength and ductility compared to hollow tubes. They are an optimum combination of the two materials in terms of structural resistance, utilising the compressive strength of confined concrete together with the high strength-to-weight ratio and tensile resistance of steel. They are also easy to construct, with the outer tube providing formwork for the inner concrete core. Many different shaped tubes are used as columns in construction, and most research to date has focussed on circular, square and rectangular tubes. In recent years attention has been paid to utilising elliptical hollow sections (EHS). This is a new structural shape which offers the additional benefits of visual appeal and potentially superior structural efficiency from the presence of major and minor bending axes. Most of the research to date on elliptical hollow sections has focussed on the behaviour of individual components without considering their interaction with other members in an assembly. The design and construction of EHS connections can be cumbersome, since the connecting components must be constructed to fit the curved EHS face. This research explores a number of new I-beam to EHS column connection arrangements, using simple, readily available components, and compares the responses of hollow and concrete-filled, stiffened and unstiffened columns. The connections were investigated through a series of experiments which were conducted in the University of Bradford. Moment-rotation relationships were recorded and compared for each connection type, showing the moment-resistance and rotational capacity of each joint. Different deformed shapes and failure modes were observed for each connection and the benefits of using concrete-infill and steel stiffeners in the columns were highlighted. The findings pave the way to develop design recommendations for these types of connections which will improve their economy and constructability, reducing material wastage and enabling more widespread use of composite EHS columns and other curved profiles in buildings.

1 INTRODUCTION

Curved, tubular columns, such as circular hollow sections (CHS) offer many advantages. The aesthetic appeal renders them a favourable choice among architects while the cross-section shape provides high strength, ductility and torsional resistance. The use of concrete infill can further enhance the response by mitigating inward local buckling, and the circular tube is the ideal shape to confine the concrete core, while also acting as formwork during the concrete pour. In recent years, hollow and concrete-filled elliptical hollow sections (EHS) have also become popular, owing to a highly appealing shape, and the presence of major and minor bending axis, which can potentially increase the structural efficiency. EHS have been utilised in many structures around the world, such as Heathrow Airport in London and Barajas Airport in Madrid. Many researchers have studied the response of hollow EHS (Chan and Gardner, 2008) and concrete-filled EHS (Yang *et al.*, 2008; Dai and Lam, 2010; Sheehan *et al.*, 2012) members but less work has been carried out on EHS connections. One notable disadvantage associated with curved columns is the complexity of connection design. Welded EHS joints have been investigated by Bortolotti *et al.* (2003) and Pietrapertosa and Jaspart (2003) for brace-chord connections in trusses. Willibald *et al.* (2006) examined EHS gusset-plate connections using branch-plates or through-plates, and concluded that the through-plate connections were approximately twice as great as the branch-plate connections. Different arrangements of concrete-filled tubular connections have also been tested by several researchers. Wang *et al.* (2009) conducted experiments on beam-column connections comprising blind bolts and concrete-filled tubular columns of different shapes. Han and Li (2010) tested concrete-filled CHS columns, connected to beams using external diaphragms, subjected to seismic loading. Previous research has provided several possible structural solutions for concrete-filled CHS connections. However, the manufacture of such connections is often complicated, requiring special additional components. Furthermore, the behaviour of concrete-filled EHS, with more complex geometry, has been far less explored to date.

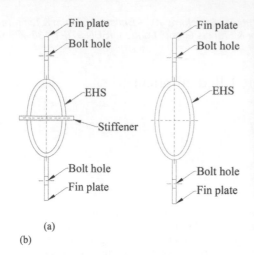

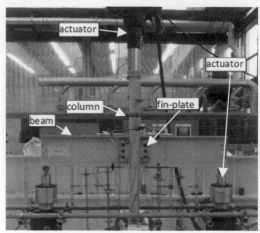

Figure 2. Typical test set-up.

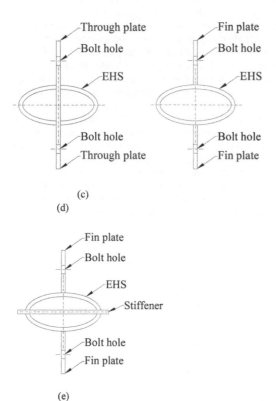

Figure 1. Plan view of Joint types: (a) Joint-A; (b) Joint-B; (c) Joint-C; (d) Joint-D; (e) Joint-E.

Hence this paper addresses the response beam to EHS column connections. Five different connection configurations have been proposed, exploring major and minor axis bending, and comparing the benefits of using concrete infill, through-plates and stiffening plates. The beams are bolted to fin/through plates, which are in turn welded to the columns, providing a connection that is relatively simple and easy to manufacture.

2 EXPERIMENTAL STUDY

2.1 *Specimen design and fabrication*

Five different joint arrangements were considered in this study as shown in Figure 1. Two specimens were tested for each joint type, one of which comprised a hollow steel EHS column and the other employing concrete infill. Joints A and B underwent bending about the major axis and Joints C, D and E were bent about the minor axis. Hollow joints were named Joint-A, Joint-B, etc. while the corresponding filled joints were Joint-AC, Joint-BC, etc. In all joints, the beams ($305 \times 127 \times 48$ UB, 900 mm in length) were connected by bolts to fin plates ($220 \times 110 \times 8$mm), which were in turn welded onto the column face. ($200 \times 100 \times 5$ EHS). The study examined the contribution of concrete infill to these connections and also the effect of stiffening plates. As shown in Fig. 1, Joints A and E employed stiffening plates running through the section in the transverse direction to the fin plates, while in Joint C, the fin-plates were replaced by a single through-plate. Each connection used three M20 Gr.8.8 bolts or Gr.10.9 bolts.

Concrete of the same target strength (C30) was used to fill all of the specimens. Two batches of concrete were used to fill the 5 specimens, which had an average 28-day strength of 37 MPa and an average test-date strength of 42 MPa.

2.2 *Testing procedure*

The test set-up is shown in Figure 2. One 250 tonne hydraulic actuator was positioned above the column and exerted a compressive force equal to 40% of the columns compressive resistance. Concentrated, upwards forces were then applied simultaneously at the beam ends using two 100 tonne hydraulic actuators, which replace the floor-slab load that would occur in a real structure. These were applied in displacement increments until failure of one/more joint components.

Curved connection pieces, as shown in Figure 3, were attached to the tops of the 100 tonne actuators. These allowed rotation of the beam ends in the plane of the testing frame, while constraining the beams against out-of plane rotation/twisting.

2.3 *Instrumentation*

The instrumentation for Joint-A is shown in Figure 4. Load cells in each of the actuators measured the applied loads, while the actuator head positions were also recorded. LVDTs L1-L4 measured the lateral deflections at the position of the top and bottom bolts

curved face

welded side plate (to prevent out of plane deflection)

Figure 3. Roller at beam end.

in each fin-plate connection, which were then used to calculate the joint rotation. LVDTs L5-L8 measured the vertical deflection of the underside of the beam, providing an alternative measure of the joint rotation and recording the deformed shape of the beams, to indicate if beam bending occurred. LVDTs L9 and L10 measured the vertical displacements at opposite sides of the plate on top of the column.

LVDTS L11-L14 measured the horizontal displacement of the column face at the top and bottom of each of the beams, to determine whether column face deformations were concave or convex at these locations. Several strain gauges were utilized to capture the response of the column face (in terms of longitudinal and hoop strains) and also to monitor the fin plate for tearing at the welded or bolted connections. Strain gauges C1 and C2 measured the longitudinal strains at the points of minimum and maximum cross-section curvature at a location above the connection, while C8 and C9 were in similar positions in a lower region of the column. C3 measured the hoop strain in the column midway between the top of the two joints. C4-7 measured the hoop and longitudinal strains in the column face at the top of and bottom of the connecting beam (on one side). C10-13 were used at the top and bottom of the fin plates on each side to monitor horizontal strains close to the welded column connection.

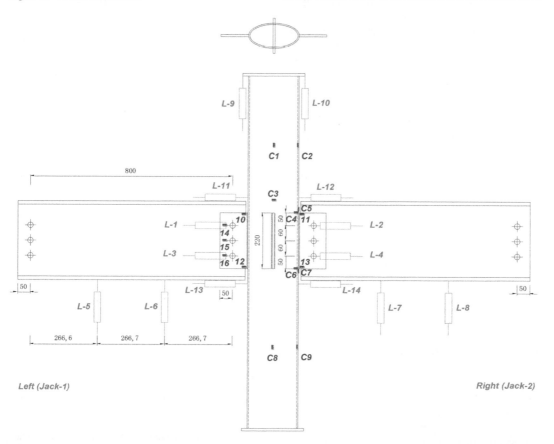

Figure 4.   Instrumentation for Joint-A.

Finally, C14-16 were positioned adjacent to the top, middle and bottom bolts on one side to measure horizontal strains in the plate at this location. These strain gauges showed the position of the centre of rotation of the joint and how it moved as the test progressed. Similar arrangements of LVDTs and strain gauges were used for the other 9 specimens.

3 RESULTS

3.1 Moment-rotation behaviour

In all test specimens, an initial gap existed between the beam and column faces, and at some stage during the test, the top flange of the beam made contact with the column face. Owing to imperfections in the test set-up, these gaps were different on the left and right hand sides of each specimen and between specimens, leading to deviations in responses. Figures 5–9 show the relationship between moment and rotation for each of the test specimens, where moment is calculated as $0.8 \times$ the load at the beam end and rotation is calculated using the displacements measured by LVDTs 1–8. The square data points indicate the moment and rotation on the left hand side and the circular data points display the data recorded from the right hand side. Hollow data points correspond to the hollow columns and filled points to the filled columns. In the case of Joint-EC, Grade 8.8 bolts were used for the initial test, but after failure of these bolts, a repeat test was carried out using Grade 10.9 bolts. Hence in this case, two sets of data are presented, with the initial test represented by the triangular data points and the final test represented by the square and circular filled points.

During the initial stages of the test, rotation was resisted by friction between the beam, fin-plate and bolts. When the force exceeded the frictional resistance, slippage occurred between the joint components. After slipping, the moment increased slowly with increasing rotation, but the slope of the curved gradually increased as the test progressed, with the joint becoming very stiff, until one or more of the connection components finally failed.

Details of the maximum moment M_u, rotation θ_u and the ratio of ultimate moment for filled and hollow columns $M_{u,filled}/M_{u,hollow}$ are provided in Table 1. Hollow specimens achieved significantly larger rotations than their concrete-filled counterparts, particularly for joints A, D and E. The use of concrete significantly enhanced the ultimate moment in all cases, most notably for Joint types B and D, where no stiffeners/through-plates were used.

3.2 Failure mode

The failure mode of Joint A is shown in Figure 10 (a). Inwards local buckling was observed on the column face near the top of the connection. In contrast to this, no local buckling was observed in the concrete-filled

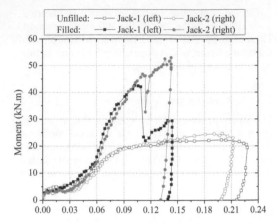

Figure 5. Moment-rotation relationship for Joint-A and Joint-AC.

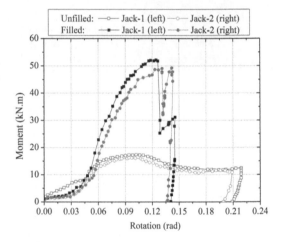

Figure 6. Moment-rotation relationship for Joint-B and Joint-BC.

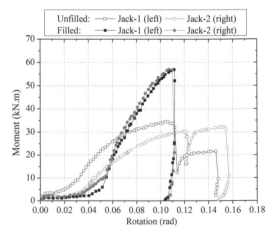

Figure 7. Moment-rotation relationship for Joint-C and Joint-CC.

82

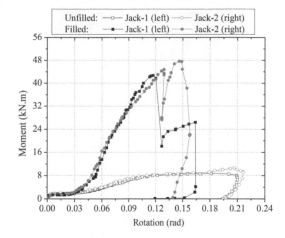

Unfilled: —○— Jack-1 (left) —○— Jack-2 (right)
Filled: —■— Jack-1 (left) —●— Jack-2 (right)

Figure 8. Moment-rotation relationship for Joint-D and Joint-DC.

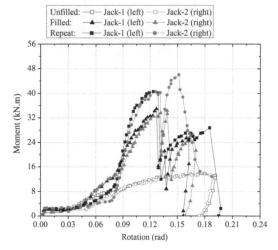

Unfilled: —□— Jack-1 (left) —○— Jack-2 (right)
Filled: —▲— Jack-1 (left) —▲— Jack-2 (right)
Repeat: —■— Jack-1 (left) —●— Jack-2 (right)

Figure 9. Moment-rotation relationship for Joint-E and Joint-EC.

Table 1. Ultimate moments and rotations for test specimens.

Specimen	M_u (kNm)	θ_u (mrad)	$M_{u,filled}/M_{u,hollow}$
Joint-A	22.3	200	–
Joint-AC	43.8	110	1.96
Joint-B	16.0	100	–
Joint-BC	49.6	120	3.10
Joint-C	30.0	110	–
Joint-CC	57.2	110	1.91
Joint-D	8.4	180	–
Joint-DC	43.6	110	5.19
Joint-E	13.3	180	–
Joint-EC	33.8	130	2.55
Joint-EC repeat	41.4	130	3.11

Joint-AC, Joint-BC exhibited little inward deformation, however some cracking occurred in the concrete core, which was revealed upon cutting the steel tube after the test, as shown in Figure 11 (d).

As expected, the moment capacity of Joint-A was higher than that of Joint-B connection owing to the enhancement provided by the stiffener plate in the minor axis direction of EHS tube. However, the mode of local buckling failure was the same for the two joints. Greater differences were observed between the stiffness of Joint-B and Joint-BC than between Joint-A and Joint-AC, since the transverse stiffener contributed significantly to the response of Joint-A. Although it had a limited effect on the stiffness, the addition of concrete did significantly increase the maximum bending moment of Joint-AC, as shown in Figure 5. Surprisingly, the resistance of the concrete-filled unstiffened joint, Joint-BC was actually higher than the stiffened Joint-AC. One possible explanation for this result is that the joint response in these cases is governed by bolt failure, and the enhanced column resistance actually led to earlier failure of the bolts.

In Joint C, the through plate prevented large local deformations but inward local buckling occurred in the hollow column, above the location of the connection, as shown in Figure 12 (a). Shear failure of the bottom bolts occurred in the final stage of the test. The right bottom bolt of connection with concrete infill, Joint-CC also failed. The in these two tests, the fin plates and bolts endured almost the whole load transferred from the beam, except for the direct compression from the top flange of the beam which led to the inward deformation of EHS tube.

In Joints D and E, local buckling occurred above the connection, but noticeable deformations also occurred in the column, adjacent to the connection, as exemplified by the images of Joints D and E presented in Figure 12 (b) and (c). In Joint-DC, inward deformation of the steel tube caused cracking in the concrete core, directly above the connection. The middle and bottom bolts on the left hand side failed during the test of Joint-DC and in addition, tearing occurred at the bottom of the fin plate on the right hand side.

equivalent, Joint-AC, as shown in Figure 10 (b). After the test, the steel tube was cut to reveal the condition of the core concrete, presented in Figure 10 (c). No cracks were observed in the core concrete for Joint AC, demonstrating the efficiency of the stiffener, concrete core and steel tube when working together. The test of Joint-AC was eventually terminated when one of the bolts failed in shear.

Severe local deformations were observed for Joint-B. The view of the failed specimen from the front is presented in Figure 11 (a). In a similar manner to Joint-A, inward local buckling occurred on the left and right faces of the column at the top of the connection, causing the cross-section to become square in shape. A plan view of this cross-section is shown in Figure 11 (b). At the bottom of the connection, inward local buckling occurred on the front and back of the column, ovalising the cross-section and increasing the aspect ratio as shown in Figure 11 (c). Similarly to

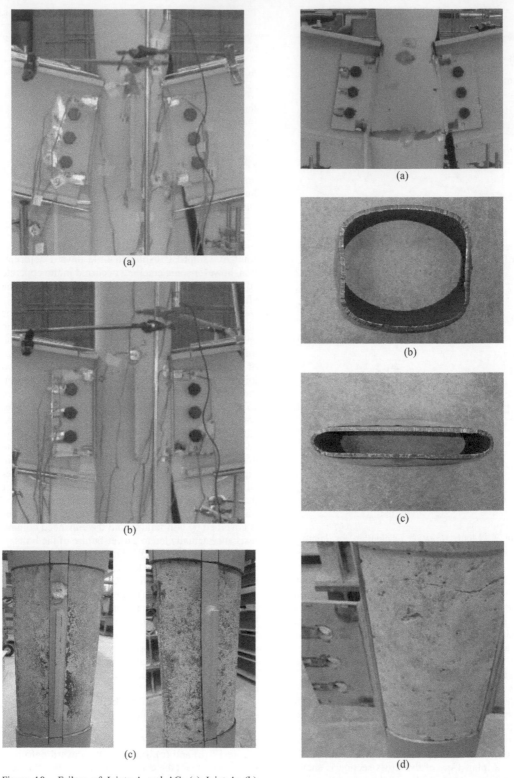

Figure 10.   Failure of Joints A and AC: (a) Joint-A; (b) Joint-AC; (c) Concrete core of Joint-AC.

Figure 11.   Failure of Joints B and BC: (a) Joint-B; (b) Plan view of Joint-B column – top of connection; (c) Plan view of Joint-B column – bottom of connection (d) Concrete core of Joint-BC.

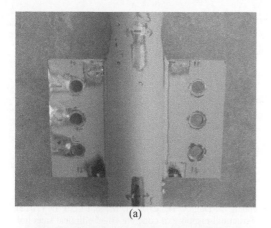

(a)

(b)

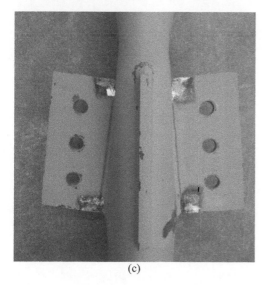

(c)

Figure 12. Deformed shapes for hollow specimens under minor axis bending: (a) Joint-C; (b) Joint-D; (c) Joint-E.

For the first test of Joint-EC, using Grade 8.8 bolts, the bottom bolts of both sides failed and hence Grade 10.9 were then adopted to repeat the experiment with the expectation of a better load-bearing capacity. Initial cracks were observed on both sides of the column at the end of the initial experiment using Grade 8.8 bolts, and the ones on right side extended during the repeated test using Grade 10.9 bolts. However, owing to the enhanced transverse stiffness provided by the concrete infill, the cracks on left side did not lead to tearing failure of column wall. Instead, the bottom and middle bolts of the left hand side failed in sequence. Cracks were observed on the concrete core when the steel wall was cut after the test.

The moment capacity of the through-plate connection, Joint-C with the unfilled column was significantly higher than that of the unreinforced connection Joint-D and the connection that was stiffened in the major axis direction, Joint-E. As stated previously, the through-plate resisted most of the shear force and bending moment transferred from the beams. The response of Joint-C demonstrated that although in an EHS tube, the stiffness in the major axis direction is higher than that in the minor axis direction, the moment capacity in the minor axis direction can still be enhanced by welding a stiffening plate in the major axis.

Although the minor axis through plate connection with concrete infill, Joint-CC, failed with a low joint rotation, it displayed a greater bending moment capacity than the other minor axis connections. However, little difference was observed between the ultimate moments of Joint-DC and Joint-EC. Similarly to the concrete-filled tubular joints under major axis bending, the unstiffened concrete-filled Joint-DC actually endured a larger bending moment than the stiffened case, Joint-EC. In both cases, the use of the stiffener prevented cracks from occurring in the concrete core.

4 CONCLUSION

A series of experiments has been carried out to examine I-beam to EHS column bolted fin-plate connections, considering both major and minor axis bending, and assessing the benefits of using concrete infill, transverse stiffeners and through-plates. For hollow column joints, the principal mode of failure was observed to be inward buckling of the column wall, at the point of contact with the top flange of the beam. The use of concrete infill was shown to mitigate inward local buckling, and consequently, failure of the concrete-filled connections occurred by shearing of the bolts.

Owing to the clearance of the bolt holes, the initial moment-rotation response was governed by friction between the joint components, and became increasingly stiff after slippage occurred.

For both major and minor axis bending, hollow column joints with transverse stiffeners endured greater bending moments than their unstiffened counterparts.

However, this increase in resistance did not occur between unstiffened concrete-filled column joints and stiffened concrete-filled joints, possibly owing to the fact that the failure of these connections occurred in the bolts.

Generally, the use of concrete infill reduced the rotational capacity of these connections, but significantly increased the ultimate bending moment, with the factor of increase ranging between 1.91 and 5.19. For hollow specimens under minor axis bending, the use of a through-plate provided a considerable increase in moment resistance.

In summary, these experiments have verified the merits of using concrete-infill, transverse stiffening plates and through-plates in I-beam to EHS column connections and has demonstrated the potential for practical, easy-to-construct connection solutions for EHS members.

REFERENCES

Bortolotti, E., Jaspart, J. P., Pietrapertosa, C., Nicaud, G., Petitjean P. D. and Grimault, J. P. (2003). Testing and modelling of welded joints between elliptical hollow sections. *Proceedings of the 10th International Symposium on Tubular Structures, Madrid:* 259–266. London: Taylor & Francis.

Chan, T. M. and Gardner, L. (2008). Compressive resistance of hot-rolled elliptical hollow sections. *Engineering Structures* 30(2): 522–532.

Dai, X. and Lam, D. (2010). Numerical modelling of the axial compressive behaviour of short concrete-filled elliptical steel columns. *Journal of Constructional Steel Research* 66(7): 931–942.

Han, L.-H. and Li, W. (2010). Seismic performance of CFST column to steel beam joint with RC slab: Experiments. *Journal of Constructional Steel Research* 66(11): 1374–1386.

Pietrapertosa, C. and Jaspart, J. P. (2003). Study of the behaviour of welded joints composed of elliptical hollow sections. *Proceedings of the 10th International Symposium on Tubular Structures, Madrid:* 601–608. London: Taylor & Francis.

Sheehan, T., Dai, X. H., Chan, T. M. and Lam, D. (2012). Structural response of concrete-filled elliptical steel hollow sections under eccentric compression. *Engineering Structures* 45: 314–323.

Wang, J.-F., Han, L. -H. and Uy, B. (2009). Behaviour of flush end plate joints to concrete-filled steel tubular columns. *Journal of Constructional Steel Research* 65(4): 925–939.

Willibald, S., Packer, J. A. and Martinez-Saucedo, G. (2006). Behaviour of gusset plate connections to ends of round and elliptical hollow structural section members. *Canadian Journal of Civil Engineering* 33(4): 373–383.

Yang, H., Lam, D. and Gardner, L. (2008). Testing and analysis of concrete-filled elliptical hollow sections. *Engineering Structures* 30(12): 3771–3781.

Tubular Structures XV – Batista, Vellasco & Lima (eds)
© *2015 Taylor & Francis Group, London, ISBN 978-1-138-02837-1*

Effect of bolt gauge distance on the behaviour of anchored blind bolted connection to concrete filled tubular structures

M. Mahmood
Department of Civil Engineering, The University of Nottingham, UK
Department of Civil Engineering, University of Diyala, Iraq

W. Tizani & C. Sansour
Department of Civil Engineering, The University of Nottingham, UK

ABSTRACT: This study focuses on investigating the effect of bolt gauge distance on the column face bending behaviour for anchored blinded bolting connections to concrete filled tubular columns. Full scale experimental tests were conducted for six samples and ABAQUS 6.13 was employed to perform a parametric study to extend the experimental data. Results show that, increasing the bolt gauge by 75% and 125% can improve the column face bending strength by 26% and 58% respectively and the initial stiffness by 22% and 44% respectively. At the ultimate resistance of the component and for all the bolt gauges that were investigated in this study, complete yielding of the tubular corners was recorded. This finding contradicts with exciting knowledge which assumes that the yield lines could extend only to the centreline of the SHS wall.

1 INTRODUCTION

Concrete filled tubular columns have high strength, high stiffness and high ductility. The concrete and the tubular section work as a complementary part to each other. The concrete provides the support to the tubular walls and improve their buckling strength. The tubular structures provide the confinement to the concrete and increase its strength. In addition, the tubular section can act as a formwork during the construction period, which can reduce the cost and construction time (Ellobody and Young, 2006). However, performing the connection to this type of construction represents a challenge due to the difficulty of access to the inside of the tubular sections.

Lindapter company patented a practical blind bolting system called HolloBolt (HB) (Figure 1,a). It's installation needs only inserting it in the pre drilled fixture and steel work and tightening it to open the sleeves inside the tubular column and perform the connection. HB has a 40% higher shear strength than the standard bolt (Occhi, 1995). However, it has a low stiffness compared with standard bolt under tensile load (Barnett et al., 2000). To overcome the problem of low stiffness, Tizani and Ridley-Ellis (2003) developed a novel Extended anchored HolloBolt (termed EHB) (Figure 1, b) fastener. The idea of the development is based on increasing the shank length and attaching an anchoring nut at the shank's end. Embedding the extended shank length and the nut inside the concrete improves the behaviour of the fastener through anchoring the EHB in the concrete. The extended shank length and

nut size are limited to the size of the tubular section and the diameter of the bolt hole. Pitrakkos and Tizani (2013) carried out an extensive experimental program to investigate the tensile behaviour of EHB under pure tension. The study confirmed that the tensile capacity of the EHB component is controlled by its internal bolt tensile capacity.

The work presented in this paper is part of a series of studies, which are investigating the behaviour of the different components of EHB connection under variety of loading conditions to provide the required knowledge for producing design guidance for EHB connections. In this study, the focus is set on investigating the effect of bolt gauge distance on the bending behaviour of the column face component for EHB connections.

2 EXPERIMENTAL TESTS

2.1 Test specimens

A series of experimental tests were conducted to investigate the effect of bolt gauge (g) on the column face bending behaviour. All the tests were performed using $300 \times 300 \times 8$ SHS and M16 EHB (dummy EHB). The shape and the dimensions of the dummy EHB are exactly similar to the installed EHB, and it was manufactured from high strength steel alloy (Elamin, 2013) to limit the failure at the column face. The minimum and maximum values of the bolt gauge are controlled by the geometry of the EHB and the tubular section.

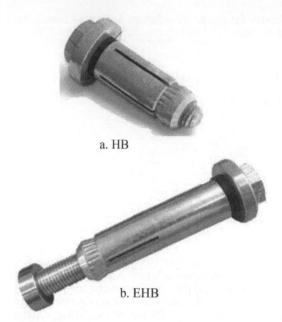

a. HB

b. EHB

Figure 1. HolloBolt (HB) and Extended HolloBolt (EHB).

As well as providing a suitable distance for concrete placement, which is not less than 20 mm according to Eurocode (CEN, 2004b).

The maximum opening width of the sleeves was calculated theoretically and it was found equal to 46.6 mm (Figure 2). Therefore, the minimum bolt gauge (g_{min}) required to satisfy the previous conditions was calculated as follows:

$$g_{min} = 2\left(\frac{46.6}{2}\right) + 20 = 66.6 \ mm$$

Similarly, the minimum edge distance between the sleeves edges and the inside wall of the tubular section should not be less than 20 mm. Therefore, Equation (1) was suggested to calculate the maximum bolt gauge (g_{max}).

$$g_{max} = b - 2t - 86.6 \tag{1}$$

where b = tubular section width (mm) and t = tubular section plate thickness (mm).

The experimental programme was carried out by testing a pair of identical samples for each bolt gauge to confirm the reliability of the results. Six samples were tested: two for bolt gauge of 80 mm (G80-1 & G80-2), two for bolt gauge 140 mm (G140-1 & G140-2) and two for bolt gauge of 180 mm (G180-1 & G180-2). Apart from the bolt gauge, all the geometrical and material properties and the test procedure were similar. The concrete mix was designed to have a compressive strength of 40 N/mm² on the day of the test. The anchored length of the EHB was 80 mm and all the samples were cut from one batch of tubular section.

The actual section dimensions and the holes size for all the tested samples are listed in Table 1. These values were used in the validation of the finite element model.

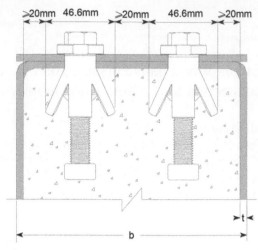

Figure 2. Maximum and minimum bolt gauge distance.

Table 1. Actual section dimensions and holes size for the tested samples.

	G80-1	G80-2	G140-1	G140-2	G180-1	G180-2
Section width (b) mm	300	300	300	300	300	300
Plate thickness (t) mm	7.98	7.98	7.98	7.98	7.97	7.97
Corner thickness (t_c) mm	9.08	9.10	9.11	9.12	9.09	9.08
Hole diameter (d) mm	25.88	25.85	25.90	25.91	25.87	25.88

2.2 Test setup and instrumentations

The test setup is presented in Figure 3. The test rig was designed to apply pull-out load on the EHBs. The loading was displacement controlled with a rate of 0.015 mm/sec. During the test, a video gauge system was employed to record the column face displacement, bolt slip and the sample movement. This system can measure a wider range of displacement compared with the linear potential meter and it was used to provide accurate results in similar studies (Mahmood et al., 2014). The sample movement was recorded to eliminate it from the column face displacement and the bolt slip. Recording the bolt slip requires attaching some fittings to the sample. A 300 mm length M16 threaded rode was attached to the end of the EHB inside the sample and it passed through a hole in the back of the sample. This rod was covered by 20 mm diameter steel tube to protect it from concrete during the casting (Figure 4).

A Digital Image Correlation (DIC) system was used to monitor the strain distribution in the column face.

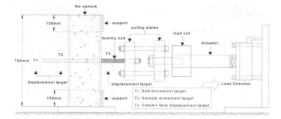

Figure 3.   Test arrangement.

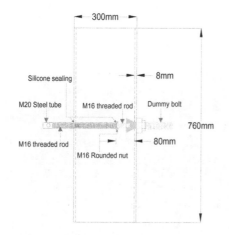

Figure 4.   Sample setup.

2.3   Test results

The specified cube concrete compressive strength in the day of the test was 39.3, 40.1 and 40 N/mm² for samples with bolt gauge of 80, 140 and 180 mm respectively. The effect of variation in the concrete strength was assumed to be neglected since the maximum difference is 0.8 N/mm². The force displacement curves for all the tested samples are presented Figure 5. Samples G80-1 and G80-2 show early nonlinear behaviour at about 66% of the yield load, whereas the linear behaviour for samples G14-1, G140-2, G180-1 and G180-2 extends up to about 80% of the ultimate strength. The early nonlinear behaviour for components with small bolt gauge might be due to the limited role of anchoring. In this study, the ultimate strength of the connection represents the peak point before the strength starts dropping. Once the ultimate strength was reached, there were some concrete crushing sounds coming from the samples. These sounds were synchronised with a drop in the strength, and this scenario was repeated for all the tested samples. However, the rate of the drop was increased with the use of large bolt gauge. A possible explanation for this behaviour is the use of small bolt gauge results in a stress concentration at the concrete between the bolts (the gauge distance). This limits the advantages of anchoring the EHB in the concrete through the concrete failure and losing the composite action with the bolts at early stage. However, with the use of large bolt gauge, the stress distributed over a wider concrete area.

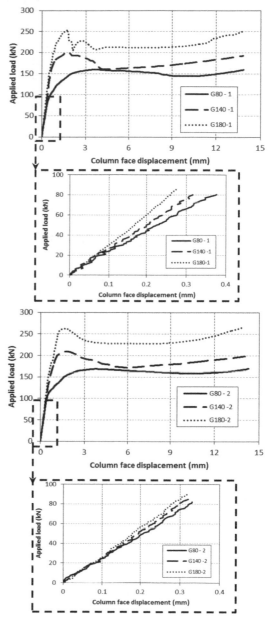

Figure 5.   Load versus column face displacement curves.

This results in delaying the concrete failure and maintaining the connection strength to much higher than that for the small bolt gauge. Therefore, the concrete failure causes sudden drop in the strength. After this drop, the connection strength starts increasing due to the membrane action in the column face.

Results indicate that, increasing the bolt gauge by 75% and 125% could improve the ultimate strength of the connection up to 26% and 58% respectively and the initial stiffness up to 22% and 44% respectively.

The previous results suggest that, increasing the bolt gauge can improve the column face bending strength

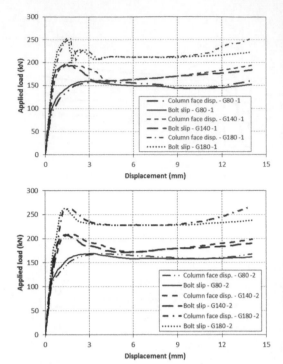

Figure 6. Column face displacement and bolt slip.

Table 2. Pull-out displacements.

Sample	Column face displacement (mm)
G80-1	11.014
G80-2	11.180
G140-1	9.340
G140-2	9.201
G180-1	8.140
G180-2	8.557

a. G80

b. G180

Figure 7. Bolt gauge effect on bolt movement.

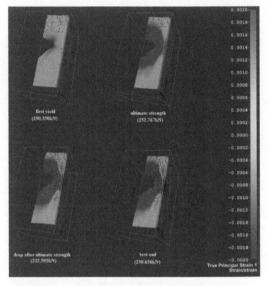

Figure 8. Strain outline sample G180-1.

more than the stiffness. The improvement comes from changing the loading distribution on the component by varying the bolt gauge. As increasing the bolt gauge, results in decreasing the shear span of the applied load which requires more work to bend the column face component.

Figure 6 shows comparison between the column face displacement and the bolt slip for all the tested samples, and the values of the pulling-out displacements are summarized in Table 2. The pull-out point is identified when the column face displacement becomes less than the bolt slip at the same load value.

It is noticeable that for larger gauge distances, the bolts tend to pull-out earlier (at a lower face displacement) than those with relatively smaller gauge distance. This may be attributed to the different in the higher load and also when the bolt starts pulling-out, it begins pushing the hole sides outward. With low values of bolt gauge, the distance between the bolts is small

and each bolt is trying to push the column face plate towards the another bolt. This increases the pressure of the holes on the sleeves and results in holding the bolts in the holes for higher displacement (Figure 7a). However, with large bolt gauge there is a long distance between the bolts so that the hole can expand easier to allow the bolt pulling-out (Figure 7b).

The DIC results show that the first yield was at the hole sides at 75% of the ultimate strength. Then it starts propagating in all directions. At the ultimate resistance of the connection, the yield pattern has a shape of half circle with a radius extend from the hole center to corner of the tubular section. Then, propagation of the yield was mainly in the column face. In contrast to the previous understanding which assumes that the yielding could only extend to the centreline of the tubular corners, this finding confirmed that the whole corner of the tubular section yields at the ultimate load. The yield outline is presented in Figure 8, only half of sample was monitored due the limited access to the column face during the test.

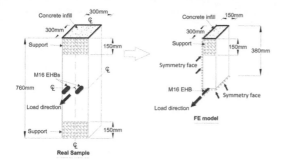

Figure 9. FE details.

3 FINITE ELEMENT MODELLING AND ANALYSIS

ABAQUS 6.13 (Dassault, 2013) was used to perform Finite Element (FE) analysis. The FE model consists of three parts: tubular section, concrete and the EHBs. To reduce the modelling and computational cost, the advantage of symmetry was considered and only quarter of the sample was modelled. The material and geometrical properties and the boundary conditions were considered to be similar to the real samples. The details of the FE model are presented in 3.1.

All the model components were modelled using (C3D8) element. This element is suitable for simulating the complete nonlinear behaviour including contact and geometrical nonlinearities (Abaqus, 2013). The contact between the three parts of the model was modelled using surface based and pair algorithm. Both normal and tangential contacts are existing in the model. The coefficient of friction between steel and concrete was considered as 0.25 (Elremaily and Azizinamini, 2001, Hu et al., 2003) and between steel and steel was considered as 0.45 .(Wang, 2012).

The analysis was performed by applying a static uniform displacement at the EHB similar to the experimental test. The increments were calculated using RIKS method. This method is powerful in predicting the nonlinear behaviour of the structures, because it has the ability for simulating the softening part in the behaviour.

3.1 Modelling of tubular section

The actual dimensions that are summarised in Table 1 were used to create the geometry of the tubular section in the model. The measured material properties reported by Mahmood et al. (2014) were used in this study. The experimental measured yield strength (f_y) was 406 N/mm^2, the ultimate strength (f_u) was 537 N/mm^2, the Young's modulus of elasticity (E_s) was 207000 N/mm^2 and the Poison's ratio was 0.3.

3.2 Modelling of concrete

The concrete geometry was modelled using the interior dimensions of the tubular section. A hole with a shape

and dimensions similar to the EHB was created at the bolt location. The concrete elastic behaviour was simulated by the Young's modulus of elasticity (E_c) and Poison's ratio. E_c was calculated using equation (2) (CEN, 2004a) and the poison's ratio was considered as 0.2.

$$E_c = 22000 \left(\frac{f_{cm}}{10}\right)^{0.3} \qquad (2)$$

where E_c = concrete Young's modulus (N/mm^2) and f_{cm} = mean value of concrete cylinder compressive strength (N/mm^2).

The plastic behaviour of the concrete was simulated using the Concrete Damage Plasticity (CDP) model. The failure mechanisms for this model are the tensile cracking and compression crushing. The concrete compression stress-strain curve was predicted using the Euro code multi-linear model for nonlinear structural analysis (CEN, 2004a). The model assumes that the concrete has a linear behaviour until 0.4f_{cm}. Then, then the behaviour was predicted using the following equations.

$$f_c = \left(\frac{k\eta - \eta^2}{1 + (k-2)\eta}\right) f_{cm} \qquad (3)$$

$$k = \frac{1.05 \, E_c \, \varepsilon_{c1}}{f_c'} \qquad (4)$$

$$\varepsilon_{c1} = \frac{0.7(f_{cm})^{0.31}}{1000} < 0.0028 \qquad (5)$$

$$\varepsilon_{c2} = 0.0035 \qquad (6)$$

$$\eta = \frac{\varepsilon_c}{\varepsilon_{c1}} \qquad (7)$$

where f_c = concrete compressive stress at any point on the stress-strain curve (N/mm^2), ε_{c1} = concrete compressive strain at the maximum stress (f_{cm}), ε_{c2} = concrete compressive strain at the end of stress-strain curve and ε_c = compressive strain in the concrete.

The following plasticity parameters were used: dilation angle is 35°, eccentricity is 0.1, the ratio of initial equibiaxial compressive yield stress to initial uniaxial compressive yield stress (σ_{bo}/σ_{co}) is 1.16, yield shape parameter (kc) is 1 and viscosity parameter (μ_o) is 0. The tensile behaviour of concrete was simulated using a bilinear model with ultimate strength of 0.1f_{cm}.

3.3 Modelling of EHB

The geometry of the EHBs is exactly similar to dummy EHB. They are assumed to have high strength, therefore, they were modelled as elastic material using Young's modulus of 390000 N/mm^2 and poison's ratio of 0.3 (Elamin, 2013).

Table 3. Experimental and numerical ultimate column face strength.

Sample	Ultimate load (kN)		
	Experimental	Numerical	Difference %
G80-1	159.665	172.092	8
G80-2	169.136		2
G140-1	200.880	2011.279	5
G140-2	208.666		1.5
G180-1	252.767	275.687	9
G180-2	262.842		5

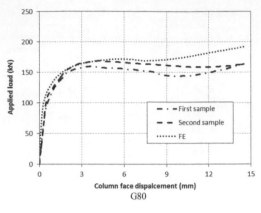

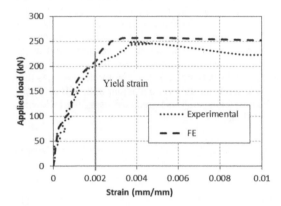

Figure 10. Load strain curves: experimental versus FE at bolt hole, sample G180-1.

4 VERIFICATION OF FINITE ELEMENT MODEL

A comparison between the experimental and FE analysis results was carried out to verify the FE model. The comparison was considered the following parameters: ultimate column face strength strain values at the bolt hole and the load versus the column face displacement behaviour. The maximum difference in the ultimate column face strength was 9% and it was recorded for sample G180-1. The experimental and numerical ultimate column face strength is listed in Table 3.

Figure 10 presents a comparison between the experimental and numerical results for the stain values at the bolt hole. Good correlation between the both results was achieved.

The numerical and experimental load versus column face displacement for all samples is plotted in Figure 11. The numerical results agreed very well with the experimental results up to the ultimate column face strength. Then, the FE model shows stiffer behaviour than experimental results due to the limited damage in the concrete model.

5 PARAMETRIC STUDY

A parametric study was conducted using the validated FE model. Sample G140-1 was used as a base

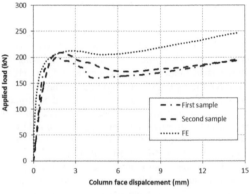

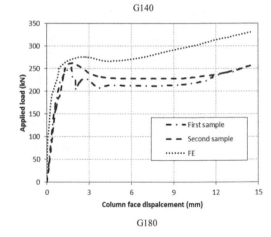

Figure 11. Load-column face displacement, experimental versus FE.

sample. Eleven models were analysed with different values of bolt gauge, ranged from 80 mm to 180 mm. The applied load versus the column face displacement curves are plotted in Figure 12. Although the bolt gauge distance was increased by constant value (10 mm), the amount of improvement in the ultimate column face strength was increased with the use of higher bolt gauge. For instance, increasing the bolt gauge from 80 mm to 90 mm improved the column face strength by %2, whereas increasing the bolt

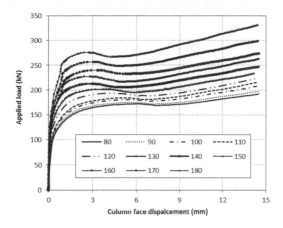

Figure 12. Effect of bolt gauge on the column face bending behaviour.

gauge from 170 mm to 180 mm enhanced the column face strength by %8. Also the connection could reach its ultimate strength at lower column face deformation with the use of large bolt gauge. For example the connection with a bolt gauge of 80 mm reaches its ultimate strength at column face displacement of 6.140 mm, while the connection with a bolt gauge of 180 mm reaches its ultimate strength at column face displacement of 2.817 mm.

6 CONCLUSIONS

This paper presented an experimental and numerical investigation of the effect of bolt gauge distance on the bending behaviour of the column face component for a novel anchored blind bolted connection to concrete filled tubular columns. The behaviour of all the investigated samples was approximately linear up to the ultimate resistance of the column face. The sharpness of the drop in the strength after the ultimate column face strength was increased with the use of large bolt gauge. The bolt gauge distance has a clear influence at the strength and stiffness of the column face. However, the use of small bolt gauge could limit the advantages of using the anchored blind bolting system. The ultimate strength of the connection could be reached with less deformation and the bolt starts bulling out at lower column face deformation with the use of large bolt gauge. The yield outline was mainly in the column face and tubular corners. Complete yielding of the corners at the ultimate resistance was recorded, contrary to standard understanding that the yielding could only extend to the centreline of the corners.

ACKNOWLEDGMENT

The authors wish to acknowledge TATA Steel and Lindapter International® for supporting work. Gratitude is expressed to Mr Trevor Mustard, of TATA Steel, and Mr Neil Gill, of Lindapter International. The first author would like to thank the higher committee for education development in Iraq for providing the chance to perform this research.

REFERENCES

Abaqus, I. 2013. ABAQUS Analysis User's Manual: Volume IV: Elements. *Abaqus, Inc.* Dassault Systèmes.

Barnett, T., Tizani, W. & Nethercot, D. 2000. Blind Bolted Moment Resisting Connections to Structural Hollow Sections. *Connections in Steel Structures IV: Steel Connections in the New Milennium*, 23–25.

CEN 2004a. Eurocode 2: Design of Concrete Structures: Part 1-1: General Rules and Rules for Buildings. *EN 1992-1-1.* British Standards Institution.

CEN 2004b. Eurocode 4: Design of Composite Steel and Concrete Structures: Part 1-1: General Rules and Rules for Buildings. *EN 1994-1-1.* CEN.

Dassault 2013. Abaqus v. 6.13 [Software]. *Dassault Systèmes Simulia Corp.* Providence, RI: Dassault Systèmes Simulia Corp.

Elamin, A. 2013. *The Face Bending Behaviour of Blind-Bolted Connections to Concrete-Filled Hollow Sections.* PhD Thesis, University of Nottingham.

Ellobody, E. & Young, B. 2006. Nonlinear Analysis of Concrete-Filled Steel SHS and RHS Columns. *Thin-walled structures*, 44, 919–930.

Elremaily, A. & Azizinamini, A. 2001. Design Provisions for Connections between Steel Beams and Concrete Filled Tube Columns. *Journal of Constructional Steel Research*, 57, 971–995.

Hu, H. T., Huang, C. S., Wu, M. H. & Wu, Y. M. 2003. Nonlinear Analysis of Axially Loaded Concrete-Filled Tube Columns with Confinement Effect. *Journal of Structural Engineering*, 129, 1322–1329.

Mahmood, M., Tizani, W. & Sansour, C. 2014. Effect of Tube Thickness on the Face Bending for Blind-Bolted Connection to Concrete Filled Tubular Structures. *International Journal of Civil, Architectural, Structural and Construction Engineering*, 8, 904–910.

Occhi, F. 1995. Hollow Section Connections Using (Hollofast) Hollobolt Expansion Bolting. *Second Interim Report 6G-16/95.* Sidercad, Italy.

Pitrakkos, T. & Tizani, W. 2013. Experimental Behaviour of a Novel Anchored Blind-Bolt in Tension. *Engineering Structures*, 49, 905–919.

Tizani, W. & Ridley-Ellis, D. The Performance of a New Blind-Bolt for Moment-Resisting Connection. TUBULAR STRUCTURES-INTERNATIONAL SYMPOSIUM-, 2003. 395–400.

Wang, Z. 2012. *Hysteretic Response of an Innovative Blind bolted Endplate Connection to Concrete Filled Tubular Columns.* PhD Thesis, University of Nottingham.

Tubular Structures XV – Batista, Vellasco & Lima (eds)
© *2015 Taylor & Francis Group, London, ISBN 978-1-138-02837-1*

Beam to concrete-filled rectangular hollow section column joints using long bolts

V.L. Hoang, J.F. Demonceau & J.P. Jaspart
University of Liege, Belgium

ABSTRACT: This paper presents a research on a specific type of unstiffened extended end-plate joint used to connect I-shaped beams to concrete-filled rectangular hollow section columns. The main idea is to use long bolts throughout the column to connect the beam end-plates, so avoiding intermediate connecting elements (e.g. a reverse U channel) or special bolts (e.g. blind bolts). However, the use of long bolts for beam-to-column connections is still rare in the construction and no design procedure exists in the Eurocodes; this justifies the present research. Firstly, a test program within a RFCS European project titled HSS-SERF "High Strength Steel in Seismic Resistant Building Frames", 2009–2013 was performed. In this project, specimens subjected to significant bending moments (and shear) or to shear only was defined. Then, analytical developments based on the component approach and aimed at predicting the joint response have been carried out; their validity is demonstrated through comparisons with the tests. Finally, design guidelines have been provided.

1 INTRODUCTION

In order to connect the beam end-plates to the rectangular hollow section columns (with or without concrete inside), the following solutions are usually adopted in the construction (Figure 1): use of special bolts (blind bolts/flowdrill connectors) or use of intermediate elements (such as reverse U channels). These solutions are adopted to overcome the difficulty of placing bolts when the column section is a closed one. In the blind-bolt/flowdrill bolt joints, the beam end-plates are directly connected to the column faces as these bolts do not need an access to the inner side of the column faces. With respect to the joints using the U channels, a U channel is welded to the column and then the beam end-plates are attached to the U channel face by "classical" bolts. Design rules for these joint configurations are not yet covered in the current Eurocode 3, part 1–8 but these kinds of joint have been widely investigated in the literature (de Silva (2003, 2008), Elghazouli (2009), France (1999a, 1999b, 1999c), Gome (1990), Huang (2013), Jaspart (1997, 2005), Mágala-Chuquitaype (2010, 2010b, 2012), Park (2012), Vandegans (1995)). However, it can be pointed out that the two above joint solutions have some disadvantages, in particular their cost and their generally low rigidity and resistance. Indeed, on one hand, the use of special bolts or of additional pieces of the reverse U channel is costly. On the other hand, the mechanical behaviour of the mentioned joint solutions are mainly governed by the column or U channel faces component, subjected to the transverse tension forces through the bolts, which presents generally a rather weak rigidity and a resistance.

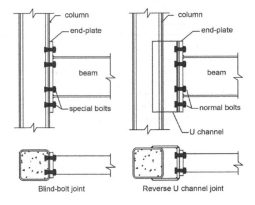

Figure 1. Blind-bolt and reverse U channel joints.

To avoid the disadvantages of the above solutions, it is proposed to use long bolts throughout the column, connecting the beam end-plates (Figure 2). Regarding the configuration, it appears that the cost of this joint may be reduced in comparison with the joints using special bolts or U channels. Moreover, the column face is not directly subjected to the tension forces through the bolts, so the rigidity and resistance of the joints may be improved. However, the use of long bolts for beam-to-column connections is still rare in the construction and no design procedure exists in the current Eurocodes.

The present paper summarizes the researches on the proposed joint configuration, from the experimental tests to the development of the design procedure. Section 2 presents the application of the component

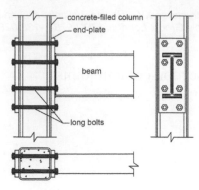

Figure 2. Proposed joint configuration.

Table 1. Design rules for the joint under bending.

No.	Components	Design rules
1	Bolts in tension	(1)
2	Beam web in tension	
3	End plate in bending	
4	Beam flange and web in compression	
5	Column in compression (joint side)	(2)
6	Column in compression (opposite side)	
7	Column panel in shear	

(1): They are basic components covered in EN1993-1-8 (§6.1.3); the design rules can be directly applied.
(2: The rules of "H-shaped column web" component in EN1993-1-8 (§6.1.3) can be used for the lateral faces of the steel tube; the rules in EN1994-1-1 may be used for the concrete core: §A.2.3.2 for the "column panel in shear" and §8.4.4.2 for the "column in compression".

method to the joint configuration; in which the additional rules needing to complete the design procedure are highlighted. Section 3 summarizes the experimental results. Section 4 devotes to the analytical developments and their validation. Section 5 is finally addressed to the concluding remarks.

2 DESIGN RULES

Let us consider a bolted extended end-plate joint using "normal" bolts and H-shaped column as a reference case, for which design rules are recommended in EN1993-1-8 and EN1994-1-1, and they are not reminded herein. These design rules may be applied to the investigated joint configuration; however the following remarks should be taken into account.

Joint under bending: Table 1 identifies the components of the investigated joint and the corresponding design rules. In principle, the resistance and stiffness of the joint under bending can be completely characterized. However, the component "column webs in tension" in the reference joint (i.e. using short bolts) should be replaced by the component "column webs in

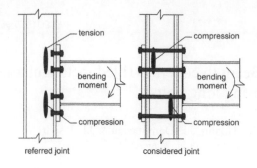

Figure 3. Column components.

Table 2. Design rules for the joint in shear.

If the preloading in the bolts, leading to friction forces between the end-plate and the column, is omitted:

No.	Components	Design rules
	End-plate in bearing	Basic components in
2	End-plate in block tearing	EN1993-1-8; the rules
3	Steel tube wall in bearing	can be directly applied.
4	Bolts in shear	Additional rules are
5	Concrete core in bearing	needed, see Section 4.

If the preloading in the bolts is considered (slip resistance): see EN1993-1.8 (§3.9.1).

Table 3. Load introduction.

Load transferred from the bolt to the steel tube:

$$F_{bolt-tube} = \min\left(nF_{tube}; \frac{A_s E}{A_s E + A_c E_c} nF_{bolt} \right)$$

Load transferred from the bolt to the concrete core:

$$F_{bolt-concrete} = \min\left(nF_{concrete}; \frac{A_c E_c}{A_s E + A_c E_c} nF_{bolt} \right)$$

Slip resistance between the steel tube and the concrete:

$$F_{tube-concrete} = F_{friction} + nF_{bolt}$$

the notations are defined in Figure 4.

compression" in the considered joint configuration as illustrated in Figure 3. Moreover, as the long bolts are used, the consideration of the preloading effect may be included in the calculation of the joint rigidity; this aspect will be detailed in Section 4.

Joint under shear and load introduction: Table 2 identifies the activated components of the joint under shear loads, while Table 3 summarizes the introduction of shear load at the joint level. Again, the bolt preloading may influence the characterization of the joint in shear. In one hand, when the preloading is taken into account, meaning that the shear load is transferred through friction between the end-plate and the steel tube faces, design rules are available in EN1993-1-8. On the other hand, if the bolt preloading is omitted, some additional rules are required for the considered joints, in particular: the resistance of the long bolts in

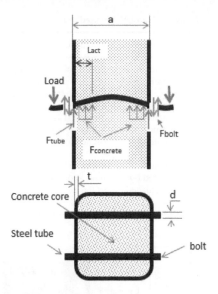

Figure 4. Behavior of long bolts and concrete-filled column under shear.

a is the width of the tube
d is the nominal diameter of the bolt
f_{yb} is the yield strength of the bolt
f_{ub} is the ultimate strength of the bolt
f_y is the yield strength of the tube
f_u is the ultimate strength of the tube
f_{ck} is the characteristic strength of the concrete
f_{cd} is the design strength of the concrete
n is the number of bolt sections in shear
t is the thickness of the tube wall
A_c is the area of the concrete core
A is the area of the steel tube cross-section
E is the Young modulus of the steel tube
E_c is the Young modulus of the concrete
F_{tube} is the bearing resistance of the tube wall [*]
F_{bolt} is the shear resistance of the bolts[*]
$F_{concrete}$ is the bearing resistance of the concrete core[*]
$F_{tube-concrete}$ is the slip resistance between the steel tube and the concrete core[*]
L_{act} is the active length of the bolts[*]
[*]: these quantities will be clarified in Section 4.

shear, the bearing resistance of the concrete core, and the steel tube-concrete slip resistance. These specific components will be investigated in Section 4.

3 EXPERIMENTAL RESULTS

Four tests on joint under bending (and shear) and six tests on joint under shear were defined and performed within HSS-SERF project "High Strength Steel in Seismic Resistant Building Frames", 2009–2013. The detail of the tests can be found in Hoang (2013), only the main points are summarized in the following.

3.1 *Joint under bending*

The aim of these tests was to prequalify the studied joints for building frames in medium to strong

Table 4. Geometrical properties and used materials of the tested specimens (Figure 5).

Tests	Column tube	Loading protocol
D1	SHS 300 × 300 × 12.5	Monotonic
D2	S460 grade	Cyclic
F1	SHS 250 × 250 × 10	Monotonic
F2	S700 grade	Cyclic

– C30/37 concrete is used for all the specimens; S355 steel is used for the beams and the end-plates; 10.9 bolts are used.
– Filet welds of 5 mm and 8 mm are used to connect the beam web and beam flanges to the end-plate, respectively.

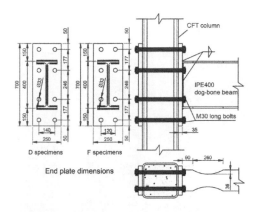

Figure 5. Geometrical properties of the specimens.

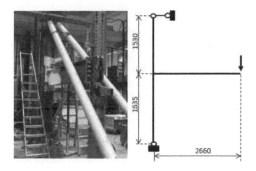

Figure 6. Test set-up used for the joint specimens.

earthquake area. Therefore, the dog-bone beam solution was used to ensure the location of the plastic hinges in the beam and so to avoid the joint yielding. Also two different steel grades (S460 and S700) were used for the column steel tubes to investigate the possibility of using high strength steel. However, only the rigidity of the joints will be presented and discussed in this section.

The geometrical properties and the used materials of the specimens are presented in Table 4 and Figure 5, while the test set-up is presented in Figure 6. The bolts were preloaded according to the combined method recommended in EN1090-1. The load – displacement (at the load application point) curves are presented in Figure 7, and one of the tested joints at

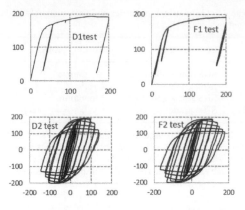

Figure 7. Load-displacement curves of the joint specimens.

Figure 8. D2 specimen at failure.

Table 5. Stiffness of the specimens (in kNm/rad).

Test	Measured stiffness	Average stiffness	k_b factor[*]
D1	154 900	149 860	≈23.0
D2	144 820	(D specimens)	
F1	113 850	113 400	≈18.0
F2	112 950	(F specimens)	

[*]ratio between the joint rigidity and the unit rigidity of the IPE400 beam.

failure is shown in Figure 8. The joint stiffness provided by the tests are reported in Table 5. From the tests, the following observations can be done:

- From the load-displacement curves, it can be seen that the rigidity of the specimens does not change during the tests, until the plastic hinges develop in the beam. It means that the bolt preloading remains until the end of the tests.
- The joints have a quite high rigidity, the coefficient k_b (ratio between the joint rigidity and the unit rigidity of the IPE400 beam) are about 23.0 and 18.0 (Table 5) for D and F configurations, respectively, with a beam span equals to 7.5 m (span coming from the reference building from which the joints were extracted). This means that the studied joints can be classified as rigid according to the criteria

Table 6. Objectives through the used test set-ups.

Set-up	Specimen	Objective
Set-up 1 (Figure 10)	T1: without nuts T4: preloaded bolts	Slip resistance between the concrete core and the steel tube with the presence of the long bolts.
Set-up 2 (Figure 10)	T2: non-preloaded bolts T5: preloaded bolts	Shear resistance of the bolts and bearing resistance of the concrete core.
Set-up 3 (Figure 10)	T3: non-preloaded bolts T6: preloaded bolts	Shear resistance of the bolts and bearing resistance of the steel tube wall.

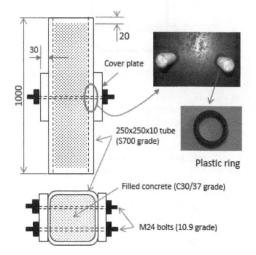

Figure 9. Tested column stubs under shear.

recommended in EN1993-1-8 ($k_b \geq 8.0$ for braced frames and $k_b \geq 25.0$ for unbraced frames), at least for braced frames.

3.2 Joints under shear

The specimen geometrical properties and materials are presented in Figure 9, more detail can be found in Hoang (2013). The specimens are made of concrete-filled rectangular column stub with a height of 1000 mm, two long bolts with a diameter of 24 mm passing through the column stub, and two cover plates representing the beam end-plates. 33 mm holes in the tube wall were made for the 24 mm bolts, and the plastic rings were used to center the bolt shanks in the holes and so, to avoid initial contacts between the bolts and the steel tubes. Different testing set-ups shown in Figure 10 were adopted for varied objectives, as presented in Table 6. Displacement transducers are used to record the displacements of the specimens, in particular: (1) the relative displacement between the steel tube and the bolts; (2) the relative displacement between the concrete and the steel tube; and (3) the relative displacement between the cover plates and the steel tube.

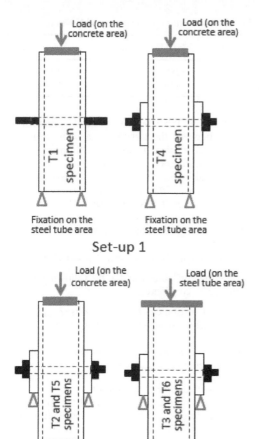

Set-up 1

Set-up 2   Set-up 3

Figure 10.   Used testing set-ups for the column stub tests.

Table 7 and Figure 11 reports the reached maximal loads and the observed failure modes of all the tested specimens. Applied load vs. displacement curves are given in Figure 12. These results will be used to propose analytical models in the next section.

4   ANALYTICAL MODELS

As mentioned in Tables 1, 2 and 3, there are some components of which the design rules should be investigated, they are dealt with in this section.

4.1   *Bolt preloading effect to the joint stiffness*

The bolt preloading has effects on the bolt stiffness itself but also on the stiffness of the column in transverse compression/tension component (Figure 13). Indeed, if the bolt preloading is omitted, the bolts and the column are two separate components while they work together if the preloading is considered as represented in Figure 13. The following equations

Table 7.   Maximal loads and failure modes of the column stub tests.

Test	Maximal load (kN)	Failure modes (Figure 11)
T1	1437	Significant slip between the concrete and the tube; the bolts are significantly deformed but not failed.
T2	1218	Small slip between the concrete and the tube; the bolt failed in shear; very small deformation of the tube in bearing.
T3	1210	Small slip between the concrete and the tube; the bolt failed in shear; significant deformation of the tube in bearing.
T4	3516	Significant slip between the concrete and the tube; the bolts are significantly deformed but not failed.
T5	1182	Small slip between the concrete and the tube; the bolt failed in shear; very small deformation of the tube in bearing.
T6	1218	Small slip between the concrete and the tube; the bolt failed in shear; significant deformation of the tube in bearing.

Note that the failure sections of the bolts are at the interface between the cover plate and the steel tube, in the unthreaded portion of the shank.

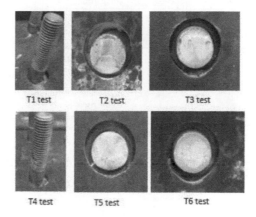

Figure 11.   Critical zones in the tested specimens.

can be used to estimate the effective stiffness of the "column + bolt" component for the two cases:

$$k_{column+bolt} = (1/k_{column} + 1/k_{bolt})^{-1} \qquad (1)$$

$$k_{column+bolt} = k_{column} + k_{bolt} \qquad (2)$$

where the preloading effect is omitted and considered, respectively.

In Eqs. (1) and (2), k_{column} and k_{bolt} are respectively the stiffness coefficients of the "column in compression/tension" component and "bolt in tension" component when they are considered in isolation, using the formulas recommended in EN1993-1-8 and EN1994-1-1.

Normally, the rigidity of the concrete-filled column in compression is much higher than the one of the long

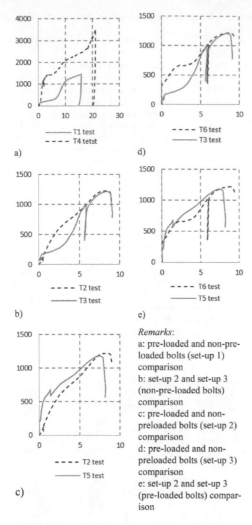

Table 8. Joint stiffness comparison (in kNm/rad).

Specimens	Eq. (1)	Eq. (2)	Tests (Table 5)
D	82986	149720	149 860
F	77383	115970	113 400

summarized in Table 8. A very good agreement is observed between the test results and the proposed model predictions taking into account the bolt preloading, with a difference of less than 3%. Moreover, a significant difference between the stiffness with and without account of the preloading is observed with a ratio between these two stiffness of 1.8 for specimens D and of 1.5 for specimens F, the stiffness with account of the preloading being the highest ones.

4.2 Joint under shear and load introduction

As mentioned in Tables 2 and 3, the following quantities should be considered:

- The shear resistance of the long bolts (F_{bolt});
- The bearing resistance of the concrete core ($F_{concrete}$);
- The friction resistance between the tube and the concrete core ($F_{tube-concrete}$).

4.2.1 Shear resistance of the bolts

The behavior of the bolts subjected to shear or to shear + bending is studied using the results from testing setups 2 and 3 (Figure 10), i.e. through tests T2, T3, T5 and T6. The maximal loads (around 1200 kN) and the failure modes (bolts in shear) are almost the same for these tests (Table 7 and Figure 11) which indicates that:

- At the ultimate state, there is no significant effect of the bolt preloading on the bolt shear resistance;
- The bending moment due to the gap between the application of the shear through the concrete core and the bolt support (equal to the tube thickness, see the testing setup 2 in Figure 10) is not significant and so, does not affect significantly the shear resistance of the bolts.

From the above observations, it seems to be reasonable to propose to use the shear resistance of bolts as given in EN1993-1-8 (§3.6.1) to predict the shear resistance of the long bolts in the investigated joints:

Remarks:
a: pre-loaded and non-pre-loaded bolts (set-up 1) comparison
b: set-up 2 and set-up 3 (non-pre-loaded bolts) comparison
c: pre-loaded and non-preloaded bolts (set-up 2) comparison
d: pre-loaded and non-preloaded bolts (set-up 3) comparison
e: set-up 2 and set-up 3 (pre-loaded bolts) comparison

Figure 12. Load-displacement curves for the column-stub tests.

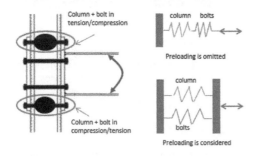

Figure 13. Effect of the bolt preloading on the joint stiffness.

bolts in tension. Therefore, it can be seen that the joint stiffness is considerably increased if the preload in the bolts is taken into account (k given by Eq. (2) is much higher than k given by Eq. (1)).

The rigidity of the components of the tested specimens were calculated, the detail can be found in Comeliau (2012) and Hoang (2014), the results are

$$F_{bolt} = \frac{\alpha_v f_{ub} A_{bolt}}{\gamma_{M2}} \quad (3)$$

the coefficient α_v is provided in EN1993-1-8, γ_{M2} is the safety factor, f_{ub} is the ultimate strength of the bolt, A_{bolt} is the area of the bolt cross-section in shear.

If formula (3) is used for the tested specimens: $F_{bolt} = 4 \times 0.6 f_{ub} A_{bolt} = 1172$ kN, where $f_{u,b} = 1008$ N/mm² (from the coupon tests); $A_{bolt} = 452$ mm² (nominal value); γ_{M2} is taken as 1.0 (to compare with the test result); and "4" is the number of bolt cross-section in shear. In comparison with the test value

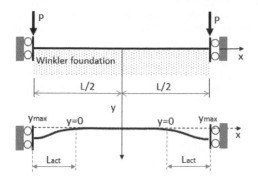

Figure 14. Active length of the bolts.

(Table 7), it can be seen that the use of the EN1993-1-8 formula to estimate the bolt strength in shear allows obtaining a good agreement with the test results.

4.2.2 Bearing resistance of the concrete core

The bearing resistance of the concrete core depends on two parameters: (1) the "active" length of the bolts taking into account their flexibility, and (2) the concrete strength taking into account the confinement effect.

Active length of the bolts

The model of a beam on an elastic foundation is used to estimate the active length of the long bolts; the Winkler foundation is adopted (Figure 14). The rotation of the two beam ends are supposed to be fully restrained due to the effects of the bolt nut and head. Two concentrated loads are considered at the two ends to simulate the action coming from the end-plates. The thickness of the steel tube wall is neglected in the model. It is proposed to define the active length as the distance from the beam end (the displacement is maximum) to the point where the displacement is considered as vanished (L_{act} in Figure 14). The following value is obtained for the active length of the bolt:

$$L_{act} = 2.5d \qquad (4)$$

Strength of the concrete core

Two effect should be taken into account when computing the strength of the concrete core under the bolts: the confinement effect due to the steel tube, and the local effect of the load. In this work, the strength of the concrete in the filled square hollow section, partially loaded, given in EN1994-1-1, §6.7.4.2(6) is proposed to be applied, in details:

$$\sigma_{c,Rd} = \left(f_{cd}(1 + \eta_{cL} \frac{t}{a} \frac{f_y}{f_{ck}}) \sqrt{\frac{A_c}{A_1}}, \quad \frac{A_c f_{cd}}{A_1}, \quad f_y \right) \qquad (5)$$

where A_1 is the loaded area under the active length of the bolts, equals to $L_{act}.d$; η_{cL} is a coefficient taking into account of the shape of the steel tube, equals to 3.5 for square sections; the other notation are defined in Figure 4.

Bearing of the concrete core

From the active length of the bolts (Eq. (4)) and the strength of the concrete (Eq. (5)), the bearing resistance of the concrete core for one shear plane can be defined:

$$F_{concrete} = L_{act} d \sigma_{Rd} = 2.5 d^2 \sigma_{Rd} \qquad (6)$$

In Eq. (6), d is the nominal diameter of the unthreaded portion of the bolts.

Eq. (6) is applied to compute the resistance of the concrete core of the tested specimens, 2380.1 kN for each specimen is obtained ($f_{ck} = 45$ N/mm^2 from the coupon tests, $f_y = 900$ N/mm^2 is used as the nominal value for the bolt). It shows that the resistance in bearing of the concrete core is higher than the bolt resistance in shear, it is in agreement with the test observations (T2 and T5 tests). However, the bearing resistance of the concrete core was not exposed through the tests and so, the so-obtained values cannot be strictly validated.

4.2.3 Steel tube – concrete slip resistance
Discussion on the test results

Tests T1 and T4 were dedicated to the characterization of the slip resistance between the steel tube and the concrete core. As can be seen in the test results reported in Table 8 and Figure 12, significant loads were reached during these tests, in particular for T4 test (3516 kN, in comparison to 1200 kN for the other tests).

During Test T4, a slip between the steel tube and the concrete occurred at a load of around 1500 kN. Then, the bolts entered into contact with the steel tubes and shear forces developed in these bolts; the system was able to sustain an additional load of around 2000 kN. However, as highlighted above, the shear resistance of the bolts has been estimated experimentally as equal to 1200 kN, which means that the bolt in shear was not the only component to support the additional load of 2000 kN.

This "over" resistance has been associated to the development of confinement effects in the concrete core, leading to high frictions between the steel tube and the concrete. The confinement effect becomes very important when the stress in the concrete is high leading to the change of Young modulus and Poisson ratio of the concrete. Noting that at the end of test T4, the normal stress in the concrete core equals to 66 N/mm^2 while the cylinder strength equals to 45 N/mm^2. A model to estimate the force in the bolts taking into account the friction has been established in Hoang (2013). It shows that even the external load of 3500 kN but the shear load in the bolts is less than 1200 kN – the ultimate value of the bolt in shear, in agreement with the test results.

Proposed model

Even a very high friction between the tube and the concrete is observed through the tests, and this phenomenon is well modelled, it is not reasonable to consider this effect in practice. Indeed, the stress in the concrete in the test is much higher than the nominal strength but this situation does not necessarily occur in practice. Therefore, the recommendation given in EN1994-1-1 (§6.7.4.2) on the friction between the concrete core and the steel tube is proposed to be applied for the investigated case, in which the confinement effect due to the present of the bolts are taken into account:

$$F_{tube-concrete} = (a-2t)b\tau_{Rd} + \mu F_{bolt}/2 + nF_{bolt} \qquad (7)$$

where a and t are defined in Figure 4; τ_{Rd} is the shear strength, which may be taken as equal to $0.4\,\text{N/mm}^2$ (see EN1994-1-1); μ is the friction coefficient, which may be taken as 0.5 (see EN1994-1-1), F_{bolt} is the shear resistance of the bolt (Eq. (3)), n is the number of bolt sections in shear.

5 CONCLUSION

A research conducted on bolted extended end-plate beam to concrete-filled rectangular hollow section column joint using long bolts was presented in this paper. It shows that the use of long bolts is a good solution to connect the beam end-plate to the concrete-filled rectangular hollow section column. It can avoid the use the intermediate elements (such as reverse U channels) or the use of special bolts (blind bolts/flowdrill connectors), leading to a saving of cost. The conducted experimental tests demonstrated the good mechanical behavior of the joints, in particular a high rigidity under bending and a high resistance in shear. Based on the component methods, some additional rules were proposed to complete the already available design rules for the investigated joints under bending, under shear and load introduction at the joint. The proposed models were validated through comparisons to the experimental results.

ACKNOWLEDGEMENTS

This work was carried out with a financial grant from the Research Fund for Coal and Steel of the European Community, within HSS-SERF project "High Strength Steel in Seismic Resistant Building Frames", Grant No. RFSR-CT-2009-00024.

REFERENCES

Comeliau L., Demonceau J.F., Jaspart J.P. 2012. Computation note on the design on bolted beam-to-column joints within HSS-SERF project. *Internal report, University of Liege.*

de Silva L.S., Neves L.F.N., Gomes F.C.T. 2003. Rotational stiffness of rectangular hollow section composite joints. *Journal of Structures Engineering* 129 (4).

de Silva L.S. 2008. Towards a consistent design approach for steel joints under generalized loading. *Journal of Constructional Steel Research* 64: 1059–1075.

Elghazouli A.Y. Mágala-Chuquitaype C., Castro J.M., Orton A.H. 2009. Experimental monotonic and cyclic behaviour of blind-bolted angle connections. *Engineering Structures* 31: 2540–2553.

EN1090-2. 2008. Execution of steel structures and aluminium structures – Part 2: Technical requirements for steel structures. *CEN, Brussels.*

EN1993-1-8. 2005. Design of steel structures. Part 1.8: Design of joints. *CEN, Brussels.*

EN1994-1-1. 2005. Design of composite steel and concrete structures. Part 1.1: General rules and rules for buildings. *CEN, Brussels.*

France J.E., Davison J.B., Kirby P.A. 1999a. Strength and rotation response of moment connections to tubular columns using flowdrill connectors. *Journal of Constructional Steel Research* 50: 1–14.

France J.E., Davison J.B., Kirby P.A. 1999b. Strength and rotation response of simple connections to tubular columns using flowdrill connectors. *Journal of Constructional Steel Research* 50: 15–34.

France J.E., Davison J.B., Kirby P.A.. 1999c. Moment-capacity and rotational stiffness of endplate connections to concrete-filled tubular columns with flowdrilled connectors. *Journal of Constructional Steel Research* 50: 35–48.

Gome FCT. 1990. Etat limite ultime de la résistance de l'ame d'une colonne dans un assemblage semi-rigide d'axe faible (in french). *Technical Report 203, University of Liege.*

Hoang V.L., Demonceau J.F., Jaspart J.P. 2014. Innovative bolted beam-to-column joints in moment resistant building frames: from experimental tests to design guidelines. In "Application of High Strength Steel in Seismic Resistant Structures", Ed. by Dubina et al. *"Orizonturi Universitare" Publishing House, Romania.*

Huang S.S., Davison B., Burgess I.W. 2013. Experiments on reverse-channel connection at elevated temperatures. *Engineering Structures* 49: 937–982.

Jaspart J.P. Pietrapertosa C., Weynand K., Busse E., Klinkhammer R. 2005. Development of a full consistent design approach for bolted and welded joints in building frames and trusses between steel members made of hollow and/or open sections: application of the component method. *CIDECT Report 5BP – 4/05, Vol 1: practical guidelines.*

Jaspart J.P. 1997. Recent advances in the field of steel joints – Column bases and further configurations for beam-to-column joints and beam splices. *Agregation Thesis, University of Liege.*

Hoang V.L. et al. 2013. High Strength Steel in Seismic Resistant Building Frames. HSS-SERF project, Deliverable D4: Prequalification tests on bolted beam-to-column joints in moment-resisting dual-steel frames. *University of Liege.*

Mágala-Chuquitaype C., Elghazouli A.Y. 2010. Behaviour of combined channel/angle connections to tubular columns under monotonic and cyclic loadings. *Engineering Structures* 32: 1600–1616.

Mágala-Chuquitaype C., Elghazouli A.Y. 2010b. Component-based mechanical models for blind-bolted angle connections. *Engineering Structures* 32: 3048–3067.

Mágala-Chuquitaype C., Elghazouli A.Y. 2012. Response and component characterization of semi-rigid connections to tubular column under axial loads. *Engineering Structures* 41: 510–532.

Park A.Y., Wang Y. C. 2012. Development of component stiffness equation for bolted connection to RHS columns. *Journal of Constructional Steel Research* 70: 137–152.

Vandegans D., Janss J. 1995. Connection between steel beam and concrete-filled R.H.S. based on the stub technique (threaded stub). In "Connections in steel structures III: Behaviour, strength and design", Ed. by Bjorhovde R., Colson A., Zandonini R.

Tubular Structures XV – Batista, Vellasco & Lima (eds)
© 2015 Taylor & Francis Group, London, ISBN 978-1-138-02837-1

Influence of outer diameter in confinement effect of CFT sections under bending

A. Albareda Valls & J. Maristany Carreras

Structures in Architecture. Polytechnic University of Catalonia (UPC), Barcelona, Spain

ABSTRACT: Concrete-filled tubes are usually used for columns and beam-columns due to their improved performance under compression, derived from the confinement effect over concrete. However, these sections are also increasingly used under bending, especially in case of large diameter pipes. Concrete in bent CFT sections becomes also confined, as a direct consequence of the circular geometry of the tube. However, it is known that the values provided by Eurocode 4 and other standards for the maximum bending strength of these sections is conservative for small diameters, but overestimated in large diameters. This is mostly due to the lack of connection between concrete and steel as the outer diameter grows, but also caused by a geometrical effect called as "ovalization" of the tube in this study. Compressed areas of the tube tend to expand outwards, while tensioned areas tend to crush concrete inwards. This phenomenon leads to a reduction of confining pressure over the core and therefore, to a reduction of the global bending strength. This study describes this effect and compares the confinement effect over the core in different specimens with different outer diameters.

1 INTRODUCTION

1.1 Concrete filled tubes under bending

Concrete Filled Tube sections (CFT) have been usually used as columns or beam-columns for their improved behavior under compression (Shimura and Okada, 2001). Lateral expansion of concrete in these sections is restricted by the steel tube, so that its compressive strength becomes clearly enhanced in most cases.

While numerous tests have been done about the compressive behavior of these sections and the confinement effect over the core, few bending experiments have been carried out involving large diameter pipes (Shanmugan and Lakshmi, 2001). Concrete of the core becomes also compressed over the neutral axis, and the circular geometry of the tube leads a uniform pressure over concrete in these parts.

Eurocode 4 about composite structures proposes a rigid-plastic analysis of the section in order to determine the maximum bending strength of the section, by using a coefficient of 0.85 for concrete (see Figure 1). This coefficient comes from a simplification of the parabola-rectangle diagram for concrete by considering also the confinement effect in those compressed areas. This coefficient grows up to 0.95 in case of the American AISC-LRFD. Even in most cases, they are conservative, for large diameter pipes these values may be overestimated (Lu and Kennedy, 1992).

This deviation of the real values with those proposed by the standards is due to the fact that few experiments have been done by using large diameter pipes. The researchers who have tested CFT sections under

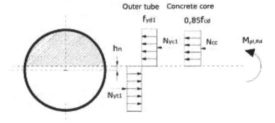

Figure 1. Rigid-plastic analysis of a CFT section under bending.

pure bending are (Chitawadagi et al., 2009), (Lu et al., 2007), (Probst et al., 2010), (Elchalakani et al., 2001), among others.

2 THE HYPOTHESIS

Derived from experiments carried out by (Elchalakani et al., 2010) and much others, we know that the maximum bending moment of CFT sections with small diameters (100 to 200 mm) is significantly higher than the values predicted by international codes or theoretical models. In the same way, we also know that this deviation decreases as the outer diameter increases. From 400 mm diameter and over, the value for the bending moment obtained from experiments are contrarily lower than the expected ones by the Standards.

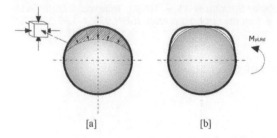

Figure 2. Effect of confinement due to geometry [a] and effect of ovalization in the compressed area of the tube [b].

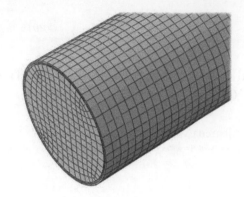

Figure 3. Detail of the mesh which has been used in the model.

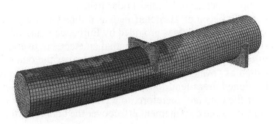

Figure 4. General view of the FE model. Half-part tests have been used.

This effect has been mostly attributed to a lack of composite action in relatively large tubes. This assumption is true, since the area of the core grows exponentially as the diameter increases, and this fact implies that tangential stresses in the interface also grow.

However, we also know that concrete in compressed areas of the core becomes confined by the steel tube and enhances its compressive strength (as in case of compressed sections). This effect, due to the influence of the circular geometry, contributes to enhance the bending strength of CFT sections with small diameters, but it does not almost exist in case of large diameters (from 400 mm and above).

This investigation studies the confinement effect over concrete in CFT sections subjected to pure bending, by using a numerical analysis and depending on the outer diameter. Confinement is usually defined depending on the D/t ratio, but it is more unusual to relate this phenomena to an absolute value of D (by considering the scale-effect).

The first hypothesis of this study is that confinement effect is influenced by the scale-effect, directly derived from geometry. Large tubes subjected to bending suffer from an "ovalization" effect in the their compressed part, independently of the D/t ratio. This effect leads to separate the tube from the core, so that confinement is usually reduced.

3 FINITE ELEMENT MODELLING

The hypothesis has been validated by using a numerical model by ABAQUS/Explicit commercial software, version 6.13.

3.1 FEM model geometry and meshing

The analysis has been carried out with solid continuum models, by using three-dimensional brick elements called C3D8 and available in ABAQUS 6.13 library. This is a three-dimensional brick element, defined by 8 nodes with three degrees of freedom at each one and without reduced integration.

Note from Figure 3 that the wall-thickness of the tube has been divided into three different elements, in order to take shell effects also into account. This is the minimum number of elements that allow to use brick instead of shell elements. The size of these elements is small enough to reproduce nonlinearities and interaction behavior in a proper way.

The mesh of the core has been automatically generated, by keeping a symmetric distribution of elements in both axes. In order to reduce the time process, half-part models have been used in most cases with symmetry conditions.

3.2 FEM material properties

3.2.1 Concrete

The damaged plasticity model for concrete (DPC), also available in ABAQUS 6.13 and based on (Lubliner et al. 1985), has been used to simulate the complex behaviour of concrete. This is a three-dimensional plasticity-based model where a damage criteria has been implemented. It is especially recommended to simulate concrete under high confined states as it is very sensitive to hydrostatic pressure.

Two hardening laws (depending on plastic strain rates) are defined separately into tension and compression. Beyond the maximum compressive or tensile yield stress, a damage process starts to degrade the stiffness matrix of the material. The damaged ratio in both states is expressed by the scalar coefficient d_c and it represents the percentage of cracking or crushing.

Using the model proposed, no discrete cracks appear beyond the maximum tensile stress due to the assumption of continuity, although they would really exist. When concrete reaches its maximum tensile

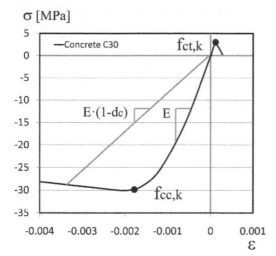

σ [MPa]

Figure 5. Uniaxial stress-strain curve used for concrete.

Table 1. Specimens tested by Elchalakani et al. (2001).

Specimen	D mm	t mm	f_c MPa	f_y MPa	E_c MPa	E_y MPa
CBC1	101.83	2.53	23.40	365	22735	2.0e5
CBC2	88.64	2.79	23.40	432	22735	2.1e5
CBC3	76.32	2.45	23.40	415	22735	2.1e5
C-NS	457.00	11.80	22.40	407	22244	2.0e5
C-S	457.00	11.80	22.40	407	22244	2.0e5

Table 2. Specimens used for validation, units in kNm.

Specimen	M_u test*	M_u FEM*	M_uFEM/$M_u$test
CBC1	11.33	11.90	1.053
CBC2	10.86	10.38	0.953
CBC3	6.92	6.55	0.946
C-NS	812.50	885.00	0.918
C-S	1239.30	1305.00	0.949

strength, its stress quickly decreases according to an linear equation.

This model allows defining the shape of the yield surface by means of the two parameters Kc and σ_{b0}/σ_{c0}: the first one describes the shape of the deviatoric plane, while the latter describes the ratio between the initial equibiaxial to the uniaxial compressive yield stress. A non-associated flow rule is used and the flow potential follows the Drucker-Praguer hyperbolic function (Richart, 1928). The dilation angle proposed is 31°, and the viscosity parameter is defined small (close to zero) in order to allow stresses outside the yield surface, and this way avoiding convergence difficulties.

The maximum tensile yield stress for concrete is considered as the 9% of the maximum characteristic compressive stress, and for the elastic Poisson's ratio, an initial value of 0.20 has been used. The initial modulus of elasticity has been considered as 33235 MPa.

3.2.2 Steel

A multilinear elastic-plastic model according to the von Mises yield criterion has been implemented for steel. A complete true stress-strain curve with hardening isotropic period has been used, obtained from uniaxial experimental tests. The Poisson's ratio has been considered 0.29 in the elastic range, and the elastic Young modulus, 210000 N/mm^2.

3.2.3 Interaction properties

Interaction between the tube and the core is always very important to guarantee the composite action of the section. The option "surface to surface" contact, available in ABAQUS, has been used in the model. Interaction properties which have been used are capable of transferring both normal and shear stresses.

The option "hard contact" has been chosen for normal stresses, and a friction coefficient according to Eurocode has been assumed for tangential stresses

(with an initial elastic slip). A constant value of 0.20 for this friction coefficient has been defined for most cases. Full connection (achieved with connectors) of specimen C-S has been defined by using a high value for the friction coefficient.

4 VALIDATION OF THE FE MODEL

In order to see if the proposed model is suitable for the analysis proposed in this study, it is necessary to validate it with different real experiments.

Three different tests of small CFT sections (diameters between 60 and 101 mm), together with two others of large pipes (457 mm) have been used for validation purpose. The first group was tested by Elchalakani et al. (2001), while the second one by Probst et al. (2010). Material and geometrical features of these specimens are shown in Table 1.

A good agreement between numerical and experimental results is observed in Table 2.

It is especially interesting to see the validity of the model for both small and large diameter groups. In case of large sections, the connection between steel and concrete becomes decisive; this is the reason why specimens C-S and C-NS have been included in the validation (with and without connectors). Probst et al. (2010) tested both specimens to see the influence of the diameter over the composite action. The obtained values are shown in Table 2.

A good agreement between numerical and experimental results can be observed in Figure 6 for section CBC1.

4.1 Validation of the friction coefficient

The friction coefficient which has been used between steel and concrete is a very important parameter

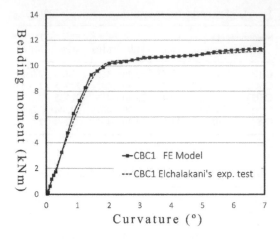

Figure 6. Validation of the model with specimen CBC1.

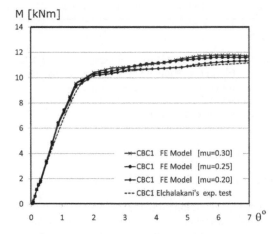

Figure 7. Comparison of FE and experimental results for CBC1, with different friction coefficients.

Table 3. Properties of the CFT Specimens proposed.

Specimen	D mm	t mm	f_c MPa	f_y MPa	D/t
C20/07-50	200.00	07.00	50.0	355.00	28.57
C40/07-50	400.00	07.00	50.0	355.00	57.14
C40/14-50	400.00	14.00	50.0	355.00	28.57
C80/14-50	800.00	14.00	50.0	355.00	57.14

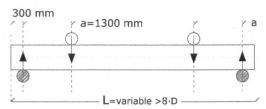

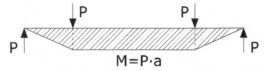

Figure 8. Scheme of the two loading point bending test used.

Table 4. Maximum and plastic bending moments.

Specimen	M_{max} FEM*	M_{pl} Eur*	M_{max}FEM/M_{pl}Eur
C20/07-50	156.05	104.50	1.49
C40/07-50	659.50	493.00	1.33
C40/14-50	1196.04	836.40	1.43
C80/14-50	3767.50	3950.00	0.95

*Values expressed in kN·m

in order to guarantee a correct transfer of stresses between steel and concrete. Different values of this coefficient have been tested to analyse section CBC1. The obtained results have been compared with the original experimental curve (see Figure 7).

It is clear that the value of the friction coefficient which more fits with the experimental curve is the same as proposed by Eurocode 4: a value of 0.20. This coefficient is more reliable for small tubes than in large tubes, since the adherence in this last group is not always guaranteed.

5 NUMERICAL STUDY

With the aim of describing the influence of the outer diameter D on the effect of confinement over the core in CFT sections, four different specimens have been analysed by using the previously described model. These specimens vary from 200 to 800 mm diameter,

all them with the same material properties, 50 MPa concrete and 355 MPa steel (see Table 3).

The previously mentioned specimens have been tested under pure bending, with a two-point bending test (see Figure 8). The length of the tubes have been defined differently, not less than 8 times the diameter.

6 ANALYSIS OF RESULTS

6.1 Analysis of the moment-deflection curves

The obtained moment-deflection curves have been compared with the values proposed by Eurocode 4 for the plastic bending moment resistance of circular CFT sections (Table 4).

From the results shown above, we can see that the Eurocode provides a set of values that are lower than those obtained by the numerical model in case of C20/07-50, C40/07-50 and C80/14-50 sections. Contrarily, in case of C80/14-50, we see that the Eurocode

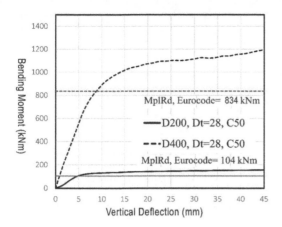

Figure 9. Maximum bending moments for sections with D/t = 28.

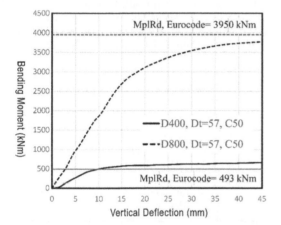

Figure 10. Maximum bending moments for sections with D/t = 57.

Table 5. Maximum and plastic bending moments.

Specimen	M_u FEM*	M_{pl} Theo*	M_u FEM/ M_{pl} Eur
C20/07-50	156.05	108.44	1.43
C40/07-50	659.50	476.18	1.38
C40/14-50	1196.04	867.48	1.37
C80/14-50	3767.50	3809.45	0.98

*Values expressed in KN·m

slightly overestimates the bending capacity of the section in a 5% more (see Figures 9 and 10).

In the following table (Table 4) we can see that the values coming from a theoretical rigid-plastic approach follow the same trend as those from the Eurocode. All sections, except for the case of C80/14-50 are underestimated by a theoretical rigid-plastic approach.

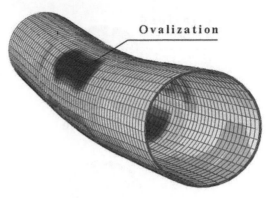

Figure 11. Effect of "ovalization" in steel tubes.

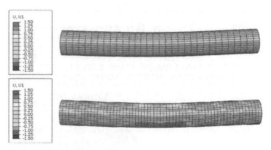

Figure 12. Lateral displacements of the tube UX for specimens C20/07-50 (top) and C40/14-50 (down).

6.2 Behavior of the steel tube

The analyses validates the initial hypothesis, in reference to the effect of "ovalization" of the steel tube. It tends to expand outwards in the compressed areas, and tends to deform contrarily in cracked areas of the core (see Figure 11). This phenomenon is independent of local buckling effects, but it clearly depends on the global diameter of the tube (this is, scale-effect). As larger is the tube, larger is this lateral deformation.

Lateral deformation that causes the effect of "ovalization" is about 1.50 mm in specimen C40/14-50, but much less in C20/07-50 (see Figure 12). This deformation follows different patterns depending on the length of the tube and the way of loading. In case of having the two points of loading very close one from the other, this deformation tends to be like infinite-shaped. In case of having the two points of loading relatively separated, this deformation looks more elliptical. These patterns of deformation could be theoretically explained by using the classical theory of elasticity of compressed shells, embedded by the two edges.

When this effect becomes significant, the steel tube becomes separated from the core, so that it is no more confined (see next Section). This phenomenon may reduce the adherence between both components too, considering that in large pipes this adherence it is already very low.

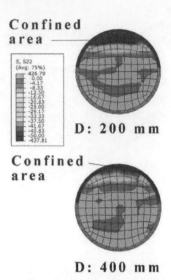

Confined area

D: 200 mm

Confined area

D: 400 mm

Figure 13. Confined areas (areas over 50 MPa are shown in dark grey) in case of specimens with a D/t ratio of 28.

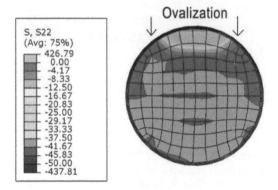

Figure 14. Consequences on confinement of "ovalization" effect.

6.3 Stress distribution in concrete core

The most important purpose of this study is to analyse the stress distribution in the concrete core, and the influence of "ovalization" of the tube over confinement effect. Concrete subjected to hydrostatic pressure enhances its compressive strength. In case of circular CFT sections, concrete of upper areas of the core becomes clearly confined, as a direct consequence of the circular geometry of the tube.

Figure 13 shows in dark grey those areas of the core which become subjected over 50 MPa (this is confined). The following figure compares the stress state of the core (longitudinal stresses) in case of 200 mm and 400 mm diameter sections, with the same D/t ratio of 28. Note that these confined areas are significantly larger in section C20/07-50 than in C40/14-50.

It is important to compare always sections with the same D/t ratio, since this implies that the compressed over cracked areas ratio in the core is similar in both cases. It is clear that, independently of the composite

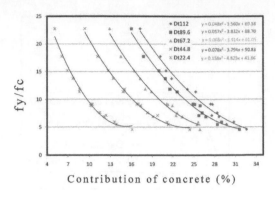

Figure 15. Contribution in % of the core to the global bending strength.

action, there is a clear scale-effect over confinement effect.

If we look at section C80/14-50, we see that the difference is even clearer. The section with a diameter of 800 mm shows the evidence of "ovalization" effect of the tube in relatively large diameters (Figure 14).

7 CONCLUSIONS

Concrete in CFT sections not only becomes confined under compressive states, but also under bending. This phenomenon takes place derived directly from the geometry of the tube (circular). Compressed concrete tends to expand laterally, but this expansion is limited by the walls of the steel tube.

Eurocode 4 for composite sections provide a methodology to determine the maximum bending moment of CFT sections, but the results obtained by this method are quite conservative in case of small diameter pipes, but truly overestimated for large diameters.

One of the reasons of this deviation is that the composite action between steel and concrete decreases as the diameter grows. In most cases, it is needed to add connectors in case of big diameter pipes to transfer stresses from one to the other component.

However, this paper shows that there is a second factor that contributes to enhance the global bending strength in small diameters, and to reduce it in case of large pipes: the "ovalization" effect of the steel tube. This phenomenon reduces the conditions of confinement of the core, and therefore, the maximum bending moment of the section.

The numerical analysis which has been carried out in this study, although it is not wide, explains the reduction of confinement effect as the absolute diameter of a section grows. This reduction is directly caused by this curious effect of "ovalization" that comes from the scale-effect.

Although the lack of connection is more important to define the ultimate bending strength, the reduction of the capacity of the core also contributes in this cause. Normally, the core in CFT sections contributes between 5 and 30% of the bending moment of the section (see Figure 15) depending on the D/t ratio and material strengths.

In case of sections with high D/t ratios, the contribution of concrete is more significant. Between a confined and unconfined concrete may be a difference of 30% or 40% of the total strength, so that in case of sections with high D/t ratios, this effect clearly reduces de bending capacity of the section.

REFERENCES

ANSI/AISC 360-10. Specification for Structural Steel Buildings. June 22, 2010. American Institute of Steel Construction, Chicago.

Chitawadagi, M.V.; Narasimhan, M.C. 2009. Strength deformation behaviour of circular concrete filled steel tubes subjected to pure bending. Journal of Constructional Steel Research; 1–10.

Cimpoeru, S.J.; Murray, N. 1993. The large deflection pure bending properties of a square thin-walled tubes. International. J. Mech. Sci. 1993;35(3-4):247–56.

Elchalakani, M.; Zhao, X.L.; Grzebieta, R.H. 2001. Concrete-filled circular steel tubes subjected to pure bending. Journal of Constructional Steel Research, 2001; 57: 1141–1168.

Eurocode 4. Design of steel and concrete structures. Part 1.1, general rules and rules for buildings. ENV 1994-1-1:1994. London

Han, L; Zhao, X. 2006. Double Skin Composite Construction. Structural Engineering Materials; 8:93–102.

Lu, F.W.; Li, S.P.; Li, D.W.; Sun, G. 2007. Flexural Behavior of concrete filled non-uni-thickness walled rectangular steel tube. Journal of Constructional Steel Research; 63: 1051–1057.

Lu, Y.; Kennedy, L. 1992. The flexural behaviour of concrete-filled hollow structural sections. Structural Engineering Report, 178.

Probst, A.D.; Kang, T.H.; Ramseyer, C.; Kim, U. 2010. Composite Flexural Behavior of Full-Scale Concrete-Filled Tubes without Axial Loads. Journal of Structural Engineer, ASCE, ; 1401–1412.

Richart, Brandtzaeg and Brown. 1928. A Study of Failure of Concrete under Combined Compressive Stresses. - [s.l.] : University of Illinois, Eng. Exp. Stn. Bull. 185.

Shanmugam, N.E.; Lakshmi, B. 2001. State of the art report on steel-concrete composite columns. Journal of Constructional Steel Research.; 57:1041–1080.

Shimura, Y.; Okada, T. 2001. Concrete Filled Tube Columns. Nippon Technical Steel Report, N° 77,78, July 1998; 57–64.

Tubular Structures XV – Batista, Vellasco & Lima (eds)
© *2015 Taylor & Francis Group, London, ISBN 978-1-138-02837-1*

Casting of composite concrete-filled steel tube beams with self-consolidating concrete

J.M. Flor, R.H. Fakury, R.B. Caldas & F.C. Rodrigues
Departamento de Engenharia de Estruturas, Universidade Federal de Minas Gerais, MG, Brasil

A.H.M. de Araújo
Vallourec Research Center Brasil, Vallourec Tubos do Brasil S.A.

ABSTRACT: This paper reports tests that have been conducted as part of a research project that aims to investigate the flexural behavior of real-scaled rectangular concrete-filled steel tube (CSFT) beams. Concrete-filling tests were carried out on two beam specimens of 12-m in length which were horizontally cast with self-consolidating concrete under two different conditions: with and without pressure. Ultrasonic inspection tests were also conducted to evaluate the efficacy of such non-destructive technique in detecting imperfections on the surface of concrete core due to casting. Finally, a total of six short column specimens were split into three groups to investigate an effective surface treatment to minimize the bond resistance between steel and concrete: oil, grease and mold release wax-based agent.

1 INTRODUCTION

In many applications steel hollow sections are concrete-filled in order to gain structural advantages (Han et al. 2014). Over the last two decades, some studies have been carried out on CFST beams (Lu & Kennedy 1994) (Elchalakani et al. 2001) (Gho & Liu 2004) (Han 2004) (Han et al. 2006) (Lu et al. 2009) (Moon et al. 2012) (Jiang at al. 2013) (Wang et al. 2014). However, previous experimental research work limited their scope to reduced-scale prototypes to investigate the flexural behavior of the composite beams. In general, specimen lengths were within 1000–2000 mm range. Besides, the concrete was cast vertically into the steel hollow section. Few researchers reported a very small amount of longitudinal shrinkage at the top of the specimens during curing. In such cases, high-strength epoxy (Han, 2009) or high-strength cement mortar (Gho & Liu, 2004) (Jiang et al. 2013) was used to fill this longitudinal gap.

Furthermore, gap between steel tube and concrete core has been recognized as a type of initial concrete imperfection in CFST circular members (Liao et al. 2011). Circumferential gaps may be caused by the concrete shrinkage and usually appears in vertical elements. On the other hand, spherical-cap gaps may occur in horizontal members due to the constructional process.

Experimental studies shall thus be conducted on the casting procedures of large-scaled CFST beams so as to promote their practicability as composite members in buildings with tubular structures. Within this context, the current study focuses on the examination of an adequate concrete mix and a feasible and efficient concrete-filling process to cast horizontally real-scaled steel tubes in the field.

This paper reports the initial stage of an ongoing experimental investigation on the flexural behavior of real-scaled rectangular concrete-filled steel tube (CFST) beams which will be tested to failure under pure bending. A total of fifteen 6000-mm-long specimens, including 12 ordinary CFST beams and 3 CFST beams with a concrete solid slab attached to the steel flange by shear connection will be tested in the program.

The main objectives of this paper are threefold. First, to report a series of concrete-filling tests performed in order to evaluate issues concerned with the feasibility and quality of casting of large scale rectangular tubes in the horizontal position in the field. Second, to describe a series of ultrasonic inspection controlled tests to check the efficacy of such non-destructive technique in detecting voids or discontinuities that may appear on the surface of the concrete core due to casting. Third, to report a series of surface treatment tests conducted aiming to identify an effective way to minimize the bond resistance at the interface between steel and in-filled concrete (see subsection 2.4 for further explanations).

2 EXPERIMENTAL PROGRAM

2.1 Material and test specimens

Steel tubes with nominal yield strength of 300 MPa were used in the tests. Ready-mix self-consolidating concrete (SCC), designed for a compressive cylinder

Table 1. Information of the tested specimens.

Specimen	Dimensions $D \times B \times t \times L$ (in mm)	Number of specimens
CF-R250 NP	$250 \times 150 \times 6.4 \times 12000$	1
CF-R250 PR	$250 \times 150 \times 6.4 \times 1000$	1
ST-R250 OL	$250 \times 150 \times 6.4 \times 1000$	2
ST-R250 GR	$250 \times 150 \times 6.4 \times 1000$	2
ST-R250 MR	$250 \times 150 \times 6.4 \times 1000$	2

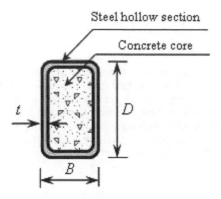

Figure 1. Rectangular concrete-filled steel hollow section.

strength at 28 days of 30 MPa and for a slump-flow of 550–660 mm, was used to cast the test specimens.

A total of 8 concrete-filled steel tube specimens were prepared. The nominal cross-section dimensions and specimen length for each test series are shown in Table 1.

The test specimens were labeled such that the test type and depth of steel tubes as well as the test condition could be identified from the label. The first two letters indicate the test type, where the prefix "CF" refers to concrete-filling tests and "ST" refers to surface treatment tests. The letter "R" refers to rectangular hollow section and is followed by the cross-sectional depth. The last two letters indicate the specific test condition, where the suffix "NP" refers to concrete-filling with no pressure, "PR" refers to concrete-filling with pressure, "OL" refers to surface treatment with oil, "GR" refers to surface treatment with grease, and "MR" refers to surface treatment with mold release wax-based agent. A cross sectional view of the specimens is illustrated in Figure 1.

2.2 Concrete-filling tests

Concrete-filling tests were conducted on two 12000-mm long specimens using different casting procedures. Both specimens were designed to have a circular, 100 mm in diameter hole on the superior flange at both ends of the steel tube. Additionally, both specimens were manufactured with a longitudinal cut throughout the entire length, splitting the tube into two halves, in order to facilitate mold releasing for

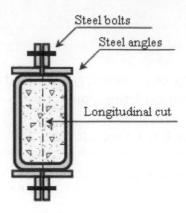

Figure 2. Details of the concrete-filling test specimen.

visual inspection of the in-filled concrete. At every meter along the length of the tube, two steel angles (L 75.6 × 75.6 × 6.3 × 75 mm) were welded both on the top and bottom flanges and fastened together with steel bolts to keep the specimen closed during casting and curing period, as shown in Figure 2. Specimens were capped at both ends with a 6.3-mm thick steel plate (250 mm × 300 mm) fastened to two steel angles (L 75.6 × 75.6 × 6.3 × 210 mm) welded on the exterior surface of the webs of the tube. During manufacture, the outer and inner surfaces of the steel tubes were painted with a protective coating to avoid corrosion. Before casting, the inner surface of the specimens was also treated with a thick coat of wax-based mold release agent proper for metallic formwork to prevent concrete from bonding to the steel surface.

Both specimens were horizontally filled with SCC without any vibration. Two different apparatus were designed to be coupled to the tube at one of its circular holes during casting: a PVC funnel and a metallic device, as shown in Figure 3. The former was used for casting one specimen under no pressure condition. The later was used for casting the other specimen under pressure condition. In this case, the concrete was pumped into the tube through a hose that was manually attached to the metallic device already fixed to the tube. After casting, the CFST specimens were kept undisturbed until they were opened for visual inspection of the in-filled concrete.

2.3 Ultrasonic inspection controlled tests

Ultrasonic inspection controlled tests were also conducted on sandwich-type models to investigate the efficacy of such non-destructive technique in detecting voids or discontinuities that may appear on the surface of the in-filled concrete casting. These models consisted of a 100-mm thick concrete layer intercalated by two 6.3-mm thick steel plates (250 mm × 300 mm). The model's components were fastened together by four steel bolts. Surface voids were simulated by fixing circular PVC caps of different diameters on one steel plate, as illustrated in Figure 4. The PVC caps had pre-defined locations according to the map shown

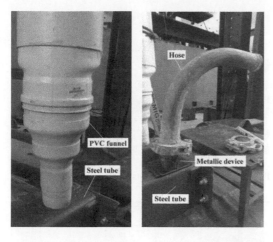

Figure 3. Concrete-filling tests apparatus: (a) PVC funnel, (b) metallic device and hose.

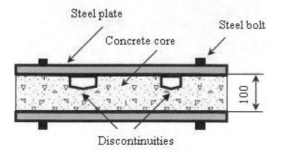

Figure 4. Ultrasonic inspection tests: model's cross-section.

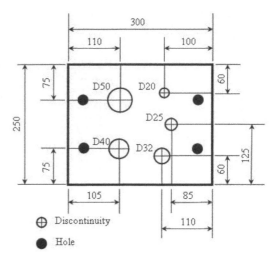

Figure 5. Ultrasonic inspection tests: discontinuities mapping.

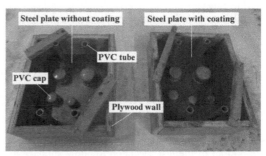

Figure 6. Ultrasonic inspection tests: preparation of models.

on Figure 5. In this figure, the labels identify the PVC caps as follow: the letter "D" refers to diameter and the following two digits indicate the diameter, expressed in mm.

Two models were prepared to investigate how surface treatment affects the efficacy of the ultrasonic inspection. In the first model, the surfaces of the steel plates were treated with protective anticorrosive paint coating. In the second model, the surfaces of the steel plates received no treatment.

A SCC concrete mix was designed to have the same properties of the ready-mix concrete used in the concrete-filling tests. One single batch was produced in the laboratory to cast both models. The PVC caps were previously fixed on the first steel plate surrounded by plywood walls and the four holes were protected with PVC tubes to avoid filling. Figure 6 shows the preparation of the models before casting. Concrete was then poured inside this box. After the curing period, the second steel plate was placed over the concrete layer. Four steel bolts were used to keep the model's components together. Finally, the model was turned up-side-down so to have the discontinuities faced upward.

Ultrasonic tests employed Olympus EPOCH XT ultrasonic flaw detector, Krautkramer MB4 SNB transducer and SAE 10W30 automotive oil as couplant.

2.4 Surface treatment tests

Six 1000-mm long CFST column specimens were manufactured with an anticorrosive paint coating applied to both inner and outer surfaces. The specimens were split into three groups of two specimens each. Three different inner surface treatments were used to investigate the most effective method to minimize the bond resistance between the steel tube and the concrete core: oil, grease and mold release wax-based agent. Based on the results, the inner surface of a series of CFST beams specimens will be treated accordingly before being subjected to pure bending tests. The purpose is to evaluate the behavior and performance of CFST beams under minimal bond resistance condition. This is believed to be the worst case in regards to flexural behavior of the composite beams.

Steel tubes were kept in an upright position for concrete casting. The bottom ends of the steel tubes were tightly covered with a polyethylene sheet which was removed prior to the test. SCC concrete was poured from the top without any vibration up to a mark below the top of the steel tube so to establish a 50-mm "gap" between the steel tube and the concrete core. After

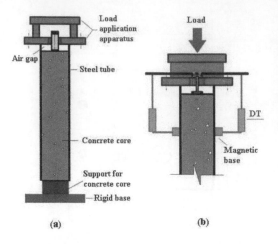

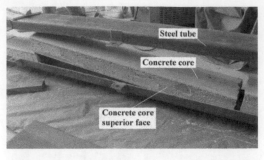

Figure 8. Concrete-filling tests: specimen mold releasing.

Figure 7. Surface treatment test setup (a) longitudinal view, (b) instrumentation.

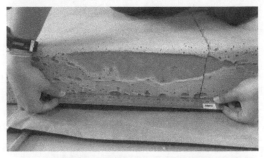

Figure 9. Concrete-filling tests: typical discontinuities on the top surface of concrete core.

casting, the CFST specimens were kept undisturbed until testing.

Displacement transducers (DTs) with a measuring range 0–100 mm were installed on two opposite faces of the specimen to measure the relative displacement (slippage) between the steel tube and the concrete core. Both DTs and a load cell were connected to a data acquisition system.

The tests were carried out under axial compression in a 500 kN capacity hydraulic actuator. The load was applied to the steel tube alone at the top of the test specimen and only the concrete core was supported at the bottom of the test specimen to allow slippage to occur. The loading process was performed under load-controlled type of loading. The tests were performed under monotonic loading and followed until the end of test. Each load interval was maintained for about 2–3 minutes. At each load increment, the load and the relative displacement readings were recorded. Figure 7 shows the test setup.

2.5 Results and discussions

2.5.1 Concrete-filling tests
The duration of each concrete-filling test, i.e., the time period between the start of concrete injection inside the tube at one end of the specimen and the start of concrete overflow at the other end, was 3 minutes, approximately. Both casting procedures were considered successful and efficient in filling completely the interior of the 12-m long tubes. Due to its lower height, ease of use and fast coupling to the tube, the metallic device showed to be an excellent solution to facilitate the injection of concrete inside the tube in the field. Additionally, SCC seems to be the most appropriate option for casting large-scaled CFST beams due to its high fluidity that allows self-consolidation without the use of any vibration and flow ability through reinforcement.

Specimens were opened approximately eight months after casting for visual inspection of the in-filled concrete. Figure 8 presents the mold releasing process. It was possible to macroscopically detect innumerous discontinuities or "gaps" of different sizes, shapes and depth on the top surface as shown in Figure 9.

The depth of the gaps was evaluated along the specimen entire length. On average, gaps were 1.97-mm and 2.99-mm deep for the specimen filled with the PVC funnel (without pressure) and the metallic device (with pressure), respectively. These observations corroborate previous literature findings regarding initial concrete imperfections due to construction process in circular CFST members.

The influence of these initial concrete imperfections originated from the horizontal casting procedure on the flexural behavior and resistance of rectangular CFST beams will be evaluated in the next phase of the present research project.

2.5.2 Ultrasonic inspection controlled tests
Ultrasonic inspection effectiveness on detecting discontinuities on the steel-concrete interface showed to be very sensitive to the surface conditions of the steel plate. As can been observed from Figures 10–11, ultrasonic inspection seems to be relatively effective for detecting gaps on the concrete core when the steel surface is not treated with any coating.

In the upcoming phase of this investigation, ultrasonic inspection experiments shall be carried out on two reduced-scaled CFST beam specimens with no

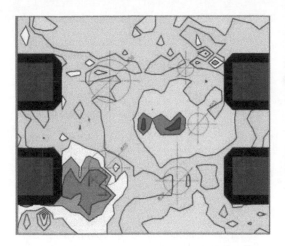

Figure 10. Ultrasonic inspection tests: results for model with coating.

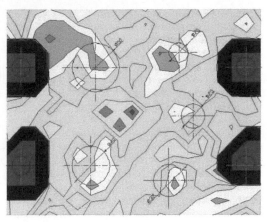

Figure 11. Ultrasonic inspection tests: results for model without coating.

Table 2. Surface treatment tests results.

Specimen	Ultimate Load (kN)	τ_{mexp} (MPa)	τ_{exp}/τ_{Rd}
ST-R250 OL 1	180.7	0.260	0.65
ST-R250 OL 2	221.7	0.319	0.80
ST-R250 GR 1	59.5	0.086	0.21
ST-R250 GR 2	56.1	0.081	0.20
ST-R250 MR 1	50.0	0.072	0.18
ST-R250 MR 2	24.1	0.035	0.09

inner surface treatment. Instead, it shall be cleaned to remove any rust and loose debris present. For preparing the specimens with gaps, PVC caps will be placed into the steel tube of CFST specimens prior to concrete casting according to a previously defined map. For one specimen, the location of the gaps shall be disclosed to the test operator prior to the experiment. For the other specimen, the efficacy of the technique shall be put into test and evaluation.

2.5.3 Surface treatment tests

Figure 12 shows the load-slippage relationship of all six specimens tested and Table 2 gives a summary of the test results.

The frictional resistance to movement along the steel-concrete interface (τ_{exp}) was determined dividing the ultimate load by the interface area. The interface area was calculated from the product of the interface length (950 mm) by the inner perimeter of the steel hollow section. The frictional resistance was also compared to the design shear resistance (τ_{rmRd}) specified by Brazilian standard ABNT NBR 8800:2008 for rectangular CFST columns, that is, 0,40 MPa. The experimental-to-design resistance ratio is also given in Table 2.

As can be seen from the table, the paint and mold release wax-based agent surface treatment provided the best results, reducing the bond resistance to values as low as 9% of the design shear resistance. By combining paint and grease, it was possible to reduce the bond resistance to 20–21% of the shear resistance provided by the Brazilian standard. However, the oiled specimens did not show satisfactory results: the paint and oil surface treatment was able to reduce the bond strength to 65–80% of the design value.

Experimental bond resistance values have been reported in the literature related to CFST columns. Rectangular ($120 \times 80 \times 5$ mm) specimens tested by Shakir-Khalil (1993) showed an average bond strength of 0.83 MPa. The author also reported an average bond strength of 0.44 MPa for rectangular

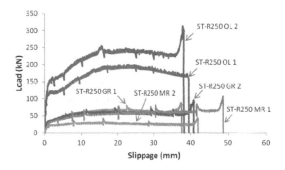

Figure 12. Surface treatment tests: load-slippage curves.

($150 \times 150 \times 5$ mm) specimens. Starossek & Falah (2009) found bond strength of 0.685 MPa for square ($150 \times 150 \times 6.3$ mm) specimens. Should the experimental results for the frictional resistance be compared to the experimental average values for the bond resistance instead of the design value, the effectiveness of the surface treatment would be even more evident.

3 CONCLUSIONS

The following conclusions can be drawn within the scope of the current study:

- Self-consolidating concrete shows to be an effective concrete mix to in-fill horizontally large-scaled rectangular steel tubes having length up to 12 m regardless the use or not of pressure during casting.
- Initial concrete imperfections, also named gaps, are expected to occur on the top surface of the concrete core whenever casting is made with rectangular CFST beams positioned horizontally.
- The influence of such gaps on the behavior of rectangular CFST beams shall be further investigated.
- Ultrasonic inspection demonstrates to be a non-destructive method of relative efficacy for detecting gaps on the concrete core when the steel surface is not treated with any coating.
- The association of paint and grease or of paint and mold release wax-based agent reduces significantly the bond resistance at the steel-concrete interface of rectangular CFST short column specimens. The frictional resistance was 20–21% of the Brazilian standard design shear resistance for the former and 9–18% for the latter.
- The association of paint and oil seems to be not effective to minimize the bond resistance at the interface between the steel tube and the concrete core of rectangular CFST short column specimens.

ACKNOWLEDGEMENT

This paper is based on the dissertation research being conducted by Flor for the degree of Doctor of Philosophy at Federal University of Minas Gerais (UFMG). Concrete-filling tests and surface treatment tests were carried out in the Experimental Structural Analysis Laboratory (LAEEs) at UFMG. Ultrasonic inspection tests were conducted at the Center of Nuclear Technology Development (CDTN). The authors are grateful to Vallourec Tubos do Brasil S.A. for providing all the steel tubes in the experimental programme. Thanks are given to Lafarge for providing the ready-mixed self-consolidating concrete. The financial support of the National Council for Scientific and Technological Development (CNPq) and Foundation for Research Support of Minas Gerais (FAPEMIG) is acknowledged.

REFERENCES

ABNT NBR 8800:2008. Design of steel and composite structures for buildings. Rio de Janeiro: Associação Brasileira de Normas Técnicas (Brazilian Association of Technical Standards); 2008 (in Portuguese).

Elchalakani, M.; Zhao, X.L. & Grzebieta, R.H. 2001. Concrete-filled circular steel tubes subjected to pure bending. Journal of Constructional Steel Research 57: 1141–1168.

Gho, W.-M. & Liu, D. 2004. Flexural behaviour of high-strength rectangular concrete-filled steel hollow sections. Journal of Constructional Steel Research 60: 1681–1696.

Han, L.-H. 2004. Flexural behaviour of concrete-filled steel tubes. Journal of Constructional Steel Research 60: 313–337.

Han, L.-H.; Li, L. & Bjorhovde, R. 2014. Developments and advanced applications of concrete-filled tubular (CFST) structures: members. Journal of Constructional Steel Research 100: 211–228.

Han, L.-H.; Lu, H.; Yao, G.-H. & Liao, F.-Y. 2006. Further study on the flexural behaviour of concrete-filled steel tubes. Journal of Constructional Steel Research 62: 554–565.

Jiang, A.; Chen, J. & Jin, W. 2013. Experimental investigation and design of concrete-filled steel tubes subject to bending. Thin-walled Structures 63: 44–50.

Liao, F.-Y.; Han, L.-H. & He, S.-H. 2011. Behaviour of CFST short column and beam with initial concrete imperfection: experiments. Journal of Constructional Steel Research. 67: 1922–1935.

Lu, H.; Han, L.-H.; Zhao, X.-L. 2009. Analytical behavior of circular concrete-filled thin-walled steel tubes subjected to bending. Thin-walled Structures 47: 346–358.

Lu, Y.Q. & Kennedy, D.J.L. 1994. The flexural behaviour of concrete-filled hollow structural sections. Canadian Journal of Civil Engineering 21(1): 111–130.

Moon, J.; Roeder, C.W.; Lehman; D.E.; Lee, H.-E. 2012. Analytical modeling of bending of circular concrete-filled steel tubes. Engineering Structures 42: 349–361.

Shakir-Khalil, H. 1993. Pushout strength of concrete-filled steel hollow sections. The Structural Engineer 71(13): 230–233 and 243.

Starossek, U. & Falah, N. 2009. The interaction of steel tube and concrete core in concrete-filled steel tube columns. In Shen, Chen & Zhao (eds.), Tubular Structures XII; Proc. 12th International Symposium on Tubular Structures, Shanghai, 8–10 October 2008. Leiden: CRC Press/Balkema.

Wang, R.; Han, L.-H.; Nie, J-G.; Zhao, X.-L. 2014. Flexural performance of rectangular CFST members. Thin-walled Structures 79: 154–165.

Tubular Structures XV – Batista, Vellasco & Lima (eds)
© 2015 Taylor & Francis Group, London, ISBN 978-1-138-02837-1

Analytical behaviour of concrete-encased CFST column to steel beam joints

W. Li, W.W. Qian & L.H. Han
Department of Civil Engineering, Tsinghua University, Beijing, China

X.L. Zhao
Department of Civil Engineering, Monash University, Melbourne, Australia
National 1000-talent professor at Tsinghua University

ABSTRACT: The concrete-encased concrete-filled steel tubular structures have been used in some high-rise buildings in China. This kind of steel and concrete composite structure has higher strength and better fire performance when compared to its concrete-filled steel tube counterpart. However, very few research has been conducted for the behaviour of concrete-encased concrete-filled steel tube to steel beam joints, which hinders the application of this kind of structures. This paper reports the analytical behaviour of concrete-encased concrete-filled steel tube to steel beam joints. Various connection details are proposed, and the corresponding numerical models are established and verified. The behaviour of different kinds of joints is compared using the verified numerical models. It is shown that for the joint which has the internal diaphragm, the strength and stiffness characteristics are the best among all joint specimens.

1 INTRODUCTION

The concrete-encased concrete-filled steel tubular (concrete-encased CFST) column consists of the core concrete, the inner steel tube and outer reinforced concrete (RC). When compared to the conventional concrete-filled steel tube (CFST), the concrete-encased CFST has higher fire resistance and durability owing to the presence of the outer RC. With the protection of outer RC, the buckling mode of the steel tube can also be improved. When compared to the traditional RC member, concrete-encased CFST can achieve higher ductility for the inner steel tube. Also, the inner tube can be designed to sustain the constructional load first, and then the outer reinforcement can be assembled and the concrete can be placed. When compared to the concrete-encased steel column which usually uses H-shape steel beam, the concrete-encased CFST is expected to have higher ductility and strength, for the inner tube can provide certain confinement to the core concrete. The concrete-encased CFST members have attracted interests of structural designers and have been used in some high-rise buildings and industrial buildings in China.

Although the concrete-encased CFST columns have many advantages over other column types, their usage has been limited due to the complexity of the beam-to-column joints. The concrete-encased CFST column can be connected to either RC beam or steel beam in practice. For the joint with RC beams, the steel reinforcements play important roles in the composite joint. However, some reinforcements may go through the steel tube which can certainly reduce the capacity of the cross section. For the joint with steel beams, the concrete-encased CFST member can easily be connected with beams by welding or other connection details while the capacity of the inner steel tube is not reduced. Another advantage of connecting with steel beams is that, the steel beams and the inner tubes can form a temporary frame to resist the constructional load. Meanwhile, the structural system consisted of concrete-encased CFST columns and steel beams is expected to have higher ductile behaviour than its counterpart with RC beams. Figure 1 shows the photo of concrete-encased CFST column to steel beam joints.

The behaviour for the concrete-encased CFST column to steel beam joints can be complicated for it has various kinds of components. Although some data is available on concrete-encased CFST column behaviour, limited work has been conducted to these composite joints. Usually the steel beam can be connected directly to the inner steel tube for the concrete-encased CFST structures, thus the joint behaviour maybe similar to that of concrete-filled steel tubular structures in some extent. In the past, extensive studies have been carried out to analyze the seismic behaviour of CFST column to steel beam joints (Alostaz & Schneider 1996, Schneider & Alostaz 1998, Nishiyama et al. 2004, Ricles et al. 2004, Li et al. 2009, Han & Li 2010, Li & Han 2011), where

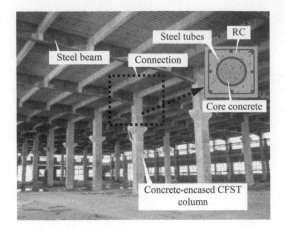

Figure 1. Concrete-encased CFST column to beam joints (Liao et al. 2014).

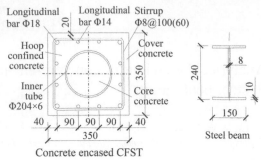

Figure 2. Cross sections of column and beam (units:mm).

some useful information could be found to help understanding the behaviour of concrete-encased CFST joints.

For the behaviour of concrete-encased CFST column to beam joints, Liao et al. (2014) has carried out experimental research on both steel beams and RC beams joints. The stiffeners were applied for the steel beam joints. The main test parameter was the axial load level on the column. The influence of the RC slab was also evaluated. The test results showed that the composite joints displayed favourable seismic performance and might be applicable in earthquake-prone regions. The authors (Qian et al. 2013) also conducted analytical investigation on the cyclic behaviour of concrete-encased CFST joints.

However, to date, there's limited research on the behaviour of concrete-encased CFST column to steel beam joints with various connection details. This paper thus presents an analytical investigation on various connection alternatives to concrete-encased CFST columns. The objective is to study the joint behaviour of different connection details for the joints with steel beams. Six different kinds of connection details were proposed. Nonlinear finite element (FE) models were developed for concrete-encased CFST column to steel beam joint under monotonic loading. Experimental results reported by Liao et al. (2014) were used to verify the numerical model. The load-displacement as well as the moment rotation relationships were simulated, and the inelastic performance of each joint was inferred from the verified FE models.

2 CONNECTION DETAILS

The concrete-encased CFST column has circular inner steel tube and square outer RC profile, since this type of composite column is most commonly used in practice and presents several detailing difficulties. The exterior joints are modeled, as it is sufficient to investigate the connection details using this joint type. All joint specimens have the same beam and column configurations. The schematic views for the cross sections of the concrete-encased CFST column and the steel beam are depicted in Figure 2.

Six connection details are proposed in this paper considering the performance and the construction ability of the joints. The connection details are as follows:

– Type I: the joint uses the external diaphragm. The joints with external diaphragm was a significant improvement than the simple welded connection details. The width of the external diaphragm is not provided for concrete-encased CFST columns in current codes of practice, therefore the width is designed according to the diameter of inner tube and positions of longitudinal bars.
– Type II: the joint uses the external diaphragm. The width of the external diaphragm could affect the casting of the outer concrete, and designers are tending to use narrow external diaphragm if the joint performance can keep the same. The width of Type II joint is only half of that of the Type I joint;
– Type III: the joint uses the 1/3 external diaphragm. The diaphragm is mainly used to transfer the tensile and compressive load to the inner tube. The diaphragm width of Type III joint is same as that of the Type I joint, while only 1/3 of the diaphragm is considered;
– Type IV: the joint uses the internal diaphragm. The internal diaphragm is commonly used in CFST column to beam joint, and joints using internal diaphragm can achieve favorable performance under both monotonic and cyclic loading;
– Type V: the joint uses vertical stiffeners. The trapezoid stiffeners are designed for the joint specimen;
– Type VI: the steel beam is directly welded onto the tube wall without any additional connection detail. Usually it is not recommended to weld steel beam directly onto the steel tube in conventional CFST structures, however, this connection method might work with the help of outer RC.

Figure 3 gives the schematic view of joint specimens. The width of the Type I joint is 30 mm, while that of the Type II joint is 15 mm. The diameter of the hole for the internal diaphragm is 100 mm. It is used for the

118

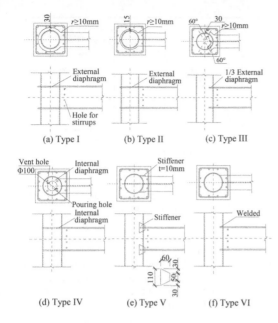

Figure 3. Configuration of connection details (units:mm).

Table 1. Material properties for steel components.

Components	t or d (mm)	f_y (N/mm²)	f_u (N/mm²)	E_s (N/mm²)
Beam flange	10	345	449	2.06×10^5
Beam web	8	345	449	2.06×10^5
Diaphragm	10	345	449	2.06×10^5
Tube	6	345	449	2.06×10^5
Stirrup	8	335	436	2.06×10^5
Longitudinal bar	14	335	436	2.06×10^5
Longitudinal bar	18	335	436	2.06×10^5

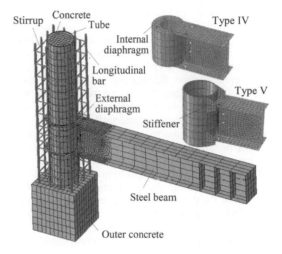

Figure 4. Schematic view of FE model.

concrete placement together with four Φ24 mm vent holes at corners. The thickness for the external and internal diaphragms is the same as that of the beam flange. The thickness of the vertical stiffener for the Type V joint is also 10 mm.

The distance between two rotation points of the column is 1500 mm, and the distance between the loading point and the column surface is 1125 mm. The moment capacity of the steel beam is 152 kNm.

3 FINITE ELEMENT MODELLING

3.1 General description

The FE models were established using the software ABAQUS, where the nonlinearity of the material, the contact between two materials and the large geometry deformation were considered.

For the material modelling, the steel was modeled by the elasto-plastic model with the combination of isotropic and kinematic hardening models. In the elastic stage, the Young's modulus and Poisson's ratio were set to be 206,000 N/mm² and 0.3, respectively. The von Mises yield criterion and an associated plastic flow rule were used. Ideal elastic-plastic models were applied for the tube and the beam, whilst the linear hardening model was applied for reinforcements. The yield and tensile strength for different steel components are shown in Table 1.

For the concrete materials, the concrete damaged plasticity (CDP) model in ABAQUS with a non-associated plastic flow rule and isotropic damage was employed. The concrete for the composite column was divided into three parts to consider the behaviour differently, i.e. the core concrete, the hoop-confined

concrete and the cover concrete, as shown in Figure 2. The stress-strain curves developed by Han et al. (2007), Han and An (2013) and Attard et al. (1996) were used for the core concrete confined by the steel tube, the hoop-confined concrete confined by stirrups and the cover concrete without confinement, respectively. The compressive strength of concrete cubes for the inner and outer concrete was 60 MPa and 40 MPa, respectively. The tensile performance of concrete was described by the fracture energy based approach. The Poisson's ratio was set to be 0.2 for concrete. Other parameters such as the dilation angle (ψ), the flow potential eccentricity (e_f), the ratio of the compressive strength under biaxial loading to uniaxial compressive strength (f_{b0}/f_c') and the stress invariant ratio on the tensile/compressive meridian (K_c) were set as 30°, 0.1, 1.16 and 2/3, respectively.

The shell element (S4R) was used to model the steel beams, the diaphragms and the stiffeners. The solid element (C3D8R) was used to model the concrete and steel tube. Steel reinforcements were modeled by the truss element (T3D2). The mesh convergence study was performed to determine the proper mesh density. Figure 4 shows the element mesh of typical composite joints.

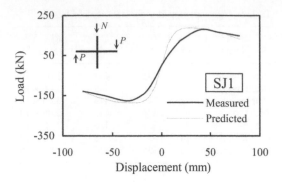

Figure 5. Comparison of measured and predicted results for concrete-encased CFST column to steel beam joint.

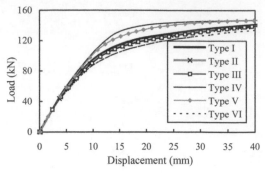

Figure 6. Load-displacement relationship for concrete-encased CFST column to steel beam joints.

The joint models had several contact relationships, i.e. the steel tube to core concrete surface, the steel tube to outer concrete surface, the diaphragm to concrete surface, the stiffener to concrete surface and the steel reinforcement to concrete pair. For the contact behaviour of the tube-concrete, diaphragm-concrete and stiffener-concrete surfaces, the surface-based interaction models were used. The Coulomb friction model and the hard-contact model were used in the tangential and normal direction, respectively. Usually the frictional factor ranges from 0.2~0.6. Previous investigation showed that the frictional factor of 0.6 is suitable for the steel tube-concrete interface (Li & Han, 2011). Therefore the frictional factor was taken as 0.6 in this analysis. For the steel reinforcements, the longitudinal bars and stirrups were embedded in the concrete elements.

The exterior joint was assumed to be pinned at both column ends while only the vertical movement was allow at the end where the axial load was applied. The beam end was allow to move in the plane. Two steps were induced to apply the loads on the joint. A constant axial load of 100kN was applied on the top of the column in the first step for all six joint specimens, and then the vertical loads was applied at the beam end. The Newton-Raphson method was used for the numerical solution.

3.2 Verification

Liao et al. (2014) conducted experimental investigation on the concrete-encased CFST column to beam joints, among which two specimens were interior joint with steel beams. The constant axial load was applied on the top of the column, and the reverse cyclic loading was applied at both beam ends. The specimens showed ductile behaviour, and the load-deformation relationships as well as other indexes for evaluating the seismic behaviour were obtained.

The specimen SJ1 was taken for the validation of the numerical model. The profile of the outer concrete was 350 mm, and the diameter of the inner steel tube was 140 mm. The axial load level on the column was approximately 0.2. More details could be found in Liao et al.(2014).

The measured average ultimate strength for the composite joint was 178kN, while the calculated one was 186kN. The load-deformation relationship obtained from the numerical model gave a reasonable prediction on the envelop curve of the cyclically loaded specimen, as shown in Figure 5.

4 ANALYTICAL BEHAVIOUR

The load-displacement (P-Δ) relationships for all six specimens are depicted in Figure 6.

For the P-Δ relationships, all curves exhibit several stages, i.e. the elastic stage, the elasto-plastic stage and the plastic hardening stage. The stiffness of the curves for the Type IV joint is 1.33×10^4 kN/m, which is the highest one among these joints. The strength for the Type IV and Type V joints is higher than that of other four joints. For the joints with external diaphragm, the joint with full diaphragm (the Type I joint) shows the highest stiffness and strength, while the differences between the Type II and Type III joints are small. The joint without any additional connection details (the Type VI joint) presents the lowest strength among these joints. The strength of the Type VI joint is 123kN when the 2% inter-storey drift (obtained by dividing the beam end displacement with the beam length) is reached, which is 14.6% lower than that of the Type IV joint.

The moment-rotation (M-θ) relationships for the six specimens are shown in Figure 7, where the joint moment M_j is obtained by multiplying the load with the force arm, and then normalized by the plastic moment resistance of steel beam $M_{b,p}$. It can be seen that similar to P-Δ relationships, M-θ relationships also have three different stages. The stiffness and strength of the internal diaphragm joint (the Type IV joint) is the highest among these joints, while that of the simple welded joint (the Type VI joint) is the lowest. The joint classification method from Eurocode 3, Part:1–8 is employed to analyze M-θ relationships tentatively. The joint whose initial stiffness of the M-θ relationship is larger than $25EI_b/L_b$ and $8EI_b/L_b$ can be regarded as the rigid one for sway and non-sway frames, respectively, where E is the elastic modulus of beam, I_b is

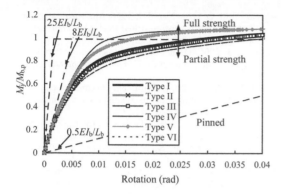

Figure 7. Load-displacement relationship for concrete-encased CFST column to steel beam joints.

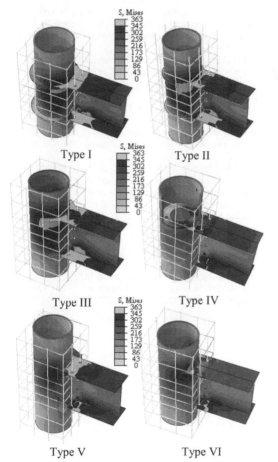

Figure 8. Stress distributions for steel tubes, diaphragms, flanges and reinforcements.

the second moment of area of beam, and L_b is the beam span. According to this classification method, all joints can be regarded as semi-rigid joints, for the initial stiffness of $M\text{-}\theta$ relationships is all close to but less than $8EI_b/L_b$. Figure 7 also indicates that, for Type IV and Type V joints, the joint moments are more than

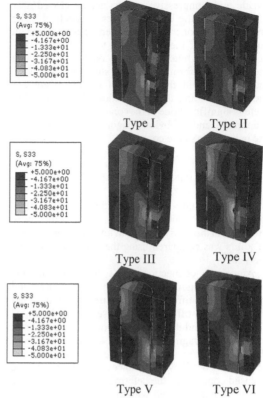

Figure 9. Stress distributions for core concrete and outer concrete.

the beam flexural strength when the rotations reach 0.02rad. Other joints could be catalogued to "full-strength" ones only when the rotation is larger than 0.03rad or more.

Figures 8 shows the stress distributions for steel components when the rotation of the joint reaches 0.02rad. For the Type I joint, the end of beam flange yields at this deformation level and 1/3 of the tensile diaphragm hardens. For the Type II joint, the average stress level of the diaphragm is higher than that of the Type I joint, while the hardening region on the tube is larger as well. For the Type III joint, the hardening region on the tube is the largest one among joints having external diaphragms. For the Type IV joint using internal diaphragm, the diaphragm helps to transfer the load to the core concrete. The stress on the tube is smaller than that of joints using external diaphragms, and the hardening regions only exist near edges of the beam end. For the Type V joint, the stress level on steel tube is lower than that of joints using external diaphragms, as the vertical stiffeners help to transfer the load to the tube more uniformly. For the Type VI joint, stress concentrations are found at edges of the beam flange, and the stress on the beam tensile flange is less than the yield stress.

Figure 9 shows the stress distributions of the core and outer concrete. It can be seen that for all specimens, tensile and compressive regions are developed near the upper and the bottom flanges, respectively. The encased outer concrete certainly enhance the strength and stiffness of the bare CFST to steel beam connection. Meanwhile, the encased concrete constrained the outward deformation of the steel tube and work together well with the interior CFST component. For the Type IV joint, the deformation of the core concrete is significant due to the constraint of the internal diaphragm, and an obvious diagonal strut is formed in the panel zone's core concrete.

5 CONCLUSIONS

Six different kinds of connection details for concrete-encased CFST column to steel beam joints were proposed in this study. The behaviour of different kinds of joints was compared using the verified numerical models.

Among six connection details, the joint using the internal diaphragm exhibited the highest stiffness and strength. The stiffness and the strength of the joint using vertical stiffeners was the second highest one. The stiffness and the strength of joints using different external diaphragms were almost the same. However, the stiffness and strength of the joint without any connection detail was the lowest one. The concrete encased constrained the deformation of the steel tube and work together well with the interior CFST component. All six joints could be classified as semi-rigid joints tentatively according to Eurocode 3.

ACKNOWLEDGEMENTS

This research is part of Projects 51208281 supported by National Natural Science Foundation of China (NSFC) and the Specialized Research Fund for the Doctoral Program of Higher Education (SRFDP) (20110002110017). The financial support is highly appreciated.

REFERENCES

Attard, M. M. & Setunge, S. 1996. Stress-Strain relationship of confined and unconfined concrete. *ACI Materials Journal*, 93(5): 432–442.

Alostaz, Y. M. & Schneider, S. P. 1996. Analytical behavior of connections to concrete-filled steel tubes. *Journal of Constructional Steel Research*, 40(2): 95–127.

Eurocode 3. 2005. Design of steel structures — Part 1–8: Design of joints. *European Committee for Standardization.*

Han, L. H. & An, Y. F. 2013. Performance of concrete-encased CFST stub columns under axial compression. *Journal of Constructional Steel Research.*

Han, L. H. & Li, W. 2010. Seismic performance of CFST column to steel beam joint with RC slab: experiments. *Journal of Constructional Steel Research* 66(11): 1374–1386.

Han, L. H., Yao, G. H. & Tao, Z. 2007. Performance of concrete-filled thin-walled steel tubes under pure torsion. *Thin-Walled Structure* 45(1): 24–36.

Li, W. & Han L.H. 2011. Seismic performance of CFST column to steel beam joints with RC slab: analysis. *Journal of Constructional Steel Research* 67(1): 127–139.

Li, X., Xiao, Y. & Wu, Y.T. 2009. Seismic behavior of exterior connections with steel beams bolted to CFT columns. *Journal of Constructional Steel Research*, 65(7): 1438–1446.

Liao, F.Y., Han, L.H. & Tao, Z. 2014. Behaviour of composite joints with concrete encased CFST columns under cyclic loading: Experiments. *Engineering Structures* 59(2): 745–764.

Nishiyama, I., Fujimoto, T. & Fukumoto, T., et al. 2004. Inelastic force-deformation response of joint shear panels in beam-column moment connections to concrete-filled tubes. *Journal of Structural Engineering ASCE*, 130(2): 244–52.

Ricles, J.M., Peng, S.W. & Lu, L.W. 2004. Seismic behavior of composite concrete filled steel tube column-wide flange beam moment connections. *Journal of Structural Engineering ASCE*, 130(2): 223–232.

Schneider, S.P. & Alostaz, Y.M. 1998. Experimental behavior connections to concrete-filled steel tubes. *Journal of Constructional Steel Research*, 45(3): 321–52.

Qian, W.W., Li, W., Han, L.H. & Zhao, X.L. 2013. Analytical behaviour of concrete-encased concrete-filled steel tubular column to steel beam joint. In: *Proceeding of the Pacific Structural Steel Conference (PSSC 2013)*, Singapore, 8 to 11 October 2013, 285–290.

Tubular Structures XV – Batista, Vellasco & Lima (eds)
© 2015 Taylor & Francis Group, London, ISBN 978-1-138-02837-1

Composite columns made of concrete filled circular hollow sections in fire case and with restrained thermal elongation – numerical and experimental analysis

T.A.C. Pires
Federal University of Pernambuco, Recife, Brazil

J.P.C. Rodrigues
Institute for Sustainability and Innovation on Structural Engineering, University of Coimbra, Portugal

J.J.R. Silva
Federal University of Pernambuco, Recife, Brazil

ABSTRACT: This paper presents the results of a numerical study on the fire behavior of Concrete Filled Circular Hollow (CFCH) columns with restrained thermal elongation. Parameters such as the slenderness of the column, load level, stiffness of the surrounding structure and steel reinforcement ratio were tested. The model was validated with the result of fire resistance tests. In addition, the paper suggests simple equations to evaluate the critical time of CFCH columns and also compare its results with the ones of simple calculation methods proposed by EN1994-1-2 (2005). The research shows critical times smaller than the fire resistances suggested in the literature for similar type of CFCH columns. The numerical model presents results in close agreement with the experimental data. The tabulated data method of EN1994-1-2 (2005) may be unsafe and the simple calculation model is conservative to evaluate the fire resistance of CFCH column in fire.

1 INTRODUCTION

Nowadays Concrete Filled Circular Hollow (CFCH) columns are largely used in construction due structural and architectural advantages such as the load bearing capacity, both at room and high temperatures, and aesthetic appearance. The fire performance of structures of this type is usually carried out by the designer with numerical or analytical methods in order to cover a wide range of situations in a faster and less cost way. Concerning the CFCH columns some researchers have already presented models to predict the thermomechanical behavior of these columns in fire, however there are not yet a consensus in some aspects. Recently researches will be discussed in next lines.

Ding & Wang (2008) presented a parametric study carried out with a finite element model developed in ANSYS to access the fire behavior of CFCH columns. The research also presented results for rectangular cross-section columns. A set of 34 columns were study varying relevant parameters such as external diameter (between 150 and 355.6 mm), length of the column (between 3810 and 5200 mm), boundary and load conditions. The thermal analysis was developed in a 2D model and the fire resistance (*i.e.* structural analysis) in a 3D model. Important features often neglected by others researchers were addressed in this study such as

the influence of an air gap and slip at the steel tube – concrete core interface, the concrete tensile behavior and the column initial imperfection on fire behavior of the CFCH columns. The main conclusions addressed for both types of columns were including or not the slipping between the steel tube and concrete core has a minor influence on the calculated fire resistance. However the slip between the steel tube and concrete core gives a better prediction of columns deflections. The tensile strength and tangential stiffness of concrete have a small influence in the fire resistance. The air gap improves the accuracy of the predictions for both temperatures and structural behavior. Finally, the initial deflection of the column has a small influence in fire resistance and the common maximum value of L/1000 should be used.

Schaumann et al. (2009) presented a 2D numerical model to study CFCH columns in fire. They compared the North American and the European recommendations for material properties of high strength concrete at high temperatures and found important differences and divergences in the numerical results for the CFCH columns filled with this material. The numerical results were compared with tested columns already present in the literature varying parameters such as external diameter (203.2, 219.1 and 273.1 mm), tube filling (plain, steel fibre and bar reinforced concrete), cross-section shape (circular and square) and concrete strength (81.7–107 MPa). Also they carried a parametric study considering a larger range of parameters values of section size, cross-section shape and slenderness of the columns. The computer program showed conservative results for the fire resistance of CFCH columns filled with steel fibre or bar reinforced concrete but overestimated it for CFCH columns filled with plain high strength concrete. According to the authors the differences arise from the numerical model that did not consider the local effects like gaping cracks and local plastic buckling.

Hong and Varma (2009) proposed a 3D finite element model to predict the behavior of concrete filled rectangular hollow columns under fire conditions. A three step sequentially coupled analysis was suggested in this research. Initially, fire growth and development was determined with the Fire Dynamics Simulator (FDS), followed by heat transfer and stress analyses, both with ABAQUS finite element program. A large sensitive analysis was carried out and relevant parameters that influence the model were pointed out, such as better steel and concrete thermal-mechanical models, geometric initial imperfections and composite interaction steel tube – concrete core. The model did not consider the evaporation and migration of the water inside the concrete core neither the heat loss in the interface steel tube – concrete core. The authors concluded that the model is useful to investigate the behavior of this type of columns and to conduct parametric studies to design guidelines for fire situation.

Finally, Espinos et al. (2010) presented a larger sensitive analysis with important improvements on modeling with Abaqus the behavior of CFCH columns in fire. The main improvements were the thermal conductance and friction model for the steel-concrete interface; thermal expansion coefficients; thermomechanical models of steel and concrete; and type of finite element employed for the reinforcement bars. The model presented better agreement for the leaner columns than the massive columns. The authors justify the fact due the major contribution of the concrete core in failure mechanism. Also the numerical results presented some divergence with high strength concrete filled columns. One of the main contributions of this research was to suggest optimal values to represent the thermal-mechanical interface between the steel tube and concrete core. This thermal resistance was set as a constant value of 200 W/m2 for the gap conductance. Also a radiative heat transfer mechanism was modeled in this interface. The mechanical interface considered the contact formulation for the normal behavior and the coulomb friction model for tangent behavior. In addition they concluded that the eurocode 4 simple calculation model may lead to unsafe results of slenderness columns. However the model was conservative when analyze eccentric loads.

A validation of the proposed models were not investigated considering the restraints of columns to their thermal elongation. The response of these columns when inserted in a building structure in fire is different from when isolated. Restraints on column thermal elongation, provoked by the building surrounding structure, plays a role on columns stability, since it induces different forms of interaction between the heated column and the cold adjacent structure. However the axial has a different role from the rotational restraining, while the former reduces the critical time and temperature of the columns, the latter increases them (Rodrigues et al. 2000).

On the other hand, these numerical approaches present some differences in relevant parameters when modeling of CFCH columns such as: the initial steel tube imperfections, steel and concrete mechanical properties in fire models, thermal expansion coefficients for steel and concrete, thermal and structural behavior at interface steel tube – concrete core and the mesh elements type and its geometric order (linear or quadratic elements). It is important to make it clear that these differences does not necessarily mean errors on the modeling procedure, but reflects the need for further studies because to represent a real fire situation of a building structure by experiments and numerical modeling is a difficult and challenging task. On this context, this work offers a contribution to area.

This paper presents a three-dimensional nonlinear finite element model to predict the behavior of CFCH columns in fire, considering its more relevant parameters and the restraints of columns to their thermal elongation that adds a rather difficult step to the problem analysis.

The main objective of this research is to present a parametric study that includes a range of practical values of load level, diameter of the column and ratio of reinforcement for the columns. Based on this study, simple equations to evaluate the critical time of the CFCH columns are proposed. Finally, a comparison between the research results and the simplified calculation methods proposed by EN1994-1-2 (2005) is presented.

2 EXPERIMENTAL TESTS

A total of 40 fire resistance tests on CFCH columns with restrained thermal elongation were carried out in the Laboratory of Testing Materials and Structures of the University of Coimbra – Portugal and the results presented already in several papers (Pires et al. 2012). The tests studied the influence of parameters in the fire resistance of the columns such as the slenderness, cross-sectional diameter, loading level, stiffness of surrounding structure, steel reinforcement ratio and degree of concrete filling inside the steel tube (completely filled or with a ring around the internal surface of the steel tube wall).

The columns were 3000 mm length with external cross-section diameter of 168.3 or 219.1 mm. The infill of the columns was plain concrete (PC) or steel bar reinforced concrete (RC) filling all the cross-section (TOT) or forming a ring around the inner wall of the steel tube (RING). The concrete ring thickness was 40 and 50 mm for the 168.3 and 219.1 mm diameter columns, respectively.

The diameter of the steel reinforcing bars was 10 and 12 mm respectively for the for the 168.3 and 219.1 mm diameter columns and were made with A500 steel grade. In all cases, 6 mm diameter stirrups, spaced 200 mm apart, were adopted. The distance of the central axis of the longitudinal rebar to the inner surface of the column wall was 30 mm in all cases. All the columns were made from circular hollow steel sections steel grade S355 and filled with concrete C25/30 resistance class. However the real mechanical properties at ambient temperature of the materials used in the fabrication of the CFCH columns were assessed and the values used in the numerical model.

The surrounding structure imposed an axial and rotational restraining to the thermal elongation of the heated column. Two values were tested: a lower stiffness (klow) that corresponded to 13 kN/mm of axial stiffness and 4091 and 1992 kN·m/rad of rotational stiffness in X1 and X2 directions, respectively, and a higher stiffness (khigh) that corresponded to 128 kN/mm of axial stiffness and 5079 and 2536 kN·m/rad of rotational stiffness in X1 and X2 directions, respectively.

The load applied was 30 and 70% of the design value of the buckling load calculated according to EN1994-1-1 (2005). Although the columns were 3.0 m tall, just 2.5 m of the specimen height was directly exposed to the furnace heating. The heating rate followed the ISO834 standard fire curve.

3 NUMERICAL MODELLING

A three-dimensional nonlinear finite element model was developed in Abaqus (2005) to simulate the behavior of CFCH columns in fire.

Continuing the research on this area, this work makes use of the main suggestions proposed by the authors presented in the first section of this paper.

The columns were simulated with an initial geometric imperfection of L/1000. The material model considered the temperature dependent formulation for thermal and mechanical properties presented in EN1993-1-2 (2005) and EN1992-1-2 (2004) for steel and concrete, respectively. For the concrete, a peak value of 2659 J/kg K corresponding to a moisture content of 4.25% determined experimentally was adopted.

A sequentially coupled thermal-stress analysis was carried out instead of fully coupled thermal-stress analysis, because the former is less time-consuming and leads to fewer problems of convergence than the latter.

The model considered the friction and thermal model for the interface steel tube – concrete core proposed by Espinos et al. (2010).

A basic axial flexural wire feature connector was applied on the top plate of the CFCH columns to simulate the restraint of the surrounding structure to thermal elongation. An axial-rotational spring was set-up at the top plate with the respective value for the stiffness of the surrounding structure.

All parts of the model (except the reinforced bars) were meshed with three-dimensional twenty node solid elements: a DC3D20 element for the thermal model and a C3D20R element for the structural model. The bars were meshed with one-dimensional three node truss elements (T3D3 element) for the structural model and with a three node heat transfer link (DC1D3 element) for the thermal model. An approximate global size of 20 mm for the element was defined and was sufficient for the accuracy of the numerical results to be good.

A sensitive analysis was then carried out in order to determine some main variables for the numerical model employed to the cases studied. The results of the fire resistance tests carried out by the authors on the CFCH columns with restrained thermal elongation and described in the last section were used to validate the model. The results presented already in several papers (Pires et al. 2014).

4 NUMERICAL DATA – DISCUSSION AND SIMPLIFIED EQUATIONS

This section presents simplified calculation equations to represent the fire performance of CFCH columns based on experimental and numerical results.

The results were obtained using Abaqus (2005), employing the numerical model already tested, in order to discuss the response of CFCH columns in fire, based on the numerical data. The parameters chosen for this study were: the load level (10%, 20%, 30%, 50%, 70% and 90% Ned), the diameter of the column (168.3 mm, 219.1 mm, 323.9 mm and 457 mm) and the ratio of steel bars reinforcement (0% – without reinforcement, 3% and 6%). These ranges include the typical values for CFCH columns, especially for the load level and steel reinforcement bars. Values outside these ranges have no meaning for practical cases.

It should be emphasized that issues such as the lack of experimental results, especially for more massive CFCH columns, the difficulty in standardizing tests in order to compare results, the use of different criteria to define the fire resistance, the limitation on numerically represent the experiments, and the accuracy of mathematical models that describe the mechanical behaviour of materials, raise challenges as to completely understanding the behaviour of CFCH columns subjected to fire. Nevertheless, the proposal of simplified calculation equations prompts discussion on some of these key points, and seeks to contribute towards future studies that will be undertaken in this field.

Figure 1 shows the calculated critical times arising from the load level for different CFCH columns with steel bar reinforcement ratios of 0%.

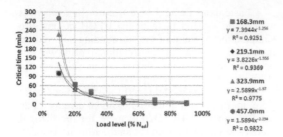

Figure 1. Critical times arising from a load level of CFCH columns with a reinforcement ratio of 0%.

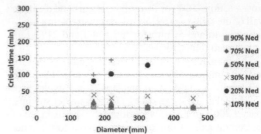

Figure 2. Critical times arising from CFCH column diameters with 3% steel reinforcement.

The results ratify the great influence of the load level on the critical times of the columns, as previously shown. The higher the load level, the lower the critical time is. However, in general, the simulated columns present low critical times (i.e. under 30 min) for load levels above 50% Ned.

A regression analysis was carried out and a relationship between the independent variables (i.e. load level η, column diameter d and reinforcement ratio ρ) and the dependent variable (i.e. critical time) was defined. Figures 1 also presents the regression functions.

The regression functions presented an acceptable approximation with the numerical data. The choice of regression function was the coefficient of determination R^2.

Therefore, the critical times of CFCH columns may be expressed by equations 1, 2 and 3, as shown below.

For the reinforcement ratio ρ = 0%

$$t_{cr} = \begin{cases} 7.3944 \times \eta^{-1.256} & if \quad d = 168.3mm \\ 3.8226 \times \eta^{-1.556} & if \quad d = 219.1mm \\ 2.5899 \times \eta^{-1.970} & if \quad d = 323.9mm \\ 1.5894 \times \eta^{-2.294} & if \quad d = 457.0mm \end{cases} \quad (1)$$

For the reinforcement ratio ρ = 3%

$$t_{cr} = \begin{cases} 6.8203 \times \eta^{-1.330} & if \quad d = 168.3mm \\ 2.9842 \times \eta^{-1.876} & if \quad d = 219.1mm \\ 1.4324 \times \eta^{-2.396} & if \quad d = 323.9mm \\ 0.7280 \times \eta^{-2.660} & if \quad d = 457.0mm \end{cases} \quad (2)$$

For the reinforcement ratio ρ = 6%

$$t_{cr} = \begin{cases} 13.804 \times \eta^{-1.027} & if \quad d = 168.3mm \\ 5.5099 \times \eta^{-1.624} & if \quad d = 219.1mm \\ 0.7399 \times \eta^{-3.117} & if \quad d = 323.9mm \\ 0.5593 \times \eta^{-3.175} & if \quad d = 457.0mm \end{cases} \quad (3)$$

where η = load level and d = cross section diameter.

The numerical results also showed that for load levels of 10% and 20% the greater the diameter of the columns is, the greater the calculated critical times are. For a load level of 30% the critical time does not change and, finally, for load levels above 50% a slight reduction in critical times is observed (Fig. 2).

The same tendency was observed for columns with reinforcement ratios of 0% and 6%.

On applying the concept of fire resistance as employed by some authors and presented in section 1, it is expected that this resistance will increase for CFCH columns with the highest diameters. The same tendency was identified when applying the critical time criteria, as defined before in this paper.

As mentioned in before, the same numerical and experimental data were used to compare these models, and showed good agreement. The discretization and boundary conditions of the columns followed the modelling described in section 3, for all computer runs. On the other hand, in order to study the standard behaviour of CFCH columns the data used for the mechanical properties of materials correspond to the theoretical standard values given in the literature while for the fire curve the ISO834 was adopted.

However, in this case, for columns with the highest diameters, as mentioned above, an increase in the critical time was not observed for the highest load levels (Fig. 2). Several comments may be made about these results.

It has been commented in the literature, as for instance by Espinos et al. (2010), that the agreement between experimental and numerical results is better for less massive columns than for massive ones. According to these authors, the error observed in these larger columns may be justified by the higher contribution of the concrete core and its more complex failure mechanisms. However, further studies are still necessary before this conclusion can be drawn.

There are two other key questions that arise from this issue. First, how large can the column diameter be with regard to applying the numerical techniques available? Second, what is, in fact, the influence of the load level on the behaviour of the largest columns?

It is important to observe that in their paper Espinos et al. (2010) analysed columns with diameters of up to 273.1 mm and load levels up to 45%. The numerical results obtained here tend to show similar behaviour for load levels of up to 30% and diameters of 168.3 mm and 219.1 mm. Comparisons between experimental and numerical results for columns with larger diameters are scarce due to the inherent difficulty of the tests. However, far from proposing a final conclusion, these values should be taken as a reference for

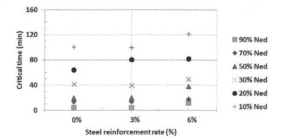

Figure 3. Critical times arising from a reinforcement ratio for CFCH columns with a diameter of 168.3 mm.

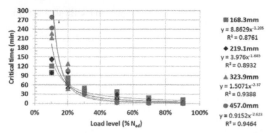

Figure 4. Critical times arising from the CFCH load level.

future research. It is also important to comment that the fire resistance adopted in the studies mentioned above differs from the critical time considered here.

Furthermore, all the experimental and numerical analyses carried out in this study have taken into account the axial and rotational restraining to thermal elongation, thus simulating the effect of the surrounding structure.

This consideration has not yet been taken into account in the research about CFCH columns in a fire situation.

Another key point: the failure mode also may justify the column behaviour shown by the numerical results. The thickness was kept constant when the diameter of the columns was increased. On analysing the deformed shapes of the columns (output from Abaqus), local buckling is more frequent for higher diameters. This may suggest the failure is sooner of CFCH columns with a steel tube thickness of 6.3 mm, a higher load level, and diameters of 323.9 and 457 mm.

Previously, the experimental (section 2) and numerical analyses (section 3) showed that the higher the diameter of the columns, the higher the volume of local buckling is. This same tendency is observed when using numerical data. The number of cases of local buckling increases in columns with diameters of 323.9 and 457 mm. In addition, it is important to comment that local buckling in columns with a reinforcement ratio of 6% is less frequent.

Steel reinforcement and the increase on its ratio in CFCH columns slightly increase the critical time of the columns calculated, especially for those with a reinforcement ratio of 6% (Fig. 3). However, this small increase does not justify great changes in the critical time of the CFCH columns due to there having been an increase in the reinforcement ratio.

Considering the above conclusions, a simplified equation (Eq. 4) independent of the steel reinforcement ratio may be addressed for CFCH columns and provides good agreement with numerical data. Figure 4 presents the regression function for this situation and its coefficient of determination R^2.

$$t_{cr} = \begin{cases} 8.8629 \times \eta^{-1.205} & if \quad d = 168.3mm \\ 3.9760 \times \eta^{-1.685} & if \quad d = 219.1mm \\ 1.5071 \times \eta^{-2.370} & if \quad d = 323.9mm \\ 0.9152 \times \eta^{-2.623} & if \quad d = 457.0mm \end{cases} \qquad (4)$$

where η = load level and d = cross section diameter.

Equations 1-4 may be a reference for future studies. It seems that further research should be carried out including into other parameters such as the thickness of the steel tube, the compressive resistance of concrete, the yield strength of the steel tube and the support boundary conditions, principally for massive CFCH columns. A wider database would probably provide a non-linear multivariate regression analysis and, consequently, a more general equation would be proposed (perhaps with greater reliability) to predict the critical time of CFCH columns. However, it is not known to the authors whether or not a similar study has already been addressed for CFCH columns with restrained thermal elongation, at least is not published, and therefore the simplified equations presented here are the first attempt to approach a parametric study for this purpose.

5 COMPARISON WITH THE SIMPLIFIED CALCULATION AND TABULATED DATA METHODS OF EN1994-1-2 (2005)

Finally, a comparison with the simplified calculation and tabulated data methods given in EN1994-1-2 (2005) is presented. This comparison showed that the tabulated data method can be unsafe for some cases and the simplified calculation method may be conservative, when compared to the results obtained from the simplified equations developed or by the experimental tests (section 2).

Two simple methods for assessing the structural behaviour of CFCH columns under fire were presented. They are: the tabulated data method and the simplified calculation model (SCM) both presented in EN1994-1-2 (2005). A comparison of both methods with the experimental and numerical results obtained for similar columns are presented in what follows.

5.1 Tabulated data method

Table 4.7 of EN1994-1-2 (2005) prescribes minimum cross-sectional dimensions, minimum reinforcement ratios and minimum axis distance for reinforcing bars of CFCH columns to reach a standard fire resistance according to their load levels η.

Table 1. Standard fire resistance and critical times for CFCH columns.

CFCH columns	Standard fire resistance	Critical time t_{cr} (min)
Section diameter: d = 168.3 mm Reinforcement ratio: ρ = 0.0% Distance of axis bars: us = 30 mm	R30	26–27
Section diameter: d = 219.1 mm Reinforcement ratio: ρ = 0.0% Distance of axis bars: us = 30 mm	R30	21–27
Section diameter: d = 168.3 mm Reinforcement ratio: ρ = 2.5% Distance of axis bars: us = 30 mm	R30	30–31
Section diameter: d = 219.1 mm Reinforcement ratio: ρ = 2.2% Distance of axis bars: us = 30 mm	R60	43–46

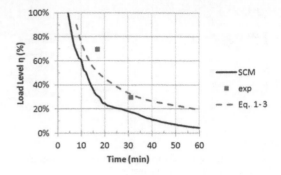

Figure 6. Critical times for SMC, experimental tests and simplified equation for a CFCH-RC column total filled with a diameter of 168.3 mm.

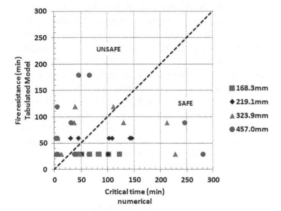

Figure 5. Tabulated method vs. numerical study of the critical time of the CFCH columns.

Table 1 presents the standard fire resistance, according to Table 4.7 of EN1994-1-2 (2005), and the experimental critical time of the CFCH columns presented in section 2. The other tested columns could not be classified in accordance with the standard fire resistance (i.e. they have fire resistances under 30min) and thus a comparison was not possible for these cases. In fact, the load level applied in the tests was slightly higher than the limits prescribed by EN1994-1-2 (2005) and a 2% tolerance was considered to enable comparison.

Three columns with a diameter of 219.1 mm are in the unsafe zone (i.e. they have critical times lower than the standard fire resistance) and lie outside the 5 min tolerance line. Two columns with a diameter of 168.3 mm are in the unsafe zone and the other two columns are close to the boundary of the safe zone.

Figure 5 plots the standard fire resistance of the columns versus their critical time as evaluated in numerical simulations. Some columns, especially the larger columns with diameters of 323.9 and 457.0 mm, presented critical times lower than the tabulated standard fire resistance, so they are positioned in the unsafe zone.

These results suggest that the tabulated data method may be slightly unsafe especially for larger columns.

5.2 Simplified calculation model

EN1994-1-2 (2005) presents a simplified calculation model (SCM) to determine the design value of the loadbearing capacity of a CFCH column in axial compression subjected to fire.

Figure 6 shows the critical times obtained by the SCM, experimental tests (load level of 30% and 70%) and simplified calculation equations (load level between 10% and 90%) for similar CFCH columns.

The temperatures measured for the columns in experimental tests were used as the first step of the simple calculation model (SMC). In the second step, the design axial buckling load $N_{fi,Rd}$ was normalized by the design value of the buckling load at room temperature N_{ed} in order to make the comparison for each load level η possible. The partial material safety factor in fire design ($\gamma_{M,fi}$) was 1.0.

The results show that the simplified calculation model for fire the resistance gives smaller results for the critical times than those obtained with experimental tests and numerical simulations. This suggests that the simplified calculation model is conservative when evaluating the design value of the resistance of a CFCH column in a fire situation.

The implementation of the SCM is not a straightforward task. In addition the proposing of a simplified mathematical model for the heat transfer problem requires an incremental solution for the mechanical problem. It can be concluded that the simplified equations proposed can be a viable alternative to determine critical times instead of using SCM as shown by Figure 6.

6 CONCLUSIONS

This paper presented a study of a three-dimensional non-linear finite element model developed with Abaqus to predict the behavior of CFCH columns and verify the influence of several parameters in fire. The restraining to thermal elongation of the column is an important parameter analysed in this paper that added a rather difficult step to the numerical modelling. The numerical model was validated with experimental results of fire resistance tests on CFCH columns with restrained thermal elongation, carried out by the authors. This research was complemented with the development of a parametric study. A range of practical values for the load level, diameter of the column and ratio of steel reinforcement was studied. Based on numerical data, simplified calculation equations to evaluate the critical time of CFCH columns with restrained thermal elongation were proposed. Finally, a comparison between the experimental results and the ones obtained with the EN1994-1-2 (2005) simplified calculation method and the tabulated data, for this type of columns, was done. The idea was to verify the reliability of these methods to assess the structural behaviour in fire of CFCH columns with axial and rotational restraining to thermal elongation.

The main conclusions from this research are commented on in what follows.

- The critical times of the CFCH columns tested in this research were smaller than those registered by some researches for similar experimental tests – however without restraining to their thermal elongation;
- The load level and slenderness of the columns had a great influence on the critical time of the columns. If one of them is reduced, the column critical time increases;
- The proposed numerical model presented results in close agreement with those obtained in fire experiments conducted on CFCH columns with restrained thermal elongation. Therefore, the numerical modelling can be considered as an option to assess the fire performance of CFCH columns;
- The numerical critical times obtained were slightly higher than the experimental ones. However, in most cases, this difference was not large, less than 5 minutes;
- The temperatures calculated with the numerical model were slightly lower than those measured in experimental tests (around 100°C);
- The numerical relative restraining forces obtained were in close agreement with those measured in the experiments. In general, the error was less than 10%;
- In general, the numerical axial deformations were higher than the experimental results. However, in most cases this difference was less than 5 mm, which is negligible given the length of the columns;
- The ratio of steel reinforcement does not justify great increases in critical times;
- The set of simplified equations proposed is a viable alternative for assessing the critical times of CFCH columns;
- The tabulated data method was shown to be slightly unsafe when compared to the results obtained in experimental and numerical simulations especially for larger columns;
- The SCM leads to safe results in comparison with the numerical simulations and experimental tests.

ACKNOWLEDGMENTS

The authors would like to thank the PhD scholarship of the first author given by Erasmus Mundus in the framework of the "Improving Skills Across Continents (ISAC)" programme, the Portuguese Foundation for Science and Technology (FCT) due to financial for the research under the framework of PTDC 65696/2006 project and, A. Costa Cabral S. A. a steel retailer and Metalocardoso S. A. steel structures builder for their support with steel tubes for the specimens.

REFERENCES

Abaqus User's manual: volumes I–III, version 6.7. Pawtucket, Rhode Island: Hibbit, Carlsson and Sorensson Inc., 2005.

Ali FA, Shepherd A, Randall M, Simms IW, O'Connor DJ, Burgess I. "The effect of axial restraint on the fire resistance of steel columns", Journal of Constructional Steel Research, 46 (1–3), pp. 305–306, 1998.

Chabot M, Lie TT. "Experimental studies on the fire resistance of hollow steel columns filled with bar-reinforced concrete", IRC Internal Report, n°628, National Research Council of Canada, Institute for Research in Construction, 1992.

Ding J, Wang YC. "Realistic modeling of thermal and structural behavior of unprotected concrete filled tubular columns in fire", Journal of Constructional Steel Research, vol. 64, pp. 1086–1102, 2008.

EN 206-1. "Concrete Part 1: Specification, performance, production and conformity". CEN – European Committee for Standardization; p. 84, 2007.

EN 1991-1-2, "Actions on structures. Part 1-2: general actions – actions on structures exposed to fire", CEN – European Committee for Standardization, 2002.

EN 1992-1-2 "Design of concrete structures. Part 1–2: general rules – structural fire design", CEN – European Committee for Standardization, 2004.

EN 1993-1-2, "Design of steel structures. Part 1–2: general rules – structural fire design", CEN – European Committee for Standardization, 2005.

EN 1994-1-1, "Design of composite steel and concrete structures. Part 1–1: general rules and rules for buildings", CEN – European Committee for Standardization, 2005.

EN 1994-1-2, "Design of composite steel and concrete structures. Part 1–2: general rules – structural fire design", CEN – European Committee for Standardization, 2005.

Espinos A, Romero ML, Hospitaler A. "Advanced model for predicting the fire response of concrete filled tubular columns", Journal of Constructional Steel Research, vol. 66, pp.1030–1046, 2010.

Hong S, Varma AH. "Analytical modeling of the standard fire behavior of loaded CFT columns", Journal of Constructional Steel Research, vol. 65, pp. 54–69, 2009.

ISO834-1. "Fire resistance tests-elements – elements of building construction – Part 1 General requirements", ISO – International Organization for Standardization, 1999.

Lie TT. "Fire resistance of circular steel columns filled with bar-reinforced concrete", Journal of Structural Engineering, vol. 120, n° 05, pp. 1489–1509, 1994.

Lie TT., ed. "Structural fire protection", Manuals and Reports on Engineering Practice, n° 78, ASCE, 1992.

Neves IC, Valente JC, Rodrigues JPC. "Thermal restraint and fire resistance of columns", Fire Safety Journal, 37, pp. 753–771, 2002.

Pires TAC, Rodrigues JPC, Rêgo Silva JJ. "Fire resistance of concrete filled circular hollow columns with restrained thermal elongation", Journal of Structural Steel Research, 77, pp. 82–94, 2012.

Rodrigues JPC, Neves IC, Valente JC. "Experimental research on the critical temperature of compressed steel elements with restrained thermal elongation", Fire Safety Journal, 35, pp. 77–98, 2000.

Schaumann P, Kodur V, Bahr O. "Fire behavior of hollow structural section steel columns filled with high strength concrete", Journal of Constructional Steel Research, vol. 65, pp. 1794–1802, 2009.

Valente JC, Neves IC. "Fire resistance of steel columns with elastically restrained axial elongation and bending", Journal of Structural Steel Research, 52 (3), pp. 310–331, 1999.

Tubular Structures XV – Batista, Vellasco & Lima (eds)
© 2015 Taylor & Francis Group, London, ISBN 978-1-138-02837-1

Buckling resistance of concrete-filled steel circular tube columns composed of high-strength materials

M. Karmazínová

Brno University of Technology, Brno, Czech Republic

ABSTRACT: The paper deals with the problems of the buckling compression of steel-concrete composite columns. The analysis of load-carrying capacity is oriented to the experimental verification and theoretical studies of the buckling resistance, especially with regards to the usage of high-strength concrete. Within the framework of this research the following basic steel-concrete columns with different cross-section configuration have been investigated: concrete-filled steel circular tubes with the diameter in the range from 133 mm to 168 mm and thickness from 4 mm to 5 mm. The buckling length of investigated columns was 3 m respecting the typical usual length used in building structures, for example. Material of investigated columns has been chosen as follows: steel grade was either S 235, or S 355; concrete strength class was in the range from C 20/25 up to C 80/90. Within the experimental research, steel tubes filled by concrete and steel tubes without concrete subjected to buckling compression force have been tested. A total of 45 test specimens have been verified: so far 27 of them have been tested and completely evaluated, another 18 specimens have been tested recently and are evaluated currently. Besides influence of high-strength materials (mainly concrete), the theoretical analysis also was oriented to actual imperfections and their influence on the buckling resistance. Within the experimental investigation, among others the material tests and their evaluation, as well as the measurement of actual geometrical imperfections, have been performed to verify the theoretical results obtained from theoretical studies.

1 INTRODUCTION

The particular problem of the load-carrying capacity of composite columns composed of steel tubes filled by high-strength concrete has been experimentally and theoretically investigated on several groups of structural members simulating, from the viewpoint of geometry and usual dimensions, typical building columns. According to this condition, the columns of 3 m length (see also Bradford et al.) and two basic circular tube cross-sections, i.e. TR Ø152/4.5 and TR Ø159/4.5, have been chosen for the experimental and theoretical analysis. The attention has been mainly paid to the influence of high-strength concrete on the buckling resistance. Steel and steel-concrete columns of the following materials have been investigated: steel grades S 235 and S 355; concrete strength classes C 20/25 and C 80/95 for the first cross-section and C 50/60 and C 70/85 for the second one, in accordance with the overview of the investigated column types presented in Table 1.

The columns with such cross-section configurations and material combinations, for which the highest ratio of the buckling capacity to full-plastic capacity has been calculated, have been chosen for the investigation. Table 1 shows several groups of specimens only, which already have been tested; next test specimens have been produced and currently are preparing for testing.

Table 1. Overview of investigated column types.

Cross-section	column group	steel grade	concrete strength class
TR Ø152/4.5	TS 1	S 235	steel only
	TS 2	S 235	C 20/25
	TS 3	S 235	C 80/95
TR Ø159/4.5	TS 4	S 235	steel only
	TS 5	S 235	C 50/60
	TS 6	S 235	C 70/85 (min)
TR Ø159/4.5	TS 7	S 355	steel only
	TS 8	S 355	C 50/60
	TS 9	S 355	C 70/85 (min)

2 THEORETICAL CALCULATIONS

2.1 Buckling load-carrying capacity

The buckling load-carrying capacity of the composite steel-concrete column, calculated using simplified methods given in EN 1993-1-1, EN 1994-1-1 (see also Kuranovas & Kvedaras 2007, Liang 2011b, Yu et al. 2008) is usually written as

$$N_b = \chi N_{pl}, \tag{1}$$

where, for the column composed of hollow steel cross-section filled by concrete, the full-plastic section load-carrying capacity (without buckling) N_{pl} is usually given in the form

$$N_{pl} = A_a \cdot f_y + A_c \cdot f_c \qquad (2)$$

with steel and concrete cross-section areas A_a, A_c, yield strength f_y and cylindrical concrete strength f_c. The reduction buckling factor χ depends on the non-dimensional slenderness $\bar{\lambda}$ known as

$$\bar{\lambda} = \sqrt{N_{pl}/N_{cr}} , \qquad (3)$$

where the critical force N_{cr} is

$$N_{cr} = \pi^2 \cdot (EI)_{eff} / L_{cr}^2 , \qquad (4)$$

the effective flexural stiffness is

$$(EI)_{eff} = E_a I_a + 0.6 \cdot E_c I_c \qquad (5)$$

with steel and concrete Young's modulus of elasticity E_a, E_c and corresponding second moments of area I_a, I_c. Then, the buckling reduction factor χ is

$$\chi = \frac{1}{\phi + \sqrt{\phi^2 - \bar{\lambda}^2}} , \qquad (6)$$

if the coefficient ϕ is

$$\phi = 0.5 \cdot \left[1 + \alpha_1 \cdot (\bar{\lambda} - 0.2) + \bar{\lambda}^2 \right], \qquad (7)$$

where the buckling curve (EN 1994-1-1 2008) of "a" with the imperfection factor value of $\alpha_1 = 0.21$ is applied for hollow cross-sections filled by concrete.

Within the particular theoretical analysis, the expected buckling load-carrying capacities have been calculated for steel and steel-concrete columns with material specifications mentioned above, using formulas (1) to (7). The calculations have been performed with the mean values of concrete cylindrical strength and mean values of steel yield strength, to obtain buckling load-carrying capacities corresponding with the predicted ultimate capacities. The capacity values have been determined for the buckling length of $L_{cr} = 3$ m, corresponding with hinges on both column ends, which is usual for the columns of building structures, and further for the pure compression, to compare the full-plastic capacity N_{pl} and the buckling capacity $N_b = \chi N_{pl}$.

Of course, there are many other methods of calculation based on analytical approaches (Bergmann et al. 1995, Lee et al. 2011, Uy et al. 2011, for example) resulting from the theory of elasticity and plasticity, or based on numerical approaches (Gupta et al. 2007, Liang 2011a, Portolés et al. 2011a).

2.2 Initial eccentricity of steel-concrete composite columns

The conception of compression member buckling covered by the normative rules for the design of steel structures arises from the model of the real members with the equivalent initial curving, which is similar to the form of the stability lost of the ideal member (see Chalupa 1976, Melcher 1977, Chalupa & Melcher 1975, for example).

Accepting the approach for steel slender columns, then for columns composed of two materials with different mechanical properties the concept of substitute (equivalent) steel cross-section can be used, that means geometrical parameters, i.e. the area A_i and the second moment of area I_i of the substitute cross-section must be applied as follows:

$$A_i = A_a + \frac{A_c}{n} , \qquad (8)$$

$$I_i = I_a + \frac{I_c}{n} , \qquad (9)$$

where n is the ratio of steel-to-concrete Young's modulus of elasticity

$$n = \frac{E_a}{E_c} . \qquad (10)$$

The maximal stress σ_{max} of steel-concrete slender column with the cross-section area consisting of steel part A_a and concrete part A_c and with yield strength f_y and concrete cylindrical strength f_c can be given by the formula (1)

$$\sigma_{max} = \sigma_0 \cdot \left[1 + m_0 \Big/ \left(1 - \frac{\sigma_0}{\sigma_{cr}} \right) \right], \qquad (11)$$

where the axial stress σ_0 is

$$\sigma_0 = \frac{N}{A_i} , \qquad (12)$$

the critical stress of the ideal member σ_{cr} is

$$\sigma_{cr} = \pi^2 \frac{E_a}{\lambda_i^2} , \qquad (13)$$

the relative initial eccentricity m_0 can be calculated

$$m_0 = \frac{e_0}{j_i} \qquad (14)$$

and the buckling strength can be given by

$$\bar{\sigma}_0 = \frac{1}{2} \left[f_y + (1 + m_0) \cdot \sigma_{cr} \right] - \sqrt{ \frac{1}{4} \left[f_y + (1 + m_0) \cdot \sigma_{cr} \right]^2 - f_y \cdot \sigma_{cr} } . \qquad (15)$$

132

Analogically to the members in tension – see (6), the condition for buckling load-carrying capacity of steel-concrete compression column can be written as

$$\frac{N}{A_i} \le \frac{\overline{\sigma_0}}{\gamma_M} = \overline{\sigma_0} \cdot \frac{f_{yd}}{f_y} = \chi \cdot f_{yd} \qquad (16)$$

with the reduction buckling factor

$$\chi = \frac{\overline{\sigma_0}}{f_y} = \frac{1}{2}\left[1 + (1 + m_0) \cdot \frac{\sigma_{cr}}{f_y}\right] -$$
$$- \sqrt{\frac{1}{4}\left[1 + (1 + m_0) \cdot \frac{\sigma_{cr}}{f_y}\right]^2 - \frac{\sigma_{cr}}{f_y}} \cdot \qquad (17)$$

Then, the relative initial eccentricity m_0 can be considered according to the form of

$$m_0 = \alpha_1 \cdot \frac{f_y}{\sigma_{cr}} = \alpha_1 \cdot \left(\frac{\lambda_i}{\lambda_{fy,i}}\right)^2, \qquad (18)$$

where the characteristics of material slenderness is

$$\lambda_{fy,i} = \pi \cdot \sqrt{\frac{E_a}{f_y}} = \pi \cdot \sqrt{\frac{E_a}{\gamma_M \cdot f_{yd}}} \cdot \qquad (19)$$

The reduction buckling factor χ in dependence on the slenderness λ can be written as

$$\chi = \frac{1}{2}\left[1 + \alpha_1 + \left(\frac{\pi}{\lambda_i} \cdot \sqrt{\frac{E_a}{f_y}}\right)^2\right] -$$
$$- \sqrt{\frac{1}{4}\left[1 + \alpha_1 + \left(\frac{\pi}{\lambda_i} \cdot \sqrt{\frac{E_a}{f_y}}\right)^2\right]^2 - \left(\frac{\pi}{\lambda_i} \cdot \sqrt{\frac{E_a}{f_y}}\right)^2} \cdot \qquad (20)$$

The maximal initial eccentricity e_0 in the middle of the member length can be given by the formula

$$e_0 = j \cdot \alpha_1 \cdot \left(\frac{\lambda}{\lambda_{fy}}\right)^2 = \alpha_1 \cdot \frac{f_y}{\pi^2 \cdot E} \cdot \frac{L_{cr}^2}{z}, \qquad (21)$$

where z is the distance of the cross-section edge from the gravity centre.

For circular tubes, the distance z is a half of the diameter, i.e. $d/2$, so the equivalent geometrical imperfection is given as

$$e_0 = \alpha_1 \cdot \frac{f_y}{\pi^2 \cdot E} \cdot \frac{2 \cdot L_{cr}^2}{d}. \qquad (22)$$

The initial geometrical imperfection can also be determined experimentally, using Southwell's line (see below). This equivalent imperfection can be utilized as the substitution of all member imperfections and

Table 2. Overview of test specimens: cross-section configuration and materials properties (mean values).

| Column group* | steel | | concrete | |
	yield strength $f_{y,m}$ [MPa]	Young's modulus $E_{a,m}$ [GPa]	cylindrical strength $f_{c,m}$ [MPa]	Young's modulus $E_{c,m}$ [GPa]
TS 1	356.4	has not	–	–
TS 2		been	27.0	32.0
TS 3**		measured	87.1	49.0
TS 4	344.4	–	–	
TS 5		209.2	56.7	36.2
TS 6			82.1	45.3
TS 7	475.4	–	–	
TS 8		199.8	56.7	36.3
TS 9			82.1	45.3

*3 test specimens in each column group TS.
**3 specimens tested: 1 test only valid – global buckling failure, 2 tests invalid – local buckling failure (see Fig. 2).

it is one of significant parameters determining the buckling curve and buckling reduction factor.

3 EXPERIMENTAL VERIFICATION OF BUCKLING LOAD-CARRYING CAPACITY

3.1 Test specimens and material properties

To verify the actual behaviour of steel and steel-concrete columns and to evaluate the contribution of concrete to the buckling capacity, the test specimens of geometrical and material parameters corresponding with ones mentioned above have been experimentally verified. The overview of test specimen groups accordant to described column types is shown in Table 2, together with the mean values of measured physical-mechanical properties.

3.2 Loading tests of steel and steel-concrete columns subjected to axial compression force

The European Standard rules are useable for steel grade up to S 460 and for concrete strength class up to C 60/75, but for higher strengths of concrete it would not be used without further verification. In the first phase of this research it was verified, that the usage of high-strength concrete is effective because it gives increasing in load-carrying capacity, which corresponds with the assumed values calculated according to the standards. The influence of particular concrete properties has been in detail investigated.

Within the experimental verification the investigation has been oriented, among others, to the comparison of the buckling load-carrying capacity of steel columns (groups TS 1, TS 4, TS 7) with the buckling capacity of steel-concrete columns using normal-strength concrete (groups TS 2, TS 5, TS 8) and high-strength concrete (groups TS 3, TS 6, TS 9). The column ends have been structurally solved so, to have being simply supported.

133

Figure 1. Illustration of test arrangement and test performance: left – tubes TR Ø152/4.5, right – tubes TR Ø159/4.5.

Figure 2. Failed specimens: typical global buckling failure – left up tubes TR Ø152/4.5, right up tubes TR Ø159/4.5; below local buckling failure occurred in two cases – tube ends unfilled by concrete.

Table 3. Buckling compression load-carrying capacities: theoretical values vs. experimental results.

Column group	buckling compression load-carrying capacity				
	theory (calculation) N_{th} [kN]		experiments N_{ex} [kN]		$\dfrac{N_{ex}}{N_{th}}$
TS 1	607	668	640	603	1.049
TS 2	870	1001	905	929	1.086
TS 3	1160	–	–	1499	1.292
TS 4	654	672	738	712	1.082
TS 5	1270	1351	1509	1408	1.120
TS 6	1486	1557	1501	1515	1.026
TS 7	788	870	850	857	1.090
TS 8	1366	1710	1710	1705	1.251
TS 9	1457	1767	1711	1755	1.197

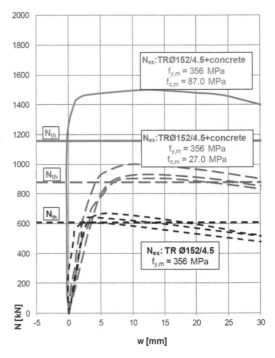

Figure 3. "N–w" diagrams: column groups TS 1, TS 2, TS 3 – TR Ø152/4.5, steel S 355, concrete C20/25, C80/95.

capacities calculated, are listed in Table 3. The theoretical values have been calculated according to the procedure mentioned above, with the actual mechanical properties, which have been measured helping the material test specimens cut from the failed specimens after loading test finishing. For the comparison, the average ratios of the experimental capacity to the calculated theoretical values are added to Table 3, too.

In Figs. 3, 4 and 5 the basic relationships of the buckling capacity N_{ex} and deflections w at the mid-length are drawn for steel columns (TS 1, TS 4, TS 7), for steel-concrete columns with normal-strength concrete (TS 2) and for steel-concrete column with different qualities of high-strength concrete (TS 3, TS 5,

The illustrations of the test arrangement and test realization are depicted in Figs. 1 and 2. The photos show the testing equipment, test specimens on the beginning loading process and failed test specimens.

The buckling load-carrying capacities obtained from the loading tests in comparison with the expected

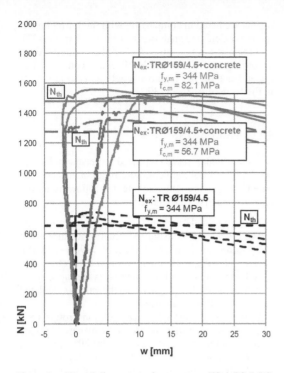

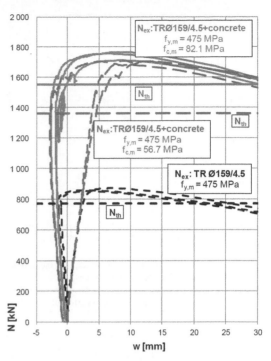

Figure 4. "N–w" diagrams: column groups TS 4, TS 5, TS 6 – TR Ø159/4.5, steel S 235, concrete C50/60, C70/85.

Figure 5. "N–w" diagrams: column groups TS 7, TS 8, TS 9 – TR Ø159/4.5, steel S 355, concrete C50/60, C70/85.

TS 6, TS 8, TS 9). For the comparison, the calculated values of the buckling load-carrying capacity are added to the graphs in Figs. 3, 4, 5, too.

From Figs. 3, 4 and 5 it is in general seen, that tubes filled by concrete of mentioned material properties show increasing in the buckling compression load-carrying capacity even up to about two times in comparison with steel columns; the actual buckling capacity of steel-concrete columns N_{ex} obtained from experiments very good corresponds with the theoretical buckling capacity N_{th} calculated by standard rules (for example Bukovská & Karmazínová 2012, Karmazínová et al. 2009, Karmazínová & Melcher 2010, 2011); experimental values are in the following range: from by 3% (TS 6) minimally, usually about by 10% (TS 2, TS 5), and maximally up to by 20–25% (TS 9, TS 8) higher than calculated values.

It is also seen, that the buckling capacities of the members with concrete C 50/60 and C 70/85 are about the same (TS 8 and TS 9 or TS 5 and TS 6), although the measured parameters of both concretes very good correspond with the assumed properties, but Young's modulus of concrete C 50/60 has been obtained much more than normally assumed.

From the comparison of experimental resistances $N_{ex,i}$ and theoretical values $N_{th,i}$ calculated according to EN 1994-1-1 for steel-concrete columns and according to EN 1993-1-1 for steel columns, using measured materials properties, the "$N_{ex} - N_{th}$" diagrams have been derived (see Fig. 6). Then the relationship between experimental and theoretical values

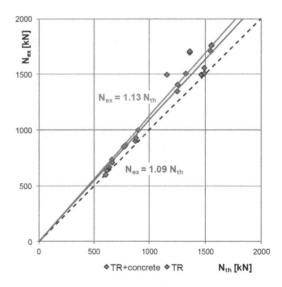

Figure 6. "N_{ex}–N_{th}" diagrams: $N_{ex} = 1.13\ N_{th}$ for steel-concrete columns; $N_{ex} = 1.09\ N_{th}$ for steel columns.

of the buckling resistance of steel-concrete columns can be derived in the form $N_{ex} = 1.13\ N_{th}$, if the variation coefficient of ratio N_{ex} / N_{th} is $v = 0.0786$. The similar relationship for steel columns has been derived in the form $N_{ex} = 1.09\ N_{th}$, which is also shown in Fig. 6, for the illustration and comparison.

135

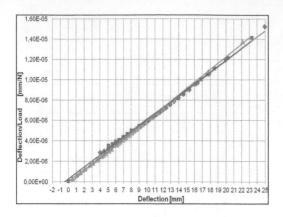

Figure 7. Illustration of Southwell's lines: specimen group TS 8 – TR Ø159/4.5, steel S 235, concrete C50/60.

Table 4. Comparison of initial imperfections: theoretical (calculated) vs. actual (experimental) values.

Column group	equivalent initial geometrical imperfection			
	theory (calculation acc. paragraph 2.2)		experiments	
	e_0 [mm]	e_0/L [−]	e_0 [mm]	e_0/L [−] (mean)
TS 1			0.473, 0.344, 0.139	1/9633
TS 2	4.478	1/686	0.973, 1.120, 1.672	1/2446
TS 3			0.070	1/43857
TS 4			1.000, 0.500, 0.500	1/3600
TS 5	3.961	1/757	0.100, 0.100, 0.100	1/30000
TS 6			0.100, 0.400, 0.600	1/8182
TS 7			0.400, 0.100, 0.400	1/10000
TS 8	5.731	1/523	0.100, 0.500, 0.300	1/10000
TS 9			0.100, 0.100, 0.100	1/30000

3.3 Actual equivalent geometrical imperfections

The experimental verification has been realized with the specimens represented by steel circular tubes and steel circular tubes filled by concrete composed of normal-strength materials and high-strength ones.

The investigated columns were simply supported on both ends with structural length of $L = 3070$ mm or $L = 3000$ mm, respectively, that the buckling length was $L_{cr} = 3070$ mm and $L_{cr} = 3000$ mm, respectively. Material properties including their combinations in particular cross-sections are evident from Tables 2 and 3.

The actual values of the initial eccentricities have been derived using Southwell's lines, examples of which are illustrated by the graphs in Fig. 7. On horizontal axis there are the values of the transverse deformation, i.e. deflection w in the mid-length of the member. On vertical axis there are the values of the ratio of deflection w to axial force N, i.e. w/N. The initial eccentricity e_0, that means the equivalent initial geometrical imperfection, is such value of the deflection, for which $w/N = 0$.

In Table 4, the actual equivalent initial geometrical imperfections derived experimentally evaluating test

results are listed in comparison with the initial imperfections calculated analytically using the formula (22), including the ratios of e_0/L. Member imperfections according to EN 1994, EN 1993 are considered in dependence on the member length as $e_0 = L/300$, or in the form of $e_0/L = 1/300$.

4 CONCLUSIONS

From particular results verified experimentally, for steel-concrete columns composed of circular steel tube filled by high-strength concrete, the following partial conclusions may be deducted:

- Buckling load-carrying capacity in compression can be calculated using the usual procedure given by European Standard rules (EN 1994-1-1 2008), like as for the column with normal concrete.
- Actual values of buckling load-carrying capacity are in average by 13% higher than the values calculated with measured properties.
- In the case of test specimens column groups TS 3 and TS 8, experimental load-carrying capacities are considerably higher, that means by 29% and 25%, compared to other cases, wherein the capacities are only slightly higher, that means about by 10% usually. In the case of the column group TS 3, one specimen only has been tested, so the result obtained cannot be taken as generally acceptable. In the case of the column group TS 8, the higher increase of the load-carrying capacity in comparison with other results can be caused by the fact, that load-carrying capacity of slender columns (non-dimensional slenderness is about 1.3 here) is more influenced by concrete modulus of elasticity E_c (within flexure stiffness $E_c I_c$), than by concrete strength (like in pure compression).
- From results of column groups TS 5 and TS 6 and also TS 8 and TS 9 it is seen, that the influence of Young's modulus is understandably more significant than the influence of concrete strength. The efficient methods how to evaluate the importance of the influence of particular parameters, can be probabilistic approaches (Beck et al. 2009), including sensitivity analysis (Kala et al. 2009, 2010) for example, which has been continuously utilized in parallel with testing.
- Actual initial equivalent geometric imperfections are much more less than theoretical values calculated by analytical formulas, as well as the values recommended in European Standards (EN 1994-1-1 2008, EN 1993-1-1 2008).

ACKNOWLEDGEMENT

The paper has been elaborated within the framework of the solution of the following projects:

- The project No. LO1408 "AdMaS UP – Advanced Materials, Structures and Technologies" supported by the Ministry of Education, Youth and Sports of the Czech Republic, under the "National Sustainability Programme I";

- The project of the specific research FAST-S-14-2544 supported by the same institution.

REFERENCES

Beck, A. T., de Oliveira, W. L. A., De Nardim, S. & El Debs, A. 2009. Reliability-based evaluation of design code provisions for circular concrete-filled steel columns, *Engineering Structures*, Vol. 31, pp. 2299–2308. ISSN 0141-0296.

Bergmann, R., Dutta, D., Matsui, C. & Meinsma, C. 1995. *Design guide for concrete-filled hollow section columns (5)*. CIDECT (Ed.) and Verlag TÜV Rheinland, Cologne.

Bradford, M. A., Loh, H. Y. & Uy, B. 2002. Slenderness limits for filled circular steel tubes, *Journal of Constructional Steel Research*, Vol. 58, No. 2, pp. 243–252. ISSN 0143-974X.

Bukovská, P. & Karmazínová, M. 2012. Behaviour of the tubular columns filled by concrete subjected to buckling compression, *Procedia Engineering*, Issue 40, 2012, pp. 68–73. doi: 10.1016/j.proeng.2012.07.057.

Chalupa, A. 1976. *Strength of slender members and webs* (in Czech language), Technical Report on Steel Structures, Vítkovice, Vol. 3, 1976.

Dundu, M. 2012. Compressive strength of circular concrete filled steel tube columns, *Thin-Walled Structures*, Vol. 56, pp. 62–70. ISSN 0263-8231.

Gupta, P. K., Sarda, S. M. & Kumar, M. S. 2007. Experimental and computational study of concrete filled steel tubular columns under axial load, *Journal of Constructional Steel Research*, Vol. 63, pp. 182–193. ISSN 0143-974X.

Kala, Z., Karmazínová, M., Melcher, J., Puklický, L. & Omishore, A. 2009. Sensitivity analysis of steel-concrete structural members, In *Proc. of the 9th International Conference on Steel-Concrete Composite and Hybrid Structures ASCCS 2009* held in Leeds, Research Publishing Services: Singapore, pp. 305–310. ISBN 978-981-08-3068-7.

Kala, Z., Puklický, L., Omishore, A., Karmazínová, M. & Melcher, J. 2010. Stability problems of steel-concrete members composed of high-strength materials, *Journal of Civil Engineering and Management*, Vol. 16(3), pp. 352–362. doi: 10.3846/jcem.2010.40.

Karmazínová, M, Melcher, J. J. & Röder, V. 2009. Load-carrying capacity of steel-concrete compression members composed of high-strength materials, In *Proceedings of the 9th International Conference on Steel-Concrete Composite and Hybrid Structures ASCCS 2009* held in Leeds, Research Publishing Services: Singapore, pp. 239–244. ISBN 978-981-08-3068-7.

Karmazínová, M. & Melcher, J. J. 2010. Buckling resistance of steel-concrete columns composed of high-strength materials, In *Proceedings of the International Colloquium on Stability and Ductility of Steel Structures SDSS'Rio 2010*, Rio de Janeiro, pp. 895–902. ISBN 978-85-285-0137-7.

Karmazínová, M. & Melcher, J. J. 2011. Design assisted by testing applied to the determination of the design resistance of steel-concrete composite columns, In *Proceedings of the 13th WSEAS International Conference on Mathematical and Computational Methods in Science and Engineering*, WSEAS, Catania, pp. 420–425. ISBN 978-1-61804-046-6.

Kuranovas, A. & Kvedaras, A. K. 2007. Behaviour of hollow concrete-filled steel tubular composite elements, *Journal of Civil Engineering and Management*, Vol. 13, No. 2, pp. 131–141. doi: 10.3846/jcem.2010.40.

Lee, S. H., Uy, B., Kim, S. H., Choi, Y. H. & Choi, S. M. 2011. Behavior of high-strength circular concrete-filled steel tubular (CFST) column under eccentric loading, *Journal of Constructional Steel Research*, Vol. 67, pp. 1–13. ISSN 0143-974X.

Liang, Q. Q. 2011a. High-strength circular concrete-filled steel tubular slender beam-columns, Part I: Numerical analysis, *Journal of Constructional Steel Research*, Vol. 67, No. 2, pp. 164–171. ISSN 0143-974X.

Liang, Q. Q. 2011b. High-strength circular concrete-filled steel tubular slender beam-columns, Part II: Fundamental behaviour, *Journal of Constructional Steel Research*, Vol. 67, No. 2, pp. 172–180. ISSN 0143-974X.

Melcher, J. 1977. The problems of analytical interpretation and differentiation of limit buckling stresses for centrically compressed members, In *Proceedings of Regional Colloquium on Stability of Steel Structures*, TU: Budapest.

Melcher, J. & Chalupa, A. 1975. Problems of imperfections and differentiation of buckling curves of compression steel members (original in Czech language), *Journal of Civil Engineering*, Vol. 11.

Portolés, J. M., Romero, M. L., Bonet, J. L. & Filippou, F. C. 2011a. Simulation and design recommendations of eccentrically loaded slender concrete-filled tubular columns, *Engineering Structures*, Vol. 33, pp. 1576–1593. ISSN 0141-0296.

Portolés, J. M., Romero, M. L., Filippou, F. C. & Bonet, J. L. 2011b. Experimental study of high-strength concrete-filled circular tubular columns under eccentric loading, *Journal of Constructional Steel Research*, Vol. 67, pp. 623–633. ISSN 0143-974X.

Portolés, J. M., Serra, E. & Romero, M. L. 2013. Influence of ultra-high strength infill in slender concrete-filled steel tubular columns, *Journal of Constructional Steel Research*, Vol. 86, pp. 107–114. ISSN 0143-974X.

Uy, B., Tao, Z. & Han, L. H. 2011. Behaviour of short and slender concrete-filled stainless steel tubular columns, *Journal of Constructional Steel Research*, Vol. 67, pp. 360–378. ISSN 0143-974X.

Yu, Q. Tao, Z. & Wu, Y. X. 2008. Experimental behaviour of high-performance concrete-filled steel tubular columns, *Thin-Walled Structures*, Vol. 46, No. 4, pp. 362–370. ISSN 0263-8231.

EN 1993-1-1 *Design of Steel Structures – Part 1-1: General Rules and Rules for Buildings*, CEN: Brussels, 2008.

EN 1994-1-1 *Design of Composite Steel and Concrete Structures – Part 1-1: General Rules and Rules for Buildings*, CEN: Brussels, 2008.

Tubular Structures XV – Batista, Vellasco & Lima (eds)
© 2015 Taylor & Francis Group, London, ISBN 978-1-138-02837-1

Time-dependent response of three-hinged CFST arches

M.A. Bradford & Y.-L. Pi
UNSW Australia, UNSW Sydney, NSW, Australia

ABSTRACT: Arches comprised of concrete-filled steel tubular (CFST) members are finding widespread use bridges. They are advantageous because the steel provides formwork to the concrete core and the steel also contributes significantly to the strength and stiffness of the member. Shallow arches, in particular, allow for concrete casting within the pre-erected lightweight steel tubular members by pumping. However, shallow arches are known to experience geometric non-linearity and classical theories have been shown to be unconservative. Whilst providing a barrier to the drying of the concrete core, it is known that CFST members do experience concrete shrinkage and creep effects, and for shallow arches these effects may be significant enough to lead to the phenomenon of creep buckling. Most studies have focused on continuous arches, but the use of pinned arches is often dictated by transportation restrictions and, despite three-pinned arches being statically determinate, their response to both limit point and bifurcative buckling is complex.

1 INTRODUCTION

CFST arches are known to experience quasi-viscoelastic behaviour because of shrinkage and creep of the concrete core, and this may precipitate creep buckling in shallow arches (Bradford et al. 2006, 2011). Most studies of the non-linear response of arches have focused on one-fold (pinned) or three-fold (fixed) indeterminate arches. However, arches are often pinned to produce a statically determinate member, due to transportation requirements and the need to eliminate the secondary effects of thermal straining.

A crown-pinned arch has bending moment distributions that are greatly different to those of pin-ended and fixed-ended arches without such a crown pin. Hitherto, most research of quasi-viscoelastic effects in CFST arches has not considered those with a crown pin, but their deformations are larger than those of pin-ended and fixed-ended arches and they may be more susceptible to non-linear deformations than pinned and fixed arches. As a result, the long-term interaction between geometric non-linearity and the creep and shrinkage of the concrete core may be particularly important in crown-pinned CFST arches, and warrants study. This paper focuses therefore on an investigation of the effects of geometric non-linearity of crown-pinned CFST circular arches in conjunction with creep and shrinkage of the concrete core on their non-linear long-term elastic in-plane behaviour under a central concentrated load (Fig. 1). Analytical solutions are derived for the non-linear deformations, internal forces and buckling loads, providing an understanding of the long-term structural response and buckling of crown-pinned CFST arches.

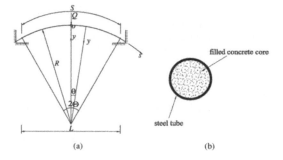

Figure 1. Crown-pinned CFST arch: (a) geometry and axes and (b) cross-section.

2 NON-LINEAR LONG-TERM ANALYSIS

Several methods have been proposed for treating creep and shrinkage of concrete, and numerous empirical methods are available for estimating the creep coefficient and shrinkage strain (Gilbert & Ranzi 2011). Among these, the age-adjusted effective modulus method is attractive because it can lead to solutions of the well-posed problem in closed form. For this, the shrinkage strain at time t is taken as

$$\varepsilon_{sh} = \left(\frac{t}{t+d}\right)\varepsilon_{sh}^*, \tag{1}$$

where $d = 35$ days for moist curing and ε_{sh}^* is the final shrinkage strain. Creep of the concrete core is accounted for using the age-adjusted effective modulus given by

$$E_{ec} = \frac{E_c}{1 + \phi(t,t_0)\chi(t,t_0)},\qquad(2)$$

where E_c is the elastic modulus of the concrete,

$$\phi(t,t_0) = \frac{(t-t_0)^{0.60}\phi_u}{10 + (t-t_0)^{0.60}}\quad t > t_0\qquad(3)$$

is the creep coefficient, t_0 the age at first loading, $\phi_u = 2.4$ (Uy 2001) the final creep coefficient,

$$\chi(t,t_0) = 1 - \frac{(1-\chi^*)(t-t_0)}{20 + (t-t_0)}\quad t > t_0\qquad(4)$$

is the aging coefficient with $\chi^* = k_1 t_0/(k_2 + t_0)$ being its final value, where $k_1 = 0.78 + 0.4\exp(-1.33\,\phi(\infty, 7))$, $k_2 = 0.16 + 0.8\exp(-1.33\,\phi(\infty, 7))$. The effective axial stiffness of the section is taken as $AE = A_s E_s + A_e E_{ec}$ and the effective flexural stiffness of the section as $EI = E_s I_s + E_{ec} I_c$, where A_s and A_c are the areas of the steel tube and concrete core respectively and I_s and I_c the second moments of these areas about the centroid respectively, and where E_s is the elastic modulus of the steel tube.

The non-linear normal strain at a point on the cross-section can be written as (Pi et al. 2011)

$$\varepsilon = \tilde{w}' - \tilde{v} + \tfrac{1}{2}\tilde{v}'^2 - y\tilde{v}''/R,\qquad(5)$$

in which $\tilde{v} = v/R$, $\tilde{w} = w/R$ v and w are the radial and axial displacements respectively, and where $(\ )' = \mathrm{d}(\ )/\mathrm{d}\theta$ (Fig. 1). Invoking the principle of virtual work produces

$$\delta\Pi = \int_0^\Theta \left[-NR(\delta\tilde{w}' - \delta\tilde{v} + \tilde{v}'\delta\tilde{v}') - M\delta\tilde{v}'' \right]\mathrm{d}\theta$$
$$- QR\delta\tilde{v}_0 = 0 \qquad \forall\,\delta\tilde{v},\delta\tilde{w},\qquad(6)$$

where N and M are the axial compressive force and bending moment given by

$$N = -AE\left(\tilde{w}' - \tilde{v} + \tfrac{1}{2}\tilde{v}'^2\right) - \varepsilon_{sh}A_c E_{ec};\ M = \frac{-EI\tilde{v}''}{R}.\quad(7)$$

Integrating Equation 6 by parts leads to the differential equations of axial and radial equilibrium as

$$-N' = 0 \quad\text{and}\quad \frac{\tilde{v}^{iv}}{\mu_e^2} + \tilde{v}'' = -1\qquad(8)$$

respectively, and to the static boundary conditions

$$\tilde{v}'' = 0 \quad\text{and}\quad \tilde{v}''' + \mu_e^2\tilde{v}' - \frac{QR^2}{2EI} = 0\quad\text{at }\theta = 0\qquad(9)$$

for singly and three-pinned arches and to

$$\tilde{v}'' = 0\quad\text{at }\theta = \Theta\qquad(10)$$

for three-pinned arches, where

$$\mu_e^2 = \frac{NR^2}{EI}\qquad(11)$$

is a time-dependent axial force parameter. Solving the differential equation for the radial deformation subject to the boundary conditions produces

$$\tilde{v} = \frac{1}{\mu_e^2}\left\{1 - \cos\mu_e\theta + \frac{\beta_e^2}{2} - \frac{\mu_e^2\theta^2}{2}\right.$$
$$\left. - \mathcal{H}(\theta)\sin\mu_e\theta\tan\frac{\beta_e}{2} + \frac{P[\mathcal{H}(\theta)\mu_e\theta - \beta_e]}{2\beta_e}\right\}\qquad(12)$$

for three-pinned CFST arches and

$$\tilde{v} = \frac{1}{\mu_e^2}\left\{\frac{1}{\cos\beta_e} - \cos\mu_e\theta - \beta_e\tan\beta_e - \frac{\mu_e^2\theta^2}{2}\right.$$
$$+ \frac{\mathcal{H}(\theta)\sin\mu_e\theta\left(\beta_e - \sin\beta_e\right)}{\cos\beta_e}\Bigg\}$$
$$+ \frac{P\left[\tan\beta_e - \beta_e + \mathcal{H}(\theta)\left(\mu_e\theta - \sin\mu_e\theta/\cos\beta_e\right)\right]}{2\mu_e^2\beta_e}\qquad(13)$$

for singly-pinned CFST arches, where $P = QR^2\Theta/EI$ is a dimensionless load, $\beta_e = \mu_e\Theta$ and where

$$\mathcal{H}(\theta) = \begin{cases} 1 & \theta > 0 \\ -1 & \theta < 0 \end{cases}\qquad(14)$$

represents Heaviside's step function. Substituting Equations 12 and 13 into Equation 7 produces

$$M = -\frac{EI}{R}\left[\cos\mu_e\theta - 1 + \mathcal{H}(\theta)\sin\mu_e\theta\tan\frac{\beta_e}{2}\right]\qquad(15)$$

for three-pinned CFST arches and

$$M = -\frac{EI}{R}\left[\cos\mu_e\theta - 1\right.$$
$$\left. + \frac{\mathcal{H}(\theta)\sin\mu_e\theta}{\cos\beta_e}\left(\sin\beta_e - \beta_e + \frac{P}{2\beta_e}\right)\right]\qquad(16)$$

for singly-pinned CFST arches.

A relationship between the applied load Q (through P) and the axial force N (through β_e) can be established by substituting Equations 12 and 13 into the first of Equations 7 and integrating it over the entire arch $\theta \in [-\Theta, \Theta]$, which leads to the transcendental quadratic equation of equilibrium

$$A_1 P^2 + A_2 P + A_3 = 0,\qquad(17)$$

where the coefficients A_1, A_2 and A_3 are given by

$$A_1 = \frac{1}{8\beta_e^4}; \; A_2 = 0; \; A_3 = \frac{\beta_e^2}{\lambda_e^2} + \frac{\beta_e - \sin\beta_e}{2\beta_e^3(1 + \cos\beta_e)}$$
$$- \frac{1}{6} + \frac{\varepsilon_{sh}A_cE_{ec}}{\Theta^2 AE} \quad (18)$$

for three-pinned arches and

$$A_1 = \frac{1}{16\beta_e^4}\left(2 + 2\sec^2\beta_e - \frac{3\tan\beta_e}{\beta_e}\right);$$

$$A_2 = \frac{1}{4\beta_e^4}\left[\frac{\beta_e\tan\beta_e(1 + \cos\beta_e) - 2}{\cos\beta_e} + 2 - \frac{\beta_e^2}{\cos\beta_e^2}\right];$$

$$A_3 = \frac{\beta_e^2}{\lambda_e^2} + \frac{1}{4\beta_e^2}\left[\frac{(\beta_e^2 - 1)\tan\beta_e}{\beta_e} + \right.$$

$$\left. \frac{1 + \beta_e^2 - 2\beta_e\sin\beta_e}{\cos^2\beta_e}\right] - \frac{1}{6} + \frac{\varepsilon_{sh}A_cE_{ec}}{\Theta^2 AE},$$

for singly-pinned arches, in which

$$\lambda_e = \frac{\Theta}{2}\frac{S}{r_e} \quad \text{with} \quad r_e = \sqrt{\frac{EI}{AE}} \quad (20)$$

is a slenderness parameter for the arch.

3 EFFECTS OF NON-LINEARITY ON LONG-TERM STRUCTURAL BEHAVIOUR

Bradford & Pi (2014) have presented a theory for the geometric linear response of a CFST arch, for which the non-linear term in Equation 5 is omitted. Figure 2 compares the linear and non-linear results for a three-pinned arch, while the counterpart results are shown in Figure 3 for a singly-pinned arch. Both figures consider a shallow arch ($f/L = 1/10$) and a deeper arch ($f/L = 1/3$). In the analyses, $Q = 0.07N_{E2}$ was applied to the shallow arch (Figs. 2a and 3a) and $Q = 0.217N_{E2}$ to the deeper arch (Figs. 2b and 3b), where N_{E2} is the second-mode flexural buckling load of a pin-ended column under uniform axial compression with the same length as that of the arch.

It can be seen that the radial displacements at the time $t = 200$ days are much larger than those at time $t = 15$ days using both linear and non-linear analyses and, in particular, that the non-linear analysis predicts significantly larger deflections than does linear analysis in the time domain.

The effects of geometric non-linearity on the long-term radial displacements are shown in Figure 4 as variations of the dimensionless long-term central displacement $v_c/v_{c,15}$ with time t for a three-pinned and singly-pinned arch, where first loading is at time $t = 15$ days and $v_{c,15}$ is the short-term deflection at that time. Because the length S of a deep arch is larger than that of a shallow arch for the same span L, for comparison, the same sustained central load $Q = 0.05N_{L,E2}$

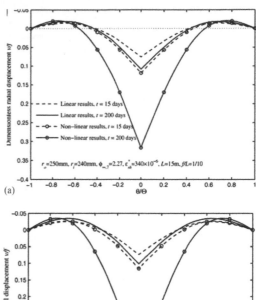

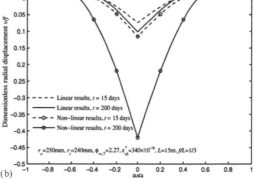

Figure 2. Radial displacements for three-pinned arch with (a) $f/L = 1/10$ and (b) $f/L = 1/3$.

was applied to both the deep and shallow arches, where $N_{L,E2} = \pi^2 EI/(L/2)^2$.

It can be seen from Figures 4a and 4b that as the time t increases, creep and shrinkage of the concrete core produce a significant long-term increase of the radial displacements of crown-pinned CFST arches under a sustained load. It can also be seen that the increases of the long-term radial displacements predicted by non-linear analysis are much larger than those predicted by linear analysis. This indicates that linear analysis may underestimate long-term increases of the radial displacements of crown-pinned CFST arches and so cannot correctly determine the long-term serviceability limit state of CFST arches.

The effects of geometric non-linearity on the long-term bending moments are shown in Figure 5 for crown-pinned arches as variations of the dimensionless bending moment $4M/QL$ with the dimensionless length θ/Θ at times $t = 15$ days and $t = 200$ days, where a central load $Q = 0.217N_{E2}$ was applied. It can be seen that the bending moments at the time $t = 200$ days are much larger than those at time $t = 15$ days from non-linear analysis of both singly and three-pinned arches. However, linear analysis does not predict any long-term changes of the bending moment for three-pinned arches because they are statically determinant, and those changes predicted by linear analysis for

141

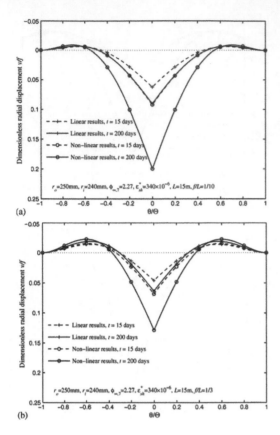

(a)

(b)

Figure 3. Radial displacements for singly-pinned arch with (a) $f/L = 1/10$ and (b) $f/L = 1/3$.

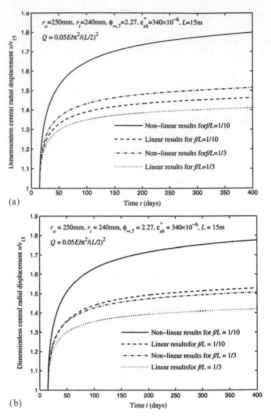

(a)

(b)

Figure 4. Linear and non-linear central displacements: (a) three-pinned arches and (b) singly-pinned arches.

singly-pinned arches, despite their being statically indeterminate, are too small to detect.

4 EFFECTS OF NON-LINEARITY ON LONG-TERM BUCKLNG

It has been demonstrated experimentally (Wang et al. 2006) that creep and shrinkage deformations may produce non-linear creep buckling of shallow concrete arches in the long term. To investigate whether crown-pinned arches also display this creep buckling behaviour, the non-linear equilibrium paths described by Equations 12, 13 and 17 are shown in Figure 6 for three and singly-pinned arches as variations of the dimensionless load $Q/(\Theta N_{E2})$ with the dimensionless radial displacement v_c/f at the times $t = 15$ days and $t = 100$ days, where f is the rise of the arch.

It can be seen from Figures 6a and 6b that when the upper limit point is reached, a further increase of the displacement is associated with a decrease of the external load and with a decrease of the axial force along the unstable path until the lower limit point is reached. Following this, as the displacement continues to increase, the external load increases again along the remote stable equilibrium path. In practice, the sustained load remains unchanged and the CFST arch

cannot follow the equilibrium path shown by the solid lines after the upper long-term limit point buckling and the lower limit point cannot be reached. When the long-term buckling occurs, the equilibrium configuration of the arch will suddenly jump from the limit point to a remote equilibrium point as shown by the horizontal dashed lines. The movement of the arch after buckling is associated with a build-up of kinetic energy, which leads to the familiar sudden and noisy snap-through phenomenon.

Because the limit points are local extrema on the non-linear equilibrium paths, differentiating Equation 17 with respect to β_e leads to the equilibrium equation between the dimensionless load P and the axial force parameter β_e at the limit points as

$$B_1 P^2 + B_2 P + B_3 = 0, \qquad (21)$$

where the coefficients B_1, B_2 and B_3 are given by

$$B_i = -\frac{\beta_e \partial A_i}{2 \partial \beta_e} \quad i = 1, 2, 3 \qquad (22)$$

Solving Equations 17 and 21 simultaneously will then produce the buckling load at the upper limit point, as well as the buckling load at the lower limit point and the corresponding axial forces as shown in Figure 6.

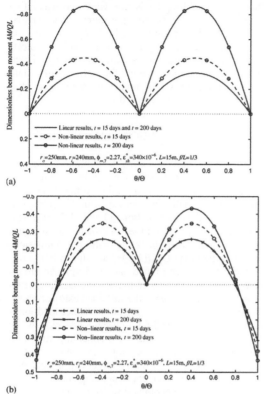

(a)

(b)

Figure 5. Bending moments along the length of deep arches: (a) three-pinned arch and (b) singly-pinned arch.

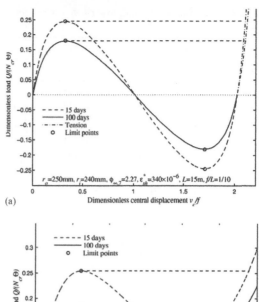

(a)

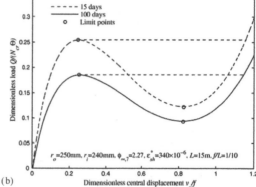

(b)

Figure 6. Non-linear equilibrium: (a) three-pinned arch and (b) singly-pinned arch.

It is also known (Pi et al. 2011) that in addition to long-term limit point buckling, pin-ended and fixed CFST arches can buckle in an antisymmetric bifurcation mode. To clarify whether this can occur with crown-pinned arches, the radial displacements in Figures 2 and 3 are worth re-inspection. The slopes of these curves can be determined by differentiating Equations 12 and 13, and it can be seen from Figures 2 and 3 that as $\theta \to 0$ from negative and positive values of θ, the slope has two limit points and the sharp kink and free rotation do not allow the symmetrical radial displacements from bifurcating into antisymmetric ones. Hence, a crown-pinned CFST arch can only buckle in a symmetric or limit point mode, and not in an antisymmetric or bifurcation mode.

The dimensionless non-linear long-term buckling loads are compared with the short-term buckling loads in Figure 7 as variations of the dimensionless buckling load $Q/(\Theta N_{E2})$ with the included angle 2Θ, where length of the arch is constant with a slenderness ratio $S/r = 100$ and the short-term and long-term buckling loads are determined at $t = 15$ days and at $t = 200$ days respectively. It can be seen that in the long term, non-linear analysis predicts a significant reduction of the in-plane buckling loads of crown-pinned CFST arches. Hence, a crown-pinned arch that satisfies the stability limit state in the short term may lose its stability

Table 1. Dimensionless buckling load $Q/(N_{E2}\Theta)$ for three-pinned arches.

2Θ	$t = 15$ days		$t = 200$ days	
Degree	L	N	L	N
15	0.5745	0.1322	0.4165	0.0855
30	0.7027	0.2186	0.5094	0.1532
60	0.7360	0.2493	0.5335	0.1791
90	0.7540	0.2566	0.5466	0.1853
120	0.7755	0.2589	0.5626	0.1872
150	0.8038	0.2599	0.5827	0.1882
180	0.8557	0.2605	0.6203	0.1886

Note: L = Linear results from ANSYS, N = Non-linear results.

in the long term and fail by creep buckling. To further illustrate the effects of geometric non-linearity on the long-term buckling loads, typical in-plane buckling loads obtained from the finite element analysis of ANSYS are compared with the corresponding non-linear results in Table 1 for three-pinned arches and in Table 2 for singly-pinned arches, where the arches are the same as those of Figure 6.

Table 2. Dimensionless buckling load $Q/(N_{E2}\Theta)$ for singly-pinned arches.

| 2Θ | $t = 15$ days | | $t = 200$ days | |
Degree	L	N	L	N
15	0.6788	–	0.4695	–
30	0.7815	0.2166	0.5578	0.1514
60	0.8398	0.2656	0.6053	0.1895
90	0.8634	0.2789	0.6241	0.2006
120	0.8870	0.2846	0.7841	0.2054
150	0.9160	0.2871	0.6632	0.2075
180	0.9526	0.2885	0.6900	0.2087

Note: L = Linear results from ANSYS, N = Non-linear results

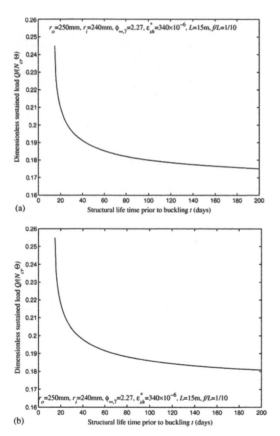

(a)

(b)

Figure 7. Prebuckling structural lifetime: (a) three-pinned arch and (b) singly-pinned arch.

Although linear analysis predicts no long-term changes of internal forces for three-pinned CFST arches and very small long-term changes of internal forces for singly-pinned CFST arches, creep and shrinkage of the concrete core reduce their bending stiffness significantly, which leads to a reduction of the linear buckling loads of crown-pinned CFST arches in the long term, as shown in Tables 1 and 2.

However, it can be seen from Tables 1 and 2 that the linear buckling loads are very unconservative when compared with their non-linear counterparts, and that the linear long-term buckling loads at time $t = 200$ days are even higher than the non-linear short-term buckling loads. This demonstrates again that linear analysis cannot predict the stability limit state for crown-pinned CFST arches correctly, and so a non-linear analysis is required.

5 NON-LINEAR STRUCTURAL LIFETIME

Because non-linear buckling is dominant, the pre-buckling structural life for the non-linear long-term buckling of crown-pinned CFST arches have been investigated, and the time until structural buckling can be determined using Equations 17 and 21. Typical values of the structural lifetime t of crown-pinned arches prior to their creep buckling with the dimensionless sustained load $Q_{sus}/(N_{E2}\Theta)$ are shown in Figures 7a and 7b for three-pinned and singly-pinned arches respectively. It can be seen that the prebuckling structural lifetime increases as the sustained load decreases, as expected. For a sufficiently low sustained load, long-term buckling of the CFST arch cannot develop, while for a high sustained load, the prebuckling structural lifetime of the arch can be quite short. It can also be seen that in the first 100 days of prebuckling structural life, the decrease of the corresponding sustained load is quite rapid, but becomes slow following this.

6 CONCLUSIONS

The effects of geometric non-linearity on the long-term elastic in-plane behaviour and buckling of crown-pinned CFST circular arches under a sustained central concentrated load have been investigated, and analytical solutions for their non-linear deformations and buckling loads were derived. It was found that geometric non-linearity influences the long-term behaviour of crown-pinned CFST arches significantly. When geometric non-linearity is not considered, much smaller long-term deformations are predicted, no long-term internal force changes are predicted for three-pinned CFST aches (as they are statically determinate) and very small long-term internal force changes are predicted for singly-pinned CFST arches (despite being statically indeterminate). The non-linear analysis has shown that the long-term deformations may be so large that they significantly reduce the reserve of the service limit state of crown-pinned CFST arches. The non-linear analysis also predicted significant long-term increases of and bending moments for three-pinned arches, which is different to that predicted by linear theory.

It was also found that the non-linear long-term buckling loads of crown-pinned CFST arches are smaller than their linearly-derived counterparts, significantly so in some cases. Hence, geometric non-linearity together with the creep and shrinkage of the

144

concrete core may reduce the reserve of the stability limit state of crown-pinned arches in the long-term. It can be concluded that non-linear analysis is needed to determine the long-term structural response and buckling of crown-pinned CFST arches accurately.

The non-linear long-term behaviour of crown-pinned CFST arches is different to that of arches without an internal pin. Firstly, in the long term, pin-ended and fixed CFST arches without an internal pin may buckle in either a symmetric limit point mode (by snap-through) or in an antisymmetric bifurcation mode, while crown-pinned arches may only buckle in a symmetric limit point mode. Secondly, non-linear buckling dominates the long-term buckling of shallow pin-ended and fixed CFST arches without internal pins, and linear buckling dominates the behaviour of these arches when they are deep. However, non-linear buckling dominates the behaviour of both shallow and deep CFST crown-pinned arches.

ACKNOWLEDGEMENTS

The work in this paper was supported by the Australian Research Council through Discovery Projects (DP120104554, DP130102934 and DP140101887) awarded to both authors and an Australian Laureate Fellowship (FL100100063) awarded to the first author.

REFERENCES

Bradford, M.A., Wang, T. & Gilbert, R.I. 2006. Coupled geometric viscoelastic non-linearities in parabolic concrete-filled steel tubular arches. *19th Australasian Conference on the Mechanics of Structures and Materials, Christchurch, New Zealand, December, 43–48*. Bath UK: Taylor & Francis.

Bradford, M.A., Pi, Y.-L. & Qu, W. 2011. Time-dependent in-plane behaviour and buckling of concrete-filled steel tubular arches. *Engineering Structures* 33(5): 1781–1795.

Bradford, M.A. & Pi, Y.L. 2014. Geometric nonlinearity and long-term behavior of crown-pinned CFST arches. *Journal of Structural Engineering, ASCE* 04014190-1 – 04014190-11.

Gilbert, R.I. & Ranzi, G. 2011. *Time-Dependent Behaviour of Concrete Structures*. London: Spon Press.

Pi, Y.-L., Bradford, MA. & Qu, W. 2011. Long-term non-linear behaviour and buckling of shallow concrete-filled steel tubular arches. *International Journal of Non-Linear Mechanics* 46(9): 1155–1166.

Uy, B. 2001. Static long-term effects in short concrete-filled steel box columns under sustained loading. *ACI Structural Journal* 98(1): 96–104.

Wang, T., Bradford, M.A. & Gilbert, R.I. 2006. Creep buckling of shallow parabolic concrete arches. *Journal of Structural Engineering, ASCE* 132(10): 1641–1649.

Earthquake resistance

Tubular Structures XV – Batista, Vellasco & Lima (eds)
© 2015 Taylor & Francis Group, London, ISBN 978-1-138-02837-1

Elliptical-hollow-section braces under cyclic axial loading

Y.M. Huai

State Key Laboratory of Disaster Reduction in Civil Engineering, Tongji University, Shanghai, China
Department of Structural Engineering, Tongji University, Shanghai, China

T.M. Chan

Department of Civil and Environmental Engineering, The Hong Kong Polytechnic University, Hong Kong

W. Wang

State Key Laboratory of Disaster Reduction in Civil Engineering, Tongji University, Shanghai, China
Department of Structural Engineering, Tongji University, Shanghai, China

ABSTRACT: In this paper, finite element analysis was conducted to simulate the cyclic behaviour of elliptical hollow section (EHS) and circular hollow section (CHS) braces with pin-ended conditions. The aims were to investigate the influence of the cross-section slenderness, member slenderness and cross-section shape on the hysteretic performance. Prior to the numerical simulations, tensile coupon tests and stub column tests had been conducted to obtain material properties, and cross-section capacities of the hot-rolled and cold-formed hollow section tubes. Pilot numerical simulation results indicated that the member slenderness affects the ultimate load; the cross-section slenderness would have significant impact on the energy dissipation capacity, and EHS specimens dissipate more energy than their CHS counterparts.

1 INTRODUCTION

Concentrically-braced frame (CBF) is one of the common lateral force resisting system due to its high lateral stiffness and load bearing capacity. As it is simpler to design than eccentrically-braced frame and moment-resisting frame, CBF is often used these days. In this seismic-resistant system, braces are the main members to dissipate the seismic energy, and subject to buckling under cycle lateral loading. However, repeated buckle would reduce the stiffness and load bearing capacity of the braces and hence, affects the overall performance of the structure dramatically. Thus, there has been significant amount of research to investigate the influence of the cross-section slenderness, member slenderness and cross-section shape on the hysteretic performance. These parameters have been codified in many codes of practice.

Nip *et al.* (2010) conducted experimental and numerical investigation on square and rectangular hollow section members under cyclic axial loading to study their hysteretic response and fracture life. They compared steel tubes with different materials including hot-rolled carbon steel, cold-formed carbon steel and cold-formed stainless steel. In addition to rectangular hollow section, CHS braces have also been examined by Takeuchi and Matsui (2011) and Sheehan and Chan (2012) carried out experimental investigation on concrete-filled and slender CHS braces. In order to get a reliable assessment on buckling, stub column

tests are commonly used to analyze local buckling of steel tube. Experimental and numerical investigations have also been conducted by researchers on EHS stub columns (Chan and Gardner, 2008), concrete-filled EHS stub columns (Bradford and Roufegarinejad, 2007; Yang *et al.*, 2008; Zhao and Packer, 2009; Dai and Lam, 2010; Sheehan *et al.*, 2012), and special cross section elliptical steel tubes (Uenaka, 2013).

In this paper, elliptical hollow section (EHS) braces under cyclic axial loading were numerically examined. Fundamental tensile coupon tests and stub column tests were first carried out and the experimental results were used to develop the Finite Element (FE) models. The developed FE models were adopted to numerically assess the influence of the cross-section slenderness, member slenderness and cross-section shape on the hysteretic performance.

2 TENSILE COUPON TETS

The EHS tube was cold-formed from the parent hot-rolled CHS tube. The parent CHS was classified as Q345b in accordance with the Chinese specifications. Two longitudinal coupons were cut from each CHS tube while three longitudinal coupons were machined from each EHS tube as shown in Figure 1. Tensile coupon tests were carried out in accordance with GB/T 228.1-2010 (ISO 6892-12009). The nominal and measured average yield stress (hot-rolled)/0.2% proof stress (cold-formed) f_y, ultimate stress f_u, modulus of elasticity E and percentage of elongation after fracture are summarized in Table 1.

As shown in Table 1, two sets of specimen labelled with "-A" and "-B" were examined under each 'CHS" and "EHS" category.

The resulting stress-strain relationships are shown in Figure 2 and correlated closely to those expected. For CHS coupons, the hot-rolled steel exhibited linear behaviour up to yield and then a yield plateau before advancing into the strain hardening region. In contrast, for EHS coupons, the cold formed material was non-linear in the initial stages and did not display plateau during yielding. Judging from the test results, the EHS coupons show a higher strength and lower ductility than CHS coupons. This was attributed by the cold-forming process.

3 STUB COLUMN TESTS

3.1 Specimens

A total of 12 stub columns was tested at Tongji University. The stub column test aimed at investigating the cross-sectional response and the effect of local buckling. All stub columns were 250 mm in length. EHS and CHS stub columns in series A and B had similar cross-section area and EHS in series A was cold-formed from the parent hot-rolled CHS in series A. The specimen details and measured geometry are

presented in Tables 2 and 3. The nomenclature adopted for naming the specimens consists of three parts. The first part of the specimen ID refers to the steel-type – C denotes hollow circular steel tube, E denotes hollow elliptical steel tube. The second part refers to different geometry of the hollow section and the final part refers to the specimen type (SC = stub column in this case).

Table 3 also shows the equivalent diameter d_e which can be calculated by Equation 1(Gardner $et al.$, 2011).

$$d_e = h\left\{1 + \left[1 - 2.3\left(\frac{t}{h}\right)^{0.6}\right]\left(\frac{h}{b} - 1\right)\right\} \qquad (1)$$

where h is the larger outer diameter (approx. 120 mm), b is the smaller outer diameter (approx. 60 mm). Geometric parameters are presented in Figure 3.

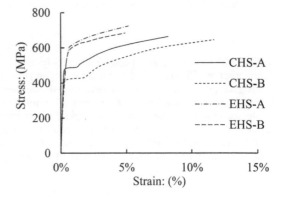

Figure 2. Stress-strain relationship.

Table 2. Measured geometric properties for CHS stub column specimens.

Specimen ID	Average measured dimensions (mm)		Area (mm²)	Maximum imperfection
	d	t	A	(mm)
C-A-SC1	95.53	5.3	1502	0.19
C-A-SC2	95.49	5.31	1505	0.234
C-A-SC3	95.51	5.36	1518	0.189
C-B-SC1	95.61	7.49	2074	0.184
C-B-SC2	95.58	7.43	2058	0.166
C-B-SC3	95.57	7.39	2047	0.071

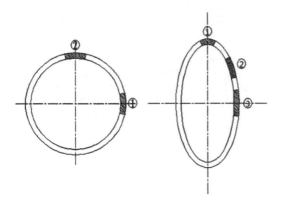

Figure 1. Location of tensile coupon.

Table 1. Nominal and measured material properties.

Specimen	Steel	Nominal material properties		Average measured material properties				
		f_y MPa	f_u MPa	E MPa	f_y MPa	f_u MPa	f_u/f_y	% Elongation after fracture
CHS-A	Hot-rolled	400	570	208000	476	616	1.29	23
CHS-B	Hot-rolled	360	530	207000	417	574	1.38	26
EHS-A	Cold-formed	400*	570*	212000	599	682	1.14	18
EHS-B	Cold-formed	388*	574*	199000	578	647	1.12	17

*Based on material properties of the parent hot-rolled CHS tube

Table 3. Measured geometric properties for EHS stub column specimens.

| EHS Specimens ID | Average measured dimensions (mm) | | | Equivalent diameter (mm) | Area (mm²) | Maximum imperfection (mm) |
	h	b	t	d_e	A	
E-A-SC1	120.37	59.98	5.42	197.66	1443	1.107
E-A-SC2	120.17	59.51	5.24	198.14	1393	0.905
E-A-SC3	120.11	59.74	5.25	198.01	1396	0.779
E-B-SC1	120.18	58.07	9.03	181.88	2271	0.506
E-B-SC2	120.31	57.47	8.85	182.8	2226	0.561
E-B-SC3	120.33	57.49	8.91	182.63	2239	0.505

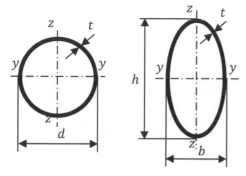

Figure 3. Geometric parameters.

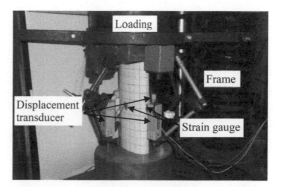

Figure 4. Test set up for stub column test.

Prior to testing, local geometric imperfections were measured along each of the four sides of the tube, in addition to the geometric measurements. For the CHS stub columns, the largest imperfection is 0.234 mm, about 5% of the thickness. While maximum imperfection of 1.107 mm (22% of the thickness) was recorded for the EHS stub columns.

3.2 Test preparation

All stub column tests were performed in a 2000 kN capacity SANS 200T testing machine as shown in Figure 4. Four longitudinal strain gauges were used at the mid-height, and four longitudinal displacement transducers were used to measure the relative distance between two flat bearing plates. The test specimens were placed in the center of the machine. For alignment purpose, the variation between the strain measurements from the average strain should be less than 5% at the alignment load which was around 10% of the yield load. After the alignment, the load was steadily increased at a constant rate of 1 mm/min, until the load reached the yield level. The loads were then maintained for 3 minutes. After that, the loads were steadily increased (through displacement control) until the load dropped to half of the predicted yield level load or the specimens were severely deformed.

3.3 Results

Most stub columns failed in a similar manner, with local buckling occurring at either the top or bottom

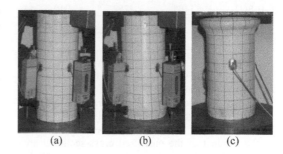

Figure 5. C-A-SC1 stub column (a) at beginning of test; (b) during test after occurrence of local buckling; (c) at end of test.

of the specimen. The hot-rolled tubes exhibited a distinct yield plateau which was not observed for the cold-formed EHS tubes. For the hollow stubs, the load decreased rapidly after the ultimate value. This can be explained by the significant changes propagated in the stub-column geometry as the test progressed. As the axial displacement increased, the stub column buckled in a ring-shaped mode at two or more locations along the length, after which serious deformation occurred at one of these rings location and it became very stiff until the end of test. Figure 5 shows the deformation history for stub column: C-A-SC1.

All failure modes of stub column CHS and EHS tubes were similar and a typical failure mode (one ring at the end) of CHS and EHS stub columns are presented in Figures 6(a) and (b) respectively. The

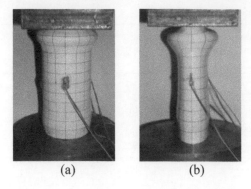

(a)   (b)

Figure 6.   Failure mode of stub column (a) CHS; (b) EHS.

Table 4.   Maximum load from hollow stub column tests.

Specimen ID	$N_{u,test}$ (kN)	$N_{pl,Rd}$ (kN)	$N_{u,test}/N_{pl,Rd}$
C-A-SC1	917	715	1.28
C-A-SC2	903	716	1.26
C-A-SC3	897	723	1.24
C-B-SC1	1200	865	1.39
C-B-SC2	1203	858	1.40
C-B-SC3	1204	854	1.41
E-A-SC1	775	864	0.90
E-A-SC2	772	834	0.93
E-A-SC3	761	836	0.91
E-B-SC1	1415	1308	1.08
E-B-SC2	1405	1282	1.10
E-B-SC3	1420	1289	1.10

maximum loads, $N_{u,test}$ from hollow stub column tests are summarized in Table 4.

In Table 4, the third column presents the maximum load predicted by BS EN 1993-1-1 (2005). The resistance of the hollow stub column was taken as the product of the measured cross-sectional area and the yield stress. The forth column presents the ratio between the maximum test load and the prediction from the Eurocode. The hot-rolled CHS stub columns significantly exceeded the predicted ultimate load from BS EN 1993-1-1 (2005). For cold-formed EHS stub columns, loads measured were lower than predicted in particular for E-A-SC1 to 3 specimens.

Load-end shortening relationship was derived for each stub column, and typical curves are presented in Figure 7. Resulting load-end shortening correlated closely to those expected. For hot-rolled CHS stub columns, they exhibited linear behaviour up to yield and followed by a plateau before advancing into the hardening region. In contrast, for cold-formed EHS stub columns, they displayed non-linear behaviour in the initial stages of the test and did not show yield plateau. Judging from the test results, the cold-formed EHS stub columns show higher strength and lower ductility than the hot-rolled CHS counterparts.

Overall, local bucking occurred earlier in the EHS stub columns. In series A, local-buckled displacements

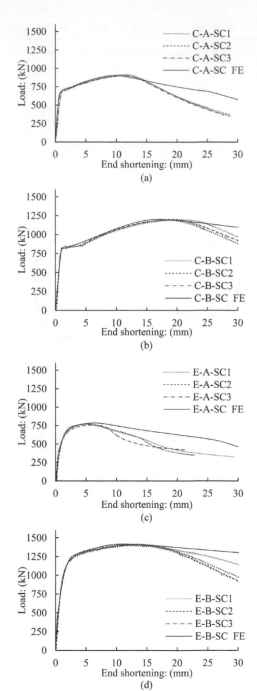

Figure 7.   Axial load – axial displacement relationship: (a) specimens: C-A; (a) specimens: C-B; (a) specimens: E-A; (a) specimens: E-B.

were approximately 4% (CHS) and 2% (EHS) of length while that of series B were about 7% (CHS) and 6% (EHS) of length. In addition, the deformability of the cold-formed EHS stub columns in series A was much lower than that of the parent CHS stub columns.

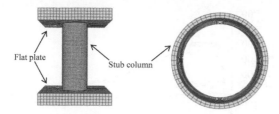

Figure 8. FE model of stub column.

4 NUMERICAL SIMULATIONS

4.1 Numerical simulation of stub column under axial load

Four FE models were developed and assessed against the experimental results. To obtain two elements in thickness direction, mesh size was designed to be equal to half of the tube thickness. C3D8R, 8-node linear integration reduced brick solid element was used for modelling the steel tube. This type of element can take into account the stress variation through thickness and it is more efficient than traditional 8-node brick element. This element will also be used in the companion numerical investigation on concrete-filled stub columns. Two thick flat plates were used to apply a uniform load and simulate loading condition in laboratory, as shown in Figure 8.

As showed in Figure 7, the numerical results correlated closely to the experimental findings. The experimental and numerical load-displacement curves overlapped with each other up to the ultimate point. In addition, deformations of stub columns were also shown in Figure 9.

4.2 Modelling of hollow steel brace under cyclic axial load

Same element, C3D8R, 8-node linear integration reduced brick solid element was employed. Each numerical brace was divided into three parts with different mesh size. The mesh size was determined by different level of stress concentration and local deformation. More refined mesh was employed in the middle region, of which the mesh size was about half of hollow section thickness while in other parts the mesh size was about twice the hollow section thickness.

Six CHS and six EHS braces were modelled to investigate the influence of the cross-section slenderness, member slenderness and section shape on the hysteretic performance.

Tables 5 and 6 show the main geometric specimens parameters, including the member length L, outside diameter d, thickness t, slenderness $\bar{\lambda}$ (0.8, 1.3 and 1.5), area A (around $1400\,mm^2$ and $2000\,mm^2$ respectively), diameter to thickness ratio (CHS: d/t; EHS: d_e/t).

Pin-ended boundary conditions were employed at the ends, to mimic the pin-ended bracing tests to be carried out at Tongji University. Figure 10 indicates

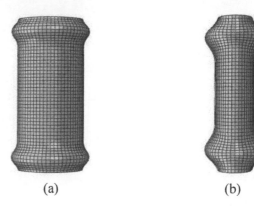

(a)   (b)

Figure 9. Deformed shape: (a) CHS stub column; (b) EHS stub column.

Table 5. Geometric parameters of CHS braces.

Specimen ID	L (mm)	t (mm)	d/t	A (mm²)	$\bar{\lambda}$
C-A-1	1950	5	19.0	1414	0.8
C-A-2	3150	5	19.0	1414	1.29
C-A-3	3675	5	19.0	1414	1.5
C-B1	1925	7	13.57	1935	0.8
C-B2	3100	7	13.57	1935	1.29
C-B3	3600	7	13.57	1935	1.5

Table 6. Geometric parameters of EHS braces.

Specimen ID	L (mm)	t (mm)	d_e (mm)	d_e/t	A (mm²)	$\bar{\lambda}$
E-A-1	2275	5	158.00	31.60	1335	0.8
E-A-2	3650	5	158.00	31.60	1335	1.29
E-A-3	4250	5	158.00	31.60	1335	1.5
E-B-1	2200	8	131.29	16.41	2060	0.8
E-B-2	3550	8	131.29	16.41	2060	1.29
E-B-3	4150	8	131.29	16.41	2060	1.51

Figure 10. End conditions and mesh sizes.

the FE pin-ended braces where cycle load was applied at node RP-1 (reference point 1).

Numerical specimens were tested under a cyclic loading protocol outlined in ECCS (1986), as shown in Figure 11. At each displacement $0.25\delta_y$, $0.5\delta_y$, $0.75\delta y$ there was one cycle and for displacement $1.0\delta_y$, $2\delta_y$, $4\delta_y$, $6\delta_y$, $8\delta_y$, and $10\delta_y$ there were three cycles, where

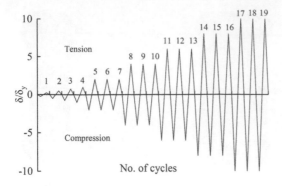

Figure 11. Cyclic loading protocol.

δ_y was the yield displacement of the specimen. At each cycle, specimens were first loaded in compression and then in tension, and the loading procedure was symmetric in compression and tension. The δ_y was obtained by multiplying the member length and values of yield strain which could be calculated by the yield stress/0.2% proof stress and the modulus of elasticity from the tensile coupon tests.

The inbuilt linear kinematic hardening material model was employed and all the material parameters were taken from the material coupon tests. By using the inbuilt analysis step 'buckle' in ABAQUS (2011) linear buckling modes could be obtained. Multiplying the coefficient of the first buckling mode by $L/1000$ aims to obtain an assumed global geometric imperfection. Then using the step 'Static-General', the numerical brace was cyclically loaded. At the same time, relationship of load and displacement and all plastic energy dissipation (ALLPD) were obtained.

4.3 Numerical results

In CBF, braces are the main members to resist the horizontal load, and buckled deformation occur under cyclic lateral loading. Besides, repeated buckle would reduce the stiffness and load bearing capacity of the braces and hence impose a significant impact on the overall performance of the structure. Thus, the reduction of maximum compression force after buckling is a significant parameter of this structural system. In Figure 12, maximum compression for the first cycle at each displacement amplitude of braces with different $\bar{\lambda}$ and geometry are plotted.

Figure 12 shows that at different member slenderness all of the compressive resistances fell to stable and similar values, though dramatic degradation is observed for braces with member slenderness of 0.8. In contrast, the reduction is more gentle for braces with member slenderness of 1.3 and 1.5. Overall, hot-rolled CHS braces and cold-formed EHS braces shared similar trend at different $\bar{\lambda}$ level.

Hysteretic performance of the braces may influence the seismic response of the system, and one of

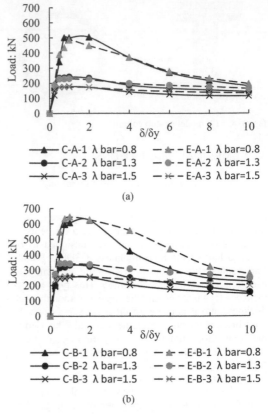

Figure 12. Maximum compression force for first cycle at each displacement amplitude: (a) specimens in series A; (b) specimens in series B;

the important indicator is the capacity of energy dissipation. The hysteretic curves of braces under cyclic loading are presented in Figures 13 and 14, which show the energy dissipation capacities of CHS and EHS braces respectively.

Figure 13 indicates that, for CHS braces, specimens with a smaller member slenderness as well as smaller diameter to thickness ratio d/t have larger enclosed area of load-displacement hysteresis loop per cycle. Specimens with smaller d/t resist larger compression and tension. Besides, comparing the hysteresis loops of first cycle and others with similar displacement amplitude, the values of compressive resistance significantly reduce as well as the area of hysteresis loops. The loops are almost coincided for the second and third cycles. The gap between first cycle and other becomes larger when $\bar{\lambda}$ level increases.

From Figure 14, same trend can be found that by increasing the member slenderness $\bar{\lambda}$ and diameter to thickness ratio d_e/t, the shape of hysteresis loops become more spread out.

However, comparing specimens C-A to E-A, EHS brace has larger enclosed area of load-displacement hysteresis loop per cycle and larger ultimate load than the CHS brace which has similar member slenderness

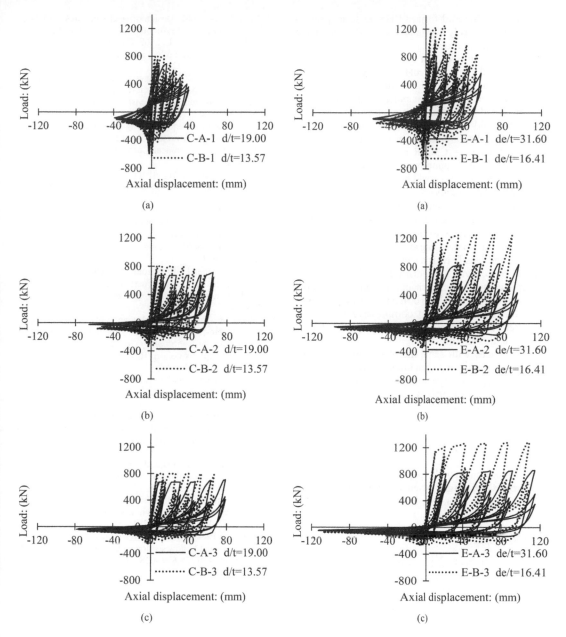

Figure 13. Load displacement hysteresis curves for CHS specimens: (a) $\bar{\lambda} = 0.8$; (b) $\bar{\lambda} = 1.3$; (c) $\bar{\lambda} = 1.5$.

Figure 14. Load displacement hysteresis curves for EHS specimens: (a) $\bar{\lambda} = 0.8$; (b) $\bar{\lambda} = 1.3$; (c) $\bar{\lambda} = 1.5$.

$\bar{\lambda}$ and cross-section area. While comparing specimens C-A to E-B, EHS brace displays larger energy dissipation capacity and larger compressive and tensile resistances than the corresponding CHS brace which has similar member slenderness $\bar{\lambda}$ and diameter to thickness ratio d/t (or d_e/t for EHS). In addition, when d/t (or d_e/t) and cross-section area are the same, because of the larger displacement in slender braces, total area of hysteresis loops of slender braces, as well as energy dissipation capacities, are similar to that in stocky short braces.

5 CONCLUSIONS

In this paper, fundamental performance of EHS stub columns and braces which were cold formed from parent hot-rolled CHS was investigated. By comparing against parent hot-rolled CHS, the following observations can be drawn:

(1) Cold formed steel material has higher strength but lower ductility than the parent hot-formed

material. For cold-formed EHS, material properties are unevenly distributed which depend on the cold-forming process.

(2) Tests on stub column under axial compression illustrated that after cold forming, local buckling occur earlier on EHS stub columns than its CHS counterparts. Comparing with the parent hot-rolled CHS stub columns, the ultimate loads on the cold-formed EHS stub columns are lower.

(3) Numerical simulation of stub columns illustrated that element – C3D8R and the corresponding mesh size were appropriate to simulate the EHS stub columns and could be adopted for the pilot numerical simulations on braces.

(4) For numerical braces with same member slenderness and cross-sectional shape, braces with smaller diameter to thickness ratio have better energy dissipation capacity, and larger compressive and tensile resistances.

(5) For braces with same diameter to thickness ratio and cross-sectional shape, increasing member slenderness, the shape of hysteresis loop become more spread out, and compressive resistances after buckle reduce. However, energy dissipation capacities remain stable.

(6) For braces with same member slenderness and similar diameter to thickness ratio, EHS braces have better energy dissipation capacities than their CHS counterparts when buckling about the major axis.

ACKNOWLEDGMENTS

The work presented in this paper was supported by the Key Projects of Natural Science Foundation of China (NSFC) through Grant No. 51038008.

REFERENCES

ABAQUS. 2011. Standard User's Manual. Hibbitt, Karlsson and Sorensen, Inc, Version 6.11.

Bradford, M.A. & Roufegarinejad, A. 2007. Elastic local buckling of thin-walled elliptical tubes containing elastic infill material. *Interaction and Multiscale Mechanics*, 1(1), 143–156.

BS EN 1993-1-1 2005. Design of steel structures, Part 1-1: General rules and rules for buildings. Milton Keynes: BSI.

Chan, T. M. & Gardner, L. 2008. Compressive resistance of hot-rolled elliptical hollow sections. *Engineering Structures*, 30(2), 522–532.

Dai, X. & Lam, D. 2010. Numerical modelling of the axial compressive behaviour of short concrete-filled elliptical steel columns. *Journal of Constructional Steel Research*, 66(7), 931–942.

ECCS 1986. Recommended Testing Procedure for Assessing the Behaviour of Structural Steel Elements under Cyclic Loads, Brussels: European Convention for Constructional Steelwork.

Gardner, L., Chan, T. M., Abela, J. M. 2011. Design recommendations for hot-finished elliptical hollow sections. *NCCI for EC, 3*.

GB/T. 2010. 228.1-2010. Tensile test of metallic material – part 1: test method at ambient temperature. *China Standard Press*

ISO, EN. 2009. 6892-1: 2009. *Metallic materials–Tensile testing-Part, 1*

Nip, K. H., Gardner, L., Elghazouli, A. Y. 2010. Cyclic testing and numerical modelling of carbon steel and stainless steel tubular bracing members. *Engineering Structures*, 32(2), 424–441.

Sheehan T. & Chan T.M. 2014. Cyclic response of hollow and concrete-filled CHS braces. Structures and Buildings, *Proceedings of Institution of Civil Engineers*, 167(SB3): 140–152. http://dx.doi.org/10.1680/stbu.12.00033.

Sheehan, T., Dai, X.H., Chan, T.M., Lam, D. 2012. Structural response of concrete-filled elliptical steel hollow sections under eccentric compression. *Engineering Structures*, 45, 314–323.

Takeuchi, T. & Matsui, R. 2011. Cumulative Cyclic Deformation Capacity of Circular Tubular Braces under Local Buckling. *Journal of Structural Engineering*, 137(11), 1311–1318.

Uenaka, K. 2014. Experimental study on concrete filled elliptical/oval steel tubular stub columns under compression. *Thin-Walled Structures*, 78, 131–137.

Yang, H., Lam D., Gardner, L. 2008. Testing and analysis of concrete-filled elliptical hollow sections. *Engineering Structures*, 30(12), 3771–3781.

Zhao, X.L. & Packer, J.A. 2009. Tests and design of concrete filled elliptical hollow section stub columns. *Thin-Walled Structures*, 47(6/7), 617–628.

Tubular Structures XV – Batista, Vellasco & Lima (eds)
© 2015 Taylor & Francis Group, London, ISBN 978-1-138-02837-1

Seismic response and damage distribution of concrete filled steel tube frame

K. Goto
Kumamoto National College of Technology, Kumamoto, Japan

ABSTRACT: Seismic response and damage of concrete filled steel tube frame (CFT frame) are calculated in relation with the column overdesign factor (r_{cb}) and its distribution. The numerical analysis method to predict the damage of CFT frame under strong ground motion is obtained by introducing the damage ratios of cracking and local buckling of both CFT column and H-section beam. CFT frames with r_{cb} distributed uniformly or not-uniformly are designed. By the use of the presented analysis method the static analysis and the seismic response analysis of these designed CFT frames have been calculated and the damages of CFT frame are obtained quantitatively. From these results it is pointed that the distribution of r_{cb} is one of the important design factor.

1 INTRODUCTION

It is well known that the column overdesign factor (r_{cb}) is an important earthquake resistant design factor of multi-story frame. The weak-beam, strong-column designed structure is commonly used in earthquake resistant design to make the frame structure collapse according to the entire failure mechanism pattern, which allows the yielding of all the beams in flexure prior to possible yielding of columns. To ensure that a frame structure collapses according to the beam-hinging pattern, the columns are generally over-designed with r_{cb}. For multi-story, multi-bay structure, because the number of beams or columns is different for each node, r_{cb} is defined for each beam-to-column connection (node) as the ratio between the sum of the column strength and the sum of the beam strength at that node, as shown in Equation 1.

$$r_{cb(k)} = \frac{\sum_i {}_c M_{ui}}{\sum_i {}_b M_{ui}} \qquad (1)$$

where $k = k$-th node, ${}_c M_{ui}$, ${}_b M_{ui} =$ the ultimate moment of the columns and beams, respectively, connected to the k-th node.

It was pointed out that the necessary value of r_{cb} to prevent the damage concentration at specific story or members is $r_{cb} = 1.5$ in Japanese design manual (Building Centre of Japan, 2008). Moreover, it was pointed out that the necessary value of r_{cb} to prevent the local buckling of concrete filled steel tube column (CFT column) under strong seismic load is $r_{cb} = 2$ (Saisho and Goto 2004). This critical value was obtained by the seismic response analyses of many CFT frames designed with r_{cb} distributed uniformly at every node. However, r_{cb} is distributed not-uniformly in real multi-story steel frame (Kawashima et al., 2006), and the distribution relation between r_{cb} and damage of building under strong ground motion is unknown.

Although CFT frame is designed with high enough r_{cb} at all nodes uniformly, the members' damage of the local buckling or the cracking may be concentrated at some stories according to other design or analysis condition, i.e., the shear strength of frame, the input ground motion and the material of members (Goto and Saisho, 2010; Goto 2012).

In this study seismic response and distribution relations between damages and r_{cb} are investigated. By the use of the presented analysis method the seismic responses of CFT frame designed under quite different design conditions are calculated and the damages of CFT frame to collapse dynamically are obtained quantitatively for design condition.

2 SEISMIC RESPONSE ANALYSIS

2.1 *Multi-story CFT frame model*

In the dynamic analysis of CFT frame, multi-story plane frame is assumed to be composed of the rigid panel zones of beam-column connection and the axially elastic members with elastic-plastic hinges at both ends as explained in Figure 1 (Saisho & Goto 2004). This assumption related to the rigid panel zone can be satisfied in most real CFT frames because the steel beam to CFT column moment connection has large shear strength. The mass of frame is concentrated in every panel zone and distributed uniformly in it. The displacement of frame can be expressed only by the

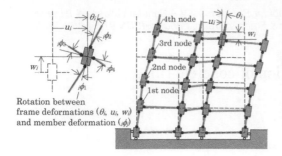

Rotation between
frame deformations (θ_i, u_i, w_i)
and member deformation ($_i\phi_j$)

Figure 1. Multi-story CFT frame model for numerical analysis.

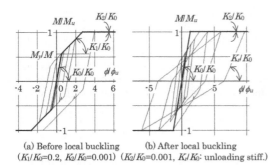

(a) Before local buckling (b) After local buckling
(K_1/K_0=0.2, K_2/K_0=0.001) (K_2/K_0=0.001, K_r/K_0: unloading stiff.)

Figure 2. Restoring force models of CFT column.

rotation (θ_i, i: number of panel zone), the horizontal displacement (u_i) and the vertical displacement (w_i) of every rigid panel zone.

The viscous damping of frame is expressed by the Rayleigh Damping in which the damping factors of the first mode (h_1) and the second mode (h_2) are assumed to be $h_1 = 0.02$ and $h_2 = 0.02$.

2.2 *Restoring force model of CFT column*

The restoring force model of elastic-plastic hinge is obtained on the basis of the dynamic loading tests of CFT column (Saisho & Goto 2001). According to the test results the non-dimensional restoring force (M/M_u) of CFT column until the local buckling of steel tube is approximated by the Tri-linear model whose skeleton curve is explained in Figure 2 (a). After the local buckling of steel tube, it is expressed by the Clough model (Clough & Johnston 1966) as shown in Figure 2 (b). The stiffness ratios at the plastic range in Figures 2 (a)–(b) are given as $K_1/K_0 = 0.2$, $K_2/K_0 = 0.001$ which are approximated on the basis of the test results (Saisho & Goto 2001). The restoring force models mentioned above are defined by the non-dimensional restoring force (M/M_u) in which the ultimate bending strength (M_u) changes at every moment by the varying axial force of CFT column. Accordingly the restoring force model can simulate the effect of varying axial force of CFT column. M_u is obtained by the generalized superposed strength method considering the confined effect (Saisho & Goto 2001).

(a) Steel tube cracking   (b) Local buckling

Figure 3. Cracking and local buckling of CFT column.

2.3 *Cracking and local buckling of CFT column*

From the cyclic loading tests of CFT column it is shown that the steel tube of CFT column cracks and fractures as shown in Figure 3 (a) when the accumulated plastic strain of steel tube becomes to be equal to the critical value ($\alpha\varepsilon_f$). From this result the cracking condition of CFT column is expressed by Equation 2 (Saisho & Goto 2001).

$$\sum \varepsilon_{TC} + \sum \varepsilon_T = \alpha\varepsilon_f \qquad (2)$$

where ε_T: the plastic tension strain of steel tube in the tension stress side. ε_{TC}: the plastic tension strain due to the local buckling deformation of steel tube in the compression stress side. $\alpha(= -0.3\rho + 5.0)$: constant expressed by the strength ratio of filled concrete to steel tube $\rho(= \sigma_c A_c / \sigma_u A_s$, A_c, A_s: sectional areas of concrete and steel tube respectively, σ_c: compression strength of filled concrete, σ_u: tensile strength of steel tube), ε_f: fracture elongation of steel tube, Σ: summation of plastic strain under cyclic load. From Equation 2, the cracking damage ratio of CFT column ($_cD_{cr}$) is expressed by Equation 3.

$$_cD_{cr} = (\sum \varepsilon_{TC} + \sum \varepsilon_T) / \alpha\varepsilon_f \qquad (3)$$

As shown in Equation 2, the local buckling of steel tube (Fig. 3 (b)) is closely related to the steel tube cracking. The local buckling condition is obtained on the basis of the upper bound theorem of the limit analysis (Saisho et al. 2004). The damage ratio of local buckling ($_cD_{lb}$) is decided by the use of critical deformation ($_c\delta_{lb}$) that corresponds to the CFT column deformation for the steel tube to buckle locally.

$$_cD_{lb} = (_c\delta_{PC} - {_c\delta_{PT}}) / {_c\delta_{lb}} \qquad (4)$$

where ($_c\delta_{PC} - {_c\delta_{PT}}$) is the amplitude of plastic deformation of CFT column. By the use of Equations 3–4, the local buckling and the steel tube cracking in the restoring force of CFT column mentioned above are decided.

2.4 *Restoring force model of H-section beam*

The H-section beam of multi-story CFT frame is also expressed by the axially elastic member with the

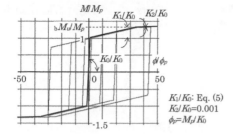

Figure 4. Restoring force model of H-section beam.

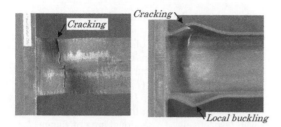

Figure 5. Cracking and local buckling of H-section beam.

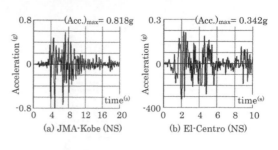

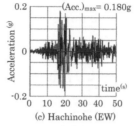

Figure 6. Time-histories of input ground motions.

elastic-plastic hinges at both ends as shown in Figure 1. The restoring force of the elastic-plastic hinge is decided by the Tri-linear model shown in Figure 4 in which the restoring force characteristics are given by the full plastic moment (M_p) and the ultimate bending strength of H-section beam ($_bM_u$). The strain hardening behavior of H-section beam affects the seismic response and collapse of CFT frame under strong ground motion. Accordingly the strain hardening of H-section beam in the model cannot be neglected. It is given by the value K_1 which is obtained by assuming H-section beam is approximated by two-flange section member.

$$\frac{K_1}{K_0} = \frac{1/y-1}{1.5y(1-y)(1+u)-1} \qquad (5)$$

where $y(=\sigma_y/\sigma_u)$ and $u(=\varepsilon_u/\varepsilon_y)$ mean the yield stress ratio and the ultimate tensile strain ratio respectively.

2.5 Cracking and local buckling of H-section beam

When strong alternating repeated load is applied to the H-section cantilever beam, flange buckles locally and after that cracks (Fig. 5) even if the fracture at the welded joint is avoided. According to the dynamic loading tests of H-section cantilever beam, the cracking fracture of H-section beam is considered as the very low-cycle fatigue behavior and assumed to be approximated by the Palmgren-Miner rule (Goto 2010, Goto & Saisho 2010). From the Palmgren-Miner rule, the damage ratio is given by the accumulation of the each cycle damage (Dowling 2007). Therefore, the cracking damage ratio of H-section beam ($_bD_{cr}$)

in each instant under random repeated load can be expressed by Equation 6.

$$_bD_{cr} = \sum_j \frac{1}{N_{fj}} \qquad (6)$$

where N_{fj} means the number of cycle to fracture under j-th cycle load and N_f is approximated by the Coffin-Manson relationship as shown in Equation 7.

$$\varepsilon_{pa} = \varepsilon_f(2N_f)^c \qquad (7)$$

where ε_{pa}: the plastic strain amplitude of H-section beam flange buckled locally (Goto & Saisho 2010), ε_f: fracture elongation, c: material modulus.

The local buckling condition of flange at the beam end is obtained on the basis of the upper bound theorem of the limit analysis (Goto 2010). The damage ratio of local buckling ($_bD_{lb}$) is decided by the use of critical deformation ($_b\delta_{lb}$).

$$_bD_{lb} = (_b\delta_{PC} - _b\delta_{PT})/_b\delta_{lb} \qquad (8)$$

where ($_b\delta_{PC} - _b\delta_{PT}$) is the amplitude of plastic deformation of H-section beam.

2.6 Input ground motion

To calculate the seismic response and damage of CFT frame, El-Centro (California 1940), Hachinohe (Aomori, Japan 1968) and JMA-Kobe (Kobe, Japan 1997) are used as input ground motion. These time histories of acceleration and acceleration response spectra are shown in Figures 6 respectively. In the calculation, the maximum velocity (V_m) of these input ground motions is amplified to $V_m = 1.0$ m/s to analyze the local buckling and cracking of the members.

159

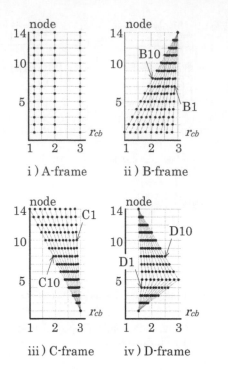

i) A-frame   ii) B-frame

iii) C-frame   iv) D-frame

Figure 7.   Distribution of r_{cb} (in case of 15-story frame).

2.7 *Design of multi-story CFT frame*

Calculated CFT frames are the 7, 10 and 15 story 3 bay frames in this study. They are designed under the following conditions.

1. The distribution of story shear strength ratio is decided by the Japanese design code. The base shear strength coefficient C_B ($=Q_1/W$, Q_1: ultimate shear force of the first story, W: weight of frame) of 7, 10, and 15 story frames are also decided by the design code and they are $C_B = 0.40$, 0.38 and 0.25 respectively. The frame strength is calculated by the limit analysis assuming the collapse mechanism of frame with plastic hinges at every beam-end and the upper and lower column-ends in the top story and the first story.

2. The column overdesign factor (r_{cb}) of every node except for the highest story is designed as shown in Figure 7. The distributions of r_{cb} are defined as follows.

   A-frame: r_{cb} are distributed uniformly.
   B-frame: r_{cb} are increased from the first node monotonically and linearly.
   C-frame: r_{cb} are decreased from the first node monotonically and linearly.
   D-frame: r_{cb} are increased from the first node to 1/3 of frame height and decreased to the highest node.

3. The strength ratio of filled concrete to steel tube (ρ) affects the restoring force characteristics of CFT column strongly (Saisho & Goto 2001). For this reason the strength ratios (ρ) of all CFT columns

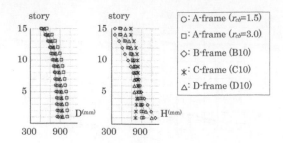

Figure 8.   Design results of 15-story frame.

are assumed to be the same and they are $\rho = 3.0$ in this study.

4. Every H-section beam of multi-story frame satisfies the critical conditions of the width-to-thickness ratio of flange (b/t_f) and web (h/t_w) and the lateral buckling parameter ($L_b h/A_f$). They are $b/t_f = 5$, $h/t_w = 71$ and $L_b h/A_f = 375$ (b: half width of flange, h: depth of beam, t_f, t_w: thicknesses of flange and web respectively, A_f: sectional area of flange, L_b: beam length).

All CFT frames are designed under the conditions mentioned above and another condition that any dimensions of steel tube and H-section are available. The story-height of every CFT frame is 4.0 m and the span lengths of outer span and inner span are 8.0 m and 6.0 m respectively. The weight of each story is 2000 kN. The yield stress (σ_y) and the tensile strength (σ_u) of steel tube and H-section beam are $\sigma_y = 340 \text{ N/mm}^2$ and $\sigma_u = 440 \text{ N/mm}^2$. The fracture elongation (ε_f) of steel tube and H-section beam is $\varepsilon_f = 0.20$. The compression strength of filled concrete (σ_c) is $\sigma_c = 60 \text{ N/mm}^2$.

Some results of designed 15-story CFT frames are shown in Figure 8 in which there are diameters (D) of CFT column and heights (H) of H-section beam at the outside span. The design conditions of CFT frames are quite different among them. But every diameter (D) of CFT column and height (H) of H-section beam is practical dimension. From these designed frames it is ascertained that the design method of multi-story CFT frame is useful.

3   RESPONSE UNDER MONOTONIC HORIZONTAL LOAD

Static analysis under monotonic horizontal load is carried out. In this analysis it is assumed that vertical load and horizontal load are applied to the center of every rigid panel zone. The distribution of the horizontal load (H_i) is decided by the Japanese design code.

The calculated load-deformation relations of 15-story frames are shown in Figure 9. The vertical axis is the load expressed by the summation of the horizontal load (ΣH_i ($=Q_1$)) and frame weight (W). The horizontal axis is the frame deformation expressed by the horizontal deformation of the top floor (U_{15}) and frame height (H_F). It is shown that the load-deformation

160

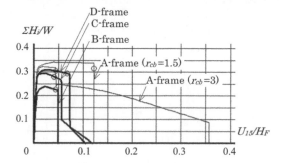

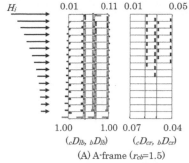

Figure 9. Distribution of r_{cb} (in case of 15-story frame).

relations greatly vary according to the distribution of r_{cb}. Especially, it is observed the maximum load of B-frame (0.24) is decreased to 68% of the design load (0.35).

The damage distributions at the point that the frame load decreased to 95% of the maximum frame load (shown by the circle in Figure 9) are shown in Figure 10. The damages are the local buckling damage ratio ($_cD_{lb}$, $_bD_{lb}$) and the cracking damage ratio ($_cD_{cr}$, $_bD_{cr}$). These damage ratios are expressed by the length of thick line perpendicular to the axis of column and beam. The numerical values in the figures explain the damage ratios of CFT columns in the first story and the highest story. In these figures the main damage occurs to the beam-end because the frame is designed by the weak-beam type. However, the damage appears to some column-ends or the damage concentration occurs at some nodes having enough large r_{cb} because of the not-uniform r_{cb} -distribution shape.

4 SEISMIC RESPONSE ANALYSIS

4.1 A-frame

The damage distributions of CFT frames under JMA-Kobe whose maximum velocity (V_m) is amplified to $V_m = 1.0\,\text{m/s}$ are shown in Figure 11. The column overdesign factors (r_{cb}) of these CFT frames are distributed uniformly. The damages of CFT frame are the story deformation angle (Δ/L), the local buckling damage ratios of CFT column ($_cD_{lb}$) and the cracking damage ratios of CFT column ($_cD_{cr}$) and H-section beam ($_bD_{cr}$). To show the effect of r_{cb}, calculations in these figures are carried out under the design condition of $r_{cb} = 1.2, 1.5, 2, 3$.

Although all CFT frames are designed under the same design conditions on the ultimate story shear strength, the strength distribution along the story and the strength ratio of filled concrete to steel tube (ρ), the CFT frames of $r_{cb} = 1.2$, many CFT columns buckle locally (($_cD_{lb})_m = 1.0$) and high $_cD_{cr}$ appear.

On the other hand, the local buckling of CFT column does not appear in the CFT frame of $r_{cb} = 3$. Therefore, $_cD_{cr}$ are low. However, $_bD_{cr}$ are high at the upper stories. From these calculated results, it is

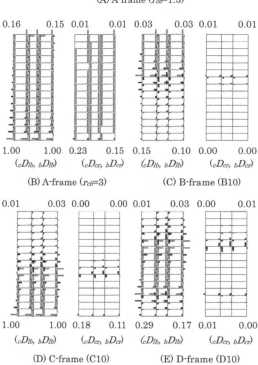

Figure 10. Distribution of r_{cb} (in case of 15-story frame).

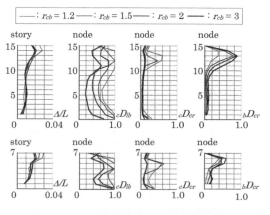

Figure 11. Damage distributions of A-frame.

161

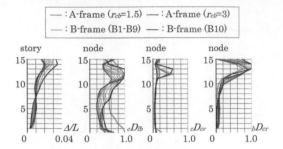

— : A-frame (r_{cb}=1.5) — : A-frame (r_{cb}=3)
— : B-frame (B1-B9) — : B-frame (B10)

Figure 12. Damage distributions of B-frame.

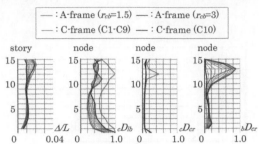

— : A-frame (r_{cb}=1.5) — : A-frame (r_{cb}=3)
— : C-frame (C1-C9) — : C-frame (C10)

Figure 13. Damage distributions of C-frame.

pointed that the damage are distributed not uniformly even if the CFT frame designed with high enough r_{cb}.

4.2 B-frame

To avoid damage concentration at specific stories, r_{cb} are increased from the first node ($r_{cb} = 1.0$–2.8) to the highest node ($r_{cb} = 3.0$) monotonically and linearly (Figure 7). Figure 12 shows the maximum damage distributions of CFT frames under JMA-Kobe ($V_m = 1.0$ m/s). From the calculated results, as the r_{cb} at the lower stories decrease, the frame response at the upper stories increases. And high damage ratios appear at the upper stories which have high r_{cb} enough. Accordingly, this r_{cb} distribution shows the dangerous distribution.

4.3 C-frame

CFT frames whose r_{cb} are distributed decreasingly from the first node ($r_{cb} = 3.0$) to the highest node ($r_{cb} = 1.0$–2.8) monotonically and linearly (Figure 7) have been analyzed. Figure 13 shows the maximum damage distributions of CFT frames under JMA-Kobe ($V_m = 1.0$ m/s). From the calculated results, damage ratios increase at the lower story which have high r_{cb} enough, and decrease at the upper story which have low r_{cb}. This results show contrary results of B-frame, and show the damage concentration at some stories. Accordingly, this r_{cb} distribution also shows the dangerous distribution.

4.4 D-frame

From the results of B, C-frames, it was pointed out that some damage concentrations appear by the distribution of r_{cb}. On the other hand, it was also pointed out that responses at some stories become small.

In this section, finding the safe r_{cb} distribution, which the damage concentration does not appear and the damages are distributed uniformly, is attempted by combining the both r_{cb} distributions.

CFT frames whose r_{cb} are distributed increasingly from the first node ($r_{cb} = 1.5$) to the (n-1)/3 node ($r_{cb} = 1.65$–3.0, in case of n-story frame) monotonically, and decreasingly from the (n-1)/3 node to the highest node ($r_{cb} = 1.5$) have been analyzed. Figures 14, 15 and 16 show the maximum damage

— : A-frame (r_{cb}=1.5) — : A-frame (r_{cb}=3)
— : D-frame (D1-D9) — : D-frame (D10)

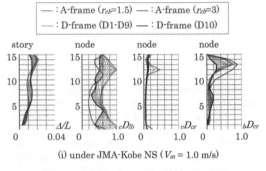

(i) under JMA-Kobe NS ($V_m = 1.0$ m/s)

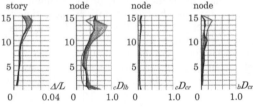

(ii) under El-Centro EW ($V_m = 1.0$ m/s)

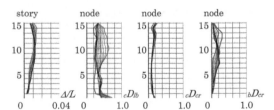

(iii) under Hachinohe EW ($V_m = 1.0$ m/s)

Figure 14. Damage distributions of 15-story D-frame.

— : A-frame (r_{cb}=1.5) — : A-frame (r_{cb}=3)
— : D-frame (D1-D9) — : D-frame (D10)

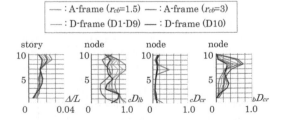

Figure 15. Damage distributions of 10-story D-frame under JMA-Kobe NS ($V_m = 1.0$ m/s).

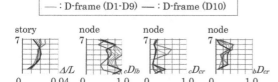

Figure 16. Damage distributions of 7-story D-frame under JMA-Kobe NS ($V_m = 1.0\,\mathrm{m/s}$).

distributions of 15, 10 and 7 story CFT frames respectively. In 15-story frame results, although the tendency for $_bD_{cr}$ at 5–10 story to become large is seen, damages are smaller than that of B and C-frame and distributed uniformly when the r_{cb} at the (n-1)/3 node is large. In 10-story frame results, although the response at middle story becomes large a little, damages are distributed uniformly. However, in 7-story frame, the local buckling occurs at 1st node, and high cracking damage ratio appears. Accordingly the safe r_{cb} distribution condition is not applicable to the low story frame.

5 CONCLUSIONS

Seismic responses and damages of many CFT frames under strong ground motions have been analyzed and it is pointed out that not only the ultimate story shear strength and the shear strength distribution but also the column overdesign factor (r_{cb}) and its distribution affect both damage ratios of the local buckling and the steel tube cracking strongly. From the calculation in this study the following points are obtained.

1. In case r_{cb} are increased from the first node to the highest node linearly (B-frame), high damage ratios appear at the upper stories because the response of frame becomes large.
2. In case r_{cb} are decrease from the first node to the highest node linearly (C-frame), high damage ratios appear at the lower stories.

3. In case r_{cb} are distributed increasingly from the first node to the (n-1)/3 node (in case of n-story frame) linearly and decreasingly from the (n-1)/3 node to the highest node (D-frame), the damage ratios become low and distributed uniformly in 10 and 15 story frames. However this r_{cb} distribution condition is not applicable to the low story frame.

REFERENCES

Building Centre of Japan, 2008, *Manual for cold-formed rectangular steel tube design and construction*, (In Japanese)

Dowling, N.E. 2007. *Mechanical Behavior of Materials*. New Jersey: Person Prentice Hall.

Kawashima T, Deguchi Y and Ogawa K, 2009, Optimum strength distribution of structural members in steel frames, *Journal of structural and construction engineering*, Vol. 635: Architectural institute of Japan, pp. 147–155. (In Japanese)

Goto K, 2012, Seismic response and damage limit of concrete filled steel tube frame, *Proceedings of the 14th International Symposium on Tubular Structures*, U.K., pp. 533–539.

Goto K, 2010, Crack fracture type of H-section cantilever beam under strong cyclic load, *Proceedings of 9th Pacific Structural Steel Conference*, China, pp. 1017–1023.

Goto K and Saisho M, 2010, Crack damage of multi-story CFT frame under strong motion, *Proceedings of the 13th International Symposium on Tubular Structures*, Hong Kong, pp. 149–157.

Saisho M and Goto K, 2001, Restoring force model of concrete filled steel tube column under seismic load, *Proceedings of the 9th Pacific Structural Steel Conference*, China, pp. 453–458.

Saisho M and Goto K, 2004, Ultimate earthquake resistant capacity of CFT-frame, *Proceedings of the 13th World Conference on Earthquake Engineering*, Paper No. 2613., Canada, pp. 1–15.

Saisho M, Kato M, and Gao, S, 2004, Local buckling of CFT-column under seismic load, *Proceedings of the 13th World Conference on Earthquake Engineering*, Paper No. 2614., Canada, pp. 1–15.

Tubular Structures XV – Batista, Vellasco & Lima (eds)
© 2015 Taylor & Francis Group, London, ISBN 978-1-138-02837-1

Seismic design of partially concrete-filled steel tubular columns with enhanced ductility

I.H.P. Mamaghani

Department of Civil Engineering, University of North Dakota, USA

ABSTRACT: This paper deals with the seismic design and analysis of partially concrete-filled steel bridge piers with enhanced ductility supporting highway bridge superstructures. The basic characteristics of the thin-walled steel tubular structures are noted and the importance of partial concrete fill in improving the strength and ductility of such structures is explained. A seismic design method for ultimate strength and ductility evaluation of concrete-filled, thin-walled, steel tubular bridge piers is presented. The application of the method is demonstrated by comparing the computed strength and ductility of some bridge piers with test results. The method is applicable for both the design of new, and retrofitting of existing, thin-walled steel tubular bridge piers. The effects of some important parameters, such as width-to-thickness ratio, column slenderness ratio, height of infill concrete, and residual stress, on the ultimate strength and ductility of thin-walled steel tubular bridge piers are presented and discussed.

1 INTRODUCTION

The use of concrete-filled steel hollow sections in structures generally leads to a more efficient and economical system for resisting severe environmental loads such as seismic forces. The objective of this paper is to address the seismic design and ductility evaluation of partially concrete-filled steel box columns that are widely used in Japanese bridge pier construction in seismic areas due to their excellent seismic-resisting characteristics such as high ductility, improved strength, and energy absorption capacity (Usami et al. 1995).

Thin-walled steel tubular columns are vulnerable to damage caused by local and overall interaction buckling during a major earthquake. A sound understanding of the inelastic behavior of thin-walled steel tubular beam-columns is important in developing a rational seismic design methodology and ductility evaluation of such structures. On the other hand, fully or partially filling the tube with concrete substantially enhances the ductility and thereby the seismic performance of the structure (Mamaghani and Packer 2002). For this purpose, a seismic design method for ultimate strength and ductility evaluation of hollow and concrete-filled thin-walled steel tubular beam-columns is presented. The method involves an elastoplastic pushover analysis and definition of failure criterion taking into account local buckling and residual stresses due to welding. The application of the method is demonstrated by comparing the computed strength and ductility of some cantilever bridge piers with test results. The effects of some important parameters, such as width-to-thickness ratio, column slenderness ratio,

and height of infill concrete, on the ultimate strength and ductility of thin-walled steel tubular bridge piers, are also presented.

2 STEEL BRIDGE PIERS

Steel bridge piers have found wide application in highway bridge systems in Japan compared with other countries, where such structures are much less adopted. In general, steel bridge piers are designed as either single columns of the cantilever type or one- to three-story frames. Steel columns in highway bridge systems are commonly composed of relatively thin-walled members of closed cross-sections, either box or circular in shape because of their high strength and torsional rigidity. Such structures differ from columns in buildings. The former are characterized by failure attributed to local buckling in the thin-walled members; irregular distribution of the story mass and stiffness; strong beams and weak columns; low rise (1–3 stories); and a need for the evaluation of the residual displacement. These make them vulnerable to damage caused by local and overall interaction buckling in the event of a severe earthquake. For example, Figure 1 shows a steel bridge pier of circular box section which suffered severe local buckling damage in the Kobe earthquake. The piers were partially filled with concrete ,and local buckling occured at the hollow portion just above the concrete-filled section. In this section, some important parameters, such as strength and ductility capacity, affecting the seismic performance of steel bridge piers are addressed and discussed.

Figure 1. Damaged partially concrete-filled steel tubular column, Kobe earthquake, 1995, Japan.

2.1 Steel sections

The cross-section of columns used in bridge piers greatly differs from those used in buildings. The main differentiating characteristics of the cross-section of bridge pier columns compared with those in buildings are:

1. the cross-section is large (about 3–4 m in bridge piers compared with 0.6–1 m in tall buildings);
2. stiffened plates are used;
3. the ratio of applied axial compressive load to the squash load of the cross-section is small (less than about 0.15, compared to 0.6–0.7 for the leeward columns of a tall building);
4. for concrete-filled sections, the ratio of the squash load of the outer steel cross-section (hollow section) to the squash load of the whole (steel plus concrete) section, $\bar{\gamma}$, is small ($\bar{\gamma}$ is about 0.2 for bridge piers and is about 0.75 for columns in buildings).

The ductility behaviour of steel bridge piers of welded box sections is mainly governed by local component members such as plates, with or without stiffeners.

2.2 Characteristic parameters

The most important parameters considered in the practical design and ductility evaluation of thin-walled steel hollow box sections are the width-to-thickness ratio parameter of the flange plate R_f and the slenderness ratio parameter of the column $\bar{\lambda}$ (Mamaghani et al. 1996a, 1997). While the former influences local buckling of the flange, the latter controls the global stability. They are given by:

$$R_f = \frac{b}{t}\frac{1}{n\pi}\sqrt{3(1-v^2)\frac{\sigma_y}{E}} \quad (\text{for box section}) \tag{1}$$

$$R_t = \frac{d}{2t}\sqrt{3(1-v^2)}\ \frac{\sigma_y}{E}\,(\text{for circular section}) \tag{2}$$

$$\bar{\lambda} = \frac{2h}{r}\frac{1}{\pi}\sqrt{\frac{\sigma_y}{E}} \tag{3}$$

in which $b =$ flange width; $t =$ plate thickness; $\sigma_y =$ yield stress; $E =$ Young's modulus; $v =$ Poisson's ratio; $n =$ number of subpanels divided by longtudinal stiffeners in each plate panel ($n = 1$ for unstiffened sections); $d =$ diameter of the circular section; $h =$ column height; and $r =$ radius of gyration of the cross section (Usami et al. 1995). The stiffener's equivalent slenderness ratio $\bar{\lambda}_s$, magnitude of axial load P/P_y, and type of stiffener material are other important parameters considered in a practical design. The parameter $\bar{\lambda}_s$ controls the deformation capacity of the stiffeners and local buckling mode, and is given by (Usami 1996):

$$\bar{\lambda}_s = \frac{1}{\sqrt{Q}}\frac{a}{r_s}\frac{1}{\pi}\sqrt{\frac{\sigma_y}{E}} \tag{4}$$

$$Q = \frac{1}{2R_f}[\beta - \sqrt{\beta^2 - 4R_f}\,] \le 1.0 \tag{5}$$

$$\beta = 1.33R_f + 0.868 \tag{6}$$

where $r_s =$ radius of gyration of a T-shape cross-section consisting of one longitudinal stiffener and the adjacent subpanel of width (b/n); $a =$ distance between two adjacent diaphragms; and $Q =$ local buckling strength of the sub-panel plate. An alternative parameter reflecting the characteristics of the stiffener plate is the stiffener's relative flexural rigidity, γ, which is interdependent on $\bar{\lambda}_s$ and obtained from elastic buckling theory (DIN-4114 1953). Thus, only $\bar{\lambda}_s$ is considered in the ductility equations.

The elastic strength and deformation capacity of the column are expressed by the yield strength H_{y0} and the yield deformation (neglecting shear deformations) δ_{y0}, respectively, corresponding to zero axial load. They are given by:

$$H_{y0} = \frac{M_y}{h} \tag{7}$$

$$\delta_{y0} = \frac{H_{y0}h^3}{3EI} \tag{8}$$

where $M_y =$ yield moment and $I =$ moment of inertia of the cross section. Under the combined action of buckling under constant axial and monotonically increasing lateral loads, the yield strength is reduced from H_{y0} to a value denoted by H_y. The corresponding yield deformation is denoted by δ_y. The value H_y is the minimum of yield, local buckling, and instability loads evaluated by the following equations (Usami, 1996):

$$\frac{P}{P_u} + \frac{0.85H_yh}{M_y(1-P/P_E)} = 1 \tag{9}$$

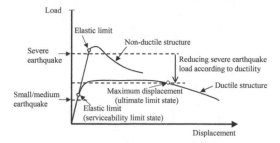

Figure 2. Schematic load-displacement response of ductile and non-ductile steel bridge piers.

$$\frac{P}{P_u} + \frac{H_y h}{M_y} = 1 \qquad (10)$$

in which P = the axial load; P_y = the squash load; P_u = the ultimate load; and P_E = the Euler load.

2.3 Loads

In bridge piers, the primary loadings may be considered as axial compressive forces (to support superstructures) with little variation and alternating bending in response to seismic loads and wind loads. Bending force is the primary section force, appearing as a stress resultant which produces repetitive axial forces in opposite, outer, thin plate elements of the cross section. These thin elements are subjected to cyclic axial compression and tension due to alternating bending moments, resulting in successive buckling until the occurance of complete sectional collapse.

3 DESIGN PHILOSOPHY

It has been widely realized in modern seismic design philosophy that a rational seismic design method should be able to explicitly evaluate the structural non-linear performance that is expected to occur under the design seismic load. Recently, for economic reasons, the design of structural restraint to severe earthquake has focused on the idea of ductility-based design, a shift away from strength-based design. This requires attention to both strength and ductility. Figure 2 illustrates a schematic load-displacement response of ductile and non-ductile steel bridge piers. As shown in this figure, design loads for severe earthquakes can be considerably reduced according to the ductility of the structure. Since bridge piers are either statically determinate or indeterminate to a low degree, an increase in strength for large displacements cannot be expected, even if local instability is disregarded. Hence, a large degradation in strength, due to local buckling at least, needs to be prevented.

3.1 Methods of improving strength and ductility

The methods to improve strength and ductility of steel bridge piers of hollow box sections can be summarized as follows:

1. limiting the width-to-thickness ratio of component plates;
2. limiting the slenderness of columns;
3. filling the tube with concrete (composite construction);
4. using the inelastic characteristics of high performance structural steel; and
5. using of a ductile cross-sectional shape.

Among these, in practice, limiting the width-to-thickness ratio of component plates and the slenderness of the column are the most suitable methods for enhancing the strength and ductility. The Japanese design guideline limits the width-to-thickness ratio parameter to $0.3 \leq R_f \leq 0.5$, the rigidity of longitudinal stiffeners to $\gamma/\gamma^* \geq 3.0$ (γ^* = optimum value of γ), the column slenderness ratio to $0.25 \leq \bar{\lambda} \leq 0.60$, and the axial compressive force ratio to $P/P_y \leq 0.20$ (Usami 1996, Fukumoto 1997). These quite conservative limitations ensure reasonable safety for steel bridge piers, even in the event of a strong earthquake.

Increasing the ductility capacity of steel bridge piers has been one of the main issues in retrofit work. In practice, piers can be retrofitted by one or a combination of the following methods:

1. filling with concrete inside steel sections from the base to an optimal level;
2. reinforcing with longitudinal stiffeners;
3. increasing the rigidity of longitudinal stiffeners;
4. adding transverse stiffeners between diaphragms;
5. placing a diaphragm over the filled-in concrete; and
6. corner reinforcement for box sections.

Among these options, the most efficient alternative for improving the seismic performance of hollow steel sections is to partially fill them with concrete.

3.2 Partially concrete-filled columns

Filling the hollow steel tube with concrete (and thereby using composite construction) is one of the most suitable available methods for improving strength and ductility because the concrete core inhibits local buckling of the thin steel tube and, at the same time, the steel tube provides a confining stress to the concrete core. The main ideas underlying partially concrete-filled bridge piers are:

1. to increase the ductility for resisting strong earthquakes, without substantially increasing the stiffness and/or strength, compared with steel bridge piers of hollow tubular sections of the same size;
2. to reduce the self weight compared with fully concrete-filled steel bridge piers.

This is advantageous because:

1. fully concrete-filled steel bridge piers can substantially increase stiffness and/or strength, but they behave in a relatively brittle (nonductile) manner and do not provide sufficient ductility to resist very strong earthquakes;
2. foundation structures supporting bridge piers become large and expensive if the self-weight and strength of the piers are increased.

The behavior of concrete-filled steel columns has been studied by several researchers, including Mamaghani & Packer (2002), Usami et al. (1995) and Usami (1996). The results of conducted tests and observed damage sustained by bridge piers in the Kobe earthquake indicate that local buckling may occur at the hollow portion just above the concrete-filled section, when the height of concrete fill is small compared to the height of the column. The following section presents and discusses a numerical procedure for the ultimate strength and ductility evaluation of partially concrete-filled thin-walled steel bridge piers.

4 PROCEDURE FOR STRENGTH AND DUCTILITY EVALUATION

The strength and ductility evaluation of partially concrete-filled, thin-walled, steel box columns of the cantilever type, modeling bridge piers, consists of the following steps:

1. Initially design the general layout of the new structure or determine the general layout of the existing structure to be retrofitted. Determine the section type and size, the height of column, material, height of concrete fill, and estimate the superstructure weight and lateral loads via an elastic seismic design method.
2. Carry out a pushover analysis to obtain the load-displacement relationship. The basic procedures in the pushover analysis utilized in this study are summarized as follows:
   (a) Based on the general layout and loading condition of the structure, establish the analytical model as shown in Figure 3 by using beam-column elements.
   (b) Set the height of concrete fill h_c (initially h_c/h may bet set to 0.25).
   (c) Choose failure criteria for both steel and concrete. The ultimate limit state corresponding to the failure criteria is used as the termination of the analysis. The adopted stress-strain relation for steel and concrete, effective failure length, and failure criteria are explained later in this section.
   (d) Apply the constant vertical load (including weight of the superstructure) and laterally push the structure until the ultimate limit state is reached.
   (e) Based on the base shear H versus top lateral displacement δ curve from the analysis,

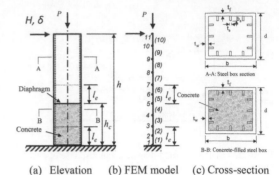

(a) Elevation (b) FEM model (c) Cross-section

Figure 3. Analytical modeling of partially concrete-filled steel box columns.

determine the ultimate strength H_u and ductility δ_u of the structure.
3. Check the seismic performance of the structure. If the capacity cannot meet the seismic demand, change the height of concrete fill and repeat the procedure starting from step 2 until the damage indices for both steel and concrete simultaneously approach unity (corresponding to optimum height of concrete fill). If the capacity still cannot meet the seismic demand, change the general layout of the structure and repeat the procedure from step 1 until the capacity meets the seismic demand. The seismic verification procedure is presented in a subsequent section.

The above procedure is easily applicable for both fully concrete-filled steel sections and steel hollow sections. It can be implemented for both the design of new steel bridge piers and the retrofit of existing bridge piers composed of thin-walled members filled with concrete.

4.1 Failure strain for steel plate members

Local buckling controls the capacity of bridge piers composed of thin-walled steel members. A number of experimental and numerical analyses have shown that the critical local buckling of thin-walled steel structures always appears in the compressive flange plates within an effective failure range (Usami et al. 1995). The ductility of isolated plates with and without longitudinal stiffeners has been investigated through extensive parametric analyses (Usami et al. 1995, Banno et al. 1998). The normalized failure strain $\varepsilon_{f,s}/\varepsilon_y$ of a plate under pure compression is given by (Usami et al. 1995).

For an unstiffened plate:

$$\frac{\varepsilon_{f,s}}{\varepsilon_y} = \frac{0.07}{(R_f - 0.2)^{2.53}} + 1.85 \leq 20.0 \tag{11}$$

For a stiffened plate:

$$\frac{\varepsilon_{f,s}}{\varepsilon_y} = \frac{0.145}{(\bar{\lambda}_s - 0.2)^{1.11}} + 1.19 \leq 20.0 \tag{12}$$

in which $\varepsilon_{f,s}$ = failure strain corresponding to 95% of maximum strength after peak load. The occurance of local buckling can be neglected when $\bar{\lambda}_s$ is smaller than 0.2.

4.2 *Failure criteria*

In a thin-walled steel bridge pier, excessive deformation tends to develop in a local part; consequently the redistribution of stress becomes unexpected. The test results indicate that the local buckling occurs near the column base (for a composite section) or in the hollow section just above the concrete fill in the range of about 0. 7b (b = breadth of the flange) or between the transverse diaphragms, if any, depending on the height of the concrete (Usami 1996). These critical local parts are defined by an effective failure length l_e, as marked in Figure 3. To establish the ultimate state, the average strain in the outer fibers of the concrete and steel segments along the length l_e, at composite and hollow parts, is recorded. The failure criterion can be described by damage indices for steel D_s and concrete D_c, which are defined by:

$$D_s = \frac{\varepsilon_{ave,s}}{\varepsilon_{f,s}} \tag{13}$$

$$D_c = \frac{\varepsilon_{ave,c}}{\varepsilon_{f,c}} \tag{14}$$

The ultimate limit state of the structure is attained when any one of these indices first reaches unity (Usami et al. 1995). Here, $\varepsilon_{ave,s}$ represents the average strain of the compressive flange for the box section and $\varepsilon_{ave,c}$ is the average strain of the most compressive outer fibre of the concrete core over corresponding effective failure lengths. $\varepsilon_{f,c}$ denotes the failure strain of concrete and is defined to be $\varepsilon_{f,c} = 0.011$. $\varepsilon_{f,s}$ denotes the failure strain of steel plate members given by Equations 11 and 12.

5 NUMERICAL ANALYSIS

The pushover analysis, discussed above, is applied to simulate the load-displacement response of the tested columns with and without stiffeners (Usami et al. 1995). An example of the analytical model is shown in Figure 3. The partially concrete-filled steel bridge pier is modeled by using beam-column elements. An elastoplastic finite element formulation for a beam-column, considering geometrical and material nonlinearities, was developed and implemented in the computer program FEAP (Zienkiewicz 1977), which was used in the analysis (Mamaghani 1996, Mamaghani et al. 1996b).

Uniaxial stress-strain relations for steel and concrete are assigned to beam-column elements, which are employed to model the steel and concrete segments. The monotonic stress-strain curve of the modified two-surface plasticity model (2SM), developed by

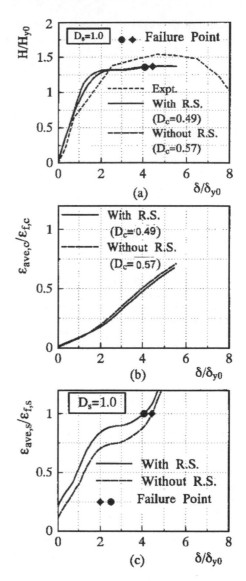

Figure 4. Comparison of test UU2 and analysis.

the author and his co-workers at Nagoya University (Mamaghani et al. 1995, Shen et al. 1995), and is employed in the analysis. The stress-strain model for the concrete material employed in the analysis accounts for the confinement effect provided by the surrounding steel plates (Uasmi et al. 1995).

The analysis accounts for the residual stresses due to welding (Mamaghani et al. 1996a). The details of the numerical analysis and comparison between test results and analytical results have been reported in the works by Mamaghani & Packer (2002) and Usami et al. (1995).

5.1 *Numerical examples*

Figure 4 shows the results for the cantilever column with $R_f = 0.664$, $\bar{\lambda} = 0.362$, and $h_c/h = 0.3$, without

169

longitudinal stiffeners obtained from tests and analysis. Figure 4a compares normalized lateral load-lateral displacement (H-δ) curves for the experiments and analysis with and without residual stresses. In this figure the envelope curve (dashed line) of the hysteretic behavior of the cyclic experiments (Usami et al. 19950) is shown for comparison. Figure 4b shows the concrete damage index $D_c = \varepsilon_{ave,c}/\varepsilon_{f,c}$ versus normalized lateral displacement. Figure 4c shows the steel damage index $D_s = \varepsilon_{ave,s}/\varepsilon_{f,s}$ versus normalized lateral displacement. As described in the previous section, the failure of the column is attained when any one of these indices first reaches unity. The failure points marked in Figure 4 correspond with the ultimate displacement of the column at failure, taking place in the hollow section just above the concrete fill level with $D_s = 1.0$ (Figure 4c), while $D_c = 0.57$ for the case without residual stress and $D_c = 0.49$ for the case with residual stress, (Figure 4b). Figure 5 compares the results for the same column with concrete fill height of $h_c = 0.5h$. In this case the failure of the column is attained in the concrete-filled section with $D_s = 1.0$, (Figure 4b), while $D_s = 0.35$ for the case without residual stress and $D_s = 0.64$ for the case with residual stress (Figure 5c).

5.2 Effects of residual stresses on strength and ductility capacity

The results in Figures 4 and 5 indicate that residual stress has the effect of reducing the stiffness of the column regardless of the height of concrete fill. It has the effect of reducing the ductility of columns in which failure occurs in hollow steel box sections (δ_u/δ_{yo} is reduced from 4.43 to 4.08 for column UU2, Figure 4a). However, the residual stress causes an increase in ductility when the governing failure mode is concrete failure for a higher h_c/h ratio (e.g., $h_c/h = 0.5$). As shown in Figure 5a, δ_u/δ_{yo} is increased from 6.36 to 6.79 for column UU3. This is attributed to the spread of plasticity along the effective failure length owing to early yielding of steel in the presence of residual stress, which in turn causes the larger deformation capacity. The results of these analyses indicate that the residual stress has almost no effect on the maximum load carrying capacity of the column. Figures 5a and 6a show that the H-δ curves obtained from analyses considering residual stress effects agree well with those of the experiments. The failure modes are the same in both the experiments and analyses indicating the accuracy of the employed analytical model. The ductility of the column δ_u/δ_{yo} is improved by 66 percent (from 4.08 to 6.79) when the height of concrete fill is increased from 30 to 50 percent of the column height ($h_c = 0.3h$ to $h_c = 0.5h$) while the change in column ultimate strength is 3.6%, which is negligibly small (H_u/H_{yo} increased from 1.36 to 1.41). This satisfies the main idea underlying partially concrete-filled thin-walled steel bridge piers, namely to increase the ductility for resisting

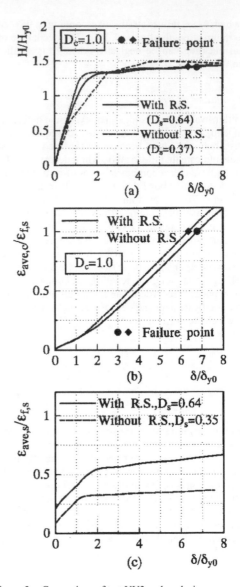

Figure 5. Comparison of test UU3 and analysis.

strong earthquakes without substantially increasing the strength.

Examples of the results obtained for a few experiments are given in Table 1. The effects of some important parameters on strength and ductility capacity of partially concrete-filled columns are summarized as follows:

1. Failure modes are the same in both the experiments and analyses indicating the accuracy of the analytical model employed; Table 1.
2. Failure occurs in hollow steel box sections for lower h_c/h ratios ($h_c/h = 0.3$); Table 1.
3. Failure occurs in concrete-filled sections for higher h_c/h ratios ($h_c/h = 0.5$); Table 1.
4. Residual stress has the effect of reducing the ductility of columns in which failure occurs in hollow

Table 1. Effects of residual stress, R_f, $\bar{\lambda}$, h_c/h on strength, ductility capacity and failure mode.

Specimen				Without R.S.		With R.S.		Without R.S.		With R.S.		Failure mode
	h_c/h	R_f	$\bar{\lambda}$	H_u/H_{y0}	δ_u/δ_{y0}	H_u/H_{y0}	δ_u/δ_{y0}	D_s	D_c	D_s	D_c	
UU2	0.3	0.664	0.362	1.36	4.43	1.36	4.08	1.0	0.57	1.0	0.49	Steel
UU3	0.5	0.664	0.362	1.41	6.36	1.41	6.79	0.35	1.0	0.64	1.0	Concrete
UU5	0.3	0.664	0.577	1.24	3.53	1.22	3.14	1.0	0.49	1.0	0.45	Steel
UU7	0.3	0.854	0.381	1.40	2.96	1.34	2.02	1.0	0.33	1.0	0.17	Steel

steel box sections ($h_c/h = 0.3$). However, the residual stress causes an increase in ductility when the governing failure mode is concrete failure ($h_c/h = 0.5$). The effect of residual stress becomes more apparent with the increase in width-to-thickness ratio R_f; Table 1.

5. The ductility capacity substantially decreases with an increase in column slenderness ratio $\bar{\lambda}$ and width-to-thickness ratio R_f, where steel failure is the governing failure mode.

6. The results from experiments and analysis suggest that the optimum ductility of partially concrete-filled, thin-walled, steel bridge piers can be attained when concrete failure is the governing failure mode with $D_c = 1.0$ and the damage sustained by steel is very close to unity. That is, the optimum seismic design and ductility capacity of such bridge piers can be achieved by arranging the bridge pier parameters such as R_f, $\bar{\lambda}$ and h_c/h to satisfy this optimum failure mode.

6 SEISMIC VERIFICATION METHOD

To verify seismic performance of a bridge pier, the following condition for supply (earthquake-resistance) capacity S provided by the structure and demand capacity D required by an earthquake should be satisfied:

$$S \geq D \qquad (15)$$

The demand capacity D for a specific earthquake motion, can be obtained from the discussed pushover analysis by determining the H-δ curve, in which the ultimate displacement of the bridge pier δ_u corresponds to the ultimate strength H_u at failure. The factored lateral displacement δ_f, can be obtained by applying a safety factor against the ultimate displacement δ_u, and its corresponding factored lateral strength H_f can be obtained from the H-δ curve. Based on the equivalence in energy absorption capacity criterion, the supply and demand capacities are given by:

$$D = k_{hc}W \qquad (16)$$

$$S = \sqrt{2\mu_f - 1}\,H_f \qquad (17)$$

$$\mu_f = \frac{H_y}{H_f}\frac{\delta_f}{\delta_y} \qquad (18)$$

where k_{hc} = seismic coefficient; W = equivalent weight of both the superstructure and pier; μ_f = factored global ductility. This condition should be met for the design of both new and retrofitted structures.

7 CONCLUSIONS

The paper covers seismic design and ductility evaluation of concrete-filled steel tubular bridge piers supporting highway bridge superstructures with enhanced energy dissipating mechanisms. The seismic design concepts and some important characteristic parameters of thin-walled steel bridge piers are presented and discussed. The main ideas underlying partially concrete-filled steel box columns are pointed out. A seismic design and an evaluation procedure are described to estimate the strength and ductility of such bridge piers through an elastoplastic pushover analysis. It is assumed that the structural ultimate limit state is attained when the compressive strain in the flange plate or concrete fill reaches their prescribed failure strains.

Using this procedure, the ultimate strength and ductility capacity of partially concrete-filled thin-walled steel bridge piers were estimated. By comparison with strength and ductility estimations obtained from tests reported in the literature, the reliability of the procedure was verified. A good agreement between the failure modes observed in experiments and pushover analysis provided proof of the assumptions of the material models, effective failure length concept, and failure criteria, on which the strength and ductility of partially concrete-filled steel bridge piers were calculated. The effects of some important parameters, such as width-to-thickness ratio, column slenderness ratio, height of infill concrete and residual stress, on the ultimate strength and ductility of thin-walled steel tubular beam-columns, were presented and discussed. The procedure can be used in seismic design and retrofit of partially concrete-filled thin-walled steel bridge piers to evaluate the ultimate strength and ductility. It can also be easily extended to the seismic design and retrofit of fully concrete-filled steel tubular columns and frames.

REFERENCES

Banno, S., Mamaghani, Iraj H.P., Usami, T. & Mizuno, E. 1998. Cyclic Elastoplastic Large Deflection Analysis of Thin Steel Plates, *Journal of Engineering Mechanics*, ASCE, USA, Vol. 124, No. 4, 363–370.

DIN-4114. 1953. *Stahbau, Stabilitatsfalle (Knickung, Kippung, Beulung), Berechnungsgrund- lagen, Richtlinien*, Blatt2. Berlin, Germany (in German).

Fukumoto, Y. 1997. *Structural Stability Design- Steel and Composite Structures*, Pergamon Press, Elsevier Science Ltd., Oxford, Great Britain.

Mamaghani, I. H. P. 1996. Cyclic Elastoplastic Behavior of Steel Structures: Theory and Experiments, *Doctoral Dissertation*, Department of Civil Engineering, Nagoya University, Nagoya, Japan.

Mamaghani, I.H.P. & Packer J.A. 2002. Inelastic Behavior of Partially Concrete-Filled Steel Hollow Sections, *4th Structural Specialty Conference of the Canadian Society for Civil Engineering*, Montréal, Québec, Canada, s71.

Mamaghani, I. H. P., Shen, C., Mizuno, E. & Usami, T. 1995. Cyclic Behavior of Structural Steels, I: Experiments, *Journal of Engineering Mechanics*, ASCE, USA, Vol.121, No. 11, 1158–1164.

Mamaghani, I. H. P., Usami, T. & Mizuno, E. 1996a. Cyclic Elastoplastic Large Displacement Behaviour of Steel Compression Members, *Journal of Structural Engineering*, JSCE, Japan, Vol. 42A, 135–145.

Mamaghani, I. H. P., Usami, T. & Mizuno, E. 1996b. Inelastic Large Deflection Analysis of Structural Steel Members Under Cyclic Loading, *Engineering Structures*, UK, Elsevier Science, 18(9), 659–668.

Mamaghani, I. H. P., Usami, T. & Mizuno, E. 1997. Hysteretic Behavior of Compact Steel Box Beam- Columns, *Journal of Structural Engineering*, Japan Society of Civil Engineers (JSCE), Japan, Vol. 43A, 187–194.

Shen, C., Mamaghani, I. H. P., Mizuno, E. & Usami, T. 1995. Cyclic Behavior of Structural Steels, II: Theory, *Journal of Engineering Mechanics*, ASCE, USA, Vol. 121, No. 11, 1165–1172.

Usami, T., Suzuki, M., Mamaghani, I. H. P. & Ge, H.B. 1995. A Proposal for Check of Ultimate Earthquake Resistance of Partially Concrete Filled Steel Bridge Piers, *Journal of Structural Mechanics and Earthquake Engineering*, JSCE, Tokyo, 525/I-33, 69–82, (in Japanese).

Usami, T. 1996. *Interim Guidelines and New Technologies for Seismic Design of Steel Structures*, Committee on New Technology for Steel Structures, JSCE, Tokyo (in Japanese).

Zienkiewicz, O. C. 1977. *The Finite Element Method*, 3rd Ed., McGraw-Hill.

Fire resistance

Tubular Structures XV – Batista, Vellasco & Lima (eds)
© 2015 Taylor & Francis Group, London, ISBN 978-1-138-02837-1

Optimal economic design of unprotected circular concrete-filled steel tubular columns at ambient temperature and under fire condition

D. Hernández-Figueirido, A. Piquer & J.M. Portolés
Dpt. Ing Mecánica y Construcción, Universitat Jaume I, Castellón, Spain

A. Hospitaler
Instituto de Ciencia y Tecnología del Hormigón, Universitat Politècnica de València, Valencia, Spain

J.M. Montalvá
Universitat Politècnica de València, Valencia, Spain

ABSTRACT: This paper presents the results of a global comparison between three types of circular columns: steel hollow sections, plain concrete-filled steel tubular columns and reinforced concrete-filled steel tubular columns, submitted to static loads and attending to fire resistance capacity and economic aspects. The analysis is focused on simple columns taking into account the models of Eurocode 3 and 4 to verify the column capacity. The automatic implementation of the algorithms allows achieving a high amount of case studies, covering the realistic possibilities of building columns: with different combinations of geometry (D,t,L), materials (fay, fck, fsy), bar reinforcement and axial load (NEd). Design curves are presented for optimum values of the thickness, diameter, sectional aspect ratio (diameter to thickness), reinforcement, resistance of the materials and fire requirements for various load level and column lengths. Thin-walled columns and high strength concrete (with the limitations imposed by Eurocode 4) provide the optimal economic solution of the problem.

1 INTRODUCTION

Steel hollow sections are commonly used in construction due to their structural efficiency and attractive appearance. Compared to conventional open sections, hollow sections have excellent strength capacities (compression, bending and torsion), lower drag coefficients, less painting area and enable a more elegant structure design, Wardenier et al. (2010).

The main steel elements' disadvantage, as compared to concrete ones, is their reduced capacity during fire situations and the necessity of protecting steel profiles to enhance structural fire resistance, with the consequent incremental cost. The use of concrete to protect steel sections against the action of fire and corrosion was the origin of the composite structures: steel sections encased in concrete, B. Uy (2003).

Concrete-Filled Steel tubular columns (CFST) were developed later. This type of mixed columns does not need any external formwork for the concrete (the steel tube is a permanent one) and the costs of erection were reduced. Besides, concrete filling of hollow steel sections enhances fire resistance while maintaining aesthetic properties, turning this structural system into an attractive and competitive alternative. The fire resistance of unprotected circular hollow steel tubular (CHT) columns is normally found to be less than 30 minutes, Twilt et al. (1996). Several authors like

Han et al. (2003a) and Lu et al. (2010), have proved experimentally that concrete filling can increase fire resistance up to 2 or 3 hours. Furthermore, CFST have other valuable properties like high load-bearing capacity, high seismic resistance and fast construction technology, Zhao et al. (2010). All these characteristics have led to an extended use of CFST columns over the last few decades in many structural engineering applications such as columns in high-rise buildings, bridge piers, industrial buildings, structural frames and supports, electricity transmission poles, etc, Han et al. (2014).

In the past, a number of experimental studies were conducted to evaluate fire performance of CFT structures but, there are no studies that seek for the optimal configuration of CFST columns subjected to fire requirements based on economical criteria.

The present research offers the optimal economical configuration of a column that fulfills engineering requirements such as load-carrying capacity and safety under the action of fire.

2 PROBLEM DEFINITION

This study is concerned with the optimal design of steel tubular columns taking into account the required

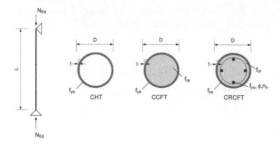

Figure 1. Geometry of the columns and notation of the problem to be analyzed.

Table 1. Summary of the variables employed and their ranges.

	Circular Hollow Tube (CHT)	Circular Concrete-Filled Steel Tube (CCFT)	Circular Reinforced Concrete-Filled Steel Tube (CRCFT)
Materials			
f_{ya} (MPa)	235, 275, 355, 420, 460		
f_{ck} (MPa)		25, 30, 35, 40, 45, 50	
f_{ys} (MPa)			400, 500
Geometry of elements			
D (mm)	21.3, 26.9, 33.7, 42.4, 48.3, 60.3, 76.1, 88.9, 101.6, 114.3, 139.7, 168.3, 177.8, 193.7, 219.1, 244.5, 273.0, 323.9, 355.6, 406.4, 457, 508, 610, 711, 762, 813, 914, 1016, 1067, 1168, 1219		
t (mm)	2, 2.5, 3, 4, 5, 6, 6.3, 8, 10, 12, 12.5, 16, 20, 25, 30		
Φ (mm)			8, 10, 12, 14, 16, 20, 25, 32, 40
n_ϕ			2, 3, 4, 5, 6, 7, 8, 9, 10, 11, 12

load bearing capacity and fire resistance for building construction. In figure 1 the different types of tubular columns studied are represented. The main aim is to give to designers some aspects for selecting the most economical structural solution. Many feasible configurations fulfill the design and fabrication constraints and designers should select the best one among all these possibilities.

Among several feasible solutions, the optimal one, in economical terms, is the cheapest one. In order to compare the different feasible solutions for each pair of axial load and column length, a cost function has been defined. For this purpose, the data cost base of BEDEC [21] has been employed. The data cost base utilized is specific for Spain at the present time. It is therefore assumed that in other countries or periods of time it will be different. However, the objective is to draw general conclusions that can be useful for any place or time.

2.1 Variables and constants

The design variables and the parameters of the problem are all the data required to define a given structure. The variables in the design of a column are the external factors (load applied and boundary conditions), sectional geometry and column length, materials and configuration employed. In this study the load applied and the length of the column are considered as constants. The optimal solutions were obtained for fixed values of L and N_{Ed}. The range of variables studied in this work is limited for the field of application of the codes considered, in this case, the European Standard Codes. Thus, Table 1 shows a summary of the variables of the problem and their ranges.

Where f_{ya} is the steel tube yielding strength, f_{ck} characteristic compressive strength of concrete (150×300 cylinder test), f_{ys} is the yield strength of reinforcing steel bars, D is the external diameter of the steel tube, t is the thickness, ϕ the diameter of the reinforcement, and n_ϕ the number of bars.

Regarding the concrete cylinder strength, the same strength can be obtained with different mixtures by varying the components employed. For this study, mixtures proposed by different authors are adopted and summarized in Table 2.

2.2 Constraints

The studied problem has several constraints:

1) Geometrical constraints for avoiding incongruences;
2) Limitations of the models that are used for verifying structural requirements (range of application of Eurocodes):
   − Requirements for member buckling resistance (Clause 6.7.3.3 (2))
   − Restrictions to avoid premature buckling, detected in the section classification. (Clause 6.7.3.2(2))
   − Restrictions of the simplified model: relative slenderness, λ; reinforcement ratio; and steel contribution ratio, δ. (Clause 6.7.1)
3) Constraints due to the fact that the proposed column does not resist the load applied and does not fulfill stability or fire resistance requirements.
   − Structural and stability capacity of steel CHT columns are verified with Eurocode 3 part 1-1 (EC3-1-1).
   − Fire resistance capacity of CHT columns is evaluated with Eurocode 3 part 1-2 (EC3-1-2).
   − Composite columns capacity are evaluated with Eurocode 4, part 1-1 (EC4-1-1).
   − Kodur's model (1995) is employed for quantifying the resistance of composite columns at high temperature.

3 ANALYSIS AND RESULTS

3.1 Hollow versus in-filled columns

Due to the fact that the number of feasible columns is quite high, yet achievable, all feasible solutions of

Table 2. Mixtures used in the study for obtaining the cost of different concrete strengths.

Authors	f_{ck} (MPa)	CEM I (kg/m³)	Sand (d′ < 4 mm) (kg/m³)	Aggregates (kg/m³)	Water (l)	Silica fume (kg/m³)	Additives (kg/m³)	d′ (mm)
Han (2003b)	25	457	1129	608	206			12
Hernández-Figueirido (2012)	30	348	1065	666	220			12
Chitawadagi (2012)	35	390	702	889	175	20	9	12
Han (2003b)	40	414	630	1170	207			15
Chitawadagi (2012)	45	410	722	885	164	20		12
Tao (2007)	50	523	581	1077	220			15

*Where d′ is the maximum diameter size of the aggregates.

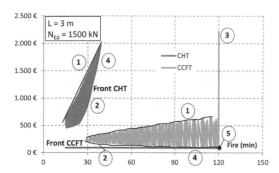

Figure 2. Cost – fire resistance graph for feasible solutions of CHT and CCFT columns with 3 m length and a load of 1500 kN.

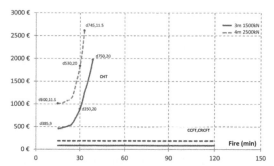

Figure 3. Cost vs fire resistance of CHT, CCFT and CRCFT columns. Pareto's frontier are represented.

CHT and CFST columns have been analyzed. In figure 2 the cost (in Euros) of all the feasible solutions are given on the vertical axis versus the fire resistance in minutes. It shows a global view of the space of feasible solutions. The configuration of 3 m in length and 1500 kN as axial load applied are the boundary conditions selected for this representation.

The lines represent the limits of the solution space and the constraints of the problem.

– Line 1 shows a constitutive condition of the problem: the stability of the column. It is the safety upper limit of the columns that satisfy the requirement of resistance and stability at ambient temperature.
– Line 2 is the limit that separates the sections that can fail by local buckling and are not included in this study.
– Line 3 is a limitation of the Kodur's formula, used for the calculus of the fire resistance of CFST columns: 120 minutes.
– The number of feasible solutions is too high to evaluate them as a whole. Then, from here on the Pareto Frontier will be represented with the objective of showing the best columns. The Pareto frontier is the envelope of minimum costs for each period in fire resistance. It is marked with number 4.

– The graph shows that in some cases it is possible to find the best possible column, that is, to find the global optimum for both objectives: fire resistance and cost. This is the case of CFST columns where there is an absolute optimum point which is better than the rest of feasible solutions. The optimum configuration is labeled in the graph with number 5.

In figure 2, the benefits of in-filling CHT columns in economical and fire resisting terms are clear. In figure 3 the Pareto frontiers for CHT, CCFT and CRCFT (cost versus fire resistance) have been presented. In order to obtain a general view of the solutions, the results of two solutions have been plotted: a 3 m long column with 1500 kN of axial load (continuous line) and 4 m long column with 2500 kN (discontinuous line). The cost reduction and the enhancement of fire resistance in composite columns are clearly confirmed. Here in after, the study will focus on the analysis of the composite columns, due to their cost-economical benefits in comparison to steel hollow columns.

3.2 Parametric study of composite columns

Parametric studies have the aim of exploring the influence of several factors on the optimal configuration

177

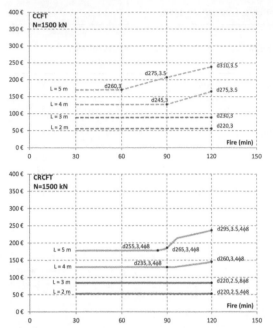

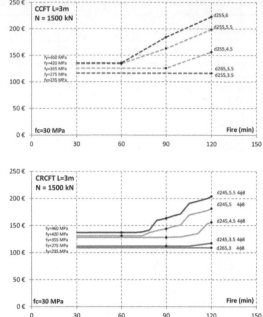

Figure 4. Analysis of the optimal solution for different column lengths.

Figure 5. Analysis of the grade of steel employed.

of composite columns. These factors are: 1) Column length, L; 2) Materials employed: f_{ya}, f_{ck}, f_{ys}; 3) Sectional geometry: D, t; 4) Influence of reinforced concrete versus plain concrete: ϕ, n_ϕ. The boundary conditions in the parametric analysis was the applied axial load: 1500 kN.

3.2.1 Length of the column

In figure 4 the Pareto frontiers of fire resistance versus the cost of columns with different lengths are represented. Obviously, the cost of a slender column is higher than a shorter one because more materials are employed and the capacity is reduced by second-order effects. There is not much difference between the results of CCFT and CRCFT columns: the cost of CRCFT columns is around 4% higher than the CCFT ones when the demanded load is 1500 kN.

In the design process of a composite column, the material selection is a fundamental aspect that affects the cost and the capacity of the element. In this section the influence of steel and concrete strength are evaluated. In figure 5, concrete strength has been set at 30 MPa, therefore the influence of the yielding limit of the steel tube can be appreciated. In all cases, regardless of the column length, load applied and plain or reinforced concrete, the economically optimal solution is reached with the lower class of steel, f_{ya} being equal to 235 MPa.

3.2.2 Materials employed in the composite columns

On the other hand, the influence of concrete strength is studied in figure 6. In this case, the fixed value for the steel tube yield limit is 275 MPa. The Pareto frontiers

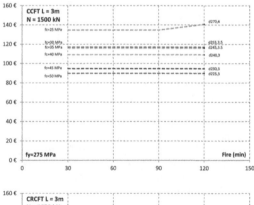

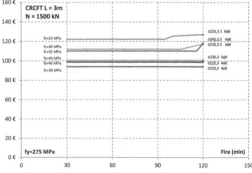

Figure 6. Analysis of the concrete strength grade employed.

have been classified by the concrete strength and the optimum solution was reached with the greatest concrete strength class allowed by the current European design code EC4, 50 MPa.

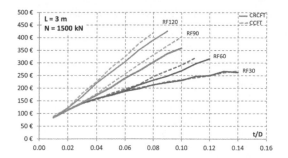

Figure 7. Ratio t/D vs cost, and the optimal configuration is obtained with thin walled sections.

To sum up, we can state that the optimum configuration is obtained with the highest concrete capacity and the minimum steel yield limit. These results can be explained due to the fact that at high temperatures the steel tube loses its resistance quickly and the applied load is transferred to the concrete core, which has lower thermal capacity and is able to sustain the required capacity for a longer period of time. Also, the steel tube confines the concrete section preventing or delaying the spalling problem of concrete.

3.2.3 *Geometry influence*

The geometry of a CFST column depends on the diameter and the thickness of the steel tube, both correlated. With the aim of determining the optimal configuration, two ratios are studied: thickness to diameter ratio and diameter to column length ratio.

In figure 7, the thickness to diameter ratio (t/D) of the CFST column is represented in front of the cost and the information is classified considering fire resistance. This graph shows that, regardless of the load, the length, the time or the reinforcement, for fixed boundary conditions there is a global optimum with a t/D ratio value around 0.01. This means that the economical optimum is always found in composite columns built with thin-walled tube sections. In the future it will be interesting to continue studying CFST columns including sections that are susceptible to failure due to local buckling, in order to obtain more economical solutions that fulfill all requirements.

4 CONCLUSIONS

The use of Concrete-Filled tubular steel columns is increasing as a structural element. This widespread use is due to their excellent structural performance such as their high load-bearing capacity, fast construction technology, ductility and large energy absorption (a desirable capacity in case of a seism) and high fire resistance without external protection.

The present research paper has investigated the economical interest in the use of CFT columns as an alternative to traditional hollow tubes. Besides, it has studied the optimal configuration of CFST columns submitted to static concentric loading, from an economical point of view. Different configurations,

lengths and loads applied were considered and the optimal cost configuration, geometry (D, t, ϕ, n_ϕ) and materials (f_{ya}, f_{ck}, f_{ys}) have been obtained in each case, depending on the fire resistance requirement. In addition, the calculus method used in the analytical study has been explained in detail.

The main conclusions of the research are summarized as follows:

1) Composite CFST columns offer lower cost solutions than CHT columns. Their high strength capacity leads to a reduction of the sections and, consequently, costs. As well, fire resistance of CFT columns without external protection is higher than steel hollow columns. CHT columns reach 30 minutes in fire exposure at a high cost while composite columns reach 120 minutes at a lower cost.

2) For the considered loads and lengths the cost solutions of CFT columns with plain or reinforced concrete filling are similar.

3) Regardless of column length and imposed load, optimal configurations were obtained for the highest concrete strength and the minor yielding limit of the steel tube. This is due to the fact that in fire situations the steel tube temperature increases rapidly, then the steel gradually loses strength and stiffness and the load is transferred to the concrete core.

4) The geometry study for the optimal solution shows that the most economical geometry is always obtained with thin-walled sections. Optimal configurations are formed by minimum thickness of the steel tube without it suffering from local buckling (a constraint of the model imposed by EC4).

ACKNOWLEDGEMENTS

The authors wish to express their sincere gratitude to University Jaume I, that has given its support to this study under their Research Promotion program (project P1-1A2013-09 entitled "Determinación de la configuración optima, geometría y materiales, de pilares mixtos tubulares rellenos de hormigón. Reducción de costes e incremento de prestaciones: presente y futuro de la construcción").

REFERENCES

Chitawadagi, M.V., Narasimhan, M.C. & Kulkarni, S.M. 2012. Axial capacity of rectangular concrete-filled steel tube columns – DOE approach. Construction and Building Materials. Vol. 24–4, pp. 585–595.

European Committee of Standardization EN 1993–1-2:2005. Design of steel structures-Part 1–2: General rules structural fire design, European committee for standardization, Brussels.

European Committee of Standardization. 2005. EN 1993-1-1: Eurocode 3: Design of steel structures – Part 1-1: General rules and rules for buildings.

European Committee of Standardization. EN 1994-1-1:2004 Eurocode 4: Design of composite steel and concrete structures Part 1-1: General rules and rules for buildings, 2004.

Han, L.H. & Yang, Y.F. 2003b. Analysis of thin-walled steel RHS columns filled with concrete under long-term sustained loads. Thin-Walled Structures. Vol. 41–9, pp. 849–870.

Han, L.H., Zhao, X.L., Yang, Y.F. & Feng, J.B. 2003a. Experimental study and calculation of fire resistance of concrete-filled hollow section columns. Journal of Structural Engineering 129(3): 346–356.

Han, L.H., Li, W. & Bjorhovde, R. 2014. Developments and advanced applications of concrete-filled steel tubular (CFST) structures: Members.Journal of Constructional Steel Research, Vol. 100, pp. 211–228.

Hernández-Figueirido D., Romero, M.L., Bonet, J.L. et al. 2012. Ultimate capacity of rectangular concrete-filled steel tubular columns under unequal load eccentricities. Journal of constructional steel research. Vol. 68, pp. 107–117.

Institut de Tecnologia de la Construcció de Catalunya. Construction products data base, banco BEDEC. http://www.itec.es/nouBedec.e/bedec.aspx [January 2011].

Kodur, V.K.R. & Lie, T.T. 1995. Experimental studies on the fire resistance of circular hollow steel columns filled with steel-fibre-reinforced concrete. NRC-CNRC Internal Report: No. 691, Ottawa, Canada.

Lu, H., Han, L.H. & Zhao, X.L. 2010. Fire performance of self-consolidating concrete filled double skin steel tubular columns: Experiments. Fire Safety Journal 45(2): 106–115.

Tao, Z., Han, L.H. & Wang, D.Y. 2007. Experimental behavior of concrete-filled stiffened thin-walled steel tubular columns. Thin-Walled Structures. Vol. 45–5, pp. 517–527.

Twilt, L., Hass, R., Klingsch, W., Edwards, M. & Dutta, D. 1996. Design guide for structural hollow section columns exposed to fire. Köln: TÜV-Verlag.

Uy, B. & Liew, R. 2003. Chapter 51, Composite Steel-Concrete Structures. The Civil Engineering Handbook, second Edition. W.F. Chen, R. Liew, CRC Press, ISBN:978-0849309588.

Wardenier, J., Packer, J.A., Zhao, X.L. & van der Vegte, G.J. 2010. Hollow sections in structural applications, Delft: Bouwenmet Staal.

Zhao, X.L., Han, L.H. & Lu, H. 2010. Concrete-filled tubular members. 1st Ed. Oxon: Spon Press.

Tubular Structures XV – Batista, Vellasco & Lima (eds)
© 2015 Taylor & Francis Group, London, ISBN 978-1-138-02837-1

Structural analysis of tubular truss in fire

J.A. Diez Albero, T. Tiainen, K. Mela & M. Heinisuo
Tampere University of Technology, Tampere, Finland

ABSTRACT: Tubular trusses with welded gap joints are internally indeterminate if they are modeled with beam elements using continuous chords and eccentricities at the joints. Trusses may be externally statically determinate or not, depending on their support conditions. The start point was a tubular truss, which was sizing and shape optimized. This means that the members and the joints can resist just the load in normal situation. Next the same truss was analyzed using material and geometrical non-linear theory with reduced load in fire allowing horizontal displacement at the other end and restraining it. The idea was to study: 1) is the normal tubular truss such that in externally determinate case the linear theory can be used in fire, 2) what happens in externally indeterminate case? The results were: 1) the linear analysis can be used in fire to determine the stress resultants and 2) the linear model is very conservative.

1 INTRODUCTION

Tubular steel trusses with welded gap joints are widely used in buildings as floor and roof girders due to their nice aesthetic appearance, good load bearing properties and cost efficiency. Fire resistance research of steel trusses and the effects of trusses to the surrounding structures stepped forward after WTC towers collapses (Quintiere ct al. 2002), (Usmani et al. 2003) and Banerjee (2013). After researches of WTC in NIST an interaction system between fire – thermal – structural analyses has been developed to enhance the fire design, including also trusses (Duthinh et al. 2008). In (Flint et al. 2007) it is shown that individual member buckling of the truss is often the first mode of the collapse but the truss will not fall after this first fail of stability. Also, the behaviors of the joints between the truss and the column, have a great influence on the final failure. Special finite elements have been formulated just for the fire design of trusses (Yang et al. 2008). In (Choi et al. 2008) a composite truss which is restrained against horizontal movements at its ends is analyzed in fire. It is shown that the truss behavior changes from bending to catenary action via progressive buckling of compressive web members. In (Sun et al. 2012) is developed a dynamic analysis method for the analysis of these kinds of collapses, including truss-frames. References (Wang et al. 2008) and Pada (2012) includes truss analyses in the cooling phase of the fire. In (Wang et al. 2011) are analyzed pre-stressed steel trusses in fire. In the present paper normal trusses formed by triangular free spaces between members are considered. The trusses with large openings, such as in Ölmez & Topkaya (2011), are outside the scope of this paper. Only planar trusses are considered here.

Tubular steel trusses in fire have been considered experimentally in (Liu et al. 2010). They found that the governing failure model was the local buckling of the brace. Using the same experiments for the validation of the finite element model they found analytically (Jin et al. 2011) that the most important parameters dealing with the fire resistance of the tubular trusses are: the wall thickness ratio τ, diameter ratio β and chord diameter/thickness ratio γ. The results demonstrate that the critical temperature of the truss can be improved significantly by the increase of the brace diameter and the wall thickness of the chord while changing the wall thickness of the brace has limited effects. Artificial neural networks are used in (Xu et al. 2013) to define the steel temperatures of the trusses in fire with the reference truss of (Liu et al. 2010).

The structural analysis of the trusses in fire includes mainly two resistance checks: members and joints. The joints of trusses to the surrounding structures, and the support conditions of the surrounding structures, are extremely important as shown in many aforementioned studies. When considering only an isolated tubular truss, as in this study, the behavior of the members and joints of the truss are important to know. The behavior of the tubular members in fire can be predicted rather well. The behavior of the welded tubular joints has been extensively studied at room temperatures, but "there is paucity of research on tubular connections under fire conditions", as is concluded in a recent study by Ozyurt & Wang (2013). Before going to the resistance checks of members and joints, the actions of the members and joints should be known.

The finite element method using beam elements is a normal way to perform the structural truss analysis, as in this study. The analysis model should follow

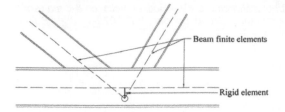

Figure 1. Local analysis model of K-joint.

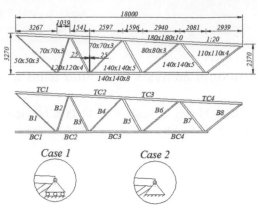

Figure 2. Tubular truss and boundary conditions at the right end, member notations.

the mass center of the members of double symmetric cross-sections. When approaching the joints, then many options are available for the local analysis models of the joints. In Boel (2010) and (Snijder et al. 2011) seven possibilities for local analysis models of welded tubular joints have been studied including semi-rigid joint models. In EN 1993-1-8 no local analysis model is given for welded tubular joints, but rules to construct that are given. It is stated that in some cases the eccentricities at the joints should be taken into account, at the ends of the braces should be hinges and the chords may be considered as the continuous beams. In this study the local analysis model for K-joint of Figure 1 is used which fulfills the requirements of EN 1993-1-8.

The use of eccentricity elements at the joints and by modeling the chords as continuous beams means that the truss is internally statically indeterminate (hyperstatic). Opposite to this is a statically determinate (isostatic) structure in which the actions of members and joints can be determined using only equilibrium conditions if the truss as a whole is externally statically determinate. Externally statically determinate means that the support reactions of the truss can be defined using only equilibrium equations. It is well-known that the displacements actions, fire elongations of the members in this case, do not have any effect to the stress resultants of the truss if it is statically determinate both internally and externally when deformations can be considered small and thus geometrical non-linearity is omitted. Also, the changes of the material properties, due to fire in this case, have no effects to the stress resultants in the statically determinate (internal and external) cases. Stress resultants, axial and shear forces and bending moments, should be known to check the resistances of the members and joints. In this paper the main focus is at the development of the stress resultants during fire, not in the final design of members and joints.

One-span symmetric welded tubular roof trusses are analyzed in fire. The truss, made of grade S355 cold-formed steel, was optimized with respect to the fabrication costs using sizing and shape optimization at the ultimate limit state load in room temperature (Mela et al. 2013). The constraints were derived from requirements for the members and for the joints appearing in the Eurocodes. This means that not much extra material is used in the members of the truss, when considering the design in the normal conditions supported externally determinate way.

The trusses are analyzed in this study under increasing gas temperature originating from the ISO 834 fire supposing elevated temperature around the members of the truss and supposing uniform constant reduced load at the top chord. The heat transfer from the gas to the steel members is calculated using EN 1993-1-2, ending up to the uniform temperature around the whole cross-sections of the members. The joints are supposed to act so, that the temperatures of the members are constant along their lengths. The truss is analyzed without fire protection so that temperature effects are the largest.

The same truss is analyzed supposing the external support conditions as statically determinate (as in the original case) and statically indeterminate so that horizontal movements at the ends of the trusses are totally restrained simulating two extreme cases. The truss is analyzed in both cases using the linear elastic theory and material and geometrical non-linear theory using beam elements and routines of the program ABAQUS/Standard. The stress resultants are recorded with respect to increasing time in fire. The details of calculations can be found in Diez Albero (2014).

The idea was to find out how well the linear elastic theory works when the structure is internally and externally statically indeterminate. The internal indeterminate condition follows from the continuous chord and from the small eccentricity elements describing the normal situation in this kind of roof trusses. The calculation in fire was carried out until convergence stop using static ABAQUS.

2 THERMAL ANALYSIS

The truss is shown in Figure 2. Welded gap joints with 50 mm gap are at each joint. Two support cases 1 and 2 are considered. The uniform vertical design load in the normal conditions at the top chord was 23.5 kN/m and in this fire case 8.2 kN/m. The load is constant in the fire analysis and the ISO fire is progressing up to the convergence stop.

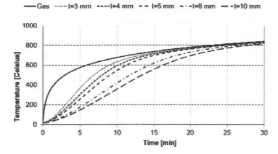

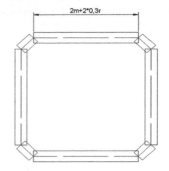

Figure 3. Temperatures of members.

Figure 4. Cross-section used in ABAQUS.

The steel temperatures can be calculated step by step, when the gas temperatures are known:

$$\Delta T_{steel} = [A_m/(V\rho c)][\varepsilon\sigma(T_{gas}^4 - T_{steel}^4) + h(T_{gas} - T_{steel})]\Delta t \quad (1)$$

where

V is the volume of the member per unit length [m³];
ρ is the density of the material [7850 kg/m³];
c is the specific heat of the member [600 Ws/(kgK)];
ΔT_{steel} [K] is the temperature change of the member during the time step Δt [s], 5 s;
ε is the emissivity of the surface of the member, 0.7;
σ is the Stefan-Boltzmann constant 5.67×10^{-8} W/(m²K⁴);
h is the convective heat transfer [25 W/(mK)];
A_m is surface area of the cross-section [m²];
Am/V in this case is $1/t$ [1/m], where t is the wall thickness of the tube in meters.

The gas temperature T_{gas} [Celsius] is using ISO 834:

$$T_{gas} = 20 + 345 \log_{10}(8\ t_{min} + 1) \quad (2)$$

where t_{min} is the time in minutes. The temperatures of members are derived using Equations (1)–(2) and they are shown in Figure 3. These temperatures were given to the ABAQUS model for the members.

3 MECHANICAL ANALYSIS

The program ABAQUS/Standard and its beam elements B3.3 were used in the analysis. Each truss member was divided into 8–10 elements (minimum element length 0.5 m) and one element was used for the eccentricities, properties as HEM1000.

The tubular cross-section was compensated using the cross-section built from eight planes as shown in Figure 4. Flat part (2m) is increased with 2*0.3r where r is the mid-line radius of the corner.

The rule shown in Figure 4 is originating from the German standard and it has shown Kukkonen & Heinisuo (2005) to give reliable area and moment of inertia for the cold-formed tubular sections. The temperatures are given in ABAQUS for the mid-points of the planar parts of the cross-section when beam in plane is

considered. A bi-linear elastic-perfectly plastic material model (without strain hardening and without decay phase) with the yield strength 355 MPa and the elastic modulus 210000 MPa and Poisson's ratio 0.3 were given for steel. In fire the reduction factors $k_{y,\theta}$ and $k_{E,\theta}$ according to EN 1993-1-2, Table 3.1 were given for the members following their temperatures in Figure 3. Thermal elongations for the members were given using EN 1993-1-2, equations 3.1a-c. Geometrically non-linear option of ABAQUS was used in the non-linear analysis. Only half of the truss was modeled due to symmetry.

4 RESULTS FOR CASE 1 AND 2 WITHOUT FIRE

The stress resultants without fire and using the linear theory are given in Table 1.

It can be seen that the shear forces and the bending moments are about the same in both cases. Instead axial forces differ considerably. The compression of the top chord is much smaller in the restrained case going to almost zero near the support (TC4). In this case the resultant of support reactions (vertical + horizontal) is almost parallel to the brace B8 meaning that this brace transfers almost all the loads from the truss to the support. Tensile forces at the bottom chord are about 10% larger in the restrained case than in the unrestrained case. The signs of the axial forces at braces B1 and B2 changes but the magnitudes are small there. The maximum deflection in Case 2 (~50 mm) was only about 2 mm smaller than in Case 1 (~52 mm).

5 RESULTS FOR CASE 1 WITH FIRE

When in Case 1 the elastic modulus is changing following the steel temperatures and elongation of steel is taken into account, but the yielding of steel is not taken into account then the deflection starts to increase with respect to time. This case is named as NLMAT/NOT YIELD. When the yielding is also allowed the case is named as NLMAT. In this case the convergence was stopped after 11.5 minutes due to large yielding at the

Table 1. Stress resultants without fire using linear theory.

N: + comp.	N [kN]	Q [kN]	M_{max} [kNm]	M_{min} [kNm]
Case 1, unrestrained				
TC1	438	16.26	7.77	−8.28
TC2	422	15.73	6.37	−8.55
TC3	353	18.33	13.16	−7.09
TC4	165	22.59	14.50	−16.38
BC1	−433	0.00	0.58	0.58
BC2	−437	0.53	1.94	0.81
BC3	−390	0.27	1.10	−0.11
BC4	−266	2.75	9.38	−3.09
B1	−5			
B2	3			
B3	36			
B4	−45			
B5	79			
B6	−112			
B7	162			
B8	−206			
Case 2, restrained				
TC1	307	15.87	6.77	−8.55
TC2	286	16.04	7.39	−8.19
TC3	206	18.36	13.43	−6.89
TC4	2	22.68	14.95	−16.11
BC1	−483	0.00	0.61	0.61
BC2	−479	0.68	2.08	0.72
BC3	−421	0.35	1.14	−0.16
BC4	−284	2.97	10.00	−3.49
B1	4			
B2	−3			
B3	44			
B4	−55			
B5	87			
B6	−124			
B7	173			
B8	−219			

Table 2. Axial forces in unstrained case.

Member	Linear, no fire	NLMAT/ NOT YIELD	NLMAT	NLMAT& NLGEO
TC1	438	436	436	432
TC2	422	419	418	394
TC3	353	352	350	300
TC4	165	166	168	146
BC1	−433	−434	−435	−437
BC2	−437	−434	−435	−428
BC3	−390	−388	−383	−345
BC4	−266	−266	−271	−238
B1	−5	0	0	8
B2	3	−1	−1	−10
B3	36	36	39	75
B4	−45	−43	−49	−71
B5	79	77	72	94
B6	−112	−110	−101	−81
B7	162	162	164	148
B8	−206	−206	−210	−179

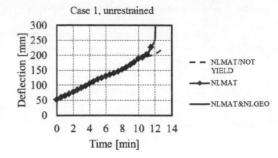

Case 1, unrestrained

Figure 5. Maximum deflection versus time, unrestrained case.

brace B6. When the material and geometrical model is used the case is called as NLMAT&NLGEO. Note, that no virtual imperfections were given in this analysis. In the case NLMAT&NLGEO the convergence stopped after 13 minutes in fire due to large global displacements, maximum displacement was about 1.5 m. Figure 5 illustrates the mid-span deflections with respect to time.

It can be seen that the deflection starts to grow at 11.5 minutes due to plastification of the brace B6. After that geometrical non-linearities govern up to 13.5 minutes.

The axial loads of the members are given in Table 2 for the unrestrained case. The values for NLMAT/NOT YIELD and NLMAT&NLGEO are given after 13 minutes fire and NLMAT/NOT YIELD are given after 11.5 minutes fire.

It can be seen that the axial forces of all members are about the same in three first columns of Table 2. This means that the linear model without fire can predict very well the axial forces in this truss up to the beginning of the failure at 11.5 minutes. When using the total non-linear theory (NLMT&NLGEO) the maximum chord forces near the mid-span of the truss are about the same as using the linear model and when approaching the support the axial forces at the chords start to decrease. In the braces B3, B4 and B5 the axial forces increase in this case and in the braces B6, B7 and B8 the axial forces decrease. This means redistribution of the forces at the braces, but it can be seen that largest axial forces of the braces are getting smaller, meaning safe design for these members if the linear model is used to determine the stress resultants. The original utilization ratios of the braces B3, B4 and B5 were rather small indicating that they can resist the increased axial loads when the geometrical non-linearities are taken into account.

The bending moments using the four theories are given in Table 3.

Moments increase at three first columns so that absolute maximum for the top chord is changing from 16.38 to 26.77 kNm and for the bottom chord from 9.38 to 9.72 kNm. The temperature of the top chord at 11.5 minutes is according to Figure 3 424°C and at the bottom chord 490°C. This means reduction factor $k_{y,\theta}$ as 0.95 and 0.80, respectively. The moment

Table 3. Bending moments for unrestrained truss.

Member	Linear, no fire	NLMAT/ NOT YIELD	NLMAT	NLMAT& NLGEO
M_{max} [kNm]				
TC1	7.77	12.84	13.96	18.34
TC2	6.37	−0.40	−1.05	−20.84
TC3	13.16	14.00	24.56	−3.11
TC4	14.50	15.56	26.77	3.72
BC1	0.58	0.88	1.07	0.57
BC2	1.94	3.26	3.39	13.72
BC3	1.10	1.71	3.90	25.94
BC4	9.38	9.43	9.72	22.71
M_{min} [kN/m]				
TC1	−8.28	−10.63	−9.30	−21.06
TC2	−8.55	−14.73	−20.37	−83.95
TC3	−7.09	−8.98	−12.43	−77.51
TC4	−16.38	−15.88	−11.75	−23.63
BC1	0.58	0.88	1.07	0.16
BC2	0.81	1.15	1.32	1.52
BC3	−0.11	1.55	1.60	10.41
BC4	−3.09	−2.49	0.10	7.95

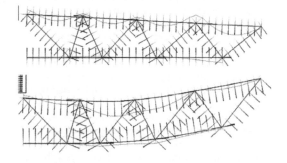

Figure 6. Bending moments for unrestrained case using linear theory without fire and NLMAT&NLGEO after 13 minutes of fire.

resistances of the chord in the normal situation are 143 and 69 kNm. So, the maximum moments at 11.5 minutes are less than 20% of the moment resistances. The moments at 13 minutes using NLMAT&NLGEO are large, but the convergence, meaning among others equilibrium, could be found there, too. The last run was done using NLMAT&NLGEO theory but adding the initial bow imperfections $L/1000$ (L is the length of the member) to the compressed braces B7, B5 and B3, but the result did not change compared to the results above (NLMAT&NLGEO). This was expected because the lowest eigenvalue (α_{cr} in the Eurocodes) was above 10 for this truss.

When considering the bending moments based on the linear theory without fire and the moments using NLMAT&NLGEO after 13 minutes of fire it can be noted that the maximum moment at the bottom chord changes from the element BC4 to BC3 see Fig 6. The eccentricity moments are not any more critical at the final phase. The top chord moments

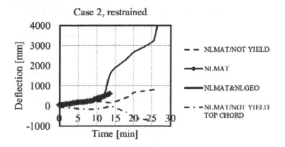

Figure 7. Maximum deflections versus time, restrained case.

change dramatically due to plastification of the braces and due to large displacements, especially at TC2 and TC3.

As a conclusion it can be stated that the linear theory can be used to predict the stress resultant for this truss in Case 1 in fire design. The axial forces of the chords are safe using the linear theory and redistribution of the axial forces at the braces is happening near the failure.

6 RESULTS FOR CASE 2 IN FIRE

The maximum deflections are given in Figure 7 for this case.

It can be seen that the deflections of the bottom chord using three theories are about the same up to about 10 minutes. The deflections NLMAT&NLGEO are a little bit smaller at this time range than the displacements NLMAT. The convergence stopped at 13.5 minutes using NLMAT, instead the final failure using NLMAT&NLGEO was at 26.5 minutes with very large deflections, about 4.1 meters.

The stress resultants after 13.5 minutes in fire are given in Table 4 for using two theories.

In this case, as expected, thermal expansions play an important role. If they are neglected in NLMAT/NOT YIELD, the stress resultants are about the same as in the linear case (see Table 1) up to 8 minutes with maximum 330 kN, and up to 13.5 minutes the maximum axial force at the top chord increase to 443 kN. But with the thermal elongation the maximum is 5499 kN (see Table 4) which is well above the resistance of the top chord after 13.5 minutes in fire. The yielding starts at the top chord TC4 leading to large plastic strains although in the beginning the axial force was near zero and convergence stops in NLMAT at 13.5 minutes. As seen in Table 4 there are also large bending moments at the top chord in this case. The axial force at the brace B4 has changed from tension 55 kN (beginning of the fire) to compression 44 kN after 13.5 minutes of fire using NLMAT. This is rather slender member.

The stress resultants using NLMAT&NLGEO after 26.5 minutes fire are given in Table 5.

Table 4. Stress resultants using two theories after 13.5 minutes in fire.

$N: +$ comp.	N [kN]	Q [kN]	M_{max} [kNm]	M_{min} [kNm]
NLMAT/NOT YIELD				
TC1	4702	58.60	121.01	−42.40
TC2	4870	34.80	8.99	−63.39
TC3	5168	18.95	4.47	−17.21
TC4	5499	19.72	−4.95	−28.57
BC1	1174	0.00	−3.26	−3.26
BC2	932	2.13	0.81	−3.48
BC3	620	2.33	8.81	0.27
BC4	320	4.72	10.45	−10.95
B1	−252			
B2	186			
B3	−237			
B4	296			
B5	−194			
B6	269			
B7	−196			
B8	247			
NLMAT				
TC1	1339	33.85	37.34	−31.96
TC2	1365	29.37	17.38	−35.09
TC3	1384	18.67	19.85	−1.13
TC4	1303	21.76	10.53	−18.20
BC1	−92	0.00	2.48	2.48
BC2	−134	0.91	5.35	3.52
BC3	−179	1.45	7.13	1.82
BC4	−141	1.08	5.09	0.20
B1	−45			
B2	32			
B3	−33			
B4	44			
B5	23			
B6	−35			
B7	86			
B8	−110			

Table 5. Stress resultants after 26.5 minutes fire using NLMAT&NLGEO.

Member	N [kN]	Q [kN]	M_{max} [kNm]	M_{min} [kNm]
TC1	−136	14.17	3.51	−12.40
TC2	−148	13.31	4.10	−7.95
TC3	−177	12.24	0.59	−6.86
TC4	−236	7.97	2.26	−1.26
BC1	−136	2.40	0.59	0.00
BC2	−135	6.53	0.88	−0.59
BC3	−116	3.98	2.47	0.76
BC4	−86	1.69	2.83	1.04
B1	4			
B2	9			
B3	26			
B4	−12			
B5	27			
B6	−23			
B7	54			
B8	−60			

It can be seen that the axial forces are totally different compared previous values. Both top and bottom chords are now in tension, meaning that catenary action is active. However, by comparing the values in Table 5 to the values of Table 1, Case 2, restrained, it can be seen that when using the stress resultants of the linear case to the design, they are very conservative, meaning that they can be used, but the result is very safe and not economical.

7 CONCLUSIONS

The research questions were: 1) is the normal tubular truss such that in externally determinate case the linear theory can be used in fire, 2) what happens in externally non-determinate case? The results were: 1) the linear analysis can be used in fire to determine the stress resultants and 2) the linear model can be very conservative.

Based on this example, the modeling of the chord as a continuous beams and using the eccentricity elements at the local joint models do not make the truss "very much" internally indeterminate, if it is externally determinate. In normal trusses where the members are in triangles, the quantities of the axial forces at the members are almost independent on the local joints models, hinge/rigid/semi-rigid, eccentricities. The failure in fire starts at buckling of members both in this study and in the referred papers. Perhaps this explains why the linear theory can be used up to the failure in this case.

If the truss is externally indeterminate the axial forces together with other stress resultants of the truss are very dependent on the support conditions to the surrounding structures and the non-linear analysis should be used. The global behavior of the truss including redistribution of the stress resultants due to progressive buckling of members in fire play the main role in this case and e.g. small eccentricity elements do not have the major role. This result is very similar to those in the referred papers, but joint eccentricities are taken into account in this study.

Externally determinate truss is used frequently in practical applications. This example shows that for those trusses the design can be completed using linear elastic theory up to the failure in fire. This is essential information for future optimization tasks of tubular trusses in fire.

Joint behavior especially under large deformations at elevated temperatures is rather open question and only few papers were found dealing joint resistance in fire. In few fire tests of trusses the buckling of the members has been critical.

ACKNOWLEDGEMENTS

The truss of this study was taken from the document D4.5 of RFCS project RUOSTE dealing case studies completed in TUT during 2013.

REFERENCES

Banerjee, D. 2013. Uncertainties in steel temperatures during fire. *Fire Safety Journal*, Volume 61: 65–71.

Boel, H. 2010. Buckling Length Factors of Hollow Section Members in Lattice Girders. *Ms Thesis, Eidhoven University of Technology*.

Choi, S-K., Burgess, I. & Plank, R. 2008. Performance in fire of long-span composite truss systems. *Engineering Structures*, Volume 30, Issue 3: 683–694.

Diez Albero, J. 2014. Structural analysis of welded tubular trusses in fire. *Ms Sci Thesis. Tampere University of Technology*, Tampere.

Duthinh, D., McGrattan, K. & Khaskia, A. 2008. Recent advances in fire-structure analysis. *Fire Safety Journal*, Volume 43, Issue 2: 16.

Flint, G., Usmani, A., Lamont, S., Lane, B. & Torero, J. 2007. Structural response of tall buildings to multiple floor fires. *J Struct Eng, ASCE*, 133: 1719–1732.

Heinisuo, M. & Kukkonen, J. 2005. Design of Cold-Formed Members Following New EN 1993-1-3. *Department of Civil Engineering, Structural Engineering Laboratory, Research Report 132*, Tampere University of Technology, Tampere.

Jin, M., Zhao, J., Liu, M. & Chang, J. 2011. Parametric analysis of mechanical behavior of steel planar tubular truss under fire. *Journal of Constructional Steel Research*, Volume 67, Issue 1: 75–83.

Liu, M., Zhao, J. & Jin, M. 2010. An experimental study of the mechanical behavior of steel planar tubular trusses in a fire. *Journal of Constructional Steel Research*, Volume 66, Issue 4: 504–511.

Mela, K., Heinisuo, M. & Tiainen, T. 2013. D4.5, *Rules on high strength steel (RUOSTE), RFSR-CT-2012-00036*, 2013.

Ozyurt, E. & Wang, Y. 2013. Resistance of T- and K-joints to tubular members at elevated temperatures. *Proceedings of Application of Structural Fire Design (Eds. Wald F. and Burgess I.), 19–12 April, Prague, Czech Republic*: 179–185.

Pada, D. 2012. Steel Skeleton Behaviour in Decaying Fire. *Licenciate Thesis,* Tampere University of Technology, Tampere.

Quintiere, J. G., di Marzo, M. & Becker, R. 2002. A suggested cause of the fire-induced collapse of the World Trade Towers. *Fire Safety Journal*, Vol. 37, No. 7: 707–716.

Snijder, H., Boel, J., Hoenderkamp, J. & Spoorenberg R. 2011. Buckling length factors for welded lattice girders with hollow section braces and chords. *Proceedings of Eurosteel, Budapest*: 1881–1886.

Sun, R., Huang, Z. & Burgess, I. 2012. The collapse behaviour of braced steel frames exposed to fire. *Journal of Constructional Steel Research*, Volume 72: 130–142.

Usmani, A., Chung, Y. & Torero, J. 2003. How did the WTC Towers Collapse? A New Theory. *Fire Safety Journal*, Vol. 38: 501–533.

Wang, P., Li, G. & Guo, S. 2008. Effects of the cooling phase of a fire on steel structures. *Fire Safety Journal*, Volume 43, Issue 6: 451–458.

Wang, X., Zhou, M. & Wang, W. 2011. Numerical Analysis of Pre-Stressed Steel Trusses Subjected to Fire Load. *Advanced Material Research*, Vols. 163–167: 799–803.

Xu, J., Zhao, J., Wang, W. & Liu, M. 2013. Prediction of temperature of tubular truss under fire using artificial neural networks. *Fire Safety Journal*, Volume 56: 74–80.

Yang, Y., Lin, T., Leu, L. & Huang, C. 2008. Inelastic post-buckling response of steel trusses under thermal loadings. *Journal of Constructional Steel Research*, Volume 64, Issue 12: 1394–1407.

Ölmez, H. & Topkaya, C. 2011. A numerical study on special truss moment frames with Vierendeel openings. *Journal of Constructional Steel Research*, Volume 67, Issue 4: 667–677.

Tubular Structures XV – Batista, Vellasco & Lima (eds)
© 2015 Taylor & Francis Group, London, ISBN 978-1-138-02837-1

Effects of truss behavior on critical temperatures of welded steel tubular truss members exposed to fire

E. Ozyurt & Y.C. Wang

School of Mechanical, Aerospace and Civil Engineering, University of Manchester, UK

ABSTRACT: This paper presents the results of a numerical investigation into the behavior of welded steel tubular truss at elevated temperatures. The purpose is to assess whether the current method of calculating truss member limiting temperature, based on considering each individual truss member and using the member force from ambient temperature analysis, is suitable. Finite Element (FE) simulations were carried out for Circular Hollow Section (CHS) trusses using the commercial Finite Element software ABAQUS. The results of the numerical parametric study indicate that due to truss undergoing large displacements at elevated temperatures, some truss members (compression brace members near the truss centre) experience large increases in member forces. Finally, this paper proposes and validates an analytical method to take into consideration the additional compression force due to large truss displacement. This is based on assuming a maximum truss displacement of span over 30.

1 INTRODUCTION

Hollow structural sections of all types are widely used in truss construction due to their attractive appearance, light weight and structural advantages. They are commonly used in onshore and offshore structures e.g. bridges, towers, stadiums, airports, railway stations, offshore platforms etc. For these structures, fire presents one of the most severe design conditions, because the mechanical properties of the steel degrade as the temperature increases.

For truss design, both the members and the joints should be checked. Truss member design at ambient temperature is relatively easy, involving mainly design checks for tension and compression resistance after performing static analysis to obtain the member forces. There is abundant amount of literature on the behavior and strength of and of truss joints at ambient temperature. Indeed, the CIDECT (2010) design guide and Eurocode EN 1993-1-8 present design equations to calculate the ambient temperature static strength of practically all tubular truss joints.

Under fire condition, the current method for truss member design involves calculating the member force using static analysis at ambient temperature and then finding the critical temperatures of the member using the ambient temperature member force. The member force – critical temperature relationship can be evaluated using design methods such as those in BS 5950 Part 8 and EN 1993-1-2. However, the member force obtained from truss static analysis at ambient temperature may not be correct at elevated temperatures due to large deformations of the truss. There has been little study to investigate how truss member forces change

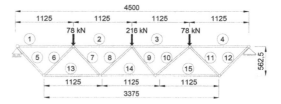

Figure 1. Test Girder B of Edwards (2004) (dimensions in mm).

at elevated temperatures and how such changes affect the member critical temperatures.

The specific scope of this paper is to investigate whether the member-based fire resistance design approach is safe, and if not, to develop a modified member-based method to take into consideration truss behavior.

2 VALIDATION OF FINITE ELEMENT MODEL

For validation, the fire tests of Edwards (2004) and Liu et al. (2010) were simulated and compared with the test results. Figure 1 and Figure 2 show the tested trusses. Failure modes and displacement-temperature curves were compared.

2.1 *Material properties*

Table 1 summarizes the member sizes and material grades. For both tests, the ambient temperature mechanical properties were based on their coupon

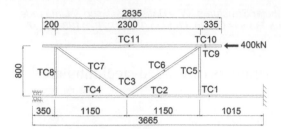

Figure 2. Test specimen SP1 of Liu et al. (2010) (dimensions in mm).

Table 1. Dimensions and ambient temperature mechanical properties of the test trusses.

Member Type	Dimensions (mm)	Young's modulus (GPa)	Yield stress (MPa)	Tensile Stress (MPa)
Girder B (Edward, 2004)				
Inner bracings (Member 7, 8, 9 and 10)	60 × 60 × 4.0	190	384.92	487.38
Outer bracings (Member 5, 6, 11 and 12)	60 × 60 × 8.0	190	279.93	436.63
Top chord (Member 1, 2, 3 and 4)	100 × 100 × 10.0	210	397.16	574.49
Bottom chord (Member 13, 14 and 15)	100 × 100 × 10.0	210	397.16	574.49
SP1 (Liu et al., 2010)				
Brace	102 × 5	202	376	559
Top chord	180 × 8	193	368	553
Bottom chord	219 × 8	196	381	565

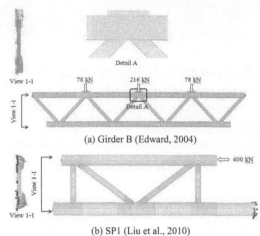

(a) Girder B (Edward, 2004)

(b) SP1 (Liu et al., 2010)

Figure 3. Finite element models of the test trusses for validation study.

a) Test observations, b) Shell elements c) Beam Elements
(Liu et al., 2010)

Figure 4. Comparison of failure modes.

results. Extensive temperature measurements of the truss members were made in both tests and the recorded temperatures were used in the numerical analysis.

2.2 *Finite element type and initial imperfection*

For the chord and brace members, ABAQUS element types S4R (4 noded shell element) or B21 (2 noded line element) may be used. In the case of modeling using shell elements, quadratic wedge solid elements (C3D15) instead of shell elements were used for the weld to allow accurate meshing of the weld geometry (Cofer et al., 1992). At the weld-tubular section interface, the brace and chord members were tied with the weld elements using the ABAQUS "tie" function with surface to surface contact. The brace and chord members were chosen as the master surface and the weld elements the slave surface. Owing to symmetry in loading and geometry, to reduce computational time, only half of the truss was modeled when using shell elements, with the boundary conditions for symmetry

being applied to the nodes in the various planes of symmetry as shown in Figure 3.

Eigenvalue buckling analysis was performed on the numerical models in order to define the possible buckling modes for compressed members in the trusses. Lanczos was chosen as eigensolver together with the request five buckling modes. Initial imperfections were included, based on the lowest buckling mode from eigenvalue analysis. The maximum initial imperfection was according to EN 1993-1-1.

2.3 *Comparison with test results*

2.3.1 *Test SP1 of Liu et al. (2010)*

The SP1 truss failed due to buckling of the diagonal brace member in compression. Figure 4 compares the observed failure mode of the truss and the simulated deformed shapes using both shell elements and line elements for the brace and chord members. The agreement between the test and simulation results is excellent. Figure 5 compares the detailed simulation

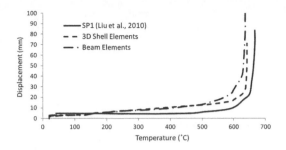

Figure 5. Comparison for displacement-temperature curves of SP1 (Liu et al., 2010).

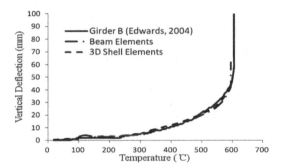

Figure 6. Comparison for displacement-temperature curves of Girder B of Edwards (1992).

displacement-temperature curves with the test results of Liu et al. (2010). From the comparison, it can be seen that both the four-noded shell elements and line elements give close prediction of the test results. The critical temperatures from the test, from the simulations using shell elements and using line elements are 666°C, 642°C and 636°C respectively. This agreement is acceptable.

2.3.2 Girder B of Edwards (1992)

Figure 6 compares the displacement-temperature curves at the centre point of the top chord member. The agreement with the test result is excellent for using both the line and shell elements.

The failure mode of the truss, obtained from the test, from the simulations using shell elements and using line elements was due to buckling of the middle diagonal compressive brace members (members 8 and 9 in Figure 1) at 606°C, 602°C and 595°C respectively. Figure 7 compares the simulated and observed failure modes. It is clear that the numerical simulation model, either using 2D line elements or 3D shell elements, is suitable for simulating the overall behavior of welded tubular trusses in fire. However, from a computational point of view, using line elements is preferable because the simulation was very fast. The line element model was used to conduct the parametric study in the next sections.

(a) Test observation (Edward, 2004)

(b) Failure mode using shell elements

(c) Failure mode using line elements

Figure 7. Comparison of failure modes for test Girder B of Edwards (1992).

3 INFLUENTIAL FACTORS ON STRUCTURAL BEHAVIOR OF TRUSS AT ELEVATED TEMPERATURE

The behavior of a truss at elevated temperatures is affected by many design factors as well as design assumptions. The parametric study in section 4 of this paper will investigate, in detail, the effects of different design factors. This section will present the results of a number of numerical investigations to examine the effects of different design assumptions. These assumptions are:

– Joints: Joints: the welded truss joint may be considered to be rigid (Figure 8a), pinned (Figure 8c) or semi-rigid (where the chord members are continuous, but the brace members are pinned, Figure 8b);
– This investigation was to assess whether restraint to some differential thermal elongation of the members due to some temperature difference would cause any difference in truss member critical temperatures.

Figure 8 illustrates the joint and boundary conditions of the trusses. Table 2 lists the truss member dimensions. The material strength of steel was $f_y = 355$ N/mm² $f_u = 443.75$ N/mm². The elastic modulus of steel was assumed to be 210 GPa. The elevated temperature stress-strain curves and the thermal expansion coefficient of steel were based on Eurocode EN-1993-1-2.

For comparison, ABAQUS simulations of the individual truss members were carried out to obtain the critical temperatures of all the members according to the current member based design method. In these analyses, the member forces were from the same as those at 20°C.

In the truss simulations, the rates of temperature rise of all the members were assumed to be in proportion to their individual critical temperatures. Whether or not

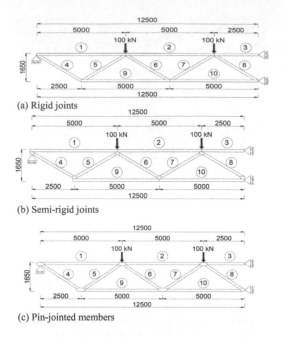

(a) Rigid joints

(b) Semi-rigid joints

(c) Pin-jointed members

Figure 8. Truss configurations used in numerical analyses (half span, dimensions in mm).

Table 2. Truss member dimensions.

Member Type	Dimensions (mm)
Bottom and top chords	Φ 323.9 × 8
Outer bracings (Member 4 and 5)	Φ 193.7 × 4
Inner bracings (Member 6, 7 and 8)	Φ 114.3 × 5

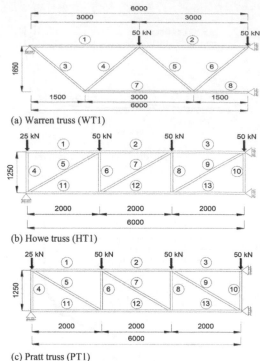

(a) Warren truss (WT1)

(b) Howe truss (HT1)

(c) Pratt truss (PT1)

Figure 9. Truss configurations used in parametric study (half span, dimensions in mm).

thermal expansion was not considered, the truss failure temperatures were almost identical. This indicates that any differential thermal elongation due to different truss members being heated to different temperatures had negligible effect on the failure temperature of the trusses. The failure temperatures of the trusses with rigid, semi-rigid and pinned joints are 535°C, 529°C and 523°C respectively.

Since assuming pinned joints gives the lowest truss failure temperature, it is suggested that this assumption may be made in fire resistant design of the truss. However, even assuming pin-joints, the member-based calculation method of using the ambient temperature forces may not be safe. For example, for the truss in Figure 8, the member-based critical temperature of member 7 of the truss, obtained numerically using ABAQUS, was 558°C. This is higher than the actual critical temperature (535°C) of the member in the rigid truss.

The reason for this difference in the critical temperature of member 7 between using the member-based method and using truss analysis is due to the increased member force in the member in truss analysis when the truss deflection is high. To confirm this, static structural analysis was performed using the deformed configuration of the truss. The initial member forces, the member force at truss failure (523°C) and the member force from static analysis based on the deformed shape of the truss were 184.4 kN, 213.6 kN and 213.6 kN respectively.

Furthermore, the ABAQUS model of member 7 was rerun, in this case using the final member forces (213.6 kN), and the failure temperature was 523°C, exactly the same as the failure temperature of the member in the pin-jointed truss.

Parametric study will investigate how the member forces change and the effects of this change on member critical temperatures. In the investigations, the truss will be assumed to be pin-jointed.

4 PARAMETRIC STUDY

Extensive numerical simulations have been conducted on Warren, Howe and Pratt trusses. Figure 9 the different types of the trusses being investigated.

4.1 Simulation results and discussions

4.1.1 Effect of span-to-depth ratio
Different span to depth ratios gives different chord axial force values and different diagonal member

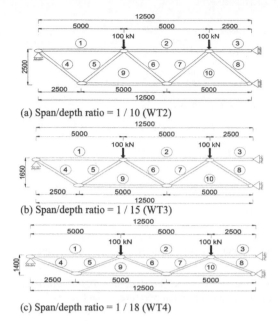

(a) Span/depth ratio = 1 / 10 (WT2)

(b) Span/depth ratio = 1 / 15 (WT3)

(c) Span/depth ratio = 1 / 18 (WT4)

Figure 10. Warren trusses with different span/depth ratios (half span, dimensions in mm).

Table 3. Effect of truss span to depth ratio on critical temperature.

		Member 7	Truss
WT2	Critical temperature	556°C	540°C
	Max. displacement	–	389 mm
	(P_θ/P_{20})	1.00	1.07
WT3	Critical temperature	558°C	523°C
	Max. displacement	–	587 mm
	(P_θ/P_{20})	1.00	1.19
WT4	Critical temperature	541°C	509°C
	Max. displacement	–	663 mm
	(P_θ/P_{20})	1.00	1.22

angles. Figure 10 illustrates the geometric configurations of the Warren trusses with different span-depth ratios. Table 3 provides a summary of the simulation results. In Table 3, P_θ/P_{20} is the ratio of the member force at elevated temperature (P_θ) from truss analysis to that at ambient temperature (P_{20}). In all cases, failure was initiated in member 7. The member based analysis gives a critical temperature of more than 30°C higher than truss analysis for truss span to depth ratio of 18.

As the span to depth ratio increases, the angles between the chord and brace members become smaller. As will be shown in Section 5 which presents a method to calculate the increased member force in compressive diagonal member, it is the vertical component of the chord force in the deformed truss that increases the compression force in the diagonal brace members.

This increase is in inverse relationship to the angle between the diagonal brace member and the chord member. A smaller angle (for large truss span to depth ratio) leads to a higher increase in the brace member force.

4.1.2 Effect of member slenderness (λ)

This case used the Warren truss shown in Figure 10b and the Pratt truss shown in Figure 9c. The influence of different slenderness values for the Warren truss ($\lambda = 89$, 78 and 62) was examined by changing the cross section of critical member 7 ($\Phi101.6 \times 6.3$, $\Phi114.3 \times 5$ and $\Phi139.7 \times 3$ respectively for the above slenderness values). In the case of the Pratt truss, the critical member was member 10 and the member size was $\phi76.1 \times 2.5$, $\phi60.3 \times 3.6$ and $\phi48.3 \times 5$ to give slenderness values of $\lambda = 48$, 62 and 81 respectively.

Because the overall geometry of the trusses was unchanged, the truss member forces were similar in all cases. However, a comparison between Warren and Pratt trusses shows that the increase in member force in the Pratt truss is greater. This is expected because the increase in the compression force (due to the vertical component of the compression chord force in the deformed position of the truss) of the brace is shared by two diagonal members in the Warren truss but resisted by only one single member in the Pratt truss.

4.1.3 Effect of applied load ratio

For this investigation, the Warren truss in Figure 10b and the Pratt truss in Figure 9c were used. Each point load was 50 kN. The member forces change similarly as discussed in the previous section. However, the effect of the same change in the member compression force is different at different load ratios. For the Pratt truss with a high load ratio (applied load = 100 kN), the reduction of truss critical temperature from the member based analysis is nearly 60°C.

4.1.4 Effect of truss span

In this investigation, the span of a Warren truss was changed as 4.5 meters, 12 meters and 25 meters but the span to depth ratio was kept constant at 20.8. Since the span to depth ratio was constant, the member forces experienced very similar changes. Due to the shallow truss depth, the increase in member force was quite high, resulting in truss critical temperatures much lower (by as much as 76°C) than that from member based analysis.

4.1.5 Effect of truss configuration

Warren, Howe and Pratt trusses of the same span, the same span-to-depth ratio and the same load level were simulated. Figure 9 shows the truss dimensions.

As expected, the more brace members are involved in sharing the load (Howe truss > Warren truss > Pratt truss), the lower increase in the compression force of each member are obtained.

The results of this investigation show that the effect of increased member force may be ignored for Howe trusses. In the case of increased member forces, Pratt truss analysis revealed a more significant effect on the failure temperature compared to Warren trusses. This effect should be considered for fire resistance analysis of both Warren and Pratt trusses when failure takes place in the middle brace members.

4.1.6 *Effect of number of braces*

For this investigation, the number of brace members was different, all other dimensions being equal. The total load was 500 kN.

Changing the number of brace members in a truss changes the maximum chord compression force. As the number of brace members increases, the chord compression force at the truss centre increases. This increase in the chord compression force results in a higher increase in the brace compression force. For truss A, which is sensibly proportioned, the brace compression force at truss failure more than doubled the initial compression force. This resulted in a truss critical temperature of more than 100°C lower than the member based result which did not include this increase in the brace force.

4.1.7 *Effect of failure location*

In the previous cases, failure of the truss was initiated in the brace near the centre of the truss. In this simulation, failure of the truss was forced to start from the brace member near the support by reducing the size of this member. A Pratt truss with a span of 100 meters was used.

When the failure is near the support, using truss analysis gave the same results as the member based analysis. This is because the increase in the brace force was very small, a result of small chord compression force and small change in the line of action in the chord compression force.

4.2 *Summary of results*

In summary, these results show that due to truss undergoing large displacements at elevated temperatures, some truss members (compression brace members near the truss centre) experience large increases in member forces. Therefore, when calculating the member critical temperatures, it would not be safe to use the member forces from ambient temperature structural analysis. Using the ambient temperature member force may overestimate the truss member critical temperature (based on truss analysis) by 100°C.

The influence of slenderness on the truss member forces was ignorable owing to the unchanged overall geometry of the trusses. Also, span of the trusses had no effect on the critical temperature as long as the span to depth ratio was constant.

As will be explained in the next section, the changes in forces in the top chord, bottom chord and tension brace members of Howe, Pratt and Warren trusses are small at high temperatures. However, the effect of the deformed shape of Pratt and Warren trusses at elevated temperatures on the compression brace members is considerable. Furthermore, the angle between

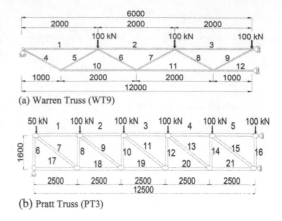

(a) Warren Truss (WT9)

(b) Pratt Truss (PT3)

Figure 11. Dimensions of trusses used for detailed comparison of additional forces (half span, dimensions in mm).

diagonal member and chord member has a significant influence on failure temperature of the trusses.

5 DEVELOPMENT OF A SIMPLE METHOD

The lower truss critical temperatures from truss analysis compared to member based analysis are a result of increased brace compression force. A method will be developed in this section to calculate this force increase. This increase is a result of the large deflection of the truss. The maximum truss deflection ranged from 0.014 to 0.03 of the truss span (span to deflection ratios of 71 and 33 respectively). In the fire protection industry, a deflection limit of span/30 is often used (BSI, 2009) to determine fire resistance. For simplicity and for safety, this value will be used in developing the calculation method.

To enable detailed comparison for forces in different truss members between calculation using the proposed analytical method and truss analysis results, results for the Warren and Pratt trusses shown in Figure 11 are used.

The critical temperature of the Warren truss was 623°C, due to buckling of brace member 9 in compression. At truss failure, the maximum truss deflection was 293 mm. For the Pratt truss, the critical temperature was 617°C, reached due to failure of member 16. At the failure temperature, the maximum truss deflection was 638 mm.

The results in Table 4 and Table 5 reveal the following trend clearly:

- Forces in the tension brace members decrease. Therefore, for these members, the current member based design method is on the safe side;
- The changes in forces in the chord members are small. Therefore, the current member based design method is acceptable;
- The large percentage change in forces occurs in the compression brace members near the centre of the truss. Away from the truss centre, the change

Table 4. Comparisons of the member forces for the Warren truss in Figure 11 (unit in kN).

Member No	ABAQUS		
	$F_{20°C}$	$F_{max,Abaqus}$	$F_{max,Abaqus}/F_{20°C}$
1	−433.0	−416.2	0.96
2	−1125.8	−1107.8	0.98
3	−1472.0	−1462.7	0.99
4	500.0	501.6	1.00
5	−500.0	−512.4	1.02
6	300.0	274.4	0.91
7	−300.0	−345.5	1.15
8	100.0	57.6	0.58
9	**−100.0**	**−167.1**	**1.67**
10	866.0	878.7	1.01
11	1385.6	1414.5	1.02
12	1558.8	1608.4	1.03

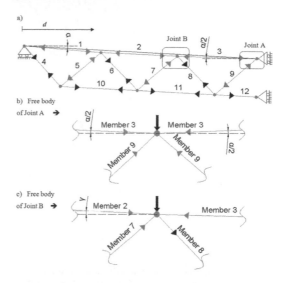

Figure 12. Deformed shape and free body diagrams for a Warren truss.

Table 5. Comparisons of the member forces for the Pratt truss in Figure 11 (unit in kN).

Member No	ABAQUS		
	$F_{20°C}$	$F_{max,Abaqus}$	$F_{max,Abaqus}/F_{20°C}$
1	−703.1	−713.0	1.01
2	−1250.0	−1267.5	1.01
3	−1640.6	−1655.3	1.01
4	−1875.0	−1891.1	1.01
5	−1953.1	−1987.0	1.02
6	−500.0	−498.8	1.00
7	834.8	851.4	1.02
8	−450.0	−453.7	1.01
9	649.3	658.1	1.01
10	−350.0	−365.4	1.04
11	463.8	460.6	0.99
12	−250.0	−284.1	1.14
13	278.2	279.1	1.00
14	−150.0	−186.0	1.24
15	92.8	92.5	1.00
16	**−100.0**	**−189.6**	**1.90**
17	0.0	23.7	–
18	703.1	732.4	1.04
19	1250.0	1278.2	1.02
20	1640.6	1660.0	1.01
21	1875.0	1892.3	1.01

in compression force in the brace members rapidly diminishes.

5.1 Maximum increase in compression brace force at centre of truss

Refer to Figure 11 which shows the deformed geometry at the joint at the centre of the truss, the increase in compression force in the compression brace members (member 9) is to resist the vertical components of the compression chords (member 3).

Assuming the maximum truss deflection is δ. The angle between the straight line drawn from the support

to the maximum deformed position of the compression chord is $\alpha = \delta/(L/2)$ where L is the total span of the truss. The angle between the deformed compression chord and the horizontal is approximately $\alpha/2$. Based on the above assumptions, the additional vertical force from one of the two chord members with the maximum compression force (chord member 3) can be calculated as:

$$F_{maximum\ chord\ compression} * \frac{\delta}{L/2} * \frac{1}{2} \tag{1}$$

Therefore, the maximum increase in compression force in the brace member (member 9) can be calculated as follows:

$$\Delta F_{truss\ centre} = \frac{F_{maximum\ chord\ compression} * \frac{\delta}{L}}{\sin\theta} \tag{2}$$

where θ is the angle between the compression brace member at the truss centre (member 9) and the horizontal.

The total compressive force in the brace member at the truss centre (member 9) is:

$$F_{truss\ centre} = F_{truss\ centre,0} + \Delta F_{truss\ centre} \tag{3}$$

5.2 Increase in compression brace force away from the centre of truss

Away from the centre of truss, the increase in force in the compression brace members decreases rapidly for three reasons: (1) the chord compression force are lower; (2) the additional vertical force from one of the two chord members with the higher compression force are shared by two brace members; (3) as shown in Figure 12c, the relative chord rotation (γ) between two

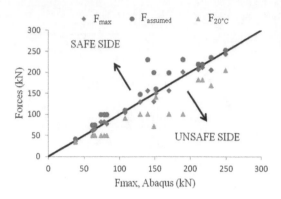

Figure 13. Comparison for forces in critical members of Warren and Pratt trusses between analytical calculations, ABAQUS simulation results and ambient temperature values.

adjacent members is much smaller than angle $\alpha/2$ to the horizontal at the centre of the truss (Figure 12b). Because the percentage increases in forces in these members are small, gross approximation is acceptable when calculating the force increase in these members. Assuming the chord compression force decreases linearly from at the truss centre to 0 at support, and assuming that the relative rotation of the chord members at each node is $\alpha/2$, the increase in compression force in compression brace members not at the centre of the truss can be approximately calculated as:

$$\Delta F_{other\ brace\ member} = \frac{1}{2} * \Delta F_{truss\ centre} * \frac{d}{L/2} \quad (4)$$

where is d is the distance from the support to the node connecting the compression brace member whose force is calculated.

For the Pratt truss, as shown in Figure 11, because there is only one vertical member at the centre, the additional compressive force is twice as given in Equation 5 and $\theta = 90°$. Therefore:

The increase in compression force in the members away at the centre can be calculated using Equation 4 for the same reasoning as for the Warren truss.

The accuracy of the proposed analytical method has been checked against all the simulation trusses which were used in the parametric study. Figure 13 compares the maximum compression brace forces at the centre of the trusses from truss analysis with those calculated using Equations 5 and 8 for two maximum truss deflections: the maximum truss deflection at failure from ABAQUS analysis and the assumed maximum truss deflection of L/30. Also, original truss member forces at 20°C (F 20°C) are included in Figure 13. The results in Figure 13 indicate that the proposed calculation method can produce very accurate results if the actual maximum deflection of truss is used. If a value of L/30 is used for the maximum deflection, the calculation results are reasonably close to the ABAQUS simulation results, and are on the safe side. In contrast, using the ambient temperature forces can greatly underestimate the member forces and produce unsafe design.

6 CONCLUSIONS

This paper has presented the results of a numerical investigation into the behavior of welded steel tubular truss at elevated temperatures.

The main conclusion of the parametric study on truss behavior was that the member-based calculation method for calculating member critical temperatures, in which the ambient temperature truss member forces are used, may not be safe.

This paper has developed an analytical method to calculate increases in truss compressive brace member forces.

REFERENCES

BSI, BS 5950: Part8: Code of Practice for the Fire Protection of Structural Steelwork. 2003: UK.

BSI, BS 476: Part 10: Guide to the principles, selection, role and application of fire testing and their outputs. 2009: UK.

CIDECT, Design Guide for Circular Hollow Section (CHS) Joints Under Predominantly Static Loading, ed. S. Edition. 2010, Verlag TUV Rheinland, Germany.

Cofer, W.F. and J.S. Jubran, Analysis of welded tubular connections using continuum damage mechanics. Journal of Structural Engineering, 1992. 118(3): p. 828–845.

Edwards, M., Fire Performance of SHS Lattice Girders. Tubular Structures V, 2004. 5: p. 95.

Liu, M.L., J.C. Zhao, and M. Jin, An experimental study of the mechanical behavior of steel planar tubular trusses in a fire. Journal of Constructional Steel Research, 2010. 66(4): p. 504–511.

Tubular Structures XV – Batista, Vellasco & Lima (eds)
© 2015 Taylor & Francis Group, London, ISBN 978-1-138-02837-1

Fire performance of innovative slender concrete filled steel tubular columns

A. Espinós, M.L. Romero, E. Serra, V. Albero & A. Hospitaler
Instituto de Ciencia y Tecnología del Hormigón (ICITECH), Universitat Politècnica de València, Spain

ABSTRACT: This paper investigates the experimental behaviour of innovative slender concrete filled steel tubular (CFST) columns under standard fire test conditions. A total of 36 fire tests were carried out in the testing facilities of AIDICO (Instituto Tecnológico de la Construcción) in Valencia, in the framework of the European Project FRISCC (Fire resistance of innovative and slender concrete filled tubular composite columns). The aim of this project was to provide a full range of experimental evidence on the fire behaviour of CFST columns of different cross-section shape, a necessary basis for the development of numerical models and simplified design methods. Four different section shapes were tested in this experimental program: circular (6 tests), square (6 tests), elliptical (12 tests) and rectangular (12 tests). Large eccentricities were used, in order to extend the current experimental database, with load eccentricity ratios up to 0.75. The study was focused on slender columns, all the specimens having a length of 3180 mm and being in most cases hinged at both ends. The steel tubes had a nominal yield strength of 355 MPa, while the concrete infill had a theoretical compressive strength of 30 MPa. The load level applied to the columns was a 20% of their load bearing capacity at room temperature. This experimental study allows understanding the influence of the cross-section shape on the fire performance of CFST columns, as well as the effect of the load eccentricity and reinforcement ratio. The test results are compared and analysed in this paper, where the influence of the different parameters used in the experimental program are studied, specially the effect of large eccentricities combined with high slenderness. Comparative graphs are given and conclusions are drawn on the basis of these results.

1 INTRODUCTION

The fire resistance of CFST columns with circular and square cross-section has been deeply studied by means of experimental investigations carried out in the last decades, as those from the research programs promoted by CIDECT (COMETUBE 1976, Grandjean et al. 1980, Kordina & Klingsch 1983), National Research Council of Canada (Lie & Chabot 1992, Chabot & Lie 1992, Kodur & Lie 1995), or those carried out at Fuzhou University (China) by Han and co-workers (Han et al. 2003), researchers from University of Seoul (Korea) (Kim et al. 2005), and the authors of this paper (Romero et al. 2011, Moliner et al. 2013). Nevertheless, there is a lack of experimental results on other section shapes, such as elliptical or rectangular sections, while large eccentricities have not been studied deeply, which motivates the need for carrying out new experimental programs in order to increase the existing experimental database.

One of the aims of the European Project FRISCC (Fire resistance of innovative and slender concrete filled tubular composite columns) is to provide a full range of experimental evidence on the fire behaviour of CFST columns, a necessary basis for the development of numerical models and simple calculation rules.

After a full review of the results of previous tests available in the literature, an extensive experimental program is designed in the framework the referred European Project. In this experimental program, four different section shapes are used: circular, elliptical, square and rectangular hollow sections filled with concrete, focusing on slender columns and large eccentricities. By means of the experimental results, the relative fire performance of the different section shapes is analysed in the paper, as well as the effect of the load eccentricity and percentage of reinforcement.

2 EXPERIMENTAL INVESTIGATION

2.1 Design of the experimental program

The experimental program consisted of a total of 36 fire tests. The parameters studied were the cross-section shape (CHS, SHS, EHS and RHS), sectional dimensions, member slenderness, load eccentricity and reinforcement ratio. Large eccentricities (relative values of 0.5 and 0.75) were applied to some of the circular and square columns, while for the elliptical and rectangular columns, relative eccentricities of 0.2 and 0.5 were applied about both minor and major axis.

The square columns were designed to have approximately the same steel area than their circular counterparts (i.e. same quantity of steel), in order to compare their effectiveness in the fire situation for the same steel usage. This was intended also for the elliptical and rectangular columns, although due to the limitations in the availability of the elliptical sections in the market, this equivalence between rectangular and elliptical columns was not possible to obtain in some of the cases.

2.2 Test setup

The tests were performed in the facilities of AIDICO (Instituto Tecnológico de la Construcción) in Valencia (Spain), using a 5×3 m furnace equipped with a hydraulic jack with a maximum capacity of 1000 kN. The load level applied to the columns was a 20% of their maximum capacity at room temperature. With this load level applied and kept constant, the ISO-834 fire curve was prescribed, with unrestrained column elongation. All the columns were tested under pinned-pinned (P-P) boundary conditions, except for the elliptical columns E1 to E6, which were designed as pinned-fixed (P-F) in order to reduce their slenderness. All the column specimens had a length of 3180 mm. For each column, two ventilation holes of 15 mm diameter were drilled in the steel tube wall at 100 mm from each column end. Steel end plates of dimensions $300 \times 300 \times 15$ mm were welded to the column ends. The data of all the tested specimens and resulting fire resistances measured in minutes are listed in Tables 1 to 4. Figure 1 presents some details of the experimental setup.

2.3 Instrumentation

In order to register the temperature evolution inside the columns during the fire tests, three layers of six thermocouples each were placed at different heights. The axial elongation of the columns was measured by means of a LVDT located outside the furnace.

2.4 Material properties

The hollow tubes used in the experimental program had a S355 steel grade, nevertheless the real strength (f_y) of steel was obtained by performing the corresponding coupon tests, and is summarized in Tables 1 to 4. Normal strength concrete (30 MPa) was used for the column infill. In order to determine the compressive strength of concrete, sets of concrete cylinders were prepared and cured in standard conditions during 28 days. All cylinder samples were tested on the same day as the column fire test. The cylinder compressive strength of all the tested specimens (f_c) can be found in Tables 1 to 4. The bar-reinforced specimens had the arrangements and geometrical reinforcement ratios (ρ) given in these tables, using 6 mm stirrups with 30 cm spacing.

Figure 1. Test setup and details of the column ends.

Table 1. Characteristics of the column specimens, circular columns.

No.	D (mm)	t (mm)	Rebar	ρ (%)	f_c (MPa)	f_y (MPa)
C1	193.7	8	$6\phi12$	2.74	36.37	359.06
C2	273	10	$6\phi16$	2.40	37.62	369.73
C3	193.7	8	$6\phi12$	2.74	43.23	359.06
C4	273	10	$6\phi16$	2.40	35.96	369.73
C5	193.7	8	$6\phi16$	4.86	35.76	359.06
C6	273	10	$8\phi20$	5.00	36.89	369.73

No.	B.C.	$\bar{\lambda}_z$	e/D	Load (kN)	Time (min)
C1	P-P	0.75	0.5	186.65	26
C2	P-P	0.54	0.5	387.46	30
C3	P-P	0.76	0	535.57	29
C4	P-P	0.53	0	882.90	72
C5	P-P	0.77	0.75	152.41	29
C6	P-P	0.56	0.5	391.53	57

3 ANALYSIS OF RESULTS

3.1 General

The typical failure mode observed in all the columns was overall buckling. Figures 3 to 6 show the evolution of the axial displacement measured at the top end of the columns versus the fire exposure time for all the tested specimens, grouped according to their section shape and dimensions.

In a qualitative comparison between all the results, it can be clearly seen that the failure mode of all the rectangular and elliptical columns was the same, with

Table 2. Characteristics of the column specimens, square columns.

No.	B (mm)	t (mm)	Rebar	ρ (%)	f_c (MPa)	f_y (MPa)
S1	150	8	4φ12	2.52	45.03	452.74
S2	220	10	4φ16+4φ10	2.80	39.72	560.25
S3	150	8	4φ12	2.52	43.15	452.74
S4	220	10	4φ16+4φ10	2.80	42.39	560.25
S5	150	8	8φ12	5.04	48.67	452.74
S6	220	10	4φ20 + 4φ16	5.15	38.84	560.25

No.	B.C.	$\bar{\lambda}_z$	e/D	Load (kN)	Time (min)
S1	P-P	0.91	0.5	161.13	26
S2	P-P	0.66	0.5	446.53	23
S3	P-P	0.90	0	404.29	32
S4	P-P	0.66	0	882.90	54
S5	P-P	0.94	0.75	133.18	29
S6	P-P	0.67	0.5	452.63	29

Table 3. Characteristics of the column specimens, rectangular columns.

No.	H (mm)	B (mm)	t (mm)	Rebar	ρ (%)	f_c (MPa)
R1	250	150	10	–	0	37.85
R2	250	150	10	4φ16	2.69	39.63
R3	250	150	10	–	0	31.96
R4	250	150	10	4φ16	2.69	36.31
R5	250	150	10	–	0	32.92
R6	250	150	10	4φ16	2.69	42.45
R7	350	150	10	–	0	38.18
R8	350	150	10	4φ16 + 4φ10	2.61	37.61
R9	350	150	10	–	0	37.33
R10	350	150	10	4φ16 + 4φ10	2.61	37.97
R11	350	150	10	–	0	39.73
R12	350	150	10	4φ16 + 4φ10	2.61	38.18

No.	f_y (MPa)	B.C.	$\bar{\lambda}_y$	$\bar{\lambda}_z$	e/H	e/B	Load (kN)	Time (min)
R1	428.269	P-P	0.53	0.82	0	0	650.80	19
R2	428.269	P-P	0.54	0.85	0	0	699.84	23
R3	428.269	P-P	0.52	0.80	0	0.2	374.67	23
R4	457.689	P-P	0.55	0.86	0	0.5	276.87	27
R5	457.689	P-P	0.54	0.83	0.2	0	456.74	24
R6	457.689	P-P	0.54	0.85	0.5	0	322.14	34
R7	473.999	P-P	0.41	0.83	0	0	928.93	30*
R8	473.999	P-P	0.40	0.86	0	0	988.78	21
R9	503.718	P-P	0.41	0.85	0	0.2	540.06	22
R10	473.999	P-P	0.40	0.86	0	0.5	383.88	25
R11	503.718	P-P	0.42	0.85	0.2	0	683.04	22
R12	503.718	P-P	0.41	0.88	0.5	0	481.36	18*

*Failed test.

only two stages in the axial displacement versus time curve: axial elongation of the column and sudden failure after the yielding of the steel tube occurs, thus not taking advantage of the contribution of the concrete

Table 4. Characteristics of the column specimens, elliptical columns.

No.	H (mm)	B (mm)	t (mm)	Rebar	ρ (%)	f_c (MPa)
E1	220	110	12	–	0	34.72
E2	220	110	12	–	0	39.11
E3	220	110	12	–	0	38.17
E4	220	110	12	4φ10	2.37	33.34
E5	220	110	12	4φ10	2.37	37.83
E6	220	110	12	4φ10	2.37	36.63
E7	320	160	12.5	–	0	37.30
E8	320	160	12.5	4φ16	2.57	41.22
E9	320	160	12.5	–	0	43.73
E10	320	160	12.5	4φ16	2.57	42.35
E11	320	160	12.5	–	0	36.49
E12	320	160	12.5	4φ16	2.57	35.53

No.	f_y (MPa)	B.C.	$\bar{\lambda}_y$	$\bar{\lambda}_z$	e/H	e/B	Load (kN)	Time (min)
E1	372.45	P-F	0.47	0.84	0	0	397.19	21
E2	347.54	P-F	0.46	0.83	0	0.18	281.84	26
E3	348.06	P-F	0.46	0.82	0	0.45	198.96	28
E4	348.06	P-F	0.46	0.84	0	0	409.63	22
E5	347.54	P-F	0.47	0.85	0	0.18	287.94	25
E6	369.71	P-F	0.48	0.87	0	0.45	204.51	26
E7	522.64	P-P	0.52	0.93	0	0	589.76	30
E8	522.64	P-P	0.54	0.97	0	0	681.94	31
E9	522.64	P-P	0.53	0.95	0	0.2	361.04	30
E10	522.64	P-P	0.54	0.97	0	0.5	249.42	37
E11	522.64	P-P	0.52	0.93	0.2	0	440.44	32
E12	522.64	P-P	0.53	0.96	0.5	0	286.54	38

core, due to the high slenderness of these specimens. Nevertheless, in some of the tests performed on circular and square columns (those with concentric load or reduced eccentricity) the curve presented four stages, with a contribution of the concrete core after the steel tube yielding, which is reflected as a plateau in this curve. See Espinos et al. (2010) for a more detailed description on the different failure modes.

In those fire tests where the eccentricity was applied about the major axis, as the one shown in Figure 2, interaction between major and minor axis was observed. This effect was especially noticeable for those columns subjected to a relative eccentricity of 0.2H.

3.2 Circular and square columns

For each geometry (circular and square), the results are presented in two different graphs, where the columns have been grouped according to their sectional dimensions. Figures 3a and 3b correspond to the circular columns, while Figures 4a and 4b correspond to the square columns. Comparing the circular columns with their square counterparts, which made use of the same quantity of steel, the fire response of the circular columns resulted more efficient. This can be seen by comparing cases C1-S1 and C5-S5, where for the same

Figure 2. Column E12 after fire test.

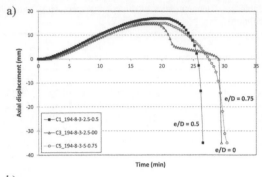

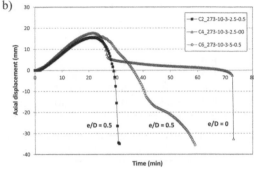

Figure 3. Results of the fire tests on circular columns: a) 193.7 × 8 mm, b) 273 × 10 mm.

fire resistance time, the circular columns sustained higher loads (with increments of 15.8% and 14.4%, respectively), or comparing cases C2-S2 (30.4% time increment with a 13.2% reduction of applied load) and C3-S3 (32.5% load increment with a 9.4% reduction in time). The more significant increase in fire resistance can be observed by comparing cases C4 and S4, where for the same load applied, the circular column achieved a 33.3% higher fire resistance time. Therefore, it can be concluded that, for the same steel usage, the circular columns present a better fire behaviour than the square columns. It is worth noting that the slenderness values of the square columns were higher in all cases.

It is also important to note that the A/V-ratio of the circular columns was lower than that of the square columns, which make them perform better in the fire situation, as they expose a lower surface to the fire for the same volume.

The influence of the load eccentricity can be studied through Figures 3b and 4b. If cases C2 and C4 are compared, it can be seen that for the same column dimensions and percentage of reinforcement, the fire resistance time was significantly reduced when applying the eccentricity (30 min), in comparison to the concentrically loaded test (72 min), having the second case 2.28 times the load applied to the first case. If the percentage of reinforcement is increased from 2.5% to 5%, with the same load eccentricity applied (C2 versus C6), the fire resistance time increases (30 min versus 57 min). Therefore, this result confirms that the reinforcement contributes to improve the fire resistance of the columns. In the case of the square columns,

the fire resistance of the concentrically loaded column (S4) was higher than that of the corresponding eccentrically loaded column (S2) (54 min versus 23 min), having the first case twice the load applied to the second case. The beneficial effect of the reinforcement can be also seen by comparing cases S2 and S6 (23 min versus 29 min increasing the reinforcement from 2.5% to 5%).

3.3 Rectangular and elliptical columns

The results of the rectangular and elliptical columns are presented in different graphs, according to their sectional dimensions. Figures 5a and 5b correspond to the rectangular columns, while Figures 6a and 6b correspond to the elliptical columns. Note that in the second series of rectangular columns (Figure 5b), the results of specimens R7 and R12 have been omitted, as these fire tests were anomalous.

The influence of the load eccentricity can be observed in these figures, for both the major and minor axis. It can be observed that, as the load eccentricity was increased, the fire resistance time also increased, which was due to the differences on the applied load. In effect, as the load level applied to all the columns was the same (20% of their theoretical maximum capacity at room temperature), the value of the load applied to the columns with higher eccentricity was lower, and therefore the resulting fire resistance time was higher. However, it results more useful to see this comparison in terms of load increment. For instance, for the elliptical tests, the load applied to the concentrically loaded

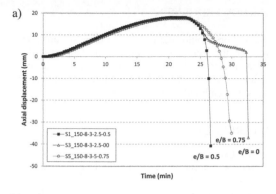

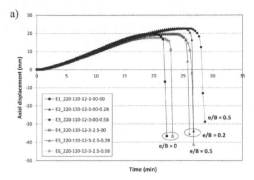

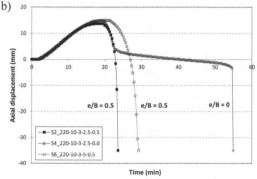

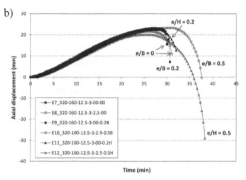

Figure 4. Results of the fire tests on square columns: a) 150 × 8 mm, b) 220 × 10 mm.

Figure 6. Results of the fire tests on elliptical columns: a) 220 × 110 × 12 mm, b) 320 × 160 × 12.5 mm.

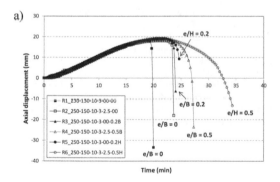

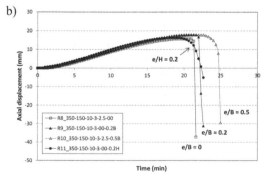

Figure 5. Results of the fire tests on rectangular columns: a) 250 × 150 × 10 mm, b) 350 × 150 × 10 mm.

columns was approximately two times the load applied to the columns with relative eccentricity of 0.5 (E1 vs E3 or E4 vs E6), while the difference in terms of fire resistance time was not proportional to the load increment (25% time difference between specimens E1 and E3 and a 15.4% time difference between specimens E4 and E6).

The effect of the eccentricity applied about the major axis can be observed by comparing case R1 against R5 and R6 (Figure 5a), with the same column dimensions and relative eccentricities of 0, 0.2H and 0.5H. As it can be seen, the fire resistance time also increases when applying increasing eccentricities on the major axis, due to the reduction in the applied load. This behaviour can be also noticed in the rectangular columns, comparing case E7 against E11 and E12 (Figure 6b). It should be noted that the load applied to the concentrically loaded columns was about two times the load applied to the columns with 0.5H relative eccentricity.

Comparing between reinforced and unreinforced specimens, it can be seen that, although the load applied to the reinforced specimens was higher, the values of their fire resistance times were similar or in some cases higher than those of the unreinforced columns (see R1/R2 or E1/E4), which confirms the favourable effect of the contribution of the reinforcing bars in the fire situation.

If the elliptical sections are compared with their rectangular counterparts (E1-R1, E4-R2, E2-R3 and

201

E6-R4), having the same load eccentricity and percentage of reinforcement, the fire resistances obtained are similar, although the loads applied to the rectangular columns were much higher than those applied to the elliptical columns (between a 30% and 70% load increment). In these cases the slenderness of all the columns was similar, whereas the steel area of the rectangular columns was about a 30% higher than that of the elliptical columns, which contributed to sustain a higher load during a similar amount of time.

If the elliptical series E7-E12 (Figure 6b) is compared against the rectangular series R7-R12 (Figure 5b), having a similar steel area (7.5% difference), the elliptical columns achieve a higher fire resistance time (44.4% higher on average, discarding cases R7 and R12), although it should be noticed that the loads applied to the rectangular columns were higher to those applied to the elliptical columns, with a 50.9% average increment. Therefore, in this case it is difficult to reach a conclusion in favour of one or other section shape. Further studies are needed for obtaining a conclusive result, which the authors will carry out in the future by means of numerical simulations.

4 SUMMARY AND CONCLUSIONS

An experimental program on slender CFST columns of different cross-section shapes subjected to elevated temperatures was carried out, comprising a total of 36 fire tests. The following parameters were studied: cross-section shape, sectional dimensions, member slenderness, load eccentricity and reinforcement ratio. Innovative columns such as those composed of elliptical sections filled with concrete were included in this study. The load eccentricity was applied to the columns on the major and minor axis, and large eccentricities with relative values up to 0.75 were used.

Comparing between the circular and square columns, it was found that, for the same steel usage, the circular columns presented a better fire performance than the square columns. Additionally, for the same column dimensions and percentage of reinforcement, the fire resistance time was significantly reduced when applying the load eccentrically. Furthermore, it was found that for the same load eccentricity, when the percentage of reinforcement was increased, the fire resistance time also increased.

Comparing between the elliptical and rectangular columns, the fire resistances obtained were similar or higher for the elliptical columns, although the loads applied to the rectangular columns were much higher than those applied to the elliptical columns, therefore in this case the results did not allow reaching a conclusion in favour of one or other section shape, further numerical studies being needed.

ACKNOWLEDGEMENTS

The authors would like to express their sincere gratitude to the European Union for the help provided through the Project RFSR-CT-2012-00025, carried out with a financial grant of the Research Programme of the Research Fund for Coal and Steel.

REFERENCES

COMETUBE. 1976. Fire resistance of structural hollow sections. Cometube research. CIDECT programme 15A. Final report.

Grandjean, G., Grimault, J.P. & Petit, L. 1980. Determination de la duree au feu des profils creux remplis de beton. CIDECT Research Project 15B–80/10. Cologne, Germany: Comité International pour le Développement et l'Etude de la Construction Tubulaire.

Kordina, K. & Klingsch, W. 1983. Fire resistance of composite columns of concrete filled hollow sections. CIDECT Research Project 15C1/C2–83/27. Cologne, Germany: Comité International pour le Développement et l'Etude de la Construction Tubulaire.

Lie, T.T. & Chabot, M. 1992. Experimental studies on the fire resistance of hollow steel columns filled with plain concrete. Internal report No. 611. Ottawa, Canada: Institute for Research in Construction, National Research Council of Canada (NRCC).

Chabot, M. & Lie, T.T. 1992. Experimental studies on the fire resistance of hollow steel columns filled with bar-reinforced concrete. Internal report No. 628. Ottawa, Canada: Institute for Research in Construction, National Research Council of Canada (NRCC).

Kodur, V.K.R. & Lie, T.T. 1995. Experimental studies on the fire resistance of circular hollow steel columns filled with steel-fibre-reinforced concrete. Internal report No. 691. Ottawa, Canada: Institute for Research in Construction, National Research Council of Canada.

Han, L.H., Zhao, X.L., Yang, Y.F. & Feng, J.B. 2003. Experimental study and calculation of fire resistance of concrete-filled hollow steel columns. *Journal of Structural Engineering* (ASCE) 129(3): 346–356.

Kim, D.K., Choi. S.M., Kim, J.H., Chung, K.S. & Park, S.H. 2005. Experimental study on fire resistance of concrete-filled steel tube column under constant axial loads. *International Journal of Steel Structures* 5(4): 305–313.

Romero, M.L., Moliner, V., Espinos, A., Ibañez, C. & Hospitaler, A. 2011. Fire behavior of axially loaded slender high strength concrete-filled tubular columns. *Journal of Constructional Steel Research* 67(12): 1953–1965.

Moliner, V., Espinos, A., Romero, M.L. & Hospitaler, A. 2013. Fire behavior of eccentrically loaded slender high strength concrete-filled tubular columns. *Journal of Constructional Steel Research* 2013; 83: 137–146.

Espinos, A., Romero, M. & Hospitaler, A. 2010. Advanced model for predicting the fire response of concrete filled tubular columns. *Journal of Constructional Steel Research* 66(8–9), 1030–1046.

Impact, blast and robustness

Tubular Structures XV – Batista, Vellasco & Lima (eds)
© *2015 Taylor & Francis Group, London, ISBN 978-1-138-02837-1*

Experimental and theoretical development for pipe-in-pipe composite specimens under impact

X. Qian & Y. Wang
Department of Civil and Environmental Engineering, National University of Singapore, Singapore

ABSTRACT: This article describes an experimental investigation on the cement composite filled pipe-in-pipe composite specimens under drop weight impact loadings. The specimens consist of two steel hollow sections with the annulus in between filled with ultra-lightweight cement composite. The experimental program examines the primary contribution of each individual material layer (including the wall of the outer steel pipe, the cement composite layer and the wall of the inner steel pipe) in resisting the impact resistance of the pipe-in-pipe composite specimens. Based on the experimental findings, the subsequent study proposes a theoretical model in predicting the impact resistance for the pipe-in-pipe specimens. The theoretical model combines a previously validated load-indentation relationship of the pipe-in-pipe specimens under static loads with a dynamic vibration model and an energy approach to estimate the load and displacement histories under the lateral drop-weight impact. The theoretical model predicts closely the impact resistance of the pipe-in-pipe specimens under the impact loads.

1 INTRODUCTION

Concrete filled or cement composite filled steel structures have emerged as a common practice to enhance the structural resistance and efficiency in recent industrial practice. Such composite structures combine effectively the high compressive strength of concrete and the tensile strength and ductility of the steel materials, and lead to economical structural solutions.

Accidental events, *e.g.*, impacts caused by dropped object, ships or other conditions, create critical threats to the safety of tubular structures, in which the composite construction becomes increasingly popular. The resistance of the composite tubes against impact loading requires an improved understanding on the mechanics involved, which provide the basis for an engineering procedure to estimate the impact resistance of the composite structures.

A number of researchers have contributed significantly to the understandings on the impact resistance of pipe structures. Wierzbicki & Suh (1988) have developed a simplified ring-generator model to estimate the load-indentation resistance of the pipes under lateral loads,

$$P = \sigma_y t_o \sqrt{2\pi\delta t_o} \qquad (1)$$

where σ_y refers to the yield strength of the steel and t_0 is the wall thickness of the steel pipe. The load-indentation relationship in Eq. (1) has become a widely recognized model in engineering practice. Equation (1) provides the theoretical basis to develop the load-indentation relationship for pipe-in-pipe composite structures (Qian et al. 2013).

The last decade has observed substantial research and development on the impact response of steel and composite tubular structures. Bambach et al. (2008) have examined the response of square hollow sections (SHSs) and concrete-filled SHSs under low-velocity, high mass impact. Their experiments showed that the strengthening effect provided by the concrete becomes negligible for compact sections. Wang et al. (2013) have reported two distinctive failure mechanisms for concrete filled steel tubes under lateral impacts, the ductile failure for specimens with high constraining effect and the brittle mechanism for specimens with low constraining effect. The previous research work on the impact response for composite pipes focuses on the single-skin type of specimens with the hollow space inside the tube member completely filled with concrete material. The experimental work on double-skin composite pipes investigates mainly the axial resistance of such composite pipes under compression (Uenaka et al. 2010, Zhao et al. 2010, Li et al. 2012).

This study examines the impact resistance of the double-skin pipe-in-pipe specimens subjected to a drop-weight impact through a combined experimental and theoretical approach.

2 EXPERIMENTAL PROGRAM

2.1 *Specimens and test setup*

The experimental program (Wang et al. 2014) includes seven double-skin pipe-in-pipe specimens, as listed in Table 1. Each specimen consists of a steel outer pipe, a steel inner pipe and the ultra-lightweight cement composite layer in the annulus between the outer and inner pipe. All specimens have a fixed length of 2 m.

The outer pipe for all specimens has the same outer diameter, *i.e.*, $D_o = 219$ mm. The first three specimens in Table 1 (CCFPIP-1 to CCFPIP-3) share the same inner pipe dimensions with $D_i = 140$ mm and $t_i = 5$ mm. This set of specimens examine the effect of the wall thickness of the outer pipe in resisting the impact loading. The second set of specimens (CCFPIP-4 to CCFPIP-6) have the same variations in the wall thickness of the outer pipe, but a different wall thickness for the inner pipe (compared to the first set of specimens). As the first six specimens have the same diameter for the inner pipe, the thickness of the cement composite material (t_c) in between the two pipes remains at a similar level, ranging from 30 mm to 35 mm. The last specimen in Table 1, CCFPIP-7, has a substantially different thickness of the cement composite layer ($t_c = 20$ mm).

Figure 1a illustrates the test configuration of a typical specimen. The indenter has a semi-cylindrical shape with a radius of 30 mm. The indenter connects to a drop weight of 1350 kg, as shown in Figure 1b. The boundary conditions on the two ends of the pipe specimens employ a roller support on one end and a pin support on the other.

The experimental procedure measures the impact load through three dynamic load cells between the indenter and the drop weight. The test setup employs a laser system, which contains two laser sources, to monitor the impact velocity of the indenter, as indicated in the Figure 1a. The impact velocity equals the vertical distance between the two laser lights divided by the time for the indenter to reach from the upper laser light to the lower laser light. As the lower laser light locates at distance of Δ_{l-p} above the top surface of the pipe, the initial impact velocity, V_o, equals,

$$V_o = \sqrt{V_p^2 + 2g\Delta_{l-p}} \qquad (1)$$

where V_p refers to the velocity measured using the laser system, and g represents the gravitational acceleration. Table 1 represents the initial impact velocity, V_o, for all specimens.

The material for the steel pipes utilize the S355 steels, while the cement composite employs the ultralight cement composite (ULCC) developed by Chia et al. (2011). Figure 2a shows the uni-axial true stress versus the true strain relationship measured using the tension specimens fabricated from the steel pipes. Figure 2b illustrates the stress-strain relationship for the ULCC measured from compression tests on concrete cubes cured for 28 days. The density of the ULCC material equals 1460 kg/m³, significantly lower than the normal strength concrete.

2.2 *Experimental results*

The impact process causes severe local plastic deformations near the impact zone, as indicated in Figure 3,

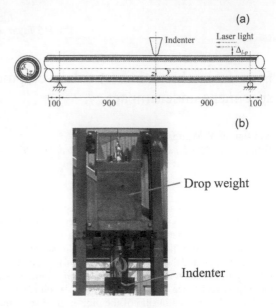

Figure 1. (a) Configuration of the double-skin specimens; and (b) the drop weight and indenter.

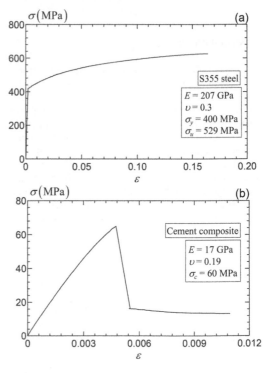

Figure 2. Materials properties: (a) uniaxial true stress-true strain curve for S355 steels; and (b) compressive stress-strain relationship for ultra-lightweight cement composite.

which shows the failure mechanism of the composite specimens after the test. Figure 3 also illustrates the cracking of the cement composite material at the bottom of the specimen, revealed after the post-test sectioning.

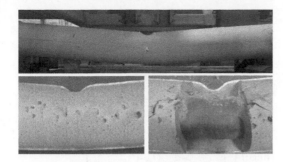

Figure 3. Typical failure mechanisms observed for the composite specimens.

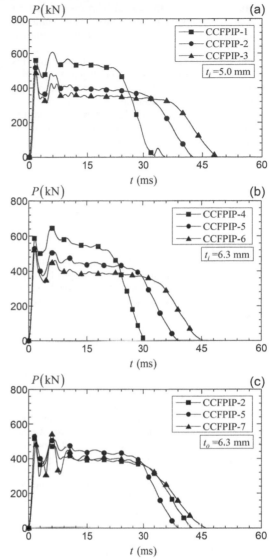

Figure 4. Impact force history for specimens: (a) with $t_i = 5$ mm; (b) with $t_i = 6.3$ mm; and (c) with $t_o = 6.3$ mm.

Figure 4 compares the impact force history for all seven specimens. The impact force increases sharply initially as the indenter strikes upon the specimen. The initial contact between the indenter and the specimen causes some vibrations in the indenter, which leads to the fluctuations observed in the impact force shown in Figure 4. After the vibration stage, the impact forces reaches a plateau value as the indenter and the pipe specimen converge towards the same speed. Once the displacement of the pipe at the indentation zone becomes larger than that of the indenter, the indenter separates from the pipe specimen. The impact force therefore decreases to zero due to the loss of contact between the indenter and the composite pipe specimen.

Figure 4a compares the impact force history for specimens with the same inner pipe but different thicknesses in the outer pipe. The wall thickness of the outer pipe imposes an apparent effect on the impact response of the specimens. The impact force increases with the wall thickness of the outer pipe. A larger impact force leads subsequently to a shorter impact duration for specimens subjected to the same initial impact momentum.

Figure 4b observes the similar impact force history for specimens with a slightly thicker inner pipe ($t_i = 6.3$ mm) than the specimens in Figure 4a. The impact forces in Figure 4b therefore show marginally larger magnitudes than those in Figure 4a.

In Figure 4c, the specimen CCFPIP-2 and CCFPIP-5 share the same outer pipe (see Table 1), and the inner pipe with a different wall thickness. The impact force history exhibits minor differences between the two specimens. Figure 4c also compares the impact response between specimens CCFPIP-5 and CCFPIP-7, which share the same outer pipe, the same thickness of the inner pipe, but a different thickness of the cement composite layer. The thickness of the cement layer does not impose a noticeable effect on the impact resistance of the composite pipe specimens.

The comparison in Figure 4 examines the effect of three material layers in the impact force resistance, *i.e.*, the outer pipe thickness, the inner pipe thickness and the cement composite thickness. The results in Figure 4 confirm that the outer pipe thickness contributes most significantly to the impact resistance, among the three material layers. The failure mechanism in Figure 3 also affirms that the outer pipe experiences significantly higher plastic deformations than the inner pipe in the indentation zone.

3 A THEORETICAL MODEL

This study aims to develop a theoretical model to predict the impact history for cement composite filled pipe-in-pipe specimens. The theoretical approach assumes a plastic hinge mechanism of the composite pipes, as shown in Figure 5.

The total displacement of the specimen at the indentation location, w_t, follows,

$$W_t = W_g + \delta \tag{2}$$

where W_g refers to the global displacement at the bottom of the pipe and δ represents the local indentation. Equation (2) separates the displacement solution of the composite pipe specimens into two parts, the global displacement and the local indentation. This allows the derivation of the global displacement, W_g, from the beam theory, and the local indentation from a load-indentation relationship.

The theoretical approach further assumes that the indenter and the pipe remain in contact after the first strike. The total displacement of the pipe, W_t, therefore, observes the conservation of momentum of the drop weight,

$$w_t(t)=V_o t-\frac{1}{m_d}\int_0^t P(\tau)(t-\tau)d\tau \qquad (3)$$

where m_d denotes the mass of the drop weight, or 1350 kg in this study.

The global displacement W_g in Eq. (2) decomposes into an elastic component, W_{ge} and a plastic component, W_{gp}. The elastic global displacement derives from Timoshenko's beam theory (Lee 1940, Goldsmith 1960),

$$w_{ge}(t)=\frac{2}{m_p}\sum_{i=1,3...}^{\infty}\frac{1}{\omega_i}\int_0^t P(\tau)\sin[\omega_i(t-\tau)]d\tau \qquad (4)$$

where m_p refers to the mass of the pipe, as listed in Table 1, and ω_i defines the ith angular frequency of the natural vibration for the beam.

The theoretical approach calculates the plastic global displacement, W_{gp}, through an energy approach. The kinetic energy transferred from the drop weight to the pipe (E_d) dissipates through three different energies, the global elastic bending of the beam (E_{ge}), the local indentation (E_δ) and the energy absorbed in the plastic hinges of the beam. The kinetic energy equals the change in the kinetic energy of the indenter before and after the impact. The elastic bending energy of the beam derives from the integration of the moment-rotation relationship along the beam. The energy absorbed in the plastic hinge equals,

$$E_{gp}=2M_p\theta=\frac{4M_p}{L_o}w_{g,p} \qquad (5)$$

Alternatively,

$$w_{gp}=\frac{\left(E_d-E_\delta-E_{ge}\right)L_o}{4M_p} \qquad (6)$$

The evaluation of the local indentation energy requires a load-indentation relationship, in order to compute the impact force history and the displacement history. This study employs the load-indentation relationship developed by Qian et al. (2013) for cement composite filled double skin pipes. The indentation resistance includes three contributions, the outer pipe, the inner pipe and the cement composite layer,

$$P=P_o+P_i+P_c \qquad (7)$$

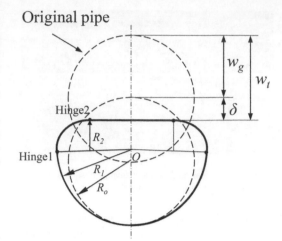

Original pipe

Figure 5. Assumed mechanisms in the cross section of the pipe specimen.

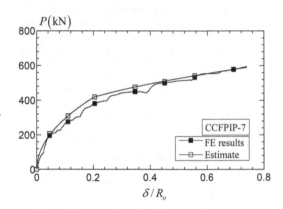

Figure 6. Validation of the load-indentation relationship for the cement composite filled pipe-in-pipe specimens.

The indentation resistance of the steel pipes (both inner and outer pipes) depends on,

$$P_i=\frac{(C_1-C_2 n)}{4}\sigma_y t_i^2\sqrt{\frac{2\pi\delta}{R_i}} \qquad (8a)$$

$$P_o=\frac{(C_1-C_2 n)}{4}\sigma_y t_o^2\sqrt{\frac{2\pi\delta}{R_o}} \qquad (8b)$$

where C_1 and C_2 are coefficients derived from the numerical analysis, t denotes the thickness of the inner or outer pipe, and R refers to the radius of the inner or outer pipe. The contribution of the cement composite to the indentation resistance derives from an effective loading area on the cement composite (Qian et al. 2013).

$$P_c=8f_c t_c\sin\beta(r+t_o+t_c) \qquad (9)$$

where f_c defines the compressive strength of the cement composite, t_c refers to the thickness of the

208

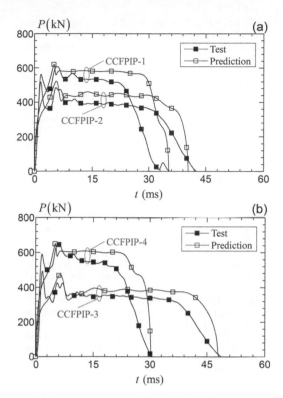

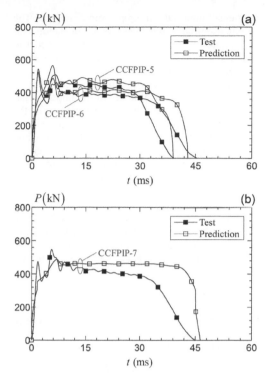

Figure 7. Comparison of the predicted impact load history with the experimental measurement: (a) for CCFPIP-1 and CCFPIP-2; and (b) for CCFPIP-3 and CCFPIP-4.

Figure 8. (a) Comparison of the predicted impact load history with the experimental measurement: (a) for CCFPIP-5 and CCFPIP-6; and (b) for CCFPIP-7.

cement composite layer, r denotes the radius of the indenter, and β represents the disperse angle of the concrete material ($\tan \beta = 0.5$ in Eurocode 2).

Figure 6 compares the load-indentation resistance measured from a static load-indentation test with the estimation using the theoretical approach described above.

The strain rate effect for the low velocity impact with the initial velocity less than 10 m/s remains insignificant as demonstrated by Firouzsalari & Showkati (2013). This study therefore does not consider the strain rate effect in the theoretical approach.

Table 2. Comparison between the impact test results and theoretical predictions.

	Prediction/Test	
Specimen	P_{max}	w_{max}
CCFPIP-1	1.02	1.11
CCFPIP-2	1.01	1.09
CCFPIP-3	0.96	1.12
CCFPIP-4	1.01	1.04
CCFPIP-5	1.04	1.15
CCFPIP-6	0.93	1.11
CCFPIP-7	0.90	1.01
Mean	0.98	1.09
CoV	0.053	0.045

4 VALIDATION OF THE THEORETICAL APPROACH

Figures 7 and 8 compare the impact load history predicted using the theoretical model described above with the experimentally measured response for all seven cement composite filled pipe-in-pipe specimens, as also listed in Table 2. The proposed theoretical approach predicts closely the maximum impact load resistance for all specimens, with a slight over-estimation of the post-peak plateau load resistance. For all specimens, the theoretical approach predicts closely the impact duration, as indicated in Figs. 7 and 8.

The comparison in Figs. 7 and 8 implies that the theoretical approach over-estimates slightly the total

impact momentum absorbed by the pipe specimens, as the theoretical model assumes zero energy loss during the impact. This over-estimation in the impact momentum leads consequently to a slight over-estimation on the maximum global displacement, w_{max}, which denotes the maximum value of w_g. Table 2 confirms the over-estimation in the global displacement.

5 CONCLUSIONS

This paper highlights the experimental study on the response of cement composite filled pipe-in-pipe specimens under lateral drop weight impact. The theoretical approach integrates a load-indentation relationship into a classical Timoshenko beam model to estimate the global response of the composite pipes specimens under impact loads.

The experimental study examines the contribution of three different material layers to the impact resistance of the composite pipes. The variation in the thickness of the outer pipe specimen introduces significant effects in the impact response of the pipes, while the changes in the inner pipe thickness and the cement composite layer thickness does not cause noticeable changes in the impact force history.

The proposed theoretical model predicts closely the maximum impact resistance of the pipes, with a slight over-estimation on the maximum global displacement experienced by the composite pipe specimens. Wang et al. (2015) have extended a similar approach to estimate the impact response of fully grouted pipe structures.

REFERENCES

Bambach, M.R., Jama, H., Zhao, X.L. & Grzebieta, R.H. 2008. Hollow and concrete filled steel hollow sections under transverse impact loads. *Eng. Struct.*, 30(10), 2859–2870.

Chia, K.S., Zhang, M.H. & Liew, J.Y.R. 2011. High-strength ultra lightweight cement composite material properties. In: *Proc. 9th Int Symp High Performance Concrete – Design, Verification & Utilization*. Roturua, New Zealand, 9–11 Aug, 2011, Primary Section A8-paper 2.

Eurocode 2. 2004. Design of concrete structures – Part 1-1: General rules and rules for buildings. CEN.

Firouzsalari, S.E. & Showkati, H. 2013. Thorough investigation of continuously supported pipelines under combined pre-compression and denting loads. *Int. J. Press. Vessels Pip.*, 104, 83–95.

Goldsmith, W. 1960. Impact, the theory and physical behavior of colliding solids. London: Edward Arnold Publishers.

Lee, E.H. 1940. The impact of a mass striking a beam. *J. Appl. Mech.*, 7: A129-38.

Li, W., Han, L.H. & Zhao, X.L. 2012. Axial strength of concrete-filled double skin steel tubular (CFDST) columns with preload on steel tubes. *Thin-Walled Struct.*, 56, 9–20.

Qian, X., Wang, Y., Liew, R.J.Y. & Zhang, M.-H. 2013. A load-indentation formulation for cement filled pipe-in-pipe composite structures. *Eng. Struct.*, under review.

Wang, Y., Qian, X., Liew, R.J.Y. & Zhang, M.-H. 2014. Experimental behavior of cement filled pipe-in-pipe composite structures under transverse impact. *Int. J. Impact Eng.*, 72, 1–16.

Wang, Y., Qian, X., Liew, R.J.Y. & Zhang, M.-H. 2015. Impact of cement composite filled steel tubes: an experimental, numerical and theoretical treatise. *Thin-Walled Struct,* 87, 76–88.

Uenaka, K., Kitoh, H. & Sonoda, K. 2010. Concrete filled double skin circular stub columns under compression. *Thin-Walled Struct.*, 48(1), 19–24.

Wang, R., Han, L.-H. & Hou, C.-C. 2013. Behavior of concrete filled steel tubular (CFST) members under lateral impact: experiment and FEA model. *J. Constr. Steel Res.*, 80, 188–201.

Wierzbicki, T. & Suh, M.S. 1988. Indentation of tubes under combined loading. *Int. J. Mech. Sci.* 30(3–4), 229–48.

Zhao, X.L., Tong, L.W. & Wang, X.Y. 2010. CFDST stub columns subjected to large deformation axial loading. *Eng Struct.*, 32(3), 692–703.

Tubular Structures XV – Batista, Vellasco & Lima (eds)
© 2015 Taylor & Francis Group, London, ISBN 978-1-138-02837-1

Field blast testing and FE modelling of RHS members

C. Ritchie & J.A. Packer
Department of Civil Engineering, University of Toronto, Toronto, Canada

M. Seica
Explora Security Ltd., London, UK
Department of Civil Engineering, University of Toronto, Toronto, Canada

X.L. Zhao
Department of Civil Engineering, Monash University, Victoria, Australia

ABSTRACT: Full-scale, blast arena testing has been performed on cold-formed rectangular hollow section (RHS) members in flexure. RHS members with width-to-thickness ratios of 15 and 24 were tested at two scaled distances. These tests were heavily instrumented to gather detailed information on the behaviour of both the RHS members and the air blast wave profile. Additionally, material testing was performed to confirm the properties of the steel. Data from the blast arena testing is used to validate predictive and numerical models. One modelling method in blast-resistant design is explicit finite element (FE) analysis. ANSYS LS-DYNA has been used to numerically model the RHS members and to subject them to blast loading profiles similar to those from the arena tests. The LS-DYNA models incorporate measured test parameters and include strain-rate-dependent properties for the steel. A comparison between data obtained from the field blast tests and the model is then presented herein.

1 INTRODUCTION

The ability to resist blast loading, whether accidental or malicious, is increasingly becoming a requirement for structures around the world. Key to achieving performance in blast-resistant design are the energy-dissipating characteristics of the chosen structural system. Tubular steel members have been cited as ideal elements for this application due to their geometry and ability to withstand large plastic deformations while maintaining stability (Astaneh-Asl 2010). In contrast, under blast loads concrete may experience brittle failure, including spalling and scabbing, leading to an increased hazard level. Existing rectangular hollow section (RHS) research has focused on small, thin-walled RHS under impact loads (Bambach et al. 2008, Remennikov et al. 2011, Hou et al. 2011) and larger RHS under contact and near-field blast loads (Remennikov & Uy 2013, Karagiozova et al. 2013). The response of full-scale RHS under far-field air blast loads is different. Accordingly, research is being performed to advance knowledge regarding the performance and design of full-scale tubular steel sections under blast loading. Two main aspects of this research are full-scale field blast testing and numerical finite element (FE) analysis.

2 BLAST ARENA TEST SERIES

Through a partnership between the Explora Foundation and the University of Toronto "Centre for Resilience of Critical Infrastructure" blast arena tests on RHS members were conducted in July 2012 and June 2013. Walker et al. have discussed the initial results and observations from these tests (2012, 2013). Ritchie et al. (2014) followed up with a Single degree of freedom (SDOF) analysis of the 2013 test series results using the Single-degree-of-freedom Blast Effects Design Spreadsheet (SBEDS). (USACE 2005). The data obtained from the 2013 test series is conducive to a detailed FE analysis, the details of which are covered in subsequent sections.

2.1 *Test site*

Two blast arena explosive tests were performed on a target containing RHS members. For the first test, the RHS targets were placed at a scaled distance of $2.7 \, \text{m/kg}^{1/3}$ and for the second test they were placed at a scaled distance of $1.9 \, \text{m/kg}^{1/3}$.

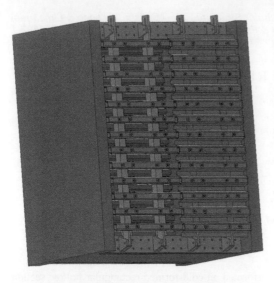

Figure 1. Computer rendering of RHS target, showing cladding removed on the left side.

Figure 2. RHS targets before test (2013 Test 2).

2.2 Rectangular hollow section targets

Two pairs of RHS120 × 120 × 5 and RHS120 × 120 × 8 members, manufactured to EN 10219 (CEN 2006) Grade S355J2H, were tested in each firing. The members thus had nominal width-to-thickness ratios of 15 and 24. The effective span of the simply-supported RHS members within the target was 3.26 m. Corrugated steel decking, common for each pair of members and strengthened with smaller RHS members, was attached to each pair of RHS to transfer the airblast loading. Four RHS "beams", with pin and slotted-end connections, were support by a concrete box reaction structure.

Figure 1 shows a computer rendering of the RHS target. One section of the steel decking is removed to illustrate the RHS members behind. The RHS target, consisting of two pairs of duplicate beams, was tested side-by-side with a similar target containing two pairs of concrete-filled RHS members, the results of which are not included in this paper. Large concrete inverted T-barriers and cubes were placed around the RHS targets to limit clearing effects (Fig. 2).

Figure 3. Instrumentation inside the RHS target.

2.3 Instrumentation

Four parameters were measured during each test by an ultra high-speed data acquisition system: free-field pressure, reflected pressure, displacement, and strain. The free-field pressure gauges were mounted on 1.5 m tall stands and set at the same standoff distance as the RHS targets. Five reflected pressure gauges were used, one on each of the outer edge of the concrete reaction structure at mid-height and three spaced vertically between the targets (Fig. 2). One displacement gauge and two strain gauges were mounted at the midspan of each RHS member on the rear "flange" (Fig. 3). Additional manual displacement measurements were taken after the tests to confirm the final shape of the RHS members.

2.3.1 Data processing

The methods used to process the raw data obtained from the instrumentation are described in detail by Ritchie et al. (2014). These methods were used to reduce the inherent "noise" in the data while accurately capturing the critical values, such as peak pressures and maximum displacements, obtained from the data.

2.4 Preliminary observations

The testing in 2012 did not produce significant plastic deformations in the RHS members (Walker et al. 2012). Therefore, the initial scaled distance in 2013 was chosen to ensure plastic deformation was developed. Indeed, after the first test there was visible plastic deformation in both sizes of RHS members. Due to the test configuration, a smaller scaled distance was chosen for the second test to facilitate larger plastic deformations in the concrete-filled RHS members. This led to significant plastic deformations in both unfilled RHS sizes (Fig. 4), with the smaller of the

Figure 4. Deformed RHS members after test (2013 Test 2).

Table 1. Average measured yield strengths.

Specimen	Flat yield strength MPa	Corner yield strength MPa
RHS120 × 120 × 5	427	563
RHS120 × 120 × 8	420	521

two reaching the maximum displacement allowed for in the targets.

At the conclusion of the test series, the steel cladding was removed to facilitate further observations. In both tests, the thinner RHS members underwent clear local buckling at midspan.

2.5 Steel material testing

Subsequent investigations were carried out at the University of Toronto to determine the mechanical and geometric properties of the two RHS sizes used. Five tensile coupons, three from the flats and two from the corners, were tested to ASTM A370 (ASTM 2009), to determine the material properties. Table 1 presents the average measured yield strengths for the two sizes.

3 EXPLICIT FINITE ELEMENT ANALYSIS

Explicit FE analysis is a common numerical method used for blast-resistant design and analysis. The high cost of field blast arena tests often makes it difficult to conduct a large number of field experiments.

Therefore, FE analyses are necessary to expand of the test results. The field test results are used to validate the numerical models, which can then be used for subsequent parametric analysis.

Numerical analysis of tubular steel elements subjected to blast and impact loading is a fairly recent research endeavour. The majority of the research in this field has been completed using one of two proprietary explicit finite element codes: ABAQUS/Explicit (Simulia 2014) and LS-DYNA (LSTC 2014). The existing research in this area has focused on RHS and circular hollow sections (CHS), unfilled and concrete-filled, subject to close-in blast and impact loading.

3.1 Previous FE investigations

In the last few years a number of FE investigations that share some similarities to the present research were undertaken. Zeinoddini el al. (2008) studied axially pre-loaded CHS members subjected to lateral impacts. Zhao et al. (2009) ran simulations on concrete-filled RHS subjected to airblast loading. Jama et al. (2009) used explicit finite analysis to evaluate the global and local deformation in a RHS member subjected to a surface charge loading. Bambach (2011) modelled unfilled and concrete-filled, stainless steel RHS members subjected to lateral impacts. Walker et al. (2011) examined lateral impact loading of cold-formed RHS members. Wang et al. (2013) conducted lateral impact tests on concrete-filled CHS members. Yousuf et al. (2013, 2014) subjected lateral, static and impact, loads to unfilled and concrete-filled stainless steel RHS members. Remennikov & Uy (2013) modelled unfilled and concrete-filled RHS members subjected to surface and near-field blast loading. Han et al. (2014) ran lateral impact loading models using drop hammer tests on concrete-filled CHS. Each of the models presented in the aforementioned papers used slightly different methods. These papers were used to inform the development of the numerical model of RHS members used in this study. The key parameters from these existing simulations performed on unfilled RHS members are summarized in Table 2.

3.2 Computer software

This investigation used ANSYS LS-DYNA to develop the numerical model (ANSYS 2013). ANSYS is a powerful pre-processor that, when paired with the LS-DYNA module, can output LS-DYNA keyword files for analysis. ANSYS has several built in LS-DYNA keywords that can be utilized for modelling. It also allows "keyword snippets" that support many other LS-DYNA keywords, providing modelling flexibility.

3.3 Numerical model

Three-dimensional CAD models of the RHS members, their supports and cladding, were adapted from the computer manufacturing model (Fig. 1). This model was reduced to a single RHS member, with the steel

Table 2. Key parameters from selected explicit finite element research on unfilled RHS.

Resource	Loading type	Maximum RHS depth mm	FEA code	RHS Element type	Strain rate dependent material model
Remennikov & Uy (2013)	near-field airblast	100	LS-DYNA	shell	N/A
Yousuf et al. (2013, 2014)	lateral impact	100	ABAQUS/Explicit	solid	Constant dynamic increase factors
Walker et al. (2011)	lateral impact	102	ANSYS Explicit	shell	Johnson-Cook
Bambach (2011)	lateral impact	50	ABAQUS/Explicit	shell	Cowper-Symonds
Remennikov et al. (2011)	lateral impact	100	LS-DYNA	shell	N/A
Jama et al. (2009)	surface blast	50	LS-DYNA	shell	Cowper-Symonds

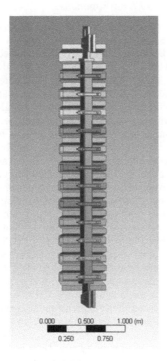

Figure 5. ANSYS DesignModeler model.

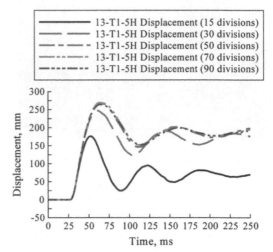

Figure 6. Mesh sensitivity analysis.

decking and steel connection details (Fig. 5). The complete model was then imported into ANSYS where it could be further prepared for meshing using the built-in DesignModeler module. A line of symmetry was produced in ANSYS to reduce the model to a single RHS member. The remaining model consisted of the main RHS member, the steel decking, the smaller RHS members (supporting the cladding in the transverse direction), the welded RHS connection details, a portion of the concrete box connections, and the bolts (Fig. 5).

3.4 *Element properties*

With the exception of the study by Yousuf et al. (2013, 2014), all previous FE studies presented in Table 2 utilized shell elements for the RHS members. Shell elements are well suited to flexural models of full-scale RHS members. Four-noded Belytschko-Tsay shell elements with five integration points through the thickness (LSTC 2014) were used in the study herein. Another 3D analysis was completed using eight-noded (brick) elements and the results confirmed that the less computationally expensive shell elements were suitable for this problem.

A sensitivity analysis was run to determine the appropriate mesh size for the RHS members. A non-uniform mesh with finer regions at the top and bottom connections and midspan was chosen. The RHS were divided into quarters lengthwise and the "bias" options were used to control the mesh size. Mesh densities, denoted by the number of divisions per quarter, are compared in Figure 6 and Table 3.

Based on the mesh sensitivity analysis, the optimal number of divisions per quarter was 50. This results in shell elements ranging from 8 to 22 mm in length.

The steel decking and smaller RHS members were also modelled using Belytschko-Tsay shell elements. The bolts, RHS connection, and the connections at the reaction points were all modelled using eight-noded solid elements with a single integration point.

Table 3. Mesh sensitivity analysis.

Number of divisions	Maximum displacement mm	Difference from previous %
15	176	N/A
30	248	+40.9
50	268	+8.06
70	269	+0.37
90	265	−1.49

Table 4. Cowper-Symonds parameters for mild steel.

Resource	$C\ s^{-1}$	p	Strain rate range s^{-1}
Cowper & Symonds (1957)	40.4	5	9.5×10^{-7} − 300
Abramowicz & Jones (1986)	6844	3.91	40–195
Yu & Jones (1991)	1.05×10^7	8.30	0.0012-140
Marias et al. (2004)	844	2.207	0.001-918
Jama et al. (2009)	844	2.207	82-403
Sun & Packer (2014)	3427	9.59	100-1000
	70926	12.43	100-1000

3.5 Material properties

For numerical modelling, the main RHS member was partitioned into flat and corner areas and the appropriate material properties were used for the respective parts of the RHS. This same method was employed by Walker et al. (2011) to account for the cold-forming of the RHS members. The RHS members were modelled using the Johnson-Cook material model (Johnson & Cook 1985) built into LS-DYNA. The Johnson-Cook material model includes terms that account for strain hardening, strain rate effects, and thermal effects. LS-DYNA provides the option to use the default Johnson-Cook logarithmic strain rate term, or use the Cowper-Symonds expression (Cowper & Symonds 1957). For this investigation the Cowper-Symonds expression was chosen for strain rate effects. The complete expression for the dynamic stress (σ) is shown in Equation 1.

$$\sigma = \left(\sigma_y + B\varepsilon_p{}^n\right)\left(1 + \left(\tfrac{\dot{\varepsilon}_p}{c}\right)^{1/p}\right)\left(1 - \left(\tfrac{T-T_r}{T_m-T_r}\right)^m\right) \quad (1)$$

where σ_y is the yield stress, B is a strain-hardening parameter, n is the strain-hardening index, $\dot{\varepsilon}_p$ is the plastic strain rate, C and p are Cowper-Symonds strain rate parameters, T is the temperature, T_r is the reference temperature, T_m is the melting point, and m is a Johnson-Cook parameter.

A review of Cowper-Symonds parameters used for mild steel is shown in Table 4 and Figure 7. A key consideration for these parameters is the strain rate for which they have been verified.

During the blast tests the strain rates measured were approximately $2\ s^{-1}$. This falls well below the range

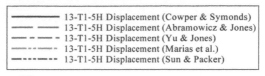

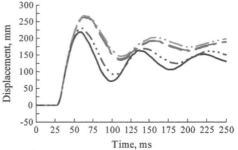

Figure 7. Cowper-Symonds parameter sensitivity analysis.

of some of the tests listed in Table 4. The majority of the tests that extend to this level are grouped at the larger maximum displacement Figure 7. Therefore, the Cowper-Symonds parameters proposed by Marais et al. (2004), which have been used for cold-formed RHS, were used for this investigation for both the flat and corner elements. Typical strain hardening parameters were found using material tests. Typical thermal properties for mild steel were used for all elements (Jama et al. 2009).

The steel decking was modelled using a tri-linear stress-strain curve to failure. The top and bottom RHS connections were modelled as elastic steel material. The remaining parts: the smaller RHS, the concrete box connection details, and the bolts were modelled as rigid materials as they sustained negligible deformations compared to the other components. An analysis was run (with elastic properties) to confirm that making these parts rigid had very little effect on the displacement-time history of the large RHS test member.

3.6 Contact properties

Three contact properties were used in the LS-DYNA model. A single surface contact with friction engaged was used for the majority of the contact surfaces between parts. The exceptions to this were the welded joint between the RHS and RHS T-connection, and the sliding contact between the RHS T-connection and the M24 bolt. The welded RHS detail was modelled using a tied contact. The sliding connection between the RHS T-connection and the bolt was modelled using a surface-to-surface contact with friction engaged. The friction coefficients used for all contacts were 0.74 for dynamic and 0.57 for static. These values are typical of mild steel-to-mild steel contact and obtained from the LS-DYNA Theory Manual (LSTC 2014).

3.7 Boundary conditions

Boundary conditions were imposed on the concrete box connection support details (fixed) as well at

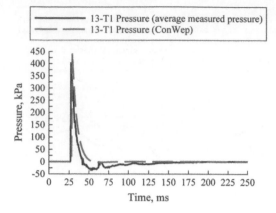

Figure 8. Typical reflected pressure-time histories (2013 Test 1).

Table 5. Global maximum displacements of select RHS members.

RHS member	Test mm	LS-DYNA mm	LS-DYNA vs. Test % difference
13-T1-5H-1	252	268	+6.3
13-T1-5H-2	259	268	+3.5
13-T1-8H-1	N/A	159	N/A
13-T1-8H-2	171*	159	−7.0
13-T2-8H-1	289*	286	−1.0
13-T2-8H-2	283*	286	+1.0

Note: *indicates data obtained from manual measurements.

Table 6. Global final displacements of select RHS members.

RHS member	Test mm	LS-DYNA mm	LS-DYNA vs. Test % difference
13-T1-5H-1	182	188	+3.3
13-T1-5H-2	177	188	+6.2
13-T1-8H-1	57	63	+10.5
13-T1-8H-2	66	63	−4.5
13-T2-8H-1	155	198	+31.0
13-T2-8H-2	139	198	+46.0

the line of symmetry to represent the reaction and connectivity conditions.

3.8 Load properties

LS-DYNA has several built in options for simulating blast loads on a structure. This includes the simple ConWep (Hyde 1988) loading, based on charge weight and standoff, and more complex fluid structure inter-action modelling. Additionally, given the availability of the test data, the measured pressures from the blast tests can be applied directly onto the model. Two load cases were considered for this study, the measured pressure-time history and the ConWep loading. The measured pressured-time histories were applied to the "flat" sections of the decking perpendicular to the charge location. A scaled version of the measured pressure (P) was applied to the "angled" sections of the decking to account for the angle of incidence (θ) of the corrugated steel decking. This was done using the same formula as ConWep (Eq. 2), which incorporates the reflected (P_r) and free-field (P_{so}) pressures. Using the peak pressure, the resulting angled pressure-time histories (P) were $0.516P_r$ for test 1 and $0.384P_r$ for test 2.

$$P = P_r\cos^2\theta + P_{so}(1 + \cos\theta - 2\cos^2\theta) \qquad (2)$$

Figure 8 gives a comparison between the measured test pressure-time and the ConWep pressure-time history of a typical element at target midheight.

For test 1 the measured pressure values were lower and resulted in a smaller impulse then the equivalent ConWep values. Test 2 showed a much better correlation between the two. For both tests, the ConWep loading does not accurately represent the negative phase of the pressure-time history. The subsequent LS-DYNA results presented use the average pressure measured by the pressure gauges for a more accurate representation.

4 RESULTS

4.1 Displacement-time histories

Typical results for the displacements are presented in Tables 5–6 and Figure 9.

The LS-DYNA models accurately replicate the maximum displacement of the RHS specimen. The final displacements produced through the LS-DYNA analysis are acceptable for the test 1 specimens but significantly larger for the test 2 specimens shown. A more accurate determination of the complete pressure profile might improve the LS-DYNA analysis under high pressure cases. A more rigorous fluid-structure interaction analysis is required to determine the true nature of the applied reflected pressure. Improving this parameter for specimen 13-T2-8H should produce more accurate results. Figure 9 illustrates that, nonetheless, the predicted displacement-time histories presented herein (based on measured pressure) are still far more accurate than displacement predictions using the ConWep loading.

Unlike a SDOF analysis, a FE analysis is also capable of capturing local effects of the RHS members. Figure 10 illustrates the typical local deformation (local buckling) seen in the thinner (5H) RHS members. The LS-DYNA model captures a similar local buckling, but it is slightly less pronounced. The LS-DYNA models also correctly determine the lack of local effects for the thicker RHS members (no local buckling).

The LS-DYNA models do a good job of capturing the stiffness of the dynamic system. Table 7 presents

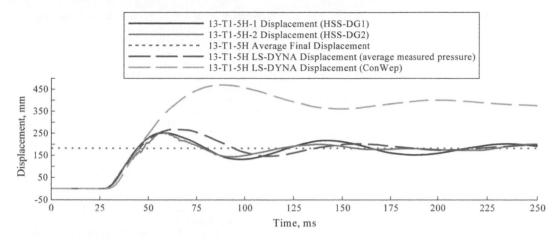

Figure 9. Typical displacement-time history (2013 Test 1) – measured and predicted by FE analysis.

(a)   (b)

Figure 10. Typical deformed RHS member (a) and corresponding LS-DYNA model (b).

Table 7. Natural periods of selected RHS members.

RHS member	Test ms	LS-DYNA ms	SDOF ms	LS-DYNA vs. Test % difference
13-T1-5H-1	100.1	91.0	64.8	−9.1
13-T1-5H-2	76.5	91.0	64.8	+19.0
13-T1-8H-2	78.7	67.0	56.1	−14.9

the test and LS-DYNA periods, as well as SDOF periods found during a previous analysis (Ritchie et al. 2014), for select RHS members.

The stiffnesses found using LS-DYNA are improved over those found using SDOF analysis, which averaged an approximate 20% difference. Based on the uncertainty in determining accurate test periods (as seen for the 13-T1-5H specimen), the LS-DYNA periods are deemed acceptable.

5 CONCLUSION

The discussion presented herein illustrates the methodology used for an explicit finite element model that can be used to predict and analyse the results of full-scale blast tests on RHS members. The LS-DYNA predicted displacement-time histories correlate reasonably well with the test data, when measured reflected pressures are used for the loading. Using the measured pressure test data as a loading rather then the ConWep charge weight and standoff loading provided a far more accurate analysis. While the ConWep loading may be a suitable tool for design, it is not recommended for analysis. However, there is a difficulty in applying a measured pressure-time relationship to a surface such as the corrugated steel decking used for the RHS targets herein. FE models can also capture more details that are important to blast-resistant design, including local deformations and strain-time behaviour. Refinements to aspects of the numerical model are being investigated to further improve the correlation with test results.

ACKNOWLEDGEMENTS

The authors are grateful for the substantial financial aid and in-kind support of the Explora Foundation to the University of Toronto "Centre for Resilience of Critical Infrastructure". Financial support has also been received from the Natural Sciences and Engineering Research Council of Canada (NSERC), the Steel Structures Education Foundation (SSEF), the Thornton Tomasetti Foundation, and Australian Research Council (ARC) Discovery Grant DP130100181. The technical advice and assistance provided by Professor David Yankelevsky, Mr. Martin Walker, and Mr. Alex Eytan are also highly appreciated.

REFERENCES

Abramowicz, W. & Jones, N. 1986. Dynamic progressive buckling of circular and square tubes. *International Journal of Impact Engineering*, 4(4): 243–270.

ANSYS. 2013. ANSYS LS-DYNA user's guide. Canonsburg, PA, USA.

Astaneh-Asl, A. 2010. Notes of blast resistance of steel and composite building structures. Steel TIPS Report. Structural Steel Educational Council. Moraga, CA, USA.

ASTM International. 2009. ASTM A370-09: Standard test methods and definitions for mechanical testing of steel products. West Conshohocken, PA, USA.

Bambach, M.R. 2011. Design of hollow and concrete filled steel and stainless steel tubular columns for transverse impact loads. *Thin-Walled Structures*. 49(10): 1251–1260.

Bambach, M.R., Jama, H.H., Zhao, X.L. & Grzebieta, R.H. 2008. Hollow and concrete filled steel hollow sections under transverse impact loads. *Engineering Structures*. 30(10): 2859–2870.

Cowper, G.R. & Symonds, P.R. 1957. Strain hardening and strain-rate effects in the impact loading of cantilever beams. Brown University, Division of Applied Mathematics.

European Committee for Standardization (CEN). 2006. EN 10219-1: Cold formed welded structural hollow sections of non-alloy and fine grain steels – Part 1: Technical delivery conditions. Brussels, Belgium.

Han, L.-H., Hou, C.-C., Zhao, X.-L. & Rasmussen, K.J.R. 2014. Behaviour of high-strength concrete filled steel tubes under transverse impact loading. *Journal of Constructional Steel Research*. 92: 25–39.

Hou, C., Han, L. & Tao, Z. 2011. Simulation of concrete-filled steel tubular members under transverse impact. Proceedings of the 2011 World Congress on Advances in Structural Engineering and Mechanics (ASEM'11). Seoul, Korea, 18–23 September 2011.

Hyde, D. 1988. Microcomputer programs CONWEP and FUNPRO, applications of TM 5-855-1, 'Fundamentals of Protective Design for Conventional Weapons' (user's guide). U.S Army Engineer Waterways Experiment Station. Vicksburg, Mississippi, USA.

Jama, H.H., Bambach, M.R., Nurick, G.N., Grzebieta, R.H. & Zhao, X.-L. 2009. Numerical modelling of square tubular steel beams subjected to transverse blast loads. *Thin-Walled Structures*, 47(12): 1523–1534.

Johnson, G.R. & Cook, W.H. 1985. Fracture characteristics of three metals subjected to various strains, strain rates, temperatures and pressures. *Engineering Fracture Mechanics*. 21(1): 31–48.

Karagiozova, D., Yu, T.X. & Lu, G. 2013. Transverse blast loading of hollow beams with square cross-sections. *Thin-Walled Structures*, 62: 169–178.

Livermore Software Technology Corporation (LSTC). 2014. LS-DYNA theory manual. Livermore, California, USA.

Marais, S.T., Tait, R.B., Cloete, T.J. & Nurick, G.N. 2004. Material testing at high strain rate using the split-Hopkinson pressure bar. *Latin American Journal of Solids and Structures*. 1(1): 319–39.

Remennikov, A, Kong, S.Y. & Uy, B. 2011. Response of foam- and concrete-filled square steel tubes under low-velocity impact loading. *ASCE Journal of Performance of Constructed Facilities*. 25(5): 373–381.

Remennikov, A. & Uy, B. 2013. Simplified modeling of hollow and concrete-filled tubular steel columns for near-field detonations. Proceedings of the 15th International Symposium on Interaction of the Effects of Munitions with Structures (ISIEMS 15). Potsdam, Germany.

Ritchie, C., Packer, J.A., Seica, M. & Yankelevsky, D. 2014. Field Blast Testing and SDOF Analysis of Unfilled and Concrete-Filled RHS Members. Proceedings of the 23nd International Symposium on Military Aspects of Blast and Shock (MABS 23). Oxford, United Kingdom, 7–12 September 2014.

Simulia. 2014. Abaqus theory manual. Providence, RI, USA.

Sun, M. & Packer, J.A. 2014. High strain rate behaviour of cold-formed rectangular hollow sections. *Engineering Structures*. 62–63: 181–192.

U.S. Army Corps of Engineers (USACE). 2005. PDC-TR 05-01: Single-degree-of-freedom blast effects design spreadsheets (SBEDS).

Walker, M., Ritchie, C., Spiller, K., Seica, M.V., Packer, J.A. & Eytan, A. 2013. Challenges and outcome of full-scale blast experimentation on structural steel members and glass elements. Proceedings of the 15th International Symposium on Interaction of the Effects of Munitions with Structures (ISIEMS 15). Potsdam, Germany, 16–20 September 2013.

Walker, M., Seica, M.V., Eytan, A. & Packer, J.A. 2012. Standards and strategies for blast testing of structures and devices. Proceedings of the 22nd International Symposium on Military Aspects of Blast and Shock (MABS 22). Bourges, France, 4–9 November 2012.

Walker, M., Seica, M. & Packer, J. 2011. Behaviour of square hollow section steel members under transverse high-impact loading. Proceedings of the 14th International Symposium on Interaction of the Effects of Munitions with Structures (ISIEMS 14). Seattle, Washington, USA, 19–23 September 2011.

Wang, R., Han, L.-H. & Hou, C.-C. 2013. Behavior of concrete filled steel tubular (CFST) members under lateral impact: Experiment and FEA model. *Journal of Constructional Steel Research*. 80: 188–201.

Yousuf, M., Uy, B., Tao, Z., Remennikov, A. & Liew, J.Y.R. 2013. Transverse impact resistance of hollow and concrete filled stainless steel columns. *Journal of Constructional Steel Research*. 82: 177–189.

Yousuf, M., Uy, B., Tao, Z., Remennikov, A. & Liew, J.Y.R. 2014. Impact behaviour of pre-compressed hollow and concrete filled mild and stainless steel columns. *Journal of Constructional Steel Research*. 96: 54–68.

Yu, J. & Jones, N. 1991. Further experimental investigations on the failure of clamped beams under impact loads. *International Journal of Solids and Structures*. 27(9): 1113–37.

Zeinoddini, M., Harding, J.E. & Parke, G.A.R. 2008. Axially pre-loaded steel tubes subjected to lateral impacts (a numerical simulation). *International Journal of Impact Engineering*. 35(11): 1267–1279.

Zhao, J.H., Wei, X.Y. & Ma, S.F. 2009. The finite element analysis for concrete filled steel tubular columns under blast load. Proceedings of the 9th International Conference on Analysis of Discontinuous Deformation, Singapore, 25–27 November 2009.

Tubular Structures XV – Batista, Vellasco & Lima (eds)
© 2015 Taylor & Francis Group, London, ISBN 978-1-138-02837-1

Behaviour of reverse channel tension zone subjected to impact loads

P. Barata, J. Ribeiro & A. Santiago
ISISE – Institute for Sustainability and Innovation in Structural Engineering, University of Coimbra, Portugal

M.C. Rigueiro
ISISE – Institute for Sustainability and Innovation in Structural Engineering, Polytechnic Institute of Castelo Branco, Portugal

ABSTRACT: When an exceptional event occurs, unforeseen and extreme loading scenarios may arise with severe intensity and complexity. The unpredictable consequences, as the collapse of the World Trade Center, have highlighted the vulnerability of steel joints under impact loading and fire. It is recognized that joints play a very significant role in the structural behaviour; hence, in the design for enhanced structural robustness, high ductility is required from joint details and improvements are demanded. An increasingly attractive solution to connect tubular hollow section elements is the use of the reverse channel; this joint exhibits high ductility through the ability to perform under catenary action allowed by the deformation of the channel's web panel. This paper presents a study evaluating the behaviour of the reverse channel's tension zone when subjected to tensile impact loading. Firstly, monotonic and impact tests are conducted in a special purpose experimental setup and secondly, a finite element model, including description of the material's strength depending on the strain rate, is validated enabling further studies of this component.

1 INTRODUCTION

Tubular sections are inherently efficient due to their closed cross-sections and improved second moment of area in the weak axis; this allows for an increased buckling resistance for column elements and avoids lateral buckling when used as beams. Therefore, buildings with an improved strength-to-weight ratio, in comparison with rather common I and H shaped cross-sections, are designed. Additionally, the rectangular and circular closed cross-sections are regarded as being aesthetical and architecturally appealing shapes. However, connecting hollow sections is often considered to be complicated and expensive, due to the lack of access to the inside to perform a bolted connection; this limits the connection of members to welding performed "on site".

An increasingly attractive option is the use of the reverse channel connection as presented in Figure 1. This connection is particularly useful to connect beams to circular or rectangular hollow-section columns, both composite or non-composite; the main feature is that it allows the use of simple conventional bolted connections, e.g., partial-depth end-plates and web cleats, by assembling the connection to the back of a channel section that is welded in advance to the column section (Hicks et al., 2002).

Recent experimental studies, on different types of connections of steel beams to concrete-filled tubular columns (Ding & Wang, 2007), have shown that the reverse channel connection appears to have the

Figure 1. Examples of the reverse channel joints.

best combination of desirable features: moderate construction cost, ability to develop catenary action and extremely high ductility through deformation of the channel's web. Their weakest feature is the reduced ability to act as rigid moment joint capable of delivering the required stiffness to a frame.

The architectural attractiveness of tubular cross-section makes them most suited for iconic buildings in which the steel structure is exposed to the public's eyesight (Figure 2); therefore, the steel members are prone to remain vulnerable and relatively unprotected in the event of unforeseen loading scenarios during the design stage, such as fire, impact or explosion, and yet be capable of preventing the buildings' progressive collapse.

Figure 2. Pavililion at Expo 92, Seville, Spain (Packer, et al., 2009).

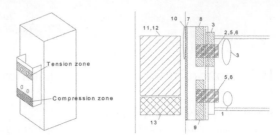

Figure 3. Reverse channel connection active components.

According to (ARUP, 2011) and (McAllister, 2002), the behaviour of connections is considered crucial to fully assess the resistance of structural steelwork for buildings in avoiding progressive collapse due to accidental loadings. A functional requirement raised in FEMA's report is: "Connection performance under impact loads … needs to be analytically understood and quantified for improved design capabilities and performance as critical components in structural frames". Additionally, a recent report presented by Arup made the following recommendation (rec. n° 26): "… *the strain rate enhancement of yield strengths in connections could still be important. It is recommended that research is undertaken to examine this effect using rate-sensitive material models*" (ARUP, 2011).

Despite real evidences, current design standards provide little guidance on this issue. Eurocode 1, Part 1.7 (EC1-1-7, 2006) considers accidental situations due to internal explosions and impact, whilst fire and earthquake hazards are considered in specific parts of the Eurocodes. Accidental actions are represented by an equivalent static force corresponding to the equivalent action affects in the structure and it is recommended to take strain rate effects in the material properties of the impactor and the structure. Eurocode 3, Part 1.8 (EC3-1-8, 2004) deals with the design of joints (primarily of its resistance and stiffness) under monotonic loading, only.

Most of the available studies on joints under accidental loadings are focused on fire (Spyrou et al., 2004) & (Wald, et al., 2006) or seismic hazard (Yorgun & Bayramoglu, 2001) & (Broderick & Thomson, 2002)). But, in the last decade, research on the study of joints behaviour under impact and blast loads is growing, especially regarding the influence of strain-rate effect in the connection response. Some examples of these studies are reported in Arup (ARUP, 2011).

2 STRUCTURAL MODEL – REVERSE CHANNEL COMPONENT

2.1 *Reverse channel design*

Currently, design guidance for a joint using the reverse channel as a connecting element is still absent from the Eurocodes (EC3-1-8, 2004); yet several research groups are studying this joint typology, thus providing guidance to its resistance calculation. Any approach to this is deemed to follow the component method established in the Eurocode (EC3-1-8, 2004); based on the individual behavior of the several components reported, the user is able derive the behavior of various joint configurations by assembling the components present (in the joint).

Following the component method, a typical reverse channel beam-to-column connection, subject to bending moment is divided into three major zones: tension, shear and compression. Figure 3 identifies the active components in each zone:

(1) beam flange and web in compression;
(2) bolts in tension;
(3) end-plate in bending;
(4) beam web in tension;
(5) bolts in bearing;
(6) bolts in shear;
(7) welds;
(8) reverse channel in bending;
(9) reverse channel in compression;
(10) column wall in bending;
(11) column side walls in shear;
(12) column side walls in transverse tension;
(13) column side walls in compression.

Most of these components are already defined and characterized in the Eurocode 3 part 1-8 (EC3-1-8, 2004). Component (10), however, is not included in the Eurocode 3 part 1-8 but has been object of study by the "Comité International pour le Développement et l'Étude de la Construction Tubulaire" – CIDECT in their Report 5BP-4/05 (Jaspart et al., 2005) and design guide (Kurobane, et al., 2004); the following failure modes for this component should be taken into account:

i) shear strength of the wall adjacent to a weld;
ii) punching shear through the tube wall;
iii) yielding of the tube wall using a yield line mechanism (Kurobane, et al., 2004).

Although similar to what may be observed in a connection to a hollow section, components (8) and (9) are not covered in these guides; the main difference is the limited length available in the reverse channel to extend the yield lines, when compared to a

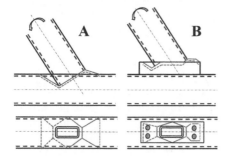

Figure 4. View of yield line length in components: a) wall in tension and wall in compression (adapted from (Packer, et al., 2009)); b) reverse channel web in tension and compression.

Figure 5. Geometry of specimens: a) connection geometry; b) Reverse channel section.

whole element, and the localized loading (and yielding) developed in the tension zone as exemplified in Figure 4.

2.2 Structural details and geometrical properties

Following the component method philosophy, it is required to provide the behaviour of both compression and tension zones; in this paper only the tension zone is addressed under monotonic and dynamic loading. The studied reverse channel tension zone specimens are built from a steel hollow section (SHS 200) cut lengthwise with 200 mm of length, with the geometry reported in Figure 5. The web of the reverse channel is bolted to the end plate through two bolts M24, grade 10.9 partial threaded, with a pitch of 85 mm. Thicknesses of 8, 10 and 12 mm are considered for the monotonic studies, while under dynamic conditions only RC-8 is assessed.

The procedure for the design of the reverse channel components has been the subject of study from Lopes and co-authors (Lopes, et al., 2013) under monotonic loading and fire hazard. Within the tension zone, two active components are identified: bolts in tension and RHS web in transverse tension. Table 1 provides the component's design values for each of studied reverse channel thicknesses; failure is driven by local yielding around the bolts.

3 EXPERIMENTAL PROGRAM – REVERSE CHANNEL COMPONENT

3.1 Test procedure and instrumentation

The test setup used in the experiments has been designed and built at the University of Coimbra and

Table 1. Summary of the design values.

	Bolts		RHS tension		
	Tension [kN]	Punching shear [kN]	Local failure [kN]	Global failure [kN]	F_{pl} [kN]
RC-8	317.2	937.3	144.2	149.1	144.2
RC-10	317.2	937.3	220.4	230.4	220.4
RC-12	317.2	937.3	310.1	328.2	310.1

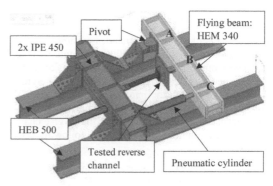

Figure 6. Layout of the experimental system.

is presented in Figure 6. It consists of a very stiff structure (shown in grey colour) anchored to the floor. The loading mechanism is based on the principle of a 2nd class lever. The yellow beam is loaded in one end (point C) by a loading device (pneumatic cylinder in red colour) and it's allowed to rotate around the pivot axis in the other end (point A). The yellow beam is supported at mid span by the reverse channel specimen (shown in green colour in the middle of the testing layout), which is limited by two additional pins in its ends. Such boundary conditions of the reverse channel assure transmission of axial tensile forces only.

During the monotonic tests, the loading device was a servo actuator with 1000 KN capacity, whilst during the dynamic tests, a tank was filled with gas, up to the maximum operating pressure of 30 MPa (300 Bar). Once the pressurized air was released into the chamber, the gas accelerated a ram with approximately 40 kg of mass that impacted the flying HEM340 beam. The dynamic equilibrium of the system is controlled by measuring the beam's acceleration and displacements, and the transient applied load.

Due to the need to control the inertia of the flying beam, the instrumentation includes accelerometers placed in the mass centre of the beam (points B) and near the load application point (point C); additionally a load cell is placed between the ram of the cylinder and the flying beam (near point C) measuring the forces discussed later in Figure 9; a laser distance gauge in the load application point (point C) and two others in the reverse channel are used to measure displacements.

Table 2. Summary of experimental test results.

Test		RC-8		RC-10		RC-12	
		#1	#2	#1	#2	#1	#2
$F_{pl,na}$	[kN]	144.2		220.4		310.1	
$F_{Rd,exp}$	[kN]	133.7	121.1	217.4	218.7	308	309.1
F_{max}	[kN]	404.7	406.5	570	570	591.2	528.6
ΔF_{max}	[mm]	58.1	53.3	44.02	66.2	53.24	37.08
$K_{e,test}$	[kN/mm]	119	131.6	195.8	166.7	267	214.3
$K_{pl,test}$	[kN/mm]	6.57	6.48	7.5	6.6	5.8	5.9

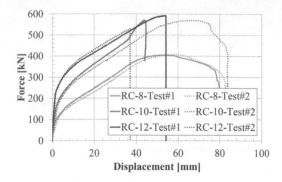

Figure 7. Reverse channel monotonic results: RC-8; RC-10; RC-12.

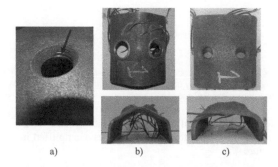

Figure 8. Deformed specimens after the test: a) bolt thread marks b) RC-8-Test#1; c) RC-12-Test#1.

3.2 Experimental programme

The experimental program includes monotonic tests and sequential impact test. With the monotonic tests the behaviour of the reverse channel tension zone is evaluated and these results are used as reference for comparison with the impact tests. The sequential impacts were applied to reach the failure and allow the measurement of the stiffness of the reverse channel (during the unloading phase). The following tests are considered:

(1) Monotonic tests of reverse channel tension zone with 8, 10 and 12 mm thickness (RC-8, RC-10 and RC-12; two tests per thickness were arranged;

(2) Sequential impact test on RC-8: rapidly applied loading of 160 Bar, followed by 185 Bar, followed by 200 Bar, and finally 240 Bar (test RC-8-D160-185-200-240);

3.3 Experimental results

3.3.1 Results under monotonic loading

Figure 7 presents the F-δ curves, which highlight the large amount of displacement capacity available after yielding occurs. Table 2 presents a summary of the test results; an increase in the elastic stiffness, plastic strength and maximum force is observed with the increase of the reverse channel thickness; a reduction in the displacement at the maximum strength (Δ_{Fmax}) as the thickness increases from 10 to 12 mm is reported. The post limit stiffness remains somewhat indifferent (Figure 7).

Figure 8 shows the deformed shape of the specimens for RC-8 and RC-12. The failure for RC-8 has been the *pull-out* of the bolts, i.e., the bolts' holes become largely deformed enabling the bolts to pass-through it. It can be observed that RC-8 exhibits both local and global plastic deformation in Figure 8 b). Oppositely, RC-12 (Figure 8 c)) exhibits only local failure in the bolt area, and bolt *pull-out* is not observed. The thread marks presented in Figure 8 a) demonstrate that the bolts are subject to both tension and shear, which would eventually lead to the bolt rupture in both RC-12 tests and in RC-10-Test#1.

3.3.2 Results under impact loading

This section concerns the evaluation of the experimental test results; both, the force applied to system (black solid line) and to the specimen (black dashed line), and also the displacement readings for the first impact of the sequential test (160 bar – dashed grey line) are presented in Figure 9. The dotted grey line represents an idealized function [f(t)] to apply the 40 mm displacement, recorded in the end of the sequential experimental test, later in the numerical model.

It should be noted that the applied force in the specimen reads a maximum of 150 kN, yet this value ought to be multiplied by 1.97 due to the relation of the 2nd lever arm spans (see the experimental setup Figure 6). The inertia effects of the "flying beam" are included, calculated with the measured accelerations in the mass center and in the point where the force is applied.

Concerning the results of the impact tests, Figure 10 depicts the results of the sequential test RC-8-D160-185-200-240. After impact #1, the specimen yields with plastic strength of 239 KN (point A, in Figure 10), which corresponds to an increase of 65% of the value obtained in the monotonic reference tests. A maximum resistance of 303 KN for a maximum displacement of 17 mm is measured. Figure 11a) depicts the corresponding deformed shape; both local and global failure can be observed after this first impact.

For both impact #2 and #3 the strength flow remains above the maximum reached in the reference monotonic test values. Finally, after application of impact#4 (240 Bar), it is observed that the strength remains above the reference test values, but the maximum displacement of 39 mm is substantially smaller. This

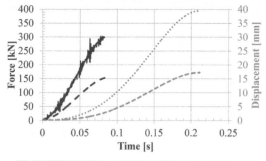

Figure 12. Parts and mesh composing the reverse channel model.

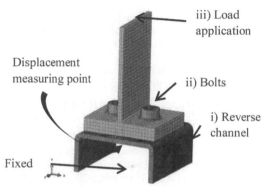

Figure 13. Model assembly and boundary conditions.

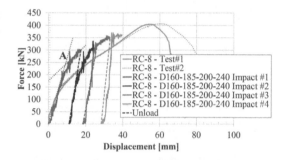

—Applied force - Pressure = 160 Bar
— Applied force on reverse channel specimen
--160 Bar - Displacement = 17 mm - Measurement
··· 240 Bar - Displacement = 40mm - Idealized

Figure 9. Shape and magnitude of the applied force vs. time & displacement reading vs. time for RC-8-D160-185-200-240 impact #1.

Figure 10. RC-8-D160-185-200-240: F-δ response.

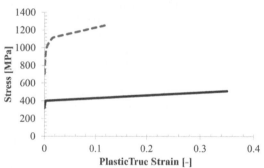

Figure 14. Reverse channel material properties.

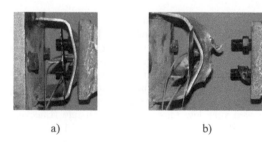

a)   b)

Figure 11. Deformed shape of RC-8-D160-185-200-240: a) after impact#1; b) after impact#4.

corresponds to a decrease of 50% of the displacement capacity of the reverse channel tension zone when compared with the monotonic reference tests. The observed failure mode is the excessive yielding around the bolt holes with bolt *pull-out*, similarly to the monotonic tests (Figure 11 b)).

4 FINITE ELEMENT MODEL

4.1 *Description of the FE model*

A simplified 3D numerical model is built with software ABAQUS (Abaqus, 2011), making use of the implicit algorithm (both static and dynamic) to establish the non-linear behaviour of the reverse channel component. The geometry presented previously is considered, hence the model composed of the parts shown in Figure 12: i) the reverse channel, ii) the bolts and iii) the rigid tensile load application part. Figure 13 presents the model assembly and boundary conditions: the reverse channel is tied to a fixed central node. Normal contact interactions include hard contact and tangential contact with a 0.2 friction coefficient following penalty formulation. The displacement measuring point is located in center of the back surface of the reverse channel.

4.2 *Material characterization*

A simplified elasto-plastic material description is considered for the reverse channel's steel and bolts

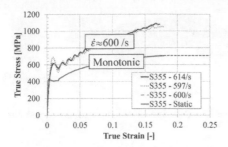

Figure 15. True stress – logarithmic strain relationship of steel under high-strain rate (approx. $600\,\text{s}^{-1}$) for $t = 15\,\text{mm}$ plate, S355 (Saraiva, 2012).

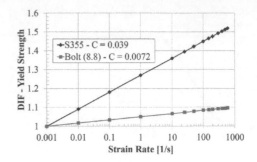

Figure 16. Dynamic increase factor (DIF) of the yield strength as a function of the strain-rate.

(Figure 14) both with $E = 210\,\text{GPa}$, and the following yield and hardening properties:

$fy_{steel} = 400\,\text{MPa}$ and $Etan_{steel} = 316\,\text{MPa}$;
$fy_{bolt,10.9} = 1100\,\text{MPa}$ and $Etan_{bolt,10.9} = 1350\,\text{MPa}$.

It is known that steel's strength is affected by the rate of imposed deformation; the effects of different strain rates on the stress-strain relationship of steel are illustrated in Figure 15 (Saraiva, 2012). These true stress-logarithmic strain curves are obtained from an experimental programme carried out at the University of Coimbra, using a Compressive Split Hopkinson Pressure Bar (SHPB) for the dynamic tests; an average strain rate around $\dot{\varepsilon} = 600\text{s}^{-1}$ is applied. Comparison against monotonic results shows that:

i) the yield and ultimate strengths (f_y, f_u) increase near 50% the results obtained under monotonic loading;
ii) the total strain on rupture (ε_{cu}) decreases, and;
iii) the elastic modulus (E) remains indifferent to the loading rate.

A simplified way to consider high strain rate enhancement in the stress-strain material law is to adopt a dynamic increase factor (DIF), given by the relation of the dynamic yield strength, σ_{dyn} to the yield strength obtained under static conditions, σ_y:

$$DIF = \frac{\sigma_{dyn}}{\sigma_y} \qquad (1)$$

Finite element models aiming to simulate the behaviour of structural elements when subject to impact loads require a constitutive law representing the behaviour of materials for a range of strain rates. The Johnson–Cook model (Johnson & Cook, 1983) is purely empirical and is able to account not only for the strain rate sensitivity but also for thermal softening behavior. This constitutive law assumes that the slope of flow stress δ_y, is independently affected by each of the mentioned variables therefore only the strain rate dependency is considered (equation 2):

$$DIF = [1 + C\,ln\dot{\varepsilon^*}] \qquad (2)$$

where: $\dot{\varepsilon^*} = \dot{\varepsilon}/\dot{\varepsilon}_0$ is the reference dimensionless plastic strain rate ($\dot{\varepsilon}_0 = 0.001\,s^{-1}$); $\dot{\varepsilon}$ is the strain rate and C is the strain rate constant.

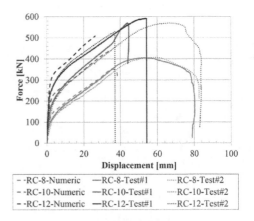

Figure 17. Reverse channel – Experimental Vs. numerical F-δ responses under monotonic loading.

Thus, based on the results from SHBT presented before and using the second term of Johnson–Cook's law (equation 2), $C_{steel} = 0.039$ for $600\,\text{s}^{-1}$ is calculated to fit the experimental data (Figure 15). The dependency on the strain rate for the bolts' material is accounted considering literature reports: impact tests on A 325 bolts recovered from the WTC debris showed very low sensitivity to strain rate (Luecke, et al., 2005), hence a maximum DIF of 1.1 is considered for the bolts, thus a value of $C_{bolt} = 0.0072$ is obtained. Figure 16 provides the applied DIF for strain rate values between 0.001 and $600\,\text{s}^{-1}$ following the Johnson–Cook law.

4.3 Validation under monotonic loading

Figure 17 presents the comparison of the experimental and numerical F-δ responses for the studied reverse channel thicknesses. The numerical model is able to approximate the behaviour of the reverse channel tension zone under monotonic load in the elastic; plastic transition and hardening phases for all thicknesses. Yet, the model may still require improvements for description of the damaged phase.

Figure 18 plots the equivalent plastic strain patterns (PEEQ) for the various RC thicknesses at large levels of displacement. It can be observed that 3

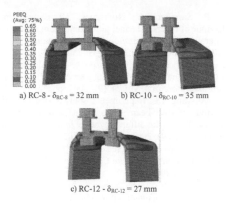

a) RC-8 - δ_{RC-8} = 32 mm   b) RC-10 - δ_{RC-10} = 35 mm

c) RC-12 - δ_{RC-12} = 27 mm

Figure 18.   Equivalent plastic strain (PEEQ) pattern.

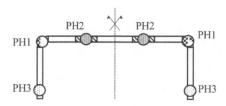

Figure 19.   Plastic hinge development.

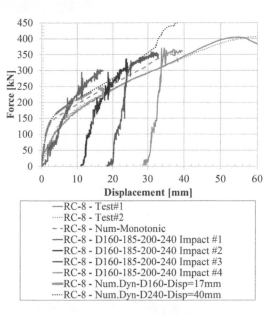

Figure 20.   Numerical vs. experimental.

different plastic hinges in each half of the reverse channel are developed. The plastic hinge formation evolves according to Figure 19: firstly, PH1 (1st plastic hinge) in the round corner and PH2 in the bolt region, start their development almost simultaneously, being responsible for the transition knee to the plastic phase in the F-δ responses. Later, a minor plastic hinge is developed in the support region – PH3 (most visible for RC-10). No plastic strains in the bolts (except for the washer) are observed.

4.4   *Validation under dynamic loading*

The loading is modelled in a simplified way, considering the displacement fields recorded experimentally (Figure 9). The procedure is to apply the recorded displacements/time function and retrieve the generated reaction forces measured in the reverse channel tension zone. Two different load levels are considered: i) considering the first impact of test RC-8 – D160-185-200-240, and ii) considering that the final displacement observed in the final of this sequential test is applied in the same time period (Figure 9).

Figure 20 compares the experimental and numerical F-δ responses; the numerical model is able to generally reproduce plastic behaviour exhibited throughout the sequential test; however it is noted a rather high disagreement with the first impact test of 165 Bar in terms of the maximum force.

4.4.1   *Analysis of the dynamic response*

Figure 21 compares the monotonic and short transient dynamic FE responses. It shows that the elastic stiffness remains unchanged, although an increase

in the plastic resistance of $170.9/153.1 = +11.6\%$ is observed due to the increase in the deformation rate and enhancement of the material's strength. Both FE analysis load schemes yield the same response. No failure is observed in FE models under dynamic loading.

The strength enhancement is analysed in Figure 22 a) through the strain rate (ER) plot for a time increment of $t = 0.1695$ seconds. In this increment, a displacement of 32 mm and a force of 371 kN are applied in the RC tension component; the plot shows strain rate mean values around 3 to 4/s in the plastic hinges; comparing this value with Figure 16, a DIF of 1.3 ought to be observed in the F-δ response; However, comparison of Figure 18 a) and Figure 22 b) shows that the plastic hinges are under developed when subject the dynamic loading; the plastic hinge evolution is, nonetheless, the same way as observed for the monotonic tests (Figure 19).

5   CONCLUSIONS

This paper concerns the tension zone of the reverse channel joint under monotonic and short transient loads; both experimental and numerical FE results are reported. Under monotonic loading, thicknesses of 8, 10 and 12 mm are considered while under dynamic loading only RC-8 is considered in a sequential impact test with 4 increasing load magnitudes.

The reference monotonic results demonstrated the high ductility capacity of this joint typology. It is found that the maximum force occurs for a 300% higher level of displacement than the plastic strength's displacement.

Under dynamic loading conditions the plastic strength is increased due to the development of strains

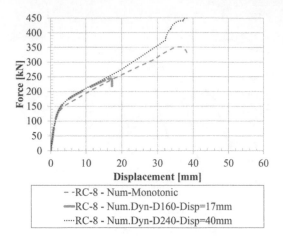

- - -RC-8 - Num-Monotonic
——RC-8 - Num.Dyn-D160-Disp=17mm
······RC-8 - Num.Dyn-D240-Disp=40mm

Figure 21. Numerical results comparison of monotonic and dynamic numerical response.

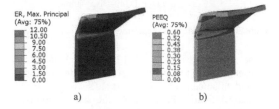

a)   b)

Figure 22. a) Strain rate (ER) plot and b) equivalent plastic strain pattern (PEEQ) for increment $F = 371$ kN; $\delta = 32$ mm (presented for a quarter of the model for RC-8-D240).

at an elevated rate; a strength enhancement of 65% has been observed in the experimental tests, whilst in the numerical results only 11.6% is noted. From the numerical analysis a mean value of strain rate around 3 to 4/s is detected; this would lead to a dynamic increase factor around 1.3, yet the plastic hinges look slightly under developed.

Concerning the displacement capacity, a harsh reduction up to 50% has been observed comparing monotonic and dynamic experimental results; the failure mode has been for both situations, the bolts *pull-out,* for RC-8. No failure was observed in the numerical models.

ACKNOWLEDGEMENTS

The authors acknowledge financial support from Ministério da Educação e da Ciência (Fundação para a Ciência e Tecnologia) under research project *PTDC/ECM/110807/2009.*

REFERENCES

Arup, Review of International Research on Structural Robustness and Disproportionate Collapse, *Department for Communities and Local Government*, October, 2011.

Barata, P., Santiago, A., Rodrigues, J.P. Experimental behaviour of reverse channel joint component at elevated temperatures, *4th International Conference on integrity, reliability & failure*, 2013, pp. 749–750.

Broderick, B., Thomson, A. The response of flush end-plate joints under earthquake loading, *Journal of Constructional Steel Research*, Vol. 58 (9), 2002, pp. 1161–1175.

Ding, J. and Wang, Y.C. Experimental study of structural fire behaviour of steel beam to concrete filled tubular column assemblies with different types of joints. *Engineering Structures*, 12(29), 2007, pp. 3485–3502.

Eurocode 1: actions on structures. Part 1-7, General actions - accidental actions, *European Committee for Standardization*, Brussels, 2006.

Eurocode 3: Design of steel structures part 1-8: Design of joints, *European Committee for Standardization*, Brussels, 2004.

Hicks, S.J., Newman, G.M., Edwards, M., Orton, A. Design Guide for Concrete Filled Columns. *Corus Tubes/ The Steel Construction Institute*, London, UK, 2002.

Jaspart, J.-P., Pietrapertosa, C., Weynand, K., Busse, E., and Klinkhammer, R. Development of a full consistent design approach for bolted and welded joints in building frames and trusses between steel members made of hollow and/or open sections. *Application of the component method. Draft final report – volume 1: Practical design guide*, Research project 5BP, CIDECT, 2005

Johnson, G.R, Cook, W.H. "A constitutive model and data for metals subjected to large strains, high strain rates and high temperatures", Proceedings of the 7th International Symposium on Ballistics, The Hague, The Netherlands, pp. 541–547, 1983.

Karlsson, & Sorensen, I.U., (2011). Abaqus Theory Manual, v.6.11.

Kurobane, et al., Design guide for structural hollow section column connections, CIDECT, 2004, LSS Verlag.

Lopes, F., Santiago, A., Simões da Silva, L., Heistermann, T., Veljkovic, M., Guilherme da Silva, J. Experimental behaviour of the reverse channel joint component at elevated and ambient temperatures, *International Journal of Steel Structures*, Volume 13(3), 2013, pp. 459–472.

Luecke, W. et al., 2005. Mechanical Properties of Structural Steels. s.l., Federal Building and Fire Safety Investigation of the World Trade Center Disaster, NIST NCSTAR 1-3D.

McAllister, T., World Trade Center building performance study: data collection, preliminary observations and recommendations; Federal Emergency Management Agency, *Federal Insurance and Mitigation Administration*, Washington, D.C., FEMA Region II, New York, 2002.

Packer, J.A., et al., "Design guide for rectangular hollow section (RHS) joints under predominantly static loading, CIDECT, 2009, LSS Verlag.

Saraiva, E., "Variação das propriedades mecânicas do aço relacionadas com problemas de impacto em estruturas", Master Thesis at University of Coimbra, in portuguese, 2012.

Spyrou, S., Davidson, J.B., Burgess, I.W. and Plank, R.J. Experimental and analytical investigation of the 'tension zone' components within a steel joint at elevated temperatures, *Journal of Constructional Steel Research*, vol. 60, 2004, pp. 867–896.

Wald, F., Simões da Silva, L., Moore, D.B., Lennon, T., Chladná, M., Santiago, A., Beneš, M., Borges, L. Experimental behaviour of a steel structure under natural fire, *Fire Safety Journal*, Vol. 41(7), 2006, pp. 509–522.

Yorgun, C., Bayramoglu, G. Cyclic tests for welded-plate sections with end-plate connections, *Journal of Constructional Steel Research*, Vol. 57(12), 2001, pp. 1309–1320.

Offshore structures

Tubular Structures XV – Batista, Vellasco & Lima (eds)
© 2015 Taylor & Francis Group, London, ISBN 978-1-138-02837-1

Tubular based support structures for offshore wind turbines

J. Müglitz
ZIS Industrietechnik GmbH, Meerane, Germany

S. Weise
WeserWind GmbH, Bremerhaven, Germany

J. Hermann & U. Mückenheim
SLV Halle GmbH, Halle/Saale, Germany

K.A. Büscher
HAANE Welding Systems GmbH, Borken, Germany

ABSTRACT: Worldwide there is a strong tendency towards power generation by wind, with both on- and offshore equipment. The welded tubular support structures currently in use are based on three design principles, mo-nopile, tripod and jacket constructions. The large diameters, the wall thicknesses up to 50 mm, the framework mass of several hundred tons, the demand of endurance strength and the high job lot of almost identical units require new solutions for design, assessment and manufacturing. This article presents a number of new developments for the manufacturers of support structures in the offshore wind industry.

1 INTRODUCTION

Suitable support structures are necessary for offshore wind turbines. For the windmill towers experience of approximately 20 years is available because of its onshore use. In contrast the support structures are virgin soil for the engineers.

Within the planned lifetime of 25 years the support structures need to withstand all the prevailing loads. These are:

- static and dynamic forces, induced by weight and wind,
- forces by wave pounding, ice drift, storm and marine growth,
- extraordinary loads, for example a collision of a boat or ship,
- corrosion.

Even though there are other designs in use, for example concrete foundations, the article only considers welded steel constructions which are founded on the sea ground.

The use of tubes instead of box sections made of welded plates has several key advantages.

These are the low approach flow resistance, the very good utilization of the material, the continuous curvature of the braces surface and its low surface area. From the geometric point of view there are nearly no limitations to the design of a tubular structure, i.e. the number and intersecting angles of the braces, which are part of the nodes. Tubular structures have a high

Figure 1. Different types of support structures (from left: Monopile, Tripod, Jacket).

freedom of design and can be adapted very closely to the existing load spectrum.

Different configurations and design principles have been developed in the last few years. At the moment it is not clear, which will assert itself on the market, but some tendencies are visible. In Fig. 1 the three fundamental principles can be seen.

The **monopile** is a single vertical tube, driven into the soil, with diameters up to 6 m. There is a tendency to fabricate and install piles with increasing diameter up to 8 m to cover the needs of increased water depth beyond 30 m and of bigger turbines beyond the mainly used 3,6 MW machines. While the steel structure can be scaled up to 10 m in diameter, aspects of geotechnical assessment, transportation and installation need to be considered as limiting factors.

Figure 2. Tripod.

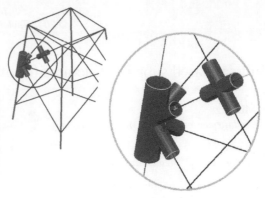

Figure 3. Jacket node types.

In contrast to the monopile there are lattice structures, which will be considered in this article, with intersecting contours between tubes, i.e. curved welding seams. They need to be cut for fitting and need to be welded in high quality.

The **tripod** (see Figure 2) is a lattice structure, with a rather low degree of decomposition. This foundation is meant for water depths of more than 30 m and turbines with 5 MW and more. It is a three-legged structure, consisting of 10 main components. It has a central tube, which is connected to three pile guides via conical braces, three upper braces and three lower braces. Tube diameters commonly range from 2.500 up to 6000 mm. The structure has an advantageous resonance frequency. Its behavior in critical situations, for example in case of a ship collision, is unproblematic. The handling and the installation on- and offshore have been developed and proven for 120 tripods installed in the German Bight of the North Sea.

The **jacket** is a lattice structure, consisting of legs and bracings with typical X, Y- and K-shaped nodes. Design variations with three, four or six legs are common.

The degree of decomposition is high, i.e. it is a slender design. It makes it possible to use tube dimensions which are similar to the pipeline industry in terms of diameter and thickness, making the purchase of the tubes easy and cost effective. The total mass of a jacket is essentially smaller, around half the mass of a tripod. This decreases the installation costs, because smaller installation crane vessels would be sufficient for the grounding of the jackets, even the use of the weather sensitive offshore lift would become feasible.

Each of the three design principles has its advantages and disadvantages (SCHAUMANN2010) and will be selected for a special project according to the application and environmental conditions. Besides pure technical aspects, questions of ecology,

certification and admission have strong influences on the decision for or against a special structure.

The long distance to the coast is typical for offshore wind farms in Germany, especially in the North Sea. AC-DC converter stations are necessary nearby the wind farms to connect them with the electricity network. These transformer stations need to be erected offshore as well. They are designed as platforms and are similar to platforms used for gas and oil extraction.

In contrast to the support structures, which will be produced in production runs of 50 ... 100 pieces, the platforms are usually unique designs.

2 DIMENSION AND DESIGN

With the standardization of the nodes of a jacket, a modular construction system is possible and useful. According to the special boundary conditions a cast or a welded node design can be a more economical solution. In the following text only welded nodes are considered.

The result of the general jacket design are two main node types. A double-K-node forms the connection of the legs with the bracings, an X-shape node forms the intersection of two diagonals (Fig. 3). The main chord at a double-K-node has typically a diameter between 900 and 1500 mm. Four nozzle stumps with diameters from 600 ... 1000 mm are connected to the main tube with an intersecting angle of around 30 to 45 degrees. The X-node is comparatively small, the intersection angle of the axes is approximately rectangular. But at X-nodes two tubes with the same diameter are intersecting (1:1 – intersection). This creates a high weld seam curvature and could cause precarious NDT test conditions. Typical wall thicknesses are between 15 and 30 mm.

The mass and the dimensions of jacket nodes make it possible to establish a prefabrication with higher production runs. The transportation of parts by trucks is possible, so the prefabrication does not need to be carried out at the place of assembly. A large amount of the welding work, especially all the welding at the three dimensionally formed intersecting contours, can be done in a preferred welding position and under common conditions of a welding shop floor. Automation by robots or CNC- machines is worth consideration. Special attention should also be paid to the geometrical accuracy of prefabricated nodes, because it is an essential precondition for efficient assembly.

For the connection of the nodes to a framework only girth welds are needed. This can be done semi-automatically using orbital welding, a technology, which is widely used and field-tested in pipeline construction. The single sided root welding for these fatigue loaded structures is a key technology and is well controlled with modern GMA-welding processes.

The tube dimensions in tripods, both diameter and wall thickness, are two to three times higher compared to jackets for comparable applications. Due to the dimensions the tubes need to be made out of sheet metal plates as rolled sections. For manufacturing the curved contours afterwards a big tube cutting machine would be necessary. Nevertheless these parts could be prefabricated separately from the final assembly lines and transported by truck.

Assembly and welding require a certain accuracy of the prefabricated brace. It also has to be taken into account that the part mass of the braces may easily exceed 50 tons.

It is a special characteristic of a tripod, that the nodes have a double side weld. Some employers value this type of weld because of the execution excellence of the welds for fatigue resistance. It should also be mentioned that the welding needs to be done manually under quite harsh conditions.

3 MANUFACTURING PROBLEMS

The new level of offshore support structures deals with the part dimensions, the quality demands and the number of identical units.

Some experience with large tubular structures from the steel building industry, especially bridge constructions, can be transformed (EULER2013). They need to be advanced because of other boundary conditions and even larger tube dimensions.

A typical fabrication task is the manufacturing of reliable T-connections which includes three-dimensionally formed intersecting contours. Full penetration welds are mainly required in offshore industry executed as a HV- or K-seam. The thermal cutting technologies, like plasma- or oxygen cutting, are standard for preparing the parts. Laser cutting is impossible because of the high wall thickness.

The longitudinal and circular seams in pipe fabrication can be welded by submerged arc welding. Bringing the powder on top of the molten pool stabilizes the pool crater. Realizing that the part has to be arranged in a flat position.

The real challenge is the assembly weld, i.e. the connection of the pre-assembled components to a support structure framework. At first a solution for the accurate positioning of the heavy duty parts is necessary.

Also the deformation of the parts under their own mass can not be disregarded.

Moving the parts into a preferred welding position is impossible. The weld has to be done in all positions, with the exception of downhill-welding technique, which is prohibited in the rules and standards (GL2010). Only the GMA- welding is suitable for the demands. Its specific properties make it less economical, but more flexible.

The authorized inspecting organizations demand a very high quality welding seam. Also small discontinuities are forbidden. Therefore all seams need to be tested non-destructively by ultrasonic testing and by dye penetrant inspection of surface cracks. The evaluation of the test results is done manually and has naturally a subjective component. Automatic, completely objective test routines will be used in the near future. Phased array and time of flight defraction (ToFD) become more and more available and first standards are published already. Nevertheless these methods are not yet state of the art in the offshore steel construction.

On the other hand high standards are demanded regarding the material properties of the weld, mentioning the hardness, the toughness and the strength of the material in the welding seam and beside the weld, the so-called heat affected zone.

All the materials used are fine grained steel grades normalized or thermomechanically treated with yield strength from 355 to 460 MPa in general. The nodes are loaded with multiaxial stress under difficult environmental conditions, so the improved toughness needs to be guaranteed to avoid brittle fracture. Therefore test temperatures are specified for charpy impact tests down to $-60°C$. The test temperature scheme is based on empirical knowledge nowadays. The fracture mechanic methodology from EN 1993-1-10 very likely to be applied to offshore wind conditions in the near future. Meeting the toughness requirements calls for careful welding process control, which involves choosing the welding process itself, the welding parameters and the consumables, observing of the preheating temperature, energy input per unit length and interpass temperature. This is only possible with the string bead technique for manual welding.

The summation of the demands recommend that an open-air assembly weld makes it even difficult to guarantee the manufacturing quality. It must take place in a closed shop floor.

4 CHARGING AND SOLUTIONS

4.1 *Welding preparation*

According to the regulations (DNV2011), (GL2007), (GL2012) valid in Germany it is not allowed to create a welding seam as a combination of but and fillet weld with a transition zone as known from the steel construction industry (AWS2010), (KOCH2001). A welding preparation with a full penetration is required.

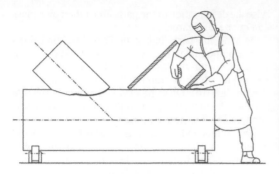

Figure 4. Manual welding from inside.

This can only be realized at nodes with smaller intersection angles by welding from the inside of the tube as well.

Starting with a diameter of around 600 mm it is possible to weld comparatively short stubs from the inside (Fig. 4). The welder stands beside the stub and needs to reach into the stub. If the diameter is a lot higher, the welder needs to get inside the tube through a manhole, well known from the vessel and shipbuilding industry.

The requirements for the manual skills of welders are already high for welding only from the outside of a tube. Special attention needs to be paid to the root pass welding without backing and sometimes without visual inspection. If also welded from inside the tube, the welding process is more tolerant, but the operating conditions become a lot more difficult because of the limited accessibility and the preheating temperature.

A new algorithm for calculating this special kind of welding preparation has been developed. Included in the special CAM-system, which is used for the manufacturing of the nodes, the welding preparation is calculated completely automatic. The calculation is based on a couple of simple rules, easy to understand by skilled workers and easy to adjust with several parameters.

Input parameters of the calculation are the seam opening angle α_{1i} inside and α_{1a} outside the tube. Both are in a range of about 45 degrees, are defined by the welding engineer and are described in the welding procedure specification (WPS). The bevel angle α_2 describes the maximal cutting angle of the special cutting technology. It is the angle between the torch axis $\underline{b}$ and the normal vector $\underline{n}$ at the tube surface:

$$\alpha_2 = \max[\text{angle}(\underline{b},\underline{n})] \qquad (1)$$

The bevel angle is influenced by the cutting process and the cutting process itself is determined by the wall thickness s of the part. Mild steel plasma cutting is limited to about 50 mm of wall thickness, whereas oxyfuel cutting has practically no limitation regarding the wall thickness. Possible bevel angles are:
Plasma cutting $\alpha_2 < =47$ deg. at s < 50 mm
Oxyfuel cutting $\alpha_2 < =60 \ldots 70$ deg. 10 < s < ∞ (practical no limit)

Figure 5. Nozzle (711 × 50) with welding preparation both outside and inside and smooth transition.

With the condition

$$\max(\alpha_{1i},\alpha_{1a}) < \alpha_2 \qquad (2)$$

at any intersecting angle and also in cases, where the axes are not intersecting but only crossing, it is possible to create a welding preparation with revolving full penetration. According to practical experiences it is recommended to enhance the seam opening angle α_{1i} inside because of the poor accessibility. Also the curving of the nozzle appears convex and so decreases the accessibility even more.

The change in orientation of the welding preparation from outside to inside can be smooth, where in a transition zone of given length the outer seam decreases and the inner seam increases. So in the transition zone a K-seam occurs. It is also possible and requested by customers to have the transition abrupt, with a non-continuous jump from the outside to the inside. In both cases the correct welding of such a seam is a big challenge also for high qualified welders.

4.2 Rolled sections with intersection contour

Tubes with diameters of more than 1500 mm and thickness above 25 mm are no longer commercial. So for tripods all the main components need to be manufactures with rolled sections made from plate material. If the section is rolled and welded with a longitudinal seam the intersecting contour can be cut on behalf of a special machine, a tube cutting machine or a robot equipment (Figure 6).

This machine is very special and therefore expensive. The rolled section and the machine need to be adjusted very accurately. In addition the rolled section needs to be struted because of its elastic deformation under gravity. For example: A rolled section without any stiffener with 3600 mm outer diameter and a wall thickness of 50 mm has an elongation of 3610 in horizontal and 3590 in vertical direction. The adjustment will never be perfect which results in higher gaps in the stage of assembly.

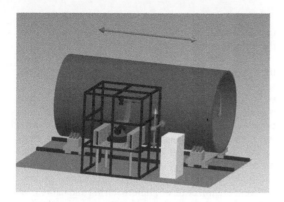

Figure 6. Robot based tube cutting machine.

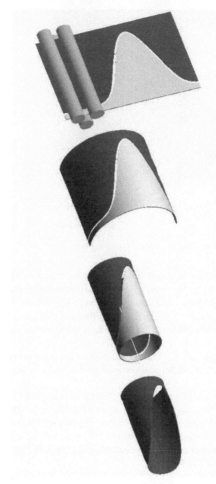

The question is: Why not change the process by cutting the intersecting contour including the welding preparation as sheet metal before rolling, as seen in Figure 7? It opens up very smart technical and commercial potentials:

- A complete process step, the cutting of the rolled sections with the tube cutting machine, is removed. The tube cutting machine is no longer necessary.
- The sheet metal parts are easy to handle with conventional equipment and to transport. They can be prefabricated in a specialized job shop and transported via truck, i.e. the job shop must not be situated at the same place where the assembly is executed.
- A tube-tube-connection, i.e. an intersection, can have a variable wall thickness, according to the load, along its elongation. From heel to toe zone the thickness increases up to 150 ... 200%. Material costs, mass and weld seam volume is reduced significantly.

Figure 7. Tube intersection, rolled from sheet metal.

The rolling of the sheet metal parts to cylindrical sections is comparable easy. The rolled segment needs to be a rectangular shape because of constant bending stiffness, so the sheet metal part is not removed after cutting but connected via bridges and is removed after the rolling process. Segments with different wall thickness are connected together with a special welding detail, a transition or sharpening, to avoid stress concentration. The transition is formed with a ratio of at least 1:4. According to the experiences in practical use it is recommend to create the transition after the rolling. Otherwise there are plastic deformations in the region of sharpening which result in higher geometric tolerances. Along the wall transition the intersecting contour gets a discontinuity, a buckle at both ends of the transition zone.

The rolling process puts on the material a cold deformation with the risk of increasing the hardess and decreasing the ductility. Significant changes of the material properties are not be expected because the deformation degree is low. It has to be checked regularly anyways. This effect is more critical if a polygonal chamfering is used instead of rolling. Then the hazard

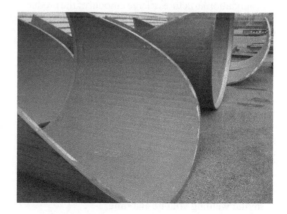

Figure 8. Chamfered tube segments.

of exceeding the approvable plasticization is extremely high.

The generally small degree of deformation results in only small plastic deformations in the rolled segment, both in circumference and in longitudinal direction.

233

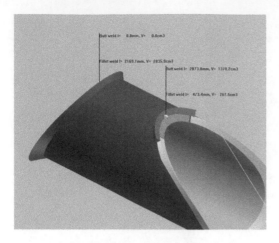

Figure 9.   Calculation of the welding volume.

The calculation is based on a 3D-model of the node which needs to be unrolled into a sheet metal part. This can be done based on the neutral axis, which is with k = 0.5 exactly in the middle of the wall thickness. With a modification of the factor k in the quantity of hundredth and some more (in first approach) linear acting compensation factors the congruence between the theoretical model and the real part is sufficiently accurate.

Some special geometric transformations are necessary. Also for this process step special software needs to be developed. According to our experiences common CAD-systems are not able to do this task in the required quality and with acceptable effort (EPA2010), (TUBECUT2014).

4.3   Welding with low stress concentration

For foundation structures a large amount of the production costs are allotted to the welding, both for tripods and jackets.

For production planning and cost calculation it is of essential significance to calculate the volume of the welding seam quickly and easily along the intersection of the tubes. Up to this point the possibilities of common CAD-systems are constricted and only partly useful, regarding the high effort for modeling. Also comparable simple nodes are impossible to parameterize. A special software module calculates the seam volume right away and communicates with the CAD-system based on STEP-data interface.

Exemplary calculations with variations of the geometrical parameters show a great economic potential of saving welding volume and time.

The specified seam volume is a theoretical one assuming full penetration. But the welding regulation includes also a demand for a certain outer weld geometry providing low stress concentration. This results in a sculptural weld to generate corner arcs, which increases the welding seam volume dramatically. Rough calculations result in a real seam volume which is twice as much as the theoretical one.

Tolerances, produced during the tube and node manufacturing process, influence the root gap as well as the welding seam volume.

In the future the replacement of the traditionally used thermal cutting techniques or the combination with milling could make sense. Although the costs of the prefabrication will increase, a decrease of the overall costs because of more accurate fittings seems possible. This effect might be achieved by measuring the chord tube and milling the fitting according to the measured real connecting shape. From the mathematical point of view there would not be any additional difficulties, but it requires a completely different organization of manufacturing and data flow.

In industrial production with the conventional manufacturing process, with the accuracy of the used semi-finished goods, with the common equipment and with the qualification of the staff, limiting the gaps between the intersecting parts of tripods to less than 12 mm is not realistic.

4.4   Seam volume and deposition rate

A replacement of arc welding techniques in the next years for the described tasks is not foreseeable.

Alternative processes, like electron beam welding, are far from practical application in this industrial sector. Indeed a higher welding speed by connecting the parts is possible, but the fitting of the weld flanks needs to be at a magnitude of 0.2 mm! Also the compliance of the material properties, especially the ductility, is not easy. Beyond that not only the nozzle but also the main tube needs machining (MUE2004).

Simple circular and longitudinal welds at rolled sections are widely mechanized and can be fabricated fast and with low costs, in high, repeatable quality by submerged arc welding (indenture number 121, 125). Cost effectiveness is measured by assimilated deposited metal and is called deposition rate. An amount of 10 to 20 kg/h is accessible with submerged arc welding process.

The seams at intersecting contours of nodes need to be fabricated by GMA-welding (indenture number 135). The deposition rate of this arc welding process is about 3 to 3.5 kg/h.

A crucial fabrication problem is the antagonism between the high quality demands on one hand and the extremely high seam volumes on the other. For example, with a tripod the connections of the three upper braces with the central tube alone need more than 1000kg of filler metal!

The manual weld with string bead technique, in overhead position, normally executed at racks or platforms, decreases the deposition rate. Technical innovations, like welding with cored wire, increases not the quantity but only the quality of the weld (WEISE2013).

In Europe highly qualified and trained welders are not available in the desired amount on the employment market. They realize a weekly average deposition rate at curved seams of significantly lower than 1 kg/h.

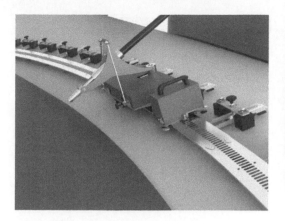

Figure 10. Robot like, rail based welding device.

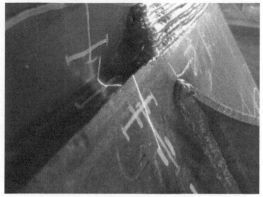

Figure 11. Weld at a tripod, generated with the robot like welding device consisting of more than 100 welding seams.

Therefore it makes sense to eliminate three-dimensionally formed welding seams completely. This is possible if cast nodes are being used. At welded nodes a complete displacement of three-dimensionally formed welding seams at the stage of prefabrication is highly recommended. This gives the preference to the design principle "Jacket".

5 POSSIBILITIES OF AUTOMATION

The antagonism between welding quality demands and cost effectiveness can be dissolved only by automation.

From the principal point of view the dimensions of jacket nodes enable the possibility of robot welding. No kinematic problems are to be expected. The one and only problem is to find a way of programming the robot movement, which guarantees at least the same quality as a manual welder with much higher power-on time of the arc. This task sounds easy but emerges as very complex. The main problems are:

- The multilayer welding seam which consists in extreme cases of more than 100 single seams.
- The continuous change of the curvature and bevel direction along the seam which results in continuous change of the welding parameters (arc currency, welding speed ...).
- The demanded welding quality without any discontinuity and also the demanded material parameters in the seam and in the head affecting zone.

Using an automatic robot path generation based on CAD-data on one hand and welding knowledge assigned into algorithms on the other hand seems promising. This can be done offline, i.e. in the office. At the robot, i.e online, by the help of camera based sensors, the path is transformed to the real geometrical conditions. This task needs an interdisciplinary collaboration of designers, software developers and welding engineers. To solve this problem great efforts are being made at the moment.

For welding the very long seams at tripod structures a rail based solution was developed (WEISE2014).

The machine, moving close to the parts surface, does not appear to be robot-like but has robot like properties. It is small, lightweight, cheap and modular and enables at least four times higher deposition rates than a manual welder (LÖFFLER2013).

6 SUMMARY AND FORECAST

Offshore-windmills are a challenge also for the manufacturer. New and unconventional solutions are necessary along the complete process chain to meet the technical demands in combination with cost effectiveness. The highest potential for improvement is found in welding.

By adjusting the design on one hand and abandoning manual welds on the other hand future demands can be fulfilled.

REFERENCES

Schaumann, P.; Lochten-Holtgreven, S.: Optionen der Stahlbauweise in der Windenergie. Tagungsband zum Deutschen Stahlbautag, 7./8.10.2010, Weimar.

Euler, M., et al. (Universität Stuttgart, Hochschule München, Hochschule der BW München, SLV Halle GmbH): Ermüdungsgerechte Fachwerke aus Rundhohlprofilen mit dickwandigen Gurten. Forschungsbericht FOSTA P815, Stuttgart, 2013.

Det Norske Veritas: Design of Offshore Steel Structures, DNV-OS-C101, 2011.

Germanisch Lloyd: IV Industrial Services 6 Offshore Technology, Hamburg, 2007.

Germanisch Lloyd: IV Industrial Services 2 Guidelines for Certification of Offshore Wind Turbines, Hamburg, 2011.

AWS D1.1/D1.1M:2010 An american national standard. American welding Society, 2010.

Koch, M., Müglitz, J.: Tubular Frameworks in Steel Construction – Welding and Cutting. Tubular Structures IX, Proceedings of the 9th International Symposium and Euroconference on Tubular Structures, Düsseldorf, Germany, 03.- 05.04.2001, A.A. Balkema Publishers, 2001.

European Patent Application EP10726876, 2010: Verfahren zur Herstellung überschwerer Rohrverbindungen, bevorzugt für Offshore-Windenergieanlagen.

Müglitz, J., Sobisch, G., Müller, U., Langrock, S.: 3D-Verbindungen- Schneiden, Fräsen, Elektronen-strahlschweißen und Prüfen.

Löffler, T.: Schweißtechnische Erprobung von schienengeführten, roboterähnlichen Maschinen mit 3D-torsionsgekrümmten Schnittkonturen, Bachelor-arbeit. Hochschule Mittweida, Germany, 2013.

6th Int. Conference Beam Technology, 26.-28.04.2004, Halle, Germany.

www.macaso.com, Software system TubeCut, 2014.

Weise, S.: Konzeption, Entwicklung, Umsetzung und Optimierung – Ein Erfahrungsbericht aus dem Stahlbau für Offshore-Gründungsstrukturen. Tagungsband DVS Congress 2013, DVS-Berichte 296, DVS Media GmbH, Düsseldorf, 2013, S.280–286.

Weise, S.; Ströfer, M.; Müglitz, J.: Schienengeführte, roboterähnliche Maschinen zum Schweißen, Schneiden und Prüfen. 14. Tagung Schweißen in der maritimen Technik und im Ingenieurbau, 2014, Tagungsband SLV Nord Hamburg, S.51–59.

Tubular Structures XV – Batista, Vellasco & Lima (eds)
© 2015 Taylor & Francis Group, London, ISBN 978-1-138-02837-1

Stinger design for *Pioneering Spirit* – the world's largest pipelay vessel

Y. Yu, J. van der Sman & J. van Lammeren
Allseas Engineering BV, Delft, The Netherlands

ABSTRACT: This paper presents the design of the *Pioneering Spirit* stinger based on the boundary conditions of the vessel. The stinger is designed for worldwide operations and will be installed in the slot between the bow sections of the vessel where the topsides lift system is located. During topsides lifting activities, the stinger will be removed from the vessel. An extensive loads on system approach has been defined for different operational modes of the stinger. The main structure of the stinger has been designed mostly with welded circular hollow sections. For the interfaces between the stinger sections and some local high stress locations, tubulars with stiffener plates or plate structures are used. The stinger starboard main hinge and one joint close to this main hinge consist of cast nodes in order to reduce the stress concentrations and hence improve the fatigue capacities. The design challenges for the stinger are discussed.

1 INTRODUCTION

1.1 *S-lay and stinger*

Allseas is a global leader in offshore pipeline installation and subsea construction. The company's applied pipelaying technique is the so-called S-lay method. The configuration of the pipe from the vessel to the seabed is shaped as an "S" (see Fig. 1).

In order to keep the pipeline in the correct "S"-shape and hence to guarantee the pipeline strain within the allowable value, the pipeline is held onboard by tensioners and supported by the stinger structure as it leaves the vessel. Moreover, the vessel must be kept in a controlled position either by dynamic positioning (DP) or by anchoring. Thus, the tensioner capacity, the stinger and the DP or anchoring capacities are three major parameters for determining the pipe lay capacity. Pipelines consist of pipe joints where each joint is a piece of pipe with a length of 12.2 m for a standard single joint. The S-lay pipe joints are welded horizontally on the production line in the vessel, which is called the

"firing line". After each completion of joint welding, the vessel moves forward or backward in the case of *Pioneering Spirit* and the pipe slides down. Step by step, the pipe is laid on the seabed.

The stinger, a space frame tubular structure used to support the pipeline and to guide it from the vessel to the seabed in a pre-designed curvature, is an indispensable part of the S-lay system. The firing line is equipped with welding, non-destructive testing (NDT) and field joint coating (FJC) stations, and occupies almost the whole length of the vessel in order to maintain efficient pipeline production. The stinger must be mounted on the bow or on the aft of the vessel in order to be in line with the firing line for easy pipe joint welding production. The stinger usually consists of two to three sections with roller boxes on each section that act as support points for the pipeline. Both the relative angle between each stinger section and stinger roller box heights can be adjusted in order to meet the requirement of the stinger radius for a specific pipelay installation project. The stinger radius is important since the theoretical pipe bending strain is equal to the pipe radius divided by the stinger radius when the pipe passes over the stinger.

The stinger structure consists of top and bottom chords, vertical, horizontal and diagonal braces of circular hollow sections welded into multi-planar tubular joints. Although the calculation of global member forces in chords and braces seems straightforward, the local joint stresses are complicated especially when dynamic wave loads and pipeline loads are applied on the stinger.

The loads applied on the stinger depend on its operational scenarios. Usually, the governing stinger operation modes include pipelaying, transit and survival conditions. During pipelaying, the stinger is

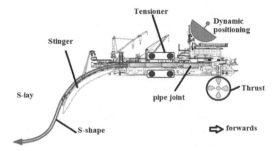

Figure 1. S-lay system.

submerged in water, and the loads on the stinger are dead weight, buoyancy, wave, current and vessel motion induced loads, and loads from the pipeline which are called roller box (RB) loads. Stinger survival is the case when weather conditions deteriorate and the stinger has to be retracted from the water to its highest position in order to avoid unacceptable wave drag and inertia loads on the stinger, and to keep the loads on the vessel within the capacity of the stinger-vessel interfaces. Stinger transit happens when the vessel sails across the ocean from one pipelay project location to another with the stinger retracted to its highest position. During survival and transit, the loads on the stinger are dead weight, wind and vessel motion induced loads. Except for dead weight, buoyancy and current loads, all other loads are dynamic loads. Wind and wave induced dynamic response of a stinger is usually negligible due to its high natural frequency, which is far from the wave or wind excitation frequency.

1.2 Allseas' pipelay vessels

Allseas operates a fleet of pipelay vessels: the DP pipelay vessels *Solitaire*, *Audacia* and *Lorelay*, and the anchor mooring barge *Tog Mor*. As the world's first DP pipelay vessel, *Lorelay* has executed various small to medium diameter pipeline projects. In 1996, she first extended the limits of S-lay installation and set a world record of pipelaying with a water depth of 1645 m. *Solitaire*, operating since 1998, is currently the largest pipelay vessel in the world and is optimized for laying medium to large diameter pipelines at high speed. In 2003 she set a world record of 9.3 km per day in pipelaying speed, and in 2006 she wrote a new world record with a water depth of 2775 m. *Audacia* is Allseas' latest pipelay vessel. Operational since 2007, she is optimised for the execution of small-to large-diameter pipeline projects. Unlike *Solitaire* and *Lorelay*, *Audacia's* stinger is located on the bow and she sails faster than other vessels. *Tog Mor* is an anchored barge and was converted for pipelay between 2001 and 2002. She operates primarily in shallow water areas.

1.3 Pioneering Spirit stinger

Allseas' dynamically positioned installation/decommissioning and pipelay vessel *Pioneering Spirit*, with an overall hull length of 382 m, width of 124 m and depth of 30 m, has been designed as a multi-purpose vessel to transport, install and remove topsides and jackets and to lay the heaviest pipelines in S-lay mode. The jacket installation and removal equipment is located at the stern of the vessel, while the topsides installation and removal equipment is located at the vessel's bow. The stinger will be fitted in the bow slot of *Pioneering Spirit*. During topsides lifting the stinger will be removed from the vessel.

By doubling the tension capacity of *Solitaire*, *Pioneering Spirit* will provide a pipe top tension capacity of 2000 tonnes and a bottom tension capacity of around 700 tonnes. The high top and bottom tension capacity of *Pioneering Spirit* provides the possibility to design a high capacity stinger, for a wide range of water depths and pipeline diameters with adjustable configurations. Based on *Pioneering Spirit's* tension capacity and initial studies about the current and future pipelay market, the total stinger length was determined to be around 198 m. As a result, *Pioneering Spirit* will surpass *Solitaire* as the largest S-lay capacity vessel in the world.

2 DESIGN BOUNDARY CONDITIONS AND CHALLENGES

As described in Section 1.3, it was decided to fit the stinger in the bow slot based on initial investigations. During design of the stinger, the decision was made to increase the width of *Pioneering Spirit* by 6.75 m towards the portside and keep the firing line location unchanged. As a result, the firing line is located 3.375 m off the vessel center line towards the starboard side. The maximum slot width is 58.75 m and the slot length is 122 m. The shape of the slot can be seen in (Fig. 2).

After a preliminary feasibility study, the concept was selected to combine a stinger transition frame (STF) with a three-section stinger. Boundary conditions for this design concept are:

- The length of the stinger transition frame (STF) is 55 m and the length of the three-section stinger is 143 m. In total, the length is 198m.
- The STF, working as an extension of the firing line with the possibility of radius adjustment, is supported at the slot end of the vessel. The forward end of the STF can either be connected by a sliding mechanism to section 1 of the stinger during pipelay and good weather transit (Fig. 2) or suspended from the topsides lifting beams (Fig. 5) to obtain maximum clearance with the waterline in bad weather conditions and to reduce wave slamming loads on the STF. Up to five field joint coating (FJC) stations will be accommodated on the STF, and should be closed from the environment with the possibility to open the roof for inline structure installations. A general view of the STF and the stinger during pipelay operations is shown in (Fig. 2).
- The stinger is supported on main hinges connected to retractable stinger hinge beams protruding from both bows, and suspension supported with redundancy winches at the forecastle decks via the stinger hang-off (SHF) tackle wires (see Fig. 2).
- The vessel-STF and vessel-stinger interfaces should not be obstacles to the topsides installation and removal activities. Thus, these interfaces should be designed such that they can be installed and removed quickly, or such that the topsides activities are not hindered and slot dimensions are not compromised.
- The STF and the stinger will be installed and removed with a dedicated barge that will also serve

as storage location when *Pioneering Spirit* is in heavy lift mode.

- In order to avoid water slamming on the stinger, sufficient stinger tip clearance to water level should be maintained for survival and transit.
- Transportation of the STF and the stinger can be done either with *Pioneering Spirit* or with a dedicated barge, depending on the project schedule. The STF and the stinger will be designed to be able to transit worldwide with *Pioneering Spirit* under heading and sea state limitations. For worldwide transit with the barge, no heading and sea state limitations will be applied.
- A quick radius adjustment system should be applied in open seas in order to increase time available for pipelay.
- The top and bottom tension capacities of *Pioneering Spirit* are 2000 tonnes and around 700 tonnes, respectively. The stinger will be designed to meet the demand of the current and future pipelay market based on *Pioneering Spirit's* capacity.
- The external pipe diameters will be within the range of 6″–68″ including concrete coating, which cover the current and future pipelay market.
- The challenges of this stinger design are the large stinger size, the restrictions in arrangements of stinger supports due to the topsides removal activities and last but not least, the relative bow deflections. By reducing stiffness due to the slot, the two bows will have relative deflections under wave and current loads especially in vertical and transversal directions. These relative bow deflections make the stinger design more complicated.

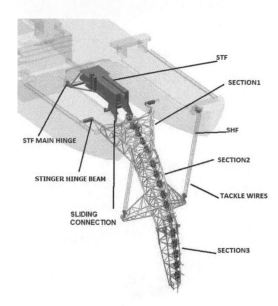

Figure 2.   General arrangement of the STF and the stinger.

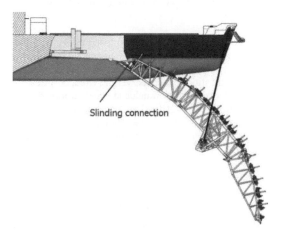

Figure 3.   STF and stinger during pipelay.

3   LOADS ON SYSTEM

Based on the boundary conditions and the stinger operational scenarios, loads on system were defined. This paper presents only some important general loads on system. For the stinger design, a more detailed loads on system was developed which is not further described in this paper.

3.1   *STF and stinger during pipelay*

With the combination of STF and stinger, the range of pipelay radii can vary from 110 m to 425 m applied for ultra-deep to shallow water pipelay projects, respectively. A general arrangement side view of the STF and stinger with a radius of 110 m is shown in (Fig. 3) as an example. An isometric view can be found in (Fig. 2).

As previously described, during pipelay the aft end of the STF is supported at the slot end of the vessel. The forward end of the STF is connected by a sliding mechanism to section 1 (see Fig. 3).

The stinger is supported on main hinges and suspension supported via the stinger hang-off (SHF) tackle wires at section 2 (see Fig. 2). The STF includes up to five work stations for application of field joint coating (FJC) on the pipe. The loads on the stinger system are dead weight, buoyancy and bow deflection imposed displacement loads at the support locations, friction at the sliding connections, Morrison drag and inertia loads due to wave and current and RB loads from the pipeline. Depending on the combination of the stinger radius and RB loads, the stinger is designed to resist environmental loads of wave heights up to Hs = 4.5 m in combination with a current speed of up to 1.5 m/s without heading control.

3.2   *STF and stinger during transit with Pioneering Spirit (manned)*

This is the case when the STF and the stinger sail across the ocean with *Pioneering Spirit*. The support boundary conditions are the same as those during pipelay, and the crew can continually work on the STF for project preparations and maintenance: the so-called

sliding connection

Figure 4. Stinger and STF during survival and transit (manned).

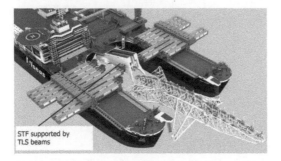

STF supported by TLS beams

Figure 5. Stinger and STF during survival and transit (unmanned).

manned or connected condition (see Fig. 4). In order to keep sufficient stinger tip clearance to water level, the stinger is retracted to its up position. The loads on the STF and stinger are dead weight, wind, bow deflections and vessel motion induced accelerations due to waves and frictions at the sliding connection. The maximum design significant wave height is Hs = 6 m for all wave headings in combination with a wind speed of 25 m/s.

3.3 STF and stinger during survival and transit with Pioneering Spirit (unmanned)

For ocean transit of the STF and the stinger with *Pioneering Spirit* in wave heights Hs higher than 6 m, the stinger keeps the same configuration as described in Section 3.2, while the STF will be cleared of all personnel and lifted up using a dedicated hang-off system suspended from the topsides lift system (TLS) beams in order to reduce wave slamming loads on the STF and hence reduce the loads on the stinger and the vessel (see Fig. 5).

Two survival design cases are included: 1) maximum wave height Hs = 8 m with pipe abandonment and recovery (A&R) cable hanging on the stinger; 2) maximum wave height Hs = 12.5 m without A&R cable. For both survival conditions, the vessel is under heading control in order to reduce vessel motion induced loads.

The loads on the STF and the stinger are dead weight, imposed displacements due to bow deflections at the stinger-vessel interfaces, wind load and vessel induced accelerations due to waves. The wind speed for transit and survival case 1) is 25 m/s and for survival case 2) is 35 m/s.

Figure 6. General overview of the barge supports.

3.4 Transit with the dedicated barge

Depending on the project schedule, ocean transit of the STF and the stinger can also be done with the dedicated stinger barge instead of using *Pioneering Spirit*. A general overview of STF and stinger on the barge is shown in (Fig. 6). The aft part of the STF is supported on the barge and the forward part of the STF is rested on the stinger sliding connection, which is the same as during pipelay. Preliminary analysis has shown that if only static loads are applied on the stinger, four supports on the stinger and two supports on the STF are enough to carry the loads. However, under the design wave transport conditions, large roll and sway movements occur. Extra supports (sea fastening supports) for these dynamic effects have to be designed for both the stinger and the STF. Loads on the stinger and the STF are dead weight, wind, accelerations due to design waves, frictions at the supports and at the STF-stinger sliding connections and barge deformations at support locations due to design waves. The maximum design wave is Hs = 12.5 m and the maximum wind speed is 35 m/s without heading control.

3.5 Radius adjustment in open sea

The stinger radius adjustment has been designed to be operated in open sea with mild weather conditions without the use of external equipment. This procedure reduces more vessel down time compared to other radius adjustment systems, such as with support barges in a sheltered area.

The radius adjustment consists of the STF part and the stinger part. For the STF part, the STF angle adjustment is obtained by sliding the STF over stinger section 1 where sliding pads are constructed.

For the stinger part, the radius adjustment consists of adjustment of the relative angle between sections 1 and 2 and that between sections 2 and 3. Sections 1 and 2 are connected with each other via portside (PS) and starboard (SB) lower hinges at the bottom plane, and via PS and SB pup pieces at the top plane of the stinger (see Figs 7 and 8 for the locations of each item).

Between sections 1 and 2, the pup pieces are connected to the so-called PS and SB "hanenkammen" (cockscombs in English) (see Fig. 8). The same principle is applied to the connections between sections 2 and 3.

The final tuning of the STF and the stinger radius is achieved by adjustment of the roller box heights. A

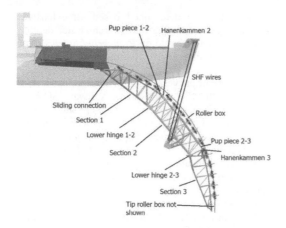

Figure 7. Explanation of the stinger parts.

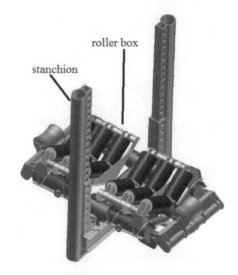

Figure 9. Height adjustable roller box structure.

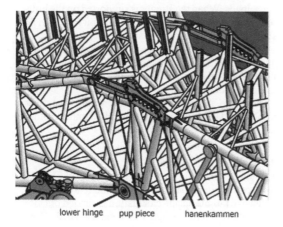

Figure 8. View of lower hinge, pup piece and hanenkammen (roller boxes are not shown).

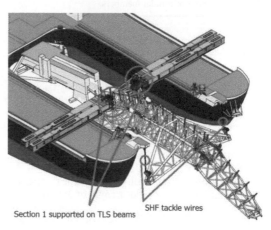

Section 1 supported on TLS beams   SHF tackle wires

Figure 10. Section 1 supported on TLS beams.

typical roller box structure is shown in (Fig. 9). The two stanchions have pin holes in height. Depending on the pipelay radius requirement, the roller box can be supported at the predefined pin locations. The stanchions are welded on the stinger main structure as shown in (Fig. 8).

During radius adjustment, the loads, if applicable, are dead weight, buoyancy, wave and wind loads and loads due to bow deflections. The maximum design wave is Hs = 2 m with heading control. The design wind speed is 10 m/s.

3.5.1 *Radius adjustment between sections 1 and 2*
When the angle between sections 1 and 2 is adjusted, section 1 is supported by the TLS beams as shown in (Fig. 10), and the pup pieces between sections 1 and 2 are disconnected. By lowering or lifting the SHF tackle wires, section 2 will rotate around the lower hinges between sections 1 and 2 until the pup pieces between sections 1 and 2 reach the desired position.

3.5.2 *Radius adjustment between sections 2 and 3*
For the adjustment of sections 2 and 3, the stinger adjustment system (SAS) cables with a fixed length

are connected from the vessel to section 3 (see Fig. 11). First the pup pieces between sections 2 and 3 are disconnected, and then the SHF tackle wires are lowered or lifted until the pup pieces reach a new desired position. As a result, the angle between sections 2 and 3 is adjusted.

4 GLOBAL AND DETAIL DESIGN

For the stinger design, a combination of Lloyd's LAME (2009) and API RP2A WSD (2007) has been applied. For tubular members and tubular joints, the static strength has been checked in accordance with API, while the detail design for plates and stiffeners has been based on Lloyd's LAME.

4.1 *Material of structure*

The materials used for tubular members, plates and castings are all Lloyd's approved in accordance with Lloyd's rules (2010) and Allseas specifications (2012).

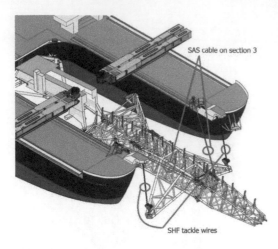

SAS cable on section 3

SHF tackle wires

Figure 11. Radius adjustment between sections 2 and 3.

Table 1. Material of tubular members.

Specified yield strength MPa	Tensile strength MPa	Thickness range mm	Grade normal material	Z-grade material
		6< =t< =25	DH46	DH46 + Z35
460	570–720	25 <t< =50	EH46	EH46 + Z35
		50 <t< =150	FH46	FH46 + Z35

The material grade of the main tubular structure is shown in Table 1.

Plates, stiffeners and roller box frames are also Lloyd's material with a yield strength of 355 MPa.

Furthermore, lower parts of the hanenkammen plates have been designed with material FH69 and casting joints have been designed with material yield strength of 355 MPa.

4.2 Global design

The stinger global design has been carried out based on the design concept and loads on system as described in the previous chapters. The design analyses have been done with the FE programme STAAD/Pro V8i. An in-house Matlab code has been used to automatically check the member strength and joint strength in accordance with API RP2A WSD. The fatigue design results used for making the stinger inspection plan are not included in this paper.

4.2.1 Loads and load combinations
The loads and load combinations depend on the stinger operational scenarios. Based on these scenarios, comprehensive loads on system were determined which are partly described in Chapter 3.

Weight and buoyancy loads can be easily determined based on the designed structure and its configuration.

RB loads were determined based on extensive dynamic pipelay analyses for different water depths and pipe diameters based on *Pioneering Spirit's* tension and thruster capacity and the possible stinger radii.

Wave, current and vessel motion induced loads were calculated with the hydrodynamic software AQWA. The three-hour time domain results were statistically analysed based on the most critical combinations of wave direction, frequency, spectrum type, current speed and current direction. Afterward, design loads were determined and the corresponding nodal load results were used as input for the structural design analyses.

Due to the flexibility of the two hulls of *Pioneering Spirit* at the bow side, relative deflections of the hulls exist in all three translational directions, especially in the transversal and vertical directions under wave loads. When the stinger is supported on the hulls, relative bow deflections result in reaction forces on the stinger.

Friction forces exist between the STF and the stinger at the sliding pads. The magnitude of the friction force is equal to the sliding pad contact force multiplied by a friction coefficient.

After the above individual basic loads were determined, load combinations were applied and structural design analysis was carried out.

4.2.2 Design principles
The stinger is a space frame structure of tubular hollow sections under static and dynamic loads. Usually, extensive fatigue design calculation is carried out after the static strength design has been finished. However, too many last stage modifications are time and labour cost consuming, and the design deadline would not be met. Thus, attention should be paid to both static and fatigue strength design right from the first design stage, which means that when the members are designed, the joint design should be considered carefully based on the knowledge of joint behaviour and the possible dynamic load variations applied on the joint. A static and fatigue load friendly joint will be with a smaller chord radius over wall thickness ratio (γ) and with a smaller brace over chord wall thickness ratio (τ) for specific nominal stresses in the chord and the braces. As a rule of thumb, it is recommended to design the connection with $\gamma \leq 10$–12.5 and $\tau \leq 0.5$ for a fatigue sensitive structure in order to limit the stress concentration factors (SCF), and hence to enhance the static strength and the fatigue life of the joint (Ermolaeva et al. 2010). If it is too difficult to reduce the joint SCF according to the above method, for example, when the chord wall thickness is too high and it will result in further negative thickness effect on fatigue life, a detailed joint design with stiffener plates or casting nodes could be applied.

Pipelay mode is one of the governing cases for the stinger design. Stinger buoyancy balances part of the structural dead weight and works as a positive effect on one hand. On the other hand, too much

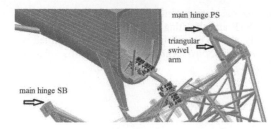

Figure 12. Triangular swivel arm.

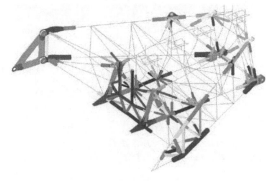

Figure 13. Detailed design of section 1.

buoyancy results in stinger uplift and thus a compli-cated heave compensation system has to be designed. This is usually avoided by selecting smaller diameter tubular members. Furthermore, large diameter tubu-lar members cause high drag loads which are one of the major loads especially in the transversal direction. Transversal loads are critical for the stinger design since there are usually no transversal supports on the stinger except at its main hinges.

4.2.3 *Design challenges*

Challenges were encountered during the design of the *Pioneering Spirit* stinger, and are mentioned mainly as follows:

– The boundary conditions
– The size and weight
– The bow deflections.

As mentioned earlier, the stinger is located in the bow slot and should not be an obstacle for the topsides lifting activities. Thus, the STF is designed to be only supported at the slot end at its aft, while at its fore end it is supported on stinger section 1 to avoid interface supports from the vessel. For the stinger, it is supported on retractable main hinge beams at the main hinges and suspension supported on the forecastles. The main hinge beams (shown in Fig. 2) will be retracted into the inside of the hulls during topsides lifting activities, and the forecastle supports are not obstacles either. Due to the slot width, wide wings were designed on stinger section 2 in order to avoid side lead when the stinger is suspended via the SHF tackle wires.

Due to its design capacity requirements and bound-ary conditions, the stinger size and weight are the largest compared with all Allseas' other stingers. This makes the design, fabrication and installation quite challenging. For installation and removal of the stinger, a dedicated barge was designed.

In order to reduce the forces on the stinger due to bow deflections especially in the transversal direc-tions, a triangular "swivel arm" was introduced at the portside of the stinger main hinge (see Fig. 12). Spher-ical ball bearings were designed on the three vertices of the triangular swivel beam. When bow transver-sal deflections at the main hinge supports occur, the stinger can rotate around the hinges of the swivel arm so that no transversal reaction force will be applied on the main hinge supports.

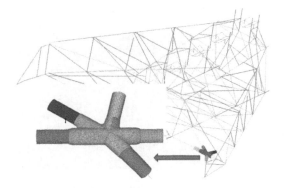

Figure 14. Casting joint at bottom plane section 1.

4.3 *Detail design*

Detailed design analyses on the stinger main hinges, swivel arm, lower hinges 1-2 and 2-3, pup pieces 1-2 and 2-3, hanenkammen (see Figs 2, 7 and 8 for item locations) and all joints with stiffeners and casting joints were carried out with the FE programme FeMap and Nastran. Two casting joints were designed: one at the stinger SB main hinge, and another close to this main hinge at the bottom plane in order to increase the fatigue capacity. Locations of the detailed FE model of section 1 are shown in (Fig. 13) as an example. Cast-ing node at bottom plane close to the SB main hinge is shown in (Fig. 14).

5 FABRICATION

Companies involved with the fabrication of the stinger and its related major structures are:

– Iemants N.V. Belgium: stinger primary structure, walkways and platforms
– China Merchants Heavy Industry (Shenzhen) China: stinger transition frame (STF)
– COSCO Shipyard (Zhoushan) China: stinger barge
– Allseas Fabrication B.V. the Netherlands: roller boxes.

The structure fabrication was performed under Lloyd's surveyor's approval. The factory fabricated

Figure 15. Section 1 under construction.

Figure 16. Sections 1 and 2 under construction.

Figure 17. Sections 1, 2 and 3 under construction.

stinger parts were transferred to the quay site and were further welded together as complete sections. Afterwards, the STF, the stinger sections, roller boxes and outfitting were installed together on the quay site and loaded out on to the stinger barge which will transport the STF and the stinger to *Pioneering Spirit*. Eventually, the STF and the stinger will be installed on *Pioneering Spirit*, which is expected to be achieved in 2015.

6 CONCLUSIONS

Design of the *Pioneering Spirit* stinger has been a challenging task due to its strict boundary conditions and high demands on capacities. After completion of installation in 2015, it will be the largest capacity pipelay stinger in the world and will make *Pioneering Spirit* suitable for ultra-deep to shallow water pipelay of small to large diameter pipelines. *Pioneering Spirit*'s first pipelay job will be to install 888 km of 32″ pipeline in the Black Sea at over 2000 m water depth.

REFERENCES

Allseas specification 2012. Pieter Schelte – general specification for special structural components, *Allseas PI-10110-024-N-F-010*

API RP2A 2007. Recommended practice for planning, designing and constructing fixed offshore platforms – working stress design, API RP2A, 21st edition

Ermolaeva, N., Yu, Y. & Zhao, L. 2010. Design and fatigue assessment of a stinger. In Young (ed.), *Tubular Structures XIII: Proc. Intern. Symo., Hong Kong, 15–17 December 2010.* Rotterdam: Bakema. 547–555

Lloyd's LAME 2009. Code for lifting appliances in a marine environment, Lloyd's Register of Shipping

Lloyd's Rules 2010. Rules for the manufacture, testing and certification of materials, *Lloyd's Register*

Tubular Structures XV – Batista, Vellasco & Lima (eds)
© *2015 Taylor & Francis Group, London, ISBN 978-1-138-02837-1*

Comparative assessment of the design of tubular elements according to offshore design standards and Eurocode 3

T. Manco, J.P. Martins & L.S. da Silva
ISISE, Civil Engineering Department, University of Coimbra, Portugal

M.C. Rigueiro
ISISE, Civil Engineering Department, Polytechnic Institute of Castelo Branco, Portugal

ABSTRACT: The objective of this paper is to perform a comparative analysis of design standards for the offshore construction, with focus on the structural design of circular steel tubular elements. It is intended to conclude about the safety level of two sets of standards: offshore *vs.* construction standards. To this end, the bearing capacity obtained by ISO 19902 and Eurocode 3 (that does not contain specific provisions for offshore structures) through part 1-1 and part 1-6 will be compared for a broad parametric study. Subsequently, a numerical analysis using ABAQUS will be carried out to evaluate the accuracy and safety level that ISO and EC3 assume. To this end, the performance of a set of individual circular steel tubular elements will be evaluated under axial compression and hydrostatic pressure.

1 INTRODUCTION

Steel circular tubular elements have a vast application in Civil Engineering. With the exception of gravity structures, the generality of offshore structures supported on the seabed is typically built with this type of elements. Due to their dimensions and high values of slenderness that offshore structures can reach, these elements are prone to buckling phenomena. In particular they are very prone to global and local buckling by axial compression and hoop buckling due to the hydrostatic pressure.

In what concerns the comparison between standards, the bibliography presents some studies which are applied to offshore structures. However, few of them deal in a comprehensive way with the differences obtained in relation to the bearing capacity of the elements, for example (Idrus et al., 2010), (HSE, 2001) and (Tuen, 2012). In spite of that, they still omit or still do not treat with the desired thoroughness some aspects deemed important like the reduced parametric variation, the superficial evaluation of hydrostatic pressure and the reduced development of combined loadings. Regarding the inclusion of Eurocode in comparisons with standards directly applied to the design of offshore structures, what can be found in the bibliography is much scarcer, for example (Tuen, 2012) and (Steck, n.d.) but once again, it is believed that there are gaps to be filled and that the study should be continued and deepened. In this respect the main limitations encountered concern: The class of cross-sections studied; the non-accounting of the hydrostatic pressure; and the limited evaluation of the interaction

between loadings in Eurocode. In an even more significant way it is verified a scarcity of studies that conclude about the safety level that each standard has inherent to itself, for example through the comparison with results obtained by the finite element method.

The purpose of this paper is to present a comparative analysis in the design of steel circular tubular elements in offshore structures (in particular fixed structures), through different regulatory approaches namely ISO 19902 and Eurocode 3, the last one does not contain specific provisions for offshore structures. It is intended to evaluate the results in terms of strength that these two sets of standards (offshore standards *vs* construction standards) present and conclude about the safety level inherent to each of the groups. Furthermore, it will be assessed if EC3 can be applied to offshore structures. This requires the application of two distinct parts of this standard: part 1-1 (EN 1993-1-1) for class 1, 2 and 3 and part 1-6 (EN 1993-1-6) for class 4 cross-sections (shell elements). Looking for more in-depth conclusions at this level, the results obtained by the expressions set out in each standard will be compared with the results obtained from the numerical study through the finite element method which was carried out using ABAQUS software under axial compression and hydrostatic pressure.

2 NUMERICAL MODELING

In this study was performed a linear elastic bifurcation analysis (LBA) (through the Lanczos algorithm available in ABAQUS) to determine the elastic critical load (R_{cr}) and the eigenmodes of the perfect cylindrical tubes. The first was used to calculate the slenderness of the element ($\lambda = \sqrt{(R_{pl}/R_{cr})}$), where R_{pl} is the yielding load. The eigenmodes were used as patterns for the geometric equivalent imperfections in the geometrically and materially nonlinear analysis with imperfections (GMNIA).

In geometrically nonlinear problems involving energy loss in the system (negative stiffness), characteristic of buckling problems, the equilibrium path (load-displacement) is not monotonic and as such resolution algorithms that can efficiently translate this behavior should be applied. One of the most efficient methods used in post-buckling problems, is the so-called Riks method (Riks, 1978), which was used here. This method modifies the load factor on each iteration, so it has to resolve simultaneously the unknowns load and displacement, using the "arc-length" technique for the resolution of the problem.

In what concerns the material, it was used a linear-elastic and an elastic-perfectly plastic behavior, respectively for the LBA and GMNIA analysis, with a modulus of elasticity (E) of 210 GPa, a Poisson coefficient (υ) of 0.3 and a yield stress (f_y) of 355 MPa. The elastic-perfectly plastic behavior is, according to the bibliography (e.g. (Rotter & Schmidt, 2008)), the behavior more often applied to this type of analysis with mild steels like the ones here considered.

The imperfections, following studies of authors such as Koiter (1945), were considered through a pattern of equivalent imperfections in the form of initial deflections perpendicular to the middle surface of the shell. These patterns are given by the eigenmodes of perfect elements through LBA's. These imperfections, being "equivalent", must cover the effects of imperfections like load eccentricities, residual stresses, etc.

The provisions given by the Eurocode 3 were used to simulate the imperfections in the numerical model. In opposition to other standards, this one leads specific indications regarding the way equivalent geometric imperfections should be considered. In the absence of measurements, equivalent geometric imperfections can be estimated according to the fabrication quality of the element. EC3-1-6 takes this into account through the consideration of 3 distinct quality classes – Class A, B and C with decreasing order of quality. For the calculation of the respective imperfection amplitudes, EC3-1-6 indicates that these should be assessed through gauges of length to the situation of meridional compression and circumferential compression.

In the absence of a direct correspondence between standards for offshore structures and the Eurocode, in relation to fabrication related imperfections, the normal fabrication class (class C) of EC3-1-6 was adopted in the numerical analysis. This is mainly due to the following reasons: *i*) according to the type of element and its usual fabrication process, it is expected that the fabrication quality is not the highest according to the parameters of the EC3-1-6, which is reserved for more sensitive structures; *ii*) it is more useful to obtain the minimum resistance given by EC3-1-6 (through the most unfavorable class), since thus it is possible to conclude that a resistance in Eurocode superior to other standards implies that this standard is the least conservative for all classes of fabrication used; *iii*) it was found that this is the curve that best fits the results from ISO 19902 for buckling phenomena due to hydrostatic pressure (not in the case of local buckling for compression, since the procedure is relatively distinctive, as will be seen).

For the global buckling phenomena induced by axial compression, imperfections with the shape of the first global eigenmode were used. To the amplitude of this imperfection, i.e. lack of linearity of the element, values of L/250, L/300 (buckling curve "a" in EC3-1-1 for elastic and plastic analysis, respectively) and L/500 (amplitude usually referred in the bibliography for hot-finished tubular elements, e.g. Ziemian (2010)) were tested.

The modeling of the cylindrical tubes was made using shell-type elements S8R5. This element has 8 nodes with 5 degrees of freedom per node, 3 translations and 2 rotations (omitted the one perpendicular to the surface). To establish the mesh size was carried out a convergence study that led to the conclusion that the element size of $5 \times 5 \, cm^2$ corresponded to the best relationship between the precision of the results and the required computational effort.

Regarding the boundary conditions, simply supported elements were considered to axial compression. This means in one end unrestrained rotations in both axles of the plane of the end cross-section and in the other end these two rotations as well as the axial translation (axial rotation was blocked). For hydrostatic pressure both ends were considered clamped, once the axial direction does not have here a significant role.

In the case of axial compression, it was applied a point load in a reference point on the axis of the element. For hydrostatic pressure was applied a constant centripetal pressure over the surface of the cylindrical tube.

For the validation of the model subjected to axial compression influenced by global buckling phenomena experimental tests from Chen & Ross (1978) were used (Table 1). For the model subjected to axial compression influenced by local buckling phenomena a series of experimental tests conducted between 1976 and 1977 were used (Ostapenko and Gunzelman, 1976), (Gunzelman and Ostapenko, 1977) and (Marzullo and Ostapenko, 1977) (Table 2). Regarding the hydrostatic pressure experimental tests from Windenburg and Trilling (1934) were used (Table 3).

Table 1. Validation for axial compression (global buckling).

D_{ext} (m)	t (mm)	L (m)	Δ_{w0} (mm)	F_{cr} (MN)	F_{pl} (MN)	λ (Num)	λ (EC3-1-1)	χ_x (Num)	χ_x (Exp)	Dif. (%)
0.380	7.80	7.60	6.00	5.98	2.81	0.685	0.701	0.925	0.910	1.48
0.380	7.80	11.0	4.40	2.88	2.81	0.988	1.014	0.846	0.877	-3.08
0.560	7.80	11.0	4.20	9.01	4.17	0.680	0.684	0.968	0.968	0.01

Table 2. Validation for axial compression (local buckling).

D_{ext} (m)	t (mm)	L (m)	Δ_{w0} (mm)	F_{cr} (MN)	F_{pl} (MN)	λ (Num)	λ (EC3-1-6)	χ_x (Num)	χ_x (Exp)	Dif. (%)
0.717	8.35	2.05	5.96	51.45	5.93	0.340	0.336	0.942	0.998	-5.57
1.787	7.17	3.03	2.99	39.25	15.11	0.620	0.617	0.766	0.814	-4.79
1.532	6.55	2.44	2.86	32.87	19.56	0.771	0.767	0.953	0.912	4.14

Table 3. Validation for hydrostatic pressure (hoop buckling).

D_{ext} (m)	t (mm)	L (m)	Δ_{w0} (mm)	P_{cr} (MPa)	P_{pl} (MPa)	λ (Num)	λ (EC3-1-6)	χ_θ (Num)	χ_θ (Exp)	Dif. (%)
0.406	1.27	0.813	0.60	0.22	1.55	2.681	2.619	0.111	0.111	-0.01
0.406	1.35	0.406	0.43	0.50	1.96	1.988	1.946	0.182	0.204	-2.13
0.406	1.30	0.102	0.17	1.96	1.77	0.950	0.888	0.663	0.637	2.57
0.406	2.74	0.406	0.44	2.96	3.82	1.140	1.082	0.513	0.520	-0.78

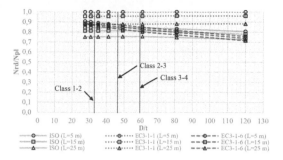

Figure 1. Normalized axial compression strength.

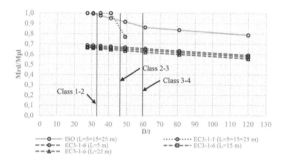

Figure 2. Normalized bending strength.

3 COMPARISON OF RESULTS

3.1 *Comparison of the strength given by ISO 19902, EC3-1-1 and EC3-1-6*

In this section, the design strengths (with partial safety coefficients) inherent to ISO 19902 and EC3 will be compared.

3.1.1 *Parametric variation*
The parametric variation used in this analysis aimed to cover all classes of cross-sections and check the limits of applicability of ISO 19902, being this the most restrictive standard (D/t < 120). The variation of geometry was obtained maintaining constant a diameter (D) of 1.5 m, varying the thickness (t) of the wall of the cross-section from 12.5 to 54.5 mm with increment of 4 mm. For which one of this sections a length (L) of 5, 15 and 25 m was established in order to differentiate the local and global buckling phenomena.

3.1.2 *Axial compression*
The effective length factor (K) was considered with the value of 1.0. The EC3-1-1 was applied to all classes of cross-sections in order to access the effect of global buckling in the larger D/t ratios, despite the fact that this standard it is not applicable in class 4.

From Figure 1 it is possible to conclude that EC3-1-1 leads to less conservative results to the global resistance in axial compression than ISO 19902 for every D/t ratios. As expected, the resistance of the longer elements is lower due to the greater susceptibility to the occurrence of global buckling, this is clearly perceptible both in ISO and EC3-1-1. The EC3-1-6 is not much influenced by this variable. With the maintenance of L/D ratio it can be seen that the variation of thickness has no impact on global buckling phenomena.

With regard to local buckling it turns out that the increasing of the D/t ratio has a greater impact in reducing the strength in EC3-1-6 that in ISO, as is noticeable by the steeper curves in the first standard. However, the lower and upper limits of the strength in these two standards for class 4 and for the lengths considered are very similar.

3.1.3 *Bending*
For calculating the bending strength EC3-1-1 uses the plastic section modulus for classes 1 and 2, and the elastic section modulus for class 3. This distinction leads to an appreciable reduction in the bearing capacity in class 3, as can be seen in Figure 2. In comparison with ISO it is noted that for class 1 and 2 cross-sections the results are quite similar to those obtained by EC3-1-1. On the other hand, for class 3 due to the reduction of resistance in EC3, ISO becomes the least conservative, and remained so up to the maximum D/t ratio presented here. In EC3-1-6 the bearing capacity was considered by converting the bending into meridional compression in the wall of the cross-section and verifying it to local buckling like the procedure carried out for axial compression.

The length of the elements has no influence on the bending strength, with the exception of EC3-1-6, where this variable has only a residual impact. In addition, this standard was applied to sections unsusceptible to local buckling, whence it can be concluded that its application to cross-sections with low slenderness leads to results considerably more conservative than the remaining standards.

3.1.4 *Axial compression + bending*
The adopted procedure consisted in fixing the value of the stress caused by the bending in 30% (Figure 3) and 60% (Figure 4) of the value of the yield stress (f_y), withdrawing the maximum value permitted to the axial

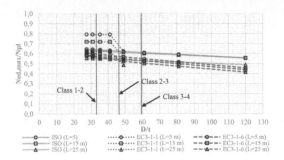

Figure 3. Normalized axial compression + bending ($\sigma_b = 0.3f_y$) strength.

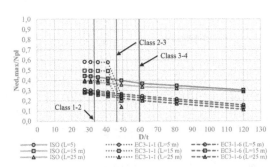

Figure 4. Normalized axial compression + bending ($\sigma_b = 0.6f_y$) strength.

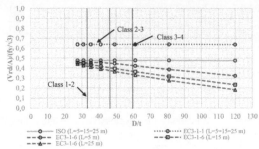

Figure 5. Normalized shear strength (without torsion).

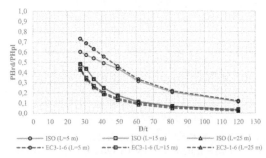

Figure 6. Normalized hydrostatic pressure strength.

load by solving numerically the interaction formulas in each standard. With this, it was intended to assess the influence that bending has in the decreasing of the axial load strength. The bending reduction factors (C_m) were obtained in ISO 19902. For a jacket leg this standard specifies the value of 0.85, having been the value used in both standards.

Analyzing Figure 3 and Figure 4 it can be concluded that EC3-1-1 gives a greater importance to the length of the elements. For class 1 and 2 cross-sections, the EC3-1-1 is less conservative than ISO. The difference is significant for the shorter element. EC3-1-1 for class 3 specifies a considerable breakdown in the strengths compared to the ones given for class 1 and 2. This reduction is mainly due to the interaction factors k which have for class 3 much more conservative values than for class 1 and 2.

3.1.5 Shear (without torsion)

In Figure 5 is presented the normalized shear strength, i.e. the stress V_{Rd}/A over the maximum permissible shear stress ($f_y/\sqrt{3}$) where it is verifiable that the differences between the shear strength given by ISO 19902 and EC3-1-1 are significant due to the differences in the calculation expressions. ISO 19902 is clearly more conservative that EC3-1-1. Furthermore, it can be concluded that shear strength, in both standards, is independent of the D/t ratio, as well as the length of the element. Due to the fact that EC3-1-6 deals with the shear similarly to other loads by buckling check, this standard establishes strengths that are

naturally dependent on the D/t ratio. The strength given by this part of the Eurocode is considerably lower than the other two standards, in particular for higher D/t ratios.

3.1.6 Hydrostatic pressure

EC3-1-1 does not take into account the hydrostatic pressure. From Figure 6 it can be seen that there is an excellent adjustment between the curves that represent the strength in both standards, especially for D/t > 60 (which marks approximately the limit of the class 4 cross-sections). Furthermore, it is clear that the length of 5 m presents a resistance considerably higher in both standards, while for the larger lengths the resistant depth shows an indifference to this variable. Where the differences begin to be noticeable ($\approx$ D/t < 60), EC3-1-6 is more on the safe side than ISO for the longest cylindrical tubes (15 and 25 m), being on the other hand, less on the safe side for the shortest tube (5 m).

3.2 Comparison with numeric results

The parametric variation used in this study fitted the type of loading in analysis. It should be noted that the reduction buckling curves established by standards, subsequently presented, do not take into account any partial safety coefficient in order to allow the comparison with the obtained numerical results.

3.2.1 Axial compression
3.2.1.1 Local buckling
In order to facilitate the comparison between numerical results and those established by ISO and EC3-1-6,

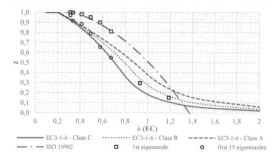

EC3-1-6 - Class C ·········· EC3-1-6 - Class B − − − − EC3-1-6 - Class A
— · — ISO 19902   □ 1st eigenmode   ○ first 15 eigenmodes

Figure 7. Local buckling – EC3-1-6 vs ISO 19902 vs GMNIA (EC3-1-6 class C imperfections) – L = 10 m.

Table 4. Local buckling by axial compression – GMNIA (EC3-1-6 class C imperfections) vs EC3-1-6 – L = 10 m.

D_{ext} (m)	t (mm)	Δ_{w0} (mm)	F_{cr} (MN)	F_{pl} (MN)	λ (Num)	λ (EC)	χ_s (EC)	1st mode		First 15 modes		
								χ_s (Num)	Dif. (%)	Mode (Num)	χ_s (Num)	Dif. (%)
1.5	30	18.75	604.92	49.18	0.285	0.313	0.922	1.000	7.73	9	0.969	4.71
1.5	25	15.63	429.82	41.13	0.309	0.335	0.905	0.997	9.19	13	0.915	1.01
1.5	15	10.55	163.92	24.84	0.389	0.413	0.838	0.980	14.16	13	0.882	4.41
1.5	10	8.62	75.01	16.62	0.471	0.492	0.760	0.949	18.87	13	0.790	2.82
1.5	7.0	7.21	37.49	11.66	0.558	0.577	0.665	0.898	23.28	13	0.652	-1.24
1.5	5.0	6.09	19.33	8.34	0.657	0.674	0.543	0.809	26.61	9	0.547	0.40
1.5	2.5	4.31	4.92	4.18	0.921	0.932	0.228	0.294	6.66	3	0.292	6.49
1.5	1.5	3.34	1.78	2.51	1.185	1.189	0.107	0.149	4.17	3	0.149	4.15

the respective curves were plotted on the same chart. To this end, once the axis of abscissas used by the two standards is distinct, it was chosen for this axis the slenderness λ calculated by EC3-1-6 taking into account the correspondence between the variables represented in the two standards.

The procedure regarding the eigenmodes to be used as imperfections consisted in search in the first 15, one with the pattern of deformation as close as possible to the gauge length indicated by EC3-1-6 and introducing it the respective amplitude. It was possible to conclude that when the pattern of the eigenmodes deviates from the one given by the gauge lengths, the differences can be considerable.

The study of the local buckling phenomena was made for lengths of 5 and 10 m. However, to economize space and by the fact that the results obtained are similar, the data of the shortest element will not be presented.

Through the observation of the chart in Figure 7, the following conclusions can be drawn. Firstly the curve of ISO 19902 is, for the slenderness in which it is employed, considerably above (i.e. it is less conservative) that any curve for the different classes of EC3-1-6. Even for the class with the highest fabrication quality (class A) in EC3, the differences are around 10%. Having this in mind it can be conjectured that the expressions of reduction to local buckling for axial compression on ISO were calibrated for quite low imperfections, according to the criteria of the EC3-1-6. However, it should be noted that by limiting its application to thicknesses ≥6 mm and ratios D/t <120, the formulas of ISO imply that the represented curve do not have applicability to the larger slenderness. In fact, for the numerical results presented for the length of 10 m, only the first three (with thicknesses of 0.03, 0.025 and 0.015 m) have applicability in ISO. This evidence has as consequence that ISO 19902 does not allow the elements to be strongly influenced by local buckling phenomena, since the value of χ is always high. Furthermore, since the partial safety factor for axial compression in ISO is considerably higher than the one in EC3-1-6 (1.18 vs 1.1) it is expectable that these differences are dimmed in the values obtained for the final strengths.

Regarding the numerical results obtained (more specifically those defined by the first 15 eigenmodes), they approach quite well the class C curve of EC3-1-6, since geometric imperfections were calculated taking into account this fabrication class. There are not, as expected, on local buckling phenomena by axial compression significant differences in the results obtained for strength in the elements with lengths of 5 and 10 m.

Curiously, the results obtained using the first eigenmode for the elements with 10 m in length fit particularly well the curve of ISO. This aspect leads to believe that the use of the first eigenmodes for longer elements leads to values close to the ones indicated by this standard, which corroborates the idea that ISO was formulated taking into account imperfections less injurious for the verification of local buckling phenomena.

Given the existence of numerical values lower than the curves given by the standards, the final strengths (using partial safety coefficients) were compared with the numerical results. This permitted to conclude that EC3-1-6 is safe for all the geometries considered, while ISO is safe only when applied in the scope of the standard. Which means that ISO is valid for D/t < 120 and t ≥ 6 mm due to its high safety partial factors, even using the imperfections indicated in EC3-1-6.

3.2.1.2 Global buckling

EC3-1-1 and ISO 19902 use, once again, different variables to define the global buckling curves. ISO uses to define the limits of this curve the yield stress reduced by local buckling phenomena, unlike EC3-1-1 which has not naturally this aspect into consideration. However, taking into account the D/t ratios used in this study for the verification of the overall buckling phenomena, ISO considers that local phenomena are negligible and therefore the verification to global buckling phenomena is done also through the yield stress. That being so, the variables on the axis of abscissas take perfectly the same meaning and they can be plotted and compared directly with the numerical results in a single chart (Figure 8).

By the fact that the effective length of the elements was considered with the value of 1.0, i.e. as simply supported elements, the shape of the imperfections for the analysis of global buckling phenomena is given by the 1st eigenmode.

From Figure 8, a good approximation between the curves given by ISO and EC3-1-1 (curve a → α = 0.21) is verified, being that the latter is slightly more conservative for the represented intermediate

Table 5. Global buckling – EC3-1-1 vs ISO 19902 vs GMNIA.

Test N.	D_ext (m)	t (mm)	D/t	L (m)	A (cm²)	F_cr (MN)	F_pl (MN)	λ (EC=ISO)	λ (Num.)	Error λ (%)
1	1.5	50	30	10	2277.7	1266.39	80.86	0.255	0.253	0.97
2	1.5	50	30	15	2277.7	608.73	80.86	0.383	0.364	5.01
3	1.5	50	30	20	2277.7	352.60	80.86	0.510	0.479	6.56
4	1.5	50	30	25	2277.7	228.83	80.86	0.638	0.594	7.30
5	1.25	50	25	25	1885.0	135.17	66.92	0.771	0.704	9.51
6	1	50	20	25	1492.3	71.01	52.98	0.973	0.864	12.63

Test N.	χχ (EC)	χχ (ISO)	Imp. = L/250				Imp. = L/300			
			Δw0 (mm)	χχ (Num)	Dif. EC (%)	Dif. ISO (%)	Δw0 (mm)	χχ (Num)	Dif. EC (%)	Dif. ISO (%)
1	0.988	0.982	40	0.945	-4.29	-3.7	33.33	0.959	-2.92	-2.33
2	0.957	0.959	60	0.918	-3.94	-4.14	50	0.935	-2.24	-2.44
3	0.921	0.928	80	0.817	-10.38	-11.03	66.67	0.844	-7.67	-8.32
4	0.875	0.887	100	0.798	-7.72	-8.9	83.33	0.830	-4.48	-5.66
5	0.812	0.835	100	0.735	-7.72	-10.01	83.33	0.766	-4.59	-6.88
6	0.685	0.737	100	0.647	-3.74	-8.97	83.33	0.678	-0.63	-5.86

Test N.	χχ (EC)	χχ (ISO)	Imp. = L/500			
			Δw0 (mm)	χχ (Num)	Dif. EC (%)	Dif. ISO (%)
1	0.988	0.982	20	0.987	-0.07	0.52
2	0.957	0.959	30	0.967	0.99	0.79
3	0.921	0.928	40	0.908	-1.27	-1.92
4	0.875	0.887	50	0.898	2.32	1.14
5	0.812	0.835	50	0.853	4.09	1.8
6	0.685	0.737	50	0.759	7.48	2.25

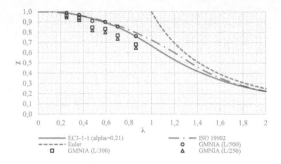

Figure 8. Global buckling – EC3-1-1 vs ISO 19902 vs GMNIA.

slenderness. In this respect, it should be noted that the slenderness value calculated by the numerical results ($\lambda = \sqrt{(f_y/\sigma_{cr})}$) begins to deviate from the ones calculated from standards (same in both) as its value increases. This aspect has only as consequence the translation, in relation to the curves, of the numerical points of greater slenderness to the left. Thus, the results presented in this chart must be examined, for the greater slenderness, taking this aspect into consideration. Despite this particularity, the relevant aspect is the value obtained for the strength of the elements through the buckling coefficient χ and this turns out to be, in general, closer between the standards and the numerical analysis when considering the value L/500 for the amplitude of imperfections. Thus, the fact that for amplitudes of L/250 and L/300 the numerical results obtained are lower than those obtained by ISO seems to indicate that the calibration of its expressions was made taking into account lower imperfections values that the ones indicated by Eurocode (through the imperfections e_0/l).

Using partial safety coefficients of ISO and EC3, the strength values given by ISO are found being more on the safe side than EC3-1-1 due to higher partial safety coefficients (1.18 vs 1.10). Moreover, the clearance between the strength in EC3-1-1 and the numerical results is very low. In fact, this standard presents some values slightly below the determined numerically (maximum −1.27%). However, if verified in parallel the strength given by EC3-1-6 (with $\gamma_{M1} = 1.1$) it turns out that this standard gives lower values than the numeric ones (and therefore safe values). This fact allows to evidence that, in some cases, even in the scope of the EC3-1-1 the resistance is being controlled (according to EC3-1-6) by local buckling.

3.2.2 *Hydrostatic pressure*

The eigenmodes used as pattern for imperfections were chosen through a similar process to the one explained

to local buckling by axial compression. That is, in the first eigenmodes (in this case the first 25) were sought the ones that more approached the deformation indicated by EC3-1-6, now through the gauge length in circumferential direction. The use of the first eigenmode was found to be too conservative, using the amplitudes defined by EC3-1-6 compared with the strength given by both standards. The amplitude of imperfections in EC3-1-6 is given by the maximum of two expressions, being one influenced primarily by the length and the other by the thickness. However, it was found that the most conditioning expression is the one that depends directly of the thickness, originating too conservative values, when this variable becomes significant. Therefore, the same tubes were also tested with the value of imperfections independently of thickness.

The verification of the circumferential hoop buckling phenomena was done to lengths of 5 and 10 m, from where it can be concluded that this variable has a significant impact reducing the strength. For the shortest element the data will not be presented.

In order to overlap the reduction hoop buckling curves of ISO 19902 and EC3-1-6 it was adopted once again as abscissa for the chart the slenderness defined by EC3-1-6. The results show that the class C curve from Eurocode fits very well the one obtained by ISO.

The numerical results show that the use of the first eigenmode results in too conservative values for the lower slenderness in any of the standards, even with the use of imperfections calculated as independent of the thickness. However, the results show that a careful choice of the eigenmodes used as imperfections leads to results close to the ones defined in the standards. Despite that, there is a small caution to the greater thickness of the longest elements in relation to ISO (Table 6). The slenderness calculated numerically deviates from the one calculated by Eurocode. The reason is due to the fact that shell elements are not the most suitable for low slender cross-section. For this reason, solid elements (C3D20R) were also used to simulate these cylindrical tubes. The slenderness calculated in this way approaches the one given by Eurocode. The difference to the strength values is not so significant. This evidence leads to the conclusion that the critical load (not so much the strength value)

Table 6. Hoop buckling – EC3-1-6 vs ISO 19902 vs GMNIA (EC3-1-6 class C imperfections) – L = 10 m.

Test N.	D_{ext} (m)	t (mm)	A (cm²)	Δw_0 (EC) (mm)	Δw_0 (EC ind. t)(mm)	P_{cr} (MPa)	P_{pl} (MPa)	$\sigma_{\theta Rcr}$ (MPa)	λ (EC)	λ (Num)
1	1.5	100	4398.2	62.50	18.56	132.52	47.33	927.65	0.537	0.619
2	1.5	50	2277.7	31.25	18.56	20.75	23.67	300.86	1.031	1.086
3	1.5	25	1158.5	18.56	18.56	4.18	11.83	123.26	1.651	1.697
4	1.5	15	699.8	18.56	18.56	1.32	7.10	65.35	2.397	2.331
5	1.5	10	468.1	18.56	18.56	0.43	4.73	31.98	3.252	3.332
6	1.5	5	234.8	18.56	18.56	0.08	2.37	11.65	5.473	5.519

				1st Eigenmode		1° Eigenmode (Imp. indep. t)			First 25 Eigenmodes (Imp. indep. t)			
Test N.	χ_θ EC	χ_θ ISO	χ_θ Num	Dif. EC (%)	Dif. ISO (%)	χ_θ Num	Dif. EC (%)	Dif. ISO (%)	Mode	χ_θ Num	Dif. EC (%)	Dif. ISO (%)
1	0.885	0.979	0.379	-50.69	-60.06	0.676	-20.94	-30.31	15	0.876	-0.97	-10.34
2	0.473	0.562	0.264	-20.93	-29.85	0.341	-13.26	-22.18	18	0.516	4.28	-4.64
3	0.183	0.176	0.192	0.89	1.63	0.192	0.89	1.63	12	0.241	5.72	6.46
4	0.087	0.084	0.098	1.12	1.47	0.098	1.12	1.47	12	0.149	6.16	6.51
5	0.047	0.051	0.054	0.71	0.38	0.054	0.71	0.38	15	0.087	3.98	3.65
6	0.017	0.018	0.021	0.41	0.29	0.021	0.41	0.29	15	0.038	2.12	2

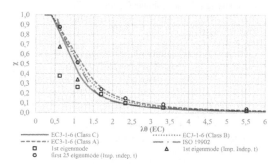

Figure 9. Hoop buckling – EC3-1-6 vs ISO 19902 vs GMNIA (EC3-1-6 class C imperfections) – L = 10 m.

Legend:
— EC3-1-6 (Class C)   ·········· EC3-1-6 (Class B)
– – – EC3-1-6 (Class A)   – · – · ISO 19902
□ 1st eigenmode   △ 1st eigenmode (Imp. indep. t)
○ first 25 eigenmode (Imp. indep. t)

Table 7. Statistical analysis of the ratio between the strengths of the standards (ISO and EC3-1-6) and the numerical results for local buckling by axial compression.

		Without P.S.C.		With P.S.C.	
	t (mm)	ISO	EC3-1-6	ISO	EC3-1-6
Mean $R_{standard}/R_{numeric}$	15-30*	1.085	0.970	0.906	0.896
	1.5-10	1.497	0.881	1.259	0.809
	combined	1.379	0.906	1.158	0.834
Standard deviation $R_{standard}/R_{numeric}$	15-30*	0.036	0.022	0.034	0.017
	1.5-10	0.272	0.187	0.230	0.174
	combined	0.298	0.162	0.254	0.151
Coefficient of variation (%)	15-30*	3.36	2.26	3.78	1.94
	1.5-10	18.20	21.29	18.27	21.54
	combined	21.63	17.86	21.89	18.09

*Where ISO has applicability.

Table 8. Statistical analysis of the ratio between the strengths of the standards (ISO and EC3-1-6) and the numerical results for global buckling by axial compression.

	Without P.S.C.		With P.S.C.	
	ISO	EC3-1-1	ISO	EC3-1-1
Mean $R_{standard}/R_{numeric}$	0.991	0.972	0.840	0.972
Standard deviation $R_{standard}/R_{numeric}$	0.017	0.041	0.015	0.041
Coefficient of variation (%)	1.75	4.19	1.75	4.19

Table 9. Statistical analysis of the ratio between the strengths of the standards (ISO and EC3-1-6) and the numerical results for hoop buckling by hydrostatic pressure.

		Without P.S.C.		With P.S.C.	
	t (mm)	ISO	EC3-1-6	ISO	EC3-1-6
Mean $R_{standard}/R_{numeric}$	25-100	1.021	0.977	0.817	0.888
	5-15	0.657	0.622	0.526	0.565
	combined	0.839	0.799	0.671	0.727
Standard deviation $R_{standard}/R_{numeric}$	25-100	0.152	0.121	0.121	0.110
	5-15	0.148	0.131	0.118	0.118
	combined	0.238	0.221	0.190	0.201
Coefficient of variation (%)	25-100	14.84	12.40	14.85	12.41
	5-15	22.50	20.99	22.48	20.94
	combined	28.36	27.61	28.35	27.60

calculated by shell elements with low slenderness (for very small D/t ratios) presents non-negligible errors, since the analysis moves away from the assumptions of the use of shell elements.

Making use of the partial safety coefficients, the normalized final strengths to hoop buckling established by ISO and EC3-1-1 are safe. In addition, ISO leads to conservative values even when applied outside the limits set by itself.

3.3 Statistical analysis

To the data referred (either presented or not) in section 3.2.1 and 3.2.2, in this section is carried out its statistical analysis. To this end, it was used the mean, the standard deviation and the coefficient of variation (ratio between the standard deviation and the mean) of the ratio between the strength of the standards (for ISO and EC3) and the strength obtained numerically. The results with and without the use of partial safety coefficients (P.S.C.) are also presented. The objective is to verify the existence of any passages from unsafe situations (mean of $R_{standard}/R_{numeric} > 1$) to safe situations (mean of $R_{standard}/R_{numeric} \leq 1$) with the use of P.S.C.

Regarding the local buckling, in Table 7 is confirmed that ISO presents on average non-conservative values (even using P.S.C.) when analyzed all the data together. For this reason, a separate analysis was done for the geometries in which ISO has applicability (thicknesses of 15, 25 and 30 mm) where it can be concluded that the standard is safe when considered the P.S.C.. For this type of buckling Eurocode gives better adjusted strengths to numerical results (lower mean and standard deviation).

Regarding the global buckling (Table 8) it is verifiable that both ISO and EC3-1-1 present results quite well adjusted to the numerical ones (mean of the ratios ≈ 1 and low standard deviations). Thus, the use of a considerable value for the P.S.C in ISO ($\gamma_M = 1.18$) implies final strengths considerably on the safe side. For EC3-1-1, by the fact that $\gamma_{M1} = 1.0$, the strength value remains unchanged with the use of P.S.C, although it is on average safe even in this way (using imperfection values of L/500).

Table 9 shows that on average, both ISO and EC3-1-6 present safe values for strengths compared to numerical results to hoop buckling. Nonetheless, it is important to highlight that once the values of strength for the slenderest cylindrical tubes can be quite low, values considerably away from the unit are expected to the ratios between the strength in standards and the numerical analysis (despite the differences between them being small, as seen in section 3.2.2). Therefore it is expected that this fact affects the statistical results when used all the data simultaneously. For this reason the statistical analysis was divided in two groups of thicknesses (the greatest: 25, 50 and 100 mm; and the smallest: 5, 10 and 15 mm). The difference of the results between these groups is quite noticeable especially in the difference of means, where it appears that the ratios are closer to the unit in thicknesses that cause larger strengths.

4 CONCLUSIONS

Below are some of the main findings arising from the preparation of this study:

– ISO leads to safe values, using imperfections defined by EC3-1-6, for local buckling phenomena by axial compression only for the field of application set out in the standard.
– The use of EC3-1-6 for the verification of local buckling phenomena by axial compression is safe beyond the limits of application of ISO (D/t < 120).
– The use of imperfection amplitudes defined in EC3-1-1, $e_0/L = 1/300$ and $1/250$, for elastic and plastic analysis, respectively, for the global buckling curve "a" was found to lead to lower numerical values than the ones defined by this standard and ISO. The amplitude value L/500 is on average the one better adjusted to the global buckling curves in both standards.
– In some cases it was found that EC3-1-6 provides more conservative values than EC3-1-1, even in classes 1 to 3. Thus, EC3 leads safe values to global buckling (for value imperfections of L/500) when both parts of EC3 are used together.
– ISO leads to safe values for global buckling.
– The strengths obtained to hydrostatic pressure by ISO and EC3-1-6 (class C) are very similar and safe even when applied to D/t > 120.

REFERENCES

ABAQUS User's manual – Version 6.11 (2011). Dassault Systèmes Simulia Corp., USA.

Chen, W. and Ross, D. (1978). "The Strength of Axially Loaded Tubular Columns – Tests of Fabricated Tubular Columns". Fritz Engineering Laboratory Report No. 393.8, Lehigh University, Pennsylvania.

DNV (2011). "Comparison of API, ISO, and NORSOK Offshore Structural Standards". Det Norske Veritas.

European Committe for Standardization (2007). "EN 1993-1-6:2007 – Eurocode 3: Design of steel structures – Part 1–6: Strength and Stability of Shell Structures". CEN, Brussels.

European Committe for Standardization (2005). "EN 1993-1-1:2005 – Eurocode 3: Design of steel structures – Part 1–1: General rules and rules for buildings". CEN, Brussels.

Gunzelman, S. and Ostapenko, A. (1977). "Local Buckling Tests on Three Steel Large-Diameter Tubular Columns", Fritz Engineering Laboratory Report No. 406.7, Lehigh University, Pennsylvania.

HSE (2001). "Comparison of tubular member strength provisions in codes and standards". Offshore Technology Report 2001/084, United Kingdom.

Idrus, A., Potty, N. and Nizamani, Z. (2010). "Tubular strength comparison of offshore jacket structures under API RP2A and ISO 19902". Perak, Malaysia.

International Organization for Standardization (2007). "ISO 19902 – Petroleum and natural gas industries – Fixed Steel Offshore Structures". Ed.1, ISO, Geneva.

Koiter, W. (1945). "On the Stability of Elastic Equilibrium", (Original in Dutch). PhD thesis, University of Delft. Delft. (Translation AFFDL-TR-70-25, Wright-Patterson Air Force Base, 1970).

Marzullo, M. and Ostapenko, A. (1977). "Tests on Two High-strength Short Tubular Columns". Fritz Engineering Laboratory Report No. 406.10, Lehigh University, Pennsylvania.

Ostapenko, A. and Gunzelman, S. (1976). "Local Buckling of Tubular Steel Columns". Proceedings, ASCE National Structural Engineering Conference, Wisconsin.

Riks, E. (1978). "A unified method for the computation of critical equilibrium states of non-linear elastic systems". Acta Technica Academiae Scientiarium Hungaricae, Tomus 87 (1–2).

Rotter, J. and Schmidt, H. (2008). "Buckling of Steel Shells – European Design Recommendations". Ed. 5, ECCS Press – P125, Brussels. 91-9147-000-92.

Standards Norway (2013). "NORSOK N-004 – Design of steel structures". Ed. 3, Standards Norway, Norway.

Steck, M. (n.d.). "Comparison of internacional regulations with regard to the archieved bearing capacity of structural members". Germanischer Lloyd, Hamburg.

Teng, J.G. and Rotter, J.M. (2004). "Buckling of thin metal shells". Ed. 1, Spon Press, London. ISBN 0-203-30160-9.

Tuen, E. (2012). "Structural resistence safety level in Eurocode versus Norsok and ISO". Master thesis, University of Stavanger – Faculty of Science and Technology, Stavanger.

Windenburg, D. and Trilling, C. (1934). "Collapse by Instability of Thin Cylindrical Shells Under External Pressure". ASME, Transactions Vol. 34, No. 11, Washington.

Yamaki, N. (1984). "Elastic Stability of Circular Cylindrical Shells". North Holland, Elsevier Science Publishers, Amsterdam.

Ziemian, R. (2010). "Guide to stability design criteria for metal structures". Ed. 6. John Wiley & Sons, Inc, New Jersey. 978-0-470-08525-7

Stainless and high strength steel structures

Tubular Structures XV – Batista, Vellasco & Lima (eds)
© 2015 Taylor & Francis Group, London, ISBN 978-1-138-02837-1

Low cycle fatigue of high-strength steel tubes with longitudinal attachments

J. Hrabowski, S. Herion & T. Ummenhofer
KIT Steel and Lightweight Structures, Karlsruhe, Germany

ABSTRACT: Longitudinal attachments are often used in crane structures to fix cables, anchors or gratings. Crane structures are exposed to very high loads but with rather low number of load cycles. In the so called low cycle fatigue (LCF) domain the use of high strength steels offer considerable advantages. Whereas the application of post weld treatment methods is still a relevant question. In the following, fatigue tests on tubes with longitudinal attachments and post weld treatment by means of TIG-dressing, high frequency hammer peening (HFHP) and grinded roots are presented.

1 INTRODUCTION

For crane structures, the demands for high load capacity and decreasing total construction weight have to be balanced. This can be achieved by using high-strength steels. Nowadays, ultra-high strength fine grained steels S1300QL with sufficient toughness and weldability are available on the market and used in highly stressed components.

At complex structures sometimes welding in such highly stressed areas cannot be avoided. Special requirements on the welded seams are made here. These can be fulfilled with the help of modern post weld treatment methods.

2 GENERAL

2.1 High-strength steel

The development of fine grained structural steels with higher and highest strengths is aimed at an increase of load capacity together with constant weight. Recently, the standardized steel grade with the highest yield strength is S960QL in EN 10025-6 (2009) with 960 MPa. But, although not covered by standards, steels with yield strengths up to 1300 MPa are produced and already used for crane structures. High-strength fine grained steels can be welded with usual methods, such as MAG and submerge welding. But filler material is only available up to yield strength of 960 MPa.

The fine grained structure of high- and ultra-high-strength steels is sensitive to high temperatures, however, so special attention has to be paid concerning working temperature and energy input during welding.

2.2 Low-cycle fatigue

Conventional standards and recommendations, such as EN 1993-1-9 (2010) and Hobbacher (2008) provide linear fatigue design S-N-curves starting at 10,000 load cycles. But fatigue test data is mainly based on tests between 100,000 and 2 million load cycles. Low cycle fatigue means usually a number of failure load cycles below 10,000 to 40,000, in which the transition to high-cycle fatigue is fluid and not clearly defined. So the question is where the line to low cycle fatigue has to be drawn and the so called Woehler-curves loose validity.

In practice often the yield strength R_e is considered to be the lower limit to low cycle fatigue (Hrabowski et al. 2011). For more detailed analyses low cycle fatigue of ductile materials can be separated from high cycle fatigue by means of the deformation criterion (Gudehus & Zenner 2000). Herein, also the influence of the stress range ratio is considered within the following formulae:

$$R_e^* = R_e \cdot (1 - R) \tag{1}$$

where R_e is the yield strength in MPa and R is the stress range ratio with

$$R = \sigma_{min} / \sigma_{max} = const. \tag{2}$$

For axial loading the maximum stress range to be reached is

$$R_m^* = R_m \cdot (1 - R) \tag{3}$$

where R_m is the tensile strength in MPa and R is the stress range ratio according to eq. (2).

For conventional structural steels material properties are given in standards, e.g. EN 10025-2 (2005). Beneath that, material properties can also be taken from inspection certificates. For ultra-high strength steels with yield strength over 960 MPa, which are not standardized, this is the only way to proceed.

In Table 1, the deformation limit R_e^* and the maximum stress range R_m^* are calculated for steels S355

Table 1. Deformation limit R_e^* and maximum stress range R_m^* for stress range ratio R = +0.1

Steel	R_e	R_m	R_e^*	R_m^*	Material properties acc.
S355	355	500	320	450	EN 10025-2 (2005)
S960	960	1060	860	950	EN 10025-6 (2009)

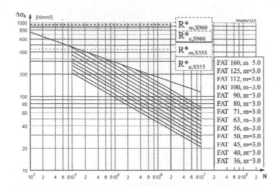

Figure 1. Fatigue resistance S-N curves according to Hobbacher (2008) with deformation limit R_e^* and maximum stress range R_m^* (R = +0.1) for S355 and S960.

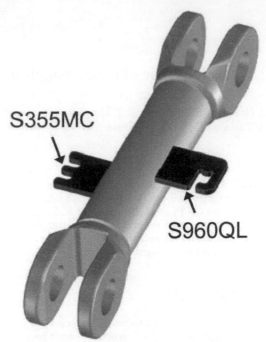

Figure 2. CHS specimen with longitudinal attachments with off-set made of S355MC and S960QL (source: Liebherr Werk Ehingen GmbH).

and S960. On the basis of the fatigue tests presented in the following, the values are given for stress range ratio R = +0.1.

Fatigue design is usually independent of the materials strength. In Figure 1 it is shown, where higher strength steels still can be advantageous for fatigue resistance.

According to Hobbacher (2008), the maximum possible fatigue S-N curve, corresponding to base material, has characteristic fatigue strength at two million load cycles of 160 MPa. Due to the low notch effect, a slope of the S-N curve of m = 5.0 is assumed. This curve is also the cut-off limit for the other FAT curves with m = 3.0. Now, in Figure 1 also the deformation limit R_e^* and the maximum stress range R_m^* (R = +0.1) for steel grades S355 as well as S960 are illustrated. The horizontal line representing the deformation limit $R_e^* = 320$ MPa for S355 (Tab. 1) cuts the S-N-curves for FAT 63 and higher between 10,000 and 60,000 load cycles. Hence, for weld details corresponding to FAT 63 and higher, failure can occur using S355 although fatigue resistance is proven. In contrast, for S960 the FAT 160 curve can be elongated below 1000 load cycles without cutting the deformation limit line, as it is shown in Figure 1.

3 EXPERIMENTAL INVESTIGATION

3.1 Fabrication of specimens and test program

The base element for the specimens is a 400 mm long hot rolled circular hollow section 101.6 × 5.6 mm made of *FineXcell® 960 ImpactFIT 40* by Vallourec

Deutschland GmbH. This high performance steel was formerly called FGS100WV and is similar to S960G1QL. Two longitudinal attachments at a time, each 80 mm long, were welded with a lateral off-set on the tubes, see Figure 2. To investigate the influence of the attachment's material, they were made of undermatching steel S355MC and matching steel grade S960QL with corresponding filler material. To be able to distinguish the attachments easily, they were formed differently as shown in Figure 2.

The clevises welded to the end of the tubes were selected according to the application in crane construction of lattice girders. The specimens were provided by the companies Liebherr-Werk Ehingen GmbH and the Salzgitter Mannesmann Forschung GmbH.

Within the test program given in Table 2, three different types of post weld treatment have been investigated: Burr grinding, TIG-dressing and high frequency hammer peening (HFHP). For HFHP the hammer tool developed by Dynatec GmbH for high frequency impact treatment (HiFIT) has been used. Reference fatigue tests on specimens in as welded condition have been also carried out. Moreover, a test series was carried out with larger tube dimensions 139.7 × 8.8 mm and attachments in as welded condition.

The post weld treatment (PWT) has been carried out according to the instructions of Haagensen & Maddox (2011). While for burr grinding and hammer peening no special demands for high strength steels have to be met, during TIG-welding the energy input per unit

Table 2. Test program.

Series	Weld condition	Tubes [mm]	No. of tests
MB2_AW	As welded	101.6 × 5.6	10
MB2_WIG	TIG-dressed	101.6 × 5.6	10
MB2_G	Grinded	101.6 × 5.6	10
MB2_HFH	HFHP	101.6 × 5.6	10
MB3_AW	As welded	139.7 × 8.8	9

Table 3. Improvement of stress range for butt welds with yield stress >350 MPa due to post weld treatment according to Haagensen & Maddox (2011).

Weld condition	Factor	FAT	FAT
As welded	1.0	63	71
TIG-dressing	1.3	80	90
Burr grinding	1.3	80	90
Hammer peening	1.5	90	100

length should not exceed 0.8 kJ/mm and the working temperature should be below 200°C.

3.2 Test-setup

The fatigue tests were carried out at the Research Center for Steel, Timber and Masonry, Karlsruhe Institute of Technology (KIT) on two different testing machines: a universal testing machine with 6.3 MN maximum load capacity by Schenck and one with 3.0 MN maximum load capacity by MFL. The tests were performed as tensile test with constant load amplitude and stress range ratio of R = +0.1.

The implementation of the specimen into the testing machine was realistically carried out with clevis pins to the connection link (Fig. 2).

3.3 Classification

The fatigue behaviour of longitudinal attachments is dominated by the stiffeners length, the shape of the stiffener and the weld execution (Puthli et al. 2006). Rectangular stiffeners with 80 mm length are classified into FAT 71 according to Hobbacher (2008) and EN 13001-3-1 (2010). In EN 1993-1-9 (2010) the 80 mm long stiffener represents a border case between FAT 63 and FAT 71. Former investigations of Ummenhofer et al. (2013) on rectangular attachments also in the low cycle fatigue regime indicate a fatigue resistance of 63 MPa at two million load cycles (Ummenhofer et al. 2013).

Haagensen & Maddox (2011) provide design S-N-curves for post weld treated steel structures. The possible improvements of stress range for different methods are listed in Table 3. Thus, for burr grinding and TIG-dressing an improvement of 2 FAT classes in comparison to the as welded condition can be applied. For high frequency hammer peened (HFHP) welds an increase of 3 FAT classes is possible, which corresponds to an improvement of stress range by a factor 1.5.

3.4 Fatigue test results

In General, cracks occurred in the heat affected zone at the weld toe as can be seen in Figures 4 to 7. As critical failure a through-thickness crack was defined. The controlling during the test was carried out via the increase of the machine's range. Herein, the transition from first crack on the surface to through-thickness

Figure 3. Specimen in testing machine with clevis bearing.

crack and even crack lengths of about 40 mm is fluently. That means that after the crack trough the wall thickness further loading is still possible.

Some of the specimens failed at the butt weld to the clevis at the tubes end, especially within the series with post weld treatment. These test results are not decisive for the investigations, so they are not considered for the further evaluation. Nevertheless, they represent a certain minimum value since the specimens offered resistance at least until the failure of the butt weld. So

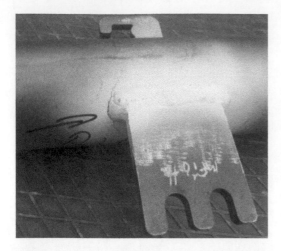

Figure 4. Specimen MB2_AW-02 with crack in as welded condition.

Figure 5. Specimen MB2_G-08 with crack and grinded weld toe.

in the S-N-diagrams they are illustrated with arrows but not considered for evaluation.

Cracks occurred on both, the S355 attachments as well as the S960 attachments with a more or less equal distribution. Only the high frequency hammer peened specimens show a tendency cracks at the S355 stiffeners.

The results of the fatigue tests are summarized in Table 4. In the first three columns the series name, the weld condition and the tubes' dimensions are given. The fourth column gives the number of tests carried out with the number of test results not considered in evaluation in parenthesis. In the fifths column the range of the numbers of cycles to failure is listed to show the rather small spectrum. In the next three columns the fatigue test results are statistically evaluated by linear regression analysis according to EN 1990 (2010) and fatigue strengths $\Delta\sigma_C$ at two million load cycles for survival probability $P_S = 95\%$ and $P_S = 50\%$ as well as the free calculated slopes are given. Since in the high cycle range between 100,000 and two million load cycles nearly no test results are available, the S-N-curves are extrapolated by far. This leads to

Figure 6. Specimen MB2_WIG-06 with crack an TIG-dressed weld.

Figure 7. Specimen MB2_HFH-04 with crack and weld toe peened with HFHP.

Table 4. Fatigue test results.

Series	PWT	Tube dimension [mm]	Number of tests (not cons.)	Load cycle range	$\Delta\sigma_C$ $P_s = 50\%$ [MPa]	$\Delta\sigma_C$ $P_s = 95\%$ [MPa]	Slope m	FAT acc. IIW
MB2_AW	As welded	101.6 × 5.6	10 (0)	5.134–56.235	60	43	2.5	71
MB2_WIG	TIG-dressed	101.6 × 5.6	10 (1)	16.969–258.786	117	95	3,1	90
MB2_HFH	HiFIT	101.6 × 5.6	10 (2)	9.622–71.239	92	79	2.9	100
MB2_G	Grinded	101.6 × 5.6	10 (3)	11.341–129.025	82	57	2.8	90
MB3_AW	As welded	139.7 × 8.8	9 (2)	14.089–74.100	75	43	2.9	71

a rather large scatter band resulting in low fatigue strengths $\Delta\sigma_C$ at two million load cycles and survival probability $P_S = 95\%$. In the last column the corresponding FAT classes according to Hobbacher (2008) and Haagensen & Maddox (2011) from Table 3 are added.

3.5 Strain measurements and crack propagation

On two samples per series strain gauge measurements have been carried out to validate the applied loads and the inner stresses and to determine the point of the initial crack.

Single strain gauges of the type FLA-3-11-3 L as well as 5s chains of the type FX-1-11-5 L are applied in a low distance to the fillet weld in the direction of the stiffener (Fig. 8).

The strain measurement is carried out both before the test beginning with static loads and during the fatigue tests with cyclic loads.

In Figure 8 the maximum strains at the crack initiation point on specimen MB2_WIG-02 with TIG-dresses weld is evaluated. The first decrease of measured strains represent the begin of crack propagation. After the crack runs through the wall thickness the strains stabilize, and the specimen can still bear further load cycles.

At the same two samples with strain gauge measurements, also the crack propagation is taken visually during the test. Here, as crack beginning the first visible crack on the surface is defined. The transition from first crack on the surface to through-thickness crack and to crack lengths of about 40 mm is smooth.

Since cracks frequently appear in several places, the cracks are distinguished in the following in primary cracks as decisive failure criterion and secondary cracks. The evaluation of crack propagation is summarized in Table 5. Herein, the crack characteristic (primary or secondary) and the crack length at the end of the test are given. In the last column of Table 5, the load cycle quotient is calculated by the number of load cycles at first visible crack N_{CI}, divided by the number of load cycles to failure N_f.

3.6 Discussion

First fatigue cracks occur after about 60% of the total lifetime of a specimen. At specimens with greater

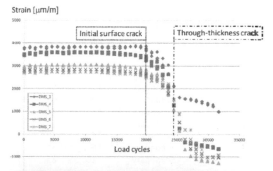

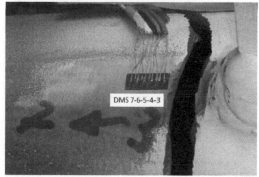

Figure 8. Strain gauge measurements specimen MB2_WIG-02.

diameters cracks appears a little earlier, at specimens with TIG-dressed welds a little later. Considering an interaction of the safety factors for action and resistance, no cracks should arise during the design life time of the component.

Crack observation and strain gauge measurement show a distinctive crack propagation phase and the components provide the possibility to transfer the loads long after crack initiation. These are qualities, which are often disputed to high-strength steel.

Figure 9 shows the S-N-diagram for tubes with longitudinal attachments in as welded condition. Series MB2 (tubes dimension 101.6 × 8.8 mm) and MB3 (tubes dimension 139.7 × 8.8 mm) are evaluated together in one S-N-curve. The larger total number of tests provides statistically ensured results. But the rather small spectrum of failure load cycles between 5,000 and 75,000 cycles requires a wide extrapolation

Table 5. Examination of crack propagation.

Specimen	Crack	Crack length [mm]	Load cycle quotient N_{CI}/N_f [%]
MB2_AW-11	Primary	33	63
MB2_AW-12	Primary	19.5	78
MB2_AW-12	Secondary	14.5	39
MB2_WIG-02	Primary	150	62
MB2_WIG-08	Primary	21	84
MB2_WIG-08	Secondary	19	66
MB2_HH-02	Primary	43	55
MB2_HH-11	Primary	44	61
MB2_HH-11	Secondary	25	61
MB2_HH-11	Secondary	20	61
MB2_HH-11	Secondary	10.5	55
MB2_G-02	Primary	25	60
MB2_G-02	Secondary	22	60
MB2_G-02	Primary	20	61
MB2_G-11	Secondary	21.5	52
MB2_G-11	Secondary	11	52
MB3_AW-02	Primary	45	68
MB3_AW-02	Secondary	17	44
MB3_AW-02	Secondary	16.5	44
MB3_AW-11	Primary	18.5	39
MB3_AW-11	Secondary	16	35
MB3_AW-11	Secondary	4	85

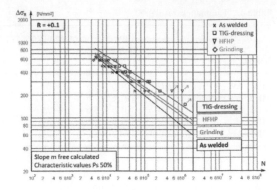

Figure 10. Comparison of S-N-Curves for fatigue test on post weld treated specimens.

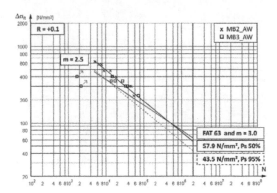

Figure 9. S-N-Curve for fatigue test on specimens in as welded condition

to the characteristic value for fatigue resistance at two million load cycles. This fact together with the resulting steep slope of the curve with m = 2.5 lead to a low value of 43.5 N/mm² for the fatigue resistance.

The curve for FAT 63 and m = 3.0 according to EN 1993-1-9 (2010) for attachments with lengths l > 80 mm is on the save side for all tests results as illustrated in Figure 9. And it cuts the S-N- curve for 95% survival probability in the low-cycle fatigue region.

Former investigations of Puthli et al. (2006) at the research Center for Steel, Timber and Masonry in Karlsruhe have shown that components with longitudinal attachments of 80 mm length are overestimated by the codes and provide a FAT class of 63. With this assumption also the improved classes for post weld treatment have to be adapted to the initial FAT 63. So according to Haagensen & Maddox (2011) FAT 80 can be

assumed for TIG-dressed and grinded welds and FAT 90 for specimens with HFHP-treatment.

In Figure 10 the test results for specimens in as welded condition are compared to the results of post weld treated specimens. To disregard the influence of the scatter, the 50%-fractiles are shown. As mentioned before, the curves are extrapolated by far from low cycle fatigue region to two million load cycles. So the values for the fatigue resistance at two million load cycles are to be handled with care and not mandatory. But on the basis of Figure 10, a comparison of the effectiveness of the different post weld treatment methods can be drawn.

So the low cycle fatigue tests on specimens with TIG-dressing achieved the most load cycles to failure and test results fulfill FAT 90. High frequency hammer peening and grinding of the weld toe perform equal during the tests. This is in contrast to the recommendations of Haagensen & Maddox (2011), where hammer peening is considered to provide the highest improvement of fatigue resistance. One possible reason can be seen in the high amplitude loads. Functionality of HFHP lies in the height of the induced residual compression stresses up to the materials yield strength. It is assumed that during very high loading the compressive stresses plasticize out. The fact that with HFHP-treatment the cracks occurs mainly on the S355 stiffener and the fillet welds with lower strength confirm this assumption.

Nevertheless, in the low-cycle fatigue regime a FAT 80 and therefore 30 % improvement of the fatigue resistance can be achieved via HFHP-treatment.

4 CONCLUSIONS

Low cycle fatigue tests on high-strength steel tubes with 80 mm longitudinal attachments have been carried out. As already figured out in former investigations (Puthli et al. 2006) the fatigue resistance of such details is overestimated by the codes and the authors recommend FAT 63.

Constant amplitude tests on post weld treated specimens resulted in the highest fatigue resistance for filet welds with TIG-dressing. When special demands concerning energy input and working temperature are met, TIG-welding is very effective and the fatigue resistance also of high-strength steel components can be improved by 50%. But TIG-welding is also expensive. HFHP-treatment and grinding of the welds can be carried out more easily and cheaper. But with higher loads the HFHP-method seams to lose efficiency due to the very high stresses. The presented investigations still provide a FAT 80 for HFHP-treated high-strength steel specimens, which corresponds to a 30% improvement as recommended for specimens with grinded weld toes.

ACKNOWLEDGEMENTS

The authors gratefully acknowledge the financial support of FOSTA (Forschungsvereinigung Stahlanwendung e.V.) for these investigations and Mr. Heise, FOSTA for administration of the research project P900. The presented research was carried out in collaboration with TU Darmstadt, System Reliability and Machine Acoustics SzM, where we thank the responsible staff members for the good teamwork. The authors also like to thank the members of the working group of the research project P900 for their support and contributions to the presented work. Special thanks go to Salzgitter Mannesmann Forschung GmbH and Liebherr Werk Ehingen GmbH for the manufacturing of the specimens.

REFERENCES

EN 1990. 2010. Eurocode: Basis of structural design, Berlin: DIN German Institute for Standardization

EN 1993-1-9. 2010. Eurocode 3: Design of steel structures – Part 1–9: Fatigue, Berlin: DIN German Institute for Standardization

EN 10025-2. 2011. Hot rolled products of structural steels – Part 2: Technical delivery conditions for non-alloy structural steels, Berlin: DIN German Institute for Standardization

EN 10025-6. 2009. Hot rolled products of structural steels – Part 6: Technical delivery conditions for flat products of high yield strength structural steels in the quenched and tempered conditions, Berlin: DIN German Institute for Standardization

EN 13001-3-1. 2013. Cranes – General Design – Part 3-1: Limit States and proof competence of steel structure, Berlin: DIN German Institute for Standardization

Gudehus, H. & Zenner, H. 2000. Leitfaden für eine Betriebsfestigkeitsrechnung: Empfehlung zur Lebensdauerabschätzung von Maschinenbauteilen, Düsseldorf: Verein zur Förderung der Forschung und Anwendung von Betriebsfestigkeits-Kenntnissen in der Eisenhüttenindustrie (VBFEh) im Verein Deutscher Eisenhüttenleute (VDEh)

Haagensen, P.J. & Maddox, S.J. 2011. IIW Recommendations on Post Weld Improvement of Steel and Aluminium Structures, IIW-Doc. No. XIII-2200r1-07, Paris: International Institute of Welding

Hobbacher, A. 2008. Recommendations for fatigue design of welded joints and components. IIW-Doc. No. XIII-2151-07/XV-1254-07, Paris: International Institute of Welding.

Hrabowski, J. et al. 2011. Low-cycle fatigue behavior of high-strength steel butt welds, Proceedings of the Twenty-First International Offshore and Polar Engineering Conference, Maui, Hawaii, USA, June 19–24, 2011, Vol. IV, pp. 282–287

Puthli, R. et al. 2006. Beurteilung des Ermüdungsverhaltens von Krankonstruktionen bei Einsatz hoch- und ultrahochfester Stähle, (Estimation of the fatigue behavior of crane structures when using high strength steels), Final Report P512 Forschungsvereinigung Stahlanwendung e.V., Düsseldorf: Verlag und Vertriebsgesellschaft mbH

Ummenhofer, T. et al. 2013. Bemessung von ermüdungsbeanspruchten Bauteilen aus hoch- und ultrahochfesten Feinkornbaustählen im Kran- und Anlagenbau, (Design of members for crane surtctures made of high-strength fine grained steels under fatigue load), Final Report P778 Forschungsvereinigung Stahlanwendung e.V., Düsseldorf: Verlag und Vertriebsgesellschaft mbH

Tubular Structures XV – Batista, Vellasco & Lima (eds)
© *2015 Taylor & Francis Group, London, ISBN 978-1-138-02837-1*

Local and local-overall buckling behaviour of welded stainless steel box section columns

H.X. Yuan & X.X. Du
School of Civil Engineering, Wuhan University, Wuhan, PR China

Y.Q. Wang & Y.J. Shi
Department of Civil Engineering, Tsinghua University, Beijing, PR China

L. Gardner
Department of Civil and Environmental Engineering, Imperial College London, London, UK

L. Yang
The College of Architecture and Civil Engineering, Beijing University of Technology, Beijing, PR China

ABSTRACT: This paper summarises the previously conducted experiments on local and local-overall buckling behaviour of welded stainless steel box section columns under axial compression. A total of thirteen stub column tests and eight local-overall interactive buckling tests were carried out. The box section test specimens were manufactured and fabricated by welding hot-rolled stainless steel plates of austenitic grade EN 1.4301 and duplex grade EN 1.4462. The material properties, welding residual stresses, initial local and global geometric imperfections were all accurately determined prior to the loading tests. The stub column specimens were tested under pure axial compression with fixed end boundary conditions, revealing the local buckling behaviour of stiffened stainless steel plates. The long columns with intermediate overall slenderness ratios were axially loaded between two pin-ended supports, which were subject to local-overall interactive buckling failure. The obtained test strengths were used to evaluate the existing design methods: the design provisions of Eurocode 3 Part 1.4, the direct strength method (DSM) and its modified form. The EN 1993-1-4 design provisions provide conservative strength predictions while the DSM slightly overpredicts the buckling capacities for the stub columns. Furthermore, both the EN 1993-1-4 design provisions and the DSM in the EN 1993-1-4 format tend to generate overpredicted interactive buckling resistances for columns of austenitic grade and underestimated resistances for those of duplex grade. The modified DSM was proved to offer more accurate strength predictions for both local and local-overall buckling resistances of welded stainless steel box section columns.

1 INTRODUCTION

Welded stainless steel sections can be fabricated by using plates with flexible dimensions and thicknesses, which helps to gain the popularity in structural skeletons compared with the cold-formed sections. The buckling behaviour of welded stainless steel sections may be different from the welded carbon steel sections due to the nonlinear material properties and the welding residual stresses. Experimental tests on welded stainless steel sections have been carried out recently, including the stub column tests on welded I-sections (Kuwamura 2003, Saliba & Gardner 2013a), shear buckling tests on plate girders (Real et al. 2007, Saliba & Gardner 2013b) and lateral-torsional buckling tests on welded I-section beams (Wang et al. 2014). Clearly, more experimental work on welded stainless steel sections involving multiple section types and alloys is required.

The current main design codes – Eurocode 3 Part 1.4 (EN 1993-1-4 2006), SEI/ASCE 8-02 (2002) and AS/NZS 4673 (2001) all provide design provisions for structural stainless steel sections. The EN 1993-1-4 provisions are suited to the design of both cold-formed and welded sections, while SEI/ASCE 8-02 and AS/NZS 4673 can only be used for cold-formed ones. Currently no design code for structural stainless steel has been published in China, though there is a substantial production surplus in stainless steel alloys. Promoting the structural use of stainless steel provides a solution to the further development of this industry, which can be stimulated by the first Chinese design code considering both cold-formed and welded sections.

This paper summarises the previously conducted experiments on local and local-overall buckling behaviour of welded stainless steel box section columns under axial compression, including thirteen stub column tests and eight local-overall interactive buckling tests. Moreover, the residual stress distributions were studied by means of the sectioning method for eight welded stainless steel box sections. The experiments form part of the fundamental work to underpin the development of the first Chinese design code for structural stainless steel. Part of the summarised research results presented in this paper has been published in international journals, hence further details can be found in these referred publications.

2 TEST SPECIMENS

2.1 Material properties

Two stainless steel alloys, namely austenitic grade EN 1.4301 and duplex grade EN 1.4462, were considered. According to the ASTM A959 (2011) designation system, these two alloys are equivalent to type 304 and 2205, respectively. The tensile coupons were prepared and tested using a 100 kN capacity universal testing machine. The accurately measured thicknesses of the hot-rolled coil plates were 6.00 mm for the two stainless steel grades.

Average measured tensile material properties from the test coupons are summarised in Table 1, where the following symbols are used: E_0 is the initial Young's modulus, $\sigma_{0.01}$, $\sigma_{0.2}$ and $\sigma_{1.0}$ are the 0.01%, 0.2% and 1% proof stresses, respectively, σ_u is the ultimate tensile stress, ε_f is the plastic strain at fracture, measured from the fractured tensile coupons as elongation over the standard gauge length, and n is the Ramberg-Osgood strain hardening coefficient. The ratios of 0.2% proof strengths given in the final column of Table 1 provide a measure of the anisotropic characteristics of the materials. A typical set of full stress-strain curves is plotted in Figure 1, revealing typical non-linear characteristics and illustrating the differences between the two stainless steel grades.

2.2 Geometric dimensions

The geometry and notation for the welded stainless steel box sections is shown in Figure 2.

Butt welds with a nominal weld size of 5 mm were adopted to fabricate the box sections. Symmetric welding sequences were introduced to alleviate the welding distortions, followed by additional straightening process using a hydraulic press.

The nominal outer sectional dimensions ranged from 100 mm to 400 mm, covering a wide range of plate width-to-thickness ratios from 14.7 to 64.9. All the constitutive plates of the specimens were cut from hot-rolled coil with an accurately measured thickness of 6.00 mm. The nominal lengths of the stub columns lay within the limits set by the Structural Stability Research Council (Ziemian 2010), at a nominal value of three times the depth of the section. The lengths of long column specimens were designed by means of the CUFSM (Li & Schafer 2010), allowing the interaction between local plate buckling and overall flexural buckling. The average measured geometric dimensions for all the test specimens, including 13 stub columns for stub column tests and 8 columns of intermediate slenderness for local-overall interactive buckling tests, were summarised in Tables 2 and 3.

The specimen labelling convention indicates the section type, the material grade and the nominal section depth. Stub column specimen R304-200, for example, is a rectangular section, of grade EN 1.4301 (304) alloy and 200 mm nominal section depth. All the test specimens for interactive buckling tests were appended with a symbol 'i'.

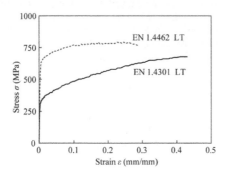

Figure 1. Full stress-strain curves from tensile coupon tests.

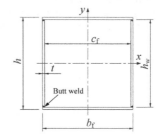

Figure 2. Geometry and notation for test specimens.

Table 1. Average measured material properties from tensile coupon tests.

Grade	t (mm)	Direction	E_0 (MPa)	$\sigma_{0.01}$ (MPa)	$\sigma_{0.2}$ (MPa)	$\sigma_{1.0}$ (MPa)	σ_u (MPa)	ε_f (%)	n	Anisotropic ratio (DT/LT or TT/LT)
1.4301	6.00	L*	188600	186.3	312.6	354.4	695.7	60.6	5.8	–
		D	194000	227.9	310.9	352.0	678.6	62.5	9.6	0.99
		T	201900	241.3	318.7	364.5	683.5	59.2	10.8	1.02
1.4462	6.00	L	193200	404.4	605.6	665.0	797.9	34.6	7.4	–
		D	191000	451.2	635.3	695.9	803.3	38.2	8.8	1.05
		T	221200	470.2	696.4	767.0	869.0	31.2	7.6	1.15

*L: Longitudinal, D: Diagonal, T: Transverse.

2.3 *Initial geometric imperfections and load eccentricity*

Prior to testing, the initial geometric imperfections in the test specimens were accurately determined by means of experimental techniques. The stub column specimens were designed to be short enough to prevent overall buckling, with the local imperfections being measured. Both the local and global imperfections in test specimens subject to local-overall interactive buckling were determined.

Schematic profiles of typical local geometric imperfection distributions measured in the box sections are plotted in Figure 3. The local imperfection magnitude w_0 of a cross-section was defined in relation to the junctions between flanges and webs.

The local imperfections of the test specimens were measured using the tool shown in Figure 4, which involved a digital linearly-varying displacement transducer (LVDT), driven by a calibrated electric guideway at a constant rate (2 mm/s), recording positional data across the section and corresponding time points. Three cross-sections of each stub column specimen – the mid-length section and two end sections, and three other cross-sections of each column for interactive buckling tests – the mid-length section and two quarter point sections, were continuously measured for each constitutive plate. The local imperfection amplitude (w_0) for each specimen was taken as the maximum

value among the three cross-sections. The measured results of the test specimens for stub column tests and local-overall interactive buckling tests are summarised in Tables 4 and 5, and the maximum value is close to $b/200$.

By means of an optical theodolite and a calibrated vernier caliper (Ban et al. 2012), the initial global geometric imperfections were measured. A schematic view and field photo of the measurement setup are shown in Figure 5. On the basis of a virtual straight line generated from the theodolite, five cross-sections – the mid-length section, two quarter point sections and two end sections, were measured along each column specimen for determining the amplitude of the initial global curvature. The maximum deviation from a straight line between the two ends of the members was taken as the global imperfection amplitude, $v_0 = \max(v_1, v_2, v_3)$.

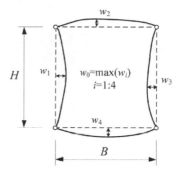

Figure 3. Schematic view of initial local geometric imperfections.

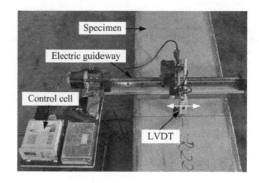

Figure 4. Measurement of local geometric imperfections.

Table 2. Measured geometric dimensions of the specimens for stub column tests.

Specimen	b_f (mm)	h (mm)	L (mm)	A (mm²)	c_f/t	h_w/t
R304-200	100.4	199.9	600.1	3459.2	31.3	14.7
R304-300	200.1	299.7	901.2	5852.8	48.0	31.3
R304-400	200.0	400.3	1201.5	7059.0	64.7	31.3
S304-130	130.3	129.8	399.7	2975.8	19.6	19.7
S304-200	200.3	200.5	600.4	4665.0	31.4	31.4
S304-300	301.3	300.7	900.9	7079.2	48.1	48.2
S304-350	350.5	350.1	1051.0	8262.8	56.3	56.4
R2205-200	100.4	200.1	601.0	3462.2	31.4	14.7
R2205-300	200.9	300.6	900.1	5873.6	48.1	31.5
R2205-400	200.2	401.3	1201.2	7072.6	64.9	31.4
S2205-130	130.5	130.3	399.5	2985.8	19.7	19.8
S2205-300	299.9	301.0	898.9	7065.7	48.2	48.0
S2205-350	349.8	350.4	1051.0	8256.8	56.4	56.3

Table 3. Measured geometric dimensions of the specimens for interactive buckling tests.

Specimen	b_f (mm)	h (mm)	A (mm²)	h/b_f	c_f/t	h_w/t	L (mm)	L_e (mm)	L_e/r_y
S304-300-i	299.8	300.0	7053.4	1.0	48.0	48.0	4497.3	4877.3	40.7
R304-300-i	200.5	300.1	5863.1	1.5	31.4	48.0	3998.0	4378.0	52.5
R304-360-i	179.8	360.2	6336.6	2.0	28.0	58.0	3998.6	4378.6	57.0
R304-400-i	200.4	400.3	7064.4	2.0	31.4	64.7	3702.7	4082.7	47.5
S2205-300-i	299.0	300.6	7050.4	1.0	47.8	48.1	4497.3	4877.3	40.7
R2205-300-i	199.7	300.0	5852.8	1.5	31.3	48.0	3996.2	4376.2	52.7
R2205-360-i	180.1	360.5	6342.8	2.0	28.0	58.1	3997.8	4377.8	56.9
R2205-400-i	199.5	400.2	7052.4	2.0	31.3	64.7	3697.5	4077.5	47.7

Table 4. Measured initial local imperfections of the stub column test specimens.

Specimen	Local geometric imperfections				w_0 (mm)
	Flange 1	Flange 2	Web 1	Web 2	
R304-200	0.14	0.23	−0.18	−0.15	0.23
R304-300	0.46	0.15	−0.22	−0.46	0.46
R304-400	−0.41	−0.42	0.35	0.42	0.42
S304-130	−0.21	−0.16	0.15	0.21	0.21
S304-200	−0.24	−0.66	0.36	0.22	0.66
S304-300	0.43	0.50	0.59	0.34	0.59
S304-350	−0.60	−0.72	0.58	0.62	0.72
R2205-200	−0.17	−0.19	0.51	0.18	0.51
R2205-300	−0.26	−0.15	0.49	0.32	0.49
R2205-400	−0.29	−0.15	1.38	0.81	1.38
S2205-130	−0.15	−0.15	0.18	0.15	0.18
S2205-300	−0.36	−0.43	0.52	0.48	0.52
S2205-350	−0.86	−0.72	0.57	0.86	0.86

Table 5. Measured initial geometric imperfection amplitudes and load eccentricities of the test specimens for interactive buckling tests.

Specimen	w_0 (mm)	v_0 (mm)	e_c (mm)	e_{Eq} (mm)	e_{Eq}/L
S304-300-i	0.62	1.58	1.75	3.33	1/1350
R304-300-i	0.58	−1.96	5.49	3.52	1/1134
R304-360-i	0.69	−0.94	11.10	10.16	1/394
R304-400-i	0.62	0.28	−4.16	−3.87	1/956
S2205-300-i	0.53	0.56	0.56	1.12	1/4021
R2205-300-i	0.65	0.72	−1.27	−0.55	1/7223
R2205-360-i	0.78	−0.11	−2.36	−2.47	1/1618
R2205-400-i	1.12	−1.42	2.29	0.86	1/4282

separated into two simple states – pure axial compression (σ_F) and pure bending (σ_M), as illustrated in Figure 6. For each cross-section, the internal axial force F and the moment M are given by Equation (1).

$$\begin{cases} F = \sigma_F A = EA\varepsilon_F \\ M = \sigma_M W_{el} = EW_{el}\varepsilon_M \end{cases} \tag{1}$$

in which ε_F and ε_M correspond to the strains generated by pure axial compression (σ_F) and pure bending (σ_M), respectively, E is the material Young's modulus, A is the cross-section area and W_{el} is the elastic section modulus. Hence the initial load eccentricity e_c can be computed from

$$e_c = \frac{M}{F} = \frac{W_{el}}{A}\frac{\varepsilon_M}{\varepsilon_F} \tag{2}$$

The obtained load eccentricities for the test specimens subject to local-overall interactive buckling are listed in Table 5. The equivalent initial eccentricity e_{Eq} of a column may be approximated as the sum of the loading eccentricity e_c plus initial geometric imperfections v_0:

$$e_{Eq} = e_c + v_0 \tag{3}$$

2.4 Welding residual stresses

The welding residual stresses in the test specimens were determined by means of the sectioning method, and the detailed experimental process were presented in a previous study (Yuan et al. 2014b). Eight welded stainless steel box sections for residual stress measurements were fabricated in parallel with the test specimens by using the same welding process. The cross-section of the test specimens were cut into pre-designed strips (as shown in Figure 7), with the relieved residual strains being determined from the changes in length. All cutting operations performed during the sectioning process of this experimental programme were conducted using an automated electric spark wire-cutting machine, with minimal heat input brought into the test specimens. The welding residual

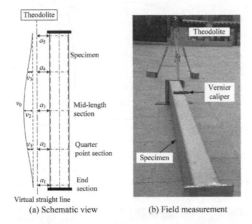

(a) Schematic view   (b) Field measurement

Figure 5. Measurement of global geometric imperfections.

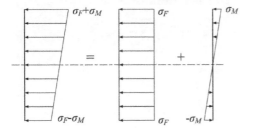

Figure 6. Stress state of a column section subjected to eccentric loading.

Based on the described method, the global geometric imperfections were measured and are summarised in Table 5. It can be observed that the overall geometric imperfections of the test specimens are generally small, with the maximum amplitude reaching only $L/2000$.

In addition, the initial load eccentricity applied to each interactive buckling test specimen about the minor axis (y axis) can be estimated by means of the strain gauge readings at the two quarter point cross-sections. The obtained axial stress state can be

Figure 7. Sliced residual stress test pieces (R304-300).

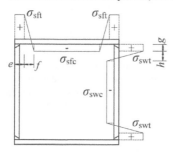

Figure 8. Residual stress pattern in welded stainless steel box sections.

Figure 9. Stub column test set-up (R2205-300).

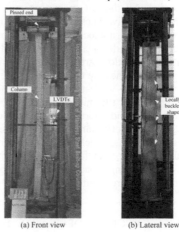

(a) Front view   (b) Lateral view

Figure 10. Test set-up for local-overall buckling tests (R2205-360-i).

stresses were computed by multiplying the relieved residual strains and the Young's modulus.

Based on the measured residual stress results, a new predictive model for determining the residual stresses in welded stainless steel box sections was proposed by referring to the ECCS predictive model (ECCS 1976) for welded carbon steel sections, with the basic distribution pattern shown in Figure 8. The key parameters of the newly proposed model are listed in Table 6, which shows lower peak tensile residual stresses, narrower peak tension zones but wider transition zones, compared with the existing models for carbon steel structural sections.

3 COLUMN BUCKLING TESTS

3.1 Stub column tests

The stub columns were axially loaded between two fixed end boundary conditions by means of a 5000 kN capacity hydraulic testing machine, with four LVDTs being symmetrically placed around each test specimen, as shown in Figure 9. Strain gauges were attached to the mid-length cross section, which helped to align the test specimens and monitoring the local plate buckling. The loading process was controlled by axial load at a rate of around 120 kN per minute until the ultimate load, followed by displacement control at a rate of around 4 mm per minute. The test specimens were unloaded once excessive end shortening and a clear decline in axial load were observed.

All thirteen test specimens were subject to local plate buckling failure. The axial load and corresponding end shortening deformation were continuously recorded, with the ultimate load $N_{u,Exp}$ and end shortening at ultimate load $\delta_{u,Exp}$ being summarised in Table 7. The stress-strain curves derived from the stub column tests and other details can be found in Yuan et al. (2014c).

3.2 Local-overall interactive buckling tests

The test set-up and instrumentation scheme for the interactive buckling tests was devised by referring to the proposals from Structural Stability Research Council (Ziemian 2010), as shown in Figure 10. The two pinned ends can rotate freely about the minor axis. Two steps were taken to align the test columns. Firstly, geometrical centreing was implemented by means of a laser, coupled with the achievement of close contact between the end plates and supports. In the second step, a preloading procedure was undertaken up to around 10% of the estimated ultimate load. The strain readings from the mid-length and quarter point cross-sections were used to assess the alignment and make further adjustments.

Table 6. Key parameters in the proposed models for residual stresses in welded stainless steel box sections.

Alloy	Ratio	$\sigma_{sft} = \sigma_{swt}$	e	f	g	h
Austenitic	$h/t(b_f/t) < 20$	$0.8\sigma_{0.2}$	0	$5t_f$	0	$5t_w$
	$h/t(b_f/t) \geq 20$	$0.8\sigma_{0.2}$	$t_w + 0.025c_f$	$5t_f$	$0.025h_w$	$5t_w$
Duplex	$h/t(b_f/t) < 20$	$0.6\sigma_{0.2}$	0	$5t_f$	0	$5t_w$
	$h/t(b_f/t) \geq 20$	$0.6\sigma_{0.2}$	$t_w + 0.025c_f$	$5t_f$	$0.025h_w$	$5t_w$

The tests were initially load controlled, at a rate of 90 kN per minute; after ultimate load, the test was switched to displacement control, at a rate of 25 mm per minute. The test was continued beyond ultimate load, enabling the post-ultimate response to be recorded. All the test specimens experienced local-overall interactive buckling. Typically, the test specimens exhibited first local buckling, followed by overall buckling. Under increasing load, and as the overall lateral deflections grew, the local buckles on the tension (or less heavily compressed) side of the member were eased while those on the other side developed further, as illustrated in Figure 10(b). The axial load versus end shortening deformation and mid-length lateral deflection curves were recorded for all test specimens, which can be found in Yuan et al. (2014a) along with further details. The ultimate test load-carrying capacities and corresponding deformation values are summarised in Table 8.

4 EVALUATION OF THE DESIGN METHODS

Based on the previously obtained test results, the EN 1993-1-4 design provisions that can predict the local and local-overall interactive buckling capacities of welded stainless steel box section columns are evaluated in this section. Meanwhile, the DSM can be used to determine the buckling resistances by considering the elastic buckling and yield capacity of the full cross-section (Becque et al. 2008, Rossi & Rasmussen 2013), with further modifications being made for welded stainless steel sections (Yuan et al. 2014a, Yuan 2014), and they are also assessed.

4.1 Overall view of the design formulae

In EN 1993-1-4, the cross-section resistance for axial compression can be determined as follows:

$$N_{c,Rd} = Af_y / \gamma_{M0} \qquad \text{for class 1, 2 and 3 cross-sections} \qquad (4)$$

$$N_{c,Rd} = A_{eff}f_y / \gamma_{M0} \qquad \text{for class 4 cross-sections} \qquad (5)$$

in which f_y is the material yield strength, taken as $\sigma_{0.2}$, γ_{M0} is the partial factor for cross-section resistance (set equal to unity in this study), and A_{eff} is the effective area of the class 4 cross-sections accounting for the local buckling, with the corresponding effective width reduction factor ρ given by Equation (6) for welded Class 4 internal elements.

$$\rho = \frac{0.772}{\lambda_p} - \frac{0.125}{\lambda_p^2} \leq 1 \qquad (6)$$

where the element slenderness λ_p is defined as

$$\lambda_p = \sqrt{\frac{f_y}{\sigma_{cr}}} = \frac{b/t}{28.4\varepsilon\sqrt{k_\sigma}} \qquad (7)$$

in which b/t is the relevant width-to-thickness ratio, k_σ is the buckling coefficient dependent on the boundary conditions and applied stress conditions, and ε is a material factor given by Equation (8):

$$\varepsilon = \sqrt{\frac{235}{f_y} \cdot \frac{E_0}{210000}} \qquad (8)$$

The current Class 3 slenderness limit for welded internal elements in pure compression is 30.7ε.

Meanwhile, the local-overall interactive buckling resistance of an axial compression member can be taken as:

$$N_{b,Rd} = \chi A_{eff}f_y / \gamma_{M1} \qquad (9)$$

in which γ_{M1} is the partial factor for member buckling (set equal to unity in this study), χ is the overall buckling reduction factor, and can be calculated from Equations (10) and (11).

$$\chi = \frac{1}{\phi + \sqrt{\phi^2 - \lambda^2}} \leq 1 \qquad (10)$$

$$\phi = 0.5[1 + \alpha(\lambda - \lambda_0) + \lambda^2] \qquad (11)$$

in which the imperfection factor α and limiting slenderness λ_0 are taken as 0.49 and 0.40 respectively for welded compression members with hollow sections undergoing flexural buckling, and the non-dimensional slenderness λ can be defined by Equation (12).

$$\lambda = \sqrt{\frac{A_{eff}f_y}{N_{cr}}} \qquad \text{for class 4 cross-sections} \qquad (12)$$

where N_{cr} is the elastic global buckling load of the member.

Table 7. Comparison of stub column test results with predicted strength values.

Specimen	$N_{u,Exp}$ (kN)	$\delta_{u,Exp}$ (mm)	EN 1993-1-4 $\frac{N_{c,Rd}}{N_{u,Exp}}$	Modified DSM $\frac{N_{cl}}{N_{u,Exp}}$	DSM $\frac{N_{cl}}{N_{u,Exp}}$
R304-200	1068.6	3.8	0.93	0.96	0.91
R304-300	1317.1	2.5	1.02	1.08	0.99
R304-400	1351.7	3.3	1.01	1.08	0.99
S304-130	1066.4	7.3	0.87	1.02	0.94
S304-200	1354.2	2.9	0.94	0.97	0.91
S304-300	1393.2	2.0	1.00	1.16	1.06
S304-350	1423.9	2.8	1.00	1.21	1.10
R2205-200	1802.5	3.2	0.93	0.97	0.89
R2205-300	2140.4	3.7	0.94	1.08	0.97
R2205-400	2320.3	5.2	0.88	1.01	0.90
S2205-130	1897.5	4.2	0.91	0.91	0.86
S2205-300	2221.5	3.9	0.93	1.17	1.05
S2205-350	2363.5	4.6	0.89	1.16	1.04
Mean	–	–	0.94	1.06	0.97
COV	–	–	0.05	0.09	0.07

The DSM utilises a continuous reduction curve for the treatment of local plate buckling (Rusch & Lindner 2001). The DSM formulae for predicting the resistances of structural stainless steel sections were presented by Rossi & Rasmussen (2013). Stresses in cross-section can grow higher than the nominal yield strength, allowing the exploitation of strain hardening capacity. The axial cross-sectional resistance N_{cl} considering local buckling can be determined from Equation (13).

$$N_{cl} = \begin{cases} \left[\left(1 - 2.11\lambda_l\right)\left(\frac{f_u}{f_y} - 1\right) + 1 \right] A f_y & \text{for } \lambda_l \leq 0.474 \\ \left[\frac{0.95}{\lambda_l^{0.8}} - \frac{0.22}{\lambda_l^{1.6}} \right] A f_y & \text{for } \lambda_l > 0.474 \end{cases} \quad (13)$$

in which $\lambda_l = \sqrt{f_y/\sigma_{cr}}$, where σ_{cr} can be obtained from CUFSM, with element interaction being considered.

The DSM formulae for predicting column flexural buckling strength was presented in the format of EN 1993-1-4 design provisions (Becque et al. 2008), and the nominal axial resistance N_{ce} can be calculated by Equation (12).

$$N_{ce} = \chi A f_y \quad (14)$$

where χ is obtained from Equations (10) and (11) by using the full cross-section area A. The local-overall interactive buckling resistance N_{cl} is given by Equation (15), taking the local plate buckling into consideration.

$$N_{cl} = \begin{cases} N_{ce} & \text{for } \lambda_l \leq 0.55 \\ \left[\frac{0.95}{\lambda_l} - \frac{0.22}{\lambda_l^2} \right] N_{ce} & \text{for } \lambda_l > 0.55 \end{cases} \quad (15)$$

in which $\lambda_l = \sqrt{N_{ce}/N_{crl}}$, where N_{crl} is the elastic critical local buckling load and may be obtained from CUFSM (Li & Schafer 2010).

Modifications to the current DSM formulae for predicting the local and local-overall interactive buckling capacities of welded stainless steel box section columns were proposed previously (Yuan et al. 2014a, Yuan 2014). Specifically, Equations (13) and (15) can be replaced by Equations (16) and (17), respectively. Meanwhile, the parameters α and λ_0, were proposed separately for the austenitic and duplex grades: austenitic grades – $\alpha = 0.49$, $\lambda_0 = 0.20$; duplex grades $\alpha = 0.34$, $\lambda_0 = 0.40$

$$N_{cl} = \begin{cases} \left[\left(1 - 2.11\lambda_l\right)\left(\frac{f_u}{f_y} - 1\right) + 1 \right] A f_y & \text{for } \lambda_l \leq 0.474 \\ \left[\frac{0.86}{\lambda_l^{0.8}} - \frac{0.17}{\lambda_l^{1.6}} \right] A f_y & \text{for } \lambda_l > 0.474 \end{cases} \quad (16)$$

$$N_{cl} = \begin{cases} N_{ce} & \text{for } \lambda_l \leq 0.474 \\ \left[\frac{0.86}{\lambda_l^{0.8}} - \frac{0.17}{\lambda_l^{1.6}} \right] N_{ce} & \text{for } \lambda_l > 0.474 \end{cases} \quad (17)$$

4.2 Comparison with the test results

The previously obtained tensile material properties were used to calculate the predicted axial design resistances. Comparisons between the test results and the three design methods – EN 1993-1-4, the DSM and the modified DSM are summarised in Table 7 and Table 8.

From Table 7 it can be observed that the EN 1993-1-4 design provisions generate conservative axial cross-section resistances with an average strength ratio of 0.94, while the DSM offers slightly overestimated strength predictions with a corresponding strength ratio of 1.06. For the modified DSM, the mean value of predicted to test strength ratios is 0.97 with a corresponding coefficient of variation (COV) of 0.07, which shows an average increase in resistance of about 3% compared with the EN 1993-1-4 predictions and a lower scatter compared with the DSM.

The mean values of predicted interactive buckling resistances over test strengths are 1.01 and 1.00 for the EN 1993-1-4 design provisions and the DSM in the EN 1993-1-4 format, while the corresponding COV values are equal to 0.11 and 0.12, respectively. It can be seen that the interactive buckling capacities of columns of grade EN 1.4301 are overpredicted while those of columns of grade EN 1.4462 are underestimated. For the modified DSM, the mean value of strength ratio is 0.94 with a corresponding COV of 0.06, revealing good strength predictions for welded sections of both austenitic and duplex grades.

5 CONCLUSIONS

The experimental research on local and local-overall buckling behaviour of welded stainless steel box section columns has been summarised in this paper. The magnitude and distribution of welding residual stresses were measured using the sectioning method. The initial geometric imperfections and load

Table 8. Comparison of interactive buckling test results with predicted strength values.

Specimen	$N_{u,Exp}$ (kN)	$\delta_{u,Exp}$ (mm)	$\Delta_{u,Exp}$ (mm)	EN 1993-1-4 $\frac{N_{b,Rd}}{N_{u,Exp}}$	DSM-EN 1993-1-4 $\frac{N_{cl}}{N_{u,Exp}}$	Modified DSM $\frac{N_{cl}}{N_{u,Exp}}$
S304-300-i	1330.1	8.1	4.4	1.04	1.11	0.97
R304-300-i	1153.3	6.9	9.6	1.03	1.05	0.91
R304-360-i	970.3	5.5	15.5	1.20	1.17	1.03
R304-400-i	1145.2	7.0	12.0	1.14	1.10	0.98
S2205-300-i	2054.1	13.1	9.7	0.94	0.99	1.01
R2205-300-i	1853.6	9.3	13.4	0.87	0.86	0.86
R2205-360-i	1673.3	8.8	23.7	0.94	0.87	0.88
R2205-400-i	1976.6	9.8	15.7	0.91	0.84	0.87
Mean	–	–	–	1.01	1.00	0.94
COV	–	–	–	0.11	0.12	0.06

eccentricities were accurately determined by means of experimental techniques. A series of tests was carried out for welded stainless steel box section columns subject to local buckling and local-overall interactive buckling. The EN 1993-1-4 design provisions, the DSM for structural stainless steel and the modified DSM formulae were evaluated based upon the obtained test strengths. It was revealed that the EN 1993-1-4 design provisions provided conservative resistance predictions while the DSM slightly overpredicted the cross-sectional resistances for the stub columns. Moreover, both EN 1993-1-4 and the DSM were found to provide overestimated interactive buckling resistances for columns of austenitic grade and underestimated resistances for those of duplex grade. The modified DSM was proved to offer more accurate strength predictions for both local and local-overall buckling resistances of welded stainless steel box section columns.

ACKNOWLEDGEMENTS

The authors are grateful for the financial support from the China Postdoctoral Science Foundation (Grant No. 2014M560626), the Specialised Research Fund for the Doctoral Program of Higher Education (Grant No. 20110002130002) and the National Natural Science Foundation of China (Grant No. 51108007).

REFERENCES

AS/NZS 4673. 2001. Cold-formed stainless steel structures. Sydney: Standards Australia.

ASTM A959. 2011. Standard guide for specifying harmonized standard grade compositions for wrought stainless steels. West Conshohocken, PA: ASTM International.

Ban, H.Y., Shi, G., Shi, Y.J. & Wang, Y.Q. 2012. Overall buckling behavior of 460 MPa high strength steel columns: Experimental investigation and design method. Journal of Constructional Steel Research 74: 140–150.

Becque, J., Lecce, M. & Rasmussen, K.J.R. 2008. The direct strength method for stainless steel compression members. Journal of Constructional Steel Research 64(11): 1231–1238.

ECCS. 1976. Manual on stability of steel structures – Part 2.2 Mechanical properties and residual stresses. 2nd Edition, Bruxelles: ECCS Publ.

EN 1993-1-4. 2006. Eurocode 3: Design of steel structures – Part 1.4: General rules–Supplementary rules for stainless steels. CEN.

Kuwamura, H. 2003. Local buckling of thin-walled stainless steel members. Steel Structures 3: 191–201.

Li, Z. & Schafer, B.W. 2010. Buckling analysis of cold-formed steel members with general boundary conditions using CUFSM: conventional and constrained finite strip methods. Proceedings of the 20th Intl. Spec. Conf. on Cold-Formed Steel Structures. St. Louis, MO. November, 2010.

Real, E., Mirambell, E. & Estrada, I. 2007. Shear response of stainless steel plate girders. Engineering Structures 29: 1626–1640.

Rossi, B. & Rasmussen, K.J.R. 2013. Carrying capacity of stainless steel columns in the low slenderness range. Journal of Structural Engineering ASCE 139(6): 1088–1092.

Rusch, A. & Lindner, J. 2001. Remarks to the direct strength method. Thin-Walled Structures 39(9): 807–820.

Saliba, N., & Gardner, L. 2013a. Cross-section stability of lean duplex stainless steel welded I-sections. Journal of Constructional Steel Research 80: 1–14.

Saliba, N., & Gardner, L. 2013b. Experimental study of the shear response of lean duplex stainless steel plate girders. Engineering Structures 46: 375–391.

SEI/ASCE 8-02. 2002. Specification for the design of cold-formed stainless steel structural members. New York: American Society of Civil Engineers (ASCE).

Wang, Y.Q., Yang, L., Gao, B., Shi, Y.J. & Yuan, H.X. 2014. Experimental study of lateral-torsional buckling behavior of stainless steel welded I-section beams. International Journal of Steel Structures 14(2): 411–420.

Yuan, H.X., Wang, Y.Q., Gardner L. & Shi, Y.J. 2014a. Local-overall interactive buckling of welded stainless steel box section compression members. Engineering Structures 67: 62–76.

Yuan, H.X., Wang, Y.Q., Shi, Y.J. & Gardner L. 2014b. Residual stress distributions in welded stainless steel sections. Thin-Walled Structures 79: 38–51.

Yuan, H.X., Wang, Y.Q., Shi, Y.J. & Gardner L. 2014c. Stub column tests on stainless steel built-up sections. Thin-Walled Structures 83: 103–114.

Yuan, H.X. 2014. Local and local-overall buckling behaviour of welded stainless steel members under axial compression. Beijing: Department of Civil Engineering, Tsinghua University.

Ziemian, R.D. 2010. Guide to stability design criteria for metal structures. 6th ed. New York: John Wiley & Sons, Inc.

Tubular Structures XV – Batista, Vellasco & Lima (eds)
© 2015 Taylor & Francis Group, London, ISBN 978-1-138-02837-1

Experimental investigation of cold-formed high strength steel tubular sections undergoing web crippling

H.T. Li & B. Young

Department of Civil Engineering, The University of Hong Kong, Hong Kong, China

ABSTRACT: This paper presents a series of tests carried out on cold-formed high strength steel tubular sections undergoing web crippling. The tests were conducted on square and rectangular hollow sections of high strength steel with measured 0.2% proof stress ranged from 679 to 1025 MPa. The measured web slenderness values of the tubular sections ranged from 8.3 to 35.6. Tensile coupon tests were performed to obtain the material properties. The web crippling tests were conducted under the End-Two-Flange (ETF) and Interior-Two-Flange (ITF) loading conditions as specified in the North American Specification and Australian/New Zealand Standard for cold-formed steel structures. The test strengths obtained from this study were compared with the design strengths calculated from the North American Specification, Australian/New Zealand Standard and European Code for cold-formed steel structures. The web crippling test results obtained from this study are valuable for the development of design rules for cold-formed high strength steel.

1 INTRODUCTION

Web crippling is a form of localized buckling that could occur at supports or points of structural members where concentrate load applied. Cold-formed steel sections, such as square and rectangular hollow sections, could fail by web crippling. The current web crippling design rules in most international specifications for cold-formed steel structures are empirical in nature, and are based on experimental investigation conducted by researchers from the 1940s onwards, such as Winter & Pian (1946), Zetlin (1955), Khan & Walker (1972), Hetrakul & Yu (1978), Yu (1981), Wing & Schuster (1982), Santaputra (1986), Bakker (1992), Bhakta et al. (1992), Prabakaran (1993), Wu et al. (1997), Gerges & Schuster (1998), Parabakaran & Schuster (1998), Beshara & Schuster (2000), Young & Hancock (2001), Wallace & Schuster (2004). However, most of the tested specimens used to formulate the empirical web crippling design equations in the international specifications for cold-formed steel structures are having the web slenderness value greater than 40 and the 0.2% proof stress (yield stress) less than 500 MPa. This is mainly due to the limitation of the cold-forming technology in the past. On the other hand, high strength steel is becoming increasing attractive for high-rise buildings and bridges due to its structural and architectural advantages. Although extensive investigations have been conducted on normal steel grade sections with 0.2% proof stress of less than 350 MPa undergoing web crippling, research on high strength steel sections undergoing web crippling is limited up to date. Web crippling tests were conducted by Santaputra et al. (1989). The tested

specimens were hat sections and built-up sections formed from high strength steel sheet with measured thickness ranged from 1.17 to 2.24 mm. Zhou & Young (2007) investigated square and rectangular hollow sections with measured web slenderness value ranged from 16.5 to 49.7 and the 0.2% proof stress ranged from 448 to 707 MPa. However, it should be noted that the tests conducted by Zhou & Young (2007) were high strength stainless steel sections instead of carbon steel sections. Therefore, it is important to investigate cold-formed high strength carbon steel sections undergoing web crippling.

The purpose of this study is to provide test data for cold-formed high strength carbon steel square and rectangular hollow sections subjected to web crippling under End-Two-Flange (ETF) and Interior-Two-Flange (ITF) loading conditions, which has been recently conducted at the University of Hong Kong. The test strengths were compared with the design strengths predicted by the current North American Specification (NAS 2012), Australian/New Zealand Standard (AS/NZS 2005) and European Code (EC3 2006) for cold-formed steel structures to study the appropriateness of the current design equations for high strength steel tubular sections undergoing web crippling.

2 EXPERIMENTAL INVESTIGATION

2.1 Test specimens

A series of tests was conducted on square and rectangular hollow sections of cold-formed high strength steel

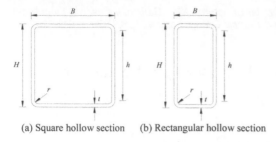

(a) Square hollow section   (b) Rectangular hollow section

Figure 1. Definition of symbols.

with nominal 0.2% proof stresses of 700 and 900 MPa. Before testing, the height (H) and width (B) of the cross-sections were measured using a Mitutoyo digital caliper, the thickness (t) was measured by a Mitutoyo digital micrometer, and the inner radius (r) and outer radius (R) were measured using Moore Wright radius gauges. The test specimens were cold-formed from flat strips into eight different sections, having the overall measured web heights ranged from 50.0 to 160.2 mm, overall flange widths ranged from 50.0 to 160.2 mm, and thicknesses ranged from 3.893 to 3.998 mm. The measured inner and outer corner radii ranged from 4.5 to 6.8 mm and 8.4 to 10.6 mm, respectively. The web slenderness ratio (h/t) ranged from 8.3 to 35.6 mm, where h is the depth of the flat portion of the web ranged from 32.8 to 142.0 mm.

The specimen length (L) was determined in accordance with the NAS (2012) and AS/NZS (2005). Generally, the clear distance from the end of the specimen to the edge of the bearing plate was designed to be 1.5 times of the overall height of the web rather than 1.5 times of the flat portion of the web, the latter being the minimum specified in the NAS (2012) and AS/NZS (2005). Tables 1 and 2 show the measured dimensions of the test specimens using the nomenclature defined in Figure 1, where H is the overall height of web, B is the overall width of flange, t is the thickness, r is the inner corner radius of the sections, and h is the flat portion of the web. The loading or reaction force was applied through bearing plates. Six pairs of bearing plates were machined to specify dimensions, and the thickness was 50 mm for all the bearing plates. The bearing plates were designed to act across the full flange widths of the sections, excluding the rounded corners. It should be noted that all flanges of the specimens were not fastened to the bearing plates during testing.

The test specimens are labelled such that the loading condition, nominal 0.2% proof stress, nominal cross-section dimensions, and the bearing length could be identified, as shown in Tables 1 and 2. For example, the labels "ETF-H50 × 100 × 4N50-R" and "ITF-V100×100 × 4N50" define the following specimens. The first three letters of the labels indicate the loading condition of End-Two-Flange (ETF) or Interior-Two-Flange (ITF) was used in the tests. The next letter shows the material of the specimens. The letter "H" indicates the nominal 0.2%

proof stress of the specimen was 700 MPa, while the letter "V" indicates the nominal 0.2% proof stress of the specimen was 900 MPa. The following symbols are the nominal cross-section dimensions $H \times B \times t$ of the specimen in millimeters ($50 \times 100 \times 4$ means $H = 50$ mm $B = 100$ mm and $t = 4$ mm; $10 \times 10 \times 4$ means $H = 100$ mm, $B = 100$ mm and $t = 4$ mm). The notation N50 indicates the bearing length was 50 mm If a test was repeated, the "−R" indicates the repeated test.

The material properties of the test specimens were determined from tensile coupon tests. Tensile coupon specimens were prepared in accordance with the ASTM (2013) using 12.5 mm wide coupon of 50 mm gauge length. The coupons were taken from the center of the face at 90 degree angle from the weld in the longitudinal direction of the untested specimens. A MTS 810 material testing machine with maximum capacity of 250 kN was used to conduct the coupon tests. Displacement control was used during the tests. Two strain gauges and a calibrated MTS extensometer of 50 mm gauge length were used to measure the longitudinal strain of the coupons. A data acquisition system was used to record the load and strain at regular intervals during the coupon tests. The static load was obtained by pausing the applied straining for 100s near the 0.2% proof stress and the ultimate strength, as suggested by Huang & Young (2014), allowing the plastic straining associated with stress relaxation to take place. Table 3 summarizes the material properties determined from the tensile coupon tests, namely, Young's modulus (E), static 0.2% proof stress ($\sigma_{0.2}$), static tensile strength (σ_u) and elongation after fracture (ε_f) based on a 50 mm gauge length. Figure 2 shows the static stress-strain curve obtained from tensile coupon test for H160 × 160 × 4 section.

2.2 Test setup and procedure

The web crippling tests of the cold-formed high strength steel square and rectangular hollow sections were conducted under ETF and ITF loading conditions specified in the NAS (2012) and AS/NZS (2005) for cold-formed steel structures, as shown in Figure 3. Two half rounds were used to simulate the hinge supports. Vertical web deformations of the specimens were obtained by the average readings of four calibrated linear variable displacement transducers (LVDTs). The test setup of the ETF loading and ITF loading are shown in Figure 4 and Figure 5, respectively. For ETF loading condition, two identical bearing plates were carefully positioned at the end of the specimens, whereas the bearing plates were positioned at the mid-length of the specimens for ITF loading condition.

A servo-controlled hydraulic testing machine was used to apply a compressive force to the test specimens. Displacement control was used to drive the hydraulic actuator at a constant speed of 0.5 mm/min. A data acquisition system was used to record the load and the readings of the LVDTs at regular intervals for all the tests.

Table 1. Measured dimensions and experimental ultimate loads for specimens under ETF loading condition.

Specimen ($H \times B \times t$)	Height H (mm)	Width B (mm)	Thickness t (mm)	Inner radius r (mm)	Length L (mm)	Web slenderness ratio h/t	Exp. load per web P_{Exp} (kN)
ETF-H80 × 80 × 4N90	80.2	80.2	3.903	4.7	210.7	16.1	75.7
ETF-H80 × 80 × 4N50	80.1	80.2	3.904	4.7	170.7	16.1	55.2
ETF-H120 × 120 × 4N120	120.6	120.7	3.929	4.5	301.0	26.4	79.6
ETF-H120 × 120 × 4N60	120.6	120.7	3.918	4.5	240.9	26.5	53.9
ETF-H160 × 160 × 4N150	160.0	160.2	3.998	5.1	389.0	35.5	72.6
ETF-H160 × 160 × 4N90	160.2	160.2	3.985	5.1	330.4	35.6	57.2
ETF-H50 × 100 × 4N90	50.1	100.3	3.976	4.6	164.1	8.3	100.0
ETF-H50 × 100 × 4N50	50.2	100.3	3.974	4.6	124.5	8.3	66.6
ETF-H50 × 100 × 4N50-R	50.1	100.3	3.967	4.6	125.3	8.3	66.7
ETF-H100 × 50 × 4N50	100.3	50.1	3.919	4.6	200.1	21.3	51.0
ETF-H100 × 50 × 4N30	100.4	50.1	3.926	4.6	180.1	21.2	38.3
ETF-H100 × 50 × 4N30-R	100.4	50.0	3.918	4.6	180.0	21.3	36.6
ETF-V80 × 80 × 4N90	80.0	80.2	3.951	6.2	211.5	15.1	87.3
ETF-V80 × 80 × 4N50	80.0	80.3	3.961	6.2	171.0	15.1	60.8
ETF-V100 × 100 × 4N90	100.1	99.9	3.951	6.8	239.4	19.9	66.6
ETF-V100 × 100 × 4N50	100.3	99.9	3.947	6.8	200.8	20.0	50.1
ETF-V120 × 120 × 4N120	120.9	120.7	3.923	6.4	300.4	25.6	77.5
ETF-V120 × 120 × 4N60	120.9	120.8	3.927	6.4	240.2	25.5	55.9

Table 2. Measured dimensions and experimental ultimate loads for specimens under ITF loading condition.

Specimen ($H \times B \times t$)	Height H (mm)	Width B (mm)	Thickness t (mm)	Inner radius r (mm)	Length L (mm)	Web slenderness ratio h/t	Exp. load per web P_{Exp} (kN)
ITF-H80 × 80 × 4N90	80.1	80.1	3.925	4.7	330.5	16.0	137.7
ITF-H80 × 80 × 4N50	80.1	80.1	3.893	4.7	291.1	16.2	120.0
ITF-H120 × 120 × 4N120	120.7	120.7	3.909	4.5	480.6	26.6	152.9
ITF-H120 × 120 × 4N60	120.7	120.7	3.93	4.5	421.3	26.4	139.4
ITF-H160 × 160 × 4N150	160.1	160.2	3.988	5.1	630.1	35.6	164.0
ITF-H160 × 160 × 4N90	160.2	160.2	3.99	5.1	571.4	35.6	148.9
ITF-H50 × 100 × 4N90	50.2	100.4	3.969	4.6	239.1	8.3	143.3
ITF-H50 × 100 × 4N90-R	50.1	100.3	3.966	4.6	240.0	8.3	144.2
ITF-H50 × 100 × 4N50	50.0	100.4	3.973	4.6	200.0	8.3	118.1
ITF-H100 × 50 × 4N50	100.3	50.1	3.927	4.6	349.0	21.2	125.1
ITF-H100 × 50 × 4N30	100.3	50.0	3.926	4.6	330.1	21.2	97.6
ITF-H100 × 50 × 4N30-R	100.4	50.1	3.923	4.6	329.1	21.2	98.9
ITF-V80 × 80 × 4N90	80.0	80.2	3.95	6.2	331.6	15.1	164.3
ITF-V80 × 80 × 4N50	80.0	80.2	3.948	6.2	290.6	15.1	140.3
ITF-V100 × 100 × 4N90	100.2	99.9	3.953	6.8	390.3	19.9	150.7
ITF-V100 × 100 × 4N50	100.2	99.9	3.96	6.8	349.6	19.9	131.1
ITF-V120 × 120 × 4N120	120.9	120.8	3.925	6.4	480.9	25.5	175.9
ITF-V120 × 120 × 4N60	120.9	120.8	3.921	6.4	421.4	25.6	150.7

2.3 Test results

The experimental web crippling loads per web (P_{Exp}) are given in Tables 1 and 2 for ETF and ITF loading conditions, respectively. Repeated tests were conducted on ETF-H50 × 100 × 4N50, ETF-H100 × 50 × 4N30, ITF-H50 × 100 × 4N90 and ITF-H100 × 50 × 4N30 specimens, and the repeated test results are close to the first test values with a difference of 0.1%, 4.3%, 0.7% and 1.3%, respectively. The small difference between the repeated tests demonstrated the reliability of the test results.

3 COMPARISON OF TESTS STRENGTHS WITH CURRENT DESIGN STRENGTHS

The web crippling design rules specified in the NAS (2012) is based on a unified equation derived by Prabakaran (1993) and Prabakaran & Schuster (1998). The unified web crippling equation is able to accommodate for various section geometries and loading conditions. Consideration is also given to whether the members are fastened to the bearing plates or not.

Table 3. Material properties obtained from tensile coupon tests.

Section	E (GPa)	$\sigma_{0.2}$ (MPa)	σ_u (MPa)	ε_f (%)
H80 × 80 × 4	214.4	735	832	11.0
H80 × 80 × 4-R	210.9	725	834	11.2
H100 × 100 × 4	214.7	715	808	12.4
H120 × 120 × 4	215.1	721	814	12.4
H160 × 160 × 4	212.4	751	829	12.3
H50 × 100 × 4	206.9	703	828	10.9
H50 × 100 × 4-R	211.3	679	820	11.5
V80 × 80 × 4	209.5	1025	1173	7.6
V80 × 80 × 4-R	214.2	997	1198	8.1
V100 × 100 × 4	205.3	971	1079	7.8
V120 × 120 × 4	206.7	976	1148	8.1
V120 × 120 × 4-R	205.4	969	1142	8.2

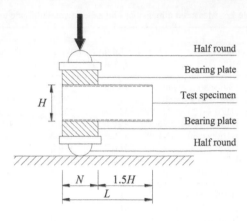

(a) End-Two-Flange (ETF) loading

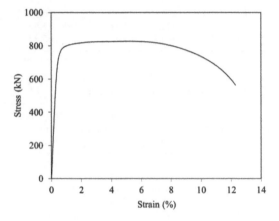

Figure 2. Stress–strain curve obtained from tensile coupon test for H160 × 160 × 4 section.

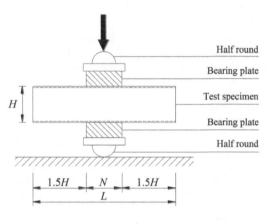

(b) Interior-Two-Flange (ITF) loading

Figure 3. Loading conditions of web crippling tests.

The AS/NZS (2005) has adopted the web crippling design rules from the 2001 edition of the NAS (2001). Although the NAS was revised in 2007 and 2012, no changes were introduced to the web crippling design for sections with single web, such as the hollow sections. Hence, the web crippling design strengths for hollow sections predicted by the NAS (2012) and AS/NZS (2005) are identical. In calculating the web crippling strengths for square and rectangular hollow sections, Table C3.4.1-2 of the NAS (2012) was used.

EC3 Part 1-3 (2006) provides design rules to calculate the web crippling strength, also known as the local transverse resistance of the web, for cross-sections with a single web and for cross-sections with two or more webs. Hollow sections are belonged to sections with two or more webs when calculating the web crippling strength according to the EC3 (2006)

The web crippling loads per web obtained from the cold-formed high strength steel square and rectangular sections were compared with the unfactored design strengths (nominal strengths) predicted using the NAS (2012), AS/NZS (2005) and the EC3 (2006). As the web crippling design rules of hollow sections are identical for the NAS (2012) and the AS/NZS (2005), hence, the web crippling strengths calculated by the NAS (P_{NAS}) are identical to those calculated from the AS/NZS ($P_{AS/NZS}$). Tables 4 and 5 show the comparison of the test strengths per web (P_{Exp}) with the nominal strengths predicted using the NAS (2012), AS/NZS (2005) and EC3 (2006) for ETF and ITF loading conditions, respectively. The nominal strengths were calculated using the measured cross-section dimensions and the measured material properties as shown in Tables 1–3.

The design strengths predicted by the current NAS (2012) and AS/NZS (2005) are unconservative for the tested square and rectangular hollow sections under the ETF and ITF loading conditions. The mean values of the tested-to-predicted load ratio are 0.64 and 0.72 with the corresponding coefficient of variation (COV) of 0.226 and 0.061 for ETF and ITF loading conditions, respectively, as shown in Tables 4 and 5. It should be noted that test strengths are as low as 45% and 63% of the design strengths predicted by the current NAS (2012) and AS/NZS (2005) for ETF and ITF loading conditions, respectively. For the EC3 (2006),

(a) End view

(b) Side view

Figure 4. Test setup of ETF loading condition.

(a) End view

(b) Side view

Figure 5. Test setup of ITF loading condition.

the design strengths are very conservative for both ETF and ITF loading conditions. The mean values of the tested-to-predicted load ratio are 2.56 and 5.56 with the COV of 0.257 and 0.127 for ETF and ITF loading conditions, respectively. It is noteworthy that the web crippling design strengths for ETF and ITF loading conditions predicted by the EC3 (2006) are identical for a given hollow section. The design equation in the EC3 (2006) does not consider the h/t ratio and using a same bearing length of 10 mm for ETF and ITF loading conditions despite the fact that sections may have different web slenderness and members may be

loaded with different bearing lengths. It has been found that 50% and 28% enhancement of the web crippling strength could be obtained by increasing the bearing length for ETF and ITF loading conditions, respectively, and the effect of increasing the bearing length for ETF loading condition is more obvious compared to the ITF loading condition.

4 CONCLUSIONS

A test program on cold-formed high strength steel square and rectangular hollow sections undergoing

Table 4. Comparison of web crippling tests strengths with current design strengths for ETF loading condition.

Specimen	Measured Ratio			Test Exp. load per web	Comparison NAS	EC3
	r/t	N/t	h/t	P_{Exp} (kN)	$\frac{P_{Exp}}{P_{NAS}}$	$\frac{P_{Exp}}{P_{EC3}}$
ETF-H80 × 80 × 4N90	1.2	23.1	16.1	75.7	0.78	3.21
ETF-H80 × 80 × 4N50	1.2	12.8	16.1	55.2	0.60	2.34
ETF-H120 × 120 × 4N120	1.1	30.5	26.4	79.6	0.83	3.30
ETF-H120 × 120 × 4N60	1.1	15.3	26.5	53.9	0.60	2.24
ETF-H160 × 160 × 4N150	1.3	37.5	35.5	72.6	0.73	2.89
ETF-H160 × 160 × 4N90	1.3	22.6	35.6	57.2	0.61	2.30
ETF-H50 × 100 × 4N90	1.2	22.6	8.3	100.0	1.00	4.22
ETF-H50 × 100 × 4N50	1.2	12.6	8.3	66.6	0.70	2.81
ETF-H50 × 100 × 4N50-R	1.2	12.6	8.3	66.7	0.70	2.82
ETF-H100 × 50 × 4N50	1.2	12.8	21.3	51.0	0.60	2.21
ETF-H100 × 50 × 4N30	1.2	7.6	21.2	38.3	0.46	1.65
ETF-H100 × 50 × 4N30-R	1.2	7.7	21.3	36.6	0.45	1.59
ETF-V80 × 80 × 4N90	1.6	22.8	15.1	87.3	0.69	3.11
ETF-V80 × 80 × 4N50	1.6	12.6	15.1	60.8	0.50	2.16
ETF-V100 × 100 × 4N90	1.7	22.8	19.9	66.6	0.57	2.47
ETF-V100 × 100 × 4N50	1.7	12.7	20.0	50.1	0.45	1.87
ETF-V120 × 120 × 4N120	1.6	30.6	25.5	77.5	0.66	2.90
ETF-V120 × 120 × 4N60	1.6	15.3	25.5	55.9	0.51	2.09
				Mean	0.64	2.56
				COV	0.226	0.256

Table 5. Comparison of web crippling tests strengths with current design strengths for ITF loading condition.

Specimen	Measured Ratio			Test Exp. load per web	Comparison NAS	EC3
	r/t	N/t	h/t	P_{Exp} (kN)	$\frac{P_{Exp}}{P_{NAS}}$	$\frac{P_{Exp}}{P_{EC3}}$
ITF-H80 × 80 × 4N90	1.2	22.9	16.0	137.7	0.70	5.77
ITF-H80 × 80 × 4N50	1.2	12.8	16.2	120.0	0.69	5.11
ITF-H120 × 120 × 4N120	1.2	30.7	26.6	152.9	0.72	6.40
ITF-H120 × 120 × 4N60	1.1	15.3	26.4	139.4	0.74	5.77
ITF-H160 × 160 × 4N150	1.3	37.6	35.6	164.0	0.73	6.56
ITF-H160 × 160 × 4N90	1.3	22.6	35.6	148.9	0.74	5.95
ITF-H50 × 100 × 4N90	1.2	22.7	8.3	143.3	0.74	6.06
ITF-H50 × 100 × 4N90-R	1.2	22.7	8.3	144.2	0.75	6.11
ITF-H50 × 100 × 4N50	1.2	12.6	8.3	118.1	0.68	4.98
ITF-H100 × 50 × 4N50	1.2	12.7	21.2	125.1	0.74	5.40
ITF-H100 × 50 × 4N30	1.2	7.6	21.2	97.6	0.63	4.21
ITF-H100 × 50 × 4N30-R	1.2	7.6	21.2	98.9	0.64	4.26
ITF-V80 × 80 × 4N90	1.6	22.8	15.1	164.3	0.74	5.86
ITF-V80 × 80 × 4N50	1.6	12.7	15.1	140.3	0.71	5.01
ITF-V100 × 100 × 4N90	1.7	22.8	19.9	150.7	0.76	5.59
ITF-V100 × 100 × 4N50	1.7	12.6	19.9	131.1	0.74	4.85
ITF-V120 × 120 × 4N120	1.6	30.6	25.5	175.9	0.80	6.57
ITF-V120 × 120 × 4N60	1.6	15.3	25.6	150.7	0.79	5.64
				Mean	0.72	5.56
				COV	0.062	0.127

web crippling was conducted. Specimens with different steel grade, web slenderness and bearing length were tested. The web crippling tests were conducted under the End-Two-Flange (ETF) and Interior-Two-Flange (ITF) loading conditions in accordance with the North American Specification (NAS 2012) and Australian/New Zealand Standard (AS/NZS 2005) for cold-formed steel structures. The web crippling test strengths were compared with the nominal strengths predicted by the current NAS (2012), AS/NZS (2005) and European Code (EC3 2006) for cold-formed steel structures. It is shown that the design strengths predicated by the current specifications are either unconservative or very conservative. Therefore, the current specifications are not capable to predict the cold-formed high strength steel square and rectangular hollow sections undergoing web crippling.

ACKNOWLEDGEMENTS

The authors are grateful to Rautaruukki Corporation for providing the test specimens. The research work described in this paper was supported by a grant from the Research Grants Council of the Hong Kong Special Administrative Region, China (Project No. HKU17209614E).

REFERENCES

American Society for Testing and Materials (ASTM). 2013. Standard test methods for tension testing of metallic materials. *E8/E8M-13a*, West Conshohocken, Penn., USA.

Australian/New Zealand Standard (AS/NZS). 2005. Cold-formed steel structures, *AS/NZS 4600: 2005*, Standards Australia, Sydney, Australia.

Bakker, M.C.M. 1992. Web crippling of cold-formed steel members. Ph.D. Thesis, Eindhoven Univ. of Technology, Eindhoven, The Netherlands.

Beshara B., and Schuster R.M. 2000. Web crippling data and calibrations of cold-formed steel members. Final Report. University of Waterloo, Waterloo, Canada.

Bhakta, B.H., LaBoube, R.A., and Yu, W.W. 1992. The effect of flange restraint on web crippling strength. *Civil Engineering Study 92-1, Cold-Formed Steel Series, Final Rep.*, Univ. of Missouri-Rolla, Rolla, Missouri, USA.

European Committee for Standardization (EC3). 2006. Eurocode 3: Design of steel structures—Part 1.3: General rules—Supplementary rules for cold-formed members and sheeting. *EN 1993-1-3*, European Committee for Standardization, Brussels.

Gerges, R.R., and Schuster, R.M. 1998. Web crippling of single web cold formed steel members subjected to end one-flange loading. *Proc., 4th Int. Specialty Conf. on Cold-Formed Steel Structures*, St. Louis, Missouri, USA.

Hetrakul, N., and Yu, W.W. 1978. Structural behavior of beam webs subjected to web crippling and a combination of web crippling and bending. *Civil Engineering Study 78-4, Final Rep.*, Univ. of Missouri-Rolla, Rolla, Missouri, USA.

Huang, Y., and Young, B. 2014. The art of coupon tests. *Journal of Constructional Steel Research*, 96: 159–175.

Khan, M.Z., and Walker, A.C. 1972. Buckling of plates subjected to localized edge loading. *The Structure Engineer*, 50(6): 225–232.

North American Specification (NAS). 2001. North American Specification for the design of cold-formed steel structural members, American Iron and Steel Institute (AISI), Washington, D.C., USA.

North American Specification (NAS). 2012. North American Specification for the design of cold-formed steel structural members, *AISI S100-12*, American Iron and Steel Institute (AISI), Washington, D.C., USA.

Prabakaran, K. 1993. Web crippling of cold-formed steel sections. M.S. Thesis, Univ. of Waterloo, Waterloo, Canada.

Parabakaran, K., and Schuster, R.M. 1998. Web crippling of cold formed steel sections. *Proc., 4th Int. Specialty Conf. on Cold-Formed Steel Structures*, St. Louis, Missouri, USA.

Santaputra, C. 1986. Web crippling of high strength of cold-formed steel beams. Ph.D. Thesis, Univ. of Missouri-Rolla, Rolla, Missouri, USA.

Santaputra, C., Parks, M.B., and Yu, W.W. 1989. Web-crippling strength of cold-formed steel beams. *J. Struct. Eng.*, 115(10): 2511–2527.

Wallace, J.A. and Schuster, R.M. 2004. Web crippling of cold formed steel multi-web deck sections subjected to end one-flange loading. *Proc., 6th Int. Specialty Conf. on Cold-Formed Steel Structures*, Univ. of Missouri-Rolla Press, Rolla, Missouri, USA.

Wing, B.A., and Schuster, R.M. 1982. Web crippling for decks subjected to two flange loading. *Proc., 6th Int. Specialty Conf. on Cold-Formed Steel Structures*, Univ. of Missouri-Rolla Press, Rolla, Missouri, USA.

Winter, G., and Pian, R.H.J. 1946. Crushing strength of thin steel webs. *Engineering Experiment Station, Bulletin No. 35*, Cornell Univ., N.Y., USA.

Wu, S., Yu, W.W., and LaBoube, R.A. 1997. Strength of flexural members using structural grade 80 of A653 steel (web crippling tests). *Civil Engineering Study 97-3, Cold-Formed Steel Series, Third Progress Report*, Univ. of Missouri-Rolla, Rolla, Missouri, USA.

Young, B., and Hancock, G.J. 2001. Design of cold-formed channels subjected to web crippling. *J. Struct. Eng.*, 127(10): 1137–1144.

Yu, W.W. 1981. Web crippling and combined web crippling and bending of steel decks. *Civil engineering study 81-2, structural series*, Univ. of Missouri-Rolla, Rolla, Missouri, USA.

Zetlin, L. 1955. Elastic instability of flat plates subjected to partial edge loads. *Journal of the Structural Division, ASCE Proceedings*, 81: 1–24.

Zhou, F., and Young, B. 2007. Cold-formed high-strength stainless steel tubular sections subjected to web crippling. *J. Struct. Eng.*, 133(3): 368–377.

Tubular Structures XV – Batista, Vellasco & Lima (eds)
© *2015 Taylor & Francis Group, London, ISBN 978-1-138-02837-1*

Behaviour of eccentrically loaded ferritic stainless steel stub columns

O. Zhao & L. Gardner
Imperial College London, London, UK

B. Rossi
KU Leuven, Belgium

B. Young
The University of Hong Kong, Hong Kong, China

ABSTRACT: This paper presents a comprehensive experimental study of ferritic stainless steel tubular cross-sections under combined loading. Two square hollow section (SHS) sizes – SHS 40 × 40 × 2 and SHS 50×50 × 2 in grade EN 1.4509 stainless steel were considered. In total, two concentric compression tests, ten uniaxial bending plus compression tests and four biaxial bending plus compression tests were carried out. The experimental results were analysed and then compared with the design strengths predicted by the current European code EN 1993-1-4 (2006) and American Specification SEI/ASCE-8 (2002), revealing undue conservatism. Recent proposals (Liew & Gardner 2014; Zhao et al. 2014a) have been made to extend the scope of the deformation-based Continuous Strength Method (CSM) to the case of combined loading. Their applicability to ferritic stainless steels was evaluated herein and substantial improvements in design efficiency were found.

1 INTRODUCTION

Cold-formed ferritic stainless steels are increasingly becoming an attractive choice in a range of engineering applications due to their unique combination of moderate material price and favourable mechanical properties. Compared to their austenitic and duplex counterparts, the ferritic grades have no or very low nickel content and thus relatively low material price. Previous studies on ferritic stainless steels are briefly reviewed herein. Hyttinen (1994) performed a series of uniaxial eccentric compression tests on tubular members to investigate the interaction buckling behaviour of ferritic stainless steel beam-columns and assess the accuracy of the European code and American Specification. Van den Berg (2000) collected previous test data on ferritic stainless steel open sections and studied the flexural-torsional buckling behaviour of I-section columns and lateral torsional buckling behaviour of lipped channel beams. A series of 48 concentric compression tests on ferritic stainless steel lipped channel sections were carried out by Rossi et al. (2010) to investigate their combined distortional and overall flexural-torsional buckling behaviour. Afshan & Gardner (2013a) conducted a comprehensive experimental programme on square hollow sections (SHS) and rectangular hollow sections (RHS), including stub column tests, beam tests and flexural buckling tests, to verify the basic structural performance of ferritic stainless steel elements. The web crippling behaviour of ferritic SHS and RHS, strengthened by CFRP and FRP, were

carefully tested and studied by Islam & Young (2012, 2013). However, to date, there have been no investigations into the cross-sectional behaviour of ferritic stainless steel sections under combined loading, and this is therefore the focus of the present study.

An experimental study was conducted at the University of Liège and Imperial College London to investigate the behaviour of ferritic stainless steel cross-sections under combined loading. The two studied cross-sections were SHS 40 × 40 × 2 and SHS 50 × 50 × 2 of grade EN 1.4509, which are class 1 and class 3, respectively, according to the slenderness limits stated in EN 1993-1-4 (2006). In total, the laboratory testing programme comprised two stub column tests, ten uniaxial eccentric stub column tests and four biaxial eccentric stub column tests. The experimental data were carefully analysed and used to assess the accuracy of the codified design provisions given in EN 1993-1-4 (2006) and SEI/ASCE-8 (2002). Furthermore, the applicability of two new design proposals (Liew & Gardner 2014; Zhao et al. 2014a), which were derived through extension of the CSM to the case of stainless steel cross-sections under combined loading, were carefully evaluated.

2 EXPERIMENTAL INVESTIGATION

2.1 *Material tests and imperfection measurements*

Prior to structural testing, tensile coupon tests and imperfection measurements were conducted. The

Table 1. Average measured tensile flat material properties.

Cross-section	E GPa	$\sigma_{0.2}$ MPa	σ_u MPa	n	$n'_{0.2,u}$
SHS $40 \times 40 \times 2$	196	499	526	6.6	4.2
SHS $50 \times 50 \times 2$	190	466	515	6.6	7.6

Table 2. Average measured tensile corner material properties.

Cross-section	E GPa	$\sigma_{0.2}$ MPa	σ_u MPa	n	$n'_{0.2,u}$
SHS $40 \times 40 \times 2$	200	639	646	7.2	–
SHS $50 \times 50 \times 2$	226	623	658	5.3	1.5

Table 3. Measured dimensions of stub column specimens.

Specimen	L mm	H mm	B mm	t mm	r_i mm	ω_0 mm
SHS $40 \times 40 \times 2$-1A	150.0	40.0	40.1	2.0	1.8	0.01
SHS $50 \times 50 \times 2$-2A	200.1	50.1	50.2	1.9	2.5	0.01

Table 4. Summary of test results for stub columns.

Specimen	N_u kN	δ_u mm	$N_u/A\sigma_{0.2}$
SHS $40 \times 40 \times 2$-1A	183.3	1.85	1.13
SHS $50 \times 50 \times 2$-2A	205.0	2.08	1.13

detailed procedures and experimental setup for the coupon tests are described by Afshan et al. (2013), while only a brief summary of the key test results is reported herein, For each cross-section, the average measured flat and corner material properties are summarised in Tables 1–2, where E is the Young's modulus, $\sigma_{0.2}$ is the 0.2% proof stress, σ_u is the ultimate tensile strength, and n and $n'_{0.2,u}$ are the strain hardening exponents, which are determined based on the 0.01% and 0.2% proof stresses, and 0.5% proof stress and ultimate tensile strength, respectively (Mirambell & Real 2000). Due to the absence of global buckling phenomena, only local geometric imperfections were measured, following the procedures of Schafer & Peköz (1998).

2.2 Stub column tests

For each cross-section, one concentric stub column test was carried out using a SCHENCK 600 kN hydraulic testing machine with fixed end platens. The measured geometric dimensions and imperfection amplitudes of the test specimens are reported in Table 3, where L is the member length, B and H are the outer cross-section width and depth, respectively, t is the material thickness, r_i is the internal corner radius and ω_0 is the measured maximum local geometric imperfection. The true end-shortening values were obtained by eliminating the elastic deformation of the end platens from the end-shortening measurements on the basis of the strain gauge readings. The modified true load–end shortening curves are presented in Figure 1, while the key test results, including the ultimate load N_u, the end shortening at ultimate load δ_u and the ratio of $N_u/A\sigma_{0.2}$ are reported in Table 4.

2.3 Combined loading tests

For each cross-section size, five uniaxial bending plus compression tests and two biaxial bending plus compression tests were conducted, with the aim of

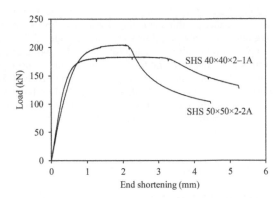

Figure 1. Load–end shortening curves for stub column tests.

investigating, the cross-sectional behaviour of ferritic stainless steel tubular sections under combined loading. Measurements of the geometric properties were performed before the end plates were welded to the specimen ends. The measured geometric properties and imperfection amplitudes are listed in Tables 5 and 7 for specimens tested under uniaxial and biaxial eccentric compression, respectively.

The combined loading tests were performed using a Zwick/Roell 600 kN hydraulic testing machine with hemispherical bearings at both ends to provide pin-ended boundary conditions. The specimens were eccentrically bolted to the hemispherical bearings and the initial eccentricities were varied to provide a range of bending moment-to-axial load ratios. Figure 2 depicts the test setup consisting of two inclinometers positioned at both ends of the specimens to measure the end rotations, four strain gauges attached to the extreme fibres of the cross-sections at mid-height to obtain the maximum and minimum longitudinal strains, and two LVDTs located along both principal axes to determine the generated lateral deflections and thus the second order bending moments (Fujimoto et al. 2004; Zhao et al. 2014b).

Tables 6 and 8 report the key experimental results for the combined loading tests, including the failure

Table 5. Measured dimensions of uniaxial bending plus compression specimens.

Specimen	L mm	H mm	B mm	t mm	r_i mm	ω_0 mm
SHS 40 × 40 × 2-1B	150.0	40.0	40.1	2.0	1.8	0.01
SHS 40 × 40 × 2-1C	150.0	40.1	40.1	2.0	1.8	0.01
SHS 40 × 40 × 2-1D	150.1	40.1	40.0	2.0	1.8	0.01
SHS 40 × 40 × 2-1E	150.0	40.0	40.0	2.0	1.8	0.01
SHS 40 × 40 × 2-1F	150.0	40.0	40.0	2.0	1.8	0.01
SHS 50 × 50 × 2-2B	200.1	50.1	50.1	1.9	2.5	0.02
SHS 50 × 50 × 2-2C	199.9	50.0	50.1	1.9	2.5	0.02
SHS 50 × 50 × 2-2D	200.0	50.0	50.1	1.9	2.5	0.02
SHS 50 × 50 × 2-2E	200.0	50.1	50.2	1.9	2.5	0.02
SHS 50 × 50 × 2-2F	200.0	50.1	50.2	1.9	2.5	0.02

Table 6. Summary of uniaxial bending plus compression test results.

Specimen	e_0 mm	N_u kN	e' mm	M_u kNm	ϕ_u deg
SHS 40 × 40 × 2-1B	8.5	119.0	0.70	1.10	1.65
SHS 40 × 40 × 2-1C	18.0	86.6	0.97	1.64	1.98
SHS 40 × 40 × 2-1D	28.4	68.9	1.10	2.03	2.49
SHS 40 × 40 × 2-1E	36.4	56.2	1.12	2.11	2.49
SHS 40 × 40 × 2-1F	48.7	47.4	1.39	2.38	3.06
SHS 50 × 50 × 2-2B	8.8	139.2	0.73	1.32	1.38
SHS 50 × 50 × 2-2C	20.4	106.2	0.77	2.24	1.45
SHS 50 × 50 × 2-2D	27.2	81.6	0.78	2.28	1.51
SHS 50 × 50 × 2-2E	40.5	68.9	0.89	2.85	1.55
SHS 50 × 50 × 2-2F	50.0	57.9	0.96	2.95	1.75

load N_u, the initial loading eccentricity e_0, the generated lateral deflection at the failure load e', the failure moment $M_u = N_u(e_0 + e')$ and the corresponding end rotation at failure ϕ_u. Note that the initial loading eccentricities reported in Tables 6 and 8 are not the nominal but the actual values, which are calculated on the basis of the strain gauge readings, following the derivation procedures of Zhao et al. (2014b). The experimental load–end rotation curves are shown in Figures 3–8. Typical local buckling failure modes from the combined loading tests are depicted in Figure 9, in which the welded end plates have been cut from the specimens after the combined loading tests.

3 COMPARISON OF TEST CAPACITIES WITH DESIGN STRENGTHS

3.1 General

In this section, the codified design provisions for stainless steel cross-sections under combined loading, as given in EN 1993-1-4 (2006) and SEI/ASCE-8 (2002), are firstly examined. Then, two new design proposals (Liew & Gardner 2014; Zhao et al. 2014a), which were derived through extension of the deformation-based Continuous Strength Method (CSM) to the case of combined loading, are fully described. The accuracy of each method is evaluated, as reported in Table 9, by

Figure 2. Combined loading test setup.

Table 7. Measured dimensions of biaxial bending plus compression specimens.

Specimen	L mm	H mm	B mm	t mm	r_i mm	ω_0 mm
SHS 40 × 40 × 2-1G	150.0	40.1	40.0	2.0	1.8	0.01
SHS 40 × 40 × 2-1H	150.1	40.0	40.0	2.0	1.8	0.01
SHS 50 × 50 × 2-2G	200.0	50.0	50.2	1.9	2.5	0.02
SHS 50 × 50 × 2-2H	200.0	50.0	50.1	1.9	2.5	0.02

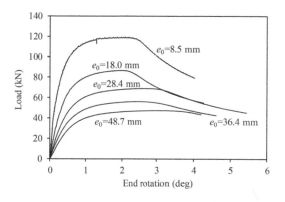

Figure 3. Load–end rotation curves for SHS 40 × 40 × 2 under uniaxial eccentric compression (Specimens: 1B to 1F).

means of the ratio of the test to predicted capacities $R_u/R_{u,pred}$, in which R_u is the distance on the N–M interaction curve from the origin to the test data point (see Fig. 10), while $R_{u,pred}$ is the distance from the origin to the intersection with the design interaction curve, assuming proportional loading. A value of $R_u/R_{u,pred}$ greater than unity indicates that the test data point lies outside the interaction curve and is safely predicted.

Table 8. Summary of biaxial bending plus compression test results.

Specimen	e_{0y} mm	e_{0z} mm	N_u kN	e'_y mm	e'_z mm	M_{uy} kNm	M_{uz} kNm	ϕ_{uy} deg	ϕ_{uz} deg
SHS 40 × 40 × 2-1G	19.0	19.0	69.9	1.12	1.12	1.41	1.41	1.92	1.92
SHS 40 × 40 × 2-1H	13.0	28.0	65.5	0.72	1.12	0.90	1.91	1.22	2.44
SHS 50 × 50 × 2-2G	14.0	25.0	86.8	0.62	0.95	1.27	2.25	0.83	1.51
SHS 50 × 50 × 2-2H	14.0	30.0	77.9	0.57	1.00	1.13	2.41	0.70	1.53

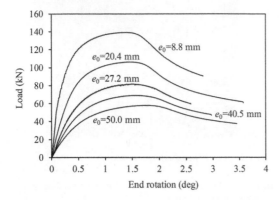

Figure 4. Load–end rotation curves for SHS $50 \times 50 \times 2$ under uniaxial eccentric compression (Specimens: 2B to 2F).

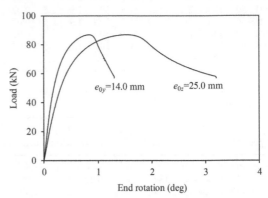

Figure 7. Load–end rotation curves for SHS $50 \times 50 \times 2$-2G under biaxial bending plus compression.

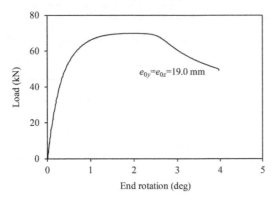

Figure 5. Load–end rotation curves for SHS $40 \times 40 \times 2$-1G under biaxial bending plus compression.

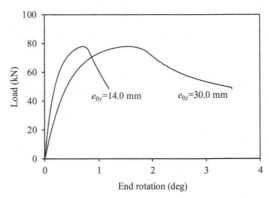

Figure 8. Load–end rotation curves for SHS $50 \times 50 \times 2$-2H under biaxial bending plus compression.

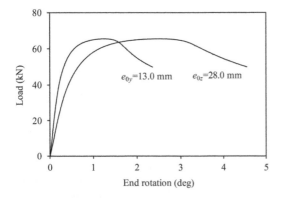

Figure 6. Load–end rotation curves for SHS $40 \times 40 \times 2$-1H under biaxial bending plus compression.

3.2 European code EN 1993-1-4 (EC3)

For cross-sectional capacity under combined loading, the current European code for stainless steel, EN 1993-1-4 (2006), adopts the same design provisions as those given in EN 1993-1-1 (2005) for carbon steel, in which the interaction formula for class 3 cross-sections is derived based on a linear elastic response, as given by Eq. (1), where N_{Ed} is the design axial load, and $M_{Ed,y} = N_{Ed}(e_y + e'_y)$ and $M_{Ed,z} = N_{Ed}(e_z + e'_z)$ are the design bending moments about the two principal axes.

$$\frac{N_{Ed}}{A\sigma_{0.2}} + \frac{M_{Ed,y}}{M_{el,y}} + \frac{M_{Ed,z}}{M_{el,z}} \leq 1 \tag{1}$$

Figure 9. Local buckling failure modes of specimens SHS 40 × 40×2-1D and SHS 40 × 40 × 2-1H.

Table 9. Comparison of the test results with design strengths predicted by different methods.

Specimen	$\dfrac{R_u}{R_{u,EC3}}$	$\dfrac{R_u}{R_{u,ASCE}}$	$\dfrac{R_u}{R_{u,csm1}}$	$\dfrac{R_u}{R_{u,csm}}$
SHS 40 × 40 × 2-1A	1.13	1.13	1.09	1.09
SHS 50 × 50 × 2-2A	1.13	1.13	1.13	1.13
SHS 40 × 40 × 2-1B	1.10	1.22	1.10	1.08
SHS 40 × 40 × 2-1C	1.09	1.26	1.10	1.07
SHS 40 × 40 × 2-1D	1.11	1.32	1.14	1.09
SHS 40 × 40 × 2-1E	1.06	1.28	1.11	1.04
SHS 40 × 40 × 2-1F	1.09	1.34	1.17	1.08
SHS 50 × 50 × 2-2B	1.25	1.18	1.12	1.09
SHS 50 × 50 × 2-2C	1.40	1.29	1.17	1.14
SHS 50 × 50 × 2-2D	1.28	1.16	1.05	1.01
SHS 50 × 50 × 2-2E	1.41	1.27	1.14	1.08
SHS 50 × 50 × 2-2F	1.39	1.24	1.12	1.04
SHS 40 × 40 × 2-1G	1.10	1.68	1.15	1.09
SHS 40 × 40 × 2-1H	1.12	1.65	1.18	1.11
SHS 50 × 50 × 2-2G	1.75	1.58	1.13	1.07
SHS 50 × 50 × 2-2H	1.71	1.54	1.11	1.05
Mean	1.26	1.33	1.13	1.08
COV	17.7%	13.6%	3.0%	3.0%

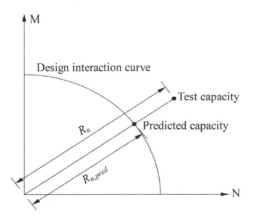

Figure 10. Definitions of R_u and $R_{u,pred}$.

The interaction formulae for class 1 and class 2 cross-sections are derived by assuming full plasticity throughout the cross-section at failure, as given by Eqs (2) and (3) for RHS under major axis bending plus compression and minor axis bending plus compression, respectively, and Eq. (4) for RHS subjected to biaxial eccentric compression, in which $M_{R,y}$ and $M_{R,z}$ are the reduced plastic moment capacities due to the existence of the axial force N_{Ed}, n is equal to $N_{Ed}/A\sigma_{0.2}$, a_w and a_f are the ratios of the web area A_w and flange area A_f to gross cross-section area A, respectively, and α_e and β_e, which are equal to $1.66/(1-1.13n^2)$, are the interaction coefficients for biaxial bending.

$$M_{Ed,y} \le M_{R,y} = M_{pl,y}\frac{(1-n)}{(1-0.5a_w)} \le M_{pl,y} \tag{2}$$

$$M_{Ed,z} \le M_{R,z} = M_{pl,z}\frac{(1-n)}{(1-0.5a_f)} \le M_{pl,z} \tag{3}$$

$$\left[\frac{M_{Ed,y}}{M_{R,y}}\right]^{\alpha_e} + \left[\frac{M_{Ed,z}}{M_{R,z}}\right]^{\beta_e} \le 1 \tag{4}$$

The accuracy of EN 1993-1-4 (2006) is assessed by comparing the experimental results with the EC3 predicted capacities. As reported in Table 9, EN 1993-1-4 results in reasonable strength predictions for class 1 cross-section SHS 40 × 40 × 2 but leads to undue conservatism for class 3 cross-section SHS 50 × 50 × 2. Overall, the mean value of $R_u/R_{u,EC3}$ ratio is equal to 1.26 with a COV equal to 17.7%.

3.3 American Specification SEI/ASCE-8

The American Specification SEI/ASCE-8 (2002) employs the same interaction formula for both the cross-sectional and global behavior of stainless steel elements under combined loading, as given by Eq. (5). However, as discussed by Zhao et al. (2014a), for a short beam-column under constant first order bending moment, the equivalent moment factors (C_{my} and C_{mz}) and the magnification factors (a_{ny} and a_{nz}) are all approximately equal to unity. Thus, Eq. (5) reduces to the linear interaction formula given by Eq. (6), in which M_{ny} and M_{nz} are the codified bending resistances calculated based on the inelastic reserve capacity according to clause 3.3.1.1 of SEI/ASCE-8 (2002).

$$\frac{N_{Ed}}{A\sigma_{0.2}} + \frac{C_{my}M_{Ed,y}}{M_{ny}\alpha_{ny}} + \frac{C_{mz}M_{Ed,z}}{M_{nz}\alpha_{nz}} = 1 \tag{5}$$

$$\frac{N_{Ed}}{A\sigma_{0.2}} + \frac{M_{Ed,y}}{M_{ny}} + \frac{M_{Ed,z}}{M_{nz}} = 1 \tag{6}$$

The mean $R_u/R_{u,ASCE}$ ratio, as given in Table 9, is 1.33 and the corresponding COV is equal to 13.6%, revealing overall more conservative but less scattered predictions by SEI/ASCE-8 (2002) than by EN 1993-1-4 (2006).

3.4 Continuous Strength Method (CSM)

The Continuous Strength Method (CSM) is a deformation-based design approach (Afshan and Gardner 2013b), which relates the strength of a cross-section to its deformation capacity and employs a bi-linear material model to consider strain hardening. The bi-linear material model in CSM was previously calibrated based on the material data from carbon steels, aluminum and stainless steels of austenitic and duplex grades, and has recently been extended to cover ferritic stainless steel grades by Bock et al. (2014). The relationship between the deformation capacity, in terms of the strain ratio $\varepsilon_{csm}/\varepsilon_y$, where ε_{csm} is the maximum attainable strain of the cross-section and $\varepsilon_y = \sigma_{0.2}/E$ is the yield strain, and the cross-section slenderness $\bar{\lambda}_p$, calculated as $\sqrt{\sigma_{0.2}/\sigma_{cr}}$, in which σ_{cr} is the elastic buckling stress of the cross-section allowing for element interaction (Schafer & Ádány 2006; Theofanous & Gardner 2012), is defined by Eq. (7), where, ε_u, which may be predicted by $\varepsilon_u = 0.6-0.6\sigma_{0.2}/\sigma_u$, is the strain at the material ultimate stress.

$$\frac{\varepsilon_{csm}}{\varepsilon_y} = \frac{0.25}{\bar{\lambda}_p^{3.6}} \leq \min\left(15, \frac{0.4\varepsilon_u}{\varepsilon_y}\right) \quad (7)$$

The strain hardening modulus E_{sh}, employed in the CSM elastic, linear hardening material model, may be calculated for ferritic stainless steel from Eq. (8).

$$E_{sh} = \frac{\sigma_u - \sigma_{0.2}}{0.45\varepsilon_u - \varepsilon_y} \quad \text{if } \frac{\varepsilon_y}{\varepsilon_u} < 0.45, \text{ else } E_{sh} = 0 \quad (8)$$

The CSM design stress σ_{csm} can be found from Eq. (9), while the CSM resistances for RHS subjected to pure compression and pure bending are respectively determined by Eqs (10) and (11) (Afshan & Gardner 2013b), where α is equal to 2 for RHS in bending about either principal axis.

$$\sigma_{csm} = \sigma_{0.2} + E_{sh}\left(\varepsilon_{csm} - \varepsilon_y\right) \quad (9)$$

$$N_{csm} = A\sigma_{csm} \quad (10)$$

$$M_{csm} = M_{pl}\left[1 + \frac{E_{sh}}{E}\frac{W_{el}}{W_{pl}}\left(\frac{\varepsilon_{csm}}{\varepsilon_y}-1\right)-\left(1-\frac{W_{el}}{W_{pl}}\right)/\left(\frac{\varepsilon_{csm}}{\varepsilon_y}\right)^\alpha\right] \quad (11)$$

Based on the assumption of a linear through-depth strain distribution and the bi-linear material model, the CSM resistances for RHS under various combined loading cases, including major axis bending plus compression, minor axis bending plus compression and biaxial bending plus compression, can be derived by integration and represented through interaction expressions, as shown in Eqs (12)–(14) (Liew & Gardner 2014), respectively

$$M_{Ed,y} \leq M_{R,csm1,y} = M_{csm,y}\left[1-\left(\frac{N_{Ed}}{N_{csm}}\right)^{\alpha_y}\right]^{1/b_y} \quad (12)$$

$$M_{Ed,z} \leq M_{R,csm1,z} = M_{csm,z}\left[1-\left(\frac{N_{Ed}}{N_{csm}}\right)^{\alpha_z}\right]^{1/b_z} \quad (13)$$

$$\left(\frac{M_{Ed,y}}{M_{R,csm1,y}}\right)^{\alpha_{csm1}} + \left(\frac{M_{Ed,z}}{M_{R,csm1,z}}\right)^{\beta_{csm1}} \leq 1 \quad (14)$$

in which N_{csm}, $M_{csm,y}$ and $M_{csm,z}$ are the CSM compression and bending (major and minor axes) resistances, which act as the end points of the design curves and are calculated according to Eqs (9)–(11), $M_{R,csm1,y}$ and $M_{R,csm1,z}$ are respectively the reduced CSM bending resistances about the major and minor axes due to N_{Ed}, of which the extent of reduction is determined by $\alpha_y = 1.2 + A_w/A$, $\alpha_z = 1.2 + A_f/A$ and $b_y = b_z = 0.8$, and α_{csm1} and β_{csm1} are the interaction coefficients for biaxial bending, which are equal to $1.75 + W_r(2n_{csm}^2-0.15)$ and $1.6+(3.5-1.5W_r)n_{csm}^2$, respectively, where W_r is the ratio of major to minor axis plastic section moduli $W_{pl,y}/W_{pl,z}$ and n_{csm} is the ratio of design axial force to CSM compression resistance N_{Ed}/N_{csm}.

The accuracy of the CSM was evaluated by comparing the test capacity with the CSM predicted capacity; the results are reported in Table 9. Overall, the CSM offers much more accurate and consistent predictions than the European code and American Specification, with the mean $R_u/R_{u,csm1}$ ratio of 1.13 and the corresponding value of COV equal to 3.0%.

3.5 Simplified CSM

The simplified CSM was proposed by Zhao et al. (2014a) for the design of austenitic and lean duplex stainless steel cross-sections under combined loading. The proposed interaction curves adopt the general format of the design interaction curves for class 1 and class 2 cross-sections in EN 1993-1-4 (2006) but the end points are anchored to the CSM compression and bending resistances (N_{csm}, $M_{csm,y}$ and $M_{csm,z}$) rather than the yield load ($A\sigma_{0.2}$) and plastic bending capacities ($M_{pl,y}$ and $M_{pl,z}$). The interaction formulae are given by Eqs (15)–(17) for RHS subjected to major axis, minor axis and biaxial eccentric compression, respectively, where $M_{R,csm,y}$ and $M_{R,csm,z}$ are the reduced CSM bending resistances about the major and minor axes, respectively, due to the presence of the axial load N_{Ed}, and $\alpha_{csm} = \beta_{csm} = 1.66/(1-1.13n_{csm}^2)$ are the interaction coefficients for biaxial bending, which are taken from Eurocode 3 but based on the CSM end points.

$$M_{Ed,y} \leq M_{R,csm,y} = M_{csm,y}\frac{(1-n_{csm})}{(1-0.5a_w)} \leq M_{csm,y} \quad (15)$$

$$M_{Ed,z} \leq M_{R,csm,z} = M_{csm,z}\frac{(1-n_{csm})}{(1-0.5a_f)} \leq M_{csm,z} \quad (16)$$

$$\left[\frac{M_{Ed,y}}{M_{R,csm,y}}\right]^{\alpha_{csm}} + \left[\frac{M_{Ed,z}}{M_{R,csm,z}}\right]^{\beta_{csm}} \leq 1 \qquad (17)$$

The quantitative evaluation of the simplified CSM is reported in Table 9. The mean $R_u/R_{u,csm}$ ratio is equal to 1.08 with the corresponding COV of 3.0%, revealing the highest accuracy and consistency in the prediction of cross-section capacity under combined loading among the four design methods.

3.6 Summary

Overall, the American Specification SEI/ASCE-8 (2002) results in the most conservative strength predictions among the four methods in the design of ferritic stainless steel cross-sections under combined loading, mainly owing to the use of linear interaction design curves. The European code EN 1993-1-4 (2006) generally leads to more accurate predictions than SEI/ASCE-8, but with increased scatter. The CSM (Liew & Gardner 2014) and simplified CSM (Zhao et al. 2014a) perform well for ferritic stainless steels and yield more accurate predictions with significantly lower scatter.

Figures 11–12 depict the uniaxial bending plus compression test results compared against the design interaction curves obtained from the four design methods, clearly indicating the CSM and simplified CSM offer substantial advantages over the codified design methods in the prediction of ferritic stainless steel cross-section resistance under combined loading, in terms of both accuracy and consistency.

4 CONCLUSIONS

A comprehensive experimental programme has been performed to investigate the cross-sectional behaviour of ferritic stainless steel sections under combined loading. A total of two stub column tests, ten uniaxial bending plus compression tests and four biaxial bending plus compression tests were conducted. The test setup and experimental procedures for each type of test have been fully described. The full load–deformation histories have been presented for all test types. Key test results including the ultimate loads and the corresponding deformation parameters at ultimate load have also been tabulated. The obtained experimental results were then utilized to assess the accuracy of four design methods, including two codified methods – EN 1993-1-4 (2006) and SEI/ASCE-8 (2002), and two deformation-based design approaches – CSM (Liew & Gardner 2014) and simplified CSM (Zhao et al. 2014a). Generally, the American Specification results in the most conservative predictions among the four methods. The European code leads to more accurate but scattered strength predictions than SEI/ASCE-8 (2002). The two deformation-based CSM design approaches are well suited to ferritic stainless steel design, yielding a much higher level of accuracy and consistency in the prediction of ferritic stainless steel

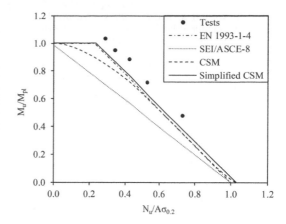

Figure 11. Comparison of uniaxial bending plus compression test results of SHS $40 \times 40 \times 2$ with the four design curves.

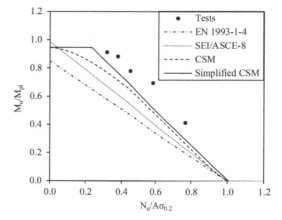

Figure 12. Comparison of uniaxial bending plus compression test results of SHS $50 \times 50 \times 2$ with the four design curves.

cross-sectional resistances under combined loading, compared to the two codified methods.

ACKNOWLEDGEMENTS

The authors would like to thank Stalatube Oy, Finland for the supply of test specimens, and Mr Max Verstraete from the University of Liege and Mr Gordon Herbert from Imperial College London for their assistance in the tests. They are also grateful to the Joint PhD Scholarship from Imperial College London for its financial support.

REFERENCES

Afshan, S. & Gardner, L. 2013a. Experimental study of cold-formed ferritic stainless steel hollow sec-tions. *Journal of Structural Engineering* 139(5): 717–728.

Afshan, S. & Gardner, L. 2013b. The continuous strength method for structural stainless steel design. *Thin-Walled Structures* 68(4): 42–49.

Afshan, S., Rossi, B. & Gardner, L. 2013. Strength enhancements in cold-formed structural sections – Part I: Material testing. *Journal of Constructional Steel Research* 83: 177–188.

ASCE. 2002. Specification for the design of cold-formed stainless steel structural members. *SEI/ASCE 8-02*, Reston, VA.

Bock, M., Gardner, L. & Real, E. 2014. Material and local buckling response of cold-formed ferritic stainless steel sections. *Journal of Constructional Steel Research*, submitted.

EC3. 2005. Design of steel structures: Part 1-1: General rules and rules for buildings. *EN 1993-1-1:2005*, Brussels, Belgium.

EC3. 2006. Design of steel structures: Part 1–4: General rules: Supplementary rules for stainless steel. *EN 1993-1-4:2006*, Brussels, Belgium.

Fujimoto, T., Mukai, A., Nishiyama, I. & Sakino, K. 2004. Behavior of eccentrically loaded concrete-filled steel tubular columns. *Journal of Structural Engineering* 130(2): 203–212.

Hyttinen, V. 1994. Design of cold-formed stainless steel SHS beam-columns. *Report 41*, University of Oulu, Oulu, Finland.

Islam, S. M. & Young, B. 2012. Ferritic stainless steel tubular members strengthened with high modulus CFRP plate subjected to web crippling. *Journal of Constructional Steel Research* 77: 107–118.

Islam, S. M. & Young, B. 2013. Strengthening of ferritic stainless steel tubular structural members using FRP subjected to Two-Flange-Loading. *Thin-Walled Structures* 62: 179–190.

Liew, A. & Gardner, L. 2014. Ultimate capacity of structural steel cross-sections under combined loading. *Structures*, in press.

Mirambell, E. & Real, E. 2000. On the calculation of deflections in structural stainless steel beams: An experimental and numerical investigation. *Journal of Constructional Steel Research* 54(1): 109–133.

Rossi, B., Jaspart, J. P. & Rasmussen, K. J. R. 2010. Combined distortional and overall flexural-torsional buckling of cold-formed stainless steel sections: Experimental investigations. *Journal of structural engineering* 136(4): 361–369.

Schafer, B. W. & Peköz, T. 1998. Computational modelling of cold-formed steel: Characterizing geometric imperfections and residual stresses. *Journal of Constructional Steel Research* 47(3): 193–210.

Schafer, B. W. & Ádány, S. 2006. Buckling analysis of cold-formed steel members using CUFSM: conventional and constrained finite strip methods. *Proc., 18th Int. Specialty Conf. on Cold-formed Steel Structures*, Orlando, USA.

Theofanous, M. & Gardner, L. 2012. Effect of element interaction and material nonlinearity on the ultimate capacity of stainless steel cross-sections. *Steel and Composite Structures* 12(1): 73–92.

Van Den Berg, G. J. 2000. The effect of non-linear stress-strain behaviour of stainless steels on member capacity. *Journal of Constructional Steel Research* 135(1): 135–160.

Zhao, O., Rossi, B., Gardner, L. & Young, B. 2014a. Behaviour of structural stainless steel cross-sections under combined loading – Part II: Numerical modelling and design approach. *Engineering Structures*, in press.

Zhao, O., Rossi, B., Gardner, L. & Young, B. 2014b. Behaviour of structural stainless steel cross-sections under combined loading – Part I: Experimental study. *Engineering Structures*, in press.

Tubular Structures XV – Batista, Vellasco & Lima (eds)
© *2015 Taylor & Francis Group, London, ISBN 978-1-138-02837-1*

CFRP strengthened cold-formed stainless steel tubular sections subjected to concentrated loading under ITF loading condition

F. Zhou, P. Huang & H. Peng
Department of Building Engineering, Tongji University, Shanghai, China

ABSTRACT: This paper presents a series of tests on carbon fiber reinforced polymer (CFRP) strengthened cold-formed stainless steel square and rectangular hollow sections subjected to web crippling. A total of 22 web crippling tests were conducted in this study under Interior-Two-Flange (ITF) loading condition. The tests were performed on five different tubular section sizes which covered a slightly wide range of measured web slenderness ratios from 18.73 to 61.74. The effects of CFRP layout, bearing length and web slenderness on the strength enhancement have been studied. A non-linear finite element model which includes geometric and material non-linearities was developed for CFRP strengthened cold-formed stainless steel tubular section subjected to web crippling and verified against the experimental results presented in this study. The debonding between CFRP and cold-formed stainless steel tubular sections was carefully modelled by using cohesive elements. It was shown that the calibrated model accurately predicted the web crippling strengths, web deformations and failure modes of the tested specimens.

1 INTRODUCTION

Cold-formed stainless steel sections often experience web crippling failure due to high local intensity of concentrated loads or reactions. In the literature, the web crippling behaviour of cold-formed stainless steel sections has been researched (Korvink et al. 1995; Zhou and Young 2006). However, these investigations did not consider any strengthening in the web of the sections. For square and rectangular hollow section members, it is often difficult to provide transverse stiffeners at loading points, especially when the stiffeners are located away from the ends of the members. Hence, web crippling strength enhancement using Carbon Fibre-Reinforced Polymer (CFRP) in localized region can be considered as an attractive solution.

This paper mainly investigated the web crippling behaviour of cold-formed austenitic stainless steel tubular sections strengthened with CFRP sheets. Firstly, a series of web crippling tests on cold-formed stainless steel tubular sections strengthened with CFRP sheets is presented. The typical load-web deformation behaviour, failure loads, failure modes and strength enhancement due to CFRP strengthening in the web of the cold-formed stainless steel tubular sections are reported. The effects of CFRP layout, bearing length and web slenderness on the strength enhancement are discussed based on the test data obtained in this study. Secondly, a non-linear finite element model which includes geometric and material non-linearities is developed to investigate the web crippling behaviour of cold-formed stainless steel tubular sections strengthened with CFRP sheets. The finite

element analysis (FEA) program ABAQUS (2008) was used for the numerical investigation. Cohesive element which can simulate the debonding between CFRP and cold-formed stainless steel tubular sections was used to model the adhesive. The developed finite element model was verified against the test results obtained in this study.

2 MATERIAL PROPERTIES

Square and rectangular hollow sections fabricated by cold-rolling from normal strength material of austenitic stainless steel type 304 have been considered in this study. Tensile coupon tests were carried out to determine the material properties of the cold-formed austenitic stainless steel tubular sections. The material properties obtained from the tensile coupon tests are summarized in Table 1, which includes the static 0.2% tensile proof stress ($\sigma_{0.2}$), static tensile strength (σ_u) and elongation after fracture (ε_f) based on a gauge length of 50 mm.

The CFRP sheet used in this study is HITEX-C300 which has a mass area ratio of 300 g/m2 and nominal thickness of 0.167mm. The main characteristics of the fibres are their strength and Young's modulus. The specified material properties provided by the supplier are listed in Table 2. The adhesive adopted is Lica-100 A/B. The properties of adhesive have significant influence on external bonded CFRP strengthening. The key mechanical properties of adhesive for strengthening structure are effective bond strength, elastic modulus

Table 1. Measured material properties of cold-formed stainless steel tubular sections.

Table 1. Measured material properties of cold-formed stainless steel tubular sections.

Test specimen $b_f \times d \times t$ (mm)	$\sigma_{0.2}$ (MPa)	σ_u (MPa)	ε_f (%)
$120 \times 120 \times 2.7$	393	734	17.3
$120 \times 120 \times 3.6$	445	760	16.8
$80 \times 80 \times 3.6$	448	758	17.2
$60 \times 120 \times 1.8$	388	761	19.2
$60 \times 120 \times 2.7$	420	751	17.1

Table 2. Material properties of CFRP sheets.

σ_u (MPa)	E_0 (MPa)
3492.5	2.6×10^5

Table 3. Material properties of adhesive.

σ_u (MPa)	E_0 (MPa)	ε_f (%)	Shear strength (MPa)
48.4	3443	1.6	19.6

and elongation. The material properties of adhesive provided by the material supplier are listed in Table 3.

3 TEST PROGRAM

3.1 Test specimens

In this study, a total of 22 web crippling tests were conducted on cold-formed stainless steel square and rectangular hollow sections strengthened with CFRP sheets. The test specimens were fabricated by cold-rolling from normal strength material of austenitic stainless steel type 304. The test specimens were subjected to Interior-Two-Flange (ITF) loading condition. The tests were performed on five different hollow section sizes as shown in Table 1. The specimens had the nominal thickness ranged from 1.8 to 3.6 mm, the nominal depth of the webs ranged from 80 to 120 mm, and the flange widths ranged from 60 to 120 mm. The measured web slenderness values h/t of the tubular sections ranged from 18.73 to 61.74. The specimen lengths (L) were determined according to the American Specification (ASCE 2002) and the Australian/New Zealand Standard (AS/NZS 2001). Generally, the clear distance from the edge of the bearing plate to the end of the member was set to be 1.5 times the overall depth (d) of the web rather than 1.5 times the depth of the flat portion of the web (h), the latter being the minimum specified in the specifications. The measured dimensions of the test specimens are shown in Table 4 using the nomenclature defined in

Fig. 1. The specimens without strengthening of CRFP were also tested for reference purpose.

3.2 Specimens labelling

The test specimens were labelled such that the type of loading condition, nominal dimensions of the specimen, bearing length and CFRP layout can be identified, as shown in Table 4. For example, the label "ITF60×120 × 1.8N60-C" defines the following specimen: The first three letters indicate the loading condition of Interior-Two-Flange (ITF). The following symbols are the nominal dimensions of the specimen in mm, where $60 \times 120 \times 1.8$ ($b_f \times d \times t$) means the flange width = 60 mm, web depth = 120 mm and thickness = 1.8 mm. The following three letters "N60" indicate the bearing length of 60mm. The following letter indicates whether CFRP was used or not, where "C" means CFRP was used, and the reference test specimen without CFRP is represented as "B".

The CFRP strengthened width for all specimens was the same length as the bearing length N, except for the second phase investigation of the effect of different widths of CFRP strengthening where additional following symbols "(N+d)" were used to indicate its CFRP strengthened width, as shown in Table 4. For example, the label "ITF120 × 120 × 2.7N60 −C(N+d)" indicates the CFRP strengthened width is N+d.

3.3 Specimen preparation

The strength of the adhesive bond is directly proportional to the quality of the surfaces to which it is mated (Fawzia 2007). In this study, the first step in specimen preparation was grinding the surface of the specimens with sand paper to remove the galvanized and other impurities. And then the outer surfaces of the cold-formed stainless steel tubular sections as well as the CFRP were cleaned with acetone. The adhesive was then applied uniformly on the CFRP sheets and then attached onto the cold-formed stainless steel surfaces. The excess adhesive and air were removed using a ribbed roller that applied on the CFRP sheets with a small amount of force. The fibre direction was along the transverse direction of the web. The thickness of the adhesive layer was maintained uniform to be approximately 1.0 mm. The test specimens were tested after 7 days of curing at room temperature. The CFRP sheets were bonded to the webs only at the middle of the specimens for ITF loading condition.

3.4 Bearing plates and CFRP widths

The load was applied through bearing plates. The bearing plates were fabricated using high strength steel of yield stress approximately 800 MPa. All bearing plates had the thickness of 50 mm and the length of 200 mm. The bearing plates acted across the full flange widths of the sections, excluding the rounded corner. The bearing length (N) was chosen to be the full-flange

Table 4. Measured specimen dimensions and test results of CFRP strengthened cold-formed stainless steel tubular sections subjected to ITF loading condition.

Specimen	b_f (mm)	d (mm)	t (mm)	P_{Exp} (kN)	$P_{Exp-C}/$ P_{Exp-B}
ITF120 × 120 × 2.7N60-B	120.3	120.4	2.860	93.21	100.0%
ITF120 × 120 × 2.7N60-C	120.1	120.6	2.885	94.71	101.6%
ITF120 × 120 × 2.7N60-C(N+d)	120.3	120.5	2.855	97.50	104.6%
ITF120 × 120 × 2.7N120-B	120.4	120.5	2.933	111.07	100.0%
ITF120 × 120 × 2.7N120-C	120.4	120.1	2.860	114.81	103.4%
ITF120 × 120 × 3.6N60-B	120.6	120.1	3.650	151.88	100.0%
ITF120 × 120 × 3.6N60-C	120.6	120.1	3.605	154.24	101.6%
ITF120 × 120 × 3.6N120-B	120.7	120.2	3.620	172.25	100.0%
ITF120 × 120 × 3.6N120-C	120.6	120.2	3.635	175.09	101.6%
ITF80 × 80 × 3.6N30-B	80.0	79.8	3.660	157.34	100.0%
ITF80 × 80 × 3.6N30-C	80.0	80.0	3.653	159.77	101.5%
ITF80 × 80 × 3.6N60-B	80.1	79.8	3.673	179.59	100.0%
ITF80 × 80 × 3.6N60-C	79.9	79.9	3.658	182.44	101.6%
ITF60 × 120 × 1.8N30-B	60.3	120.2	1.840	34.19	100.0%
ITF60 × 120 × 1.8N30-C	60.4	120.2	1.843	35.18	102.9%
ITF60 × 120 × 1.8N60-B	60.6	120.0	1.830	39.38	100.0%
ITF60 × 120 × 1.8N60-C	60.7	120.1	1.873	40.10	101.8%
ITF60 × 120 × 1.8N60-C(N+d)	60.4	120.2	1.801	41.55	105.5%
ITF60 × 120 × 2.7N30-B	60.6	120.1	2.745	80.23	100.0%
ITF60 × 120 × 2.7N30-C	60.5	120.1	2.728	81.79	101.9%
ITF60 × 120 × 2.7N60-B	60.6	120.1	2.730	86.94	100.0%
ITF60 × 120 × 2.7N60-C	60.5	120.1	2.715	89.89	103.4%

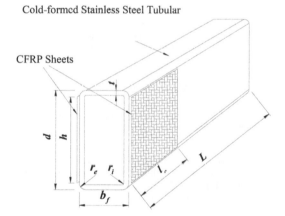

Figure 1. Definition of symbols.

Figure 2. Photograph of test setup.

and half-flange widths of the section. The width of CFRP sheet for strengthening is identical to the bearing length N. As mentioned before, additional CFRP width of N+d was also used to study the effects of CFRP width on CFRP strengthening against web crippling. The flanges of the specimens were not fastened to the bearing plates during the tests.

3.5 Web crippling tests

The web crippling tests were carried out under ITF loading condition which is specified in the American Specification (ASCE 2002) and Australian/New Zealand Standard (AS/NZS 2001). Fig. 2 shows the photographs of the test setup.

For ITF loading condition, two identical bearing plates with half round of the same width were positioned at the mid-length of the specimens. Hinge supports were simulated by two half rounds. The specimen was seated between the two bearing plates during the test. A servo-controlled hydraulic testing machine was used to apply a concentrated compressive force to the test specimen. Displacement control was used to drive the hydraulic actuator at a constant speed. Three transducers were used to record the web deformations of the specimens. The web deformations of the specimens were obtained by the average readings of the three transducers measured between the two bearing plates.

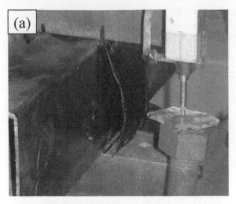

(a) Adhesion failure

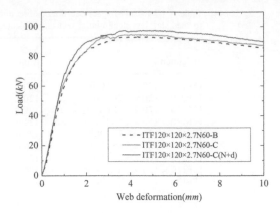

Figure 4. Comparison of CFRP length effect on load-web deformation behaviour for specimen ITF120 × 120 × 2.7N60.

(b) Combination of adhesion and cohesion

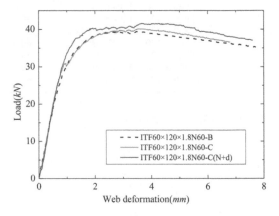

Figure 5. Comparison of CFRP length effect on load-web deformation behaviour for specimen ITF60 × 120 × 1.8N60.

(c) Tear of CFRP sheets

Figure 3. Failure modes of CFRP strengthened cold-formed stainless steel specimens.

3.6 *Test results and discussion*

In this study, three main failure modes, namely adhesion failure, combination of adhesion and cohesion failure and tear of CFRP sheets were observed as shown in Fig. 3. The adhesion failure was found at physical interface between the adhesive and the adherents. It depends on the surface characteristics of adherent such as the roughness and other factors. The cohesion failure is fully control by the adhesive properties. As the CFRP sheets used were unidirectional fabrics, the tear of CFRP sheets were also observed.

Almost of the CFRP strengthened specimens failed due to debonding. Generally, debonding initiated from the end of the CFRP sheet that experienced high interfacial stresses developed in the region. The debonding was propagated gradually towards the mid-height of the webs (Islam and Young 2011). The CFRP sheet strengthened webs were not able to resist the applied load when the propagation of the cracks had reached a certain limit (Islam and Young 2013). The typical load-web deformation behaviour of CFRP strengthened cold-formed stainless steel specimens are shown in Figs. 4 and 5.

A series of tests was conducted on three different cold-formed stainless steel specimens ETF60 × 120 × 1.8N60, ITF120 × 120×2.7N60 and ITF60 × 120 × 1.8N60 to investigate the effects of different widths of CFRP sheets on the strength enhancement. Two kinds of CFRP strengthened widths of N and N+d were considered for each specimen, where N was the bearing length and d was the overall depth of the section. The load-web deformation curves are plotted in Figs. 4 and 5 for ITF120 × 120 × 2.7N60 and ITF60 × 120 × 1.8 N60, respectively. It is evident that the web crippling capacity of the CFRP strengthened cold-formed stainless steel tubular sections increased by increasing the width of the CFRP sheet for ITF60 × 120 × 2.7N60 and ITF60 × 120 × 1.8N60. Compared to the reference test, when the CFRP width increases from N to N +d, the web crippling strengths increases from 1.6% to 4.6% for ITF120 × 120 × 2.7N60. For ITF60 × 120 × 1.8N60, the web crippling strengths increases from 1.8% to 5.5% with CFRP strengthened width increases from N to N+d. It is shown that the increase of CFRP strengthened width provides some improvement on strengthening of the cold-formed stainless steel tubular sections for ITF loading condition.

4 NUMERICAL INVESTIGATION

The non-linear finite element software package ABAQUS (2008) was used to simulate the CFRP strengthened cold-formed stainless steel tubular sections subjected to web crippling. Five main components, i.e. the bearing plate, cold-formed stainless steel tubular section, CFRP sheet, adhesive and the interfaces between the bearing plate and the cold-formed stainless steel tubular section, between the stainless steel section and the adhesive, as well as between the adhesive and the CFRP sheet have been carefully considered in the finite element modal (FEM). In the FEM, the measured cross-section dimensions and material properties were used.

The bearing plates were modelled using discrete rigid plates. The cold-formed stainless steel tubular section and the CFRP sheet were modelled using the S4R shell elements. The adhesive was modelled using cohesive elements which can simulate the initiation and the propagation of damage leading to eventual failure at the bonded interface. The constitutive response of cohesive element is defined by using a traction-separation description in ABAQUS (2008). The finite element mesh used in the model was investigated by varying the size of the elements to provide both accurate results and less computational time. The typical finite element mesh of the CFRP strengthened cold-formed stainless steel square hollow section under the ITF loading condition is shown in Fig. 6.

The top bearing plate was restrained against all degrees of freedom, except for the translational degree of freedom in the Y direction and the rotation about the X-axis. The bottom bearing plate was restrained against all degrees of freedom, except for the rotation about the X-axis, as shown in Fig. 6.

The "surface to surface contact" option provided in ABAQUS was used for the interaction between the bearing plate and the cold-formed stainless steel tubular section. Bearing plate was treated as master surface and cold-formed stainless steel tubular section as a slave surface. The contact pair allows the surfaces to separate under the influence of a tensile force. However, the two contact surfaces are not allowed to penetrate each other.

The interface between the cold-formed stainless steel tubular section and the adhesive as well as the interface between the adhesive and the CFRP sheet were all modelled using tie constraints. The "tie constraint" ties two surfaces together, which make the translational and rotational degrees of freedom the same as for the pair of surfaces.

The loading method used in the finite element analysis was identical to that used in the tests. The displacement control method was used for the analysis of the CFRP strengthened cold-formed stainless steel square and rectangular hollow sections subjected to web crippling. Transverse compressive load was applied to the specimen by specifying a displacement to the reference point of the discrete rigid plate that modelled the bearing plate. Generally, a displacement of 10 mm was specified, and the displacement is equivalent to the web deformation of the specimen.

The measured stress-strain curves for flat portions of the cold-formed stainless steel tubular sections were used in the finite element analysis. In addition, the predicted stress-strain curves of the corner portions for the same series were also used. The material properties of corner portions are different from those of flat portions. The equation (1) proposed by Ashraf et al. (2006) was used to obtain the yield stress of the corner portion of cold-formed stainless steel tubular section.

$$\sigma_{0.2}^c = \frac{1.88\sigma_{0.2}}{\left(\dfrac{r_i}{t}\right)^{0.194}} \tag{1}$$

where $\sigma_{0.2}$ is the yield stress of flat portion, r_i is the internal corner radius and t is the thickness of cold-formed stainless steel tubular section.

The modified Ramberg-Osgood equation (Rasmussen 2003) was used to predict the stress–strain curves of the corner portions. The material behaviour provided by ABAQUS allows for the multi-linear stress-strain curve to be used. The first part of the multi-linear curve represents the elastic part up to the proportional limit stress with measured Young's modulus and Poisson's ratio equal to 0.3. Both the measured stress–strain curves of the flat portions and the predicted stress–strain curves of the corner portions were converted to true stresses and logarithmic strains due to large in-elastic strains involved in the post buckling analysis. The true stress (σ_{true}) and plastic true strain (ε_{true}^{pl}) were specified in ABAQUS (2008).

The CFRP sheet was treated as a linear elastic material. The material properties of CFRP shown in Table 2 were used in the FEM. The elastic behaviour of the adhesive was defined using the command *Elastic, type = TRACTION. Knn, Kss and Ktt are the elastic stiffness of the normal and the two shear directions, respectively, which are calculated as follows (Fernando 2010):

$$K_{nn} = \frac{E_a}{T_0}, \quad K_{ss} = K_{tt} = 3\left(\frac{G_a}{T_0}\right)^{0.65} \quad (2)$$

where T_0 is the initial thickness of the adhesive layer, E_a is the tensile elastic modulus of the adhesive, and G_a is the shear modulus of the adhesive.

The damage initiation behaviour was defined using the command *Damage Initiation, criterion = QUADS. The damage evolution behaviour was defined using the command *Damage Evolution, type = ENERGY, mixed mode behaviour = POWER LAW, power = 1. The mode I fracture energy was assumed to be equal to the tensile strain energy. The interfacial fracture energy

of Mode II can be calculated as follows (Fernando 2010):

$$G_f^s = 628t_a^2 R^2 \ (N / mm^2 \cdot mm) \quad (3)$$

where t_a is the adhesive thickness in mm and R is the tensile strain energy of the adhesive which is equal to the area under the uniaxial tensile stress (in MPa) – strain curve.

The effects of imperfections and residual stresses on web crippling behaviour of cold-formed stainless steel tubular sections were found to be negligible (Gardner and Nethercot 2004) and thus ignored in this study.

5 VERIFICATION OF FINITE ELEMENT MODEL

In the verification of the finite element model (FEM), a total of 22 CFRP strengthened cold-formed stainless steel square and rectangular hollow sections subjected to web crippling were analysed. A comparison between the experimental results and the finite element results was carried out. The main objective of this comparison is to verify and check the accuracy of the finite element model. The comparison of the test results (P_{Exp}) with the web crippling strengths (P_{FEA}) per web predicted from the FEA is shown in Table 5 for ITF loading condition. It can be seen that good agreement has been achieved between both results for all specimens. The failure modes obtained from the tests were also verified by the FEM, as shown in Fig. 6. It is shown that the web crippling strengths and the failure

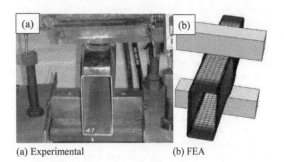

Figure 6. Comparison of experimental and FEA failure modes for specimen ITF60 × 120 × 1.8N60-C: (a) Experimental; (b) FEA.

Table 5. Comparison of experimental and numerical results for ITF loading condition.

Specimen	P_{Exp} (kN)	P_{FEA} (kN)	P_{Exp}/P_{FEA}
ITF120 × 120 × 2.7N60-B	93.21	94.52	0.986
ITF120 × 120 × 2.7N60-C	94.71	100.27	0.945
ITF120 × 120 × 2.7N60-C(N+d)	97.50	99.34	0.981
ITF120 × 120 × 2.7N120-B	111.07	110.57	1.005
ITF120 × 120 × 2.7N120-C	114.81	110.17	1.042
ITF120 × 120 × 3.6N60-B	151.88	154.76	0.981
ITF120 × 120 × 3.6N60-C	154.24	154.92	0.996
ITF120 × 120 × 3.6N120-B	172.25	175.54	0.981
ITF120 × 120 × 3.6N120-C	175.09	184.27	0.950
ITF80 × 80 × 3.6N30-B	157.34	154.49	1.018
ITF80 × 80 × 3.6N30-C	159.77	157.77	1.013
ITF80 × 80 × 3.6N60-B	179.59	184.20	0.975
ITF80 × 80 × 3.6N60-C	182.44	189.34	0.964
ITF60 × 120 × 1.8N30-B	34.19	35.14	0.973
ITF60 × 120 × 1.8N30-C	35.18	35.79	0.983
ITF60 × 120 × 1.8N60-B	39.38	40.44	0.974
ITF60 × 120 × 1.8N60-C	40.10	41.80	0.959
ITF60 × 120 × 1.8N60-C(N+d)	41.55	43.16	0.963
ITF60 × 120 × 2.7N30-B	80.23	76.93	1.043
ITF60 × 120 × 2.7N30-C	81.79	77.56	1.055
ITF60 × 120 × 2.7N60-B	86.94	90.49	0.961
ITF60 × 120 × 2.7N60-C	89.89	93.47	0.962
Mean, P_m			0.987
COV, V_p			0.031

modes reflect good agreement between the experimental and finite element results. Hence, the FEM closely predicted the behaviour of CFRP strengthened cold-formed stainless steel square and rectangular hollow sections subjected to web crippling.

6 CONCLUSIONS

A test program on CFRP strengthened cold-formed stainless steel tubular sections subjected to web crippling has been presented. The web crippling tests were performed under Interior-Two-Flange (ITF) loading condition. It is shown that the test specimens were mainly failed in three failure modes, namely the adhesion, combination of adhesion and cohesion as well as tear of CFRP sheets. It is also shown that the increase of CFRP strengthened width can provide some improvements on strengthening of the cold-formed stainless steel tubular sections.

A non-linear finite element model which includes geometric and material non-linearities has been developed for CFRP strengthened cold-formed stainless steel tubular sections subjected to web crippling. The debonding between the CFRP and the cold-formed stainless steel tubular section was carefully modelled using cohesive element. The developed model was verified against the experimental results obtained in this study. The finite element model can accurately predicted the behaviour of CFRP strengthened cold-formed stainless steel tubular sections subjected to web crippling.

ACKNOWLEDGEMENTS

The research work described in this paper was supported by "the Chinese National Natural Science Foundation (Project No. 51108337)" and Shanghai Municipal Natural Science Foundation (Project No. 11ZR1439400).

REFERENCES

ABAQUS. 2008. *ABAQUS Standard User's Manual.* 2007. Hibbitt, Karlsson and Sorensen, Inc. Vols. 1-3, Version 6.8, USA.

American Society of Civil Engineers (ASCE). 2002. *Specification for the design of cold-formed stainless steel structural members.* SEI/ASCE-8-02, Reston, Virginia.

Ashraf M, Gardner L, Nethercot DA. 2006. Finite element modelling of structural stainless steel cross-sections. *Thin-Walled Structures* 44(10): 1048–1062.

Australian/New Zealand Standard (AS/NZS). 2001. *Cold-formed stainless steel structures.* AS/NZS 4673:2001, Standards Australia, Sydney, Australia.

Fawzia S, Al-Mahaidi R, Zhao XL, et al. 2007. Strengthening of circular hollow steel tubular sections using high modulus CFRP sheets. *Construction and Building Materials* 21(4): 839-845.

Fernando D. 2010. *Bond behaviour and debonding failures in CFRP-strengthened steel members.* PhD, Hong Kong Polytechnic University, Kowloon.

Gardner L, Nethercot DA. 2004. Numerical modeling of stainless steel structural components – A consistent approach. *Journal of Structural Engineering, ASCE* 130(10): 1586–1601.

Islam S, Young B. 2011. FRP strengthened aluminium tubular sections subjected to web crippling. *Thin-Walled Structures* 49(11): 1392–1403.

Islam S, Young B. 2013. Strengthening of ferritic stainless steel tubular structural members using FRP subjected to Two-Flange-Loading. *Thin-Walled Structures* 62: 179–190.

Korvink SA, Van den Berg GJ, Van der Merwe P. 1995. Web crippling of stainless steel cold-formed beams. *Journal of Constructional Steel Research* 34(2): 225–248.

Rasmussen K. 2003. Full-range stress-strain curves for stainless steel alloys. *Journal of Constructional Steel Research* 59(1): 47–61.

Zhou F, Young B. 2006. Cold-formed stainless steel sections subjected to web crippling. *Journal of Structural Engineering, ASCE* 132(1): 134–144.

Tubular Structures XV – Batista, Vellasco & Lima (eds)
© *2015 Taylor & Francis Group, London, ISBN 978-1-138-02837-1*

Tests on ferritic stainless steel simply supported and continuous SHS and RHS beams

I. Arrayago, E. Real & E. Mirambell
Department of Construction Engineering, Universitat Politècnica de Catalunya, UPC, Barcelona, Spain

ABSTRACT: Development of efficient design guidance for stainless steel structures is key for the spreading of this corrosion-resistant material by considering both nonlinear behavior and strain hardening into predicting expressions, together with allowing the consideration of moment redistribution in indeterminate structures. With the aim of analyzing the bending moment redistribution capacity in ferritic stainless steel beams (RHS and SHS), an experimental programme is presented. The tests contribute to the assessment of EN 1993-1-4 (2006) and classical and new plastic design methods available in the literature to indeterminate stainless steel structures, not currently allowed for stainless steels. Additional tests results reported by other authors are also studied. The analysis indicated that cross-sectional classification limits seem to be too optimistic for ferritic stainless steels and further research is needed for the extension of plastic design to these grades, although promising predictions of ultimate loads are obtained for austenitic and lean duplex stainless steels.

1 INTRODUCTION

The advantages offered by stainless steels for construction, such as excellent corrosion resistance, good mechanical properties and fine aesthetic appearance, have been widely spread. However, and in order to make stainless steel a competitive material, the specific mechanical features of this material need to be also considered.

Various metallic alloys such as stainless steel present a nonlinear stress-strain relationship, even for low strain values, together with an important strain hardening. Hence, this special behavior needs to be taken into account when proposing design expressions. In a similar way to EN1993-1-1 (2005) for carbon steel, current European design guidance for stainless steel, EN1993-1-4 (2006), considers four cross-sectional classes depending on their local buckling susceptibility, and a different resistance is assigned to each class. EN1993-1-1 (2005) permits plastic design for carbon steel cross-sections classified as Class 1, whose rotation capacity is the one required for plastic analysis. Nevertheless, no plastic design is allowed for stainless steel elements in EN 1993-1-4 (2006) despite the high ductility characterizing stainless steels. These realities, together with the fact that strain hardening effects are not considered when indeterminate structures are designed, leads to overconservative predictions.

Although tests on continuous stainless steel beams have already been conducted in austenitic and lean duplex stainless steels (Real and Mirambell (2005) and Theofanous et al. (2014)) with the aim of assessing the moment redistribution capacity of stainless steel beams and the possibility of incorporating plastic design, no experimental results of ferritic stainless

steels are yet available as far as the authors know. Hence, the objective of the continuous beam tests on ferritic stainless steel hollow elements (RHS and SHS) was to understand the behavior of indeterminate ferritic stainless steel structures and the redistribution capacity of the beams. Furthermore, four-point bending tests were also conducted in the same cross-sections in order to utilize them in the analysis of indeterminate structures.

Additionally, a new design method based on the Continuous Strength Method (CSM) for indeterminate structures, proposed by Gardner et al. (2011) and Theofanous et al. (2014), was also assessed with the conducted tests.

2 EXPERIMENTAL PROGRAMME

Five different rectangular and square hollow sections (RHS and SHS) were analyzed as simply supported and continuous beam tests: S1-80 × 80 × 4, S2-60 × 60 × 3, S3-80 × 40 × 4, S4-120 × 80 × 3 and S5-70 × 50 × 2. All the tested specimens were made from ferritic stainless steel grade EN 1.4003 and were cold-rolled and seam welded.

2.1 Tensile tests and measured dimensions

All the specimens tested in this experimental programme were produced in ferritic stainless steel grade EN 1.4003. Some tensile tests were conducted (see Figure 1) in several coupons extracted from the cross-sections in order to obtain information about the real stress-strain behavior of the tested materials. Two flat (F) and two corner (C) specimens were extracted from

Figure 1. Tested flat coupon, testing of S3-F coupon and tested corner coupon.

Table 1. Average tensile test results for flat and corner specimens.

Section	E [GPa]	$\sigma_{0.2}$ [MPa]	σ_u [MPa]	ε_u [%]	ε_f [%]
S1 – F	174	521	559	8.2	21.7
S1 – C	145	577	645	1.1	7.9
S2 – F	187	485	505	6.8	20.9
S2 – C	167	555	587	1.0	10.1
S3 – F	182	507	520	3.6	21.0
S3 – C	129	558	601	1.0	7.0
S4 – F	177	430	490	12.6	27.1
S4 – C	180	540	583	1.0	10.1
S5 – F	180	418	480	13.8	26.8
S5 – C	131	552	575	1.1	6.5

Table 2. Average measured dimensions for simply supported (L_1) and continuous (L_2) beams.

Section	H [mm]	B [mm]	t [mm]	R_{ext} [mm]	L_1 [mm]	L_2 [mm]
S1	80.3	79.9	3.8	7.2	1700	3200
S2	60.2	60.1	2.9	6.3	1700	3200
S3	79.9	39.8	3.9	7.1	1700	3200
S4	119.8	79.9	2.9	7.1	1700	3200
S5	70.1	49.8	2.0	4.3	1700	3200

Table 3. Weighted tensile material properties for beam specimens.

Section	E [GPa]	$\sigma_{0.2}$ [MPa]	σ_u [MPa]	ε_u [%]
S1	174	539	587	5.83
S2	187	509	533	4.81
S3	182	529	554	2.45
S4	177	453	509	10.09
S5	180	449	502	10.83

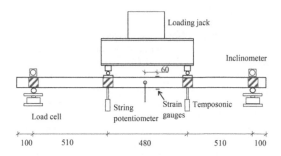

Figure 2. Schematic diagram of the test setup for four-point bending tests. Dimensions in mm.

each studied cross-section and tested in accordance with ISO 6892-1 (2009) provisions.

The average key material parameters are presented in Table 1, where E is the Young's modulus, $\sigma_{0.2}$ is the 0.2% proof stress, σ_u is the ultimate tensile stress of the material, ε_u is the corresponding ultimate strain and ε_f is the strain at fracture. E values for corner coupons were found to be unusually low, which might be caused by some eccentricities introduced when gripping the specimens.

Average values of the measured dimensions of the tested beam elements are presented in Table 2, where H is the height of the cross-section, B the width, t represents the thickness and R_{ext} the external radius. L_1 is the total length of simply supported elements, while L_2 is the length of continuous beams.

Considering the measured cross-sectional dimensions and the tensile properties corresponding to flat and corner coupons, weighted tensile properties were calculated (see Table 3). The weight of each zone (flat/corner) was calculated according to the part of the total area that represents in the whole cross-section. However, Young's modulus values corresponding to flat coupons have been considered.

2.2 Simply supported beam tests

Eight simply supported beams were tested under four-point bending loading conditions in order to determine the bending moment and rotation capacities of every studied cross-section, considering both major (denoted as Mj) and minor (Mi) bending axes for the RHS.

The adopted test configuration is presented in Figures 2 and 3: the load was applied as two line loads at a distance of 510 mm from both supports, being separated by 480 mm.

Both support reactions were measured in order to verify the symmetry of the system, and the deflections at the midspan and loading sections were also measured using a string potentiometer and two temposonic transducers respectively. The deflection measurements were also used for the determination of the curvature of the specimens at each load step. In addition, two inclinometers recording end rotations were placed at the support sections. Strain-gauges were also placed

Figure 3. General view of the failed S3-Mi specimen subjected to four-point loading conditions.

Figure 4. Local buckling at S5-Mi specimen subjected to four-point loading conditions.

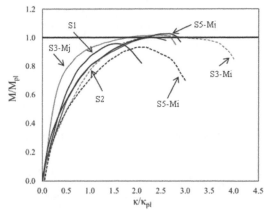

Figure 5. Normalized moment-curvature diagrams for simply supported beam tests.

Table 4. Summary of experimental results for simply supported beams.

Section	F_u [kN]	δ_u [mm]	M_u [kNm]	M_u/M_{el}	M_u/M_{pl}	R
S1	66.1	42.4	16.9	1.18	0.96	–
S2	27.2	59.6	6.9	1.23	1.00	1.4
S3-Mj	43.2	63.8	11.0	1.36	1.02	1.8
S3-Mi	26.3	104.4	6.7	1.26	1.01	2.1
S4-Mj	64.1	16.3	16.3	1.03	0.84	–
S4-Mi	48.6	22.5	12.4	0.97	0.83	–
S5-Mj	19.2	48.0	4.9	1.26	1.03	1.9
S5-Mi	13.9	49.9	3.5	1.09	0.94	–

at a distance of 60mm from the midspan section, measuring the extreme tensile and compressive strains. Both the loading and support sections were stiffened by inserting wooden blocks in order to prevent web crippling and ensure that the section failed under pure bending moment. The specimens were tested in a 1000kN MTS hydraulic machine under displacement control at a rate of 2 mm/min.

The specimens failed by local buckling of the compressed flange at loading sections since web crippling was prevented, as it is shown in Figure 4. Simply support test results are presented in Figure 5, where normalized ultimate moments are shown against normalized curvatures. No curves are presented for section S4 as some problems during data acquisition made curvature calculations impossible, even if ultimate loads were recorded.

Table 4 also presents experimental results for the simply supported beams: ultimate loads F_u and the corresponding midspan deflections δ_u are presented, together with the ultimate bending moment M_u reached. The comparison of the bending capacities against elastic (M_{el}) and plastic (M_{pl}) bending moment capacities is also gathered, and finally, the rotation capacity R is provided for those beams showing a M_u/M_{pl} ratio greater than 1.

The rotation capacity R has been calculated according to Equation 1, where κ_u is the curvature at which

the moment-curvature curve falls below M_{pl} on the descending branch, and κ_{pl} is the elastic curvature corresponding to M_{pl} at the ascending branch. Curvatures have been calculated according to Equation 2, where u_2 is the midspan deflection and u_{av} is the average vertical displacement at the loading points. L represents the length between the loading points.

$$R = \kappa_u / \kappa_{pl} - 1 \qquad \kappa = \frac{8 \cdot (u_2 - u_{av})}{4 \cdot (u_2 - u_{av})^2 + L^2} \qquad (1) \text{ and } (2)$$

Table 4 shows that the only cross-section not reaching the elastic bending capacity, and therefore, experimentally classified as Class 4, is S4-Mi. For sections S1, S4-Mj and S5-Mi, ultimate bending capacities lay between the elastic and plastic bending capacities, so they can be classified as Class 3. A minimum rotation capacity of R > 3 is usually adopted for guaranteeing moment redistribution in carbon steel cross-section, and the same limit is required for stainless steels, since no specific limit is provided, as noted in Theofanous et al. (2014). Therefore, the remaining cross-sections might be considered Class 2 as they reach the plastic moment capacities, but the rotation capacity is lower than 3.

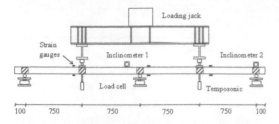

Figure 6. Schematic diagram of the test setup for continuous beam tests. Dimensions in mm.

Figure 7. General view of five-point bending test setup.

2.3 Continuous beam tests

Eight double span continuous beam tests were also conducted in order to determine their redistribution capacity. The objective of these tests was to assess whether plastic design, which is not currently allowed for stainless steel structures, is applicable to ferritic stainless steel cold-formed beams. All specimens were 3200 mm long and were tested in a two span structural configuration, with identical span lengths of 1500 mm. The test configuration is presented in the scheme shown in Figure 6: the beams were placed over two spans, subjected to two concentrated loads applied at the midspan section of each span. All support reactions were measured by load cells in order to evaluate the reaction redistribution during the tests, midspan deflections were recorded by two temposonic transducers and rotations were also measured by inclinometers at the outer support section of the right span and at a distance of 250 mm from the internal support.

Extreme tensile and compressive strains were measured by several strain gauges at a distance of 60 mm from the loading sections and the internal support. All the loading and support sections were stiffened with wooden blocks in order to prevent web crippling in these sections. The specimens were tested in a 1000 kN MTS hydraulic machine under displacement control at a rate of 2 mm/min, and failed by local buckling of the compressed flange at the internal support and at the loading sections (see Figure 7).

Experimental results for continuous beam tests are summarized in Table 5, where ultimate loads F_u and the corresponding midspan deflections δ_u are presented, together with the measured reaction at the interior

Table 5. Summary of experimental results for continuous beams.

Section	F_u [kN]	δ_u [mm]	R_u [kN]	θ_u^1 [rad]	θ_u^2 [rad]
S1	119.5	24.6	79.8	0.047	0.025
S2	51.7	29.1	34.0	0.053	0.038
S3-Mj	84.2	23.5	56.1	0.048	0.025
S3-Mi	52.4	47.4	34.6	0.068	0.047
S4-Mj	106.5	11.4	69.5	0.022	0.010
S4-Mi	87.4	16.7	58.7	0.029	0.012
S5-Mj	34.4	20.6	22.5	0.038	0.025
S5-Mi	26.7	27.8	17.6	0.055	0.033

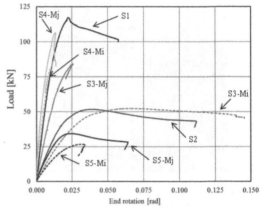

Figure 8. Load-end rotation curves for continuous beam tests.

support, R_u. The rotations corresponding to F_u at a distance of 250 mm from the internal support θ_u^1 and at the outer support section θ_u^2 are also provided. Test results are presented as load curves plotted against end rotations at the outer supports for continuous beam tests in Figure 8.

3 ASSESSMENT OF EXISTING DESIGN METHODS

3.1 EN1993-1-4 and plastic design

The predicted ultimate capacities of simply supported beams can be easily determined by calculating the bending moment resistances.

Codified expressions for the determination of the bending moment capacity according to EN1993-1-4 (2006) depend on section classification and can be summarized by Equation 3: for cross-sections classified as Class 1 or 2, the plastic bending capacity (considering $\beta = 1$) needs to be applied. For Class 3 sections, the elastic bending capacity is determined by considering $\beta = W_{el}/W_{pl}$, and finally, for

Class 4 cross-sections, effective properties need to be considered through $\beta = W_{eff}/W_{pl}$.

$$M_{c,Rk} = \beta \cdot W_{pl} \cdot \sigma_{0.2} \qquad (3)$$

where W_{pl} is the plastic modulus, W_{el} is the elastic modulus, W_{eff} is the effective modulus and $\sigma_{0.2}$ is the 0.2% proof stress.

Concerning continuous beams, EN1993-1-4 (2006) considers that the collapse of the beam is reached with the formation of the first plastic hinge at the central support, based on elastic calculations, without allowing any redistribution. This method usually provides too conservative predictions of the ultimate loads. Although no plastic design is currently available in EN1993-1-4 (2006), the classical plastic design approach with rigid-plastic material codified in EN1993-1-1 (2005) is also analyzed in this paper in order to determine its accuracy and the possibility of recommending its inclusion in EN1993-1-4 (2006) for stainless steel.

3.2 New design methods

The development of accurate design methods or expressions that do not provide overconservative resistance predictions is doubly important when metallic materials with high strain hardening levels, such as stainless steel, are considered. Therefore, new design methods, based on deformation capacity instead of considering discrete cross-sectional capacities, are of interest. The Continuous Strength Method (CSM), developed and presented in Gardner (2008), is a new design method which considers the deformation capacity of a cross-section ε_{csm} in terms of its slenderness through Equations 4 and 5.

$$\frac{\varepsilon_{csm}}{\varepsilon_y} = \frac{0.25}{\overline{\lambda}_p^{3.6}} \text{ but } \frac{\varepsilon_{csm}}{\varepsilon_y} < \min\left(15, \frac{0.4\varepsilon_u}{\varepsilon_y}\right) \qquad (4) \text{ and } (5)$$

where ε_y is the elastic strain for the proof stress corresponding to a 0.2% plastic strain, $\overline{\lambda}_p$ is the relative cross-sectional slenderness and ε_u the ultimate strain.

Expressions and work examples of the determination of the predicted ultimate capacities are widely developed in Afshan and Gardner (2013). The expression predicting the bending capacity of a cross-section according to CSM is given by Equation 6.

$$\frac{M_{csm}}{M_{pl}} = 1 + \frac{E_{sh}}{E}\frac{W_{el}}{W_{pl}}\left(\frac{\varepsilon_{csm}}{\varepsilon_y} - 1\right) - \left(1 - \frac{W_{el}}{W_{Pl}}\right) \cdot \left(\frac{\varepsilon_{csm}}{\varepsilon_y}\right)^{-2} \qquad (6)$$

where W_{el} is the elastic modulus, W_{pl} is the plastic modulus, E_{sh} is the strain-hardening modulus and E is the Young's modulus. For the definition of the strain hardening modulus, Afshan and Gardner (2013) proposed some expressions for austenitic and duplex stainless steels, but recent research works by Bock

et al. (2014) demonstrated the need of a new equation for ferritic grades, given by Equation 7.

$$E_{sh} = \frac{\sigma_u - \sigma_{0.2}}{0.45\varepsilon_u - \varepsilon_y} \qquad \text{if} \quad \frac{\varepsilon_y}{\varepsilon_u} < 0.45 \qquad (7a)$$

$$E_{sh} = 0 \qquad \text{if} \quad \frac{\varepsilon_y}{\varepsilon_u} > 0.45 \qquad (7b)$$

In order to accurately predict the collapse loads of stocky stainless steel continuous elements, considering moment redistribution and strain-hardening, Theofanous et al. (2014) assessed the applicability of the new design method developed by Gardner et al. (2011), the CSM for indeterminate structures. The method assigns the full CSM cross-sectional resistance to the critical plastic hinge and allows a degree of strain-hardening for the rest of the hinges. The rotation demand of each hinge is calculated through Equation 8 where θ_i is the relative rotation derived from kinematics considerations for the collapse mechanism considered, h_i is the section height at the considered location and $(\varepsilon_{csm}/\varepsilon_y)_i$ is the corresponding normalized strain ratio at the ith hinge.

$$\alpha_i = \frac{\theta_i \cdot h_i}{\left(\varepsilon_{csm}\middle/\varepsilon_y\right)_i} \qquad (8)$$

The critical hinge is that showing the greatest rotation capacity demand relative to the deformation capacity of the cross-section, and the rest of rotation demands are calculated according to Equation 9. The collapse load is calculated through the virtual work principle as in conventional plastic design.

$$\left(\frac{\varepsilon_{csm}}{\varepsilon_y}\right)_i = \frac{\alpha_i}{\alpha_{crit}}\left(\frac{\varepsilon_{csm}}{\varepsilon_y}\right)_{crit} \leq \left(\frac{\varepsilon_{csm}}{\varepsilon_y}\right)_{i,limit} \qquad (9)$$

Sufficient deformation capacity for moment redistribution to occur is guaranteed by ensuring a minimum rotation capacity of $R > 3$. Nevertheless, this criterion should be revised when stainless steels are considered, since the plastic moment capacity of those cross-sections is not clear. Gardner et al. (2011) propose a new criterion, based in deformation capacity, in order to guarantee that a cross-section is capable of moment redistribution in indeterminate structures: minimum value of $\varepsilon_{csm}/\varepsilon_y = 3$ for I-sections and 3.6 for box sections.

3.3 Assessment of design methods and discussion

The assessment of the existing design methods for simply supported and continuous stainless steel beams is presented herein: EN1993-1-4 and CSM.

For simply supported beams, ultimate bending moment predictions have been calculated according to Equation 3, considering the cross-sectional classification currently codified in EN1993-1-4 and the revised

Table 6. Assessment of existing design methods for simply supported beams.

Section	EN1993-1-4 $M_{c,Rk}/M_u$	Revised limits $M_{c,Rk}/M_u$	CSM M_{csm}/M_u	$\varepsilon_{csm}/\varepsilon_y$
S1	0.85	1.04	1.04	3.5
S2	1.00	1.00	1.00	5.5
S3-Mj	0.98	0.98	1.01	15.0
S3-Mi	0.79	0.99	0.99	3.2
S4-Mj	0.95	0.98	1.09	1.6
S4-Mi	0.81	0.84	–	0.3
S5-Mj	0.80	0.97	0.93	1.9
S5-Mi	0.79	0.82	–	0.5
Mean	0.87	0.95	1.01	–
COV.	0.106	0.081	0.051	–

Table 7. Assessment of existing design methods based on elastic analysis for continuous beams.

Section	EN1993-1-4 Class	$F_{h1,EN}/F_u$	Revised limits Class	$F_{h1,revEN}/F_u$	CSM $F_{h1,csm}/F_u$
S1	3	0.85	1	1.04	1.04
S2	2	0.95	1	0.95	0.95
S3-Mj	1	0.92	1	0.92	0.94
S3-Mi	3	0.72	1	0.90	0.90
S4-Mj	4	1.04	2	1.31	1.18
S4-Mi	4	0.81	4	0.84	–
S5-Mj	4	0.80	1	0.98	0.94
S5-Mi	4	0.74	4	0.77	–
Mean		0.85		0.96	0.98
COV.		0.127		0.167	0.112

slenderness limits proposed by Gardner and Theofanous (2008). CSM predictions have been determined by calculating the relative slenderness according to Equation 10.

$$\bar{\lambda}_p = \sqrt{\sigma_{0.2}/\sigma_{cr}} \tag{10}$$

where σ_{cr} is the critical buckling stress calculated by the provisions of EN1993-1-4 (2006) for the most slender plate element. The assessment of the three methods is studied in Table 6 by presenting predicted-to-experimental bending moment ratios, where it is shown that the new slenderness limits seem to be too optimistic for the experimental results of the conducted tests. Current limits codified in EN1993-1-4 (2006) provide safe but conservative predictions of bending resistances. $\varepsilon_{csm}/\varepsilon_y$ ratios for each tested cross-section are also presented in order to determine whether the CSM for indeterminate structures is applicable.

Concerning continuous beam tests, the assessment of the different design methods presented before is studied in Tables 7 and 8, by providing the predicted-to-experimental collapse load ratios. In Table 7 the predicted ultimate loads have been calculated through elastic calculation, when the first plastic hinge at the central support is formed, F_{h1}. Ultimate bending moment capacities according to expressions codified in EN1993-1-4, but considering both EN1993-1-4 (2006) cross-sectional classification and the revised limits, together with CSM provisions have been considered. For high slenderness values (S4-Mi and S5-Mi) this method is not applicable, so the resistances calculated according to CSM have not been considered in Tables 7, 8 and 9.

Table 8 presents the assessment of the existing design methods, but allowing for plastic design when cross-sections are classified as Class 1, considering the two available classifications, $F_{u,EN}$ and $F_{u,revEN}$, together with the CSM for indeterminate structures presented before, $F_{u,csm}$, when $\varepsilon_{csm}/\varepsilon_y$ ratios (see Table 6) indicate that it is applicable. It is important to highlight that for the tests presented in this

Table 8. Assessment of existing design methods for continuous beams allowing for plastic design.

Section	EN1993-1-4 Class	$F_{h1,EN}/F_u$	Revised limits Class	$F_{h1,revEN}/F_u$	CSM $F_{h1,csm}/F_u$
S1	3	0.85	1	1.18	1.17
S2	2	0.95	1	1.08	1.08
S3-Mj	1	1.04	1	1.04	1.06
S3-Mi	3	0.72	1	1.66	1.02
S4-Mj	4	1.04	2	1.31	–
S4-Mi	4	0.81	4	0.84	–
S5-Mj	4	0.80	1	1.11	–
S5-Mi	4	0.74	4	0.77	–
Mean		0.87		1.11	1.06
COV.		0.145		0.258	0.068

paper, the rotation capacity demand on the three plastic hinges, calculated by Equation 8, is the same, so the full CSM resistance is assigned to all hinges. This method, therefore, is equivalent to considering classical plastic design but with the bending moment capacity determined according to CSM instead of M_{pl}, for this particular test setup. The differences in load predictions between Tables 7 and 8, especially for S3, owe to the consideration of global plastic analysis and M_{pl} instead of elastic analysis, with M_{el}.

If elastic analysis is assumed, the best ultimate predictions are obtained, as shown in Table 7, when the revised class limits are considered, although some unsafe predictions are found (S1, S4-Mj). However, Table 8 demonstrates that the consideration of plastic design for the tested specimens classified as Class 1 overestimates the collapse loads, as revised classification limits proposed by Gardner and Theofanous (2008) are too optimistic for ferritic stainless steels, which show considerably lower ductility and therefore, lower rotation capacity.

Figure 9 presents the measured experimental load-end rotation curves for continuous beam tests, where loads have been normalized by the collapse loads determined according to conventional plastic design,

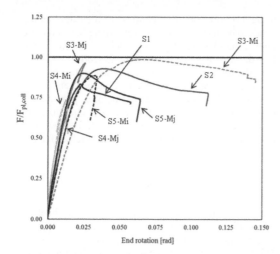

Figure 9. Normalized load-end rotation experimental curves for continuous beam tests.

$F_{pl,coll}$. This figure demonstrates that the consideration of plastic design overestimates the capacity of all the tested beams, since none of them can be considered a Class 1 cross-section, as it has been established in Section 2.2.

3.4 Additional experimental results from literature

Additional test data on stainless steel continuous beams reported in the literature is also presented and included in this section. Two span continuous beam tests in EN 1.4301 four austenitic RHS and two H-sections were conducted and reported in Mirambell and Real (2000) and Real and Mirambell (2005) under different test setups. Recent continuous beam tests have also been reported in Theofanous et al. (2014), where EN1.4301/1.4307 austenitic RHS and SHS and EN1.4162 lean duplex H-sections were tested. More detailed information about the experimental programmes and test results can be found in the original publications. Since the previously conducted continuous tests consider different stainless steel families and cross-section types than the ones presented in this paper, this additional experimental data has been included in the analysis.

Table 9 presents the assessment of the aforementioned design methods for stainless steel indeterminate structures for the experimental results gathered from the literature. The two cross-sectional classifications have been considered, the one currently codified in EN1993-1-4 (2006) and the one revised by Gardner and Theofanous (2008). In both cases, plastic design has been considered for Class 1 cross-sections, while elastic calculations have been conducted for the rest of the cross-sections, with the corresponding moment resistance: M_{pl} for Class 2 sections, M_{el} for Class 3 and M_{eff} for Class 4 cross-sections. Finally, $F_{u,csm}$ considers the ultimate load predictions according to the new CSM method for indeterminate structures when applies.

Table 9. Assessment of existing design methods allowing for plastic design for additional test data.

| Section | EN1993-1-4 | | Revised limits | | CSM |
	Class	$\frac{F_{u,EN}}{F_u}$	Class	$\frac{F_{u,revEN}}{F_u}$	$\frac{F_{u,csm}}{F_u}$
RHS and SHS					
$80 \times 80 \times 3$	4	0.78	2	0.98	–
$80 \times 80 \times 3$	4	0.72	2	0.91	–
$120 \times 80 \times 3$	2	0.56	1	0.71	0.75
$120 \times 80 \times 3$	2	0.61	1	0.78	0.82
$50 \times 50 \times 3$	1	0.75	1	0.75	0.91
$50 \times 50 \times 3$	1	0.76	1	0.76	0.78
$60 \times 60 \times 3$	2	0.70	1	0.78	0.91
$60 \times 60 \times 3$	2	0.73	1	0.82	0.95
$100 \times 100 \times 3$	4	0.79	4	0.82	–
$100 \times 100 \times 3$	4	0.80	4	0.83	–
$60 \times 40 \times 3$	1	0.54	1	0.54	0.70
$60 \times 40 \times 3$	1	0.54	1	0.54	0.70
$40 \times 60 \times 3$	1	1.03	1	1.03	0.85
$40 \times 60 \times 3$	1	1.07	1	1.07	0.75
H-sections					
$200 \times 140 \times 6 \times 6$	4	0.66	4	0.67	–
$200 \times 140 \times 8 \times 6$	4	0.64	3	0.64	–
$200 \times 140 \times 10 \times 8$	1	0.80	1	0.80	0.92
$200 \times 140 \times 12 \times 8$	1	0.72	1	0.72	0.92
$200 \times 140 \times 6 \times 6$	4	0.55	4	0.56	–
$200 \times 140 \times 8 \times 6$	4	0.57	3	0.57	–
$200 \times 140 \times 10 \times 8$	1	0.85	1	0.85	0.83
$200 \times 140 \times 12 \times 8$	1	0.81	1	0.81	0.89
$100 \times 100 \times 8 \times 8$	1	0.89	1	0.89	1.14
$100 \times 100 \times 8 \times 8$	Failed by early lateral torsional buckling				
Mean		0.73		0.78	0.86
COV.		0.198		0.191	0.135

According to what is observed in Table 9, the best design approach for austenitic and lean duplex elements is the CSM for indeterminate structures, even if it is not applicable to all cases. However, EN1993-1-4 (2006) expressions, with the revised classification limits and allowing for plastic design, do also provide considerably good results while keeping calculations simple.

The predicted collapse loads calculated by classic plastic design $F_{pl,coll}$ have been normalized by the experimental ultimate load for all the test results, regardless of the cross-sectional classification, and presented in Figure 10.

As it can be appreciated, the capacity of the most slender specimens is overpredicted when plastic design is considered, while the prediction gets more accurate for cross-sections showing an intermediate slenderness. For the stockiest cross-sections, where the effect of strain hardening is more influential, the consideration of an elastic-perfectly plastic material results in overconservative predictions. A similar behavior is observed for the analyzed stainless steel families, although a more extensive database would be necessary to extract more general conclusions, as for ferritic stainless steel continuous beams the adoption of plastic design still needs to be studied.

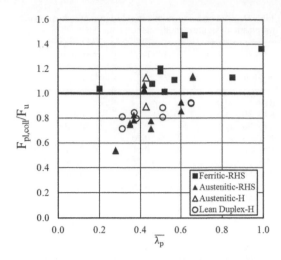

Figure 10. Assessment of the classic plastic design method.

Table 10. Assessment of design methods allowing plastic design.

Section	$F_{u,EN}/F_u$	$F_{u,revEN}/F_u$	$F_{u,csm}/F_u$
Ferritic RHS	0.87	1.11	1.06
Austenitic RHS	0.72	0.79	0.82
Austenitic H	0.89	0.89	1.14
Lean duplex H	0.70	0.70	0.89
Mean	0.77	0.86	0.90
COV.	0.197	0.277	0.152

Table 10 summarizes the assessment of the considered design methods for the different stainless steel families and cross-sectional types analyzed in this work, and the global results are in line with the conclusions extracted from Table 9. However, it is important to note that the application of plastic design to ferritic cross-sections needs to be more deeply analyzed, as class limits for Class 1 cross-sections have been found to be too optimistic for both classifications considered in this paper.

4 CONCLUSIONS

Some experimental results on ferritic stainless steel simply supported and continuous beams are presented in this paper in order to understand the behavior of indeterminate structures and the redistribution capacity of these cross-sections. Additionally, test results reported by other authors have also been included in the analysis. Experimental ultimate loads have been compared to calculated capacities according to different design methods.

These comparisons highlighted that EN1993-1-4 (2006) expressions are the most appropriate approach for the prediction of ultimate loads in RHS ferritic stainless steel beams. The revised class limits seem to be too optimistic for ferritics, so their application needs to be deeply analyzed. Nevertheless, when more ductile stainless steel grades, such as austenitic and lean duplex, are considered, the revised cross-sectional classification limits provide more accurate EN1993-1-4 (2006) ultimate capacity estimations, and CSM provisions for indeterminate structures provide the best predictions of the collapse loads. Therefore, an extensive finite element analysis would be now necessary in order to extend this study to stockier ferritic cross-sections.

ACKNOWLEDGEMENTS

This experimental programme was possible thank to the funding from the Ministerio de Economía y Competitividad (Spain) under the Project BIA 2012-36373. The first author would like to acknowledge the financial support provided by the Secretaria d'Universitats i de Recerca del Departament d'Economia i Coneixement de la Generalitat de Catalunya. The authors would also like to mention Acerinox for their special support and trust.

REFERENCES

Afshan S, Gardner L. The continuous strength method for structural stainless steel design. Thin-Walled Structures 2013(68): 42–49.

Bock M, Gardner L and Real E. (2014). Material and local buckling response of ferritic stainless steel sections. Thin-Walled Structures (submitted, under revision).

EN ISO 6892-1. 2009. (2009). Metallic materials – Tensile testing – Part 1: Method of test at room temperature. Brussels: European Committee for Standardization (CEN); 2009.

European Committee for Standardization Eurocode 3. (2006). Design of steel structures. Part 1–4: General rules. Supplementary rules for stainless steels. Brussels, Belgium; 2006.

European Committee for Stan dardization. (2005). EN 1993-1-1. European Committee for Standardization Eurocode 3. Design of steel structures. Part 1–1: General rules and rules for buildings. Brussels, Belgium.

Gardner L. The Continuous Strength Method. Proceedings of the Institution of Civil Engineers – Structures and Buildings. 2008; 161(2): 127–133.

Gardner L and Theofanous M. (2008). Discrete and continuous treatment of local buckling in stainless steel elements. Journal of Constructional Steel Research 64, 1207–1216.

Gardner L, Wang FC, and Liew A. (2011). Influence of strain hardening on the behaviour and design of steel structures. International Journal of Structural Stability and Dynamics, 2011; 11(5):855–75.

Mirambell E and Real E. (2000). On the calculation of deflections in structural stainless steel beams: an experimental and numerical investigation. Journal of Constructional Steel Research 54(4), 109–133.

Real E and Mirambell E. (2005). Flexural behaviour of stainless steel beams. Engineering Structures, 28(6), 926–934.

Theofanous M, Saliba N, Zhao O, and Gardner L. (2014). Ultimate response of stainless steel continuous beams. Thin-Walled Structures 83, 115–127.

Tubular Structures XV – Batista, Vellasco & Lima (eds)
© *2015 Taylor & Francis Group, London, ISBN 978-1-138-02837-1*

Behaviour of bridging mega-truss tubular structures fabricated from S690 steel plates: Part I – Structural aspects

S.P. Chiew, K.W. Cheng & C.K. Lee

School of Civil and Environmental Engineering, Nanyang Technological University, Singapore

ABSTRACT: Mega-trusses with stack-up structures have captured the attention of structural engineers and designers in recent years due to a growing demand for more usable land space and horizontal connectivity between high-rise buildings. Such new structural forms are able to intensify and optimize land use by creating space for functional buildings such as offices and commercial facilities over a mega bridging structure which could be located over busy expressways or roads. This study is the first of a two-part series focusing on the structural challenges including the selection of a suitable structural form, the suitable use of external post-tensioning to facilitate the stack-up stage construction and vibration mitigation due to structural response to external loadings such as construction and/or expressway-induced vibration and wind. The selection of a suitable structural form is simulated numerically based on the following criteria: (i) material usage, (ii) type and cost of foundation required, (iii) deflection control and (iv) aesthetics. The use of high-strength steel (HSS) with a yield strength of 690 MPa for the main structural members to reduce the total weight of the mega-truss will be explored. The technique of external post-tensioning, which is an active strengthening system involving the introduction of external forces to the structural members using high strength tendons, will be employed to further control the deflection and to facilitate the stack-up stage construction. Finally, a preliminary vibration analysis of the mega-truss for its fundamental frequencies and modes shapes will be presented.

1 INTRODUCTION

Since the independence of the island state, Singapore underwent rapid development and urbanization. In the past few decades especially, both the population and population density have increased dramatically. It is of little doubt that land optimization and urban development – space demand for sustainable urban living will increase rapidly in Singapore. In general, similar to many other major cities in the world, Singapore followed a very similar course of urbanization: (1) formation of skyline mainly consists of many "pen-shape" skyscrapers, (2) gradual extension of the Central Business District with the demolishment of old and historical buildings, (3) development of new "satellite" towns with rapid increase in residual building height and population per unit land area. However, due to the very limited size of the country, such traditional development phase, especially paradigm of "intensification by height" which tends to generate more and more crowded living conditions and overloading the energy and transportation infrastructures of the city, is unlikely to meet the expectations of most residents in terms of living standard and to deliver a long term sustainable urban living solution. Unfortunately, such problems will be further aggravated with the increasing average heights of the skyline of the city.

To address the above challenges and to meet the objectives of sustaining Singapore's long-term growth through creating new space cost-effectively and developing a livable, sustainable and resilient city, the critical turning point is to give up the traditional paradigm of 1D (vertical) intensification approach and moves towards the more effective and sustainable 2D (both vertical and horizontal) intensification approach (Figure 1). While intensification in the vertical direction is familiarly linked to those ubiquitous pen-shaped skyscrapers, extension into the second dimension (horizontal) requires the relatively less appreciated and new concept of bridging building, whose merit and potential has only been partially explored to a very limited extent thus far.

While bridges have been built for many millennia to span geographical obstacles such as river, there are only a few historical examples that a bridge bears other functions such as to provide residential or commercial spaces. The main reason for the absence of functional bridging buildings in the past was largely due to shortcoming of contemporary construction materials, structural engineering knowledge (both analysis and design) and building technology limitations as well as safety concerns. As a result, the main function of most bridges with buildings is still largely remained as a bridge, i.e. to provide linkage across physical obstacles. However, with the advancement of

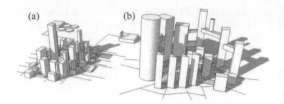

Figure 1. (a) 1D intensification vs (b) 2D intensification.

Figure 2. 3D Model of the Functional Bridging Building (FBB) with dimensions.

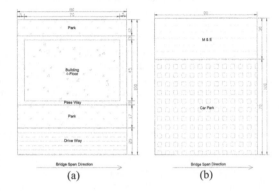

Figure 3. Land usage of the proposed FBB. (a) shows the upper deck and (b) shows the basement.

bridge engineering, structural engineering knowledge and the use of modern building construction technology as well as advanced building materials, it is now possible to build functional bridging structures by combining vertical and horizontal developments. As a result, both the problems of space creation and improving the livability in compact precincts could be solved simultaneously.

The next step is to take this technology further. It should be emphasized that the main novelty of the current study is to fully extend those "other functions" of bridges to such an extent that they will be at least as important as the function of providing linkage. As a more sustainability friendly approach for the intensification of low-density developments, this study is to 'marry' existing practices of bridge engineering and structural engineering, and develop new and safe technologies for design and construction for an innovative building called "Functional Bridging Building" (FBB). This paper addresses some challenges the FBB is going to face, and proposes a prototype system comprising bridging mega-truss and stack-up structures incorporating the use of some innovative technologies regarding high strength steel material and external post-tensioning.

2 PROTOTYPE FOR FBB

2.1 Dimensions and land usage

The FBB is just a general name for structures that are able to perform functions of both bridges and buildings at the same time, while its form can vary according to the requirements. The aim of the current study is to propose a viable and cost-effective FBB over one of the important expressways in Singapore. Besides the required connection for both sides of the expressway, buildings are to be built on top of the structure as part of land intensification and creation.

In this study, it is expected that the mega-truss will span 80m in the longitudinal direction and 100m in the transverse direction with a depth of 5m. Underneath the mega-truss runs an expressway with a carriageway of 80m and with an headroom of 5m. Furthermore, above the mega-truss a 0.2m thick concrete slab and a 4-storey stacked-up building with a height of 16m will be built (Figure 2).

The "land" usage plan of the FBB is shown in Figure 3. As can be seen from (a), the 4-storey building will be sandwiched by two parks. On its left a driveway

with a width of 20m will provide access to the FBB. The space below the deck will be utilized as a carpark to fully exploit the space provided by the depth of the mega-truss. The size of the carpark will be 80 m × 70 m while a section of the below deck space will be reserved for M&E purposes.

2.2 Design actions

Steel trusses normally used as roof in commercial and industrial buildings usually bear and transfer no loads except their self-weight. That is different from the mega-truss in this study which would be subjected to significant permanent and imposed loads transferred from the stack-up building, the pedestrian loads transferred from the walkways, and other accidental load. Therefore, to design the FBB as a bridge system seems to be more appropriate.

The design loadings for the mega-truss are depicted in Table 1 (BSI 2002). The design loadings are divided into two categories – those above the deck and those below. The intended use of the stack-up building is for light industry and research laboratories with possible vibration-sensitive equipment ($q_k = 7.5\,\mathrm{KN/m^2}$). In calculating the dead load ($8.6\,\mathrm{KN/m^2}$) for the stack-up building, the following information is assumed:

- Number of storeys: 4
- Storey height: 4 m

Table 1. Design loadings on the FBB.

Location	Section	Dead loads g_k(KN/m^2)	Live loads q_k(KN/m^2)
Above	Stack-up building	8.6	7.5
Deck	Side walk	3.84	5.0
	Green belt	30	5.0
	Deck & driveway	Calculated directly in software	Traffic loads selected in software
Below	Truss	Calculated directly in software	
Deck	M&E		1.5
	Car park	Calculated directly in software	2.5

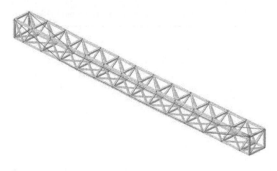

Figure 5. Two transfer trusses at the left edge of the mega-truss are shown connected to each other through top and bottom transverse members and diagonal bracings.

Figure 6. Each transfer truss is a warren truss strengthened by external post-tensioning. Red line depicts profile of external pre-stressed tendon.

Figure 7. Stiffened members are highlighted in red.

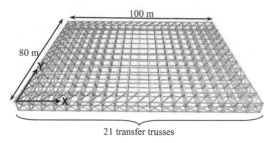

Figure 4. The mega-truss consists of 21 transfer trusses spaced along 100 m in the transverse direction.

- Partition wall area: 3% of total
- Columns & beams area: 4% of total
- Floor thickness: 0.12 m
- Density of normal concrete for beams and columns: 24 KN/m^3
- Density of lightweight concrete for partition walls: 16 KN/m^3

The green belt consists of a 0.2 m concrete floor plate and a 1.5 m deep soil layer. The density of the soil layer is assumed to be 16.8 KN/m^3. The loadings for the driveway are rather standard and are calculated directly in our structural analysis software.

2.3 Structural model and section properties

The 3-D structural model for the mega-truss consists of 21 mega transfer trusses spaced along 100 m in the transverse direction. The center-to-center spacing of the trusses is 5 m and each truss spans 80 m in the longitudinal direction (Figure 4).

Each transfer truss has a depth of 5 m and consists of a top chord, a bottom chord and vertical and diagonal web members (Figure 6). The transfer trusses are connected to each other by transverse members and additional diagonal bracings are also added at the edges (Figure 5). One critical factor is the use of high strength steel and composite materials to reduce the

weight of the structures so that it is both physically viable and economically cost effective to construct the functional bridge building over a large span at elevated height. Currently, the focus of the material side is on the application of the reheated, quenched and tempered structural steel plate, RQT-S690, which comply with the EN 10025-6 grade S690 specification (BSI 2004). It has nominal yield strength of 690 N/mm^2, and tensile strength between 790 N/mm^2 and 930 N/mm^2. However, the ductility of this material of about 15% is only half of the values of normal strength structural steel. The top and bottom chords are constructed out of S690 high-strength steel while the web members are fabricated with standard S355 mild steel. The material aspects of this project will be studied in detail in the Part II of this study.

For all the steel members, square hollow sections (SHS) are used but the dimensions vary from truss to truss depending on the applied loadings on the deck. Table 2 shows the dimensions of the sections for different regions of the mega-truss. The lower half of the table depicts the dimensions of the stiffened members which utilize larger sections for certain parts of the truss (Figure 7).

2.4 External post-tensioning

Due to the weight of the stack-up building, external post-tensioning is necessary to control any excessive deflection. The external post-tensioning technique has long been an attractive option for design loads enhancement or strength and serviceability problems correction in both new and existing bridge and building structures. It is a method commonly used to strengthen

Table 2. Dimensions of square hollow sections (SHS).

Region member	Stack-up building	Green belt	Driveway
Top chord	750*750*50	500*500*40	500*300*20
Bottom chord	500*500*36	400*400*36	300*300*10
Diagonal web	400*400*26	400*400*20	300*300*16
Vertical web	450*450*32	500*500*40	500*300*16
Stiffened top chord	900*900*64	550*550*54	400*400*34
Stiffened bottom chord	550*550*60	500*500*38	400*400*12.5
Stiffened web member	500*500*46	400*400*34	350*350*12.5

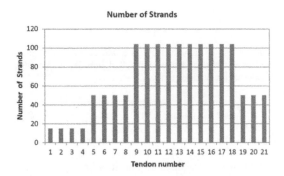

Number of Strands

Figure 8. Number of strands required for each truss. The largest number of strands is needed for the trusses located beneath the stack-up building.

reinforced concrete structures with high-strength pre-stressing steel strands. The benefits post-tensioning brings to a structure include large clear span, high carrying capacity and great system stiffness (Nordin 2005).

For the FBB model, the external post-tensioned tendon is anchored at the two ends of the top chord and has two turning points (deviators) connected to the bottom chord of every girder. The distance between the two deviators is approximately 24.6 m (Figure 6). This particular profile is selected based on consideration of several competing profiles and selecting the one which yields the lowest deflection/material consumption ratio.

For this study, the strand type Y1860S7 with a diameter of 15.7 mm and a nominal strength of 1860 MPa is selected. The number of strands required for each truss varies according to its load density and is decided by limiting the strand tensile stress to be within the range from 1350 MPa to 1550 MPa (Figure 8). This is to ensure the strands are optimally utilized such that the stress does not exceed the yield strength.

3 MODELLING RESULTS

3.1 Ultimate limit state

The load combination for the ultimate limit state according to BS EN 1990 Table A1.2(b) (BSI

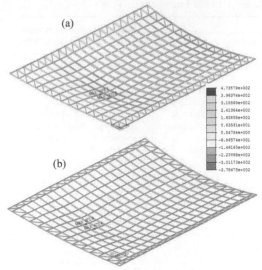

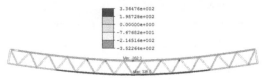

Figure 9. Contour plots of the combined (axial + bending) stresses (in MPa) for both (a) the upper deck and (b) the lower deck. The maximum tensile stress is indicated in (b) and the maximum compressive stress is indicated in (a). All steel members are fabricated with S690 high-strength steel.

Figure 10. Contour plot of the combined (axial + bending) stresses (in MPa) for a particular transfer truss. It can be seen clearly that the top chord is subjected to compressive stress whereas the bottom chord is subjected to tensile stress.

2005) and BS EN 1993-1-11, cl. 5.3(1) (BSI 2006) is defined as $1.35 \times$ (Dead Load) $+ 1.5 \times$ (Live Load) $+ 1.0 \times$ (Post Tension Load). Figure 9 shows the combined (axial + moment) stress (in MPa) for the mega-truss. In Figure 9, negative stress values indicate compression whereas positive values indicate tension. It is clear that the top deck beam members are predominantly in compression whereas the lower deck members are predominantly in tension (Figure 9). The maximum tensile stress experienced by the lower deck members is around 474 MPa and the maximum compressive stress experienced by the upper deck members is around 379 MPa. It should be noted that all the members are fabricated with S690 high-strength steel.

For the stresses experienced by the web members, Figure 11 shows the combined (axial + moment) stress (in MPa) for the web members which are fabricated with S355 mild steel. It can be seen that the web members experience compression and tension alternately along the longitudinal direction. The maximum tensile stress is around 266 MPa while the maximum compressive stress is around 304 MPa. The shear stresses

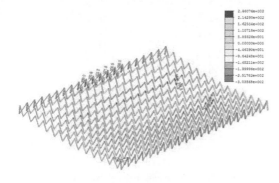

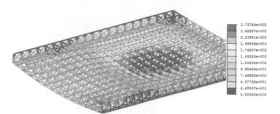

Figure 13. Contour plot of the deflection (in mm) for the mega-truss under SLS.

Figure 11. Contour plot of the combined (axial + bending) stresses (in MPa) for the web members The maximum tensile and compressive stresses are indicated. All steel members are fabricated with S355 mild steel.

Table 3. Dominant modes in the global x-direction (transverse).

Mode	Effective modal mass (%)	Effective modal mass (cumulative)	Natural frequency (Hz)
1	79.37	79.37	0.414
9	14.92	94.29	1.322

Table 4. Dominant modes in the global y-direction (longitudinal).

Mode	Effective modal mass (%)	Effective modal mass (cumulative)	Natural frequency (Hz)
15	17.96	17.96	2.044
36	41.05	59.01	3.046

Figure 12. Contour plot of the deflection (in mm) for the mega-truss under ULS.

have also been examined and it is found that the maximum shear stress for the S355 web members is around 14.3 MPa while that for the S690 members (upper and lower decks) is around 106 MPa. Finally, the maximum tensile stress for the pre-stressing strand is found to be 1531 MPa (less than the nominal yield strength of the tendon).

Figure 12 shows the deflection of the mega-truss under ULS. Not surprisingly, the maximum deflection of around 663 mm occurs near the center where the stack-up building is located.

Table 5. Dominant modes in the global z-direction (vertical).

Mode	Effective modal mass (%)	Effective modal mass (cumulative)	Natural frequency (Hz)
2	69.26	69.26	0.717
5	6.91	76.17	0.880

3.2 Serviceability limit state

According to Eurocode BS EN 1990 (BSI 2005), Serviceability Limit state (SLS) is the design state such that the structure remains functional for its intended use subject to routine loading.

3.2.1 Deflection

The load combination for the serviceability limit state (BS EN 1990: cl. A1.4) is defined as $1.0 \times$ (Live Load). Figure 13 shows the deflection of the mega-truss under SLS. Once again, the maximum deflection of around 274 mm occurs near the center where the stack-up building is located. However, it should be noted that this deflection is greater than the maximum allowable of 222 mm under SLS. For the external pre-stressing tendons, it is found that the maximum tensile stress is about 157 MPa.

3.2.2 Vibration

A free-vibration analysis is conducted for the mega-truss with the effects of post tensioning ignored. A total of 50 modes are included in the analysis. The dominant modes in each of the three spatial directions together with their effective modal masses and natural frequencies are given in Tables 3, 4 and 5, respectively.

Figure 14 shows the mode shapes for the dominant modes in the global x-direction (transverse direction). For the lateral direction, according to BS EN 1990: cl. A2.4.3.2 (BSI 2005), if the natural frequency is smaller than 1.5 Hz, then a verification of the pedestrian comfort criteria should be performed. From Table 3, it can be observed that both the dominant modes have

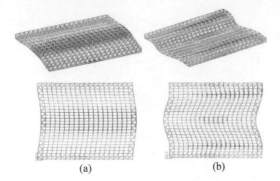

(a)   (b)

Figure 14.   Dominant modes in the global x-direction (transverse direction). Top row shows isometric views of the two modes (a) 0.414 Hz and (b) 1.322 Hz. Bottom row shows the top-down views.

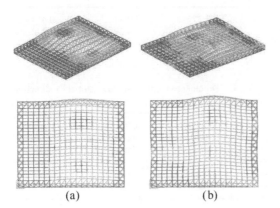

(a)   (b)

Figure 15.   Dominant modes in the global y-direction (longitudinal direction). Top row shows isometric views of the two modes (a) 2.044 Hz and (b) 3.046 Hz. Bottom row shows the top-down views.

frequencies smaller than 1.5 Hz. Therefore, we conclude that lateral stiffening of the mega-truss is needed in the global x-direction in order to raise the natural frequencies above the threshold of 1.5 Hz.

Figure 15 shows the mode shapes for the dominant modes in the global y-direction (longitudinal direction). From Table 4, both dominant modes exhibit natural frequencies above the threshold of 1.5 Hz However, BS EN 1990: cl. A2.4.3.2 (BSI 2005) also mentions that for lateral frequencies between 1.5 Hz and 2.5 Hz, a verification of the comfort criteria *may* be specified for the particular project. Referring to Table 4, the first mode has a frequency of 2.044 Hz which falls within the specified range. In this case, an increase in the stiffness in the longitudinal direction may be needed. From Figure 15, we see that the mode shapes result from the relative movement between the upper and lower decks. Therefore, there may be a need to reduce such relative movement by increasing the stiffness of the web members connecting the upper and lower decks, though this is not as critical as for the vibration in the transverse x-direction.

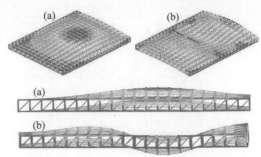

Figure 16.   Dominant modes in the global z-direction (vertical). Top row shows isometric views of the two modes (a) 0.717 Hz and (b) 0.880 Hz. Bottom row shows the front view (into y-direction).

Finally, Figure 16 shows the mode shapes for the dominant modes in the global z-direction (vertical). For vibration in the vertical direction, according to BS EN 1990: cl. A2.4.3.2 (BSI 2005), if the natural frequency is smaller than 3.0 Hz, then a verification of the pedestrian comfort criteria should be performed. From Table 5, both dominant modes exhibit frequencies less than 1 Hz, which necessitate considerable stiffening of the mega-truss in the vertical direction.

4   CONCLUSION

This paper presents the preliminary design and analysis of a steel mega-truss structure for the purpose of supporting a stack-up building on its deck. Several challenges which have been addressed include (1) external post-tensioning, (2) material selection, (3) deflection and (4) vibration control. It has been demonstrated that the existing design needs further improvement and refinement in the following areas:

1. Excessive deflection in SLS.
2. Natural frequencies in the lateral transverse and vertical directions need to be raised to meet Eurocode specifications.

In fact, it is conceivable that the above two issues are intertwined and interdependent. For example, a decrease in the vertical deflection by using steel members with a larger stiffness may also lead to an improvement in the vibration characteristics of the structure, and vice versa. The current research to date has focused on the basic structural form of the mega-truss. Future research will focus on resolving the above two major challenges.

ACKNOWLEDGEMENT

This research is supported by the Singapore Ministry of National Development Research Fund on Sustainable Urban Living (Grant No. SUL2013-4).

REFERENCES

BSI, 2005. *Eurocode 0: Basis of structural design.* London: BSI.

BSI, 2002. *Eurocode 1: Actions on structures – Part 1-1: General actions – Densities, self-weight, imposed loads for buildings* London: BSI.

BSI, 2006. *Eurocode 3: Design of steel structures – Part 1-11: Design of structures with tension components.* London: BSI.

BSI, 2004. *Hot rolled products of structural steels: part 6 technical delivery conditions for flat products of high yield strength structural steels in the quenched and tempered condition*, BS EN 10025-6, in, British Standards Institution, London.

Nordin, H. 2005. Strengthening structures with externally prestressed tendons: literature review. Luleå tekniska universitet, Luleå. Technical report/Luleå University of Technology, no. 2005:06.

Tubular Structures XV – Batista, Vellasco & Lima (eds)
© 2015 Taylor & Francis Group, London, ISBN 978-1-138-02837-1

Behaviour of bridging mega-truss tubular structures fabricated from S690 steel plates: Part II – Material aspects

K.W. Cheng, S.P. Chiew & C.K. Lee

School of Civil and Environmental Engineering, Nanyang Technological University, Singapore

ABSTRACT: Mega-trusses with stack-up structures have captured the attention of structural engineers and designers in recent years due to a growing demand for more usable land space and horizontal connectivity between high-rise buildings. Such new structural forms are able to intensify and optimize land use by creating space for functional buildings such as offices and commercial facilities over a mega bridging structure which could be located over busy expressways or roads. This study is the second of a two-part series focusing on the material aspects pertaining to reheated, quenched and tempered high-strength steel (RQT-S690) which has a strength-to-weight ratio almost twice that of mild steel (S355). A major challenge confronting the use of RQT-S690 in modern structures is its potential lack of deformation capacity, thereby leading to premature brittle fracture under as-weld conditions. This study experimentally investigates the effect of post-weld heat treatment (PWHT) on the residual stresses, ductility and ultimate strength of welded plate-to-plate joints. Test results in the form of global load-displacement curves at the braced ends and residual stresses at certain crucial locations close to the weld toe demonstrate that PWHT may be effective in reducing the residual stresses and enhancing the deformation capacity without significantly sacrificing the ultimate strength.

1 INTRODUCTION

Structural steel has been used in construction for more than a century. Responding to market demand for high strength construction materials, quenched and tempered high strength steel plates with yield strength more than 690 MPa were developed in the 1960s (Bjorhovde 2004). Notwithstanding its distinct advantage in achieving a higher load carrying capacity, research demonstrates that high strength steel suffers from insufficient deformation capacity and susceptibility to heat as compared to mild steel (Bjorhovde et al. 2001, Može & Beg 2011, Ricles et al. 1998, Coelho & Bijlaard 2007, Bhadeshia & Honeycombe 2006).

Welding, as one of the most important tools available to engineers in their efforts to reduce the production and fabrication costs, is of great economic importance for steel structures. In addition, greater freedom in design is also achieved by the proper use of welding. However, welding may lead to problems for the heat treated high strength steel which is inherited from the heat-vulnerable microstructures (Bhadeshia & Honeycombe 2006). Research has shown that the amount of residual stresses in welded quenched and tempered steel structures are high (e.g. Lee et al. 2012, Wang et al. 2012) and the deterioration of mechanical properties in the heat affected zone (HAZ) including strength, hardness and toughness is inevitable (Mohandas 1999).

This study investigates the effect of post-weld heat treatment (PWHT) on the tensile behavior of reheated,

quenched and tempered steel (RQT-S690) plate to plate joints fabricated by manual shield metal arc welding method. Joints with angles 45° and 90° and with different thicknesses are tested. The effects of PWHT on the residual stress, ductility and ultimate strength of the welded RQT-S690 steel plates will be quantified.

2 MATERIAL, SPECIMENS AND TEST SETUP

2.1 Material and specimens

Reheated, quenched and tempered high strength steel (RQT-S690) plates are used in this study. RQT-S690 has nominal yield strength of 690 N/mm², and tensile strength between 790 N/mm² and 930 N/mm² and elongation capacity around 15%. These RQT steel plates comply with the EN 10025-6 grade S690 specification (BSI 2004), which is approximately equivalent to the ASTM A514 steel (ASTM 2009). The mechanical properties of RQT-S690 acquired from standard coupon tensile test are shown in Table 1 in comparison with a randomly selected hot rolled rectangular hollow section manufactured to grade S355J2H. Although the ductility of RQT-S690 is sacrificed during the hardening processes, the extraordinarily high strength offers better serviceability under elastic stage.

The specimens of coupon are cut from 8 mm thick RQT-S690 HSS plates in the longitudinal direction. The configuration is designed according to EN 10025-5. The specimens are fabricated by welding two plates with dimensions of 440 mm × 150 mm × 8 mm (as

Table 1. Material properties of test specimens.

	$f_{2.0,n}$ (MPa)	$f_{u,n}$ (MPa)	E_n (GPa)	Elongation
S355J2H	461.6	535.1	204.5	34.2%
RQT-S690	817.1	849.8	201.3	14.7%
% difference	77.0%	58.8%	1.6%	57.0%

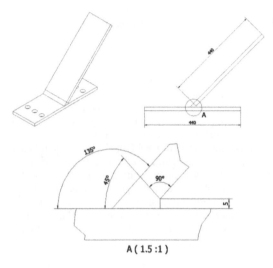

Figure 1. Configuration of the 45° joint.

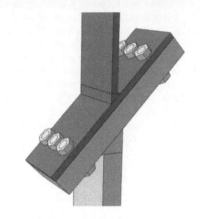

Figure 2. Tensile test setup.

well as 12 mm and 16 mm). The joints are designed according to the AWS structural welding code (AWS 2008). Two types of specimens are studied, i.e. RQT-S690 plate-to-plate joints in 45° and 90°. The configuration of the 45° joint is shown in Figure 1. Three bolt holes are drilled at both sides of the chord to fix the specimens in the test rig. The center-to-center distance between two rows of bolt holes is 290 mm.

In fact, it is expected that these two types of joints will be failed differently: the 45° specimens would fail due to tension in the chord plate, while the chord plate of the 90° specimen is actually taking bending loads. Nevertheless, the two types of loading are supposed to lead to tensile failure either at the weld toe or the bolt area.

2.2 Tensile test

The tensile tests are carried out in an INSTRON 8506 servo-hydraulic test machine with maximum loading capacity of 200 T. To fix the specimen into the uniplanner test machine, two types of support joints made of S355 with thickness of 45 mm were fabricated according to the same configuration. The specimens are fixed into the support joints by 6 high strength hexagon bolts in grade 10.9HR, M24, as shown in Figure 2. As the test progresses, the load-displacement relationship is monitored. A constant loading rate of 1 mm/min is applied until failure.

2.3 Post-weld heat treatment

PWHT is normally applied to mild steel weldment to remove residual stress, restore deformations during welding or improve the load-carrying capability in the brittle fracture temperature range of service. In fact, the beneficial effects of PWHT are not primarily due to reduction of residual stresses, but rather improvement of metallurgical structure by tempering and removal of aging effects. This process is widely accepted as beneficial for mild steel weldment since the microstructure, i.e. the mixture of pearlite and proeutectoid ferrite formed at temperatures above normal PWHT range, would be little altered unless the time of heating is prolonged or a too high treatment temperature is employed (Stout 1987). However, PWHT may introduce unpredictable changes into the microstructure of hardened steel weldment, which is extremely complicated and normally very sensitive to heat. This is why PWHT for quenched and tempered steel as well as cold work hardened steel is forbidden by AWS (clue 3.14, AWS 2008).

In order to fulfill the heat treatment task, the laboratory heating furnace Nabertherm LH216 with a maximum heating capacity of 1200°C is employed. The external and internal dimensions (width × depth × height) of the oven are 900 × 900 × 1200 mm and

Figure 3. Oven.

$500 \times 500 \times 700$ mm, respectively, as shown in Figure 3. Inside the furnace, the specimens are simply supported in the oven by three ceramic bars to release the thermal expansion or contraction during the heat treatment. Besides the thermometer measuring the temperature in the furnace, a thermocouple is attached to the specimens to monitor their inner temperatures, guaranteeing the temperature to be within the range of the designed values $\pm 3°$C below $600°$C and $\pm 5°$C from $600°$C to $1000°$C.

3 RESULTS

3.1 Stress-strain behaviour for PWHT HSS

To study the residual strength of RQT high strength steel after PWHT, a PWHT residual strength test is conducted. Standard coupon specimens are first gradually heated and cooled in the laboratory furnace and the residual strength is tested in ambient temperature condition. In this study, the targeted elevated or treatment temperatures are $400°$C, $600°$C, $800°$C, $900°$C, and $1000°$C. A constant heating time rather than constant heating speed is set herein, since different parts suffer to different extents in heat, yet the same period in fire conditions. After the preset temperature is reached, a ten-minute duration is maintained for the temperature to be stabilized and uniformly distributed. Subsequently, the specimens are cooled down naturally in the furnace until $300°$C when they are moved out and continue to cool in air. $300°$C is chosen as the critical temperature since steel's properties would not change below it according to the elevated temperature test results.

The stress-strain curves of RQT-S690 after heated and cooled down as shown in Figure 4. The temperatures in the legend refer to the highest temperature that the specimens attain in the furnace. Figure 4 shows that the mechanical properties are stable below $400°$C or above $900°$C. Otherwise in general, as the temperature rises, the yield and ultimate strengths of HSS keep

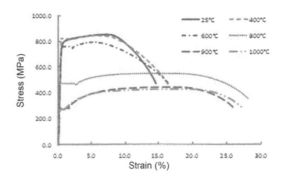

Figure 4. Stress-strain curves of RQT-S690 after heated to different temperatures.

Table 2. Parameters for PWHT.

Specimen Thickness (mm)	Maintaining time (minutes)			
	$600°$C		$570°$C	
	AWS	Actual	AWS	Actual
8	20	15*	38.4	20
12	30		57.6	
16	40		76.8	

*15min is the minimum holding time required by AWS. 90° T specimens only went through the 570°C post weld heat treatment.

decreasing with increasing ductility. When temperature reaches $800°$C, the largest elongation is obtained. $1000°$C is considered to be the end of this series test, since the difference between stress-strain curve of $900°$C and $1000°$C almost only exist in ductility.

Through the post-weld heat treatment study of the standard coupon test, it is found out that RQT-S690 high strength steel tends to turn back to the softer but more ductile parent steel after exposure to high temperatures. $600°$C is the critical temperature above which significant loss of strength occurs, with minor losses occurring between $400°$C and $600°$C. Therefore, $600°$C is chosen as the heat treated temperature used in following plate-to-plate joints post weld heat treatment.

Based on the PWHT procedure for normal strength steels specified by the AWS structural steel welding code, PWHT at $600°$C and a sub-critical temperature $570°$C are designed for RQT-S690. However, as it was demonstrated that maintaining the temperature at $600°$C for 10 minutes is sufficient to introduce noticeable changes to the mechanical properties of RQT-S690 base metal (Chiew et al. 2014), reduced maintaining time is used due to consideration of the fact that RQT-S690 is more vulnerable to heat compared to normal strength steel.

3.2 Residual stress study for PWHT-600 joints

PWHT is carried out for the 45° plate-to-plate joints. It takes one and a half hours to heat the specimens

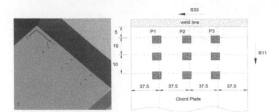

Figure 5. Strain gauge scheme on chord plate.

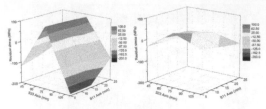

Figure 7. Residual stress for 12 mm specimen in S33 Direction: (a) as-weld; (b) PWHT-600.

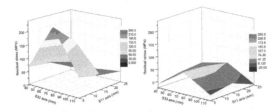

Figure 6. Residual stress for 12 mm specimen in S11 Direction: (a) as-weld; (b) PWHT-600.

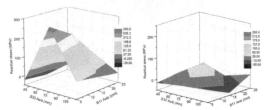

Figure 8. Residual stress for 16 mm specimen in S11 Direction: (a) as-weld; (b) PWHT-600.

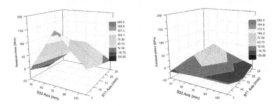

Figure 9. Residual stress for 16 mm specimen in S33 Direction: (a) as-weld; (b) PWHT-600.

from 25°C to 600°C and 15 min for the temperature inside the specimens to stabilize (see Table 2). A short maintaining time is chosen due to consideration of deterioration in the mechanical properties and embrittlement of quenched and tempered steel caused by long time tempering. Subsequently, the specimens are held within the furnace to be cooled slowly. To verify the relieving of residual stresses by heat treatment, two set of specimens (12 mm and 16 mm thickness) are chosen for the analysis of residual stress.

The standard ASTM hole-drilling method (ASTM 2008) is employed to measure residual stress in this study. This method can identify in-plane residual stresses by removing localized stresses and measuring strain relief in the boundaries of the drilled hole. Following ASTM, the drilling is limited to 2 mm from the surface of specimen by 8 successive drilling steps. Therefore, it causes relatively little damage to the specimen and allows localized residual stress measurements. In the designed locations, a hole is drilled in the center of the special strain gauge rosette, and then the released strain is recorded for further analyzing the residual stresses.

A special type of strain rosette FRAS-2 is used to measure the released strain of the specimen during drilling. As the residual stresses caused by welding are confined to areas near the weld, 9 points on the chord plate are chosen to capture the residual stress distribution near the weld toe. The strain gauge scheme is shown in Figure 5. For each point, both longitudinal (S11) and transversal (S33) residual stresses are measured. The test results of each point are shown in Figures 6–8.

Residual stress is induced during welding due to the highly localized heating and subsequent uneven cooling, non-linear material properties and thermal expansion/shrinkage. It could be seen from Figure 6(a) that the gradient of residual stress changes drastically

along both S11 and S33 directions. The transversal residual stresses distribution is generally symmetrical to the center plane perpendicular to the weld line. The values of residual stresses in the center plane are positive, which means the material is in tension. The stresses decrease steadily as it move toward the edges from the center plane. The distribution of longitudinal stress seems to be significantly affected by the distance to the weld line. In general, the stresses far away from the weld line are lower than those close to it. Figure 6(b) shows the residual stress distribution in longitudinal direction of the duplicate specimen heated to 600°C. It could be seen that the level of residual stresses after PWHT-600 is greatly decreased, while maintaining the general trend in distribution. The peak stress value drops to 60 MPa from 236 MPa.

The transversal residual stresses (in S33 direction) of both as-weld and PWHT-600 specimens are shown in Figure 7. It could be seen in Figure 7(a) that both the starting and ending positions of the weld line bear extreme compression with the starting position in greater compression, which seems due to the multi-pass welding with reaction stresses from extra distortion and subsequent mismatch as welding is being carried out. After heat treatment, with the stress redistribution and microstructural transformation, the compressive stresses are significantly reduced (Figure 7b).

Figure 8 shows the residual stress distribution perpendicular to weld line (S11 direction) for 16 mm 45° joints. The residual stress distribution in Figure 8(a) is similar to that in Figure 6(a) in that the center plane contains the largest tensile stress. At point P2 (5 mm from weld toe), the maximum residual stress for the specimen without heat treatment reaches 285 MPa. Although there is a singular point which might be caused by measuring error, it could be seen from Figure 8(b) that the residual stresses are reduced to a low level. The heat treated effect is significant for the 16 mm specimens.

3.3 Load-displacement behaviour

The first batch of tests only includes the 45° Y specimens in three different thicknesses (see Table 2). The employed PWHT methods did not change the failure type of those specimens, but did bring in certain changes to the load carrying capacity or the global ductility. All specimens, except for those in 8 mm, exhibited failure mainly due to the brittle fracture at the weld toe. Generally, although PWHT at 600°C for 20 minutes (PWHT-600) improved the deformation ability, or the global ductility of the specimens to some extent, the load carrying capacity at the same displacement level is compromised. However, an encouraging improvement on the maximum load could be seen for the 8 mm and 16 mm specimens, which motivate further tests for PWHT at a lower temperature (PWHT-570).

Compared to the effects of PWHT-600, the effects of PWHT at 570°C (PWHT-570) seems to be much gentler. For most of the loading time, the PWHT-570 just shared the same curve with the AW until the final stage. For the 8 mm specimens, only a small decrease in the final strength could be observed. The weld connection proves to be stronger than the load carrying capacity of the bolt area. Since the strength deterioration of the base material during PWHT is predicted according to the literature (Chiew et al. 2014), the slight decrease in the global load carrying capacity is still acceptable.

On the other hand, the PWHT-570 produced encouraging results in the 16 mm specimens. First, the load-deformation (L-D) curve of the PWHT went much further than that of the as-welded (AW) in terms of maximum load. Second, the failure mode of the PWHT started to change. For the AW specimen, the first cracks in the weld toe are followed immediately by a through thickness fracture (TTF). However, the first surface cracks appear much later in the PWHT-570 specimen and the specimen was still able to carry more load after that. Although the test ends at about 850 KN because the bolts fail, the displacement after the first cracks reached more than 15% of the total and the L-D curve still has the potential to go higher. Since the design shear load resistance of the bolts are supposed to be much higher than the weld connection, the

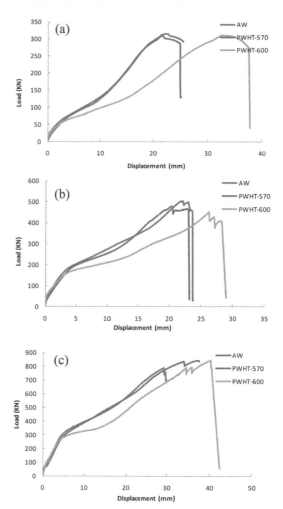

Figure 10. Load-displacement curves for the 45° Y-joints of thickness (a) 8 mm. (b) 12 mm and (c) 16 mm.

PWHT for the 16 mm is considered to be successful hence.

Unfortunately, the PWHT-570 seems to be too much for the 12 mm specimens. Earlier failure is triggered by the fracture of the HAZ. The evidence is that the first cracks are followed by a long almost-horizontal stage in the LD curve until the final TTF. We surmise that the notorious grain boundary growth in the microstructure by overheating may be blamed for the embrittlement of the HAZ (Kaplan & Murray 2008).

As stated before, during to the geometries of the specimens, the loading applied to the 45° Y and 90° T specimens are different. However, not much difference in the failure modes are observed. The weld connection of the 8 mm specimen was strong enough to bear loads until the bolt area failed. Therefore, it is reasonable to suppose that PWHT-570 only manages to lower down the ultimate strength in a negligible extent. The focus here is still given to the 16 mm specimen. It can be clearly seen that the specimens are still able to carry loads after the crack initiated in the HAZ for

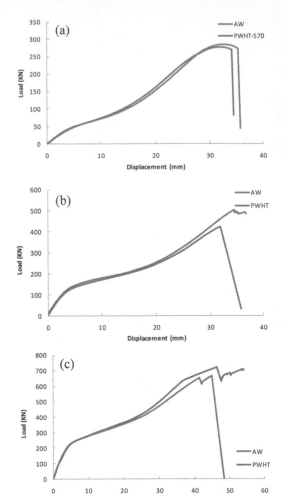

Figure 11.   Load-displacement curves for the 90° T-joints of thickness (a) 8 mm. (b) 12 mm and (c) 16 mm.

a long time. The test for PWHT-570 ends when the L-D curve reaches 700 KN again after cracking, due to safety considerations. Checking the specimen after the test reveals that the HAZ is only partially cracked at the middle of the weld toe, while necking is formed at the bolt area which is the sign of the failure type shifting, as shown in Figure 12. Accordingly, it can be confirmed that PWHT-570 is effective in improving the crack resistance in the HAZ and help avoid brittle failure of the 16 mm 90° T specimens.

However, the embrittlement of overheating seems to be exacerbated in the 12 mm specimen. In contrast to the 45° Y specimens, the 12 mm 90° T specimen failed suddenly in the form of TTF without any noticeable clue in the L-D diagram.

4   CONCLUSION

This paper presents the experimental investigations of the effect of heat treatment on the material properties

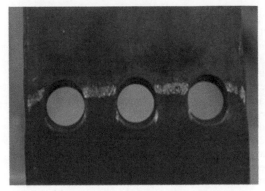

Figure 12.   Failure modes of PWHT-570: cracks in the HAZ (top) and necking at the bolt area (bottom).

of reheated, quenched and tempered high strength steel RQT-S690, as well as residual stress of plate-to-plate joints. The coupons are heated to high temperatures, which range from 400°C to 1000°C, and cooled down at ambient temperature. It is demonstrated that 600°C is the critical temperature above which serious deterioration of strength will occur, although above 85% of the nominal strength is still retained at that temperature.

The standard ASTM hole-drilling method is employed to measure the value of residual stress at a critical region of the joints. The experimental results indicate that while large tensile or compressive residual stresses could appear near the weld toe, the post weld heat treatment could significantly reduce the magnitude of these residual stresses.

The potential improvement in the ductility of welded plate to plate joints by PWHT techniques is also investigated. In total, 15 plate-to-plate specimens in different shapes or thicknesses are heat treated and subsequently tested under uniaxial tension test. Test results confirmed that PWHT could be useful in improving the load carrying capacity and deformation ability of the 16 mm plate-to-plate joints, and avoid sudden brittle fracture. However, overheating could be detrimental to the joints as the 12 mm specimens have demonstrated to be weakened after the same PWHT procedures. Finally, PWHT shows almost no influence on the 8 mm specimens, since they fail due to insufficient load resistance of the cross section capacity at the bolt area rather than at the weld connection.

ACKNOWLEDGEMENT

This research is supported by the Singapore Ministry of National Development Research Fund on Sustainable Urban Living (Grant No. SUL2013-4).

REFERENCES

ASTM E837-08, 2008. *Stardard test method for determining residual stresses by hole-drilling strain-gage method*, ASTM international, West Conshohocken, PA 19428–2959, United States.

ASTM, 2009. *Standard specification for high-yield-strength, quenched and tempered alloy steel plate, suitable for welding*, ASTM International, West Conshohocken, United States.

AWS, 2008. *Structural Welding Code–steel,* American National Standards Institute, Miami.

Bhadeshia, H.K.D.H. & Honeycombe, R.W.K. 2006. *Steels: microstructure and properties*, 3th ed., Elsevier Science & Technology, Oxford, United Kingdom.

Bjorhovde, R. 2004. Development and use of high performance steel, *Journal of Constructional Steel Research*, 60: 393–400.

Bjorhovde, R., Engestrom, M., Griffis, L., Kloiber, L. & Malley, J. 2001. *Structural steel selection considerations, Reston (VA) and Chicago (IL)*: American Society of Civil Engineers (ASCE) and American Institute of Steel Construction (AISC).

BSI, 2004. *Hot rolled products of structural steels: part 6 technical delivery conditions for flat products of high yield strength structural steels in the quenched and tempered condition*, BS EN 10025-6, in, British Standards Institution, London.

Chiew, S.P., Zhao, M.S. & Lee, C.K. 2014. Mechanical properties of heat-treated high strength steel under fire/post-fire conditions, *Journal of Constructional Steel Research*, 98: 12–19.

Girão Coelho, A.M. & Bijlaard, F.S.K. 2007. Experimental behaviour of high strength steel end-plate connections, *Journal of Constructional Steel Research*, 63: 1228–1240.

Kaplan, C.K. & Murray, G. 2008. *Thermal, Metallurgical and Mechanical Phenomena in the Heat Affected Zone*, John Wiley & Sons.

Lee, C.K., Chiew, S.P. & Jiang, J. 2012. Residual stress study of welded high strength steel thin-walled plate-to-plate joints, Part 1: Experimental study, *Thin-Walled Structures*, 56: 103–112.

Mohandas, T., Madhusudan Reddy, G. & Satish Kumar, B. 1999. Heat-affected zone softening in high-strength low-alloy steels, *Journal of Materials Processing Technology*, 88: 284–294.

Može, P. & Beg, D. 2011. Investigation of high strength steel connections with several bolts in double shear, *Journal of Constructional Steel Research*, 67: 333–347.

Ricles, J.M., Sause, R. & Green, P.S. 1998. High-strength steel: implications of material and geometric characteristics on inelastic flexural behavior, *Engineering Structures*, 20: 323–335.

Stout, R.D. 1987. *Weldability of steels*, fourth ed., Welding research council, New York.

Wang, Y.-B., Li, G.-Q. & Chen, S.-W. 2012. The assessment of residual stresses in welded high strength steel box sections, *Journal of Constructional Steel Research:* 76, 93–99.

Tubular Structures XV – Batista, Vellasco & Lima (eds)
© *2015 Taylor & Francis Group, London, ISBN 978-1-138-02837-1*

Compressive behavior of innovative hollow long fabricated columns utilizing high strength and ultra-high strength tubes

F. Javidan, A. Heidarpour & X.L. Zhao
Department of Civil Engineering, Monash University, Melbourne, VIC, Australia

J. Minkkinen
SSAB, Hämeenlinna, Finland

ABSTRACT: A comprehensive experimental study is carried out on the compressive behaviour of high strength innovative hollow columns. In terms of geometric appearance, the innovative square section of columns is the produced by welding steel tubes with nominal diameter-to-thickness ratio of 23.78 to mild steel plates with a nominal width-to-thickness ratio of 70. The tubes are of three types of mild, high strength (HS) and ultra-high strength (UHS) steel with nominal tensile strength of 890 MPa and 1380 MPa, respectively. The connection of high strength tubes to the sides of plates causes a satisfactory increase in the ductility and improves local buckling behaviour of each mild steel plate whilst the restraint provided by plates enhances the resistance of each tube against global buckling. This effective interaction between the elements leads to taking advantage of the high strength of tubes and, simultaneously, keeping the weight of the section as low as possible. Experimental outcomes of steel columns containing tubes are compared with similar columns without tubes, showing a significant increase in the peak compressive load and ductility.

1 INTRODUCTION

High strength (HS) and ultra-high strength (UHS) steel circular tubes are widely used in industrial productions due to their unique geometry and material characteristics such as strength, energy absorption, weight etc. The most wide spread utilization of high strength steel has been in automobile manufacturing, for instance its application in propeller shafts, suspension parts and doors under impact and crush conditions (Link and Grimm, 2005; Shunsuke et al., 2004; Tanabe et al., 1995; Vankuren and Scott, 1977). Accordingly, the above mentioned specifications of high strength steel can be proposed as an exceptional alternative for conventional structural material. The use of this material can lead to a great deal of reduction in the mass of consumed material and increase in the overall load bearing capacity. This is due to the fact that high strength steel and mild steel have proven to possess similar weights per equal volume. Published researches can be found on the employment of high strength steel circular tubes in structures and structural elements. Namely, cantilever roof structures (Gangopadhyay et al., 1990), stub columns consisting of high strength steel tube (Yang et al., 2010; Suzuki et al., 1984), gusset-plate welded connections in structural steel hollow sections (Ling et al., 2007a; Ling et al., 2007b) and CFRP strengthened butt-welded ultra-high strength circular steel tubes (Jiao and Zhao, 2004b). In terms of material properties, fracture-mechanical characterization

of high-strength steel tubes adopting numerical and experimental data was done with respect to the specimen geometry and also thermo-mechanical conditioning during the production of tubes (Wiedemeier et al., 2010). In terms of cost, a comparison among high strength steel tubes and normal steel tubes in realistic possibilities of building columns, considering the strength, stability and stiffness conditions, clearly indicated that using high strength steel in structures is more economic (Long et al., 2011). From another economic point of view, the use of high strength material leads to the consumption of less steel weight and longer life span of structure. Specifically, UHS steel material utilized in this study exhibits an increase of 2.5 times of life span than that of standard mill tubes (RAEX®).

In compressive loading conditions, HS and UHS tubes experience failure modes due to the brittle behavior of the material as well as global buckling depending on the geometry and slenderness. In fact, sections consisting of pure HS and UHS steel have lower ductility compared to mild steel which has an undermining effect on the global performance of the whole structure, especially under specific loading conditions such as seismic loading. Therefore, an innovative section consisting of HS or UHS tubes welded to mild steel plates has been proposed and investigated in previous studies (Zhao et al., 2004), having the benefits of utilizing a high strength steel material with highly acceptable compressive behavior in addition to having a convenient ductile behavior. Furthermore, the closed geometry of the proposed section consisting of circular tubes exhibits convenient performance under compression in terms of local buckling (Javidan et al., 2014). The behavior of short lengths of these innovative sections was investigated under various conditions such as compression loading and fire (Heidarpour et al., 2014; Rhodes et al., 2005; Van Binh et al., 2004). Other research works can be found investigating specifications of welded HS and UHS elements such as the effect of welding on the high strength material in the vicinity of weld known as the heat affected zone (HAZ). This has previously been the area of attention of a number of studies (Mazzina et al., 2013; Jiao and Zhao, 2004a; Shi et al., 2013).

The present research work, as opposed to previous literature available focusing only on stub columns, experimentally investigates the behavior of long innovative specimens under static compression. A comprehensive comparison is drawn out on the effect of tube material of mild, high strength and ultra-high strength steel incorporated in the section. On top of strength, material properties and economic specifications of high strength steel in general, this material is proven to be more sustainable compared to conventional steel. This study introduces the innovative sections in a comprehensive structural point of view and compares it with conventional welded columns.

2 SPECIMEN SPECIFICATIONS

2.1 Geometric specifications

The Innovative Fabricated Column (IFC) section welded from steel plates and tubes is shown in Figure 1. The plate elements used for IFC are mild steel with the width of 210 mm and thickness of 3 mm. Tubes with nominal outer diameter of 76.1 mm and thickness of 3.2 mm are used having three different material properties. In total, 14 specimens are tested in different lengths of one and two meters including individual tubes, conventional welded box sections and fabricated innovative columns. Geometric specifications of all test specimens are shown in Table 1. As for specimen labeling, **S** represents **s**ingle tube, **WB** presents conventional **W**elded **B**ox and **IFC** shows the initials of **I**nnovative **F**abricated **C**olumn. The letters that comes after the dash indicated the material type of the tube where mild tube, high strength tube and ultra-high strength tube are presented as MT, HST and UHST, respectively. The number at the end of each specimen label refers to the length of column. Slenderness ratios are calculated for each column by dividing the ratio of effective length to radius of gyration.

2.2 Fabrication, welding and imperfections

In the proposed fabricated section, fillet weld is used to connect the plates and tubes to each other. This welding configuration is more convenient in terms of cost and performing only on the outside of column. The welding

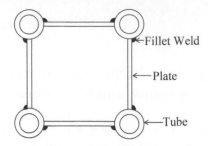

Figure 1.   IFC section geometry.

Table 1.   Geometric specifications of test specimens.

Specimen label	Length (mm)	Slenderness ratio
S-MT1	1000	19.4
S-HST1	1000	19.4
S-UHST1	1000	19.4
WB-1	1000	8.4
S-MT2	2000	38.8
S-HST2	2000	38.8
S-UHST2	2000	38.8
WB-2	1000	16.7
IFC-MT1	1000	2.3
IFC-HST1	1000	2.3
IFC-UHST1	1000	2.3
IFC-MT2	2000	4.5
IFC-HST2	2000	4.5
IFC-UHST2	2000	4.5

wire is 2.4 mm of the category AWS A5.9 ER2209. Type of welding is gas tungsten arc welding (G.T.A.W) with a 0.2% proof stress and tensile strength equal to 560–620 MPa and 800–835 MPa, respectively. The existence of imperfections has a direct effect on the compressive behavior of members and provision are made in design standards regarding the maximum value allowed for imperfections of different members in welded elements. Therefore, the out-of-flatness and out-of-square of the column section in the plates and tubes and the length deviations of each fabricated section is obtained using a precise measuring system. The center of each plate and tube in one end of the column is considered as a reference point for that member and measurements are done accordingly throughout the length. Besides, the distance of the reference points of plates and tubes are also measured relative to each other. These values meet the standard requirements and are also considered in the numerical modelling.

2.3 Material properties

Tension material tests are conducted for each steel type following the guidelines of ASTM-E8M-04 (Officials and Materials, 2004) and AS1391(Standards, 2007). The coupon configuration and dimensions for plates and tubes are shown in Figure 2. The longitudinal specimens extracted from tube samples have the required 90 degrees angle from the tube weld as outlined in

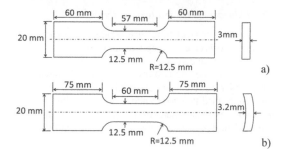

Figure 2. Coupon specimen configuration and dimensions; a) plate b) tube.

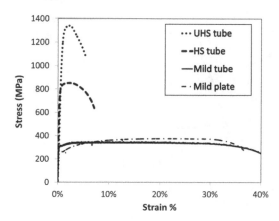

Figure 3. A stress strain curve for each material type.

Table 2. Mechanical properties of material.

Material	Elastic modulus (GPa)	Proof stress 0.2% (MPa)	Ultimate strength (MPa)	Ultimate strain
Mild plate	207	265	376	23%
Mild tube	198	305	342	11%
HS tube	215	785	841	2.5%
UHS tube	206	1216	1397	2.1%

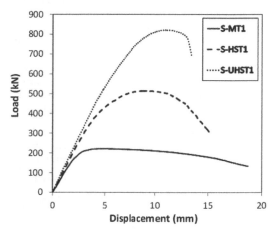

Figure 4. Load-displacement curve of 1 m single MT, HST and UHST specimens.

above standards. All tests were done in a Shimadzu tensile test machine with a constant grip displacement rate of 0.5 mm/min. Data acquisition is followed using both strain-gauges and the laser extensometer data to have sufficient understanding of yield and post-yield behavior. Due to the occasion of early failure in strain gauges, the laser extensometer provides the material results until failure. Stress versus strain curves read from the laser extensometer data of the average test specimen from the mild steel plate and three different steel tubes are shown in Figure 3. Along with the strength increase in HS and UHS materials, the ductility experiences a significant decrease. The stress values are obtained from precise measurement of each individual test coupon. The average of mechanical properties of mild steel plate, mild steel tube, high strength steel tube and ultra-high strength steel tube extracted from each of the test specimens are presented in Table 2. Yield strength has been determined based on proof strength using the offset method with a lay-off equal to 0.2% and the ultimate strain is the equivalent strain at ultimate stress.

3 EXPERIMENTAL TESTS

Testing equipment used in all column compression tests is a 5000 kN capacity Amsler machine with fixed ends and clamped conditions. Strain gauges were installed on column surfaces and Linear Variable Differential Transformers were used for reporting the displacements of the loading plate. Static loading was applied with a rate equal to 0.5 mm/min.

3.1 Single tubes

For further overview of the behavior of the tubes incorporated in the fabricated columns in terms of strength and ductility, all single tubes with different material properties and lengths have been individually tested subjected to static compression. For the aim of avoiding the member failure taking place in the vicinity of loading plates, fixed rings with the thickness of 25 mm were placed on both ends. Figures 4 and 5 compare the load versus displacement curves of the three types of columns with one and two-meter lengths, respectively. The peak load undergone by MT, HST and UHST with one meter length is 221.9, 514.5 and 821.6 kN, respectively and that for two-meter lengths is 204.5, 367.6 and 583.6 kN, respectively. The failure modes of the three specimens with one and two-meter lengths are also shown in Figure 6. The dominant failure mechanism of mild steel single tube is global buckling. However, with changing the material of steel from mild to higher strength steel, the ductility of failure experiences a noticeable difference. It is obvious from Figure 6 for one and two-meter specimens that the global buckling bending curve reduces in HST and UHST compared to MT. The ratio of maximum stress generated in single tubes to the ultimate material strength of each single tube specimen is calculated. The maximum stress is obtained from the load bearing capacity of each tube divided by the actual cross-sectional area. This ratio for MT, HST and UHST is equal to 0.83, 0.8, 0.77 for one meter and 0.76, 0.58 and 0.54 for two-meter tubes. These numbers indicate that ductility of tube has an effect on the compressive behavior reducing its capacity compared to the actual material capacity. It can also be understood that with the increase in the length of single tubes, this strength reduction is more obvious which reaches to about half the actual strength of the material. In such circumstances, it is highly beneficial to utilize the high strength of material in structures in a way that the brittle failure and buckling effects are reduced compared to the actual material capacity. This performance of high strength steel has been considered when it is incorporated in the proposed innovative sections.

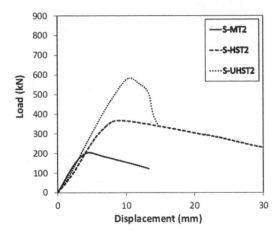

Figure 5. Load-displacement curve of 2 m single MT, HST and UHST specimens.

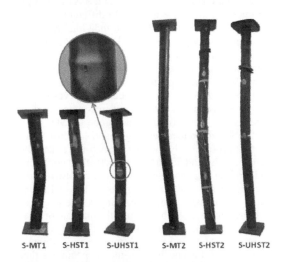

Figure 6. Failure mechanism comparison of one and two-meter single MT, HST and UHST specimens.

3.2 Fabricated one and two- meter columns

Four one-meter steel tubes are welded to corners of four mild steel plates to shape the innovative fabricated section as introduced in section 2.2. These columns are fabricated in one and two meters lengths. Behavioral variations resulted from material alteration of steel tubes from mild steel to HS and UHS steel tubes in one and two-meter specimen column groups are shown in Figures 7 and 9, respectively, in which the load versus

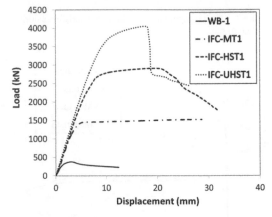

Figure 7. Load-displacement curve of one meter IFC column specimens.

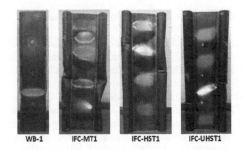

Figure 8. Failure mode of one meter IFC column specimens.

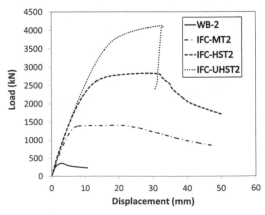

Figure 9. Load-displacement curve of two meters IFC column specimens.

displacements curves of each column is presented. Coupled with the strength rise, post peak behavior exhibits a difference in both one and two-meter column types when varying the tube materials. The failure mechanisms of three types of columns in two lengths of one and two meters are shown in Figures 8 and 10 respectively. Considering two very important specifications among all three tested columns of same length, the cross-sectional area and weight is kept constant. Therefore, the increase in the load bearing capacity of columns consisting of higher strength steel is significantly convenient in both design and economic point of view. Peak load undertaken by each of the test specimens is quantitatively discussed and comprehensively investigated in section 4.

4 RESULTS AND DISCUSSIONS

With the aim of having a realistic comparison among the behavior of the one meter innovative column with a member of the same cross sectional area, the superposition of four individual single tubes in addition to a welded box column is obtained. All individual plates and tubes incorporated in the superposition have similar geometries as the elements utilized in the innovative

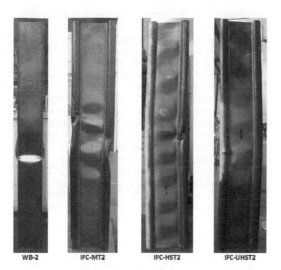

WB-2   IFC-MT2   IFC-HST2   IFC-UHST2

Figure 10. Failure mode of two meters IFC column specimens.

section. The results of compressive strength of single tubes, welded box sections, superpositions and IFC elements of three different tube steel material and two different lengths are presented in Table 3. The effect of global buckling in reducing the load bearing capacity of columns with the increase in height is observed from the results of the sixth column of the table. However, in case of fabricated sections, imperfections play a more important role in the behavior of columns within the category of one and two meters length. For instance, the maximum imperfection measured in the one meter fabricated column consisting of UHST was 1.32 times more than that of the two-meter one which justifies the higher compressive strength in the later. The IFC columns show a significantly higher strength compared to the welded conventional box sections. Calculating and interpreting the ratio of IFC compressive strength to the superposition of similar elements, shows extreme benefit when using the IFCs especially in higher lengths and consisting of higher strength steel tubes. This is due to the fact that the single tubes of HST and UHST, as mentioned in section 3.1, experience a failure with a quite higher brittleness. Consequently, welding the tubes to corners of mild steel plates provides a superior interaction among the two elements helping the high strength tubes take advantage of their high strength material. As presented in the last column of Table 3, IFC-HST2 and IFC-UHST2 show more that 50% beneficial behavior.

The load versus displacement curve of the proposed innovative column and the superposition of individual elements is extracted. Both above mentioned curves in addition to the compression test results of single tube and conventional welded box column for high strength and ultra-high strength steel materials are shown in Figure 11 and 12, respectively. The ductility of the IFC sections has also been taken into consideration against the ductility of conventional welded box section. Ductility factors (Park, 1989) calculated as the ratio of ultimate displacement (D_u) to yield displacement (D_y-which is the displacement where yielding first occurs in the system) are equal to 5.2 and 4.1 for IFC-HST1 and IFC-UHST1, respectively. Despite the slight ductility reduction of IFC-HST against IFC-UHST which is obvious because of the brittle nature of the material, both of these specimens show an advance in ductility compared to the conventional box section which has a ductility factor equal to 3.

Table 3. Compressive strength of all specimens.

Tube material	Length (meters)	Single Tube (kN)	Welded box (kN)	Superposition (kN)	IFC (kN)	Ratio: IFC to Superposition
Mild	1	221	378	1266	1523	1.20
	2	204	361	1179	1411	1.19
HST	1	514	378	2436	2909	1.19
	2	367	361	1831	2836	1.55
UHST	1	821	378	3665	4051	1.10
	2	583	361	2695	4126	1.53

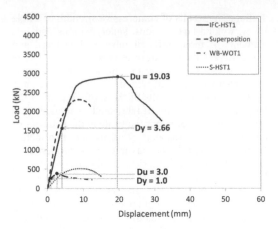

Figure 11. Load bearing capacity and ductility of one meter IFC-HST comparison against one meter single tube, welded box and superposition specimens.

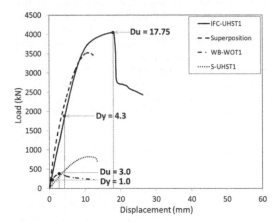

Figure 12. Load bearing capacity and ductility of one meter IFC-UHST comparison against one meter single tube, welded box and superposition specimens.

5 CONCLUSIONS

The outcome of this research work gives rise to the introduction of an innovative steel column taking advantage of the superior material properties of high strength and ultra-high strength steel materials in construction. To avoid the undesirable brittleness of the higher strength steel material, the proposed fabricated innovative section is welded to mild steel plates to advance the overall structural behavior to a more ductile system. The following conclusions are made based on limited experimental data:

- Utilizing high and ultra-high strength steel materials in construction will result in a significant increase in strength of the load-bearing elements while increasing the sustainability of the structure in terms of material consumption, life span, recyclability etc.
- Based on test results, sections fabricated from high strength steel tubes advance the load bearing limit

of the column to around two times the strength of a similar section consisting of mild steel tubes. This strength improvement will grow to a ratio of almost three times in the case where ultra-high strength steel tubes are utilized.
- Comparing the innovative column test outcomes to the quantitative superposition of four single tubes and a conventional box section with similar geometries and equal cross-sectional area shows a benefit of up to more than 50% in two meter columns of high and ultra-high steel tubes.
- Comparing ductility of the high strength innovative column with conventional welded box columns made of similar plate dimensions shows a satisfactory increase.

ACKNOWLEDGEMENTS

This project was supported by (i) Australian Research Council through Discovery Projects DP1096454 and DP130100181 awarded to the second and third authors; and (ii) SSAB steel company.

REFERENCES

Gangopadhyay, K. K., Guha, D. & Reddy, P. V. N. 1990. Salt Lake Stadium roof, Calcutta, with high strength steel tubes. *Structural engineer London,* 68(20), pp 397–404.
Heidarpour, A., Cevro, S., Song, Q.-Y. & Zhao, X.-L. 2014. Behaviour of stub columns utilising mild-steel plates and VHS tubes under fire. *Journal of Constructional Steel Research,* 95(0), pp 220–229.
Javidan, F., Heidarpour, A., Zhao, X.-L. & Minkkinen, J. 2014. The compressive behaviour of innovative hollow long columns consisting of mild steel plates and tubes. *7th European conference on steel and composite structures (EUROSTEEL 2014).* Naples, Italy.
Jiao, H. & Zhao, X.-L. 2004a. Tension Capacity of Very High Strength (VHS) Circular Steel Tubes after Welding. *Advances in Structural Engineering,* 7(4), pp 285–296.
Jiao, H. & Zhao, X. L. 2004b. CFRP strengthened butt-welded very high strength (VHS) circular steel tubes. *Thin-Walled Structures,* 42(7), pp 963–978.
Ling, T. W., Zhao, X. L., Al-Mahaidi, R. & Packer, J. A. 2007a. Investigation of block shear tear-out failure in gusset-plate welded connections in structural steel hollow sections and very high strength tubes. *Engineering Structures,* 29(4), pp 469–482.
Ling, T. W., Zhao, X. L., Al-Mahaidi, R. & Packer, J. A. 2007b. Investigation of shear lag failure in gusset-plate welded structural steel hollow section connections. *Journal of Constructional Steel Research,* 63(3), pp 293–304.
Link, T. M. & Grimm, J. S. 2005. Axial crash testing of advanced high strength steel tubes. *SAE Technical Papers.*
Long, H. V., Jean-François, D., Lam, L. D. P. & Barbara, R. 2011. Field of application of high strength steel circular tubes for steel and composite columns from an economic point of view. *Journal of Constructional Steel Research,* 67(6), pp 1001–1021.
Mazzina, R., Gomez, G., Solano, M., Perez, T. & Lopez, E. Study on weldability of high strength steel for structural applications. ASM Proceedings of the International Conference: Trends in Welding Research, 2013. 208–216.

Officials, A. A. S. H. T. & Materials, A. S. T. 2004. E8M-04 Standard Test Methods for Tension Testing of Metallic Materials (Metric)1. ASTM International.

Park, R. 1989. Evaluation of ductility of structures and structural assemblages from laboratory testing. *Bulletin of the New Zealand National Society for Earthquake Engineering,* 22(3), pp 155–166.

RAEX®, R. *Demand more from wear-resistant steels* [Online]. RUUKKI RAEX®REFERENCES. Available: http://www.ruukki.com.

Rhodes, J., Zhao, X. L., Van Binh, D. & Al-Mahaidi, R. 2005. Rational design analysis of stub columns fabricated using very high strength circular steel tubes. *Thin-Walled Structures,* 43(3), pp 445–460.

Shi, G., Jiang, X., Zhou, W., Chan, T.-M. & Zhang, Y. 2013. Experimental investigation and modeling on residual stress of welded steel circular tubes. *International Journal of Steel Structures,* 13(3), pp 495–508.

Shunsuke, T., Koji, S. & Akio, S. 2004. High strength steel tubes for automotive suspension parts - High strength steel tubes with excellent formability and forming technology for light weight automobiles. *JFE Technical Report,* 4), pp 32–37.

Standards, A. 2007. AS 1391-2007 Metallic Materials-Tensile testing at ambient temperature. Standards Australia.

Suzuki, T., Ogawa, T., Motoyui, S. & Kato, M. Plastic local buckling strength of high strength structural steel tubes. 1984. 227–233.

Tanabe, H., Anai, I., Miyasaka, A. & Tanioka, S. 1995. HAZ softening-resistant high-strength steel tubes for automobile propeller shafts,

Van Binh, D., Al-Mahaidi, R. & Zhao, X. L. 2004. Finite element analysis (FEA) of fabricated square and triangular section stub columns utilizing very high strength steel tubes. *Advances in Structural Engineering,* 7(5), pp 447–457.

Vankuren, R. C. & Scott, J. E. 1977. Energy absorption of high-strength steel tubes under impact crush conditions. *SAE Technical Papers.*

Wiedemeier, B., Sander, M., Džugan, J., Richard, H. A. & Peters, A. Fracure-mechanical characterization of high-strength steel tubes. 18th European Conference on Fracture: Fracture of Materials and Structures from Micro to Macro Scale, 2010.

Yang, L. Y., Li, Z. L., Wei, L., Kang, D. C., Ma, Z. Y. & Duan, B. 2010. An investigation of high strength steel tubes' ultimate load capacity under axial compression. *Xi'an Jianzhu Keji Daxue Xuebao/Journal of Xi'an University of Architecture and Technology,* 42(2), pp 201–204+210.

Zhao, X. L., Van Binh, D., Al-Mahaidi, R. & Tao, Z. 2004. Stub column tests of fabricated square and triangular sections utilizing very high strength steel tubes. *Journal of Constructional Steel Research,* 60(11), pp 1637–1661.

Tubular Structures XV – Batista, Vellasco & Lima (eds)
© 2015 Taylor & Francis Group, London, ISBN 978-1-138-02837-1

Innovative corrugated hollow columns utilizing ultra high strength steel tubes

M. Nassirnia, A. Heidarpour & X.L. Zhao
Department of Civil Engineering, Monash University, Melbourne, VIC, Australia

J. Minkkinen
Ruukki Metals Oy, Hämeenlinna, Finland

ABSTRACT: Individual tubular structures are weak in compression from buckling point of view. However, this limitation could be resolved if they are utilized in conjunction with other kind of structural elements. In the current study, the improvement induced by using Ultra High Strength (UHS) tubes in innovative corrugated columns is investigated. UHS tubes used at the corners have nominal yield stress of 1300 MPa. Three different corrugated plate configurations are introduced and fabricated so that the effect of corrugation parameters such as angle of corrugation and height of corrugated plate are experimentally investigated. The results prove the superior performance of proposed innovative columns under compression compared to the accumulated capacity of similar column without tubes and four individual tubes. Also, it is shown that using the tubes at the corner not only enhances the capacity of innovative corrugated columns compared to that of the column without tubes, but also it will enhance the performance of individual UHS tubes under compression.

1 INTRODUCTION

Aoki and Ji (2000) developed an innovative way of increasing the capacity of steel columns without sacrificing other properties such as significantly adding weight. The idea was to weld steel tubes to the apexes of square and triangular steel columns. It was concluded that the plates provide lateral restraint to the steel tubes which delay the buckling, and the total load supported by the fabricated sections was greater than the load obtained from summation of the capacities of individual components. This structural solution was further developed by Zhao et al. (2004) and Heidarpour et al. (2014) by replacing the mild steel tubes with very high strength (VHS) steel, and also recently by Heidarpour et al. (2013) by replacing the mild-steel tubes with stainless steel tubes. The welded plates as lateral supports of steel tubes can be replaced by self-strengthened plates like corrugated plates. Nassirnia et al. (2014) studied the fundamental behaviour of square welded columns comprising corrugated plates for further development of plate-tube interaction configuration.

Many construction materials such as steel, aluminum, etc., from material point of view, fall into isotropic category. However, certain materials display direction-dependent properties; consequently, these materials are referred to as anisotropic. In anisotropic materials stressed in one of the principal directions, the lateral deformations in the other principal directions could be smaller or larger than the deformation in the direction of the applied stress depending on the material properties. For a general anisotropic material, the matrix of material constants, because of symmetry, contains 21 independent constants. This means that all the strains are coupled to all the stresses. Some materials such as wood, plywood, delta wood, and fiber-reinforced plastics, etc., fall into this category. These materials possess natural anisotropy. Besides plates made of anisotropic materials, a number of manufactured plates made of isotropic materials also may fall into the category of anisotropic plates such as corrugated and stiffened plates, etc. Such a type of anisotropy is referred to as structural anisotropy (Ventsel & Krauthammer 2001).

2 PROPOSED INNOVATIVE COLUMNS

UHS tubes are not recommended to be used in structures individually; however, as proposed in this research, they can be utilized as structural elements. Proposed innovative sections are fabricated from corrugated plates which are welded to UHS tubes at the corners. In this configuration, even though the section is still ductile, thanks to UHS tubes, load carrying capacities of these columns are significantly high compared to conventional sections.

2.1 Corrugated plates

The corrugated plates, known as self-strengthened plates, are regularly produced from flat plates. The

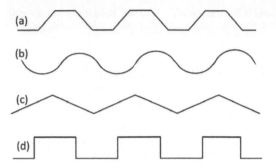

Figure 1. Versatile corrugation profiles: a) Trapezoidal, b) Sinusoidal, c) Triangular, d) Rectangular.

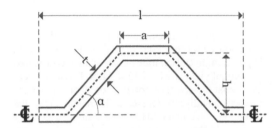

Figure 2. A corrugated module dimension parameters.

Table 1. Different corrugated plate types (dimensions).

	α (°)	a (mm)	h (mm)	l (mm)
Type I	45	20	15	70
Type II	75	25.7	15	59.42
Type III	75	18.44	22	48.67

corrugations increase the bending strength of the plate in the direction perpendicular to the corrugations, but not parallel to them. The profile of a corrugated plate may take several shapes: sinusoidal (approximated by tangent arcs), trapezoidal, triangular, or rectangular as shown in Figure 1.

The shape of profile has little influence on the performance characteristics of a self-strengthened plate; while we shall be concerned primarily with the depth of corrugation, plate thickness, and wavelength (Giovanni 1982). The most common profiles used are trapezoidal and sinusoidal; thus the former shape is chosen in this research.

2.2 Trapezoidal corrugated plates

As mentioned before, the strength characteristics of corrugated plates mainly depend on the dimension ratios of the profile parameters as shown in Figure 2.

In the current study, three different types of profiles with versatile parameters, described as Type I, Type II, and Type III in Table 1, have been defined in order to examine the effect of geometry ratios in fabricated innovative columns.

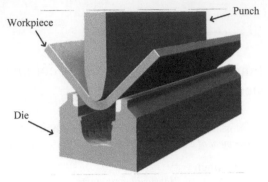

Figure 3. Press-brake forming machine.

Each corrugated plate is 3 mm thick ($t = 3$ mm) and has 3 modules of corrugations.

2.3 Corrugation process

Nowadays the corrugation process is carried out using the process of roll forming. This modern process is highly automated to achieve mass production with low costs. However, in the current study, due to using thick plates corrugated plates are produced through press-brake method. In this method which is a common mechanism of cold forming process, individual folds and corners are created from originally flat plate punched between a v-block and a die (see Figure 3). In general, cold working concludes with decrease in ductility and increase in yield and ultimate strength around folded regions (Karren 1967).

2.4 UHS steel tubes

UHS steel tube is produced from a durable high-strength structural material where Carbon, Silicon, Manganese, Phosphorus, and Sulphur form its main chemical compositions. The hollow tube designed for cost-effective applications, is 3 mm thick with nominal outer diameter of 76.1 mm. Utilizing individual tube as structural elements is discouraged due to its low performance under compression; however, with regard to high hardness and tensile strength properties, it shows outstanding performance while used in the configuration proposed in this paper. Despite their high hardness and strength, the weldability of UHS tube is good.

2.5 Welding procedure

In the current study, trapezoidal corrugated plates and UHS tubes are welded together to form an integrated square column. The fillet weld used in this work is produced based on Gas Tungsten Arc Welding (G.T.A.W) procedure, utilizing ER2209 wire. The welding process was completed by Crossline Engineering Pty Ltd. A schematic view of innovative fabricated column with corner tubes is depicted in Figure 4.

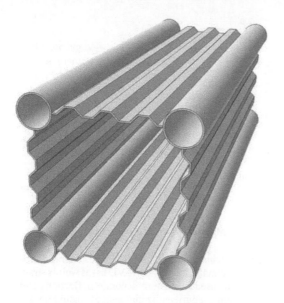

Figure 4. An innovative fabricated corrugated column.

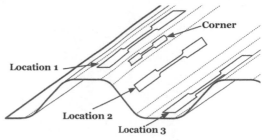

Figure 5. Location dependent tensile coupons.

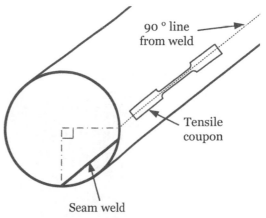

Figure 6. UHS tube tensile coupon location.

3 EXPERIMENTS

Along with the experimental results of innovative corrugated columns with corner tubes, another column called control column which is consisting of unfolded flat plates welded to the UHS tubes at corners is also fabricated to investigate the effect of corrugation on the compressive strength. In addition, the experimental results are compared with those of corrugated columns with no tubes at the corners.

The corrugated plates are folded from flat mild steel plates. The virgin steel plate is chosen from Grade 250 steel. It is noted that for UHS steel tubes, the nominal tensile strength is 1250 MPa.

3.1 Structural elements material properties

In order to find out the mechanical properties of the materials used in the proposed innovative columns, tensile tests are conducted on the coupons taken from corrugated plates and UHS tubes as described below.

For corrugated plates, material properties of folded regions are enhanced during cold forming process. The intensity of the enhancement depends on the location where the coupon specimen is taken from. Therefore, four types of coupon specimens, as shown in Figure 5, are taken from the nominated locations as follows: Location 1 (top flat face) and 3 (bottom flat face) specimens which are taken from horizontal strips at top and bottom strips, respectively; Location 2 (side flat face) specimens which are taken from inclined strips, and corner specimens which are taken from the edge fillet produced during folding process. The corner test specimens are prepared in accordance with the Australian standard (Standards Australia AS1391 2007) while the rest of specimens are made based

on ASTM standard (ASTM E8 2013). The location-based mechanical properties obtained from tensile test have been averaged from three specimens taken from the same location. 36 tensile tests were conducted on the coupon specimens taken from all three types of the corrugated plates. Moreover, three tensile tests were carried out on three coupon specimens taken from the virgin material. The results show that corner material properties enhancement may affect the mechanical properties of the flat regions. Therefore, it is proposed to extend corner properties to adjacent flat regions, approximately at a distance of t away from each side (Karren 1967).

On the other hand, for the current UHS tubes, material properties of the tube depending on the distance from the weld line may vary. Therefore, the tensile coupons are taken from a location at approximately 90° from the weld line, as recommended by standard (ASTM E8 2013).

Both curved end sides of coupons taken from UHS tubes are flattened without heating (no touch on gauge length) to be tested under tension in conjunction with flat grips.

3.2 Material scale experiments

In Figure 7, a close view of a coupon test setup under active test is shown. The displacement control method is produced by the Instron 4204 testing machine available in Civil Engineering Laboratory,

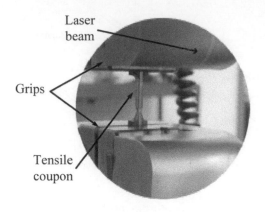

Figure 7. Tensile coupon test setup.

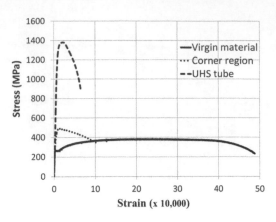

Figure 9. Material properties comparison.

UHS tube   Innovative column

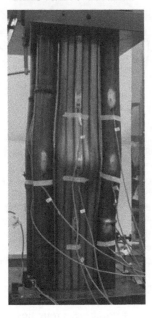

Figure 8. Individual tube and fabricated column under compression.

Monash University. Even though two strain gauges are attached at the centre of each coupon side, a non-contact MTS Laser extensometer is also utilized to capture strain values, particularly once strain gauges are failed over higher deformations.

3.3 *Large scale experiments*

UHS steel tubes are tested under compression loading either individually or fabricated in innovative corrugated column. For individual tube tests, 2 collars are used as support for both ends of tube.

Figure 8 demonstrates individual UHS tube and Type III innovative fabricated column under compression. The tests are conducted using a 5000 kN Amsler

Testing Machine. In the compression tests, linear variable differential transformers (LVDTs) as well as strain gauges are used to capture accurate deflections and local strains. Fourteen strain gauges including six strain gauges at middle column and four strain gauges at a distance of 250 mm from either top or bottom end, are attached on each corrugated column whilst twelve gauges are attached to the control column. All strain gauges are attached to a data acquisition system and the data is recorded at points on the outer surface of the plate. For the displacement control compression tests, the rate is adjusted to be 0.5 mm/min.

4 RESULTS AND DISCUSSION

The fabricated innovative columns proposed in this research are 1-meter long. The boundary conditions of supports are assumed to be clamped.

Test result data of tensile coupons shows that Young's modulus and yield stress for the virgin material are 209 GPa and 262 MPa, respectively. For UHS tubes, 0.2% proof stress and ultimate tensile stress are 1250 MPA and 1450 MPa, respectively.

4.1 *Material properties comparison*

Figure 9 shows the stress-strain curves for the coupons taken from UHS tubes, virgin material and also corner of corrugated columns. It is worth noting that the stress-strain curves for the coupons taken from locations 1, 2 and 3 of the corrugated plates are similar to that of virgin material.

It can be seen from Figure 9 compared to virgin mild steel, the UHS tubes are very brittle, but have higher ultimate tensile stress.

4.2 *Individual UHS tube*

The compression test results of 1m individual UHS tubes show that they can carry up to 820 kN axial force. The load versus strain curve of the tube is drawn in Figure 10.

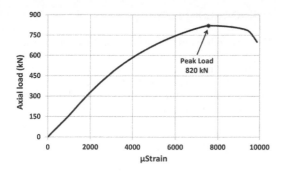

Figure 10.   Load vs. strain curve for UHS steel tube.

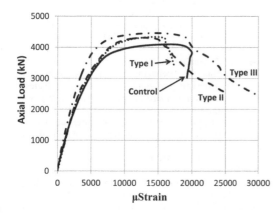

Figure 11.   Experimental load vs. strain curves (Innovative columns with corner tubes).

Table 2.   Column case number definition.

	Description
Case 1	Control column
Case 2	Type I innovative column
Case 3	Type II innovative column
Case 4	Type III innovative column
Case 5	Conventional welded column equivalent to case 2
Case 6	Conventional welded column equivalent to case 3
Case 7	Conventional welded column equivalent to case 4

4.3   Innovative fabricated column

Figure 11 illustrates the experimental curves of load versus strain during compression tests for corrugated columns as well as control column with corner tubes. The axial strain was measured from strain gauges and LVDTs. Since the strain gauges might not be able to give accurate strain values after buckling, the strain values from the strain gauges are only used for curve plotting before the occurrence of column buckling. After that, the axial strain was calculated from the measured axial shortening divided by the overall length (L) to evaluate the post-buckling behaviour of the test specimen (Tao, Han & Wang 2008).

It can be seen that energy absorption of the columns is descending from Type III to Type I, and then control test. It is seen that depending on the nature of application where a hollow section is required, conventional columns could be replaced by suitable corrugated columns which have higher capacity for absorbing energy. It is worth mentioning that this development is currently investigating in Civil Engineering Department at Monash University.

Also, as shown in Figure 11, by comparing the area surrounded under each curve up to peak point, it is clearly understood that corrugation has increased energy absorption and ductility, notably in Type III column.

Comparing mechanical behaviour of innovative corrugated columns with UHS tubes with those of corrugated columns without corner tubes recently reported (Nassirnia et al. 2014) shows that innovative columns with tubes have dramatically higher performance. For instance, the compressive load which is carried by Type I column with tubes is 5.5 times greater than that of the same corrugated column without corner tubes. In addition, incorporating UHS tubes with flat side plates in a column (control column with corner tubes) can increase the compressive load capacity 11 times greater than that of similar column without corner tubes.

4.4   Weight and cost analysis

As discussed earlier, experimental results demonstrate excellent mechanical performance of proposed innovative columns with corner UHS tubes compared to conventional hollow columns.

For industrial purpose, the weight and cost comparison of proposed sections compared to that of conventional sections might be a point of interest. Thus, seven different cases are studied as listed in Table 2. All column cases in this table are assumed to be 1 m long.

Cases 1 to 4 are the columns examined and tested in the current study. Cases 5, 6, and 7 are referred to corresponding conventional square hollow columns with the same load carrying capacity and coverage area to that of cases 2, 3, and 4, respectively. In order to find out desirable plate thickness in cases 5 to 7 for which 'the aforementioned criteria are satisfied, the regulations and formula in Australian standard is utilized (Standards Australia AS4100 1998). The calculations show that equivalent columns should be 10.5 mm thick, around 3.5 times thicker than plate thickness in innovative columns.

By comparing fabrication cost of corrugated column with corner tubes (cases 2-4) and thick conventional columns (cases 5-7), it can be concluded that

Generally, global buckling mode is the governing failure mode of the single UHS tubes. However, by incorporating them at the corners of corrugated column, this issue is resolved such that the strength of the ultra-high strength materials is fully utilized.

around 20% of raw material cost of column can go for the corrugation process (leading to the same fabrication cost), whilst innovative columns are much lighter.

5 CONCLUSIONS

This paper has discussed the post-buckling behaviour of the innovative corrugated columns with UHS corner tubes. The test experiments were conducted not only on the large scale, but on the material scale as well.

Three different types of corrugated plates were introduced. The plates were welded to UHS tubes at the corners to form a square hollow section. Another hollow column similar to innovative columns but with flat plates (called control column) was considered to show the performance of proposed sections.

From experiment results, it can be concluded that proposed innovative columns can boost bearable axial loading by 10% comparing to control columns. The innovative corrugated columns with tubes can carry axial load up to 5 times higher than that of corrugated columns without UHS tubes carry.

In addition, the loading capacity of innovative columns with corner tubes is higher than the accumulated capacity of individual elements consisting of a corrugated column and 4 individual UHS tubes, which shows superior performance of proposed sections in resolving buckling limitation of UHS tubes.

Apart from mechanical performance of innovative columns, the proposed columns are competitive in terms of cost and weight compared to the sections already available in the civil market.

ACKNOWLEDGMENT

The research work presented in this paper was supported by Australian Research Council through Discovery Projects (DP1096454 and DP130100181) awarded to the second and third authors. The authors wish also to thank Ruukki Corporation in Finland for providing steel materials.

REFERENCES

ASTM E8 2013, 'Standard Test Methods for Tension Testing of Metallic Materials', *Annual book of ASTM standards*.

Giovanni, D 1982, *Flat and corrugated diaphragm design handbook*, vol. 11, CRC Press.

Karren, K 1967, 'Corner properties of cold-formed steel shapes', *Journal of the Structural Division*, vol. 93, no. 1, pp. 401–32.

Nassirnia, M, Heidarpour, A, Zhao, XL & Minkkinen, J 2014, 'Stability Behavior of Innovative Fabricated Columns Consisting of Mild-Steel Corrugated Plates', in *7th European Conference on Steel and Composite Structures*, Naples.

Standards Australia AS1391 2007, 'Metallic materials – Tensile testing at ambient temperature', *New South Wales, Australia*.

Standards Australia AS4100 1998, 'Steel Structures', *New South Wales, Australia*.

Tao, Z, Han, L-H & Wang, D-Y 2008, 'Strength and ductility of stiffened thin-walled hollow steel structural stub columns filled with concrete', *Thin-Walled Structures*, vol. 46, no. 10, pp. 1113–28.

Ventsel, E & Krauthammer, T 2001, *Thin plates and shells: theory: analysis, and applications*, CRC press.

Tubular Structures XV – Batista, Vellasco & Lima (eds)
© *2015 Taylor & Francis Group, London, ISBN 978-1-138-02837-1*

New experimental determination of fatigue strength of tubular truss joints in steel grades up to S690

S. Herion & T. Ummenhofer
Research Centre for Steel, Timber and Masonry, Karlsruhe Institute of Technology KIT, Germany

M. Veselcic
CCTH – Competence Center for Tubes and Hollow Sections, Karlsruhe, Germany

F. Zamiri & A. Nussbaumer
Swiss Federal Institute of Technology Lausanne, ICOM – Steel Structures Laboratory, Lausanne, Switzerland

ABSTRACT: In this paper research work on welded butt weld connections between hot-rolled seamless tubes and steel castings are presented. Comparison is made between different weld preparations for the butt welds including different steel grades, repair methods, the influence of residual stresses and a study on large scale trusses. The steel grades investigated include S550QH as well as S690QH to complement previous test results on lower steel grades. These investigations are intended to lead towards the development of a concept for improving the fatigue life for bridges using such end-to-end connections.

1   INTRODUCTION

The outcome of two research projects funded by CIDECT and FOSTA (a national steel research funding organization in Germany) were presented at previous ISTS conferences, at ISTS-12 in Shanghai (2008) by Veselcic et al. and at ISTS-13 in Hong Kong (2010) by Nussbaumer et al.

Due to the very convincing results from these projects an extension of the work was carried out by a research group consisting of ICOM of EPFL Lausanne, the Competence Center for Tubes and Hollow Sections (CCTH), Karlsruhe, the Büro für Ingenieurarchitektur Dietrich in Traunstein and the Karlsruhe Institute of Technology KIT. The project was called *"P816 - Optimal application of hollow sections and cast steel nodes in bridge buildings with the usage of steel S355 up to S690"* and, as the previous project, sponsored by the German Forschungsvereinigung für Stahlanwendung FOSTA in Düsseldorf. The fatigue classes in both preceding projects were determined on the basis of a lower bound of the experimental dataset due to the little number of available results. To obtain more results and thus to enlarge the dataset for statistical evaluation, additional tests on previously investigated steel grades S355J2H and S460NH were carried out.

Additionally, new cast to CHS girth welded test specimens made of hot-rolled seamless hollow sections of steel grades S550QH and S690QH and the matching cast steel material were produced. The specimens were tested under 4-point bending as well as within complete truss girders. With this enlargement of the fatigue test database the influence of several parameters such as steel grade and tube thickness can be assessed.

This paper summarizes the whole project and presents the test results and the re-evaluated S-N-curves. One major finding is the confirmation of the relative independence of steel grade (from both CHS and cast steel) on the fatigue strength of end-to-end connections in tubular truss joints.

All these investigations will help allow for safer and more economical design of bridges made of thick-walled hollow sections and cast nodes.

2   RESEARCH PROJECT

2.1   *Overview*

Based on the previous project the new research work was planned as a combined work of Karlsruhe Institute of Technology (KIT) and EPFL Lausanne. With this it was possible to investigate a lot of different aspects on test specimens under fatigue loading of tube to tube connections between cast steel joints and steel hollow sections.

Not all aspects of the research project are included in this paper, namely: architectural aspects of tubular bridges which were studied by the office of engineering-architecture Richard Dietrich, "integral planning" of bridges made of tubular steel structures which was enhanced and time-tested and the economical aspects of material, welding, non-destructive-testing, etc. which were also studied.

In this way, the economic value for the erection of such structures can be optimized and therefore a realization in practice will be promoted.

3 ANALYSIS OF BUTT WELDED JOINTS

3.1 Overview

Butt welds are used in bridge constructions in multifaceted varieties as a connection of two chords or between cast steel joint and steel tube. These connections are butt welds with or without weld backing. All welds can be carried out as single side welds only. Within the scope of this project different variants for butt weld designs are investigated.

Since the chemical as well as the mechanical-technological properties of the used cast steel parts are nearly the same as for the hot-rolled hollow sections, all following assumptions are not only valid for butt welds between hollow sections and cast steel parts but furthermore for butt welds connecting hollow sections also.

The tests carried out on butt-welded specimen have been performed in close cooperation between Karlsruhe Institute of Technology (KIT) and the Center of competence for tubes and hollow sections (CCTH). On the basis of the test results derived from project P591(2010) and in close collaboration with EPFL Lausanne and the involved participants of the industry in the monitoring group of the project the research program for the examination of the butt welds was proposed. The objective was to reduce the previously investigated variants to the decisive ones and maintain a wide spectra of possible butt welded joints to perform an economic feasibility study with the different variants.

In the first part the testing program with the different variant is presented. Subsequently the chosen materials are given and the fabrication of the test specimens is described. Special attention is placed on the non-destructive examinations. Furthermore, repair welding on different test specimens is considered. One of the main aspects of the project was on the investigation of high strength steels. These test results are illustrated with a detailed analysis.

Furthermore, FE-calculations concerning the influence of the different wall-thicknesses and geometries have been performed. However, these results are not part of this paper.

3.2 Testing program

Detailed information on the manufacturing processes was presented in former publications mentioned above. Details on quality levels and welding parameters mentioned in this section are described in detail in Veselcic et al. (2006, 2007, 2009).

For the research projects steels according to Table 1 are considered for the tests. With the use of high strength steel, a reduction of the member thickness can be realized in practice. To ensure good weldability, the cast quality is chosen according the previous projects. For the welding, pre-heating was only used for the steel grade S690. For all other steel grades pre-heating could be omitted, which entails a significant cost reduction.

Altogether, four butt weld solutions have been condensed from the previous project (see figure 1). The testing programme for all four solutions was identical.

A summary of the section dimensions of all tests on butt welds is given in Table 2. The influencing factors

Table 1. Steel Grades for the Hollow Sections and corresponding Cast Steels.

Hollow Sections	Standard	Cast Steel	Standard
S355J2H	EN 10210	G20Mn5(V)	EN 10293
S460NH	EN 10210	G10MnMo V 6-3	EN 10293
S460NH	EN 10210	G10Mn7V	EN 10293
S550QH	prEN 10210	G10Mn7V	EN 10293
S690QH	EN 10210	G10MnMo V 6-3	EN 10293

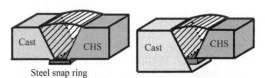

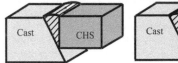 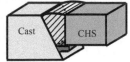

Solution 2 : backing using steel ring

Solution 3 : beveled, with tack welded backing steel ring (t = 3 to 4 mm)

Solution 5 : no backing ring and not beveled

Solution 6 : with tack welded backing steel ring and not beveled

Figure 1. The four different variants of butt weld solutions studied.

Table 2. Testing Program for all four solutions.

Outside diameter Ømm	Cast steel wall thickness mm	CHS wall thickness mm	No. of tests	Steel grade
193.7	20/30*	20	2	S460
			2	S550
			2	S690
298.5	30/40 *	30	2	S690

* depending on chosen solution.

studied were the diameter and thickness of the tubular beams and the detailing of the CHS-castings butt welds. In solution 2 the cast steel wall thickness is chosen to be the same size as that of the hollow section to be welded to it. For the other solutions 10 mm is added to the cast steel components.

All fatigue test specimens were made out of two CHS and one cast steel member, forming tubular beam specimens. The tests were carried out under 4-point bending as shown in Figure 2. Altogether 32 tests have been performed on high strength steel.

After welding the test specimens, an ultrasonic inspection of the different variants is carried out to detect any defects. However, for practical applications, it should be pointed out that with a multitude of regulations for ultrasonic inspection, the necessary requirements should be specified at an early stage of a project to avoid subsequent disagreements between the client and contractor and to maintain quality standards.

For the investigations three main aspects were taken into account.

The first aspect was an extension on the existing data concerning steel grades with $f_y = 355$ MPa and $f_y = 460$ MPa. For this, additional to the tests of specimens made of S460, further tests with steel grades S550 and S690 have been performed. These steel grades were also realized for the cast steel parts. With this a large scope, also with respect to future constructions could be covered.

Secondly, failed beam specimens from the preceding project were repaired and examined again. Further tests on the available specimens can lead to an extended fatigue life and higher detail categories. Additionally, the fatigue resistance of the repaired weld should be

Figure 2. Test rig for 4-point bending fatigue tests on beam specimens diameters 193.7 and 298.5 mm.

investigated. With repair welding, the life expectancy of bridge structures can be extended.

3.3 Tests on repair welded beam specimens

On test specimens previously tested repair welding should be conducted to study the influence of a repaired weld. Furthermore, as always two welds were simultaneously tested with only one failed in the tests, another test result can be obtained for the weld that not failed yet.

Prior to the selection of the specimens to be repair welded all suitable specimens have been ultrasonically tested. Based on this the specimens with small defects in the still intact weld could be excluded Only the specimens with none or small defects are taken into account for repair. Otherwise the weld would crack before the repair welded weld has any significant cycles to fatigue. The failed weld was tested with magnetic particle testing to obtain the full length of the existing crack. For repairing the crack was grind to provide a good quality.

After repair welding the fatigue tests were continued. The parameters used previously have not been changed. In total, 4 fatigue tests (with 4 repaired welds and 4 still intact welds) were performed. The test frequency was between 20 Hz and 30 Hz (depending on the stiffness of the specimen) with a load ratio $R = Q_{min}/Q_{max}$, of 0.2.

For each test, further applied cycles to fatigue the specimen was determined. The tests were stopped when either the repaired weld or the previously intact weld cracked. For two test specimen a crack occurred in both welds. For these specimens the repaired weld featured lower cycles to fatigue as for newly welded welds.

Both other test specimens showed a crack only in the previously intact weld. The endured cycles to fatigue was more than 50% higher than previously obtained. Die test results for all specimens are presented in Table 3. A proposal concerning further fatigue life of repair welds could not be derived based on only two test results. For this further tests are necessary.

3.4 Tests in high strength steel

For all fatigue tests specimens with two welds have been investigated. All fatigue tests have been stopped, when a crack through the wall-thickness (through crack) was visually observed. All failures for tests observed started from inside at the weld root with the

Table 3. Test results on beam specimens after repair welding.

specimen	previous cycles of fatigue	Repaired weld loading S_R	Intact weld cycles of fatigue	cycles of fatigue	crack
V3-B1	1.193.693	170 MPa	232.365	1.426.058	Both
V5-B1	1.838.046	170 MPa	255.735	2.093.781	Both
V5-B2	209.471	233 MPa	153.069	362.540	Existing weld
V6-B1	434.024	170 MPa	226.188	660.212	Existing weld

crack propagating in the weld area or the heat affected zone.

In total, 32 fatigue tests (with 64 welds altogether) were carried out at load ratios $R = Q_{min}/Q_{max}$, of 0.2.

3.5 Evaluation of test results

All fatigue classes determined in the evaluation are only suggested as a lower bound of the experiments due to the number of available testing data. With more experimental results using the existing and only partially broken test specimens.

Since the specimens contained two butt welds, the results are biased towards the weaker joint, allowing a conservative safety margin for the evaluations. Also, since in the bending tests only part of the joint is subjected to the highest stress ranges and the weld root is less loaded, the expected fatigue life is obviously higher than in the case of tension tests, where the weakest joint part will fail first.

The failure criteria used for determining S-N curves is through-thickness cracking. As expected, the bending tests give good fatigue strengths, with characteristic nominal stress range values at 2 million cycles, $\Delta\sigma_C$, ranging between 100 and 120 N/mm2 and a fatigue slope coefficient m = 5.

The fatigue test results are summarized in Table 4. Here not only the test results obtained in this project are described but furthermore they are compared with the previous results.

It is notable that for welding the high strength steel the correspondent filler material is chosen respectively. For some tests however the filler material used had lower yield strength. In an evaluation of all specimens no negative effect of the yield strength was observed. The evaluation of the specimens with filler material of lower yield strength leads to a higher detail category in comparison to the specimens with adequate yield strength. Therefore the filler material had only a small or nearly no influence on the test results.

According to Eurocode 3, part 1-9 (2005), only solution 2 (see Table 4, grey marked section) can directly be classified. It is a transverse butt weld, full penetration, between curved plates (EN1993-1-9, Table 8.3, detail 14), with a fatigue strength or corresponding detail category of 71 N/mm2 and m = 3 (with size effect reduction if t > 25 mm). Note that these joints cannot be classified using EN1993-1-9, Table 8.6 (hollow sections), detail 3, since here the allowable range is only for t < 12.5 mm. All tests and statistical analyses show higher slopes, usually closer to m = 5.

Solution 2 and 5 in both projects give comparable results. With this no difference in steel quality concerning the fatigue life is observed. The other solution 3 and 5 give slightly lower results (approximately 5%).

In a first analysis, the results on the different steel grades were pooled together, i.e. it was agreed that there was no significant difference in fatigue behavior between the different steel grades. Also, all specimen sizes could be pooled together as no significant size effect was noticed apart from the effect of additional bending due to misalignment. (see also Nussbaumer et al. 2013).

However, in a second analysis the test results have been evaluated according to their steel grades (S355, S460, S550 and S690). With this all test results for one steel grade have been pooled together for evaluation. The parameters of this evaluation are presented in table 5.

It stands out that in the statistical evaluation the difference of the 95% survival probability compared to the mean stress for the steel grades S355, S460 and S690 is roughly 20%, with a minimum characteristic value of approx. 110 N/mm², m = 5.4. In the case of S550 the difference is nearly 40%. This could possibly be attributed to defects in the welding. Furthermore the small number of tests has to be taken into account. The slope ended up with m = 3.6 which also gives the

Table 4. Test results for all solutions on butt welds.

		Project P591		Project P816		
Solution		lower boundary	Number of tests	lower boundary	Additional test results	Total test results
		$\Delta\sigma_C$ and m		$\Delta\sigma_C$ and m		
1		122 (m=5)	4	-	-	4
2		119 (m=5)	14	120 (m=5)	8	22
3		128 (m=5)	4	120 (m=5)	9	13
4		130 (m=5)	4	-	-	4
5		113 (m=5)	14	105 (m=5)	10	24
6		100 (m=5)	12	100 (m=5)	9	21

Table 5. Parameters of the evaluation for different steel grades.

Steel grade	calculated slope m	95% survival probability	Difference between mean stress to 95% survival probability	Stress range S_R as lower boundary of test results
		[N/mm²]	in %	[N/mm²]
S355	6,5	109,8	17	100
S460	5,4	111,1	23	110
S550	3,6	72,8	38	105
S690	6,8	126,5	18	110

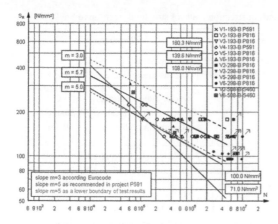

Figure 3. S-N curve of complete test results.

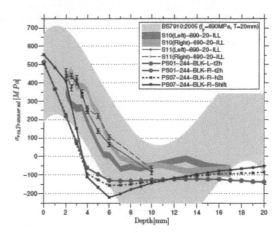

Figure 4. Transverse residual stresses from measurements and computations.

impression that there have been some problems with the welding. In all other cases the slope is between m = 5.4 and 6.8.

Of particular concern is the lower boundary of the test results. The stress variation range is lower for the steel grade 550 compared to the other steel grades. This is due to the higher variation.

A complete evaluation of all test results is presented in figure 3. In total, there are 108 individual test results put together for the evaluation. For all test specimens there are two test results available – one concerning the failed weld and one for the non-failed weld. In cases where both welds have failed both test results have been considered.

Assuming a slope of m = 5 a lower boundary for all test results can be derived which leads to a detail category of 100. This graph is almost identical with the graph 95% survival probability with a slope of m = 5.7 determined in the statistical evaluation.

4 STUDY ON LARGE-SCALE TRUSSES AND THEIR WELDING RESIDUAL STRESSES

4.1 Testing program

Since one crucial point differentiating small (in this case beam specimens) and large-scale tests (truss girders) is the residual stress field present in the fatigue cracking region, emphasis is put on measuring them and accounting for them. Thus the experimental program consisted in two separate studies:

– determination of residual stresses in K-joints made out of S690 (using neutron diffraction facilities at ILL, Grenoble), in order to compare with those measured on S355 K-joints,
– fatigue tests on 2 large-scale truss girder with K-joints in S 690 and including matching cast tube parts, in order to verify the fatigue behavior observed on the beam test specimens.

4.2 Results on residual stresses

The largest measured residual stresses in K-joints made out of S690 were about 60% of the yield stress in the direction transverse to the weld bead. Compared with previous measurements on S355, the residual stress field is found not to be a function of the yield stress, which is in opposition with BS 7910. An example of the measurement results is shown in Figure 4. As can be seen, the measurements in the transverse to the weld bead, for the most part, fall within the band from BS 7910 and also follow its trend given in function of the depth from the surface. The longitudinal components of the stresses show the same trend, the radial being of less interest (the values in this direction were relatively low).

As shown on Figure 4, welding simulations with a simplified nb of passes, but considering phase transformations, were also carried out and gave values and a trend globally similar to the measured values. Also, the values at the left and right toes were found to be very similar, even if influenced by the start position.

4.3 Trusses fatigue tests

Similar tube sizes and truss size as in the previous studies were used, namely upper and lower chords were 193.4 × 20 mm circular hollow sections and braces were 101.6 × 8 mm. The trusses were 9.66 m in length and 1.97 m in height. Keeping the size also provided a more convenient comparison between the specimens. In particular, the eccentricity ratio was kept close to the values used in specimens S5 to S7 tested during P591 (P591, 2010). In addition to the K-joints and next to them, cast nodes were introduced. They were made like tubes parts in order to get butt joints between them and the tubes. They were 20 and 30 mm thick and made out of G10MnMoV6-3.

A three-point load-controlled bending test setup similar to experiments previously done was employed. The tests were carried out at constant amplitude

Table 6. Summary of proposed new S-N curves for tubular bridge construction made of structural steels up to S690.

Detail category*	Detail	Description	Requirements
$\Delta\sigma_{Cb} = 112$ $\Delta\sigma_{Ct} = 71**$ $(m=5)$		Butt-joint between CHS and cast steel with backing made of ceramic elements or a steel ring	Weld must be UT controlled. The cast steel meets a quality level V1S1 at its ends, with NDT control on surface cracks. Tube slenderness: $\gamma \leq 5.0$ Minimum wall thickness $T \geq 20$mm Maximum wall thickness $T \leq 60$mm
$\Delta\sigma_{Cb} = 112$ $\Delta\sigma_{Ct} = 71**$ $(m=5)$		Butt-joint between CHS and cast steel with different thicknesses: - beveled, with tack welded backing steel ring ($t = 3$ to 4 mm) or ceramic backing - with TIG root pass as backing	Weld must be UT controlled. The cast steel meets a quality level V1S1 at its ends, with NDT control on surface cracks. Tube slenderness: $\gamma \leq 5.0$ The thickness variation should not exceed $T_1/T_2 > 0.6$
$\Delta\sigma_{Cb} = 100$ $\Delta\sigma_{Ct} = 71**$ $(m=5)$		Butt-joint between CHS and cast steel with different thicknesses and CHS not beveled, with or without backing	Account for secondary bending correction using $k_f = \left(1 + \dfrac{6e}{T_1} \dfrac{T_1^{1.5}}{T_1^{1.5}+T_2^{1.5}}\right)$ with $T_2 > T_1$ Minimum wall thickness $T \geq 20$mm Maximum wall thickness $T \leq 60$mm
$\Delta\sigma_{C,hs} = 80\ (m=3)$	Welded K-joint		For chord wall thicknesses T > 16 mm, account for size effect using Equation (2)

* expressed in nominal stress range, $\Delta\sigma_C$ (index b for bending or t for tension), or hot spot stress range, $\Delta\sigma_{C,hs}$

** detail category 71 was temporarily predefined as a lower bound. Conservative estimation because a lot of test specimens did not fail

and with a load ratio R equal to 0.1. The loading frequencies, were selected as 1.8 Hz and 1.3 Hz for S10 and S11, resp. Cyclic loading was stopped at regular intervals to repeat the static tests and crack detection. After the failure of the first joint in each truss, the tests were stopped temporarily to carry out the repair operation and then the cyclic loading continued. A rapid repair method using post-tensioning – which did not require dismounting the truss from the test platform– was successfully applied to the trusses. Truss S11 was tested first; the test lasted 516000 cycles. Afterwards, truss S10 was tested with a lower load range up to 1949000 cycles.

Nominal stresses in truss joints were numerically evaluated using a beam element structural model. From previous studies, the following modeling procedure was found to give most accurate results:

– Nodal eccentricities are simulated by rigid links
– For section properties of brace elements in the connection region, the real section is used instead of rigid section properties.
– Brace moments are evaluated at brace-to-chord surface intersection.

Measurements of the strains were also carried out. Experimentally determined were generally in agreement with the numerical values. Hot-spot stresses at the joints were calculated by multiplying axial and bending stresses in the joint by relevant stress concentration factors (SCF). However, the formulas from the CIDECT recommendations were not used since geometric parameters of test trusses are outside the application range for SCF tables given by Zhao et al. [CIDECT recommendations, 2000]. Instead, SCF tables provided in publication [VSS578 2004] were used for calculation of the SCF. These tables were the result of an extensive parametric study at a parameter range more suitable for bridge application. structures ($0.5 \leq \beta \leq 0.7$, $4 \leq \gamma \leq 12$, and $0.3 \leq \tau \leq 0.7$, with realistic consideration of nodal eccentricities in FE models). Using these SCF, the values of the hot spot stresses determined from the nominal ones are the closest to the real values (extrapolated at hot spot from strain meas.).

For the K-joints, no difference was observed, nor in the behavior neither the fatigue strength, between S690 trusses and S355 trusses previously tested. Thus, the same fatigue category can be used for both steel grades. The trusses systematically failed from the K-joints, i.e. for the load combination applied, the K-joints have a lower fatigue strength compared to the cast to tube butt joints. Thus only run-outs but no failures resulted form the fatigue tests on the cast to tube butt joints. Subsequent NDE by KIT on some of the joints using phased-array method confirmed that no fatigue crack initiated in these joints. A couple of the run-outs are below but near to the curve $\Delta \sigma Ct = 71$, $m = 5$, which is a logic confirmation that for these load levels and number of cycles no fatigue cracks should be found.

5   SUMMARY AND CONCLUSIONS

Experimental studies were initiated with the aim of establishing design recommendations for the fatigue behavior of end-to-end connections. Especially the previously derived detail categories for material S355 and S460 were examined using higher steel grades up to S690.

To obtain this, four variants of butt-welded connections had been determined to be investigated on high strength steel. Furthermore, repair welding has been performed on selected test specimens. All fatigue tests have been carried out either as 4-point bending tests on beam specimens or 3-point bending on the trusses. The results obtained on beam specimens have thus been verified on large scale truss girder tests. All experimental data have been evaluated according Eurocode 3 and using S-N curves with a slope $m = 5$.

The following conclusions can be drawn from this program combined with the previous ones:

- For the beam specimens, some difference between steel grades was observed. For the steel grades S355, S460 and S690, this difference in the 95% survival probability compared to the mean stress was roughly 20%, with a minimum characteristic value of approx. $110 \, N/mm^2$, $m = 5.4$. Only S550 gave worse results, but number of results is very limited so not conclusive.
- There was no difference in results for trusses made of S355 and S690. For all tests there was a failure on the K-joints. No failure was observed from the butt welded joint, the run-outs being in agreement with results from the beam tests. There is thus a lower fatigue strength of the K-joint compared to butt welded joint.
- For tubular bridge constructions made of CHS and cast steels, the obtained fatigue classifications for butt-welds are also applicable to high strength steels up to S690, a summary of the detail categories is given in table 6 as a proposal for design recommendations and regulations.
- The K-joints made out of S690 showed residual stresses about 60% of the yield stress in the direction transverse to the weld bead. The absolute values are close to previous measurements on S355, thus the residual stress field is found not to be a function of the yield stress, which is in opposition with BS 7910.

In this paper, structural hollow sections in bridge construction are shown to meet the project goals: functionality, aesthetics, durability, and economic viability.

ACKNOWLEDGMENTS

The authors gratefully acknowledge the financial support of FOSTA (Forschungsvereinigung Stahlanwendung e.V.) and CIDECT for these investigations. The authors would also like to thank the members of the working group for the research project P816 "Optimal application of hollow sections and cast steel nodes in bridge building with the usage of steel S355 up to S690". Thanks are also addressed to the companies who participated in this project, ROPROTEC GmbH, Maurer Söhne, Schmitt Werkstoffprüfung, Friedrich Wilhelms-Hütte, Vallourec Deutschland GmbH, Brütsch/Rüegger AG (Schweiz), Zwahlen et Mayr S.A. (Schweiz).

REFERENCES

Veselcic, M., Herion, S., Puthli, R. (2003). "Cast steel in tubular bridges – New applications and technologies," Proceedings of the 10th International Symposium on Tubular Structures; Swets & Zeitlinger, Lisse, Netherlands.

MacDonald, K.A. and Maddox, S.J., (2003). New guidance for fatigue de-sign of pipeline girth welds, Engineering Failure Analysis, Elsevier science Ltd., No. 10, pp. 177–197.

Veselcic, M., Herion, S., Puthli, R. (2006). "Selection of butt-welded connections for joints between tubulars and cast steel nodes under fatigue loading," Proceedings of the 11th International Symposium on Tubular Structures; Taylor and Francis, London, United Kingdom.

Schumacher, A. and Nussbaumer, A., (2006). Experimental study on the fatigue behavior of welded tubular k-joints for bridges. Engineering Structures, Vol. 28, pp. 745–755.

Herion, S. (2007). "Guss im Bauwesen," Sonderdruck aus Stahlbau Kalender 2007, Ernst & Sohn.

Veselcic, M., Herion, S., Puthli, R. (2007). "Cast Steel and Hollow Sections – New Applications and Technologies," Proceedings of the 17th International Offshore and Polar Engineering Conference (ISOPE-2007), Lisbon, July 1–7, 2007.

Herion, S., Veselcic, M., Puthli, R. (2007). "Cast steel – new standards and advanced technologies" 5th International Conference on Advances in Steel Structures, Singapore.

Haldimann-Sturm, S., Nussbaumer, A., (2008). Determination of allowable defects in cast steel nodes for tubular bridge applications, International Journal of Fatigue, Elsevier Ltd., vol. 30, pp. 528–537.

Nussbaumer, A. and Borges, L., (2008). Size effects in the fatigue behavior of welded tubular bridge joints, Materialwissenschaft und Werkstofftechnik, Wiley interscience, Vol. 39, No 10, Oct., pp. 740–748.

M. Veselcic, S. Herion, R. Puthli, (2008) Selection of butt-welded connections for joints between tubulars and cast steel nodes under fatigue loading, Proceedings of the 12th International Symposium on Tubular Structures; Taylor & Francis, London, United Kingdom.

Borges, L., (2008). Size effects in the fatigue behaviour of tubular bridge joints, Ph.D. Thesis, Thesis EPFL no 4142, Swiss Federal Institute of Technology (EPFL), Lausanne.

Lotsberg, I., (2009). Stress concentrations due to misalignment at butt welds in plated structures and at girth welds in tubulars, Int. Journal of Fatigue, vol. 31, No. 8–9, pp. 1337–1345.

Schumacher, A., Costa Borges, L., Nussbaumer, A., (2009). A critical examination of the size effect correction for welded steel tubular joints, Int. J. of Fatigue, Elsevier Ltd., vol. 31, pp. 1422–1433.

SNF, (2009). Swiss National Fund, Modelling of micro- and macro-structural size effects in the fatigue of welded tubular structures, grant no 200021-112014, 2006-2009.

Veselcic, M., Herion, S., and Puthli, R., (2009). Selection of butt welding methods for joints betweeen tubular steel and steel castings under fatigue loading, Proceedings Tubular Structures XII, held in Shanghai 8-10 oct. 2008, eds. Shen, Chen and Zhao, CRC press, Taylor & Francis group, London, pp. 499–506.

P591, (2010). Wirtschaftliches Bauen von Straßen- und Eisenbahnbrücken aus Stahlhohlprofilen, Final report, Project FOSTA P591, Forschungsvereinigung Stahlanwendung e. V., Düsseldorf, Germany.

Nussbaumer, A, Herion, S., Veselcic, M., Dietrich, R., (2010). "New S-N curves for details in bridges with steel truss tubular superstructure," Proceedings of the 13th International Symposium on Tubular Structures; CRC Press/Balkema, Netherlands

VSS578 (2004). Schumacher, A., Sturm S., Walbridge, S., Nussbaumer, A. and Hirt, M. A., Fatigue design of bridges with welded circular hollow sections, Research mandate 88/98, Swiss federal highway administration, EPFL, ICOM, Lausanne (earlier published as report ICOM 489E, 2003).

EN 287: 2004. Qualification test of welders – Fusion welding – Part 1: Steels.

EN 1369. 1997. Founding, Magnetic particle Inspection.

EN 1370. 1997. Founding, Surface roughness inspection by visual tactile comparators.

EN 1371-1. 1997. Founding, Liquid penetrant inspection – Part 1: Sand, gravity die and low pressure castings.

EN 1559-1. 1997. Founding, Technical conditions of delivery.

EN 1559-2. 1997. Founding, Technical conditions of delivery, Part 2: Additional requirements for steel castings.

DIN 1681. 1985. Steel castings for general purposes; technical delivery conditions.

EN 1993-1-1. 1993. Design of steel structures – Part 1-1: General rules and rules for buildings.

EN 1993-1-9. 2005. Design of steel structures – Part 1-9: Fatigue.

EN 10210. 2006. Hot finished structural hollow sections of non-alloy and fine grain steels.

EN 10293. 2005. Steel castings for general engineering uses.

EN 12680. 2003. Founding – Ultrasonic inspection – Part 1: Steel castings for general purposes.

EN 12681. 2003. Founding – Radiographic inspection.

EN 17182: Steel castings with improved weldability and toughness for general purposes, DIN, Mai 1992.

EN 17205. 1992. Quenched and tempered steel castings for general purposes.

EN ISO 5817. 2003. Welding – Fusion-welded joints in steel, nickel, titanium and their alloys (beam welding excluded) – Quality levels for imperfections.

EN ISO 9692. 2004. Welding and allied processes – Recommendations for joint preparation.

AD HP 5/3 and App.1. 2002. Manufacture and testing of joints – Non-destructive testing of welded joints.

CIDECT Report – Project 7W. 2007. Fatigue of End-To-End CHS Connections, Sixth Interim Report, University of Karlsruhe.

CIDECT, 2009. Project 7W, Fatigue of end-to-end connections, University Karlsruhe, 2006-09.

CIDECT recommendations, 2000. Zhao, X. L., Herion, S., Packer, J. A. and al. Design guide for circular and rectangular hollow section joints under fatigue loading. CIDECT, Comité International pour le développement et l'étude de la construction tubulaire, guide no 8, TüV-Verlag Rheinland, Köln.

EN 10210, 1997. Hot finished structural hollow sections of non-alloy and fine grain structural steels, Part 1: technical delivery requirements and Part 2: tolerances, dimensions and sectional properties, Brussels, CEN.

EN 10293, 2005. Steel castings for general engineering uses, Brussels, CEN.

EN 1993-1-9, 2005. Design of steel structures, Part 1.9: fatigue strength of steel structures, Brussels, CEN.

DIN 17182, 1992. General-purpose steel castings with enhanced weldability and higher toughness; technical delivery conditions, Deutsches Institut für Normung E.V., Berlin.

Connections

Tubular Structures XV – Batista, Vellasco & Lima (eds)
© 2015 Taylor & Francis Group, London, ISBN 978-1-138-02837-1

Stress concentration factors of square bird-beak SHS T-joints under brace axial loading

B. Cheng, Q. Qian, J.C. Zhao & Z.A. Lu
Shanghai Jiao Tong University, Shanghai, China

X.L. Zhao
Monash University, Melbourne, Australia

ABSTRACT: The stress concentration factors of square bird-beak square hollow section (SHS) welded joints under brace axial forces are numerically studied in this paper. Square bird-beak joints considered have T-shapes and are simply supported at their chord ends. Refined finite element models are developed to obtain the strain/stress concentration factors (SNCFs/SCFs) of square bird-beak joints with various dimensions. The FE models are validated by comparing with the experimental data. Critical locations of high stress concentrations are identified. The influences of three major non-dimensional parameters, i.e., brace/chord width ratio β, chord wall slenderness ratio 2γ, and brace/chord wall thickness ratio τ, on the stress concentration factors (SCFs) of square bird-beak T-joints are revealed on the basis of numerous parametric studies.

1 INTRODUCTION

Bird-beak joint, an innovative type of tubular constructions composed of square hollow sections (SHS), is generated by simply rotating the members of a conventional SHS joint at 45° about theirs longitudinal axes. Therefore, there are two types of bird-beak SHS T-joints, that is, square bird-beak joint generated by rotating only chord and diamond bird-beak joint generated by rotating both chord and brace, as shown in Fig. 1. These innovative connections obviously have more elegant configurations, and would assist in relieving lateral wind loads and provide higher structural resistances. The actual reason for the better mechanical behaviors was supposed to be that the bird-beak orientation would assist in transferring forces between members by natural in-plane action rather than by bending of the walls.

With regard to the fatigue behavior of bird-beak joints, Ishida (1992) carried out the earliest fatigue tests of bird-beak T-joints with the load case of brace axial force being considered; Keizer (2003) and Keizer et al. (2003) investigated the stress concentration factors of diamond bird-beak joints under brace axial force in his master degree thesis; Tong et al. (2014) presented an experimental study on stress concentration factors for diamond bird-beak T-joints under axial force and in-plane bending on the brace; Cheng et al. (2014) carried out the tests to determine the strain concentration factors of square bird-beak T-joints under chord and brace axial forces. Existing fatigue researches about bird-beak joints are far from systematic and complete.

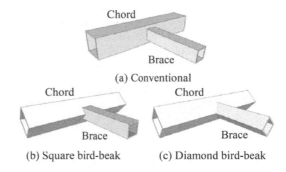

Figure 1. SHS-SHS welded T-joints.

Therefore, this study numerically investigates the stress concentration characteristics of square bird-beak T-joints. The load case of brace axial force is considered. Refined finite element models of both conventional and innovative joints, whose meshes are highly consistent with the strain extrapolation rules, are established. Critical locations of high stress concentrations are identified. The influences of three major non-dimensional parameters on SCFs of square bird-beak T-joints are obtained on the basis of parametric finite element analysis.

2 SHS T-JOINTS AND FE MODEL

2.1 *Dimensions*

Similar with conventional T-joints, three types of non-dimensional parameters, that is, brace/chord width

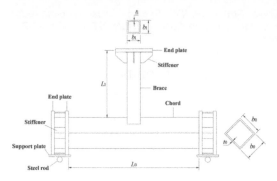

Figure 2. Dimensions of square bird-beak T-joints.

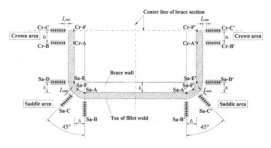

Figure 3. Arrangement of strip gauges at hot spots (top view).

ratio $\beta = b_1/b_0$, chord wall slenderness ratio $2\gamma = b_0/t_0$, and brace/chord wall thickness ratio $\tau = t_1/t_0$, are mainly considered for square bird-beak T-joints. Here, b_0 and t_0 represent the sectional width and wall thickness of the chord; b_1 and t_1 correspond to the sectional width and wall thickness of the brace, as shown in Fig. 2.

2.2 Hot spots

For the new-type bird-beak joints, the junction configurations which contain the so-called crown areas and saddle areas, as used for CHS-CHS tubular joints, are much more complicated. By referring to the findings from Cheng et al. (2014) and adopting the similar criteria used for conventional joints (van Wingerde 1992, Zhao et al. 2000), four crown hot lines (i.e., Cr-B and Cr-C on the chord and Cr-F and Cr-A on the brace) and six saddle hot lines (i.e., Sa-B, Sa-C and Sa-D on the chord and Sa-E, Sa-F and Sa-A on the brace) were selected, as shown in Fig. 3.

For a hot line, the stress/strain at weld toe was calculated from the stresses/strains within a fixed region away from the weld toe (i.e., the extrapolation region) by the use of an extrapolation approach. In the present research, the boundaries of extrapolation regions followed the rules provided in CIDECT Design Guide No. 8 (Zhao et al. 2000), that is, each extrapolation region started from L_{min} away from the weld toe and had a length of t, where t represents the sectional wall thickness of the member on which the strains were extrapolated, and L_{min} takes the greater value of $0.4t$

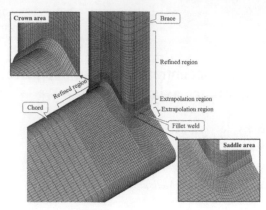

Figure 4. Finite element model of square bird-beak T-joint (half-structure).

and 4 mm, as shown in Fig. 3. Furthermore, quadratic extrapolation, as suggested by Cheng et al. (2014), was also used in this research.

2.3 Finite element model

A general-purpose finite element program ANSYS 12 was used to obtain the strain/stress concentration factors (SNCFs/SCFs) of SHS T-joints. The analyses were based on the small deformation and the material elasticity, i.e., neither geometrical nor material nonlinearity was considered. Solid elements (SOLID95) with 20 nodes and three degrees of freedom for each node were used. Geometries of fillet welds were accurately considered in FE models. In order to take into account the complexities of stress distributions near the junction, the meshes in these areas were refined. The element sizes within the selected extrapolation regions were exactly specified to be 2 mm.

Fig. 4 demonstrates the typical finite element meshes of the half-structure of a square bird-beak T-joint. The selected mesh-controlling parameters had been validated by comparing to the half-size refined meshes, i.e., the elements within the extrapolation regions had uniform lengths of 1 mm. Errors in SNCF were found to vary between 0.1–4% for all hot spots considered, indicating that the established FE meshes should be accurate enough for the study.

3 NUMERICAL ANALYSIS RESULTS

3.1 Comparison with experimental data

To further verify the validity of FE models, experimental SNCFs of nine orthogonally designed square bird-beak SHS T-joints (Cheng et al. 2014) were employed for the comparison. The measured sectional widths, wall thicknesses, member lengths as well as fillet weld sizes of specimens were introduced into the FE models. It can be seen from Fig. 5 that in most occasions the FEM strain concentration factors (i.e., $SNCF_{FEM}$) are slightly greater than those

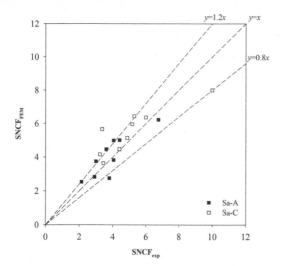

Figure 5. Comparison of experimental and FEM SNCFs.

obtained from the test (i.e., $SNCF_{exp}$). The average value of $SNCF_{FEM}/SNCF_{exp}$ ratios goes to 1.15 and 1.07, respectively for chord spots and brace spots. The coincidence between the test and the simulation could be regarded as fairly good.

3.2 Comparison of hot spot locations

By also employing the nine square bird-beak T-joints tested by Cheng et al. (2014) as objects, the comparison of SNCF magnitudes at all considered hot spots were made. It can be found that four of the nine square bird-beak joints have maximum chord SNCFs at saddle spots Sa-C, and the other five joints have maximum chord SNCFs at saddle spots Sa-D. As for the brace spots, there are four joints whose maximum brace SNCFs occurred at saddle spots Sa-A, and the maximum brace SNCFs of the other five joints occurred at saddle spots Cr-F. The SCF differences between saddle spots Sa-C and Sa-D or between saddle spots Sa-A and Sa-F or between crown spots Cr-A and Cr-F are not much evident.

In comparison, the maximum chord SNCFs of seven joints are greater than the corresponding maximum ones in the brace, with the differences between them being 12–40%. While for the other two joints where highest stress concentrations occurred in the braces, the maximum chord SNCFs are as large as 90% of the maximum brace SNCFs. Therefore, it can be concluded that the chord of a square bird-beak joint is more prone to higher stress concentrations than the brace, which is highly consistent with experimental results.

3.3 Principal stress versus the stress perpendicular to weld toe

There are two types of stresses, i.e., the principal stress and the stress normal to weld toe, to be used for the definition of hot spot stress. The stresses normal to

weld toe are adopted by the AWS (2010), while the IIW (2008) and many researchers (van Wingerde 1992; van Wingerde et al. 1997; Packer and Wardenier 1998) use the principal stresses by stating that these stresses are usually perpendicular to the weld toes for simple connections.

We compared the differences between these two stresses by analyzing the FEM results of some square bird-beak joints with typical non-dimensional parameters, and it shows that the differences depended on the hot line locations. $SCF_{principal}$ at those hot lines normal to the weld toes (i.e., Cr-C, Cr-B, Cr-F, Sa-D, Sa-C, Sa-B, Sa-A) were up to 3% larger than the corresponding $SCF_{perpendicular}$. It can therefore be considered that, for square bird-beak joints, the differences between principal stress based SCFs and normal-to-weld stress based SCFs are insignificant.

3.4 Relationship between SNCF and SCF

Relationships between SNCF and SCF of conventional tubular joints have been investigated in existing research (van Delft et al. 1987, Frater 1991). It was found that SNCF could be converted to SCF by multiplying a coefficient of 1.2 for CHS joints or 1.1 for RHS joints. This relationship, which depends on the joint configuration, the local position concerned as well as the loading case, can be deduced on the basis of Hooke's Law in linear elasticity and then be expressed as follows (Sing et al. 1999)

$$SCF = \left(\frac{1+v \dfrac{\varepsilon_{//}}{\varepsilon_{\perp}}}{1-v^2} \right) \cdot SNCF \qquad (1)$$

in which, v is the Poisson's ratio, $\varepsilon_{//}$ is the strain component parallel to the weld toe, and ε_{-} is the strain component perpendicular to the weld toe.

The SNCF-to-SCF relationships of square bird-beak T-joints subject to brace axial force were investigated in this research by the use of finite element method. Fig. 6 shows the numerical SNCFs and SCFs at hot spots where the principal strains/stresses were almost perpendicular to the weld toes. By fitting these data using the least square method, the relationship between SNCF and SCF of square bird-beak T-joints can be approximately written as the following

$$SCF = 1.15 \cdot SNCF \qquad (2)$$

4 PARAMETRIC STUDIES

4.1 Weld sizes

In the parametric numerical analysis where more cases of non-dimensional parameters were considered, the prequalified weld details for PJP tubular joints provided in AWS Code (2008) were adopted, and the weld sizes within different junction areas (i.e., the crown

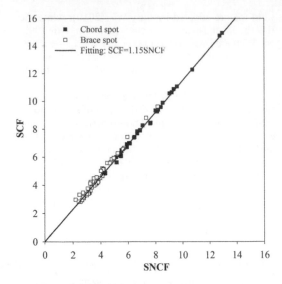

Figure 6.   Comparison of SNCFs and SCFs.

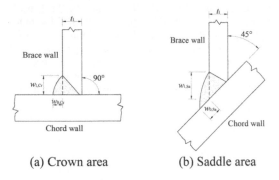

(a) Crown area   (b) Saddle area

Figure 7.   Weld details used for parametric analysis.

areas where the welded plates were perpendicular to each other and the saddle areas where the oblique angles between two plates were near or equal to 135°) were determined as follows

- For the crown areas in Fig. 7(a)
    Weld length projected on the brace $w_{1,Cr} = t_1$
    Weld length projected on the chord $w_{0,Cr} = t_1/2$
- For the saddle areas in Fig. 7(b)
    Weld length projected on the brace $w_{1,Sa} = t_1 \cdot \sqrt{2}$
    Weld length projected on the chord $w_{0,Sa} = (t_1 \cdot \sqrt{2})/2$.

4.2   Influences of non-dimensional parameters

Fig. 8–10 demonstrate the typical variations of chord and brace SCFs of square bird-beak joints under brace axial force, as functions of three non-dimensional parameters considered. The following can be concluded:

- *Case of variable β.* Both chord SCFs at spots Sa-C and brace SCFs at spots Sa-A first increase and then decrease as β grows from 0.3 to 1.3, with the brace/chord width ratios corresponding to the curve

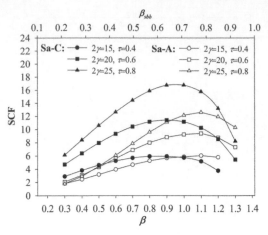

Figure 8.   Variations of chord and brace SCFs with β, under brace axial force.

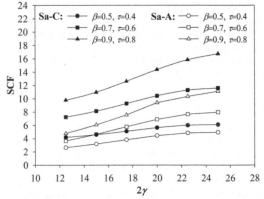

Figure 9.   Variation of chord and brace SCFs with 2γ, under brace axial force.

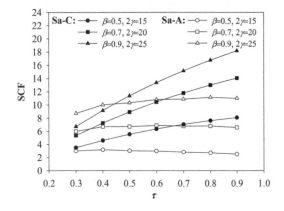

Figure 10.   Variations of chord and brace SCFs with τ, under brace axial force.

peaks (i.e., β_{pk}) taking values around 0.9 (for chord) or 1.1 (for brace). The slopes of curves for $\beta > \beta_{pk}$ are much greater than those symmetric ones for $\beta < \beta_{pk}$. Comparatively speaking, a greater 2γ or

346

τ is more likely to produce higher rates of increases or decreases in SCFs with increasing β.

- *Case of variable 2γ.* Both chord SCFs and brace SCFs are improved as 2γ is enhanced from 12.5 to 25, with the rates of increase within the middle variable zones where 2γ take values between 15 and 20 being of the highest. Comparatively, the larger β or τ is, the higher the rate of increase in SCF becomes.
- *Case of variable τ.* Increases in chord SCF with enlarging τ are evident, with the rates of increase being slightly depressed as τ goes greater. Variations in brace SCF with varying τ are relatively insignificant, with both descending and declining tendencies with increasing τ being observed for various cases of β and 2γ.

4.3 *Definition of brace/chord width ratio*

For a square bird-beak joint, the brace/chord width ratio β could reach a maximum value of 1.4 with the brace wall being exactly supported by the chord corners. In other words, the brace of a square bird-beak joint could have a larger width than the chord, which is unrealistic for conventional joints. In case of brace axial force, peak points are always observed from the SCF-β curves of the two types of joints, that is, SCF decreases from the peak value as β is decreased from β_{pk} to 0 or is increased from β_{pk} to the maximum value β_{max}, as shown in Fig. 8. The primary difference lies in that β_{pk} of square bird-beak joints are much larger than the conventional ones. Therefore, a new definition of brace/chord width ratio based on the contact length between brace and chord walls could be introduced for square bird-beak joints so that the brace/chord width ratio β_{max} of two joint types are unified to 1.0.

In the new definition, the normalized brace width takes the contact length on the chord walls (i.e., $\sqrt{2} \cdot \frac{b_1}{2}$) instead of original b_1. The normalized brace/chord width ratio of square bird-beak joints (β_{sbb}) can thus be calculated as

$$\beta_{sbb} = \left(\sqrt{2} \cdot \frac{b_1}{2} \right) \Big/ b_0 = \frac{\sqrt{2}}{2} \cdot \frac{b_1}{b_0} = 0.707\beta \qquad (3)$$

The normalized definition of brace/chord width ratio is convinced to have a better reflection to the natural features of SCF variations. The SCF variations with β_{sbb} have also been provided in Fig. 8, where the peak β_{sbb} under brace axial forces are observed to be within 0.6–0.8.

5 CONCLUSIONS

This paper investigated the stress concentration factors of square bird-beak T-joints subject to brace axial forces. Based on the finite element analysis, the following can be concluded for the new-type joints:

(1) Stresses normal to the weld toes can be used for the definition of hot spot stress.

(2) SCFs can be converted from SNCFs by multiplying a coefficient of 1.15.

(3) The stresses within crown and saddle areas are of the highest. The maximum chord SCFs appear at the saddle spots Sa-C or Sa-D, while the maximum brace SCFs belong to the saddle spots Sa-A or Cr-F.

(4) Both chord SCFs and brace SCFs are generally improved as any one of three non-dimensional parameters (β, 2γ or τ) is enlarged. The exceptions are that an increase in large β (e.g., $\beta > 0.9$) may significantly depress the chord SCFs, and that the increase of τ may also slightly reduce the brace SCFs.

(5) The normalized brace/chord width ratio $\beta_{sbb} = 0.707\beta$ has a better reflection to the natural features of SCF variations of square bird-beak joints.

ACKNOWLEDGEMENTS

The present research was undertaken with support from the special project of Science and Technology Research and Development Plan funded by the Ministry of Railway of P.R.C. (no. J2011G002).

REFERENCES

American Welding Society (AWS). (2010). Structural welding code-steel, ANSI/AWS D1.1/D1.1M:2010, American Welding Society, Miami.
Ansys 12 [Computer software].
Cheng, B., Qian, Q., and Zhao, X. L. (2014). "Tests to Determine stress concentration factors for square Bird-Beak SHS joints under chord and brace axial forces." ASCE's Journal of Structural Engineering; DOI: 10.1061/(ASCE)ST.1943-541X.0001095.
Frater G.S. (1991). "Performance of Welded Rectangular Hollow Structural Section trusses." PhD thesis, University Toronto, Canada.
International Institute of Welding (IIW). (2008). Recommended fatigue design procedure for welded hollow section joints-part 1: Recommendations, part 2: Commentary, XV-1035-99, Cedex, France.
Ishida, K. (1992). "Experimental research on fatigue behavior of diamond bird-beak joint." Proc., Symp. on Structural Engineering, Architectural Institute of Japan, Tokyo (in Japanese).
Keizer, R. (2003). "Stress concentration factors in diamond bird beak T-joints." M.S. Thesis, Delft Univ. of Technology, Delft, Netherlands.
Keizer, R., Romeijn, A., and Wardenier, J. (2003). "The fatigue behaviour of diamond bird beak T-joints." Proc., Int. Symp. on Tubular Structures, Vol. 10, M. A. Jaurietta, et al., eds., E&FN Spon, London, 303–310.
Packer, J. A., and Wardenier, J. (1998). "Stress concentration factors for non-90° X-connections made of square hollow sections." Can. J. Civ. Eng., 25(2), 370–375.
Sing, P. C., Chee, K. S., and Nai, W. W. (1999). "Experimental and numerical stress analyses of tubular XT-joints." J. Constr. Engrg.,ASCE, 125, 1239–1248.
Tong, L. W., Fu, Y. G., Liu, Y. Q., and Zhao, X. L. (2014). "Stress concentration factors of diamond bird-beak SHS

T-joints under brace loading." Thin-Walled Struct., 74(1), 201–212.

van Delft, D.R.V., Noordhoek, C., and Da Re M.L. (1987). "The Results of the European Fatigue Tests on Welded Tubular Joints Compared with SCF Formulas and Design Lines." Proc., Steel in Marine Structures, Delft, The Netherlands, 565–577.

van Wingerde, A. M. (1992). "The fatigue behaviour of T- and X-joint made of square hollow section." HERON, 37(2), 1–180.

van Wingerde, A. M., Packer, J. A., and Wardenier, J. (1997). "SCF formulae for fatigue design of K-connections between square hollow sections." J. Constr. Steel Res., 43(1), 87–118.

Zhao, X. L., et al. (2000). "Design guide for circular and rectangular hollow section joints under fatigue loading." Comité International pour le Développement et l'Etude de la Construction Tubulaire, TÜV-Verlag, Köln, Germany.

Tubular Structures XV – Batista, Vellasco & Lima (eds)
© *2015 Taylor & Francis Group, London, ISBN 978-1-138-02837-1*

Experimental evaluation of the directional strength increase for fillet welds to rectangular hollow sections

J.A. Packer, M. Sun & P. Oatway
University of Toronto, Toronto, Canada

G.S. Frater
Canadian Steel Construction Council, Markham, Canada

ABSTRACT: A series of laboratory experiments on fillet-welded connections between rectangular hollow sections (RHS) and end-plates is presented. A total of 21 specimens, with the welds designed to be the critical element, were tested to failure by application of axial tension loading to the RHS members. Complete mechanical and geometric properties of all materials were obtained and weld fracture was the governing failure mode for all test specimens. An evaluation of the simple design methods in steel codes for fillet welds to RHS members was thus performed, to ascertain both their accuracy and level of safety (or structural reliability). It was found that, even for such rigid connections, restrictions need to be placed on the application of the "directional strength enhancement factor" for fillet welds which is currently permitted in certain steel design codes and standards.

1 INTRODUCTION

Contemporary codes, standards, specifications and design guides for steel hollow structural section connections (Wardenier et al. 2008; Packer et al. 2009; Packer et al. 2010; ISO 2013) usually acknowledge that there are two weld design philosophies available: (i) design for the capacity of the attached branch member (which satisfies any loading scenarios), or (ii) design for the actual forces in the attached member (a "fit-for-purpose" approach). The latter approach involves a consideration of the effective properties of the weld group, since all of the weld length may not be effective, because the base to which the weld is attached is flexible in a typical hollow section connection. Method (i) enables a pre-qualified weld size to be readily specified, but Method (ii) generally allows for weld "downsizing" and hence is popular. ANSI/AISC 360-10 (AISC 2010) has adopted this Method (ii) in Section K for welded connections to RHS by specifying various weld effective lengths for different connect types and loading situations.

The fillet weld design clauses in the current American (AISC 2010; AWS 2010) and Canadian (CSA 2013a; CSA 2014) steel and welding standards permit the use of a so-called "directional strength enhancement factor" for fillet welds, which enables the weld strength to be enhanced when the direction of loading is non-parallel to the axis of the weld. This has been shown to be a generally unsafe concept, when applied to RHS-to-RHS fillet-welded connections and used in conjunction with current AISC 360-10 Chapter K weld effective lengths/properties, because target structural

reliability levels are not achieved (McFadden & Packer 2014). Thus, AISC does not permit the use of this "directional strength enhancement factor" when the "effective length method" of AISC 360 is used for proportioning welds in hollow section connections. One of the purposes of this investigation, presented herein, was to determine if the "directional strength enhancement factor" was even applicable to fillet welds between a RHS branch and a rigid base, where all of the weld length would be effective.

The adoption of the fillet weld "directional strength enhancement factor", or "sinθ factor", into North American codes and standards, was based on experimental research on predominantly lap splice connections where the connection was loaded in shear (Miazga & Kennedy 1989; Lesik & Kennedy 1990). Fillet welds in many RHS connections, on the other hand, have the branch member loaded in tension or bending. Furthermore, fillet welds to hollow section members are inherently one-sided (welding is only performed on one side of the hollow section wall), thus the weld joint will be subject to a local eccentricity that – with tension loading in the attached wall – will produce tension at the root of the weld (see Fig. 1). CSA W59 (2013a) Clause 4.1.3.3.2 even states that …"Single fillet and single partial joint penetration groove welds shall not be subjected to bending about the longitudinal axis of the weld if it produces tension at the root of the weld". EN 1993-1-8 (CEN 2005) Clause 4.12 states that such local eccentricity, producing tension at the root of the weld, should be taken into account, but it specifically notes that …"Local eccentricity need not be taken into account if a weld is

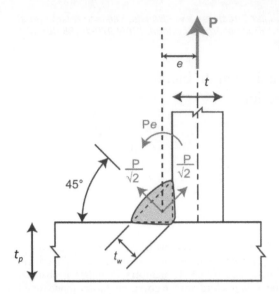

Figure 1. Eccentrically-loaded fillet weld under axial tension (adopted from McFadden and Packer 2013).

taken as the product of the nominal stress of the weld metal (F_{nw}) and the weld effective throat area (A_w) with a resistance factor ($\phi_w = 0.75$) applied. The nominal strength (R_n) is determined as follows:

$$R_n = F_{nw}A_w \tag{1a}$$

$$F_{nw} = 0.60X_u \tag{1b}$$

where X_u = ultimate strength of weld metal.

Alternatively, for a linear weld group with a uniform leg size, loaded through the centre of gravity (i.e. all weld elements are in a line or are parallel hence having the same deformation capacity under the applied load), the provisions of Section J2.4(a) permit the use of the directional strength enhancement factor (Eq. (1c)) for the calculation of the nominal stress of the weld metal (F_{nw}). It shall be noted that Chapter K of AISC 360 (2010) does not permit the use of such a "sin θ factor" when the "effective length method" is used for proportioning welds in hollow section connections.

$$F_{nw} = 0.60X_u \left(1.00 + 0.50 \sin^{1.5} \theta\right) \tag{1c}$$

where θ = angle of loading measured from the weld longitudinal axis (in degrees).

As a special application of Section J2.4(a), Section J2.4(c) gives provisions for concentrically loaded connections with weld elements of multiple orientations (longitudinal and transverse to the direction of applied load). According to Lesik and Kennedy (1990), the deformation capacity of fillet welds decreases as the angle of loading increases (i.e. maximum deformation capacity for weld element loaded longitudinally; minimum deformation capacity for weld element loaded transversely). As a result of such incompatibility, the transverse weld prevents the longitudinal weld from reaching its full capacity before failure of the joint takes place. Thus, Section J2.4(c) specifies that the nominal strength (R_n) of concentrically loaded joints with both longitudinal and transverse fillet welds be determined as the greater of Equations 2a and 2b.

$$R_n = R_{nwl} + R_{nwt} \tag{2a}$$

$$R_n = 0.85R_{nwl} + 1.5R_{nwt} \tag{2b}$$

where R_{nwl} = total nominal strength of longitudinally loaded fillet welds; R_{nwt} = total nominal strength of transversely loaded fillet welds with F_{nw} calculated by Equation 1b.

used as part of a weld group around the perimeter of a structural hollow section". The basis for this Eurocode waiver is unknown. AWS D1.1 Section 2.6.2 (2010) states that, in the design of welded joints, the calculated stresses shall include those due to eccentricity caused by alignment of the connected parts, size and type of welds, but this Section pertains to connections which are "non-tubular". These code clauses thus recognize that eccentric loading on a fillet weld, causing tension at the weld root, may result in reduced weld capacity.

Another factor influencing the strength of fillet welds is the amount of weld penetration into the base metals at the root. Small fillet welds and large fillet welds tend to have the same degree of root penetration; for large multi-pass welds the root penetration is generally determined just by the root pass. In laboratory testing, then, the strength of small fillet welds will be aided proportionally more than for large fillet welds, by the root penetration. Since most laboratory tests on weld-critical joints involve relatively small welds (because the failure mode of weld fracture must be achieved), one must bear in mind that the results achieved are likely more optimistic compared to results from large-weld tests.

2 REVIEW OF RELEVANT CODES AND STANDARDS

2.1 *ANSI/AISC 360 (AISC 2010)*

In Section J of AISC 360, unless over-matched weld metal is used, the design strength ($\phi_w R_n$) of a single fillet weld is based on the limit state of shear rupture along the plane of the weld effective throat. It is

2.2 *CAN/CSA S16 (CSA 2014)*

Similar to AISC 360 (2010), unless over-matched weld metal is used, the Canadian steel structures design standard specifies that the design strength ($\phi_w R_n = 0.67R_n$) of a single fillet weld be evaluated based on the limit state of shear rupture along the plane of the weld effective throat. The nominal strength of a fillet-welded joint is the sum of the nominal strengths of weld elements with different orientations.

$$R_n = 0.67 A_w X_u \left(1.00 + 0.50 \sin^{1.5}\theta\right) M_w \qquad (3a)$$

$$M_w = \frac{0.85 + \theta_1/600}{0.85 + \theta_2/600} \qquad (3b)$$

where θ & θ_1 = angle of loading (in degrees) of the weld element under consideration; θ_2 = angle of loading (in degrees) of the weld element in the joint that is nearest to 90°; M_w = strength reduction factor to consider for the variation in deformation capacity of weld elements with different orientations (similar to the "0.85" factor in Equation 2b.

3 TEST SPECIMENS AND GEOMETRIC MEASUREMENTS

A total of 21 RHS-plate-RHS welded connections, with angles between RHS and plate of 60° and 90° (see Fig. 2), were designed and fabricated to be weld-critical under the applied axial tension loads. By welding the RHS to a rigid base the effect of any flexibility of the surface on which the fillet weld lands is removed, and all of the fillet weld length can hence be considered effective. The RHS designation, angle between RHS and plate, and plate thickness (t_p) of all connection specimens are listed in Table 1. All RHS were cold-formed to CAN/CSA G40.20/40.21 Grade 350W (2013b).

A matching electrode with a nominal tensile strength of 490 MPa was used for all fillet welds. The weld sizes were selected to satisfy the minimum requirements in Table J2.4 of AISC 360 (2010). Prior to testing, all test welds were ground (see Fig. 3) into a triangular shape using a hand held grinder so that the weld leg sizes, as well as the theoretical effective throat thicknesses, could be measured externally with accuracy using a fillet weld gauge (see Fig. 4). The weld cross-sectional dimensions were measured externally at numerous locations around the weld perimeter. The averages of these externally measured theoretical throat thicknesses were used for strength calculation to ensure that the connections would be weld-critical during testing.

After testing and noting the actual load at failure, where possible, each connection was cut normal to the longitudinal axis of the welds at eight locations around the perimeter of the RHS (two cuts per side). After surface polishing, the eight cross-sections were subjected to macroetch examinations in accordance with ASTM E340-06 (2006) using a 10% nital etchant solution to observe the weld profile. These cross-sections were then scanned and input into computer programs so that the weld cross-sectional dimensions, especially

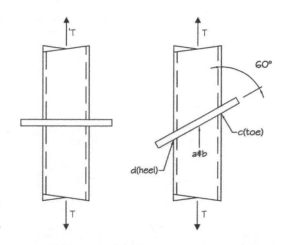

Figure 2. Connection specimens.

Figure 3. Typical weld after grinding.

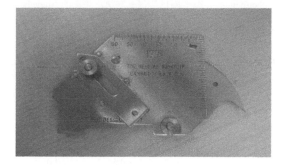

Figure 4. Fillet weld gauge.

the effective throat thickness (the shortest distance from the root to the face of the weld), can be measured with accuracy. The throat thickness of a fillet weld was taken as the height of the largest triangle that can be inscribed within the fusion faces and the weld surface (see Fig. 5). The internal measurements by macroetch examinations agreed well with the external measurements using the fillet weld gauge. The averages of the measured effective throat thicknesses (t_w) are listed in Table 1.

Since the current design requirements of AISC 360 (2010) and CSA S16 (2014) are based only on the limit state of shear rupture along the plane of the weld effective throat, the measured values of the weld leg sizes are not reported herein. The total lengths of weld (l_w in Table 1) were determined based on the RHS perimeters, considering the rounded corners.

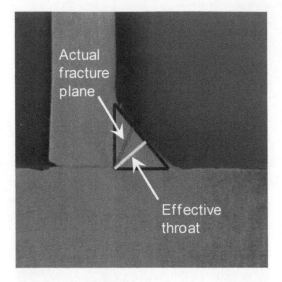

Figure 5. Example of fillet-weld throat measurements from the macroetch examinations.

4 EXPERIMENTAL INVESTIGATION

4.1 *Tensile coupon tests*

Tensile coupons were machined, from the RHS (flat faces away from the weld seam) and from the intermediate plates, and tested in accordance with ASTM A370 (2013) to determine the material properties of the base metals. The averages of the measured yield stresses (F_y and F_{yp}, determined by the 0.2% strain offset method) and ultimate strengths (F_u and F_{up}) are listed in Table 2.

To determine the material properties of the as-laid weld metals, all-weld-metal tensile coupons were created in accordance with AWS D1.1 (2010). The averages of the measured yield stresses (F_{yw}) and ultimate strengths (X_u) of the as-laid weld metals are listed in Table 2.

It was found that the measured F_y and X_u values are generally much higher than the nominal values.

4.2 *Instrumentation*

For all connection specimens, a group of four strain gauges (Group A) were mounted on the four faces of the RHS well above the intermediate plate. The purpose of these strain gauges was to measure any strain difference between the opposite RHS faces during testing, thereby monitoring any unintentionally induced bending moments. Typical load-strain curves of the four sides of an RHS are shown in Figure 6. It can be

Table 1. Measured geometric properties.

Specimen No.	RHS Designation	Angle between RHS and Plate	t_p (mm)	l_w (mm)	Average of t_w (mm)
1	127 × 127 × 8.0	90°	25.0	481	3.7
2	127 × 127 × 8.0	90°	25.0	481	5.6
3	178 × 178 × 8.0	90°	25.0	481	4.7
4	178 × 178 × 8.0	90°	25.0	481	5.7
5	127 × 127 × 13.0	90°	25.0	668	7.0
6	127 × 127 × 13.0	90°	25.0	668	7.4
7	178 × 178 × 13.0	90°	25.0	668	6.1
8	178 × 178 × 13.0	90°	25.0	668	7.0
9	127 × 127 × 9.5	90°	19.0	475	3.5
10	127 × 127 × 9.5	90°	19.0	475	3.2
11	127 × 127 × 9.5	90°	19.0	475	2.8
12	127 × 127 × 9.5	90°	19.0	475	4.4
13	127 × 127 × 9.5	90°	19.0	475	4.3
14	127 × 127 × 9.5	90°	19.0	475	4.8
15	127 × 127 × 9.5	90°	19.0	475	7.1
16	127 × 127 × 9.5	90°	19.0	475	8.1
17	127 × 127 × 9.5	90°	19.0	475	10.1

Specimen No.	RHS Designation	Angle between RHS and Plate	t_p (mm)	l_w (mm) at a & b*	at c*	at d*	Average of t_w (mm) at a & b*	at c*	at d*
18	127 × 127 × 9.5	60°	19.0	137	119	119	5.0	4.2	7.3
19	127 × 127 × 9.5	60°	19.0	137	119	119	3.8	3.5	4.1
20	127 × 127 × 9.5	60°	19.0	137	119	119	4.4	4.8	8.4
21	127 × 127 × 9.5	60°	19.0	137	119	119	7.2	5.6	10.2

*Locations a, b, c and d are indicated in Figure 2.

Table 2. Measured material properties.

Specimen No.	RHS*		Plate		Weld Metal	
	F_y (MPa)	F_u (MPa)	F_{yp} (MPa)	F_{up} (MPa)	F_{yw} (MPa)	X_u (MPa)
1, 2, 3 & 4	412	478	383	563	563	619
5, 6, 7 & 8	380	489	383	563	563	619
9, 10, 11 & 19	426	500	351	558	634	687
12, 13, 14, 15, 16, 18, 20 & 21	426	500	351	558	641	739

*The material properties of RHS were determined by testing tensile coupons taken from the flat faces.

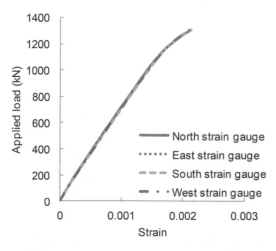

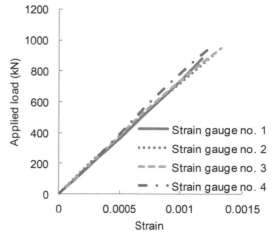

Figure 6. Typical load-strain curves from four strain gauges on four sides of RHS (Group A strain gauges, see Section 4.2).

Figure 7. Typical load-strain curves from four strain gauges on one side of RHS (Group B strain gauges, see Section 4.2 and Figure 8).

deduced that no bending moment was applied to the connection. Hence, the specimen was loaded in pure tension during testing.

In order to further ensure that the weld elements were uniformly loaded along their lengths, another group of eight strain gauges (Group B) were placed on two adjacent faces of the RHS just above the intermediate plate for all connection specimens (see Fig. 8). Typical load-strain curves at four different locations of one side of an RHS are shown in Figure 7. It can be deduced that the weld was uniformly loaded during testing.

4.3 Connection tests

All connection specimens were tested using a MTS universal testing machine with a capacity of 2650 kN. The connection specimens were loaded to failure in a quasi-static manner to eliminate any strain rate effect. All connection specimens failed by weld rupture (see Fig. 8). The actual loads at failure are listed in Table 3.

5 RESULTS AND DISCUSSIONS

The analysis of test data has been performed using the measured weld effective throat size (i.e. the minimum distance between the root of the fillet weld and the face

Figure 8. Specimen at failure.

of the triangular weld profile), which is the weld theoretical or effective throat size that a designer would use in calculations. This corresponds to the yellow line shown in the example of Figure 5, even though the typical fracture plane through the weld is generally closer to the RHS fusion face (as indicated by the red line in Figure 5) and has a longer failure line. This measured weld throat size was multiplied by the weld length to obtain the weld area, where the weld length was taken as the RHS perimeter considering the rounded corners

Table 3. Comparison of actual strengths and nominal strengths per design specifications.

Specimen No.	Actual Strength (kN)	Nominal Strength (kN)			
		AISC 360-10 without "sinθ factor"	AISC 360-10 with "sinθ factor"	CSA S16-14 without "sinθ factor"	CSA S16-14 with "sinθ factor"
1	831	652	978	728	1092
2	1166	1008	1511	1125	1688
3	1235	843	1265	942	1412
4	1311	1015	1522	1133	1700
5	2433	1724	2586	1925	2888
6	2574	1836	2754	2050	3075
7	2525	1521	2281	1698	2547
8	2302	1739	2609	1942	2913
9	1020	685	1028	765	1148
10	960	627	940	700	1049
11	840	548	822	612	918
12	1140	927	1390	1035	1552
13	1200	906	1358	1011	1517
14	1207	1011	1516	1129	1693
15	1494	1495	2243	1670	2505
16	1578	1706	2559	1905	2858
17	1788	1978	2966	2208	3312
18	1131	1214	1633	1321	1919
19	982	802	1070	871	1262
20	1270	1230	1681	1344	1960
21	1534	1707	2292	1858	2697

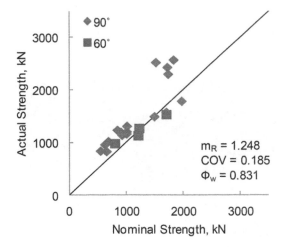

$m_R = 1.248$
$COV = 0.185$
$\Phi_w = 0.831$

Figure 9. Comparison of actual strengths and nominal strengths per AISC 360-10 without directional strength enhancement factor.

(l_w in Table 1). Using this weld length provides a more scientific evaluation of the true "sinθ effect", although it is recognized that most designers would simply calculate the weld length from H_b and B_b dimensions (especially if the branch was inclined). The H_b and B_b approach always produces a longer weld length, thus exaggerating the real weld length and giving a higher predicted strength, which is unconservative.

The fracture strengths for these welded joints are then compared to predictions by existing steel design codes/specifications (AISC 2010 and CSA 2014), using measured geometric and mechanical properties for the RHS and weld. Thus, it is then possible to determine if an appropriate safety index (or safety margin) is achieved, as expected for brittle elements (welds) in structural design codes, both with and without application of the fillet weld directional strength increase. The strengths of the four weld elements in the 60° connections (Specimen Nos. 18–21) were calculated individually since they were loaded in a different manner. All predicted nominal strengths are listed in Table 3.

The predicted nominal strengths (R_n) of the test welds per AISC 360 without any directional strength enhancement factor are compared to the actual strengths at failure in Figure 9. For the 90° connections, the nominal strengths of the test welds were determined using Equations 1a and 1b. For the 60° connections, the nominal strengths of the test welds were computed using Equation 2a. In this case, R_{nwl} is for the oblique welds at locations a and b (see Fig. 2), based on their real oblique weld lengths. Hence, all the "sinθ effects" are omitted.

The predicted nominal strengths of the test welds per AISC 360 with the directional strength enhancement factor are compared to the actual strengths at failure in Figure 10. For the 90° connections, the nominal strengths of the test welds were determined using Equations 1a and 1c. For the 60° connections, the nominal strengths of the test welds were computed using Equation 2b with the 1.5 factor for R_{nwt}. Equations 1a and 1b are used to calculate R_{nwt} since the directional strength increase is considered by the 1.5 factor already. Equations 1a and 1c are used to calculate R_{nwl}

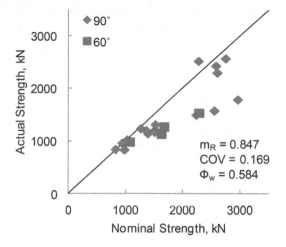

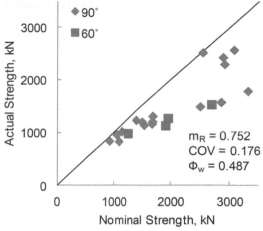

Figure 10. Comparison of actual strengths and nominal strengths per AISC 360-10 with directional strength enhancement factor.

Figure 12. Comparison of actual strengths and nominal strengths per CSA S16-14 with directional strength enhancement factor.

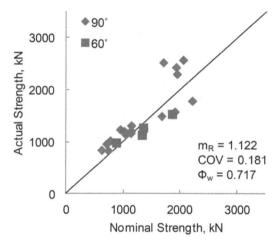

Figure 11. Comparison of actual strengths and nominal strengths per CSA S16-14 without directional strength enhancement factor.

to account for the directional strength increase factor for the 60° oblique welds. Also, R_{nwl} is multiplied by a 0.85 factor (similar to the M_w factor in Equation 3a per CSA S16) to account for the difference in deformation capacity between the oblique and transverse weld elements. (In theory, the 0.85 factor should really be higher since the 0.85 applies to longitudinal welds).

Similarly, the predicted nominal strengths of the test welds per CSA S16 with and without the "sinθ factor" are computed using Equations 3a and 3b, and are compared to the actual strengths in Figures 11 and 12.

To assess whether adequate safety margins are inherent in the correlations shown in Figures 9–12, one can check to ensure that the minimum safety indices of $\beta = 4.0$ (per Chapter B of the AISC Specification Commentary) and $\beta = 4.5$ (per Annex B of CSA S16) are achieved, using a simplified reliability analysis in

which the resistance factor ϕ_w is given by (Fisher et al. 1978; Ravindra and Galambos 1978):

$$\phi_w = m_R \exp(-\alpha\beta\text{COV}) \qquad (4)$$

where m_R = mean of the ratio: (actual strength)/(nominal strength); COV = associated coefficient of variation of this ratio; and α = coefficient of separation taken to be 0.55 (Ravindra and Galambos 1978). The calculated m_R, COV and ϕ_w values are shown in Figures 9–12.

For predicted nominal strengths per AISC 360 without the directional strength enhancement factor (Fig. 9), the application of Equation 4 produced $\phi_w = 0.831$ which exceeds 0.75 as specified by AISC 360, hence on the basis of the available experimental evidence, the predictions can be deemed adequately conservative. Similarly, the predicted nominal strengths per CSA S16 without the directional strength enhancement factor (Fig. 11) can be deemed adequately conservative as well since the calculated $\phi_w = 0.717$ exceeds 0.67 as specified by CSA S16.

However, the predicted nominal strengths per AISC 360 and CSA S16 with the directional strength enhancement factors (Figs. 10 and 12) are unsafe, since the calculated ϕ_w values (0.584 and 0.487, respectively) are lower than the corresponding specified resistance factor values (0.75 and 0.67, respectively).

6   CONCLUSIONS

In this study, a total of 21 RHS-plate-RHS weld critical connections were tested to failure under axial tension loading. The design methods in AISC 360-10 and CSA S16-14 for fillet welds to RHS members were evaluated by comparing the actual strengths of the fillet welds to the predicted nominal strengths. It was found that without the use of the $(1.00 + 0.50\sin^{1.5}\theta)$ directional strength enhancement term, an adequate level of safety can be achieved. However, the use of such a directional strength enhancement factor leads to unsafe predictions, hence restrictions need to be placed on the application of it for fillet welds to RHS members in these steel design codes.

NOTATION

e	eccentricity
l_w	total length of weld
m_R	mean of ratio: (actual strength)/ (nominal strength)
t	wall thickness of RHS
t_p	thickness of intermediate plate
t_w	effective throat thickness of weld
A_w	effective throat area of weld
B_b	overall width of RHS branch member
COV	coefficient of variation
F_{nw}	nominal stress of weld metal
F_u	ultimate strength of RHS
F_{up}	ultimate strength of plate
F_y	yield stress of RHS
F_{yp}	yield stress of plate
F_{yw}	yield stress of weld metal
H_b	overall height of RHS branch member
M_w	strength reduction factor to consider for the variation in deformation capacity of weld elements with different orientations
P	applied force
RHS	rectangular hollow section
R_n	nominal strength
R_{nwl}	total nominal strength of longitudinally loaded fillet welds
R_{nwt}	total nominal strength of transversely loaded fillet welds (without "sin θ" factor applied)
X_u	ultimate strength of weld metal
α	coefficient of separation
β	safety (reliability) index
ϕ_w	resistance factor for weld metal
θ	angle of loading measured from the weld longitudinal axis for fillet weld strength calculation (in degrees)
θ_1	angle of loading (in degrees) of the weld element under consideration;
θ_2	angle of loading (in degrees) of the weld element in the joint that is nearest to 90°

REFERENCES

AISC 2010. Specification for structural steel buildings. *ANSI/AISC 360-10*, Chicago, USA.

ASTM 2006. Standard test method for macroetching metals and alloys. *ASTM E340-06*, West Conshohocken, USA.

ASTM 2013. Standard test methods and definitions for mechanical testing of steel products. *ASTM A370-13*, West Conshohocken, USA.

AWS 2010. Structural welding code – Steel, 22nd. ed., *ANSI/AWS D1.1/D1.1M:2010*, Miami, USA.

CEN 2005. Eurocode 3: Design of steel structures – Part 1-8: Design of joints. *EN 1993-1-8:2005*, Brussels, Belgium.

CSA 2013a. Welded steel construction (metal arc welding). *CSA W59-13*, Toronto, Canada.

CSA 2013b. General requirements for rolled or welded structural quality steel/structural quality steel. *CAN/CSA-G40.20-13/G40.21-13*, Toronto, Canada.

CSA 2014. Design of steel structures. *CSA S16-14*, Toronto, Canada.

Fisher, J. W., Galambos, T. V., Kulak, G. L. & Ravindra, M. K. 1978. Load and resistance factor design criteria for connectors. *Journal of Structural Division*, American Society of Civil Engineers, 104(9):1427–1441.

ISO 2013. Static design procedure for welded hollow section joints – Recommendations. *ISO 14346:2013 (E)*, Geneva, Switzerland.

Lesik, D. F. & Kennedy, D. J. 1990. Ultimate strength of fillet welded connections loaded in plane. *Canadian Journal of Civil Engineering*, 17(1): 55–67.

McFadden, M. R. & Packer, J. A. 2013. *Effective weld properties for RHS-to-RHS moment T-connections*. Phase 1 report to the American Institute of Steel Construction, University of Toronto, Toronto, Canada.

McFadden, M. R. & Packer, J. A. 2014. Effective weld properties for hollow structural section T-connections under branch in-plane bending. *Engineering Journal*, American Institute of Steel Construction, 51(4):(pp. pending)

Miazga, G. S. & Kennedy, D. J. 1989. Behaviour of fillet welds as a function of the angle of loading. *Canadian Journal of Civil Engineering*, 16(4):583–599.

Packer, J. A., Wardenier, J., Zhao, X. L., van der Vegte, G. J. & Kurobane, Y. 2009. Design guide for rectangular hollow section (RHS) joints under predominantly static loading, Design Guide No. 3, 2nd. ed., CIDECT, Geneva, Switzerland.

Packer, J. A., Sherman, D. R. & Leece, M. 2010. Hollow structural section connections, Steel Design Guide No. 24, AISC, Chicago, USA.

Ravindra, M. K. & Galambos, T. V. 1978. Load and resistance factor design for steel. *Journal of Structural Division*, American Society of Civil Engineers, 104(9):1337–1353.

Wardenier J., Kurobane, Y., Packer, J. A., van der Vegte, G. J. & Zhao, X. L. 2008. Design guide for circular hollow section (CHS) joints under predominantly static loading, Design Guide No. 1, 2nd. ed., CIDECT, Geneva, Switzerland.

Tubular Structures XV – Batista, Vellasco & Lima (eds)
© *2015 Taylor & Francis Group, London, ISBN 978-1-138-02837-1*

Investigation of weld effective length rules for RHS overlapped K-connections

K. Tousignant & J.A. Packer

Department of Civil Engineering, University of Toronto, Toronto, Canada

ABSTRACT: A laboratory-based test program was conducted to assess the performance of welds in rectangular hollow section (RHS) overlapped K-connections. A 10 m span, simply-supported, RHS Warren truss was fabricated with nine connections designed to be weld-critical with varied key factors that affect weld strength in RHS connections: branch member overlap, chord wall slenderness, and branch-to-chord width ratio. By means of a point load applied to strategic panel points, sequential fracture of the welds to the overlapping branches was obtained. The normal strains adjacent to the welds around the RHS perimeter and branch axial loads were measured. By using the mechanical and geometrical properties of the welds and RHS, the structural reliability (or safety index) of the existing AISC specification provisions for weld effective lengths in RHS overlapped K-connections was determined. The existing AISC 360-10 formulae were found to be conservative, and hence more liberal recommendations are proposed.

1 INTRODUCTION

It is well known that the flexibility of rectangular hollow section (RHS) connections results in the non-uniform loading of the welds around a joint. Most international design recommendations have thus required that welds to branches be designed to develop the yield capacity of the member, in order to resist any arrangement of loads in the member. This requirement is almost exclusively based on old recommendations from the International Institute of Welding (IIW) (IIW 1989).

Designing welds to branches to develop the yield strength of the member is justifiable when there is low confidence in the design forces, or if plastic stress redistribution is required in the connection. This results in relatively large weld sizes and may be excessively conservative in some situations. In any respect, it has been shown that there is no definitive international agreement for how to proportion a fillet weld in order to develop the capacity of a member (McFadden et al. 2013), and thus many connections designed using this approach, despite merit, have applied larger-than-necessary welds.

Extensive laboratory tests at the University of Toronto, both on isolated RHS connections and complete trusses, have led to the development and use of a more modern, "fit-for-purpose" approach – which has become internationally recognized as an alternative to the former approach for welding tubular structures (IIW 2012, ISO 2013). By designing welds to resist the actual load in the branch member, smaller, more appropriate weld sizes typically result. It has been found that in order to account for the non-uniform loading of the weld perimeter due to the flexibility of RHS connections using a "fit-for-purpose" approach, it is necessary to use effective weld properties, including the effective length of welds, whereby portions of the weld which do not contribute to the overall resistance of the joint are discounted.

Recommendations have already been made for the effective length of welds in axially-loaded T-, Y-, X-, and gapped K-connections (Frater & Packer 1992a,b, Packer & Cassidy 1995), and these have been adopted by the American Institute of Steel Construction (AISC) in a separate section in their latest (2010) specification (ANSI/AISC 360-10: Section K4) on "Welds of Plates and Branches to RHS". As no information was available, at the time, for the effective length of welds in axially-loaded RHS overlapped K-connections, speculative formulae were agreed upon by the AISC 360 HSS technical committee and released in the same specification. While thought to be conservative, there is an absence of experimental evidence to support their validity.

A laboratory-based test program was hence conducted to assess the performance of welds in RHS overlapped K-connections in order to validate or modify these speculative aspects of the existing AISC specification. This paper reports on the findings.

2 EXPERIMENTATION

2.1 Scope

Nine RHS overlapped K-connections, with all web members at 60° to the chord, were designed into a 10 m span Warren truss, fabricated from RHS made to CAN/CSA G40.20/G40.21 (CSA 2013) Class C and ASTM A1085 (ASTM 2013) using a semi-automatic flux-cored-arc-welding (FCAW) process with a CO_2 shielding gas. The truss replicated in-situ member continuity and deflection effects (Frater & Packer 1992b), and hence was a way to ensure that the correct boundary conditions were taken into account for the connection tests. As the strength and behaviour of RHS overlapped K-connections depends on the overlap (O_v), branch-to-chord width ratio (β), and chord wall slenderness (B/t) (Packer & Henderson 1997), one of the truss chord members was selected to produce a more rigid connection ($\beta = 0.71$ and $B/t = 14.0$), and the other was selected to be more flexible ($\beta = 0.50$ and $B/t = 26.7$). The amount of overlap was also varied, so as to cover a broad range of connection parameters. The truss layout is shown in Figure 1.

The experiments focused on the welds to tension-loaded overlapping branches, which did not "pin down" the adjacent overlapped branch. This would not be the case if the overlapped branch was the tension member. The latter is a more complicated case that would be difficult to investigate experimentally. The properties of the nine test connections are shown in Table 1.

2.2 Geometric and mechanical properties of the as-laid welds

Correct input for the geometric and mechanical properties of the joints is a prerequisite for scientific analysis of the experimental data, and hence great lengths were taken to very accurately obtain this information for each RHS connection.

Such welded joints, when overlapped, are comprised of three distinct weld regions (Fig. 2): a longitudinal 90° fillet-welded region (side a), a 60° fillet-welded region (sides c and d) and a longitudinal partial-joint-penetration (PJP) flare-bevel-groove-welded region, formed by the butt joint of a matched-width connection between the overlapping branch members (side a'). In the latter region, the deposition of sound weld metal to the bottom of the flare is hindered by bridging of the weld pud-

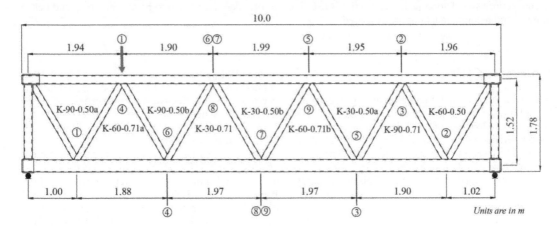

Figure 1. Truss layout and loading sequence.

Table 1. Measured properties of nine RHS overlapped K- (test) connections.

Test	RHS web member			RHS chord member						
	$B_b \times H_b \times t_b$ mm × mm × mm	A_b^* mm²	F_{yb}^{**} MPa	$B \times H \times t$ mm × mm × mm	A^* mm²	F_y^{**} MPa	O_v %	β	B/t	P_a^{***} kN
K-90-0.50a	127.0 × 127.0 × 7.78	3625	412	254.4 × 254.4 × 9.24	8804	387	90	0.50	27.5	1233
K-90-0.50b							90			1275
K-60-0.50							60			594
K-30-0.50a							30			765
K-30-0.50b							30			740
K-90-0.71	127.0 × 127.0 × 7.78	3625	412	178.7 × 178.7 × 12.53	7776	380	90	0.71	14.3	1139
K-60-0.71a							60			972
K-60-0.71b							60			861
K-30-0.71							30			1055

Note: $\theta_i = \theta_j = 60°$; and $B_{bj}/t_{bj} = 16.3$ for all connections.
*Cross-sectional areas determined by cutting a prescribed length of RHS, weighing it, and then using a density of 7850 kg/m³ to calculate its cross-sectional area.
**Yield strength of all RHS determined from tensile coupon tests performed according to ASTM A370 (2009).
***Force in overlapping web member at weld fracture.

dle between the surfaces of the two branches (Packer & Frater 2005). Thus, the PJP weld throat is highly variable. To establish consistency in the measurements, a 6 mm root gap with backing was designed into the weld detail resulting in a complete penetration (CP) detail along these weld elements. The detail was qualified in accordance with Clause 4.13 of AWS D1.1 (AWS 2010) for complete joint penetration butt joints in tubular connections and verified by ultrasonic weld inspection; the throat dimension (t_w) was hence measured according to Equation 1:

$$t_w = t_{bi} - d \qquad (1)$$

where t_{bi} = thickness of the overlapping branch member; and d = greatest perpendicular dimension measured from a line flush to the overlapping branch member surface to the weld surface.

Fillet weld sizes were determined by making a negative mould of each test weld at numerous locations along its length then machining it normal to the longitudinal axis of the root and measuring the legs

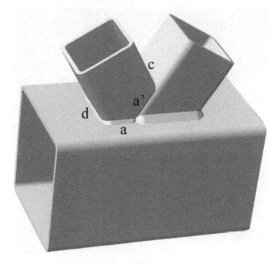

Figure 2.   Weld element regions.

and throat dimension (at each position). The effective throat was taken as the minimum distance between the root and the face of the diagrammatic weld. Over 180 weld dimensions were taken for the nine connections and the average measured values for the throats are shown in Table 2.

Mechanical properties of the as-laid welds were determined by tensile coupon tests as specified by AWS D1.1 (AWS 2010). The average yield stress (by 0.2% strain offset) was 563 MPa and the average ultimate strength (F_{EXX}) was 619 MPa with 27.5% elongation at rupture. The measured ultimate strength was 28% stronger than the nominal strength of the AWS E71T-1C electrode used.

2.3   Instrumentation and loading strategy

A single, carefully-planned, point load was applied by a universal testing machine to a strategic truss panel point resulting in a distribution of axial forces which accentuated the load in a predetermined branch member, and hence the weld to it. Since the experiments were intended to examine weld behaviour, failure was designed to always occur in the weld, instead of by some connection failure mode. After rupture occurred in one of the nine critical test welds, the location of the point load was altered to cause failure at another joint. The truss was thus translated, rotated 180° and/or inverted and the supports were re-positioned. The test order was according to the sequence shown in Figure 1.

Linear-strain gauges (SGs), oriented along the longitudinal axis of the truss members, were used to measure axial loads in branch members and to record the strain distribution around the branch members adjacent to the welded test joint. These were positioned, for the latter purpose, centred 25 mm away from the weld toe to avoid the strain concentrations caused by the notch effect (Packer & Cassidy 1995). The actual weld fracture loads (P_a) were obtained from a pair of SGs located in-plane and at mid-height of the truss, in the constant stress region (Mehrota & Govil 1972), on opposite faces of the branch members.

Table 2.   Measured weld throat and predicted fracture load for test joints according to existing AISC 360-10 provisions for weld effective lengths in RHS overlapped K-connections.

| Test | Measured weld throat dimension | | | | | | P_n^* |
	a mm	a′ mm	b mm	b′ mm	c mm	d mm	kN
K-90-0.50a	3.45	3.18	3.12	3.58	3.76	4.27	833
K-90-0.50b	4.60	3.45	3.81	3.66	3.84	4.27	872
K-60-0.50	2.67	3.81	2.39	3.56	3.86	4.22	613
K-30-0.50a	3.35	4.60	2.95	3.96	4.34	3.63	384
K-30-0.50b	3.28	4.29	3.05	4.29	4.01	3.53	380
K-90-0.71	3.18	3.89	3.18	3.81	3.63	3.84	886
K-60-0.71a	3.99	3.51	3.86	3.12	3.78	3.86	648
K-60-0.71b	3.43	3.56	3.23	3.76	3.84	3.96	661
K-30-0.71	4.57	4.93	3.40	4.78	4.27	3.78	461

Note: b and b′ are analogous to a and a′, but on the opposite side of the overlapping branch.
*Nominal predicted fracture load according to the existing AISC 360-10 specification provisions.

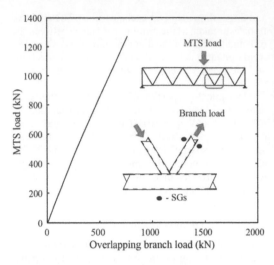

Figure 3. Typical point load versus branch load magnitude relationship (in test K-30-0.50a).

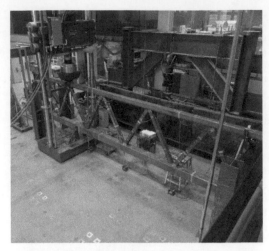

Figure 4. Testing arrangement.

The elastic loads were thus calculated according to Equation 2:

$$P_a = A_b E \varepsilon \qquad (2)$$

where A_b = cross-sectional area of the branch, determined by weighing the cross-section; E = elastic modulus of the RHS, determined by tensile coupon tests in accordance with ASTM A370 (ASTM 2009); and ε = average strain measured on opposite faces of the RHS.

Figure 3 shows the relationship between the applied MTS load and the load in the branch member in test K-30-0.50a (measured by SGs). By virtue of a constant slope (indicating a linear variation in average strain), it can be seen that the member itself remained elastic throughout the entire load range.

All of the welds to the overlapping branch members failed in a brittle manner, by fracture along a plane through the weld, which occurred simultaneously at all locations around the branch perimeter. The laboratory testing arrangement – from the first stage of the loading sequence – and a typical weld fracture – from test K-30-0.50a – are shown in Figures 4 and 5, respectively.

Figure 5. Weld fracture in test K-30-0.50a.

3 COMMENTS ON RESULTS

Graphs of load versus strain for the transverse and side welds of test K-30-0.50a are shown in Figure 6. It is seen that the magnitude of strain decreases as a function of the distance from the toe of the connection, which is caused by differences in the relative stiffness of the chord ($\beta = 0.50$ and $B/t = 27.5$) and the overlapped branch ($\beta = 1.00$ and $B_{bj}/t_{bj} = 16.3$) that result in the latter attracting more load. The magnitude of strain along the branch transverse faces decreases towards the mid-wall locations (SGs 1 and 13) – except

for the latter stages of stress redistribution at the toe of the connection – and the variation is expectedly more pronounced along the flexible heel of the connection (Fig. 6(c)), with much of the weld remaining in compression for the entire load range.

A study of the distributions of strain around the branch members adjacent to the test welds in all nine overlapped K-connections showed that the longitudinal welds are completely effective in resisting the load, whereas the transverse welds are only partially effective for certain connection parameters and generally become less effective as the β-ratio decreases, as the overlap decreases, and as the chord wall slenderness value increases. These are the same trends observed in the formulation of the existing AISC specification provisions.

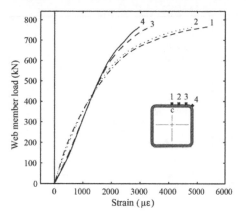

(a) Strain across transverse toe weld.

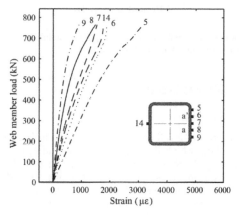

(b) Strain along longitudinal fillet and PJP welds.

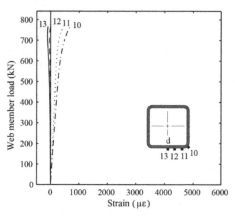

(c) Strain across transverse heel weld.

Figure 6. Load versus strain relationship in test K-30-0.50a.

4 EVALUATION OF AISC 360-10

4.1 Existing provisions for weld effective lengths in RHS overlapped K-connections

According to AISC 360-10, the available strength of welds to axially-loaded RHS branches is based on the limit state of shear rupture along the plane of the weld effective throat in accordance with Equation 3:

$$P_n = F_{nw} t_w l_e \tag{3}$$

where $F_{nw} =$ nominal strength of weld metal; $t_w =$ weld effective throat around the perimeter of the branch; and $l_e =$ effective length of fillet and groove welds.

An LRFD resistance factor, φ, equal to 0.75 and 0.80, applies for fillet welds and PJP groove welds, respectively.

In Table J2.5 of AISC 360-10, F_{nw} is specified as $0.60F_{EXX}$ for both fillet and PJP welds. In the case of the former, it implies that the failure mode is by shear rupture on the effective throat; however, for PJP groove welds, it is an arbitrary reduction factor – to account for the issues that arise from bridging of the weld puddle between surfaces. Thus, to evaluate the current design provisions, a more suitable term of $1.00F_{EXX}$ is used for F_{nw} for groove welds.

The formulae for l_e are given in section K4 of AISC 360-10 and are as follows:

– When $25\% \leq O_v < 50\%$:

$$l_{e,i} = \frac{20_v}{50} \left[\left(1 - \frac{O_v}{100}\right)\left(\frac{H_{bi}}{\sin\theta_i}\right) + \frac{O_v}{100}\left(\frac{H_{bi}}{\sin(\theta_i+\theta_j)}\right) \right] + b_{eoi} + b_{eov} \tag{4}$$

– When $50\% \leq O_v < 80\%$:

$$l_{e,i} = 2\left[\left(1 - \frac{O_v}{100}\right)\left(\frac{H_{bi}}{\sin\theta_i}\right) + \frac{O_v}{100}\left(\frac{H_{bi}}{\sin(\theta_i+\theta_j)}\right) \right] + b_{eoi} + b_{eov} \tag{5}$$

– When $80\% \leq O_v \leq 100\%$:

$$l_{e,i} = 2\left[\left(1 - \frac{O_v}{100}\right)\left(\frac{H_{bi}}{\sin\theta_i}\right) + \frac{O_v}{100}\left(\frac{H_{bi}}{\sin(\theta_i+\theta_j)}\right) \right] + B_{bi} + b_{eov} \tag{6}$$

where $i =$ subscript used to refer to the overlapping branch; $j =$ subscript used to refer to the overlapped branch; $H_b =$ overall height of the branch member measured in the plane of the connection; and $\theta =$ included angle between the branch and the chord ($=60°$ for all test connections).

The terms b_{eoi} and b_{eov} are empirically derived from laboratory tests (Davies & Packer 1982) and quantify the effective widths of weld to the branch face, normal to the plane of the connection:

$$b_{eoi} = \frac{10}{B/t}\left(\frac{F_y t}{F_{ybi} t_{bi}}\right) B_{bi} \leq B_{bi} \tag{7}$$

$$b_{eov} = \frac{10}{B_{bj}/t_{bj}}\left(\frac{F_{ybj} t_{bj}}{F_{ybi} t_{bi}}\right) B_{bi} \leq B_{bi} \tag{8}$$

where $B =$ overall width of the chord, normal to the plane of the connection; $B_b =$ overall width of the branch, normal to the plane of the connection; $t =$ wall thicknesses of the chord; $t_b =$ wall thicknesses of the branch; $F_y =$ yield stress of the chord; and $F_{yb} =$ yield stress of the branch.

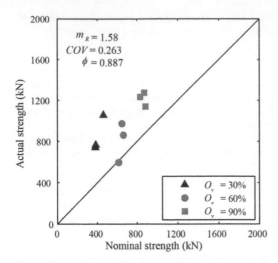

Figure 7. Correlation of the existing AISC 360-10 provisions with the test results.

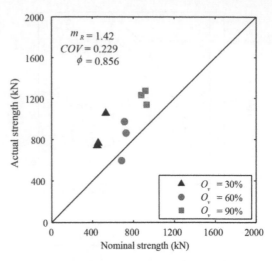

Figure 8. Correlation of the proposed modified AISC 360-10 provisions with the test results.

AISC 360-10 also currently limits the values of $b_{eoi}/2$ and $b_{eov}/2$ through a notwithstanding clause, which states,

"When $B_{bi}/B > 0.85$ or $\theta_i > 50°$, $b_{eoi}/2$ shall not exceed $2t$ and when $B_{bi}/B_{bj} > 0.85$ or $(180° - \theta_i - \theta_j) > 50°$, $b_{eov}/2$ shall not exceed $2t_{bj}$."

Thus, for the overlapped K-connections tested, the upper limits of $b_{eoi} = 4t$ and $b_{eov} = 4t_{bj}$ apply.

4.2 Safety level implicit in AISC 360-10

In order to assess whether adequate or excessive safety margins are inherent, one can check to ensure that a minimum safety index ($\beta+$) of 4.0 (as currently adopted by AISC 360-10 per Chapter B of the Specification Commentary) is achieved using a simplified reliability analysis in which the resistance factor (φ) is given by Equation 9 (Fisher et al. 1978, Ravindra & Galambos 1978), below:

$$\phi = m_R \cdot \exp(-\alpha\beta^+ COV) \qquad (9)$$

where m_R = mean of the ratio of actual element strength to predicted nominal element strength; COV = associated coefficient of variation of the ratio of actual element strength to predicted nominal element strength; and α = coefficient of separation taken to be 0.55 (Ravindra & Galambos 1978).

The implied resistance factor, φ, is equal to 0.887 for the existing AISC specification provisions and is larger than the necessary resistance factors for fillet welds and PJP groove welds (0.75 and 0.80, respectively) indicating an excessive level of safety for the current AISC 360-10 formulae. Figure 7 shows the correlation of the existing AISC 360-10 predicted nominal strengths with the experimental results.

5 RECOMMENDATION

5.1 Background

By means of 12 full-scale experiments on isolated T-connections which also showed excessive safety in the current AISC 360-10 formula for the effective elastic section modulus for in-plane bending for RHS moment T-connections, McFadden & Packer (2013, 2014) proposed a change to the requirement restricting the effective widths of welds to the branch face from two times the chord wall thickness ($2t$) to a more reasonable limit of $B_b/4$.

Their proposal increases the effective length of the transverse weld elements in most RHS connections and was shown to be applicable to the formulae for the effective length of welds in axially-loaded RHS T- and X- (or Cross) connections.

5.2 Proposal

Since the same pattern is observed for RHS overlapped K-connections, it is proposed that the existing formulae for the effective length of welds be modified in the same manner, by changing the requirement,

"When $B_{bi}/B > 0.85$ or $\theta_i > 50°$, $b_{eoi}/2$ shall not exceed $2t$ and when $B_{bi}/B_{bj} > 0.85$ or $(180° - \theta_i - \theta_j) > 50°$, $b_{eov}/2$ shall not exceed $2t_{bj}$."

to:

"When $B_{bi}/B > 0.85$ or $\theta_i > 50°$, $b_{eoi}/2$ shall not exceed $B_{bi}/4$ and when $B_{bi}/B_{bj} > 0.85$ or $(180° - \theta_i - \theta_j) > 50°$, $b_{eov}/2$ shall not exceed $B_{bi}/4$."

This change produces the correlation with the test data given by Figure 8.

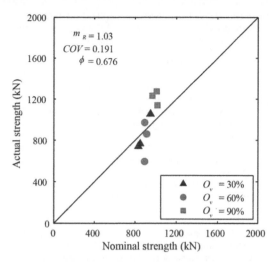

$m_R = 1.03$
$COV = 0.191$
$\phi = 0.676$

$O_v = 30\%$
$O_v = 60\%$
$O_v = 90\%$

Figure 9. Correlation without effective length rules.

5.3 Safety level implicit in recommendation

The implied resistance factor, φ, is equal to 0.856 for the recommended modification to the existing AISC specification provisions, which is still larger than the necessary resistance factors for fillet welds and PJP groove welds. More importantly, using these modified AISC provisions for RHS overlapped K-connections results in consistency for the aggregate recommended design rules for welds in RHS connections, including axially-loaded RHS T- and X- (or Cross) connections, moment T-connections, and overlapped K-connections.

It is worth noting that if no effective length rules are applied, and the total weld length was used to determine the strength of the welded joint to the overlapping branch, then the correlation with the test data shown in Figure 9 results. The implied resistance factor, φ, is equal to 0.676 which is less than the necessary resistance factors for fillet and PJP welds, illustrating that such an approach provides an insufficient safety margin.

6 CONCLUSIONS

Based on a series of careful laboratory tests on weld-critical in-situ RHS overlapped K-connections, it was shown that the existing AISC specification provisions for the effective length of welds (in such connections) can be safely modified to increase the effective length of the transverse weld elements. The proposed change is consistent with already-existing recommendations to change the AISC specification provisions for the effective length of welds in axially-loaded RHS T- and X- (or Cross) connections, and moment T-connections.

In the data analysis presented herein, the so-called "fillet weld directional strength increase" factor has not been applied when determining predicted strengths.

ACKNOWLEGEMENTS

Financial support for this project was provided by the American Institute of Steel Construction (AISC), the Natural Sciences and Engineering Research Council of Canada (NSERC), and the Steel Structures Education Foundation (SSEF). Fabrication was provided, as a gift in-kind, by Walters Inc., Hamilton, Canada and the hollow structural sections used were donated by Atlas Tube, Harrow, Canada.

NOTATION

A	cross-sectional area of the chord
A_b	cross-sectional area of the branch
B	overall width of the chord
B_b	overall width of the branch
B_{bi}	overall width of the overlapping branch
B_{bj}	overall width of the overlapped branch
COV	coefficient of variation of the ratio of actual element strength to predicted nominal element strength
E	Young's modulus of RHS
F_{EXX}	ultimate strength of weld metal
F_{nw}	nominal strength of weld metal
F_y	yield stress of the chord
F_{ybi}	yield stress of the overlapping branch
F_{ybj}	yield stress of the overlapped branch
H_b	height of the branch
H_{bi}	height of the overlapping branch
O_v	branch overlap
P_a	actual weld fracture load
P_n	nominal predicted weld fracture load
b_{eoi}	effective width of weld to the branch face adjacent to the chord
b_{eov}	effective width of weld to the branch face adjacent to the overlapped branch
d	greatest perpendicular dimension measured from a line flush to the overlapping branch member surface to the weld surface
l_e	effective length of fillet and groove welds
m_R	mean of the ratio of actual element strength to predicted nominal element strength
t	wall thicknesses of the chord
t_b	wall thickness of the branch
t_{bi}	wall thicknesses of the overlapping branch
t_{bj}	wall thicknesses of the overlapped branch
t_w	weld throat dimension
α	coefficient of separation
β	branch-to-chord width ratio (B_{bi}/B)
β^+	safety index
ε	average strain measured on opposite faces of the RHS
θ_i	included angle between the overlapping branch and the chord
θ_j	included angle between the overlapped branch and the chord
ϕ	LRFD resistance factor

REFERENCES

American Institute of Steel Construction (AISC) 2010. ANSI/AISC 360-10. Specification for structural steel buildings. Chicago, IL, USA.

ASTM International 2009. A370-09. Standard test methods and definitions for mechanical testing of steel products. West Conshohocken, PA, USA.

ASTM International 2013. ASTM A1085-13. Standard specification for cold-formed welded carbon steel hollow structural sections (HSS). West Conshohocken, PA, USA.

American Welding Society (AWS) 2010. ANSI/AWS D1.1/D1.1M:2010. Structural welding code – steel, 22nd ed. Miami, FL, USA.

Canadian Standards Association (CSA) 2013. G40.20-13/G40.21-13. General requirements for rolled or welded structural quality steel. Toronto, Canada.

Davies, G. & Packer, J.A. 1982. Predicting the strength of branch plate – RHS connections for punching shear. *Canadian Journal of Civil Engineering* 9(3): 458–467.

Fisher, J.W., Galambos, T.V., Kulak, G.L., & Ravindra, M.K. 1978. Load and resistance factor design criteria for connectors. *Journal of the Structural Division, American Society of Civil Engineers* 104(9): 1427–1441.

Frater, G.S. & Packer, J.A. 1992a. Weldment design for RHS truss connections, I: Applications. *Journal of Structural Engineering, American Society of Civil Engineers* 118(10): 2784–2803.

Frater, G.S. & Packer, J.A. 1992b. Weldment design for RHS truss connections, II: Experimentation. *Journal of Structural Engineering, American Society of Civil Engineers* 118(10): 2804–2820.

International Institute of Welding (IIW) 1989. Doc XV-701-89. Design recommendations for hollow section joints – predominantly statically loaded, 2nd ed. Paris, France.

International Institute of Welding (IIW) 2012. Doc XV-1402-12. Design recommendations for hollow section joints – predominantly statically loaded, 3rd ed. Paris, France.

International Standards Organization (ISO) 2013. ISO 14346. Static design procedure for welded hollow-section joints – recommendations. Geneva, Switzerland.

McFadden, M.R. & Packer J.A. 2013. Effective weld properties for RHS-to-RHS moment T-connections. Phase 1 project report to American Institute of Steel Construction, Toronto, Canada: University of Toronto.

McFadden, M.R. & Packer, J.A. 2014. Effective weld properties for hollow structural section T-connections under branch in-plane bending. *Engineering Journal, American Institute of Steel Construction* 51(4).

McFadden, M.R., Sun, M., & Packer, J.A. 2013. Weld design and fabrication for RHS connections. *Steel Construction* 6(1): 5–10.

Mehrotra, B.L. & Govil, A.K. 1972. Shear lag analysis of rectangular full-width tube connections. *Journal of the Structural Division, American Society of Civil Engineers* 98(1): 287–305.

Packer, J.A. & Cassidy, C.E. 1995. Effective weld length for HSS T, Y, and X Connections. *Journal of Structural Engineering, American Society of Civil Engineers* 121(10): 1402–1408.

Packer, J.A. & Frater, G.S. 2005. Recommended effective throat sizes for flare groove welds to HSS. *Engineering Journal, American Institute of Steel Construction* 42(1): 31–44.

Packer, J.A. & Henderson, J.E. 1997. Hollow structural section connections and trusses – a design guide, 2nd ed. Toronto, Canada: Canadian Institute of Steel Construction.

Ravindra, M.K. & Galambos, T.V. 1978. Load and resistance factor design for steel. *Journal of the Structural Division, American Society of Civil Engineers* 104(9): 1337–1353.

Tubular Structures XV – Batista, Vellasco & Lima (eds)
© 2015 Taylor & Francis Group, London, ISBN 978-1-138-02837-1

Experimental investigation of built-in replaceable links in external diaphragm connection between steel I-beam and CHS column

M. Khador
The University of Warwick, Coventry, UK

T.M. Chan
The Hong Kong Polytechnic University, Hong Kong, China

ABSTRACT: This paper investigates experimentally the cyclic behaviour of an external diaphragm connection between steel I-beam and circular hollow section column. The proposed joint includes two diaphragm plates welded to the outer circumference of the column and bolted to the beam flanges with two tapered cover plates. The cover plates were integrated in the connection to act as replaceable links after seismic actions. A series of full-scale laboratory experiments of the joint was conducted to investigate its energy dissipation and hysteretic response. The cover plates used in the experimental work had the same geometry but differed from each other on steel grade, bolt-holes size, use of stiffeners and bolts' preloading force. The results confirmed that the main energy dissipation fuse in these connections was yielding in the reduced section areas of the cover plates. Connection slip created a second fuse of energy dissipation when bolts preloading force was properly controlled.

1 INTRODUCTION

Tubular columns own many structural and architectural features that can make them more favourable than open-section columns in steel moment-resisting frames (Kurobane 2004 & Wardenier et al. 2010). A research study conducted by Alostaz & Schneider (1997) revealed that connections made by welding open-section beams directly to the outer skin of tubular columns were not suitable for seismic applications because they exhibited, under cyclic loading, significant distortions of the tube wall, which prevented plastic hinging of the beam and led to brittle failures of the weld. Different configurations of I-beam to tubular column connections incorporating stiffening plates at the levels of beam flanges have since been investigated, including internal, through and external diaphragm connections (Cheng & Chung 2003, Chen et al. 2004 & Wang et al. 2011). Because internal and through diaphragm connections discontinue the column at the connection level and require extensive welding work that could result in higher risks of weld defects, different arrangements of externally stiffened connections to tubular columns have been examined e.g. T-shaped stiffeners, angle stiffeners and diaphragm plates (Ting et al. 1991, Shin et al. 2008 & Wang et al. 2008).

Schneider & Alostaz (1998) carried out an investigation on the use of external diaphragm plates (DPs) welded to the outer circumference of a tubular column. The experimental cyclic results indicated high local stress levels in the DPs at the re-entrant corners

adjacent to weld. These stress concentrations resulted in fractures in both diaphragm plates, and the fractures propagated through the plates into the column wall before plastic hinging of the beam. Schneider & Alostaz (1998) recommended shifting the beam further from the column and reducing the sharpness of the re-entrant corners between the DPs and the beam to allow for better stress distribution in the DPs, better stress flow around the column and keep high strains within the beam flanges.

A numerical parametric study on a single-sided I-beam to circular hollow section (CHS) column joint under monotonic and cyclic loadings was carried out by Sabbagh et al. (2013) to establish joint details that eliminate the structural deficiencies identified above. To avoid fracture in the I-beam flanges due to local buckling failure and to protect the primary members of the joint under seismic actions, a joint detailing was proposed, which involved shifting the beam away from the column face, using tapered cover plates (TCPs) to connect the I-beam flanges to the diaphragm plates and bolting the I-beam web to a single-sided web stub plate welded to the column face. The TCPs were integrated in the joint to act as a sacrificial energy dissipation fuse by designing them to reach full plasticity before the I-beam reached its design plastic bending resistance in order to create a multi-fuse energy dissipative mechanism in the joint under seismic actions. Yielding in TCPs was the main energy dissipation mechanism accompanied by bolts slippage and yielding in the beam's section located right after the connection as two extra energy dissipation fuses.

In this paper, the work of Sabbagh et al. (2013) is extended experimentally. A series of full-scale laboratory experiments of an external diaphragm joint was conducted in the Structures Laboratory at the University of Warwick to examine its seismic performance and the influence of certain parameters on the overall hysteretic response and energy dissipation of the connection. Efficient choice of the geometry and material properties of the different joint components was crucial to achieve a moment-resisting frame that can be classified in the Damage Control Structural Performance range, as identified in FEMA-356 (2000), for which post-seismic repair operations are required yet minimal to provide quick reoccupancy of the building. To achieve this, the TCPs were deigned to dissipate most of the seismic energy and act as sacrificial components that can be easily replaced after a seismic event whilst the I-beam and the rest of the joint components remain within their elastic range. Because of the influence of the replaceable links on the overall seismic performance of the joint, different types of them were investigated in the experimental program, and selected results are presented herein.

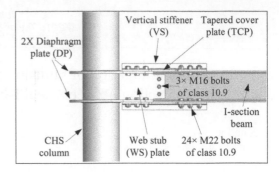

Figure 1. Front view of the test specimen.

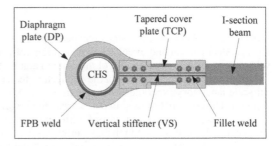

Figure 2. Top view of the test specimen.

2 EXPERIMENTAL SET-UP

2.1 Test specimens

The proposed single-sided joint included two diaphragm plates shop-welded to the outer circumference of a circular hollow section column using full-penetration butt (FPB) welds and field-bolted to the I-beam flanges with two tapered cover plates. A single-sided web stub (WS) plate was welded to the CHS column and bolted to the I-beam web with three M16 bolts of class 10.9. The specimens of the tests were identical except for their tapered cover plates that were bolted to the I-section beam from one end and to the diaphragms plates from the other end using 24 × M22 bolts of class 10.9. The I-beam (UKB: 203 × 133 × 30) was used in this joint and was designed to be 2 m long. The CHS column (244.5 × 10) with the height of 2 m was used in all tests, and its boundary conditions were designed to be hinges to represent the inflection points at the mid-height of columns in a moment frame. The arrangement and components of the beam to column joint after assembly are illustrated in Figures 1–2.

The geometry of the diaphragm plates, illustrated in Figure 3, was chosen efficiently to allow for a smooth stress flow around and into the CHS column to eliminate any stress concentration in them or distortion of the column wall.

The web connection was designed to transfer shear forces from the I-beam to the column face efficiently. The four parameters investigated in the experiments were all related to the TCPs and their bolts, whilst the rest of the joint components were kept similar in all tests. Figure 4 shows the general geometry of the TCPs. The reference used for tests' titling represents the values of the examined parameters and follows the format

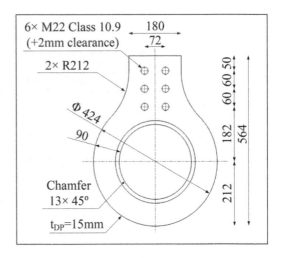

Figure 3. Geometry of the DPs. Dimensions in mm.

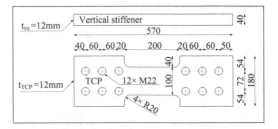

Figure 4. Geometry of the TCPs and their stiffeners when applicable. Dimensions in mm.

Table 1. Tests titling according to the type of TCPs.

Test Title	Steel grade G	Size of holes H	Use of stiffeners S	Bolt preload BP
1 S235-OSH-NS-FP	S235	Oversized	No	Full
2 S355-OSH-NS-FP	S355	Oversized	No	Full
3 S235-OSH-NS-HP	S235	Oversized	No	Half

G-H-S-BP; where G is the steel grade of the TCPs and it is either S235 or S355, H represents the size of the TCPs bolt-holes that were either standard round holes or oversized holes, S reflects the use or absence of stiffeners in TCPs and BP is the preloading force acting on the TCPs bolts that were either full-preloaded in line with Eurocode 3: BS EN 1993-1-8 (2005) for slip-resistant connections where $F_{p,Cd} = 0.7 \times f_{ub} \times A_s$ or half-preloaded $F_{p,Cd} = 0.35 \times f_{ub} \times A_s$ where f_{ub} and A_s are the ultimate strength and the tensile stress area of the bolt respectively. This paper investigates the influence of two out of the four parameters detailed above, which are the TCPs steel grade and bolts preloading, through the results of three experiments. The investigation of the influence of the use of stiffeners was presented in a previous publication by the authors (Khador & Chan 2014). Table 1 presents the type of the TCPs used in each of the tests reported in this paper.

2.2 Material properties

All the joint components were of steel grade S355 except the TCPs that were either S235 or S355 as detailed above. Tensile coupon tests were carried out in the Material Laboratory at the University of Warwick in accordance with BS EN ISO 6892-1 (2009) to determine the actual material properties of the steel that composes the different components of the test specimens.

2.3 Test rig

A test rig was designed and built in the Structures Laboratory at the University of Warwick to transfer the loads applied to test specimens to the strong floor safely. A 100kN hydraulic actuator, connected to a reaction frame, was used to apply vertical displacement-controlled loading to the specimen's I-beam. A loading mechanism consisting of swivel hinges was designed to accommodate the rotations of the beam whilst maintaining the verticality of the actuator. Two more hinges were designed and connected to the ends of the specimen's column to provide the column's boundary conditions assumed at the design stage. Another reaction frame was built to restrain the column at both ends and transfer the forces to the strong floor. The two reaction frames were connected together with horizontal and diagonal bracing to provide out-of-plane stability. Lateral supports were designed and built on both sides of the beam to

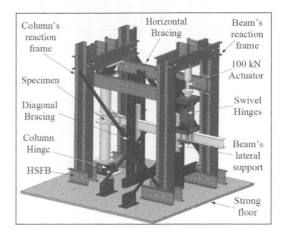

Figure 5. Schematic arrangement of the test rig.

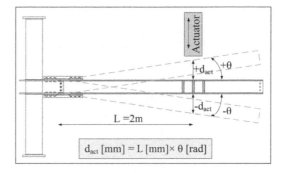

Figure 6. Loading protocol.

avoid any lateral torsional buckling during testing. The schematic of the test rig is shown in Figure 5.

2.4 Loading protocol

The loading protocol adopted for this experimental study was based on the provisions of AISC (2010) for qualifying beam-to-column moment joints in special and intermediate moment frames. The cyclic loading was conducted by controlling the vertical deflection of the beam section located under the actuator as illustrated in Figure 6. Table 2 presents the proposed drift angles for the different cycles and the equivalent actuator displacements.

2.5 Instrumentations

Electrical post-yield strain gauges were used at the expected locations of plastic hinges, stress concentration regions and other areas of interest. Eight displacement Transducers (DTs) were installed horizontally at different locations of the test specimen to measure primary slips and displacements. Electronic inclinometers (INCs) were also used in the tests to measure angle of rotation at five different locations of the specimen and calculate the accurate rotation of the connection. All instrumentations were calibrated prior

Table 2.	Loading protocol.	
Cycle number	Drift angle rad	Actuator displacement mm
1–6	±0.00375	±7.5
7–12	±0.005	±10
13–18	±0.0075	±15
19–22	±0.01	±20
23–24	±0.015	±30
25–26	±0.02	±40
27–28	±0.03	±60
29–30	±0.04	±80

Table 3. Summary of the experimental results.

	Test Title	Maximum moment kN.m	Maximum rotation mrad	Dissipated energy kN.m.rad
1	S235-OSH-NS-FP	88.1	21.1	12.76
2	S355-OSH-NS-FP	99.6	20.1	11.41
3	S235-OSH-NS-HP	65.9	25.8	12.76

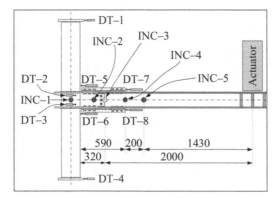

Figure 7. Instrumentations. Dimensions in mm.

to and after conducting the tests. Figure 7 shows the locations of the DTs and INCs on a schematic drawing of the test specimen.

3 EXPERIMENTAL RESULTS

3.1 Hysteretic response

Tests 1–2 were conducted up to the end of the 30th cycle of the loading protocol described in section 2.4 of this paper, whereas Test–3 was stopped at the end of its 29th loading cycle due to technical challenges related to the laboratory facilities. When comparing the results of two tests, they were considered up to the maximum loading cycle that was reached in both of them. The results of all tests revealed that yielding was limited to the reduced section areas of the TCPs whereas all the other joint components showed elastic response throughout testing as intended at the design stage. The main response parameters obtained from the three reported tests are presented in Table 3, and their hysteresis plots are illustrated in Figures 8–10. In these graphs, the moments of the connection M_{conn} were calculated at the beam's connected end and were then normalised by the design elastic bending resistance of the beam $M_{b,el,y,Rd} = 117.46\,kN.m$. The connection rotations θ_{conn} were obtained by subtracting the measurements of INC–2 from those of INC–3

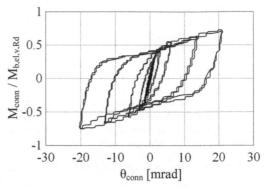

Figure 8. Moment–rotation curve of Test–1: S235-OSH-NS-FP.

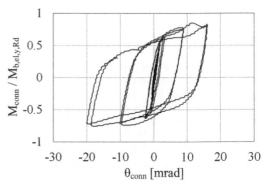

Figure 9. Moment–rotation curve of Test–2: S355-OSH-NS-FP.

to exclude the elastic deformations of the column web panel and the web stub from the evaluation of the rotation capacity of the connection. It can be noted from Figures 8–10 that the connection exhibited generally stable hysteretic behaviour in the three tests.

3.2 Energy dissipation

Accumulated energy dissipation values at the end of each loading cycle of the three tests were obtained by calculating the areas enclosed by the moment–rotation hysteresis loops, see Figure 11. The total energy dissipation values at the end of the three tests are presented in Table 3. It can be seen from Figure 11 that the energy dissipation values of the connection were very small in the three tests and the difference between them

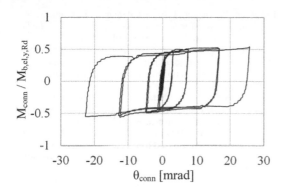

Figure 10. Moment–rotation curve of Test–3: S235-OSH-NS-HP.

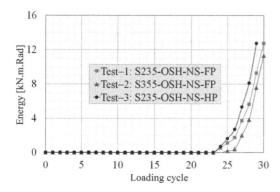

Figure 11. Accumulated energy dissipation for Tests 1–3.

was negligible up to cycle 24 after which a dramatic increase is noticed in the three tests due to the onset of yield in the TCPs, slip in the connection or both.

3.3 Steel grade comparison

Both pairs of the tapered cover plates used in the specimens of Test–1 and Test–2 were unstiffened (NS), had the same size of bolt holes (OSH) and the same preloading force acting on their bolts (FP). The two tests differed only in the steel grade of their TCPs and hence comparing their results demonstrated the influence of TCPs steel grade on the overall hysteretic behaviour of the connection. The use of S355–TCPs resulted in a stronger connection than the one with S235-TCPs. However, comparing strain measurements in both tests revealed that there were higher stress demands imposed on the I-beam and the diaphragm plates when the steel grade of the TCPs was higher. Furthermore, the maximum rotation of the connection with S355–TCPs was 5% less than its S235 counterpart. The last observation was interpreted further in Figure 11 that shows higher accumulated energy dissipation rates for the connection with S235–TCPs than the one with S355–TCPs in all the inelastic loading cycles.

3.4 Bolts preloading force comparison

Both pairs of TCPs used in the specimens of Test–1 and Test–3 were unstiffened (NS), of the same steel grade (S235) and had the same size of bolts' holes (OSH) i.e. they were identical in all parameters except the preloading force acting on their bolts and hence comparing their results demonstrated the influence of bolts preloading value on the overall performance of the connection. It was found that reducing the bolts' preloading force in Test–3 to 50% of that in Test–1 caused the stress demands imposed on the I-beam and the DPs to be lower in Test–3 by 12% and 11% respectively, and improved the maximum rotation of the connection by 18% due to the activation of connection slip. This resulted in higher energy dissipation rates in Test–3 than that in Test–1 as can be seen from Figure 11 and allowed the connection rotation to exceed the minimum rotation capacity required for plastic hinge regions in medium ductility class (DCM) structures, which is 25 mrad.

4 CONCLUSIONS

A series of full-scale laboratory experiments were carried out on an external diaphragm connection to CHS column incorporating different types of tapered cover plates which were designed to act as replaceable links in the connection after seismic actions whilst the rest of the joint components keep intact. The results of three tests were reported in this paper highlighting the effect of using different steel grades of TCPs and controlled bolts preloading force on the overall hysteretic response of the connection and its energy dissipation.

The results presented in this paper confirmed that the main energy dissipation fuse in these connections was yielding in the reduced section areas of the TCPs; and that using higher steel grades of TCPs imposed higher stress demands on the beam and the diaphragm plates and dissipated less energy. Reducing the preloading force acting on the TCPs bolts to half the value recommended in Eurocode 3: Part 1–8 (2005) for slip-resistant connections created a second fuse for energy dissipation that was slip activation in the oversized bolt-holes of the connection, which subsequently led to higher rates of energy dissipation.

ACKNOWLEDGMENTS

The authors are grateful to the Engineering and Physical Science Research Council (EP/1020489/1) for the project funding and Damascus University for the doctoral scholarship awarded to the first author. Financial support from TATA Steel, the School of Engineering at the University of Warwick and the technical support from the Structures Laboratory at the University of Warwick are gratefully acknowledged. The authors wish also to thank Dr Alireza Bagheri Sabbagh for his advice and contribution to this research.

REFERENCES

Alostaz, Y. M. & Schneider, S. P. 1997. Analytical behavior of connections to concrete-filled steel tubes. *Journal of Constructional Steel Research* 40(2): 95–127.

ANSI/AISC 341-10. 2010. *Seismic provisions for structural steel buildings*. American Institution of Steel Construction (AISC).

BS EN 1993-1-8. 2005. *Eurocode 3: Design of steel structures – Part 1-8: Design of joints*. British Standards Institution (BSI).

BS EN ISO 6892-1. 2009. *Metallic materials – Tensile testing – Part 1: Method of test at ambient temperature*. British Standards Institution (BSI).

Chen, C. C., Lin, C. C. & Tsai, C. L. 2004. Evaluation of reinforced connections between steel beams and box columns. *Engineering Structures* 26(13): 1889–1904.

Cheng, C. T. & Chung, L. L. 2003. Seismic performance of steel beams to concrete-filled steel tubular column connections. *Journal of Constructional Steel Research* 59(3): 405–426.

FEMA-356. 2000. *Prestandard and commentary for the seismic rehabilitation of buildings*. The American Society of Civil Engineers and the Federal Emergency Management Agency (FEMA).

Khador, M. & Chan, T. M. 2014. Structural behaviour of external diaphragm connection between steel I-section beam and circular hollow section columns under cyclic loading. *Proceedings of the 7th European Conference on Steel and Composite Structures (EuroSteel), Napoli, 10–12 September 2014.*

Kurobane, Y., Packer, J. A., Wardenier, J. & Yeomans, N. 2004. *Design guide for structural hollow section column connections*. Tüv-Verlag. (ed.). Köln: CIDECT.

Sabbagh, A. B., Chan, T. M. & Mottram, J. T. 2013. Detailing of I-beam-to-CHS column joints with external diaphragm plates for seismic actions. *Journal of Constructional Steel Research* 88: 21–33.

Schneider, S. P. & Alostaz, Y. M. 1998. Experimental behavior of connections to concrete-filled steel tubes. *Journal of Constructional Steel Research* 45(3): 321–352.

Shin, K. J., Kim, Y. J. & Oh Y. S. 2008. Seismic behaviour of composite concrete-filled tube column-to-beam moment connections. *Journal of Constructional Steel Research* 64(1): 118–127.

Ting, L. C., Shanmugam, N. E. & Lee, S. L. 1991. Box-column to I-beam connections with external stiffeners. *Journal of Constructional Steel Research* 18(3): 209–226.

Wang, W. D., Han, L. H. & Uy, B. 2008. Experimental behaviour of steel reduced beam section to concrete-filled circular hollow section column connections. *Journal of Constructional Steel Research* 64(5): 493–504.

Wang, W., Chen, Y., Li, W. & Leon, R.T. 2011. Bidirectional seismic performance of steel beam to circular tubular column connections with outer diaphragm. *Earthquake Engineering & Structural Dynamics* 40: 1063–1081.

Wardenier, J., Packer, J. A., Zhao, X. L. & van der Vegte, G. J. 2010. *Hollow sections in structural applications*. Geneva: CIDECT.

Tubular Structures XV – Batista, Vellasco & Lima (eds)
© *2015 Taylor & Francis Group, London, ISBN 978-1-138-02837-1*

Structural behaviour of T RHS joints subjected to chord axial force

A. Nizer
PGECIV – Civil Engineering Post-Graduate Program, UERJ – State University of Rio de Janeiro, Brazil

L.R.O. de Lima, P.C.G. da S. Vellasco, S.A.L. de Andrade, E. da S. Goulart & A.T. da Silva
Structural Engineering Department, UERJ – State University of Rio de Janeiro, Brazil

L.F. da C. Neves
INESCC, Civil Engineering Department, University of Coimbra, Portugal

ABSTRACT: This paper presents an experimental and numerical analysis of the chord normal stresses influence on the "T" tubular joints behaviour. The experimental program consisted of six tests: two without chord axial forces; two with compressive chord forces; and the remaining two with tensile chord forces. These results were compared to numerical results obtained with the aid of the ANSYS 12.0 software, and to the design provisions present in the: Eurocode 3; NBR 16239:2013 Brazilian standard and ISO 14346. This assessment indicated that the results obtained for chords without axial forces and with compressive forces, in both experiments and numerical analyses, were safe when compared to the values predicted by all the investigated standards, where the best fit was obtained by ISO 14346. On the other hand, the results associated to cases with chord tension forces proved to be conservative.

1 INTRODUCTION

An increase in the use of tubular structural elements in Brazil such as the example depicted in Figure 1 was observed in recent years as a consequence of their aesthetical and structural advantages. A review of the analytical approaches available to predict the structural response of these structures indicated the major importance of their connections. Many examples in nature show the optimum properties of the tubular shape when loaded in compression, torsion and bending in all directions. Furthermore, the section closed shape without sharp corners reduces the surface area to be protected and extends the corrosion protection life (Wardenier *et al.*, 2010a).

Extensive work has been performed to address the technological and design issues of these structures around the World in the last decades (Rondal *et al.*, 1992, Wardenier, 2000). Several international design codes and recommendations explicitly cover the design of tubular joints (Wardenier, 2000, EN 1993-1-8, 2010, IIW, 1989 and IIW, 2009). Last year, two codes focusing on tubular joints design were updated with the recent advances in this field (ISO14346, 2013, NBR16239, 2013).

The joints between the tubular members are a critical issue, and many studies indicate that further research is needed, especially for some particular joint geometries. Over the last few years many studies have been performed in Brazil to better understand the

Figure 1.   Cidade Nova Subway Station, Rio de Janeiro, Brazil.

tubular joints behaviour aiming to provide Brazilian structural engineers with an extensive background to their structural design and response.

In the past, some experimental results on the influence of the chord stresses over the brace load capacity were published. These results showed that compressive chord stresses considerably reduce the joint resistance. However, very few studies indicated the same reduction in load carrying capacity for joints under the action of chord tensile forces. This trend is reflected by: EN 1993-1-8 (2010) and NBR 16239 (2013) design codes. At this point it is necessary to observe that the second edition of the CIDECT design guide for RHS joints (Packer et al., 2009) and ISO 14346 (2013) preconize some joint resistance reduction for both cases, i.e., tensile and compressive chord stresses. Additionally, Lipp & Ummenhofer (2014) based on experimental and numerical results, proposed a reduction of the CHS joint design capacity to consider the presence of tensile chord loads.

These were the main motivations for the present investigation where experiments and numerical simulations centred the influence of tension and compression chord stresses on RHS chord to SHS braces T joint geometry were performed (Nizer, 2014).

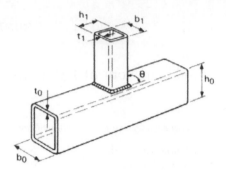

Figure 2. T-joint main geometrical parameters (Packer et al., 2009).

2 TUBULAR JOINTS BACKGROUD

Traditionally, design rules for hollow sections joints are based on either plastic analysis or on a deformation limit criteria. The use of plastic analysis to define the joint ultimate limit state is based on a plastic mechanism corresponding to the assumed yield line pattern. Typical examples of these approaches can be found on Packer et al. (2009), Cao et al. (1998), Packer (1993) and Kosteski et al. (2003).

One interesting investigation dealing with RHS T tubular joints was published by Korol & Mirza (1982), focusing on a numerical FE model with shell elements that indicated a simultaneous increase of the joint resistance with the increase of the geometrical parameters β and γ (Figure 2). This study identified the need for establishing a deformation limit failure criteria for these connections. Packer et al. (1989) also observed the same trend of connection resistance increase and developed a failure path theory to estimate the plastic load capacity of connections.

Zhao & Hancock (1993), Cao et al. (1998), Zhao (2000), and Kosteski et al. (2003) also observed that both the joint's resistance and initial stiffness increase with a simultaneous increase of the geometrical parameters β and γ (Figure 2). Wardenier et al. (2010b) and Zhao et al. (2010) presented the latest recommendations for tubular joints design concerning the IIW (2009) improvements.

When deformation limit criteria are used to obtain the tubular joint resistance, these are associated to the out-of-plane deformation of the loaded chord face, and the ultimate limit state is established as a maximum deformation of this component in that direction. These deformation limit criteria are used because the joint stiffness does not vanish but may maintain considerable values due to membrane effects for slender chord faces (i.e., with a large width to thickness ratio) and after complete yielding by bending.

The deformation limit proposed by Lu et al. (1994) may be used in the evaluation of axial or bending loads for joints subjected to bending and axial forces. The joint resistance is based on the comparison of the deformation at the chord-brace intersection for two loads levels: the ultimate resistance, N_u, which corresponds to a chord out-of-plane displacement of

Table 1. Limits for geometrical parameters – T tubular joints – chord yield.

EN 1993-1-8 (2010) and NBR 16239 (2013)	ISO14346 (2013)
$0,25 \leq \beta = b_1/b_0 \leq 0,85$	$\beta \geq 0,1 + 0,01b_0/t_0$ but $0,25 \leq \beta \leq 0,85$
$10 \leq 2\gamma = b_0/t_0 \leq 35$ and class 1 or 2	$2\gamma = b_0/t_0 \leq 40$ and class 1 or 2
b_1/t_1 e $h_1/t_1 \leq 35$ and class 1 or 2	b_1/t_1 e $h_1/t_1 \leq 40$ and class 1 or 2
$0,5 \leq h_0/b_0 \leq 2,0$	$0,5 \leq h_0/b_0 \leq 2,0$
$0,5 \leq h_1/b_1 \leq 2,0$	$0,5 \leq h_1/b_1 \leq 2,0$

$\Delta_u = 3\%$ b_0, and the serviceability limit, N_s, that corresponds to an out-of-plane displacement of $\Delta_s = 1\%$ b_0. For a joint with a peak load (N_{peak}) with an associated deformation smaller than 3% b_0, this peak load is considered to be the joint resistance. If the peak load (N_{peak}) is associated to a deformation greater than 3% b_0, the ratio N_u/N_s should be considered. If the ratio N_u/N_s is less than 1.5, the joint design should be based on the ultimate limit state, N_u. Conversely, if N_u/N_s is greater than 1.5, the joint design is controlled by the serviceability limit multiplied by 1.5, i.e., $1.5 \times N_s$.

At this point it is fair to mention that Zhao et al. (2010) performed an evaluation of this criterion and the authors concluded that a better agreement with experimental results was obtained using only the 3% deformation limit criterion.

3 DESIGN CODES

According to EN 1993-1-8 (2010), NBR 16239 (2013), CIDECT Design Guide (Packer et al., 2009) and ISO14346 (2013), several geometrical parameters must be verified prior to the evaluation of the joint resistance. These parameters are presented in Figure 2, where b_0 and t_0 and b_1 and t_1 represent the width and the thickness of the chord and the brace, respectively. The limits for these geometrical parameters are presented in Table 1, for all investigated design codes, related to the chord yield failure.

Table 2. Design resistances of SHS brace to RHS chord T joints.

EN 1993-1-8 (2010) and NBR 16239 (2013)	$N_{1,Rd} = \dfrac{1}{\gamma_{M5}} \dfrac{k_n f_{y0} t_0^2}{(1-\beta)\,sen\theta_1} \left(\dfrac{2\beta}{sen\theta_1} + 4\sqrt{1-\beta}\right)$

$k_n = 1,3 - \dfrac{0,4}{\beta} n \leq 1,0$ if $n > 0$ (chord in compression)

$k_n = 1,0$ if $n \leq 0$ (chord in tension)

$n = \dfrac{N_{0,Ed}}{A_0 f_{y0}} + \dfrac{M_{0,Ed}}{W_{el,0} f_{y0}}$

ISO14346 (2013)

$F_i^* = Q_u Q_f \dfrac{f_{y0} t_0^2}{sen\theta_1}$

$Q_u = \dfrac{2\eta}{(1-\beta)\,sen\theta_1} + \dfrac{4}{\sqrt{1-\beta}}$

$Q_f = (1 - |n|)^{C_1}$ with $C_1 = 0,6 - 0,5\beta$ if $n < 0$ (chord in compression) and

$C_1 = 0,10$ if $n \geq 0$ (chord in tension)

$n = \dfrac{F_0}{F_{pl,0}} + \dfrac{M_0}{M_{pl,0}}$

Design equations for chord yield in design codes consist of a strength function for the non-dimensional joint parameters β and γ, and an additional function for the chord use (n_p), the chord yield stress f_{y0} and the chord thickness t_0. Table 2 summarizes the equations dealing with the chord face failure mode for the T joints studied in this work according to EN 1993-1-8 (2010), NBR 16239 (2013) and ISO14346 (2013).

Table 3. Experimental programme – mechanical parameters ($E = 200$ GPa, $v = 0.3$, $\beta = 0.71$ and $2\gamma = 35$).

Test	Chord use (n)	f_{y0} [MPa]	f_{u0} [MPa]	Chord load [kN]
TN01N0	0	361.9	418.6	0
TN02N0	0	361.9	418.6	0
TN03N50+	+0.5	361.9	418.6	306.9
TN04N70+	+0.7	361.9	418.6	429.6
TN06N50−	−0.5	361.9	418.6	306.9
TN05N70−	−0.7	361.9	418.6	429.6

4 EXPERIMENTAL PROGRAMME

The experimental programme for SHS brace to RHS chord T joints, summarized in Table 3, consisted of six prototypes made with ASTM-A36 steel grade. The chords were rectangular hollow sections RHS $140 \times 80 \times 4$ with a 1000 mm length, while the braces were square hollow sections SHS $100 \times 100 \times 3$ with a 400 mm length (Fig. 3). The adoption of these sections correspond to non-dimensional parameters of $\beta = 0.71$ and $2\gamma = 35$. The prototypes were instrumented with strain gauges and LVDTs to obtain the full joints behaviour (Nizer, 2014).

In the first two tests, only the braces were axially loaded in compression using a 3000 kN Universal Lousenhausen test machine. In one test the chord ends was free to rotate (TN01N0), while in another the chord was restrained for rotation around its longitudinal axis (TN02N0). In the other four tests the prototypes were subjected to axial tension or compression forces in the first load step using a hydraulic jack with +50% (TN03N50+), +70% (TN04N70+), −50% (TN06N50−) and −70% (TN05N70−) of the chord member axial yield capacity. During this first load step, the braces were not connected to the test machine to avoid bending of the braces caused by elongation of the chord due to first load step. In the second load step, the axial load was kept constant while the braces were axially loaded in compression.

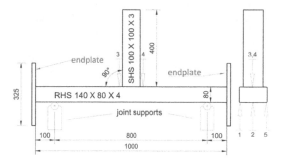

Figure 3. T Joint geometrical properties and LVDTs layout.

For the tests where compression forces acted on the chords, a hydraulic jack coupled with a load cell was used between the chord endplate and the reaction frame (Fig. 4 and Fig. 6a). For the tests where the chords were loaded in tension, a hydraulic jack coupled with a load cell was introduced outside the reaction frame while four DWIDAG bars connected the chord endplate (Fig. 5 and Fig 6b).

The measurement of the mean chord indentation was performed with five LVDTs: two placed at the chord top face, near the brace weld (3 and 4) and three

Figure 4. Experimental layout – chord in compression.

Figure 5. Experimental layout – chord in tension.

used at the chord bottom face to compensate the beam displacements (1, 2 & 5) (Fig. 3).

5   NUMERICAL MODEL

The numerical model developed for this study adopted four-node thick shell elements (SHELL181) with six degrees of freedom per node available in the Ansys Element Library (Ansys 12, 2010), therefore considering bending, shear and membrane deformations. The finite element mesh was more refined near the welds, where the stress concentration is more likely to occur, and as more regular as possible, with well-proportioned elements to avoid numerical problems. Figure 7 presents an overview of the developed finite element model. The welds were modelled with shell elements as well, according to Lee & Wilmshurst (1997) – see Figure 8. The weld size was modelled according to the mean measured weld size (5mm). Convergence studies were also performed to define the numerical model optimum mesh size.

Axial loading was applied by displacing the nodes at the right-hand end of the chord, while the right-hand end was left restrained to simulate the experimental tests boundary conditions. The brace loading was also applied using displacements at the top nodes. This procedure simulated the test set-up previously described and that was used to calibrate this numerical model. A full nonlinear analysis was performed considering material and geometrical nonlinearities. The stress *versus* strain curves obtained from the tensile coupons tests were adopted in the numerical model (Nizer,

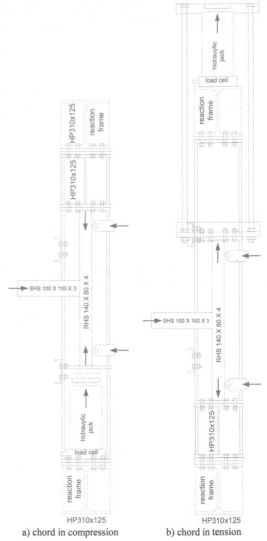

a) chord in compression   b) chord in tension

Figure 6.   Load application scheme.

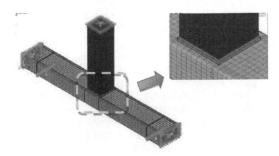

Figure 7.   T joint numerical model and boundary conditions.

2014) using the *von Mises* yield criterion. The geometrical nonlinearity was implemented according to the updated Lagrangian Algorithm.

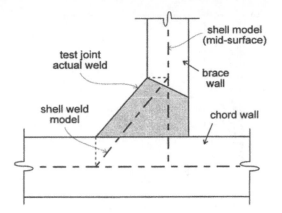

Figure 8. Weld model – shell elements (Lee & Wilmshurst, 1997).

6 RESULTS

6.1 *Experimental versus numerical*

The experimental load *versus* displacement curves depicted in Figure 9, were obtained from the difference between LVDTs readings located at the top and bottom chord faces (Fig. 3).

As previously referred, the chord boundary conditions affect the joint global behaviour, fact that was also cnfirmed with the comparison of the results from tests TN01N0 and TN02N0. For the first test, where the chord end rotations were free, the load corresponding to 3% of chord indentation was equal to 84.47 kN, while for the case where the rotations were restrained, this load was 90.1 kN.

Observing Figure 9, it can be noticed that normal chord stresses directly affect the global response of the investigated SHS to RHS T joints ($\beta = 0.71$ and $2\gamma = 35$). For the tests where tension loads were applied to the chord, an increase of the joint resistance was observed. On the other hand, when compressive axial loads were applied to the chord, a decrease of the joint resistance was noticed. For the test TN05N70− (n = −0.7) a premature decrease of the joint resistance associated to an elevated axial force level corresponding to a chord use ratio of 0.7 was also observed. This fact may be related to the β value and it will be discussed in the next sections of this paper together with the numerical results.

Figures 10–15 present the comparison of experimental and numerical load *versus* displacement curves. The curves of non-preloaded (n = 0) chords, show that the numerical results reached greater values of resistance when compared to their experimental counterparts. Residual stresses from the welding process may occur in the SHS to RHS T-joints. Generally, these stresses increase the flexibility of the load–displacement behaviour (Lipp & Ummenhofer, 2014). This influence is not included in the numerical models because the welding process before brace loading has not been simulated. The existence of residual

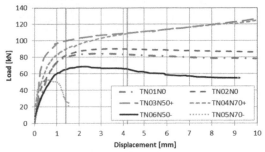

Figure 9. Load *versus* displacement curves.

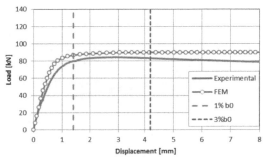

Figure 10. Load *versus* displacement curves – TN01N0.

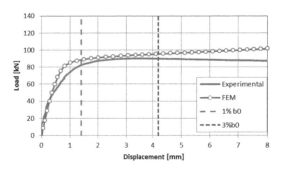

Figure 11. Load *versus* displacement curves – TN02N0.

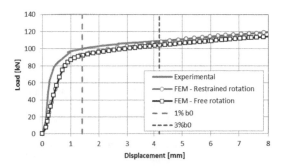

Figure 12. Load *versus* displacement curves – TN03N50+.

stresses may be one reason for the observed differences between experimental and numerical results. Numerical results for the chord stresses were obtained for free chord end and for fully restrained chord end.

375

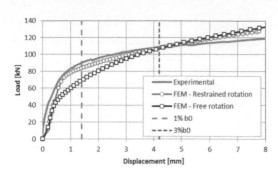

Figure 13. Load *versus* displacement curves – TN04N70+.

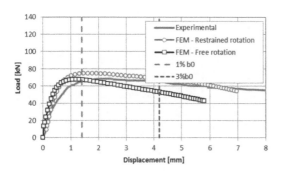

Figure 14. Load *versus* displacement curves – TN06N50–.

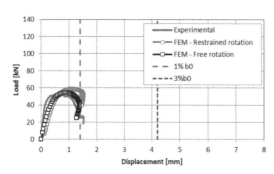

Figure 15. Load *versus* displacement curves – TN05N70–.

For the cases where tensile stresses were applied to the chord, the numerical results showed load *versus* displacement curves in a lower boundary when compared to their experimental counterparts. On the other hand, when compressive chord stresses were considered, the numerical results presented higher values than their experimental counterparts.

Aiming to investigate the influence of the parameter β over the joint global response, eight different geometries were considered in the numerical model corresponding to β values of 0.35, 0.42, 0.50, 0.57, 0.67, 0.71, 0.78 and 0.85. The load *versus* displacement curves for these cases is depicted in Figure 16. The β value of 0.71 corresponds to the experimental test TN05N70. It is clear in Figure 16 that varying the β value strongly affects the joint global response. For the cases of $\beta \leq 0.57$, peak loads were observed beyond a chord indentation corresponding to 1% b_0. On the

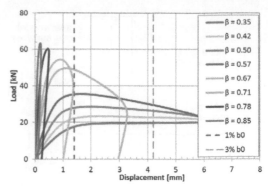

Figure 16. Load *versus* displacement curves – evaluation of the β value influence.

Figure 17. Numerical and experimental deformed shapes – compressive chord stress (n = −0.7) and $\beta = 0.71$.

other hand, for the cases where $\beta > 0.57$, peak loads were observed before the chord indentation of 1% b_0. Figure 17 presents the deformed shapes obtained from experimental and numerical analyses. It is quite easy to spot from the numerical curves the occurrence of local buckling in the chord top face justifying the abrupt decrease in the load *versus* displacement curve for this case. This occurred due the superposition of normal stresses caused by the chord compressive stress and the bending on the chord top face due to brace loading.

Table 4. Summary – experimental *versus* numerical results.

Test	Chord use (n)	N_{EXP} [kN]	N_{FEM} [kN]	N_{FEM}/N_{EXP}
TN01N0	0	84.5	89.98	1.06
TN02N0	0	90.1	95.32	1.06
TN03N50+	+0.5	109.2	107.95	0.99
TN04N70+	+0.7	109.0	107.77	0.99
TN06N50−	−0.5	68.6	67.15	0.98
TN05N70−	−0.7	52.9	54.41	1.03

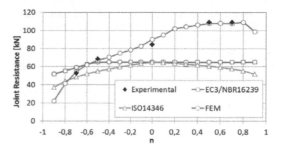

Figure 18. Comparison of experimental, numerical and design codes results.

Table 5. Summary of comparison – experimental, numerical and design codes results.

Test	N_{EXP} [kN]	N_{FEM} [kN]	N_{EC3} and N_{NBR} [kN]	N_{ISO} [kN]
TN01N0	84.5	89.98	65.06	65.06
TN02N0	90.1	95.32	65.06	65.06
TN03N50+	109.2	107.95	65.06	60.70
TN04N70+	109.0	107.77	65.06	57.68
TN06N50−	68.6	67.15	65.06	54.98
TN05N70−	52.9	54.41	59.07	48.56

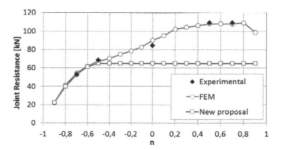

Figure 19. Comparison of experimental, numerical and new proposal design equation.

Similar conclusions concerning normal stress distribution along the thickness in CHS-X joints subjected to tensile chord stresses were obtained by Lipp & Ummenhofer (2014).

Table 4 presents a summary of the comparison between experimental and numerical results. The joints resistances were obtained using the 3% deformation limit criteria from Lu *et al.* (1994) or peak load in the cases when this occurs before a 3% b_0 indentation. As observed, a good agreement between experimental and numerical results was obtained with a maximum 6% error.

6.2 Numerical results versus design codes

Table 5 and Figure 18 present the comparison between experimental, numerical and design codes results.

In the numerical and experimental results, for the tests where tension loads were applied to the chord, an increase of the joint resistance was verified. On the other hand, a decrease of the joint resistance was observed when compression loads were applied to the chord.

The increase of the joints resistance is not considered in the codes. In EN 1993-1-8 (2010) and in NBR 16239 (2013), no reduction of the joint resistance is proposed when tension loads are applied to the chord. According to these two codes, application of compression loads to the chord results in reductions in the joint resistance for the cases where $\beta \geq 0.5$. On the other hand, ISO14346 (2013) prescribes a joint resistance reduction when tension or compression loads are applied to the chord.

Comparing the numerical results to the resistance values obtained from the design codes, for the cases where compression loads were applied to the chord, ISO14346 (2013) presented better results for $0.6 \leq \beta \leq 0.8$. For the cases with $\beta > 0.8$, neither EN 1993-1-8 (2010), NBR 16239 (2013) nor ISO14346 (2013) produced good results. So, for the SHS to RHS tubular joint studied in this work ($\beta = 0.71$ and $2\gamma = 35$), to perform a safe design, a new equation for the case where compression loads are applied to the chord for $n \leq -0.5$ is proposed:

$$k_n = -3.6|n|^2 + 3.4|n| + 0.2 \leq 1 \qquad (1)$$

The results obtained using this new equation are plotted in Figure 19, and it should be stressed that the joint resistance is not reduced where tension loads are applied to the chord, but only for compressive chord loads and if $n \leq -0.5$. The better agreement between experimental and numerical results, when chord compressive loads are applied, and using the proposed equation is depicted in Figure 19.

7 CONCLUDING REMARKS

This paper presented an experimental and numerical analyses of the influence of chord normal stresses on "T" tubular joints behaviour. The experimental program consisted of six tests, being two without chord axial forces, two with compressive chord forces and the remaining two with tensile chord forces.

According to the experimental and numerical results, it may be concluded that normal chord stresses directly affect the global response of the investigated SHS to RHS T joints ($\beta = 0.71$ and $2\gamma = 35$). For the tests where tension loads were applied to the chord, an increase in the joint resistance was verified. On the other hand, when compressive axial load was applied to the chord, a reduction of the joint load carrying capacity was observed. For the test TN05N70− (n = −0.7) a premature reduction of the joint resistance associated to an elevated axial force level (corresponding to a chord utilization ratio of 0.7) was observed.

Both EN 1993-1-8 (2010) and NBR 16239 (2013), do not prescribe a reduction of the load carrying capacity when tension loads are applied to the chord. Alternatively these two codes, for cases where compression loads are applied to the chord and $\beta \geq 0.5$, prescribe reductions in the joint resistance. On the other hand, ISO14346 (2013) includes a joint resistance reduction if tension or compression loads are applied to the chord.

A new equation for the case where compression loads are applied to the chord for $n \leq -0.5$ was proposed based on EN 1993-1-8 (2010) and NBR 16239 (2013) formulation. The obtained results, presented in Figure 19, showed an improved performance of the proposed equation when compared to experimental and numerical results.

ACKNOWLEDGEMENTS

The authors would like to thank CAPES, CNPq and FAPERJ for the financial support to this research program. This work has also been partially supported by the Portuguese Foundation for Science and Technology under project grant[s] PEst-OE/ EEI/UI308/ 2014.

REFERENCES

ABNT NBR 16239 (2013) Design of steel and composite structures for buildings using hollow sections, Associação Brasileira de Normas Técnicas, São Paulo, Brazil (*in portuguese*).

Ansys 12.0 ®, 2010 ANSYS – Inc. Theory Reference.

Cao, J.J., Packer, J.A., Young, G.J. (1998), Yield line analysis of RHS connections with axial loads, *Journal of Constructional Steel Research*, 48: 1–25.

EN 1993-1-8 (2010) Eurocode 3 – Design of steel structures – Structures – Part 1–8: Design of joints. CEN, ECCS, Brussels.

IIW (1989) Design recommendation for hollow section joints—Predominantly statically loaded, 2nd Edition, International Institute of Welding, Document XV-701-89, Cambridge, U.K.

IIW (2009), Static design procedure for welded hollow section joints – Recommendations. 3rd Edition, International Institute of Welding, Sub-commission XV-E Annual Assembly, Singapore, IIW Doc. XV-1329-09.

ISO 14346 (2013) Static design procedure for welded hollow-section joints – Recommendations, International Organization for Standardization, Switzerland.

Kosteski, N., Packer, J.A., Puthli, R.S. (2003), A finite element method based yield load determination procedure for hollow structural section connections, *Journal of Constructional Steel Research*, 59: 427–559.

Korol R., Mirza F. (1982) Finite Element Analysis of RHS T-Joints. *Journal of the Structural Division*, ASCE. 108:2081–2098.

Lee, M.M.K, Wilmshurst, S.R., 1997, Strength of Multiplanar Tubular KK-Joints Under Antisymmetrical Axial Loading. Journal of Structural Engineering, 123 (6): 755–764.

Lipp, A., Ummenhofer, T. (2014) Influence of tensile chord stresses on the strength of circular hollow section joints. Steel Construction – Design and Research, ECCS, 7: 126–132.

Lu L.H., de Winkel G.D., Yu Y., Wardenier J. (1994) Deformation limit for the ultimate strength of hollow section joints. Proceedings of the 6th International Symposium on Tubular Structures, Melbourne, pp. 341–347.

Nizer, A. (2014) Analysis of the influence of chord normal stresses on the behavior of hollow sections connections. Master Dissertation, PGECIV – Post Graduate Program in Civil Engineering, UERJ, Brazil (in Portuguese).

Packer J., Morris G., Davies G. (1989) A limit states design method for welded tension connections to I-section webs. *Journal of Constructional Steel Research*. 12: 33–53.

Packer, J.A., (1993), Moment Connections between Rectangular Hollow Sections, *Journal of Constructional Steel Research*, 25: 63–81.

Packer, J. A., Wardenier, J., Zhao, X.-L., Vegte, G. J. van der, Kurobane, Y. (2009). *Design guide for rectangular hollow section (RHS) joints under predominantly static loading*. 2nd Edition, Construction with hollow steel sections, No. 3, CIDECT.

Rondal J., Wurker, K.G., Wardenier J., Dutta D., Yeomans N. (1992) Structural Stability of Hollow Sections, CIDECT.

Wardenier J. (2000) Hollow Sections in Structural Applications, CIDECT.

Wardenier, J., Packer, J.A., X.-L. Zhao and van der Vegte, G.J. (2010a) Hollow sections in Structural Applications, CIDECT.

Wardenier, J., van der Vegte, G.J., Packer, J.A. and Zhao, X.-L. (2010b) Background of the New RHS Joint Strength Equations in the New IIW (2009) Recommendations, 13th. International Symposium on Tubular Structures, Hong Kong, China.

Zhao X., Hancock G. (1993) Plastic Mechanism analysis of T-joints in RHS subject to combined bending and concentrated force. Proceedings of the Fifth International Symposium on Tubular Connections held at Nottingham, UK, E & FN Spon, London, pp. 345–352.

Zhao X.-L. (2000) Deformation limit and ultimate strength of welded T-joints in cold-formed RHS sections. Journal of Constructional Steel Research. 53:149–165.

Zhao, X.-L., Wardenier, J., Packer, J.A., van der Vegte, G.J., (2010) Current static design guidance for hollow-section joints, Structures and Buildings 163, SB6: 361–373.

Tubular Structures XV – Batista, Vellasco & Lima (eds)
© *2015 Taylor & Francis Group, London, ISBN 978-1-138-02837-1*

Influence of tensile chord stresses on the strength of CHS X-joints – Experimental and numerical investigations

A. Lipp & T. Ummenhofer

Karlsruhe Institute of Technology, Research Center for Steel, Timber and Masonry, Germany

ABSTRACT: The current version of the Eurocode for the design of joints does not account for reduction of brace load capacity for welded hollow section joints with tensile preloading of the chord. However, the more recent 2nd edition of the CIDECT design guides or the ISO standard for welded hollow section joints includes a strength reduction due to tensile chord stresses. This paper presents results of experimental and numerical investigations on the influence of tensile chord stresses to the capacity of circular hollow section X-joints. A newly proposed chord load function for circular hollow section joints subjected to tensile chord stresses is the conclusion of this study.

1 INTRODUCTION

Light lattice structures made of steel hollow profiles are applied widely in structural engineering because of constructional and aesthetical advantages. The main fields of application are bridge constructions, onshore- and offshore-wind energy plants, crane constructions but also buildings with large spans such as exhibition halls and sports stadiums.

Unstiffened welded hollow section joints exhibit a very complex structural behaviour. For this reason, researchers began to study about 50 years ago the strength and distribution of stiffness, the material properties and further characteristics of joint types. In numerous experimental tests have many different failure modes been observed. Furthermore, researches identified a large number of parameters which influence the load capacity of hollow section joints. One of these parameters is the chord pre-loading. Already in the 1960's, experimental studies on the influence of chord preloading on the brace load capacity were published. The results of these early studies have been that compressive chord stresses considerably reduce the strength of the joint. It was supposed for tensile chord stresses that these stresses only lead to an insignificant reduction of the ultimate brace load capacity due to stabilisation of the chord wall by tensile stresses. However, no limitations of chord deformation in the ultimate and serviceability limit state were considered at that time. More recent studies, considering different deformation criteria, indicate that there is a limitation of joint strength (because of governing deformation aspects) when the chord is subjected to tensile stresses although not being as severe as for compressive stresses.

2 CURRENT SITUATION

The brace load capacity of unstiffened welded CHS and RHS joints depend on numerous parameters. One important parameter is the utilisation of the chord due to axial chord stresses. The external elastic pre-load ratio n_p and the maximum plastic pre-load ratio n are the two common definitions of the chord pre-load ratio.

n_p is the elastic chord utilisation (based on the elastic limit state) due to axial stresses without stresses from the force component(s) of the brace(s) parallel to the chord. The plastic chord utilisation n (based on the plastic limit state) is calculated with a linear interaction relationship between the axial force and the bending moment. Figure 1 illustrates the definition of both pre-load ratios using the example of a K-joint without an external bending moment M_{0p} acting in the chord. In the case of an X-joint with an axially loaded chord, the

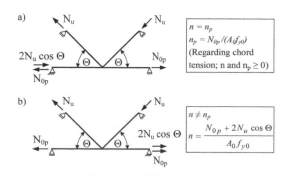

Figure 1. Definitions of pre-load ratios n_p and n.

preload ratio n_p is similar to n because no force components of the braces parallel to the chord are transferred into the chord.

Togo (1967) already documented in the 1960s that compressive chord stresses reduce the brace load capacity significantly based on experimental investigation on CHS X-joints with preloaded chords, Another main conclusion has been that the influence of tensile chord stresses on the ultimate brace load capacity has a minor effect. Other early studies, summarized by the authors (2012), agreed with this opinion. However, at that time no limitation of the chord indentation in the ultimate and serviceability limit states existed.

In the current version of the European code EN 1993-1-8 (2010) which is the version from 2005 with incorporated amendment from 2009, there is also no reduction of the brace load bearing capacity when the chord is subjected to tensile pre-loading. The reduction factor due to chord preloading is unity for tensile chord stresses.

The chord load function according to the 1st edition of the CIDECT design guide No. 1 (Wardenier et al., 1991) for all types of CHS joints, which has been adopted into EN 1993-1-8 (2010), has the following form:

$$f(n_p) = 1 + 0.3n_p - 0.3n_p^2 \quad \text{chord compress. } (n_p < 0)$$

$$f(n_p) = 1 \quad \text{chord tension } (n_p \geq 0)$$

Whereas the function for all CHS joints according to the 2nd edition of the CIDECT design guide (Wardenier et al., 2008), adopted into ISO 14346 (2013), is as follows:

$$f(n) = (1 - |n|)^{C_1}$$

with the coefficient:

$C_1 = 0.45 - 0.25\beta$ for T-, Y- and X-joints with compressive chord stresses $(n < 0)$

$C_1 = 0.25$ for K-joints with compressive chord stresses $(n < 0)$

$C_1 = 0.20$ for all joint types with tensile chord stresses $(n \geq 0)$

There exists a reduction of joint stiffness and a limitation joint strength (caused by governing deformation aspects) due to tensile chord stresses as stated in topical numerical studies, carried out for instance by Liu et al. (1998, 2004), van der Vegte et al. (2001a, 2001b, 2003), Choo et al. (2003a, 2003b) and Wardenier et al. (2007). It has to be noted, that different deformation criteria, for example the $0.03d_0$-deformation limit (where the local indentation of the chord is limited to 3% of the outer diameter of the chord) by Lu et al. (1994) or the displacement limit by Yura et al. (1981), have been considered in all recent surveys. The authors (2012) discussed the results of these investigations in

detail. Based on numerical studies, the 2nd edition of the CIDECT design guides for CHS (Wardenier et al., 2008) and RHS joints (Packer et al., 2009) include a reduction of joint capacity due to tensile chord stresses.

3 EXPERIMENTAL RESEARCH

3.1 General notes

Experimental data on the effect of tensile chord stresses on hollow section joints is very limited. Within the scope of the recent CIDECT research project 5CC (Ummenhofer and Lipp, 2013), two uniaxial pre-tests on stub CHS specimens (see chapter 3.2) and eight tests on CHS X-joints with different preload ratios have been performed (see chapter 3.3).

The non-dimensional geometric parameters β (ratio between the outer diameter of the brace and the chord) and 2γ (slenderness of the chord = ratio between the outer diameter and the thickness of the chord) of the specimens have been defined on the basis of a numerical study carried out by van der Vegte et al. (2003). The reduction with the highest magnitude of brace load capacity occurred for $\beta = 0.48$ and $2\gamma = 25.4$ for CHS X-joints. Therefore, these parameters have been used for the experiments. The chords for all specimens were made of the identical, hot finished circular hollow section according to EN 10210 (2006) in S355. For this reason, all specimens have the same material properties.

Tensile coupon tests have been carried out from the chord material. The mean yield stress f_{y0} of the test coupons was 458 N/mm^2 and the mean tensile strength f_{u0} was 603 N/mm^2.

3.2 Pre-tests on stub CHS specimens

In order to ensure the defined tensile preload ratios in the biaxial tests on CHS X-joints, two pre-tests on stub CHS test specimens with the same end-plates and branch plates as the CHS X-joints (see chapter 3.3) have been carried out. The test specimens (with a length of the CHS of 400 mm, end-plates and branch plates with a thickness of 25 mm) have been loaded in tension axially until fracture using a servo-controlled hydraulic jack. The displacement of the movable crosshead of the testing device has been recorded in pre-tests. Further strains into the longitudinal direction of the test specimens have been measured with four strain gauges per specimen in the middle of the CHS.

Figure 2 shows the specimens after testing. Both specimens fractured in a ductile manner in the base material of the circular hollow section approximately in the middle of the CHS. After testing there haven't been observed any cracks in the area of the welds between the CHS and the end-plates, respectively the branch plates. A plot of the tensile stresses in the CHS against the longitudinal strains in the centre of one test specimen is shown in figure 3. The tensile stresses have been calculated with the measured dimensions of the CHS of the test specimen and the actual test force. A good agreement between the stress-strain-curves from the tensile coupon tests and the pre-tests on the stub CHS specimens has been observed.

3.3 Tests on CHS X-joints

The experimental programme on CHS X-joints consisted of eight test specimens divided into two series (series 1 and series 2) and is summarised in table 1. For all eight test specimens, the chords were made of the

Figure 2. Stub CHS specimen after ductile.

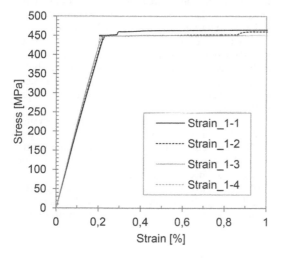

Figure 3. Experimental stress-strain-curve from one pre-test (Ummenhofer and Lipp, 2013).

identical circular hollow section. Series 1 consisted of two uniaxial loaded tests without tensile pre-loading of the chord member (CHS_X_1 and CHS_X_2). The braces were axially loaded in compression. The braces have been loaded with a hydraulic jack. Series 2 are biaxial tests and consist of six test specimens. The test specimens were subjected to axial tensile preload forces in the first load step with different utilisations of the chord of 60% (CHS_X_8), 75% (CHS_X_7), 90% (CHS_X_3 and CHS_X_4) and respectively 100% (CHS_X_5 and CHS_X_5) of the yield stress of the chord material.

In the second load step, the preload was maintained by the testing device and the braces of the test specimens were axially loaded in compression using a hydraulic jack.

For the experiments on CHS X-joints a test device with a bending table has been used (see figure 4). On the bending table, sliding carriages have been installed. On these sliding carriages, two frames with high bending stiffness have been mounted to avoid deflections at the ends of the braces perpendicular to the longitudinal axis of the braces. Due to this installation, the ends of the braces were just able to move on the longitudinal brace axis. Furthermore, friction forces between the frames and the bending table were minimized using the sliding carriages. The branch plates at the ends of the chord of the CHS X-joint were clamped into the testing device. Therefore, the chord was fixed against rotation around its longitudinal axis. The vertically oriented chord was at first loaded axially in tension with a defined preload. The braces have not been connected with the frames yet to avoid bending of the braces caused by elongation of the chord due to tensile preloading. When the defined preload was reached, the force was maintained by utilization the servo-controlled hydraulic jack of the test device. Then, the ends of the braces have been connected with the frames using screws. After this procedure, the braces have been loaded in compression using a horizontally oriented hydraulic jack until the final displacement level.

The measurement of twice the mean indentation of the chord ($=2 \cdot \delta_{1,mean}$) has been carried out with two magnetic position sensors (see figures 4 and 5). One sensor has been positioned at the left side, another one

Table 1. The experimental research programme, nominal dimensions, material properties and non-dimensional joint parameters.

Test specimen	Loading case	n [−]	Nominal dimensions [mm]				Non-dimensional parameters [−]				Steel grade EN 10210-1	f_{y0} [MPa]
			d_0	t_0	d_1	t_1	α	β	2γ	τ		
CHS_X_1	uniaxial	0	101.6	4.0	51.0	4.0	11.81	0.50	25.4	1.0	S355	355
CHS_X_2		0										
CHS_X_3	biaxial	0.9	101.6	4.0	51.0	4.0	11.81	0.50	25.4	1.0	S355	355
CHS_X_4		0.9										
CHS_X_5		1										
CHS_X_6		1										
CHS_X_7		0.75										
CHS_X_8		0.6										

Figure 4. Test rig for CHS X-joints (Ummenhofer and Lipp, 2013).

Figure 5. Test specimen deflection $2 \cdot \delta_{1,mean} \approx 50\,mm$ (CHS_X_7).

at the right side of the test specimen. The inside width between the chord and the clamps for the sensors was 75 mm when the braces were unloaded.

From this configuration it becomes evident that the measured deflections $2 \cdot \delta_{1,mean}$ also include shortenings of the braces. However, these deformations are negligible proportionally to the chord indentations. At the end of the brace, opposite to the horizontally oriented hydraulic jack, a load cell has been positioned. This load cell was screwed via an end-plate with the brace (see figure 4, right brace). Figure 5 shows a strongly deformed X-joint after testing. During the

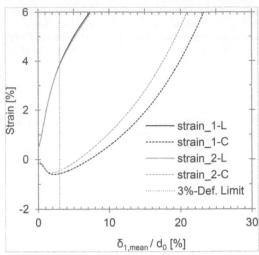

Figure 6. Experimental strain vs. chord indentation – specimen CHS_X_3: L = longitudinal; C = circumferential (Ummenhofer and Lipp, 2013).

loading procedure of the braces, strains in the chord into the longitudinal direction of the chord (strain gauges 1-L and 2-L) and into circumferential direction of the chord (strain gauges 1-C and 2-C) of one specimen with a preload ratio of 90% (CHS_X_3) have been recorded. The strain gauges have been applied on the surface of the chord in the middle between both braces. Figure 6 shows the results of the strain measurement. The strains are plotted against indentation of the chord.

During the loading procedure of the braces, strains in the chord into the longitudinal direction of the chord (strain gauges 1-L and 2-L) and into circumferential direction of the chord (strain gauges 1-C and 2-C) of one specimen with a preload ratio of 90% (CHS_X_3) have been recorded. The strain gauges have been applied on the surface of the chord in the middle between both braces. Figure 6 illustrates the results of the strain measurement. The strains are plotted against indentation of the chord.

Figure 7 shows the load displacement curves of all CHS X-joint tests. The axial brace load N_1 is divided by the measured yield stress of the chord material f_{y0} and the squared thickness of the chord t_0^2. The indentation of the chord $\delta_{1,mean}$ has been normalised on the outer diameter of the chord d_0.

4 FEM-SIMULATION OF THE EXPERIMENTS

Finite element analyses have been carried out to simulate the experiments on the X-joints. The analyses of the joints were performed using the static procedure of the program system. Furthermore, geometric and material nonlinearities have been considered in the finite element analyses.

Three dimensional 20-noded elements with three translatory degrees of freedom per node and reduced

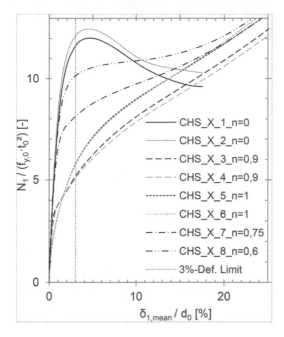

Figure 7. Experimental load-displacement curves of the CHS X-joints (Ummenhofer and Lipp, 2013).

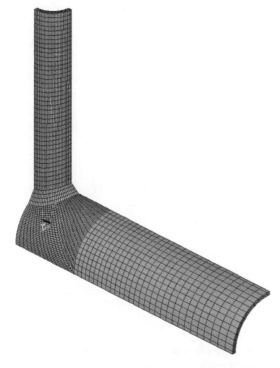

Figure 8. Meshing of the finite element model (Ummenhofer and Lipp, 2013).

integration behaviour were used for the simulation of welded CHS X-joints. These elements show a quadratic displacement behaviour. Convergence studies were performed to define the mesh size of the numerical model. Four layers of solid elements were applied over the thickness of the chord wall and three layers over the thickness of the brace wall. The elements of the chord were approximately cubic in the intersection area of the chord and the brace. Figure 8 shows the meshing of the numerical model.

Within the scope of the research project tensile coupon tests of the chord wall were carried out. The measured nominal stress-strain curves from the tensile coupon tests were then converted into true stress-logarithmic strain curves. The relationship between nominal stress and true stress or rather nominal strain and logarithmic strain is as follows:

$$\sigma_{true} = (1 + \varepsilon) \cdot \sigma$$

$$\varepsilon_{\ln} = \ln(1 + \varepsilon)$$

where σ and ε are the measured nominal stress and strain from the tensile coupon tests.

The converted stress-strain curves were used in the finite element analyses. Therefore, a multilinear isotropic hardening model was used for large strain analyses of the CHS X-joints. This model uses von Mises' yield criteria coupled with an isotropic hardening assumption. The Young's modulus and the Poisson's ratio used in the finite element models was $210000 \, N/mm^2$ or rather 0.3.

For the elements of the weld, the same material properties were used as for the elements of the hollow section. The weld geometry was considered in the finite element analysis because of the significant influence on the strength and stiffness of CHS X-joints as shown by van der Vegte (1995). In this study, the same 20-noded elements were used for the welds as well as for the hollow sections. The weld size was modeled according to the mean measured weld size. Figure 9 shows a comparison of the experimental and numerical load-displacement curves of the non-preloaded (n = 0) and the fully preloaded (n = 1) joints. The numerical load-deformation curves of the joints with n = 0 and n = 1 are clearly in a better agreement with the experimental curves compared to the curves with n = 0.6, 0.75 and 0.9. To clarify this aspect further research is required.

5 CONCLUSION AND RECOMMENDATIONS

Both experimental and numerical data achieved in the CIDECT research project 5CC (Ummenhofer and Lipp, 2013) show, that tensile chord stresses strongly influence the load-deformation characteristic of the investigated CHS X-joint ($\beta = 0.5$ and $2\gamma = 25.4$). This relationship becomes evident in figure 7. A reduction of the joint stiffness occurs due to tensile chord preloads that depend on the present preload ratio. Therefore, the magnitude of limitation of the joint strength strongly depends on the deformation limit used. Furthermore, experimental and numerical load-deformation curves show, that tensile chord stresses do

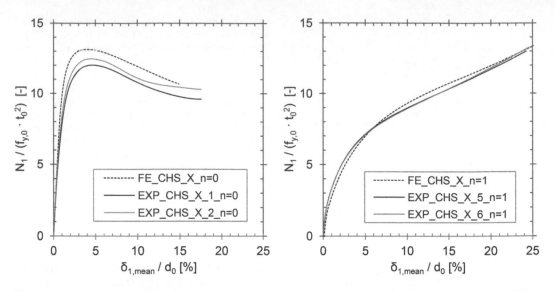

Figure 9. Comparison of experimental and numerical load-displacement curves with $n = 0$ and $n = 1$ (Ummenhofer and Lipp, 2013).

not influence the ultimate brace load capacity if chord indentations are not considered. All tensile preloaded joints ($n > 0$) reached the same compressive axial brace load as those joints not preloaded ($n = 0$). This observation is in agreement with early experiments on CHS X-joints carried out by Togo (1967). However, the pre-loaded joints showed large deformations ($\delta_1 \approx 20$ to $25\% \cdot d_0$, see figure 7) and large strains (see figure 6) in this loading state. Therefore, the authors of this article recommend the use of the $0.03d_0$-deformation limit by Lu et al. (1994) in order to avoid large joint deformations under service loads and cracking in the ultimate limit state. During the experiments, no cracks have been observed, but the tests have been carried out at room temperature and nominal chord thicknesses of 4 mm. Numerous parameters, such as low temperature, thick walled chords and high strain rates, can result to a reduction of joint ductility.

The use of the $0.03d_0$-deformation limit causes a distinctive limitation of joint strength due to tensile chord stresses. Figure 10 illustrates the effect of tensile chord preloading on the capacity of the investigated X-joints based on the $0.03d_0$-deformation limit and a comparison with the CIDECT function from 2008 (Wardenier et al., 2008). Experimental data with $n > 0$ have been normalised on the mean value of the experimental data with $n = 0$. Both experimental and numerical joint strength for a pre-load ratio $n = 1$ is not zero as characterised by the CIDECT function. For this reason a new chord load function $f(n)$ for CHS joints is proposed here. The design formula for uniplanar CHS X-joints according to EN 1993-1-8 (2010) bases on Togo's ring model calibrated with experimental data. Togo's simple ring model simplifies a CHS X-joint into a ring with a diameter $d_0 - t_0 \approx d_0$, a thickness t_0 and an effective width w_e. The brace loads are applied as line loads in the four saddle points of the joint. Six plastic hinges (or yield lines, each with a length of

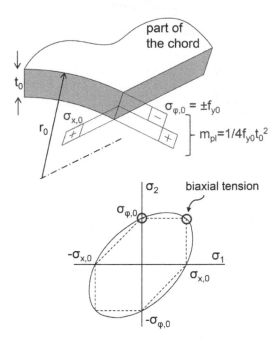

Figure 10. Superposition of longitudinal and circumferential stresses in the chord member (Lipp and Ummenhofer, 2014).

the effective width of the ring), where stresses act into circumferential direction of the chord, arise in the ultimate limit state. Tensile and compressive stress blocks of the plastic moment m_{pl} are shown in figure 10.

Using of the von Mises or Tresca yield criterion, yielding of the chord material occurs, when the longitudinal and also the circumferential tensile stresses reach the yield stress f_{y0} (see figure 10). Thus, one half of the chord thickness can still be taken into account for

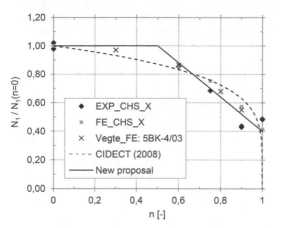

Figure 11. Effect of tensile chord preloading an a CHS X-joint with $\beta = 0.5$ and $2\gamma = 25.4$ based on 0.03d-deformation limit (Ummenhofer and Lipp, 2013).

longitudinal tensile chord stresses without reduction of the brace load capacity. Against this background the newly proposed chord load function f(n) for tensile chord stresses is a bilinear function with no strength reduction for n smaller than 50% (see figure 11):

$$f(n) = 1.6 - 1.2n \leq 1$$

For full preloading, the brace load capacity is still 40% of the non-preloaded joint. In general, the function is in a good agreement with experimental and numerical data. It has to be noted, that the experimental data points with preload ratios of 90% are below the proposed function. This circumstance does not result into joint collapse, but slightly increased indentations and strains in the serviceability and ultimate limit state. The new chord load function has been developed for CHS X-joints but can also be used for other CHS joint types. This approach is on the safe side because the effect of tensile chord stresses is most significantly for X-joints as shown by van der Vegte (1995).

NOTATION

CHS   circular hollow section
RHS   rectangular or square hollow section
m_{pl} plastic moment capacity per length unit (Togo's ring model)
$M_{pl,i}$ plastic moment capacity of member i
$N_{i,Rd}$ design value of the joint resistance, expressed in terms of the internal axial force in the brace member i (i = 1 or 2)
$N_{pl,i}$ axial yield capacity of member i
d_i external diameter of CHS member i
f_{u0} tensile strength of the chord
f_y yield stress
f_{y0} yield stress of the chord
g' gap between brace members in a K or N joint divided by the chord wall thickness

n   plastic stress ratio in the chord $= N_0/N_{pl,0} + M_0/M_{pl,0}$
n_p elastic prestress ratio in the chord $= \sigma_{p,Ed}/f_{y0}$
w_e effective width of the ring (Togo's ring model)
r_i external radius of CHS member i
σ nominal stress
σ_{true} true stress $= (1 + \varepsilon)\sigma$
$\sigma_{p,Ed}$ maximum compressive stress in the chord excluding the stress due to the components parallel to the chord axis of the axial forces in the braces
t_0 chord wall thickness
β ratio of the mean diameter or width of the brace members to that of the chord $= d_1/d_0$ (for CHS-T, -Y and -X joints) $= (d_1 + d_2)/2d_0$ (for CHS-K and -N joints)
ε nominal strain
ε_{ln} logarithmic strain $= \ln(1 + \varepsilon)$
γ half external diameter to thickness ratio of the chord $(2\gamma = d_0/t_0)$
γ_M partial safety factor on joint resistance
θ_i included angle between brace member i (i = 1 or 2) and the chord

REFERENCES

EN 10210. 2006. Hot finished structural hollow sections of non-alloy and fine grain steels – Part 1: Technical delivery conditions; – Part 2: Tolerances, dimensions and sectional properties.

EN 1993-1-8. 2010. Eurocode 3: Design of steel structures. Part 1-8: Design of joints.

ISO 14346. 2013. Static design procedure for welded hollow section joints – Recommendations.

Choo, Y. S., Qian, X. D., Liew, J. Y. R., Wardenier, J. 2003a. Static strength of thick-walled CHS X-joints – Part I: New approach in strength definition. *Journal of Constructional Steel Research*, 59, pp. 1201–1228.

Choo, Y. S., Qian, X. D., Liew, J. Y. R., Wardenier, J. 2003b: Static strength of thick-walled CHS X-joints – Part II: Effect of chord stresses. *Journal of Constructional Steel Research*, 59, pp. 1229–1250.

Lipp, A., Ummenhofer, T. 2014. Influence of tensile chord stresses on the strength of circular hollow section joints. *Steel Construction*, 7(2), pp. 126–132.

Liu, D. K., Wardenier, J. 1998. Effect of axial chord force in chord members on the strength of RHS uniplanar gap K-joints. *Proc. 5th Pacific Structural Steel Conference, Seoul.*

Liu, D. K., Wardenier, J., Vegte, G. J. van der. 2004. New chord stress functions for rectangular hollow section joints. *14th International Offshore and Polar Engineering Conference, Toulon.*

Lu, L. H., Winkel, G. D. de, Yu, Y., Wardenier, J. 1994. Deformation limit for the ultimate strength of hollow section joints. *Proceedings 6th International Symposium on Tubular Structures, Melbourne*, Balkema, Rotterdam, pp. 341–347.

Packer, J. A., Wardenier, J., Zhao, X.-L., Vegte, G. J. van der, Kurobane, Y. 2009. *Design guide for rectangular hollow section (RHS) joints under predominantly static loading.* 2nd Edition, Construction with hollow steel sections, No. 3, CIDECT.

Togo, T. 1967. Experimental Study on Mechanical Behaviour of Tubular Joints. *Dissertation* (in Japanese), Osaka University.

Ummenhofer, T., Lipp, A. 2012. Comprehensive discussion of the results of available research on the effect of tensile chord stresses on the load capacity of unstiffened, welded hollow section joints. Final report CIDECT 5CA-7/12.

Ummenhofer, T., Lipp, A. 2013: New chord load function. Final Report CIDECT 5CC-6/13.

Vegte, G. J. van der, Makino, Y., Choo, Y. S., Wardenier, J. 2001a. The influence of chord stress on the ultimate strength of axially loaded uniplanar X-joints. *Proc. 9th International Symposium on Tubular Structures, Düsseldorf*, pp. 165–174.

Vegte, G. J. van der, Makino, Y. 2001b. The effect of chord stresses on the static strength of CHS X-joints, *Memoirs of the Faculty of Engineering*, Kumamoto University, Vol. 46, No. 1, pp. 1–24.

Vegte, G. J. van der, Liu, D. K., Makino, Y., Wardenier, J. 2003. New chord load functions for circular hollow section joints. Final report CIDECT (revised) 5BK-4/03.

Vegte, G. J. van der 1995. The static strength of uniplanar and multiplanar tubular T- and X-joints. *Doctoral Dissertation*. Delft University of Technology, Delft University Press.

Wardenier, J., Kurobane, Y., Packer, J. A., Dutta, D., Yeomans, N. 1991. *Bemessung und Berechnung von Verbindungen aus Rundhohlprofilen unter vorwiegend ruhender Beanspruchung*. 1. Auflage, CIDECT Handbuch, Teil 1, Köln, Verlag TÜV Rheinland.

Wardenier, J., Vegte, G. J. van der, Liu, D. K. 2007. Effect of chord loads on the strength of RHS uniplanar gap K joints. Final report CIDECT 5BU-7/07.

Wardenier, J., Kurobane, Y., Packer, J. A., Vegte, G. J. van der, Zhao, X.-L. 2008. *Design guide for circular hollow section (CHS) joints under predominantly static loading*. 2nd Edition, Construction with hollow steel sections, No. 1, CIDECT.

Yura, J. A., Zettlemoyer, N., Edwards, I. F. 1981. Ultimate capacity of circular tubular joints. *Journal of the Structural Division*, American Society of Civil Engineers.

Tubular Structures XV – Batista, Vellasco & Lima (eds)
© 2015 Taylor & Francis Group, London, ISBN 978-1-138-02837-1

Tension testing of welds for X-joints with CHS branches to SHS chord

W. Wang
State Key Laboratory of Disaster Reduction in Civil Engineering, Tongji University, Shanghai, China

Q. Gu & J.J. Wang
Department of Structural Engineering, Tongji University, Shanghai, China

ABSTRACT: Experiments were conducted to investigate the axial tensile behavior of welds for tubular X-joints with CHS branches to square hollow section (SHS) chord. Five non-rigid and three rigid welded joints were tested under monotonic loading. Geometry of weld was measured by the use of negative mold. The strain distribution, the failure modes and the strength of welds was studied. The test results indicate the uneven strain distribution in non-rigid joints, in contrast to the uniform strain distribution in rigid joints. The strength of welds for non-rigid joints shows a significant reduction compared with rigid joints. Finally the prediction formula for effective length of fillet welds is proposed, which shows good agreement with test results.

1 INTRODUCTION

Single-layer reticulated shells have become a particularly popular choice for the large-span roof systems of steel structures because of their light weight, appealing architectural appearance and rapid erection. Tubular sections are a common selection for primary load-carrying members of this type of onshore structures. In the practical applications, the sections are profiled and welded to form unstiffened X-joints. Research work has been done to study the behavior of unstiffened joints, including the tubular joint rigidity (Wang & Chen, 2005) and the flexural behavior of tubular joints (Chen & Wang, 2003), as well as the behavior of tubular joints under cyclic loading (Wang & Chen 2007, Wang et al. 2012a,b). The fillet weld, because of its inexpensive cost and simple operation in weld proceeding, is generally adopted in welded joints. The weld is vitally important to ensure the strength of the joints, thus there are currently two design methods for fillet welds, namely pre-qualified method and fit-for-purpose method (McFadden & Packer, 2013). Pre-qualified method requires that the weld be proportioned to develop the yield strength of the connected branch wall at all locations around the branch. The design provision for fillet weld, based solely on the thickness of the branch, is generally conservative and results in relatively large weld size. Fit-for-purpose method requires that the weld be designed to resist the applied load in the branch. The design method takes the weld mechanical properties into consideration and results in appropriate weld size.

When the applied axial loads transfer from the branch to the chord in tubular joints, the connection deformation must be considered caused by the flexibility of the connection. Therefore the proper mode for loading distribution of welds should be studied when fit-for-purposed method is used to judge the strength of welds. Frater (1992) and Packer (1995) explored the strain distribution of welds for RHS joints by series of experimental research and proposed the effective length of welds. The AISC (2010) "Specification" adopts the effective length and puts forward the design formula of welds for RHS joints under axial tension. For tubular joints with CHS branches to SHS chord, nevertheless there are currently few research work. Thus axial tensile behavior testing is conducted in this paper to investigate the strain distribution, the failure modes and strength of welds for this type of joints.

2 EXPERIMENTAL PROGRAM

2.1 Specimens

8 X-joint specimens were tested, including 5 non-rigid joints and 3 rigid welded joints. The non-rigid joints referred to directly welded joints, varying in parameters included the width ratio β and the joint angle θ (see Fig. 1). The rigid joints referred to butt joints with a stiffener plate. The square chord was replaced by the plate in rigid joints to eliminate flexible deformation of the welds, in compare with the non-rigid joints. The details of the specimens are shown in Figure 1. The properties of the specimens are shown in Table 1. For the name of each specimen, the signal 'X' or 'GX' respectively stands for the non-rigid or rigid joints and the number indicates the value of θ. The notation '-1' and '-2' respectively corresponds to two different values of β. For all specimens, the length of the branch was about 3 times of its diameter to allow the uniform

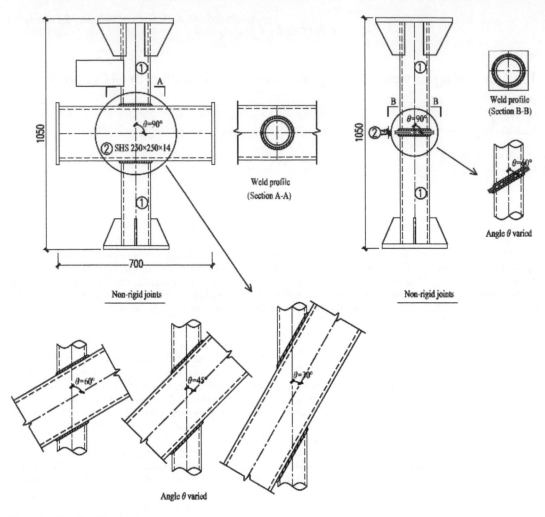

Figure 1. Details of the specimens.

Table 1. Geometrical characteristics of specimens.

Specimen	Section ① (mm)	Section ② (mm)	θ	β	Expected h_f (mm)
X90-1	CHS 133 × 10	SHS 250 × 250 × 14	90°	0.53	6
X90-2	CHS 180 × 10	SHS 250 × 250 × 14	90°	0.72	6
X60-1	CHS 133 × 10	SHS 250 × 250 × 14	60°	0.53	6
X45-1	CHS 133 × 10	SHS 250 × 250 × 14	45°	0.53	6
X30-1	CHS 133 × 10	SHS 250 × 250 × 14	30°	0.53	6
GX90-1	CHS 133 × 10	−180 × 14	90°	–	6
GX90-2	CHS 180 × 10	−230 × 14	90°	–	6
GX60-1	CHS 133 × 10	−210 × 14	60°	–	6

loading transmission from the branch to the connection region. Similarly, the length of the chord was about 3 times of its width.

All specimen members were made of Q345B steel. Fillet welds were used for all X-joints in accordance with the Chinese Code for welding of steel structures GB50661-2011. The material quality testing for steel and the weld metal was carried out before the experiment and the measured material properties were shown in Table 2.

As the behavior of welds was the main focus and the tests were meant to be weld-critical, the expected

Table 2. Measured material properties.

Sections	Material	f_y (MPa)	f_u (MPa)	f_y/f_u	Elongation at fracture (%)
CHS133 × 10	Q345B	348	529	0.66	26
CHS180 × 10	Q345B	375	563	0.67	27
SHS250 × 250 × 14	Q345B	325	401	0.81	24
Weld deposit	E50 eletrode	416	500	0.83	27

Table 3. Measured average effective thickness h_e for each joint (in mm).

Specimen	X90-1	X90-2	X60-1	X45-1	X30-1	GX90-1	GX90-2	GX60-1
U	5.0	5.1	6.0	6.1	7.7	5.0	4.4	7.1
L	5.5	4.5	6.1	6.7	8.2	4.7	4.6	6.4

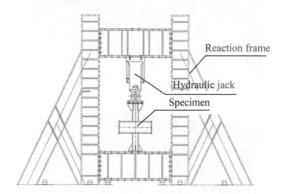

Figure 2. Overview of testing arrangement.

leg size (h_f) for each fillet weld, namely the leg length along the chord and the branch, was specified as 6 mm with reference to pre-analysis. To gain the actual geometric properties of welds, the silicone rubber impression was used to shape the negative mold of welds for each specimen. The effective thickness (h_e) of each slice was measured and given in Table 3. Labels 'U' and 'L' in the table stood for the upper part and the lower part of the X joints respectively.

2.2 Test setup and loading procedure

The overall view of testing arrangement is shown in Figure 2. All the specimens were tested to failure in a static manner under axial load applied in the branch. The main content in the process of test measuring included the relative deformation in the connection and the strain distribution of welds around the intersection line. The strain distribution was measured by strain gauges secured to the branch about 15 mm away from the weld toe to avoid the strain concentrations. The typical measuring arrangement is shown in Figure 3.

3 TEST RESULTS AND DISCUSSIONS

3.1 Failure modes

For non-rigid joints, initial crack could be seen at the weld toe on the chord. The initial crack grew fast to the weld toe nearby when the load increased and finally fractured through the wall of the chord. For rigid joints, sudden fracture happened at the connections when the load reached maximal value.

The failure modes in the test include: (1) fracture in the wall of chord along the weld toe (CF); (2) fracture of weld metal (WF); (3) fracture of the material in the heat affected zone of the branch (BHAZF). The different failure modes are shown in Figure 4. Mode CF is generated when the material in the heat affected zone of the chord is unable to deform itself together with the wall of the chord. This mode can be seen in all non-rigid joints. Mode WF takes place when the weld metal is not capable to transmit the axial load, resulting in the brittle rupture. The cause of mode BHAZF is the weakness of the connection between the weld and the heat affected zone in the branch. Generally, mode WF and BHAZF can be seen in rigid joints.

3.2 Curves of load versus relative deformation

Curves of load versus relative deformation for each specimen are shown in Figure 5. The yield strength (P_{by}) and the ultimate strength (P_{bu}) calculated using measured branch dimensions and material properties are marked in the figure. In addition, the ultimate deformation $\Delta_{Lu} = 0.03B$ (Lu et al. 1994) is also marked to study the deformation growth in non-rigid joints. Significant deformation could be seen in non-rigid joints, indicating that the joints fail when the relative deformation is remarkable. The fracture of each non-rigid joint happens when the relative deformation exceeds Δ_{Lu} except specimen X90-2. For rigid

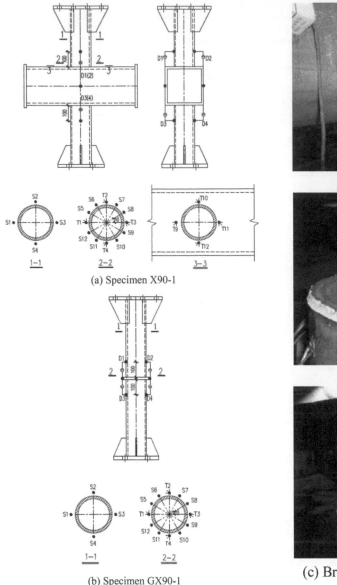

(a) Specimen X90-1

(b) Specimen GX90-1

Figure 3.   Instrumentation arrangements.

(a) Chord failure (CF)

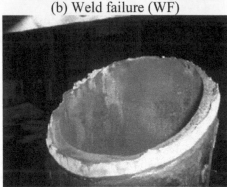

(b) Weld failure (WF)

(c) Branch's HAZ failure (BHAZF)

Figure 4.   Failure modes of the specimens.

joints, the specimens fail at tiny deformation since there is a low deformation capacity in these joints.

3.3   Curves of load versus strain around circumference

Load versus strain curves of welds for the upper part of X90-1 and GX90-1 are shown in Figure 6 and Figure 7, as the representative of non-rigid and rigid joints respectively. The strain for measuring point 'a' and 'b' in the Figure 6 is the equivalent strain calculated by strain rosette. The yield strain of the branch calculated by measured material properties is also marked as ε_y.

It is seen in Figure 7 that there is uniform strain distribution in rigid joints. Instead, the strain distribution is uneven in non-rigid joints as shown in Figure 6. For welds at specific positions like measuring point '3', '4' and '5', there is a dramatic strain growth. In contrast, for welds at other positions like measuring point '1' and '7', there is little strain growth.

3.4   Loading efficiency and effective length of welds

Figure 6 shows the load versus strain distribution relationship of welds for non-rigid joints. The difference of the strain growth indicates the discrepancy in loading transmission for local weld in different position. The slope of the load-strain curve for each measuring

390

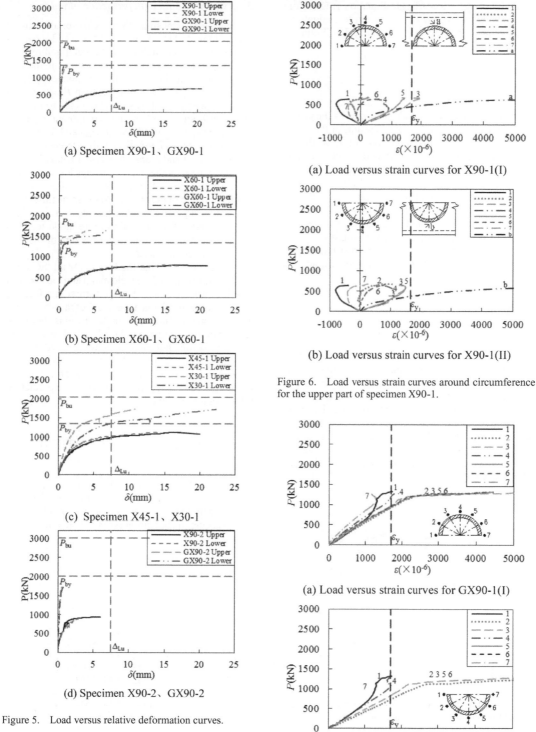

(a) Specimen X90-1、GX90-1

(b) Specimen X60-1、GX60-1

(c) Specimen X45-1、X30-1

(d) Specimen X90-2、GX90-2

Figure 5.　Load versus relative deformation curves.

(a) Load versus strain curves for X90-1(I)

(b) Load versus strain curves for X90-1(II)

Figure 6.　Load versus strain curves around circumference for the upper part of specimen X90-1.

(a) Load versus strain curves for GX90-1(I)

(b) Load versus strain curves for GX90-1(II)

Figure 7.　Load versus strain curves around circumference for the upper part of specimen GX90-1.

point is studied first when the joint is in elastic loading, marked as k_i. The minimal absolute value of the slope happens at the measuring point with the fastest strain growth, marked as k_m. The local loading efficiency factor of welds η is thus defined to quantify the uneven strain distribution according to Equation 1. Therefore, η is considered to provide a criterion to evaluate the strain growth, as well as the loading distribution of welds.

$$\eta = \frac{k_i}{k_m} \qquad (1)$$

The generalized abscissa X is defined by the projection of the intersection line onto the axis of the chord. Therefore, the generalized abscissa value of the left and right endpoints of the intersection line is respectively -1 and 1. The relationship between η and X (see Figure 8) thus presents how the weld behaves in the process of loading transmission. It can be seen that the curves shape like the mountains, with peaks at the center and decrease along both sides, which correspond to the law of the path of loading transmission.

To look into the uneven loading distribution, the effective length factor χ is defined by Equation 2 according to the η-X curves.

$$\chi = \frac{\sum \eta dl}{L_w} \qquad (2)$$

where dl is the length of segment of the weld if the intersection line is divided into several parts and can be calculated by mathematical characteristics, and L_w is the geometric length of the intersection line. Those segments when η is negative are excluded. The result of effective length factor computed by experimental results is shown in Table 4. Therefore, the uneven loading distribution of welds around the intersection line is equivalent to the uniform distribution along the effective length of χL_w.

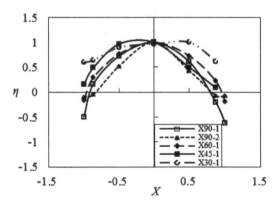

Figure 8. The local loading efficiency factor of welds around the intersection line for non-rigid joints.

3.5 Axial tensile strength prediction of welds

The analysis of axial strength of welds for all joints is listed in Table 5. Most joints failed in a sudden manner with fracture through the weld metal and/or base metal. P_{max} refers to the maximal axial load in the test. The symbol '>' in the table refers to situation that the joint has potential strength higher than P_{max} given in the table since the joint was unable to load to fracture because of limitation of the test equipment capacity. The plastic capacity of the chord (P_{ju}) and the punching capacity of the chord (P_{su}) with reference to IIW (2012) "recommendations" are also included in the table. It can be seen that P_{max} is always lower than the ultimate branch strength P_{bu}, eliminating the possibility of the fracture of the branch. At the same time, P_{max} is lower than P_{su} while higher than P_{ju} for non-rigid joints, validating that the main failure mode for non-rigid joints is mode CF due to the fracture of the heat affected zone of the chord.

It should be noted that the failure for all the joints includes the fracture of the weld metal and the material in the heat affected zone. For the former, the fracture load reflects the mechanical behavior of the weld itself. For the latter, the ultimate strength of welds should be higher than the maximal axial load in the test since the heat affected zone fails before the weld. Nevertheless, P_{max} is considered the ultimate strength of welds in this paper regardless of the distinction mentioned above, indicating a certain degree of safety for the prediction of the strength of welds.

4 PARAMETRIC ANALYSIS OF EFFECTIVE LENGTH FACTOR

To study the effective length factor in general conditions for non-rigid joints, parameter analysis is done by finite element method. The parameters include θ, β, τ and γ, varying in the common range listed in Table 6.

Table 5. Analysis of the strength of welds.

Specimen	P_{ju} (kN)	P_{su} (kN)	P_{by} (kN)	P_{bu} (kN)	P_{max} (kN)	Failure Mode
X90-1	406	860	1345	2044	684	CF
X90-2	635	1164	2003	3007	945	CF
X60-1	489	1092	1345	2044	806	CF
X45-1	641	1539	1345	2044	1125	CF
X30-1	1040	2823	1345	2044	>1729	CF
GX90-1	/	/	1345	2044	1323	WF
GX90-2	/	/	2003	3007	>1727	Not fail
GX60-1	/	/	1345	2044	1644	WF BHAZF

Table 4. The effective length factor χ of welds for non-rigid joints.

Specimen	X90-1	X90-2	X60-1	X45-1	X30-1
Computed by Experimental results	0.385	0.316	0.458	0.458	0.583
Computed by the fitting formula	0.372	0.353	0.420	0.500	0.670

Table 6. Parameters of joints.

Parameter	Value			
θ	90°	60°	45°	30°
β	0.4	0.6	0.8	
τ	0.4	0.7	1	
γ	5	10	20	30

The elastic analysis for a total of 144 finite element models is carried out.

In the finite element analysis, the stress distribution of the wall of the branch about 15mm away from the weld toe is considered to replace that of the welds. The local loading efficiency factor can be calculated by Equation 3.

$$\eta = \frac{\sigma_i}{\sigma_{max}} \qquad (3)$$

where σ_i is the axial stress along the branch in positions mentioned above in a joint, and σ_{max} is the maximum among σ_i in different positions in this joint.

The η-X curves for several joints are shown in Figure 9. It can be concluded that the parameters γ and θ have a great influence on the shape of the curve. The reason is that the parameter γ reflects the stiffness of the chord, while parameter θ affects the length and feature of the intersection line.

According to Equation 2, the result of the effective length factor for 144 finite element models is studied and the fitting formula for χ is proposed as Equation 4. The comparison of the value of χ computed by the experiments results and the fitting formula is listed in Table 4, which shows good agreement.

$$\chi = \frac{1.61}{(\sin\theta)^{0.85}} \cdot \frac{1}{3.05 + \beta} \cdot \frac{1}{(\gamma - 4.61)^{0.13}} \qquad (4)$$

5 CONCLUSIONS

Based on the tension testing of eight X-joints with CHS branches to SHS chord, the strain distribution, the failure modes and the strength of welds was studied, and some conclusions can be draw as follows:

(1) The axial strength of welds for non-rigid joints is much lower than corresponding rigid joints. It is the result of the uneven loading distribution in non-rigid joints, as well as the fracture of the material in heat affected zone due to its brittleness. The uneven load distribution results in high stress concentrations but the deformation capacity is not sufficient to redistribute the stresses and to allow a complete plastic stress redistribution.
(2) The curves of local loading efficiency factor η versus the generalized abscissa X of welds shape like the mountains, with peaks at the center and decrease along both sides, which correspond to the law of the path of loading transmission.

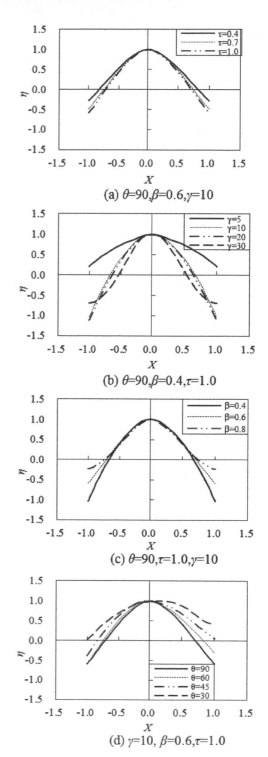

(a) θ=90,β=0.6,γ=10

(b) θ=90,β=0.4,τ=1.0

(c) θ=90,τ=1.0,γ=10

(d) γ=10, β=0.6,τ=1.0

Figure 9. The η-X curves for joints of various parameters.

(3) The formula for predicted the effective length of fillet welds is proposed and shows good agreement with test results.

ACKNOWLEDGMENTS

The research presented in this paper was supported by the Natural Science Foundation of China (NSFC) through Grant No. 51378380.

REFERENCES

American Institute of Steel Construction, Specification for structural steel buildings, Chicago, USA, 2010.

Chen, Y.Y., Wang, W., 2003. Flexural behavior and resistance of uni-planar KK and X tubular joints, *Steel & Composite Structures,* 3(2), 123–140.

Frater, G.S., Packer, J.A., 1992. Weldment design for RHS truss connections. I: Applications, *Journal of Structural Engineering.* 118(10), 2784–2803.

Frater, G.S., Packer J.A., 1992. Weldment Design for RHS Truss Connections. II: Experimentation, *Journal of Structural Engineering,* 118(10), 2804–2819.

GB50661-2011, Code for welding of steel structures, China Architecture & Building Press, Beijing, 2011

International Institute of Welding. Static design procedure for welded hollow section joints – recommendations, 3rd edition, Paris, France, 2012.

Lu, L.H., de Winkel, G.D., Yu, Y., Wardenier, J., 1994. Deformation limit for the ultimate strength of hollow section joints, *Proceedings of the 6th International Symposium on Tubular Structures, Tubular Structures VI*: 341–347 Rotterdam: Balkema.

McFadden, M.R., Packer, J.A., 2013. Weld design and fabrication for RHS connections, *Steel Construction*, 6(1), 5–10.

Packer, J.A., Cassidy, C.E., 1995. Effective weld length for HSS T, Y, and X connections, *Journal of Structural Engineering,* 121(10), 1402–1408.

Wang, W., Chen, Y.Y., 2005. Modelling and classification of tubular joint rigidity and its effect on the global response of CHS lattice girders, *Structural Engineering and Mechanics,* 21(6), 677-698.

Wang, W., Chen, Y.Y., Zhao, B.D., 2012. Effects of loading patterns on seismic behavior of CHS KK-connections under out-of-plane bending, *Journal of Constructional Steel Research,* 73, 55–65.

Wang, W., Chen, Y.Y., Meng, X.D., Leon, R.T., 2012. Behavior of thick-walled CHS X-joints under cyclic out-of-plane bending, *Journal of Constructional Steel Research,* 66(6), 826–834.

Wang, W., Chen, Y.Y., 2007. Hysteretic behaviour of tubular joints under cyclic loading, *Journal of Constructional Steel Research*, 63(10), 1384–1395.

Tubular Structures XV – Batista, Vellasco & Lima (eds)
© 2015 Taylor & Francis Group, London, ISBN 978-1-138-02837-1

Influence of the vent hole shape on the strength of RHS K-joints in galvanized lattice girders. A numerical study

M.A. Serrano-López, C. López-Colina, J. Díaz-Gómez & F. López-Gayarre
Department of Construction, University of Oviedo, Gijon, Spain

G. Iglesias-Toquero
Grupo Condesa, Vitoria, Spain

ABSTRACT: The hot-dip galvanizing process is a common technique to avoid corrosion in steel lattice girders. Vent holes are required when structural hollow sections are used as chord and brace members in these galvanized structures. In this situation some concern arises among designers due to the possible strength reduction of the joint as a consequence of the section's reduction caused by the vent holes. To the authors' best knowledge just some recent research has studied this problem. In that study, circular holes were considered in a position that was good for the structural performance of the joint, but avoids the total recovering of the fluid zinc. Some different shapes have been considered in this new proposal in an attempt to extend the research and to find a better configuration of the vent holes. Results of ultimate load capacities allow pointing out some conclusions to help designers to take their decisions properly.

1 INTRODUCTION

Hot-dip galvanizing (HDG) that is a process of coating fabricated steel, is probably the most cost effective protection method against corrosion for structural hollow sections (SHS). Although zinc has been used in construction for more than 150 years to protect steel from corrosion there is an ever increasing use of the term 'galvanizing' to describe the coating of steel with zinc. So it is important to clarify misconceptions as commented by Glinde (2013). In the hot-dip galvanizing process, individual components or full structures depending on the size of the kettle containing the molten zinc, are chemically cleaned through degreasing, pickling and fluxing before being immersed in the bath of molten (450°C) zinc of at least 98% as it is recommended by the American Galvanizers Association (2014). The steel components are lowered into the kettle at an angle that allows air to escape from tubular shapes, and the zinc to flow into, over, and through the entire piece (Fig. 1).

While immersed in the kettle, the iron in the steel metallurgically reacts with the zinc to form a series of zinc-iron intermetallic layers not only creating a barrier between the steel and the environment, but also cathodically protecting the steel. This creates a bonded coating that provides complete coverage of the steel component both internally and externally for hollow sections. As the intermetallic layers are harder than the base steel, they also provide a high resistance to abrasion. The coating grows perpendicular to the surface due to the diffusion process in the galvanizing kettle, ensuring that corners and edges have at least equal thickness to flat surfaces assuring a long maintenance-free service life. As hot-dip galvanizing provides maintenance-free performance for 75 years or more in most environments as it is stated by Cornish (2013), its Life-Cycle Cost (LCC) is almost always the same as its initial cost. Consequently, when analyzing LCC, hot-dip galvanizing has advantages over other protection systems.

When the piece to be galvanized is a lattice girder with hollow sections as chords and braces, at least the brace members have to be drilled to provide ventilation holes near their connection with the chords. Vent holes are necessary to allow both, the immersion of the structure in the zinc bath and to recover the remaining liquid to be reused. Therefore on the one hand holes have to be big enough to efficiently accomplish the both above mentioned necessities, but on the other hand there is a risk associated with the reduction in the joint resistance if the size of the vent hole is too large. To date there have been no reported joint failures recognized as a consequence of the vent holes presence in the joint, but some concern in fact exists among designers due to the absence of guidance or recommendations in the codes of design as EN-1993-1-8 (2005) and ANSI/AISC 360-10 (2010), in the standards for technical delivery conditions EN-10219-2 (2006) and EN-10210-2 (2006) or in the design guides for structural hollow sections Packer et al. (2009), Dutta el al. (1998) and Packer & Henderson (1997). Probably this

Figure 1. Galvanization of a RHS K-joint.

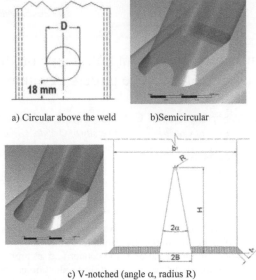

a) Circular above the weld   b)Semicircular

c) V-notched (angle α, radius R)

Figure 2. Geometries of different vent holes.

lack of knowledge is reducing unjustifiably the use of this kind of structures despite of its overall efficiency.

With the aim of facing this problem some research has been conducted previously under a pair of projects funded by the CIDECT (Comité International pour le Development et l'Etude de la Construction Tubu-laire) and the EGGA (European General Galvanizing Association). Iglesias & Landa (2007) proposed some recommendations about the size, shape and position of vent holes. Later, Serrano et al. (2011) through a research work, including numerical simulation and a large experimental program on full scale joints, inves-tigated if the recommended holes had any adverse effect on the resistance of several joint configurations. Although some failure modes were identified depend-ing on the joint configuration, the main conclusion was that holes had little adverse influence on the joint resistance.

The above mentioned project focused in circular holes with a position that was good for the structural performance of the joint, but avoided the total recover-ing of the fluid zinc. Therefore some different shapes and positions have been considered in this new pro-posal in an attempt to extend the research and to find the best configuration of the vent holes.

2 NUMERICAL MODEL

To carry out a deep analysis allowing to determine the best geometry of vent holes in terms of joint efficiency and liquid recovering, a refined numerical model was developed. The FE software Ansys Workbench V.14 academic research was used.

Four different models of RHS K-joints were consid-ered in this new work. The first one does not include

any vent hole and was mainly used to be compared with experimental test results allowing to validate the FE model. Then three more models as an extension of the previously validated and including three different vent holes were developed. Figure 2 shows them. The first one (Fig. 2a) includes a circular hole away from the weld and in the same position considered in the previous research. The other two that allow the total recovering of liquid after galvanization were a semi-circular and a V-notched holes cutting off the weld between braces and chord. The three holes present the same filling area that is referred in any case to the area of the circular hole with a diameter D. As it is recom-mended by Iglesias & Landa (2007) this diameter is the 25% of the X-section of any brace. The diameter of semicircular hole is therefore $\sqrt{2}xD$. In case of V-notched hole, the cutting off in the weld is lower than in the semicircular configuration but some new variables of design appears (Fig.2c). The maximum opening $2B$ and the height of the notch H is calculated for the angle α, the radius R and the filling area that depends on D. In the simulation for this research a radius of 1 mm and an angle of 10° were considered.

The selected elements in the FE model were a four-noded shell 181 with six degrees of freedom at each node. This element has proved to be adequate for numerical simulation of thin walled hollow sections by López-Colina et al. (2010–2011) and Del Coz et al. (2012). To adjust the element size to necessities of simulation, a finer mesh area with element size of 3 mm was considered inside a sphere of 150 mm in diameter centered in the point of axis concurrence. Furthermore, two refinements were done in the welds and hole borders.

A parametric model allowed to adapt easily the model to every individual joint. Figure 3 shows a

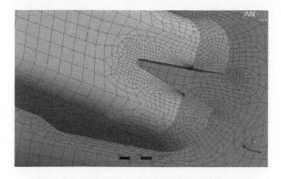

Figure 3. Mesh of V-notch hole configuration and weld.

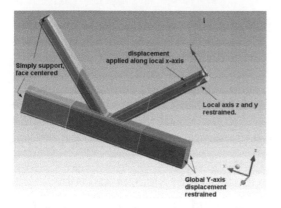

Figure 4. Global boundary conditions in FE Models.

Table 1. RHS members and actual material properties on FE.

Gap (g/b$_0$)	Joint members (bxhxt)	f$_y$ [N/mm^2]	Tangent modulus [N/mm^2]
Test 1 0,10	Chord 258x258x8.4	289	1110
	Comp. Brace 99.4x99x6.0	337	802
	Tension Brace 103x99x5.2	317	932
Test 2 0,43	Chord 100x100x3.8	261	963
	Braces 40x40x4.0	279	996
Test 3 0,15	Chord 100x100x4.1	394	1291
	Braces 60x60x4.2	377	1813
Test 4 0,15	Chord 99.8x99.8x4.3	283	907
	Braces 59.7x59.7x4.4	320	958
Test 5 0,10	Chord 102x102x3.9	363	1440
	Braces 64x64x3.3	369	1350
Test 6 0,10	Chord 102x102x4.2	321	1550
	Braces 64x64x4.0	349	729

*g (gap). b$_0$ (chord width). b, h, t (width, height, thickness).

meshed model for the V-notched geometry and figure 4 shows the adopted global boundary conditions.

The compression load applied in the tests by means of a jack on the end of the compressed brace member was simulated by small steps of displacement introduced on the end of this brace. Meanwhile the other brace was in tension due to its boundary conditions that include a pinned end for this second brace. One end of the chord was pinned while the other was left free in the load plane but its displacement was restrained in the perpendicular axis (see fig. 4). These global boundary conditions ensure that axial forces were transmitted to the members. The compression load over the brace was increased gradually until the collapse of the joint was reached.

3 VALIDATION OF NUMERICAL MODEL

To validate the numerical model some experimental results from tests carried out on RHS K-joints under two CIDECT projects at the University of Delft by Wardenier et al. (1976–1978) were used. Six tests on symmetrical K-joints with a brace angle of 45° and a variable gap between the braces were selected to be compared with the model. The nominal steel grade was S275 hot finished for tests 1, 2, 5 and 6 but cold finished for tests 3 and 4. The material properties for the joint members were obtained from coupon

tests. In the numerical model these values were later corrected to take into account the actual material properties of the steel instead of the engineering values from the standard tests. The material behavior was simulated assuming an isotropic bilinear hardening. Table 1 presents the RHS members and the gap for each joint together with the main material properties considered in the FE model. The tangent modulus allows to simulate the steel behaviour after yielding. The welds were modelized considering the same material properties than the corresponding welded brace member. Although some gaps were smaller than those in EN-1993-1-8 (2005) they were selected to avoid eccentricity of the axial force applied on the brace member. In any case the tests were prior to the code formulation and they are just used in this research for validation of the numerical model.

In order to validate the model not just regarding the joint resistance but as well regarding the joint stiffness, force-displacement curves were registered by the simulation and compared with those from the experiments in test 1. Figure 5 plots one of these curves showing very good agreement between test and simulation. The displacement, as indentation, was measured on point 1 (figure 5) for the compression brace member. The dotted line in figure 5 indicates the maximum

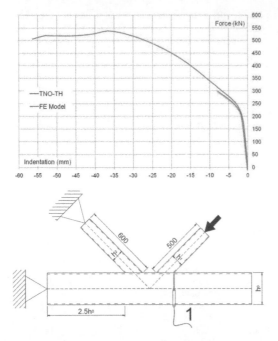

Figure 5. Force-indentation curve in compression brace.

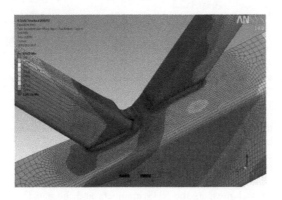

Figure 6. Von Mises stress with large deformations.

Table 2. Maximum load from test and numerical model.

Joint	2	3	4	5	6
Test	102	215	172	198	193
Model	105.9	232.1	185.3	204.3	198.6

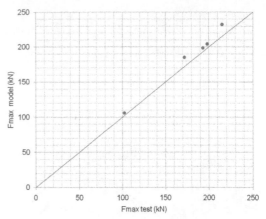

Figure 7. Model vs. test for validation.

compression load reached by the compression brace member in the test. The simulated joints in the validation process did not include any hole to reproduce the tested joints. To reduce the computation time and taking advantage of the symmetry just half of the K-joint was modelized. Figure 6 shows the Von Mises stress for the half joint modeled after large deformations. The plastification started in the brace members just above the weld and it was spread later to the chord face.

The validation in case of tests 2 to 6 was carried out by comparison of the maximum load reached with the model *vs* those on the tests. Table 2 presents these results for comparison. Deviations were always under 8% with a mean value of 5.1%. Figure 7 shows these values with very small differences supporting the validation of the model.

4 EXTENSION OF NUMERICAL STUDY

Once the model was validated it was extended to consider other K-joint configurations combined with the three types of ventilation holes above mentioned (circular, semicircular and V-notched). A series of 8 symmetrical K gap joints, involving RHS as chords and SHS as braces with different sizes and angles of 45°, 55° and 60° were studied. So it was possible to evaluate the influence of several geometrical variables together with the shape of the vent hole in the joint resistance. Chord and braces dimensions, brace angles and the adopted gaps to avoid any eccentricity in the joint, are presented in Table 3. Four different possibilities including or not different shapes of vent holes were analyzed for every joint. So a total number of 32 cases were studied.

A nominal steel grade S275 was considered assuming a bilinear behaviour of the steel with a tangent modulus of $1330 \, \text{N/mm}^2$ in this part of the research. A non-linear analysis was carried out taking into account geometric non-linearities, non-linear material properties and non-linear buckling.

The results for the maximum load supported by the joint and the maximum load when it is considered the criteria of an indentation of 3% of the chord width (b_0) were registered. Table 4 presents the maximum load reached by all the combinations K-joint/vent-hole for an indentation of 3% of b_0 because it was the minimum value of both criteria in all cases. Nevertheless the failure mode was not conclusive for joints 1 to 5 because ultimate loads based on 3% indentation and those based on the maximum load supported by the joint were very close with differences lower than 5%.

Table 3. Joint dimensions in the extended model.

Joint	Chord RHS $(b_0 x h_0 x t_0)$	Braces SHS $(b_1 x t_1)$	β (b_1/b_0)	γ $(b_0/2t_0)$	Gap [mm]
KB1	60x80x5	30x3	0.5	6	37.6
KB2	60x80x5	40x3	0.66	6	23.4
KC3	60x80x4	30x3	0.5	7.5	23.4
KB4	60x80x5	30x3	0.5	6	19.4
KB5	80x100x5	40x3	0.5	8	43.4
KC6	80x100x5	40x4	0.5	8	43.4
KC7	80x100x5	40x5	0.5	8	43.4
KC8	60x110x5	30x3	0.5	6	28.8

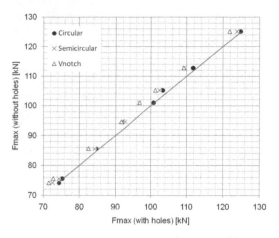

Figure 8. Comparison of ultimate loads based on 3%b_0.

Table 4. Ultimate joint resistances based on 3%b_0 [kN].

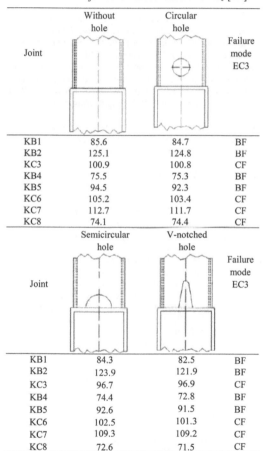

Joint	Without hole	Circular hole	Failure mode EC3
KB1	85.6	84.7	BF
KB2	125.1	124.8	BF
KC3	100.9	100.8	CF
KB4	75.5	75.3	BF
KB5	94.5	92.3	BF
KC6	105.2	103.4	CF
KC7	112.7	111.7	CF
KC8	74.1	74.4	CF

Joint	Semicircular hole	V-notched hole	Failure mode EC3
KB1	84.3	82.5	BF
KB2	123.9	121.9	BF
KC3	96.7	96.9	CF
KB4	74.4	72.8	BF
KB5	92.6	91.5	BF
KC6	102.5	101.3	CF
KC7	109.3	109.2	CF
KC8	72.6	71.5	CF

Joint KB1 with semicircular hole

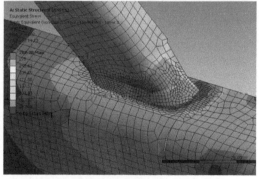

Joint KC8 with V-notched hole

Figure 9. Samples of failure modes.

Also the predicted failure mode BF (brace failure) or CF (chord plastification) according to formulation in EN-1993-1-8 (2005) is pointed out in the table.

According with table 4, if joints with vent holes are compared with the same K-joint without the hole, differences are very small. The registered deviations of mean values regarding the joint without the hole were lower than 1% for circular holes, just 2.2% for semicircular holes and only 3.4% in case of V-notched holes.

Figure 8 shows in the same graphic these three comparisons. So neither the hole presence nor the shape and position of considered holes did not imply a significant strength reduction.

The influence of some other geometrical variables in the joint resistance like the wall thickness, the β ratio or the brace angle was also analyzed in this study. It could be seen that an increase in the brace thickness t_1 not only implied the expected increase in the load

399

supported by the joint but also a different failure mode that changed from local buckling in the brace member for the smaller thickness to chord plastification for the thicker brace walls.

The brace angle effect, confirmed other previous research by Serrano et al. (2012) that suggested a lower joint resistance for higher brace angles. For a brace angle of 60° the predicted failure mode also changed from BF to CP. As well a reduction in the chord thickness produced a change again in the failure mode. Figure 9 shows a pair of samples of the failure modes. The first one shows a brace failure in a joint with a semicircular hole while the second corresponds to a typical chord failure in a V-notched hole specimen.

5 CONCLUSIONS

A FE model has been developed to analyze the behavior of RHS K-joints when the brace members include a ventilation hole for the galvanization process. The model was successfully validated by means of experimental tests and it was extended to other joint configurations. The work was focussed in joints prone to a failure by local buckling in the brace member at the joint location (effective width failure at the compression brace) and chord plastification. From this research some conclusions may be drawn:

- The developed FE model is adequate to simulate the joint behaviour and to predict their strength and stiffness as it was proved by the good agreement with the experimental results.
- In the numerical simulation two different failure modes were observed depending on the geometrical ratios of brace and chord but never depending of the shape of the vent hole. In several cases both failure modes appear almost simultaneously in a kind of combined failure mode.
- In case of brace failure, as local buckling is away from vent holes there is not influence of the hole presence in the brace member.
- When the ratio of thickness led to a chord failure by plastification neither a significant influence of the hole presence was observed.
- Based on the small differences on the joint behavior for the considered shapes of vent holes it is concluded that it does not matter which type of these holes are used.
- A recommendation for designers is that they may choose any of the considered vent hole configurations just taking care of the suggested filling areas.

ACKNOWLEDGEMENTS

The authors would like to express deep gratitude to the IEMES Research Group at Oviedo University.

As well, the authors wish to acknowledge the financial support provided by the Spanish Ministry of Economy and Competitiveness to the Research Project with the reference BIA2013-43177-P. This project has been co-financed with FEDER funds.

Furthermore they would also like to thank Swanson Analysis Inc. for their assistance with the use of ANSYS University program.

REFERENCES

AISC. 2010 *Specification for Structural Steel Buildings. ANSI/AISC 360-10*. American Institute of Steel Construction. Chicago. Il.

American Galvanizers Association. 2014. *Galvanize lt! Online seminar: 'Sustainable development & hot-dip galvanizing'.* http://www.galvanizeit.org/sustainable-development-and-hot-dip-galvanizing. Accessed 25 March 2014).

Cornish R. 3013. *American Galvanizers Association. 'General Galvanizing in North America-Association, key metrics and markets'.* 9th Asia Pacific General Galvanizing Conference, Singapore. September 2013.

Del Coz Díaz JJ, Serrano-López MA, López-Colina C, and Alvarez Rabanal F. 2012. *Effect of vent hole geometry and welding on the static strength of galvanized RHS K-joints by FEM and DOE*. Engineering Structures 41: 218–233.

Dutta D, Wardenier J, Yeomans N, Sakae K, Bucak Ö and Packer JA, (1998) *Design guide 7: For fabrication, assembly and erection of hollow section structures*. TÜV-Verlag. 1998.

EN-1993-1-8. 2005. *Eurocode 3: Design of steel structures. Part 1-8: Design of joints*. European Committee for Standardization.

EN-10219-2. 2006. *Cold formed welded structural hollow sections of non-alloy and fine grain steels. Tolerances, dimensions and sectional properties*. European Committee for Standardization.

EN-10210-2. 2006. *Hot finished structural hollow sections of non-alloy and fine grain steels. Part 1: Technical delivery conditions; Part 2: Tolerances, dimensions and sectional properties*, European Committee for Standardization.

Glinde H. 2013. *Galvanizing explained. The misunderstood process*. Hot Dip Galvanizing Magazine. Galvanizers Association. Issue 1.

Iglesias G & Landa P. 2007. *Recommendations for holes needed due to galvanization process*. Final report project 14B. UPV-EHU, ICT, CIDECT.

López-Colina C, Serrano MA, Gayarre FL, Suárez FJ. 2010. *Resistance of the component 'lateral faces of RHS' at high temperature*. Engineering Structures 32 (4): 1133–1139.

López-Colina C, Serrano MA, Gayarre FL, and del Coz JJ. 2011. *Stiffness of the component 'lateral faces of RHS' at high temperature*. Journal of Constructional Steel Research 67 (12): 1835–1842.

Packer JA, Wardenier J, Zhao XL, Vegte, GJ, van der and Kurabane Y. 2009 *Design guide 3, 2nd Edition: For rectangular hollow section (RHS) joints under predominantly static loading*. LSS Verlag.

Packer JA & Henderson JE. 1997. *Hollow Structural Section Connections and Trusses, A Design Guide*. Canadian Institute of Steel Construction.

Serrano-López MA, López-Colina C and Iglesias G, 2011. *Static strength of RHS K-joints in which brace members may be affected by vent holes for truss Hot-Dip Galvanizing*. Final report project 5BX. Univ Oviedo, ICT, CIDECT.

Serrano-López MA, López-Colina C, Lozano A, Iglesias G and González J. 2012. *Influence of the angle in the strength of RHS K-joints in galvanized lattice girders*. Proceedings of the 14th ISTS. Tubular Structures XIV. 113–120.

Wardenier J. & De Koning C.H.M. 1976. *Rig comparison tests for welded joints*. Project 5S-76_33.TNO, The University of Delft, CIDECT.

Wardenier J. & Stark J.W.B. 1978. *The static strength of welded lattice girder joints in structural hollow sections*. Final report project 5Q/78/4. TNO, The University of Delft, CIDECT.

Tubular Structures XV – Batista, Vellasco & Lima (eds)
© 2015 Taylor & Francis Group, London, ISBN 978-1-138-02837-1

Design of hollow section joints using the component method

J.P. Jaspart
ArGEnCo Department, University of Liège, Belgium

K. Weynand
Feldmann + Weynand GmbH, Aachen, Germany

ABSTRACT: The component method is a design approach for the characterization of the mechanical properties of structural joints. Initially, the component method has been developed for joints between open sections. However, the design of joints between tubular hollow sections follows a different approach. A joint is considered as a whole when determining its resistance(s). Those design rules are based on simple theoretical mechanical models and they are then fitted through comparisons with experimental tests. As a consequence, their field of application is often restricted to the domain for which the rules have been validated. Under the umbrella of CIDECT, a project is being carried out to develop a unified design approach for steel joints independent of the type of section by extending the field of application of the component method. To achieve this objective, rules recommended for hollow section joints have to be "converted" into a component format. Ways and means for the development of such a unified design approach are presented. The paper presents also design rules for some hollow section joints as examples.

1 INTRODUCTION

1.1 Objective

The so-called component method is a design approach for the characterization of the mechanical properties of structural joints. Initially it has been developed for joints between open sections. The application of this design approach for joints in tubular construction has been initiated in CIDECT. As an outcome of first projects, e.g. Jaspart et al. 2005, the possibility to extend this method to hollow section joints has been demonstrated and the advantage of its use has been pointed out:

- Unified approach for all joints
- Applicability to any constitutive material(s) and combinations of profiles (tubular sections, open sections, rolled, cold-formed, hot-finished, built-up sections …)
- Applicability to any connection system (bolts, rivets, welds …) and connecting devices (plates, cleats, gussets, splices …)
- Applicability to joints in any structural system (frames, trusses, masts …)
- Applicability to joints under static loading but also fire, earthquake, fatigue, exceptional events, etc.

Furthermore, in the last years CIDECT has carried out some gap analysis (Jaspart et al. 2011) in order to highlight further research and development needs. Resulting from this work, CIDECT has asked the authors of this paper to reformat the contents of EN 1993-1-8 chapter 7 (Hollow section joints) into a "component format" compatible with the one used in the EN 1993-1-8 chapter 6 (Structural joints connecting H or I sections) so as to profit from the above-listed advantages and to allow later on an easy application to joints between members with tubular and open sections.

1.2 Design approaches

In the last decades, much research work has been devoted in all parts of the world to the design of joints in steel structures. All these efforts led to the development of new connection types and new design rules which have been progressively implemented in design recommendations and national or international codes (e.g. Packer et al. 1992, Wardenier et al. 1991).

These actions have been extended by other ones aimed at preparing design handbooks for practitioners as well as appropriate simplified design tools, so facilitating the transfer of new technologies to the constructors and designers.

A quick look at the outcome of these different initiatives is enough to understand that two separate ways have been followed by researchers as far as the type of connected members is concerned:

- joints between tubular hollow sections
- joints between hot-rolled or built-up I or H sections

This divergence results from two different design approaches which are briefly described below.

1.2.1 CIDECT approach (failure mode approach)

The design rules for joints between tubular hollow sections are based on simple theoretical mechanical models and they are then fitted through comparisons with results of experimental and numerical investigations. As a consequence, their field of application is often restricted to the domain for which the rules have been validated.

For different failure modes observed in the experimental tests, the design formulae generally give a resistance value for the joint as a whole, for specific loading cases and specific joint configurations; this restricts the field of application and hence the freedom to modify the joint detailing. It has also to be mentioned that usually no information is provided with regard to the stiffness or the ductility of the joints.

1.2.2 Component method approach

In the case of joints between open sections, a new theoretical approach based on the so-called "component method" has been developed. It allows a theoretical determination of the resistance, stiffness and ductility properties based on mechanical models. It may be applied to a wide range of joint configurations and connection types in steel structures and has been extended to composite steel-concrete joints. Research activities are in progress in view of its application to timber joints or joints between concrete precast elements, but also to various loading situations (fire, earthquake, impact, robustness ...). Besides its theoretical background, the component method provides the user with information about the relative stiffness, resistance and ductility of all the constitutive parts of the joints (called components), so allowing:

– to show how the stiffness and resistance are "distributed" between the joint components
– to derive the actual failure mode
– to assess the related level of ductility
– to give indications on how to stiffen or strengthen the joint in an easy and economic way when needed

1.3 The component method

1.3.1 A basis for a unified approach

In Europe both design approaches are reflected by Eurocode 3 in the latest version of EN 1993 Part 1.8. For "mixed" joint configurations, i.e. where hollow and open sections are connected together, EN 1993 Part 1.8 provides some guidelines for welded connections but not for bolted connections, i.e. structural solutions with mixed types of sections are difficult to apply for practitioners as they are not directly covered in the design rules. Rules for some "mixed" bolted connections were covered by the ENV version (prestandard) of Eurocode 3, but they were not included in the final EN version. Those rules are of course found in the various CIDECT design guides (e.g. Packer et al. 2009, Wardenier et al. 2008).

The present paper reports on the development of a unified design approach for steel joints independent of the type of connected elements by extending the field

of application of the component method to tubular joints.

The component method is a three step procedure which may be defined as follows:

– identification of the constitutive individual components of the joint
– determination of the stiffness/resistance properties of all these components by using appropriate design formulae
– combination or "assembly" of the single components so as to derive the stiffness/resistance properties of the whole joint

The properties of joints to be evaluated in practice strongly depend on the type of global frame analysis and design process which is followed by the designer; for instance:

– for an elastic analysis combined with an elastic verification of the members and joints, the stiffness and the elastic resistance of the joints should be derived
– for an elastic analysis combined with a plastic verification of the members and joints, the stiffness and the plastic resistance are required
– for a rigid-plastic analysis, only the plastic resistance and the rotational capacity of the joints will have to be evaluated

This approach is very comprehensive and, as already said, the objective of the ongoing project is to extend it to joints involving other types of sections and, in particular, hollow sections. Practically speaking, this requires:

– to define and characterize the relevant components;
– to verify that the available "assembly" procedures (which are based on general principles like equilibrium, compatibility of displacements, ...) are general enough to be considered as independent of the actual nature of the constitutive components and are therefore still relevant.

The field of application of this unified design procedure is wide as the number of components may be enlarged to cover any new joining solution that could be proposed by designers or fabricators. This is one of the main advantages of the procedure. But it has to be recognized that its practical application may sometimes, when the number of components become significant, be rather long and cumbersome. That is why, for daily practice, the user will favor practical design tools much more in line with his request for efficiency. Amongst the practical design tools allowing a quick and easy characterization of the joints, software or sometimes design sheets and tables of standardized joints appear to be the most efficient. Many already exist and are used in practice for joints with open sections (e.g. SCI/BCSA 2011).

1.3.2 Basic joints components

As explained above, any joint is considered as a set of individual components. Therefore, in agreement with the principles of the component method, the first step of the work will consist in establishing the list of components required to cover the present scope of chapter 7 of EN 1993 Part 1.8. Not only the geometrical configuration of the studied joints will have to be taken into consideration, but also the type of loading to which the joint is subjected (axial forces, bending moments, shear forces, combinations of axial forces and bending moments, ...). In a next step, design resistance formulae for each of the constitutive individual components in shear, tension or compression will have to be derived from chapter 7 of EN 1993 Part 1.8 through an appropriate "conversion" process to be specified.

1.3.3 *Assembly – Determination of joint properties*

Once the components will be identified and characterized, the way on how to assemble the latter will have to be derived.

To assemble the components means to express the fact that the forces acting on the whole joint distributes amongst the constitutive components in such a way that:

- the internal forces in the components are in equilibrium with the external forces applied to the joint
- the resistance of a component is nowhere exceeded
- the deformation capacity of a component is nowhere exceeded

As far as the resistance of the whole joint to external forces is concerned, the fulfilment of these three rules is enough to ensure that the evaluated design resistance is smaller than the actual joint resistance.

2 FIELD OF APPLICATIONS

In its section 7.1, EN 1993-1-8 provides the scope of chapter 7 and its general field of application. Within these limits, the clauses provided by chapter 7 may be used for the evaluation of the design resistance of the hollow section joints.

In section 7.2, EN 1993-1-8 explains that six failure modes can be identified for any hollow section for joints between hollow and open sections and basically that the design resistance of a joint may be evaluated as the minimum of the six design resistances corresponding to these six failure modes.

Further to these quite general statements, EN 1993-1-8 points out that not all these six failure modes are always relevant and therefore that simplification may be achieved, when possible, by disregarding failure modes which could be recognized as irrelevant. In order to turn this conclusion into reality, EN 1993-1-8 specifies successively for:

- welded joints between CHS members (section 7.4)
- welded joints between CHS or RHS brace members and RHS chord members (section 7.5)
- welded joints between CHS or RHS brace members and I or H section chords (section 7.6)
- welded joints between CHS or RHS brace members and channel section chord members (section 7.7)

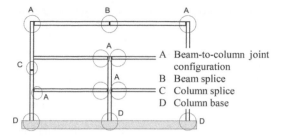

Figure 1. Different types of joint configurations in a building frame.

A Beam-to-column joint configuration
B Beam splice
C Column splice
D Column base

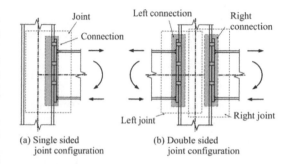

Figure 2. Joints and connections.

(a) Single sided joint configuration
(b) Double sided joint configuration

The specific limited fields of application in which some of the failure modes may be disregarded, so reducing the number of computations to be achieved by the designer.

For the "remaining" (relevant) failure modes to be checked, formulae are then provided, but again with some more further "local" limitations on joint geometry. These ones reflect usually the domain in which the design rule has been validated.

3 GENERALITIES ABOUT THE CONVERSION PROCESS

3.1 *Definitions of connection, joint and joint configuration in EN1993-1-8 chapter 6*

Chapter 6 of EN 1993-1-8 addresses the design of joints between members with open cross-sections and applies in priority to building frames; the latter consist of beams and columns, usually made of H or I shapes that are assembled together by means of connections. These connections are between two beams, two columns, a beam and a column or a column and the foundation (Figure 1).

A connection is defined as the set of the physical components which mechanically fasten the connected elements. One considers the connection to be concentrated at the location where the fastening action occurs, for instance at the beam end/column interface in a major axis beam-to-column joint. When the connection as well as the corresponding zone of interaction between the connected members are considered together, the wording joint is used (Figure 2a).

Depending on the number of in-plane elements connected together, single-sided and double-sided joint configurations are defined (Figure 2). In a double-sided configuration (Figure 2b), two joints – left and right – have to be considered.

The definitions illustrated in Figure 2 are also valid for other joint configurations and connection types.

To point out the need for distinguishing connections from joints, reference is here made to one-sided and double-sided joint configurations illustrated in Figure 2.

In these ones, the left (if any) and right connections are subjected to the forces respectively in the left (if any) and right beams while the shear force applying to the column web panel results from the equilibrium of all the forces acting on it. The shear force in the panel V_{wp} may be evaluated as follows:

$$V_{wp} = \frac{M_{b1} - M_{b2}}{z} - \frac{V_{c1} - V_{c2}}{2} \qquad (1)$$

Another formula to which it is sometimes referred, i.e.:

$$V_{wp} = \frac{M_{b1} - M_{b2}}{z} \qquad (2)$$

is only a rough and conservative approximation of Eq. (1). In both formulae, z is the lever arm of the resultant tensile and compressive forces in the connection(s) and M_{b1}, M_{b2}, V_{c1} and V_{c2} are the acting bending moments and shear forces in right (b1) and left (b2) beams and lower (c1) and upper (c2) columns respectively. Both above-mentioned formulae are given in EN 1993-1-8 chapter 5.3 (Modelling of beam-to-column joints).

As a direct outcome, it is concluded that the connection response depends only on the forces acting in the corresponding beam and therefore that it is the same in both joint configurations; but also that the response of the joints is influenced by that of the column web panel in shear which depends on how the joint configuration is loaded. For its computation, the designer may decide, according to EN 1993-1-8: (i) to model separately the column web panels and the connections or (ii) to model the joints.

A similar concept applies to minor axis beam-to-column joints (beam connected to the web of the column) or to 3-D joint configurations (beams connected to the flange and the web of the column).

In daily practice, it is used to refer to the "joint" concept and not to model separately the column webs panels and the connections and, in order to facilitate the application, so-called "transformation parameters β" are defined. These ones allow to easily account for the influence of the column web panels on the response of the joints. Chapter 5 of EN 1993-1-8 provides the user with accurate or approximate values of these β parameters.

3.2 Application to EN 1993-1-8 chapter 7

Even if it is not explicitly said, EN 1993-1-8 chapter 7 refers also to the concept of joints, and therefore implicitly to the concept of transformation factors.

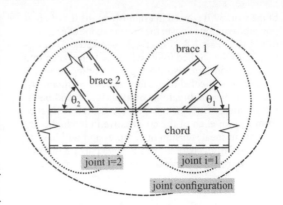

Figure 3. Example of a joint configuration including two joints.

In fact, in chapter 7, a joint configuration is constituted of as many joints as there are braces connected to the chord (see Figure 3). The verification of the resistance of the joint configuration is achieved through the successive check of the resistance of all the constitutive joints. For symmetry reasons, the number of joints to check may be reduced (as an example, for DK joints in Table 7.15 of EN 1993-1-8 where the four actual joints reduce to two).

This means that, in a later stage of revision of the Eurocodes, a rewording of section 5.3 of EN 1993-1-8 chapter 5 could be easily achieved so as to extend its scope from "beam-to-column joints" to "joints" and so to cover the whole scope of EN 1993-1-8.

3.3 Design checks for joint configurations involving tubular members

As already said above, to check the resistance of a joint configuration consists in checking successively the resistance of all its constitutive joints. And for the verification of a joint named i, the following general formula is provided in chapter 7:

$$\frac{N_{i,Ed}}{N_{i,Rd}} + \frac{M_{ip,i,Ed}}{M_{ip,i,Rd}} + \frac{M_{op,i,Ed}}{M_{op,i,Rd}} \leq 1,0 \qquad (3)$$

In this expression:

- $N_{i,Ed}$, $M_{ip,i,Ed}$ and $M_{op,i,Ed}$ are the forces acting in the brace of the considered joint i, respectively the design axial force, the design in-plane moment and the design out-of-plane moment;
- $N_{i,Rd}$, $M_{ip,i,Rd}$ and $M_{op,i,Rd}$ are the design resistances of the considered joint i, respectively the design axial resistance, the design in-plane moment resistance and the design out-of-plane moment resistance.

For welded joints between CHS members, formula (3) is substituted by the following one:

$$\frac{N_{i,Ed}}{N_{i,Rd}} + \left[\frac{M_{ip,i,Ed}}{M_{ip,i,Rd}} \right]^2 + \frac{M_{op,i,Ed}}{M_{op,i,Rd}} \leq 1,0 \qquad (4)$$

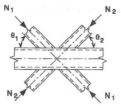

Additional check: $N_{1,Ed} \sin\theta_1 + N_{2,Ed} \sin\theta_2 \leq N_{x,Rd} \sin\theta_x$

where $N_{x,Rd} \sin\theta_x = \max\left\{ \left|N_{1,Rd} \sin\theta_1\right|, \left|N_{2,Rd} \sin\theta_2\right| \right\}$

Figure 4. Additional design check to cover interactions.

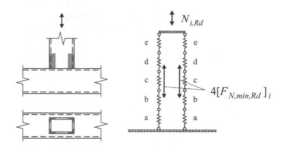

Figure 5. Example of joint representation by means of the component approach.

Obviously, possible interactions between joints belonging to the same joint configuration may occur and should be considered in the verification procedure. This fact is well recognized in EN 1993-1-8 chapter 6 and chapter 7.

In chapter 6, most of the relevant interactions are considered at the component level while in chapter 7, these ones lead more generally to additional checks for the whole joint configuration. An example is provided in Figure 4 for a DK joint.

Through the conversion process of chapter 7 to a component style, it will be seen that some of these "joint configuration" extra checks may be turned into a "component" modified check, so decreasing the number of verifications to be effectively achieved, but without changing the resistance of the joint configuration presently provided by the application of chapter 7.

3.4 Design checks for joints involving tubular members

For each joint belonging to a joint configuration, Eq. (3) has to be applied, what requires first to evaluate its three individual resistances $N_{i,Rd}$, $M_{ip,i,Rd}$ and $M_{op,i,Rd}$.

To derive these individual resistances, reference is made to the component approach. Its application is described in the following sections, first for a particular joint and then in more general terms.

3.4.1 Component approach – particular example

According to the component method, the joint may be conceptually considered as a system of springs, each of these ones representing a specific component. Figure 5 illustrates the concept for a T joint between RHS profiles under axial force. In this graph, the notations "a" to "e" relate to the relevant active components which may be identified as follows:

a. (chord) face in bending
b. (chord) side wall(s) in tension or compression
c. (chord) side wall(s) in shear
d. (chord) face under punching shear
e. (brace) flange or web(s) in tension or compression

In Figure 5, the axial forces is seen to be carried out from the brace to the chord through four loading zones located at the corners of the brace sections. This

assumption, once it is agreed, has to be respected all along the design process. For each brace cross-section type, the number and the location of the load transfer zones will have to be carefully defined.

In this particular case, the design resistance of the joint under axial force may be derived as:

$$N_{i,Rd} = 4 \cdot \left[F_{N,\min,Rd} \right]_i \tag{5}$$

where $[F_{N,\min,Rd}]_i$, the minimum resistance of the active components (a)...(e) for joint i under axial force, is expressed as:

$$\left[F_{N,\min,Rd} \right]_i = \min \left[F_{a,N,Rd} ; F_{b,N,Rd} ; ... ; F_{e,N,Rd} \right]_i \tag{6}$$

The procedure may be extended similarly to all loading situations and so, at the end, the design resistances of the joint under $N_{i,Ed}$, $M_{ip,i,Ed}$ and $M_{op,i,Ed}$ are given by:

$$N_{i,Rd} = 4 \cdot \left[F_{N,\min,Rd} \right]_i \tag{7}$$

$$M_{ip,i,Rd} = 2 \cdot \left[F_{M_{ip},\min,Rd} \right]_i z_{ip} \tag{8}$$

$$M_{op,i,Rd} = 2 \cdot \left[F_{M_{op},\min,Rd} \right]_i z_{op} \tag{9}$$

where $[F_{N,\min,Rd}]_i$, $[F_{M_{ip},\min,Rd}]_i$ and $[F_{M_{op},\min,Rd}]_i$, respectively the minimum resistance of the active components for joint i under axial force, the minimum resistance of the active components for joint i under in-plane moment and the minimum resistance of the active components for joint i under out-of-plane moment are expressed as:

$$\left[F_{N,\min,Rd} \right]_i = \min \left[F_{a,N,Rd} ; F_{b,N,Rd} ; ... ; F_{e,N,Rd} \right]_i \tag{10}$$

$$\left[F_{M_{ip},\min,Rd} \right]_i = \min \left[F_{a,M_{ip},Rd} ; F_{b,M_{ip},Rd} ; ... ; F_{e,M_{ip},Rd} \right]_i \tag{11}$$

$$\left[F_{M_{op},\min,Rd} \right]_i = \min \left[F_{a,M_{op},Rd} ; ... ; F_{e,M_{op},Rd} \right]_i \tag{12}$$

where z_{ip} is the relevant lever arm under in-plane moment and z_{op} is the relevant lever arm under out-of-plane moment.

These lever arms result from the assumption made on the location of the loading transfer zones within the joint.

3.4.2 *Component approach – generalization*

The conversion of EN 1993-1-8 chapter 7 into a "component style" requires therefore, for each joint in each joint configuration covered in the normative document, to:

– identify the active components
– derive the resistance of these active components
– assemble the components

Successively for each individual loading situations (axial force, in-plane bending moment and out-of-plane bending moment).

When this is achieved, the resistance of each joint may then be checked through Eqs. (3) or (4).

In terms of assembly, it may be shown that the above-mentioned Eqs. (7) to (9) may be generalized to all joints under the scope of EN 1993-1-8.

4   FROM FAILURE MODES TO COMPONENTS

Six failure modes are listed in chapter 7 of EN 1993-1-8 to be possibly relevant to determine the resistance of hollow section joints:

– Chord face failure (yielding);
– Chord side wall failure (yielding and/or instability);
– Chord shear failure (yielding and/or instability);
– Chord punching shear;
– Brace failure;
– Local buckling failure (brace or chord).

In the ongoing CIDECT project, five failure modes have been considered so far. In full agreement with the component method principles, five corresponding components are identified:

a. Chord face in bending;
b. Chord side wall(s) in tension or compression;
c. Chord side wall(s) in shear;
d. Chord face under punching shear;
e. Brace flange and web(s) in tension or compression component.

The names of these components are chosen so as to respect the terminology used in EN1993-1-8 chapter 6. These basic components are illustrated in Table 1. Local buckling failure will be investigated in a later step.

5   COMPONENT DESIGN RESISTANCE FORMULAE

5.1   *"Chord face in bending" component*

The chord face in bending component (a) for a CHS chord is here selected to present the way on how the contents of EN 1993-1-8 Chapter 7 is "transferred", without any change in the level of design resistance, into a component format.

Let's consider first the case of a X joint made of a CHS chord and two IPE braces, the latter ones being subjected to compression forces N_i^*.

Table 1.   Illustration of basic components (from EN 1993-1-8).

| a: Chord face in bending |
| b: Chord side wall(s) in tension/compression |
| c: Chord side wall(s) in shear |
| d: Chord face under punching shear |
| e: Brace flange and web(s) in tension or compression |

According to the well-known Ring Model initially developed by Togo (1967):

– It is assumed that most of the loading is transferred at the saddles of the brace, since the chord behaves most stiffly at that part of the connection perimeter.
– The load N_i^* in the brace can be divided into two loads of $0.5 \cdot N_i^*$ perpendicular to the chord at the saddles (at a distance noted $c_1 d_1$).
– These loads are transferred by an effective length B_e of the chord.
– The load $0.5 \cdot N_i^*$ is considered as a linear distributed load over the effective length B_e.
– To take into account the influence of the axial load in the chord, a chord stress function k_p is used. This one has been determined by experiments and numerical investigations.
– The width ratio β is defined as equal to d_1/d_0.
– Neglecting the influence of axial and shear stresses on the plastic moment resistance $m_{pl,Rd}$ of the chord per unit length allows, if it is assumed that $d_0 - t_0 \approx d_0$, to derive an evaluation of the design resistance N_{Rd} of the joint as follows:

$$N_{Rd} = \frac{8B_e/d_0}{1-c_1\beta} k_p m_{pl,Rd} \tag{13}$$

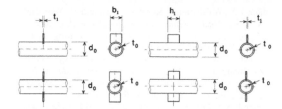

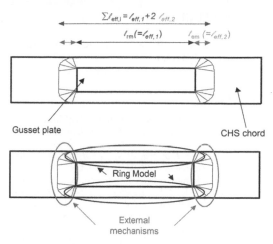

Figure 6. Different "load introduction" situations in a welded T and X joint respectively (left: gusset plate perpendicular, right: gusset plate longitudinal) (from EN 1993-1-8).

Figure 7. Yield line pattern and effective lengths.

- The effective length B_e has been determined experimentally and depends on the β ratio. An average value is $B_e = 2.5d_0 \div 3.0d_0$

In reality, much more complex yield patterns develop in the chord face when it is subjected to tension/compression transverse forces; and these ones differ significantly according to the specific way on how the transverse load is transferred to the CHS chord. From that point of view, the four situations illustrated as examples in Figure 6 have each individually to be considered as particular ones and specific design resistance formulae have consequently to be suggested.

In EN 1993-1-8, for instance, the following design resistances are suggested (see Figure 6):

Gusset plate welded perpendicular to CHS chord:

$$N_{Rd} = k_p f_{y0} t_0^2 \left(4 + 20\beta^2\right) / \gamma_{M5}$$
$$= 4k_p \left(4 + 20\beta^2\right) m_{pl,Rd} \tag{14}$$

Gusset plate welded longitudinal to CHS chord:

$$N_{Rd} = 5k_p f_{y0} t_0^2 \left(1 + 0,25\eta\right) / \gamma_{M5}$$
$$= 20k_p \left(1 + 0,25\eta\right) m_{pl,Rd} \tag{15}$$

Whatever is the case, the general format of the equations is as follows:

$$N_{Rd} = \left(\sum \bar{l}_{eff,i}\right) \cdot k_n \cdot m_{pl,Rd} \tag{16}$$

where $\sum \bar{l}_{eff,i}$ designates the total equivalent (effective) length coefficient characterizing the yield line pattern actually developing in the chord face. The meaning of $\sum \bar{l}_{eff,i}$ may be highlighted through the following interpretation of the basic Ring Model formula:

$$N_{Rd} = \frac{8 \cdot B_e/d_0}{\left(1 - c_1 \cdot \beta\right)} \cdot m_{pl,Rd} \cdot k_p$$
$$= \sum l_{eff,i}/d_0 \cdot m_{pl,Rd} \cdot k_p \tag{17}$$
$$= \sum \bar{l}_{eff,i} \cdot m_{pl,Rd} \cdot k_p$$

In Figure 7, for an joint between a CHS chord and a longitudinal gusset plate, the actual yield lines developing in the chord are illustrated; moreover, it shows how the different zones of the yield line pattern contribute individually to $\sum l_{eff,i}$. In this figure, the thickness t_1 of the gusset has been voluntary exaggerated.

From Figure 7 it can be seen that the total effective length is the sum of the effective length of the "inner part" (Ring model) and the external part. For compatibility with the typical definitions of components in chapter 6 of EN 1993-1-8 it is suggested to subdivide the axial force acting on the gusset is into four equal forces, in this particular case two forces at the same location. The distance between the two pair of forces is equal to h_1. Therefore one has:

$$N_{Rd} = 4 \cdot \left[F_{a,Rd}\right] \tag{18}$$

The design resistance of the "chord face component in compression" (or of the "chord face component in tension") may finally be expressed as follows:

$$F_{a,Rd} = \left(0,5\bar{l}_{eff,1} + \bar{l}_{eff,2}\right) k_p m_{pl,Rd} \tag{19}$$

With Eqs. (14) to (19) the values for $\bar{l}_{eff,1}$ and $\bar{l}_{eff,1}$ can be derived. More details can be found in Weynand et al. (2014).

5.2 Other components

For the other components, a similar procedure is used:

- Select an analytical expression representing the physics of the studied phenomenon;
- Compare the EN 1993-1-8 to the selected analytical expression and so calibrate effective length(s) or the effective width(s) according to the case.

ACKNOWLEDGEMENT

Financial support for the work reported in this paper is provided by CIDECT.

REFERENCES

EN 1993-1-1. 2005. *Eurocode 3: Design of Steel Structures – Part 1-1: General rules and rules for buildings*, Brussels: CEN.

EN 1993-1-8. 2005. *Eurocode 3: Design of Steel Structures – Part 1-8: Design of joints*, Brussels: CEN.

Jaspart, J.-P.; Pietrapertosa, C.; Weynand, K., Busse, E.; Klinkhammer, R. 2005. *Development of a full consistent design approach for bolted and welded joints in building frames and trusses*, CIDECT project 5BP, report 5BP-4/05, Final report.

Jaspart, J.-P.; Herion, S.; Weynand, K., Welch, S. 2011. *CIDECT Gap Analysis*, CIDECT internal report.

SCI/BCSA. 2011. *Joints in Steel Construction: Simple Joints To Eurocode 3*, Publication P358, www.steel-sci.com.

Packer, J. A.; Wardenier, J.; Kurobane, Y.; Dutta, D.; Yeomans, N. 1992. *Design Guide for rectangular hollow section (RHS) joints under predominantly static loading*. Köln: Verlag TÜV Rheinland GmbH.

Packer, J. A.; Wardenier, J.; Zhao, X.-L.; van der Vegte, G. J.; Kurobane, Y. 2009. *Design Guide for rectangular hollow section (RHS) joints under predominantly static loading*. 2nd edition, www.cidect.org/en/Publications.

Weynand, K.; Jaspart, J.-P., Zhang, L. 2014. *Component method for tubular joints*, CIDECT project 16F, report 16F-5/14, 1st interim report.

Togo, T. 1967: Experimental study on mechanical behaviour of tubular joints. PhD thesis, Osaka University, Japan, (in Japanese).

Wardenier, J.; Kurobane, Y.; Packer, J.A.; Dutta, D.; Yeomans, N. 1991. *Design Guide for circular hollow section (CHS) joints under predominantly static loading*. Köln: Verlag TÜV Rheinland GmbH.

Wardenier, J.; Kurobane, Y; Packer, J. A.; van der Vegte, G. J.; Zhao, X.-L.. 2008. *Design Guide for circular hollow section (CHS) joints under predominantly static loading*. 2nd edition, www.cidect.org/en/Publications.

Tubular Structures XV – Batista, Vellasco & Lima (eds)
© 2015 Taylor & Francis Group, London, ISBN 978-1-138-02837-1

Numerical study of through plate-to-CHS connections

A.P. Voth
Read Jones Christoffersen Ltd., Toronto, Canada

J.A. Packer
Department of Civil Engineering, University of Toronto, Toronto, Canada

ABSTRACT: Branch plate-to-circular hollow section (CHS) connections are often chosen as an easy-to-fabricate and cost-effective option. Branch plates can be slotted through the CHS and welded to two opposing faces to produce a through plate connection, where strengthening is required; however, design equations are not available at present due to limited research. The results of a finite element study, presented here, indicate that the behaviour of through plate-to-CHS connections closely match the sum of branch plate-to-CHS connection behaviour in tension and compression for a given geometric configuration. A partial design strength function, shown to be valid for a wide range of connection geometries, which is the sum of existing design recommendations for branch plate-to-CHS connections in tension and compression, is proposed for through plate-to-CHS connections.

1 INTRODUCTION

Branch plate-to-hollow structural section (HSS) connections are often chosen as an easy-to-fabricate and cost-effective option making them ideal for both enclosed framed structures and for exposed steel work, including cable-stayed roof systems and tubular arch bridges. The strength of the hollow section chord can be low depending on connection geometry, particularly for heavily loaded tension-only plate-to-CHS connections, often requiring strengthening or stiffening techniques.

Some of the more common strengthening techniques include using ring stiffeners on either the outside or the inside of the CHS as annular ring plates, and concrete or grout filling of the chord. Significant research has been completed on both of these strengthening methodologies whereby plastification of the CHS chord is restrained which thus increases the connection resistance (e.g. Alostaz & Schneider 1996; Packer 1995; Zhao & Packer 2009; Lee & Llewelyn-Parry 2005; Willibald 2001). Another widely used stiffening and strengthening method is a through plate-to-HSS connection where the branch plates can be slotted through the HSS and welded to two opposing faces (see Fig. 1).

An experimental program for through plate-to-CHS connections, that was conducted at the University of Toronto (Voth 2010; Voth & Packer 2012a), indicated that through plate connection behaviour is comprised of two independent mechanisms: one in compression and one in tension, that occur on opposite sides of the CHS chord during load application. Preliminary examination of this behaviour also illustrated

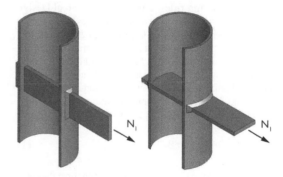

Figure 1. Longitudinal (left) and transverse (right) through plate-to-CHS connection types.

that the capacity of a through plate connection is approximately the summation of the capacities of a branch plate-to-CHS connection tested in tension and a branch plate-to-CHS connection tested in compression, provided that connection geometry for the branch connection and the through connection were similar. The dual mechanism for plate-to-CHS connections differs from the approach that has been previously applied to plate-to-RHS connections, where one yield line mechanism is present on both the top and bottom connection faces. For RHS connections this results in a through plate connection having approximately double the ultimate capacity of a branch plate connection – regardless of branch or through plate connection loading sense – tension or compression (Kosteski 2001; Kosteski & Packer 2001, 2003). As there is a limited set of experimental and numerical results for through plate-to-CHS connections, a study that explores an expanded range of connection geometries was needed.

A numerical parametric study has been carried out for T-type though plate-to-CHS connections to examine the influence of various geometric parameters on connection capacity. This study had the aim of combining tension and compression branch plate-to-CHS connection capacity functions to develop through plate-to-CHS connection capacity. The following describes the numerical finite element study and proposed expressions for the ultimate limit state.

1.1 Design resistance of T-type branch plate connections

Branch plate-to-CHS connection resistance is determined by using the lower of two limit states: chord plastification and chord punching shear, given that both the branch plate and the weld are adequately designed and are non-critical. The two limit states are highly dependent on connection geometry (including the orientation and dimensions of the both the branch plate and chord member). Axially loaded T-type branch plate-to-CHS connection design guidelines (CIDECT Design Guide No. 1, 2nd Edition, by Wardenier et al. 2008, IIW design rules, 2012 and AISC Steel Design Guide No. 24 by Packer et al. 2010) were developed by adapting existing CHS-to-CHS design guidelines to a limited set of experimental results for branch plate-to-CHS (Washio et al. 1970, Akiyama et al. 1974) using regression analysis (van der Vegte et al. 2008). The chord plastification connection resistance, expressed as an axial force in the branch member, is given by CIDECT (Wardenier et al. 2008) as:

$$N_1{}^* = Q_u Q_f f_{y0} t_0{}^2 / \sin\theta_1 \qquad (1)$$

where Q_u is a partial design strength function that predicts connection resistance without chord axial stress, and Q_f is a chord stress function that reduces connection resistance based on chord normal stress influence. These functions, along with chord punching shear expressions, are summarized in Table 1.

An extensive experimental and numerical study was conducted by Voth & Packer (2012a, 2012b and 2012c) for X- and T-type branch plate-to-CHS connections in an effort to re-evaluate the current CIDECT chord plastification partial design strength function, Q_u (Table 1). The numerical study of T-type plate-to-CHS connections, which consisted of approximately 100 connection geometries (Voth & Packer 2012c), concluded that the behaviour of branch plate connections tested under plate axial compression load varied significantly with respect to connections under branch plate axial tension load.

As the capacity of the tension-only connections was found to be underutilized under the current CIDECT design guidelines (Wardenier et al. 2008), regression analysis of the numerical results was undertaken and

Table 1. CIDECT design resistance of T-type branch plate-to-CHS connections under axial load (Wardenier et al. 2008).

	Transverse plate	Longitudinal plate
Chord plastification	$Q_u = 2.2(1 + 6.8\beta^2) \cdot \gamma^{0.2}$ $Q_f = (1 - |n|)^{C_1}$ where for chord compression stress $(n < 0)$, $C_1 = 0.25$ for chord tension stress $(n \geq 0)$, $C_1 = 0.20$	$Q_u = 5(1 + 0.4\eta)$ $n = \frac{N_0}{N_{pl,0}} + \frac{M_0}{M_{pl,0}}$
Punching shear	$N_1* = 1.16b_1 t_0 f_{y0}$ when $b_1 \leq d_0 - 2t_0$	$N_1* = 1.16h_1 \frac{t_0 f_{y0}}{\sin^2\theta_1}$

Range of validity:
Compression chords must be class 1 or 2, but also $2\gamma \leq 50$
Tension chords must be $2\gamma \leq 50$
Transverse plate: $0.4 \leq \beta \leq 1.0$; longitudinal plate: $1 \leq \eta \leq 4$
$f_{y1} \leq f_{y0}, f_y/f_u \leq 0.8, f_{y0} \leq 460\,\text{MPa}$

Note: θ_1 is the angle of the force acting on the plate.

new partial design strength functions were developed as follows (Voth 2010):

$$Q_{u,90°,C} = 2.9\zeta(1 + 3\beta'^2)\gamma^{0.35} \qquad (2)$$

for transverse T-type branch plate in compression,

$$Q_{u,90°,T} = 2.6\zeta(1 + 2.5\beta'^2)\gamma^{0.55} \qquad (3)$$

for transverse T-type branch plate in tension,

$$Q_{u,0°,C} = 7.2\zeta(1 + 0.7\eta') \qquad (4)$$

for longitudinal T-type branch plate in compression and,

$$Q_{u,0°,T} = 10.2\zeta(1 + 0.6\eta') \qquad (5)$$

for longitudinal T-type branch plate in tension. A lower bound reduction factor of $\zeta = 0.85$ was used based on regression analysis of the numerical results, for application to limit states design.

2 RESEARCH PROGRAM AND FINITE ELEMENT CONNECTION MODELLING

To investigate the increase in connection capacity of through plate-to-CHS connections, when compared to their branch plate counterparts, the same 52 connection geometries used for T-type branch plate- to-CHS connections were initially studied numerically. The high ultimate capacity of the through plate-to-CHS connections proved difficult to capture without causing non-converged solutions and chord end failure. As a result, only 18 of the previous T-type configurations produced results to the point of connection ultimate capacity: five transverse and 13 longitudinal through plate-to-CHS connections. The resulting parametric

Table 2. Effective chord length parameter[a] for longitudinal T-type through plate-to-CHS connections.

t_0 (mm)	2γ	Nominal depth ratio, $\eta = h_1/d_0$					
		0.2	0.6	1.0	1.5	2.0	2.5
11.10	19.74	4	4	4	4		
7.95	27.56	8	8	8	8	8	8
4.78	45.84	12	12	12			

[a]Effective chord length parameter, $\alpha' = 2l_0'/d_0$.

Table 3. Effective chord length parameter[a] for transverse T-type through plate-to-CHS connections.

t_0 (mm)	2γ	Nominal depth ratio, $\beta = b_1/d_0$		
		0.2	0.4	0.6
7.95	27.56	8	8	
4.78	45.84	12	12	8

[a]Effective chord length parameter, $\alpha' = 2l_0'/d_0$.

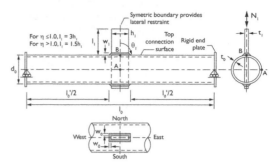

Figure 2. Parametric longitudinal T-type through plate-to-CHS connection configuration.

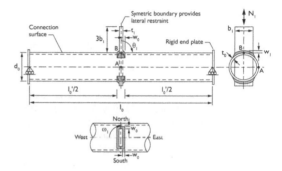

Figure 3. Parametric transverse T-type through plate-to-CHS connection configuration.

numerical finite element research study thus consisted of 18 connections, modelled by varying values of β from 0.2 to 0.6, η from 0.2 to 2.5 and 2γ from 19.74 to 45.84, as outlined in Table 2 and Table 3 with effective chord length parameter ($\alpha' = 2l_0'/d_0$) indicated. Note that transverse connections with values of $\beta > 0.8$ are improbable, depending on chord wall thickness, as sufficient space for the plate to pass through the chord is required. As the behaviour of these connections is the same for both through plate tension and compression loads, the connections were only tested in through plate tension load.

All connections were modelled with fillet welds, and the effect of the weld size on connection behaviour was incorporated by converting both β and η into effective values (β' and η'). The values of η' used in this study are 0.32, 0.72, 1.12, 1.62, 2.12 and 2.62 and the values of β' used are 0.32, 0.51 and 0.69. A plate thickness (t_1) of 19.01 mm and chord diameter (d_0) of 219.1 mm were used for all numerical models and were constructed with the geometry given in Figures 2 and 3. Where the branch plate was considered critical, the yield strength of the plate (f_{y1}) was increased to provide substantial resistance so that connection behaviour would govern.

2.1 Finite element connection modelling

The numerical analysis was carried out using the same general characteristics and methods as described by Voth (2010) and Voth & Packer (2012b and 2012c) for T-type branch plate-to-CHS connections. These previously established and validated non-linear finite element modelling techniques used are summarized herein.

Finite element models were constructed and analyzed using commercially available software: ANSYS

11.0 (ANSYS Inc. 2007). Both geometry and measured material properties, including chord end conditions and fillet weld details, were replicated within the FE model. Eight-node solid brick elements (SOLID45), each with three translational degrees of freedom per node and reduced integration with hourglass control, were used for each connection model along with three chord through-thickness elements. Uniform mesh density was used, except in locations where large deformations and/or peak stress concentrations leading to cracking and eventually fracture occurred where increased mesh density was used to better capture this behaviour; typically at joint locations between the branch plate and CHS chord. Symmetrical boundary conditions and only one quarter of the connection were modelled where geometry, restraint and loading were symmetrical. A non-linear time step analysis was used incorporating non-linear material properties, large deformation allowance and full Newton-Raphson frontal equation solver.

Multi-linear true stress-strain curves, converted from tensile coupon results of both plate, Grade 300W (CSA 2004) and CHS, ASTM A500 Grade C (ASTM 2010), material were used for FE material properties until the point of coupon necking. The post-necking behaviour was determined by an iterative method developed by Matic (Matic 1985) and modified by Martinez-Saucedo et al. (2006), using FE modelling of experimental coupons directly. The plate material had a yield and ultimate strength of 326 MPa and

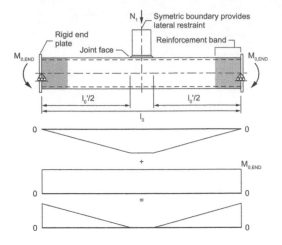

Figure 4. Member end loading to exclude chord normal stress at joint face.

505 MPa respectively, and the CHS material had a yield and ultimate strength of 389 MPa and 527 MPa, with the weld material being given the same properties as the plate. A failure criterion was imposed to emulate material ultimate fracture whereby the "death feature" of an element was activated using a previously determined maximum equivalent (von Mises) strain value of $\varepsilon_{ef} = 0.2$ (Voth 2010; Voth & Packer 2012b, 2012c). Once the maximum equivalent strain value was reached within an element, the stiffness and the stress in that element were reduced to near-zero allowing the element to freely deform.

T-type connections were modelled in three-point bending where, for a one-quarter model, the chord end was supported by a roller at the chord neutral axis (see Figs. 2 & 3) and tested using displacement controlled loading. To prevent an unstable condition, lateral restraint was provided by the symmetric boundary. A high equilibrium-induced chord bending moment (or chord normal stress) at the joint connecting face was induced, which is undesirable for determining connection behaviour without chord stress or design recommendations without a chord stress influence function (Q_f). To remove this chord normal stress due to bending at the joint face, counteracting in-plane bending moments ($M_{0,END} = N_1 l_0'/4$) were applied to the rigid chord end plates (see Fig. 4) thus allowing the chord normal stress influence function (Q_f) to remain independent to the partial design strength function, Q_u (refer to Table 1).

In-plane bending moment applied to the chord end, however, produces two additional problems that must be addressed. First, connections with high ultimate capacity due to geometric configuration (e.g. thick chords or high plate widths) or longer connection lengths produce higher end moments that may exceed the yield capacity of the chord, as the applied end moment ($M_{0,END}$) is a function of the applied connection load (N_1) and the effective chord length (l_0'). A band of elements with higher yield strength at the

CHS chord end (see Fig. 4 – "reinforcement band") was used to prevent chord end failure prior to connection capacity. If reinforcement was required, the band width was determined on an individual connection basis depending on predicted connection capacity and chord length. Second, to prevent non-convergent results which are possible with load-controlled analysis (predominantly for connections loaded in compression during periods of significant plastification and deformation), displacement-controlled analysis was used. Displacement-controlled analysis makes the calculation of the applied end moment difficult as the branch plate load for a given applied displacement is not known until the end of each time step. The applied end moment required for application at the start of each time step was determined by predicting the branch load using a Taylor series approximation and load information from the previous time step, in combination with an end-of-time-step correction and a small displacement rate based in part on the connection load-deformation curve slope.

3 PARAMETIC STUDY RESULTS AND COMPARISON TO BRANCH PLATE-TO-CHS CONNECTIONS

The load-deformation curve for each longitudinal and transverse T-type through plate-to-CHS connection was determined with the connection deformation defined as the change in distance between point A in Figures 2 and 3 and a point at the crown of the CHS chord (point B in Figs. 2 and 3). From these curves the connection ultimate capacity ($N_{1,u}$) was determined as the minimum of: (i) the load at a deformation of 3%d_0, $N_{1,3\%}$, if this deformation preceded the deformation at $N_{1,max}$, (ii) the maximum connection load, $N_{1,max}$ (the global maximum load) and (iii) branch plate yielding. For all connections the load at a deformation of 3%d_0 governed the connection capacity. Figures 5 and 6 show the normalized ultimate load [$N_{1,u}/(f_{y0}t_0^2)$] as a function of β' or η' for all 2γ values for longitudinal and transverse through plate connections. The numerical results are compared to the current CIDECT chord plastification function for T-type branch plate-to-CHS connections (Wardenier et al. 2008), Q_u, calculated using effective geometric properties.

For both longitudinal and transverse through plate-to-CHS connections, the current CIDECT design equations (Wardenier et al. 2008) presented in Table 1 for T-type branch plate-to-CHS connections do not come close to predicting connection ultimate capacity (see Figs. 5 and 6). As the CIDECT design equations for branch plate-to-CHS connections were not intended for through plate-to-CHS connections, the difference in capacity is understandable.

To determine if the numerical results are reasonable and applicable to a wider range of connection geometries, Figure 7 examines the ratio of $Q_{u,Through}$, the normalized connection capacity [$N_{1,u}/(f_{y0}t_0^2)$] for through plate connections and $Q_{u,Branch(Tens+Comp)}$, the

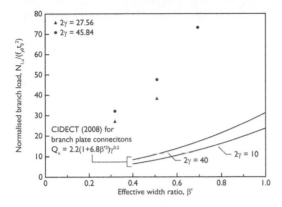

Figure 5. Parametric FE results for transverse T-type through plate-to-CHS connections.

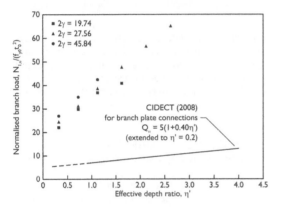

Figure 6. Parametric FE results for longitudinal T-type through plate-to-CHS connections.

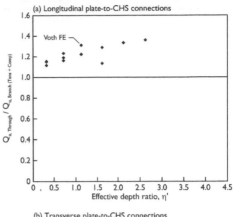

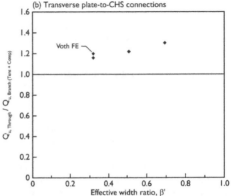

Figure 7. Comparison of through plate and the summed branch plate connection capacities (all determined by finite element analysis).

summation of the normalized connection capacities $[N_{1,u}/(f_{y0}t_0^2)]$ for T-type branch plate connections in tension and compression (previously tested by Voth 2010 and Voth & Packer 2012b, 2012c). The strength partial function (Q_u) is equivalent to the normalized connection capacity, $N_{1,u}/(f_{y0}t_0^2)$ in this case as the chord stress function (Q_f) and inclination angle term ($\sin\theta_1$) are both equal to unity.

For all longitudinal and transverse through plate-to-CHS connections, Figures 7(a) and (b), with a mean and coefficient of variation of 1.22 and 6.32% for longitudinal and 1.22 and 3.82% for transverse, show that the summation of tension and compression T-type branch plate-to-CHS connection capacities under-predicts through plate connection capacity by approximately 20%. Though the through plate-to-CHS connection ultimate capacity is close to the summation of tension and compression branch plate-to-CHS connection capacities, the numerical FE through plate connections do not directly follow the conclusions from the experimental program (Voth & Packer 2012a).

4 DESIGN RECOMMENDATION DEVELOPMENT FOR T-TYPE THROUGH PLATE CONNECTIONS

There are no established guidelines or theoretical models to use as a basis for the development of design recommendations for T-type through plate-to-CHS connections. The general behaviour of a through plate connection, however, has been established to be the approximate addition of tension and compression branch plate-to-CHS connection behaviours. As potential design expressions for both longitudinal and transverse T-type connections, with tension and compression plate loads exist, the goal of a design expression for T-type through plate connections is to add the corresponding branch plate tension and compression expressions. In this regard, the design strength partial function (Q_u) for through plate connections should ideally be equal to the expression for branch plate-to-CHS connections tested in tension, $Q_{u,T}$, plus the expression for branch plate-to-CHS connections tested in compression, $Q_{u,C}$ or

$$Q_u = Q_{u,90°,C} + Q_{u,90°,T} \tag{6}$$

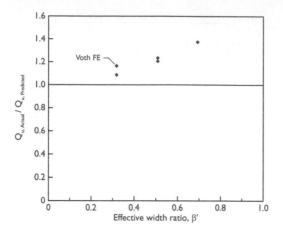

Figure 8. Comparison of potential design strength partial function, for transverse T-type through plate-to-CHS connections.

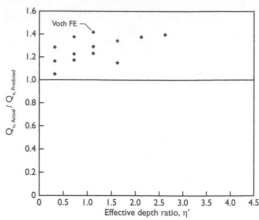

Figure 9. Comparison of potential design strength partial function, for longitudinal T-type through plate-to-CHS connections.

for transverse through plate connections and

$$Q_u = Q_{u,0°,C} + Q_{u,0°,T} \tag{7}$$

for longitudinal through plate connections where $Q_{u,90°,C}$, $Q_{u,90°,T}$, $Qu_{u,0°,C}$ and $Q_{u,0°,T}$ are Equations 2 to 5 respectively.

For transverse through plate-to-CHS connections, the actual design strength partial function, $Q_{u,Actual}$ (which is equivalent to the normalized connection capacity, $N_{1,u}/(f_{y0}t_0^2)$) to predicted design strength partial function, $Q_{u,Predicted}$ (Eq. 6) ratio is plotted against the effective width ratio, β', in Figure 8 as a graphical means of evaluating how well the potential design strength partial function captures transverse through plate connection behaviour. In this comparison the ζ term in Equations 2 and 3 is set to unity. The proposed equation has a mean of 1.21 and a coefficient of variation of 7.82% with respect to the numerical data presented. As there are limited transverse through plate-to-CHS connection results, these values are not statistically significant and are only given as an indication of the suitability of Equation 6.

Similarly, for longitudinal through plate-to-CHS the $Q_{u,Actual}$ to $Q_{u,Predicted}$ ratio is plotted against the effective width ratio, η', in Figure 9, as a graphical means of evaluating how well the potential design strength partial function captures longitudinal through plate connection behaviour. In this comparison the ζ term in Equations 4 and 5 is set to unity. The proposed equation has a mean of 1.27 and a coefficient of variation of 8.54% with respect to the numerical data presented. As with transverse through plate connections, Equation 7 under-predicts the longitudinal through plate connection ultimate capacity (see Figure 9).

The compression and tension branch plate-to-CHS connection capacities, and therefore potential design expressions, when summed, under-predict the parametric FE results for through plate-to-CHS connections possibly because of changes in chord length

and changes in yield band between the FE models for the two connection types. The expressions (Equations 6 & 7) do, however, give a good lower bound approximation for connection capacity and can be reasonably adopted to describe through plate connection behaviour.

5 PARAMETRIC NUMBERICAL STUDY CONCLUSIONS AND RECOMMENDATIONS

Through plate-to-CHS connections were numerically analysed varying values of β from 0.2 to 0.6, η from 0.2 to 2.5 and 2γ from 19.74 to 45.84 to determine connection behaviour as well as to develop design guidelines. From the 18 numerical FE analyses, proposed design strength partial functions (Q_u) presented in Equations 6 and 7, provide an adequate correlation with the numerical finite element results. The proposed design strength partial functions, for both transverse and longitudinal through plate connections, are hence summarized below:

For transverse through plate connections:

$$Q_u = Q_{u,90°,C} + Q_{u,90°,T}$$

$$Q_{u,90°,C} = 2.9\zeta(1+3\beta'^2)\gamma^{0.35}$$

and

$$Q_{u,90°,T} = 2.6\zeta(1+2.5\beta'^2)\gamma^{0.55}$$

For longitudinal through plate connections:

$$Q_u = Q_{u,0°,C} + Q_{u,0°,T}$$

$$Q_{u,0°,C} = 7.2\zeta(1+0.7\eta')$$

and

$$Q_{u,0°,T} = 10.2\zeta(1+0.6\eta')$$

The reduction factor, ζ, is analogous to a LRFD resistance factor and can be taken as 0.85, as previously recommended for T-type branch plate connections.

NOTATION

A_i = cross-section area of member i
b_1, b_1' = nominal, effective branch width $(b_1' = b_1 + 2w_0)$: 90° to chord longitudinal axis
d_0 = external diameter of CHS member
f_u = ultimate stress
f_{yi} = yield stress of member i
h_1, h_1' = nominal, effective branch depth in plane with chord longitudinal axis
i = denotes member (i = 0 for chord, i = 1 for branch)
l_0, l_1 = chord length, branch length
M_0 = chord bending moment
$M_{0,END}$ = applied in-plane bending moment
$M_{pl,0}$ = chord plastic moment capacity
N_i = axial force in member i
$N_{1,3\%}$ = branch load at 3% d_0 connection deformation
$N_{1,u}$ = connection ultimate limit state capacity
N_1^* = connection resistance expressed as an axial force in branch member
$N_{pl,i}$ = yield capacity for member i $(= A_i f_{yi})$
Q_f = chord stress influence function
Q_u = partial design strength function,
$Q_{u,R}$ = partial design strength function from experimental or numerical results
$Q_{u,p}$ = proposed partial design strength function
t_i = thickness member i
w_0, w_1 = measured weld leg length along chord, branch
α, α' = chord length parameter $(\alpha = 2l_0/d_0)$, effective chord length parameter $(\alpha' = 2l_0'/d_0)$
β, β' = nominal, effective connection width ratio $\beta = b_1/d_0, \beta' = b_1'/d_0, \beta = t_1/d_0$ for longitudinal
γ = chord radius-to-thickness ratio $(\gamma = d_0/2t_0)$
ε_{ef} = maximum equivalent strain
ζ = reduction factor
η, η' = nominal, effective branch member depth-to-chord diameter ratio $(\eta = h_1/d_0$ and $\eta' = h_1'/d_0$ for longitudinal; $\eta = t_1/d_0$ for transverse)
θ_1 = included inclination angle between branch and chord

REFERENCES

Akiyama, N., Yajima, M., Akiyama, H. & Ohtake, A. 1974. Experimental study on strength of joints in steel tubular structures. *J. Society of Steel Construction*, (10)102, 37–68 (in Japanese).

Alostaz, Y.M. & Sdf chneider, S.P. 1996. Analytical behavior of connections to concrete-filled steel tubes. *J. of Constructional Steel Research*, 40(2): 95–127.

ASTM. 2010. *Standard specification for cold-formed welded and seamless carbon steel structural tubing in rounds and shapes, ASTM A500/A500M-10*. ASTM International, West Conshohocken, USA.

ANSYS Inc. 2007. ANSYS. ver. 11.0, Cononsburg, PA, USA.

CSA. 2004. *General requirements for rolled or welded structural quality steel/structural quality steel, CAN/CSA-G40.20-04/G40.21-04*. CSA, Toronto, Canada.

IIW. 2012. *Static design procedure for welded hollow section joints: recommendations*. IIW Doc. XV-1402-12, IIW, Paris, France.

Kosteski, N. 2001. *Branch plate to rectangular hollow section connections*. Ph.D. thesis, University of Toronto, Toronto, Canada.

Kosteski, N. & Packer, J.A. 2001. Experimental examination of branch plate-to-RHS member connection types. *Proceedings of the 9th International Symposium on Tubular Structures*. Dü sseldorf, Germany, A.A. Balkema: 135–144.

Kosteski, N. & Packer, J.A. 2003. Longitudinal plate and through plate-to-hollow structural section welded connections. *Journal of Structural Engineering*, 129(4): 478–486.

Lee, M.M.K. & Llewelyn-Parry, A. 2005. Strength prediction for ring-stiffened DT-joints in offshore jacket structures. *Engineering Structures*, 27(3): 421–430.

Martinez-Saucedo, G., Packer, J.A. & Willibald, S. 2006. Parametric finite element study of slotted end connections to circular hollow sections. *Engineering Structures*, (28)14, 1956–1971.

Matic, P. 1985. Numerically predicting ductile material behavior from tensile specimen response. *Theoretical and Applied Fracture Mechanics*, (4)1, 13–28.

Packer, J.A. 1995. Concrete-filled HSS connections. *J. of Structural Engineering*, American Society of Civil Engineers (ASCE), 121(3): 458–467.

Packer, J.A., Sherman, D.R. & Lecce, M. 2010. Hollow structural section connections. *Steel Design Guide No. 24*, American Institute of Steel Construction (AISC), Chicago, USA.

van der Vegte, G.J., Wardenier, J., Zhao, X.-L. & Packer, J.A. 2008. Evaluation of new CHS strength formulae to design strengths. *Proc. 12th intern. Symp. on tubular structures*. Shanghai, China, Taylor & Francis Group, 313–322.

Voth, A.P. 2010. *Branch plate-to-circular hollow structural section connections*. Ph.D. thesis, University of Toronto, Toronto, Canada.

Voth, A.P. & Packer, J.A. 2012a. Branch plate-to-circular hollow structural section connections. I: Experimental investigation and finite-element modeling. *J. of Structural Engineering*, 138(8): 995–1006.

Voth, A.P. & Packer, J.A. 2012b. Branch plate-to-circular hollow structural section connections. II: X-Type parametric numerical study and design. *J. of Structural Engineering*, 138(8): 1007–1018.

Voth, A.P. & Packer, J.A. 2012c. Numerical study and design of T-type branch plate-to-circular hollow section connections. *Engineering Structures*, 41: 477–489.

Wardenier, J., Kurobane, Y., Packer, J.A., van der Vegte, G.J. and Zhao, X.-L. 2008. *Design guide for circular hollow section (CHS) joints under predominantly static loading, 2nd Edition*. Geneva, Switzerland: CIDECT.

Washio, K., Kurobane, Y., Togo, T., Mitsui, Y. & Nagao, N. 1970. Experimental study of ultimate capacity for tube to gusset plate joints – part 1. *Proc. annual conf. of the AIJ*. Japan.

Willibald, S. 2001. The static strength of ring-stiffened tubular T- and Y-joints. *Proc. 9th Int. Symp. on Tubular Structures*. Dü sseldorf, Germany, A.A. Balkema, pp. 581–588.

Zhao, X.-L. & Packer, J.A. 2009. Tests and design of concrete-filled elliptical hollow section stub columns. *Thin-Walled Structures*, 47(6/7): 617–628.

Tubular Structures XV – Batista, Vellasco & Lima (eds)
© 2015 Taylor & Francis Group, London, ISBN 978-1-138-02837-1

A new design equation for side wall buckling of RHS truss X-joints

J. Becque
The University of Sheffield, Sheffield, UK

T. Wilkinson
The University of Sydney, Sydney, Australia

ABSTRACT: The paper presents a new design methodology for equal width RHS X-joints failing by side wall buckling. In the new approach a slenderness parameter is defined based on the elastic local buckling stress of the side wall, obtained from a Rayleigh-Ritz approximation. When verified against a limited number of tests, the proposed design equation yields excellent results.

1 INTRODUCTION

The scope of this paper covers SHS and RHS truss connections of the X-type where the brace members and the chord member have equal width. The governing failure mode of these connections in compression is chord side wall failure.

In this paper we adopt the established CIDECT nomenclature, where b_0 and b_1 represent the chord width and the brace width, respectively, h_0 and h_1 are the chord height and the brace height, respectively, and t_0 and t_1 refer to the thicknesses of the chord wall and the brace wall, respectively.

It has been known for some time that the current CIDECT design rules (Packer et al. 2009) for chord side wall failure are quite conservative, and more so as the chord wall slenderness h_0/t_0 increases (e.g. Becque & Wilkinson 2011). The current rules are based on a design model which isolates a vertical strip in the chord side wall and designs it as a column. While defendable because of its simplicity, this approach obviously ignores the two-dimensional character of the side wall buckling as a plate.

The aim of the paper is to present an alternative design equation for chord side wall buckling, equally simple in its application, but founded on a rational plate buckling model and verified against experimental data.

2 DESIGN PHILOSOPHY

The design process of an RHS truss typically starts with a structural analysis under various load combinations in order to determine the governing internal forces. These internal forces consist mainly of tensile or compressive forces, accompanied by secondary moments which can typically be considered negligible

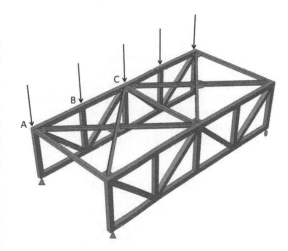

Figure 1. Sample RHS truss.

as long as the joint eccentricities are within prescribed values and the brace members are sufficiently slender. The actual design procedure then follows two steps:

1. sizing of the brace and chord members of the truss as tension or compression members.
2. a check of the connection capacities accounting for all possible failure modes using the CIDECT rules.

It is evident that the design of compressive truss members under (1) requires the determination of an effective length. As an example, we consider the truss in Figure 1 under the loading shown, with particular focus on the top chord. Seen the arrangement of the top bracing, the top chord needs to be designed as a column spanning between points A and C with out-of-plane flexural buckling being the governing failure mode. The implicit assumption in carrying out this check, however, is that the X-joint at B remains 'sound'.

Indeed, if local buckling were to occur in the chord side wall at B, this would introduce a weak link in the column A-C which would greatly reduce its out-of-plane flexural buckling capacity. It is well known that when local buckling occurs, the loss in compressive stiffness of a plate is immediate and severe (e.g. Marguerre 1937, Hemp 1945). The system could then be likened to a 'Shanley column' (Shanley 1947), albeit one where localized geometric nonlinearity rather than localized material nonlinearity (or possibly a combination of both) would be the cause of the central weak link. Because the design philosophy outlined above under (1-2) has no way of accounting for this type of local-global interaction buckling, we necessarily have to limit the design capacity of an X-joint to its side wall buckling load and neglect any post-buckling capacity. This philosophy is consistent with the current CIDECT rule for side wall failure based on flexural buckling of a 'column strip'. However, it does not condone the widespread practice of determining the capacity of an X-joint as the minimum of the peak load or the load corresponding to the 3% b_0 deformation limit (Lu et al. 1994) from a test on an isolated connection. Any argument that buckling of the side wall will lead to a rapid increase in side wall deformations and that, therefore, the load corresponding to a deformation of 3% b_0 will be somewhat representative of the buckling load, is quickly invalidated by experimental evidence. Out of the five tests X1-X5 conducted at the University of Sheffield and described in the next section, four of them reached the full peak load before even reaching the 3% b_0 side wall deformation and in no case was the 3% limit load representative of the buckling load.

3 EXPERIMENTAL PROGRAMME

Although an abundance of experimental results on equal-width RHS X-joints is available in literature, the recorded data typically includes the peak load and (in most cases) the load corresponding to the 3% b_0 deformation limit, while the load at which buckling of the side wall is first observed routinely remains unreported. A limited experimental programme was therefore conceived at the University of Sheffield encompassing five tests on equal-width SHS 90° X-joints with varying chord wall slenderness h_0/t_0. The nominal as well as the measured cross-sectional dimensions of all specimens X1-X5 are reported in Table 1. All specimens were made of hot finished 100x100 SHS, while the wall thicknesses of the chord and the brace members were varied from 3 mm to 8 mm. The overall dimensions of a typical test specimen are shown in Figure 2. A MIG welding procedure was used with W46_2_3Si1 wire ($f_y = 460$ MPa; $f_u = 600$ MPa). A simple 5 mm (X1), 8 mm (X2-X4) or 10 mm (X5) fillet weld was used to connect the top and bottom faces of the chord to the brace members, while the side walls were connected to the brace members using a butt weld with a 30° bevel on the brace ends (Figure 2).

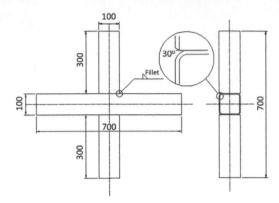

Figure 2. Specimen dimensions.

A 2000 kN test machine was used to apply a compressive load to the connection between fixed end conditions. A uniform introduction of the load into the brace members was ensured by the presence of a plate mounted on a spherical hinge underneath the ram, which made an even contact with the specimen before locking into place when the load was applied. All specimens were instrumented with two LVDTs (Linear Voltage Differential Transducers) positioned on the underside of the above mentioned plate to measure the axial shortening of the specimen, and another two LVDTs placed at the centers of the chord side walls on either side of the connection to measure the side wall displacements (Fig. 3).

The material grade was S355H (to EN10210-1 2006) for all SHS. Tensile coupons were cut from left-over pieces of the SHS segments used to fabricate the chord members and the average 0.2% proof stress was found to be $f_y = 364$ MPa. A typical stress-strain curve is shown in Figure 4. Figure 5 shows all five specimens after testing.

The limited database of five tests thus obtained was augmented with another four experiments reported by Becque & Wilkinson (2011) on equal-width X-joints. These additional data pertain to connections made of grade C450 cold-formed tube and include rectangular as well as square chord members. The tests, which will be labeled X6-X9 in this paper, generally exhibit larger h_0/t_0 ratios (some even outside the range of applicability of the current CIDECT rules) and include much larger section sizes (up to RHS 400x300x8) than those included in X1-X5. Consequently, the resulting database X1-X9 contains a more balanced mix of geometries and material properties. The measured dimensions of X6-X9 are listed in Table 1, while a typical C450 stress-strain curve is shown in Figure 4. The average 0.2% proof stress of the C450 material was measured to be 466 MPa.

The availability of the load vs. axial shortening data and the load vs. side wall displacement data for all tests X1-X9 allowed for an accurate determination of the actual side wall buckling load. Indeed, local buckling causes an abrupt change in the axial stiffness of the side wall (Marguerre 1937, Hemp 1945), while being

Figure 3. Test set-up with buckled specimen.

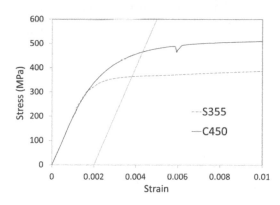

Figure 4. Stress-strain curves of S355 and C450 materials.

Figure 5. X1-X5: failed shapes.

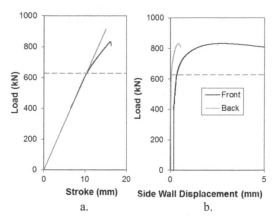

a.   b.

Figure 6. X7: a. load vs. stroke of the ram; b. load vs. side wall displacements.

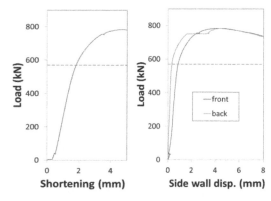

Figure 7. X5: a. load vs. axial shortening; b. load vs. side wall displacements.

accompanied by a disproportionate increase in the side wall displacements. This change in axial stiffness is unmistakable when buckling occurs in the elastic range and allows a precise pinpointing of the local buckling load. An example is given in Figure 6a/b for specimen X7. In the case where side wall buckling occurs in the elasto-plastic range, the change in axial stiffness as a result of buckling is usually not as abrupt, since it is typically interwoven with the loss of stiffness resulting from gradual material yielding. However, the rather sudden increase in side wall displacements associated with buckling typically still allows a reasonably accurate determination of the local buckling load in this case, as illustrated in Figure 7a/b for specimen X5.

The side wall buckling loads were determined for all specimens X1-X9 and are listed in Table 1 as $P_{b,test}$.

4   THEORETICAL MODEL

We now attempt to build a representative theoretical model by representing the chord side wall by a plate with thickness $t = t_0$, which extends to infinity on both sides (Fig. 8). The plate is assumed to be made of a linear elastic and homogeneous material. The loads and boundary conditions are idealized as follows:

Table 1. Test specimen dimensions.

Test	Chord –	Brace –	h_0 mm	b_0 mm	t_0 mm	$r_0^{(1)}$ mm	b_1 mm	h_1 mm	t_1 mm	$r_1^{(1)}$ mm
X1	SHS100x100x3	SHS100x100x3	100.27	100.52	2.92	6.2	100.22	100.33	3.21	6.2
X2	SHS100x100x4	SHS100x100x4	100.14	100.36	3.84	11.5	100.37	100.19	3.72	11.5
X3	SHS100x100x5	SHS100x100x5	99.80	100.25	4.89	12.7	100.08	99.90	4.85	12.7
X4	SHS100x100x6	SHS100x100x6	99.61	99.63	5.80	12.1	99.76	99.66	5.79	12.1
X5	SHS100x100x8	SHS100x100x8	99.70	99.89	7.92	15.1	100.12	99.64	7.98	15.1
X6	RHS250x150x5	SHS150x150x5	250.00	149.77	5.00	17.7	150.10	150.10	4.76	11.4
X7	SHS150x150x6	SHS150x150x6	150.18	150.23	5.86	14.1	150.48	150.35	5.86	14.7
X8	RHS350x250x10	SHS250x250x10	350.40	250.70	9.94	27.0	248.50	249.00	9.94	26.6
X9	RHS400x300x8	SHS300x300x8	400.00	300.00	7.92	22.7	300.30	300.30	7.97	22.3

(1) r = outside radius.

Table 2. Test results and predicted capacities.

Test	h_0/t_0 ($=2\gamma$) –	f_y MPa	$P_{b,test}$ kN	P_{CIDECT} kN	$P_{CIDECT}/0.9P_{b,test}$ –	P_{cr} kN	σ_{cr} MPa	λ –	$P_{cr}/P_y = \sigma_{cr}/f_y$ –	P_{pred} kN	$P_{pred}/P_{b,test}$ –
X1	34.3	363.6	124	60.3	0.54	121	206	1.327	0.585	109	0.88
X2	26.1	363.6	216	124.4	0.64	275	357	1.009	0.772	215	1.00
X3	20.5	363.6	325	219.2	0.75	569	582	0.790	0.916	320	0.99
X4	17.2	363.6	393	310.8	0.88	952	824	0.664	0.935	395	1.01
X5	12.6	363.6	565	541.2	1.06	2414	1529	0.488	0.984	558	0.99
X6	50.0	463.3	260	83.3	0.36	243	162	1.691	0.374	229	0.88
X7	25.6	450.8	628	316.8	0.56	652	370	1.104	0.791	547	0.87
X8	35.3	467.6	1270	535.3	0.47	1364	276	1.303	0.549	1228	0.97
X9	50.5	480.9	670	252.2	0.42	604	127	1.946	0.293	576	0.86
				Avg.	0.63					Avg.	0.94
				St. dev.	0.23					St.dev.	0.062

1. It is assumed that the load p transferred from the brace side wall is uniformly distributed over the brace width h_1. The total load carried by the connection (two side walls) is then given by:

$$P = 2ph_1 = 2\sigma t\, h_1 \qquad (1)$$

where the stress $\sigma = p/t$.

2. The plate is hinged along the longitudinal edges. This is obviously a conservative assumption neglecting any restraint provided by the chord top and bottom faces and by the welded connection with the brace member.

The critical buckling stress of the chord side wall is then approximated using a Rayleigh-Ritz approach i.e. by substituting a 'test' function representing the deformed shape in the energy potential. An ideal candidate to represent the longitudinal shape of the buckle is the exponential Gauss function as it approaches zero very fast when moving towards infinity and thus captures the localized nature of the failure mode very well. When also adopting a half-sine wave solution in the transverse direction (across the height of the chord wall), the proposed deformed shape is expressed by the following multiplicative function:

$$w = \Delta\cos\left(\frac{\pi y}{h_0}\right)e^{-2Bx^2} \qquad (2)$$

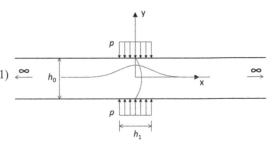

Figure 8. Idealized model.

In the above equation, w is the out-of-plane displacement of the plate, while Δ and B are (presently undetermined) parameters. Δ determines the amplitude of the displacements, while B is related to the length of the buckle. The Gauss function is prominently featured in statistics and from the study of the Gaussian (normal) distribution it is known that only 0.27% of the points in the distribution are more than three standard deviations removed from the average. From a comparison between Eq. (2) and the standard expression of the Gaussian distribution:

$$f(x,\mu,s) = \frac{1}{s\sqrt{2\pi}}e^{-\frac{(x-\mu)^2}{2s^2}} \qquad (3)$$

422

(where μ is the average and s is the standard deviation), an approximate length of the buckle is given by:

$$L_b = 6s = \frac{3}{\sqrt{B}} \tag{4}$$

The elastic strain energy U contained in the deformed shape of the plate is given by (e.g. Timoshenko & Gere 1961):

$$U = \frac{D}{2} \int_{x=-\infty}^{x=\infty} \int_{y=-h_0/2}^{y=h_0/2} \left\{ \left(\frac{\partial^2 w}{\partial x^2}\right)^2 + \left(\frac{\partial^2 w}{\partial y^2}\right)^2 \right.$$
$$\left. + 2v\left(\frac{\partial^2 w}{\partial x^2}\right)\left(\frac{\partial^2 w}{\partial y^2}\right) + 2(1-v)\left(\frac{\partial^2 w}{\partial x \partial y}\right)^2 \right\} dx\,dy \tag{5}$$

In the above equation, D is the flexural rigidity of the plate, given by:

$$D = \frac{Et^3}{12(1-v^2)} \tag{6}$$

where E is the modulus of elasticity and v is the Poisson's ratio. Substitution of Eq. (2) into Eq. (5) requires computation of the following integrals:

$$\int_{-\infty}^{\infty} e^{-2Bx^2} dx = \sqrt{\frac{\pi}{2B}} \tag{7}$$

$$\int_{-\infty}^{\infty} x^2 e^{-4Bx^2} dx = \frac{1}{16B}\sqrt{\frac{\pi}{B}} \tag{8}$$

$$\int_{-\infty}^{\infty} x^4 e^{-4Bx^2} dx = \frac{3}{128B^2}\sqrt{\frac{\pi}{B}} \tag{9}$$

and eventually leads to:

$$U = \frac{\Delta^2 D}{2}\sqrt{\frac{\pi}{B}}\left(3B^2 h_0 + B\frac{\pi^2}{h_0} + \frac{\pi^4}{4h_0^3}\right) \tag{10}$$

On the other hand, the potential energy of the applied stresses is given by:

$$V = -\frac{\sigma t}{2} \int_{x=-h_1/2}^{x=h_1/2} \int_{y=-h_0/2}^{y=h_0/2} \left(\frac{\partial w}{\partial y}\right)^2 dx\,dy \tag{11}$$

or, after substituting Eq. (2) into Eq. (11):

$$V = -\frac{\Delta^2 \sigma t \pi^2}{4h_0} \int_{x=-h_1/2}^{x=h_1/2} e^{-4Bx^2} dx \tag{12}$$

The remaining integral can only be expressed as a series:

$$V = -\frac{\Delta^2 \sigma t \pi^2}{4h_0}\left[h_1 - \frac{h_1^3 B}{3} + \dots\right] \tag{13}$$

Only the first term in the series is retained, so that:

$$V = -\frac{\Delta^2 \sigma t \pi^2}{2}\left(\frac{h_1}{h_0}\right) \tag{14}$$

Neglecting the higher order terms is acceptable, provided that:

$$\frac{h_1^3 B}{3} \ll h_1 \quad \text{or} \quad \frac{h_1^2 B}{3} \ll 1 \tag{15}$$

It will be shown at a later stage (once an expression for B has been determined) that this is indeed a reasonable assumption.

We now set the derivatives of the total energy $U + V$ with respect to B and Δ equal to zero:

$$\frac{\partial(U+V)}{\partial B} = 0 \tag{16}$$

$$\frac{\partial(U+V)}{\partial \Delta} = 0 \tag{17}$$

The calculations eventually result in simple equations:

$$B = \left(\frac{\sqrt{10}-1}{18}\right)\left(\frac{\pi}{h_0}\right)^2 = \frac{1.186}{h_0^2} \tag{18}$$

and:

$$\sigma_{cr} = 1.346\frac{\pi^2 E}{12(1-v^2)}\frac{t^2}{h_0 h_1} \tag{19}$$

For $E = 200$ GPa and $v = 0.3$, Eq. (19) becomes:

$$\sigma_{cr} = (243.10^3)\frac{t^2}{h_0 h_1} \quad \text{(MPa)} \tag{20}$$

The critical buckling load of the connection is then given by:

$$P_{cr} = 2th_1\sigma = 486.6\frac{t^3}{h_0} \quad \text{(kN)} \tag{21}$$

The condition in Eq. (15) can now be evaluated and leads to:

$$\frac{h_1}{h_0} < 1.6 \tag{22}$$

Given that the chord member is typically the larger member compared to the braces (or at most of equal size), h_1/h_0 is usually sufficiently small to satisfy Eq. (22) and, consequently, Eq. (15).

Using Eq. (4), the length of the buckle is now estimated to be:

$$L_b = \frac{3}{\sqrt{B}} = 2.76 h_0 \tag{23}$$

423

5 DESIGN

Based on the critical stress determined in Eq. (18) it is possible to define a side wall slenderness λ as follows:

$$\lambda = \sqrt{\frac{f_y}{\sigma_{cr}}} = \frac{0.00203}{t}\sqrt{f_y h_0 h_1} \qquad (f_y \text{ in MPa}) \quad (24)$$

The dimensionless buckling load of the connection P_b/P_y can then be plotted against the slenderness λ for all the experimental results (Fig. 9), where the yield load P_y is given by:

$$P_y = 2f_y h_1 t \quad (25)$$

Eq. (19) for the elastic buckling stress of the connection can then be plotted in this diagram as:

$$\frac{P_{cr}}{P_y} = \frac{\sigma_{cr}}{f_y} = \frac{1}{\lambda^2} \quad (26)$$

P_{cr} has also been calculated for each of the tested connections and the results are listed in Table 2. It is seen that there is an excellent agreement between the curve depicting elastic buckling and those test results for which σ_{cr} is sufficiently below f_y. The fact that such a close correlation was obtained is, to some extent, surprising. On the one hand, some conservative assumptions have been made in the model: the flat width of the side wall has been exaggerated by neglecting the rounded corners, and any restraint along the longitudinal edges exerted by the remainder of the connection has been neglected. On the other hand, some of the load does not enter the side wall directly from the brace wall above (or below), but instead flows through the chord top and bottom faces, thus causing additional bending in the side wall as a result of the eccentricity. The model also assumes a perfectly flat plate, while the real chord wall inevitably contains imperfections. It seems that all these effects, beneficial or detrimental, have a rather 'lucky' way of opposing and balancing each other, leading to a perfectly useable model.

To extend our model to the inelastic range we draw on the work by Bleich (1952), who proposed the following differential equation to describe buckling of a simply supported inelastic plate under uniaxial compression:

$$E_t \frac{\partial^4 w}{\partial x^4} + 2\sqrt{E_t E}\frac{\partial^4 w}{\partial x^2 \partial y^2} + E\frac{\partial^4 w}{\partial y^4} = -\sigma_x \frac{12(1-v^2)}{t^2}\frac{\partial^2 w}{\partial x^2} \quad (27)$$

where E_t is the tangent modulus and E is the elastic modulus. Although Bleich's equation is based on a semi-rational approach and more theoretically sound models have been developed (e.g. Becque 2010), it has the advantage of leading to rather simple equations. Indeed, it follows from the structure of Eq. (27)

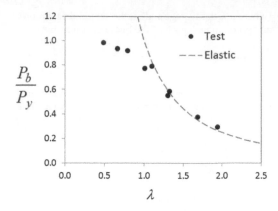

Figure 9. Comparison between test and model (elastic range).

that the inelastic buckling stress can be obtained from the corresponding buckling stress of an elastic plate by multiplying with a 'plasticity reduction factor' η, given by:

$$\eta = \sqrt{\frac{E_t}{E}} \quad (28)$$

Based on Eq. (19), we therefore propose the following equation for the inelastic buckling stress of the chord side wall:

$$\sigma_b = 1.346 \frac{\pi^2 \sqrt{EE_t}}{12(1-v^2)}\frac{t^2}{h_0 h_1} \quad (29)$$

or, with $E = 200$ GPa and $v = 0.3$:

$$\sigma_b = 544\sqrt{E_t}\,\frac{t^2}{h_0 h_1} \qquad (E_t \text{ in MPa}) \quad (30)$$

The tangent modulus E_t can thereby be obtained from a Ramberg-Osgood representation of the material stress-strain curve:

$$E_t = \frac{f_y E}{f_y + 0.002nE\left(\dfrac{\sigma}{f_y}\right)^{n-1}} \quad (31)$$

where n is a parameter characterizing the roundness of the stress-strain curve. Using the measured values: $n = 14$, $E = 200$ GPa and $f_y = 364$ MPa for the S355 material, and: $n = 18$, $E = 200$ GPa and $f_y = 466$ MPa for the C450 material, Eq. (30) can be plotted (Figure 10) and compared to the experimental data. Excellent agreement is observed.

Eq. (30) is simple in form, elegant, accurate and covers the whole slenderness range with one equation. However, it has the important drawback that it is iterative in nature. Indeed, the tangent modulus E_t has to be calculated at the buckling stress σ_b. In order to

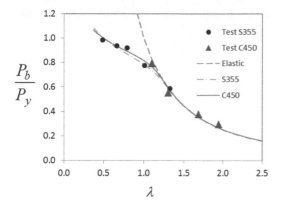

Figure 10.   Inelastic buckling predictions.

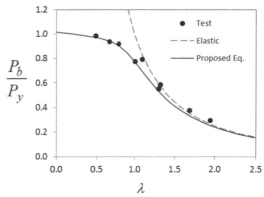

Figure 11.   Proposed design equation.

eliminate this disadvantage a new design equation is proposed, which more closely resembles the current CIDECT practice of referring in the design for side wall buckling to the equations for column buckling (e.g. EN1993-1-1 2006):

$$P_b = \chi P_y \qquad (32)$$

with:

$$\chi = \frac{1}{\phi + \sqrt{\phi^2 - \lambda^2}} \leq 1.0 \qquad (33)$$

$$\phi = \frac{1}{2}\left[1 + \alpha(\lambda - 0.2) + \lambda^2\right] \qquad (34)$$

where P_y and λ are determined by Eqs. (25) and (24), respectively, and the value of the imperfection factor α should be taken as 0.08. This value of α was obtained by visually curve-fitting the above equations to the experimental data (Figure 11). While it is seen that good agreement between Eqs (32-34) and the experiment is obtained, an even more suitable and reliable value of α in the context of design could be obtained by performing a reliability analysis to EN1990 (2002). While this is planned for the next stage of the research, to date this analysis has not been carried out yet.

Table 2 lists the ratios of the capacity predicted by Eq. (32) to the experimental result. An average ratio of 0.94 was obtained with a standard deviation of 0.06. Eq. (32) strongly outperforms the current CIDECT design rule for side wall buckling, which over the same data set features an average ratio of predicted to measured capacity of 0.63 with a standard deviation of 0.23. In this comparison, the CIDECT prediction was divided by a factor of 0.9 to eliminate the implicit safety factor (the CIDECT design rules provide factored joint resistances which already include a safety factor; Packer et al. 2009). Importantly, it is also seen that the CIDECT rule does not offer a consistent margin of safety, but is more conservative for side walls with high h_0/t_0 values. It should also be noted that the applicability of the current CIDECT rule is limited to

an h_0/t_0 ratio of 40, while the proposed rule is based on experimental data including sections with h_0/t_0 ratios of up to 50.

6   CONCLUSIONS

A new design equation is proposed for equal width RHS X-joints. The approach is founded on a rational analysis of an infinitely long plate subject to a localized distributed load. A Rayleigh-Ritz approximation is used to obtain the critical elastic buckling stress, which is subsequently used in the definition of a slenderness parameter. The new design equation is compared to experimental results including X-joints made of S355 and C450 steel and containing SHS and RHS of widely varying sizes and wall slenderness values. An excellent agreement between the proposed equation and the experimental results was observed, with an average ratio of predicted to measured capacity of 0.94 and a standard deviation of 0.06.

REFERENCES

Becque, J. 2010. Inelastic Plate Buckling. *ASCE Journal of Engineering Mechanics*, 136(9), 1123–1130.
Becque, J. & Wilkinson, T. 2011. Experimental investigation of the static capacity of grade C450 RHS T and X truss joints. In L. Gardner (ed.) *Tubular Structures XIV*, CRC Press/Balkema, Leiden, The Netherlands.
Bleich, F. 1952. *Buckling Strength of Metal Structures,* McGraw-Hill, New York.
European Standard EN10210-1. 2006. *Hot finished structural hollow sections of non-alloy and fine grain steels*, European Committee for Standardization, Brussels, Belgium.
European Standard EN1990. 2002. *Eurocode – Basis of structural design*, European Committee for Standardization, Brussels, Belgium.
European Standard EN1993-1-1. 2005. *Eurocode 3: Design of steel structures. General rules and rules for buildings*. European Committee for Standardization, Brussels, Belgium.
Hemp, W.S. 1945. The Theory of Flat Panels Buckled in Compression. *Aeronautical Research Council, Reports and Memoranda*, No. 2178.

Lu, L.H., de Winkel, G.D., Yu, Y., & Wardenier, J. 1994. Deformation limit for the ultimate strength of hollow section joints. *Tubular Structures VI*, Balkema, Rotterdam, The Netherlands, 341–347.

Marguerre, K. 1937. The apparent width of the plate in compression. *Luftfahrtforschung* 14 (3).

Packer, J.A., Wardenier, J., Zhao, X.-L., van der Vegte, G.J., & Kurobane, Y. 2009. *Design guide for rectangular hollow section (RHS) joints under predominantly static loading*, 2nd edition, Verlag TUV Rheinland, Köln, Germany.

Shanley, F. 1947. Inelastic Column Theory. *Journal of the Aeronautical Sciences*, 14(5), 261–276.

Timoshenko, S.P. & Gere, M. 1961. *Theory of elastic stability*. 2nd edition, Dover Publications, New York, USA.

Tubular Structures XV – Batista, Vellasco & Lima (eds)
© *2015 Taylor & Francis Group, London, ISBN 978-1-138-02837-1*

Performance of non-diaphragm joint of H-beam to RHS column with partially thickened wall

Y. Chen, L. Zhang & W. Jiao
State Key Laboratory of Disaster Reduction in Civil Engineering, Tongji University, Shanghai, China

ABSTRACT: The non-diaphragm joint (NDJ) connecting H shaped beams and cold-formed RHS columns is studied. The NDJ specimens with different RHS tube wall thickness in joint zone are designed. The monotonic loading tests under symmetrical and asymmetrical bending moment conditions are carried out. The stiffness and resistance of the joints are experimentally investigated and compared with the prediction equations. The features of stress distribution and local deformation in joint zone are analyzed by test results and numerical computation to determine the proper length of thickened wall in joint zone. The study shows that the joint can work well when the thickened tube wall extends to half of the beam depth out of the joint zone, and the joint stiffness and moment resistance can be predicated by proposed or available equations.

1 INTRODUCTION

The common used constructional details for joint of H-beams and RHS columns in frame structure are those with the internal or external diaphragm in the column. However, welding internal diaphragm needs onerous work during manufacturing process and the existence of internal diaphragm makes it difficult to fill in concrete inside the steel tube in concrete filled tubular (CFT) column when the tube size is small. On the other hand, the external diaphragm is unfavorable in many cases due to its protruding configuration. So, the non-diaphragm joint (NDJ) becomes one of choices for frame structure. Except from CFT column with small tube size, NDJ can be applied to low or middle rise building frames where small tubes are able to meet the requirement for column stiffness and capacity.

The initial stiffness and moment resistance are fundamental issues for design of NDJ. The previous research revealed that the stiffness of NDJ is quite different from that of the stiffened joint of H-beams and RHS column. Based on the experimental study on the welded box section columns (Morida 1989, Akiyama 1996), AIJ recommendation (2001) suggests equations to evaluate NDJ stiffness. The authors put forward a progression solution without consideration of the strength effect which is introduced in AIJ equations (Zhang, 2012). As for the joint capacity, Lu and Wardenier (1998) studied the failure modes of NDJ under shear and bending moment according to the test on specimens made of hot-rolled tubes, and proposed formulae for computation of resistance.

The previous researches reveal that both initial stiffness and moment resistance of NDJ closely rely on the wall thickness of RHS tube in joint zone. For keeping the necessary stiffness and resistance of NDJ, however, thickening the tube wall along whole column length is not always economic. Thus partially thickened wall in joint zone is more favorable. In this paper, NDJ of H-beams and RHS columns with partially thickened wall is studied. A series of specimens were tested under monotonically bending loads, and then theoretical evaluation and numerical analysis were followed. The proper geometrical sizes, thickness and length of the enhanced tube in joint zone, are finally suggested based on the consideration of joint stiffness, resistance and the economic balance.

2 EXPERIMENTAL PROGRAM AND THE RESULTS

2.1 Test specimen

Seven cruciform beam-column joint specimens including one with internal diaphragm (MSD115 in Table 1) and other six without any diaphragm were designed. The beams were welded onto the surface of the column. The specimen details are shown in Figure 1 and the geometrical parameters are listed in Table 1. Each specimen is identified by three letters and three numbers. The first letter M denotes the monotonic loading, the second letter S or U denotes the vertical loads on two side cantilever beam ends symmetrical or asymmetrical, and the third letter indicates the joint with diaphragm (D), without diaphragm but with thickened joint zone (T) or non-enhanced (N), respectively. The first two digital numbers identify the type of tube size and H beam size, while the third one presents the axial force ratio on the top of columns. Here, 0 is corresponding to no axial force, and 3 and 5 to the ratio of 0.3 and 0.5. Referring to Figure 1, all the symbols for the section configuration may be known.

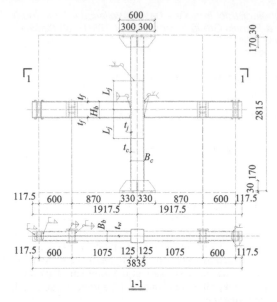

Figure 1. Test specimen.

Table 2. Mechanical properties of steel.

Coupon Code	Steel Grade	f_y MPa	f_u MPa	δ %
250 × 8-a	Q345	471.8	565.0	21.0
250 × 8-b	Q345	419.6	588.0	35.0
250 × 10-a	Q345	600.0	614.7	17.0
250 × 10-b	Q345	527.6	576.7	24.0
250 × 16-a	Q345	635.4	669.0	17.0
250 × 16-b	Q345	420.0	621.0	30.5
P-6-c	Q235	332.1	565.8	35.0
P-6-d	Q345	405.0	548.0	29.2
P-8-c	Q235	282.6	555.2	36.0
P-8-d	Q345	364.0	527.0	14.6

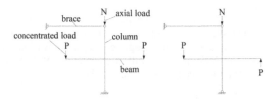

(a) Symmetric loading.　　　　(b) Asymmetric loading.

Figure 2. Schematic of test loading.

In the Table 1, the non-dimensional parameters are defined as $2\gamma = B_c/t_j$, $\beta_1 = B_b/B_c$, $\beta_2 = H_b/B_c$. Here, t_j is the tube wall thickness at joint zone.

In specimen MST110, MST113 and MUT123 the partial thickened tube portion with 16 mm thickness was welded by butt weld to column tube portion with 8 mm thickness. Since the corner radius of the tube section is the function of wall thickness, the butt weld was required to shape the interface of two portions smoothly.

The steel grade of the tube is Q345 (as the nominal yield stress 345 N/mm²), and the beams in specimen MUN115 and MUN330 are Q345, while others are Q235 (as the nominal yield stress 235 N/mm²). The mechanical properties of the steel material, yield stress f_y, tensile strength f_u and elongation δ are listed in Table 2. For the coupon of tubes, the suffix 'a' or 'b' denotes the sample cut from corner or flat part. For the coupon of steel plates, the suffix 'c' or 'd' denotes the steel grade Q235 or Q345.

2.2 Testing program

Figure 2 shows a schematic of the test set-up. Two spherical bearings were placed at the top and the bottom of the column and another two spherical bearings were placed at the two ends of the beam to achieve the expected restraint as perfect pin-boundary conditions. During tests, a constant axial load as planed was first applied on the column top, and then a pair of vertical concentrated loads was applied to free ends of the two side cantilever beams in the way symmetrical or asymmetrical.

Joint rotation was measured and it was specified as the one caused by the local flexural deformation of tube wall. For each specimen, 8 transducers numbered

D8 to D15 were arranged to measure the local deformation as shown in Figure 3. Corresponding to the axial line of column, a pair of T type steel rod accessories were stuck on to the steel tube above and beyond the upper and low flange of the beam, respectively. At two wings of the accessory, the 'front group' of transducers D8 to D11 was linked to detect the total deformation of the tube wall. At the same levels as 'front group', the 'side group' transducers D12 to D15 were fixed on the center of the column side wall to measure the sidesway of the axial of the column. By subtracting the measured data of front group from the side group, the local flexural deformation of tube wall can be obtained.

2.3 Failure mode of specimens and the loading-deformation curves

Three failure modes were observed by the tests. (1) Beam failure is the dominant mode for specimen MST113, MST110 and MSD115. The fixed beam end developed full plastic deformation accompanied with obvious local buckling. After the large deformation developed, the cracking in tensioned beam flange happened in all three specimens, causing the different ultimate levels. (2) Severe plastic deformation occurs at both fixed beam end and joint zone, as observed in specimen MUN220 and MUT123. The beam flange buckled first and the residual bending deformation of tube wall at joint zone was seen in MUN220. However, for MUT123, the local buckling deformation at the tube of column portion with 8 mm thickness

Table 1. Geometrical parameters of test specimen.

Specimen Code	$B_c \times t_c$ mm	t_j mm	L_j mm	$H_b \times B_b \times t_w \times t_f$ mm	2γ	β_1	β_2
MSD115	250×8	8	–	$350 \times 125 \times 6 \times 8$	31.3	0.5	1.4
MST110	250×8	16	550	$350 \times 125 \times 6 \times 8$	15.6	0.5	1.4
MST113	250×8	16	550	$350 \times 125 \times 6 \times 8$	15.6	0.5	1.4
MUN220	250×16	16	–	$400 \times 150 \times 6 \times 8$	15.6	0.6	1.6
MUT123	250×8	16	550	$400 \times 150 \times 6 \times 8$	15.6	0.6	1.6
MUN115	250×8	8	–	$350 \times 125 \times 6 \times 8$	31.3	0.5	1.4
MUN330	250×10	10	–	$350 \times 175 \times 6 \times 8$	25.0	0.7	1.4

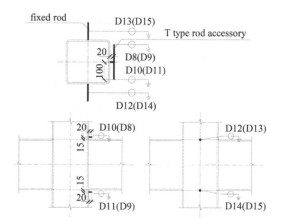

Figure 3. Arrangement of transducers for measuring joint local rotation.

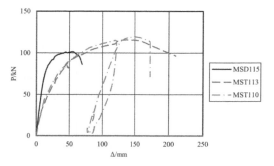

a.   Load vs. displacement at beam free end.

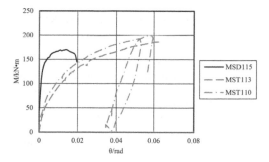

b.   Moment vs. joint rotation.

Figure 4. Performance of specimens MSD115, MST113 and MST110.

near butt weld was detected when the load reached near the ultimate. Axial compressive stress in column, relative thin wall of column portion and the uneven interface between thick and thin wall tube are three factors for MUT123 different from MUN220, though the tube thickness in joint zone are the same. (3) Plastic deformation at joint zone develops distinguished before beam failures. For specimen MUN115, the front tube wall to which beam was welded deformed out of its plane severely. For specimen MUN330, the side tube wall of joint zone developed large shear deformation together with the out-of-plane deformation of front tube wall. To understand the reason of these failure modes, we should notice the geometrical size of the specimens. The tube wall of joint zone in specimen MUN330 is thicker than that in MUN115, and also with larger β_1 than that in MUN115. These facts mean that the beam of MUN330 can bear relatively large load, and transfer stress from front wall to side wall more directly and efficiently than the behavior of MUN115.

Figure 4–6 exhibit curves of the beam load at free end versus corresponding displacement in which the rigid body movement of the specimen has been removed, and beam end moment versus joint rotation curves. The peak values in the curves are defined as the ultimate bearing capacity. To make the figures clear,

for each specimen only the data corresponding to left side beam is doted.

Figure 4 reveals that the NDJ could reach the strength equal to or even higher than that of stiffened joint (Specimen MSD115). The relatively soft flexural stiffness of the tube wall of NDJ delays the fracture of beam flange compared with the one of stiffened joint specimen. On the other hand, it shows big difference of stiffness between internal diaphragm specimen and NDJ specimens. The fact indicates that to raise the joint initial stiffness for NDJ to fully rigid level is difficult. Figure 5 shows that the two specimens, MUN220 with 16mm tube thickness in whole length of column and MUT123 with 16mm tube thickness only in joint zone with the length of 2.75 times of beam depth, exhibits almost same behavior before ultimate.

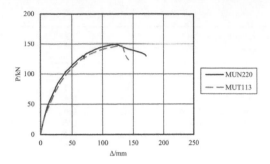

a. Load vs. displacement at beam free end.

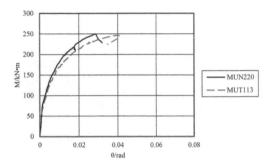

b. Moment vs. joint rotation.

Figure 5. Performance of specimens MUN220 and MUT113.

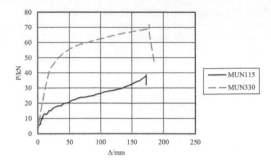

a. Load vs. displacement at beam free end.

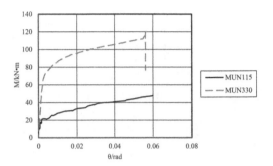

b. Moment vs. joint rotation.

Figure 6. Performance of specimens MUN115 and MUN330.

Thus, to make a portion of thickened tube in joint zone is possible and practical. The behavior of the last two specimens shown in Figure 6 tells the difference in initial stiffness and quite large different levels in resistance. Though the difference of tube wall thickness is the main reason resulting in their different performances, the axial force ratios and the non-dimensional parameters shown in Table 1 should also be taken into account.

3 EVALUATION OF THE JOINT STIFFNESS

3.1 *Equations for joint stiffness in elastic stage*

Without internal or outer diaphragm, it may be hard for NDJ to behave as fully rigid beam-column joint. Therefore, the evaluation of the flexural stiffness is important.

To evaluate the joint stiffness by equations is not easy. The previous researches have accumulated the test data (Lu 1997, Fujita 1999, Harada 2000), however, the available equations adopted by AIJ (2001) based on limited experimental results and the achieved numerical regression seem not conceptually reasonable because in the formulae for computation of elastic or initial joint stiffness, variables as M_{pj}, the yield moment of the tube wall, and, M_{pb}, the full plastic moment of the beam are used. Recently, the authors suggested a solution for the elastic stiffness of NDJ

by using trigonometric series (Zhang, 2012). In the following analysis, the Zhang-solution is used.

3.2 *Evaluation of joint stiffness by test results and theoretical expectation*

In Table 3 the measured joint stiffness from 6 pieces of NDJ specimens is compared with the prediction value by the Zhang-solution. There are mainly two ways to evaluate the elastic joint stiffness through the moment-rotation curves obtained by test measurement. One is the initial stiffness K_{ini}, and the other is secant stiffness K_s, as shown in Figure 7. However, the measured value of initial stiffness in test strongly relies on the sensibility of the sensors to the tiny deformation at the beginning stage, and how many data should be taken is always a problem. In this study, secant stiffness corresponding to the one-third of ultimate moment, denoted as K_{sT}, is used to evaluate the test specimen.

The K_P in Table 3 is the theoretical predication value which is based on Zhang-solution derived from trigonometric series, while K_{P10} is the sum of the first ten items in the solution and K_{P2} is the sum of the first two items. By the comparison, we can see that the prediction of Zhang-solution gives satisfied answer.

More important issue for structural design is to determine the classification of NDJ a rigid or semi-rigid joint in frame structure. According to Eurocode (CEN 2005), the classification of the joint should relate to the frame system which is sidesway one or not,

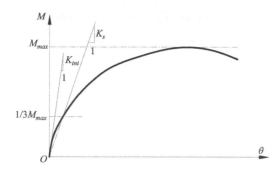

Figure 7. Definition of elastic stiffness of NDJ.

Table 3. Elastic stiffness of NDJ.

Specimen Code	K_{sT} kN.m.rad^{-1}	K_{P10}/K_{sT}	K_{P2}/K_{sT}	Joint type
MST110	1.31E+04	1.19	1.30	semi-rigid
MST113	1.25E+04	1.25	1.36	semi-rigid
MUN220	5.79E+04	0.56	0.60	semi-rigid
MUT123	2.90E+04	1.11	1.20	semi-rigid
MUN115	2.84E+03	0.87	0.96	semi-rigid
MUN330	1.59E+04	0.50	0.44	semi-rigid

and the ratio of joint stiffness with the beam rigidity. Since the beam rigidity relies on the inertial moment of beam section and beam span, several frame models for which the tested beam and column sections are valid are constituted. Suppose the ratio of steel beam depth to the span in the ordinary frame changes from 1/12 to 1/20, we can set the beam span to proportion the beam depth of specimens. From these frame models the joint classification is judged as these listed in Table 3. We can see the elastic joint stiffness of specimens in the tests could not meet the requirement of fully rigid.

3.3 Parametrical study for effect of the wall thickness and thickened tube portion length on joint stiffness

By commercial software ANSYS, the FE solid elements were used to establish the assemblage joint model to do the parametrical study. The beams and the flat part of the columns were simulated by SOLID 45, while the rest of the columns were simulated by SOLID 95. For the analysis, geometrical non-linearity switch was turned on to consider the relatively thin wall deformation even in elastic stage and the material model used the true stress-true strain relationship. The validity of the model was checked by the authors' test and previous test data. Two basic RHS column sections were adopted, one with the size of 250×8 mm and the other 300×10 mm, while the beam section as tested specimen MST110 was adopted. Part of computation results are shown in Figure 8. The vertical coordinate in the figures is the relative joint stiffness, which indicates that the joint stiffness of the FE model with various geometrical parameters are non-dimensioned by the one of the assemblage with the

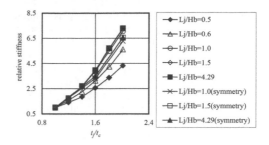

a. Effect of the wall thickness.

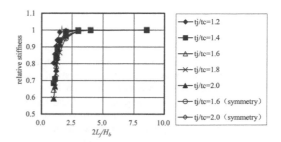

b. Effect of the thickened portion.

Figure 8. Effect of the wall thickness and thickened portion on joint stiffness.

geometrical parameters $t_j/t_c = 1.0$ and $L_j/H_b = 4.29$. The horizontal coordinates in Figure 8a is the relative thickness of thickened tube in joint zone, and in Figure 8b is the relative length of thickened tube portion to the beam depth. By the figures, it is clear that the joint stiffness shall increase non-linear with the increase of the joint wall thickness, but when the thickened tube length exceeds two times of beam depth, the joint stiffness increases limited, and when the length is large than three times of beam depth, it is almost no further contribution to the joint stiffness.

Both by the numerical study (Fig. 8a) and by deduction of Zhang-solution, it seems possible to make the tested specimen joint become rigid. For examples, to test specimen MST110, when the tube wall thickness at joint zone up to 30 mm or 25 mm respectively to the frames models supposed in Section 3.2, the elastic joint stiffness can be judged as rigid. In the same way, the tube wall thickness of specimen MUN220 at joint zone should be up to 26 mm or 22 mm. However, the greater thickness leads to more consumption and material cost of steel and the difficulty of cold-forming processing.

4 EVALUATION OF THE JOINT RESISTANCE

4.1 Equations for resistance

For assemblage of beam-column joint together with portions of connected structural members, there is failure possibility for each part of the assemblage. In the experience conducted by this study, plastic hinge at beam end, local flexural failure of the front wall at joint zone as well as plastic share deformation were observed. The available equations recommended by design codes or specifications can be used to evaluate the resistance of test specimens against those failure modes. In Equation (1) through (3), however, the resistance is presented in the form of critical load on the end of free beam.

The resistance corresponding to the failure due to plastic hinge on fixed beam end at column face can be derived as Equation (1).

$$P_{bp} = \frac{B_b t_f f_{yf} H_b + 0.25(H_b - 2t_f)^2 t_w f_{yw}}{L - 0.5B_c} \quad (1)$$

where, L = beam length from loading position to the central line of column; f_{yf} = yield stress of beam flange; and f_{yw} = yield stress of beam web.

The resistance corresponding to the local flexural failure of the front tube wall at joint zone is based on the equations which are derived from yield line model adopted by AIJ (2001) recommendation.

$$P_{cy} = \frac{0.25t_j^2 f_{yj}(B_c - t_j)(H_b - t_j)}{L - 0.5B_c}\left(\frac{4x + H_b - t_f + w_{tf}}{x^2}\right.$$
$$\left. + \frac{2}{H_b - t_f - w_{tf}}\right) + \frac{2(H_b - t_f)}{L - 0.5B_c} \cdot \frac{t_f f_{yf}}{B_c - t_j} \quad (2)$$
$$\cdot [x - \frac{0.5(B_c - t_j - B_b)(B_c - t_j)}{2x}]^2$$

where, f_{yj} = yield stress of column; w_{tf} = the sum of the beam flange thickness and the welding foot size; x=variable which can be solved by Equation (3).

$$4t_f f_{yf} x^4 - t_j^2 f_{yj}(B_c - t_j)^2 x - (B_c - t_j)^2$$
$$\cdot [0.5t_j^2 f_{yj}(H_b - t_f + w_{tf}) + (\frac{B_c - t_j - B_b}{2})^2 t_f f_{yf}] = 0 \quad (3)$$

The resistance corresponding to share failure of side wall is computed by the shear strength formula derived from the joint zone with internal or outer diaphragm. However, in Equation (4) the volume of two side walls in the equivalent joint zone is multiplied by a reduction factor of 0.9. Such failure mode happens under asymmetrical load condition.

$$P_{yv} = \frac{2.4H_b B_c t_j}{2L - B_c} f_{vj} \quad (4)$$

where, f_{vj} = the yield stress by shear.

4.2 Comparison of predicted resistance with test data

Table 4 lists the predicted resistance by Equation (1) to (4) and the capacity of specimens by tests. In the table,

Table 4. Comparison of resistance between prediction and test data.

Specimen Code	P_{bp} kN	P_{cy} kN	P_{yv} kN	P_{uL} kN	P_{uR} kN
MSD115	**92.2**	–	–	107.2	101.5
MST110	**92.2**	100.3	–	114.3	115.8
MST113	**92.2**	100.3	–	115.1	119.7
MUN220	**124.8**	127.0	278.0	153.7	136.6
MUT123	**124.8**	127.0	278.0	147.2	147.2
MUN115	116.5	**31.0**	121.6	50.4	38.7
MUN330	146.9	**70.8**	190.9	71.7	81.3

the symbol P_{uL} and P_{uR} refer to the tested ultimate load on the left and right side respectively.

For specimen MST110 and MST113 the resistance prediction matches the test result well. This fact means it possible to make NDJ a full strength joint as MSD115 which was stiffened by internal diaphragm. For specimen MUN220 and MUT123, the prediction shows that front wall failure should follow the beam failure immediately. It is what we observed in test. The local buckling of beam flange occurred first in test as mentioned in Section 2.3, and two failure modes appeared in the test. For the last two specimens, the resistance prediction correctly forecasts the failure mode of front wall in joint zone, but the mechanism of shear failure which occurred in specimen MUN330 should be further studied.

4.3 Parametrical study for effect of the wall thickness and thickened tube portion length on joint resistance

Using the same FE model established in Section 3.3 and the bi-linear material model with isotropic hardening, parametrical study for the effect of the wall thickness and thickened tube portion length on joint resistance is carried out. The computation results are shown in Figure 9. The vertical coordinate in the figures is the relative joint resistance non-dimensioned by the resistance of the assemblage with non-dimensional parameter $t_j/t_c = 1.0$ and $L_j/H_b = 4.29$. By the figures, it is understood that the joint resistance shall increase with the increase of the joint wall thickness, but when the thickened tube length is large than two times of beam depth, it is little contribution and when the length exceeds three times of beam depth, there is no further contribution, as we have looked at Figure 8.

5 ANALYSIS OF LOCAL STRESS AND DEFORMATION IN NDJ

5.1 Configuration of analysis model

In order to determine the proper length of thickened tube portion in joint zone, the local stress distribution and deformation of tube wall are investigated by the FE analysis. Five sections along the column axis are

432

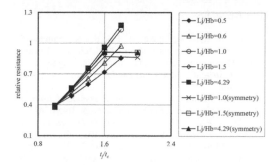

a. Effect of the wall thickness.

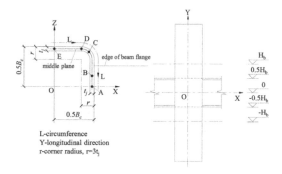

b. Effect of the thickened portion.

Figure 9. Effect of the wall thickness and thickened portion on joint resistance.

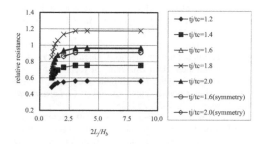

L-circumference
Y-longitudinal direction
r-corner radius, r=3t_j

Figure 10. Observation sections and points of the FE model.

selected as shown in Figure 10 and in each section five typical points on the mid-line are selected. Symmetric and asymmetrical load on two side beam ends are considered.

5.2 Results and discussion

The analysis results exhibit similar characters in all models for their local stress distribution and deformation. The results of a joint assemblage model with column size of 300×16 mm and beam size of $350 \times 125 \times 6 \times 8$ mm are picked out to demonstrate the phenomenon. Figure 11 shows the distribution of normal stress σ_l in the direction of perimeter regularized by yield stress f_y. Figure 12 shows the distribution

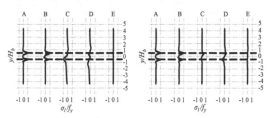

Figure 11. Distribution of normal stress in the direction of perimeter.

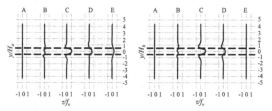

Figure 12. Distribution of shear stress in the direction of perimeter.

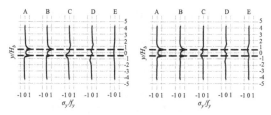

Figure 13. Distribution of normal stress in the direction of column axis.

of shear stress τ regularized by yield shear stress f_v, and Figure 13 shows the distribution of regularized normal stress in the direction of the column axis. Figure 14 shows the out of wall plane deformation.

By this investigation, it is known that the peak stress at tube wall reaching yielding level shall be within the scope corresponding to the beam depth, and the stress decreases distinctly out of the scope of two times of beam depth. The local deformation presents a similar tendency. So the length of partial thickened tube can be taken as two times of beam depth.

6 CONCLUSION

By the experimental and numerical study, we make the following conclusions:

1. To thicken the tube wall in joint zone, both the initial stiffness and moment resistance of the joint can be increased effectively. To guarantee the fully plastic moment capacity of connected beam by the thickening is relatively easy, but to acquire a fully rigid joint as internal or external diaphragm joint shall be difficult unless the tube wall is thickened in very large thickness or the connected beam is relatively slender.

Height position		Asymmetry	Symmetry
	H_b		
	$0.5H_b$		
	0		
	$-0.5H_b$		
	$-H_b$		

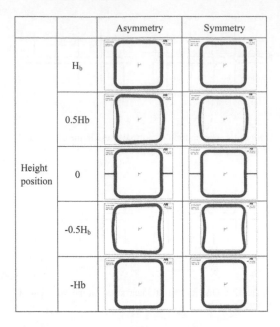

Figure 14. Distribution of out of tube wall deformation.

2. To balance the joint performance and economic benefit, the thickened tube portion can be taken the length within the two to three times of the beam depth.

3. To evaluate the stiffness of NDJ, Zhang-solution provides enough accuracy for engineering application. To predict the moment resistance of NDJ, equations adopted by AIJ (2001) is proved in good agreement with test results.

ACKNOWLEDGEMENT

The study was financially supported by Baosteel.

REFERENCES

Akiyama, H., Oh, S., Otake, F., Fukuda, K., Yamada S. 1996. General moment-rotation characteristics of beam-to-RHS-column connections without diaphragms. *Journal of Structure and Construction Engineering*, 484: 131–140.

Architectural Institute of Japan. 2001. *Recommendations for Design of Connections in Steel Structures*, Tokyo, AIJ.

CEN. 2005. *EN 1993-1-8. Eurocode 3: Design of steel structures, part 1.8: Design of joints*. Brussels.

Fujita K., Morita K., Yokoi K., Furusawa K. 1999. Structural performance of H-Beam-and-RHS column sub-assemblage models reinforced by increasing column thickness. *Steel Construction Engineering*, 6(24): 39–54.

Harada Y., Morita K., Yokoi K., Fujita K. 2000. Estimation of elastic-plastic behavior of H-beam-to-RHS column connection without diaphragm. *Steel Construction Engineering*, 7(26): 59–72.

Lu L. H. 1997. The static strength of I-beam to rectangular hollow section column connections. *Doctor Dissertation*. Delft, Netherlands: Delft University of Technology.

Lu, L., Wardenier, J. 1998. The ultimate strength of I-beam to RHS column connections. *Journal of Constructional Steel Research*, 46(1–3): 250–251.

Morida, K., Ebato, K., Watanabe, S. 1989. Research on the effect of diaphragm on the connection between box-column and H-shaped beam. *Journal of Structural and Constructional Engineering*, 338: 100–110.

Zhang, L. & Chen, Y. 2012. Progression solution of elastic stiffness of no-diaphragm joint connecting RHS tube and H-shaped beam. *Engineering Mechanism*, 29(8): 87–93.

Tubular Structures XV – Batista, Vellasco & Lima (eds)
© 2015 Taylor & Francis Group, London, ISBN 978-1-138-02837-1

Analysis of the possible failure modes in CSH bolted sleeve connections

L.R. Amparo & A.M. Sarmanho
Post-Graduate Program in Civil Engineering, Federal University of Ouro Preto, Ouro Preto, MG, Brazil

A.H.M. de Araújo
V&M do Brasil S. A., Belo Horizonte, MG, Brazil

J.A.V. Requena
Faculty of Civil Engineering, State University of Campinas, Campinas, SP, Brazil

ABSTRACT: Steel circular hollow sections (CHS) have good resistance under tension, compression and torsion loads, considered isolated or combined. Recent researches have been carried out to evaluate the behavior and resistance of these structures. A current use of steel hollow sections is roof trusses, a kind of structure that demands standard procedures and fast assemblage. This work presents an experimental study of a new type of connection called sleeve, developed to attend the assembly process. The connection consists of two tubes internally connected through a third tube with a smaller diameter with passing bolts arranged in line and cross. This connection makes easier the structure assemblage and is aesthetically advantageous if compared to usual flanged connection used in hollow sections. An experimental program of the sleeve connection with bolts in line (along the tube) and crossed at 90° was developed. In this work were tested 3 prototypes with staggered bolts to complement previous research and 12 prototypes with crossed bolts. The parameters considered in investigation were tube diameter and thickness and number of bolts. Experimental results lead to identification of failure mode and the resistance of sleeve connection. The analysis of theoretical and experimental data revealed that the dominant failure mode was bending in the bolts, and indicated a reduction in connection resistance that needs to be quantified. For the prototypes with staggered bolts, the analysis also let the development of an expression for the determination of flexural failure in the bolts. The results have demonstrated the viability of this new type of connection, but indicated that all possible failure modes need to be carefully checked.

1 INTRODUCTION

Steel circular hollow sections (CHS) have good resistance under tension, compression and torsion loads, considered isolated or combined. Recent researches have been carried out to evaluate the behavior and resistance of these structures. Many researchers have been performed in order to develop new techniques and propose solutions for proper use of these sections as seen in works done by Munse and Chesson (1963), Beke and Kvocak (2008), Martinez-Saucedo and Packer (2009), Freitas and Requena (2009), Mayor (2010), Vieira et al (2011), Silva (2012) Amparo et al (2014) and others.

To optimize the assembly of systems of trussesby tubular profiles of circular section it was proposed a kind of connection between tubular profiles called "sleeve". The connection was developed to join tubular bars and trusses modules for ease in transportation and assembly.

Figure 1 shows the sleeve connection. The sleeve connection is constituted by two tubes of the same

Figure 1. Sleeve connection with aligned bolts (Vieira, 2011).

diameter connected internally by a third tube with smaller diameter and bolts.

This work consists of a theoretical and experimental analysis of the connection, where the experimental

Figure 2. General scheme of the experiment.

Table 1. Identification of prototypes series.

Series	No. of prototypes	No. of bolts	Outer tube (diameter × thickness) mm	Inner tube (diameter × thickness) mm
LA	1	3	73.0 × 5.5	60.3 × 5.5
	2	4	73.0 × 5.5	60.3 × 5.5
CB	4	3	76.1 × 3.6	60.3 × 3.6
	3	4	76.1 × 3.6	60.3 × 3.6
CC	2	5	88.9 × 5.5	73.0 × 5.5
CD	3	6	88.9 × 5.5	73.0 × 5.5

results will be analyzed to assess the connection behavior and failure modes.

2 EXPERIMENTAL PROGRAM

To represent the connection, the prototypes are formed by an outer tube and an inner tube (sleeve) with aligned bolts. The prototypes were tested under tension in a controlled servo-hydraulic press with 200 kN of capacity. An LVDT (Linear Variational Displacement Transducer) and electrical resistance strain gages were used for the test instrumentation. Figure 2 shows a general scheme of the experiment.

The analyzed prototypes have geometric properties variations: thickness and diameter, therefore they were divided into different series, as shown in Table 1. Each series had a variation in the number of bolts. The prototypes of the series LA have staggered bolts and CB, CC and CD have crossed bolts.

The tubes were defined according to the availability lab. The tubes (outer and inner) of each series had a different steel resistance as shown in Table 2.

The gross area and net area were calculated considering the tubular section as a rectangular prismatic bar (rectangular plate). This adaptation can be found in Amparo (2014).

The effective net area (A_e) is calculated according the Equation 1:

$$A_e = C_t A_n \qquad (1)$$

Table 2. Steel resistance of tubes of prototypes.

	Steel Resistance			
	Outer tube		Inner tube	
Series	Yielding f_y (MPa)	Ultimate f_u (MPa)	Yielding f_y (MPa)	Ultimate f_u (MPa)
LA	399.5	539.5	381.0	479.0
CB	386.0	545.0	424.0	535.0
CC	375.0	474.0	399.5	539.5
CD	375.0	474.0	350.0*	450.0*

*Nominal value.

Table 3. Theoretical values to failure modes.

	Modes of failure		
	Y.G.S	F.N.A	S.B.
Prototype	N_t (kN)	N_t (kN)	F_v (kN)
LA-3-X3Y2	360.8	378.2	313.5
LA-4-X3Y2	360.8	378.2	418.0
CB-3-X2Y3	271.9	259.3	313.5
CB-4-X2Y3	271.9	259.3	418.0
CC-5-X2Y3	466.0	498.0	522.5
CD-6-X2Y3	408.2	415.4	627.1

where C_t is the shear lag coefficient; and A_n is the net area.

The value considered to shear lag coefficient (C_t) it was equal 1.0 for the calculating of the theoretical values.

The failure modes were calculated according Brazilian code ABNT NBR 8800:2008, using A_e as shown in Equation 1. The failure mode yielding of the gross section (Y.G.S. – N_t) was calculated according the Equation 2, the fracture through the effective net area (F.N.A. – N_t) according Equation 3 and the shear failure of bolt (S.B. – F_v) according Equation 4:

$$N_t = A_g f_y \qquad (2)$$

$$N_t = A_e f_u \qquad (3)$$

$$F_v = 0,4\, A_b\, f_{ub} \qquad (4)$$

where A_g is the gross area; f_y is the yielding resistance; A_e is the effective net area; f_u is the ultimate resistance; A_b is the bolt gross area; and f_{ub} is the ultimate resistance of the bolt.

Table 3 shows the theoretical values for these failure modes.

After experimental analysis, it was observed, in some prototypes, that the bolts showed a flexural mode. In some cases the limit state of bolt bending is dominant, and not the yielding of the gross section nor the fracture through the effective net area. More details are showed in Amparo et al (2014) and Amparo (2014), including one expression to bending failure of bolts.

Table 4. Experimental maximum load capacity and average load.

Prototype	Experimental maximum load capacity (kN)	Average load (kN)
LA-3-X3Y2	425.7	440.4
LA-4-X3Y2	448.7	
	446.9	
CB-3-X2Y3	247.1	244.5
	248.6	
	241.6	
	250.8	
CB-4-X2Y3	244.2	
	232.8	
	246.4	
CC-5-X2Y3	400.7	413.0
	425.3	
CD-6-X2Y3	483.0	492.6
	515.3	
	479.4	

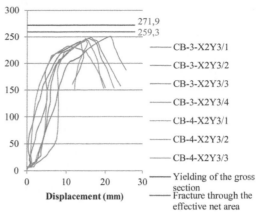

(a) prototypes of group B

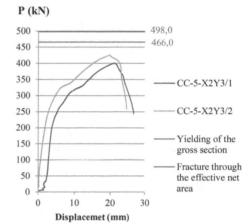

(b) prototypes of group C.

Figure 4. Curves of tension load (P) versus displacement, prototypes of group B and C.

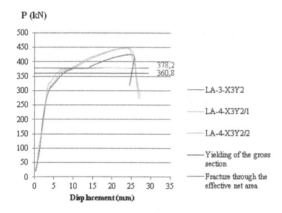

Figure 3. Curves of tension load (P) versus displacement, prototypes of group A.

3 RESULTS

Table 4 shows the experimental maximum load capacity and the average load of the experimental maximum load capacity of all prototypes of each group.

Figure 3 shows the curves of tension load (P) versus displacement for 3 prototypes of the series A. It was possible to observe that the load capacity of these prototypes was above the theoretical yielding gross section and the fracture through the effective net area (Table 3). The theoretical values of group A use the shear lag coefficient C_t equal 1.0. We observe a linear stretch followed by changes in curves slope, characterizing different failure modes of the connection.

The experimental results show that the first change in curve slope is the failure mode of bolt bending.

The average load corresponding to the fracture of the prototypes of the group A was 440.4 kN (Table 4), 16.45% higher than the expected theoretical value (Table 3). Changes in the experimental curve slopes

characterize the different failure modes. In this group, the bending in bolts was the first failure mode of the connection, observed experimentally.

In the prototypes of the series B and C, which have sleeve connection with crossed bolts, the experimental load failure mode in the net area had lower value than the theoretical load, as showed in Figure 4(a) and (b). The Figure 4 shows the curves of tension load (P) versus displacement for prototypes of series B and C, respectively.

The prototypes of the group B had the average value load corresponding to the fracture equal to 244.5 kN (Table 4). This value is 10.08% lower than the expected theoretical value (Table 3). The prototypes of the group C the average value load corresponding to the fracture was equal to 413.0 kN (Table 4), 17.07% lower than the expected theoretical value.

The prototypes of group B and C had a different behaviour compared with prototypes of the group A,

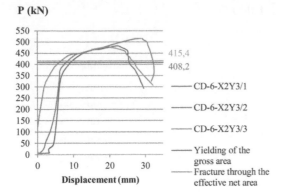

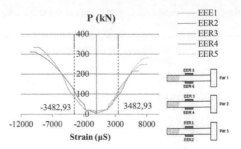

(a) bolts of prototype LA-3-X3Y2

Figure 5. Curves of tension load (P) versus displacement.

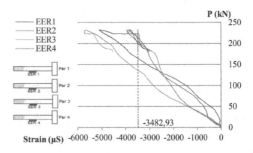

(b) bolts of prototype CB-4-X2Y3/2

all experimental curves of tension load (P) versus displacement were below the theoretical values. In these prototypes it was used of the value 1.0 as the shear lag coefficient. So, it is necessary to adopt a value that reduces the net area in an effective net area, improving the results.

The curves of tension load (P) versus displacement for prototypes of series D, which have sleeve connection with crossed bolts, is showed in Figure 5.

The average load corresponding to the fracture of the prototypes of the series D was 492.6 kN (Table 4), 18.58% higher than the expected theoretical value (Table 3).

To evaluate the forces/stresses distribution on the connection bolts due to bending, the bolts were instrumented with strain gauge (EER). Figure 6a shows the curves of tension load (P) versus strain obtained from the results of the strain gauges located in the upper and bottom regions of the bolts for the series A prototype with 3 bolts.

Figure 6a, b, c and d shows curves the curves of tension load (P) versus strain obtained from the results of the strain gauges located in the upper and bottom regions of the bolts for one prototype of each series (CB, CC and CD).

From Figure 6 it is noted that the bolt strain exceeds the strain level related to the material yield, which corresponds to 3482,93 μS or 3092,68 μS according to the length of bolts. This limit is shown in Figure 6 graph.

All other prototypes also showed the same behavior. Given this behavior it is possible to characterize the failure mode of bending the bolts.

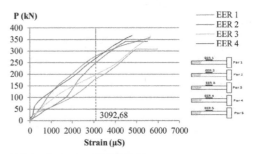

(c) bolts of prototype CC-5-X2Y3/2

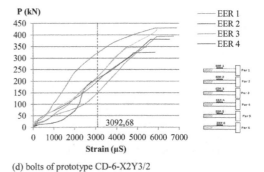

(d) bolts of prototype CD-6-X2Y3/2

Figure 6. Curve of tension load versus strain.

4 PROPOSED FORMULATIONS

From the analysis of the experiments results we noticed the presence of a limit state on the connection not foreseen in previous theoretical analyzes (Vieira, 2011). According to this experimental analysis, an equation has been proposed to represent the limit state of bolt bending for sleeve connection, Amparo (2014).

The connection resistance force due to the failure mode of bolt bending to aligned bolts, is defined by Equation 5:

$$F_b = \frac{f_{y,b} \, W_b \, \pi \, D}{2 \, d \, x} \qquad (5)$$

Table 5. Theoretical and experimental results.

Prototype	$F_{b,exp}$ (kN)	F_b (kN)	$F_b/F_{b,exp}$
LA-3-X3Y2/1	308.6	237.7	0.770
LA-4-X3Y2/1	309.9	237.7	0.767
CB-3-X2Y3/1	131.1	114.7	0.874
CB-4-X2Y3/1	148.3	152.9	1.031
CC-5-X2Y3/1	197.8	229.9	1.162
CD-6-X2Y3/1	301.2	275.9	0.916

Table 6. Shear lag coefficients (C_t).

Prototypes	Number of bolts	C_t by NBR	Experimental C_t
CB	3	0.76	0.95
CB	4	0.84	0.92
CC	5	0.86	0.83
CD	6	0.88	1.18

where $f_{y,b}$ is the bolt yield stress; W_b is the bolt elastic section modulus; D is the diameter of the inner tubes (sleeve); d is the diameter of the hole; and x is the lever arm, obtained by Equation 6:

$$x = \frac{(D_{ext} - e_{ext}) - (D_{int} - e_{int})}{2} \qquad (6)$$

where D_{ext} and D_{int} are the external diameters of the outer and inner tubes, respectively; e_{ext} and e_{int} are the thickness of the outer and inner tubes, respectively.

The connection resistance force due to the failure mode of bolt bending to crossed bolts, is defined by Equation 7:

$$F_b = \frac{f_{y,b}\, W_b\, \pi\, D}{5\, d\, x} \frac{n}{2} \qquad (7)$$

where $f_{y,b}$ is the bolt yield stress; W_b is the bolt elastic section modulus; D is the diameter of the inner tubes (sleeve); n is the number of bolts; d is the diameter of the hole; and x is the lever arm, obtained by Equation 6.

Table 5 shows a comparison between the experimental results and the results obtained by the proposed Equation 5 and Equation 7. Table 5 also shows a ratio between the theoretical and experimental value. We notice that the theoretical values are consistent according to experimental values. Just one type of prototype of each series is presented in the Table 5.

From Table 5 and Table 3 we observe that the limit state of bolt bending is dominant in the design, being important its consideration and evaluation. We notice that the theoretical results are consistent according to the experimental ones. However the Equation 7 need to be improved. In addition, a broader assessment of the behavior of this connection (crossed bolts) for other prototypes with larger variations of the geometric properties must be made.

In the prototypes of group B and C, the use of the value 1.0 as the shear lag coefficient did not provide good results for the failure mode fracture through the effective net area. So, it is necessary to adopt a value that reduces the net area in an effective net area for crossed bolts, improving the results.

By comparison the experimental result with the results given by the Equation 8. This equation was proposed by AISC (2005) and ABNT NBR 8800:2008 for a slotted tube connected to a gusset plate, adapted to sleeve connections.

$$C_t = 1 - \frac{e_c}{l_c} \qquad (8)$$

where C_t is the shear lag coefficient; e_c is the connection eccentricity and l_c is the length of connection.

In Table 6 is shown the experimental shear lag coefficient determined dividing the average value experimental load corresponding to the fracture by the net area and the ultimate stress of the inner tube.

The use of the nominal value for ultimate resistance in prototypes of series CD can explain the difference between theoretical and experimental values.

5 CONSIDERATIONS

The work carried out was suitable to evaluate the new connection for hollow sections, called sleeve connection, and its probable limit states. According to the experimental data obtained, it was possible to observe the presence of a new limit state not foreseen in the connection theoretical design, named bolt bending. The experimental results analysis allowed the proposal of an expression to design the connection for the limit state of bolt bending. The methodology used in equations proved to be consistent, showing good results compared with experimental results.

The international and national standards do not yet include the sleeve connections, the results and the analysis also proved the need and suitability of a formulation for the calculation of the shear lag coefficient to crossed bolts. Thus, it was proposed one equation based in AISC/NBR to calculate the C_t. The results showed good correlation with the experimental results.

The analysis of theoretical and experimental data revealed that the dominant failure mode was bending in the bolts, and indicated a reduction in connection resistance that needs to be quantified. The analysis also let the development of an expression for the determination of flexural failure in the bolts. The results have demonstrated the viability of this new type of connection, but indicated that all possible failure modes need to be carefully checked.

ACKNOWLEDGMENTS

The authors are grateful for the support from FAPEMIG, CNPQ, CAPES and Vallourec do Brazil.

REFERENCES

ABNT NBR 8800:2008: Projeto de Estrutura de Aço e de Estrutura Mista de Aço e Concreto de Edifícios, 2008. Associação Brasileira de Normas Técnicas, Rio de Janeiro.

American Institue of Steel Construction – AISC. 2005. Specification for Structural Steel Buildings. Manual of Steel Construction, Chicago.

Amparo, L. R., 2014. Análise Teórico-Experimental de ligações tipo luva em perfis tubulares com parafusos em linha e cruzados. Dissertação de Mestrado. Universidade Federal de Ouro Preto.

Amparo, L. R., Freitas, A. M. S., Requena, J. A. V., Araújo, A. H. M., 2014. Theoretical and experimental analysis of bolted sleeve connections for circular hollow sections. Eurosteel, Napoles, Italy.

Beke, P., Kvocak, V., 2008. Analysis of joints created from various types of sections. Eurosteel, Graz, Austria.

Freitas, A. M. S., Requena, J. A. V., 2009. Ligações em estruturas metálicas tubulares. In: Kripka, M., Chamberlaiɳ, Z.M (Org.). UPF Editora, 221p. cap. 1, p. 7–29. (Novos Estudos e pesquisas em construção metálica).

Martinez-Saucedo, G., Packer, J.A. , 2009. Static Design Recommendations for Slotted End HSS Connections in Tension. Journal of Constructional Steel Research, 135 (7), 797–805.

Mayor, I. S., 2010. Análise Teórica-Experimental de ligações tipo K e KT composta por perfis tubulares de seção retangular e circular. Dissertação de Mestrado. Universidade Federal de Ouro Preto.

Munse, W. H., Chesson, E. Jr., 1963. Riveted and bolted joints: net section design. Journal of Structural Division, ASCE, v. 89, p. 107–126.

Silva, J. M., 2012. Análise teórica-experimental de ligações tipo tubulares tipo "luva". Dissertação de mestrado. Universidade Federal de Ouro Preto.

Vieira, R. C., Vieira, R. F., Requena, J. A. V., and Araújo, A. H.M., 2011. Numerical Analysis of CHS bolted sleeve connections. Eurosteel, Budapest, Hungary.

Tubular Structures XV – Batista, Vellasco & Lima (eds)
© *2015 Taylor & Francis Group, London, ISBN 978-1-138-02837-1*

RHS beam-to-column connection: Experimental analysis of innovative bolted typology

G.B. dos Santos & E. de M. Batista
Civil Engineering Program, COPPE, Federal University of Rio de Janeiro, Brazil

A.H.M. de Araújo
Vallourec Research Center Brasil

ABSTRACT: The main purpose of the present research is to investigate the behavior of innovative beam-to-column bolted connection of steel rectangular hollow sections – RHS. The typology of the connection was conceived to combine simplicity in fabrication and erection with structural effectiveness. The connection components are two cleat plates butt-welded or inserted through the RHS column, internally connected to the webs of the RHS beam with top and bottom horizontal lines of bolts. In addition, the tested connections were divided in two concepts concerning the bolts tightening: (i) regular manual tightening and (ii) pre stressed bolts by controlled tightening. The former allows usual non-friction connection and the later permitted to investigate friction-type. The experimental results permitted to identify the connections bending moment-rotation behavior and comparison with usual concepts of rigid, semi-rigid or flexible classification, in order to support structural design procedures.

1   INTRODUCTION

Tubular steel construction is growing in Brazil due to its favorable geometric properties for structural strength and stiffness in bending and torsion if compared with open hot rolled and welded sections (Wardenier, 2010). In addition, aesthetical and architectural advantages must be considered, as well as high durability and less corrosion maintenance.

Nowadays, a recent Brazilian standard were created to support the design of tubular structures but only some connections where described (ABNT, 2012).

The design of connections is a key point of the steel construction, considering manufacturing and erection procedures, with important impact in the final costs. The concept of the connections is also an important feature for the structural behavior of the members and the global performance of the structure.

Figure 1 shows the Bending moment-Rotation classification of the steel connections according to both the Eurocode (EN, 2005) and the Brazilian code (ABNT, 2008), where Zone 1, 2 and 3 are related respectively to rigid, semi-rigid and flexible connections, according to its initial stiffness $S_{j,ini}$ given by Equations 1 to 3, in which: $k_b = 25$ since $K_b/K_c \geq 0,1$, otherwise the system is to be taken as semi-rigid; $K_b = I_b/L_b$ and $K_c = I_c/L_c$ are the ratio of moment of inertia and lengths for beams and columns, respectively.

Zone 1: $S_{j,ini} \geq k_b E K_b$ (1)

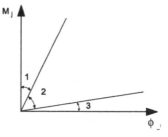

Figure 1.   Bending moment-rotation classification of the connections.

Zone 2: $k_b E K_b \geq S_{j,ini} \geq 0.5 E K_b$ (2)

Zone 3: $S_{j,ini} < 0.5 E K_b$ (3)

The present investigation is aimed at describing the experimental behavior of the innovative beam-to-column bolted connection shown in Figure 2. Planed cruciform tests allowed identification of the bending moment-rotation response for the following conditions: (a) Type 1 RHS beam-to-column bolted connection with internal (hidden) cleat plate; (b) Type 2 RHS beam-to-column connection with internal cleat plate and additional flange opening of the beam to make it easier bolting during erection. The former was conceived with nuts previously welded to the internal cleat plates in order to allow installing bolts in field;

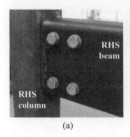
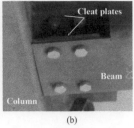

(a)   (b)

Figure 2.   (a) Connection Type 1; (b) Connection Type 2 with opening in the web flange and cleat plates crossing through the column.

Table 1.   Cruciform tests of RHS beam-to-column bolted connections.

Test Id	Type of connection	Bolts tightening method
1	Type 1 – NO/WF	Nut rotation control
2	Type 1 – NO/NF	Simple tight
3	Type 2 – WO/WF	Nut rotation control
4	Type 2 – WO/NF	Simple tight

NO: no opening, WO: with opening, WF: with friction, NF: no friction.

the latter permits usual in field positioning of nuts and bolts.

In addition, the cleat plate shown in Figure 2 was conceived in two different ways: (a) butt-welded to the column or (b) inserted and crossing through the column cross-section.

Similar connections are reported in the literature (Kurobane, 2004) with no specifications or procedures for its design. Taking into account that aligning beam and column webs would produce stiffer connection the above-cited authors show RHS with the same widths and cleat plate positioned externally to both beam and column. The concept of the present investigation was to produce improved appearance of the connection typology by introducing internal cleats plate as above cited and illustrated in Figure 2.

The present RHS beam-to-column connection was pre designed based on its local behavior, following regular code prescriptions (ABNT, 2008): (i) initial yielding of the beam, (ii) shear strength of the bolts with or without pretension, (iii) initial yielding of the plate cleats including the holes elongation by contact with bolts, (iv) strength of the fillet welding.

Finally, Type 1 and 2 connections were tested for both conditions: (i) connection without friction and (ii) friction-type connection.

2   EXPERIMENTAL PROGRAM

2.1   *Cruciform test procedures*

The experimental program was conceived to identify the connections bending moment-rotation behavior, following the above-cited concept of rigid, semi-rigid or flexible classification, according to Figure 1 and Equations 1 to 3. Four cruciform beam-to-column connection prototypes were designed and tested, as shown in Table 1: (i) Type 1 connection without friction and friction-type connection; (ii) Type 2 connection without friction and friction-type connection. Friction-type connection was accomplished by controlling bolts tightening, as describe further.

The experimental results issued from the displacement transducers and clinometers allowed identification of the onset of the nonlinear behavior of the connection, promoted by slipping between the beam

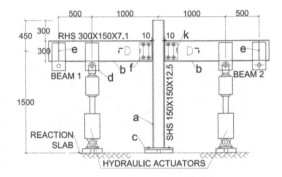

Figure 3.   Experimental setup of the cruciform tests.

Figure 4.   Cruciform Test 1 layout (Type 1 – NO/WF).

and cleat plates. In addition, in order to give an appraisal of the stress distribution, strain measurements were performed with (i) longitudinal strain gages in the bottom flange of the beams and (ii) laterally in the columns, and (iii) rosette strain gages placed close to bolts.

Figures 3 and 4 illustrate an overall view of the experimental setup of the cruciform tests of the connections, in which 1000 mm span was adopted for the applied transversal loadings (note 1.0 or 1.5 m loading span could be adopted for the tests). All the tests were performed with the help of a couple of displacement controlled servo hydraulic actuators (246 kN capacity each).

Figures 5 and 6 present the bolted beam-to-column connection with cleat plate, respectively, butt-welded and crossing through the column.

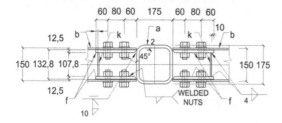

Figure 5. Beam-to-column bolted connection with butt-welded cleat plate.

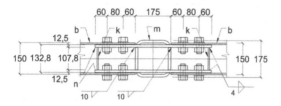

Figure 6. Beam-to-column bolted connection with cleat plate crossing through the column.

Table 2. Mechanical properties of the steel (for the components Id see Figs. 6 and 7).

Member	Label Id	f_y (MPa)	f_u (MPa)	Final elongation (%)
SHS Column $175 \times 175 \times 12.7$ mm	a/m	423.0	541.0	41.0
RHS Beam $300 \times 150 \times 7.1$ mm	b	473.0	602.0	34.0
Connection plate cleats 12,5 mm	f/n	442.0	566.0	25.0

The installation of the bolts followed practical recommended rules for pretensioned tight (ABNT, 2008) with the help of nuts rotation control (usually applied in field erection of steel structures), letting friction-type.

The steel mechanical properties displayed in Table 2 were obtained by standard direct tensile tests.

2.2 Experimental measurements

Experimental data was recorded from the following instrumentation planning, according to Figure 7: (i) HA1 and HA2 are the servo controlled hydraulic actuators with tension load capacity of 246 kN, (ii) couples of horizontal displacement transducers DT1-DT2 and DT3-DT4 are able to record relative rotation between beam and column, (iii) clinometers CL1 to CL3 are also aimed at measuring the relative rotation between beam and column, (iv) DT5 and DT6 record the beams' vertical displacements, (v) strain gages rosettes R1X to R4X are placed as close as possible of the bolts in order to follow localized strain with influence of contact between bolts and the web plate of the beam, (vi) single strain gages E1 and E2 are able to record deformations in the lateral surface of the column, (vii) strain gages E3 to E8 were planned to measure bottom beam deformations and are able to give information about stress transference between beam and plate cleats in the connection region.

2.3 Bolting for friction type connection

Friction-type connection was assured by nut rotation control according to the Brazilian code (ABNT, 2008). This is a practical method based on the measurement of the nut rotation and, for the present case of 22 mm diameter bolts, it is recommended to control turn-of-nut tightening up to 1/3 complete tur with final pretensioning force limited to 70% of the tensile strength of the bolt (125 kN for the case of 22 mm diameter A325 bolt). Bolting for friction-type connection followed the steps: (i) firmly tightening with usual wrench showed in Figure 8a; (ii) complete controlled tightening with extended wrench showed in Figure 8b in order to apply extra torque limited to 1/3 turn, as mentioned before.

In addition, more accurate measurements of the bolts pretensioning were performed with special axial tension bolts strain gages (Kyowa, model KFG) for Test 3 (Type 2 – WO/WF). These instrumented bolts were previously submitted to controlled direct tensile test in order to define its load factor constant (each of them transformed in tensile load cells), which enabled recording the pretensioning forces during the above mentioned tightening process for friction-type connection.

The recorded results in Figure 9 illustrate the different methods of bolts tightening: (i) one-step of normal tightening (wrench Figure 8a), for which we show Testing #1 and #2 in Figure 9 (unfortunately instrumented bolt no. 2 was loosen during the process and its results are only available for Testing #1); (ii) Testing #3 pretensioning the bolts for friction connection directly with the extended wrench in Figure 8b; (iii) two steps of pretensioning in Testing #4, as recommended and described before, according to code prescription (ABNT, 2008).

Although these were not extensive measurements of the pretensioning of the bolts, the following observations were issued from the experimental results in Figure 9: (i) normal tightening Testing #1 and #2 resulted in average tensile forces equal to 42.6 kN, around 24% of the tensile strength of the bolts $F_{TR} = 178$ kN; (ii) Testing #3 applied pretensioning in one step only resulted in tensile forces ranging between 44 and 70 kN, only 25–39% of the tensile strength of the bolts, (iii) Testing #4 aimed at achieving full pretensioning confirmed the friction connections were tested with appropriate pretensioning, ranging from 86 to 107 kN, respectively 49 to 60% of the tensile strength of the bolts, respecting the limitation of 70% of the bolts tensile strength.

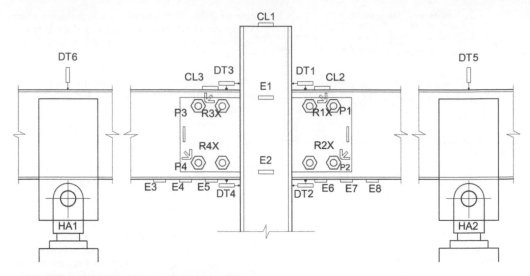

Figure 7. Instrumentation plan (beam without opening).

(a)

(b)

Figure 8. (a) Normal wrench, (b) extended wrench.

3 BEAM-TO-COLUMN CONNECTION BEHAVIOR

3.1 Behavior of the beams

Measurements with strain gages positioned in the bottom flange of the RHS beams confirmed their linear elastic behavior along all the tests, as can be observed in Figure 10 for the Test 1. These results just confirm that the nonlinear behavior of the connections was never affected by the deformations of the beams.

3.2 Behavior of the columns

Strain gages measurements in the sides of the columns, as shown in Figure 11, indicate elastic deformations with minor nonlinear contribution. One may conclude that column deformation is not a main contribution to the nonlinear behavior of the tested connections.

3.3 Bending moment-rotation response M-θ

The bending moment-rotation response of the connections was recorded with the help of both displacement

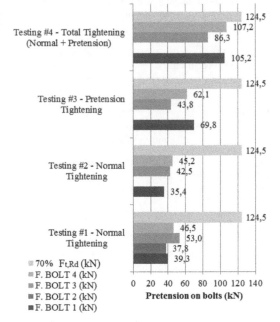

Figure 9. Axial tension loads obtained from strain gage measurements of the bolts (Test 3 beam-to-column connection).

transducers DT and clinometers CL. Relative rotation between beam and column, θ_i, obtained from DT's records was computed with Equation 4, in which: d_i is the recorded displacement of DTi, 150 mm is half height of the beam and h_i is the distance between the point of measurement of DTi and the top of the beam (see Fig. 7). In this condition, four relative rotations, $\theta_1 - \theta_4$, were recorded during the tests: θ_1 and θ_2 related to the right side beam, θ_3 and θ_4 for the left side beam, according to Figure 7. In this condition, it was admitted

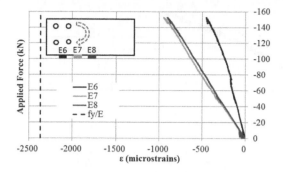

Figure 10. Measured strains at the bottom side of the right beam from Test 1 (according to Fig. 7).

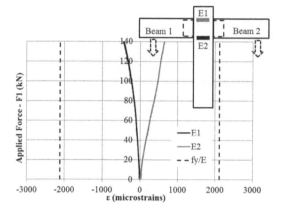

Figure 11. Measured strains at external surface of column during Test 1.

that the center of rotation of the beam was placed at its mid height.

$$\theta_1 = atan[d_1/(150 + h_1)] \qquad (4)$$

In addition, rotation measurements from clinometers CL1 to CL3 may confirm the accuracy of relative rotations recorded from DT's. Although these clinometers are analogic devices (not able to perform digital data acquisition records), the obtained results are in good agreement with those from displacement transducers, as will be confirmed in the next session.

3.3.1 Friction type connection

Figures 12 and 13 show the results of the Test 1 (Type 1 – NO/WF), and Test 3 (Type 2 – WO/WF). These are tests performed following the same methodology, the only differences are concerned to the connection configurations: (i) Test 1 is related to "no opening" in the flanges of the beam and cleat plate but welded to the column (see Fig. 5), (ii) Test 3 is related to "with opening" in the top flange of the beam and cleat plates are inserted and crossing through the column cross-section (see Fig. 6).

It was included in Figures 12 and 13 the limits for both rigid and flexible connections, according to Equations 1 and 3, taking into account beam length L_b equal

Table 3. Bending moment-rotation stiffness: experimental and estimated results related to Test 3 ($S_{,ini}$: 10^3 kNm/rad).

Beam	DT	$S_{,ini}$	$S_{,ini}$ Aver.	$S_{,ini}$ Eq. 1 and 3			
Right	DT1	16.7	17.0	L_b (m)			
				6	7	8	Class.
	DT2	17.3		≥18.5	≥15.9	≥13.9	Rigid
				<1.2	<1.0	<0.9	Flex.
Left	DT3	18.4	18.8	The same			
	DT4	19.3					

to 6, 7 and 8 m. The bending moment-rotation response of Test 1 in Figure 12 includes rotation measurements from both the displacement transducers and from the clinometers, from which one may observe accurate agreement between these results.

Figure 13 also shows the results of two unloading steps during Test 3, confirming the original linear elastic bending moment-rotation behavior for both right and left beam-to-column connections.

These results indicate the proposed beam-to-column connection exhibits rigid bolted connection, if attached including friction condition. The recorded bending moment-rotation response developed rigid connection classification up to approximately 40 kNm. After this initial elastic linear behavior, the connection displays nonlinear path because the friction effect is broken and relative displacements between the beam and cleat plates are released. Slipping displacements promote low stiff bending moment-rotation response until the bolts are able to transfer the loading effects directly to the cleat plates by contact. Finally, the connection develops a third step of its bending moment-rotation behavior, with increased stiffness and clear nonlinear response due to the development of bearing failure, conducting to localized plastic deformation around the holes. In addition, large deformations of the cleat plates develops.

The described first step – related to a rigid connection behavior – is concerned to practical design purposes and corresponds to Serviceability Limit State. It is clear from the experimental results that both types of arrangements, with and without the opening in the flange of the beam, can be classified as rigid beam-to-column connections (limited to approximately 40 kNm applied bending moment).

Finally, Table 3 presents the computed experimental results of the initial stiffness of the tested friction-type connections, which are compared with the estimated results based on Equations 1 and 3. These results confirm the classification of the bending moment-rotation response of the tested connections.

3.4 Non-friction type connection

Figure 14 shows the results of the Test 4 (Type 2 WO/NF) that differs on bolts tightening method from Tests 1 and 3, as previously explained.

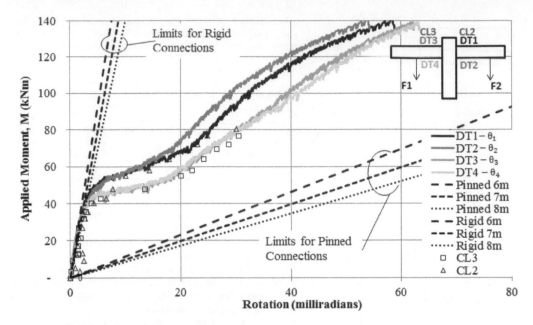

Figure 12. Applied bending moment M *versus* relative rotation θ_i for Test 1.

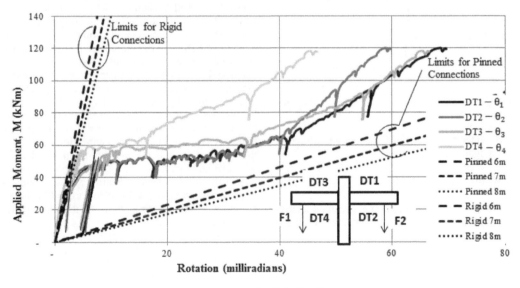

Figure 13. Applied bending moment M *versus* relative rotation θ_i for Test 3.

These results clearly indicate typical flexible behavior from the very beginning of the test for the left beam-to-column bolted connection (θ_3 and θ_4). Low friction effect allowed relative slipping between the beam and the cleat plate for applied bending moment above 5 kNm. The bolted connection in the right side of the cruciform arrangement revealed higher friction effect, up to around 13 kNm. Anyway, both connections in Test 4 exhibited rigid behavior up to very low applied bending moment (as expected, much lower than observed for the friction-type connections presented in the last session), followed by clear flexible behavior. In this condition, one may conclude the

tested non-friction connection must be consider as a flexible connection.

Test 2 (NO/NF) presented similar results of Test 4 and confirms the above cited considerations.

4 FINAL REMARKS

The present investigation permitted to evaluate innovative arrangement for steel RHS beam-to-column connection and the following conclusions are drawn:

- Both arrangements, with or without openings, allowed easy bolting. Anyway, for practical

446

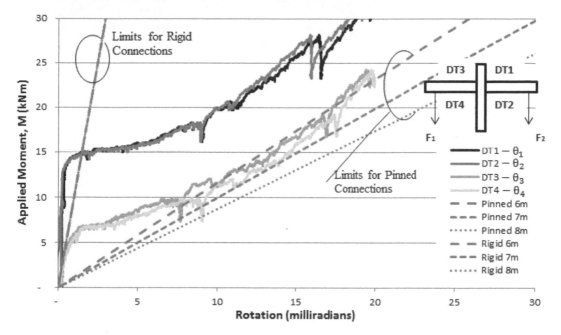

Figure 14. Applied bending moment m *versus* relative rotation θ_i for Test 4.

purposes, beams with opening could be considered a better solution, avoiding extensive welding of the internal nuts in the cleat plates.

- Top and bottom arrangement of bolts in the web of the beam were originally addressed to obtain stiff connection. This conception was actually achieved with friction-type connection.
- The tested non-friction connection proved to be useless for beam-to-column bending moment transmission purpose. The conclusion is that it does not matter the kind of bolts arrangement – in vertical or horizontal lines – if friction is not mobilized; this will be a flexible bolted connection.
- Measurements of the bolts pretensioning with the help of strain gages proved the applied method of tightening was efficient enough to promote appropriate friction between the webs of the beams and the cleat plates.
- Even if the prescribed code limits for the bolts tightening is not fully achieved, corresponding to (i) nut rotation control and/or (ii) 70% of its tensile strength, since tightening is strongly promoted with the help of extended wrench one may obtain efficient friction connection, as revealed by the experimental results.
- Measurements also showed the nonlinear bending moment-rotation behavior of the connections was achieved by (i) slipping movement between the beam and the cleat plates, (ii) the bolts-holes contact and (iii) large deformation of the cleat plates, with no contribution of the beams (always elastic) and minor contribution of localized deformation in the (plates) lateral elements of the column.

- Finally, additional experimental investigation would be useful to observe the behavior of the connection under reversal bending moments.

ACKNOWLEDGEMENTS

The authors are grateful for the support of Vallourec Tubos do Brasil during the definition of structural conception of the connections as well as for delivery of all tubular members. In addition, the authors are thankful for the support of Brafer Construções Metálicas, which manufactured all the connections prototypes.

REFERENCES

EN 1993-1-8. 2005: Design of steel structures – Part 1-8: Design of connections. CEN, Brussels.
ABNT NBR 8800: 2008. Design of steel and composite structures for buildings. Associação Brasileira de Normas Técnicas, Brazil.
ABNT NBR 16239: 2012, Design of steel and composite structures with tubular profiles. Associação Brasileira de Normas Técnicas, Brazil.
Kurobane, Y., Packer, J.A., Wardenier, J. & Yeomans, N. 2004. Design Guide for Structural Hollow Section Column Connections. CIDECT (ed.), TÜV-Verlag GmbH, Köln.
Wardenier, J. Packer, J.A, Zhao, X.–L. & van der Vegte, G.J. 2010. Hollow Sections in Structural Applications, CIDECT (ed.), Geneva, Switzerland.

Tubular Structures XV – Batista, Vellasco & Lima (eds)
© *2015 Taylor & Francis Group, London, ISBN 978-1-138-02837-1*

Finite element simulations of 450 grade cold formed K and N joints

M. Mohan & T. Wilkinson
University of Sydney, Australia

ABSTRACT: This paper addresses the numerical research work conducted on higher grade C450 tubular connections subjected to static loads in N and K configuration. These results have been used to verify the need for reduction factors imposed by CIDECT on higher grade steel. Finite element simulations have first been benchmarked against a set of previously conducted 4 K and 4 N joint test results. The failure mechanism along with 3% deformation limits, first fracture and ultimate failure load have been correlated against experimental results. The limiting loads for N joints are governed by fracture and, at fail-safe loads, are within 13% of test results. Simulations in K joints are governed by notional 3% deformation limits; loads predicted are lower and within 2% to 18% below test results. Material limits of f_u at 495 MPa and f_y at 450 MPa are generally found to be adequate without the additional correction factor imposed by CIDECT joint capacities.

1 INTRODUCTION

CIDECT Design Guides stipulate material strength and geometric limitations on the applicability of many of their design rules. The reasons are twofold: higher strength materials often have less ductility, and experience greater deformations before yield; and more slender sections may experience plate local buckling failures not necessarily considered in the background research. Packer et al. (2009) extended the applicability to yield strengths up to 455 MPa, but lays down a requirement to use the minimum of [$0.8f_u, f_y$] plus an additional correction factor of 0.9 based on extra deformation for high strength steel, rather than f_y alone, leading to a reduction of the order of 10–15% in the design capacity.

Experimental and parametric numerical investigations examining the applicability and need for these additional reduction factors to Grade C450 rectangular tubular joints were undertaken by the University of Sydney. In the framework of CIDECT program, 27 tests of joints in T, X, N and K configurations were conducted. Experimental and analytical investigations of T and X joints have been previously reported in the ISTS XIV conference by Becque and Wilkinson (2012) and Mohan and Wilkinson (2012).

Experimental results corresponding to joint tests in K and N configuration has been conducted by Yao and Wilkinson (2015). These tests have been represented by finite element models and the feasibility of eliminating the additional factor required by CIDECT on C450 steel forms the first part of this paper. Results from models using probabilistic bounds to measured f_y/f_u in the material stress strain curves are then provided. In the case of the popular K joints, effects

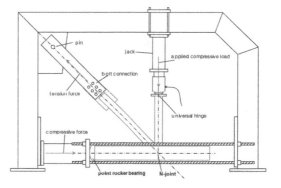

Figure 1. N test set up.

of preload and lateral constraints in the chord are compared against CIDECT joint strengths.

2 EXPERIMENTAL PROGRAM

Tests on 6 K (with gaps) and 6 N joints with some of the geometric parameters outside of CIDECT limits were conducted by Yao and Wilkinson (2015). Full details are in that paper, but for ease of reference a diagrammatic representation of the N test is shown below in Figure 1. The variety of joints failed by different modes such as tension brace failure, chord side wall buckling, chord shear, and chord face plastification. Punching shear was only observed once.

Table 1 summarises the nominal chord and brace sizes adopted in these tests along with the geometrical parameters calculated from actual measurements.

Table 1. Test configurations.

Test ID	Chord/brace size	Geometry
N1	300x300x8/	Ov = 27.5%
	150x150x5,	β i = 0.50
	200x200x6	β j = 0.66
N2	200x200x6/	Ov = 35.7%
	125x125x5,	β i = 0.62
	150 x150x6	β j = 0.74
N3	250x150x5/	Ov = 48.7%
	100x100x5,	β i = 0.40
	150 x150x5	β j = 0.60
N4	150x100x5/	Ov = 49.1%
	100x100x5,	β i = 0.68
	100x100x5	β j = 0.67
N5	250x250x6/	Ov = 63.3%
	125x125x5,	β i = 0.50
	200 x200x5	β j = 0.80
N6	150x150x5/	Ov = 91.5%
	75x75x5,	β i = 0.50
	100 x150x5	β j = 0.67
K1	200x200x6/	g = 58.7
	100x100x5	β1 = 0.50
		β2 = 0.50
K2	300x300x8/	g = 122.6
	125x125x5	β1 = 0.41
		β2 = 0.42
K3	100x150x5/	g = 79.7
	100x50x5	β1 = 1.00
		β2 = 1.00
K4	100x100x5/	g = 28.6
	100x50x5	β1 = 0.99
		β2 = 0.99
K5	100x200x5/	g = 93.8
	75x75x6	β1 = 0.75
		β2 = 0.75
K6	150x250x5/	g = 111.4
	150x100x4	β1 = 1.00
		β2 = 1.00

Figure 2. N: FEA model.

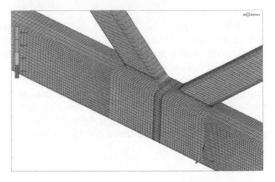

Figure 3. K: FEA model.

have been developed using MSC Patran and analysed using MSC.Marc. The mesh has been modified to a full model in parametric studies that take into account eccentricities in loading and boundary conditions. The details of a typical representation of the chord, brace and welds in an N and a K joint with higher order hex20 elements are provided in Figures 2 and 3. In each of these models, the butt welds are assumed to have 100% penetration and the size of the welds to have a minimum footprint. The models do not take into account the larger footprint following the repair of the welds in joints N2 to K6.

3 BENCHMARKING

To verify the applicability and the need for the additional factors outlined in the CIDECT code for C450 steel, numerical studies were performed on 4N (N1, N3, N5 and N6) and 4K (K1, K2, K3, K4) joints listed in Table 1. The finite element mesh methodology required was validated by a number of variations. The study establishes the need for hexahedral elements to take into account the three dimensional features at the joint, use of material models including damage constitutive model based on material coupon tests, boundary conditions that closely represent physical tests and solver parameters such as large strain, arc length, follower force effects for advanced nonlinear analyses. A brief outline of N and K joint methodology is provided in this section and a more detailed discussion is available in CIDECT project reports by Mohan et al. (2013, 2014).

3.1 Methodology

Finite element models incorporating longitudinal symmetry and hexahedral elements to represent the joints

3.2 Material

All elements in the joints were defined by true stress strain curve derived from coupons tests reported by Yao et al. (2013). A number of test results are summarised in Figure 4. The choice of this stress strain curve is also justified for the weld material Autocraft LW1-6 in argon shielding gas as it is reported to have a typical weld material yield, tensile strengths of 450 and 550 MPa and elongation of 29% (CIGWELD, 2008). These characteristics are very similar to the parent material of tubular hollow sections.

3.3 Material parameter identification for Lemaitre model

The engineering material stress strain curve with an f_y of 456 MPa and f_u/f_y of 1.13 represents nominal values. This curve converted to true stress strain curve ($f_{u\ true}/f_{y\ true} = 1.2$) superimposed with other true strain curves obtained from coupon test results is

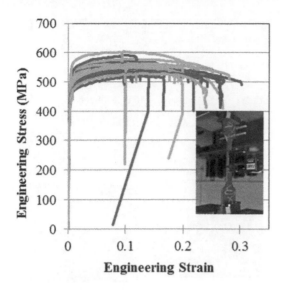

Figure 4.   Material coupon: Engineering, true stress-strain.

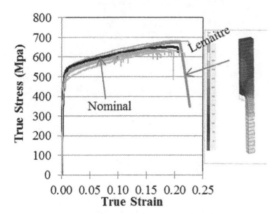

Figure 5.   Lemaitre damage model: FEA vs test.

Table 2.   Lemaitre damage model parameters.

Lemaitre damage parameters	Value
Critical damage	0.054
Maximum stress tensile test [MPa]	637
Damage resistance parameter	1.0
Equivalent strain at maximum stress	0.15

as shown in Figure 5. These data were used to define parameters for the Lemaitre damage model according to the continuum damage mechanics of Lemaitre and Chaboche (1990). The FEA model for the coupon test and the correlation between coupon test results and the measured test results are provided in Figure 5. The damage parameters for the material stress-strain curve derived from FEA simulations of a coupon test conducted on 300x8 RHS are provided in Table 2.

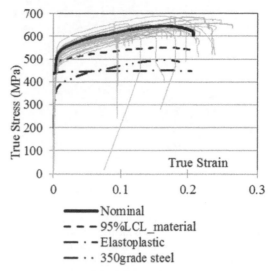

Figure 6.   Statistical variations in material parameters.

3.4   *Statistical variations in material data*

Statistical analyses of the coupon tests have been carried out. A derived stress strain curve with f_u of 550 MPa corresponds to 95% lower confidence limits. This material stress strain curve, nominal stress strain curve and a more widely used elastoplastic material stress strain curve along with other coupon test results are provided in Figure 6. Parametric studies for all three curves with appropriate Lemaitre damage parameters have been conducted for 4 N and 4 K joints and their effects on joint capacities are provided in this paper.

3.5   *Loads and boundary conditions*

Two sets in order to take into account boundary condition variations are used in simulations. Half symmetry model results are for idealised symmetric loads and boundary conditions. These are mainly reported for N joints, whereas, with K joints due to larger variations in joint capacity results from ineffective lateral restraints at the brace ends full models have also been used and reported.

4   BENCHMARKING N JOINT TESTS

Four N joint configurations were analysed for brace forces adopting large strain, follower force effects, geometric/material nonlinearity and derived parameters for Lemaitre damage material model (Table 2). The simulated joint behaviour for deformation, modes of failure and load capacities is discussed in this section.

4.1   *Observed vs simulated failure modes*

N1, N3 and N5 simulations predict fracture as the mode of failure which correlates well with test results. The premature fracture at 320 kN for N1 joint at the weld observed in the tests was due to the lack of fusion

451

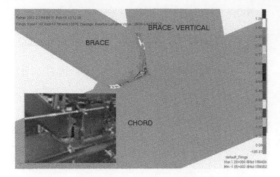

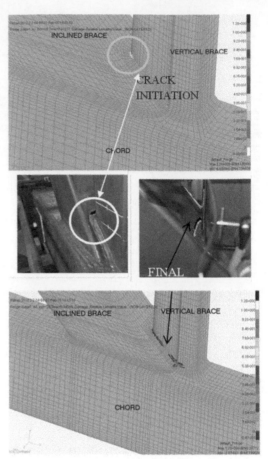

Figure 7.   N3 Fracture: Simulations at 502 kN vs test.

between the weld and parent metal; simulations for this joint with welds idealized as full penetration welds predict fracture at 546 kN. Simulated fracture locations for joints N1, N3 and N5 are similar and fracture scenarios for joints N3 and N5 are provided in Figures 7 and 8. Joint N5 exhibits punching shear failure in tests following onset of fracture and plateauing of loads. Post fracture behavior in joint N5 is not captured in simulations and both test and simulation results are provided in Figure 8. Onset of fracture and post fracture failure patterns for joints N3 and N6 in simulations and tests correlate well. The simulated onset of fracture loads predicted in joints is within 11% of tests and simulated failure mode at ultimate load capacity of the joints is within 13% of test results. The limiting fail safe loads with the exception of N5 are 14–27% larger than unfactored CIDECT design loads.

4.2   Observed vs simulated load deformation curve

The displacements monitored through lvdts during tests at the chord brace interface are compared against simulated elastic, inelastic domains for N joints. Joints N1, N3 and N5 are shown in Figures 9 to 11. As previously discussed, apart from joint N1 where early weld failure was observed, all other joints exhibit good correlation in elastic, inelastic response of the load deformation behaviour along with a good prediction on the ultimate load capacity of the joints. This correlation of results with meta models confirms the need to capture details of the joints with 3d elements. Further, the methodology confirms that the choice of parent material stress strain curve with f_y of 437 MPa and f_u/f_y of 1.13 is appropriate for all elements in the joint.

4.3   Material variations

With good validation of N tests with FEA results using meta models as outlined in section 4.2, bounds for joint capacities for possible material variations were determined. The material variations include 95% LCL material properties and elastoplastic properties discussed in section 3.4. An additional variation considered was to retain the ultimate stress, yield values of material tests but further reduce the ductility by scaling the stress strain curve to values of ultimate strain

Figure 8.   N5 Onset of fracture and beyond.

values of 0.1 and 0.05%. The results of these parametric variations as a ratio of unfactored CIDECT strength estimates at the onset of fracture are provided and compared against experimental results in Table 3. For an engineering f_u/f_y of 1.13, the minimum factor of 0.86 for N5 in Table 3 at first fracture loads are seen in the HAZ of the weld. It should also be noted that N5 has slenderness ratios just outside of CIDECT limits. Based on both numerical and experimental studies, it can be argued that under a fail-safe design (fracture allowed), load capacities in excess of 17% of first failure loads are observed and with improvements in weld design there is further scope to increase the capacity of the joint N5 without failure. Reductions of 0.9 are therefore not required and can be extended to an engineering f_u/f_y of 1.13. For other N joints, engineering f_u/f_y of 1.1 equates to unfactored CIDECT joint capacities calculated using f_y (and not factored $0.8f_u$).

4.4   Load reversal

In the current design rules, the joint capacities determined from design equations do not differentiate between compression or tension loading of overlapped and overlapping brace. In order to verify if the

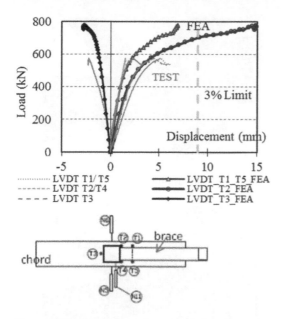

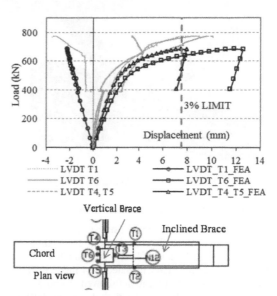

Figure 11. Joint N5: Load deformation response.

Figure 9. Joint N1: Load deformation response (Location of LVDTs shown in plan view).

Table 4. Tension vs compression in N joints.

| Compression (C)/ | N_s/N_{CIDECT} | | | |
Tension (T)	N1	N3	N5	N6
C	1.22	1.32	0.87	1.11
T	1.2	1.21	0.72	0.95
Test/CIDECT	0.72	1.33	0.8	1.25

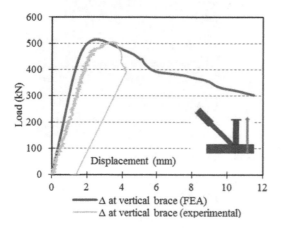

Figure 10. Joint N3: Load deformation response.

Table 3. Material stress-strain variations in N joints.

| Engineering f_u/f_y | True f_y/f_u | Ultimate Strain | N_s/N_{CIDECT} | | | |
			N1	N3	N5	N6
1.13	450/635	0.17	1.22	1.32	0.86	1.12
1.1	450/550	0.17	1.05	1.04	0.73	0.93
1	435/450	0.17	1.00	0.96	0.68	0.84
1.13	450/635	0.05	1.12	1.16	0.64	0.91
1.13	450/635	0.1	1.11	1.23	0.76	0.99
Test/CIDECT			0.72	1.33	0.80	1.25

compression in vertical brace is the governing case or results in lower joint capacities in the tests, numerical simulations for joints N1, N3, N5, N6 have been conducted with the vertical brace reversed to tensile loads whilst maintaining the rest of the boundary conditions. Simulations reveal that the first damage occurrence is at the same location, the load capacities being up to 9% less than the compressive load case. The ratio of loads from simulations when first fracture is observed to the unfactored load capacities determined as per CIDECT equations is provided in Table 4. These ratios confirm with conservative methods (such as adopting localized first fracture locations, smaller weld sizes) that the current unfactored CIDECT rules are valid for material stress strain curves with an engineering f_u/f_y of 1.13. Further, 3 of the 4 joint tests under the criteria of no fracture were acceptable for f_u/f_y of 1.13.

5 BENCHMARKING K JOINTS

Results of simulations of four K joints K1, K2, K3 and K4 are compared against experimental and CIDECT joint capacities in this section. The FEA models take into account geometrical measurements of the joint. These include thicknesses, corner radii, imperfections in the chord and brace members along with gaps between brace members. To err on the conservative side, the weld geometry in FEA simulations are modelled with a smaller foot print and do not reflect the larger weld dimensions of the actual joint.

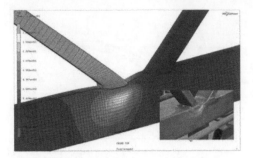

Figure 12. K1 Joint residual deformation.

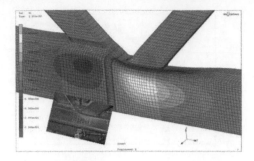

Figure 13. K3 Joint deformation.

5.1 *Observed vs simulated failure modes*

Design resistance calculated using CIDECT guide-lines predicts chord face or chord wall buckling as the mode of failure in K1, K2, K3 and K4 joints. The observed and FEA simulated first mode of fail-ure is chord face plastification followed by chord wall buckling and correlates well with those obtained using CIDECT equations. Deformation plots exhibiting per-manent deformation in the chord face and chord wall at the completion of tests for joints K1, K3 are shown in Figures 12 and 13. The only observed fracture mode in the tests was the punching shear fracture in K4 joint test (inset in Figure 14). With a gap of 29 mm, the distance between braces is nearly closed following welding. The crack observed was in the heat affected zone of the tensile brace and this mode was not suc-cessfully predicted in simulations; FEA shear stress values being less than those corresponding to material coupon test results. The probable cause for the lack of prediction may be due to residual stresses in the joint after welding/forming.

5.2 *Observed vs simulated load deformation response*

Load – deformation curves comparing measured and FEA predictions representing slender and equal width K joints at lvdt locations between braces at gaps are provided in Figures 15 to 17. The local indentation at load capacity of these joints with a partial safety fac-tor of 1.1 varies between 2.5% b_0 to 6.5% b_0 and is demonstrated in Figure 17. Chord face plastification is the first mode of failure and at 3% deformation, sim-ulations are conservative, predicted loads being lower and with a largest difference of 18% of test results. The FEA ultimate loads are determined by a decrease in applied loads similar to those observed in tests. Pre-dicted ultimate loads are greater and within 2% to 25% over test results. Load results from FEA at 3% defor-mation range from $1.1P_y$ to $1.8P_y$. The experimental ultimate load capacities of the joints are 1 to 2 times the loads predicted at 3% deformation limits.

5.3 *Material variations*

In all four K joint tests, the bounds for joint capac-ities for possible material variations discussed in

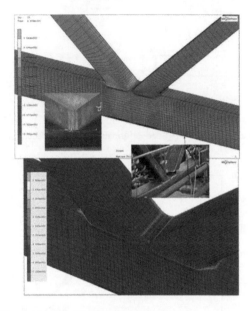

Figure 14. K4: Principal and shear stresses.

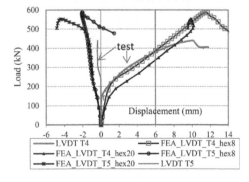

Figure 15. K1: Load deformation response.

Section 3.4 were carried out. These include 95% LCL in ultimate stress values and elastoplastic properties in material input data. The ratio of loads at 3% defor-mation results to unfactored CIDECT strength are compared against experimental results in Table 4. The largest gap between current CIDECT equations and at notional 3% deformation limits of test /FEA results is observed in joint K2. The limiting material strains or

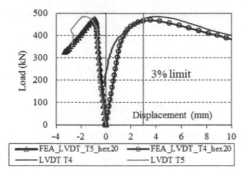

Figure 16.   K3: Load deformation response.

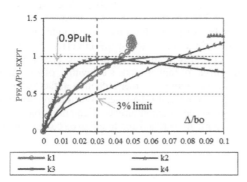

Figure 17.   K Joints summary: Load as a function of ultimate load vs deformation as a function of chord width.

Table 5.   Material stress-strain variations in K joints.

Engineering fu/fy	True fy/fu	Ultimate Strain	Ns/NCIDECT			
			K1	K2	K3	K4
1.13	450/ 635	0.17	0.9	0.6	1.1	1.2
1.1	450/550	0.17	0.8	0.5	1.1	1.2
1	435/450	0.17	0.8	0.5	1	1.1
1.4	350/495	0.17	0.8	0.4	1.4	1.1
Test/CIDECT			1.1	0.7	1.1	1.3

buckling in the chord face gap or evidence of cracking observed in the parametric FEA study occurs well beyond the 3% deformation limits in this joint. Conservative FEA predictions estimate loads of 487 kN for 95% LCL material characteristics defined by engineering f_y/f_u of 1.1. This equates to deformation limits of 5.5% b_0 and N_s/N_{CIDECT} of 0.9 corresponding to tab K2 of 0.5 in Table 5. Similar increases in N_s/N_{CIDECT} values are also observed in K1 joint. The results from these parametric studies confirm that additional reductions to f_y as $0.9f_u$ are not required up to at least engineering f_u/f_y of 1.13.

5.4   Statistical variation in joint capacities

In order to determine the deviations in load capacities for each of the joints, controlled changes to the FEA models K1 to K4 were incorporated varying one parameter at any point of time. These include

Table 6.   Statistical variations in K joint tests.

Test ID	3% Deformation				
	Test	CIDECT	MEAN FEA	COV FEA	RATIO
K1	380	355	334	2.6	0.94
K2	397	592	315	3.73	0.53
K3	474	141	453	2.9	3.21
K4	440	274	435	4.2	1.59

$$Ratio = \frac{Mean(Pfea, 3\%)}{Ppred, CIDECT}$$

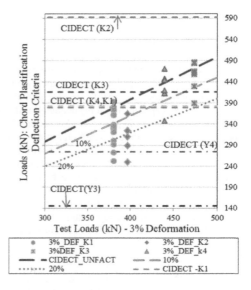

Figure 18.   Influence of boundary conditions, material property variations, perfect/imperfect geometry and modelling parameters on 3% deformation estimates.

interchanging pinned and free supports at chord ends; stiffness changes to the lateral restraints at the ends of braces (representing sinking lateral brace); imperfect vs perfect geometry; effects of numerical modelling parameters (hex8 vs hex20 elements); material property variations. A statistical compilation of these results as against test and unfactored CIDECT values is provided in Table 6.

All four joint studies have the largest scatter in results for inadequate chord axial/lateral restraints. At least one end of the chord needs to have adequate axial stiffness. Further, lateral restraints at the brace/chord ends are required in order for the joint capacities derived using CIDECT equations to be adequate.

In Figure 18, FEA capacities obtained from various parametric iterations for joints K1, K2, K3 and K4 are plotted against experimental test results. These are compared against predicted CIDECT unfactored joint capacities as K or Y joints. Capacities 10% and 20% lower than calculated CIDECT capacities are also provided to compare against calculated CIDECT resistance.

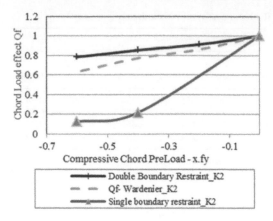

Figure 19. K2: Chord preload.

Joints K3 and K4 with β equal to 1 are required by CIDECT equations to be designed as Y joints. Both simulations and test results suggest that there is enough reserve capacity and it is adequate for these joints to be designed as K joints. Higher chord slenderness ratio has the lowest safety margins at 3% deformation limits. For K2 (smaller brace width and higher chord slenderness), both experimental and FEA derived 3% deformation values are at least 33% (18%) less than CIDECT unfactored (factored) joint capacity. Assuming loads with deformation at 5.5% b_0 is more representative of limiting material strain; this will result in an increase in loads of 25% or more.

5.5 Effect of axial loads on joint capacities

The detrimental(/positive) effects of compressive (/tensile) prechord load on joint capacities for joints K1 to K4 were considered. Figure 19 shows the results for K2. The analyses confirm that the current Q_f factors are generally adequate. Beyond a prechord compressive load of 0.4f_y, the current Q_f factors for joints K3 and K4 were found to be non-conservative.

6 CONCLUSIONS

Modes of failure observed in N joint tests are fracture at the brace and are captured well in both location and magnitude by the Lemaitre damage models. The predicted onset of fracture in N joints is within 11% of experimental results whereas the predicted final failure loads are within 13%. Post fracture response (exception N5) is correctly predicted.

In a fail-safe (fracture allowed) approach to N joint tests, for material stress strain curves with a 95% probability in f_u (f_u of 495 MPa; $f_u/f_y = 1.1$), no additional factor of 0.9 is required. FEA results correlate with CIDECT derived capacities for material with f_u of 520MPa. The only exception was N5 with slenderness limits outside of current CIDECT guidelines. CIDECT values in this joint are significantly higher than both test and simulated results and are close to ultimate failure loads.

Deformation governs the first failure mode of K joint tests and for widely accepted 3% deformation limits, results from simulations were conservative being smaller than test results and with a maximum difference of 18% of test results. The ultimate failure in simulations was larger and varied from 2% to 25% of test results.

Accurate representation of boundary conditions, particularly for K joints is required; inadequate lateral/longitudinal constraints at truss connection ends can reduce the capacity of joints by about 20%. The effect of the lateral/longitudinal constraints in a Warren truss is being studied in global/local truss analyses and will be reported separately.

K joint simulations with stress strain curves representing 95% probability in f_u, with an extension in 3% notional deformation limit for slender joints indicate that the additional factor of 0.9 in current CIDECT joint capacities is not required. Elasto plastic stress strain curves of higher grade steel with f_y of 435 MPa and $f_u/f_y = 1$ require further reduction in factors.

The current CIDECT reduction factors to take into account the detrimental effects of compression loads in the chord are verified and found to be conservative.

REFERENCES

IIW. 2009. Static design procedure for welded hollow section joints – Recommendations. 3rd edition, International Institute of Welding, Sub-commission XV-E, Annual Assembly, Singapore, IIW Doc. XV-1329-09.

CIGWELD. 2008. Welding consumables pocket guide.

Becque J. and Wilkinson, T., 2012, Experimental investigation of the static capacity of grade C450 RHS T and X truss joints, Tubular Structures XIV, pp. 177–184, Gardner (Ed), Balkema.

Becque, J., Wilkinson, T. and Syam, A. 2011. Experimental investigation of X and T truss connections in C450 cold-formed rectangular hollow sections, CIDECT report 5BV.

Lemaitre, J. and Chaboche, J.L. 1990. Mechanics of solid materials. Cambridge University Press, 1990.

MSC.Marc 2013, 2104. MSC.Software Corporation.

Packer, J.A., Wardenier, J., Zhao, X.-L., van der Vegte, G.J., Kurobane, Y. 2009. Design guide for rectangular hollow section (RHS) joints under predominantly static loading. Second edition, Construction with hollow steel sections, No. 3, Verlag TUV Rheinland, Köln, Germany.

Mohan, M. and Wilkinson, T. 2012 Fea of T and X joints in grade C450 steel, Tubular Structures XIV, pp. 185–194, Gardner (Ed), Balkema.

Mohan, M., Wilkinson, T. and Syam, A. 2013. Connections in high strength cold-formed rectangular hollow sections; part2 – FEA, CIDECT project report, 5BV 1/13.

Mohan, M., Wilkinson, T. and Syam, A. 2014. Connections in high strength cold-formed rectangular hollow sections; part3- FEA, CIDECT project report, 5BV 1/13.

Yao, E.Z., Wilkinson, T., Xu, Fadhli, M.A., Chen, L and Syam, A. 2013. Connections in high strength cold-formed rectangular hollow sections, CIDECT report 5BV 1/13.

Yao, E.Z. and, Wilkinson, T. 2015. Experimental investigation of the static capacity of grade C450 RHS K and N truss joints, Tubular Structures XV, Balkema.

Tubular Structures XV – Batista, Vellasco & Lima (eds)
© 2015 Taylor & Francis Group, London, ISBN 978-1-138-02837-1

Axially loaded K joints made of thin-walled rectangular hollow sections

O. Fleischer
CCTH – Center of Competence for Tubes and Hollow Sections, Karlsruhe, Germany

R. Puthli & T. Ummenhofer
Karlsruhe Institute of Technology (KIT), Karlsruhe, Germany

J. Wardenier
Delft University of Technology, Delft, The Netherlands

ABSTRACT: The application range of EN 1993-1-8 (EC3) limits the ratio of the section width or height to the wall-thickness of rectangular hollow sections to b/t or h/t $\leq$ 35. For sections subjected to compressive loads the additional limitation of the cross-section class 1 or 2 may even give a further reduction of these ratios. In the product standards EN 10210-2 for hot-finished hollow sections and especially EN 10219-2 for cold-finished hollow sections, there are many sections that are outside the application range. Therefore, the use of these sections is only permissible in many European countries by experimental and/or numerical verifications and expert advice, followed by acceptance for individual cases. The use of such sections in steel structures is therefore normally avoided. In this paper a design concept is presented, which allows joints with slender chords to be considered in design. Additionally, the design concept allows for joints with small gaps.

1 INTRODUCTION

In addition to the limitation of the maximum ratio of the section width or height to the wall-thickness to b/t or h/t $\leq$ 35, EN 1993-1-8 (EC3) provides a minimum and maximum gap size for design of K and N joints. In addition to the minimum gap size $g_{w,min} = t_1 + t_2$ possible for welding, which is mandatory for fillet welds to form a proper connection, it is necessary to satisfy a minimum $g_{min} = 0.5 \cdot (1 - b_i/b_0) \cdot b_0$ and a maximum gap size $g_{max} = 1.5 \cdot (1 - b_i/b_0) \cdot b_0$. This is because the design concept is based on nearly similar stiffness of the gap and the part of the chord flange situated between the braces and the chord side-wall. It should be mentioned, that EC3 uses the width ratio β for the determination of minimum and maximum gap sizes. Since the width ratio of K and N joints takes account as the brace heights $\beta = (b_1 + b_2 + h_1 + h_2)/(4 \cdot b_0)$ this is strictly correct only for symmetrical joints with square brace sections. However, the part of the chord flange situated between the braces and the chord side-wall is $0.5 \cdot (1 - b_i/b_0) \cdot b_0$, therefore strictly, the ratio of the brace to chord width b_i/b_0 should be used instead of the ratio β. The maximum gap size is additionally limited by the maximum permittable joint eccentricity $e/h_0 \leq 0.25$ in EC3.

To extend the application range of EC3, experimental and numerical investigations on K joints with a chord slenderness of $30 \leq 2\gamma \leq 55$ have been carried out at the Research Center for Steel, Timber and Masonry of the Karlsruhe Institute of Technology

(KIT). Additionally, the joints are provided with gap sizes down to the minimum gap size what allows welding, $g_{e,min} = 4 \cdot t_0$. The small brace wall-thickness t_i, which results from the small wall-thickness t_0 of the slender chords and the limitation of the wall-thickness ratio to $\tau = t_i/t_0 = 1$ requires a minimum gap size $g_{e,min}$ that is higher than the minimum gap size due to welding $g_{w,min}$ in EC3 to allow a proper welding with fillet welds, as normally used for joints with a brace thickness of $t_i \leq 8$ mm.

With the results of the experimental investigations, the design models of EC3 for chord face failure, punching shear failure and brace failure of K and N joints with gap are used as the basis and adjusted accordingly in the extended parameter range by modifying their parameters e.g. the effective lengths for punching shear or brace failure.

In addition to experimental investigations, extensive numerical parameter studies have been carried out. These are based on a numerical model which is validated with results of experimental investigations (Fleischer 2014). In these parameter studies, the influence of the gap size on the joint resistance is analyzed and applied to the extended application range, with a gap function in the semi-empirical design equation of EC3 for chord face failure of K and N joints.

Based on the experimental, numerical and analytical investigations, an approach is presented, which also allows the design of joints with a high chord slenderness $35 < 2\gamma \leq 55$ and with gap sizes between $4 \cdot t_0 \leq g \leq g_{max}$.

2 BASIC CONSIDERATIONS

2.1 Ultimate joint resistances of experimental and numerical investigations

Chord face failure cannot be observed visually in the experimental and the numerical investigations. But since in the more recent design methods the ultimate joint resistances $r_e = N_{i,u}$ are determined according to the 3% deformation limit (Lu et al. 1994), which limits the indentation of the brace into the chord flange to 3% of the chord width b_0, this method is also used here. However, punching shear failure, brace failure and chord web failure can clearly be observed in the experimental investigations and the failure loads $r_e = N_{i,max}$ are used as ultimate joint resistances in the evaluations of these failure modes.

The numerical model does not take cracking into account, so that punching shear failure or brace failure cannot be identified as governing failure modes. Therefore, only chord face failure is investigated in the numerical investigations. However, since a limitation of the joint resistance due to a load peak at a smaller indentation than 3% of the chord width b_0 does not occur for the slender sections considered here, the joint resistances $r_n = N_{i,u}$ are always determined with the 3% deformation limit (Lu et al. 1994) and these other failure modes are not considered to govern for the parametric range investigated.

2.2 Statistical evaluation of analytical resistances

Except for chord web failure, which is evaluated together with chord face failure (Section 4.2), EC3 includes design equations for K and N joints with gap for the directly observed failure modes (punching shear and brace failure) as well as for chord face failure determined with the 3% deformation limit. Based on the standardized procedure of EN 1990 (Fleischer & Puthli 2006), joint resistances $r_t = N_{i,Rd}$ are calculated with the respective design equations of EC3 using measured dimensions and material properties and compared with experimental joint resistances r_e. The numerical joint resistances r_n are compared with resistances calculated with the respective design equations of EC3 as well but using nominal dimensions and characteristic yield strengths.

By comparing calculated joint resistances r_t and experimental r_e or numerical r_n joint resistances the deviations of the mean b and the coefficients of variation (CoV) of the error terms V_δ are determined. By taking account of the coefficients of variation of the basic variables V_{xi} and the CoV of their combination V_{rt}, the CoV of the design model V_r and of the number of tests n the coefficient ξ_c is obtained, which reduces the particular analytical model r_t to its characteristic level $r_k = \xi_c \cdot r_t$. The design equations $r_d = \xi_c / \gamma_m \cdot r_t$ are then obtained by using a partial safety factor γ_m. For chord face failure, a partial safety factor of $\gamma_m = 1.1$ is used. Since for punching shear and brace failure cracking due to low deformation capacity limits the joint resistance an increased partial safety

factor of $\gamma_m = 1.25$ is applied for the determination of the design equations of these failure modes.

Also included in the coefficient ξ_c is the recalculation from mean (measured) to nominal values, which allows for the use of the nominal dimensions and the characteristic yield strength in design.

Applying the reduction to design level globally to the analytical models, $r_d = \xi_c / \gamma_m \cdot r_t$ results in modification of the design equations of EC3, which include the reduction to design level in parameters as e.g. the effective widths for punching shear or for brace failure. Following this approach, parameters are determined for the extended parameter range in such a way that the global reductions of the semi-empirical models are obtained as $\xi_c / \gamma_m = 1.00$.

2.3 Influence of yield strength on analytical resistances

The lower ductility of high strength steels and the limitation of the joint deformations requires for joints made of high strength steels a reduction of the calculated joint resistances r_t if steels with a yield strength of $f_y > 355$ N/mm² are used (Liu & Wardenier 1993, Noordhoek & Verheul 1998, Fleischer et al. 2009).

In the evaluation of the experimental results, the design resistances r_t calculated with measured dimensions and material properties for chord face failure and for punching shear failure are reduced to $0.8 \cdot r_t$ if the measured yield strength of the chord is 460 N/mm² $\leq f_{y0} \leq 700$ N/mm² and to $0.9 \cdot r_t$ if 355 N/mm² $< f_{y0} \leq 460$ N/mm². For the calculated resistances of brace failure, the measured yield strength of the failing brace is the governing factor for the reduction. Thus, in the evaluation of the experimental results the design resistances r_t of brace failure calculated with measured dimensions and material properties are reduced to $0.8 \cdot r_t$ if the measured yield strength of the failing brace is 460 N/mm² $\leq f_{yi} \leq 700$ N/mm² and to $0.9 \cdot r_t$ if 355 N/mm² $< f_{yi} \leq 460$ N/mm².

Joint resistances based on nominal dimensions and characteristic yield strengths, e.g. the mean joint resistances r_m for chord face failure (Wardenier 1982) or joint resistances r_t used in the evaluation of the numerical investigations, do not need to be reduced for joints made of steels with characteristic yield strengths of $f_y \leq 355$ N/mm². If high strength steels are used and the characteristic yield strength is 355 N/mm² $< f_y \leq 460$ N/mm² the joint resistances are reduced to $0.9 \cdot r_t$ (EC3). Although no higher strength steels are investigated it is mentioned for completeness that joints resistances of joints made of steels with a characteristic yield strengths of 460 N/mm² $< f_y \leq 700$ N/mm² have to be reduced according to EN 1993-1-12 to $0.8 \cdot r_t$.

2.4 Influence of max. chord load on analytical resistances

In the statical system of Figure 1 for the experimental and the numerical investigations, the symmetrical

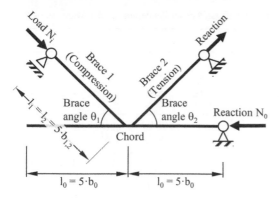

Figure 1. Statical system for the experimental and numerical investigations.

joint geometries with ignoring joint eccentricities, the compression loaded brace N_i gives a chord subjected to compressive loading $N_0 = 2 \cdot N_i \cdot \cos \theta_i$ (Fig. 1).

Depending on the ratio of the chord stress σ_0 and the yield strength of the chord $n = \sigma_0/f_{y0}$ a reduction k_n of the joint resistance for chord face failure could occur (Eq. 1).

$$N_{i,Rd} = \frac{8.9 k_n f_{y0} t_0^2 \sqrt{\gamma}}{\sin \theta_i} \left(\frac{b_i + h_i}{2b_0} \right) \qquad (1)$$

with the width b_0, wall-thickness t_0, slenderness 2γ and yield strength f_{y0} of the chord; the width b_i and height h_i and angle θ_i of the braces and the chord stress reduction factor k_n.

In the determination of the joint resistances for chord face failure $r_t = N_{i,Rd}$ the reduction k_n according to the regulations of EC3 is considered.

3 NUMERICAL INVESTIGATIONS

3.1 General

For the numerical investigations, IDEAS is used as pre-processor and ABAQUS as solver for the Finite Element (FE) analyses.

Initial studies carried out at Karlsruhe (Sarada et al. 2002, Fleischer & Puthli 2003) show that the FE analyses lead to good results if shell elements with 4 or 8 nodes and reduced integration are used. Since using 3-dimensional elements (solid elements) require a high mesh refinement due to the limited number of integration points in the thickness direction, 4 noded shell elements with reduced integration (S4R) having 9 (Simpson) integration points in thickness direction are preferred to solid elements, in order to save the CPU-time and disk space. The elements are arranged in the mid-surfaces of the sections. The welds are modelled as fillet welds, also using shell elements. Symmetry conditions are used to reduce the degrees of freedom (DOF), so that only half of the joint is modeled. Details of the discretisation of the joints and the welds can be found in Fleischer & Puthli (2003).

For the analysis, an axial compressive force N_i is prescribed at the end node of the compression brace (Fig. 1). The force is applied in increments. Modified Newton-Raphson technique with Riks arc length method is used in ABAQUS for the iterative-incremental solution procedure.

3.2 Parameter studies

The joints in the parameter study are modelled using the nominal dimensions. The outer and inner corner radii r_0 and r_i of the sections are taken from EN 10219-2, which depend on the wall-thickness.

Up to the yield strength f_y a linear elastic material ($E = 210,000 \, N/mm^2$; Poisson's ratio $\nu = 0.30$) is assumed. Subsequently, linear isotropic hardening in the plastic range up to the ultimate stress f_u at uniform strain A_{gt} is considered. The material properties are taken as S355H as in EN 10219-1 ($f_y = 355 \, N/mm^2$; $f_u = 470 \, N/mm^2$). Since no information about the uniform strain is available, A_{gt} is assumed to be 10% as a safe estimate, based on experience. Geometrical non-linearity (large displacements) is also taken into account.

Only symmetrical K joints ($b_1 = b_2 = b_i$, $h_1 = h_2 = h_i$ and $\theta_1 = \theta_2 = \theta_i$) with gap made of cold formed RHS according to EN 10219 with brace angles of $\theta_i = 45°$ are investigated. The outer dimensions of the chord sections are always $b_0 \times h_0 = 300 \times 200$ mm. For different chord slendernesses 2γ the wall-thickness t_0 is varied accordingly. While the height of the brace sections is always taken as $h_i = 100$ mm, the brace widths b_i vary.

To investigate the influence of the gap size on the joint resistance in detail, numerical models with gap sizes from $g_{e,min} = 4 \cdot t_0$ to the max. gap size for K joints g_{max} (increment $\Delta g = 2 \cdot t_0$) are analyzed for joints with a chord slenderness of $2\gamma = 30$, 40 and 50 and width ratios $\beta = 0.32$, 0.42 and 0.50 ($b_i/b_0 = 0.30$, 0.50 and 0.67). Additionally, models with a varying chord slenderness $35 \leq 2\gamma \leq 55$ (increment $\Delta 2\gamma = 5$) and width ratios $0.42 \leq \beta \leq 0.52$ ($0.50 \leq b_i/b_0 \leq 0.70$, increment $\Delta b_i/b_0 = 0.05$) are numerically analyzed for the smallest investigated gap size $g_{e,min} = 4 \cdot t_0$, the smallest g_{min} and the largest g_{max} permitted gap sizes of EC3 (Tab. 1).

3.3 Evaluation

The numerical joint resistances r_n for chord face failure are evaluated with reference to resistances $r_t^* = N_{i,Rd}^*$ calculated with nominal dimensions and characteristic yield strengths in the EC3 design equation of K and N joints with gap. Additionally, a gap function $f(g')$ is included to take account of the influence of the non-dimensional gap size $g' = g/t_0$ on the joint resistance (Eq. 2).

$$N_{i,Rd}^* = \frac{8.9 k_n f_{y0} t_0^2 \sqrt{\gamma}}{\sin \theta_i} \left(\frac{b_i + h_i}{2b_0} \right) f(g') \qquad (2)$$

459

Table 1. Min. g'_{min} and max. g'_{max} non-dimensional gap sizes for the investigated chord slenderness 2γ and width ratios β.

	2γ											
	30		35		40		45		50		55	
β	$\dfrac{g_{min}}{t_0}$	$\dfrac{g_{max}}{t_0}$	$\dfrac{g_{min}}{t_0}$	$\dfrac{g_{max}}{t_0}$	$\dfrac{g_{min}}{t_0}$	$\dfrac{g_{max}}{t_0}$	$\dfrac{g_{min}}{t_0}$	$\dfrac{g_{max}}{t_0}$	$\dfrac{g_{min}}{t_0}$	$\dfrac{g_{max}}{t_0}$	$\dfrac{g_{min}}{t_0}$	g_{max}
0.32	4	..16*	–		4	..20*	–		4	..26*	–	
0.42	4	..16*	–		4	..20*	–		4	..26*	–	
0.50	4	..14*	–		4	..20*	–		4	..24*	–	
0.42	–	–	8.8	26.3	–	–	11.3	33.8	12.5	37.5	13.8	41.3
0.44	–	–	7.9	23.6	9.0	27.0	10.1	30.4	11.3	33.8	12.4	37.1
0.47	–	–	7.0	21.0	8.0	24.0	9.0	27.0	10.0	30.0	11.0	33.0
0.49	–	–	6.1	18.4	7.0	21.0	7.9	23.6	8.8	26.3	9.6	28.9
0.52	–	–	5.3	15.8	6.0	18.0	6.8	20.3	7.5	22.5	–**	–**

Note: g_{max} is the smaller of the max. allowed gap size in EC3 and the gap size resulting from the max. Permitted eccentricity $e/h_0 \leq 0.25$.
*additional joints between $g_{e,min}/t_0 = 4$ and g_{max}/t_0 (increment $\Delta_{g/t0} = 2$) and joints with $e/h_0 = 0$: $g = 5.86$ ($2\gamma = 30$), 7.81 ($2\gamma = 40$) and 9.76 ($2\gamma = 50$).
**not analyzed, since $b_i/t_i > 35$ or cross section class of brace > 2.

with the width b_0, wall-thickness t_0, slenderness 2γ and yield strength f_{y0} of the chord; the width b_i and height h_i and angle θ_i of the braces; the reduction factor k_n and the gap function $f(g')$.

For adopting the design equation in EC3, it is necessary that the gap function does not tend towards zero for large gap sizes. Additionally, the gap function should be simple to use. Due to these basic limitations, the basic shape of the gap function $f(g')$ is specified as in Equation 3.

$$f(g') = \frac{r_n}{r_t} = A_0 + \frac{A_1}{1+g'} \qquad (3)$$

with the numerical resistances r_n, the calculated resistances r_t, the non-dimensional gap size g' and the coefficients A_0 and A_1.

The gap function $f(g')$ is obtained by the comparison of the numerical and the calculated joint resistances of EC3 for chord face failure (Eq. 3). The coefficients A_0 and A_1 of the gap functions $f(g')$ are determined by non-linear regression analyses.

In Figure 2 to Figure 4 the gap functions obtained for width ratios of $\beta = 0.32$, 0.42 and 0.50 and for a chord slenderness of $2\gamma = 40$ and 50 are given typically. The remaining functions from the numerical investigations can be found in Fleischer (2014).

The gap functions are influenced by both the width ratio β and the chord slenderness 2γ. Including these influences would result in a complex function what would hamper practical use.

Therefore, the gap function is determined as a safe estimate based on the min. of the numerical to design resistances r_n/r_t as a lower bound envelope. For the determination of the gap function only joints with a chord slenderness of $2\gamma > 35$ and joints having eccentricities within the permitted range – $0.55 \leq e/h_0 \leq 0.25$ in EC3 are taken into account (Fig. 5).

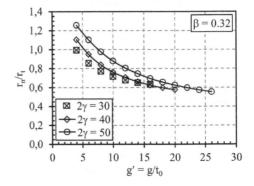

Figure 2. Gap functions for a chord slenderness $2\gamma = 40$ and $2\gamma = 50$ of joints with a width ratio of $\beta = 0.32$.

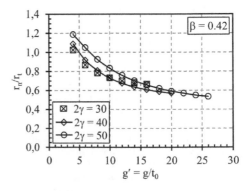

Figure 3. Gap functions for a chord slenderness $2\gamma = 40$ and $2\gamma = 50$ of joints with a width ratio of $\beta = 0.42$.

By reducing the coefficient A_0 in Equation 3 in such a way that the global reduction of the analytical model (Eq. 2) is obtained as $\xi_c/\gamma_m = 1.00$, the gap function for design is determined by using a coefficient

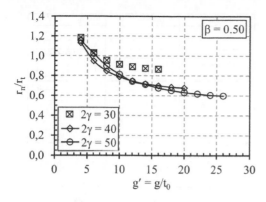

Figure 4. Gap functions for a chord slenderness $2\gamma = 40$ and $2\gamma = 50$ of joints with a width ratio of $\beta = 0.50$.

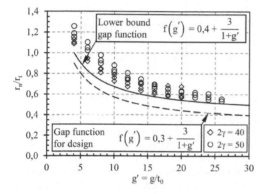

Figure 5. Lower bound gap function and gap function for design based on min. r_n/r_t-ratios.

of $A_0 = 0.32 \approx 0.3$ and thus avoid changing the main design equation in EC3 for K and N joints with gap.

4 EXPERIMENTAL INVESTIGATIONS

4.1 General

In the framework of the European research project "Design rules for cold-formed hollow sections" (Salmi et al. 2006) experimental investigations on 47 axially loaded, symmetrical K joints with gap have been carried out at the Research Center for Steel, Timber & Masonry, KIT. The specimens are made of cold-finished RHS according to EN10219 in grade S355 J2H (40 tests) as well as in grade S460 MLH (7 tests). Details on the joint geometry, the test set-up and the test results are given in Fleischer & Puthli (2003, 2006).

4.2 Chord face failure

Since the deformation of the chord webs occurs simultaneously with the chord flange, the failure loads $N_{i,max}$ where chord web deformation (chord web failure) is observed are in good agreement with the deviation of

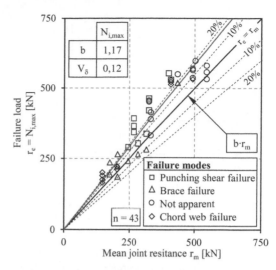

Figure 6. Comparison of the mean joint resistances r_m and the failure loads of the tests for the observed failure modes.

the mean joint resistances $b \cdot r_m$ obtained from the comparison of the mean joint resistances r_m and the failure load $N_{i,max}$ of the tests (Fig. 6).

The mean joint resistances r_m in Figure 6 are calculated with the design equation for chord face failure (Eq. 1) using nominal dimensions and characteristic yield strengths and an additional factor of 10.9/8.9 (Wardenier 1982).

As for K and N joints with geometries in the application range of EC3, separate evaluations for chord face failure and for failure where chord web deformation is observed are not required and a common evaluation of the results of both failures based on joint resistances determined with the 3 % deformation criterion (Lu et al. 1994) is carried out.

The evaluation is based on resistances r_t calculated with measured dimensions and yield strengths in the design equation for chord face failure of EC3 (Eq. 1). Without including the influence of the gap size, a global reduction factor of $\xi_c/\gamma_m = 0.76$ has to be used in Equation 1 (Fig. 7) for the determination of design resistances $N_{i,Rd}$ for the investigated joints with parameters outside the application range of EC3. Since the evaluation is based on ultimate loads determined with the 3% deformation limit and not on failure loads and due to including joints with a high chord slenderness and small gap sizes, the reduction ξ_c/γ_m is as expected smaller than 1.

To improve this reduction the design value of a gap function for the min. gap size $g_{e,min}$ is determined with a linear regression analysis to $f(g' = 4) = 0.82$ (Fig. 7).

Since the global reduction of EC3 for chord face failure (Eq. 1) without taking account of the influence of the gap size is $\xi_c/\gamma_m = 0.76$ (Fig. 7) and the design value of the gap function based on numerical investigation for the smallest investigated gap size is $f(g' = 4) = 0.9 > 0.76$ (Fig. 5) the gap function has to be limited to $f(g') \leq 0.76$. This limitation is determined

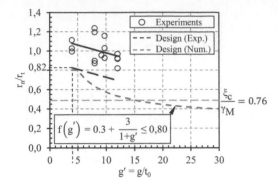

Figure 7. Linear gap function and its design value for the smallest gap size $g_{e,min} = 4 \cdot t_0$ in the experimental investigations.

with the statistical evaluation of a relatively small number of test results and may therefore be conservative. Therefore, the limitation is improved by considering a linear gap function in the experimental results for chord face failure. Here, a value for the smallest investigated gap size of $f(g') \leq 0.82 \approx 0.80$ is determined (Eq. 4).

$$f(g') = 0.3 + \frac{3}{1+g'} \leq 0.8 \qquad (4)$$

with the non-dimensional gap size g'.

However, for large gaps sizes the joint resistances based on the results of the experimental investigations are considerably higher than those based on the results of the numerical investigations. Large gap sizes are therefore not critical for the limitation of the gap function $f(g')$ for design (Fig. 7).

Therefore, design resistances $N_{i,Rd}^*$ for chord face failure for the investigated joints are determined with the design equation of EC3 for chord face failure including the influence of the gap size (Eq. 2). The gap function for design is given in Equation 4.

4.3 Punching shear and brace failure

For K joints with small gap sizes $4 \cdot t_0 \leq g \leq g_{min}$ the increased stiffness in the gap results in reduced effective lengths for both punching shear and brace failure. Therefore, the evaluation of the experimentally determined failure loads $N_{i,max}$ require large reductions of the respective design equations of EC3. For punching shear failure, a global reduction of $\xi_c/\gamma_M = 0.44$ and a global reduction of $\xi_c/\gamma_M = 0.51$ for brace failure are determined. It is assumed that the respective analytical models can be used in the extended parameter range as well, but the effective lengths have to be modified accordingly. Based on experimental observations, the following method is proposed for the determination of the reduced effective length for punching shear failure (Fig. 8).

The brace width b_i near the gap is assumed to be effective completely for all gap sizes

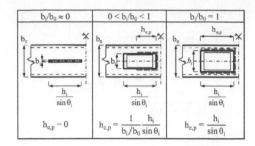

Figure 8. Eff. height $h_{e,p}$ for punching shear failure.

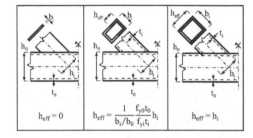

Figure 9. Eff. height h_{eff} for brace failure.

$4 \cdot t_0 \leq g \leq g_{min}$ and for chord with a slenderness of $35 < 2\gamma = b_0/t_0 \leq 55$. However, the brace walls are effective partially (Wardenier et al. 2010) depending upon the ratio of the chord to brace width b_i/b_0. For very small width ratios $b_i/b_0 \approx 0$ (only theoretically), the brace walls hardly contribute at the load transfer, for large width ratios $b_i/b_0 \approx 1$ the brace walls are effective completely. Although punching shear failure is no longer possible for joints with $b_i/b_0 > 1 - 1/\gamma$ – the width ratio β stated in EC3 is incorrect for K and N joints and b_i/b_0 should be used instead, as stated in the introduction – it is assumed due to simplification that for a width ratio of $b_i/b_0 = 1$ the brace walls contribute completely. For intermediate width ratios $0 < b_i/b_0 < 1$ the effective height for punching shear failure $h_{e,p}$ is obtained by linear regression analysis (Fig. 8).

Taking account of the shear resistance of the chord flange and the load component perpendicular to the chord flange the design resistance for punching shear failure $N_{i,Rd}$ is calculated using Equation 5.

$$N_{i,Rd} = \frac{t_0 f_{y0}}{\sqrt{3}} \left(b_i + \frac{2 f_{hep}}{b_i/b_0} \frac{h_i}{\sin \theta_i} \right) \qquad (5)$$

with the width b_0, wall-thickness t_0 and yield strength f_{y0} of the chord; the width and angle θ_i of the brace b_i and the reduction factor f_{hep}.

The reduction factor $f_{hep} = 0.62 \approx 0.6$ for the effective height for punching shear failure $h_{e,p}$ is determined in such a way that no further reduction of Equation 5 to the design level is required and $\xi_c/\gamma_m = 1.00$.

In principle the same procedure is used for the determination of the effective height h_{eff}. However, the the wall-thickness of the chord t_0 and of the braces t_i and their yield strengths f_{y0} and f_{yi} are considered additionally. Assuming ineffective brace walls for small width ratios $b_i/b_0 \approx 0$ (only theoretically) and fully effective brace wall for high width ratios $b_i/b_0 \approx 1$ the effective height for brace failure h_{eff} for intermediate width ratios $0 < b_i/b_0 < 1$ is obtained by linear regression analysis (Fig. 9).

For the effective brace cross-section the complete brace width b_i near the gap is considered additionally. Since the complete brace load N_i affects brace failure the design load is obtained from Equation 6.

$$N_{i,Rd} = t_i f_{yi} \left(b_i + \frac{2 f_{heff}}{b_i/b_0} \frac{f_{y0}}{f_{yi}} \frac{t_0}{t_i} h_i \right) \qquad (6)$$

with the width b_i, wall-thickness t_i and yield strength f_{yi} of the brace; the width b_0, wall-thickness t_0 and yield strength f_{y0} of the chord and the reduction factor f_{heff}.

The reduction factor $f_{heff} = 0.59 \approx 0.6$ of the effective height h_{eff} is also determined in such a way that no further reduction of Equation 6 to the design level is required ($\xi_c/\gamma_m = 1.00$).

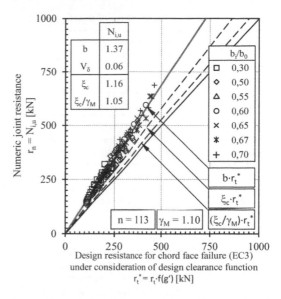

Figure 10. Calculated design resistance r_t^* vs. numerical joint resistances r_n for chord face failure.

4.4 Chord shear failure

For K joints with chords having smaller heights than widths $h_0 < b_0$ chord shear failure may govern, especially for joints with larger gap sizes. Since the specimen of the experimental investigations have only small gap sizes $g \leq 12 \cdot t_0$, shear failure does not occur in the experimental investigations. Therefore, no statistical evaluation can be carried out and the design equations of EC3 are used for the determination of the joint resistance $N_{i,Rd}$ for chord shear failure (Eq. 7) and for the remaining chord resistance for axial loadings $N_{0,Rd}$ (Eq. 8).

$$N_{i,Rd} = f_{y0} A_v / \left(\sqrt{3} \sin \theta_i \right) \qquad (7)$$

$$N_{0,Rd} = f_{y0} \left[(A_0 - A_v) + A_v \sqrt{1 - \left(\frac{V_{Ed}}{V_{pl,Rd}} \right)^2} \right] \qquad (8)$$

with the yield strength f_{y0}; the shear area A_v, the cross-section area A_0 and the shear capacity $V_{pl,Rd}$ of the chord; the shear load V_{Ed} and the brace angle θ_i.

The effective shear area A_v (Eq. 9) is thereby determined based on plastic analyses (Wardenier 1982).

$$A_v = (2h_0 + \alpha b_0) t_0 \qquad (9)$$

where:

$$\alpha = \sqrt{1 / \left(1 + \frac{4g^2}{3t_0^2} \right)}$$

with the height h_0, width b_0 and wall-thickness t_0 of the chord, the coefficient α and the gap size g.

5 GENERAL EVALUATION

With the presented design equations for chord face failure (Eq. 2) using the gap function $f(g')$ given in Equation 4, punching shear failure (Eq. 5) and brace failure (Eq. 6) and taking account of chord shear failure according to EC3, the design resistances of the numerically investigated joints are calculated and compared with their numerically determined ultimate loads (Fig. 10). Joints for which punching shear or brace failure based on the modified design equations or chord shear failure in EC3 give the smallest design resistance, are not included in the evaluation for chord face failure. Therefore, the partial safety factor for chord face failure of $\gamma_m = 1.1$ (see Section 2.2) is used for the determination of the global reduction factor ξ_c/γ_m for the related design equation (Eq. 2).

Slightly conservative design resistances are determined $\xi_c/\gamma_m = 1.05$ (Fig. 10).

6 SUMMARY AND CONCLUSIONS

A design concept for K and N joints with gap made of RHS with a chord slenderness of $35 < 2\gamma \leq 55$ is presented for sections with a yield strength of $f_y \leq 460 \, N/mm^2$. Additionally small gap sizes from $g \geq 4 \cdot t_0$ are also considered, so that bending due to joint eccentricity and secondary effects may in a number of cases not have to be considered.

Also, the design of K joints with small width ratios from $\beta \geq 0.3$ is possible for joints with a high chord slenderness $35 < 2\gamma \leq 55$. Compliance of the additional limitation of EC3 $\beta \geq 0.01 + 0.01 \cdot b_0/t_0$ can be waived. However, the max. width ratio has to be limited due to the investigated parameter range to $\beta = 0.67$.

463

Table 2. Extended application range of design proposal.

Brace sections (i = 1, 2) b_i/t_i and h_i/t_i		Chord sections b_0/t_0 and h_0/t_0	Chord and brace sections h_0/b_0 and h_i/b_0 b_i/b_0^*		Joint geometry g θ_i	
Compr.	Tension					
≤ 35 and Class 1 or 2	≤ 35	> 35 and ≤ 55	≥ 0.5 and ≤ 2.0	≥ 0.30 and ≤ 0.70	$\geq 4t_0$ and g_{max}^{**}	$\geq 30°$

Note: *width ratio $0.30 \leq \beta \leq 0.67$.
**the smaller of the max. allowed gap size of EC3 and the gap size resulting from the max. permitted eccentricity $e/h_0 \leq 0.25$.

For K and N joints with gap within the extended application range (Tab. 2) design resistances for chord face failure are determined based on the design equation in EC3 for chord face failure and additionally taking account of the gap size by a gap function (Eq. 2). The gap function $f(g')$ is verified by numerical and experimental investigations (Eq. 4). The design resistances for punching shear failure (Eq. 5) and brace failure (Eq. 6) are determined with effective heights $h_{e,p}$ and h_{eff} considering the reduced effective lengths due to the increased stiffness of the gap for small gap sizes. Additionally, chord shear failure according to EC3 has also to be considered as usual for the joint (Eq. 7 and Eq. 8).

7 FUTURE WORK

Since the strain distribution in the braces has not been measured in the experimental investigation and therefore no information is available on the secondary bending moments, it has to be verified if the secondary bending moments included in the experimental and numerical investigations are applicable in all practical cases. This can be done for example, by numerical investigations of a complete lattice girder having N joints and ratios of the brace lengths to brace heights of approximately $l_i/h_i \approx 6$ (EC3), as these would be expected to give the highest secondary bending moments.

Additionally, investigations on joints having higher width ratios and extending the gap function to joints inside the application range of EC3 would be desirable.

REFERENCES

EN 10210:2006. Hot finished structural hollow sections of non-alloy and fine grain steels – Part 1: Technical delivery conditions & Part 2: Tolerances, dimensions and sectional properties. Brussels: CEN.

EN 10219:2006. Cold formed welded structural hollow sections of non-alloy and fine grain steels – Part 1: Technical delivery conditions & Part 2: Tolerances, dimensions and sectional properties. Brussels: CEN.

EN 1990:2002. Eurocode – Basis of structural design. Brussels. CEN.

EN 1993-1-8:2005. Eurocode 3: Design of steel structures – Part 1-8: Design of joints. Brussels: CEN.

EN 1993-1-12:2007/AC:2009. Eurocode 3 – Design of steel structures – Part 1-12: Additional rules for the extension of EN 1993 up to steel grades S 700. Brussels: CEN.

Fleischer, O. & Puthli, R. 2003. RHS K joints with b/t ratios and gaps not covered by Eurocode 3. In: Tubular structures X. Ed. by M.A. Jaurrieta, A. Alonso & J.A. Chica. Lisse: Balkema, S. 207–215.

Fleischer, O. & Puthli, R. 2006. Evaluation of experimental results on slender RHS K-gap joints. In: Tubular structures XI. Ed. by J.A. Packer & S. Willibald. London: Taylor & Francis, S. 229–236.

Fleischer, O. & Puthli, R. 2009. Extending existing design rules in EN 1993-1-8 (2005) for gapped RHS K-joints for maximum chord slenderness (b_0/t_0) of 35 to 50 and gap sizes g to as low as $4t_0$. In: Tubular structures XII. Ed. by X.Z. Zhao, Y.Y. Chen & Z.Y. Shen. Boca Raton, London: CRC Press, S. 293–301.

Fleischer O., Herion, S. & Puthli, R. 2009. Numerical investigations on the static behaviour of CHS X-joints made of high strength steels. In: Tubular structures XII. Ed. by X.Z. Zhao, Y.Y. Chen & Z.Y. Shen. Boca Raton, London: CRC Press, pp. 597–605.

Fleischer, O., Puthli, R. & Wardenier, J. 2010. Evaluation of numerical investigations on static behaviour of slender RHS K-gap joints. In: Tubular structures XIII. Ed. by B. Young. Boca Raton, Fla.: CRC Press, S. 75–83.

Fleischer, O. 2014. Axial beanspruchte K-Knoten aus dünnwandigen Rechteckhohlprofilen. Berichte zum Stahl- und Leichtbau, Versuchsanstalt für Stahl, Holz und Steine. Karlsruhe: KIT Scientific Publishing.

Lu, L.H., de Winkel, G.D., Yu, Y. & Wardenier, J. 1994. Deformation limit for the ultimate strength of hollow section joints. In: Tubular structures VI. Ed. by P. Grundy, A. Holgate & B. Wong. Rotterdam: Balkema, S. 341–347.

Liu D.K. & Wardenier, J. 1993. Effect of the yield strength on the static strength of uniplanar K joints in RHS. IIW Doc. XVE-04-193.

Noordhoek C. & Verheul, A. 1998. Static strength of high strength steel tubular joints. CIDECT Report 5BD-9/98. Delft: Delft University Press.

Salmi, P., Kouhi, J., Puthli, R., Herion, S., Fleischer, O., Espiga, F., Croce, P., Bayo, E., Goñi, R., Björk, T., Illvonen, R. & Suppan, W. 2006. Design rules for cold-formed structural hollow sections: Final report 21973. Luxembourg: Publications Office of the European Union.

Sarada, S., Fleischer, O. & Puthli, R. 2002. Initial Study on the Static Strength of Thin–Walled Rectangular Hollow Sections (RHS) K-Joints with Gap. In: Proceedings of the 12th ISOPE Conference. Ed. by J. Wardenier, T. Yao, R. Ayer, R.H. Knapp & J.S. Chung. S.l.: ISOPE, S. 26–33.

Wardenier, J. 1982. Hollow section joints. Delft: Delft University Press.

Wardenier, J., Packer, J.A., Zhao, X.-L. & Vegte, G.J. van der. 2010. Hollow sections in structural applications. Zoetermeer: CIDECT & Bouwen met Staal.

Tubular Structures XV – Batista, Vellasco & Lima (eds)
© 2015 Taylor & Francis Group, London, ISBN 978-1-138-02837-1

T joints with chords made of triangular hollow sections

O. Fleischer
CCTH – Center of Competence for Tubes and Hollow Sections, Karlsruhe, Germany

S. Herion & T. Ummenhofer
KIT – Karlsruhe Institute of Technology, Karlsruhe, Germany

D. Ungermann, B. Brune & P. Dissel
TU Dortmund University, Dortmund, Germany

ABSTRACT: Many investigations on the load carrying behavior of directly welded hollow section joints made of RHS have been carried out in the past. In comparison to joints having conventionally arranged brace and chord sections, increasing joint resistances are observed for joints having square hollow chord sections rotated by 45° (Bird-Beak) as well as for joints having square hollow brace sections rotated by 45° additionally (Diamond-Bird-Beak). Analyzing the failure modes of the latter joints result in the use of newly developed triangular hollow sections. By replacing the square chords by triangular hollow sections the advantages of the Bird-Beak and of the Diamond-Bird-Beak joints are combined with a minimized material consumption. But these joints offer not only advantageous bearing properties due to a high joint rigidity but also interesting aspects in an aesthetic point of view.

1   INTRODUCTION

Directly welded joints with chords made of rectangular hollow sections (RHS) rotated by 45° (BB: Bird-Beak joints, Fig. 1b) as well as joints which braces made of RHS rotated by 45° additionally (DBB: Diamond-Bird-Beak joints, Fig. 1d) offer significant higher joint resistances compared to the design resistnaces of EN 1993-1-8 $N_{1,Rd}$ of joints with conventional aligned RHS chords and braces (Fig. 1a). This has already been observed in various investigations carried out in the last two decades e.g. Ono et al. (1991, 1993 and 1994), Ishida et al. (1993), Davies et al. (1996), Owen et al. (1996).

The failure of BB (Fig. 1b) and DBB (Fig. 1d) joints made of RHS is mainly characterized by forming plastic hinges at the corners of the chords allowing its deformation (Fig. 2a). The deformation is prevented by a tie member connecting the corners (Fig. 2b). This results in the newly developed cold-formed triangular hollow section (THS, Fig. 2c), which replaces the square chord (Fig. 1c, Fig. 1e).

Joints with a chord made of THS combine the advantage of the increasing joint resistance with a reduced material consumption. So far THS are made by welding plates. In addition to the expensive fabrication, no standard procedures or recommendations for the determination of design resistances of joints with chords made of THS exist. Therefore, their use requires extensive and costly experimental and/or

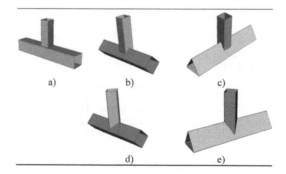

Figure 1.   T joints with varying members and alignments: a) Conventional RHS joint (upper row, left); b) BB joint made of RHS (upper row, middle); c) BB joint with THS chord and RHS brace (upper row, right); d) DBB joint made of RHS (lower row, middle); e) DBB joint with THS chord and RHS brace (lower row, right).

numerical investigations to attain the acceptance for each individual case.

For architects THS offers innovative and interesting scopes in modern design approaches. This is confirmed by several individual steel structures in which THS has already been used. Producing THS by cold-forming makes it available in significant quantities and allows a more common use.

In order to obtain a first insight in the load carrying behavior of T joints with chords made of THS 18 tests

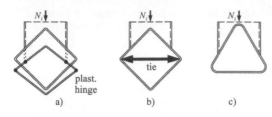

Figure 2. a) Deformed BB joint (left); b) imaginary tie member (middle); c) BB joint with THS chord (right).

Figure 4. Bridge at Cala Caldana, Menorca, Spain.

Figure 5. Civil Justice Centre, Manchester, Great Britain.

Figure 3. Roof truss of the International Airport "Suvarnabhumi", Bangkok, Thailand.

have been carried out in the framework of a German research project (Ungermann et al. 2010). Additionally, two tests with conventional T joints made of RHS are performed to investigate their different structural behavior.

2 APPLICATIONS WITH TRIANGULAR HOLLOW PROFILES

Although no design procedures for joints with chords made of THS exist THS are already used in steel structures. Checks e.g. of sufficient joint resistance are therefore carried out with experimental and/or numerical investigations. Due to no hot- or cold-formed THS are produced so far, the sections are created by welding (Fig. 7a).

The roof trusses of the International Airport "Suvarnabhumi" in Bangkok, Thailand consist of three-dimensional lattice girders. The top and the bottom chords of the girders are formed of THS to which braces made of RHS are directly welded (Fig. 3).

A bridge in Cala Galdana located at the Spanish island of Menorca is a composite structure with two parallel arcs intersected by longitudinal beams. The arcs are also made of triangular sections (Fig. 4).

The façade of the Civil Justice Centre in Manchester, England is supported by sixty metre high triangular atrium columns all suspended from the 11-storey atrium roof (Fig. 5, left). T joints transmitting the loads of the floors, cantilevering 15 m from the building's columns (Fig. 5, right).

The glazed roof of the city train station "Reinoldikirche" in Dortmund, Germany is supported

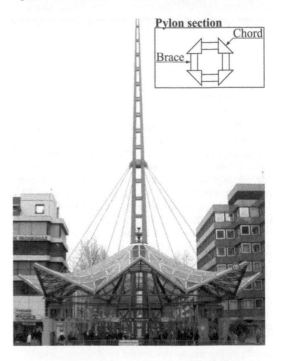

Figure 6. Pylon of city train station "Reinoldikirche", Dortmund, Germany.

by a pylon (Fig. 6). This pylon is a multi-part member, whose vertical chords are made of THS. The sections consist of three welded plates forming an isosceles triangle with RHS braces directly welded in between (see pylon section in Fig. 6).

The examples presented here show that triangular hollow sections offer interesting scopes for design

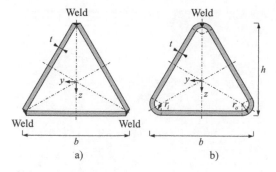

Figure 7. a) THS made of three welded plates (left); b) cold-formed THS (right).

and allow architects to design innovative and modern structures.

3 STATE OF THE ART

3.1 General

In the past the load carrying behavior of DBB T joints made of RHS has been investigated experimentally by Ono et al. (1991, 1993 and 1994). These investigations describe in detail the observed failure modes and models for the determination of the ultimate joint strength are developed by Ono et al. (1991, 1993 and 1994) and Ishida et al. (1993). Axially loaded DBB K joints made of RHS are investigated and formulae for the determination of the ultimate joint strength are developed, even for fatigue loaded K joints (Ishida et al. 1993). For BB T joints made of RHS finite element investigations have been carried out by Davies et al. (1996) and a comparison of the behavior of BB T joints made of RHS and conventional RHS and CHS systems are carried out (Owen et al. 1996).

3.2 Fabrication of THS and section properties

The newly developed cold-formed THS is an equilateral triangle with rounded corners. The basic dimensions and their denotations are given in Figure 7b. While for THS made out of welded plates three longitudinal welds are needed (Fig. 7a), for cold-formed THS only one longitudinal weld is necessary. Not only the reduced welding costs are advantageous, but also intersecting welds can be avoided for welded joints.

The outer and inner corner radii r_o and r_i correspond to the wall-thickness dependent radii given in the technical delivery standard for cold-finished hollow sections EN 10219-2 (Eq. 1). The minimum inside corner radii according to EN 1993-1-8 guarantees sufficient ductility if fully aluminium killed steels are used (Wardenier et al. 2010) and micro cracking due to bending is avoided.

$$r_o = \begin{cases} 2t & t \leq 6 \text{ mm} \\ 2.5t & 6 < t \leq 10 \text{ mm} \\ 3t & t > 10 \text{ mm} \end{cases} \quad ; r_i = r_o\text{-}t \quad (1)$$

with the wall-thickness t.

Table 1. Material properties of the band steel and section properties of the THS 160 × 6 mm based on nom. dimensions.

Measured yield strength	f_y	443 N/mm²
Measured tensile strength	f_u	546 N/mm²
Ratio of yield to tensile strength	f_y/f_u	0.81
Cross-section area	A	28.15 cm²
Shear area	A_z	11.53 cm²
Plastic shear area	$A_{pl,z}$	16.46 cm²
Moment of inertia	I_y	617.7 cm⁴
Plastic section modulus	$W_{pl,y}$	116.0 cm³

For the experimental investigations a THS is fabricated by voestalpine Krems by bending band steel in grade S355 J2 and welding the bent plate at a corner longitudinally (Fig. 7b). The THS offers a nominal width of b = 160 mm and a nominal thickness of t = 6 mm. These dimensions are also used for the notation of THS, here THS 160 × 6 mm. The measured yield and tensile strength of the bend steel and the section properties of the THS 160 × 6 mm and are given in Table 1.

Since the THS is produced by cold-forming, it is assumed that the permitted geometrical tolerances of EN 10219-1 are also applicable for the THS.

4 EXPERIMENTAL INVESTIGATIONS

4.1 General

The experimental investigations have been carried out at the Building Research Institute of the University of Dortmund under the direction of the Department for Steel Constructions.

4.2 Pilot tests

To obtain a first insight into the different load carrying behavior of conventional T joints made of RHS (Fig. 1a) and BB T joints having a THS chord and a RHS brace (Fig. 1c), two pilot tests of each have been carried out. The measured dimensions of the chord and the brace sections and the resultant width ratios $\beta = b_i/b_0$ as well as the wall-thickness ratios $\tau = t_i/t_0$ are given in Table 2.

To avoid bending of the chord ($L_0 = 1200$ mm) the T joints have been elastically supported (Fig. 8). Compressive loads are applied at the top of the braces.

The measured deflections of the load-introductions are given in Figure 9. For the dimensions and the parameters of specimens A1, A2, B1 and B2 in Figure 9 see Table 2.

The load-deflections of joints having a RHS chord are affected by membrane effects for both investigated joints. Therefore, the failure loads are attained only at deflections far above the limitation of the 3% deformation criterion of Lu et al. (1994), which limits the indentation of the brace into the chord to 3% of the chord width b_0. Due to the large indentations of

Table 2. Measured dimensions of the chord and braces and resulting geometrical parameters of the pilot tests.

No.	Chord Width, Height b_0, h_0 [mm]	Chord Wall-thickness t_0 [mm]	Brace Width, Height b_0, h_0 [mm]	Brace Wall-thickness t_0 [mm]	Parameter Width ratio β	Parameter Wall thickness ratio τ	Result Joint resist. $N_{i,u}$ [kN]
A1	202	5.6	100	4.0	0.50	0.7	97*
A2	202	5.6	151	5.6	0.75	1.0	229*
B1	269	3.9	100	4.0	0.37	1.0	204**
B2	269	3.9	151	5.6	0.56	1.4	311**

Note: *maximum indentation of 6 mm.
**maximum indentation of 8 mm.

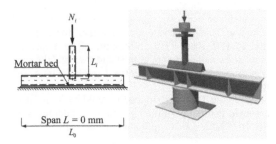

Figure 8. Schematic test set-up (left) of tests with elastic supported chord section and arragment in testing device.

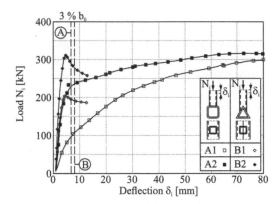

Figure 9. Load deflection curves for conventional T joints made of RHS and BB joint with THS chords and RHS braces.

the brace into the chord face chord face failure is the governing failure (Fig. 10).

However, the resistances of both joints with a THS chord are attained before the applied load has to be limited by the 3% deformation criterion. Since these joints offer a higher initial stiffness higher joint resistance than for joints with a RHS chord are obtained.

4.3 Test programme

BB T joints with a brace alignment of 0° (Fig. 1c) and DBB T joints with a brace alignment of 45° (Fig. 1e) were experimentally investigated. The chords of

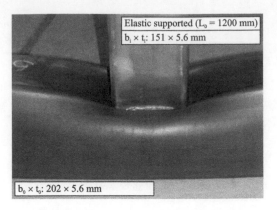

Figure 10. Chord face failure of T joint made of RHS.

Table 3. Geometry of test specimens and failure loads of tests.

Brace dimension (SHS) $b_i \times t_i$ [mm]	Brace alignment** [°]	Chord length L_0 [mm]	Width ratio β	Wall-thickness ratio τ	Brace slenderness d_i/t_i	Maximum axial load $N_{i,u}$ [kN]
90 × 6.3*	0	800	0.56	1.05	14.3	571
90 × 6.3*	45	800	0.56	1.05	14.3	625
60 × 5*	0	800	0.38	0.83	6.0	341
60 × 5*	45	800	0.38	0.83	6.0	461
120 × 7.1	0	800	0.75	1.18	16.9	366
90 × 6.3	45	800	0.56	1.05	14.3	353
60 × 5	0	800	0.38	0.83	12.0	254
60 × 5	45	800	0.38	0.83	12.0	292
120 × 7.1	0	1600	0.75	1.18	16.9	164
90 × 6.3	0	1600	0.56	1.05	14.3	165
90 × 6.3	45	1600	0.56	1.05	14.3	165
60 × 5	0	1600	0.38	0.83	12.0	163
60 × 5	45	1600	0.38	0.83	12.0	156
120 × 7.1	0	2400	0.75	1.18	16.9	105
90 × 6.3	0	2400	0.56	1.05	14.3	107
90 × 6.3	45	2400	0.56	1.05	14.3	104
60 × 5	0	2400	0.38	0.83	12.0	103
60 × 5	45	2400	0.38	0.83	12.0	106

Note: *Test with elastically supported chords.
**0° refers to BB, 45° to DBB T joints.

the specimens were made of the THS 160×60 mm as already described in Section 3.2.

Since the test specimens should cover geometrical parameters in a practical range, three RHS in grade S355 J2H with a nominal width of $b_i = 120$, 90 and 60 mm and a nominal wall-thicknesses of $t_i = 7.1$, 6.3 and 5 mm were used for the braces. The geometry of the specimens and the determined failure loads are summarized in Table 3.

To investigate the influence of bending in the chord on the joint resistance (M-N interaction) tests with chord lengths of $L_0 = 800$, 1600 and 2400 mm were carried out. To exclude influences of the load introduction on the joint the lengths of the braces were taken to $L_i = 5b_i$.

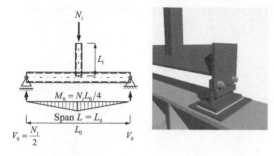

Figure 11.   Test set-up – Simply supported specimen.

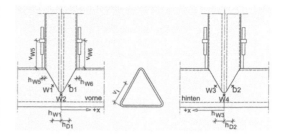

Figure 12.   Position of the measuring points.

4.4   Test set-up

All tests were performed in a hydraulic press (Schenk-Hydropuls), which offers a maximum capacity of 630 kN and a maximum stroke of the hydraulic cylinder of ±125 mm. The braces of the joints were loaded by compressive loads, applied in the centroidal axis of the brace using a spherical bearing to prevent end bending.

For the determination of the axial joint resistance, T joints with a $L_0 = 800$ mm long chord were used only. By placing a mortar bed between the chord and the support beam the chords were elastically supported and bending of the chord is eliminated. The test set-up for a T joint with an elastic supported chord is sketched in Figure 8.

For the determination of joint resistances influenced by bending of the chord sections the ends of the chord sections were simply supported (Fig. 11).

The very stiff supporting beam accommodates the bearing loads with almost no deformations. The use of the beam allows short phases of reconstruction, as for the specimens with different chord lengths only the lower supports has to be moved.

Local deformations were recorded using linear voltage displacement transducers (LVDT). By mounting the LVDT to the supporting beam the displacement measurement is insensitive to external influences. Additionally, the indentations of the load cell were not measured concurrently. The use of draw-wire displacement sensors allows for a conversion to absolute displacement coordinates. In Figure 12 the positions of the LVDT, indicated by the points "W", and the positions of the strain gauges D1 and D2 are given. At each position two strain gauges were applied. For the exact positions of the equipment see Ungermann et al. (2010).

Additionally, the cylinder stroke was measured and the applied load was recorded by a load cell.

4.5   Evaluation of test results

The joint resistances $N_{i,u}$ are compared with the load capacity of the chord. Local effects and stability phenomena are not taken into account for the

determination of the plastic bending M_{pl} capacity (Eq. 2) and the shear capacity V_{pl} (Eq. 3) of the THS.

$$M_{pl} = W_{pl,y} f_u = 63.33 \text{ kNm} \qquad (2)$$

$$V_{pl} = A_{pl,z} f_u / \sqrt{3} = 518.9 \text{ kN} \qquad (3)$$

with the plastic section modulus $W_{pl,y}$, the plastic shear area $A_{pl,z}$ and the tensile strength f_u.

In the tests no clearly defined load level occurs, which indicate first yielding. Therefore, the plastic bending M_{pl} (Eq. 2) and the shear V_{pl} (Eq. 3) capacities are calculated using the tensile strength f_u of the THS (Table 1). The comparison of the THS failure loads and the ultimate loads $N_{i,u}$ of the tests is considerably more accurate and thus more meaningful.

The chords of the elastic supported specimens were not loaded by bending, hence the ultimate load $N_{i,u}$ equates to twice of the THS shear capacity $N_{i,u} = 2 \cdot V_{pl}$.

The chords of the simple supported specimens were loaded by both, shear and bending (Fig. 11). Thus the reduction of the bending capacity M_{pl} due to shear (M-V interaction) has to be considered if the shear force V_0 is approaching the shear resistance V_{pl}. Following EN 1993-1-1 the reduced bending capacity M_0 of the single span centrally loaded static system is determined (Eq. 4).

$$M_0 = \left[1 - \left(N_i / V_{pl} - 1 \right)^2 \right] M_{pl} \qquad (4)$$

$$N_{i,u} = 4 M_{pl} / L \qquad (5)$$

with the loading of the brace N_i and the shear V_{pl} and plastic bending capacity M_{pl} of the THS and the span $L = L_0$.

However, for the simple supported tests with a span of $L \geq 800$ mm the M-V interaction can always be neglected and the ultimate loads $N_{i,u}$ of the simple supported specimens are calculated with Equation 5.

4.6   Results

In Figure 13 the experimentally determined ultimate loads $N_{i,u}$ are given in reference to the spans L. It should be mentioned that tests with elastic supported chords have a chord length of $L_0 = 800$ but their span is $L = 0$ mm. Additionally, the predicted ultimate loads

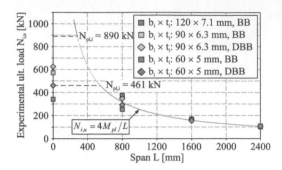

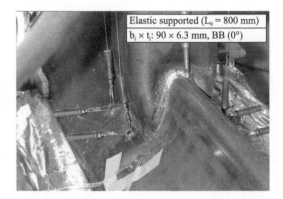

Figure 13. Comparison of calculated failure loads of the THS and the experimentally determined failure loads.

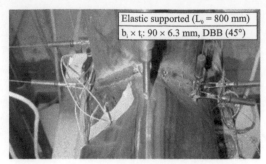

Figure 15. Local failure DBB T-joint.

ultimate loads are predicted. Because the ratio of the brace width to the span b_i/L increases with decreasing spans L, this effect gets distinctive for the smallest investigated span of $L = 800$ mm.

For the elastic (span $L = 0$ mm) and the simple supported specimens with a span of $L = 800$ mm significantly increased joint resistances were observed for DBB joints in comparison to BB joints. For DBB joints the increasing deformations of the THS webs result in a load transfer by the intersection at the corner of the THS and therefore the THS is mainly loaded by membrane stresses (Fig. 15).

For BB joints the load transfer takes place at the intersection of the brace and the THS webs, thus local bending occurs (Fig. 14).

Figure 14. Local failure BB T-joint.

of the chord (Eq. 5) and of the braces $N_{pl,i} = f_y \cdot A_i$ are given in Figure 13.

Especially for tests with elastic supported specimens local stress concentrations at the joint result in a local failure. This failure is characterised by local bending of the THS webs resulting in large deformations of the THS webs and of the braces (Fig. 14). Since local effects limit the joint resistances of the elastic supported specimens, the experimental ultimate loads are considerably lower than the calculated chord resistances (Fig. 13). Only for the DBB joint with a width ratio of $\beta = 0.38$ yielding of the brace was the governing failure mode.

The simple supported specimens fail due to exceeding the plastic bending capacity M_{pl} of the chord, local effects at the joints were not observed. This results in a good agreement of the experimental and the predicted ultimate loads, especially for tests with spans of 1600 and 2400 mm (Fig. 13).

Simple supported specimens with a width ratio $\beta = 0.38$ gave smaller ultimate loads than predicted, although this overestimation was not as pronounced as for elastic supported specimens. However, tests with higher width ratios $\beta = 0.56$ and $\beta = 0.75$ gave even higher predicted ultimate loads than the experimentally determined ultimate loads. The brace allows transferring stresses in the compressed zone and therefore contributes to the plastic section modulus. Since this effect is not considered in the calculations, lower

5 ARCHITECTURAL ASPECTS

In addition to the examination of the structural and the deformation behavior of joints a new construction method of lattice girders using THS sections has been investigated. The architectural investigations have been carried by the chair of Building Theory of TU Dortmund under the direction of Prof. Nöbel.

With the help of architectural and constructive analyzes various types of optimized girders with THS were designed and subsequently used as examples in the architectural design of a market hall. This included the architectural and constructional design of the structural systems, the representation of the spatial impression of the hall as well as the development of joint details to the adjacent components.

When designing a lattice girder made of traditional hollow sections, it is possible to use the basic shape for the chord and the braces and to adapt the section dimensions according to their loading. However, the basic shape of the used sections doesn't change usually. For girder made of THS only, joints were created having an unsatisfactory design due to their complexity. Therefore, only girders with THS chords and RHS braces are investigated here. The studies focused on lattice girders with diagonal braces only and on girders with vertical and diagonal braces. Additionally, a Vierendeel truss was investigated. In this paper, the results of the lattice girder with diagonal braces (Fig. 16) made of RHS connected at an angle of about

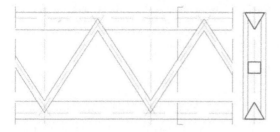

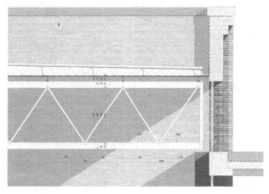

Figure 16. Truss for halls with large spans and minor loading.

Figure 17. Visualization of the hall interior.

Figure 18. Detailing of the truss.

60° to the top and bottom chords made of THS are presented exemplarily. For more details it is referred to Ungermann et al. (2010) and Ummenhofer et al. (2011).

The visible faces of THS chords appear to be larger than for RHS chords having equivalent load capacities. Additionally, the braces may be performed in smaller dimensions due to the increased joint rigidities. This results in an emphasis on the horizontal chords of the girder as in an appealing design (Fig. 17).

For the roof structure open sections are used on which a timber cladding is placed visible from within. The roof structure is bolted to the top chord by plates adapted to the geometry of the THS (Fig. 18).

The girders can be used in rooms with high creative demands on the quality of the architecture. This applies for the investigated medium spans basically loaded by dead weight, wind and snow. Among others, these girders are suitable for single storey warehouse buildings and for all hall structures like sport, exhibition and industrial halls. Architecturally, they are feasible for structures with longer spans e.g. exhibition halls or stadiums or even for railway stations and bridges.

6 CONCLUSIONS

Ultimate joint resistances of compressive loaded T joints with THS chords and RHS braces were determined experimentally. The influence of chord bending on the joint resistance was considered by testing specimens with elastic supported chords and by 3-point bending tests with varying spans. The ultimate loads of the tests were compared with predicated failure loads, which are determined based on the plastic resistances of the THS.

For spans of $L_0 = 1600$ and $2400\,mm$ the experimental joint resistances are in a good agreement to the calculated resistances. The governing failure of these tests was always plastification of the THS, local joint failure was not observed for specimens with these chord lengths. For spans of $L = 800\,mm$ deviations of the experimental ultimate loads to the calculated resistances were observed.

For the elastic supported tests with a span of $L = 0\,mm$ significantly smaller resistances than predicted were observed. Since for these tests local failure of the joints was governing, the shear resistance of the THS couldn't be reached.

For joints made of the same chord and brace sections but having different brace alignments, DBB T joints give higher joint resistances than BB T joints.

Although the position of the longitudinal weld of the THS is varied in reference to the braces, no influence on the ultimate load was detected in the experimental investigations.

The results presented here can be seen as a starting point for further research work. On the one hand, the new developed triangular hollow sections offer advantageous load carrying properties of joints with THS chords due to the high joint rigidity. On the other hand THS could offer the opportunity for interesting aesthetic aspects. Also from an economic point of view THS are an interesting alternative solution.

The presented architectural and structural studies can be used to present the new THS components to builders and architects. They offer architects a first indication of possible applications and of the integration of the components into an overall design with high demands on the architectural quality.

Due to the promising results of the research project (Ummenhofer et al. 2011) a subsequent research project is planned.

ACKNOWLEDGEMENTS

The authors want to thank the German Federation of Industrial Research Associations "Otto von Guericke" (AiF), the German Federal Ministry for Economic Affairs and Energy (BMWA) for their financial support, the Research Association for Steel Application (FOSTA) for their assistance and voestalpine Krems for providing the materials and producing the test specimens.

The paper is written in remembrance of Prof. Nöebel who sadly dies in 2013, shortly after the finalization of the research project.

REFERENCES

EN 1993-1-1:2005: Eurocode 3: Design of steel structures – Part 1-1: General rules and rules for buildings. Brussels: CEN.

EN 1993-1-8:2010. Eurocode 3: Design of steel structures – Part 1.8: Design of joints, German Version of EN 1993-1-8:2005 + AC:2009. Brussels: CEN.

EN 10219:2006. Cold formed welded structural hollow sections of non-alloy and fine grain steels – Part 1: Technical delivery conditions & Part 2: Tolerances, dimensions and sectional properties. Brussels: CEN.

Lu, L.H., G.D. de Winkel, Y. Yu, J. Wardenier. 1994. Deformation limit for the ultimate strength of hollow section joints. In: Tubular structures VI. Ed. by P. Grundy, A. Holgate and B. Wong. Rotterdam: Balkema, pp. 341–347.

Ono, T., M. Ivata, K. Ispida. 1991. An experimental study of joints of new truss system using rectangular hollow sections. In Tubular Structures IV. Ed. by J. Wardenier. and D. Dutta. Delft: Delft University Press, pp. 344–353.

Ono, T., M. Iwata, K. Ispida. 1993. Local failure of joints of new truss systems using rectangular Hollow Sections subjected to in-plane bending moment. In: Tubular Structures V. Ed. by M.G. Coutie, and G. Davies. Nottingham: E & F.N. Spon, pp. 503–5100.

Ono, T. 1994. Local failure of joints of new truss systems using rectangular Hollow Sections subjected to out-of-plane bending moment. In: Tubular Structures VI Ed. by P. Grundy, A. Holgate and B. Wong. Rotterdam: Balkema, pp. 441–448.

Ishida, K., T. Ono, M. Iwata. 1993. Ultimate strength formula for joints of a new truss system using Rectangular Hollow Sections. In: Tubular Structures V. Ed. by M.G. Coutie, and G. Davies. Nottingham: E & F.N. Spon, pp. 511–518.

Owen, J.S.; G. Davies, R.B. Kelly. 1996. A comparison of the behavior of RHS bird beak T-joints with normal RHS and CHS systems. In: Tubular Structures VII. Ed. by Farkas and Jámmai. Rotterdam: Balkema, pp. 173–180.

Davies, G., J.S. Owen, R.B. Kelly. 1996. Bird-beak T-joints in square hollow sections: a finite element investigation. In: Proceedings of the 6th ISOPE Conference. Los Angeles: ISOPE pp. 22–27.

Ungermann, D., B. Brune, P. Dissel, W.A. Noebel, O. Schmidt, S. Herion, O. Fleischer. 2010. Bestimmung der Tragfähigkeit von geschweißten T-Knoten aus (kaltgeformten) Hohlprofilen mit dreiecksförmigen Querschnitten. Schlussbericht AIF Forschungsvorhaben 15379 N, Stahlbau Verlagsgesellschaft mbH, Düsseldorf.

Ummenhofer, T., S. Herion, D. Ungermann, B. Brune, P. Dissel, O. Fleischer. 2011. T-Knoten mit Gurtstäben aus dreieckigen Hohlprofilen. In: Stahlbau Volume 80, Issue 7. Berlin: Ernst & Sohn, pp. 492–501.

Wardenier, J., J.A. Packer, X.-L. Zhao, G.J. van der Vegte. 2010. Hollow sections in structural applications. Zoetermeer: CIDECT, Bouwen met Staal.

Tubular Structures XV – Batista, Vellasco & Lima (eds)
© 2015 Taylor & Francis Group, London, ISBN 978-1-138-02837-1

Reduction of fillet weld sizes

O. Fleischer & S. Herion

CCTH – Center of Competence for Tubes and Hollow Sections, Karlsruhe, Germany

ABSTRACT: EN 1993-1-8 permits the use of steels with yield strengths $f_y \leq 460\,N/mm^2$. Since the weld sizes for hollow section joints given in the 1st edition of CIDECT Design Guide 3 (Packer et al. 1992) can only be used for yield strengths $f_y \leq 355\,N/mm^2$, they are adapted in the actual revision of CIDECT Design Guide 1 (Wardenier et al. 2008). However, the welds sizes of welds which automatically resist the design loads $N_{i,Rd}$ of the joints increase considerably with the yield strength. Therefore, design charts are developed in the framework of a CIDECT project (Fleischer et al. 2012), which allow for an efficient weld design of hollow sections joints. EN 1993-1-12 even permits the use of high strength steels with $f_y \leq 700\,N/mm^2$ but no information is given in CIDECT Design Guide 1. Therefore, a proposal for the weld sizes of joints made of steels with $f_y \leq 700\,N/mm^2$ is given additionally.

1 INTRODUCTION

In the past the design of welds for hollow section joints could easily be carried out by using the recommended weld sizes given in the 1st edition of CIDECT Design Guide 3 (Packer et al. 1992). For joints made of steels with a yield strength of $f_y \leq 355\,N/mm^2$ the recommended weld size based on EN 1993-1-8 is approximately equal to the brace thickness $a \approx t_i$. Since this weld size considers the axial load capacity of the brace $N_{i,pl}$ it resists the design loads $N_{i,Rd}$ of the joints. So, the welds have not to be checked in static analyses. Since EN 1993-1-8 permits the use of higher strength steels with yield strengths up to $f_y \leq 460\,N/mm^2$ for hollow section joints, the weld sizes are adapted accordingly. Therefore, the actual revision of CIDECT Design Guide 1 (Wardenier et al. 2008) includes recommendations for weld sizes (pre-qualified welds) of joints made of higher strength steels with yield strength up to $f_y \leq 460\,N/mm^2$. However, the welds sizes of welds which automatically resists the design loads $N_{i,Rd}$ of the joints increase considerably with the yield strength. Large weld sizes result in high welding and high straightening costs after welding. A reduction of the weld sizes proposed in CIDECT Design Guide 1 (Wardenier et al. 2008) can reduce production costs of hollow section constructions and would therefore be desirable.

Based on the simplified method of EN 1993-1-8 and the axial load capacity of the braces, the actual weld lengths are considered to reduce the weld sizes. But since the calculation of the weld sizes is hampered due to high calculation efforts to obtain the actual weld lengths, design charts help the engineer to design the welds quick and efficiently. With these charts the design of welds can be carried out as easy as the use of the pre-qualified weld sizes given in Design Guide 1 (Wardenier et al. 2008).

Furthermore, EN 1993-1-12 even permits the use of high strength steels with yield strengths of $460\,N/mm^2 \leq f_y \leq 700\,N/mm^2$. Since CIDECT Design Guide 1 (Wardenier et al. 2008) gives no recommendations with regard to the weld sizes of joints made of such high strength steels; a proposal is given to be able to do an efficient weld design for joints made of high strength steels with yield strengths $460\,N/mm^2 \leq f_y \leq 700\,N/mm^2$ as well.

The results presented in this paper are based on a CIDECT research project (Fleischer et al. 2012).

2 STATE OF THE ART

2.1 *Directional method of EN 1993-1-8*

The directional method of EN 1993-1-8 divides the loads transferred by a unit length of a weld into components parallel $\tau_{\parallel}$ and perpendicular $\sigma_{\parallel}$ to the longitudinal axis of the weld and normal $\sigma_{\perp}$ and perpendicular $\tau_{\perp}$ to the effective throat area of the weld (Figure 1).

For the determination of the effective weld area A_w (Eq. 1) the fillet weld is considered to be concentrated in the weld root.

$$A_w = l_{eff}\, a \qquad (1)$$

with the throat thickness a and the eff. length of the weld l_{eff}.

With the effective weld area A_w (Eq. 1) and the stress components, which are assumed to be distributed

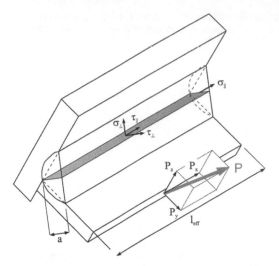

Figure 1.   Stresses in the effective throat area of a fillet weld.

constantly over the weld cross section (Figure 1) the equivalent stress σ_v is given by Eq. 2.

$$\sigma_v = \sqrt{\sigma_\perp^2 + 3\left(\tau_\perp^2 + \tau_\parallel^2\right)} \qquad (2)$$

with the components parallel $\tau_\parallel$ to the longitudinal axis and the components normal $\sigma_\perp$ and perpendicular $\tau_\perp$ to the eff. throat area of the fillet weld.

The weld capacity is sufficient if the following two criteria (Eq. 3 and Eq. 4) are fulfilled:

$$\sigma_v \leq \frac{f_u}{\gamma_{M2}\,\beta_w} \qquad (3)$$

$$\sigma_\perp \leq 0.9\,\frac{f_u}{\gamma_{M2}} \qquad (4)$$

with the tensile strength of the weaker section f_u; the correlation coefficient β_w and the partial safety factor of sections failed by cracking under tension loading γ_{M2}.

To prevent the weld from cracking due to normal stresses, the stress component normal to the effective throat area of the weld $\sigma_\perp$ is limited to 90 % of the tensile strength of the weaker section f_u (Eq. 4).

The coefficient of correlation β_w obtained from experimental work take account for a overmatching weld deposit for steels with yield strengths $f_y \leq$ 355 N/mm^2. Depending on the weaker section of the connected members, the values of β_w are given in EN 1993-1-8, Table 4.1 (Table 1).

The coefficient of correlation β_w are not fixed, but can be adopted in the National Appendices (NA) of the European member states. For example, German National Appendix DIN EN 1993-1-8/NA reduces the correlation factor for steels with a yield strength of $f_y = 420$ N/mm^2 to $\beta_w = 0.88$, for steels with a yield strength $f_y = 460$ N/mm^2 to $\beta_w = 0.85$, what gives considerably smaller weld sizes for joints made of these steel grades.

Table 1.   Correlation coefficients β_w for fillet welds acc. to EN 1993-1-8, Table 4.1.

Standards and steel grades			Correlation coefficient
EN 10025 (Sheets)	EN 10210-1	EN 10219-1	β_w
S235/S235 W	S235 H	S235 H	0.80
S275	S275 H	S275 H	0.85
S275 N/NL	S275 NH/NLH	S275 NH/NLH	
S275 M/ML		S275 MH/MLH	
S355/S355 W	S355 H	S355 H	0.90
S355 N/NL	S355 NH/NLH	S355 NH/NLH	
S355 M/ML		S355 MH/MLH	
S420 N/NL	[1]	S420 MH/MLH	1.0 (0.88[1])
S420 M/ML			
S460 N/NL	S460 NH/NLH	S460 NH/NLH	1.0 (0.85[1])
S460 M/ML		S460 MH/MLH	
S460 Q/QL/QL1			

Note: [1]S420 NH/NLH included in EN10210-1 but no correlation coefficient given in EN 1993-1-8, Tab. 4.1.
[2]DIN EN 1993-1-8/NA categories JR, J0, J2 and K2 not listed separately.

2.2   Simplified method of EN 1993-1-8

The load capacity of fillet welds is assumed to be sufficient, if at every location along the weld, the resultant load per unit length of the effective fillet weld area $F_{w,Ed}$ is smaller than its design resistance per unit length $F_{w,Rd}$. Independently of the orientation of the load acting on the effective fillet weld area A_w, the load capacity of the weld per unit length $F_{w,Rd}$ can be determined with Eq. 5.

$$F_{w,Rd} = f_{vw,d}\,a \quad \text{where} \quad f_{vw,d} = \frac{f_u/\sqrt{3}}{\beta_w\,\gamma_{M2}} \qquad (5)$$

with the design shear resistance $f_{vw,d}$ and throat thickness a of the weld; the tensile strength of the weaker section f_u; the correlation coefficient β_w and the partial safety factor for sections failed by cracking under tension loading γ_{M2}.

For the determination of the design shear resistance $f_{vw,d}$ the coefficient of correlation β_w has to be used (Eq. 5). This coefficient is based on results of former experimental investigations and is given in EN 1993-1-8 for various steel grades (Table 1).

The design load of the effective weld area per unit length $F_{w,Ed}$ is determined by the design capacity of the brace $N_{i,pl}$ and the weld length l_{eff} (Eq. 6):

$$F_{w,Ed} \leq N_{i,pl}/l_{eff} \qquad (6)$$

with the brace design capacity $N_{i,pl}$ and the weld length l_{eff}.

Since the connection has to offer sufficient deformation and rotation capacity to allow for a redistribution of stresses, the design resistance of the brace cross-section $N_{i,pl,Rd}$ has to be taken into account for

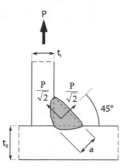

Figure 2. Distribution of the loads in a one-sided fillet weld.

the determination of the weld size, although only the load capacity of the joint $N_{i,Rd}$ has to be transmitted by the weld into the chord section. Due to the unequal stiffness distribution along the perimeter of the joint, the use of the design resistance $N_{i,Rd}$ of the joint for the determination of the weld size would require the consideration of the stress concentrations and secondary effects to avoid cracking of the weld.

Based on considerations previously made (Eq. 5 and Eq. 6) and on the partial safety factors for resistance for cross-sections $\gamma_{M0} = 1.0$ and of resistance of cross-sections in tension to fracture $\gamma_{M2} = 1.25$ (both acc. to EN 1993-1-1, 6.1) the minimum weld size is determined with Eq. 7.

$$a \geq \frac{N_{i,pl,Rd}}{f_{vw,d}} = 2.17 \, \beta_w \frac{f_{yi}}{f_u} \frac{A_i}{U_i \, K} \tag{7}$$

with the design capacity of the brace $N_{i,pl,Rd}$; the design shear resistance of the weld $f_{vw,d}$; the sectional area A_i, circumference U_i and yield strength f_{yi} of the brace; the tensile strength of the weaker section f_u; the correlation coefficient β_w and the factor K to take account for the actual length of the weld.

2.3 CIDECT design guide

The weld sizes given in the CIDECT Design Guide 1 (Wardenier et al. 2008) are based on the directional method of EN 1993-1-8 (section 2.1).

Especially for welded hollow section joints the determination of the stress components, which are necessary for the utilization of the directional method of EN 1993-1-8 is complex for brace angles deviating from $\theta_i = 90°$. Therefore, "universal" weld sizes are calculated assuming a brace angle of $\theta_i = 90°$. For the effective length of the weld l_{eff} the circumference of the brace U_i is used. Elongations of the weld due to a distortion caused by the width ratio β are neglected. With these simplifications, the section can be regarded as its sheet blank, welded on a plate (Figure 2) and the stress components can be determined easily.

Due to the unequal stiffness distribution along the intersection of the brace and the chord and the resulting stress concentrations, the weld should be able to

transmit the design capacity of the brace member P (Eq. 8).

$$P = \frac{f_{yi} \cdot t_i \cdot l}{\gamma_{M0}} \tag{8}$$

with the yield strength f_{yi} and thickness of the brace t_i; the length of the sheet blank l and the partial safety factor for resistance for cross-sections γ_{M0}.

Therefore, the stress components $\sigma_\perp$, $\tau_\perp$ ($\tau_\| = \sigma_\| = 0$) and the equivalent stress σ_v are calculated (Eq. 9 and Eq. 10):

$$\sigma_\perp = \tau_\perp = \frac{P/\sqrt{2}}{A_w} = \frac{f_{yi} \, t_i}{\sqrt{2} \, a \, \gamma_{M0}} \tag{9}$$

$$\sigma_v = \sqrt{2} \frac{f_{yi} \, t_i}{\gamma_{M0} \, a} \tag{10}$$

with the design capacity P of the sheet blank; the yield strength f_{yi} and thickness of the brace t_i; the throat thickness a and the eff. throat area A_w of the weld; the partial safety factor for resistance for cross-sections γ_{M0}; the component normal $\sigma_\perp$ and perpendicular $\tau_\perp$ to the eff. throat area of the weld and the equivalent stress σ_v.

Taking into account the two criteria (Eq. 3 and Eq. 4) the minimum throat thickness of a fillet weld is determined to:

$$a \geq \begin{cases} \sqrt{2} \, \beta_w \dfrac{f_{yi} \, \gamma_{M2}}{f_u \, \gamma_{M0}} t_i & (1) \\[2mm] \dfrac{5}{9} \sqrt{2} \dfrac{f_{yi} \, \gamma_{M2}}{f_u \, \gamma_{M0}} t_i & (2) \end{cases} \tag{11}$$

with the coefficient of correlation β_w; the yield strength f_{yi} and thickness t_i of the brace, the tensile strength of the weaker section f_u, the partial safety factors for resistance for resistance of cross-sections in tension to fracture γ_{M2} and for resistance for cross-sections γ_{M0}.

Due to the second criterion of Eq. 11 is only governing for $\beta_w \lesssim 0.79$ the weld size can always be determined with criterion (1) of Eq. 11 for the simplifications used here.

Since the design load of the brace $N_{i,Rd}$ is always smaller or equal to its load capacity $N_{i,pl,Rd}$, the weld sizes obtained by this simplification automatically resist the design loads and therefore the welds have not to be checked in static calculations.

As can be seen from Table 2 the weld size increases with the yield strength to the tensile strength ratio f_{yi}/f_u. Especially for high strength steels with a yield strength $f_y \geq 460 \, N/mm^2$ this results in large throat thicknesses for the fillet welds.

The marginal deviations of the weld sizes calculated with Eq. 11 and the weld sizes given in CIDECT Design Guide 1 (Wardenier et al. 2008) are explained by minor divergences of the material properties used for the determination of the weld sizes.

Table 2. Comparison of calculated weld sizes and weld sizes given in the CIDECT Design Guide 1 (Wardenier et al. 2008).

Steel grade	Strength				Weld size	
	Yield f_{yi} N/mm^2	Tensile f_u N/mm^2	Corr. coeff. f_{yi}/f_u	β_w	Eq. 11	CIDECT
S235 H	235[1]	360[1]	0.65	0.80	0.92·t_i	0.92·t_i
S275 H	275[1]	430[1]	0.64	0.85	0.96·t_i	0.96·t_i
S275 NH/NLH	275[2]	390[2]	0.71	0.85	1.06·t_i	0.96·t_i
S275 NH/NLH	275[3]	370[3]	0.74	0.85	1.12·t_i	0.96·t_i
S275 MH/MLH	275[3]	360[3]	0.76	0.85	1.15·t_i	0.96·t_i
S355 H	355[1]	510[1]	0.70	0.90	1.11·t_i	1.10·t_i
S355 NH/NLH	355[2]	490[2]	0.72	0.90	1.15·t_i	1.10·t_i
S355 NH/NLH	355[3]	470[3]	0.76	0.90	1.20·t_i	1.10·t_i
S355 MH/MLH	355[3]	470[3]	0.76	0.90	1.20·t_i	1.10·t_i
S420 NH/NLH	420[2]	540[2]	0.78	1.00	1.37·t_i	1.42·t_i
				(0.88[5])	(1.21·t_i)	–
S420 MH/MLH	420[3]	500[3]	0.84	1.00	1.48·t_i	1.42·t_i
				(0.88[5])	(1.31·t_i)	–
S460 NH/NLH	460[2]	560[2]	0.82	1.00	1.45·t_i	1.48·t_i
				(0.85[5])	(1.23·t_i)	–
S460 NH/NLH	460[3]	550[3]	0.84	1.00	1.48·t_i	1.48·t_i
				(0.85[5])	(1.26·t_i)	–
S460 MH/MLH	460[3]	530[3]	0.87	1.00	1.53·t_i	1.48·t_i
				(0.85[5])	(1.30·t_i)	–
S690[4]	690	770	0.90	1.00	1.58·t_i	–

Note: [1] EN 1993-1-1, Tab. 3.1, t ≤ 40 mm.
[2] EN 1993-1-1, Tab. 3.1 for hot finished hollow sections, t ≤ 40 mm.
[3] EN 1993-1-1, Tab. 3.1 for cold formed hollow sections, t ≤ 40 mm.
[4] S690 $Q/QL/QL1$ not included in CIDECT Design Guide 1 (Wardenier et al. 2008), yield f_{yi} and tensile f_u strength acc. to EN 1993-1-12, Tab. 1, t ≤ 50 mm.
[5] German National Appendix DIN EN 1993-1-8/NA categories JR, J0, J2 and K2 not listed separately.

Table 3. Eff. lengths of joints made of CHS and RHS.

Joint		$\theta_i \leq 50°$	$50° < \theta_i < 60°$	$\theta_i \geq 60°$
RHS	T, Y, X	$2h_i/\sin\theta_i + 2·b_i$	interpolate	$2h_i/\sin\theta_i + b_i$
	K/N[1]	$2h_i/\sin\theta_i + b_i$	interpolate	$2h_i/\sin\theta_i$
CHS	T, Y, K/N[1]	$2·\pi·r·K^2$ (valid for all brace angles)		

Note: [1] with gap.
[2] factor K taking account for the brace angle θ_i and the distortion due to the diameter ratio β (Eq. 16).

2.4 American code AWS

In order to be comparable to EN 1993-1-8 and since the Load and Resistance Factor Design (LRFD) method is the more modern design approach, weld design based on the LRFD method is considered here only.

According to AWS D1.1 the effective lengths l_{eff} of T, Y and K/N joints with gap made of RHS or CHS are taken into account for the design of welds (Table 3).

Due to differences in the relative flexibility of the chord loaded normal to its surface and the braces carrying membrane stresses parallel to its surface, the

Table 4. Relevant loads for the design of welds.

CHS	RHS
	Effective width failure[1,2,3] if $\beta > 0.85$[1,3]
	Plastic chord wall failure[1,2]
Brace member yield strength[1,2]	
Punching shear failure[1,2]	

Note: [1] T, Y and X joints.
[2] K/N joints with gap.
[3] omitted for joints made of square sections with equal width.

transfer of load across the weld is highly non-uniform and local yielding can be expected before the connection reaches its design load. To prevent failure of the weld and to ensure ductile behaviour of the joint, the minimum weld sizes provided in T, Y, X, and K/N joints with gap made of CHS have to be capable of transmitting the lesser of the brace member capacity and the local strength of the chord determined by punching shear failure (Table 4).

For T, Y, X made of RHS with a width ratio of $\beta > 0.85$ and for K/N joints with gap made of RHS the effective width capacity has to be considered at the design of welds, except of joints made of square hollow sections and with equal widths of the chord and the braces. Additionally, the lesser of the brace member capacity and the local strength of the chord due to web and punching shear failure has to be taken into account (Table 4).

2.5 Swedish code BSK99

In the Swedish code for steel construction BSK 99 the specified minimum strength of the weld deposit f_{euk} is considered for the determination of the design strength of the fillet weld f_{wd}. Additionally, three safety classes characterising a low, medium and high potential of danger to the building and to life and health of people are taken into account (Eq. 12).

$$f_{wd} \geq \begin{cases} \dfrac{\varphi\sqrt{f_{uk}\,f_{euk}}}{1.2\,\gamma_N} & f_u < f_{euk} \quad (1) \\[3mm] \dfrac{\varphi\,f_{euk}}{1.2\,\gamma_N} & f_u \geq f_{euk} \quad (2) \end{cases} \qquad (12)$$

with the tensile strength of base material f_{uk} and of the weld deposit f_{euk}; the factor $\varphi = 0.9$ for fillet welds and the partial safety factor $\gamma_N = 1.0$, 1.1 and 1.2 for safety classes I, II and III.

The stresses are calculated as in EN 1993-1-8 (Figure 1). The strength $F_{R\alpha}$ of the weld stressed by $\sigma_\perp$ and/or $\tau_\perp$ is determined depending on the angle between the root area and the force (Eq. 13).

$$F_{R\parallel} = 0.6\,a\,l_{eff}\,f_{wd} \quad ; \quad F_{R\alpha} = \dfrac{a\,l_{eff}\,f_{wd}}{\sqrt{2+\cos 2\alpha}} \qquad (13)$$

with the throat thickness a of the fillet weld, the eff. length of the weld l_{eff}, the angle between weld root area and load resultant α and the design strength of the weld f_{wd}.

As in EN 1993-1-8 welds subjected to both, transversal and longitudinal loading, interaction has to be considered.

3 DEVELOPMENT OF CHARTS FOR THE DETERMINATION OF THE WELD SIZE

3.1 General

To be able to do a quick and effective weld design, design charts for the determination of weld sizes are prepared. These charts are based on the simplified method of EN 1993-1-8 (section 2.2). The resulting weld sizes are limited to the weld sizes obtained from the directional method of EN 1993-1-8 for a brace angle of $\theta_i = 90°$ and omitting the effect of an increasing weld size due to the width ratio. Additionally, a set of charts for each steel grade is prepared.

The consideration of the actual weld length, taken as the intersection between the brace and the chord section, results in weld sizes which depend on the brace geometry, the brace angle and the material properties including the correlation factor β_w. The influences of the joint geometry are taken into account by the ratio of the brace sectional area to the brace perimeter A_i/U_i, the effects of an increasing weld length due to the joint geometry by the factor K. Additionally, the yield strength of the brace f_{yi}, the tensile strength of the weaker section f_u and the correlation factor β_w have to be considered for the determination of design weld sizes.

Therefore, the determination of the weld size is not possible in 2-dimensional design charts including all influencing parameters. Since 3-dimensional charts are not applicable in practice, contour plots are used for the preparation of the charts. Additionally, it is not possible to include the influence of the brace angle and the material properties, so that charts for discrete brace angles considering only one material have been developed (Fleischer et al. 2012).

3.2 Axially loaded joints made of CHS

The ratio of the brace cross-section to its circumference A_i/U_i what is expressed by a function of the brace slenderness $\gamma_i = d_i/(2t_i)$ is given by Eq. 14.

$$\frac{A_i}{U_i} = (1 - 2\gamma_i^{-1})t_i \qquad (14)$$

with the diameter d_i, thickness t_i and slenderness γ_i of the brace.

For axially loaded joints made of CHS the perimeter of the weld is considered to be effective completely.

Based on this, the length of the weld l_{eff} at the theoretical root position is taken as the intersection length of the brace and the chord section (Figure 3).

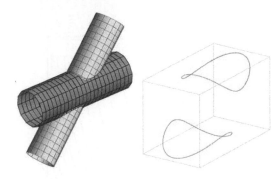

Figure 3. Exemplary intersection of brace and chord sections of an X-joint.

As can be seen from Figure 3 the length of the weld l_{eff} for joints made of CHS is influenced by the angle θ_i between the brace and the chord and the diameter ratio $\beta = d_i/d_0$ of the joint. The length of the weld l_{eff}, which is increased due to these influences in comparison to the brace perimeter U_i, is taken into account by the factor $K = l_{eff}/U_i$. acc. to AWS D1.1, Part D (Eq. 15).

$$K = \frac{\dfrac{1}{2\pi}\dfrac{1}{\sin\theta_i} + \dfrac{1}{3\pi}\dfrac{3-\beta^2}{2-\beta^2}}{+3\sqrt{\left(\dfrac{1}{2\pi}\dfrac{1}{\sin\theta_i}\right)^2 + \left(\dfrac{1}{3\pi}\dfrac{3-\beta^2}{2-\beta^2}\right)^2}} \qquad (15)$$

with the diameter ratio of the joint β and the brace angle θ_i.

For the determination of the normalized weld sizes a/t_i of joints made of CHS the geometry of the brace has to be considered additionally by the ratio of the brace cross-section to the brace perimeter A_i/U_i (Eq. 14). Thus the normalized weld sizes a/t_i of joints made of CHS depend not only on the factor K as a function of the brace angle θ_i and the diameter ratio β but also on the brace slenderness γ_i. Additionally, the yield strength of the brace f_{yi} and the tensile strength of the weaker section of the joint f_u and the correlation factor β_w influence the weld size.

Exemplarily the determination of the weld size of a joint made of CHS of grade S355 with a brace angle of $\theta_i = 45°$ is shown in Figure 4. Taking into account a diameter ratio of $\beta = 0.80$ and a brace slenderness of $\gamma_i = 10$ the weld size is determined to a $= 1.02 \cdot t_i$, what is about 92% of the weld size according to CIDECT Design Guide 1 (Wardenier et al. 2008), which gives $a = 1.10 \cdot t_i$.

If the joint is made of S460 under consideration of a correlation coefficient of $\beta_w = 0.85$ acc. to German National Annex DIN EN 1993-1-8/NA the required weld size is determined to a $= 1.13 \cdot t_i$ what is only about 76% of the weld size a $= 1.48 \cdot t_i$ given in CIDECT Design Guide 1 (Wardenier et al. 2008) for joints made of S460.

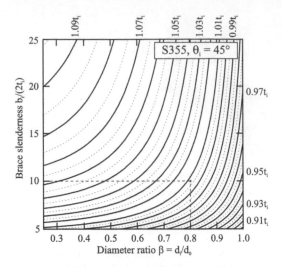

Figure 4. Exemplarily normalized weld sizes for CHS joints made of S355, brace angle $\theta_i = 45°$.

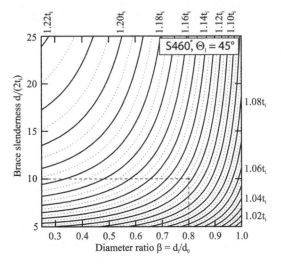

Figure 5. Exemplarily normalized weld sizes for CHS joints made of S460, brace angle $\theta_i = 45°$ acc. to German National Annex DIN EN 1993-1-8/NA.

3.3 Axially loaded joints made of CHS braces and a RHS chord

For joints with CHS braces welded on rectangular chord the width ratio $\beta = d_i/b_0$ influences the length of the weld l_{eff} only for high width ratio, so that the brace exceeds the corner radii of the chord section. Since the intersection is thereby distorted only partially the influence of the width ratio is small. Additionally, the consideration of this influence would result in different weld sizes for hot and cold finished sections. Therefore, the influence of the width ratio on the weld length of joints with CHS braces and a RHS chord is neglected. The factor K (Eq. 16) already used in EN

1993-1-8 for the determination of the design resistance for punching shear failure is used to calculate the actual length of the weld l_{eff}.

$$K = \frac{1 + \sin \theta_i}{2 \sin \theta_i} \qquad (16)$$

with the brace angle θ_i.

For the determination of the normalized weld size a/t_i, the influence of the brace geometry already used for joints made of CHS (Eq. 14) can be used and the weld sizes can be calculated by Eq. 7.

3.4 Axially loaded joints made of RHS

For joints made of RHS the length of the weld l_{eff} is influenced by the brace angle θ_i and the width ratio $\beta = b_i/b_0$. The elongation of the weld due to the width ratio β only occurs if the brace exceeds the corner radii of the chord section. Additionally, this influence only partially distorts the corner radii of the brace section to ellipses. Therefore, the influence of the width ratio is small and negligible for joints made of RHS.

For the simplified method of EN 1993-1-8 the actual length of the weld l_{eff} is considered to be effective completely. To avoid complex equations and different weld sizes for cold formed and hot finished sections, the RHS section is considered with a simplified geometry, neglecting the corner radii for the determination of the cross-section A_i, the circumference U_i and for the determination of the factor K. The ratio of the brace cross-section to its circumference A_i/U_i is then given by Eq. 17.

$$\frac{A_i}{U_i} = \frac{t_i(h_i + b_i - 2t_i)}{h_i + b_i} = \frac{\eta + 1 - \gamma_i^{-1}}{\eta + 1} t_i \qquad (17)$$

with the height h_i, the width b_i and the thickness t_i of the brace; the brace slenderness γ_i and the ratio of the brace height to the brace width η.

The factor K takes account for the elongation of the perimeter U_i due to the brace angle θ_i and the ratio of the brace height to brace width η (Eq. 18).

$$K = \frac{h_i/\sin \theta_i + b_i}{h_i + b_i} = \frac{\eta/\sin \theta_i + 1}{\eta + 1} \qquad (18)$$

with the height h_i and the width b_i of the brace; the ratio of the brace height to the brace width η and the brace angle θ_i.

For a joint made of RHS made of S355 having a brace angle of $\theta_i = 45°$, a brace slenderness of $\gamma_i = 10$ and a brace with a height to width ratio of $\eta = h_i/b_i = 1.4$ the weld size is determined to $a = 1.05 \cdot t_i$ (Figure 6). As already seen for the joint made of CHS a slight reduction of the weld size of CIDECT Design Guide 1 (Wardenier et al. 2008) of approximately 5% is determined.

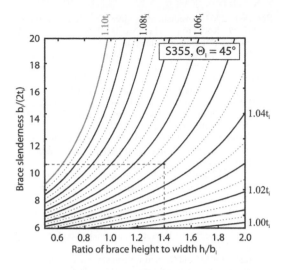

Figure 6. Exemplarily normalized weld sizes for RHS joints made of S355, brace angle $\theta_i = 45°$.

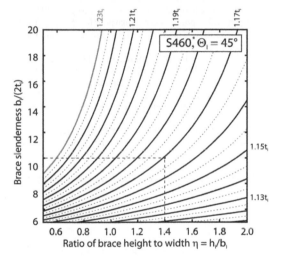

Figure 7. Exemplarily normalized weld size for CHS joints made of S460, brace angle $\theta_i = 45°$ acc. to German National Annex DIN EN 1993-1-8/NA.

If the joint is made of S460 and the correlation factor is considered acc. to German national Annex DIN EN 1993-1-8/NA $\beta_w = 0.85$, the necessary weld size is determined to $a = 1.17 \cdot t_i$ (Figure 7). This is only about 79% of the recommended weld size of CIDECT Design Guide 1 (Wardenier et al. 2008).

3.5 Axially loaded joints with longitudinal or transversal plates

For joints with longitudinal plates welded on RHS or CHS chords, neither the brace angle θ_i nor the width ratio β influences the weld length if the plate thickness is small compared to the outer dimensions of the hollow section. Neglecting the transversal welds at the

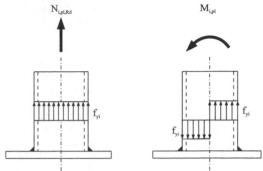

Figure 8. Plastic stress distributions for axial loading and bending.

end of the plates, the weld sizes can be determined by the directional method of EN 1993-1-8 (Eq. 19).

$$a \geq \frac{1}{2}\sqrt{2}\beta_w \frac{f_{yi}}{f_u}\frac{\gamma_{M2}}{\gamma_{M0}}t_i \qquad (19)$$

with the correlation coefficient β_w; the thickness t_i and yield strength f_{yi} of the brace; the tensile strength of the weaker section f_u; the partial safety factors of resistance for resistance of cross-sections in tension to fracture γ_{M2} and for resistance for cross-sections γ_{M0}.

Since plates are welded on both sides, half of the weld sizes given in Table 2 are sufficient.

3.6 Joints in bending

Since the weld sizes of axially loaded joints are designed for a fully plastic stress distribution, which will occur for joint loaded in bending by the plastic moment $M_{i,pl}$ as well (Figure 8) the design charts and the weld sizes given in this paper can be used for joints in bending too.

4 SUMMARY

The calculation of the throat thickness of a fillet weld with the directional method of EN 1993-1-8 requires the consideration of stress components located in the throat plane of the weld. For hollow section joints the consideration of theses stress components can get very complex, especially for brace angles different from $\theta_i = 90°$. Additionally, the consideration of the actual weld length complicates the determination of the throat thickness. Therefore, a brace angle $\theta_i = 90°$ and an actual weld length which is equal to the perimeter of the brace is assumed for the weld sizes given in the CIDECT Design Guide 1 (Wardenier et al. 2008). Due to differences in the relative flexibilities of the chord and the braces, the transfer of loads across the weld is highly non-uniform and local yielding can be expected before the connection reaches its design load. To prevent a prematurely failure of the weld and to ensure ductile behavior of the joint, the welds connecting the braces to the chord should be designed to have sufficient resistance for non-uniform stress distributions and sufficient deformation capacity to allow for the redistribution of forces. Therefore, the design of welds in EN 1993-1-8 is based on the axial load capacity of the braces. If the design of welds would be based on the load capacity of the joints, stress concentrations and secondary effects may have to be considered. These requirements result in large weld sizes, especially for high strength steels. Since large weld sizes result in high welding costs and high straightening costs after welding, a reduction of the proposed weld sizes in CIDECT Design Guide 1 (Wardenier et al. 2008) would be desirable.

Based on the directional and the simplified design methods of EN 1993-1-8 charts are prepared, which allow for the quick and efficient determination of weld sizes for various joints and steel grades (Fleischer et al. 2012). Exemplarily, charts of joints made of CHS and RHS for the steel grades S355 and S460 are given here and resulting weld sizes are compared with weld sizes of CIDECT Design Guide 1 (Wardenier et al. 2008).

Additionally, the design approaches of the AISC and AWS used in the USA and the BSK99 used in Sweden are specified. In the American specifications, the design of welds can be based on the load capacity of the joint and the actual lengths of the welds are considered. Other than in EN 1993-1-8, the strength of the weld deposit is taken into account by the American and the Swedish regulations for the design of welds. In the European regulations the strength of the weld deposit is included in the correlation factor.

Comparing the results of this paper with the design approaches of other codes, a further reduction of the sizes seems to be possible. For this, additional theoretical but also experimental investigations will be necessary.

REFERENCES

AWS. 2010. Structural Welding Code – Steel. ANSI/AWS D1.1/D1.1M. 22th Edition. Miami. FL: American Welding Society.

Boverket. 2003. Swedish regulations for steel structures, BSK99. Karlskrona: The Swedish National Board of Housing, Building and Planning.

EN 1993-1-1:2010. Eurocode 3: Design of steel structures – Part 1.1: General rules and rules for buildings. German Version of EN 1993-1-1:2005 + AC:2009. Brussels: European Committee for standardization (CEN).

EN 1993-1-8:2010. Eurocode 3: Design of steel structures – Part 1.8: Design of joints, German Version of EN 1993-1-8:2005 + AC:2009. Brussels: CEN.

EN 1993-1-8/NA:2010. National Annex – Nationally determined parameters – Eurocode 3: Design of steel structures – Part 1-8: Design of joints. Brussels: CEN.

EN 1993-1-12:2007/AC:2009. Eurocode 3 – Design of steel structures – Part 1-12: Additional rules for the extension of EN 1993 up to steel grades S 700. Brussels: CEN.

EN 10210:2006. Hot finished structural hollow sections of non-alloy and fine grain steels – Part 1: Technical delivery conditions & Part 2: Tolerances, dimensions and sectional properties. Brussels: CEN.

EN 10219:2006. Cold formed welded structural hollow sections of non-alloy and fine grain steels – Part 1: Technical delivery conditions & Part 2: Tolerances, dimensions and sectional properties. Brussels: CEN.

Fleischer O. and S. Herion. 2012. Reduction of weld sizes – Development of design charts. Final Report CIDECT Project 5BY. Comité International pour le Développement et l'Étude de la Construction Tubulaire (eds). Karlsruhe: KoRoH GmbH.

Packer J. A., J. Wardenier, Y. Kurobane, D. Dutta and N. Yeomans. 1992. Design Guide 3: For rectangular hollow section (RHS) joints under predominantly static loading. Comité International pour le Développement et l'Étude de la Construction Tubulaire (eds). Köln: TÜV – Verlag GmbH.

Wardenier J., Y. Kurobane, J.A. Packer, G.J. van der Vegte and X.-L. Zhao. 2008. Design Guide 1, 2nd edition: For circular hollow section (CHS) joints under predominantly static loading: Comité International pour le Développement et l'Étude de la Construction Tubulaire (eds). Dortmund: LSS Verlag.

Tubular Structures XV – Batista, Vellasco & Lima (eds)
© 2015 Taylor & Francis Group, London, ISBN 978-1-138-02837-1

Welding simulation of tubular K-joints in steel S690QH

F. Zamiri
Chalmers University of Technology, Gothenburg, Sweden

J.-M. Drezet & A. Nussbaumer
École Polytechnique Fédérale de Lausanne (EPFL), Lausanne, Switzerland

ABSTRACT: Residual stress state in planar tubular K-joints, in the chord within the gap region between the two braces, is studied using numerical weld modelling. The motivation comes from past full-scale fatigue tests on tubular trusses made of various steel grades with sizes typical of bridge trusses, which shows that the cracking occurs at the hot spots located in this region. Residual stress field characterization is needed in order to assess its role in fatigue cracking, especially for the case of cracks occurring on the compression brace side. Comparison between residual stress field in a Y-joint and a K-joint is made to assess the significance of restraining effect in the gap region. Phase transformations during welding and cooling down are determined and their impact on the final residual stress state is evaluated. Computed residual stresses are compared to the neutron diffraction measurements. Transient thermal field and cooling times substantially affect phase transformations. Therefore, their accurate reproduction in the analysis is important.

1 INTRODUCTION

Tensile welding residual stresses keep cracks open and thus lead to shorter fatigue lives even if welded components are subjected to, partially or fully, compressive external applied loads. Associated with welding residual stresses, distortions that result are also responsible for lower structural performance. Due to these detrimental effects on fatigue life of the structure, more realistic estimation of these stresses is necessary.

Conventional assumption for welding residual stresses is that they are of the order of yield strength of parent material. In general, through thickness residual stress distributions given by the codes, e.g. BS7910 (2005), are linear functions of yield strength. Residual stress measurements in tubular K-joints made of high-strength low-alloy (HSLA) steel by Zamiri (2014) have also shown peak residual stress values lower than yield stress. The objective of this paper is to explain the origin of these lower residual stresses using welding simulation of tubular K-joints for a steel S690.

Indeed, residual stresses can be estimated either by measurements or by numerical simulation of manufacturing process (rolling, machining, welding, etc.). Analytical prediction of residual stresses started by Rosenthal (1946) and Rykalin (1974). In early 1970s, Ueda applied finite element method for thermal stress analyses (Goldak and Akhlaghi 2005). Progresses in computational tools allowed for increasing the complexity of numerical models by implementing improved material laws, use of 3D models in lieu of 2D models, and incorporation of metallurgical transformations (Lindgren 2001).

The main source of welding residual stresses is hindered contraction of weld bead and heat affected zone by the surrounding parent material during cooling down stage. However, other important processes, namely metallurgical transformations, take place during the cooling down phase; these contribute significantly to the final state of residual stresses. The role of transformation strains has been emphasized by several authors, including Easterling (1992), Nitschke-Pagel & Wohlfhart (1992), and Voss et al. (1997).

Various concurrent processes happening during welding and subsequent cooling down, can be categorized into three major interacting fields: thermal domain, mechanical domain, and microstructure domain. Figure 1 schematically shows main interactions between thermal, mechanical and microstructural domains that occur during welding. Dark arrows show dominant effects while dotted arrows signify less important effects that generally are considered in the computations only implicitly. Temperature field affects both residual stress field and microstructural field, but the inverse effects are commonly considered as secondary. This assumption leads to de-coupling of the thermo-mechanical analysis into a staggered procedure.

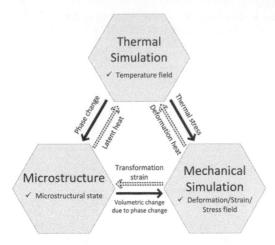

Figure 1. Interaction of temperature, mechanical, and microstructure fields during the welding and simplifications for welding simulation, after Radaj (2003). Bold arrows show the dominant effects and dotted arrows indicate factors that are not explicitly implemented in simulation.

Table 1. Welding parameters for K-Joints.

Welding process	MAG 136
Number of welding passes	8
Consumable	OK Tubrod 15.09
Preheat temperature [°C]	120
Maximum interpass temperature [°C]	250
Arc power [kW]	6.0–6.4
Average welding speed [mm/s]	7.4
Gross heat input energy [kJ/mm]	0.81–0.86
Arc efficiency [%][†]	78

[†]Based on values given by Grong (1997).

2 MODELLING OF WELDING

The K-joint modeled here was part of a fatigue tested truss. Welding parameters and temperature history were registered at the time of fabrication of the truss for one of its K-joints. Welding parameters are reported in Table 1. After the fatigue test finished, two of the K-joints were cut out of the truss and residual stress field at their gap region was evaluated using neutron diffraction method. The two K-joints were placed as attachments on the truss chords with no loading on their braces to prevent fatigue cracking in those joint. This was intentional so that as-welded residual stress would not change while the welding parameters, welding position, and welder's technic were similar to the rest of the K-joints in the truss. The welding temperature and residual stress measurements are compared to the results of finite element thermal and mechanical analyses in the following sections.

2.1 Geometry and FE mesh

The geometry is created with the method explained by Costa Borges (2008) and with dimensions mentioned

Table 2. Member sizes and non-dimensional geometric parameters of the studied K-joint.

Nominal dimensions		Non-dimensional parameters	
Chord	193.4 × 20 mm	β (d_1/d_0)	0.53
Brace	101.6 × 8 mm	γ ($d_0/2t_0$)	4.84
Eccentricity	38 mm	e/d_0	0.20
θ	60°	τ(t_1/t_0)	0.40

*Nominal angle between the chord and the braces.

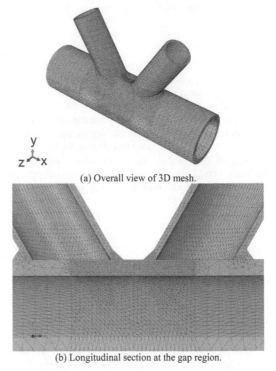

y
z x

(a) Overall view of 3D mesh.

(b) Longitudinal section at the gap region.

Figure 2. Finite element mesh of the K-joint model.

in Table 2. Weld bead is divided into three parts (or passes) such that cross section area of weld pass number 1 is 20% of the total weld bead cross section, and cross section of weld passes 2 and 3 are 40% of the total cross section each. The length of the chord and braces in the joint are taken large enough to allow for reproducing the cooling times of the welded parts similar to the actual welding.

Although the geometry has two symmetry planes, the heat loading is not symmetric Therefore, full symmetry conditions do not hold.

The geometry is discretized into 250,000 first order (linear) tetrahedral solid elements. The generated mesh for the K-joint is shown in Figure 2. Global element size is 16 mm. This is reduced to 5 mm in the vicinity of weld bead region. For the region of interest (i.e. gap region), element size is refined even more, up to 2 mm, to capture the residual stress profile with sufficient resolution. A convergence study is performed to investigate the sufficiency of mesh size. The effect of simulating only one brace (Y joint) is also investigated in this study.

2.2 Material properties

Two approaches in modeling material behavior are available (Goldak & Akhlaghi 2005). In the first approach, multi-phase steel material is considered as homogeneous and bulk material thermo-physical properties are given as the analysis input. Majority of the simulations in literature, e.g. Brickstad & Josefson (1998), are carried out using this method. For this approach, there are techniques to take into account metallurgical effects in the analysis, for example by modifying coefficient of thermal expansion (Deng 2009). The second approach predicts the behavior of heterogeneous metallic material based on the contributions from its various microstructure constituents by using mixture rules. Generally linear mixture rules are used. With this approach, firstly the evolution of micro-structure during thermal cycle is evaluated. Knowing the phase fractions at each step of transient analysis, physical properties of the material is evaluated for that step. Generally, phase-based material properties approach is applied only for the mechanical analysis step, as is the case in this study.

When metallurgical effects are taken into account, the second modelling method, per-phase material properties, can yield more accurate results than first method, bulk material properties. The drawback of the phase-based approach is that considerably more material input data are required for the material model. Temperature-dependent mechanical properties for each phase and kinetics of phase transformation are required as input data. Both modeling approaches are utilized in this study. Bulk material properties are used in the models without phase transformation. For the model with phase transformation effects included, per-phase material data are used. Identical material properties are assumed for parent metal and weld metal. Barsoum (2008) and Dai and coworkers (2010) have reported acceptable simulation results using this assumption.

2.2.1 Thermal properties

Thermo-physical properties of steel S690QH were measured and reported by Mertens & Lecomte-Beckers (2012) and are used in this study. Temperature-dependent specific heat capacity (c_p) is shown in Figure 3. The values used here fit into the range of values given by Radaj (shaded area in Figure 3) and are in general agreement with the material input data used by other researches. Lindgren (2007) has emphasized the importance of latent heat of solidus—liquidus transformation (the peak at about 1500°C) in the welding simulation, which is considered in the present study. Temperature-dependent thermal conductivity λ in the range of 2.52×10^4 to 1.00×10^5 µW/mm/K (for liquid phase) is used. Detailed values are given in Zamiri (2014). For

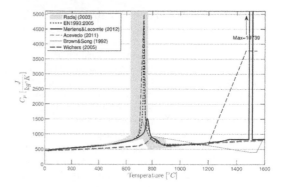

Figure 3. Specific heat capacity values from Richter (1973), EN1993 (2005), Acevedo et al. (2013), Brown & Song (1992), Wichers (2006), and Mertens & Lecomte-Beckers (2012).

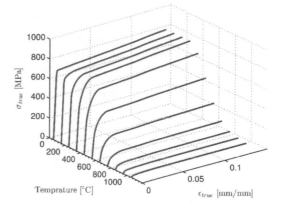

Figure 4. Eurocode 3 part 1-2 (2005) model for temperature-dependent stress–strain curve.

phase-based analysis, Morfeo uses the same thermal properties for all the constituents.

2.2.2 Mechanical properties

Temperature-dependent non-viscous mechanical properties are used. Hot tensile tests conducted on steel specimens and were in good agreement to recommended values of part 1-2 of Eurocode EN1993 (2005). Therefore, stress-strain curves given by Eurocode are used for the bulk material properties (Figure 4). For the phase-based model, yield stress of different constituents (bainitie, martensite, ferrite-pearlite, and austenite) are taken from literature as summarized in Zamiri (2014).

Coefficient of thermal expansion of steel material was measured by Mertens & Lecomte-Beckers (2012). For phase-based model, coefficients of thermal expansion of various phases are measured from Gleeble dilatometry tests and experimental formulas for lattice parameters of constituents.

2.2.3 Metallurgical properties

Kinetics of phase change can be shown best by means of continuous cooling transformation (CCT) diagrams like the one shown in Figure 5 for S690QL. CCT diagram should be read by following individual cooling curves and reading their intersection with microstructure lines (thick lines in Figure 5) to estimate volume fraction of each constituent at the end of transformation. Cooling time from 800°C to 500°C ($t_{8/5}$), which is the temperature range that austenite decomposition occurs, is a commonly accepted index for representing thermal conditions in welding of low alloy steels (Grong 1997). Observing CCT diagrams, high cooling rates (or short cooling times) lead to a martensitic structure. This corresponds to welds with low heat input. On the other hand, a weld with a large heat input will cool down slowly and the final microstructure will be bainitic-ferritic.

For the phase-based modeling approach, phase fractions are calculated for each element based on the CCT data and transient temperature history at the end of thermal analysis step. Having the phase fractions and using a linear mixture rule, the program calculates mechanical properties of steel at the given location. These properties are then used in the mechanical analysis step. The effects of volume change due to martensitic transformation are taken into account by the change in the coefficient of thermal expansion. Initial composition of studied steel is considered as 84% bainitie and 16% martensite.

2.3 Heat source

Input heat flux is calculated from the following equation:

$$q = \eta UI \quad q = \eta UI \tag{1}$$

Where q = net heat flux [J], η = arc efficiency, U = arc electric tension [V], and I = arc current [A]. Arc efficiency accounts for heat losses by radiation and convection in the arc region and molten pool. With the values from Table 1, net heat flux is equal to q = 4.8 kW and considering 8 weld passes, total heat power input is Q = 38.4 kW. Total net heat power deposited per unit length of weld is Q/v = 5.19 kJ/mm.

Due to the complex shape of multipass weld bead and individual weld passes, simplification of welding procedure is considered in the form of lumping the weld passes together as explained by Radaj (2003). Total heat flux is divided into three parts, with the first equivalent weld pass (resembling root pass) having a share of 20% of total heat flux, while 40% of total heat flux is assigned to the 2nd and 3rd equivalent passes. Another distribution, assigning 33.3% of total heat flux to each weld pass yields higher residual stresses and is used for some of the models (see Table 3). Figure 6 shows how the lumped weld passes compare to the actual welding passes.

From Fourier law for heat conduction, it derives that cooling time $t_{8/5}$ for the case of moving heat source is a function of heat flux per unit length of weld (i.e. q/v ratio) (Radaj 2003). Since the power (q) of a lumped pass is greater than an individual weld pass, the torch speed (v) needs to be augmented proportionally in order to achieve a realistic estimation of cooling times in simulations. The augmented torch speed is applied to two of the models in this study.

Heat source shape is considered either as a cylindrical heat source with uniform heat intensity or Goldak's double ellipsoid model (Goldak & Akhlaghi 2005). Both heat sources are calibrated such that they correctly reproduce the fusion zone and heat affected zone.

Weld torch trajectory for each weld pass is considered at the center of external face of that weld pass. The through-depth axis of the heat source is perpendicular to the weld face, which is in good agreement

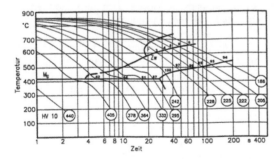

Figure 5. Continuous cooling transformation (CCT) curves for S690QL steel (Seyffarth *et al.* 1992).

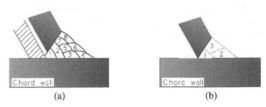

(a)　　　　　　　　　(b)

Figure 6. 8-pass weld (a) is reduced to three equivalent passes (b). Cross section of root pass is 20% of total weld cross section and two subsequent passes are 40% of total weld cross section each.

Table 3. Summary of analyzed models and their parameters.

Model	Torch speed	Material properties	Power distribution	Start/stop location	Heat source shape
K-244-BLK-N-h2t	Normal	Bulk	20%+40%+40%	heel—toe	Cylindrical
Y-244-BLK-N-h2t	Normal	Bulk	20%+40%+40%	heel—toe	Cylindrical
Y-244-BLK-N-sh	Normal	Bulk	20%+40%+40%	h—t shifted	Cylindrical
Y-333-BLK-A-sh	Augmented	Bulk	30%+30%+30%	h—t shifted	Cylindrical
Y-333-NOL-VC-A-sh	Augmented	Phase-based	30%+30%+30%	h—t shifted	Double ellipsoid

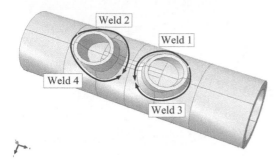

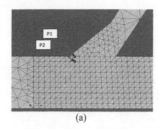

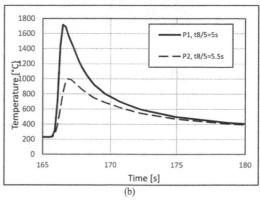

Figure 7. Sequence of welding passes. Welding start and stop points are shifted from crown toe and crown heel locations.

with Stockie's (1998) assumption of weld trajectory being coincident with bisector of the dihedral angle.

For the start/stop points, two variants are considered: (a) the heat source moves from crown toe towards crown heel, (b) the start points are shifted from crown locations to a location between saddle point and crown points as shown in Figure 7. The second variant is in coordination with the actual welding procedure, and is recommended by CIDECT (Zhao *et al.* 2000). The reason for choosing two variants is to assess the effect of start/stop points on residual stress distributions.

2.4 *Thermal initial and boundary conditions*

Initial temperature of whole welded piece is considered equal to the pre-heating temperature (120°C). Ambient temperature is assumed as 20°C. Morfeo takes into account heat flux due to convection while heat loss by radiation is not considered. Therefore, temperature-dependent film coefficient for convection heat flux is adjusted to account for the effect of heat loss by radiation as well.

2.5 *Analysis procedure*

Morfeo/Welding (2012) software is used for finite element analyses. It is a software package dedicated to manufacturing simulation tasks and can perform transient thermal-metallurgical-mechanical transient analyses.

Weld metal deposition is modeled by *quiet element* approach. All the elements in the weld bead region are available in the FE mesh from the beginning of analysis, but have near-zero material properties. Once the heat source reaches these elements, they are 'activated' by assigning real material properties to them. Element activation was used only for the mechanical analysis step.

Table 3 summarizes various models that are investigated here. Different parameters that are changed between models include shape of the joint (K- vs Y-joint), Torch speed (normal or augmented in analysis according to section 2.3), material properties, interpass power distribution, welding start/ stop location, and shape of the heat source shape.

Figure 8. Temperature time history for points P1 and P2 (a) located in fusion zone and HAZ, respectively (crown toe). Only part of time history with the highest peak temperature is shown.

3 RESULTS

3.1 *Thermal analysis results*

Estimated temperature time histories for two points located in fusion zone (FZ) and heat affected zone (HAZ) at the crown toe are shown in Figure 8. Accurate estimation of cooling time $t_{8/5}$ is crucial for correct evaluation of microstructure transformations. The values calculated here as $t_{8/5} = 5$ to 5.5 seconds, are in good agreement with value of $t_{8/5} = 3$ seconds, estimated from semi-analytical formulas given in Seyffarth et al. (1992). The calculated cooling times are short enough for formation of martensite to take place, according to CCT diagram of Figure 5.

The maximum registered temperature at measurement points located 5 mm from the weld toe at crown toe location was 350°C which is reproduced in all models with good accuracy.

Both FE models with normal and augmented torch speed give realistic sizes for the heat affected zone and fusion zone, when compared to weld macrograph. Figure 9 shows the estimated FZ and HAZ shapes for the three weld passes of model with augmented torch speed, compared to weld macrograph. The size of FZ is slightly underestimated by the model.

3.2 *Residual stress results*

Transversal component of residual stresses, i.e. residual stresses perpendicular to the weld line, are reported

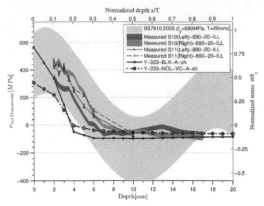

Figure 11. Comparison of effect of phase transformations on measured residual stress profiles. The wide band shows the range given by BS7910 (2005).

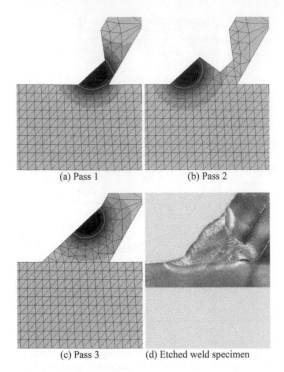

(a) Pass 1   (b) Pass 2

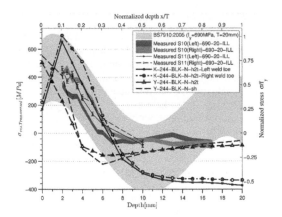

(c) Pass 3   (d) Etched weld specimen

Figure 9. Estimated FZ and HAZ from model Y-333-NOL-VC-A-sh (double ellipsoid heat source, augmented weld torch speed) compared to weld macrograph at crown toe location. Contour lines indicate 650°C and 1500°C).

Figure 10. Comparison of computed transverse residual stress profiles for K-Joint and Y-joint, together with measurement data and value range suggested by BS 7910 (2005).

in this section. Transversal component is of more interest because of its impact fatigue cracking.

Figure 10 shows the comparison transverse residual stress field at the gap region for K-joint and Y-joint models. At close-to-surface locations (Depth < 3 mm), K-joint model considerably overestimates residual stresses compared to measurements. Study of residual stress build up at the end of each

welding pass during simulation showed that the high input energy of the lumped passes on each brace affected a large part of the gap and when the effects from the weld lines of the two braces were superposed, the model would report high residual stress values. However, this was not the case for Y-joint model, since only one brace was modeled. Overall, Y-joint model with shifted start/stop locations worked better with the lumped pass weld model and gave more acceptable shape for residual stress profiles; albeit it underestimated transverse residual stresses in some locations. This model was selected as basis for subsequent analyses.

The effect of considering volumetric changes due to phase transformations on residual stress state is shown in Figure 11. As explained before, augmented speed models gave more accurate cooling times and were chosen for analysis of models with phase transformations.

Calculated transversal peak residual stresses are below yield stress. Model with phase-based material data (Y-333-NOL-VC-A-sh) reports smaller peak residual stresses compared to the other model.

Y-joint models slightly underestimate the location of minimum $\sigma_{transversal}$ profile at $0.3T_{ch}$ (T_{ch} being chord thickness) below weld toe, compared to $0.35T_{ch}-0.4T_{ch}$ from measurements. BS 7910 gives sinusoidal through-thickness residual stress profiles for transverse direction. It predicts that $\sigma_{transversal}$ will increase at higher depths (measured from chord surface) once it reaches its minima at approximately mid-thickness. However, for the case of $\sigma_{transversal}$, both calculated and measured residual stresses show only a small increase of stress with depth after the minimum is reached.

4 CONCLUSIONS

Three-dimensional weld modeling of tubular K-joint is carried out in this study in order to evaluate residual stress field in the gap region of the K-joint. A staggered thermal-metallurgical-mechanical analysis scheme is used. Effect of volumetric changes due to martensitic transformation are taken into account. The computed residual stress profiles are compared to experimental measurements.

Comparing welding residual stress measurements in steel S690QH with a similar study on normal grade steel S355J2H by Acevedo(2012) shows that increase in residual stresses in HSLA is not proportional to the increase in yield strength of material. This is mainly associated with the solid state phase transformations that take place during the cooling down phase of these steels. Low heating power input in fine-grain HSLA steels causes rapid cooling, especially at the regions close to the surface. This leads to formation of martensite phase with a positive volume change that cancels part of thermal contraction strains.

Comparison of simulation results for residual stress field in the K-joint and corresponding Y-joint showed that Y-joint model yields more acceptable residual stress values, when lumped pass heat source model is used.

Accurate reproduction of transient thermal field and cooling times is crucial when phase transformation effects are included in the model. For this reason, lumping of welding passes is a difficult decision. Since the last welding passes have more impact on the final residual stress field, it is advisable for subsequent simulations that the last two or three passes be modeled according to actual welding conditions (i.e. without lumping).

ACKNOWLEDGMENTS

This research was part of the project P816 "Optimal use of hollow sections and cast nodes in bridge structures made of S355 and S690 steel", supervised by the Versuchsanstalt für Stahl, Holz und Steine at the Technische Universität Karlsruhe, which was supported financially and with academic advice by the Forschungsvereinigung Stahlanwendung e. V. (FOSTA), Düsseldorf. The authors gratefully acknowledge their support. Thanks are also addressed to Zwahlen & Mayr S.A. (Switzerland) who contributed to the project by fabrication of steel trusses. Neutron diffraction measurements were carried out at the Institut Laue-Langevin, Grenoble, France. Their contribution is sincerely acknowledged.

REFERENCES

Acevedo, C., Drezet, J.M., & Nussbaumer, A., 2013. Numerical modelling and experimental investigation on welding residual stresses in large-scale tubular K-joints. *Fatigue & Fracture of Engineering Materials & Structures*, 36 (2), 177–185.

Acevedo, C., Evans, A., & Nussbaumer, A., 2012. Neutron diffraction investigations on residual stresses contributing to the fatigue crack growth in ferritic steel tubular bridges. *International Journal of Pressure Vessels and Piping*, 95, 31–38.

Barsoum, Z., 2008. Residual stress analysis and fatigue assessment of welded steel structures. KTH, Stockholm, Sweden.

Brickstad, B. & Josefson, B.L., 1998. A parametric study of residual stresses in multi-pass butt-welded stainless steel pipes. *International Journal of Pressure Vessels and Piping*, 75 (1), 11–25.

Brown, S. & Song, H., 1992. Finite element simulation of welding of large structures. *Journal of engineering for industry*, 114 (4), 441–451.

BS 7910, 2005. *Guide to methods for assessing the acceptability of flaws in metallic structures*. British Standards Institution.

Costa Borges, L.A., 2008. Size effects in the fatigue behaviour of tubular bridge joints (Thesis No. 4142). EPFL, Lausanne.

Dai, H., Francis, J.A., & Withers, P.J., 2010. Prediction of residual stress distributions for single weld beads deposited on to SA508 steel including phase transformation effects. *Materials Science and Technology*, 26 (8), 940–949.

Deng, D., 2009. FEM prediction of welding residual stress and distortion in carbon steel considering phase transformation effects. *Materials & Design*, 30 (2), 359–366.

Easterling, K.E., 1992. *Introduction to the physical metallurgy of welding*. 2nd ed. Oxford; Boston: Butterworth Heinemann.

EN1993, 2005. *Eurocode 3: Design of steel structures – Part 1-2: General rules – Structural fire design*. Brussels: European Committee for Standardization.

Goldak, J.A. & Akhlaghi, M., 2005. *Computational welding mechanics*. Springer Verlag.

Grong, Ø., 1997. *Metallurgical Modelling of Welding (2nd Edition)*. Maney Publishing.

Lindgren, L.-E., 2001. Finite element modeling and simulation of welding part 1: Increased complexity. *Journal of Thermal Stresses*, 24 (2), 141–192.

Lindgren, L.E., 2007. *Computational Welding Mechanics: Thermomechanical and Microstructural Simulations*. CRC Press.

Mertens, A. & Lecomte-Beckers, J., 2012. *Rapport d'essais: Caractérisation thermophysique de 2 chantillons d'acier*. Liège, Belgium: Université de Liège-Department A&M-Science of Metallic Materials (MMS).

MORFEO, 2012. *v1.7.5 User's Manual*. Gosselies, Belgium: Cenaero.

Nitschke-Pagel, T. & Wohlfahrt, H., 1992. Residual stress distributions after welding as a consequence of the combined effect of physical, metallurgical and mechanical sources. *In: Mechanical Effects of Welding*. Springer, 123–134.

Radaj, D., 2003. *Welding residual stresses and distortion: Calculation and measurement*. DVS-Verlag.

Richter, F., 1973. Die wichtigsten physikalischen Eigenschaften von 52 Eisenwerkstoffen. *Stahleisen-Sonderberichte, Duesseldorf: Verlag Stahleisen*, 8.

Rosenthal, D., 1946. The theory of moving sources of heat and its application to metal treatments. *In: Transactions of ASME*. 849.

Rykalin, N.N., 1974. Energy Sources Used for Welding. *Soudage et Techniques Connexes l (12)*, 471–485.

Seyffarth, P., Meyer, B., & Scharff, A., 1992. *Großer Atlas Schweiß-ZTU-Schaubilder*. Deutscher Verlag für Schweißtechnik DVS-Verlag GmbH.

Stockie, J.M., 1998. The geometry of intersecting tubes applied to controlling a robotic welding torch. *Mapel Tech*, 19 (2), 2.

Voss, O., Decker, I., & Wohlfahrt, W., 1997. Consideration of microstructural transformations in the calculation of residual stresses and distortion of larger weldments. *In*: H. Cerjak, ed. *Mathematical modelling of weld phenomena 4*. Presented at the Numerical Analysis of Weldability 4, 584–596.

Wichers, M., 2006. Schweißen unter einachsiger, zyklischer Beanspruchung Experimentelle und numerische Untersuchungen(Welding under uniaxial cyclic loads – Experimental and numerical research). Universitätsbibliothek Braunschweig.

Zamiri, F., 2014. Welding Simulation and Fatigue assessment of Tubular K-joints in High Strength Steel. École Polytechnique Fédérale de Lausanne, Lausanne, Switzerland.

Zhao, X.L., Herion, S., Packer, J.A., Puthli, R.S., Sedlacek, G., Wardenier, J., Weynard, K., Van Wingerde, A.M., & Yeomans, N.F., 2000. *Design guide for circular and rectangular hollow section joints under fatigue loading*. CIDECT, Comité International pour le Développement et l'Etude de la Construction Tubulaire. Köln, Germany: TÜV-Verlag.

Tubular Structures XV – Batista, Vellasco & Lima (eds)
© 2015 Taylor & Francis Group, London, ISBN 978-1-138-02837-1

Experimental investigation of the static capacity of grade C450 RHS K and N truss joints

Z. Yao
Former research associate, The University of Sydney, Australia

T. Wilkinson
The University of Sydney, Sydney, New South Wales, Australia

ABSTRACT: The results of an experimental program on K and N truss connections in C450 cold-formed RHS are presented. The aim of the program was to study the effect of the higher material yield strength and reduced f_u/f_y ratio of C450 compared to C350 steel on the various possible failure modes. At the same time the program also included RHS with wall slenderness values outside the current CIDECT geometric limits, since these slender cross-sections typically form part of the range of cold-formed RHS which are commercially on offer. With a reasonable level of confidence it can be concluded that for failure modes associated with buckling, instability, yielding and deformation (side wall buckling, face plastification and brace capacity in compression), it is recommended that the modifying factors of min[f_y, $0.8f_u$] and the extra factor of 0.9 are not required for these AS/NZS 1163 – C450 grade tubes; while for failure modes associated with fracture or ductility or liable to brittle failure modes (brace tension and punching shear) the reduction modifying factors of min[f_y, $0.8f_u$] and the extra factor of 0.9 are should remain for these AS/NZS 1163 – C450 grade tubes.

1 INTRODUCTION

This paper presents the results of an experimental program carried out at the University of Sydney with the aim of studying the static capacity of K and N truss joints in grade C450 cold-formed rectangular hollow sections (RHS). A total of 12 connections were tested, including 6 K joints and 6 N joints. A paper at the last ISTS (Becque and Wilkinson 2012) reported on the experimental investigations of T & X joints as part of this project.

The driving factors behind the research were twofold. First, the large amount of material cold-working received during the manufacturing process of RHS results in an increased material yield stress f_y, while the f_u/f_y ratio is typically reduced. It is thereby noted that, in what follows, the term 'yield stress' should be understood to represent the 0.2% proof stress, while the symbol f_u represents the ultimate tensile strength. The research aimed to investigate the effects of the altered material properties on the connection capacity and to (re)assess the need for reduction factors in the design equations when designing C450 K and N connections.

Second, advances in manufacturing techniques have allowed RHS with wall thicknesses of up to 16 mm to be produced by cold-rolling. However, upon inspection of commercial C450 RHS catalogues it is clear that a significant number of the sections on offer fall outside the geometric restrictions of CIDECT Design Guide 3 (Packer at al. 2009) in terms of their wall slenderness.

The program aimed to include these more slender sections and gather experimental data on their connection behaviour and capacity.

2 BACKGROUND

While having evolved over many decades, the CIDECT design guidelines for truss connections have typically been presented with certain restrictions on their applicability. These restrictions themselves have evolved in synch with the design rules and apply to certain geometric parameters of the connections, as well as to the material properties. Earlier versions of CIDECT Design Guide 3 (Packer et al. 1996) limited the material yield stress f_y to 355 MPa, while also imposing the requirement that f_u/f_y had to exceed 1.2. While these restrictions have been maintained in the recently published AISC Design Guide 24 for Hollow Structural Section Connections (Packer et al. 2010), the latest version of the CIDECT Design Guide 3 (Packer et al. 2009) (which is also in accord with the recommendations of the International Institute of Welding, IIW 2009) has extended the range of applicability of the design rules to yield strengths of up to 460 MPa. However, it stipulates to use the minimum of f_y and $0.8f_u$ instead of f_y in the design equations and also imposes an additional reduction factor of 0.9 on the connection capacity. Eurocode 3 EN1993-1-12 (2007) equally imposes a reduction factor of 0.9 for S460.

When researching the origins of the reduction factor of 0.9, it appears that most of the justification for it originates from experimental work specifically conducted on K gap connections. Kurobane (1981) was first to demonstrate that the ultimate capacity of CHS K gap connections in S460 is in relative terms 18% lower compared to the same joints in S235. Noordhoek et al. (1996) similarly found that CHS K gap connections of S460 had lower joint capacity factors than S235, even when an effective yield stress of $0.8f_u$ was used. However, Puhtli et al. (2010) carried out tests on CHS X-joints in S460 and observed that for nearly all connections tested, the experimentally determined capacity exceeded the capacity calculated without the factor of 0.9. The experiments were followed by numerical analyses which indicated that, while there is some justification for the inclusion of a reduction factor, the current value of 0.9 is conservative for X joints. The parametric studies, however, also revealed a dependence of the results on the f_u/f_y ratio. This dependence is the result of punching shear failures and effective width failures in tension being governed by the tensile strength f_u, while the design equations are based on the yield stress f_y. With respect to RHS connections, Mang (1978) conducted some early research on high strength S690 K-joints and observed a relative reduction in strength of 1/3 compared to S235 joints. Liu & Wardenier (2004) carried out further numerical work on RHS K gap connections in S460 and noticed a 10–16% reduction in capacity compared to S235 joints. Their research considered ultimate capacities, as well as the 3% deformation limit of Lu et al. (1994).

3 MATERIAL TESTS

Tensile coupons were tested according to the AS/NZS1391 (1991) specifications. The coupons were cut from the same segments of RHS tube as the members of the actual test specimens. For each RHS one coupon was taken from the middle of the face opposite the weld face and one coupon was taken from the middle of a face adjacent to the weld face, as illustrated in Figure 1. All coupons were 20 mm in width and were tested at a strain rate of 5×10^4/s in a 300 kN capacity MTS Sintech machine. It should be noted that the RHS employed in the tests were obtained from two different sources: sizes up to $200 \times 200 \times 6$ were manufactured in Australia by OneSteel Australian Tube Mills, while the larger sizes were imported from Japan. While all sizes are sold in Australia as C450 conforming to AS/NZS 1163 (2009), it is recognized that sections from these two origins might generally exhibit somewhat different material properties. Table 1 lists the values of f_y and f_u obtained in the coupon tests. 'Static' lower-bound values are reported, obtained by halting the test near the 0.2% proof stress and near the ultimate load and allowing the load to settle for about 2 minutes, as illustrated by the example provided in Figure 2.

It is seen from Table 1 that the material in the face opposite the weld generally exhibits a slightly higher

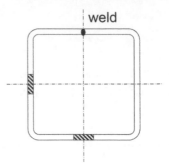

Figure 1. Location of the test coupons.

Table 1. Tensile coupon results.

		Adjacent to weld		Opposite weld	
Section	Source*	f_y MPa	f_u MPa	f_y MPa	f_u MPa
$150 \times 100 \times 4$	AUS	367.9	438.5	416.2	458.6
$150 \times 100 \times 4$	AUS	412.3	455.1	438.8	467.1
$300 \times 300 \times 8$	JAP	447.8	518.3	465.7	526.2
$300 \times 300 \times 8$	JAP	485.1	527.6	484.2	535.3
$75 \times 75 \times 5$	AUS	427.0	462.9	444.9	466.0
$75 \times 75 \times 5$	AUS	453.6	472.8	464.3	479.9
$150 \times 100 \times 5$	AUS	448.1	525.1	458.5	525.6
$200 \times 200 \times 5$	AUS	448.1	521.6	470.5	526.8
$100 \times 100 \times 5$	AUS	420.5	466.0	441.5	470.6
$75 \times 75 \times 6$	AUS	492.7	549.2	545.7	559.8
$150 \times 150 \times 5$	AUS	411.6	513.9	432.1	519.0
$100 \times 50 \times 5$	AUS	495.6	567.3	496.5	556.7
Average JAP		466.5	523.0	474.9	530.8
Average AUS		437.7	497.2	460.9	503.0

*JAP = Japanese origin; AUS = Australian origin.

yield stress than the material in the face adjacent to the weld, while the tensile strengths in both faces are similar. This can be explained by the larger amount of work-hardening undergone by the face opposite the weld during the fabrication process.

With respect to the static values, for the Australian made sections, an average yield stress f_y of 433 MPa was found, in combination with a tensile strength f_u of 475 MPa. Consequently: $f_u/f_y = 1.10$. For the Japanese sections, on average $f_y = 471$ MPa and $f_u = 527$ MPa, and thus $f_u/f_y = 1.12$. The materials fail the CIDECT requirement that $f_u/f_y > 1.2$.

Clause 10.5.4 of AS/NZS 1163:2009 states:

"Prior to tensile or impact testing, the test pieces shall be aged by heating to a temperature between 150°C and 200°C for not less than 15 min."

However, the majority of the coupons in this program were non-strain aged coupons, because the aim of these tensile tests was to determine the stress-strain behaviour of the C450 material for the purpose of the finite element modeling of the N- and K-joint

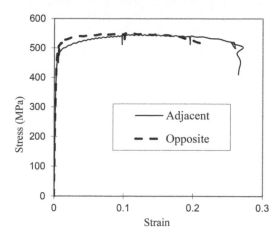

Figure 2. Results of RHS200 × 200 × 6 coupon tests.

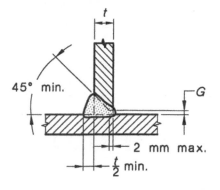

Figure 3. Compound weld (taken from AS/NZS 1554.1 2000).

tests. This ensured consistency with the properties of the material in the actual test specimens at the time of testing since the full scale tests naturally involve non-strain aged material.

Nevertheless, in order to measure the impact of strain aging on the material properties, 8 coupons were artificially strain aged in accordance with the aforementioned procedure in AS/NZS 1163:2009. A fan-forced oven with a measured temperature of 180°C–190°C was used. Specimens were lifted to the center of the oven in order to ensure good air circulation and thus even heating.

Of particular importance is the f_u/f_y ratio. For the Australian made sections, an average yield stress f_y of 435 MPa was found, in combination with a tensile strength f_u of 511 MPa. Consequently, $f_u/f_y =$ 1.18. For the Japanese made sections, on average $f_y = 459$ MPa and $f_u = 537$ MPa, and thus $f_u/f_y = 1.17$. The materials narrowly fail the CIDECT requirement that f_u/f_y has to exceed 1.2.

4 WELDING

All connections were welded according to AS/NZS 1554.1 (2000) by a certified welder. Gas metal arc welding with W503 electrode wire was selected for all welds and Argon UN1006 was used as a shielding gas. Before welding the brace members were purged using Argon UN1956. Weld failures were outside the scope of the project and undesirable, since the aim of the research was to investigate the implications of using grade C450 on the connection capacity. Therefore, full penetration butt welds with superimposed fillet welds were selected whenever possible. Figure 3, taken from AS/NZS 1554.1 (2000), shows the prequalified weld detail which was used. The decision to select a compound weld was reinforced by findings that it is difficult to obtain full penetration at the root of the weld (Wardenier et al. 2009), a conclusion which was also drawn from slicing through and visually inspecting two practice connections.

It needs be noted that a weld repair procedure was carried out after the testing of specimen N1 which showed failure in the root run of welds due to lack of fusion between the weld metal and the parental metal. In order to avoid this type of undesirable failure in the subsequent tests, WTIA (Welding Technology Institute of Australia) was requested to (i) examine of the welding problem of the failed welds, (ii) assist in developing repair methodology and procedure, (iii) supervise the practice for weld repair, and (iv) establish prequalification record (WQR) and welding procedure specification (WPS).

5 TESTS ON K AND N CONNECTIONS

5.1 Choice of sections

A total of 12 tests were carried out, including 4 K joints and 6 N joints. An overview of the test program is provided in Table 2. Keeping in mind the driving factors behind the research, which are on the one side the increased yield stress but reduced f_u/f_y ratio of the C450 steel and on the other side the increased wall slenderness of many of the cold-formed products, the experimental program was designed to include two types of sections in a balanced way:

1. sections with a nominal yield stress of 450 MPa which satisfied the geometric constraints of the CIDECT design guidelines (Design Guide 3, Packer et al. 2009). These tests investigated the effect of the altered material properties only.
2. sections with a nominal yield stress of 450 MPa which, due to their increased cross-sectional slenderness, fell outside the scope of the CIDECT design guidelines. These tests are highlighted in Table 2.

A wide range of geometries were included in the test program, with section sizes ranging from 75 × 75 × 5 to 400×400×16. The geometric parameters β, 2γ, τ and α are listed in Table 2 for each connection. The parameter β is equal to the ratio of the width of the brace member to the width of the chord member, 2γ is

Table 2. Test results.

Test ID	Chord/brace size	Geometry	Predicted failure mode	Observed failure mode	P_pred,1 (kN)	P_pred,2 (kN)	P_test,buckle (kN)	P_test,ult (kN)	P_test,3% (kN)	Ratio 1	Ratio 2
N1	300 × 300 × 8/ 150 × 150 × 5, 200 × 200 × 6	$Ov = 27.5\%$ $\beta i = 0.50$ $\beta j = 0.66$	Local yielding of tension brace	Rupture of welds of tension brace	370	432	–	573	Not reached	1.55	1.33
N2	200 × 200 × 6/ 125 × 125 × 5, 150 × 150 × 6	$Ov = 35.7\%$ $\beta i = 0.62$ $\beta j = 0.74$	Local yielding of tension brace	Local yielding of tension brace	356	418	–	648	Not reached	1.82	1.55
N3	250 × 150 × 5/ 100 × 100 × 5, 150 × 150 × 5	$Ov = 48.7\%$ $\beta i = 0.40$ $\beta j = 0.60$	Local yielding of tension brace	Local yielding of tension brace	293	365	–	505	Not reached	1.72	1.38
N4	150 × 100 × 5/ 100 × 100 × 5, 100 × 100 × 5	$Ov = 49.1\%$ $\beta i = 0.68$ $\beta j = 0.67$	Local yielding of tension brace	Local yielding of tension brace	334	424	–	489	Not reached	1.46	1.15
N5	250 × 250 × 6/ 125 × 125 × 5, 200 × 200 × 5	$Ov = 63.3\%$ $\beta i = 0.50$ $\beta j = 0.80$	Local yielding of comp. brace	Punching shear between comp. and tension braces	579	680	–	774	712	1.23	1.14
N6	150 × 150 × 5/ 75 × 75 × 5, 100 × 150 × 5	$Ov = 91.5\%$ $\beta i = 0.50$ $\beta j = 0.67$	Local yielding of comp. brace	Section yielding and column buckling of comp. brace	349	440	–	551	Not reached	1.58	1.25
K1	200 × 200 × 6/ 100 × 100 × 5	$g = 58.7$ $\beta 1 = 0.50$ $\beta 2 = 0.50$	Chord face plastification	Chord face plastification (+ chord side wall buckling)	296	355	–	440	380	1.28	1.07
K2	300 × 300 × 8/ 125 × 125 × 5	$g = 122.6$ $\beta 1 = 0.41$ $\beta 2 = 0.42$	Chord face plastification	Chord face plastification + comp. brace local buckling (+ chord side wall buckling)	487	592	~600	641	397	0.82	0.67
K3	100 × 150 × 5/ 100 × 50 × 5	$g = 79.7$ $\beta 1 = 1.00$ $\beta 2 = 1.00$	Chord side wall buckling	Chord side wall buckling	125	141	485	485	474	3.79	3.36
K4	100 × 100 × 5/ 100 × 50 × 5	$g = 28.6$ $\beta 1 = 0.99$ $\beta 2 = 0.99$	Chord side wall buckling	Local yielding of tension brace	231	274	–	449	440	1.90	1.61
K5	100 × 200 × 5/ 75 × 75 × 6	$g = 93.8$ $\beta 1 = 0.75$ $\beta 2 = 0.75$	Chord face plastification	Chord face plastification + chord side wall buckling	206	240	>400	428	365	1.77	1.52
K6	150 × 250 × 5/ 150 × 100 × 4	$g = 111.4$ $\beta 1 = 1.00$ $\beta 2 = 1.00$	Chord side wall buckling	Compression brace local buckling	96	107	~350	464	Not reached	4.83	4.34

$P_{pred,1}$ = Predicted capacity using the minimum value of f_y and $0.8f_u$ and an additional reduction factor of 0.9; Ratio 1 = $\min(P_u, P_{3\%})/P_{pred,1}$.
$P_{pred,2}$ = Predicted capacity using only f_y without an additional reduction factor of 0.9; Ratio 2 = $\min(P_u, P_{3\%})/P_{pred.}$.

the ratio of the chord width to the chord thickness, τ is the ratio of the thickness of the brace member to the thickness of the chord member, and α is the angle included between the axes of the brace member and the chord member. Table 2 also indicates whether the connection was loaded in tension (T) or in compression (C).

Four test specimens (N1, N2, N3 and N4) were designed to fail in local yielding (i.e. effective width failures) of tension brace. Among those, (i) N1 aimed to study the effect of slender compression brace (ii) N3 aimed to investigate the effect of slender chord top wall, and (iii) N4 aimed to study the effect of equal width of brace members with Ov = 50%.

Two specimens (N5 and N6) were designed to fail in local yielding (i.e. effective width failures) of compression brace. Among them, N5 was to study the effect of small overlapping member width to overlapped member width ratio and also the effect of slender chord walls.

Three specimens (K1, K2 and K5) were designed to fail in chord face plastification. Among those, K2 aimed to study the effect of small brace width to chord width ratio and also the effect of relatively slender chord walls, and K5 was to study the effect of slender chord side walls and also the effect of large gap size.

Three specimens (K3, K4 and K6) were designed to fail in chord side wall buckling. As all the three specimens had equal width brace and chord members, the gap size limitation in CIDECT design rules was inevitably violated so that they needed to be also designed as Y joints. The joint K4 aimed to study the effect of small gap size, while K6 aimed to study the effect of increase slenderness of chord and compression brace member.

Overall, tests were conducted over a wide range of (i) β values, spanning from 0.40 to 1.0, (ii) wall slenderness values, ranging from 20 to 50, (iii) overlap ratios for N-joints, spanning from 26% to 91%, and (iv) gap sizes for K-joints, covering from 29.3 mm to 123.2 mm.

The range of failure modes and geometries included in the experimental program was aimed at allowing a thorough verification and calibration of the finite element models.

Detailed shop drawings were developed for all 12 N- and K-joints. They are included in Appendix C.

Moreover, during the detailed design of the test specimens, decisions had to be made on appropriate values of the L/b_0 ratios of the RHS chord members, where L is the length of the RHS chord and b_0 is its width. The L/b_0 ratios needed to be sufficiently large to resemble conditions in actual trusses and to ensure that omission of the material beyond the L/b_0 limit did not significantly affect the connection behaviour and strength. Some research has been carried out on this topic for CHS (van der Vegte and Makino 2008), suggesting that an L/d_0 ratio of 10 for the chord is adequate for the majority of the cases considered and can be reduced below 10 in many cases, depending on the CHS wall thickness and the β ratio. For non-CHS, the

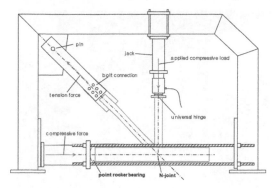

Figure 4. N test set up.

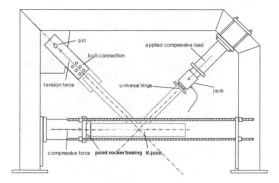

Figure 5. K joint set up.

issue is mostly considered outstanding. Some preliminary finite element studies were carried out within the context of this test program to ensure the chosen L/b_0 ratios were adequate.

A strong frame with a 1000 kN jack was used to test the smaller size X-joints in compression (X1, X2, X3, X5, X7 and X8). The set-up is illustrated in Figure 4. The specimens were tested between universal hinges, which were fitted onto the $320 \times 320 \times 32$ mm end plates of the specimen. This test configuration was chosen because it ensured a centered entry of the load into the specimens. The hinges allowed for end rotations to develop, mimicking the flexibility of the omitted parts of the brace.

5.2 Test method

A reaction frame, mounted with a 1000 kN jack, was particularly manufactured in order to test the N- and K-joints. Figures 4, 5 & 6 show various views and representations of the tets set up. In short, the perimeter SHS tubes were connected together by being welded to a 32 mm thick plate between every two segments of SHS. To ensure weld quality, full penetration butt welds with superimposed fillet welds were selected wherever possible. Beveling of the ends of each tube was carried out in order to facilitate the welding procedure. A $300 \times 300 \times 8$ SHS was welded to a 30 mm plate at each end and then bolted to one of the columns

493

Figure 6. K joint test.

Figure 7. Chord face deformation (N5).

Figure 8. Chord side wall buckling (K3).

of the frame. This tube served as the support of the chord of the N- and K-joints.

Two bolts were used to connect the column base plate to the reaction rail. This is because four high-strength Macalloy bars were used to connect the two columns of the frame, thus leaving the frame self-balanced during testing. Therefore, the base of the frame did not have to bear friction and was hence treated as a roller. Pinned boundary condition was provided to the tension brace by bolting it to two splice plates which were further connected to the corner of the frame by a shear pin. Such as connection ensured that the tension brace was free to rotate about the shear pin.

Universal hinges were provided between the jack and the compression brace of the joint. The spherical part of the hinge was bolted to a $320 \times 320 \times 32 \, \text{mm}$ plate, which was tack welded to the end of the compression brace member. The socket fitting the spherical part of the hinge was set into a plate, which was bolted to the underside of the jack. All those pinned boundary conditions of the joint ensured a centered entry of the load into the specimens, thus mimicking the situation in the actual truss which only carried axial force through individual members.

In order to eliminate any out-of-plane displacements and failure modes, a lateral restraint system was designed and installed. Two sets of lateral restraint

assemblies were provided, restraining two positions along the chord length.. In general, at the location of each restraint system, the chord side walls were clamped by two teflon plates, which were fixed onto two threaded rods. The threaded rods were then fixed to the two vertical RHS welded to the horizontal RHS. The horizontal RHS was further bolted to the reaction rail. Moreover, to effectively restrain any lateral slip, a steel plate was bolted to both ends of the horizontal RHS and pushed tightly against the concrete floor, thus providing sufficient friction against slip in the out-of-plane direction of the joint.

5.3 Experimental observations

Testing was performed in a pseudo static mode. Regular pauses in the application of increasing load were carried out to observe the formation of deformation or buckles.

Figures 7–9 illustrate various failure modes encountered in the test specimens: plastification of the chord face (Fig. 7, showing N5), chord side wall buckling (Fig. 8, showing K3) and tension failure of the brace (Fig. 9, showing N4).

A full summary of the observed failures is given in Table 2.

Figure 9.   Tension brace failure (N4).

6   TEST RESULTS: DISCUSSION

Table 2 lists the ultimate loads P_u obtained in the tests. Consideration was also given to a deformation limit which, in accordance with Lu et al. (1994), was set at 3% of the chord width. The corresponding load $P_{3\%}$ is listed in Table 2.

The current CIDECT design guidelines (CIDECT Design Guide 3, Packer et al. 2009) were used to obtain a prediction of the connection capacity. It is thereby noted that some of the connections tested (in particular, the connections highlighted in Table 2) did not satisfy the geometric constraints imposed by the CIDECT rules in terms of the wall slenderness values of the chord and/or brace members. While the CIDECT rules must not be applied to these connections, a prediction is nevertheless given. Predictions were obtained in two different ways:

1. by applying the design rules of CIDECT Design Guide 3 (Packer et al. 2009) as stated, i.e. with limiting f_y to $0.8f_u$ and applying an extra modification factor of 0.9. Actual values of f_y and f_u, as determined from coupon tests, were used. This prediction is listed as $P_{pred,1}$ in Table 2.
2. the CIDECT rules valid for $f_y = 355$ MPa were applied, without modification, to the C450 connections. Actual values of f_y as determined from coupon tests, were used. This prediction is listed as $P_{pred,2}$ in Table 2.

In either case it was taken into account that the CIDECT rules yield *factored* capacities, implicitly including a partial safety factor of 0.9. The factor of 0.9 was divided away to allow an objective comparison between the experiment and the prediction.

In Table 2, Ratio1 indicates the ratio of the experimentally determined capacity (including the 3% deformation limit) to the predicted capacity $P_{pred,1}$. Ratio2 is the ratio of the experimentally determined capacity (again including the 3% deformation limit) to the prediction $P_{pred,2}$.

6.1   Reliability analysis

Reliability analysis was performed on the test results, based on the First Order Second Moment (FOSM) method described by Ravindra & Galambos (1978). The full details are in Wilkinson et al. (2013).

Reliability analysis is a statistical procedure that combines the mean and standard variation of the experimental results to sets of predictions with other variability parameters associated with load, material and geometric variation. The study considered predictions either:

- CIDECT predictions including the modifications for fy > 355 MPa
- CIDECT predictions without the modifications for fy > 355 MPa

There are several design equations, covering different failure modes, in the CIDECT design guides. Obviously, in any given experiment, a connection usually only fails in one mode. Hence, any one experiment can only meaningfully contribute one value of P_{exp}/P_{pred} towards assessing the reliability of the equation applying to a given failure mode. However, if an experiment fails in a certain mode, it can be deduced that the experimental capacity of the other failure modes is no less than the value for the actual failure mode, and it is possible to obtain a value P_{exp}/P_{pred} that can be used in the reliability analysis of the non-occurring failure mode.

One failure mode that was not well investigated was chord punching shear. No occurrences of punching shear were observed in the K & N tests, and only one example of chord punching shear occurred experimentally (T3) in the companion paper, and it was not possible to extrapolate possible conclusions from the other experimental results.

The reliability analysis showed that for ductile failure modes, such as side wall buckling, chord face plastification etc) that there was an acceptable level of reliability when comparing the experimental results to the CIDECT predictions that did not incorporate the strength penalties for high strength steel. For brittle failure modes, an acceptable level of reliability was only achieved when the reduction factors were included.

7   CONCLUSIONS

There are a large number of joint configurations and failure modes. While the test schedule had been planned and designed to activate as many of these failure modes as possible, some did not always occur. It is difficult to obtain a large range of statistically reliable results from such a small sample size.

Nonetheless, the following conclusions can be made with a reasonable level of confidence:

1. For failure modes associated with buckling, instability, yielding and deformation (side wall buckling, face plastification and brace capacity in compression), it is recommended that the modifying factors of $\min[f_y, 0.8f_u]$ and the extra factor of 0.9 are not required for these AS/NZS 1163 – C450 grade tubes.

2. For failure modes associated with fracture or ductility or liable to brittle failure modes (brace tension and punching shear) the reduction modifying factors of $\min[f_y, 0.8f_u]$ and the extra factor of 0.9 are should remain for these AS/NZS 1163 – C450 grade tubes.

ACKNOWLEDGMENT

The authors would like to extend their sincere gratitude towards CIDECT and OneSteel Australian Tube Mills for their financial patronage of the program. The donation of the materials used in the experiments by the latter sponsor is also greatly appreciated.

REFERENCES

AS/NZS 1163 2009. "Cold-formed structural steel hollow sections." Australian Standard/New Zealand Standard, Standards Australia, Sydney, Australia.

AS/NZS 1391 1991. "Methods for Tensile Testing of Metals." Australian Standard/New Zealand Standard 1391:1991, Standards Australia, Sydney, Australia.

AS/NZS 1554.1 2000. "Structural steel welding: Part 1: Welding of steel structures." Australian Standard/New Zealand Standard, Standards Australia, Sydney, Australia.

Becque J and Wilkinson, T, 2012, "Experimental investigation of the static capacity of grade C450 RHS T and X truss joints", *Tubular Structures XIV, pp. 177–184*, Gardnr (Ed), Balkema.

Kurobane, Y. 1981. "New developments and practices in tubular joint design." *IIW Doc. XV-488-81* and *IIW Doc. XIII-1004-81*.

Liu, D.K., & Wardenier, J. 2004. "Effect of the yield strength on the static strength of uniplanar K-joints in RHS (steel grades S460, S355 and S235)." *IIW Doc. XV-E-04-293*, Delft University of Technology, Delft, the Netherlands.

Lu, L.H., de Winkel, G.D., Yu, Y., & Wardenier, J. 1994. "Deformation limit for the ultimate strength of hollow section joints." *Proceedings of the 6th International Symposium on Tubular Structures*, Melbourne, Australia, Tubular Structures VI, Balkema, Rotterdam, The Netherlands, pp. 341–347.

Mang, F. 1978. "Untersuchungen an Verbindungen von geschlossenen und offenen Profilen aus hochfesten Stählen." AIF-Nr. 3347. Universität Karlsruhe, Germany.

Packer, J.A., Wardenier, J., Zhao, X.-L., van der Vegte, G.J., & Kurobane, Y. 2009 *Design guide for rectangular hollow section (RHS) joints under predominantly static loading.* Second edition, CIDECT series "Construction with hollow steel sections" No. 3, Verlag TUV Rheinland, Köln, Germany.

Puthli, R., Bucak, O., Herion, S., Fleischer, O., Fischl, A., & Josat, O. 2010. "Adaptation and extension of the valid design formulae for joints made of high-strength steels up to S690 for cold-formed and hot-rolled sections." *CIDECT report 5BT-7/10* (draft final report), Germany.

Wardenier, J., Packer, J.A., Choo, Y.S., van der Vegte, G.J, & Orton, A. 2009. "Axially loaded T and X joints of elliptical hollow sections" *CIDECT report 5BW-6/09*.

Tubular Structures XV – Batista, Vellasco & Lima (eds)
© 2015 Taylor & Francis Group, London, ISBN 978-1-138-02837-1

Through-bolts to control ovalization of CHS T-joints under brace member compressive loads

M.A. Mohamed
Structural Engineering Department, Ain Shams University, Cairo, Egypt

A.A. Shaat
The German University in Cairo – Associate Professor (on leave) Ain Shams University, Egypt

E.Y. Sayed-Ahmed
The American University in Cairo – Professor (on leave), Ain Shams University, Egypt

ABSTRACT: Joints have usually been the challenging aspect for the design of tubular structures. The stiffness of hollow section joints usually comprises an important limitation for the designer. Therefore, increasing the joint stiffness by controlling the local ovalization at the vicinity of joints is expected to improve the economic aspect of tubular structures. In this research, a numerical study is carried out to investigate the efficiency of bracing the circular chord member of T-joints against ovalization by through-bolts in the radial direction when subject to brace member compression loads. The strengthening technique is based on. A parametric study is carried-out to investigate the effects of the number of bolts and spacing between bolts, chord diameter and brace-to-chord diameter ratio (β). The study revealed that the gain in joint stiffness is directly proportional to the number of bolts. The efficiency of this strengthening technique increases with small chord diameters and large β-ratios.

1 INTRODUCTION

Different strengthening techniques for joints of tubular structures have been developed over the past decades such as the "can" method, ring stiffeners, doubler plates, collar plates, concrete filling and through-bolts. This study is concerned with the through-bolts strengthening technique, where holes are drilled on both sides of the chord member and bolts are inserted from one side and anchored at the other.

Bains (1983) and Bradfield et al. (1994) were among the first to use through-bolts in rectangular hollow section (RHS) chord members. It was reported that joint capacity was increased by up to 18%. Zhao (1999) suggested inserting a wooden block inside the RHS to prevent the inward side wall buckling in addition to the through-bolt that prevents outward side wall buckling.

Aguilera et al. (2012) tested experimentally the effects of the number and pattern of bolts, as well as the web height-to-wall thickness ratio (h/t) of RHS chord members. The percentage gain in strength using 15 bolts increased from 3.1% to 29% for h/t ratios ranging from 34 to 65. The study showed that the number and pattern of bolts did not have a major effect on the strength gain for the selected bolts.

Sharaf & Fam (2013) conducted a parametric study using numerical analyses to investigate the effect of chord depth-to-width ratio, brace breadth-to-chord depth ratio, chord span-to-depth ratio, welding the

through bolts to chord's web, and the boundary conditions of the chord. The results showed that welding the through bolts affected the ductility of the joint significantly. The strength gain in the joint with welded bolts is 32%.

Mohamed et al. (2013) adopted a finite element model (FEM) to study the effect of using this technique on the strength gain of circular hollow chord members of T-joints. A maximum gain in strength of 41% is recorded. The study revealed the importance of having a central bolt under the brace member centerline to maximize the joint strength. It was also concluded that the larger the brace diameter-to-chord diameter ratio (β), the more efficient the through bolts technique in strengthening CHS T-joints.

In this study, the impact of using the same strengthening technique on the expected ovalization of CHS chord members of T-joints is numerically investigated. The chord ovalization (Δ) is defined as the difference between the vertical deformations of points A and B, as shown in Figure 1.

2 FINITE ELEMENT MODELING AND VERIFICATION

The finite element method provides a convenient method for structural modeling this task due to the versatility and the great advancement in its adaptation

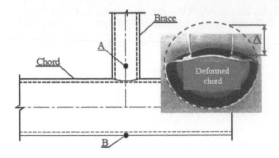

Figure 1. Undeformed and deformed shapes of CHS T-joint.

Table 1. Material properties.

Material	Yield strength (F_y) MPa	Young's modulus (E) GPa	Tangent modulus $(E_t = E/100)$ GPa
CHS	350	209	2.09
Bolts	894	209	2.09

Table 2. Verification of FEM results.

Sr.	$P_{Exp.}$ (kN)	P_{FEM} (kN)	$P_{FEM}/P_{Exp.}$	Reference/Description
1	312.0	311.90	1.00	Zhao & Hancock (1991).
2	156.0	160.11	1.02	Unstrengthened RHS T-joints
3	101.0	106.52	1.05	subjected to brace combined
4	381.0	367.29	0.96	compression and bending
5	238.0	237.74	0.99	actions.
6	305.1	312.28	1.02	Choo et al. (2005).
7	543.0	557.22	1.02	Unstrengthened CHS T-joints.
8	407.8	395.14	0.97	
9	131.0	142.59	1.09	Aguilera et al. (2012).
10	169.0	168.32	0.99	RHS T-joints strengthened
11	164.0	159.17	0.97	using through bolts.
12	162.0	164.70	1.02	
13	164.0	162.87	0.99	
14	165.0	153.94	0.93	
15	281.0	281.80	1.00	
16	294.0	295.76	1.00	
17	448.0	436.80	0.98	
18	462.0	459.35	0.99	
		Average	1.00	
		S.D.	0.03	

to computer use. A brief description of the elements adopted to model the CHS T-joints, material model and verification of the model is discussed in the following sections.

2.1 Element types and material modeling

The structural shell element (SHELL181) is used to model the chord and brace members. SHELL181 is a four-node element with six degrees of freedom at each node. This element is suitable for analyzing thin to moderately-thick shell structures and well-suited for large rotation and large strain nonlinear applications. Shell thickness option is applied to account for the bending stiffness. Element size is chosen equal to 25 mm and is kept constant for the whole specimen to avoid any mesh-dependent sensitivity of the model results.

The 3D spar (LINK180) is used to model the through bolts. LINK180 is a uniaxial tension-compression element with three degrees of freedom at each node: translations in the nodal x, y, and z directions. As in a pin-jointed structure, no bending of the element is considered. Plasticity, rotation, large deflection, and large strain capabilities are included. To allow for any inward movement, tension only option is chosen.

The stress-strain curves of the analyzed specimens are modeled using a bilinear isotropic hardening plasticity model. Table 1 lists the values of yield strength, Young's modulus and tangent modulus (E_t) for both the CHS steel material and through bolts.

2.2 Model verification

The results of the proposed FE model are verified against 18 experiments of T-joints. These experimental studies were performed by Zhao and Hancock (1991),

Choo et al. (2005) and Aguilera et al. (2012). The finite element program ANSYS (14) is adopted for the pre-processing, equation solution and post-processing of the model results.

Table 2 shows the experimental results of the previously mentioned studies against the FEM results of the current study. The table indicates good correlation with minor discrepancy of an average of value of 1.00 for the strength ratio between P_{FEM}/P_{Exp} and a standard deviation of 0.03. As such, it can be concluded that the finite element model provides realistic results and can capture the real behavior of the specimens. The model is considered suitable for the parametric study that follows.

3 PARAMETRIC STUDY

A parametric study is conducted using the verified FE model. The study is performed on CHS T-joints strengthened using through bolts and subjected to brace compression load. The effects of the through-bolts on the reductions in chord ovalization and, hence, the increases in joint stiffness are investigated.

3.1 Model description

In order to build-up the finite element model for the parametric study, the T-shape of the joints is developed, as shown in Figures 2 and 3. The brace and the chord member are assumed to be welded using a full penetration weld through the entire thickness of the member. Therefore, full contact between elements is modeled without any additional elements to simulate the welding material.

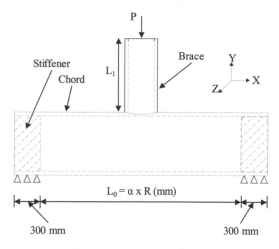

Figure 2. Schematic drawing of the T-joint.

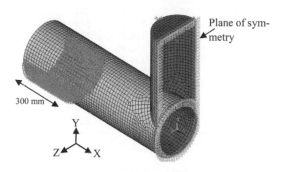

Figure 3. Finite element modeling of specimens.

Supports for the T-joints are modeled all over the nodes of the chord lower half for a distance of 300 mm, as shown in Figure 3. All nodes at this portion are prevented from translation in both the vertical (Y) and transverse (Z) directions at both ends of the chord diameter. Inner vertical plate stiffeners of 12 mm thickness are provided at both ends and penetrated inside the chord for a distance of 300 mm to avoid any premature failure at the specimen ends. The load is applied as a point load on a thick plate at the top end of the brace member to ensure uniform load distribution. The length of the brace member is kept constant at 500 mm. All brace members are chosen with thick walls ($t_1 = 11.13$ mm) to avoid premature brace failure.

3.2 Parameters

A total of 156 specimens are considered in this parametric study. They are classified into six groups. Each group consists of 26 specimens of the same chord and brace cross sections, as listed in Table 3.

The radius-to-thickness ratio of chord members ($\gamma = d_0/2t_0$) is kept as close as possible for the two chord sizes used in this study (i.e. $\gamma = 28.03$ for $d_0 = 356$ mm and $\gamma = 27.40$ for $d_0 = 610$ mm). The chord

Table 3. Groups of the parametric study.

Group	Chord	Brace	β
1	356*6.35	324*11.13	0.91
2	356*6.35	273*11.13	0.77
3	356*6.35	219*11.13	0.62
4	610*11.13	559*11.13	0.92
5	610*11.13	508*11.13	0.83
6	610*11.13	406*11.13	0.67

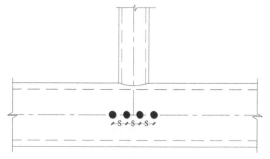

Figure 4. Arrangement of through-bolts.

length-to-chord radius ratio ($\alpha = 2L_0/d_0$) is taken equal to 9.00 (i.e. $L_0 = 1600$ mm for $d_0 = 356$ mm and $L_0 = 2742$ mm for $d_0 = 610$ mm) to ensure that the local stresses at the support locations do not overlap with the stresses at the T-joint.

The number of through bolts (n) ranges from one to five bolts. The arrangement of the bolts is always symmetric around the joint centerline, as shown in Figures 4. Different spacing values (s) between the through bolts along the centerline of the chord are considered (25 mm, 50 mm, 75 mm, 100 mm, 150 mm and 200 mm).

4 RESULTS AND DISCUSSION

Table 4 to Table 9 present the results in terms of the stiffness of each specimen along with its percentage gain compared to the stiffness of the unstrengthened specimen. The stiffness K (kN/mm) is calculated as the slope of the load-ovalization curve (i.e. $K = P/\Delta$), where P is the load at 3% of the maximum load of the specimen and Δ is the corresponding ovalization. The results listed in Table 4 to Table 9 clearly indicate that using the through-bolts technique enhances the joint stiffness and reduces its ovalization. Figure 5 shows the percentage gain in stiffness versus the number of through bolts for all specimens in groups 1 to 6 compared to the counterpart unstrengthened specimen. It is clear that the percentage gain in stiffness is greatly affected by the number of bolts, spacing between bolts, chord diameter and β ratios. The effects of each of these parameters are discussed in the following sections.

499

Table 4. Results of Group 1 specimens ($d_0 = 356$, $d_1 = 324$).

No. of bolts	Spacing (mm)	Stiffness (kN/mm)	% gain in Stiffness
0	–	99.10	N/A
1	–	161.35	63
2	25	185.63	87
2	50	192.89	95
2	75	196.99	99
2	100	199.83	102
2	150	200.73	103
2	200	225.83	103
3	25	202.38	104
3	50	217.01	119
3	75	224.02	126
3	100	229.58	132
3	150	231.49	134
3	200	229.58	132
4	25	217.00	119
4	50	237.43	140
4	75	248.03	150
4	100	254.86	157
4	150	254.85	157
4	200	248.03	150
5	25	231.47	134
5	50	257.21	160
5	75	269.71	172
5	100	275.05	178
5	150	272.36	175
5	200	257.22	160

Table 6. Results of Group 3 specimens ($d_0 = 356$, $d_1 = 219$).

No. of bolts	Spacing (mm)	Stiffness (kN/mm)	% gain in Stiffness
0	–	49.05	N/A
1	–	91.12	86
2	25	106.36	117
2	50	112.16	129
2	75	115.32	135
2	100	117.51	140
2	150	119.80	144
2	200	119.80	144
3	25	117.50	140
3	50	127.21	159
3	75	134.13	173
3	100	138.65	183
3	150	141.84	189
3	200	140.23	186
4	25	127.2	159
4	50	141.84	189
4	75	150.49	207
4	100	156.20	218
4	150	158.21	223
4	200	152.35	211
5	25	135.58	176
5	50	154.25	214
5	75	166.76	240
5	100	171.39	249
5	150	169.04	245
5	200	160.26	227

Table 5. Results of Group 2 specimens ($d_0 = 356$, $d_1 = 273$).

No. of bolts	Spacing (mm)	Stiffness (kN/mm)	% gain in Stiffness
0	–	67.53	–
1	–	127.67	89
2	25	151.75	125
2	50	159.61	136
2	75	163.86	143
2	100	165.32	145
2	150	166.80	147
2	200	165.31	145
3	25	169.86	152
3	50	183.34	171
3	75	190.9	183
3	100	194.92	189
3	150	196.99	192
3	200	192.89	186
4	25	183.33	171
4	50	203.49	201
4	75	215.33	219
4	100	220.45	226
4	150	223.11	230
4	200	212.84	215
5	25	199.11	195
5	50	223.11	230
5	75	237.41	252
5	100	240.49	256
5	150	234.41	247
5	200	217.86	223

Table 7. Results of Group 4 specimens ($d_0 = 610$, $d_1 = 559$).

No. of bolts	Spacing (mm)	Stiffness (kN/mm)	% gain in Stiffness
0	–	166.29	N/A
1	–	222.78	34
2	25	257.17	55
2	50	260.42	57
2	75	263.77	59
2	100	264.90	59
2	150	266.03	60
2	200	267.19	61
3	25	277.37	67
3	50	285.94	72
3	75	292.74	76
3	100	296.25	78
3	150	303.54	83
3	200	303.54	83
4	25	299.09	80
4	50	312.88	88
4	75	321.80	94
4	100	326.45	96
4	150	334.48	101
4	200	332.84	100
5	25	316.5	90
5	50	335.14	102
5	75	347.74	109
5	100	354.41	113
5	150	361.34	117
5	200	357.85	115

Table 8. Results of Group 5 specimens ($d_0 = 610$, $d_1 = 508$).

No. of bolts	Spacing (mm)	Stiffness (kN/mm)	% gain in Stiffness
0	–	129.83	N/A
1	–	189.26	46
2	25	223.59	72
2	50	226.88	75
2	75	229.41	77
2	100	232.00	79
2	150	234.65	81
2	200	234.64	81
3	25	247.85	91
3	50	257.15	98
3	75	263.75	103
3	100	267.18	106
3	150	270.69	108
3	200	269.51	108
4	25	268.34	107
4	50	283.12	118
4	75	291.14	124
4	100	296.75	129
4	150	301.08	132
4	200	299.62	131
5	25	287.07	121
5	50	307.08	137
5	75	318.18	145
5	100	324.87	150
5	150	328.33	153
5	200	323.17	149

Table 9. Results of Group 6 specimens ($d_0 = 610$, $d_1 = 406$).

No. of bolts	Spacing (mm)	Stiffness (kN/mm)	% gain in Stiffness
0	–	88.43	N/A
1	–	134.08	52
2	25	159.53	80
2	50	162.33	84
2	75	164.65	86
2	100	166.42	88
2	150	168.86	91
2	200	169.48	92
3	25	177.95	101
3	50	185.08	109
3	75	190.42	115
3	100	194.42	120
3	150	197.74	124
3	200	198.59	125
4	25	192.79	118
4	50	204.74	132
4	75	212.27	140
4	100	218.28	147
4	150	223.56	153
4	200	223.56	153
5	25	206.56	134
5	50	207.5	135
5	75	233.72	164
5	100	239.78	171
5	150	244.86	177
5	200	243.57	175

4.1 Effect of number of through bolts

Figure 5 shows that the percentage gain in joint stiffness is directly proportional to the number of through bolts. The maximum gain in stiffness (256%) is recorded for a specimen of a β-ratio equal to 0.77 (Group 2) and strengthened with five bolts spaced at 100 mm.

4.2 Effect of spacing between through bolts

The percentage gain in joints stiffness, listed in Table 4 to Table 9 and plotted in Figure 5, show that the effect of bolts spacing is associated with the number of bolts. The maximum gain in stiffness is usually recorded at a spacing value that decreases with increasing the number of bolts. For example, the best values of spacing for group 1 specimens are 200 mm, 150 mm, 100 and 100 for specimens strengthened with 2, 3, 4 and 5 bolts, respectively. This phenomenon is, however, valid for all investigated specimens groups 1 to 6.

4.3 Effect of chord diameter

The effect of changing the chord diameter on the joint stiffness enhancement is presented in this study. It is found that the maximum gain in stiffness is achieved at smaller diameter chord members. It should, however, be noted that this result may change should stronger bolts are used with large chord diameters.

4.4 Effect of β-ratio

The effect of changing the β-ratios on the joint strength enhancement while maintaining the same chord diameter is investigated. Figure 6 shows the percentage gain in stiffness versus different β-ratios for 75 mm spacing between bolts. Generally, the percentage gain in stiffness tends to decrease by approaching a β-ratio of 1.

4.5 Failure modes

All unstrengthened T-joint specimens experienced a common failure mode where a plastic zone is developed exactly under the centerline of the brace member and spreads in the vicinity of the joint depending on its β-ratio. The high stress zones are indicated by the Von-Misses plots, as shown in Figure 7. It is noted that as the β-ratio gets smaller, the failure area under the brace member gets smaller and is shifted upwards towards the brace member, as shown in Fig. 7a to Fig. 7c.

For strengthened specimens with small chord diameters, plastic zones are developed around the bolts. These plastic zones are developed in a separate manner for large spacing between bolts (Fig. 7d) but merged in one continuous zone along the line of bolts for small spacing values. In the case of small β values, an additional line of failure was developed above the bolts' line, as shown in Figure 7e.

501

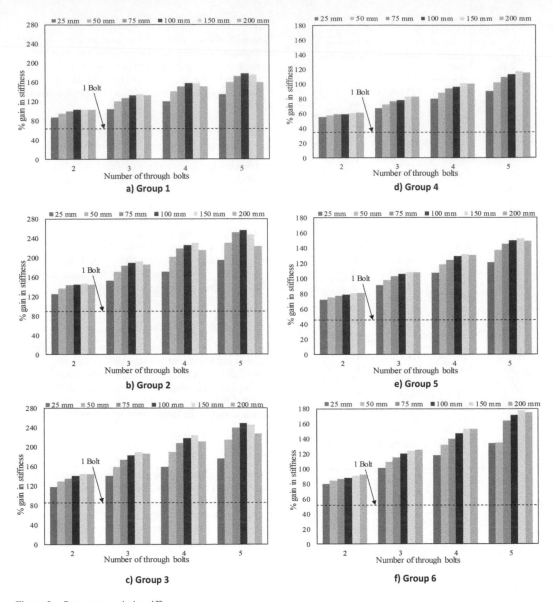

Figure 5. Percentage gain in stiffness.

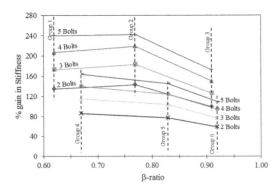

Figure 6. Effect of β-ratio on the gain in stiffness for (spacing = 75 mm).

For strengthened specimens with large chord diameters, failure occurred due to rupture of the bolts. Consequently, the stresses of the chord member around the bolts locations were lower than the surrounding areas, as shown in Figure 7f.

5 CONCLUSIONS

In this study, a numerical investigation is carried out to evaluate the use of through-bolts technique in increasing the stiffness of CHS T-joints. A finite element model is developed and a parametric investigation is carried out to examine the effects of the number of bolts, spacing between bolts, chord diameter

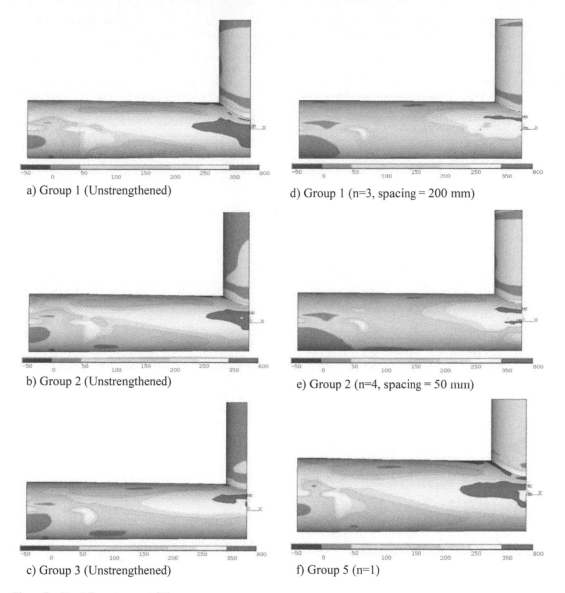

a) Group 1 (Unstrengthened)

d) Group 1 (n=3, spacing = 200 mm)

b) Group 2 (Unstrengthened)

e) Group 2 (n=4, spacing = 50 mm)

c) Group 3 (Unstrengthened)

f) Group 5 (n=1)

Figure 7.   Von-Mises stresses at failure.

and brace-to-chord diameter ratio (β). The following conclusions are drawn:

1. The percentage gain in stiffness is directly proportional to the number of through bolts. The maximum recorded gain in stiffness is 256% for specimen with a β-ratio of 0.77 strengthened with five bolts spaced at 100 mm.
2. The smaller the chord diameter, the more effective the strengthening technique for the same configuration of through bolts.
3. The smaller the brace-to-chord diameter (β) ratio, the stiffer the CHS T-joints strengthened using through bolts.

REFERENCES

Aguilera, J., Shaat, A. & Fam, A., 2012. Strengthening T-Joints of Rectangular Hollow Steel Sections against Web Buckling under Brace Axial Compression using Through-Wall Bolts. *Journal of Thin-Walled Structures*, 56(7), pp. 71–78.
ANSYS., 2011. Release 14 user's manual. Swanson analysis system.
Bains, S., 1983. *Investigating into the effects of through bolting on rectangular hollow steel beams in flexure.* s.l.: University of Sussex.
Bradfield, C., Morrel, P. & Ibrahim, A., 1994. *Improvements in the flexural capacity of rectangular hollow sections by through bolting.* Rotterdam, Balkema, pp. 109–114.

Choo, Y. et al., 2005. Static Strength of T-Joints Reinforced with Doubler or Collar Plates. *Journal of Structural Engineering*, 131(1), pp. 119–128.

Mohamed, M., Shaat, A., Sayed-Ahmed, E., 2013. Finite Element Modelling of CHS T-Joints Strengthened using Through-bolts. The Fifth International Conference on Structural Engineering, Mechanics and Computation. September 2–4, Cape Town, South Africa, pp. 1335–1340.

Sharaf, T. & Fam, A., 2013. Finite element analysis of beam-column T-joints of rectangular hollow steel sections strengthened using through-wall bolts. *Thin-Walled Structures*, Volume 64, pp. 31–40.

Zhao, X., 1999. Partially stiffened RHS sections under transverse bearing force. *Thin-Walled Structures*, Volume 35, pp. 193–204.

Zhao, X. & Hancock, G., 1991. T-joints in rectangular hollow sections subjected to combined actions. ASCE, Journal of Structural Engineering, 117(8), pp. 2258–2277.

Tubular Structures XV – Batista, Vellasco & Lima (eds)
© 2015 Taylor & Francis Group, London, ISBN 978-1-138-02837-1

Experimental study on static behavior of multi-planar overlapped CHS KK-joints

X.Z. Zhao, S.S. Han, K.H. Hu, Y.Y. Chen & A.H. Wu
Tongji University, Shanghai, China

ABSTRACT: In the past decade, considerable experimental and numerical studies have been carried out on uniplanar overlapped CHS K-joints in order to investigate the behavior of the joints and to establish formulae for their static strength calculation. However, little substantial information on multi-planar overlapped CHS KK-joints considering the construction process and loading patterns is available for tubular connections. In this study, four static tests on multi-planar CHS KK-joints with in-plane-overlapped braces (KK-IPOv joints) under axial loading were carried out. Among these four specimens, factors including different welding situations of hidden seam (presence or absence of hidden welds), presence of vertical stiffener for avoiding partial overlapping of one brace member onto another, as well as loading patterns (symmetric or anti-symmetric axial loading) were varied to study their effects on the joint behavior. Based on the tests, nonlinear finite element model is established and validated for further parametric studies. Results from both the tests and the FEM show that the welding situation of hidden seam and the presence of vertical stiffener have some effects on the stress distribution and failure mechanism, but the static ultimate capacity of the multi-planar overlapped CHS KK-joints is not affected significantly. Thus the absence of hidden welds and vertical stiffener are more suitable for the convenience of construction. The stress distribution on the chord and the failure modes differ significantly between symmetric and anti-symmetric loadings and the ultimate capacity of the multi-planar overlapped KK-joint has decreased by 17.8% when subjected to anti-symmetric loading.

NOMENCLATURE

D: Chord diameter
T: Chord thickness
d: Brace diameter
t: Brace thickness
g_l: Longitudinal gap between braces
g_t: Transverse gap between braces
e_l: In-plane eccentricity
e_t: Out-of-plane eccentricity
θ: Angle between brace axis and chord axis
Φ: Out-of-plane angle between planes in which braces lie

1 INTRODUCTION

Multi-planar CHS KK-joints, the basic components of three-dimensional triangle trusses, have been widely used in large-span structures. Multi-planar CHS KK-joints can be characterized into four types according to their geometric configurations: gap joints (Figure 1a), KK-OPOv joints with out-of-plane-overlapped braces (Figure 1b), KK-IPOv joints with in-plane-overlapped braces (Figure 1c), and KK-Ov joints with both in and out of plane braces overlapped (Figure 1d). Overlapped joints are often used in practice due to the limitation of available sizes of steel pipes as well as the requirement of spacial structural

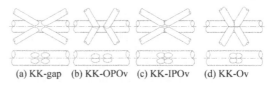

(a) KK-gap   (b) KK-OPOv   (c) KK-IPOv   (d) KK-Ov

Figure 1. Classification of multi-planar KK joints.

configuration. This paper focuses mainly on the static behaviour of KK-IPOv joints.

For overlapped joints, it is inconvenient to weld the hidden seam of overlapped braces during construction. It is thus preferable in terms of the construction process to leave the hidden seam unwelded given its effect on the joint capacity may be neglected. Existing studies on hidden seam mainly focus on uniplanar overlapped CHS K-joints. Zhao et al. (2006) tested ten overlapped CHS K-joints with or without the hidden seams welded under monotonic loading; Wu et al. (2008) studied the effects of hidden seam absence on the behavior of overlapped CHS K-joint using inite element method; Wang (2007) carried out both experimental study and numerical analysis on three specimens of two types of overlapped CHS KK-joints. However, studies on multi-planar overlapped CHS KK-joints are still insufficient. Without fully understanding and sound evidence of the effect of hidden seam absence on joint capacity, designers choose to weld each brace member to a vertical stiffener in order to avoid partial overlapping of one brace member onto another.

(Packer et al. 2009). However, this design increases the inconvenience of construction. In addition, the influence of vertical stiffener on KK-joint behavior has not been studied or verified experimentally.

For multi-planar KK-joints, gravity load within a truss gives rise to symmetric axial loads, while horizontal loads acting on trusses may cause anti-symmetric axial loads. Makino et al. (1994) initiated a research on KK-Gap joints under anti-symmetric loading. Nine specimens were tested and local deformation of chord surface between the compressive braces forming a diagonal ridge was observed. Two main types of joint failure mode by chord wall bending were identified. Based on the test results and finite element analysis results from Lee (1997), Makino et al. (1995) proposed two strength formulae for estimating the ultimate capacity of gapped KK-joints under anti-symmetric loading. Lee et al. (1997) conducted a parametric study and regression analysis with 40 finite element models and developed two sets of strength formulae for gapped KK-joints under anti-symmetric loading. Results from both the tests and the numerical analyses manifested that KK-joints behaved quite differently under symmetric and anti-symmetric loadings, and the ultimate strength was generally lower under the latter. For KK-IPOv joints under anti-symmetric loading, no experiments have been conducted in the literature.

In this study, static tests on axially loaded KK-IPOv joints were carried out. Among the four specimens, factors including different welding situations of hidden seam (welded or unwelded), presence of vertical stiffener for avoiding partial overlapping of one brace member onto another, as well as loading patterns (symmetric or anti-symmetric axial loading) were varied to study their effects on the joint behavior. Based on the tests, nonlinear finite element models were established and validated for further parametric studies.

2 EXPERIMENTAL PROGRAM

2.1 Test specimens

Configuration of KK-IPOv joints is illustrated in Figure 1. The vertical stiffener is adopted only for IPOv-HC and anti-symmetric loading is shown in brackets. Geometric parameters are identical for the four specimens (Tables 1–2).

Details on construction factors and loading patterns of specimens are listed in Table 3. Through braces 3 and 4 (see Figure 2) are in compression while lap braces 2 and 1 in tension for both IPOv-N and IPOv-W, but the hidden seam is unwelded in specimen IPOv-N and welded in IPOv-W. Vertical stiffener in the transverse direction was used in specimen IPOv-HC. Symmetric loading pattern was

Table 1. Geometric parameters.

D mm	T mm	d mm	t mm	g_l mm	g_t mm	e_l mm	e_t mm	θ °	Φ °
325	10	168	8	−49.4	50.1	−30	0	62	80

Table 2. Non-dimensional parameters.

β	γ	τ	ζ_d	ζ_t	e_l/D	e_t/D
0.517	16.25	0.8	−0.152	0.154	−0.092	0

Table 3. Details of test specimens.

Specimen	Through braces	Hidden seam	Vertical stiffener	Loading pattern
IPOv-N	3, 4	Unwelded	–	Symmetric
IPOv-W	3, 4	Welded	–	Symmetric
IPOv-HC	–	–	Presence	Symmetric
IPOv-N-A	3, 1	Unwelded	–	Anti-symmetric

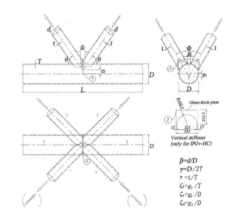

Figure 2. Configuration of KK-IPOv joints under symmetric loading (anti-symmetric loading).

applied to specimens IPOv-N, IPOv-W, and IPOv-HC. For specimen IPOv-N-A, the hidden seam was unwelded and anti-symmetric loading was applied; through braces are compressive braces 3 and 1 while lap braces are tensile braces 2 and 4.

Specimens used for tensile coupon tests were cut from the tubes used to fabricate the joints. The test results obtained are listed in Table 4. The member thickness measured before tests are listed in Table 5.

2.2 Test setup

The specimens were tested using the multi-functional spherical reaction facility consisting of two hemispheres (Figure 3) with the capability of applying loads for all kinds of multi-planar joints. The diameter of the inner space is about 6 m. One end of the chord was fixed to the reaction facility and the other end was

Table 4. Tensile coupon test results.

Specimen norminal size	Grade of steel	Yielding strength f_y (MPa)	Tensile strength f_u (MPa)	Yield ratio f_y/f_u	Elongation δ (%)
Chord ($\Phi 325 \times 10$)	Q345B	349	555	0.63	28.1
Brace ($\Phi 168 \times 8$)	Q345B	386	563	0.69	29.5
Stiffener	Q235B	239	368	0.48	30.2

Table 5. Measured member thickness.

Specimen	Brace 1	Brace 2	Brace 3	Brace 4	Chord	Stiffener
IPOv-N	8.25	8.25	8.25	8.25	10.72	–
IPOv-W	8.35	8.35	8.35	8.35	10.72	–
IPOv-HC	8.25	8.25	8.25	8.25	10.72	9.52
IPOv-N-A	8.25	8.25	8.25	8.25	10.72	–

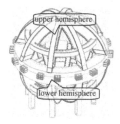

Figure 3. Spherical supporting facility.

Figure 4. Sliding end.

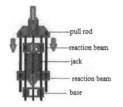

Figure 5. Tension reversing device.

Figure 6. Installation.

allowed to slide in a horizontal sleeve (Figure 4). Four hydraulic jacks sharing one oil-way were used to apply identical loads on four braces respectively. Two compressive braces were directly loaded with jacks while tension-reverse devices were used to apply loads on tensile braces (Figure 5).

Load was applied initially step by step with a 100 kN increment for each step. After the appearance of the inflection point on the compressive brace load-deformation curves, load was applied continuously.

2.3 *Measurement arrangement*

In order to obtain the actual loads in the chord and braces, 36 strain gauges were attached on the

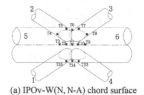

(a) IPOv-W(N, N-A) chord surface

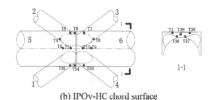

(b) IPOv-HC chord surface

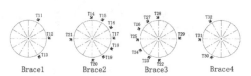

(c) Brace surface

Figure 7. Arrangement of rosette strain gauges.

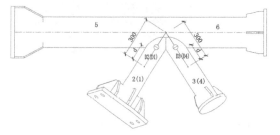

Figure 8. Arrangement of displacement gauges.

chord/brace surfaces along the length of each member; To trace the stress distribution, rosette strain gauges were applied around the intersecting curves of chord and braces, as shown in Figure 7; The deformation of chord wall under axial brace loadings was measured with four wire displacement gauges, as shown in Figure 8.

3 EXPERIMENTAL RESULTS

3.1 *Test phenomenon and failure modes*

For specimen IPOv-N, when load had reached 665 kN, the outside saddle spot T7 on the chord started to yield. After that, spots T15, T13, and T6 around the outside Intersection Point (IP) and saddle spot T5 yielded successively. The region around the outside IPs and outside saddle spots yielded first and developed with increasing load. When the load reached 1200 kN, most spots had yielded and the compressive brace load-deformation curves had inflexed, but with no obvious deformation observed on the chord surface. Load was

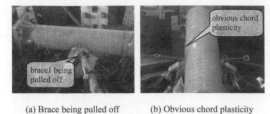

(a) Brace being pulled off   (b) Obvious chord plasticity

Figure 9. Test phenomenon and failure mode of IPOv-N.

(a) Brace being pulled off   (b) Crack at IP

Figure 10. Test phenomenon and failure mode of IPOv-W.

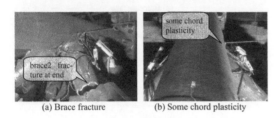

(a) Brace fracture   (b) Some chord plasticity

Figure 11. Test phenomenon and failure mode of IPOv-HC.

(c) Brace and gap region   (d) Side surface of chord

Figure 12. Test phenomenon and failure mode of IPOv-N-A.

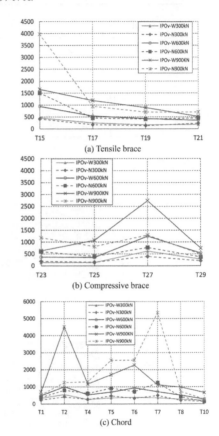

(a) Tensile brace

(b) Compressive brace

(c) Chord

Figure 13. Comparison of strain distribution for IPOv-W and IPOv-N.

applied continuously after 1200 kN. When it reached 1567 kN, cracks initiated from welds connecting the compressive brace 4, tensile brace 1 and the chord, propagated along the intersecting line and eventually caused brace 1 being totally pulled off at the throat of the welds and the cross section of the brace (Figure 9a), accompanied with obvious plastic deformation on the chord surface (Figure 9b). No obvious deformation and cracks were observed on other braces.

For specimen IPOv-W, the region around the outside IPs and outside saddle spots yielded first and developed with increasing load, similar to IPOv-N. When load reached 1393 kN, welds connecting tensile brace 2, compressive brace 3 and the chord started to crack and finally brace 2 was totally pulled off at the throat of welds, with no obvious plastic deformation on the chord surface (Figure 10a). Crack initiated at the outside IP of the compressive brace 4, tensile brace 1 and the chord (Figure 10b) and propagated along the welds between brace 1 and the chord.

For specimen IPOv-HC, the region around the outside IPs and outside saddle spots yielded first, similar to IPOv-N and IPOv-W. When load reached 1675 kN, the tensile brace 2 fractured at the stiffening end (Figure 11a), with certain plastic deformation on the chord surface (Figure 11b). No obvious deformation was observed on other braces or the vertical stiffener.

For specimen IPOv-N-A, when load reached 500 kN, spots T1 and T2 on the chord around the inner IP yielded first. T10, T32 around the inner IP yielded successively. The gap region around the inner IPs yielded first and then enlarged with load increase. This was different from specimens under symmetric loading. When the load reached 1100 kN, most spots had yielded and the compressive brace load-deformation curves had inflexed, but no obvious deformation was observed on the chord surface. Load was applied continuously after 1100 kN. When the load reached 1275 kN, welds connecting the tensile brace 2, compressive brace 3 and the chord cracked continuously and brace 2 was totally pulled off at the throat section of the welds (Figure 12a). Some plastic deformation

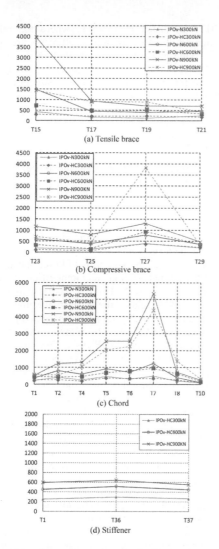

(a) Tensile brace

(b) Compressive brace

(c) Chord

(d) Stiffener

Figure 14. Comparison of strain distribution for IPOv-N and IPOv-HC.

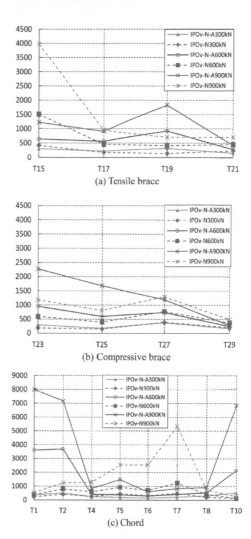

(a) Tensile brace

(b) Compressive brace

(c) Chord

Figure 15. Comparison of strain distribution for IPOv-N-A and IPOv-N.

was seen on the side surface of the chord (Figure 12b) while more obvious plastic deformation was observed in the transverse gap region (Figure 12a). No obvious deformation developed on other braces.

3.2 Strain distribution

The equivalent strain $\varepsilon_i (\times 10^{-6})$ distribution near the intersecting curves of braces and chord is shown in Figures 13–15.

It can be seen that: (1) for the three specimens subjected to symmetric loading, gauges at T15 and T27 around the outside IP measured larger strain than other positions on braces; for specimens IPOv-N and IPOv-HC, the outside saddle spot T7 measured the largest strain on the chord; for specimen IPOv-W, T6 around the outside IP and the saddle spot T2 in the transverse gap between two tensile braces have larger strain than

other measure spots on the chord. (2) the vertical stiffener experienced small strain and no yielding occurred when load reached 900 kN. (3) for the three specimens subjected to symmetric loading, the strain magnitude and distribution regularity are similar. ence on several measure spots (T2, T7). (4) strain distribution showed significant difference between symmetric and anti-symmetric loadings. For specimens under symmetric loading, spots around the outside IP and the saddle spots have larger strain on the chord; on braces, spots around the outside IP have larger strain. For the specimen under anti-symmetric loading, spots in the transverse gap and spots around the inner IP have larger strain on the chord; on braces, spots around the inner IP have larger strain.

3.3 Load-deformation curves

Brace load-deformation curves are illustrated in Figure 16. Load N is the internal force in the braces, which

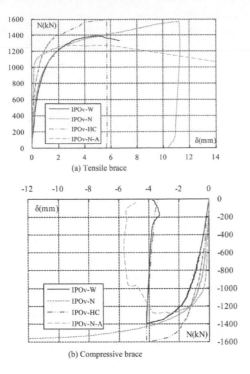

(a) Tensile brace

(b) Compressive brace

Figure 16.   Load-deformation curves.

is the actual load applied by the jacks; deformation δ is the deformation of the chord wall under tension/compression of respective brace in the direction of the original brace axis, calculated from the measured value of the wire displacement gauges deducting the elastic deformation of the braces. Tensile force in the brace is nominated to be positive and compressive force negative.

It can be seen that: (1) different specimens showed similar trend at both the elastic stage and the plastic stage. (2) for IPOv-W with welded hidden seam, the curve under tension agreed well with that of IPOv-N. (3) for specimen IPOv-HC, the peak load is slightly higher than that of IPOv-N. (4) for specimen IPOv-N-A, the initial stiffness is larger than other specimens but the peak load is the lowest among the four specimens.

3.4   Ultimate strength

Two potential failure criteria are considered and checked for each joint to get the ultimate strength: (1) the peak load on the brace load-deformation curve. (2) load corresponding to an allowable deformation, i.e., 3% of the chord diameter. The ultimate strength of joints is defined as the lowest load in all braces obtained from (1) and (2). The ultimate strength of each joint, relative value compared with IPOv-N, ultimate strength of the brace, joint efficiency, as well as comparison with predictions from EN1993 (2005), CIDECT (2008),

Li (2012) are listed in Table 6, where the relative value = (ultimate strength – ultimate strength of IPOv-N)/ultimate strength of IPOv-N × 100%; Joint efficiency = ultimate strength/brace strength; Ratio = the formula prediction/ultimate strength. Li's Equation is

$$N_{KK} = 0.9 N_K$$

$$N_K = \left(\frac{32.9}{\lambda + 25.2} - 0.084 \right) A_b f_{,b}$$

where A_b = cross section area of the brace; f_{yb} = yielding strength of the brace;

$$\lambda = \beta^{ov} \gamma \tau^{0.8 - ov}$$

where Ov = lap ratio

Tested geometric parameters and yielding strength are used in the formulae.

It can be seen from Table 6 that (1) for specimen IPOv-W, the ultimate strength and joint efficiency are lower than that of IPOv-N. This decrease is caused by premature crack of welds of the tensile brace, implying the importance of weld quality for overlapped joints. (2) for specimen IPOv-HC, the ultimate strength is slightly higher than that of IPOv-N. The joint efficiencies of both IPOv-HC and IPOv-N are close to 1.0, which means failures of the joint and the brace member almost occurred at the same time. This can be confirmed by failure modes: cross section of the tensile brace fractured in IPOv-HC while both welds and cross section of the tensile brace failed in IPOv-N. (3) The ultimate strength of IPOv-N-A is 17.8% lower than that of IPOv-N. That is because the load transfer mechanism is different between symmetric and anti-symmetric loadings. When subjected to symmetric loading, forces in two planes do not interact at the transverse gap region. The ultimate strength relies on brace strength and chord bending capacity in the longitudinal direction. However, when subjected to anti-symmetric loading, the transverse gap region on the chord is subjected to shear caused by diagonal compression and tension. The strength of joints relies on the bending capacity of the chord wall in both the transverse and the longitudinal direction. Chord plasticity rather than brace failure is the decisive factor for joint failure, which is why the joint efficiency is also low under anti-symmetric loading.

The three strength formulae for multi-planar KK-joints mentioned in this paper all take the form of multiplying the uniplanar K-joint strength formula by a multi-planar coefficient. Calculation results are all conservative compared with test results. CIDECT and Li's formulae have better prediction accuracy than EN1993. CIDECT formula is based on the effective length model, which reflects brace failure mode, the most common failure mode observed in overlapped joints. Li's formula is based on joint efficiency theory, which also reflects brace failure mode. However, EN1993 formula is based on the chord plasticity failure mode of gap joints, which is not applicable to overlapped joints. Thus, EN1993 formula has larger deviation from test results.

Table 6. Ultimate strength and comparison with formula predictions.

Specimens	Ultimate strength kN	Relative Value	Brace strength kN	Joint efficiency	N_{EN} kN	Ratio	N_{CIDECT} kN	Ratio	N_{Li} kN	Ratio
IPOv-N	1551	0	1598	0.97	802	0.52	1201	0.77	1183	0.76
IPOv-W	1393	−10.2%	1617	0.86	802	0.58	1215	0.87	1194	0.86
IPOv-HC	1598	3.1%	1598	1.00	802	0.50	1201	0.75	1183	0.74
IPOv-N-A	1275	−17.8%	1598	0.79	802	0.63	1201	0.94	1183	0.93

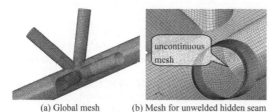

(a) Global mesh   (b) Mesh for unwelded hidden seam

Figure 17.   FE mesh map for IPOv-N.

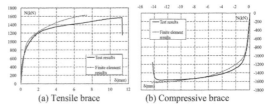

(a) Tensile brace   (b) Compressive brace

Figure 18.   Comparison of brace load-deformation curves for IPOv-N in test and finite element analysis.

4   FINITE ELEMENT ANALYSIS

4.1   *Finite element model*

Elasto-plastic large displacement analysis was carried out on all specimens tested using the commercial FE program ABAQUS to obtain the stress distribution and bearing capacity of the joints. Compared with models including welds, models without welds have enough accuracy and are used for analysis. The S4R linear shell element is used and final mesh map was shown in Figure 17.

4.2   *Comparison with experimental results*

(1) *Load-deformation curves*
Figure 18 illustrates that the brace load-deformation curves obtained from test and finite element analysis agree well for IPOv-N. For specimen IPOv-W, the peak load on the experimental curve is lower than the finite element analysis value, but the two curves showed similar trend before cracking. For specimens IPOv-HC and IPOv-N-A, good agreement between test and FE results is achieved generally.

(2) *Ultimate strength*
Table 7 lists the comparison of ultimate strength between test and finite element analysis results. Except for IPOv-W, other specimens showed good agreement between test and finite element analysis results, which validates the finite element models generated.

(3) *Stress distribution*
Figure 19 illustrates the contour plot of Von Mises stress and plastic zone of joints under symmetric and anti-symmetric loadings in the ultimate state. Specimens IPOv-N and IPOv-HC have similar distribution with IPOv-W. The stress distribution regularity in finite element analysis is in accordance with the strain distribution regularity in tests (Section 3.2).

Table 7.   Comparison of ultimate strength in tests and FEM.

Specimens	FEM results kN (1)	Test results kN (2)	Deviation % [(1)–(2)]/(2) × 100%
IPOv-N	1567.8	1551.1	1.1
IPOv-W	1572.2	1392.5	12.9
IPOv-HC	1647.6	1598.7	3.1
IPOv-N-A	1350.2	1275.2	5.9

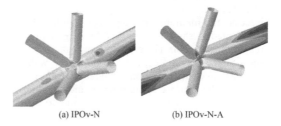

(a) IPOv-N   (b) IPOv-N-A

Figure 19.   Von Mises stress distribution in FEM.

5   DISCUSSION

5.1   *Effects of hidden seam*

Previous research indicates that for uniplanar overlapped K-joints with through compressive braces (CW joints), hidden seam absence has little influence on ultimate strength of joints and the maximum decrease of ultimate strength is only 7%. For multi-planar KK-joints considered in this study, which are all CW joints, Table 7 shows that the ultimate strength of IPOv-W is slightly higher than that of IPOv-N, which is in accordance with K-joints. Moreover, Figure 13 indicates that strain magnitude and distribution regularity are similar for the two specimens. It can be concluded that welding situation of hidden seam has some effects on the stress distribution and failure mechanism, but the static ultimate capacity of the multi-planar overlapped CHS KK-joints is not affected significantly.

5.2 *Effect of vertical stiffener*

Vertical stiffener is used to avoid partial overlapping of one brace member onto another, but with a disadvantage of additional inconvenience of construction. Both test and FEM results indicate that: the ultimate strength of IPOv-HC is slightly higher than that of IPOv-N; Strain magnitude and distribution regularity are similar for the two specimens. It is thus evident that vertical stiffener in the transverse direction has some effects on the stress distribution and failure mechanism, but the static ultimate capacity of the multi-planar overlapped CHS KK-joints is not affected significantly.

5.3 *Effect of loading patterns*

Different load transfer mechanism under symmetric and anti-symmetric loadings results in different stress distribution on the chord and different failure modes. When subjected to symmetric loading, the compression and tension in two K planes balance in the transverse direction and have resultant forces in the longitudinal direction. Forces in two planes do not interact at the transverse gap region and multi-planar KK-IPOv joints behave like uniplanar overlapped K-joints. The ultimate strength relies on brace strength and chord bending capacity in the longitudinal direction. However, when subjected to anti-symmetric loading, the compression and tension in two K planes form resultant forces in the transverse direction and the transverse gap region on the chord is subjected to shear force caused by diagonal compression and tension. Therefore, the strength of joints relies on the bending capacity of the chord wall in both the transverse and the longitudinal direction. Consequently, the ultimate strength under anti-symmetric loading is much lower than that un-der symmetric loading, which illustrates the importance of taking loading patterns into consideration in the design formula.

6 CONCLUSIONS

(1) The welding situation of hidden seam and the presence of vertical stiffener have some effects on the stress distribution and failure mechanism, but the static ultimate capacity of the multi-planar overlapped CHS KK-joints is not affected significantly. Thus the absence of hidden welds and vertical stiffener are more suitable for the convenience of construction.

(2) The stress distribution on the chord and the failure modes differ significantly between symmetric and anti-symmetric loadings. The ultimate capacity of the multi-planar overlapped KK-joint has decreased a lot when subjected to anti-symmetric loading.

(3) To avoid premature cracking, particular attention should be paid to the dimension and quality of welds. Strengthening of welds is recommended in practice.

REFERENCES

BS EN 1993-1-8, 2005. Eurocode3: Design of Steel Structures, Part 1-8: Design of joints. Standard, British Standard Institution.

Lee, M.M.K. & Wilmshurs, S.R. 1997. Strength of multiplanar tubular KK-joints under antisymmetric axial loading. *Journal of Structural Engineering* 123: 755–764.

Li, M. 2012, Research on Failure criteria and ultimate capacity equation for uniplanar overlapped circular hollow section K-joints. Dissertation, Tongji Univ., Shanghai, China.

Makino, Y. & Kurobane Y. 1994. Tests on CHS KK-joints under anti-symmetric loads. Tubular Structures VI, Grundy, Holgate & Wong (eds): 449–456. Rotterdam: Balkema.

Makino, Y. & Kurobane Y. et al. 1995. Proposed ultimate capacity equations for CHS KK-joints under anti-symmetric loads. Proceedings of the Fifth International Offshore and Polar Engineering Conference, Netherlands: The Hague.

Packer, J.A. et al. 2009. Design guide for rectangular hollow section (RHS) joints under predominantly static loading, CIDECT Design Guide No. 3, 2nd Edition, CIDECT, Geneva, Switzerland.

Wang, G.N. 2007. The experimental and numerical investigations on overlapped circular hollow section KK-joints. Dissertation, Tongji Univ., Shanghai, China.

Wardenier, J. et al. 2008. Design guide for circular hollow section (CHS) joints under predominantly static loading, CIDECT Design Guide No. 1, 2nd Edition, CIDECT, Geneva, Switzerland

Wu, J. 2008. Bearing capacity of overlap K-joints with hidden weld absence. *Journal of Tianjin University* 41(2): 226–232.

Zhao, X.Z. et al. 2006. Experimental study on static behavior of unstiffened, overlapped CHS K-Joints. *Journal of Building Structures* 27(4): 23–29.

Tubular Structures XV – Batista, Vellasco & Lima (eds)
© 2015 Taylor & Francis Group, London, ISBN 978-1-138-02837-1

Tubular joints with welder-optimized CJP-equivalent welds under high-cycle fatigue loading

X. Qian

Department of Civil and Environmental Engineering, National University of Singapore, Singapore

P.W. Marshall

MPH Systems Engineering, Houston, Texas, USA
Department of Civil and Environmental Engineering, National University of Singapore, Singapore

ABSTRACT: This article presents the experimental results of a recent joint industry project on the fatigue performance of tubular joints fabricated using a new type of welding detail, namely the welder-optimized complete joint penetration (CJP) weld, which aims to enhance the quality control and workmanship requirement for conventional CJP welds widely practiced for tubular joints. The experimental program includes two types of joint specimens: the hollow section X-joints and X-joints with the chord member filled with ultra-high strength concrete. This study compares the measured stress-concentration factors at the hot-spot locations along the brace-to-chord intersection of the hollow section specimens and those for the concrete-filled specimens. The experimental instrumentation recorded the fatigue crack initiation and propagation histories for the major fatigue cracks observed near the hot-spot positions. The experimental findings reveal that the fatigue performance of the tubular joints fabricated by the welder-optimized CJP-equivalent welds depends primarily on fatigue failure at the weld toes, similar to tubular joints welded by conventional CJP welds. The fatigue life measurements of the X-joint specimens fabricated with the welder-optimized CJP-equivalent welds follow life estimations by S-N curves originally developed for circular hollow section joints with complete joint penetration welds. The presence of the ultra-high strength concrete within the chord member reduces the hot-spot stresses at the weld-toes along the brace-to-chord intersection. The fatigue life measured for the concrete-filled specimens follows essentially a similar S-N curve as that for unfilled hollow section joints.

1 INTRODUCTION

The connections between tubular members have traditionally deployed different types of welding and joining schemes, e.g., the complete-joint penetration (CJP) welds widely practiced in offshore jacket-type structures, welds with a backing-plate in European bridge structures (Schumacher et al. 2001), and cast steel nodes (Nussbaumer et al. 2006). The complex topology of the connection geometry imposes a stringent requirement on the strength of the welds, which leads naturally to CJP welds as a convenient solution. However, the detailed quality control procedure of CJP welds requires premium workmanship to be delivered by 6GR welders (AWS 2010). To enhance the quality control of the welds, especially near the root of the welds where visual inspection is practically infeasible for tubular joints, Qian et al. (2009) have proposed a new set of welding details, inspired by the welding configuration of CJP welds and the welds with a backing plate, as shown in Figure 1, termed as the enhanced partial joint penetration welds (Qian et al. 2013b) or welder-optimized CJP welds (Marshall et al. 2013).

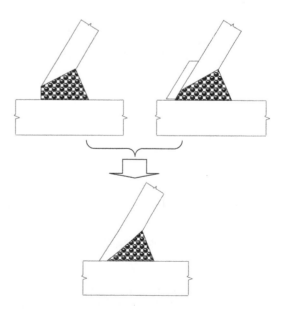

Figure 1. The idea of welder-optimized CJP-equivalent welds.

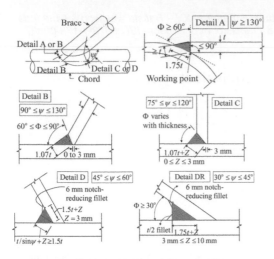

Figure 2. Details of the welder-optimized CJP welds. (Marshall et al. 2013).

Marshall et al. (2013) present a comprehensive discussion on the welder-optimized CJP-equivalent welds, including the rationale, fabrication details and design considerations. Figure 2 illustrates details of the welder-optimized CJP-equivalent welds corresponding to different local dihedral angles (ψ) along the brace-to-chord intersection.

The welder-optimized CJP-equivalent welds employ a part of the brace wall thickness as an inherent backing plate for the welds and thus provide enhanced quality control compared to conventional CJP welds, as shown in Figures 1 and 2. The volume and effective thickness of the weld material follow the strength requirement of the tubular joint. The unfused weld root, however, poses a potential threat of the root failure under fatigue loading. A joint-industry project (JIP) was, therefore, launched at the National University of Singapore to examine the fatigue performance of tubular joints using welder-optimized CJP-equivalent welds. Previous papers have reported the progress of the joint industry project, focusing on the effect of weld-toe grinding treatment (Marshall et al. 2012), the potential for root cracking (Qian et al. 2013b), and the overloading effect on the fatigue performance of these joints (Qian et al. 2013c).

This paper summarizes the experimental results for all the fatigue tests and compares the fatigue performance of tubular X-joints using welder-optimized CJP-equivalent welds, with and without the ultra-high strength concrete in the chord member.

2 EXPERIMENTAL PROGRAM

2.1 Specimens

The experimental program includes five large-scale X-joint specimens as listed in Table 1. The non-dimensional parameters used in Table 1 include the brace-to-chord diameter ratio ($\beta = d_1/d_0$), the chord

Table 1. Configuration of the X-joint specimens.

Specimen	d_0 (mm)	β	γ	τ	ΔP (kN)
J1-1	750	0.54	15	0.5	210
J1-X	750	0.54	15	0.5	210
J1-2	750	0.54	15	0.5	160
J2-1	750	0.54	15	1.0	180
J2-2	750	0.54	15	1.0	210

radius-to-wall thickness ratio ($\gamma = d_0/2t_0$), and the brace wall-to-chord wall ratio ($\tau = t_1/t_0$). The first three specimens (J1 series) have a hollow chord member, while the last two specimens (J2 series) include a chord member filled with ultra-high strength concrete. The brace-to-chord intersection has a fixed angle of 60° for all specimens. The materials for both the chord and brace utilize cold-rolled S355 steel CHSs with a Young's modulus of 203 GPa and a Poisson's ratio of 0.3. The ultra-high strength concrete in the specimen J2-1 and J2-2 had a 28-day compressive strength of 185 MPa, a Young's modulus of 70 GPa and a Poisson's ratio of 0.19.

Both specimens J1-X and J2-2 underwent post-weld burr-grinding treatment at the weld toe with a 3-mm grinding radius.

2.2 Test set-up and instrumentation

Each specimen experienced two constant-amplitude cyclic tests, using the set-up shown in Figure 3. The cyclic test applies a vertical load on top of the loading fixture attached to the top end of the chord member (see Figure 3). This vertical load generates an in-plane bending on the X-joint specimen shown in Figure 3. The first fatigue test was terminated when the measured crack depth, by an Alternating Current Potential Drop (ACPD) approach, reached 80% of the wall thickness. The experimental procedure then involved re-testing the same specimen, after rotating it by 180° in the plane of the joint.

Figure 3 denotes the position along the brace-to-chord intersection through an angular parameter, ρ. A zero ρ angle refers to the top crown point, while $\rho = 180°$ corresponds to the bottom crown point as indicated in Figure 3.

All 10 fatigue tests employed a five-element strip gauge to measure the uniaxial hot-spot strain (ε_{hs}). The hot-spot stress σ_{hs} thus follows (ARSEM 1987),

$$\sigma_{hs} = 1.15\varepsilon_{hs}E \tag{1}$$

In addition to the hot-spot strain, the experimental procedure detects the initiation and propagation of the fatigue cracks using the ACPD device. The experimental study recorded two fatigue lives for each fatigue test: the fatigue crack initiation life and the total fatigue life. The current study defines the fatigue crack initiation life as corresponding to a fatigue crack size of about 0.5 mm in depth, while the total fatigue life refers to the number of cycles at the end of the fatigue test.

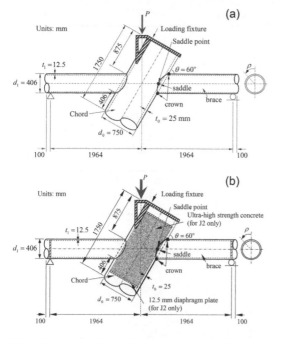

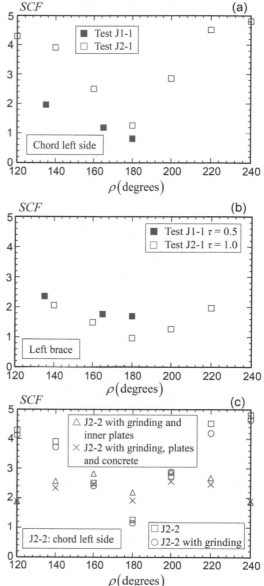

Figure 3. Set-up of the fatigue test: (a) hollow joint specimens; and (b) joint specimens filled with ultra-high strength concrete.

3 EXPERIMENTAL RESULTS

3.1 *Stress concentration factors*

The experimental study deployed the five-element uniaxial strip gauge at a distance of $0.4t$ from the weld toe to measure the hot-spot strain, where t refers to the thickness of the member. The spacing between the individual elements within each strip gauge remained fixed at 2 mm. The hot-spot strain was derived therefore from a linear extrapolation of the measured strain values over the extrapolation zone, from $0.4t$ to $0.4t + 8$ mm.

Figure 4 compares the measured stress concentration factor (SCF) values calculated from the hot-spot stress using Equation 1 along the brace-to-chord intersection of different joint specimens. The stress concentration factors follow from,

$$SCF = \frac{\sigma_{hs}}{\sigma_{nom}} \qquad (2)$$

where σ_{nom} refers to the nominal stress in the brace member. For X-joints under in-plane bending, σ_{nom} denotes the maximum elastic bending stress in the brace wall, at its intersection with the chord surface.

Previous experimental studies (Qian et al. 2013b, 2013c) demonstrate that the SCF values near the hot-spots in the chord member exceed those in the brace.

Figure 4a compares the SCF values measured from the specimen J1-1 and J2-1, the former with a τ ratio of 0.5 and the latter with a τ ratio of 1.0. The SCF values in Figure 4a and 4b refer to the values measured prior to filling the chord with ultra-high strength concrete.

Figure 4. Measured SCF values for different specimens: (a) left side of the chord in J1-1 and J2-1; (b) left brace of J1-1 and J2-1; and (c) different preparation stages of J2-2.

An increase in the brace wall thickness significantly increases the SCF values near the weld toe of the chord members, as shown in Figure 4a. The maximum SCF becomes doubled in the J2-1 specimen with a thicker brace wall, compared to the maximum SCF value in the specimen J1-1. In contrast, the SCF values at the hot-spots along the brace weld toe remain similar in magnitudes for the two X-joint specimens with different τ ratios, as demonstrated in Figure 4b, but the effect of τ is positive. The specimen with a thinner brace wall shows a slightly higher SCF value at the hot-spots near the brace weld-toe.

Table 2. Fatigue lives for all tests.

Specimen	N_i (cycles)	N_{cp} (cyles)	N_t (cycles)
J1-1 (test 1)	70,000	133,000	203,000
J1-1 (test 2)	26,000	133,000	159,000
J1-X (test 1)	175,000	310,000	485,000
J1-X (test 2)	295,000	305,000	600,000
J1-2 (test 1)	102,000	328,000	430,000
J1-2 (test 2)	168,000	310,000	478,000
J2-1 (test 1)	32,000	411,000	443,000
J2-1 (test 2)	37,000	373,000	410,000
J2-2 (test 1)	182,000	578,000	760,000
J2-2 (test 2)	104,000	876,000	980,000

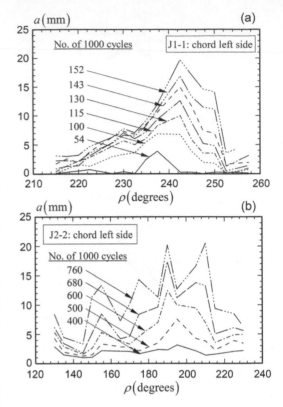

Figure 5. Fatigue crack-front profiles in the chord for: (a) J1-1; and (b) J2-2.

Figure 4c compares the SCF values at four different stages of the specimen preparation for the specimen J2-2: 1) the welded joint; 2) the welded joint with post-weld burr grinding; 3) the burr-ground joint with inner diaphragms prior to filling with concrete; 4) the burr-ground joint with ultra-high strength concrete in the chord. The square symbols in Figure 4c represent the SCF values measured for the hollow specimen prior to any weld toe burr grinding. These SCF values nearly always remain the highest compared to the SCF measurement at later stages of the specimen preparation. There was noticeable ovalization of the hollow chord, because it was only 2.5 diameters long.

After applying the 3-mm radius burr grinding at the weld toe in the chord member, the SCF values (indicated by the circular symbols in Figure 4c) decrease slightly. The maximum SCF value, measured at $\rho = 240°$, decreases by about 4%.

Before the chord member was filled with the ultra-high strength concrete, the specimen has an inner plate inserted to hold the concrete material inside the chord member. The inner plates were connected to the internal surface of the chord member by fillet welds. The presence of the inner plates stiffens the chord member and leads to a smooth distribution of the SCF values along the brace-to-chord intersection (indicated by triangles in Figure 4c). The maximum SCF decreases significantly by about 40%. Meanwhile, the hot-spot stresses near the bottom crown point increase with the presence of the inner plates.

Infilling the chord member with ultra-high strength concrete causes marginal reduction in the measured SCF values compared to the X-joint specimen with the inner plates. The SCF measurements at different stages of the specimen preparation presented in Figure 4c demonstrate that internal reinforcement causes a pronounced change in the stress distribution along the brace-to-chord intersection, which reduces the stress concentration factors effectively near the critical hot-spot locations.

3.2 Fatigue crack initiation and propagation

This experimental study defines crack initiation as when the maximum crack depth reaches 0.5 mm, measured using ACPD. Table 2 shows the fatigue crack initiation life (N_i), the fatigue crack propagation life

(N_{cp}) and the total fatigue life (N_t) for all 10 fatigue tests.

Table 2 demonstrates that the weld-toe burr grinding significantly enhances the fatigue crack initiation life for both the hollow section joint (J1-X) and the concrete-filled joint (J2-2), although burr grinding does not cause significant changes in the measured SCF values, as reflected in Figure 4c. Burr grinding also enhances the fatigue crack propagation life, compared to the fatigue tests for J1-1 specimens and those for J1-X specimens.

Figure 5 illustrates the fatigue crack-front profiles measured along the brace-to-chord intersection. Experimental observations indicate that fatigue crack initiation occurs at multiple locations along the brace-to-chord intersection and that these coalesce into a major crack. The maximum crack depth observed in Figure 5 agrees consistently with the maximum SCF values measured in the experiment (shown in Figure 4).

Figure 6 illustrates the multiple cracks observed in both the joint specimens without burr grinding at the weld toe and those with burr grinding at the weld toe. Qian et al. (2013a) have reported the effect of adjacent cracks on the fatigue crack propagation life. The presence of adjacent cracks with overlapping crack faces often slows down the crack propagation life.

Figure 7 compares the fatigue crack size (a) propagation for different cracks in different specimens:

Figure 6. Multiple cracks observed along the brace-to-chord intersection: (a) for specimens without burr grinding at the weld toe; (b) for specimens with burr grinding at the weld toe.

J1-1 without ultra-high strength concrete in the chord member (in Figure 7a), J2-1 with ultra-high strength concrete in the chord member but without burr grinding at the weld toe (in Figure 7b), and J2-2 with ultra-high strength concrete in the chord and burr grinding at the weld toe (in Figure 7c). In all three specimens, the most critical crack locates on the left side of the chord member. The hollow specimen without the ultra-high strength concrete in the chord member shows the fastest crack propagation among all three specimens. For both specimens with ultra-high strength concrete in the chord (J2-1 and J2-2), the application of the burr grinding at the weld toe in J2-2 has a longer crack propagation life, despite being subjected to a slightly higher load (and hence stress) range, as indicated in Table 1.

3.3 Comparison with design S-N curve

Figure 8 compares the fatigue lives measured in the test and two types of design S-N curves, i.e., the nominal stress S-N curve (in Figure 8a) and a hot-spot stress S-N curve (in Figure 8b). The fatigue lives in Figure 8 correspond to both the initiation of fatigue cracks (represented by filled symbols in Figure 8) and the total life, including fatigue crack propagation until crack depth at about 70~80% of the wall thickness (represented by the hollow symbols in Figure 8).

Compared to the nominal stress S-N curve (the AWS DT curve in Figure 8a), the fatigue lives corresponding to the end of the fatigue tests (with the crack depth at 80% of the wall thickness) for specimens without the ultra-high strength concrete in the chord well exceed the estimation by the AWS DT curve (AWS 2010). Similarly, the fatigue crack initiation life

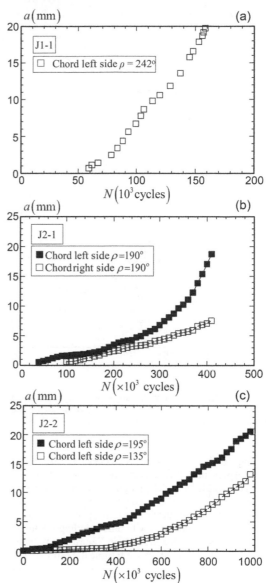

Figure 7. Fatigue crack propagation at critical front locations: (a) in specimen J1-1; (b) in specimen J2-1 with ultra-high strength concrete in the chord; and (c) in specimen J2-2 with ultra-high strength concrete in the chord and burr grinding at the weld toe.

at a crack depth of about 0.5 mm for the specimens without the ultra-high strength concrete exceeds the life estimated using the nominal stress S-N curve. It should not be surprising that some of the tests fall on the unsafe side of AWS-DT criteria, as this curve does not consider the relevant connection geometry parameters.

Compared to the hot-spot S-N curve in API (2000), the fatigue lives for the hollow specimens at both fatigue crack initiation and the end of the fatigue tests exceed the life estimation by the API X curve, as shown

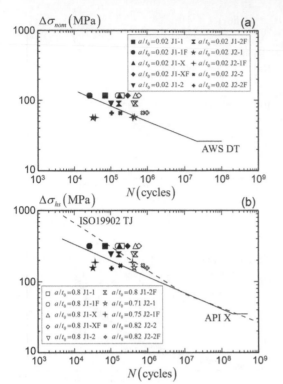

Figure 8. Comparison of the fatigue test data with design S-N curves: (a) nominal stress S-N curve in AWS (2010); and (b) hot-spot stress S-N curve in API (2005) and ISO 19902 (2007).

in Figure 8b. The hot-spot S-N curve in ISO 19902 (2007) presents a less conservative life estimation than does the API X curve. The fatigue life estimated by ISO 19902 TJ curve thus exceeds the fatigue crack initiation life for most of the hollow joint specimens, except for the hollow specimen with burr grinding treatment at the weld toe (J1-X and J1-XF in Figure 8b). With the fatigue crack propagation, the fatigue life estimated by the ISO 19902 TJ curve is less than the fatigue lives for all hollow specimens at the end of the fatigue test.

Marshall and Thang (2014) have proposed a new hot spot fatigue curve which falls in between API and ISO.

Due to the effective reduction in the stress-concentration factors by the presence of the ultra-high strength concrete in the chord member, the fatigue crack initiation lives for the four tests on the specimens with the ultra-high strength concrete in the chord member (J2-1, J2-1F, J2-2 and J2-2F) remain lower than the life estimations by the nominal stress S-N curve and the hot-spot API X curve, as demonstrated in Figure 8. Compared to the ISO 19902 TJ curve, all the fatigue crack initiation lives for the concrete-filled specimens remain lower than the S-N curve estimation.

At the end of the fatigue test, the fatigue lives for all concrete-filled specimens exceed the life estimated using the API X curve. However, two fatigue tests (J2-1 and J2-2) show fatigue lives lower than the estimation by the ISO 19902 TJ curve. These two tests also

demonstrate lower fatigue lives than those estimated using the nominal stress S-N curves, i.e., the AWS DT curve (AWS 2010).

It might be interesting to speculate about the influence of the increased ratio of punching shear to shell bending, whereby there is a difference in behavior between the strain gauge strip and the weld toe in grouted specimens, than is typical for ungrouted specimens.

4 CONCLUSIONS

This study compares the fatigue performance of hollow section X-joints and concrete-filled X-joints subjected to constant-amplitude brace in-plane bending. The entire experimental program included 10 fatigue tests on five specimens, two of which contained a chord member filled with the ultra-high strength concrete.

The presence of the internal reinforcing schemes in the chord member, for example, by filling the chord using ultra-high strength concrete or by welding internal diaphragm plates, reduces the hot-spot stresses near the weld toe in the chord member along the brace-to-chord intersection.

The decrease in the hot-spot stresses leads to an increased predicted fatigue life, as estimated by the hot-spot S-N curves. Compared to the fatigue life estimation by the design S-N curve, however, the predicted fatigue lives for the concrete-filled specimens are the same for unfilled joints. The small differences in total life correlation with AWS DT, API X and ISO 19902 TJ curves are mainly due to the different lives taken by each code.

Root crack initiation occurs only in one of the five specimens (J1-X), in which the weld-toe fatigue life has been substantially enhanced by repeated burr grinding. Root cracking did not initiate in any of the thick concrete-filled specimens, as confirmed in the post-test sectioning.

REFERENCES

American Petroleum Institute (API). 2000. Recommended Practice for Planning, Designing and Constructing Fixed Offshore Platforms, 21st Ed, API RP 2A WSD, 2000.

American Welding Society (AWS). 2010. Structural Welding Code – Steel, AWS D1.1/D1.1M:2010. 22nd edition.

Association de Recherche sur les Structures Métalliques Marines (ARSEM). 1987. Design guides for offshore structures – welded tubular joints, Éditions Technip, Paris.

International Organization for Standardization (ISO). 2007. ISO 19902:2007 Petroleum and National Gas Industries – Fixed Steel Offshore Structures. 2007.

Marshall, P.W., Qian, X., Nguyen, C.T. & Petchdemaneengam, Y. 2013. Welder-optimized CJP-equivalency welds for tubular connections. Weld World, 57, 569–579.

Marshall, P. W. & Thang, V. 2014. Radical Proposals for Hot Spot Stress Design. Steel Construction, 7(2), 84–88.

Nussbaumer, A., Haldimann-Sturm, S.C., Schumacher, A. 2006. Fatigue of bridge joints using welded tubes or cast steel node solutions. Proc. 11th Int. Symp. Tubular

Structures, Tubular Structures XI, Toronto, Canada, 31 Aug-02 Sep, 2006.

Pijpers, R. J. M., Kolsteim, M. H., Romeijn, A. & Bijlaard, F. S. K. 2009. Fatigue strength of hybrid VHSS-cast steel welded plates. *Proc. Nordic Steel Constr. Conf.*, Malmö, Sweden, 2–4 Sep, 2009.

Qian, X., Marshall, P.W., Cheong, W.K.D., Petchdemaneegam, Y. & Chen, Z. 2009. Partial joint penetration plus welds for tubular joints: fabrication and SCFs. *Proc. 62 Annual Assembly and Inter. Conf. Int. Institute Welding*. Singapore, 12–19 July, 2009.

Qian, X., Nguyen, C.T., Petchdemaneengam, Y., Ou, Z., Swaddiwudhipong, S. & Marshall, P.W. 2013a. Fatigue performance of tubular X-joints with PJP+ welds: II – Numerical investigation. *J. Constr. Steel. Res.*, 90, 49–59.

Qian, X., Petchdemaneengam, Y., Swaddiwudhipong, S., Marshall, P.W., Ou, Z. & Nguyen, C.T. 2013b. Fatigue performance of tubular X-joints with PJP+ welds: I – experimental study. *J. Constr. Steel. Res.*, 90, 49–59.

Qian, X., Swaddiwudhipong, S., Nguyen, C.T., Petchdemaneengam, Y., Marshall, P. & Ou, Z. 2013c. Overload effect on the fatigue crack propagation in large-scale tubular joint. *Fat. Fract. Eng. Mat. Struct.*, 36, 427–438.

Schumacher, A., Nussbaumer, A & Hirt, M. A. 2001. Fatigue behavior of welded circular hollow section (CHS) joints in bridges. *Proc. 9th Int. Symp. Tubular Structures, Tubular Structures IX*, Dusseldorf, Germany, 3–5 Apr, 2001.

Tubular Structures XV – Batista, Vellasco & Lima (eds)
© *2015 Taylor & Francis Group, London, ISBN 978-1-138-02837-1*

Assessment and representation of ductile fracture failure for welded tubular joints

X. Qian

Department of Civil and Environmental Engineering, National University of Singapore, Singapore

ABSTRACT: This paper first presents a ductile tearing assessment for a high-strength steel circular hollow section X-joint subjected to brace in-plane bending based on the option 3 approach outlined in the BS7910. The experimental program includes a careful monitoring of the crack growth process in the X-joint specimen made of S690 steels, with a detailed examination of the material fracture resistance through procedures prescribed in ASTM E-1820. The ductile tearing assessment demonstrates that the existing generic failure assessment curves (the option 1 and option 2 curves) lead to an un-conservative over-prediction of the load level causing unstable fracture in the joint, compared to the option 3 curve and the experimental results. With an engineering procedure, this paper further integrates a previously proposed eta-approach in estimating the crack driving force and the experimentally measured fracture resistance curve in estimating the load-deformation response of the welded tubular joints. The fracture-based representation of the tubular joint behavior assumes, conservatively, a complete unloading beyond the validity limit of the fracture resistance curve, implying unstable fracture failure in the joint. The load-deformation curve estimated using this approach agrees closely with the experimental results.

1 INTRODUCTION

Unstable fracture failure following the ductile crack extension in welded tubular connections often imposes a critical threat to the safety of tubular structures under extreme loading conditions. Existing engineering design is predicated on strength requirements under ultimate limit state conditions. The assessment of potential fracture failure remains a material requirement in structural design codes.

Investigations (Energo Engineering 2007) on the recent extreme environment actions on tubular offshore platforms revealed that fracture failure became a major failure mechanism in different structural components. Engineering assessments of ductile fracture failure require a detailed examination of the local component level, coupled with appropriate representation of fracture failure at the global structural level.

This study addresses the fracture assessment of welded tubular connections using both the generic and geometry-specific failure assessment curves described in BS7910 (2013). With the validated assessment procedure at the component level, this paper summarizes an approach to integrate the fracture failure representation, through a load-deformation relationship modified by the material fracture resistance measurement, in the global analysis of tubular frames. The validation of the proposed approach employs the reported experimental work on large-scale 2-D frames in the BOMEL joint industry project (Bolt 1994).

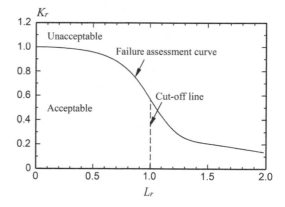

Figure 1. Option 1 FAD curve in BS7910 (2013).

2 FAILURE ASSESSMENT DIAGRAMS

The failure assessment diagram (FAD) approach examines two competing failure mechanisms, namely plastic collapse failure and fracture failure, in a structural component, as shown in Figure 1.

The vertical axis of the FAD in Figure 1, K_r, denotes the fracture ratio,

$$K_r = \frac{K_I}{K_{Ic}} \qquad (1)$$

where K_I refers to the mode I crack driving force, while K_{Ic} defines the material fracture toughness. The horizontal axis in Figure 1, L_r, denotes the load ratio,

$$L_r = \frac{P}{P_{uc}} \qquad (2)$$

where P_{uc} corresponds to the limit load of the structural component with the cracked geometry.

The FAD concept derives originally from the relationship between the applied stress and the crack driving force for a crack in an infinite plate, based on the strip-yield model (Dowling & Townley 1975, Harrison et al. 1976).

As the failure assessment diagram describes the competition between fracture failure and plastic collapse failure, the shape of the FAD depends on the elastic-plastic crack-driving force,

$$K_r = \sqrt{\frac{J_{le}}{J_{ep}}} \qquad (3)$$

where J_{le} denotes the linear-elastic J−integral value and J_{ep} refers to the elastic-plastic J-value often computed from a finite element analysis. The elastic-plastic J−value in Equation 3 exhibits both geometry and material dependence for a structural component.

Extensive numerical investigations (Milne et al. 1988) on small-scale plate type specimens have led to a geometry and material independent FAD expression,

$$K_r = \left(1 + 0.5 L_r^2\right)^{-0.5} \left[0.3 + 0.7 \exp\left(-0.6 L_r^6\right)\right] \qquad (4)$$

Equation (4) has subsequently become the option 1 failure assessment curve in BS7910 (2013). For applications with a known stress-strain curve, the failure assessment curve [based on numerical investigations reported by Milne et al. (1988)] follows,

$$K_r = \left(\frac{E\varepsilon_{ref}}{L_r \sigma_y} + \frac{L_r^3 \sigma_y}{2 E \varepsilon_{ref}}\right) \qquad (5)$$

where σ_y defines the material yield strength, and ε_{ref} refers to the true train at the corresponding true stress level ($\sigma_{ref} = L_r \sigma_y$), inferred from a material true stress-true strain curve. The option 2 failure assessment curve in BS7910 (2013) follows essentially Equation 5. BS7910 (2013) also provides a geometric and material specific failure assessment curve for a detailed assessment of a structural component based on Equation 3.

A number of previous investigations (Burdekin 2002, Yang et al. 2007, Lie & Yang 2009, Qian 2013) have examined the failure assessment diagram approach for welded tubular joints.

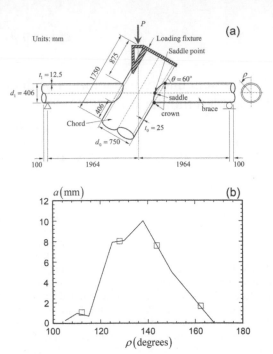

Figure 2. (a) Configuration of the X-joint under residual strength test; and (b) crack-front profile for the fatigue crack in the brace member.

3 FAILURE ASSESSMENT OF TUBULAR JOINTS

3.1 Option 1 FAD

Qian et al. (2014) reported residual strength tests on fatigue-cracked circular hollow section X-joints subjected to brace in-plane bending. This study selects one of the joint specimens with a fatigue crack at the weld toe in the brace member. Figure 2a shows the configuration of the X-joint specimen. The angle ρ in Figure 2a denotes the position along the brace-to-chord intersection when viewed from the right side of the specimen. Figure 2b illustrates the crack-front profile of the fatigue crack in the left brace member. The deepest crack front location occurs at $\rho \approx 140°$. The crack area equals about 18.9% of the brace-to-chord intersection area. The X-joint utilizes the cold-rolled S355 steel CHSs for both the chord and brace members, with a Young's modulus of 203 GPa and a Poisson's ratio of 0.3.

Figure 3 illustrates a half-symmetric model to compute the linear-elastic crack driving force required in the option 1 failure assessment procedure in BS7910 (2013). The FE model ties a local crack-front mesh to a global continuous mesh through the mesh-tying procedure, previously validated by Qian et al. (2005a, 2005b). The crack-front profile in the FE model follows the experimentally measured profile shown in Figure 2b.

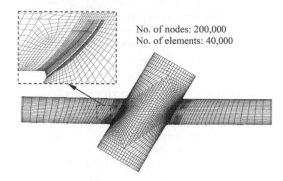

No. of nodes: 200,000
No. of elements: 40,000

Figure 3. A half-symmetric FE model for the X-joint specimen.

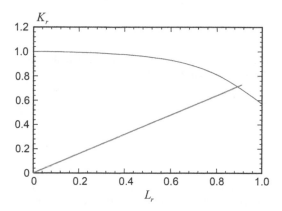

Figure 4. Failure assessment diagram and the load path for the X-joint specimen.

The material fracture test following the procedures in ASTM E-1820 (2013) reveals a fracture toughness of $J_{Ic} = 156\,\mathrm{kJ/m^2}$ (Ou 2013) The experimental work by Qian et al. (2014) showed that the plastic collapse load for the un-cracked joint equals 978 kN. Figure 4 compares the load path and the option 1 failure assessment curve in BS7910 (2013).

The intersection between the load path and the failure assessment curve in Figure 4 corresponds to an estimated failure load level of 757 kN, which remains slightly higher than the experimentally measured failure load of 744 kN.

3.2 Option 3 FAD

The option 1 FAD represents a generic failure assessment curve without the input of the structural geometry and material properties. The option 1 FAD therefore aims to provide a conservative assessment on the structural component for generic applications. With the availability of the material properties, the option 2 FAD unlocks a certain degree of conservatism in the option 1 FAD and therefore often encompasses a larger safe area than does the option 1 FAD. The option 3 FAD in Equation 3, in contrast, includes the geometric and material specific J-solutions and provides the most accurate FAD for a given structural component. This

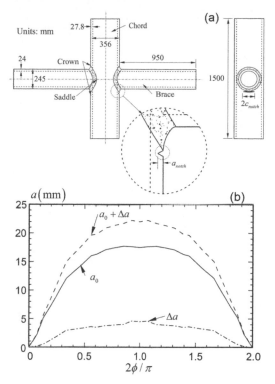

Figure 5. (a) Configuration of the X-joint made of S690 steels; and (b) the crack-front profiles.

section illustrates the assessment of option 3 FAD on a high strength steel (S690) X-joint under brace in-plane bending, reported by Qian et al. (2013a). The chord and brace members consist of hot-finished seamless pipes (EN 10210:2006) Figure 5a shows the geometric configuration of the X-joint, which entails a machined notch in the chord member near the crown point. Figure 5b illustrates the front profile of the fatigue crack and the crack-front profile at the end of the ductile tearing just prior to the unstable fracture failure. The measured fracture resistance from the standard compact tension, C(T), specimen has a toughness value (K_{Ic}) of $246\,\mathrm{MPa \cdot m^{0.5}}$ and a maximum value (K_{max}) of $306\,\mathrm{MPa \cdot m^{0.5}}$ corresponding to the 1.5 mm offset method in ASTM E-1820 (2013). Figure 6 shows the uniaxial true stress versus true strain relationship for the brace and chord materials.

Figure 7a compares the failure assessment of S690 X-joints using three different failure assessment curves in BS7910 (2013). The option 1 and option 2 failure assessment curves demonstrate a clearly larger safe area than does the option 3 FAD, implying an un-conservatism in the generic option 1 and option 2 FADs for this tubular X-joint assessment.

Figure 7a includes the load paths for two crack configurations, the fatigue pre-crack (a_0) and the crack profile at the end of a ductile crack extension $(a_0 + \Delta a)$. The load path for the fatigue pre-cracked joint utilizes the fracture toughness value K_{Ic}, while

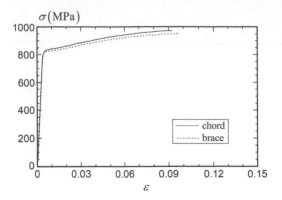

Figure 6. Uniaxial true-stress versus true strain for the S690 steels.

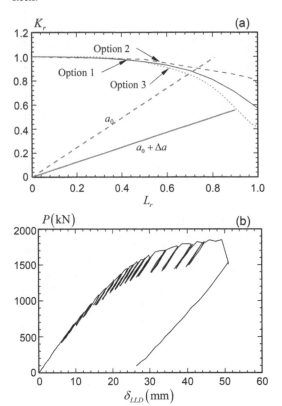

Figure 7. (a) Assessment of high-strength steel X-joint using the FADs; (b) the load versus the load-line displacement measured during the test.

the load path for the X-joint with ductile crack extension employs the maximum fracture resistance K_{max}. The intersection between the load path for the joint specimen with a_0 and the option 3 FAD shows a failure load of about 1511 kN. This implies that the crack extension may occur at this load level. The intersection of the load path for the X-joint with an extended crack-front profile and the option 3 failure assessment curve corresponds to a load level of 1863 kN, very close to the experimentally measured failure load at

which unstable fracture occurs. Figure 7b shows the load versus the load-line response measured from the experiment. The joint experiences unstable fracture failure beyond the peak load of 1864 kN.

4 FRACTURE REPRESENTATION FOR TUBULAR JOINTS IN FRAME ANALYSIS

Qian et al. (2013b) have proposed a joint formulation for tubular X- and K-joints under brace axial loads. Their joint formulation estimates the load-deformation evolution for tubular joints with severe plastic deformations along the brace-to-chord intersection. The non-dimensional load ($\bar{P}$) and the non-dimensional displacement ($\bar{\delta}$) follow the relationship below,

$$\bar{P} = C\bar{P}_u(1 - A[\ln(1 + B\bar{\delta}) - 1/\sqrt{A}]^2) \qquad (6)$$

where the non-dimensional load, $\bar{P}$, and the non-dimensional displacements, $\bar{\delta}$, are

$$\bar{P} = \frac{P\sin\theta}{\sigma_y t_0^2} \qquad (7)$$

$$\bar{\delta} = \frac{\delta}{d_0} \qquad (8)$$

In Equations 7 and 8, σ_y refers to the material yield strength, d_0 represents the outer diameter of the chord, t_0 denotes the wall thickness of the chord, and θ defines the brace-to-chord intersection angle.

The coefficients A and B in Equation 6 are derived from a regression analysis based on an extensive finite element study. Both A and B are functions of the joint geometric parameters.

The above joint formulation has not, however, considered the effect of ductile crack extension and subsequent unstable fracture failure on the load-deformation response of the tubular joint. The ductile crack extension increases the crack area, and therefore decreases the load resistance of the joint by,

$$P_{u,crack} = P_u\left(1 - \frac{A_{crack}}{L_w t_0}\right) \qquad (9)$$

where A_{crack} refers to the crack area and L_w denotes the length of the welds measured along the brace-to-chord intersection (Qian 2013). Upon unstable fracture failure, the load resistance of the joint decreases to zero.

The incorporation of Equation (9) into the existing strength formulation faces another critical challenge in defining the crack initiation, i.e., the deformation level at which the ductile crack extension initiates. Classical fracture mechanics theory defines the initiation of ductile tearing by when the crack driving force (measured by the J-integral) reaches the material fracture toughness [often measured by J_{Ic} values measured using the standard procedure in ASTM E-1820

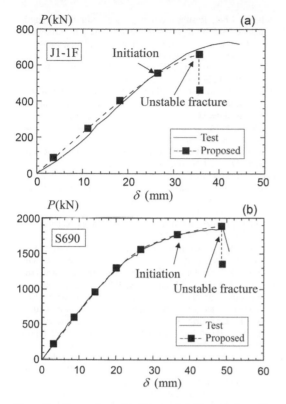

Figure 8. Comparison of the fracture joint formulation with the load-deformation responses measured during the test.

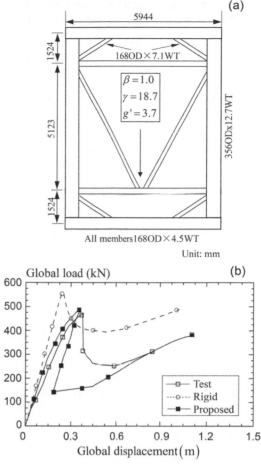

Figure 9. (a) Configuration of the 2-D tubular K-frame (Bolt 1994); and (b) comparison of the frame test and the frame analysis using two different joint formulations.

(2013)]. The integration of ductile crack extension in the load-deformation relationship requires a relationship between the load-deformation response and the crack driving force, which is often derived from an elastic-plastic, large-deformation analysis.

Zhang & Qian (2013) have proposed an η-approach to compute the elastic-plastic crack driving force directly from the area under the load-deformation relationship for tubular X- and K-joints,

$$J = \frac{\eta_{pl}}{A_{lig}} \int_0^\Delta P d\Delta \qquad (10)$$

where A_{lig} refers to the effective ligament area along the brace-to-chord intersection in resisting the brace load, and η_{pl} is a non-dimensional parameter. The η-approach originates from the η-approach proposed by Rice et al. (1973) for standard fracture specimens. This approach has subsequently become the basis to measure the fracture resistance curve ($J-R$ curve) using the single specimen, unloading compliance method in the material testing standard (ASTM E-1820 2013).

The closed-form load-deformation response in Equation 6 allows a direct evaluation of the crack driving force using Equation 10. The combination of Equations 6 and 10 allows the determination of the deformation level at which ductile crack extension initiates (at $J = J_{Ic}$). Beyond this deformation level, the proposed fracture joint formulation then integrates

Equation 9 into Equation 6 to quantify the evolution of the ductile crack extension in the joint strength. As the load further increases, the crack driving force calculated using Equation 10 escalates. The joint formulation assumes that unstable fracture occurs when the J-value in Equation 10 reaches the maximum value measured by the 1.5 mm offset line in the material J-R curve as described in ASTM E-1820 (2013).

Figure 8 compares the load-deformation responses based on Equations 6 and 9 with the load-deformation responses measured for the two X-joint specimens discussed in Section 3. The results in Figure 8 confirm the accuracy of the joint formulation at the component level. The validation of the joint formulation in the structural analysis requires calibration against the frame tests.

Figure 9a shows the configuration of one of the large-scale tubular frame tests reported by Bolt (1994). The single-bay K-frame, containing a weak K-joint, experiences a horizontal load applied to the top member. The experimental study reported a fracture failure of the K-joint initiated at the crown point. Unstable fracture failure of the K-joint leads to significant loss in the frame strength in resisting the horizontal load.

Figure 9b compares the load-deformation response of the frame measured from the test with prediction from frame analyses using two different joint formulations. The rigid joint approach (USFOS 2001), which remains the typical approach in the engineering frame analysis, assumes a fixed connection between the incoming members. The frame analysis implements the fracture joint formulation through a user-defined spring element (Qian et al. 2013b). Figure 9b shows that the rigid joint assumption over-estimates both the initial frame stiffness and the peak resistance of the K-frame. The peak load in the frame analysis with the rigid joint assumption corresponds to buckling of the compression brace, instead of fracture failure in the K-joint. The fracture joint formulation, in contrast, predicts closely the frame stiffness, the peak capacity of the frame, and the deformation at which unstable fracture takes place.

5 CONCLUSIONS

This paper compares the different failure assessment curves in BS7910 for circular hollow section X-joints and presents an approach to integrate the fracture representation in tubular joints in a frame analysis.

Compared with the experimental results, the generic failure assessment diagrams (options 1 and 2 in BS7910) provide an un-conservative assessment for the X-joints subjected to brace in-plane bending reported in this study. In contrast, the option 3 FAD provides the most accurate assessment for the welded circular hollow section X-joints.

To incorporate fracture failure in an existing load-deformation response for tubular joints, this study employs an η-approach to allow calculation of the crack driving force directly from the area under the load-deformation curve of the tubular joint. The initiation of the ductile crack extension thus occurs when the calculated crack driving force reaches the material fracture toughness measured from standard fracture specimens. The crack extension gradually decreases the load resistance of the joint and the subsequent unstable fracture failure leads to the complete loss of joint strength. A comparison of the fracture joint formulation agrees closely with the experimentally measured load-deformation response at both the joint level and the frame level.

REFERENCES

American Society of Testing and Materials (ASTM) International. 2013. ASTM E1820 Standard test method for measurement of fracture toughness.

Bolt, H. 1994. Results from large scale ultimate strength tests of K-braced jacket frame structures, In: Off-shore Technology Conference, Houston, Texas, USA, 1994.

British Standard Institute. 2013. Guide to methods for assessing the acceptability of flaws in metallic structures. British Standard Publication, BS7910:2013.

British Standard Institute. 2006. Hot finished structural hollow sections of non-alloy and fine grain steels – part 1: technical delivery conditions. BS EN 10210-1:2006.

Burdekin, F. M. 2002. The fracture behaviour of a welded tubular joint – an ESIS TC1.3 round robin on failure assessment methods Part III: UK BS7910 methodology, Eng. Fract. Mech. 69(10), 1119–1127.

Dowling, A. R. & Townley, C. H. A. 1975. The effects of defects on structural failure: a two-criteria approach. Int. J. Press. Vessels Pip., 3, 77–137.

Energo Engineering, Inc. 2007. Assessment of fixed offshore platform performance in hurricanes Katrina and Rita. MMS Report, MMS project No. 578.

Harrison, R. P., Loosemore, K. & Milne, I. 1976. Assessment of the integrity of structures containing defects. CEGB report R/H/R6, Central Electricity Generating Board, UK.

Lie, S. T. & Yang, Z. M. 2009. BS7910:2005 failure assessment diagram (FAD) on cracked circular hollow section (CHS) welded joints, Adv. Steel Constr. 5(4), 406–420.

Milne, I., Ainsworth, R. A., Dowling, A. R. & Stewart, A. T. 1988. Background to and validation of CEGB report R/H/R6-revisino 3. Int. J. Press Vessels Pip., 32, 105–196.

Ou, Z. 2013. Fracture and Failure Assessment of Fatigue-Cracked Circular Hollow Section X-Joints. PhD Thesis, National University of Singapore.

Qian, X. 2013. Failure assessment diagrams for circular hollow section X- and K-joints. Int. J. Press. Vessels Pip., 104, 43–56.

Qian, X., Dodds, R. H. Jr. & Choo, Y. S. 2005a. Mode mixity for circular hollow section X joints with weld toe cracks. J. Offshore Mech. Arct. Eng. ASME, 127(3), 269–279.

Qian, X., Dodds, R. H. Jr. & Choo, Y. S. 2005b. Elastic-plastic crack driving force for tubular X joints with mismatched welds. Eng. Struct., 27(9), 1419–1434.

Qian, X., Jitpairod, K., Marshall, P., Swaddiwudhipong, S., Ou, Z., Zhang, Y. & Pradana, M. R. 2014. Fatigue and residual strength of concrete-filled tubular X-joints with full capacity welds. J. Constr. Steel. Res., 100, 21–35.

Qian, X., Li, Y. & Ou, Z. 2013a. Ductile tearing assessment of high-strength steel X-joints under in-plane bending. Eng. Fail. Anal., 28, 176–191.

Qian, X., Zhang, Y. & Choo, Y. S. 2013b. A load-deformation formulation for CHS X- and K-joints in push-over analysis. J. Constr. Steel Res., 90, 108–119.

Rice, J., Paris, P. C. & Merkle, J. G. 1973. Some further results of J-integral analysis and estimates. ASTM STP, 536. 231–245.

USFOS. USFOS course manual. Marintek SINTEF group; 2001

Yang, Z. M., Lie, S. T. & Gho, W. M. 2007. Failure assessment of cracked square hollow section T-joint, Int. J. Press. Vessels Pip. 84(4), 244–255.

Zhang, Y. & Qian, X. 2013. An eta-approach to evaluate the elastic-plastic energy release rate for weld-toe cracks in tubular K-joints. Eng. Struct., 51, 88–98.

Tubular Structures XV – Batista, Vellasco & Lima (eds)
© *2015 Taylor & Francis Group, London, ISBN 978-1-138-02837-1*

Application of the Weibull stress approach to the prediction of brittle fracture originating from defects at the ends of groove-welded joints

T. Iwashita
National Institute of Technology, Ariake College, Fukuoka, Japan

K. Azuma
Sojo University, Kumamoto, Japan

ABSTRACT: A series of tests, which reproduced brittle fracture starting from the region around the ends of complete joint penetration groove welded joints, was conducted. The joint models under study were portion of welded beam flange-to-through diaphragm joints in beam-to-column connections. Some artificial defects were installed in the welded joints prior to testing. This paper describes the effects of plastic constraint on brittle fracture from the welded joints with the defects. The Weibull stress approach was used for evaluating the occurrence of brittle fracture and was found to predict well the test results, compared with a conventional prediction method using the fracture toughness, J_c.

1   INTRODUCTION

The 1995 Kobe Earthquake revealed that the plastic deformation capacity of steel building frames can be significantly reduced by cracks originating from various types of discontinuities in their beam-column joints. Post-earthquake investigations demonstrated that weld defects, many of which can be considered as sharply notched discontinuities, can easily lead to brittle fracture (Kinki AIJ 1997; Kurobane 2002).

Beam-column joint models, extracted from representative connections, were tested by applying bending loads to reproduce high tensile stresses at the ends of groove-welded joints in directions perpendicular to those joints. Defects were intentionally produced in the welded joints prior to testing. Test results showed that the shape of the resulting weld defect influenced the occurrence of brittle fractures, which significantly affected the fracture moment of the joint models. To predict the occurrence of brittle fractures from such defects, it is important to consider the effect of plastic constraint at the crack tips, which can exhibit a much greater fracture toughness than the base material (as determined by such conventional testing procedures as single-edge notched bend testing, in which the notch root is placed under high plastic constraint). In addition, levels of plastic constraint have been shown to vary with defect shape and location (Iwashita et al. 2006). This paper focuses on the effects of plastic constraint on the brittle fracture of groove-welded joint specimens with defects.

The Weibull stress was used to consider the effects of a loss of plastic constraint on the fracture toughness of notched specimens. It was found to predict those effects well (Iwashita et al. 2012). The validity of considering Weibull stress to predict brittle fracture from cracks, even under low plastic constraint, is discussed in this research. On this basis, the authors predicted the occurrence of brittle fracture from the type of defect present in joint models and verified the applicability of Weibull stress in predicting brittle fracture from such defects.

2   SPECIMENS AND TESTING PROCEDURES

2.1   Specimen

The test specimens adopted for this research were designed to represent the connection of a square hollow section column to a wide flange beam, with through-diaphragms at the beam flange positions as shown in Figure 1. This connection is the most commonly used type in Japan. The shaded region shows the

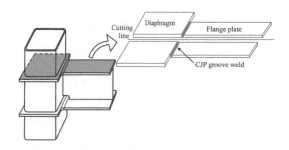

Figure 1.   Test model for groove welded joints.

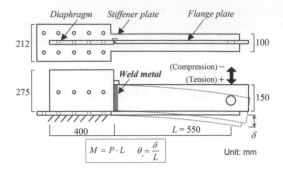

Figure 2. Specimen configuration.

$$M = P \cdot L \qquad \theta = \frac{\delta}{L}$$

Unit: mm

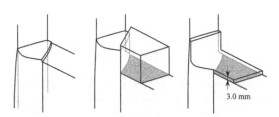

Figure 3. End of CJP weld.

portion treated by the test model. All plates composing the specimens were of Japanese Industrial Standard Grade SN490B structural steel. The flange plate was 25 mm thick, and the diaphragm was 32 mm thick. The welded plate of Figure 1 was cut along an axis of symmetry, and then a stiffener plate was welded to the cut section as shown in Figure 2. The stiffener plate was 25 mm thick and was used to prevent out-of-plane deformation of the flange plate during testing. The stiffener also provided a means of attaching the specimens. Four types of material were prepared with which to configure the specimens: A-flange, A-diaphragm, B-flange, and B-diaphragm. From these were configured two types of specimens: specimen type A, composed of an A-flange and A-diaphragm; and specimen type B, composed of a B-flange and B-diaphragm. The specimens were joined by CJP groove welds. Flux tabs and steel tabs were used in this research, as shown in Figure 3 (a) and (b). The use of flux tabs, a type of ceramic weld dam, is unique to Japanese fabricators. Steel tabs form notches between the tab and a flange plate, as shown in Figure 3 (b); the shaded region indicates a notch, and this induces a stress concentration. Then, as shown in Figure 3 (c), tabs were cut off from two specimens.

Three types of defects were prepared in this research as shown in Figure 4:

(a) Through crack: fatigue cracks were produced at weld toes on the diaphragm plate side. The cracks ran through the full plate thickness.
(b) Surface crack: fatigue cracks were produced at the weld toes on either the diaphragm plate side or the flange plate side. The cracks did not run through the full plate thickness.

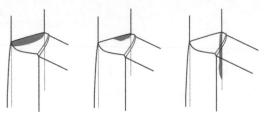

Figure 4. Defect type.

Table 1. Material tensile properties.

Material	Yield strength (MPa)	Ultimate strength (MPa)	Maximum Uniform strain (%)
A-flange	352	567	18.1
A-diaphragm	354	567	18.6
A-weld metal	461	594	14.1
B-flange	383	579	15.3
B-diaphragm	370	566	15.0
B-weld metal	463	605	15.1

Table 2. Results of Charpy impact tests and CT tests.

Material	νE_{shelf} (J)	νE_0 (J)	$\nu E_{at\ test\ temp.}$ (J)	J_c (N/mm)
A-flange	177	165	50	129
A-diaphragm	215	145	26	107
A-HAZ_flange	212	165	59	–
B-flange	188	109	20	90
B-diaphragm	169	49	10	87
B-DEPO	194	119	32	–

νE_{shelf}: Shelf energy.
νE_0: Energy absorbed at 0°C.
$\nu E_{at\ test\ temp.}$: Energy absorbed at test temperature (−45°C for Material A or −30°C for Material B).
J_c: Critical fracture toughness in terms of the J integral at the test temperature.
DEPO: Deposited weld metal.

(c) Internal crack: an area of incomplete fusion was provided within the weld root on the diaphragm plate side.

Also considered as defects for the purpose of this paper are notches formed by steel tabs (Figure. 3 (b) and (c)), as these too can cause large stress concentrations.

2.2 Material properties

Engineering stress and strain data were obtained with tensile coupon samples made from the same material as the test specimens. These data are summarized in Table 1. Table 2 shows the results of Charpy impact tests and compact tension (CT) tests. The samples exhibited a relatively high absorbed energy at 0°C. Experimental tests were next performed at the low

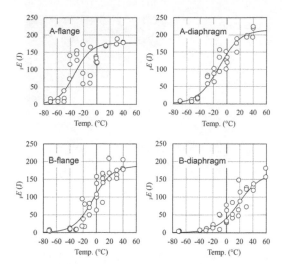

Figure 5. Temperature dependence of Charpy absorbed energy.

temperatures ($-45°C$ for material A and $-30°C$ for material B) to ensure that brittle fractures would occur in the welded joints. The test temperatures were decided by using the results of Charpy impact tests as shown in Figure 5. In addition, the authors confirmed from the CT tests that brittle fractures occurred at each test temperature.

Tensile coupon testing was performed at these low temperatures. The results are shown in Table 2. CT tests were performed per BS 7448 (BSI 1991) to measure fracture toughness at the two low temperatures ($-45°C$ and $-30°C$).

2.3 Test set-up and loading sequences

The specimens were affixed to a strong reaction frame with high-strength bolts on the diaphragm side as shown in Figure 2. The flange plate, together with a stiffener plate, is loaded as a cantilever. A lateral bracing system was provided at the tip of the flange plate. A load was applied to the tip of the cantilever by a hydraulic ram, and tests were conducted at a ram head displacement rate of 0.1 mm/s. The flange plate was thus subjected to a bending moment created by cantilever action. This loading is different from the loading applied to the beam flanges of steel moment connections in buildings during an actual earthquake. This said, the brittle fractures obtained in these tests did reproduce what was observed after the Kobe earthquake. Also, one advantage of this type of loading is that specimens can be fabricated and tested at low cost. Specimens were subjected to either monotonic or cyclic loading. Cyclic loading consisted of several cycles in the elastic range followed by cantilever rotations of θ_p, $2\theta_p$, $3\theta_p$, …, with two positive (tension) and two negative (compression) displacements at each amplitude until failure. In the case of monotonic loading, the load was increased in the positive direction until failure. The displacement, δ, at the loading point

was measured along the loading axis. The bending moment, M, and the rotation, θ, of the cantilever are defined by the equations shown in Figure. 2. We define θ_p as the elastic component of the cantilever rotation at the full plastic moment, M_p. The specimens were tested at the previously stated low temperatures. This was done because the materials used for specimens showed high toughness (Table 2), but the toughness of steel used in construction varies widely (occasionally transition temperatures are even higher than 0°C). To reduce specimen temperature to $-45°C$ or $-30°C$, the specimens were encased in an insulated box (at all areas other than the loading point) and cooled with dry ice.

3 TEST RESULTS

Mode I brittle fracture was observed in all the specimens. Its occurrence was similar to that of the brittle fracture found to have propagated from the ends of CJP groove-welded joints during the Kobe Earthquake. Figure 6 shows the failure modes in the cases of brittle fractures from the diaphragm side and the flange plate side. Test results for monotonic and cyclic load cases are summarized in Table 3. Figure 7 shows specimen fracture surfaces together with defect size and ductile crack size.

A look at the results from monotonic loading reveals that θ_{max} values for through-cracks are quite small. This is a result of extreme premature brittle fracture—the through-cracks produced a significant reduction in plastic deformation capacity. Surface-cracked specimens also experienced premature brittle fracture before reaching maximum strength. In these cases, the θ_{max} values of surface-cracked specimens were larger than those of the through-cracked specimens. An internally cracked specimen (B-F-I) shows a large moment and rotation. The θ_{max} value of a steel tab specimen (A-S-N) is small compared with that of the defect-free A-F specimen, even though the A-S-N specimen had no natural defects. The A-S-N specimen was subjected to large strain concentrations at the corner of its weld toe (flange plate side) because it had been given a notch like that shown in Figure 3 (b). These large strain concentrations induced ductile crack propagation at small deflections, which may produce premature brittle fracture.

A look at the results under cyclic loading reveals that the specimens exhibited less strength and smaller deformation than under monotonic loading. Cyclic loading acted to degrade material toughness and induce premature brittle fracture.

4 FINITE-ELEMENT ANALYSIS

4.1 Modeling methodology

The defects were modeled as double nodes on elements forming a defect boundary. The FEA models employ 8-noded brick elements under a reduced integration technique. The minimum element dimension

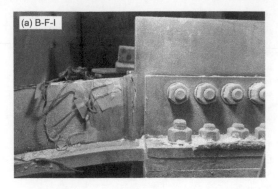
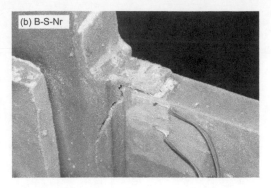

Figure 6. Failure mode. (a) Fracture from diaphragm side; (b) fracture from flange plate side.

Table 3. Test results of monotonic and cyclic loadings.

Specimen	Defect	Type	Position	M_{max} (kNm)	θ_{max} (rad)	Cycle at M_{max}	η for monotonic η_s for cyclic
A-F	None	–	–	185	0.238	–	23.0
A-F-T-1	F	T	D	91.3	0.010	–	0.2
A-F-T-2	F	T	D	111	0.028	–	1.1
A-F-S	F	S	D	135	0.059	–	4.4
A-S-N	S	N	F	141	0.053	–	4.7
B-F	None	–	–	171	0.129	–	10.1
B-F-T-1	F	T	D	100	0.012	–	0.1
B-F-T-2	F	T	D	116	0.017	–	0.5
B-F-I	L	I	D	168	0.135	–	9.5
B-S-Nr	S	Nr	F	175	0.141	–	9.4
A_C-F	None	–	–	145	0.047	$4\theta_p(+2)$	4.5
A_C-F-I	L	I	D	152	0.044	$4\theta_p(+1)$	4.7
A_C-S-N	S	N	F	135	0.033	$3\theta_p(+1)$	2.3
B_C-F	None	–	–	171	0.062	$5\theta_p(+1)$	5.6
B_C-F-I	L	I	D	156	0.039	$4\theta_p(+1)$	2.9
B_C-S-Nr	S	Nr	F	162	0.051	$4\theta_p(+2)$	3.4

η: Plastic deformation capacity.
η_s: Plastic deformation capacity for skeleton curves.

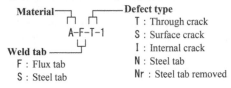

Material
Weld tab
F : Flux tab
S : Steel tab

A–F–T–1

Defect type
T : Through crack
S : Surface crack
I : Internal crack
N : Steel tab
Nr : Steel tab removed

is 0.03 mm, which is applied to the crack tip region. Figure 8 shows a typical mesh for modeling the test specimens. The plasticity of the materials was defined by the von Mises yield criterion. Stress–strain data at the relevant test temperatures were obtained through standard tensile coupon testing. Any damage to the material due to cyclic loading was ignored in this research. The specimens subjected to cyclic loading were analyzed by assuming that monotonic loads were applied and then analyzing the M-θ curves resulting from cyclic loading and conversion to skeleton curves.

4.2 FEA results

M-θ curves for the FEA models were compared with those obtained from actual testing, whereupon the analysis results were found to be comparable to the test results. Figure 9 shows moment vs. J-integral curves of type A specimens. The plotted marks on each M-J curve indicate fracture initiation as determined by moments obtained from the tests. The J values at brittle fractures are scattered and distant from the J_c line.

Different types of defects are subject to different levels of plastic constraint at the crack tips (Iwashita et al. 2006). When considering an occurrence of brittle fracture, it is therefore important to investigate the relationship between the J-integral and plastic constraint. Plastic constraint can be related to the stress triaxiality, T_s. Figure 10 shows T_s-J curves for each specimen model. T_s was taken as a peak value found below the blunted crack tips with FE analysis models.

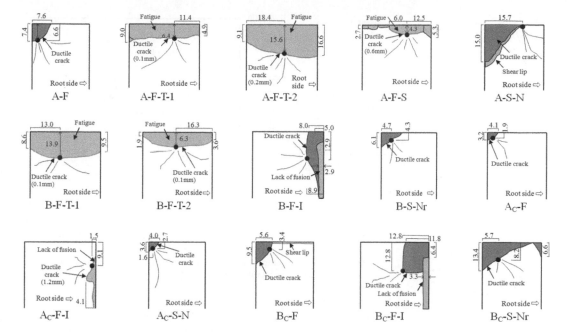

Figure 7. Fracture surfaces. *Black circles indicate fracture points.

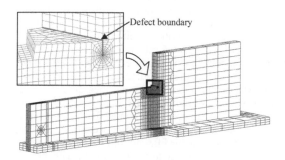

Figure 8. Finite element analysis model and defect locations.

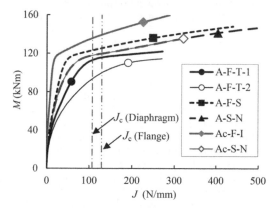

Figure 9. Moment vs. J-integral curves (Material A).

Plotted marks on each T_s-J curve represent specimen fracture points. The fracture points show a tendency for J at fracture to be small when stress triaxiality is high, and for J at fracture to be large when stress triaxiality

is low. The Ac-F-I model is an exception. The J value of this specimen at fracture is relatively small despite small stress triaxiality. This is because the damage to material toughness due to cyclic loading might induce premature brittle fracture. It is well known that cyclic loading deteriorates material fracture toughness. In addition, cyclic loading causes work-hardening that also might induces premature brittle fracture because a higher stress occurs around defect tips. The A-S-N model shows that despite low values of stress triaxiality, brittle fracture occurred at an early stage ($\theta = 0.053$). The development of the J-integral, which is taken as a crack driving force, is very large, and thus ductile crack growth for the A-S-N specimen before fracture was also quite large (ductile crack growth = 15.7 mm). Large crack growth increases the chance of encountering hard particles and may lead to premature brittle fracture despite low stress triaxiality. Taken as a whole, Figure 10 indicates that for the prediction of brittle fracture, it is very important to consider the effects of plastic constraint on fracture toughness.

5 APPLICABILITY OF WEIBULL STRESS

5.1 Weibull stress

Measurements of fracture toughness for cleavage fractures tend to show large scatter, especially in the ductile-to-brittle transition region. This scatter is caused by a large variability in the size and extent of weak spots from one specimen to another. Here, a local approach was proposed by Beremin (1983) for statistically evaluating the behavior of fracture

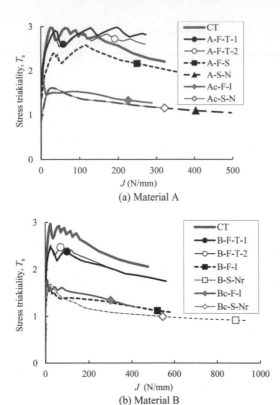

(a) Material A

(b) Material B

Figure 10. Stress triaxiality vs. J-integral curves.

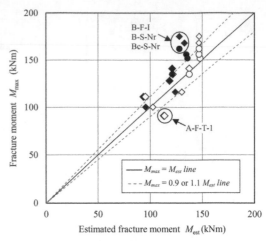

Figure 11. Relation between fracture moment and estimated fracture moment.

toughness. This approach introduces a stress parameter σ_W, termed the Weibull stress, to represent fracture resistance in place of such conventional fracture toughness parameters as K_c, J_c, and δ_c. The Weibull stress is defined as

$$\sigma_W{}^m = \frac{1}{V_0} \int_{V_p} \sigma_1^m dV \qquad (1)$$

Here, $\sigma_1 =$ maximum principal stress, $m =$ Weibull slope, and $V_p =$ fracture process zone. The quantity V_0 is a unit volume and is usually set to 1 mm^3, and authors followed it. The Weibull slope, m, which is a shape parameter, characterizes the scatter of the Weibull stress and varies between 10 and 50 for structural steel. Typically, m has a low value in the case of low-level, broadly distributed fracture toughness. The m value is defined as 20 in this paper according to other research (e.g., Beremin 1983; Mudry 1987). Weibull stress was found to not vary for maximum principal stresses ranging from $2\sigma_y$ and $3\sigma_y$; the fracture process zone, V_p, is taken to be the region in which the maximum principal stress at the crack tip is $3\sigma_y$ or greater, because lower stresses do not contribute to the development of the Weibull stress. The fracture process zone covers the actual fracture points in the tests.

5.2 Weibull stress prediction

The critical Weibull stress, σ_{Wc} is obtained from CT testing and FEA. Then, the estimated fracture moments are obtained from the critical Weibull stress, and the FEA results of each specimen model are obtained as the moments at the critical Weibull stress.

Figure 11 shows the relation between M_{max}, fracture moments obtained from experimental tests, and M_{est}, fracture moments estimated by σ_{Wc}. The estimated fracture moments for the measured J_c were also plotted in Figure 11 and were obtained as the moments at measured J_c. Estimated fracture moments of specimens under cyclic loading were calculated as well (although, in this FEA, we did not consider damage to the material due to cyclic loading). This figure shows that the prediction of brittle fracture by measured J_c is conservative and, with regards to specimens with a large fracture moment (especially B-F-I, B-S-Nr and Bc-S-Nr), provides a very safe assessment. These specimens showed large fracture moment and rotation capacity and had low plastic constraint with internal cracks or steel tab removals. For both monotonic and cyclic loading specimens, fracture moments estimated by σ_{Wc} are closer to the $M_{max} = M_{max}$ line than the fracture moments estimated with measured J_c. This result indicates that the Weibull stress approach may be effective for assessing brittle fractures originating from various defects on beam-column joints. The results for specimen A-F-T-1, however, show unrealistic results. This may be because of variation in material toughness. The size and location of critical microstructural features dictate fracture toughness, and there may thus be considerable scatter in cleavage toughness (Anderson 1994). To more fully examine the effects of such scatter on the occurrence of brittle fracture, further research is needed (the number of test specimens examined here is insufficient for this purpose). Cyclic loading effects also need to be examined in future research. Figure 11 seems to show, also, a

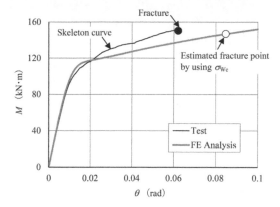

Figure 12. Moment vs. rotation skeleton curves for cyclic loading.

reasonable improvement for cyclic loading. As shown in Figure 12, however, we were not able to satisfactorily simulate the actual behavior of the specimen under cyclic loading with FE analysis because we did not consider the cyclic loading effect in this research and handled the M-θ curve for cyclic loading (Figure 12) as a skeleton curve. Thus and as also apparent in Figure 12, θ_{max} at the fracture moment can be significantly different from θ at the estimated fracture point as obtained from σ_{Wc}. This difference could be due to material damage during cyclic loading.

6 SUMMARY

The joint models represented a portion of welded beam flange-to-through diaphragm joints in beam-to-column connections. The test results show that (a) the defect type has an effect on plastic deformation capacity due to the occurrence of brittle fractures and (b) the defect type also has an effect on plastic constraint at crack tips. The fracture points exhibit a tendency for J at fracture to be small when the stress triaxiality is high, and for J at fracture to be large when the stress triaxiality is low. The Weibull stress approach was used to consider the effects of a loss of plastic constraint on fracture toughness for welded joint specimens. The estimated fracture moments by σ_{Wc} were closer to the fracture moments compared with those by measured J_c. Such findings suggest the possibility that Weibull stress could serve as an effective means for assessing brittle fracture initiation from pre-existing defects on beam-column joints. It should be noted that the FE analyses have not reproduced ductile crack growth and cyclic loading. Further research in this area is necessary to apply FE analysis to a consideration of the effects of crack growth and cyclic loading.

REFERENCES

AIJ Kinki. 1997. *Full-scale test on plastic rotation capacity of steel wide-flange beams connected with square tube steel columns*. Committee on Steel Building Structures, The Kinki Branch of Architectural Institute of Japan, Osaka, Japan. (in Japanese)

Iwashita, T., Kurobane, Y. & Azuma, K. 2006. Assessment of risk of brittle fracture for beam-to-column connections with weld defects. *Proceedings of the 11th International Symposium and IIW International Conference on Tubular Structures*, Quebec City, Canada: 601–609.

Kurobane, Y. 2002. Connections in tubular structures. *Progress in Structural Engineering and Materials* 4(1): 35–45.

Iwashita, T. & Azuma, K. 2012. Effect of plastic constraint on brittle fracture in steel: Evaluation using toughness scaling model. *Journal of Structural Engineering* 138(6): 744–752.

British Standards Institution (BSI) 1991. Fracture mechanics toughness tests – Part 1: Method for determination of KIc, critical CTOD and critical J values of metallic materials. *BS 7448-1*, London, UK.

Beremin, F. M. 1983. A Local Criterion for Cleavage Fracture of a Nuclear Pressure Vessel Steel. *Metallurgical Transactions A* 14(11): 2277–2287.

Mudry, F. 1987. A Local Approach to Cleavage Fracture. *Nuclear Engineering and Design* 105: 65–76.

Anderson, T.L. 1994. *Fracture Mechanics: Fundamentals and Applications, Second Edition*. CRC Press, Chapter 3 and 5, Boca Raton, Fl, USA.

Tubular Structures XV – Batista, Vellasco & Lima (eds)
© 2015 Taylor & Francis Group, London, ISBN 978-1-138-02837-1

Recent research developments in China on fatigue behaviour of welded joints of concrete-filled tubular trusses

L.W. Tong & K.P. Chen
State Key Laboratory for Disaster Reduction in Civil Engineering, Tongji University
Department of Structural Engineering, College of Civil Engineering, Tongji University, Shanghai, China

X.L. Zhao
Department of Civil Engineering, Monash University, Clayton, VIC, Australia

ABSTRACT: Concrete-filled tubular steel trusses have been increasingly used in arch bridges in China in order to enhance the stability of arch trusses against buckling. This composite truss is usually made of hollow section braces and concrete-filled hollow section chords by welded connections between them. The fatigue problem of welded concrete-filled tubular joints is a new topic which needs to be investigated. This paper presents the current developments in China, particularly at Tongji University, of research on the fatigue behaviour of welded concrete-filled tubular joints. The research topics cover static experiments, finite element analysis and formulae for hot spot stress at the connection between a hollow section brace and concrete-filled hollow section chord, as well as the fatigue strength experiments. The types of composite joints investigated deal with T-type and K-type connections between CHS (circular hollow section) braces and concrete-filled CHS chords, as well as T-type and Y-type connections between CHS braces and concrete-filled SHS (square hollow section) chords. It is concluded that welded concrete-filled hollow section joints have a lower hot spot stress concentration and then higher fatigue strength than welded hollow section joints. The hot spot stress concept used in CIDECT guideline for welded hollow section joints can be adopted for the fatigue assessment of concrete-filled hollow section joints by means of an experimental S-N curve for the composite joints.

1 INTRODUCTION

A considerable amount of research on the fatigue behaviour of welded joints in steel tubular trusses, for over more than thirty years in the world, has resulted in fruitful achievements (Zhao & Tong 2011). The fatigue strength design guidelines based on the hot spot stress concept have been proposed by CIDECT and IIW (Zhao et al. 2001).

Tubular trusses filled with concrete in the chord members have better comprehensive performance and are quite suitable for large span arch truss struc-tures. The concrete in the chord members improves both the bearing capacities of whole trusses and welded joints. More than 100 large-span arch truss bridges with concrete-filled tubes have been con-structed so far in China, in which the largest span is up to 460 meters. The fatigue problem of welded joints of Concrete-Filled Circle Hollow Sections (CFCHS) subjected to daily repeated vehicle loads is a new research topic.

For high-cycle fatigue problems within the elastic range, experimental study has indicated that the di-rect influence of the concrete filling in chord mem-bers is to improve the stiffness distribution of welded Circular Hollow Section (CHS) joints signif-icantly, and then to change their stress distribution and fatigue behavior (Wang et al. 2013). When fatigue failure occurs in a

Table 1. Recent research on concrete-filled tubular joints.

Research topic	Concrete-filled CHS joints	Concrete-filled CHS-to-SHS joints
SCF tests	T-joint and K-joint	T-joint
FE analysis for SCF	T-joint and K-joint	T-joint and Y-joint
Study on Formula for SCF	T-joint and K-joint	T-joint and Y-joint
Fatigue strength tests	T-joint	T-joint

CFCHS welded joint, cracks initiate at the weld toe at the maximum hot spot stress first, then propagate at the weld toe along intersection line. This fatigue behaviour is completely the same as that of the CHS welded joint. Therefore, the hot spot stress approach used for the fatigue assessment of CHS welded joints is completely suitable for the CFCHS welded joints.

In recent years, a series of research projects on the fatigue behaviour of CFCHS welded joints has been conducted at Tongji University in China. The de-tailed research topics are listed in Table 1. This paper gives a brief introduction of the relevant achievements to promote developments in this field.

2 TESTS ON HOT SPOT STRESS CONCENTRATION FACTOR

It is essential to have a good knowledge of the hot spot stress distribution or stress concentration factor (SCF) distribution along the intersection line between the chord and brace in order to understand the fatigue behaviour of a CFCHS welded joint. As shown in Figure 1a, b, SCF testing of ten CFCHS T-joint specimens were carried out under axial compressive force, tensile force and in-plane bending in the brace respectively. Concrete with different strength grades was considered (Wang et al. 2013). As shown in Figure 1c, SCF testing for seven CFCHS K-joints subjected to axial force in the brace was carried out (Sun 2001). As shown in Figure 1d, SCF testing for seven T-joint specimens made of CHS brace and CFSHS chord were carried out under axial compressive force, tensile force and in-plane bending in the brace respectively (Yang 2001). The maximum SCF_{max} in the joint versus non-dimensional geometric parameters (β, τ, γ) of the CFCHS T-joint and infilled concrete strength grades, subjected to axial compressive and tensile force in the brace, are shown in Figure 2. At same time, a comparison is made with the unfilled CHS T-joint.

All the test results from the concrete-filled tubular joints show that:

(1) Compared to CHS joints with the same geometry parameters, the SCFs of CFCHS joints significantly decrease, and the distribution of hot spot stress becomes more uniform, especially for the loading condition of axial force in the brace.
(2) When the brace of a CFCHS joint is subjected to axial tension, the SCFs are greater than those where the brace is subjected to axial compression. This is attributed to the trend of separation between chord wall and concrete, which increases the deformation of the chord wall so as to make force transfer nonuniform.
(3) The effects of non-dimensional geometric parameters on the SCF of CFCHS joints are basically the same as those on the SCF of CHS joints.
(4) The concrete strength grade does not exert a significant influence on SCF. The main influence of the concrete is to improve the stiffness distribution within a joint, thus reducing the stress concentration effectively.

3 FINITE ELEMENT ANALYSIS AND FORMULAE FOR SCF

It is essential to carry out tests because of the complexity of CFCHS joints. However, due to funding limitations, the tests are not able to cover all kinds of parameters to be investigated. Therefore, finite element analysis is an approach for further investigation in order to develop design formulae for SCFs.

As shown in Figures 3–5, finite element models were established with solid elements using the software ANSYS and ABAQUS for the test specimen

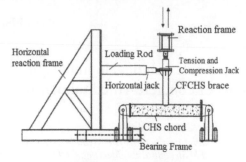

(a) Test setup for T-joints

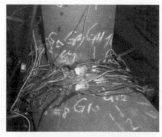

(b) Concrete-filled CHS T-joint

(c) Concrete-filled CHS K-joint

(d) Concrete-filled CHS-to-SHS T-joint

Figure 1. Tests on hot spot stress concentration factor.

of CFCHS T-joints, CFCHS K-joints and concrete-filled CHS-to-SHS T-joints, to calculate the hot spot stress distribution of these joints (Jing 2008, Wang 2008, Yang 2012). The calculated SCF_{max} of CFCHS

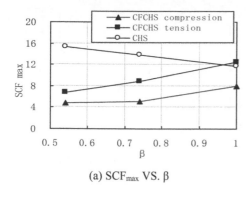

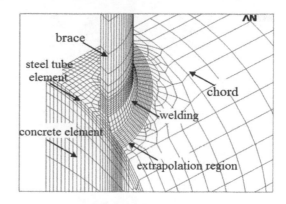

(a) SCF$_{max}$ VS. β

Figure 3. FE model for Concrete-filled CHS T-joint.

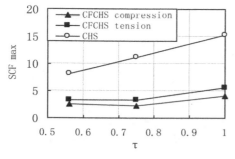

(b) SCF$_{max}$ VS. τ

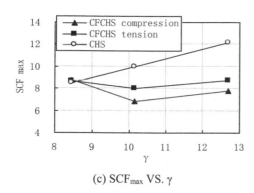

(c) SCF$_{max}$ VS. γ

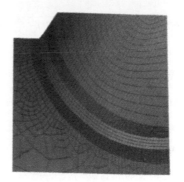

Figure 4. FE model for Concrete-filled CHS K-joint.

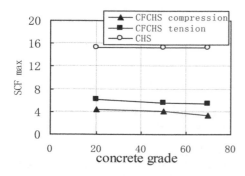

(d) SCF$_{max}$ VS. concrete strength grade

Figure 2. Effect of non-dimensional geometric parameter (β, τ, γ) and concrete cube strength grade on SCF$_{max}$ of Concrete-filled CHS T-joint.

K-joints and the comparison with the tested SCF$_{max}$, are shown in Figures 6 and 7. Their good agreement indicated that these finite element models adopted are reliable.

A parametric study has been carried out using the finite element models verified. A great number of concrete-filled tubular joints with different geometric parameters were calculated, to investigate their effects on SCF, as shown in Table 2 (Jing 2008, Wang 2008, Yang 2012).

Multivariate regression analysis was carried out for the large number of SCF data obtained from FEM. SCF formulae for different loading conditions were proposed, which can be used to calculate hot spot stress for fatigue assessment of concrete-filled tubular joints.

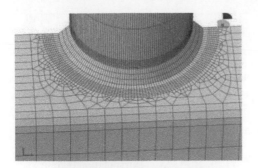

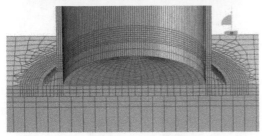

Figure 5. FE model for Concrete-filled CHS-to-SHS T-joint.

The SCF formulae for CFCHS K-joints is given in Equations (1) to (6) and Figure 7.

For the crown toe of chord at compressive side

$$SCF = 42.26(30.49 - 95.76\beta + 82.1\beta^2)\beta^{2.52}(2\gamma)^{-0.2}\tau^{0.67}\sin(\theta)^{2.37}$$
(1)

For the saddle toe of chord at compressive side

$$SCF = 8.06(5.06 - 12.47\beta + 10.41\beta^2)\beta^{0.87}(2\gamma)^{-0.09}\tau^{0.74}\sin(\theta)^{1.02}$$
(2)

For the crown toe of chord at tensile side

$$SCF = 7.16(7.24 - 21.49\beta + 18.29\beta^2)\beta^{1.61}(2\gamma)^{0.29}\tau^{0.77}\sin(\theta)^{1.05}$$
(3)

For the saddle toe of chord at tensile side

$$SCF = 5.51(6.96 - 19.03\beta + 14.89\beta^2)\beta^{1.19}(2\gamma)^{0.26}\tau^{1.04}\sin(\theta)^{1.15}$$
(4)

For the crown toe of brace

$$SCF = 6.00(4.10 - 10.61\beta + 9.16\beta^2)\beta^{0.56}(2\gamma)^{-0.05}\tau^{-0.13}\sin(\theta)^{1.50}$$
(5)

For the saddle of brace

$$SCF = 0.98(4.61 - 9.95\beta + 6.23\beta^2)\beta^{0.45}(2\gamma)^{0.25}\tau^{0.43}\sin(\theta)^{0.43}$$
(6)

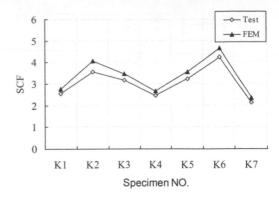

(a) SCF_{max} at chord saddle at tension area.

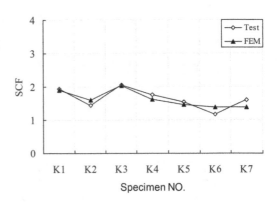

(b) SCF_{max} at brace saddle at tension area.

Figure 6. Comparison of SCF_{max} between FEM and test for Concrete-filled CHS K-joint.

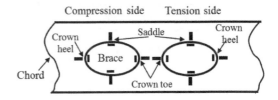

Figure 7. The location for SCF formulae for concrete-filled CHS K-joints.

4 TESTS ON FATIGUE STRENGTH

Fatigue tests on ten CFCHS T-joints and eight concrete-filled CHS-to-SHS T-joints were carried out respectively. Their test setup and fatigue failures are shown in Figure 8 and Figure 9. The results of all the fatigue tests reveal that: (1) The fatigue crack initiated exactly at the weld toe where the hot spot stress was a maximum in the joint, then propagated at the weld toe on the chord side of intersecting lines, through the thickness of the chord and finally the joint lost its static strength. (2) The concrete-filled tubular joint had a higher fatigue strength or longer fatigue

Table 2. Parametric study for SCF of Concrete-filled tubular joints.

| Joint type | Concrete-filled CHS T-joint | | Concrete-filled CHS-to-SHS T-joint | |
	T-joint	K-joint	T-joint	K-joint
Range of non-dimensional geometric parameter	$0.2 < \beta < 1$ $15 < 2\gamma < 64$ $0.2 < \tau < 1$	$0.3 < \beta < 0.6$ $24 < 2\gamma < 60$ $0.25 < \tau < 1$	$0.35 < \beta < 0.9$ $12.5 < 2\gamma < 25$ $0.25 < \tau < 1$	$0.35 < \beta < 0.9$ $12.5 < 2\gamma < 25$ $0.25 < \tau < 1$
Angle between brace and chord	/	$\vartheta = 45°$	/	$\vartheta = 30°, 50°, 70°$
Loading in the brace	Axial force In-plane bending	Axial force	Axial force In-plane bending	Axial force In-plane bending
Number of Case study	144	80	160	56

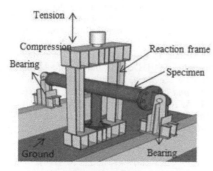

(a) Test setup

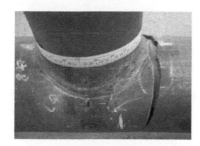

(b) Fatigue cracking

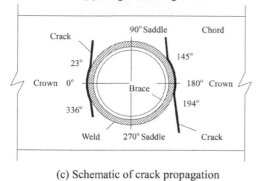

(c) Schematic of crack propagation

Figure 8. Fatigue test on concrete-filled CHS T-joint.

life than the pure tubular joint with the same geometric parameter. The S-N curves in CIDECT guidelines (Jing 2008, Wang 2008, Yang 2012) used for tubular joints are not appropriate for the fatigue assessment of concrete-filled tubular joints.

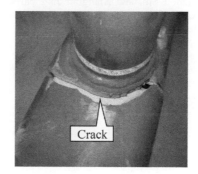

(a) Test setup.

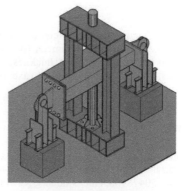

(b) Fatigue cracking

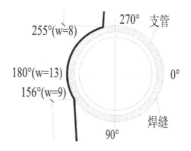

(c) Schematic of crack propagation.

Figure 9. Fatigue strength test on concrete-filled CHS-to-SHS T-joint.

5 FUTURE RESEARCH WORK

For the fatigue behaviour of welded joints of concrete-filled tubular trusses, there are many aspects that needs to be investigated further, at least dealing with: (1) Welded CFSHS joints, (2) SCF formulae for other types of joints (including multi-planar joints), (3) More fatigue tests are needed to provide more reliable S-N curves for fatigue assessment because of the limited number of test specimens to date. (4) Numerical simulation and prediction of fatigue life, based on fracture mechanics, should be developed to upgrade fatigue analysis of steel structures.

NOTATION

β is the ratio of brace diameter to chord diameter (or width), τ is the ratio of chord diameter (or width) to thickness, γ is the ratio of brace wall thickness to chord wall thickness, and ϑ is the angle between axes of brace and chord.

ACKNOWLEDGEMENTS

The research in the paper was supported financially by the Natural Science Foundation of China through the grant No. 50478108. Ph.D students, Ke Wang and Delei Yang, and Master's students, Jun Zhu, Chuanqi Sun, Weizhou Shi, and Xiangbin Jing participated in the parts of the research. The authors gratefully acknowledge all their contributions.

REFERENCES

Jing, X.B. 2008. Research on stress concentration factor of concrete-filled circular hollow section K-joints. *Thesis for Master's Degree*, Shanghai: Tongji University.

Sun, C.Q. 2008. Experimental study on hot spot stress and loading capacity of concrete-filled circular hollow section K-joints. *Thesis for Master's Degree*, Shanghai: Tongji University.

Wang, K. 2008. Study on hot spot stress and fatigue strength of welded CHS-to-CFCHS T-joints. *Thesis for Doctoral Degree*, Shanghai: Tongji University.

Wang, K., Tong, L.W., Zhu, J., Shi, W.Z., Mashirie, F.R. & Zhao, X.L. 2013. Fatigue behavior of welded T-joints with a CHS brace and CFCHS chord under axial loading in the brace. *Journal of Bridge Engineering* 18(2): 142–152.

Yang, D.L. 2012. Fatigue behavior and design method of welded truss CHS-to-CFCHS T-joints and Y-joints. Thesis for Doctoral Degree, Shanghai: Tongji University.

Zhao, X.L., Herion, S., Packer, J.A., Puhtli, R.S., Sedlacek, G., Wardenier, J., Weynand, K., van Wingerde, A.M. & Yeomans, N.F. 2001. Design guide for circular and rectangular hollow section welded joints under fatigue loading. *CIDECT and TUV-Verlag*, Koln, Germany.

Zhao, X.L. & Tong, L.W. 2011. New development in steel tubular joints. *Advances in Structural Engineering – An International Journal*, 14(4): 699–715.

Tubular Structures XV – Batista, Vellasco & Lima (eds)
© 2015 Taylor & Francis Group, London, ISBN 978-1-138-02837-1

Fatigue behaviour and detailing of slotted tubular connection

C. Baptista
IST Lisbon, GRID Engineering, Portugal

L. Borges
Structurame, Geneva, Switzerland

S. Yadav & A. Nussbaumer
ICOM – Steel Structures Laboratory, EPFL, Lausanne, Switzerland

ABSTRACT: Tubular elements are used in steel structures not only for bracing of other members but also as cross girders diaphragms. They are a competitive solution both due to the lightweight and structural efficiency. Slotted tube-to-plate connections are the natural and most common choice. However, fatigue design of the abrupt transition and differences in the stiffness corresponding to the welded connection, is a critical part of the design. Although such a fatigue connection is classified in different standards (e.g. EN1993-1-9, IIW recommendations), no guidance is given on the specific detailing. In this paper, the existing fatigue datasets for this connection are reviewed. An assessment of the different fatigue details constituting the connection (welds end, hole, etc.) using FEM models and the notch stress method is then made. The results are compared with the experiments and recommendations are issued in order to achieve the best fatigue performance.

1 INTRODUCTION

1.1 *Detail category (value at 2 million cycles)*

In slotted tubular connections, the tube may either be a circular (CHS) or a rectangular (RHS) hollow section, as shown in figure 1. Regarding the connection, its fatigue behaviour can be influenced by: the detailing of the welds between the tube and the plate, the gusset plate geometry, the tube chamfering, and probably the geometrical ratios of the connected parts (diameter/thickness of CHS, RHS, thickness ratio of plate and tube). For this or any other similar connections, the detail categories differ from standard to standard and not all consider the same influencing parameters.

EN1993-1-9:2005 classifies it (table 8.6, detail 2) for circular hollow sections (CHS) as CAT 71 (or CAT 63 for chamfered angles higher than 45°), but only for tubes with wall thicknesses less or equal to 12.5 mm; no limit on the tube diameter is given. The connection includes open holes at the end of the slit, a feature that facilitates fabrication and avoids having a non-detectable crack starting from the root of the weld. The recommendation for such a feature originates from Profs. J.W. Fisher and A. Pense, following a failure in a crane boom without warning; with open holes, fatigue cracking is visible if it develops (Fisher, 2009). However, the fatigue strength of the connection with open holes as CAT 71 cannot be taken as granted, as indeed no tests in high cycle fatigue were carried out on this detail.

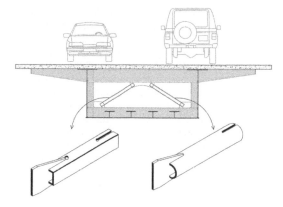

Figure 1.   Illustration of slotted tubular connections with CHS or RHS (each tube can be with or without hole).

In the IIW recommendations, a slotted CHS tube welded to plate is classified as a function of the tube diameter and plate thickness:

− FAT 63 ($d \leq 200$ mm, $t_p \leq 20$ mm)
− FAT 45 ($d > 200$ mm, $t_p > 20$ mm)

No open holes are shown and no requirement on the chamfered angle is given.

DNV recommended practice (DNVGL-RP-0005, 2014) identifies this connection as tubular joint with gusset plate. It recommends the use of FEA and hot spot stress method to design against fatigue this detail.

DNV differentiates between circular and rectangular hollow sections and also considers the closing of the tube end with plates to prevent water from getting in (sealing plates). The classification in hot spot stress is 90 for the end of the weld of the plate on the tube, which is often the critical spot.

BS7608, AISC and CIDECT design guide 8 (fatigue) do not have this detail categorized. Instead, in many codes, a "tube gusseted connection" detail is included, see section 3.2.

To summarize, the category and limits of application are confusing for this detail, not to mention the lack of guidance for computing the relevant stress in some codes. Only the tube chamfering angle, some geometrical limits of the connected parts (tube diameter and plate thickness) and the presence of open holes are differently specified. It is surprising that the detailing of the welds between the tube and the gusset plate is not part of the specifications.

1.2 Notch stress method

Finite element methods of increasing complexity (called local approaches) and devoted to analysing details under fatigue have emerged over the past years. In order to use only one S-N curve, the preferred method today is the effective notch stress method (Radaj 1990) (Fricke 2008). By modelling the local detail geometry, one can consider all global and local stress concentration effects within the model. The S-N curve provides then the intrinsic fatigue resistance of a weld toe or root with its microstructural imperfections. Thus, the method relies on the definition of a reference resistance fatigue curve, differentiating only between the weld types (E. Niemi et al. 2006). In order to get a finite and mesh independent peak stress value, all weld toes and roots of the connection are modelled with a fictitious notch radius of 1 mm, see fig. 2. This fictitious notch radius has to be added to the actual notch radius, which is usually assumed to be zero in a conservative way. Therefore it is recommended to assume generally $r_{ref} = 1$ mm for design purposes. The result is an effective notch stress or, if related to the nominal stress, a fatigue notch factor K_f (Fricke 2008). This approach together with a design reference fatigue curve of FAT 225 (225 MPa at 2 million cycles and m = 3) for steel, has been first included in the IIW Recommendation in 1996 (Hobbacher, 1996), valid for all weld types. The application to welded plate details (failing from the toe or the root) has extensively been investigated in particular for the ship industry (Fricke et al. 2002, Fricke 2012).

2 ANALYSIS OF FATIGUE DATASETS FROM LITERATURE

2.1 Analysis in nominal stress of existing tests

Only a few experimental works have been carried out by different authors on this connection and most tests

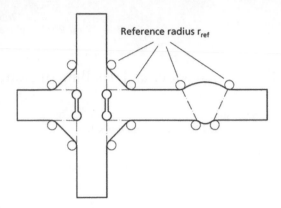

Figure 2. Applications of reference radii for the calculation of notch stresses (Hobbacher 1996).

Table 1. Categories, details and description according to (Mang et al, 1987).

Fatigue Category	Detail	Remarks

63	(61)	Tube-Plate connection
		Slotted tube end connected to a plate
		Limits:
		    Plate thickness 20mm
		    Tube diameter 200mm

45	(60)	Tube-Plate connection
		Slotted tube end connected to a plate
		Limits:
		    Plate thickness 20mm
		    Tube diameter 200mm

were on CHS connections. The results from (Zirn, 1975), (JSSC 1969), (Harada 1982) and (Uchino 1974) can be obtained from a report issued by University of Karlsruhe (Mang et al, 1987). Table 1 is adapted from this publication and shows that to be conservative for slotted cases, a category 45 should be used, with m = 3. The categories in this table and their limits correspond to the ones found in IIW, contradicting the values given in the Eurocode (and with different limits).

Reanalysing the data of the last two authors (Harada and Uchino), we came to the conclusion that the results are very scattered and no detailed information on the test specimens could be found. The JSSC test results are more homogeneous but provide a relatively low bound and again no detailed information is available. Thus the largest, most coherent and complete dataset is still the one by Zirn from his original report (Zirn 1975). Concentrating on Zirn's, it can be noticed that most cracks started from the weld toe at the end of the gusset plate welded to the tube. All tests were carried out on CHS tubes with $d = 88.9$ mm and plates $t_p \leq 20$ mm. The end of the tube was systematically closed with sealing plates. Figure 3 shows the experimental results separated into two groups depending upon the chamfered angle of the tube. Comparing the test results with the curves from the Eurocode (CAT 71 for $\alpha \leq 45°$ and 63 for $\alpha > 45°$), one notices that they

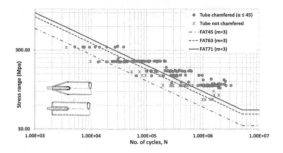

Figure 3. S-N plot in nominal stress of slotted tubular connections from Zirn, with code curves (m = 3).

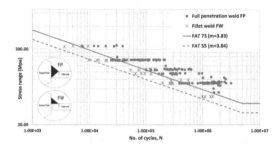

Figure 4. S-N plot in nominal stress of slotted tubular connections from Zirn, differentiating between weld types.

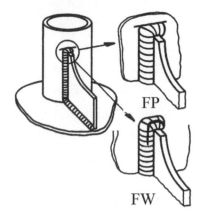

Figure 5. Illustration of tube gusseted connections (BS 7608, 1993).

his tests correspond to a characteristic valure at 2 million cycles of 36 MPa, m = 3, the lowest of all detail categories.

All of the above leads us to put into question the detail category, the differentiating parameter (chamfered angle) and the beneficial effect of an open hole (as shown in the Eurocode).

3 FATIGUE ASSESSMENT OF THE DIFFERENT INDIVIDUAL DETAILS

3.1 Longitudinal attachment

Before analysing the complete connection, it is important to understand the fatigue behaviour of the different details composing the connections, starting with longitudinal attachment, then with the influence of a hole.

A *tube gusseted connection* as shown in figure 5, is included in many codes. This connection is similar to a *slotted tube to plate* with the end of the loaded longitudinal attachment welded onto a tube. In this case cracking occurs in the tube – at the weld toe or in the weld end. However, there are two major differences regarding the stress field in the connection:

– the attachments and the tube are both welded to an endplate, the tube continues to carry its load share to the endplate,
– the tube is not slotted, thus the stress path between the tube and the attachments is smoother, closer to an unloaded longitudinal attachment.

Both these differences lead us to think that this detail shall have a better fatigue resistance then the slotted tubular connection. All the same, it is clearly shown that the plate geometry has strong influence and shall be made with a radius transition, which is not specified in the Eurocode for the slotted tubular connection.

DNV (DNV 2014) has detail categories varying from 90, with full penetration welds (FP), down to 71, with fillet welds (FW). The (BS7608 1993) has

are both unrepresentative and un-conservative. The characteristic curves (mean minus 2 std dev.) determined by linear regression were, respectively, CAT 80 with m = 4.26 and CAT 60 with m = 3.91. The difference that can be observed with the chamfering angle vanishes when using a slope m = 3 as shown in figure 3. Higher slopes, often close or equal to m = 5, are often found when analysing fatigue data on tubular connections. The IIW category FAT 45 is, on the contrary, conservative, but not the FAT 63 which should be valid for $d \leq 200$ mm, $t_p \leq 20$ mm. Different steel grades were tested but no significant difference could be seen when plotting the data.

Figure 4 shows the experimental results separated into two groups depending upon the weld type. In this case, the difference is clearer, showing that this parameter is more important. We believe the reason lies in the higher stress concentration at the weld end in the case of fillet welds. Full penetration welds allow for a smoother stress flow from the tube to the plate, and vice-versa. Note that the slopes of the regression lines are again close to m = 4 leading to CAT 73 for full penetration welds, and CAT 55 for fillet welds. Fixing the slope to m = 3 gives CAT 56 for full penetration welds, and CAT 45 for fillet welds.

More recently, Liu carried out a few fatigue experiments on this connection made with RHS and with slot gaps between the gusset plate and the tube (Liu 2006). He also found that the crack initiates from the beginning of the gusset plate, close to the slot gap, which is sharper than an open hole. Not surprisingly,

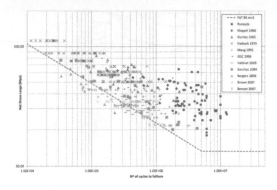

Figure 6. S-N plot in nominal stress of all results on plates with holes.

detail categories ranging from 68 for FP and failure in the tube at the weld toe, down to 43 for weld throat failure in fillet welded gusseted connections (FW), as illustrated in figure 5. DNV has higher categories but requires that the design stress include the stress concentration factor due to the overall form of the joint, which will lower the categories. In BS 7608, the design stress has to include any local bending stress adjacent to the weld end. This last point has more importance in this detail then in the slotted one since with the slot the plate is on both sides, there is less local bending. Overall, the codes do not have significantly higher categories for this detail compared to the slotted tubular connection, but the weld type is the main influencing parameter.

The notch stress method is directly applicable to this detail, but was outside of the scope of this study. Instead, the focus was directly set on modelling the complete slotted tubular connection as presented in the next section.

3.2 Plate with hole in nominal stress

The experimental results collected from literature for this detail, exemplify tests carried out in several leading European and American laboratories from 1960 to 2007. They were carefully selected in order to limit the scope to experiments typical to steel bridges. Only tests above $1 \cdot 10^4$ cycles, performed in plates under axial loading, and with the method of fabrication and crack path clearly specified were considered. The resulting data set is plotted in figure 6.

Klöppel (Klöppel, K., Weihermüller 1960) performed tests in plates with drilled holes at 4 different stress ratios (R) and showed a strong dependence on that parameter. The regression analysis shows a slope close to $m = 5$, which indicates a significant fatigue life spent in initiation.

Gurney (Gurney 1965) carried out fatigue tests on mild and high strength structural steel plates with a hole drilled and reamed. A remarkable dependence on steel type was observed with a slope also close to $m = 5$.

In (Haibach 1975) fatigue tests were reported on plates with drilled and reamed holes. Some dependence on R-ratio was observed but not on steel type.

Results presented in (Mang et al. 1991) refer to holes made by flame or plasma cut. Each plate had 3 holes spaced longitudinally at 150 mm. Plate thickness of 12, 20 and 30 mm were investigated not showing any pronounced size effect. No effect was found for the steel grade (St 37, St 52 and St E690) which indicates that initiation life is small due to plasma and flame cut rugosity. Regression analysis shows slopes close to $m = 3$ indicating that the fatigue life is fully in propagation. It is reported that flame cut results in a 30% reduction in fatigue strength compared to drilling.

Results from (JSSC 1995) are included in the background from Eurocode 3 and also here but no information is available regarding specimen dimension or steel type. Thus these results will not be used in the notch stress re-analysis.

Tests on galvanized plates with punched or drilled holes of 15 mm are presented in (Valtinat & Huhn 2003). Tests on punched holes revealed that the cracks started at the edge where the punch goes in, propagating as a corner crack until the crack reached the edge on the opposite surface. This represented almost the entire life, i.e. the propagation as an edge crack (trough thickness) was very fast. The specimens with drilled holes showed cracks mainly initiated in the mid thickness of the plate, propagating as a semi-elliptical crack. The results for punched holes show a poorer fatigue performance than the drilled ones. The authors further conclude that galvanizing has also a negative effect on the fatigue resistance.

Further tests on plates with punched or drilled holes are reported in (Sánchez et al. 2004). Again punched hole specimens were shown to have much lower fatigue life than the drilled ones. Punched holes behave with slope around $m = 3$ while drilled ones with a slope around 5. The fatigue crack for the punched specimens initiated at the transition point between the shear band and the tearing zones resulting from the punching process.

Fatigue tests in plates with plasma-cut holes were performed in (Bergers et al. 2006). Tests were done on steels S 460 MC, S 960 QL, S 1100 QL but no difference was found with the increase of steel grade. The results plot with a slope $m = 3$, indicating the predominance of propagation life due to the plasma cut surface defects. All cracks were reported to start in the inner edge of the hole.

Tests with holes done by punching, drilling or reaming are reported in (Bennett et al. 2007). The tests with drilled or reamed holes were all stopped around 2 million of cycles and considered run-outs. The fatigue performance of the punched ones was again very poor compared to drilled or reamed.

The results reported in (Brown et al. 2007) in punched and drilled holes are not included in the notch analysis because the hole on the finite width plates was

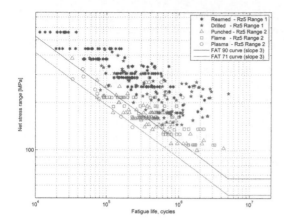

Figure 7. S-N plot in nominal net stress with classification in relation to the fabrication method of the hole.

not symmetrical and no clear information about each crack path is given.

Holes in plates are classified in Eurocode 3 as CAT 90 for the **net cross-section** nominal stress based on 3 sets of results (JSSC 1995; Valtinat & Huhn 2003; Mang et al. 1991). However the re-analysis shows this FAT category to be non-conservative. Many authors report the results in terms of **gross cross-section** nominal stress, thus the results were converted using:

$$\Delta\sigma_{gross} = \Delta\sigma_{net} \cdot \left(\frac{w}{w-d}\right) \qquad (1)$$

where w is the plate width and d is the hole diameter. The re-analysis of the data set shows that the fabrication method of the hole is the most remarkable parameter in plotting the fatigue results, see figure 7. Two different groups of holes were identified and regression analysis performed:

- **Quality Group 1**: Drilled and Reamed holes (Rz5 Range 1) – **FAT90**
  (i.e. FAT 145 with slope m = 3.03 and characteristic FAT 97 with m = 3)
- **Quality Group 2**: Punched, Flame and Plasma cut holes (Rz5 Range 2) – **FAT71**
  (i.e. FAT 91 with slope m = 2.62 and characteristic FAT 69 with m = 3)

Note: Rz5 is the ten-point roughness i.e., the average height from the five highest peaks and five lowest valleys over a given length (EN 9013).

According to EN 1090-2, holes may be formed by any process (drilling, punching, laser, plasma or flame cut), with the requirement that the local hardness and quality of the cut surface be checked. For Execution Class 4 (EXC4), punching without reaming is not allowed. Local hardness values were measured by (Mang et al. 1991; Valtinat & Huhn 2003) for punched, flame and plasma cut holes for example and they were inside the allowable range (380HV10 for steel up to S460 and 450HV10 for steel up to S690). Also residual stresses were measured but the values close to the hole edge were low. The most

important parameter distinguishing the 2 groups is the surface roughness in the hole, which may eliminate the initiation life, and introduce severe notches that act like initial fatigue cracks. The limits of roughness are defined according to EN 1090-2 for the Execution Class 4 (EXC4) as Range 2 quality for the mean height profile (Rz5) according to EN 9013, which means a maximum value of Rz5 = 40 + (0.8*t[mm]) [μm], which for the maximum plate thickness t = 30 mm, leads to 64 μm. As per the data set, surface quality was classified as Class 1 according to DIN 2310 in plasma cuts (Mang et al. 1991) which for t = 12 mm gives Rz = 65 μm. We recommend thus that for EXC4, structures subjected to fatigue loads, the Range 1 be adopted Rz5 = 10+(0.6*t[mm]) [μm].

The characteristic fatigue curves that represent the lower bound of the data set results from figure 7 may also be obtained by a linear elastic fracture mechanics approach, accounting for the initiation life in the Quality Group 1. A review of similar models is described in (Radaj et al. 2006) and has been adopted for plates with holes in low cycle fatigue by (Sehitoglu 1983) or for several high-strength steel details by (Pijpers 2011). The resistance of steel to fatigue crack initiation includes nucleation and short crack growth. The initiation life is modelled with a method described in (Haibach 2006), details can be found in (Baptista 2014). The initiation period is completed when the microcrack growth is no longer dependent on the microstructure or surface conditions, and thus, the crack growth resistance of the material starts to control the crack growth. The size of the microcrack at the transition stage is thus material dependant. We adopted the *Range 2* limit Rz5 = 40 + (0.8 × t[mm]) [μm], which for the maximum plate thickness *t* = 30 mm, leads to 64 μm.

Linear Elastic Fracture Mechanics is used for the propagation stage. Two types of cracks have to be considered to propagate the initial crack to a through-thickness crack. Observations of the failure surfaces of cracked specimens have shown that cracks propagate in this stage as corner crack or semi-elliptical cracks. Tests on punched holes revealed that the cracks started at the edge where the punch goes in, propagating as a corner crack until the crack reached the edge on the opposite surface. The specimens with drilled holes showed cracks mainly initiating in the mid-thickness of the plate, propagating as a semi-elliptical crack. In the final propagation stage, LEFM is used to propagate a through-thickness edge crack. For modelling the total life, two cases are compared:

- **Quality Group 1 (QG1)**: Initiation, propagation first as semi-elliptical and final propagation as through-thickness crack.
- **Quality Group 2 (CG2)**: Propagation first as corner crack and final propagation as through-thickness crack.

The stress intensity factor expressions used for the corner crack, semi-elliptical and through-thickness crack are detailed in (Baptista 2014). The results for

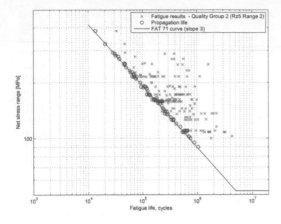

Figure 8. Comparison in nominal stress between experimental data set and model for Quality Group 2, only propagation stage considered.

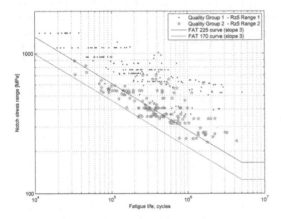

Figure 9. Comparison in notch stress between experimental data set and characteristic reference fatigue curves proposed.

CG2 are shown for example in figure 8. The cloud of points is close to the characteristic curve 71 for QG2, which shows that the model is able to properly reproduce the lower bound of the quality group. The same can be said for QG1 and characteristic curve 90. The proposed detail categories are thus confirmed to be FAT 90 for QG1 (drilled and reamed holes) and FAT 71 for QG2 (Punched, Flame and Plasma cut holes).

3.3 Plate with hole in notch stress

The data set is transformed into notch stress by considering the stress concentration factors corresponding to the plate and hole geometries (Peterson 1997). The resulting plot is shown in figure 9. The proposed characteristic curves for each of the two quality groups are also drawn. It can be seen that for QG1, the use of a FAT 225, m = 3, is appropriate. This is not the case for QG2, where only a FAT 170 should be used.

For more complex details, the net cross-section nominal stress to gross stress ratio is not uniquely defined, i.e. there is no direct geometric relation between the hole and elements composing the connection. The stress concentration factor varies with the many geometrical parameters, so that different K_f can be found for the same level of gross or net cross-section nominal stress. Thus it is only possible to determine a stress value with a FEA coupled with a *Hot Spot* or *Notch Stress* analysis.

It should be mentioned that while the analysis of the entire data set does not show any influence of the steel grade, a remarkable difference is noticed for the results of QG1, which indirectly indicates once again the importance of the initiation life in drilled and reamed holes.

3.4 Complete connection

The FEA was carried out on a series of models representing the slotted tubular connections tested by Zirn, see table 2, using the software ABAQUS. In all the tests reexamined and models, the tube is a CHS 88.9 × 5 and the attached plate thickness is 15 mm. Zirn specimens have sealing plates at the end of the tube. To see the influence of this feature, since the cost of preparing/welding these plates is not negligible, both cases were modelled. The differences in the specimens concern the type of weld, the plate and tube chamfer. Another feature, not tested by Zirn, was added to the geometry in a series of models, the presence of an open hole at the end of the slot. The hole radius was chosen to be 17.5 mm (half the plate thickness plus twice the weld leg length). A couple of cases with elongated holes were also made. In total, with the different parameters included in the models, the number of runs was 34.

All model results, given as K_f values, can be found in Table 2. An important limitation in the models is related to the initiation sites considered. At this time, since the main potential crack location observed in the tests is the cracking in the tube at the weld toe, and in one series of models where the open hole was added (and considered the new potential crack location), only these two were considered (i.e. the crack from the root or in the plate are not considered). The stress concentration factors (related to the nominal stress in the tube) for the different models (at the two details) are summarized in table 2. Note that the quality group requirement (QG1) for the open hole is important as the notch stress curve depends on it.

4 COMPARISON AND DISCUSSION

From the K_f values given in Table 2, one can deduce that for the same geometry, fillet welded connections have higher K_f compared to full penetration ones. The connections with sealing plates have a better resistance (i.e. a lower K_f) then those without them, the sealing plates make the tube carry some load over to the plate without overstressing the attachment ends in the tube.

Table 2. FEA geometries and resulting notch stress.

Details of the modeled geometries				Notch stress concentration factor, K_f					
				With sealing plate			Without sealing plate		
				Without hole	With hole		Without hole	With hole	
Tube	Gusset plate	Weld type (a=4 mm)	Specimen designation	at end of gusset plate	Max	Mean	at end of gusset plate	Max	Mean
		FW	L1	4.80	3.85	3.60	5.40	4.00	3.70
		FP	L1	**3.10**	3.95	3.74	3.20	3.85	3.60
		FP	L2*	3.60	3.50	3.35	3.70	3.70	3.50
		FP	L2	**3.40**	3.50	3.37	3.70	4.30	4.15
		FW	L3	5.50	4.00	3.72	5.90	4.80	4.55
		FP	L3	4.20	3.85	3.60	4.20	4.50	4.30
				With elongated hole					
		FP	L1	3.10	**3.08**	3.00	3.20	2.90	2.85
		FP	L2	3.40	**2.75**	2.72	3.70	3.45	3.42

Table 3. Fatigue classes in nominal stress deduced from notch analysis, cases with sealing plates.

Case	FEA w/o hole	FEA w/hole **Quality Group 1**	FEA w/hole **Quality Group 2**
L1_FW	47	**58**	44
L1_FP	**73**	57	43
L2*_FP	**64**	64	48
L2_FP	**66**	64	48
L3_FW	41	**56**	43
L3_FP	**54**	58	44
elongated			
L2*_FP_el	73	**73**	55
L2_FP-el	66	**82**	62

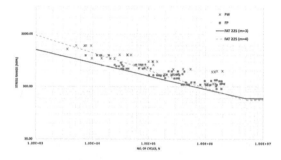

Figure 10. Comparison in notch stress between experimental data set and characteristic reference fatigue curves proposed.

For the cases with holes, note that both max value anywhere along the hole edge (not chamfered in the models) as well as the average value are given. The authors have used the max. values herein to be conservative, one could however argue that the average is more representative of well executed holes. The two models with elongated holes show it is an effective solution: the crack potential location remains inspectable in the hole edge, while the Stress Concentration Factor is lower than a circular hole.

Since the studied geometries are very close, a direct comparison between the different cases can be made in nominal stress (in the tube), taking into account the proper reference notch fatigue curves for the calculated notch factors; in other words, one can divide the FAT value by K_f. The reference notch fatigue curves, corresponding to characteristic values, are FAT 225 for the welds and FAT 170 for the holes, because usually the slot and hole are made by the same fabrication method and correspond to QG2. The results are shown in Table 3. It can be seen that Full Penetration welds may be considered in a CAT63 (except for geometry L3). In these connections, introducing the hole is even detrimental to the fatigue resistance to the whole connection, unless elongated holes in QG1 are adopted. In geometry L3, the chamfering of the plate is seen to be excessive and leads to an increase of the concentration at the weld in the tube. For Fillet Welded connections, CAT 45 is suited while the hole may increase the resistance to CAT 56 if holes are made in QG1.

Finally, one can plot Zirn results in notch stress, see Figure 10. It shows the capacity of the notch method to describe different notch cases as the difference between PP and FW is diluted in a global notch scatter band. One can note that the slope m = 4 is more adequate and, logically, confirms the analysis on figure 4 in nominal stress.

5 DETAIL RECOMMENDATIONS

The connection types with full penetration welds have a better fatigue resistance and thus should be promoted. The presence of sealing plates is also beneficial. Regarding the chamfering of the tube, it has limited influence on the fatigue resistance as long as full penetration welds are used. The same can probably be said for the chamfering of the plate, but was not fully studied; the chamfering of the part of the plate inside the tube is of more significance.

Open holes can be a good solution for tolerance reasons and NDT in order to avoid a non-detectable crack starting from weld root. But if the connection contains open holes, it is recommended to adopt the quality group 1, which corresponds to a maximum roughness of $Rz5 = 10 + (0.6*t[mm])$ [μm]. If only the quality group 2 is used because it may not be economical to ream holes after cutting them together with the slit, or to drill them separately from the slot, then a solution could be to put an elongated hole in the stress direction instead of a circular one. If one uses an elongated hole with a length double its width, the improvement can be expressed by using $0.81 \cdot K_f$ (or by multiplying the fatigue class with $1/0.81$) which approx. compensates for cut quality group 2 instead of 1.

6 CONCLUSION

The following conclusions can be drawn:

- For plates with holes, only drilled and reamed holes (Rz5 Range 1) can achieve FAT 90 in nominal net stress, and FAT 225 in notch stress (Quality Group 1). For Quality Group 2, the values are FAT 71 in nominal net stress, and FAT 170 in notch stress.
- The fatigue categories in Eurocode 3, part 1.9, for the slotted tubular connection should be updated in terms of fatigue strength, requirements and drawing of potential crack location(s).
- The connection types with full penetration (FP) welds have a better fatigue resistance and the weld type, which is one main influencing parameter, should be considered in the codes (may be instead of chamfering angle of the tube). This study shows they could be classified in CAT 63 for FP and CAT 45 for FW.
- The presence of sealing plates is beneficial to the fatigue resistance. So is the presence of the hole, but only for fillet welded plates.
- The notch method (FAT 225) is able to include the difference between FP and FW notch cases.

REFERENCES

Fisher, J. W. 2009, June 4. Prof., Lehigh Univ., USA, Private communication.

DNVGL-RP-0005. 2014, June. RP-C203: *Fatigue design of offshore steel structures*, http://www.dnvgl.com.

Mang, F., Bucak, & Klingler, J. (1987). *Wöhlerlinien katalog für Hohlprofilverbidungen*. Universität Karlsruhe: Studiengesellschaft für Anwendungstechnik von Eisen und Stahl e.V.

Liu, Y., Dawe, J., & Li, L. (2006). Experimental study of gusset plate connections for tubular bracing. *Journal of Constructional Steel Research, 62*, 132–143.

Zirn, R. (1975). *Schwingfestigkeitsverhalten geschweißter Rohrknotenpunkte und Rohrlaschenverbindungen*. Stuttgart: University of Stuttgart.

Fricke, W. et al. 2002. Comparative fatigue strength assessment of a structural detail in a containership using various approaches of classification societies. Marine Structures, 15(1), pp. 1–13.

Fricke, W. (2008). Guideline for the fatigue assessment by notch stress analysis for welded structures. International Institute of Welding.

Fricke, W. 2012. IIW Guideline for the Assessment of Weld Root Fatigue. International Institute of Welding.

Hobbacher, A. (1996). Fatigue Design of Welded Joints and Components. Cambridge (UK): Abington Publishing, 1st ed.

Radaj, D. (1990). Design and Analysis of Fatigue Resistant Structures. Abington Publishing, Cambridge (UK).

Niemi, E., Fricke, W. & Maddox, S.I. (2006). Fatigue analysis of welded components, Woodhead Publishing Limited.

Radaj, D., Sonsino, C. & Fricke, W. (2006). *Fatigue assessment of welded joints by local approaches*, CRC.

Baptista, C. (2014). A reanalysis of the fatigue strength of steel plates with holes, EPFL report N° 201786, ICOM-EPFL, Lausanne, 2014.

Bennett, C.R., Swanson, J.A. & Linzell, D.G., 2007. Fatigue Resistance of HPS-485W Continuous Plate with Punched Holes. Journal of Bridge Engineering, 12(1), pp.98–104.

Bergers, J. et al., 2006. Beurteilung des Ermüdungsverhaltens von Krankonstruktionen bei Einsatz hoch- und ultrahochfester Stähle. Stahlbau, 75(11), pp.897–915.

Brown, J.D. et al., 2007. Evaluation of Influence of Hole Making Upon the Performance of Structural Steel Plates and Connections.

Gurney, T., 1965. Some Exploratory Fatigue Tests on Notched Mild and High Tensile Steels. British Welding Journal, 12(9), p.457/61.

Haibach, E., 1975. EUR5357 – Fatigue investigation of higher strength structural steels in notched and in welded condition.

JSSC, 1995. Fatigue design recommendations for steel structures J. S. of S. Construction., ed.

Klöppel, K., Weihermüller, H., 1960. Dauerfestigkeitsversuche mit Schweißverbindungen aus St. 52. Stahlbau, 29(5), p.129/37.

Mang, F., Bucak, O. & Obert, K., 1991. Investigation into the fatigue life of specimens with flame-cut holes and plasma cutting of steel. In New Advances in Welding and Allied Processes: Proceedings of the International IIW Conference. Beijing, China: IIW.

Sánchez, L., Gutiérrez-Solana, F. & Pesquera, D., 2004. Fatigue behaviour of punched structural plates. Engineering Failure Analysis, 11(5), pp.751–764.

Valtinat, G. & Huhn, H., 2003. Festigkeitssteigerung von Schraubenverbindungen bei ermüdungsbeanspruchten, feuerverzinkten Stahlkonstruktionen. Stahlbau, 72(10), pp.715–724.

Tubular Structures XV – Batista, Vellasco & Lima (eds)
© 2015 Taylor & Francis Group, London, ISBN 978-1-138-02837-1

Performance of tube-based moment connections under cyclic loads

D. Wei, J. McCormick & M. Hartigan
University of Michigan, Ann Arbor, MI, USA

M. Fadden
University of Louisiana at Lafayette, Lafayette, LA, USA

ABSTRACT: Seismic steel moment frames rely on beam plastic hinging and moderate deformation in the panel zone of the connection to dissipate input energy from a seismic event. Although the majority of these frames utilize wide flange beams with either wide flange or hollow structural section (HSS) columns, there has been a recent increase in interest in the development of tube-based moment frame systems. To take advantage of the potential benefits of a tube-based moment frame, an external diaphragm plate connection and a through plate connection with an HSS $305 \times 203 \times 9.5$ beam and an HSS $254 \times 254 \times 15.9$ column are experimentally tested under large cyclic loads. The findings indicate that these reinforced connections can achieve stable beam plastic hinging. More innovative HSS-to-HSS moment connections utilizing beam endplates and collars to minimize field welding also are explored through a detailed finite element study. The findings provide essential data on the force transfer mechanisms and detailing requirements for these connections.

1 INTRODUCTION

1.1 Background

Low to mid-rise steel moment frames are often used in earthquake prone regions because of their light weight and ability to resist lateral forces. Traditionally these moment frames are designed based on a strong column-weak beam design philosophy in which the beams are expected to undergo the majority of the inelastic deformation during a large earthquake. The performance of seismic steel moment frames is largely dependent on the ability of the beam to undergo ductile plastic hinging with limited degradation due to local buckling or fracture. During the 1994 Northridge earthquake in the United States and the 1995 Hyogo-Nanbu (Kobe) earthquake in Japan, significant damage to steel moment frames occurred largely due to non-ductile fracture of the welded connections. This behavior led to extensive studies to understand the behavior of the connections and improve their detailing. However, the majority of the research focused on wide flange beam to wide flange column or wide flange beam to hollow structural section (HSS) column connections. Limited studies considered connections with other types of steel members, such as HSS, which has led to missed opportunities to further improve upon steel moment frame performance under seismic loads.

HSS provide many beneficial properties such as a large strength to weight ratio that can reduce the seismic weight of structures; high torsional stiffness, which can limit the need for lateral bracing; and good

bending and compression capacity. Their square and rectangular shapes also have the potential for use in modular and rapid construction applications. In Japan and Europe, studies of moment frame systems have exposed the potential benefits of HSS column to wide flange beam moment connections in seismic regions. One of the early studies of moment connections with HSS members is that by Giroux et al. (1977) in which rigid connections utilizing wide flange beams and tubular columns were experimentally tested. Tabuchi et al. (1988) investigated the cyclic behavior of rectangular and round hollow structural section columns connected to wide flange beams considering various column diameter-to-thickness ratios and axial load ratios. Korol et al. (1993) also proposed a novel connection using high strength blind bolts to connect wide flange beams to HSS columns. Although a number of studies have been conducted on seismic connections with HSS columns, connections utilizing HSS beams and HSS columns have rarely been considered.

Two studies to consider tube-based moment connections are those by Rao and Kumar (2006) and Kumar and Rao (2006) who experimentally tested and conducted finite element studies of a bolted rectangular hollow section beam to column moment connection. This connection used a channel member to transfer forces from the beam flange directly to the column sidewall, avoiding the use of internal diaphragms in the column. With the choice of a suitable channel connector, the moment connections exhibited stable hysteresis loops, adequate ductility, and good energy dissipation capacity. Fadden and McCormick (2012a) carried out 11 different HSS beam cyclic bending tests that showed that the depth-thickness and width-thickness ratios play a significant role in determining whether stable plastic hinging will occur in an HSS under pure bending. A further finite element study of HSS beam members under pure bending provided limiting values for the width-thickness and depth-thickness ratios to achieve 80% of their plastic moment capacity out to 0.04 rad. of rotation (Fadden and McCormick 2014a). This finding suggests that provided the correct HSS beam member is chosen, stable plastic hinging that meets current seismic design requirements can be achieved. To further consider the use of tube-based seismic moment connections, adequate configurations and detailing requirements must be determined and the connection's behavior characterized under large cyclic loads.

1.2 Objective

To address the lack of adequate connection configurations and detailing requirements for tube-based seismic moment connections, an experimental and finite element study is undertaken considering HSS-to-HSS connections under large cyclic loads. An experimental investigation of the behavior of an external diaphragm plate and a through plate connection is conducted at full scale to determine whether ductile limit states can be achieved and intermediate and special steel moment frame requirements met. A more innovative connection utilizing beam endplates and steel collars is then considered through a detailed finite element analysis. The thickness of the beam endplates and collars is varied to determine the optimal parameters for the connection. The findings suggest that both connections have the potential for use in low-rise seismic moment frame systems.

2 REINFORCED TUBE-BASED CONNECTIONS

2.1 Connection design

The use of diaphragm and through plates within tube-based moment connections is necessary to prevent excessive column face plastification that is often seen when beam width to column width ratios are less than 0.85 (Packer and Henderson 1997) and to prevent brittle fracture of the weld at the corner of the connection under cyclic loads when an HSS beam is directly welded to the face of an HSS column. Kamba and Tabuchi (1994) have shown that the use of internal and external diaphragm plates can address the deformation problems associated with the flexibility of an unstiffened HSS column face for HSS column to wide flange beam connections, but consideration of HSS-to-HSS seismic moment connections is limited.

To address this limitation and explore the viability of reinforced HSS-to-HSS seismic moment

Table 1. Connection component properties.

		Connection	
		Ext. Diaphragm	Through Plate
Column	size	HSS 254 × 254 × 15.9	HSS 254 × 254 × 15.9
	b/t	12.4	12.6
	h/t	12.9	13.0
	$Z_{meas} \times 10^3$ (mm³)	1178	1176
Beam	size	HSS 305 × 203 × 9.5	HSS 305 × 203 × 9.5
	b/t	18.6	18.4
	h/t	30.0	30.1
	$Z_{meas} \times 10^3$ (mm³)	859	866
Plate	Length (mm)	508	457
	Thickness (mm)	19	22.2

connections, a design procedure is developed considering the limiting parameters of HSS in cyclic bending (Fadden and McCormick 2012a,2014a); results from a detailed finite element study (Fadden and McCormick 2012b); and design philosophies outlined in the CIDECT Design Guides (Kurobane et al. 2004, Packer et al. 2009), the AISC Specification (2010a), and the AISC Seismic Provisions (2010b). The design procedure ensures that either beam plastic hinging or reinforcing plate yielding will occur prior to weld fracture allowing the plastic moment capacity of the beam to be reached. Further details of the design procedure can be found in Fadden et al. (2015).

Two external connections are considered, one that utilizes an external diaphragm plate which wraps around the column and a second that utilizes a through plate in which the HSS column was cut into three segments and then reattached with the through plates in between the segments. Table 1 provides details of the connection parameters including the member sizes and measured width-thickness ratios (b/t), depth-thickness ratios (d/t), and plastic section moduli (Z_{meas}). The reinforcing plate length refers to the length of the plate from the column face to its termination point along the length of the beam. Both the diaphragm plates and through plates are connected to the HSS column using complete joint penetration groove welds. A flare bevel groove weld is used along the length of the beam between the reinforcing plates and the beam, while a transverse fillet weld connects the end of the reinforcing plates with the top and bottom of the HSS beam. To ensure that a non-ductile weld failure does not occur, the fillet weld sizes are 19 mm and 17.5 mm for the external diaphragm plate and through plate connections, respectively. Figure 1 shows a schematic of the two connections.

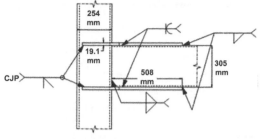

(a) External Diaphragm Plate Connection

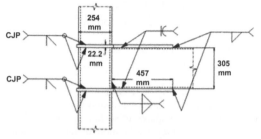

(b) Through Plate Connection

Figure 1. Schematic of the (a) external diaphragm plate and (b) through plate connections. The connection represents an exterior, one-way beam-column connection in a moment frame system.

2.2 Experimental setup and loading

The experimental setup is designed to represent a prototype tube-based seismic moment frame that has 3.7 m floor heights and 6.4 m bay widths. An exterior connection is tested where the column is half the floor height above and below the connection and the beam is half the bay width. The ends of the members are pinned to simulate the inflection points in the prototype frame. A photo of the test setup is shown in Figure 2 where the column is horizontal and the beam is vertical to allow for ease of testing. A hydraulic actuator is used to apply displacements to the beam tip using the loading protocol specified by the American Institute of Steel Construction for pre-qualification of seismic moment connections (AISC 2010b). Tip displacements are applied at a loading rate of 12.7 mm/min up to rotations of 0.07 rad.

The HSS beams and columns meet ASTM A500 Gr. B. or Gr. B/C standards. Three coupon specimens are taken from the flats of the beam and column of the external diaphragm connection, but not from the through plate connection since its members are from the same batch of material. The average yield strength and tensile strength is measured to be 426 MPa and 487 MPa for the beam and 425 MPa and 462 MPa for the column, respectively. The reinforcing plates are made from ASTM A36 carbon steel, which has a minimum specified yield strength of 248 MPa and a minimum tensile strength of 400 MPa. Materials tests on the plates are not conducted as the plates are not expected to undergo significant yielding. Further details on the experimental testing of the reinforced

Figure 2. Photographs of the test setup and connection after testing. The column is shown horizontal and the beam vertical.

connections and the results of these tests also can be found in Fadden et al. (2015).

2.3 Experimental test results

Intermediate and special moment frame system connections are expected to be able to undergo 0.02 rad. and 0.04 rad. of rotation while maintaining at least 80% of their moment capacity. To evaluate the ability of the two reinforced connections to meet these criteria, the moment-rotation behavior of the connections is considered. The moment at the centerline of the column is calculated as the applied load at the beam tip divided by the length of the beam plus half the column depth, while the rotation at the center of the connection is calculated as the horizontal displacement of the beam tip divided by the length of the beam plus half the column depth. The beam plastic moment that is used to normalize the measured moment is calculated as the experimentally measured yield strength multiplied by the experimentally measured plastic section modulus ($M_{p,beam} = F_y Z_{meas}$). The beam plastic moment of the external diaphragm plate and through plate connections are 366 kN-m and 369 kN-m, respectively.

The moment-rotation plots for the two connections are shown in Figure 3a. Both connections developed full hysteresis loops with degradation of the moment capacity not occurring until after the first 0.04 rad. cycle. The maximum moment achieved by the external diaphragm plate connection is 482 kN-m, which is approximately 32% larger than the plastic moment capacity of the beam. Likewise, the maximum moment achieved by the through plate connection is 443 kN-m, which is approximately 20% larger than the plastic moment capacity of the beam. These findings suggest that intermediate and special moment frame requirements can be met using reinforced tube-based moment connections provided that the connection is detailed such that weld fracture does not control. After reaching their maximum moments, both connections show degradation in moment capacity with continued

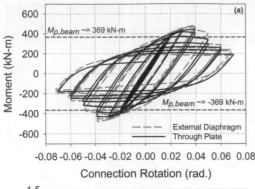

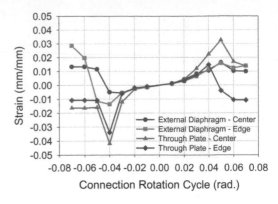

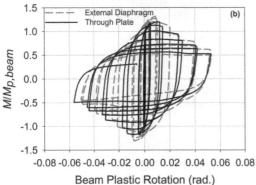

Figure 3. (a) Moment-rotation hysteresis and (b) normalized moment-beam plastic rotation hysteresis for the external diaphragm plate and through plate connections.

Figure 4. Beam strain gage measurements with respect to connection rotation.

cycling to larger rotation levels. This degradation is the result of local buckling in the beam member, which eventually leads to the initiation of fracture in the HSS beams. Fracture starts at the corner of the HSS beam due to the cold working that is imparted during the HSS manufacturing process (Fadden and McCormick 2012) and propagates toward the center of the web and flange of the beam. However, the connections are able to reach 0.07 rad. of rotation prior to terminating testing. At this point, the moment capacity of the external diaphragm plate and through plate connections has decreased to 164 kN-m and 182 kN-m.

The reinforced connections undergo desirable ductile plastic hinging of the HSS beam member prior to local buckling, which is consistent with the current strong column-weak beam design philosophy for moment frame systems. This ductile plastic hinging of the beam can be observed in Figure 3b where the normalized moment is plotted with respect to beam plastic rotation. Nearly all of the plastic rotation occurs in the beam member with the external diaphragm connection reaching a maximum beam plastic rotation of 0.053 rad. (i.e. 99% of the measured inelastic rotation) and the through plate connection reaching a maximum beam plastic rotation of 0.056 rad. (i.e. 100% of the measured inelastic rotation) during the 0.07 rad. cycle. However, this large level of plastic deformation in the beam member does lead to local buckling that limits force transfer to other components of the connections

and may decrease the efficiency of the connections. A more balanced connection design where the panel zone partially participates in sustaining the inelastic deformation demands may be more efficient and help to increase the number of cycles prior to fracture of the beam member. Further, the utilization of a beam member with a lower width-thickness ratio also will help to inhibit local buckling.

To understand the flow of forces through the HSS beam and into the reinforcing plate, strain gages are placed on the top face of the beam at its center and 25.4 mm from its edge. Along the length of the beam, the strain gages are located 50.8 mm away from the end of the reinforcing plate. Figure 4 provides the measured strain values with respect to connection rotation cycle. In general, strains are larger at the center of the beam suggesting that most of the tension and compression force is transmitted through the center of the fillet weld at the end of the reinforcing plate. It is also observed that the yield strain is surpassed during the 0.02 rad. cycle at the center of the beam and during the 0.03 rad. cycle at the edge of the beam eventually leading to full plastification of the beam section. At rotation levels greater than 0.05 rad. the strain values decrease due to the onset of local buckling further away from the column face than where the strain gages are located. Strains are also measured in the panel zone and on the face of the column on the opposite side to where the beam is connected. These strain measurements generally remained below 0.005 for the two reinforced connections, which further shows that almost all of the inelastic deformation occurred in the beam members.

3 TUBE-BASED COLLAR CONNECTION

3.1 Concept

The experimental tests of the reinforced tube-based connections show that HSS-to-HSS seismic connections are viable. However, the external diaphragm plate and through plate connections can be labor intensive and require a significant amount of field welding. To alleviate the amount of field welding that is necessary

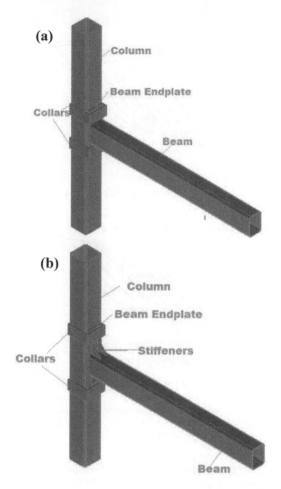

(a)

Column

Beam Endplate

Collars

Beam

(b)

Column

Beam Endplate

Stiffeners

Collars

Beam

Figure 5. Welded collar connection (a) without endplate stiffeners and (b) with endplate stiffeners.

and possibly increase construction speeds, welded collar HSS-to-HSS moment connections are considered. Figure 5 provides a schematic of the tube-based collar connection.

The welded collar moment connection consists of two steel collars that are slipped over the column member and the beam endplate to connect the beam to the column. The beam endplate and lower collar (i.e. below the beam) can be shop-welded. A gap is left between the column face and the collar to provide a location for the bottom of the beam endplate to slip into in the field. The upper collar can then be slipped down the column and over the top of endplate and welded in the field. As a result, only fillet welds are required in the field, as opposed to the complete joint penetration welds required by the tested reinforced connections, which can make construction easier and more economical. However, care must be given to provide enough tolerance to allow the beam endplate to easily slip into place between the bottom collar and column face. The beam endplate and collars also must be properly sized so as to prevent significant column face plastification or undesired brittle weld failure. To further control

the load transfer mechanism and location of inelastic deformation, stiffeners can be used in conjunction with the beam endplate connection.

3.2 Finite element models

To study the behavior of the tube-based collar connections under large cyclic loads and determine the effects of different parameters on the performance of the connection, finite element models of the connection are constructed in Abaqus CAE 6.12. As with the reinforced connections, the beam members are HSS $305 \times 203 \times 9.5$ and the column members are HSS $254 \times 254 \times 15.9$. The finite element model represents an exterior connection in a seismic moment frame with bay widths of 5.8 m and bay heights of 3 m. Half of the height of the column (1.5 m) is modeled above and below the connection with the end boundary conditions taken as pins to represent the inflections points in the column above and below the connection. Half the length of the beam member (2.9 m) also is modeled with displacements applied to its free end to simulate the cyclic loading of an earthquake.

The beam and column are modeled with S4R shell elements. To improve the accuracy and efficiency of the model, the beam mesh is divided into two sections. The region adjacent to the column face that spans 1.2 m along the beam uses 12.7 mm square elements, while for the rest of the beam the element sizes gradually increase from 12.7 mm square elements to 127 mm square elements at the displaced end. The column mesh is divided into three sections. The central portion of the column, where the connection is located, spans 1.5 m and has 12.7 mm square elements. The other two equal length regions that span 1.2 m have square element sizes that increase from 12.7 mm to 127 mm at the pinned end. The collars, beam endplate, and stiffeners are modeled with C3D8R solid elements that are 6.35 mm square. The interfaces between the beam endplate and beam, beam endplate and column face, stiffeners and beam flange, stiffeners and beam endplate, collar and beam endplate, and collar and column are fixed to each other using tie constraints to represent the welded connections. Weld failure is not considered in this finite element study.

Actual material properties from previously tested coupon specimens (Fadden and McCormick 2014) are converted to true stress and true strain and then input and applied to the flats and corners of the HSS beam and column models. The experimentally measured values of yield strength and tensile strength are 411 MPa and 496 MPa for the flats and 517 MPa (75.0 ksi) and 583 MPa for the corners. The collar, beam endplate, and stiffeners are modeled using a bilinear behavior based on the required yield strength (248 MPa) of ASTM A36 steel.

To simulate cyclic loading under earthquake excitation, the same loading protocol as is used for the experimental tests of the reinforced connections is applied to the finite element models with the exception that only two cycles are applied at each drift level.

Table 2. Collar connection parameters and values.

Parameter	Values
Endplate thickness, t_e	6.35 mm, 12.7 mm, 19.1 mm
Collar thickness, t_c	6.35 mm, 12.7 mm, 19.1 mm
Endplate stiffeners	Yes or No

The applied drift levels increase from 0.00375 rad. to 0.06 rad.

For the parametric study, beam endplate thickness, beam endplate stiffeners, and collar thickness are considered leading to ten different configurations. The values of each considered parameter are shown in Table 2. The no endplate stiffener case is only considered for one connection configuration where the endplate thickness and collar thickness are both 12.7 mm. The other nine connections utilized endplate stiffeners to provide a more uniform transfer of force from the beam to the endplate.

3.3 Parametric study results

To see the effect of the different parameters on the behavior of the tube-based collar connections, the moment-rotation results are plotted in Figure 6 for the collar connections with stiffeners that have the same endplate and collar thicknesses. The moment-rotation for the collar connection without stiffeners, which has a 12.7 mm thick endplate and collar, also is plotted. In these plots, the moment was normalized by the plastic moment capacity of the beam member, M_p. The plastic moment capacity is calculated as $M_p = F_y Z$, where F_y is the experimentally observed value of the yield strength taken from the flat of an HSS and used in the material model. The four moment connections shown produce stable hysteretic behavior. For the moment connections with stiffeners, the maximum normalized moments are 0.93, 1.03, and 1.11 for collar and beam endplate thicknesses of 6.35 mm, 12.7 mm and 19.1 mm, respectively. These results indicate that an endplate and collar thickness of at least 12.7 mm is required in order to fully develop the plastic moment capacity of the beam member which is necessary to achieve a strong column-weak beam behavior. For the moment connection without stiffeners, the maximum normalized moment was only 0.92 compared to 1.03 for the same connection with beam endplate stiffeners. This showed the effectiveness of the stiffeners in providing a better force transfer mechanism between the beam and the column.

Figure 7 provides the normalized moment-beam plastic rotation plots for the four previously discussed connections. The results suggest that as the beam endplate and collar become thinner, less inelastic behavior in the beam occurs. Use of the thickest beam endplate and collar (19.1 mm) leads to a maximum beam plastic rotation of 0.025 rad. when the connection is cycled to 0.06 rad., while the connection with the thinnest

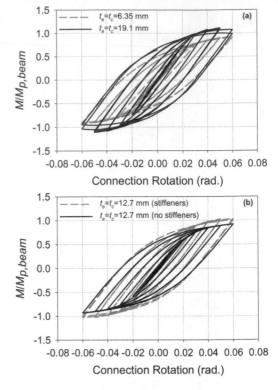

Figure 6. Normalized moment-connection rotation plots for collar connections considering (a) endplate and collar thickness and (b) the use of endplate stiffeners.

beam endplate and collar (6.35 mm) results in a maximum beam plastic rotation of only 0.006 rad. There is a further decrease in beam plastic rotation when no stiffeners are used with the beam endplate. The 12.7 mm thick beam endplate and collar connections had maximum beam plastic rotations of 0.011 rad. and 0.0004 rad. with and without stiffeners, respectively. This decrease can largely be attributed to larger deformations in the beam endplate and collar compared to the beam itself when no stiffeners are used or when a thinner beam endplate and collar are used. In general, the connection with the 19.1 mm thick beam endplate and collars shows desirable plastic hinging in the beam member based on the full hysteretic loops that are observed.

A study of the sources of inelastic rotation provides a better understanding of the overall connection behavior and whether the connection meets current strong column-weak beam design standards with the majority of the inelastic behavior occurring in the beam member. In considering the sources of inelastic rotation, the plastic rotation associated with the beam, column, column face and panel zone is determined. All of the connections that utilized stiffeners (i.e. all but one connection that is studied) have a similar distribution of inelastic behavior between the beam, column, column face, and panel zone, with the majority of the inelastic behavior occurring in the beam member. On average,

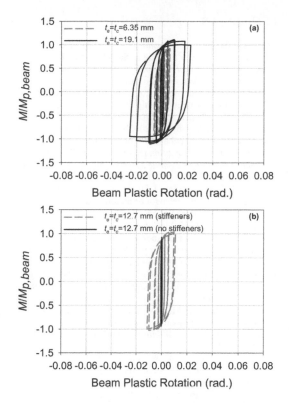

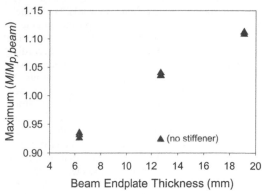

Figure 8. Normalized maximum moment-beam endplate thickness for all ten collar connections.

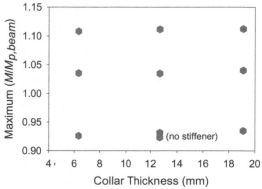

Figure 9. Normalized maximum moment-collar thickness for all ten collar connections.

Figure 7. Normalized moment-beam plastic rotation plots for collar connections considering (a) endplate and collar thickness and (b) the use of endplate stiffeners.

the connections with beam endplate and collars of the same thickness have 90.2% of their overall plastic rotation at the 0.06 rad. cycle associated with the beam member. The second largest contribution to the plastic rotation at the 0.06 rad. cycle is the panel zone with an average value of 5.68%. The percentages are based on the sum of the plastic rotations associated with the four considered components (beam, column, column face, and panel zone), which are 0.007 rad., 0.011 rad., and 0.021 rad. for the connection with 6.35 mm, 12.7 mm, and 19.1 mm thick beam endplate and collars, respectively. In general, the contribution of the panel zone, column, and column face increased with a decrease in beam endplate thickness and collar thickness, but all of the stiffened connections have the majority of inelastic behavior occurring in the beam member. The connection without stiffeners (12.7 mm thick beam endplate and collars) has a relatively small total plastic rotation (0.002 rad.) with the beam plastic rotation providing the smallest percentage of this total and the panel zone providing the largest percentage. In general, the unstiffened endplate results in much larger column, column face, and panel zone contributions than are desirable.

Figures 8 and 9 look specifically at the effects of beam endplate thickness and collar thickness on the moment capacity of the tube-based collar connections. For the collar connections with stiffeners,

the beam endplate thickness has more influence on the moment capacity of the connection than the collar thickness. An increasing trend in the maximum normalized moment is seen as the beam endplate thickness increases, regardless of the collar thickness. No clear trend is seen with respect to the collar thickness. The results suggest that the beam endplate is critical in determining the behavior of the beam and location of inelastic behavior in the connection. The collar connection without stiffeners had the least moment capacity, indicating that the beam was not fully utilized. This finding further suggests that beam endplate stiffeners may be necessary in order for these collar connections to meet current seismic design standards.

4 CONCLUSIONS

An experimental study is undertaken to evaluate the performance of two reinforced tube-based connections. The connections consist of HSS-to-HSS members with either an external diaphragm plate or a through plate to more efficiently transfer forces from the beam flanges to the column sidewalls. The connections are designed such that beam plastic hinging or reinforcing plate yielding occurs prior to non-ductile weld fracture. The connections are cycled according to a loading protocol to simulate the expected loading during a far-field earthquake and the results are evaluated in terms of the connection moment rotation behavior, beam plastic hinge behavior, and strain levels in the beam member. A second set of seismic moment connections is also considered that utilize a beam endplate and collars to connect the HSS members. Ten different tube-based collar connections are considered through a detailed finite element analysis where they are cycled through a similar loading protocol to that of the reinforced connections. The effect of beam endplate thickness and collar thickness on the cyclic behavior of the connection along with the distribution of inelastic rotations within the connection is considered in evaluating the connection.

The tube-based reinforced and collar connections are each able to achieve stable plastic hinging of the beam member. For the reinforced connections, adequate width-thickness and depth-thickness ratios of the beam are required to inhibit the onset of local buckling and the occurrence of fracture due to cycling until after the formation of the plastic hinge in the beam member. With the addition of the reinforcing plates, the transfer of forces mainly occurs at the middle of the beam flange where the reinforcing plate terminates along the length of the beam. For this reason, the fillet weld at the end of the reinforcing plate must be properly detailed to prevent non-ductile weld fracture. For the collar connections, beam endplate stiffeners ensure that the majority of the inelastic behavior occurs in the beam member and larger endplate thicknesses allow the plastic capacity of the beam to be reached. The beam endplate thickness is shown to control the moment capacity of the connection more than the collar thickness. Overall, both sets of connections show promise for use in low-rise moment frames.

ACKNOWLEDGEMENT

This work is supported by the American Institute of Steel Construction through the Milek Faculty Fellowship. The views expressed herein are solely those of the authors and do not represent the views of the supporting agency.

REFERENCES

AISC. 2010a. *Specification for Structural Steel Buildings.* Chicago: American Institute of Steel Construction.
AISC. 2010b. *Seismic Provisions for Structural Steel Buildings.* Chicago: American Institute of Steel Construction.
Fadden, M. & McCormick, J. 2012a. Cyclic quasi-static testing of hollow structural section beam members. *J. Struct. Eng. ASCE*, 138(5): 561–570.
Fadden, M. & McCormick, J. 2012b. Finite element study of the cyclic flexural behavior of hollow structural sections. *Proc. of the 7th International Conference on Behaviour of Steel Structures in Seismic Areas. Santiago, Chile, 9–11 January 2012.*
Fadden, M. & McCormick, J. 2014. Finite element model of the cyclic bending behavior of hollow structural sections. *J. Constr. Steel Res.*, 94: 64–75.
Fadden, M., Wei, D. & McCormick, J. 2015. Cyclic testing of welded HSS-to-HSS moment connections for seismic applications. *J. Struct. Eng. ASCE*, 141(2), 04014109, 10.1061/(ASCE)ST.1943-541X.0001049.
Girouz, Y.M & Picard, A. 1977. Rigid framing connections for tubular columns. *Canadian Journal of Civil Engineering*, 4(2): 134–144.
Kamba, T. & Tabuchi, M. 1995. Database for tubular column to beam connections in moment-resisting frames. *International Institute of Welding Commission XV, IIW Doc. XV-893-95.*
Korol, R.M., Ghobarah, A., & Mourad, S. 1993. Blind bolting W-shape beams to HSS columns. *J. Struct. Eng. ASCE*, 119(12): 3463–3481.
Kumar, S.R.S & Rao, D.V.P. 2006. RHS beam-to-column connection with web opening – experimental study and finite element modeling. *J. Constr. Steel Res.*, 62(8): 739–746.
Kurobane, Y., Packer, J.A., Wardenier, J., & Yeomans, N. 2004. *Design Guide for Structural Hollow Section Column Connections, CIDECT Design Guide No. 9*, CIDECT and TÜV-Verlag GmbH: Köln, Germany.
Packer, J.A. & Henderson, J.E. 1997. *Hollow structural section connections and trusses: a design guide.* Ontario: Canadian Institute of Steel Construction.
Packer, J.A., Wardenier, J., Zhao, X.L., van der Vegte, G.J., & Kurobane, Y. 2009. *Design Guide for Rectangular Hollow Section (RHS) Joints under Predominantly Static Loading, CIDECT Design Guide No. 3 (2nd. Edition)*, CIDECT: Geneva, Switzerland.
Rao, D.V.P. & Kumar, S.R.S. 2006. RHS beam-to-column connection with web opening – parametric study and design guidelines. *J. Constr. Steel Res.*, 62(8): 747–756.
Tabuchi, M., Kanatani, H., & Kamba, T. 1988. Behaviour of tubular column to H-beam connections under seismic loading. *Proc. of the 9th World Conference on Earthquake Engineering, Tokyo-Kyoto, 2–9 August 1988.*

Tubular Structures XV – Batista, Vellasco & Lima (eds)
© *2015 Taylor & Francis Group, London, ISBN 978-1-138-02837-1*

Effect of the secondary bending moment on K-joint capacity

T. Björk, N. Tuominen & T. Lähde
Lappeenranta University of Technology, Finland

ABSTRACT: Current design guides for tubular structures are based on an assumption that the joint of a K-truss can be analyzed by assuming that the brace members are pin-jointed to the chord member. This simplified approach seems to work in practice for mild or low strength steels. However, the secondary bending moment occurring in the brace members can cause additional stresses also in the joints. In this study, the secondary bending moment of the brace member of a K-joint was investigated. The secondary moment is also caused by an uneven support-effect (USE) of the brace member on the chord flange. The magnitude of the USE is highly dependent on the joint geometry. This phenomenon was studied using analytical approaches, experimental tests and numerical calculations by nonlinear FEA. The results for the different analytical approaches agreed quite well and demonstrated the importance of the USE when high strength steels are used.

1 INTRODUCTION

Based on the current design guide Eurocode (EC3, 2005) trusses made of tubular cross-sections can be analyzed using a simple approach in which the brace members are assumed to be fixed as pin-ended members on the chord member, which are approached as continuous elements. This approach implies an assumption that the joint is loaded only by axial forces and not by any bending moments. The method has been shown to work in experimental K-joint tests and the approach has also been seen to work in practice. Adoption of this analytical technique simplifies strength calculations in design of tubular cross-sections.

The approach is further based on the assumption that even if a bending moment exists in the brace member, the joint and adjacent members have sufficient plastic deformation capacity to sustain the additional stresses, mainly due to the fact that the secondary loads are displacement controlled and consequently have limited deformation.

Although this simplified method seems to work safely in practice with trusses made of low and mild strength steels, the question arises whether the assumptions underlying the approach are valid also for joints made of high strength steels, as presented for example by Sedlacek & Müller (2001). Current design rules EC3 (2007) do not set any special requirements for consideration of secondary moments for strength classes up to steel grade S700. However, EC3 (2005) sets a reduction factor of 0.9 for joints made of steel grades that are higher than S355 but less than S460 and EC3 (2007) sets a reduction factor of 0.8 for grades higher than S460 up to S700.

2 GOAL

The goal of the work is to determine the effect of the secondary moment due to USE (uneven support-effect) on the capacity of K-joints.

3 MATERIALS

The effect of the secondary moment on the capacity of K-joints was investigated using K-joints made of cold-formed rectangular hollow sections (CFRHS). The measured cross section dimensions of the test tubes and mechanical properties of the steel grades used are listed in Table 1, and the measured chemical compositions are presented in Table 2, respectively.

4 SECONDARY MOMENT OF A K-JOINT

The calculated capacity of a K-joint is based on the axial load of the brace member, but the joint can be subjected also to moment loading. A bending moment of the joint can occur for several reasons, as illustrated in Figure 1.

Bending moments acting in the plane of the tubular truss are mainly caused by four phenomena:

a) An external load is applied to chord or brace members causing a primary bending load in the joint;

b) A secondary bending moment occurs due to a large displacement of the truss and because of semi-rigid joints between the chord and brace members instead of pin-ended brace members;

c) Eccentricity of the joints means that the centerlines of the brace-members do not meet the centerline of the chord in the joint;

Table 1. Steel grades, cross section values and mechanical properties of the test tubes.

Tube size $b \times h \times t$ [mm]	Steel grade	Width b [mm]	Height h [mm]	Thickness t [mm]	Yield strength f_y [mm]	Ultimate strength f_u [mm]	Elongation A_5 [mm]
$140 \times 220 \times 8$	S700	140.2	220.7	7.80	719	820	15
$80 \times 80 \times 4$	S700	80.4	80.4	3.94	864	878	13
$150 \times 150 \times 6$	S960	150.6	150.6	5.92	1006	1156	10
$80 \times 80 \times 4$	S960	80.4	80.4	4.00	1176	1220	7

Table 2. Chemical composition from material certificates [%].

Tube size	Steel grade	C	Si	Mn	P	S	Al	Nb	V	Cu	Cr	N	Ti	Mo	Ni	B
$220 \times 140 \times 8$	S700	0.06	0.19	1.08	0.009	0.003	0.03	0.082	0.01	0.02	0.05	0.005	0.11	0.01	0.04	0.0004
$80 \times 80 \times 4$	S700	0.06	0.44	1.04	0.008	0.003	0.03	0.051	0.04	0.20	0.06	0.006	0.10	0.01	0.04	0.0004
$150 \times 150 \times 6$	S960	0.10	0.22	1.04	0.010	0.001	0.03	0.002	0.01	0.02	0.10	0.005	0.03	0.12	0.07	0.0019
$80 \times 80 \times 4$	S960	0.10	0.21	1.01	0.007	0.003	0.03	0.001	0.01	0.01	0.12	0.006	0.03	0.12	0.05	0.0021

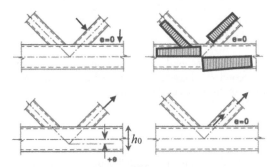

Figure 1. In-plane bending moments of a K-joint.

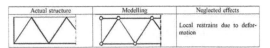

Figure 2. Potential problem involved in truss analysis of high strength steels by Sedlacek & Müller (2001).

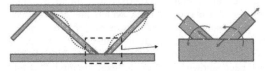

Figure 3. Secondary bending moment due to uneven support effect of the chord flange.

d) An eccentricity effect due to uneven support of the brace member on the chord face.

As mentioned earlier, according to current design codes EC3 (2005) the bending moment can be ignored as a secondary phenomenon when the load carrying capacity of the joint is defined, assuming the joint eccentricity is inside allowed limits (that is, $-0.55 < e/h_o < 0.25$). For the case of an external load causing primary bending, the bending moment can also be ignored, because EC 3 allows the brace member to be analysed as a pin-ended member independent of the local rotation stiffness of the joint.

The role of secondary moments has been investigated previously by, for example, Marshall (1992), Zhang & Niemi (1993) and Coutie (1993), but no studies specifically concerning high strength steels are available. Consequently, it is beneficial to investigate if the assumption that the secondary bending moment can be disregarded is valid also for joints made of high strength steels.

The topic is illustrated in Fig. 2 and was originally suggested by Sedlacek & Müller (2001) as one of the crucial research theme involved in the safe use of high strength steels.

This study investigates theoretically, numerically and experimentally the case where the secondary moment is caused by an USE of the brace member. The USE is a result of the boundary conditions of the brace member in the K-joint not being symmetrical about the neutral axis of brace member. In typical K-joints, the gap side is stiffer than the heel side because the space for flange deformation is more limited in the gap side than in the heel side. The phenomenon is illustrated in Fig. 3.

The bigger the difference in chord flange stiffness between the gap and heel side of the brace member, the more remarkable the USE. Consequently, small β-ratio ($=b_1/b$) together with small gap-ratio (g/t) increases the USE. On the other hand, a small gap ratio is a desired value, if a high loading capacity of the joint is sought. Due to the USE, a secondary bending moment exists in addition to the axial load. Consequently, the joint and the adjacent brace member must be designed for a combination of both axial and bending loads. The load carrying capacity of joints subjected to brace member axial loadings is well defined in current design codes EC3 (2005). However, the load carrying capacity of K-joints subjected to a bending moment is not included in current design codes.

5 THEORETICAL CAPACITY OF THE K-JOINT

The axial load capacity of the brace member in a K-joint can be calculated according to current design codes using a simple semi-empirical formula (1) that ignores the effect of the gap (EN 2005).

$$N_{i,Rd} = \frac{8.9 k_n f_{y0} t_0 \sqrt{\gamma}}{\sin\theta} \left(\frac{b_1 + b_2 + h_1 + h_2}{4b_0} \right) \quad (1)$$

where:
$\gamma = b_0/2t_0$
k_n = a coefficient considering the axial compressive membrane stress of the chord member
f_y = the yield strength of the chord member
t = the thickness of the chord member
θ = the angle between the brace and chord member
b_1 = the width of the brace member
h_1 = the height of the brace member
b = the width of the chord member.

However, the gap is a very important parameter for the capacity of the joint. Decreasing the gap size increases the load carrying capacity of the joint, until a certain limit, but decreases the deformation capacity, and vice versa. Packer (1980), Niemi (1982, 1986), Partanen & Björk (1993), Fleischer & Puthli (2009) have presented more sophisticated models for joint capacity in its limit state that take into account the gap effect.

The moment capacity of the brace member in the K-joint is dependent either on the plastic bending capacity of the member itself or on the joint capacity, which can be defined based on the yield line mode, as illustrated in Fig. 4. The presented mechanism is valid for a symmetrical K-joint with equal outer dimensions and joining angles θ of the brace members. The bending moments of the brace members are equal. The mechanism of the heel side corner is assumed to be a simple circular fan corner instead of a more sophisticated logarithmic or elliptical corner proposed by Sanezuk & Jaeger (1963). The brace members are so close to each other that their yield line mechanisms coalesce.

The following simplifications are used in location of the plastic hinges and mechanism involved:

$b = b_0 - t_0$
$\beta = (b_1 + 2k_1 a)/b_0$
$h = h_1/\sin\theta + (k_2 + k_3)a \quad (2)$
$g = g_0 - 2 k_3 a$

where:
b = the width of the chord member

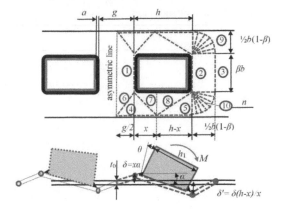

Figure 4. Yield line pattern for moment capacity of the brace member in a K-joint.

t = the wall thickness of the flange in a chord
b_1 = the width of the brace member
h_1 = the height of the brace member
θ = the joining angle of the brace member to chord
g = the nominal distance between the tension and compression brace members on the chord face
g = the distance between the plastic hinges in the gap
a = the throat thickness of the weld in the joint
$k_{1,2,3}$ = shape factors ($\approx 0 \ldots 1$) considering the weld effect on the location of the plastic hinges.

The virtual work of each hinge (i) can be calculated

$$W_i = m_p l_i a_i = \frac{f_y t_0^2}{4} l_i a_i \quad (3)$$

where:
m_p = the plastic moment of the hinge
l_i = the length of a hinge i
α_i = the rotation angle of a hinge i
f_y = the yield strength of the chord member.

The total internal work of the plastic hinges can be calculated from the terms in Table 3, obtaining the sum:

$$W_{in} = \sum_{i=1}^{10} W_i = \frac{f_y t_0^2}{2} \left\{ \begin{array}{l} b + \pi h + \dfrac{2h(\beta b + 2h)}{b(1-\beta)} + \\ \left(\dfrac{b}{g} - \dfrac{2\beta b + 8h - 2g}{b(1-\beta)} - \pi \right) x + \\ \dfrac{8}{b(1-\beta)} x^2 \end{array} \right\} \alpha \quad (4)$$

The outer work of the system is:

$$W_o = M\alpha \quad (5)$$

The place of the rotation axis (x) can be fixed by defining the minimum energy of the system:

$$\frac{dW}{dx} = \frac{d(W_{in} - W_0)}{dx} =$$

$$\frac{16}{b(1-\beta)} x + \frac{b}{g} - \frac{2\beta b + 8h - 2g}{b(1-\beta)} - \pi = 0 \quad (6)$$

559

Table 3. Work of plastic hinges.

ID	n_i	$m_{p0.i}$	l_i	α_i	W_i
1	1	$\frac{f_y t_0^2}{4}$	βb	$\left[\frac{1}{x}+\frac{2}{g}\right]\delta$	$\frac{f_y t_0^2}{4}\left[\frac{\beta b}{x}+\frac{2\beta b}{g}\right]\delta$
2	1	$\frac{f_y t_0^2}{4}$	βb	$\left[\frac{1}{x}+\frac{2(h-x)}{xb(1-\beta)}\right]\delta$	$\frac{f_y t_0^2}{4}\left[\frac{\beta b}{x}+\frac{2\beta(h-x)}{x(1-\beta)}\right]\delta$
3	1	$\frac{f_y t_0^2}{4}$	βb	$\left[\frac{2(h-x)}{xb(1-\beta)}\right]\delta$	$\frac{f_y t_0^2}{4}\left[\frac{2\beta(h-x)}{x(1-\beta)}\right]\delta$
4	2	$\frac{f_y t_0^2}{4}$	$\frac{g}{2}+x$	$\left[\frac{2}{b(1-\beta)}\right]\delta$	$\frac{f_y t_0^2}{4}\left[\frac{2g+4x}{b(1-\beta)}\right]\delta$
5	2	$\frac{f_y t_0^2}{4}$	$h-x$	$\left[\frac{2(h-x)}{xb(1-\beta)}\right]\delta$	$\frac{f_y t_0^2}{4}\left[\frac{4(h-x)^2}{xb(1-\beta)}\right]\delta$
6	2	$\frac{f_y t_0^2}{4}$	$\frac{1}{2}\sqrt{g^2+b^2(1-\beta)^2}$	$\left[\frac{\frac{g}{b(1-\beta)}+}{\frac{b(1-\beta)}{g}}\right]\frac{\delta}{l_6}$	$\frac{f_y t_0^2}{4}\left[\frac{2g}{b(1-\beta)}+\frac{2b(1-\beta)}{g}\right]\delta$
7	2	$\frac{f_y t_0^2}{4}$	$\frac{1}{2}\sqrt{4x^2+b^2(1-\beta)^2}$	$\left[\frac{\frac{2x}{b(1-\beta)}+}{\frac{b(1-\beta)}{2x}}\right]\frac{\delta}{l_7}$	$\frac{f_y t_0^2}{4}\left[\frac{4x}{b(1-\beta)}+\frac{b(1-\beta)}{x}\right]\delta$
8	2	$\frac{f_y t_0^2}{4}$	$\frac{1}{2}\sqrt{4(h-x)^2+b^2(1-\beta)^2}$	$\left[\frac{\frac{2(h-x)^2}{bx(1-\beta)}+}{\frac{b(1-\beta)}{2x}}\right]\frac{\delta}{l_8}$	$\frac{f_y t_0^2}{4}\left[\frac{4(h-x)^2}{xb(1-\beta)}+\frac{b(1-\beta)}{x}\right]\delta$
9	2	$\frac{f_y t_0^2}{4}$	$\frac{\pi b(1-\beta)}{4}$	$\left[\frac{2(h-x)}{xb(1-\beta)}\right]\delta$	$\frac{f_y t_0^2}{4}\left[\frac{\pi(h-x)}{x}\right]\delta$
10	2	$\frac{f_y t_0^2}{4}$	$\frac{nb(1-\beta)}{2}$	$\left[\frac{\pi(h-x)}{nxb(1-\beta)}\right]\delta$	$\frac{f_y t_0^2}{4}\left[\frac{\pi(h-x)}{x}\right]\delta$

This obtains the solution:

$$x=\frac{1}{16}\left[8h+2\beta b-2g-b\left(\frac{b}{g}-\pi\right)(1-\beta)\right] \quad (7)$$

The moment capacity of the joint can now be calculated:

$$M=\frac{f_y t_0^2}{2}\left\{\begin{array}{l} b+\pi h+\dfrac{2h(\beta b+2h)}{b(1-\beta)}+ \\[2mm] \left(\dfrac{b}{g}-\dfrac{2\beta b+8h-2g}{b(1-\beta)}-\pi\right)x+ \\[2mm] \dfrac{8}{b(1-\beta)}x^2 \end{array}\right\} \quad (8)$$

$$M=\frac{f_y t_0^2}{2}\left\{\begin{array}{l} b+\pi h+\dfrac{2h(\beta b+2h)}{b(1-\beta)}+ \\[2mm] \dfrac{1}{32b(1-\beta)}\left(8h+2\beta b-2g-b\left(\dfrac{b}{g}-\pi\right)(1-\beta)\right)^2+ \\[2mm] \dfrac{1}{16}\left(\dfrac{b}{g}+\dfrac{2g-2\beta b-8h}{b(1-\beta)}-\pi\right)\cdot \\[2mm] \left(8h+2\beta b-2g-b\left(\dfrac{b}{g}-\pi\right)(1-\beta)\right) \end{array}\right\} \quad (9)$$

6 EXPERIMENTAL TESTS

Two experimental tests with K-joints were carried out: BSAK1 and BSAK12. The test specimens are

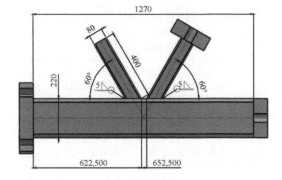

Figure 5. Test specimen.

illustrated schematically in Fig. 5 and the nominal dimension of the joint members presented in Table 4.

The test set up is illustrated schematically in Fig. 6a and the principle of the loading and boundary conditions for the specimen is given in Fig. 6b. In the BSAK12-test, the chord member had a tensile preload of 54% of the nominal yield capacity of the chord tube.

In the experiments, the load of the tensile brace was increased quasi-statically until failure occurred. The deformations were measured by displacement transducers separately for the gap and heel sides of the tension brace members enabling the axial and rotation deformations at the joint to be distinguished. The location of the displacement transducers is seen in Fig. 6a. The load of the tension brace member was measured by load cell. Two strain gauges were installed in the gap next to the weld between the brace and chord members; one on the chord side and one on the brace member.

Table 4. Dimension of the test specimens and FE-analysis joints.

Model	Note	Chord $h_0 \times$ [mm]	$b_0 \times$	t_0	Braces $h_1 \times$ [mm]	$b_1 \times$	t_1	Joint parameters α [mm]	β –	e [mm]	g [mm]	θ [°]
BSAK1	S700 Test + FEA	220	140	8	80	80	4	5	0.57	−4.0	30	60
BSAK2	S700	220	140	8	50	50	4	5	0.36	−3.7	65	60
BSAK3	S700	220	140	7	80	80	4	5	0.57	−4.0	30	60
BSAK4	S700	220	140	6	80	80	4	5	0.57	−4.0	30	60
BSAK5	butt weld in gap	220	140	8	80	80	4	5	0.57	−30.0	30	50
BSAK6	butt weld in gap	220	140	8	80	80	4	5	0.57	−45.2	30	40
BSAK7	S700	220	140	8	80	80	4	5	0.57	22.0	60	60
BSAK8	S700	220	140	8	80	80	4	5	0.57	47.9	90	60
BSAK9	S700	220	140	8	70	70	4	5	0.50	−3.5	30	63
BSAK10	S700	140	140	8	80	80	4	5	0.57	36.0	30	60
BSAK11	S700	220	140	8	80	80	4	5	0.57	0.0	34.6	60
BSAK12	S960, pretension $0.54f_y$ Test + FEA	150	150	6	80	80	4	6.5	0.53	95.9	105	60

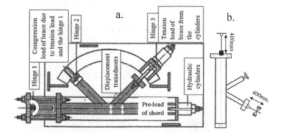

Figure 6. a) Test set up and b) real position, loading and boundary conditions.

The other two strain gauges were installed in the tensile brace member to measure the nominal stress of brace member. The results of the tests are seen in Figure 7 and the failures in Figure 8.

7 FEA ANALYSES

Nonlinear finite element analyses were carried out to compare theoretical values for the load capacity of the K-joints with experimental results. The FE-model used consisted of half of the joint, as presented in Fig. 9. The boundary conditions and the loading simulated the experimental tests and behavior of the truss, as illustrated in Figures 3 and 6b.

ABAQUS 6.13 software was used to solve the geometrical and material nonlinear behavior of the K-joint. C3D20R-type solid elements were used in the critical gap area of the joint as illustrated in Fig. 10a. The true stress-strain-material model used presented tri-linear behavior, as shown in Fig. 10b.

Fig.11 presents the deformations of the gap and heel sides of the brace member as measured in the experimental tests and illustrated also in Fig. 13.

Fig. 12 illustrates the local deformations of the gap area in both joints at the maximum calculated load level of the specimen.

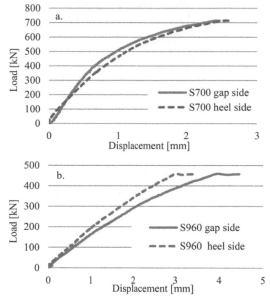

Figure 7. Load –displacement-behavior of a tension brace in a K-joint made of a) S700 and b) S960 steel.

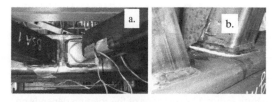

Figure 8. Failure of K-joints made of a) S700 and b) S960 steel.

8 DISCUSSION

The rotation of the brace member can be determined from differences in the displacements between the gap

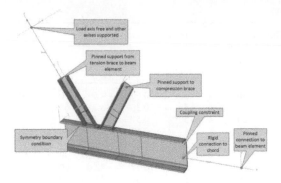

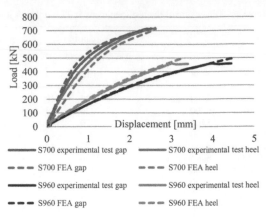

Figure 9.   Boundary conditions and loading of specimen in FEA.

Figure 11.   Load versus deformation of the brace member.

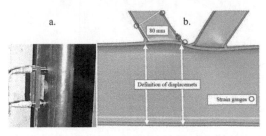

Figure 12.   Local deformations of the joints made of a) S700 (BSAK1) and b) S960 (BSAK12).

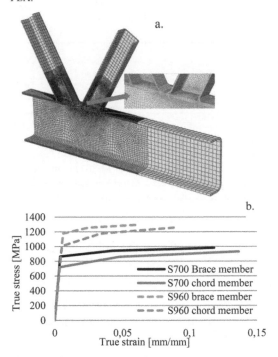

Figure 10.   a) Mesh in the gap area of the joint and b) material models for S700 and S900 steels.

Figure 13.   Definition of the joint deformations a) experimentally and b) from FEA.

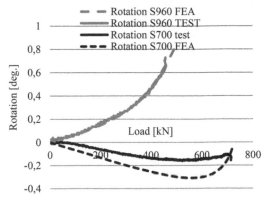

Figure 14.   Load-rotation of the brace member in K-joints.

and heel sides. The measuring system considers the rotation of the chord member equally in experimental tests and FE-analyzes, as illustrated in Fig. 13.

The rotation versus axial brace load is given in Fig. 14. Although the secondary moment increases the stresses on the gap side of the tensile brace member, the brace member rotation can be in the opposite direction in both joints. The direction of the rotation depends on the differences in stiffness between the gap and heel sides and on the amount of joint eccentricity.

Fig. 15 presents the stress distribution in the brace member at the weld toe and at a distance of 80 mm from the heel. The strain gauges were located at the same cross-section measuring the nominal stresses of the tensile brace member as illustrated in Fig. 13. The

stresses from FEA are the membrane stresses in the longitudinal direction of the brace member.

The secondary moment capacity of the brace member is dependent on the plastic moment capacity of the brace member itself or the joint, as described in section 5 above. The plastic capacity of the brace member

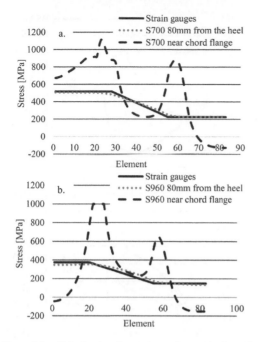

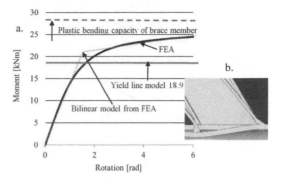

Figure 15. Calculated and experimental stresses along the brace member perimeter in a) S700 and b) S960 joints.

is the plastic modulus W_p of the cross section multiplied by the real yield strength f_y of the brace member $(80 \times 80 \times 4)$:

$$M_p = W_p f_y = 33070 \text{mm}^3 \cdot 864 \text{MPa} = 28.6 \text{ kNm}$$

The plastic capacity of the joint can be calculated for the BSAK1- joint using the following material properties and dimensions: $f_y = 719$ MPa, $t = 7.8$ mm, $h = 92.4$ mm, $b = 132.4$ mm, $g = 30$ mm, $\beta = 80.4/132.4 = 0.5735$. According to Fig. 12a the calculation is based on an assumption that the welds have only a minor effect on the location of the plastic hinges and thus on the joint capacity. Consequently for BSAK1 is set $k_1 = k_2 = k_3 = 0$ in Eq. (2). However, in BSAK12-joint the welds have much bigger effect on the places of the plastic hinges as illustrated in Fig.12b and therefore $k_1 = k_2 = k_3 = 1$ in Eq. (2). The place for the rotation line x is calculated from Eq. (7) and is obtaining for BSAK1:

$$x = \frac{1}{16} \begin{bmatrix} 8 \cdot 92.84 + 2 \cdot 0.5735 \cdot 132.4 - 2 \cdot 30 \\ -132.4 \left(\frac{132.4}{30} - \pi \right)(1 - 0.5735) \end{bmatrix} = 47.67 \text{ mm}$$

The moment capacity M of the joint from Eq. (8) is:

$$M = \frac{719 \cdot 7.8^2}{2} \left\{ \begin{array}{l} 132.4 + \pi \cdot 92.84 + \dfrac{2 \cdot 92.84 \cdot (0.5735 \cdot 132.4 + 2 \cdot 92.84)}{132.4 \cdot (1 - 0.5735)} \\ + \left(\dfrac{132.4}{30} - \dfrac{2 \cdot 0.5735 \cdot 132.4 + 8 \cdot 92.84 - 2 \cdot 30}{132.4 \cdot (1 - 0.5735)} - \pi \right) \cdot \\ 47.67 + \dfrac{8}{132.4 \cdot (1 - 0.5735)} \cdot 47.67^2 \end{array} \right\}$$

$M = 18.9 \text{ kNm}$

Figure 16. a) Pure moment – rotation of the BSAK1-joint and b) joint behavior from FEA.

Fig. 16a shows the behavior of the joint under pure moment loading. The plastic mechanism of the chord face is given in Fig. 16b, where the both brace members are subject to similar (secondary) bending moments and consequently referring to yield line model presented in Fig. 4.

Fig 17 presents the development of the stresses due to secondary moment versus the axial load in the brace member at a distance of 80 mm from the joint (defined along the heel side). The dotted line describes the interaction of axial and bending stresses based on the elastic upper limit capacity of the brace member.

If axial load and bending moment are imposed on the joint simultaneously, EC3 suggests the interaction formula:

$$\frac{N}{N_{i,Rd}} + \frac{M_s}{M} \leq 1. \tag{10}$$

For the case of ultimate loading from test BSAK1:

$N = 715$ kN ultimate axial load of the brace member

$N_{i,Rd} = 773$ kN axial load capacity of the K-joint from Eq. (1) without the reduction factor for the material:

$M_s = 3.2$ kNm secondary bending moment from strain gauges extrapolated to the joint

$M = 18.9$ kNm moment capacity of the K-joint from Eq. (9).

Applying these values in Eq. (10) obtains a total performance level for the BSAK1 joint: $0.93 + 0.17 = 1.1$, and for the BSAK12 joint: $0.54 + 0.39 = 0.94$, respectively. This result means that the BSAK1 joint has a safety factor of 1.1, based on the measured yield strength of the material, but the BSAK12 joint does not reach the theoretical capacity. The reason for this finding is that the welding effects on the HAZ near the weld are more remarkable in the joint made of S960 than in the joint made of S700. In addition, the axial preload of the chord member in the BSAK12 joint may have a capacity-reducing effect with the high eccentricity level e/h_0. The eccentricity of the joint BSAK12 is 96 mm and

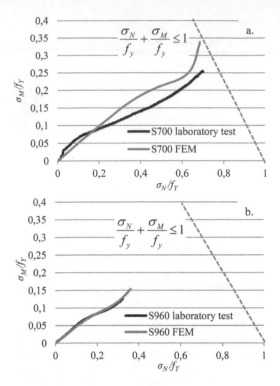

Figure 17. Relative bending stresses versus axial stresses in the brace members of a) BSAK1 and b) BSAK12.

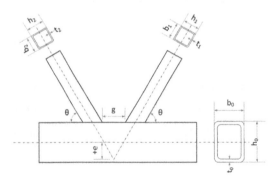

Figure 18. Parameters of the K-joint.

the limit where the effect of the eccentricity can be neglected is $0.25h_0 \leq 37.5$ mm.

Based on results for BSAK1, it is seen that secondary bending stresses can occur without any significant joint eccentricity. In order to find out the effect of joint parameters on the secondary moment, a number of different K-joints were analyzed by FEA. The investigated joint geometries are given in Fig. 18 and the dimensions in Table 4

The results of the analysis are presented in Fig. 19, where the rotation behavior versus axial load is given in Fig. 19a and the relative secondary bending stress versus axial stress in Fig. 19b. The stresses are defined using the same method as in Fig. 17 and using the same S700 material models.

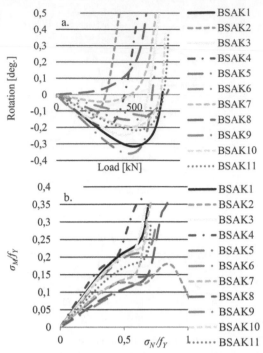

Figure 19. a) Rotation and b) secondary bending stresses of S700 K-joints.

The results show that the stresses due to secondary bending can be remarkable even if the eccentricity of the joint is insignificant. Also the displacements are so large in this case that they have effect on secondary bending moment.

9 CONCLUSIONS

Based on analytical, numerical and experimental investigations done in this study, the following conclusions can be drawn:

- The secondary bending stresses of K-joints due to the uneven support effect (USE) decrease joint capacity and should be considered if high strength steels are used;
- Other significant effects may exist, like eccentricity of the joint softening due to welding, which might have a remarkable effect on the capacity of the joint;
- The analytically, numerically and experimentally defined USE- results match quite well;
- The USE is sensitive to joint geometry and especially gap size, but more investigation with high strength steels are needed, experimentally and theoretically, in the future

ACKNOWLEDGEMENTS

Support for this work has been provided by Finnish Metals and Engineering Competence Cluster

(FIMECC) BSA-program and RFCS-Ruoste project. Tested materials were provided by SSAB Europe. We would like to thank all the participating groups.

REFERENCES

Coutie M. G. & Saidani M. 1993. Secondary moments in RHS lattice girders. Tubular Structures V. Nottingham

EN 1993-1-8. 2005. Eurocode 3: Design of Steel Structures, Part 1–8, Design of Joints

EN 1993-1-12 + AC. 2007. Eurocode, Design of Steel Structures, Part 1–12 Additional Rules for the Extension of EN 1993 up to Steel Grade S700

Fleischer, O. & Puthli R. 2009. Extending existing rules in EN 1993-1-8 (2005) for gapped RHS K-joint for maximum chord slenderness (b/t) of 35 to 50 and gap size g to as low as 4t0. Tubular Structures XII, Shanghai

Marshal P. 1992. Design of Welded Tubular Connections: Basis and Use of AWS Code Provision. Amsterdam: Elsevier

Niemi E. 1982. The effect of gap size on static strength of K-joints in rectangular hollow section trusses- a modified yield line approach, IIW Doc. XV-82-011

Niemi E. 1986. On the deformation capacity of rectangular hollow section K-joints. IIW Doc. SC-XV-E-86-107

Packer J. A., Davis G. & Coutie M G. 1980. Yield strength of gapped joint in rectangular hollow section trusses. Proc. Institution of Civil Engineering, Part 2

Partanen, T. & Björk T. 1993. On convergence of yield line theory and experimental test capacity of RHSK- and T-joints. Tubular Structures V. Nottingham

Sanezuk, L. & Jaeger, T. 1963. Grenztragfestigkeit- Theorie der Platten. Berlin

Sedlacek, G. & Müller, C. 2001. High strength steels in steel construction. Niobium: Science & Technology, International Symposium Niobium. TMS: The Minerals, Metal & Material Society, Orlando

Zhang, Z. & Niemi, E. 1993. Studies of the behavior of RHS gap K-joints by non-linear FEM. Tubular Structures V. Nottingham

Structural behaviour of cross-sections and members

Tubular Structures XV – Batista, Vellasco & Lima (eds)
© 2015 Taylor & Francis Group, London, ISBN 978-1-138-02837-1

Assessment of Eurocode 9 slenderness limits for elements in compression

M. Su
Department of Civil Engineering, The University of Hong Kong, Hong Kong
Department of Civil and Environmental Engineering, Imperial College London, London, UK

B. Young
Department of Civil Engineering, The University of Hong Kong, Hong Kong

L. Gardner
Department of Civil and Environmental Engineering, Imperial College London, London, UK

ABSTRACT: Cross-section classification is one of the key concepts for structural design in metallic materials. Current design specifications for aluminium alloys, such as Eurocode 9 (EC9), provide clear definitions and discrete design treatments for different class categories. The purpose of the present paper is to re-evaluate the yield (Class 3) slenderness limits for internal elements and outstands on the basis of substantial, recently generated experimental data on compressed aluminium alloy cross-sections. In this study, approximately 350 stub column test results have been gathered; the section types include square and rectangular hollow sections, channels and angles. The members were extruded from a variety of aluminium alloy tempers with a wide range of yield and ultimate strengths. Following analysis of the available data, the existing slenderness limits in EC9 were assessed and new slenderness limits were proposed. In conjunction with the proposed Class 3 limits, corresponding effective thickness formulae have also been established for Class 4 sections. The suitability of the proposed limits has been demonstrated through reliability analysis for both the conventional design methods in EC9 as well as the alternative methods given in Annex F of EC9. The revised limits enable more precise predictions of capacity.

1 INTRODUCTION

Section classification addresses the susceptibility of a cross-section to local buckling and defines its appropriate design resistance (Gardner & Theofanous, 2008). Eurocode 9 (EC9) (2007) adopts the cross-section classification concept, which is based on elastic-perfectly plastic material behaviour, excluding the effect of strain hardening. Given the neglect of the non-linear material stress-strain relationship, as well as the interaction between elements, in favour of simplicity, the existing slenderness limits for compressed aluminium alloy cross-sections are generally conservative. In the present study, the classification criteria in EC9 are re-assessed against approximately 350 stub column results, and the conservatism in the existing limits is illustrated in this paper.

As part of cross-section design, effective thickness formulae are used for slender (Class 4) elements. Specifically, a reduction factor for local buckling is applied to the full thickness of the slender element, in order to reduce it to an effective value. The effective thickness concept is adopted for aluminium alloy section design in EC9 (2007), while the effective width method is employed in EC3 (2006) for slender steel sections. Clearly, the effective thickness formulae should be compatible with the Class 3 limits i.e. when

the slenderness of an element is equal to the Class 3 limit, the local buckling reduction factor should be equal to unity, but should reduce thereafter.

In this paper, the available stub column test results were firstly collected and summarised. The section classification framework in EC9 is then explained, and following detailed comparisons, it is shown that revised slenderness limits are required, and are duly proposed. Reliability analysis is performed to show the suitability of the proposed limits.

2 DATA COLLECTION

Stub column test results on different aluminium alloy tempers and a wide range of cross-section slenderness were collected herein. The combined data pool of aluminium alloy stub columns includes a total of 346 results, with both closed and open sections: 110 square and rectangular hollow sections (SHS/RHS) (Mazzolani et al., 1996, 1997, 1998; Langseth & Hopperstad, 1997; Landolfo, 1999; Hassinen, 2000; Mennink, 2002; Zhu & Young, 2006, 2008), 203 plain channel sections (Mazzolani et al., 2001; Mennink, 2002) and 33 angle sections (Mazzolani et al., 2011). The average measured cross-sectional dimensions and material properties can be found in the cited papers and are used in this study. The specimens covered the four classes of cross-sections in accordance with EC9 (2007).

3 EC9 CLASSIFICATION

3.1 General

In terms of aluminium alloy structural design, EC9 (2007) is one of the key international specifications providing quantitative section classification criteria. The class of an element relates to the susceptibility to local buckling. EC9 (2007) defines four classes of cross-sections, while the American (2010) and Australian/New Zealand (1997) specifications classify cross-sections into three categories according to their failure modes: yielding (equivalent to Class 1 and 2 in EC9), inelastic buckling (equivalent to Class 3 in EC9) and elastic buckling (equivalent to Class 4 in EC9). Only Class 3 slenderness limits for elements under uniform compression are considered herein. There are two principal reasons for the variation in the slenderness limits among different specifications: the first reason relates to the pool of available data utilised in the code development; the second reason relates to the different regional practices in terms of structural reliability (Law & Gardner, 2009).

According to EC9, cross-sections are assigned to one of four classes according to their susceptibility to local buckling, as determined by comparing a slenderness parameter to codified class limits. The classification of a cross-section depends on the most slender element; i.e. the classification of an element is individually assessed based on its width-to-thickness b/t ratio, independently of other constituent elements in a cross-section.

The slenderness measure used in EC9 is β/ε, which includes the flat width-thickness ratio b/t and the yield stress f_y, as given by Equation (1):

$$\frac{\beta}{\varepsilon} = \frac{b/t}{\sqrt{250/f_y}} \tag{1}$$

Slenderness limits are given in Table 5.2 of EC9, and repeated in Table 1 of the present paper. These limits depend on the way in which the elements are supported (either one edge supported as outstands or two edges supported as internal elements), the stress distribution (uniform compression or varying stresses), and the forming process (heat-treated or non-heat-treated). Since the influence of stress distribution has been considered when calculating the slenderness parameter, the same class limits are adopted for both uniform compression and flexural elements. Table 1 also presents slenderness limits from EN1993-1-3 (2006) for carbon steel, EN1993-1-4 (2006) for stainless steel and harmonized limits proposed by Gardner & Theofanous (2008) for both materials. However, due to the differences in the Young's Modulus E, the Poisson's ratio v and the coefficient that is used to normalize the yield stress between steel and aluminium

Table 1. Existing and proposed slenderness (β/ε) limits for cross-section classification

References	Material	Class 3 β/ε limits	
		Internal	Outstand
EC9	Aluminium-Class A	22	6
	Aluminium-Class B	18	5
EN1993-1-1	Carbon Steel	24	8
EN1993-1-4	Stainless Steel	18	7
Gardner and Theofanous (2008)	Carbon Steel & Stainless Steel	21	8
Proposed herein	Aluminium-Class A & B	22	6

alloys, the aforementioned limits have been converted to a common basis, using the characteristic values for aluminium alloy adopted in EC9 (2007), as shown in Table 1.

3.2 Cross-sections subjected to compression

For cross-sections under pure compression, the key concern is whether the section is fully effective. Those cross-sections that can attain the yield load are considered to be 'fully effective', and are defined in EC9 (2007) as being either of Class 1, Class 2 or Class 3, while Class 4 sections fail by local buckling prior to yielding. Hence, the design capacities for non-slender (Classes 1, 2 and 3) sections are the yield limit Af_y, and that for slender (Class 4) sections become $A_{eff}f_y$, where A is the gross area, A_{eff} is the effective area calculated based on an effective cross-section containing elements of effective thickness. A local buckling factor ρ_c is employed to reduce the full thickness t to the effective thickness t_{eff} in any parts wholly or partly in compression.

The class limits in EC9 (2007) are presented individually for different aluminium alloys, categorised as either Class A or B materials in Table 3.2 of EC9. In order to assess the accuracy of the current class limits, the normalized experimental ultimate loads N_u/Af_y are plotted against the slenderness β/ε of the most slender element in the cross-section in Figures 1–3. Please note that all the data points in Figures 1–3 are calculated based on the measured geometrical and material properties The data in Figures 1–3 generally followed the anticipated trend whereby sections with greater slenderness have lower normalised capacity. Figure 1 shows the data for internal elements of Class A material, while Figure 2 presents the data for internal elements of Class B material; the corresponding Class 3 limits of $\beta/\varepsilon = 22$ and $\beta/\varepsilon = 18$ are indicated in both figures. Figures 1 and 2 show that the existing limit of 22 for Class A aluminium alloys may be

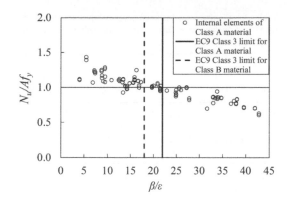

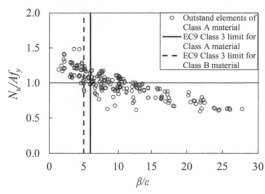

Figure 1. Comparisons between stub column results and EC9 Class 3 boundaries for internal elements of Class A material.

Figure 3. Comparisons between stub column results and EC9 Class 3 boundaries for outstands of Class A material.

4 EFFECTIVE THICKNESS FORMULAE

Compression members with Class 4 cross-sections may buckle locally prior to reaching their yield load. For stub columns of slender proportions, the normalized capacity N_u/Af_y can be deemed to be equal to the local buckling factor ρ_c by disregarding the fully corner area, as illustrated in Equation (2).

$$\frac{N_u}{Af_y} = \frac{A_{eff} f_y}{Af_y} = \frac{A_{eff}}{A} \approx \rho_c \qquad (2)$$

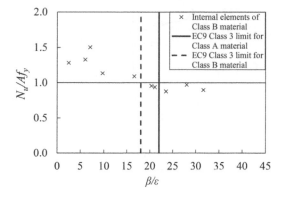

Figure 2. Comparisons between stub column results and EC9 Class 3 boundaries for internal elements of Class B material.

The calculation of the local buckling factor is based on the effective thickness formulae; one of the compatibility requirements for the formulae is that when the slenderness of an element is equal to the Class 3 limit, its local buckling factor ρ_c should be equal to unity. For aluminium alloy internal elements, it was proposed that the Class 3 limit for Class B material is harmonized with the limit for Class A material ($\beta/\varepsilon = 22$). For consistency, it is also proposed that the effective thickness formula for Class A material, specified in Clause 6.1.5 of EC9 (see Equation (3)), be adopted for Class B alloys. The normalised compression resistances N_u/Af_y versus the slenderness parameter β/ε are plotted in Figures 4 and 5, together with the curve from the effective thickness formula. It is demonstrated in Figures 4 and 5 that Class 4 sections of both Class A and B material can be accurately represented by the same design curve, given by Equation (3).

safely applied to Class B aluminium alloys, although the number of test results for Class B material is rather limited at this stage. It is therefore recommended to harmonise the limits for both Class A and B materials at $\beta/\varepsilon = 22$ for internal elements. The compression test data for outstands elements of Class A material are plotted in Figure 3, together with the corresponding Class 3 limit $\beta/\varepsilon = 6$ for Class A alloys and $\beta/\varepsilon = 5$ for Class B alloys. No compression tests have been found on cross-sections with outstands of Class B materials in the literature. It is suggested to retain the Class 3 limit at $\beta/\varepsilon = 6$ for outstands of Class A material, as indicated in Figure 3.

Table 1 summarises existing Class 3 limits for compressed elements, as found both in current design codes and in the literature. It reveals that the Class 3 limits for aluminium alloys are fairly consistent with carbon steel (EN1993-1-3, 2006), stainless steel (EN1993-1-4, 2006) and the proposed harmonised limits of Gardner & Theofanous (2008). The newly proposed Class 3 limits are seen to define accurate boundaries between sections with capacities greater than the yield limit and sections with capacities smaller than the yield limit.

$$\rho_c = \frac{32}{(\beta/\varepsilon)} - \frac{220}{(\beta/\varepsilon)^2} \quad \text{for} \quad \beta/\varepsilon > 22 \qquad (3)$$

For outstand elements, it was proposed herein that the existing Class 3 limit in EC9 ($\beta/\varepsilon = 6$) is retained for Class A material. Hence, the effective thickness formula (Equation (4)) currently given in Clause 6.1.5 of EC9 (2007) is assessed against more than 200 compression test results. The curve (Equation (4)), which is shown Figure 6, may be seen to provide a reasonable and safe fit to the test data. It is suggested to retain

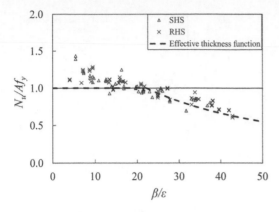

Figure 4. Relationship between N_u/Af_y and β/ε for internal element of Class A material.

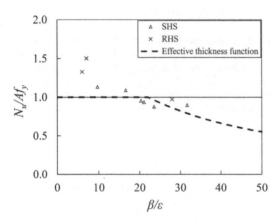

Figure 5. Relationship between N_u/Af_y and β/ε for internal element of Class B material.

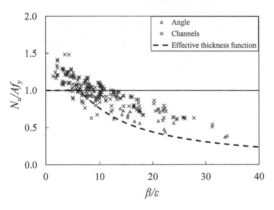

Figure 6. Relationship between N_u/Af_y and β/ε for outstands of Class A material

the current effective thickness formula for outstand elements of Class A alloys.

$$\rho_c = \frac{10}{(\beta/\varepsilon)} - \frac{24}{(\beta/\varepsilon)^2} \quad \text{for} \quad \beta/\varepsilon > 6 \qquad (4)$$

Table 2. Comparison between stub column results (346 data) and Eurocode 9 design strengths with existing and proposed limits

Stub columns		Existing limits	New limits in EC9 framework
N_u/N_{EC9}	Mean Value	1.18	1.12
	COV	0.168	0.113
$N_u/N_{EC9-AnnexF}$	Mean Value	1.16	1.10
	COV	0.169	0.112

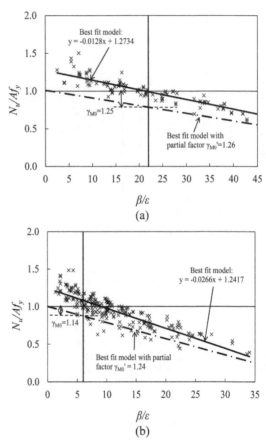

Figure 7. Determination of γ_{M0} for the best fit model for (a) internal elements and (b) outstands in EC9.

5 COMPARISON OF RESULTS

The design approaches in EC9 (2007) were used together with the existing and proposed class limits, as well as the effective thickness formulae, to predict the capacities of 346 stub columns. The design accuracy with the usage of the existing and proposed class limits is shown in Table 2. By observing the results in Table 2, the predicted capacities are found to be more accurate when adopting the newly proposed class limits. When considering the design method in the main text of EC9 (2007), the mean values of N_u/N_{EC9} were found to be 1.12 for the proposed limits, which is better than that with the existing class limits (mean value=1.18). Meanwhile, the coefficients of variation (COV) also reduce when using the proposed limits. The high COV values in Table 2 indicate shortcoming with the existing slenderness limits and design model. In Annex F of EC9 (2007), an alternative design method is provided for compressed sections, which takes into account strain hardening of aluminium alloys. Similar improvements are achieved by adopting the proposed Class 3 limits when considering the full experimental data set, though note that the presented comparisons retain the current Class 1 limit.

6 RELIABILITY ANALYSIS

In order to verify the suitability of the proposed Class 3 limits and the corresponding effective thickness formulae, reliability analyses were performed against the 346 compression tests results. The reliability analyses were performed in accordance with the criteria set out in EN1990 (2002). The parameters adopted in the statistical analyses were taken from Annex D of the AA (2010): the mean values and COVs for material properties and fabrication variables are taken as $M_m = 1.10$ (for behaviour governed by the yield stress) and 1.00 (for behaviour governed by the ultimate stress), $F_m = 1.00$, $V_M = 0.06$ and $V_F = 0.05$. The target partial safety factor γ_{M0} is recommended to be 1.10 in EC9 (2007). The reliability analysis for internal elements was performed against both Class A and B alloys, as shown in Figure 7(a). A least squares regression fit to the available data set is plotted in Figure 7(a), which is then scaled down by the required safety factor $\gamma'_{M0} = 1.26$ obtained from the conventional reliability analysis. The resulting partial factor γ_{M0} for the Class 3 limit for internal elements under uniform compression ($\beta/\varepsilon = 22$) was found to be 1.25, greater than 1.10. However, the proposed Class 3 limit is the same as the current codified value in EC9, and hence considered to be acceptable based on previous experience. Following the same procedures, the evaluated safety factor γ_{M0} of 1.14 was found for outstands, as illustrated in Figure 7(b). The evaluated safety factor obtained from reliability analysis for outstands is slightly greater than target safety factor of 1.10. Meanwhile, the Class 3 limits and the corresponding effective thickness formulae suggested herein are consistent with that in the existing codes. The harmonised class limits and effective thickness formulae may therefore be deemed acceptable and sharing a similar reliability level as existing classification limits.

7 CONCLUSIONS

Eurocode 9 (2007) is widely used for aluminium alloy design and provides cross-section classification criteria for cross-sections subjected either partly or fully to compression. All relevant compression experimental results have been gathered, including square and rectangular hollow sections, angles and channels of both Class A and B materials. The feasibility of harmonising the Class 3 limits for Class A and B aluminium alloys was investigated. Following discussion of the codified treatment of local buckling in EC9, a comprehensive assessment of the existing and proposed cross-section classification limits was performed. A harmonised Class 3 limit of 22 was proposed for internal elements and the existing Class 3 limit for outstands of Class A material was suggested to remain. Corresponding effective thickness formulae were also suggested. Comparisons of results were conducted for the design approaches in EC9 together with the existing and proposed class limits. Furthermore, reliability analyses have been performed to assess the Class 3 limits according to EN1990 (2002) against 346 compressed section results. The results illustrated that the harmonised limits are beneficial to the cross-section capacity design. Given the more accurate and consistent predictions resulted from the proposed limits, more efficient design can be anticipated in the future. Further investigations into classification for cross-sections in bending are on-going.

ACKNOWLEDGEMENT

The research work described in this paper was supported by a grant from The University of Hong Kong under the seed funding program for basic research.

REFERENCES

Aluminum Association (AA). 2010. Aluminum design manual. Washington, D.C.

Australian/New Zealand Standard (AS/NZS). 1997. Aluminum structures part 1: Limit state design. AS/NZS 1664.1:1997, Standards Australia, Sydney, Australia.

European Committee for Standardization (EN 1990), 2002. Eurocode: Basis of Structural Design. EN 1990-2002, CEN.

European Committee for Standardization (EC3). 2006. Eurocode 3: Design of steel structures—Part 1-3: General rules – Supplementary rules for cold-formed members and sheeting. EN 1993-1-3:2006, CEN.

European Committee for Standardization (EC3). 2006. Eurocode 3: Design of steel structures—Part 1-4: General rules – Supplementary rules for stainless steels. EN 1993-1-4:2006, CEN.

European Committee for Standardization (EC9). 2007. Eurocode 9: Design of aluminum structures—Part 1-1: General rules—General rules and rules for buildings. BS EN 1999-1-1:2007, CEN.

Gardner, L. & Theofanous, M. 2008. Discrete and continuous treatment of local buckling in stainless steel elements. Journal of Constructional Steel Research 64(11): 1207–1216.

Hassinen, P., 2000. Compression strength of aluminium columns – Experimental and numerical studies. In D. Camotim, D. Dubina & J. Rondal (eds.) Proceedings of the 3rd International Conference on Coupled Instabilities of

Metal Structures, CIMS'2000, Lisbon, 21–23 September, 2000. pp. 241–248, Imperial College Press, London, UK.

Landolfo, R., Piluso, V., Langseth, M. & Hopperstad, O.S., 1999. EC9 provisions for flat internal elements: comparison with experimental results. In P. Mäkeläinen and P. Hassinen (eds.) Light-Weight Steel and Aluminum Structures, ICSAS' 99, Espoo, 20–23 June, 1999, pp. 515–522. Elsevier Science Ltd, UK.

Langseth, M. & Hopperstad, O. S., 1997. Local buckling of square thin-walled aluminum extrusions, Thin-Walled Structures 27(1):117–126.

Law, K. H. & Gardner, L., 2009. Unified slenderness limits for circular hollow sections. In S.L. Chan (eds.) Proceeding of 6th International Conference on Advances in Steel Structures, Hong Kong, 16–18, December, 2008, pp. 293–300. World Scientific Publishing, Singapore.

Mazzolani, F.M., Faella, C., Piluso, V. & Rizzano, G. 1996. Experimental analysis of aluminum alloy SHS-members subjected to local buckling under uniform compression. In Proceeding of 5th International Colloquium on Structural Stability, pp. 475–488, SSRC, Rio de Janeiro, Brazil, 5–7 August, 1996. The Council, USA.

Mazzolani, F. M., Faella, C., Piluso, V. & Rizzano, G. 1997, Local buckling of aluminum alloy RHS-members: experimental analysis. In Proceeding of XVI Congresso C.T.A., Italian Conference on Steel Construction, Ancona, 2–5 October, 1997, pp. 1–12.

Mazzolani, F.M., Faella, C., Piluso, V. & Rizzano, G. 1998. Local buckling of aluminum members: experimental analysis and cross-sectional classification. Department of Civil Engineering, University of Salerno, Italy

Mazzolani, F. M., Piluso, V. & Rizzano, G., 2001, Experimental analysis of aluminum alloy channels subjected to local buckling under uniform compression. In Proceeding of Congress C.T.A., Italian Conference on Steel Construction ACS, Milano, Italy, 26–28 September, 2001, pp. 1–10.

Mazzolani, F.M., Piluso, V., & Rizzano, G., 2011, Local buckling of aluminum alloy angles under uniform compression. Journal of Structural Engineering, ASCE 137(2): 173–184.

Mennink, J., 2002. Cross-Sectional Stability of Aluminium Extrusions: Prediction of the Actual Local Buckling Behaviour. PhD thesis, The Netherlands.

Schafer, B.W. & Peköz, T., 1998. Direct strength prediction of cold-formed steel members using numerical elastic buckling solutions. In Proceeding of 14th Int. specialty conference on cold-formed steel structures, Missouri, 15–16 October, 1998, pp. 69–76. The University of Missouri-Rolla, USA.

Su, M., Young, B. & Gardner, L., 2013. The continuous strength method for aluminium alloy design, Advanced material research, 2457(742): 70–75.

Su, M., Young, B. & Gardner, L., in press. Testing and design of aluminium alloy cross-sections in compression., Journal of Structural Engineering, ASCE.

Zhu, J.H. & Young, B., 2006. Tests and design of aluminum alloy compression members. Journal of Structural Engineering, ASCE, 132(7): 1096–1107.

Zhu, J.H. & Young, B., 2008. Behavior and design of aluminum alloy structural members. Advanced Steel Construction 4(2): 158–172.

Zhu, J.H. & Young, B., 2009. Design of aluminum alloy flexural members using direct strength method. Journal of Structural Engineering, ASCE, 135(5): 558–566.

Tubular Structures XV – Batista, Vellasco & Lima (eds)
© *2015 Taylor & Francis Group, London, ISBN 978-1-138-02837-1*

Design strength of LDSS flat oval stub column under pure axial compression

K. Sachidananda & K.D. Singh

Civil Engineering Department, Indian Institute of Technology Guwahati, India

ABSTRACT: This paper presents numerical modeling of LDSS (Lean Duplex Stainless Steel) flat oval hollow section stub columns under pure axial compression using the commercial finite element software, Abaqus (2009). A parametric study of the flat oval hollow cross-section has been carried out by varying t (thickness), r (radius of curvature) and l (flat plate length), keeping w (flat plate spacing) constant at 300 mm, to assess their effects on the load bearing capacity of LDSS stub columns. It has been found that the strength gain per unit increase in area by, 1) increasing t, increases with increasing r and decreases with increasing l; 2) increasing l, increases with increasing t; while the decrease in strength per unit decrease in area by increasing r decreases with increasing t. It has been found that the current code of EN 1993-1-4 (2006) can be used for the prediction of load carrying capacity for compact/thick ($w/t \leq 40$) LDSS flat oval sections whilst it is shown to be unconservative for slender/thin ($w/t > 40$) sections. DSM (Direct Strength Method) for cold formed steel, on the other hand, is found to give good predictions for LDSS flat oval stub column.

1   INTRODUCTION

Over the past recent decades, cold formed stainless steel have become very popular in the construction industry for its numerous advantages e.g. high corrosive resistance, ductility, aesthetic appearance, high strength, smooth and uniform surfaces etc., and for a long time, among stainless steels, austenitic stainless steel with nickel content of ~8%–11% was the most sought after. But, the introduction of a new type of stainless steel *viz.*, Lean Duplex Stainless Steel (LDSS) particularly EN 1.4162 with lower nickel content of ~1.5% and improved strength to weight ratio compared to conventional austenitic, ferritic stainless steel, has changed the way of treating stain less steel as an expensive option for the construction industry. Indeed, a relatively cheaper and stronger LDSS material can be considered as a very promising alternative (e.g. Gardner, 2005, Patton and Singh, 2012, EN 10088-4, 2009). In the steel construction industry, in addition to the traditional open sections e.g. I, channel, plate girder etc., various close sections like square, rectangular, oval, circular, elliptical etc. have become more visible, primarily due to structural efficiency owing to higher moment of inertia. Amongst all sections mentioned above, flat oval section is one of the newest sections with aesthetic look. The research on oval/ellipitical hollow section for structural engineering application (i.e. civil engineering structures) is limited, for example, Gardner and Ministro (2004), Theofanous *et al.*, (2009), Zhu and Young (2011, 2012), Silvestre and Gardner (2011). However, extensive analytical studies on the elastic buckling of oval/elliptical hollow section work under axial compression was initiated in the 1950s (Marguerre, 1951,

Kemper and Chen, 1969, Hutchinson, 1968, Feinstein *et al.*, 1971) due to the requirements of aerospace industry for lighter and stronger structures (Chan, 2007). Apparently, Zhu and Young (2011, 2012) are the first to report on cold formed 'flat oval' steel section columns under axial compression through both experimental and non-linear finite element analyses. Design strengths based on experimental and numerical studies were compared with North American (NAS), Australian/New Zealand (AS/NZ) and European specification (EC3), and also with Direct Strength Method (DSM). Based on the reliability analyses it was concluded that, the reliability indices based on the above-mentioned specifications and DSM were found to be greater than the target values. DSM, which was developed for certain cold formed steel sections *viz.*, Z-, C-, Hat section etc. (but specifically not developed or not prequalified for flat-oval sections) was found to give conservative values, and showed that it can be used for the design of oval sections. In the present study, an attempt has been made to systematically study the parametric effect of cross-section on LDSS flat oval under the action of pure axial compression, using the commercial Finite Element (FE) software, Abaqus (2009). The FE parametric results are then compared with European code (EN 1993-1-4, 2006) and DSM to check its applicability for flat oval LDSS stub column.

2   FINITE ELEMENT MODELLING

In the present FE study, fixed ended FOHS (flat oval hollow section) column under pure axial compression is considered. Typical cross-section, FE

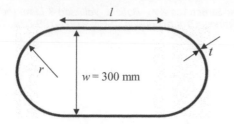

Figure 1. Flat oval hollow section.

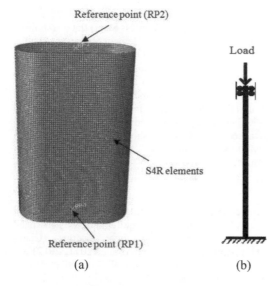

(a)　　　　　(b)

Figure 2. Typical (a) FE mesh, (b) boundary conditions of LDSS flat oval hollow column.

Table 1. Stub column dimensions (Theofanous and Gardner, 2009).

Specimen	L (mm)	B (mm)	H (mm)	t (mm)	r_i (mm)
80 × 80 × 4-SC2	332.2	80	80	3.81	3.6

L = Length, B = Width, H = Depth, t = thickness, r_i = internal corner radius

Table 2. Compressive flat material properties (Theofanous and Gardner, 2009).

Cross-section	E (MPa)	$\sigma_{0.2}$ (MPa)	$\sigma_{1.0}$ (MPa)	Compound R-O coefficients n	$n_{0.2,1.0}$
80 × 80 × 4-SC2	197200	657	770	4.7	2.6

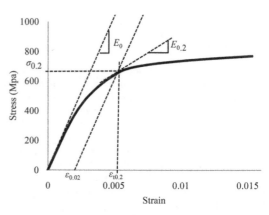

Figure 3. Experimental stress-strain curve of LDSS material Grade EN 1.4162 (Theofanous & Gardner 2009).

mesh, boundary conditions are presented schematically in Figures 1 and 2 respectively. The FE modelling approach followed is in line with those reported by Gardner and Ministro (2004), Ashraf et al. (2006), Theofanous and Gardner (2009), Patton and Singh (2012). The bottom part is fixed while allowing the top loaded part of the column as in Figure. 2. Reference points (RP1 and RP2) are provided to define the boundary conditions of the column. The column ends are constraints through kinematic coupling available in Abaqus (2009). At the loaded end all degree of freedom were restrained except for vertical translation. Four-noded doubly curved shell (S4R) elements have been utilized to discretise the models. Typical mesh size is of the dimension 11.5 mm × 11.5 mm with an aspect ratio of 1 was arrived after mesh convergence study. The lowest eigen value was used as initial geometric imperfection to perturb the geometry of the column. The scaling of the imperfection amplitude is taken as 1% of plate thickness as recommended in the literature (e.g. Chan and Gardner, 2008, Theofanous and Gardner, 2009). The FE modeling approach has been validated with an experimental result reported by Theofanous and

Gardner (2009) on LDSS square hollow column subjected to similar conditions. The geometric details of the validated square LDSS column-80 × 80 × 4-SC2 (Theofanous and Gardner 2009) is given in Table 1. Table 2 shows the material properties proposed by Gardner and Ashraf (2006) (modified version of original Ramberg–Osgood, 1943) which is used in deriving the stress-strain curve of LDSS material for the present models. Poisson's ratio is taken as 0.3. The LDSS stress (σ) – strain (ε) curve (Figure 3), consists of two material models viz., Ramberg-osgood (1943) upto 0.2% proof stress ($\sigma_{0.2}$) and Gardner and Ashraf (2006) from $\sigma_{0.2}$ to $\sigma_{1.0}$. The strength enhancement in the corner regions by cold forming-process is neglected. The residual stresses is also neglected in the model as it has negligible effect on the ultimate load (P_u) and load shortening behavior (Ellobody and Young, 2005). The material model shown in Figure 3 is used as input parameters to Abaqus (2009), by

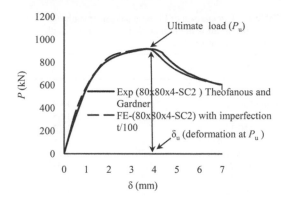

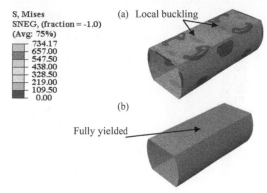

Figure 4. Variation of load with axial displacement for stub column (SHC 80 × 80 × -SC2).

Figure 6. Typical Von-Mises stress (superimposed on deformed shape) of (a) *l*300*r*300*t*3, (b) *l*300*r*300*t*20 at ultimate load.

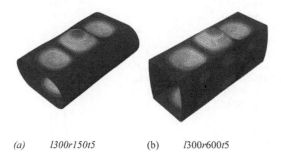

(a) *l*300*r*150*t*5   (b)   *l*300*r*600*t*5

Figure 5. First eigen buckling modes for flat oval.

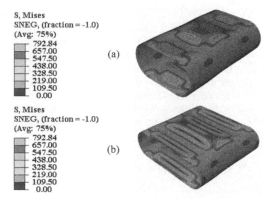

Figure 7. Typical Von-Mises stress (superimposed on deformed shape) of (a) *l*300*r*150*t*5, (b) *l*700*r*150*t*5 at ultimate load.

converting it into true stress – (σ_{true}) and true plastic strains ε^{pl}_{true} using the following equations (1) and (2).

$$\sigma_{true} = \sigma_{nom}(1 + \varepsilon_{nom}) \tag{1}$$

$$\varepsilon^{pl}_{true} = \ln(1 + \varepsilon_{nom}) - \frac{\sigma_{true}}{E_o} \tag{2}$$

where σ_{nom} and ε_{nom} are engineering stress and strain respectively.

The variation of load-axial displacement of the experimental of Theofanous and Gardner (2009) and FE results of stub column is shown in Figure 4. A good agreement can be seen between the present FE and experimental results, thus validating the FE current modeling approach. Similar FE modeling approach is then followed in the present analyses of flat oval hollow columns.

3 PARAMETRIC STUDY OF LDSS FLAT OVAL HOLLOW STUB COLUMN

A total of 77 FE models of flat oval column have been analysed by varying the flat length ($l = 300$ to 700 mm), curvature radius ($r = 150$ to 750 mm) and thickness ($t = 3$ to 20 mm). The specimens are labeled with complete details of its cross-section like *l*300*r*150*t*10 where *l*300 refer to flat length (*l*) of 300 mm, *r*150 refer to radius (*r*) of curvature as

150 mm and *t*10 refer to thickness (*t*) of plate as 10 mm, keeping constant width (*w*) between the flat plates as 300 mm. The effect of the change in cross-sectional area due to increase in flat length, curvature radius and thickness has been studied. The FE modeling followed the same pattern as discussed above. It is assumed that same material properties has been used both at the flat and curve section of flat oval section. For providing imperfection in flat oval section, the first eigen buckling modes (typical shape in Figure 5) has been used with the amplitude of 1% plate thickness.

3.1 Deformed shapes at P_u (ultimate load)

Typical von-misses stress (superimposed on deformed shape) at ultimate load of *l*300*r*300, for two thicknesses of $t = 3$ mm and 20 mm are shown in Figure 6, where it is shown that the thicker ($t = 20$ mm) model depicted a fully yielded section indicative of the whole section becoming effective while the slender section ($t = 3$ mm) showed local buckling leading to reduced load carrying capacity. The effect of the increase in flat length is shown in Figure 7, in which flat plate length has been increased from 300 to 700 mm. It can

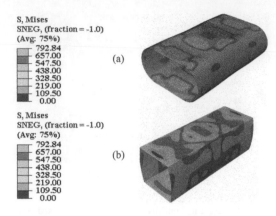

S, Mises
SNEG, (fraction = -1.0)
(Avg: 75%)

| 792.84 |
| 657.00 |
| 547.50 |
| 438.00 |
| 328.50 |
| 219.00 |
| 109.50 |
| 0.00 |

(a)

S, Mises
SNEG, (fraction = -1.0)
(Avg: 75%)

| 792.84 |
| 657.00 |
| 547.50 |
| 438.00 |
| 328.50 |
| 219.00 |
| 109.50 |
| 0.00 |

(b)

Figure 8. Typical Von-Mises stress (superimposed on deformed shape) of (a) $l300r150t5$, (b) $l300r600t5$ at ultimate load.

be seen from Figure 7 that von-mises stress distribution at the curved section is similar, but an extended local buckling commensurating with longer flat length is seen in $l700r150t5$ model (see Figure 7b). Figure 8 shows the effect on von-mises stress distribution for two values of r i.e. 150 mm and 600 mm. It is seen from Figure 8 that smaller value of r promotes better load carrying capacity (more surface area has yielded at ultimate load).

4 DESIGN CONSIDERATION

4.1 Comparison with code (EN 1993-1-4)

It may be noted that the current European code, EN 1993-1-4 (2006) is not very particular on the design provision of hollow sections consisting of both flat and curve portions e.g. flat oval sections, however Australian/New Zealand code of cold formed steel structures suggests that for $w/t < 40$, the curved sections should be considered as fully effective. Hence, the proposal by Zhu and Young (2011, 2012) to consider curved portion of the section to be fully effective (i.e. gross area of the curved section is considered for load calculation) whilst only effective area has been considered for the flat portions as per the provisions of the code. Comparison of FE and EN 1993-1-4 (2006) are shown in Tables 3 and 4 for thick/compact sections i.e. $w/t \leq 40$, for $r = 150$–750 mm, $t = 7.5$–20 mm and $l = 300$–700 mm; and for thin/slender sections i.e. $w/t > 40$, for $r = 150$–750 mm, $t = 3$–5 mm, and $l = 300$–700 mm, respectively.

Typical comparison P_u versus t, between FE and EN 1993-1-4 for thin section ($t = 3$ to 5 mm) and thick section ($t = 7.5$ to 20 mm) have been shown in Figures 9(a) and 9(b), for $l300r150$ specimen. It can be seen from Figures 9(a) and 9(b) that, predictions from EN 1993-1-4 are conservative, in general, for thick sections, whilst for thin sections they are unconservative. The reliability analysis of the flat oval sections has been calculated based on ASCE 8-02 (2002). The resistance factor has been taken as 0.91 as recommended by EN

Table 3. Comparison of FE with design strengths (EN and DSM) for variation in r (curvature radius), l (flat length) and t (thickness), ($w/t \leq 40$).

Specimen	P_{FEA}/P_{EN}	P_{FEA}/P_{DSM}
$l300r150t7.5$	1.03	1.15
$l300r150t10$	0.99	1.00
$l300r150t12.5$	1.11	1.05
$l300r150t15$	1.16	1.12
$l300r150t17.5$	1.15	1.15
$l300r150t20$	1.20	1.19
$l300r300t7.5$	1.14	1.07
$l300r300t10$	1.23	1.06
$l300r300t12.5$	1.28	1.19
$l300r300t15$	1.27	1.25
$l300r300t17.5$	1.24	1.27
$l300r300t20$	1.25	1.28
$l300r450t7.5$	1.10	1.03
$l300r450t10$	1.23	1.06
$l300r450t12.5$	1.26	1.17
$l300r450t15$	1.26	1.24
$l300r450t17.5$	1.25	1.28
$l300r450t20$	1.26	1.29
$l300r600t7.5$	1.06	0.99
$l300r600t10$	1.20	1.03
$l300r600t12.5$	1.25	1.16
$l300r600t15$	1.25	1.24
$l300r600t17.5$	1.24	1.27
$l300r600t20$	1.26	1.29
$l300r750t7.5$	1.01	0.96
$l300r750t10$	1.21	1.07
$l300r750t12.5$	1.26	1.16
$l300r750t15$	1.24	1.23
$l300r750t17.5$	1.25	1.28
$l300r750t20$	1.25	1.29
$l400r150t7.5$	0.93	1.30
$l400r150t10$	0.98	1.02
$l400r150t12.5$	1.06	1.02
$l400r150t15$	1.14	1.05
$l400r150t17.5$	1.16	1.08
$l400r150t20$	1.16	1.12
$l500r150t7.5$	0.93	1.08
$l500r150t10$	0.99	1.04
$l500r150t12.5$	1.05	1.01
$l500r150t15$	1.13	1.03
$l500r150t17.5$	1.20	1.06
$l500r150t20$	1.20	1.07
$l600r150t7.5$	1.08	1.24
$l600r150t10$	1.14	1.17
$l600r150t12.5$	1.17	1.13
$l600r150t15$	1.16	1.07
$l600r150t17.5$	1.19	1.05
$l600r150t20$	1.19	1.03
$l700r150t7.5$	1.12	1.30
$l700r150t10$	1.18	1.22
$l700r150t12.5$	1.21	1.16
$l700r150t15$	1.19	1.09
$l700r150t17.5$	1.19	1.06
$l700r150t20$	1.19	1.02
Mean	1.16	1.13
COV	0.08	0.09
Reliability index, β	2.88	3.40

Table 4. Comparison of FEA with design strengths (EN and DSM) for variation in curvature radius, flat length and thickness ($w/t > 40$).

Specimen	P_{FEA}/P_{EN}	P_{FEA}/P_{DSM}
l300r150t3	0.71	1.24
l300r150t4	0.78	1.16
l300r150t5	0.92	1.23
l300r300t3	0.85	1.19
l300r300t4	0.97	1.23
l300r300t5	1.05	1.15
l300r450t3	0.70	0.97
l300r450t4	0.86	1.04
l300r450t5	1.01	1.10
l300r750t3	0.52	0.74
l300r750t4	0.64	0.78
l300r750t5	0.74	0.81
l400r150t3	0.71	1.43
l400r150t4	0.78	1.23
l400r150t5	0.85	1.18
l500r150t3	0.72	1.36
l500r150t4	0.80	1.29
l500r150t5	0.86	1.21
l600r150t3	0.73	1.36
l600r150t4	0.81	1.27
l600r150t5	0.97	1.35
l700r150t4	0.91	1.46
l700r150t5	1.00	1.42
Mean	0.82	1.18
COV	0.16	0.17
Reliability index, β	1.28	3.03

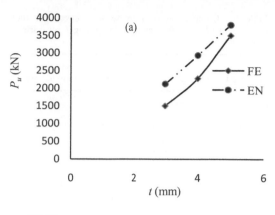

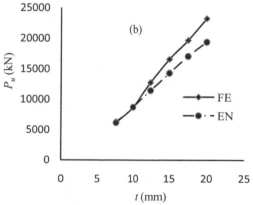

Figure 9. Comparison of FE (*l300r150*) and EN 1993-1-4 for (a) *l300r150* for $t = 3$ to 5 mm (thin sections) and (b) *l300r150* for $t = 7.5$ to 20 mm (thick sections).

Table 5. Comparison of increase in strength per unit increase in cross-section area by increasing t for various values of r.

Increase in thickness	Increase in strength per unit increase in area
l300r150t3 to *l300r150t20*	87.07%
l300r300t3 to *l300r300t20*	93.39%
l300r450t3 to *l300r450t20*	95.47%
l300r750t3 to *l300r750t20*	97.16%

1993-1-4 (2006). A target reliability index (β) value of 2.5 has been defined as the lower limit. The reliability analyses showed the reliability index (β) values of 2.88 (COV = 0.08) and 1.28 (COV = 0.16) for thick and thin sections respectively (see Tables 3 and 4), thus indicating that EN 1993-1-4 is reliable ($\beta \geq 2.5$) for thick sections but unreliable ($\beta < 2.5$) for thin sections.

4.2 Comparison with DSM

The results from Direct Strength Method (DSM) (North American Specification, Schafer, Zhu and Young etc.) for cold formed steel structures have been compared with the FE results of flat oval section. The following DSM equations (3)–(5) have been considered to compute the unfactored design strength (P_{DSM}):

$$P_{DSM} = \min(P_{ne}, P_{nl}) \qquad (3)$$

$$P_{ne} = \begin{cases} (0.658^{\lambda_c^2})P_y & \text{for } \lambda_c \leq 1.5 \\ \left(\dfrac{0.877}{\lambda_c^2}\right)P_y & \text{for } \lambda_c \leq 1.5 \end{cases} \qquad (4)$$

$$P_{nl} = \begin{cases} P_{ne} & \text{for } \lambda_l \leq 0.776 \\ \left[1 - 0.15\left(\dfrac{P_{crl}}{P_{ne}}\right)^{0.4}\right]\left(\dfrac{P_{crl}}{P_{ne}}\right)^{0.4}P_{ne} & \text{for } \lambda_l > 0.776 \end{cases} \qquad (5)$$

where,

$$P_y = f_y A, \quad \lambda_c = \sqrt{P_y/P_{cre}}; \quad \lambda_l = \sqrt{P_{ne}/P_{crl}}$$

A = Gross cross-section area.
f_y = Material yield strength taking 0.2% proof stress ($s_{0.2}$).
$P_{cre} = \pi^2 EA/(l_e/r_y)^2$, critical elastic buckling load in flexural buckling.
P_{crl} = Critical elastic local buckling load.
E = Young's modulus.
l_e = Column effective length.
r_y = Radius of gyration of gross cross-section about the minor y-axis of buckling.

Table 6. Comparison of increase in strength per unit increase in cross-section area by increasing t for various values of l.

Increase in thickness	Increase in strength per unit increase in area
$l300r150t3$ to $l300r150t20$	87.07%
$l400r150t3$ to $l400r150t20$	82.21%
$l500r150t3$ to $l500r150t20$	78.27%
$l600r150t3$ to $l600r150t20$	71.56%

Table 7. Comparison of increase in strength per unit increase in cross-section area by increasing l for various values of t.

Increase in flat length	Increase in strength per unit increase in area
$l300r150t4$ to $l700r150t4$	13.17%
$l300r150t12.5$ to $l700r150t12.5$	18.93%
$l300r150t17.5$ to $l700r150t17.5$	22.12 %
$l300r150t20$ to $l700r150t20$	24.55%

Table 8. Comparison of decrease in strength per unit decrease in cross-section area by increasing r for various values of t.

Increase in curvature radius	Decrease in strength per unit decrease in area
$l300r150t4$ to $l300r750t4$	69.98%
$l300r150t7.5$ to $l300r750t7.5$	66.32%
$l300r150t15$ to $l300r750t15$	51.30%
$l300r150t17.5$ to $l300r750t17.5$	47.60%

P_{nl} = nominal axial strength for local buckling as well as interaction of local and overall buckling.
P_{ne} = nominal axial strength for flexural buckling.
P_{DSM} = nominal axial strength for local buckling.

The P_{crl} values were obtained from the FE analyses. The FE results are then compared with the DSM predictions for both thick and thin sections in Tables 3 and 4 respectively. Also, the variation of P_u/P_{ne} (or) P_{DSM}/P_{ne} with λ_l are shown in Figure 10, for both FE and DSM. Reliability analyses results of the DSM, following similar procedure as mentioned before is presented in Tables 3 and 4 for thick and thin sections, considering resistance factor as 0.8 (Zhu and Young, 2012). It can be observed from Tables 3 and 4 that reliability index (β) values are 3.40 (COV = 0.09) and 3.03 (COV = 0.17) for thick and thin sections respectively, indicating that DSM is more reliable than EN 1993-1-4(2006), while predicting the present FE results.

The increase in column strength per unit increase in cross sectional area, as a result of increase in t (3–20 mm), for r = 150–750 mm and l = 300–600 mm, increase in l (=300–600 mm) for t = 4–20 mm, are presented in Tables 5, 6 and 7. While the decrease

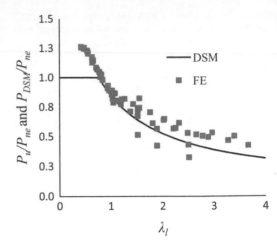

Figure 10. Comparison of FE and DSM results.

in strength per unit decrease in area, for increasing r (=150–750 mm) and t (=4–17.5 mm) is presented in Table 8.

From Table 5 it can be observed that the increase in load capacity per unit material area increment as a result of ~566% increase in t (from t = 3 mm) are ~87%, 93%, 95% and 97% for r = 150 mm, 300 mm, 450 mm, and 750 mm respectively, for $w = l = 300$ mm. Thus it can be seen the strength gain per unit increase in area (by way of increasing t) increases with increasing r or as the curvature becomes flatter, i.e. thickness increase is more beneficial for flatter curvatures. This may be because, although at higher thickness, the whole cross section becomes effective in resisting the load (leading to same strength per unit area), at lower thickness, flatter (or higher r value curve) sections produce lower load capacity, due to cross-sectional ineffectiveness.

Table 6 shows that the increase in load capacity per unit material area by increasing thickness of ~566% from t = 3 mm, along with change in flat length are of the order ~ 87%, 82%, 78% and 71% for l = 300 mm, 400 mm, 500 mm, and 600 mm respectively, for w = 300 mm and r = 150 mm. This indicates that the strength gain per unit increase in area (by way of increasing t) decreases with increasing l or as the flat length increases i.e. thickness increase is not beneficial for longer flat lengths. This may be related to thinner sections becoming ineffective at higher values of flat length (l), while at higher thickness, same strength per unit area would be reached, as the whole cross sections become effective.

For increasing in flat length, l from 300 mm to 700 mm i.e. 133% increase, the increase in load capacity per unit material area increment are seen to be ~13% 19%, 22% and 24% for t = 4 mm, 12.5 mm, 17.5 mm, and 20 mm respectively (w = 300 mm and r = 150 mm) (Table 7). This suggests that the strength gain per unit increase in area (by way of increasing l) increases with increasing t or as the thickness increases i.e. increase in flat length is beneficial for higher thickness. At higher values of t, as the section approaches a fully effective condition, more participation of flat length section in increasing load capacity occurs.

Table 8 shows the decrease in load capacity per unit material area decrement as a result of increase in r from 150 mm to 750 mm i.e. 400 % increase are found to be $\sim$70%, 66%, 51% and 48% respectively, for $t = 4$ mm, 7.5 mm, 15 mm and 17.5 mm respectively ($w = l = 300$ mm). Thus, it can be inferred that the decrease in strength per unit decrease in area (by way of increasing r) decreases with increasing t or as the thickness increases i.e. increase in curvature radius is not beneficial for higher thickness. This may be because, at lower values of t, increase in r (i.e. making it flatter curve) have relatively more pronounce effect in decreasing the load capacity due to enhance ineffectiveness in cross-section.

The comparison of FE with EN 1993-1-4(2006) for $w/t \leq 40$ shows the reliability index greater than 2.5 showing the applicability of the code while for $w/t > 40$, $\beta < 2.5$ showing unreliability nature of EN 1993-1-4. The DSM show its applicability by giving $\beta > 2.5$ for all thickness. Figure 10 between FE and DSM show its applicability for all sections of flat oval column.

5 CONCLUSION

Parametric study of fixed ended LDSS flat oval hollow stub column section for variation in flat length, curvature radius and thickness have been studied using Abaqus (2009). Based on the parameters considered in the FE analyses, following conclusions are drawn:

1) Strength gain per unit increase in area by 1) increasing t, increases with increasing r and decreases with increasing l; 2) increasing l, increases with increasing t; while the decrease in strength per unit decrease in area by increasing r decreases with increasing t.
2) Increase in load capacity per unit material area increment as a result of increase of $\sim$ 566% increase in t (from $t = 3$ mm) is $\sim \geq 87\%$ for all radii of curvatures considered, while it is $\sim \geq 71\%$ for increase in flat length, which shows that the increase in thickness has more positive effect on column strength.
3) EN 1993-1-4 (2006) is found to be reliable for thick flat oval section ($w/t < 40$) but unconservative for thin slender section, $w/t > 40$.
4) DSM is found to be applicable for design of flat oval stub column for both thick and thin sections.

REFERENCES

Abaqus (2006). Hibbitt, Karlsson & Sorensen, Inc. Abaqus/ Standard user's manual volumes I–III and ABAQUS CAE manual. Version 6.9-EF1, Pawtucket, USA.

Ashraf, M., Gardner, L. & Nethercot, D. A. (2006). Finite element modelling of structural stainless steel cross-sections. *Thin-Walled Structures*, 44, 1048–1062.

ASCE 8-02. (2002). Specification for the design of cold-formed stainless steel structural members. *SEI/ASCE-8-02*. New York: ASCE 8-02, *Society of Civil Engineering*.

Chan, T.M. (2007). Structural behavior of elliptical hollow sections. PhD thesis, *Imperial College London*.

Chan, T. M. & Gardner, L. (2008). Compressive resistance of hot-rolled elliptical hollow sections. *Engineering Structures*, 30, 522–532.

EN 1993-1-4. Eurocode 3 (2006): Design of steel structures-Part 1.4: General rules-Supplementary rules for stainless steel. CEN.

EN 10088-4. Stainless steels_part 4 (2009): Technical delivery conditions for sheet/plate and strip of corrosion resisting steels for general purposes. CEN.

Ellobody, E. & Young, B. (2005). Structural performance of cold-formed high strength stainless steel columns. *Journal of Constructional Steel Research*, 61, 1631–1649.

Feinstein, G., Erickson, B. & Kempner, J. (1971). Stability of oval cylindrical shells. *Experimental Mechanics*, 11, 514–520.

Gardner, L. (2005). The use of stainless steel in structures. *Progress in Structural Engineering and Materials*, 7, 45–55.

Gardner, L. & Ashraf, M. (2006). Structural design for non-linear metallic materials. *Engineering Structures*, 28, 926–934.

Gardner L., Ministro A. (2004). Testing and numerical modelling of structural steel oval hollow sections, 04-002-ST, London, *Department of civil and environmental engineering*, Imperial college.

Hutchinson, J. W. (1968). Buckling and Initial Post buckling Behavior of Oval Cylindrical Shells Under Axial Compression. *Journal of Applied Mechanics*, 35, 66–72.

Kempner, J. & Chen, Y. N. (1969). Post buckling of an axially compressed oval cylindrical shell. In: Hetenyi, M. & Vincenti, W. (eds.) *Applied Mechanics*. Springer Berlin Heidelberg.

Margueerre, K. (1951). Stability of the cylindrical shell of variable curvature. NASA *Technical Memorandim*, 1302

Patton, M. L. & Singh, K. D. (2012). Numerical modeling of lean duplex stainless steel hollow columns of square, L-, T-, and +-shaped cross sections under pure axial compression. *Thin-Walled Structures*, 53, 1–8.

Ramberg, W. & Osgood, W.R. (1943). Description of stress-strain curves by three parameters. *Technical note* No 902, Washington, DC: National advisory committee for aeronautics.

Silvestre, N. & Gardner, L. (2011). Elastic local post-buckling of elliptical tubes. *Journal of Constructional Steel Research*, 67, 281–292.

Theofanous, M., Chan, T. M. & Gardner, L. (2009a). Flexural behaviour of stainless steel oval hollow sections. *Thin-Walled Structures*, 47, 776–787.

Theofanous, M., Chan, T. M. & Gardner, L. (2009b). Structural response of stainless steel oval hollow section compression members. *Engineering Structures*, 31, 922–934.

Theofanous, M. & Gardner, L. (2009). Testing and numerical modelling of lean duplex stainless steel hollow section columns. *Engineering Structures*, 31, 3047–3058.

Zhu, J. & Young, B. (2011). Cold-Formed-Steel Oval Hollow Sections under Axial Compression. *Journal of Structural Engineering*, 137, 719–727.

Zhu, J. & Young, B. (2012). Design of cold-formed steel oval hollow section columns. *Journal of Constructional Steel Research*, 71, 26–37.

Tubular Structures XV – Batista, Vellasco & Lima (eds)
© 2015 Taylor & Francis Group, London, ISBN 978-1-138-02837-1

A new design method for hollow steel sections: the Overall Interaction Concept

J. Nseir, E. Saloumi, M. Hayeck & N. Boissonnade
University of Applied Sciences, Fribourg, Switzerland

ABSTRACT: The Overall Interaction Concept (O.I.C.) stands as a new design approach meant to improve the way the carrying capacity of steel profiles is predicted. First, the paper presents the results from extensive experimental tests on 55 steel hollow sections (cold-formed and hot-finished square and rectangular sections, as well as hot-rolled circular sections) aimed at characterizing the influence of section slenderness on the carrying capacity. The test campaign was also aimed at providing an experimental reference to assess numerical FE models which were shown to be in very close agreement with the experimental tests. Consequently, extensive parametric studies were performed on hot-rolled hollow sections, targeting the characterization of the onset of local buckling with respect to many parameters, such as cross-section shape, different types of simple and combined loading situations, several steel grades and various cross-section dimensions and thicknesses so as to cover "plastic" to "slender" responses of the sections.

1 INTRODUCTION

The present paper is related to the stability, resistance and design of steel hollow section members. More precisely, the behaviour of hollow sections is investigated through a large experimental campaign aiming at improving the way the performance and the carrying capacity of tubular members are actually characterized, through the development of an original "Overall Interaction Concept" (O.I.C.) (Figures 1 and 2). Based on the resistance and stability interaction, the O.I.C. further incorporates the effects of imperfections (non-homogenous material, residual stresses, out-of-straightness…) through the derivation of adequate "interaction curves" used to accurately predict the real behaviour of structural elements. The proposed concept is powerful enough to i) increase accuracy and simplicity through a sound and effective basis, ii) deal with the effects of non-linear material behaviour and local/global instabilities, and iii) advance consistency with the possibility of straightforwardly deal with any load case, including combined ones.(S.T.S.S. 2012, b, Boissonnade et al. 2013, c, Boissonnade et al. 2011).

It is suggested within the O.I.C. to enlarge the field of application of the well-known slenderness-related approach through the use of a generalized relative slenderness λ_{rel}, in which R_{RESIST} represents the factor by which the actual loading has to be multiplied to reach the *resistance* limit (i.e. no instability), while R_{STAB} is the factor used to reach the elastic buckling load. (*Instability* limit, i.e. allowable stress is infinite).

The research investigations reported in this paper are relative to a comprehensive test series that aim at providing an experimental reference and assessment

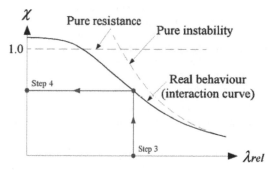

Figure 1. Resistance-instability interaction.

$Step\ 1: R_{RESIST}$ *Load ratio to reach the 'resistance' limit*

$Step\ 2: R_{STAB}$ *Load ratio to reach the 'instability' limit*

$Step\ 3: \lambda_{rel} = \sqrt{\dfrac{R_{RESIST}}{R_{STAB}}}$

$Step\ 4: \chi = f^{\circ}(\lambda_{rel})$

$Step\ 5: R_{REAL} = \chi \cdot R_{RESIST}$

Figure 2. Application steps of the O.I.C.

to the proposed approach. This experimental campaign comprised some 55 cross-sectional tests, as well as preliminary measurements of material properties, residual stresses and geometrical imperfections (Section 2);

Section 3 then presents a comparison between the test results and the predictions of purposely-derived FE models. A parametric study performed on hot-rolled hollow sections is presented in section 4 so as to investigate the adequacy of the Overall Interaction Concept.

2 EXPERIMENTAL INVESTIGATIONS

An experimental program was carried out on a wide variety of tubular cross-sectional shapes (RHS, SHS, CHS) with different fabrication processes and various dimensions (thus local plate slenderness) in order to investigate the influence of local buckling on the plastic, elastic-plastic or slender cross-section response of hollow sections. The main aim of this test campaign was to provide an experimental reference to assess numerical FE models. The testing program comprised 55 tests involving twelve hot-rolled, hot-finished or cold formed square rectangular and circular sections.

Accurate measurements of geometrical imperfections were performed, and tensile tests were carried out to determine the material stress-strain behaviour. Stub column tests were also performed for all different cross-section types. As for the main cross-sectional tests, six different load cases (LC) were distinguished; mono-axial or bi-axial bending with axial compression load cases were considered through the application of eccentrically-applied compression forces. Different M/N ratios have been adopted, in order to vary the distribution of stresses on the flanges and webs, and the following load cases have finally been considered:

- LC1: pure compression N;
- LC2: major-axis bending M_y (50%) + axial compression N (50%)[1];
- LC3: bi-axial bending M_y (33%) + M_z (33%) + axial compression N (33%);
- LC4: minor-axis bending M_z (50%) + axial compression N (50%);
- LC5: bi-axial bending M_y (25%) + M_z (25%) + axial compression N (50%);
- LC6: bi-axial bending M_y (10%) + M_z (10%) + axial compression N (80%).

2.1 Material property tests

The stress-strain behaviour of the tested specimens was captured through 55 tensile tests. For each of the eight square hollow sections (SHS) and rectangular hollow sections (RHS) parent elements, four necked coupons were cut from each flat face. In addition, two straight corner coupons were manufactured and tested for each of these sections in order to investigate the increase in strength in the cold-formed corners and to confirm uniform properties in the hot finished corners. Stresses

---

[1] The percentage between brackets indicats the relative level of axial force N_{Ed}/N_{pl} in case of a compression and $M_{Ed,y}/M_{y,el}$ or $M_{Ed,z}/M_{z,el}$ in case of major and minor axis bending respectively.

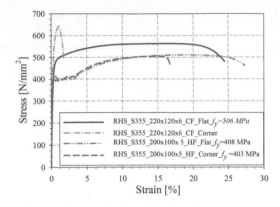

Figure 3. Stress-strain curves from flat and corner regions of a cold formed and a hot-formed sections.

were evaluated through the actual cross-section of each coupon measured before testing. However, for the corner coupons, the area was also determined by combining weight, initial length and density. Typical stress-strain curves measured from hot-finished and cold-formed material are shown in figure 3.

The highest stress level reached in corner coupons of cold-formed sections was 15% to 20% higher than the corresponding flat coupons' highest stress level. Such results caused by high cold-work in the corner regions were accompanied by a loss of ductility in a way that none of the corner coupons exceeded strains higher than 5%.

2.2 Measurement of Residual stresses

The strip-cutting method has been used to measure both flexural stresses and membrane residual stresses. It consists in a destructive technique relying on the measurement of strains triggered by the release of residual stresses after the cutting of small strips within the cross-section; material relaxation generates either elongation or shortening of the strips due to membrane stresses and a curvature due to flexural stresses, which are linked to the level of residual stresses prior to cutting.

Figures 4 and 5 display examples of the obtained residual stresses patterns for a hot-finished square section (membrane residual stresses) and a hot-rolled circular section (flexural residual stresses). In the latter, the flexural residual stresses are seen to be high due to uneven cooling between the inner and outer faces of the tube, whereas the flexural stresses were shown to be negligible in the hot-finished section compared to the membrane residual stresses. (Cidect report, 2014a)

2.3 Measurement of geometrical imperfections

Measurements of geometrical imperfections were made by means of an aluminium perforated bar containing 9 equally-spaced variable displacement transducers (LVDTs) which was displaced sideways on each specimen's plate in order to get 3D geometrical plate

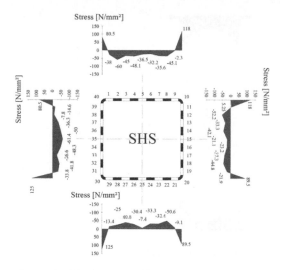

Figure 4. Membrane residual stress measurements, SHS200 × 200 × 6.3, Hot-finished.

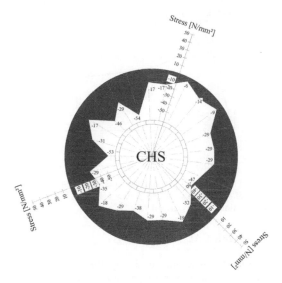

Figure 5. Flexural residual stress measurements, CHS159 × 6.3, Hot rolled.

representations; after measuring the 4 faces of a specimen, all information have been gathered into a finite element shell model that contains the measured local geometrical imperfections. An example of such an imperfect shape is shown below (magnified for sake of clarity).

2.4 Stub column tests

Twelve stub column tests were also performed for each cross-section type in order to determine the average stress-strain relationship over the complete cross-section. The length of each stub column was chosen so as to be about three times the height of the cross section, to avoid global buckling. Two strain gauges have been glued at mid-height of all the elements after

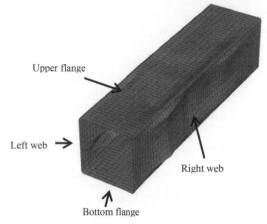

Figure 6. 3D amplified imperfect shape of the specimen SHS200 × 200 × 6.3 (amplification factor: 15).

Figure 7. Stub columns failure modes.

polishing and cleaning the surface not only to ensure that compression was kept concentrically-applied but also to check the load displacement behaviour of the specimen in the elastic range so the corresponding Young's modulus can be assessed. The stub column resulted in an average stress-strain curve of the actual profile and were performed using a 5000 kN hydraulic testing machine. Four LVDTs were used in order to record the average end-shortening behaviour.

Typical failure for stocky sections occurred with a whole cross-section yield with local buckling at the ends of the specimens ("elephant-foot failure"), whilst for slender sections, local buckling was located at the middle of the specimen. Examples of failure modes are shown in Figure 7. Detailed results are available in Cidect report (2014a).

2.5 Cross-sectional tests

The mono-axial and the biaxial-bending with axial compression test configurations were obtained through applying compression eccentrically; end-plates had been welded to the profiles with various eccentricities, according to the load case under consideration.

The response of each specimen has been carefully monitored and recorded, in view of a comparison with FE models. The end plates and the loading plates had respectively a thickness of 20 mm and 60 mm and the

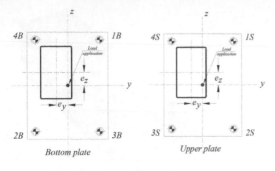

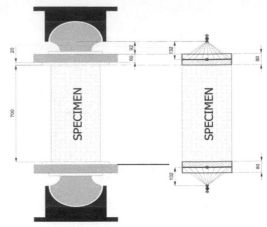

Figure 8. Upper and bottom example of endplates with double eccentricities (1B, 2B, 3B and 4B represent the bottom LVDTs positions and 1S, 2S, 3S and 4S represent the upper LVDTs positions)

Figure 10. Modelling assumptions.

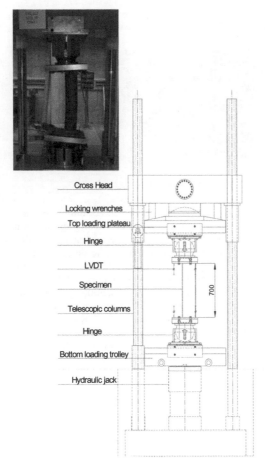

Figure 9. Cross-section test set-up.

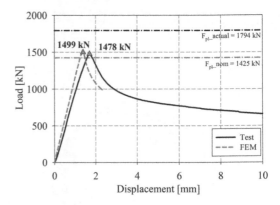

Figure 11. Numerical vs. experimental load displacement curves of specimen LC1_RHS_250 × 150 × 45_HF, Class 4 cross-section, $F_{exp}/F_{FEM} = 0.98$

3 NUMERICAL VALIDATION

In order to represent accurately the experimental behaviour of the specimens, a suitable FE-model has been developed. Specimens have been modelled with shell elements on a regular mesh, and endplates were modelled through rigid plates having an equivalent thickness of 80 mm and represented with elements that remain elastic during loading. The plates' stiffness allowed an even distribution of the applied load at the ends of the sections and prevented the cross-sectional deformation at both ends while allowing free rotations. Rigid truss elements were used to simulate the spherical hinges at both ends. All trusses were connected to the 80 mm thick end plates nodes and to the centroid of the hinge. Compression was applied through the centroid of the hinge, and load cases with major axis and/or biaxial bending were obtained through the application of an axial force acting at the centroid of the hinge with the corresponding eccentricities measured from the tests, see Figure 10.

Averaged material stress-strain behaviour including strain-hardening effects along with measured local

loading was applied evenly on the ends of the specimen (constant bending moment). Measurements were made for axial shortening / elongation and end plates rotations at both extremities. All cross-section tests have been carried out in a testing machine of 3000 kN capacity. An illustrative example of an endplate with double eccentricities and the general set-up are shown in Figure 8 and 9 respectively.

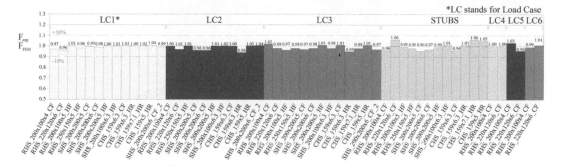

Figure 12. FE peak loads vs. experimental loads.

imperfections for each specimen were accounted for in the simulations. Measured membrane stresses were also introduced for the hot-finished profiles, whereas both (measured) flexural and membrane stresses were considered for cold-formed profiles.

Experimental and numerical peak loads were compared and numerical simulations were seen to represent the real behaviour quite accurately. Figure 11 provides an example of load-displacement curves and a graphical summary of the ratio between the ultimate loads obtained with FE-simulations to the experiments is presented in Figure 12, in which the horizontal dashed lines indicate a deviation of $+/-10\%$ between both sources. It can be seen that all numerical predictions are in excellent accordance with the test results, the maximum difference being 6% only.

4 PARAMETRIC STUDIES ON HOT-ROLLED HOLLOW SECTIONS – APPLICATION OF THE O.I.C.

4.1 Cross-sections and parameters considered

Extensive numerical calculations have been carried out for several profiles and steel grades. The profiles have been chosen so that they cover most of the class ranges according to EN 1993-1-1, CEN (2005). Firstly, 296 tubular geometries picked up from the European catalogue were considered along with 156 rectangular cross sections and 140 square cross sections. Secondly, an additional set of invented sections was analyzed. This was done in order to better visualize more distributed results along higher slenderness, since the European proposed sections would be covering only a limited range of cross-section slenderness. Thus, the newly proposed sections have been derived with respect to more severe h/b and b/t ratios; 5 values of h/b ranging from 1.0 (square sections) to 3.0 (highly rectangular ones), have been considered: $h/b = 1, 1.5, 2, 2.5$ and 3. For each h/b proposed value, b/t values varying from 15 to 115 with a step of 2 have been considered for the load cases of pure compression and major-axis bending, and values going from 15 to 115 with a step of 4 for the load cases of minor-axis bending and combined compression and bi-axial bending.

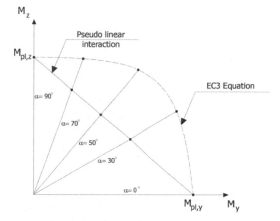

Figure 13. Selection of load cases for $N + M_y + M_z$ combined situations.

Besides, the following sets of parameters have been considered for these sections:

- 3 different steel grades: S235, S360, S690;
- Different loading conditions:

  - Pure compression;
  - Major-axis bending;
  - Minor-axis bending;
  - Combined compression and biaxial bending.

For the combined load cases, a distinction has been made within loading situations, by means of the degree of biaxiality, i.e. the M_y/M_z ratio; this ratio was varied on the basis of α angles of 0, 30, 50, 70 and 90 degrees between plastic capacities $M_{pl,y}$ and $M_{pl,z}$, as Figure 13 shows. As for the level of relative compression $n = N_{Ed}/N_{pl}$, 6 values were adopted from $n = 0$ (i.e. no compression, biaxial bending $M_y - M_z$ situation) to $n = 0.9$ (case where compression is predominat – 90% –, together with $M_y - M_z$ bending mements). The following designation adopted herein to refer to the load case under consideration is as follows:

nx_α

where x refers to the relative level of compression (expressed in percentage), and α is the angle representing the degree of biaxiality in degrees. For example,

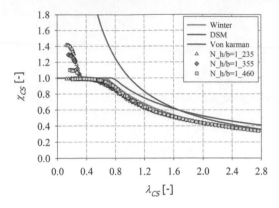

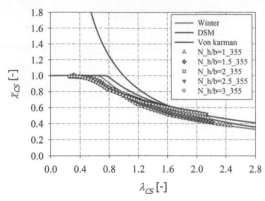

Figure 14. FE results for square sections under pure compression, represented in function of different yield stresses.

Figure 15. FE results for square and rectangular sections under compression, varying aspect ratios, S355.

n50_30 refers to a combined load case of $0.5N_{pl}$ with a degree a biaxiality having an angle of 30 degrees.

It has to be noted that the loading was applied proportionally with N, M_y and M_z being increased all together for all of the combined load cases. Consequently, 22 000 results have been calculated.

As already explained, the proposed generalized slenderness is based on the calculation of "R-factors" ("load multipliers"). Although their calculation does not raise particular difficulties in simple load cases, it may appear much more delicate under biaxial bending and compression. Consequently, R_{RESIST} was determined by means of Materially Nonlinear Analysis (MNA) calculations using the non-linear numerical software FINELg, but could also be determined by means of classic but simplified EN 1993-1-1, CEN (2005) plastic interaction formulae, while R_{STAB} was determined through Linear Buckling Analysis (LBA) calculations.

In the following subsections, the influences of yield stress, cross section shape and load cases are briefly analyzed with respect to results represented in an O.I.C. format, i.e. χ values plotted on the y axis, as a function of the generalized cross-section slenderness λ_{CS}, on the x axis.

4.2 Influence of yield stress

Figure 14 proposes the obtained numerical results for SHS in compression. One may notice a relatively really small dispersion in the results. Smooth and clear tendencies may be observed in other situations as well, e.g. RHS or different load cases. The influence of the yield stress is only pronounced as a general trend, for small relative slenderness values when $\lambda_{CS} < 0.4$. In other words, an important level of over-strength due to strain hardening effects is observed for low steel grades. Such λ_{CS} value might be safe to consider for design but still need to be investigated, especially for members. Results are presented along with the Winter, DSM and Von Karman curves, (Von Karman 1932, b, Schafer 2006).

4.3 Influence of cross section shape

For the simple compression case represented in Figure 15, it is shown that rectangular hollow sections ($h/b > 1$) reach higher relative section resistance compared to square hollow sections possessing the same relative slenderness, particularly in the slender range. The level of restraint offered by the narrow faces of the rectangular section to the wider ones is therefore shown to provide an increased cross section resistance through stress redistributions once local buckling develops in the more buckling-prone plates. Consequently, the cross-section resistance is increased with the h/b ratio, and square sections consequently exhibit the lowest resistance to compression ($h/b = 1.0$) owing to simultaneous buckling of the constitutive plates.

For major axis bending cases, the opposite is shown in Figure 16: the square hollow sections are seen to achieve higher relative bending resistances than the rectangular ones possessing the same cross-section slenderness, particularly in the slender range. The load case type plays a delicate role and decisive one for the structural behavior of elements. In contrary to the compression case, the compressed flanges in the strong axis bending load cases find themselves in need to a great restraint from the webs which in turn have higher slenderness in rectangular sections compared to square ones.

Consequently, the restraint provided by the webs to the flanges will be greater in the case of square sections, delaying thus the onset of local buckling. This is pronounced for slender sections, where failure occurs largely within the elastic material range.

For stocky sections, failure will be achieved at higher strains, where plasticity leads the structural behavior, reducing the detrimental restraint brought to the flanges.

4.4 Influence of load case

Finally, concerning the load case impact, the influence of the relative axial compression is obvious but the

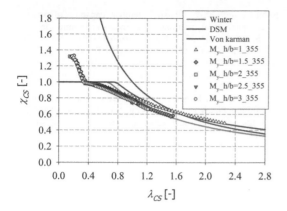

Figure 16. FE results for square and rectangular sections under major axis bending moment, varying aspect ratios, S355.

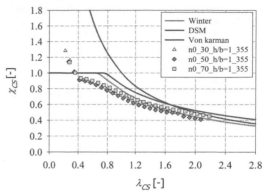

Figure 18. FE results for square sections under combined load cases with a varying level of bi-axialty (S355).

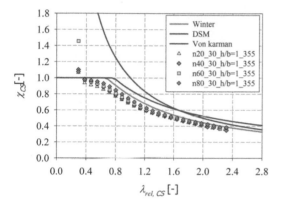

Figure 17. FE results for square sections under combined load cases with a varying level of axial forces (S355).

results on Figure 17 still show a small dispersion and clear tendencies may be emphasised.

Figure 18 finally presents results for which no axial compression is present, i.e. biaxial bending situations; it allows evidencing the influence of the degree of biaxiality in square sections. Ideally, very close tendencies should be observed, and a quite limited scatter is expected for $\alpha = 30$ and $\alpha = 70$. As can be seen in Figure 18, differences appear, especially for large λ_{CS} values. This is to be attributed to the initial geometrical imperfections (i.e. inward buckles in the flanges and outward buckles in the webs or vice-versa), which can be shown to lead to slightly different structural responses, depending on the outward-inward buckling pattern.

5 CONCLUSIONS

In the present paper, an experimental test program on rectangular, square and circular hollow sections of grade S355 structural steel was reported. Hot-finished, hot-rolled and cold-formed stub columns as

well as cross-section tests with various load cases (compression, compression + major axis-bending, compression + weak-axis bending, compression + biaxial bending) were described. Moreover, the measurements of material constitutive laws (tensile tests), imperfection and residual stress measurements were reported.

Besides, a numerical model was developed to simulate the experimental tests and excellent agreement between both sources demonstrated the appropriateness of the FE models to accurately represent the real behaviour of hollow structural shapes under combined loading.

Consecutive extensive FE parametric studies on hot-rolled hollow sections were presented and contributed to evidence the potential for the Overall Interaction Concept to become a reliable and practical alternative to the current well-known design rules, and in particular regarding resistance-instability interactions to account for the interaction between resistance and instability effects. The results demonstrated that, although being based on simple principles and with straightforward application steps, the O.I.C. may appear as an accurate and consistent approach, and serve as a basis for the next generation of standards and practical tools, especially in the frame of an increasing use of high steel or ultra-high steel grades.

REFERENCES

Boissonnade 2011. "The Concept of Cross-section Classes in Eurocode 3 into Question", *ECCS TC8 meeting*, Lisbon, June 3rd.

Boissonnade, Nseir & Saloumi 2013. "The Overall Interaction Concept: an Alternative Approach to the Stability and Resistance of Steel Sections and Members". *Proceedings of the Annual Stability Conference*, Structural Stability Research Council, St. Louis, Missouri, April 16–20.

CEN (Comité Européen de Normalisation) 2005. "Eurocode 3: Design of Steel Structures, Part 1–1: General rules and rules for buildings (EN 1993-1-1)", Brussels.

Cidect report 2014a, Project "HOLLOPOC" – Hollow profiles Overall Concept – 1st interim report, Part A:

cross-sectional resistance – Experiemental tests and validation of FE models.

Cidect report 2014b, Project "HOLLOPOC" – Hollow profiles Overall Concept – 1st interim report, Part B: cross-sectional resistance – FE parametric studies on hot-formed RHS and SHS shapes.

Kettler 2008. "Elastic-plastic cross-sectional resistance of semi-compact H- and hollow sections". *PhD Thesis*, Graz Technical University.

Schafer, 2006. "Designing Cold-Formed Steel Using the Direct Strength Method." 18th International Specialty Conference on Cold-Formed Steel Structures.

Semi-Comp. 2007. "Plastic member capacity of semi-compact steel sections – a more economic design (Semi-Comp)". 2007. *Final report (01/01/06 – 30/06/07)* – RFCS – Steel RTD (Contract RFS-CR-04044).

S.T.S.S. 2012. "Research project STSS – Simple Tools Sell Steel" (http://www.ims.org/2012/02/stss-simple-tools-sell-steel/).

Von Karman, T., E. Sechler, and L. Donel. 1932. Strength of Thin Plates in Compression. Society of Mechanical Engineers – Transactions – Applied Mechanics, 54(2): 53–56.

Tubular Structures XV – Batista, Vellasco & Lima (eds)
© 2015 Taylor & Francis Group, London, ISBN 978-1-138-02837-1

Experimental characterization of the rotational capacity of hollow structural shapes

E. Saloumi, M. Hayeck, J. Nseir & N. Boissonnade
University of Applied Sciences, Fribourg, Switzerland

ABSTRACT: Present paper details bending tests on rectangular hollow sections (RHS) and square hollow sections (SHS). The tests primarily aimed at providing data on ultimate load carrying capacities as well as rotations and curvatures in the inelastic range at plastic hinges locations. These bending tests were designed to investigate reserve capacities and plastic redistribution at the plastic-compact border ("class 2–3 border" in Eurocode 3) with the intention of characterizing the demand vs. availability equation in terms of rotational capacity. They consisted in series on 2.60 m to 4.8 m span tests, under various configurations: simply-supported with either 3-point or 4-point bending static systems, or propped cantilever configurations with mid-span and outer loaded point loads. Several cross-section shapes and dimensions (both external dimensions and thicknesses) were varied. The observed behaviour and results show i) highly variable rotational capacities, and ii) that in most cases plastic analysis isn't applicable even for class1 sections (plastic sections).

1 INTRODUCTION

Hollow profiles may be produced by many different technological processes including welding plates together, hot-rolling, cold-forming etc…Steel hollow sections are becoming more used in a large range of structural applications for their static, esthetic and functional advantages. Indeed, hollow steel beam doesn't exhibit lateral torsional buckling thus when such beams are bent, overall buckling does not occur, and the only buckling mode which can take place is local buckling.

Plastic design is based on the elastic-plastic behaviour of the material (Boeraeve & Lognard 1993) and on the assumption that flexural members should have sufficient plastic deformation capacities while maintaining the plastic moment Mp in order to develop collapse mechanism resulting from sequential plastification (Bruneau et al. 2011). To allow full redistribution of bending moments in the structure, hinges must be able to rotate without losing the bending capacity of the sections. In Eurocode 3, CEN (2005), to insure sufficient rotational capacity, width to thickness ratios are determined. For a class 1 cross-section, the plastic moment resistance can be developed with sufficient rotation capacity to allow redistribution of moments in the steel frames and the section is expected to develop a rotation capacity of 3 before the onset of local buckling. The local buckling limits which are in use were derived, mainly in the sixties and seventies, from simple plate-models (Haaijer & Thurlimann 1958, Daali & Korol 1995, Kato 1989).

In the practice of plastic design of structures, ductility is defined as the capacity of a structure to undergo

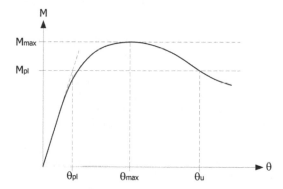

Figure 1. Generalized moment-rotation curve.

deformations after reaching its initial yield without any significant reduction in ultimate strength. The rotation capacity is defined as follow:

$$R = \frac{\theta_u - \theta_{pl}}{\theta_{pl}} \qquad (1)$$

The purpose of this research is to provide experimental data to serve as a base for future parametric studies and to investigate the accuracy of the current design codes. A more detailed knowledge of the current available criteria will contributed to a better understanding of the phenomenon thus to a more realistic and optimized design. Several experimental studies demonstrated that some sections, although classified as class 1 section according to EC3, can experience insufficient plastic rotation capacity than those defined in the codes for plastic design while others may have reserve in capacity. In addition width-to-thickness ratio limitations for flange and for web are prescribed independently of each other. Moreover, EC3 classification refers to the consideration of a very few number of parameters, selected only between material and cross-section factors, the very important member characteristics being ignored.

These considerations highlight the need to develop effective and accurate formulations to predict the rotation capacity of hollow beams in order to overcome the inconsistencies of the present codes, which is based only on the limitation of local slenderness for flange and web while many other governing parameters for the rotation capacity such as the steepness of the moment gradient etc... are disregarded (Kuhlmann 1989).

2 SCOPE OF THE RESEARCH

2.1 Test program, description and fabrication

The test program undergone at the University of Applied Sciences Western Switzerland, consisted in bending tests on hollow sections (either rectangular or square hollow sections) in order to provide an experimental data on the inelastic behavior of such members: ultimate load carrying capacity, rotation capacity ...

Four different test setup configurations were conducted: 3-point and 4-point bending static systems with a span length of 2.6 m, propped cantilever configurations with mid-span and outer loaded point loads with a span length of 4.8 m. Six different cross-sections were tested all corresponding to a class1 section according to Eurocode 3 classification. Tested beams were fabricated using the hot-formed process with a nominal yield stress $f_y = 355$ N/mm^2.

Profiles were received at the laboratory of structural engineering as 6 m long members; 400 mm part was kept for tensile tests; and two 2.8 m long profiles were cut for the simple supported configuration while a 5 m beam was reserved for the propped cantilevers configuration. The specimens' lengths were chosen long enough so that the failure mode would occur by bending and not by shear. The dimensions of each specimen were measured prior to testing.

2.2 *Material properties*

Hot formed structural steel members usually exhibit uniform material properties around the entire cross section due to their fabrication process. The stress strain curve for these profiles typically displays a sharply defined yield point and a yield plateau followed by strain hardening (Gardner et al. 2010a, b, Liew et al. 2014). For each of the tested specimens, four tensile coupons were extracted from each flat face. The coupons were 270 mm length and tested under a constant strain rate of 2.5 mm/min. The test setup with some coupons at failure is shown in Figure 2, and

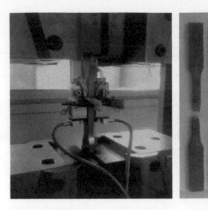

Figure 2. Tensile test rig and some tested coupons.

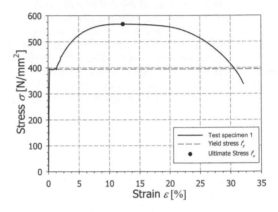

Figure 3. Stress strain curve for RHS_150×100×8_SS coupon.

Figure 3 represents a typical stress strain response for a tested coupon.

Tablen 1 Summarizes the test program and shows the geometric dimensions of all the tested profiles and their main material properties: Young's modulus E, tensile yield strength f_y and ultimate yield strength f_u.

3 THREE POINT BENDING TESTS

Six beams were tested under the 3 point bending configuration. The experimental setup shown in Figure 4 consisted of a simply supported beam on a 30 mm diameter roll. Loading was applied by means of two hydraulic jacks used to generate a concentrated force using two threaded bars connected to a loading beam. Loading was introduced to the tested specimen with half round loading point and through a 40 mm thick and 50 mm wide plate to avoid high level of stress concentration.

Many transducers were used to monitor the beam response: 1) Load cells were located under each support and under the jacks to record the support reaction and the loading force respectively. 2) Inclinometers were fixed at both ends of the beam to measure the beam end rotations. 3) Linear variable displacement

Table 1. Test program matrix.

Section reference #	$h_{measured}$ [mm]	$b_{measured}$ [mm]	$t_{measured}$ [mm]	L [mm]	f_y [N/mm^2]	f_u [N/mm^2]	E [N/mm^2]	Test configuration
RHS_150 × 100 × 8_SS_3P	149.60	99.94	8.35	2600	389	558	197649	Simply supported –
RHS_180 × 80 × 4.5_SS_3P	179.35	78.52	4.80	2600	389	541	194288	3 point bending
RHS_150 × 100 × 5_SS_3P	148.97	99.17	5.26	2600	427	583	215136	
RHS_220 × 120 × 6.3_SS_3P	217.55	120.75	6.40	2600	396	537	206337	
SHS_180 × 6.3_SS_3P	179.59	179.59	6.58	2600	389	531	211737	
SHS_180 × 8_SS_3P	179.44	179.44	7.89	2600	387	537	206441	
RHS_150 × 100 × 8_SS_4P	149.48	99.86	8.16	2600	389	558	197649	Simply supported;
RHS_180 × 80 × 4.5_SS_4P	179.59	79.71	4.81	2600	389	541	194288	4 point bending
RHS_150 × 100 × 5_SS_4P	149.13	99.48	5.13	2600	427	583	215136	
RHS_220 × 120 × 6.3_SS_4P	219.40	120.86	6.42	2600	396	537	206337	
SHS_180 × 6.3_SS_4P	179.68	179.68	6.68	2600	389	531	211737	
SHS_180 × 8_SS_4P	179.39	179.39	7.91	2600	387	537	206441	
RHS_180 × 80 × 4.5_PR_C	179.19	79.06	4.76	4800	388	530	202301	Propped cantilever;
RHS_150 × 100 × 5_PR_C	148.78	99.49	5.20	4800	403	552	215523	centrally loaded
RHS_220 × 120 × 6.3_PR_C	219.10	120.45	6.51	4800	392	536	202859	
SHS_180 × 6.3_PR_C	179.57	179.57	6.72	4800	389	533	202064	
SHS_180 × 8_PR_C	179.30	179.30	7.94	4800	386	536	213897	
RHS_180 × 80 × 4.5_PR_O	178.96	79.45	4.63	4800	385	543	208390	Propped cantilever;
RHS_220 × 120 × 6.3_PR_O	219.03	120.66	6.51	4800	395	536	211259	off-centrally loaded
SHS_180 × 6.3_PR_O	179.55	179.55	6.53	4800	385	528	194114	

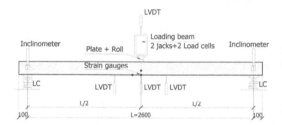

Figure 4. Test setup of the 3-point bending beam.

Figure 5. Deformed shape a RHS_150 × 100 × 5_SS_3P.

transducers were positioned along the beam to record the beam deflection. 4) Strain gauges were fixed on the tension flange to measure both its deformation and its curvature. Loading was carried out under displacement control and all readings were taken using an electronic data acquisition system. Figure 5 displays the deformed shape of the specimen RHS_150 × 100 × 5.

All six beams were tested until failure. In most cases local buckling occurred before beams reached their plastic moment except for the case of the specimen RHS_150 × 100 × 8 for which strain hardening was reached and the test had to be aborted before unloading due to high deformations and experimental limitations.

Figure 6 presents the moment-rotation curve of three tested beams in which M_{pl} is the plastic moment calculated from measured cross sections properties, θ_y is the yield rotation at the beam ends and is calculated when the middle cross section first reach the elastic moment. Figure 7 represents the total loads versus deflection for these specimens where P_{pl} is the plastic collapse load corresponding to M_{pl} and v is the deflection of the beam at mid-span. According to the plotted curves, it is clear that all represented beams failed prior to reaching their plastic capacity.

It was experimentally observed that the onset of local buckling was much localized due to the loading introduction that induced high level of load concentration. Hence, even with the loading applied through a 40 mm thick plate, loading was not uniformly distributed on the area of the plate but was applied on the plate extremities in contact with the corners edges. This can explain why beams failed prematurely by reaching 90% of the plastic moment M_{pl} – while being all class 1 in bending – and with an ultimate deflection of 33 mm. The RHS_150 × 100 × 8_SS_3P – that possess a very stocky section – was not influenced by the loading introduction and reached a 139 mm deflection at maximum loading.

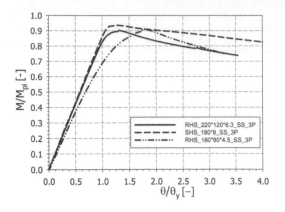

Figure 6. Normalized moment – rotation curve of specimens.

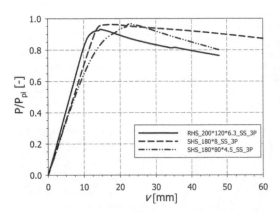

Figure 7. Normalized load – deflection curve of specimens.

Table 2. Collapse results for the 3-point bending tests.

section reference #	M_{ult}/M_{pl} [–]	P_{ult}/P_{pl} [–]	θ_u [°]	v_u [mm]
RHS_150 × 100 × 8_SS_3P	1.21	1.26	7.27	138.9
RHS_180 × 80 × 4.5_SS_3P	0.92	0.97	1.56	22.8
RHS_150 × 100 × 5_SS_3P	0.91	0.97	1.95	33.3
RHS_220 × 120 × 6.3_SS_3P	0.90	0.93	1.31	14.6
SHS_180 × 6.3_SS_3P	0.95	0.98	1.35	17.6
SHS_180 × 8_SS_3P	0.94	0.96	1.30	17.2

Table 2 summarizes the non-dimensional ultimate moment M_{ult}/M_{pl}, ultimate load P_{ult}/P_{pl} and their corresponding rotation θ_u and deflection v_u for all tested beams under the 3-point bending configuration.

4 FOUR POINT BENDING TESTS

The 4-point bending test setup differs from the 3-point bending test by the insertion of a spreader beam over the tested specimen in order to apply equivalent loads on both loading points located at quarter length of the

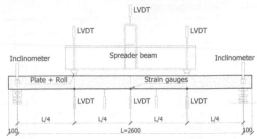

Figure 8. Test setup of the 4-point bending beam.

Figure 9. Deformed shape of a 4-point bending beam.

hinged supports. LVDT and strain gauges has been placed under the loading points and at mid span to record the beam response accurately as shown in Figure 8. Load cells were placed under both supports and hydraulic jacks; inclinometers were positioned at the beams' ends.

Six beams were tested under the 4-point loading configurations, Figure 9 shows SHS_180 × 6 at failure.

During testing, the beams' deflection remained symmetric until the peak load was reached, and a local buckling failure mode occurred at either the right of left loading point. The failure mode became more pronounced in the post buckling unloading phase leading to an increased unsymmetrical deflection shape as shown in Figure 10 for RHS_180 × 80 × 4.5.

Figure 11 represents moment versus beam end rotations; the divergence between the two curves at the loading points highlights the occurrence of local buckling.

The ultimate bending moment and the peak load didn't reach the plastic collapse load or the plastic moment for all tests except for the RHS_150 × 100 × 8 specimen who attained strain hardening but loading was stopped before reaching the peak load due to large deformations; Table 3 summaries test results for all the tested specimens.

Yield rotation θ_y is calculated when the middle cross section segment first reaches the elastic moment and plastic collapse load P_{pl} is computed when the beam attain its plastic capacity.

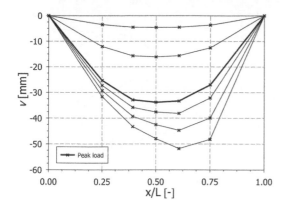

Figure 10. Deflected shape of RHS_180 × 800 × 4.5.

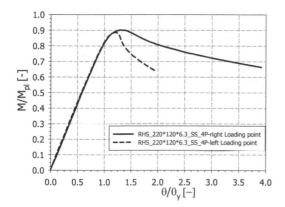

Figure 11. Moment – rotation curve of RHS_220 × 120 × 6.3.

Table 3. Collapse results for the 4-point bending tests.

section reference #	$M_{ult}/$ M_{pl} [–]	$P_{ult}/$ P_{pl} [–]	θ_u [°]	v_u [mm]
RHS_150 × 100 × 8_SS_4P*	1.37	1.41	9.09	149.2
RHS_180 × 80 × 4.5_SS_4P	0.93	0.96	1.99	27.0
RHS_150 × 100 × 5_SS_4P	0.95	0.99	2.95	47.2
RHS_220 × 120 × 6.3_SS_4P	0.90	0.93	1.31	14.0
SHS_180 × 6.3_SS_4P	0.92	0.95	1.39	16.6
SHS_180 × 8_SS_4P	0.97	0.91	1.60	20.6

* Specimen didn't reach failure

For the 3-pt bending and 4-pt bending configurations, the theoretical plastic collapse load is identical: $P_{ult} = 4 \times M_{pl}/L$. The main difference in these configurations is the steepness of the moment gradient. For the 4-pt bending beams, local buckling is supposed to occur somewhere between the loading points where the moment is constant, but experimentally it occurs at the left or right loading point due to high level of stress concentration. Local buckling was only pronounced and localized in the vicinity of the 50 mm thick plate. Therefore, the load introduction may have influenced

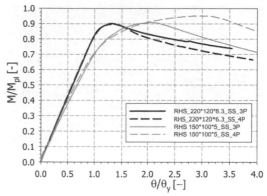

Figure 12. Comparison between 3-pt and 4-pt configurations of RHS 220×120 × 6.3 and RHS_150 × 100 × 5.

Table 4. Comparison of ultimate bending moments between 3-pt and 4-pt bending configurations.

Section	M_u/M_{pl}		
	3-pt bending	4-pt bending	divergence [%]
RHS_150 × 100 × 8	1.21	1.37	11.51
RHS_180 × 80 × 4.5	0.92	0.93	1.74
RHS_150 × 100 × 5	0.91	0.95	4.57
RHS_220 × 120 × 6.3	0.90	0.90	−0.10
SHS_180 × 6.3	0.95	0.92	−3.33
SHS_180 × 8	0.94	0.97	3.27

the beam's response, potentially explaining the lower results.

Figure 12 shows a comparison between the moment rotation curve of the 3-pt bending and the 4-pt bending configuration for the RHS_220×120 × 6.3 and the RHS_150 × 100 × 5 and Table 4 summarize all results.

5 PROPPED CANTILEEVER-CENTRALLY LOADED

Five propped cantilever of 4.8 m span length were tested with the loading being applied at mid-span. Specimens have been fixed to a braced column by welding a 30 mm thick plate to the beam's end and then bolting it by 8 10.9 M24 bolts. The plate was chosen to be thick enough so that it can be considered as perfectly rigid and a full penetration weld was performed.

Strain gauges were fixed on the tension flange; one was placed on the fixed-end 5 cm away from the plate due to the presence of weld and another one was placed at mid-span. The inclinometer was attached to the hinged end to measure the beam end rotation and a load cell was placed under the hinged support to measure the support reaction. Loading was introduced in the same way of the simply supported beams and two

Figure 13. Deformed shape of a propped cantilever centrally loaded.

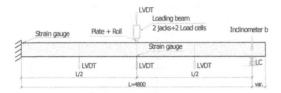

Figure 14. Test setup of the propped cantilever centrally loaded.

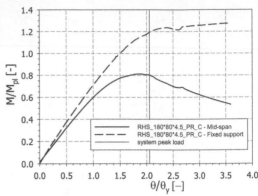

Figure 15. Normalized moment – rotation of RHS_180 × 80 × 4.5.

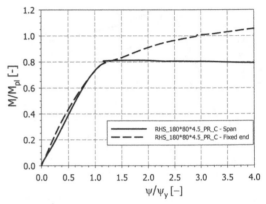

Figure 16. Moment – curvature of RHS_180 × 80 × 4.5_PR_C.

Table 5. Collapse results for the centrally loaded propped cantilever.

Section reference #	$M_{ult/load}/M_{pl}$ [–]	$M_{ult,fixed}/M_{pl}$ [–]	P_{ult}/P_{pl} [–]	θ_u [°]
RHS_180 × 80 × 4.5_PR_C	0.8	1.27	0.95	2.05
RHS_150 × 100 × 5_PR_C	1.0	1.16	1.02	2.99
RHS_220 × 120 × 6.3_PR_C	0.9	1.21	0.88	2.24
SHS_180 × 6.3_PR_C	0.9	1.15	0.88	1.34
SHS_180 × 8_PR_C	0.9	1.26	0.94	3.16

load cells were placed under the jacks to record the loading force. LVDT were placed at mid-span and at quarter span length to measure the beam deflection. The test setup is shown in Figure 14 below.

Plastic moment was first reached at the fixed support with the formation of a plastic hinge and the bending moment was then redistributed to the middle span until the plastic moment and the peak load were reached.

θ_y is calculated when the fixed-end section first reaches the elastic moment. The system collapse load is calculated based on virtual work analysis and collapse mechanism with the assumption of having rigid-perfectly plastic hinges of zero length and bending moment is computed based on the elastic method.

The moment rotation curves are plotted using the rotation given by the inclinometer at the hinged end. Results are shown for SHS_180 × 6.3 in Figure 15. As it is expected, it is shown that, as the test progresses, fixed end moment are higher than at mid-span. System peak load is reached with premature local buckling at mid-span (before reaching the plastic moment). At the fixed end, greater moment is reached due to welding restraints.

Figure 16 displays the normalized moment-curvature with the yield curvature $\psi_y = M_{el}/EI$ calculated from measured dimensions and material properties.

Table 5 summarizes normalized span moments and fixed-end moment along with the system peak load for all the 5 tested specimens.

6 PROPPED CANTILEEVER- OFF CENTRALLY LOADED

Three propped cantilever of 4.8 m span length were tested with loading applied at one third length from the hinged support. Test setup is shown in Figure 17 & 18. Arrangements of the fixed end, hinged end and

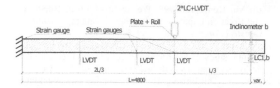

Figure 17. Test setup of the propped cantilever off-centrally loaded.

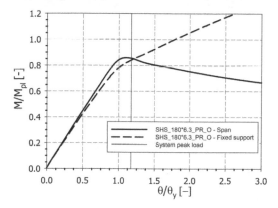

Figure 18. Deflected shape of a propped cantilever off-centrally loaded.

Figure 19. Normalized moment – rotation of RHS_180 × 80 × 4.5.

loading introduction were performed similarly to the centrally loaded cantilever.

Theoretically, span first reach the plastic moment and then failure is attained by the fixed-end reaching its plastic collapse as it is shown in Figure 19 according to the Moment-rotation graph. Span moment is higher than the fixed-end moment until the system peak load is reached after which the span moment decreases and the fixed-end moment increases to reach the plastic moment.

System peak load occurred at variable span displacement ranging between 20 mm and 60 mm, the deflection of the 3 propped cantilever off-centrally loaded are plotted against the normalized total load in Figure 20. The deflection v is measured at the loading point.

Table 6 summarizes normalized span moments and fixed-end moment along with the system peak load and corresponding end rotation θ_u for the 3 tested specimens.

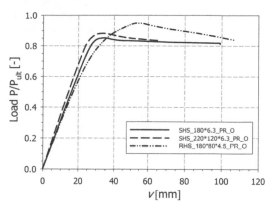

Figure 20. Normalized total load – span displacement.

Table 6. Collapse results for the off-centrally loaded propped cantilever.

Section reference #	$M_{ult/load}/M_{pl}$ [–]	$M_{ult,,fixed}/M_{pl}$ [–]	P_{ult}/P_{pl} [–]	θ_u [°]
RHS_180 × 80 × 4.5_PR_O	0.86	1.25	0.85	1.88
RHS_220 × 120 × 6.3_PR_O	0.90	1.03	0.88	1.48
SHS_180 × 6.3_PR_O	0.89	1.30	0.95	1.17

7 CONCLUSION AND FUTURE WORK

Experimental work evaluating the available rotation capacity of a section is presented. A test program has been carried out including 4 different configurations, 6 cross-sections dimensions, S355 yield strength and hot formed production route.

Tensile tests were performed for all sections and their geometrical dimensions were measured.

The results indicated, concerning the simple supported test configurations, that five out of six sections, although classified as class 1 according to EC3, experienced insufficient plastic rotation capacity. Moreover, while comparing 4-point and 3-point bending, which differ by the moment gradient, results were scattered and no clear tendency could be defined. As for the propped cantilever, the fixed section showed an increase in strength due to welding restraints while the span section did not reach its plastic moment capacity.

Following the experimental campaign, additional steps must be undertaken to further analyze these results and to be able to propose, in a further step, better formulations for plastic design.

First, the test program will be modeled on the nonlinear Finite Element Modeling software FINELg in order to validate the numerical model. In a second step, a parametric study will be launched while varying a set of parameters including the moment gradient, test configuration, yield strength, production route, the parametric study would therefore constitutes a base for the development of a more accurate and well-designed formulae to predict the rotation capacity of hollow beams.

REFERENCES

Boeraeve, P., Lognard, B. 1993. Elasto-plastic Behaviour of steel Frame works, *Journal of Constructional Steel Research* 27 (lgg3) 3–21.

Bruneau, M., Uang, C. & Sabelli, R. 2011. Ductile design of steel structure – Second edition – Chapter 14

CEN (Comité Européen de Normalisation) 2005. "Eurocode 3: Design of Steel Structures, Part 1–1: General rules and rules for buildings", Brussels.

CEN (Comité Européen de Normalisation) 2005. "Eurocode 3: Design of Steel Structures, Part 1–5: Design of plated structures (EN 1993-1-1)", Brussels.

Daali, M., Korol, R. 1995. Prediction of Local Buckling and Rotation Capacity at Maximum Moment, *Journal of Constructional Steel Research* 32, 1–13.

Gardner, L., Saari, N., Wang, F. 2010. Comparative experimental study of hot-rolled and cold-formed rectangular hollow sections, *Thin-Walled Structures,* 48:495–507.

Haaijer, G., Thiirlimann, B. 1958. On inelastic buckling in steel, *Journal of Engineering Mechanics Division, Proceedings of ASCE*, 84(EM2) Paper 1581.

Kato, B. 1989. Rotation Capacity of H-Section Members as Determined by Local Buckling, *Journal of Constructional Steel Research* 13, 95–109.

Kuhlmann, U. 1989. Definition of Flange Slenderness Limits on the Basis of Rotation Capacity Values, *Journal of Constructional Steel Research* 14, 21–40.

Liew, A., Boissonnade, N., Gardner, L., Nseir, J. 2014. Experimental tests on hot-rolled rectangular hollow sections, *Proceedings of the Annual Stability Conference SSRC* Pages: 33–51.

Tubular Structures XV – Batista, Vellasco & Lima (eds)
© 2015 Taylor & Francis Group, London, ISBN 978-1-138-02837-1

Effects of cyclic loading on occurrence of brittle fracture in notched specimens

T. Iwashita
National Institute of Technology, Ariake College, Fukuoka, Japan

K. Azuma
Sojo University, Kumamoto, Japan

ABSTRACT: Evaluating the occurrence of brittle fracture under cyclic loading is a critical issue for structures, particularly those that have experienced a major earthquake, because such loading can degrade material toughness and induce such fracture. This paper describes the effects of cyclic loading on the occurrence of brittle fracture. Notched specimens were tested under monotonic loading and other two types of cyclic loading (constant amplitude loading, and monotonic loading after constant amplitude cyclic loading). It was found that as deflection amplitude increases, cumulative plastic deformation prior to brittle fracture decreases in approximately inverse proportion although cumulative plastic deformation values show wide scatter. Another finding is that cyclic history adversely affects ductility (material toughness). Finally, our results suggest a possibility that a method proposed in this paper can be used to evaluate the occurrence of brittle fracture under various loading conditions.

1 INTRODUCTION

For structural engineers, a notable feature of the Kobe Earthquake was the occurrence of brittle fractures in welded joints at beam ends within multistory moment-resisting frames. These fractures were found to be concentrated in regions around beam bottom-flange groove welds. A post-earthquake investigation (Kinki Branch of the Architectural Institute of Japan 1997) demonstrated that cracking can be retarded by improving beam cope profiles, although also mentioned was a possibility that the ends of complete joint penetration groove welds are particularly susceptible to weld defects (lack of fusion, lack of penetration, slag inclusions). Such weld defects can lead to brittle fracture because, in effect, they act as sharp-notch discontinuities.

Our research group has worked on methods for predicting crack-induced brittle fracture. We found that two methods—the Weibull stress approach and the toughness scaling model (TSM)—enable us to predict the occurrence of brittle fracture more accurately than such conventional fracture toughness metrics as K_c and J_c (Iwashita & Azuma 2012a). It is also noted that the Weibull stress approach is effective for predicting brittle fracture initiated from both sharp cracks and notches (Iwashita et al. 2013). Note, however, that our group confirmed the effectiveness of the Weibull stress approach and TSM under only monotonic loading. It is important to also consider brittle fracture under cyclic loading, as such loading can degrade material toughness and induce premature brittle fracture in structures

exposed to major earthquakes (Iwashita & Azuma 2012a, 2012b).

This paper focuses on the effects of cyclic loading on the occurrence of brittle fracture. Notched specimens were tested under monotonic loading and two types of cyclic loading—constant amplitude cyclic loading (at one of three amplitudes), and monotonic loading after constant amplitude cyclic loading. Also, to reveal any deleterious effect of cyclic history on material toughness, three types of parameters were considered for each type of cyclic loading. Tested under each loading were three to five specimens. Test results were interpreted mainly in terms of ductility amplitude, which is related to deflection amplitude, and cumulative ductility, which is related to cumulative plastic deformation. This paper describes the effects of cyclic loading on the occurrence of brittle fracture in notched specimens based on, for all the specimens, a comparison of the relationship between ductility amplitude and cumulative ductility. Our findings suggest that a method proposed in this paper could be utilized to evaluate the occurrence of brittle fracture under various loading conditions.

2 SPECIMENS AND TESTING PROCEDURES

2.1 Specimen

All test specimens are of the same geometry (i.e., all were machined to the same dimensions; Figure 1). Each specimen has four semicircular notches (root

Table 1. Material tensile properties.

Yield strength (MPa)	Ultimate strength (MPa)	Maximum Uniform strain (%)	$_vE_0$ (J)	VE_{-20} (J)
344	596	11.9	19.9	11.1

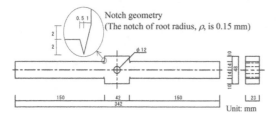

Figure 1. Specimen geometry (dimensions).

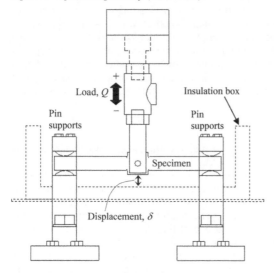

Figure 2. Positions of load application.

radius ρ: 0.15 mm) that are symmetrically located around a loading pinhole at its center. All are 23.0 mm thick.

2.2 Material properties

The samples were machined from JIS SM490A welding structural steel plate (JIS G3106). Tensile and Charpy impact test results for this material are compiled in Table 1, where $_vE_0$ indicates absorbed energy at 0°C and $_vE_{-20}$ that at −20°C (some tests were performed at this low temperature to ensure brittle fracture). As apparent in the table, this material had a low fracture toughness at −20°C. Further research is recommended to examine the influence of medium to high fracture toughnesses.

2.3 Test setup and loading sequences

In total, 31 specimens were tested under three types of loading: monotonic loading (M), constant amplitude

Figure 3. Test apparatus.

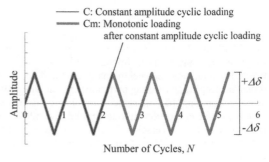

Figure 4. Loading patterns.

Table 2. Summary of specimens.

Specimen	Number of specimens	Amplitude, +/−$\Delta\delta$ (mm)	$\Delta\delta/\delta_p$	Number of cycles, N
M	3	–	–	–
C2.1	5	3.0	2.1	–
C2.6	5	3.7	2.6	–
C3.4	3	4.8	3.4	–
Cm2.1-1	4	3.0	2.1	1
Cm2.1-2	5	3.0	2.1	2
Cm2.6-1	5	3.7	2.6	1

M: Monotonic loading
C: Constant amplitude cyclic loading
Cm: Monotonic loading after constant amplitude cyclic loading
$\Delta\delta$: +/−Displacement amplitude
δ_p: Elastic displacement at the full plastic load, Q_p, (obtained by the General Yield Point method)

cyclic loading (C), and monotonic loading after constant amplitude cyclic loading (Cm). Figure 2 shows loading position. The dashed line indicates an insulated box, which we used together with a mixture of ethanol and dry ice to maintain a low test temperature. Pin supports were provided at both ends of the specimens. Displacement at the loading point was measured along the loading axis. Loading was applied to the center of the specimens by hydraulic ram, with its head moving at a displacement rate of 2.0 mm/min.

Figure 3 is a photograph of the test apparatus. Two of the three types of cyclic loading patterns are conceptually illustrated in Figure 4. For C specimens, constant amplitude cyclic loading was applied to failure. For Cm specimens, monotonic loading was increased until failure after one or two cycles of constant amplitude cyclic. Table 2 shows a summary of specimens and loading types.

3 METHODOLOGY

Test results are interpreted in terms of ductility amplitude μ and cumulative ductility η. Test data are arranged in the manner of Kuwamura & Takagi (2001). For cyclic loading, average ductility amplitude μ_p is defined as

$$\mu_p = \frac{\Delta\delta_p}{\delta_p}$$

Here, $\Delta\delta_p$ is the average displacement amplitude (half-amplitude) and δ_p is the elastic displacement at the full plastic load, Q_p, which is obtained by the general yield point method. A definition of symbols is shown in Figure 5. Because of a slight difference in $\Delta\delta_p$ by cycle during cyclic loading, $\Delta\delta_p$ was calculated as average displacement amplitude as follows:

$$\Delta\delta_p = \frac{\Delta\delta_{p1} + \Delta\delta_{p2} + \cdots + \Delta\delta_{pn}}{n}$$

in the case of Figure 5. η_p is cumulative ductility until fracture and is defined as follows:

$$\eta_p = \frac{\Delta\delta_{p1} + \Delta\delta_{p2} + \cdots + \Delta\delta_{p,n+1}}{\delta_p}$$

where $n + 1 = 5$ in the case of Figure 5. η_p for the Cm loading pattern (Figure 6) is also calculated in the same way.

η for monotonic loading is defined as η_{pM}:

$$\eta_{pM} = \frac{\delta_{pM}}{\delta_p}$$

where δ_{pM} as shown in Figure 5.

We cannot directly compare (cumulative) ductility η under cyclic loading and under monotonic loading. This said, we will attempt a comparison by assuming that monotonic loading is a subcategory of cyclic loading; that is, we assume that monotonic loading is cyclic loading under which the specimen fractures during its first cycle because of a large displacement amplitude. This allows us to use η_{pM} as a measure of both ductility amplitude and cumulative ductility for the purposes of this paper.

Load-displacement curves obtained by a monotonic load testing are known to be fairly close to load-displacement skeleton curves constructed from the load-displacement curve obtained from constant amplitude cyclic load (Cm) tests (see Figure 7). Thus, ductility according to monotonic loading (M) testing will be compared with that of skeleton curves obtained by Cm testing.

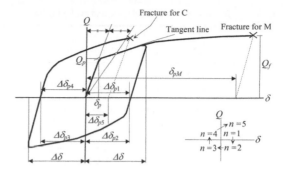

Figure 5. Definition of symbols for monotonic (M) and constant amplitude cyclic (C) loading.

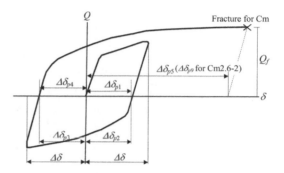

Figure 6. Definition of symbols for monotonic after constant amplitude cyclic (Cm) loading.

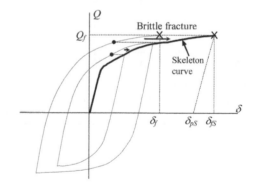

Figure 7. Skeleton curve concept, relationship δ_f and δ_{fS}.

4 TEST RESULTS

4.1 Test results

Brittle fracture was observed in all specimens upon increasing loading. Table 3 shows the test results for each specimen. As mentioned above, every specimen has four notches. When a sample is under load, two of the notches will be in tension and two in compression. Brittle fracture could potentially occur from either of the notches in tension. In actuality, brittle fracture was observed to occur from one of those notches in all specimens. Figure 8 shows an example of a specimen after brittle fracture. From Figure 9, we see that brittle

Table 3. Test results under monotonic and cyclic loadings.

Specimen	Q_p (kN)	δ_p (mm)	Q_f (kN)	δ_f (mm)	n	μ_p	η_p	δ_{pM}, δ_{pS} (mm)	$\eta_{pM}, \eta pS$	η_{p-1st}	η_{p-2nd}
M_1	23.0	1.43	33.3	11.5	–	–	–	9.4	6.6	–	–
M_2	23.0	1.43	32.6	11.5	–	–	–	9.4	6.6	–	–
M_3	22.7	1.45	32.5	11.0	–	–	–	9.0	6.2	–	–
Average	22.9	1.44	32.8	11.3	–	–	–	9.3	6.4	–	–
C2.1_1	22.7	1.46	−14.2	−0.2	10	0.88	9.3	–	–	–	–
C2.1_2	22.9	1.44	26.0	2.6	4	0.94	4.9	–	–	–	–
C2.1_3	23.1	1.42	27.2	2.8	24	0.86	21.9	–	–	–	–
C2.1_4	23.0	1.44	−26.7	−2.6	22	0.88	20.0	–	–	–	–
C2.1_5	23.0	1.43	27.5	3.0	28	0.87	25.6	–	–	–	–
Average	22.9	1.44	8.0	1.1	18	0.89	16.4	–	–	–	–
C2.6_1	23.1	1.47	−29.5	−3.6	6	1.33	9.2	–	–	–	–
C2.6_2	22.7	1.44	28.4	3.2	8	1.31	11.4	–	–	–	–
C2.6_3	23.2	1.47	−28.1	−2.9	14	1.28	18.6	–	–	–	–
C2.6_4	22.8	1.45	−29.4	−3.6	6	1.36	9.4	–	–	–	–
C2.6_5	22.7	1.44	−28.7	−3.6	14	1.30	19.4	–	–	–	–
Average	22.9	1.45	−17.5	−2.1	10	1.32	13.6	–	–	–	–
C3.4_1	23.1	1.42	30.4	4.4	4	2.09	10.2	–	–	–	–
C3.4_2	22.6	1.47	−27.9	−2.5	6	1.97	12.3	–	–	–	–
C3.4_3	22.7	1.41	−29.2	−3.4	6	2.09	13.6	–	–	–	–
Average	22.8	1.44	−8.9	−0.5	5.3	2.05	12.0	–	–	–	–
Cm2.1-1_1	22.8	1.46	29.5	4.9	4	0.97	6.0	4.0	2.7	2.9	3.0
Cm2.1-1_2	23.0	1.45	31.8	8.0	4	0.96	8.0	6.7	4.6	2.9	5.1
Cm2.1-1_3	23.1	1.44	32.6	8.8	4	0.94	8.5	7.3	5.1	2.8	5.6
Cm2.1-1_4	23.2	1.43	33.6	10.5	4	0.95	9.7	9.0	6.3	2.9	6.8
Average	23.0	1.44	31.9	8.0	4	0.96	8.0	6.7	4.7	2.9	5.1
Cm2.1-2_1	23.1	1.42	30.4	5.3	8	0.92	9.7	4.2	2.9	6.5	3.3
Cm2.1-2_2	23.0	1.43	32.9	8.7	8	0.92	12.1	7.4	5.2	6.5	5.6
Cm2.1-2_3	23.0	1.43	30.8	5.7	8	0.93	10.1	4.5	3.2	6.5	3.5
Cm2.1-2_4	23.0	1.43	28.8	3.7	8	0.92	8.6	2.6	1.8	6.5	2.2
Cm2.1-2_5	22.7	1.46	29.9	4.5	8	0.90	9.0	3.4	2.3	6.3	2.7
Average	23.0	1.43	30.6	5.6	8	0.92	9.9	4.4	3.1	6.5	3.4
Cm2.6-1_1	23.6	1.46	32.6	6.3	4	1.36	8.4	5.3	3.6	4.1	4.2
Cm2.6-1_2	23.3	1.47	32.6	8.3	4	1.42	9.9	7.4	5.1	4.3	5.6
Cm2.6-1_3	22.9	1.44	32.2	7.9	4	1.46	9.9	7.1	4.9	4.4	5.5
Cm2.6-1_4	22.9	1.44	30.8	5.6	4	1.46	8.4	4.9	3.4	4.4	4.0
Cm2.6-1_5	23.2	1.48	32.3	7.2	4	1.39	9.0	6.3	4.3	4.2	4.8
Average	23.2	1.46	32.1	7.1	4	1.42	9.1	6.2	4.3	4.3	4.8

fracture was initiated around the center of the specimen in the through-thickness direction. In all specimens, a small (less than 1.0 mm) ductile crack was found to have initiated from the tip of the notch prior to brittle fracture.

4.2 Effect of cyclic history

Example load-displacement curves are shown in Figure 10. The cross marks in the figure indicate fracture points. In the skeleton curves, displacement at fracture was significantly less for the Cm specimens than for the M specimens. As also shown in Table 3, Cm2.1-2 specimens, which were subject to additional cycling (two cycles of cyclic loading), and Cm2.6-1 specimens, which were subjected to a greater amplitude, have lower η_{pS} as compared with Cm2.1-1 specimens, although the η_{pS} values are widely scattered. This means that additional cycling has an adverse effect on ductility (material toughness).

It was found that as deflection amplitude increases, cumulative plastic deformation prior to brittle fracture decreases in approximately inverse proportion. Also, whereas cumulative plastic deformation values show much scatter, a plot of amplitude versus cumulative plastic deformation shows that average cumulative plastic deformation under constant amplitude loading follows a regression line (Figure 11). Another finding is that the scatter of cumulative plastic deformation values is narrower at higher deflection amplitudes.

4.3 Consideration of effect of cyclic history with different loading type

Generally, the ductility η_{pM} of monotonically loaded specimens is not directly comparable to the cumulative

Figure 8. Specimen after fracture.

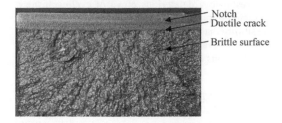

Notch
Ductile crack

Brittle surface

Figure 9. Fracture surface.

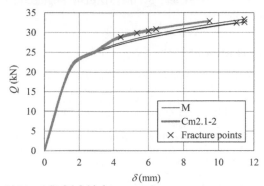

δ (mm)

(a) M and Cm2.1-2 (skeleton curves)

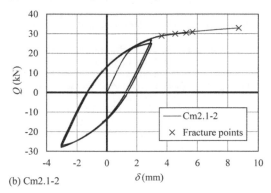

δ (mm)

(b) Cm2.1-2

Figure 10. Examples of load-displacement curves.

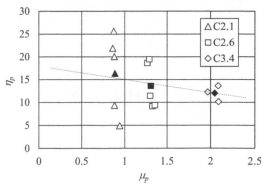

μ_p

Figure 11. Cumulative ductility versus ductility amplitude obtained from C testing (the filled symbols indicate the average η_p).

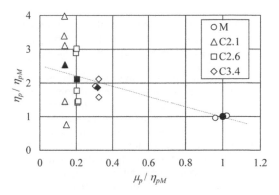

μ_p / η_{pM}

Figure 12. Cumulative ductility versus ductility amplitude obtained from C and M testing (the filled symbols indicate the average η_p).

ductility η_{pM}, of cyclically loaded specimens. However, we will attempt a comparison by assuming that monotonic loading is a subcategory of cyclic loading; that is, we assume that monotonic loading is cyclic loading under which the specimen fractures during its first cycle because of a large displacement amplitude. Here, we thus assume that cumulative ductility η_{pM} of the M specimens can be expressed in terms of either ductility amplitude μ_p or cumulative ductility η_{pM}. Figure 12 is plot of cumulative ductility versus

ductility amplitude, with both divided by cumulative ductility η_{pM}. Included are M specimens, which here are used a baseline. The same tendency previously explained in relation to Figure 11 is also observed in Figure 12, even though the latter figure includes monotonic loading results. Here, it is possible that the high degree of scatter notwithstanding, the regression line expresses the occurrence of brittle fracture with various loading types.

To examine this possibility, we add the arranged results for Cm specimens to the plot of Figure 12 to arrive at Figure 13. The results for Cm specimens are divided into two steps. The first step corresponds to the constant amplitude cyclic loading stage ($n = 1$ to 3 in the case of Figure 6), with the first step having cumulative ductility calculated as η_{p-1st} as shown in Table 3. This could be interpreted as "stored damage" with respect to material toughness. The second step corresponds to the monotonic loading stage after constant amplitude cyclic loading ($n \geq 4$ in the case of Figure 6). Second-step cumulative ductility η_{p-2nd} can also be calculated by subtracting η_{p-1st} from η_p. Here, we use the μ_p and η_{p-2nd} values of the Cm specimens (Table 3) as, respectively, the first-step and second-step ductility amplitude (i.e., the values on the horizontal axis

603

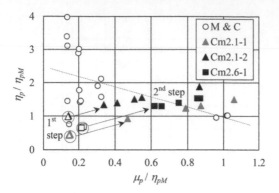

Figure 13. Cumulative ductility versus ductility amplitude with arranged Cm results (1st step plots: $x = \mu_p/\eta_{pM}$; $y = \eta_{p-1st}/\eta_{pM}$, 2nd step plots: $x = \eta_{p-2nd}/\eta_{pM}$; $y = (\eta_{p-1st} + \eta_{p-2nd})/\eta_{pM}$).

of Figure 13). More specifically, the axis values are calculated as follows—first-step plots: $x = \mu_p/\eta_{pM}$; $y = \eta_{p-1st}/\eta_{pM}$; and second-step plots: $x = \eta_{p-2nd}/\eta_{pM}$; $y = (\eta_{p-1st} + \eta_{p-2nd})/\eta_{pM}$.

As shown in Figure 13, the second-step plots fall across the regression line, with considerable scatter. This suggests a possibility of using this approach to evaluate the occurrence of brittle fracture for specimens under various loading conditions. Admittedly, there remains a problem with scatter. Further research will be needed to closely consider this scatter problem and, by extension, establish an adequate method to evaluate the occurrence of brittle fracture.

5 SUMMARY

This paper focused on the effects of cyclic loading on the occurrence of brittle fracture. Notched specimens were tested under monotonic loading and two types of cyclic loading. We found that as ductility amplitude increases, cumulative ductility prior to brittle fracture decreases in approximately inverse proportion. Also, although cumulative ductility values do show wide scatter, a plot of amplitude versus cumulative ductility nonetheless reveals that average cumulative ductility under constant amplitude loading follows a regression line. Another finding is that the scatter of

cumulative ductility values is narrower at higher ductility amplitudes. We also found that in a skeleton curve for specimens tested under monotonic loading after constant amplitude cyclic loading, ductility is small compared with that for specimens tested under monotonic loading alone. Furthermore, we revealed that a higher number of cycles and greater amplitude lead to a deterioration in cumulative ductility. We attribute this result to a deleterious effect of cycling history on material toughness. Finally, our results suggest the possibility that the method proposed in this paper can be used to evaluate the occurrence of brittle fracture under various loading conditions.

ACKNOWLEDGMENTS

This research was supported in part by JSPS Grants-in-Aid for Scientific Research (Grant Number: 25820276) and by the MAEDA Engineering Foundation.

REFERENCES

AIJ Kinki. 1997. *Full-scale test on plastic rotation capacity of steel wide-flange beams connected with square tube steel columns.* Committee on Steel Building Structures, The Kinki Branch of Architectural Institute of Japan, Osaka, Japan. (in Japanese)

Iwashita, T., Kobayashi, R. & Azuma, K. 2013. Assessment of brittle fracture for single edge notched bend specimens with different machined-notch depth. *ASME 2013 32nd International Conference on Ocean, Offshore and Arctic Engineering* 3: Paper No. OMAE2013-11144

Iwashita, T. & Azuma, K. 2012a. Effect of plastic constraint on brittle fracture in steel: Evaluation using toughness scaling model. *Journal of Structural Engineering* 138(6): 744–752.

Iwashita, T. & Azuma, K. 2012b. Brittle Fracture Initiating at Ends of CJP Groove Welded Joints with Defects: Effect of plastic constraint on brittle fracture. *Journal of Structural and Construction Engineering (Transaction of AIJ)* 77 (671): 105–112. (in Japanese)

Kuwamura, H. & Takagi, Naoto. 2001. Verification of similitude law of pre-fracture hysteresis. *Journal of Structural and Construction Engineering (Transaction of AIJ)* (548): 139–146. (in Japanese)

Tubular Structures XV – Batista, Vellasco & Lima (eds)
© *2015 Taylor & Francis Group, London, ISBN 978-1-138-02837-1*

The effect of steel strip on the quality of cold-formed hollow sections

P. Ritakallio
Ruukki Metals Oy (since September 2014: SSAB Europe Oy)

ABSTRACT: The quality of steel strip is essential for tube quality. Conventional C-Mn steels are commonly used for hollow sections. These steels are susceptible to strain ageing, thus questioning their low temperature ductility in welded structures. To ensure good low temperature characteristics Ruukki shifted to thermomechanically-rolled steels. This study compares hollow sections from the early 1990s with current Ruukki hollow sections made of thermomechanically-rolled fine grain steels. Ageing is noticeable in conventional S355J2H hollow sections. In cold-formed areas, flat faces and in corners there is a risk that the toughness does not fulfill 27 J/−20°C after welding. Ageing in Ruukki double grade S420MH/S355J2H hollow sections is still noticeable, but virtually insignificant. After ageing, the transition temperature T_{40J} remains around −70°C on the wide face, around −50°C in corners and around −40°C on the narrow face. There is only minor risk of toughness not conforming to the EN 10219 requirement of 40 J/−20°C after welding.

1 INTRODUCTION

Structural tubes for steel construction are manufactured in various locations around the world and to a variety of standards, by either a hot-finishing or seamless process or – more commonly – by cold-forming. Economic and environmental constraints favour cost competitiveness and higher strength. The cold-forming route is naturally economical and can mostly adapt to higher strength too.

The manufacture of good-quality cold-formed hollow sections consists of three crucial steps:

A. Steelmaking – ladle refining – slab casting

– appropriate chemistry of the steel

B. Hot rolling – controlled cooling and coiling

– appropriate microstructure and mechanical characteristics of the flat steel

C. Tube manufacturing – cold forming – welding – shaping

– appropriate dimensions, properties and suitability for shop fabrication, downstream processing and structural use

Hollow section quality is the sum of these steps. It is a result of the adequate processing of the steel and manufacturing of the tube. Consequently, the quality of the raw material, appropriate steel strip, is an essential requirement for success.

Conventional ferrite pearlite C-Mn steels are a commonly used material for hollow sections. These steels have poor formability and are also susceptible to strain ageing, a well-known characteristic of ferritic steels. Cold-formed corner areas, that are subsequently heated to temperatures in the range of 100–500°C, are susceptible to loss of toughness due to strain ageing in the vicinity of the welds. This loss of toughness is mainly manifested in an increase in the ductile to brittle transition temperature (Saarela et al. 1995).

Welding is a primary method of joining the components in steel structures. Ductility is one of the primary features of reliable structural materials and components, and one of the underlying principles of Eurocode 3 (EN 1993-1-1). Due to the susceptibility to strain ageing, the reliability of cold-formed hollow sections is sometimes questioned. The main concerns are related to inadequate low-temperature ductility and deformation capacity of welded joints, and loss of toughness due to strain ageing in the vicinity of the welds, especially in the cold-formed corner area. Consequently, for example, Eurocode 3 part EN 1993-1-8 includes restrictions on welding in the corner area.

Recent studies (Kosteski et al. 2003, Feldmann et al. 2012, Stranghoener et al. 2012, Eichler et al. 2012) have revealed large scatters in the low temperature ductility of cold-formed hollow sections from different suppliers in different markets. It is well known that there are both good quality products with excellent Charpy-V toughness, and products with substandard quality, too. We are dealing with a diversity of the quality.

According to Packer & Chiew (2010), many manufacturing specifications include critical shortcomings which impact negatively on the performance of cold-formed structural tubes, and the products are often not fully appreciated in various statically loaded or dynamically loaded applications. Most of the production standards are quite liberal with regard to the chemistry, hot-rolling and formability of the steel material, corner profile of rectangular hollow sections and low temperature ductility.

Table 1. Low temperature characteristics of the steel strip and cold-formed square EN 10219—S355J2H hollow sections (Soininen 1996).

Dimension			Steel strip			Flat face				Flat face Aged				Corner			Corner Aged		
H mm	B mm	T mm	US_{KV} J/cm²	T_{US} °C	T_{27J} °C	Test face	US_{KV} J/cm²	T_{US} °C	T_{27J} °C	Face [mm]	US_{KV} J/cm²	T_{US} °C	T_{27J} °C	US_{KV} J/cm²	T_{US} °C	T_{27J} °C	US_{KV} J/cm²	T_{US} °C	T_{27J} °C
100	100	8,0	250	−20	−65	H	230	0	−50	100	240	0	−30	250	0	−40	250	20	−30
100	100	10,0	200	−20	−50	H	225	0	−20	100	240	20	−10	250	10	−35	219	10	−15
150	150	8,0	230	−40	−80	H	188	−10	−70	150	220	20	−65	203	10	−65	172	−30	−55
180	180	10,0	230	−20	−70	H	200	−10	−60	180	220	0	−60	250	0	−55	200	0	−35
200	200	12,5	230	−5	−55	H	225	−20	−35	200	230	20	−15	213	0	−30	225	10	−15

As shown by Packer & Chiew, the European harmonized standard EN 10219 provides quite thorough specifications for structural tubes. All the specifications refer to measurements and testing of ready-made tube. Alternative steel materials are specified and include various conditions, like "as rolled", "normalized or normalized rolled" and "thermomechanically rolled". The strength levels include steel grades up to S460 and the impact toughness requirements include

– non-alloy steels 27 J/−20°C or 27 J/−30°C (27 J equals 35 J/cm²)
– fine grain steels 40 J/−20°C or 27 J/−50°C (40 J equals 50 J/cm²)

In addition EN 10219, Ch. 6.2, specifies the external corner profile, and even the internal corner profile, as follows: "The internal corners of square and rectangular hollow sections shall be rounded".

As noted above, the design rules, such as Eurocode 3, include restrictions on welding in the cold formed corner area. Despite this, product standards rarely contain requirements on welding and ageing.

The purpose of this paper is to highlight the impact of welding and ageing on the low temperature ductility of rectangular cold-formed hollow sections according to EN 10219 and the influence of the steel material on the quality of cold-formed structural tubes. In order to minimize the number of variables, this study is restricted to hollow sections made of non-alloy steels and thermomechanically rolled fine grain steels manufactured by Ruukki.

2 LOW TEMPERATURE TOUGHNESS OF COLD-FORMED RECTANGULAR HOLLOW SECTIONS

The Charpy-V impact test is widely used to characterize the low temperature toughness of steel materials. A complete picture of the material can be obtained by testing the material at a temperature range covering the whole range from fully ductile behaviour (upper shelf energy) over the transition temperature range down to fully brittle behaviour (lower shelf energy). This kind of testing provides the following characteristics:

– upper shelf energy US_{KV} [J/cm²]
– upper shelf temperature T_{US} [°C]

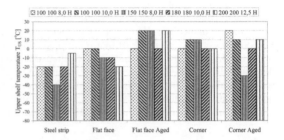

Figure 1. CharpyV upper shelf temperature T_{US} of the steel strip and cold-formed square EN 10219—S355J2H hollow sections (Soininen 1996).

– transition temperature T_{27J} or T_{40J} [°C]
– (Below T_{27J} or T_{40J} the Charpy-V energy is less than 27 or 40 [J]).

As mentioned above, the standard EN 10219, for example uses, T_{27J} or T_{40J} [°C] as acceptance criteria. This paper considers the values of US_{KV} and T_{US}, but the main focus is on the transition temperature T_{27J} or T_{40J}.

The low temperature toughness of cold-formed rectangular hollow sections has been subject to several studies (Dagg et al. 1989; Soininen 1996; Kosteski et al. 2003; Puthli & Herion 2005; Ritakallio 2010 & 2012; Ritakallio & Bjork 2014). Feldmann et al. (2012) made a detailed analysis of the results included in the previous studies. All these studies, except Soininen (1996), focused on ready-made hollow sections. Consequently, the results cannot be correlated to the quality of the steel material.

Soininen (1996) studied the evolution of properties from steel strip to cold-formed finished square EN 10219 hollow sections, including even corners and strain ageing, and the testing of structural performance of welded X-joints. The hollow sections were supplied by Ruukki during the first half of the 1990s and represented state-of-the-art technology at that time. Table 3 shows the chemistry of the steels. The low temperature characteristics are reproduced in Table 1 and in Figures 1 & 2.

According to Soininen, Table 1 and Figures 1 & 2

– cold forming affects T_{US} on flat face from −15 to +30°C and in corners from +5 to +50°C

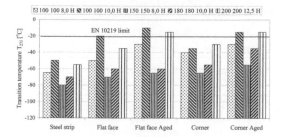

Figure 2. Charpy-V transition temperature T_{27J} of the steel strip and cold-formed square EN 10219—S355J2H hollow sections (Soininen 1996).

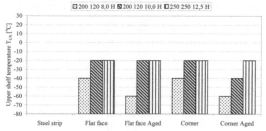

Figure 3. CharpyV upper shelf temperature T_{US} of rectangular Ruukki double grade S420MH/S355J2H hollow sections (Ritakallio & Bjork 2014).

- ageing further increases T_{US} on flat face from ± 0 to $+40°C$ and in corners from -40 to $+20°C$
- cold forming affects T_{27J} on flat face from $+10$ to $+30°C$ and in corners from $+15$ to $+25°C$
- ageing further increases T_{27J} on flat face from ± 0 to $+20$ and in corners from 10 to $+20°C$
- as a whole cold forming and ageing increases T_{27J} on flat face from $+10$ to $+40°C$ and in corners from $+25$ to $+40°C$

The effect on ageing is noticeable in the hollow sections studied by Soininen in the mid-1990s. The study revealed the occasional risk, after ageing in the flat face and corner area of the rectangular hollow sections, of low temperature toughness, which does not conform to the EN 10219 requirement for T_{27J} after welding, Figure 2.

It is worth noting that cold deformation and ageing has a rather similar impact on the flat face and corner. Due to cold deformation and/or ageing, the increase of T_{27J} in the corner is only marginally higher than in the flat face. Contrary to common belief, this study indicates that the specimen from the flat face is quite well representative for the whole section.

In order to be certain that even the corner areas have the required low temperature toughness after welding, Ruukki decided on tighter quality requirements for the Charpy-V testing of structural tubes. Since 1998, Ruukki has supplied EN10219-1&2 grade S355J2H hollow sections with Charpy-V requirement 35 J/cm² at $-40°C$.

The supply of the steel material, steelmaking and hot rolling, was adapted to the tighter needs of tube manufacturing through gradual transition to micro alloyed and thermomechanically-rolled steels. Introduction of a new type of steel raised several questions concerning the structural behaviour of welded joints and the influence of HAZ, etc. Considerable testing was conducted and a number of benefits of the new types of steel were verified over time (Ritakallio & Bjork 2014). Since 2002, Ruukki's entire production of structural tubes has been based on micro alloyed thermomechanically-rolled fine grain steels.

The CIDECT project 1A (Puthli & Herion 2005) focused on welding in cold-formed areas of rectangular hollow sections. The main goals of this research were, on the one hand, to determine the requirements for the reliable welding of cold-formed structural hollow section connections on the basis of strain ageing caused by welding and, on the other hand, to provide recommendations for the extension of the existing design rules for welding in cold-formed corner areas of rectangular hollow sections. The study provides evidence that in hollow sections made of fine grain steels or thermomechanically-rolled steels, the low temperature toughness on the flat face and in corners is similar, and the effect of ageing remains marginal.

In 2010, Ruukki introduced a new standard quality "Ruukki double grade" which conforms to EN 10219 grades S420MH and S355J2H with a Charpy-V requirement of 50 J/cm² at $-40°C$. The steel material for Ruukki double grade hollow sections is a micro alloyed thermomechanically-rolled and controlled cooled and coiled low carbon steel. The chemical analysis complies with the requirements of EN 1993-1-8, EN 1993-1-8:2005/AC for welding in the corners, the carbon equivalent is typically $C_{eqv} \approx 0.30$ and the ASTM grain size is in the range of 11 to 13. Ruukki double grade hollow sections exhibit good low temperature ductility and structural performance (Ritakallio 2012).

In 2013, the impact of ageing on the Charpy-V toughness of rectangular Ruukki double grade hollow sections was verified using three production samples: $200 \times 120 \times 8$ mm, $200 \times 120 \times 10$ mm and $250 \times 250 \times 12.5$ mm (Ritakallio & Bjork 2014). The hollow sections were tested both in ordinary cold-formed conditions and artificially aged at $250°C$ for 30 minutes. The Charpy-V test specimens were taken longitudinally on the wide face as specified in EN 10219 and in the centre of the corner. Table 3 shows the chemistry of the steels. The low temperature characteristics are shown in Table 2 and the effect of ageing on T_{US} and T_{40J} is visualized in Figures 3 & 4.

Table 2 and Figures 3 & 4 show that both upper shelf temperature T_{US} and the transition temperature T_{40J} are rather uniform and the effect of ageing is zero or minor. Ageing increases T_{40J} from $+10$ to $+20°C$ only in the corners. The transition temperature T_{40J} in the flat face is at a very low level, in the order of $-70°C$. In the corners, the transition temperature remains even after ageing at a low level of $-50°C$. These results are very consistent with the CIDECT study (Puthli & Herion 2005).

Table 2. Low temperature characteristics of the steel strip and rectangular Ruukki double grade S420MH/S355J2H hollow sections.

Dimension			Steel strip			Flat face				Flat face Aged				Corner			Corner Aged		
H mm	B mm	T mm	US_{KV} J/cm²	T_{US} °C	T_{27J} °C	Test face	US_{KV} J/cm²	T_{US} °C	T_{27J} °C	Face [mm]	US_{KV} J/cm²	T_{US} °C	T_{27J} °C	US_{KV} J/cm²	T_{US} °C	T_{27J} °C	US_{KV} J/cm²	T_{US} °C	T_{27J} °C
200	120	8,0	n.a.	n.a.	n.a.	H	240	−40	−100	200	220	−60	−100	240	−40	−100	210	−60	−90
200	120	10,0	n.a.	n.a.	n.a.	H	300	−20	−70	200	270	−20	−70	300	−20	−70	230	−40	−50
250	250	12,5	n.a.	n.a.	n.a.	H	240	−20	−70	250	270	−20	−85	240	−20	−60	290	−20	−50
200	100	10,0	260	−40	−70	B	290	0	−60	100	275	−20	−40						
200	100	10,0	260	−40	−70	H	310	−20	−75	200	300	−20	−80*	300	−60	−80*	300	−20	−80*
250	150	12,5	260	−60	−75	B	300	−20	−55	150	320	0	−45	270	−40	−70	300	0	−55
250	150	12,5	260	−60	−75	H	300	−40	−65	250	n.a.	n.a.	n.a.	270	−40	−70	300	0	−55

*Lower than reported value

Table 3. Chemistry of the steels.

Reference	Steel grade	Dimension mm	C %	Si %	Mn %	P %	S %	Al %	Nb %	V %	Ti %	N %	CEV*
Soininen 1996	S355J2H	100 × 100 × 8	0,12	0,16	1,37	0,013	0,006	0,033	0,001	n.a	n.a	n.a	0,35
Soininen 1996	S355J2H	100 × 100 × 10	0,09	0,18	1,25	0,016	0,008	0,029	0,014	n.a	n.a	n.a	0,30
Soininen 1996	S355J2H	150 × 150 × 8	0,14	0,16	1,37	0,016	0,007	0,036	0,002	n.a	n.a	n.a	0,37
Soininen 1996	S355J2H	180 × 180 × 10	0,13	0,18	1,36	0,014	0,005	0,035	0,001	n.a	n.a	n.a	0,36
Soininen 1996	S355J2H	200 × 200 × 12,5	0,12	0,16	1,45	0,014	0,005	0,043	0,001	n.a	n.a	n.a	0,36
Ruukki 2013	S420MH	200 × 120 × 8	0,07	0,19	1,43	0,011	0,003	0,026	0,023	0,007	0,017	0,0038	0,33
Ruukki 2013	S420MH	200 × 120 × 10	0,07	0,19	1,41	0,010	0,003	0,035	0,022	0,011	0,016	0,0046	0,32
Ruukki 2013	S420MH	250 × 250 × 12,5	0,07	0,17	1,40	0,013	0,006	0,028	0,023	0,012	0,014	0,0037	0,32
Ruukki 2014	S420MH	200 × 100 × 10	0,09	0,18	1,42	0,013	0,003	0,029	0,016	0,010	0,013	0,0051	0,34
Ruukki 2014	S420MH	250 × 250 × 12,5	0,06	0,19	1,40	0,012	0,003	0,034	0,027	0,011	0,016	0,0047	0,31

*$CEV = C + Mn/6 + (Cr + Mo + V)/5 + (Cu + Ni)/15$

Soininen (1996) studied square hollow sections with flat faces of equal size. The above-mentioned study (Ritakallio & Bjork 2014) used specimens taken from the wide flat face. The degree of deformation of the wide flat face is lower than deformation on the narrow flat face.

In order to better understand the evolution of Charpy-V toughness of rectangular Ruukki double grade hollow sections, additional testing was done in 2014. This study included both the steel material and two different hollow sections: 200 × 100 × 10.0 mm and 250 × 150 × 12.5 mm. During tube manufacturing, steel material samples were taken from the middle of the coil by cross cutting approximately one metre test piece, re-joining the coil ends by welding and then taking a test piece of the finished tube adjacent to the cross welds. Consequently, the test materials originated from the same area of the coil. The steel material and the hollow sections were tested both in ordinary delivery condition and artificially aged at 250°C for 30 minutes. The Charpy-V testing of hollow sections included a wide face, narrow face and corner. The notch location or orientation was varied in order to map out the possible impact of notch orientation.

The dimensions conformed to EN 10219 requirements with corner regions:

- 200 × 100 × 10.0 mm: $24.7 \leq C \leq 28.3$ mm (specified $20.0 \leq C \leq 30.0$ mm)

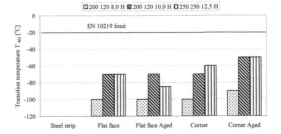

Figure 4. CharpyV transition temperature T_{40J} of rectangular Ruukki double grade S420MH/S355J2H hollow sections (Ritakallio & Bjork 2014).

- 250 × 150 × 12.5 mm: $33.1 \leq C \leq 36.1$ mm (specified $30.0 \leq C \leq 45.0$ mm)

Table 3 shows the chemistry of the steels. The grain size of the steels was ASTM 11.6 in the 200 × 100 × 10 mm tube and ASTM 12.1 in the 250 × 150 × 12.5 mm tube. Figures 5–12 show the Charpy-V impact toughness transition curves. The low temperature characteristics derived from these diagrams are included in Table 2 and the effect of cold-forming and ageing on T_{US} and T_{40J} from strip to hollow section is visualized in Figures 13 & 14.

Table 2 and Figures 5–14 show that

- the anisotropy in the steel material is almost zero

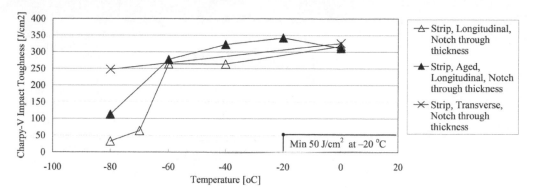

Figure 5. Charpy-V impact toughness of thermomechanically-rolled steel strip for manufacturing Ruukki double grade S420MH/S355J2II hollow sections 200 × 100 × 10 mm.

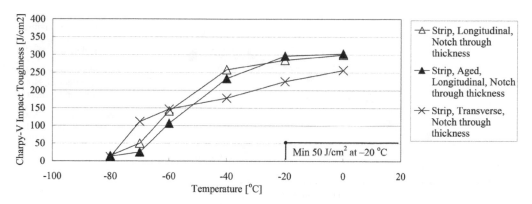

Figure 6. Charpy-V impact toughness of thermomechanically-rolled steel strip for manufacturing Ruukki double grade S420MH/S355J2H hollow sections 250 × 150 × 12.5 mm.

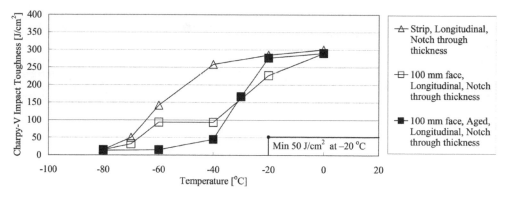

Figure 7. Charpy-V impact toughness of Ruukki double grade S420MH/S355J2H hollow section 200 × 100 × 10 mm.

- ageing has no impact on T_{US} and T_{40J} of the steel material
- notch through thickness is apparently the most conservative notch orientation on the hollow section flat face
- notch orientation in the hollow section corner has no impact on the transition temperature
- cold forming impacts T_{US} on the wide face +20°C and narrow face +40°C, but has no impact in the corner

- ageing has no consistent impact T_{US} on the flat faces, but in corners an increase +40°C can be observed
- cold forming does not impact T_{40J} on wide face and in corner, but on the narrow face an increase from +10 to +20°C can be observed
- ageing does not impact T_{40J} on wide face, but on the narrow face and in the corner of the 250 × 150 × 12.5 mm section an increase from +10 to +20°C can be observed

609

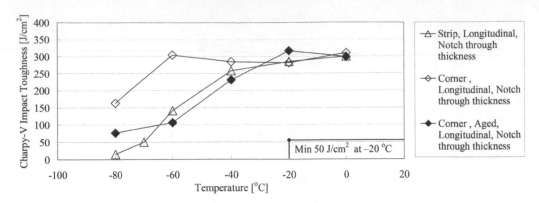

Figure 8. Charpy-V impact toughness of Ruukki double grade S420MH/S355J2H hollow section 200 × 100 × 10 mm.

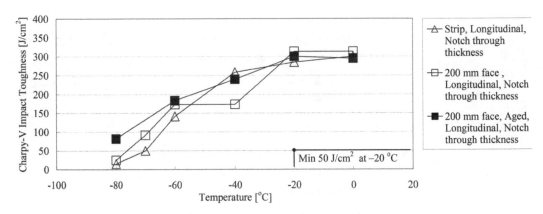

Figure 9. Charpy-V impact toughness of Ruukki double grade S420MH/S355J2H hollow section 200 × 100 × 10 mm.

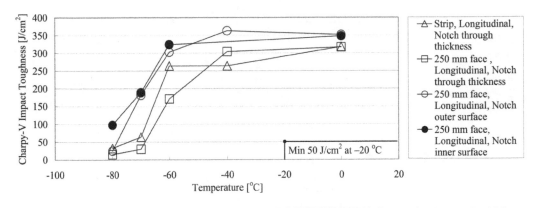

Figure 10. Charpy-V impact toughness of Ruukki double grade S420MH/S355J2H hollow section 250 × 150 × 12.5 mm.

− as a whole, cold forming and ageing impacts T_{40J} on the wide face and in corners in the range of −10 to +20°C and in the narrow face +30°C.

The impact of ageing is still noticeable in Ruukki double grade hollow sections made of thermomechanically-rolled steel strip. After ageing the transition temperature T_{40J} remains at a low level, in corners around −50°C, and on the narrow face around −40°C or below. That is notably below the EN 10219 requirement 40 J/−20°C (50 J/cm²/−20°C) for fine

grain steels, Figure 14. The risk of low temperature toughness which does not conform to EN 10219 requirements after welding is minor. In contrast to the findings of Soininen (1996), the impact of ageing is virtually insignificant.

3 SUMMARY

The quality of hollow sections is a result of the adequate processing of the steel and manufacturing of the

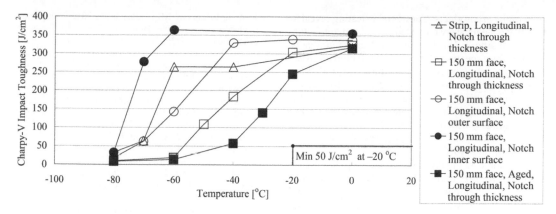

Figure 11. Charpy-V impact toughness of Ruukki double grade S420MH/S355J2H hollow section 250 × 150 × 12.5 mm.

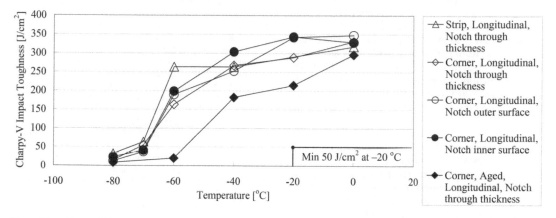

Figure 12. Charpy-V impact toughness of Ruukki double grade S420MH/S355J2H hollow section 250 × 150 × 12.5 mm.

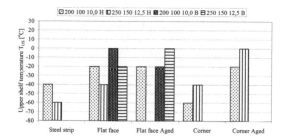

Figure 13. Charpy-V upper shelf temperature T_{US} of the steel strip and rectangular Ruukki double grade S420MH/S355J2H hollow sections.

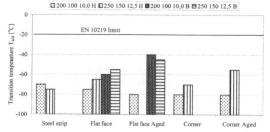

Figure 14. Charpy-V transition temperature T_{40J} of the steel strip and rectangular Ruukki double grade S420MH/S355J2H hollow sectio.

tube. The quality of the raw material, appropriate steel strip, is an essential requirement for success.

The steel materials studied by Soininen (1996) were conventional C/Mn-steels with variable chemistry and with little or no micro alloying, Table 3. The hot-rolling process was typical for that time. The steel strips had variable transition temperatures with T_{27J} ranging from −80 to −50°C, Figure 2. Cold forming increased the transition temperature T_{27J} on the flat face and in corners from +10 to +30°C. Consequently, the hollow sections had variable transition temperatures with T_{27J} ranging both on the flat face

and in corners from −70 to −20°C, Figure 2. Ageing further increased the transition temperature T_{27J} from ±0 to +20°C. Consequently, after ageing, the hollow sections had a transition temperature T_{27J} ranging both on the flat face and in corners from −55 to −10°C. As a whole, some of the products had very good low temperature toughness, but occasionally after ageing the low temperature toughness on the flat face or in corner did not conform to the EN 10219 requirement of 27 J/−20°C.

At the end of the 1990s, thermomechanically-rolled, controlled-cooled and coiled fine grain steels became

611

state of the art products. Considerable testing revealed that these steels are well suited for tube manufacturing (Ritakallio & Bjork 2014) and Ruukki gradually adopted this type of steels for the manufacture of structural hollow sections. Since 2002, the entire production of structural tubes has been based on micro alloyed, thermomechanically-rolled fine grain steels. In 2010, Ruukki introduced a new standard quality "Ruukki double grade", which conforms to EN 10219 grades S420MH and S355J2H with a Charpy-V requirement of 50 J/cm^2 at $-40°C$.

Thermomechanically-rolled steel strip for manufacturing Ruukki double grade hollow sections has a transition temperature T_{40J} of around $-70°C$. Cold-forming has only a minor impact on the low-temperature toughness of this type of steel strip. Consequently, Ruukki double grade hollow sections have a transition temperature T_{40J} both on the wide face and in corners from -80 to $-65°C$ and on the narrow face from -60 to $-65°C$, Figure 14. The impact of ageing on Ruukki double grade hollow sections is still noticeable, but virtually insignificant. After ageing, the transition temperature T_{40J} remains at a low level, around $-70°C$ or below on the wide face, around $-50°C$ in corners and around $-40°C$ or below on the narrow face, Figures 4 & 14.

In contrast to the findings of Soininen (1996), the impact of ageing in Ruukki double grade hollow sections is virtually insignificant. The risk of having a low temperature toughness which does not conform to the EN 10219 requirement of 40 J/$-20°C$ after welding is minor. Advances in steelmaking and hot rolling enable reliable and versatile Ruukki double grade S420MH/S355J2H cold-formed hollow sections for welded structures with good low temperature ductility, even in the cold-formed corner area.

REFERENCES

Dagg, H. M., Davis, K., Hicks, J. W. 1989. Charpy Impact Tests on Cold Formed RHS manufactured from Continuous Cast Fully Killed Steel. *Proceedings of the Pacific Structural Steel Conference, Australian Institute of Steel Construction, Queensland, Australia, 1989.*

Eichler, B., Feldmann, M., Sedlacek, G. 2012. Zähigkeitsdargebote kaltgefertigter Hohlprofile bei tiefen Temperaturen. *Stahlbau 81(2012), Heft 3, S. 181–189.*

Feldmann, M., Eichler, B., Kühn, B., Stranghöner, N., Dahl, W., Langenberg, P., Kouhi, J., Pope, R., Sedlacek, G., Ritakallio, P., Iglesias, G., Puthli, R. S., Packer, J. A., Krampen, J. 2012. Choice of steel material to avoid brittle fracture for hollow section structures. *EUR 25400 EN, European Commission Joint Research Centre Scientific and Policy Report for the evolution of Eurocode 3, 2012.*

Kosteski, N., Packer, J. A., Puthli, R. S. 2003. Notch Toughness of Cold-formed Hollow Sections. *Report 1B-2/03, CIDECT, Geneva, Switzerland, 2003.*

Packer, J.A. & Chiew, S.P. 2010. Production standards for cold-formed hollow structural sections. Young Ben (ed.), *Tubular Structures XIII–13th International Symposium on Tubular Structures (ISTS13), Hong Kong 2010.*

Puthli, R. S., Herion, S. 2005. Welding in Cold-Formed areas of rectangular Hollow Sections. *Report 1A-1/05, CIDECT, Geneva, 2005.*

Saarela, M., Porter, D., Laitinen, R., Myllykoski, L., Uimonen, K. 1995. Modern High Strength Structural Steel Plates for Simpler Fabrication and Improved Reliability, *Nordic Steel Construction Conference 1995.*

Soininen, R. 1996. Fracture behaviour and assessment of design requirements against fracture in welded steel structures made of cold formed hollow sections. *Lappeenranta University of Technology, Research Papers, 1996*

Stranghoener, N., Krampen, J., Lorenz, C. 2012: Impact Toughness Behaviour of Hot-Finished Hollow Sections at Low Temperatures. *Proceedings of the Twenty-second (2012) International Offshore and Polar Engineering Conference, Rhodes, Greece, June 17–22, 2012.*

Ritakallio, P.O. 2010. Ruukki Cold-formed Hollow Sections – Grade S355J2H – Random Samples. *Private communication, Rautaruukki Corporation, Test data, April 12, 2010.*

Ritakallio, P. O. 2012. Coldformed high-strength tubes for structural applications. *Steel Construction, 5 (2012), No. 3, pp. 158-167.*

Ritakallio, P. O. & Bjork, T. 2014. Low temperature ductility and structural behaviour of cold-formed hollow section structures; the progress during the last two decades. *Steel Construction, 7 (2014), No. 2, pp. 107-115.*

Tubular Structures XV – Batista, Vellasco & Lima (eds)
© *2015 Taylor & Francis Group, London, ISBN 978-1-138-02837-1*

Considerations in the design and fabrication of tubular steel transmission structures

R.M. Slocum

Vice President of Engineering, Trinity Meyer Utility Structures, LLC

ABSTRACT: Tubular steel transmission structures enjoy widespread use throughout the United States and Canada for support of high voltage electrical transmission lines. With their smaller footprint, the tubular steel transmission structures require less area, less vegetation maintenance, less site disturbance, fewer spoils and overall, they are a more environmentally responsible option when compared to the traditional steel lattice transmission towers. Due to this and other inherent advantages, there has been a dramatic increase in the demand for tubular steel transmission structures on larger projects, with increased structure heights and heavier loads. In this paper and presentation, the audience will hear about the opportunities and challenges encountered in the design and fabrication of the various types of tubular steel transmission structures.

1 INTRODUCTION

It would be overly ambitious and a bit presumptuous for one person to attempt to record the entire history and completely detail all the methods for the design and fabrication of tubular steel transmission structures. Therefore, this paper will attempt a less ambitious endeavor. The primary goal of this paper is to introduce the audience to tubular steel transmission structures and formulate the real possibility that tubular steel transmission structures are a viable alternative to steel lattice structures and that they can possibly improve the quality of life, reduce the environmental impact and promote a more sustainable approach to transmission line design.

Legend has it, the history of tubular steel transmission structures in the United States began in 1958 when a man named Roy Meyer, owner of Meyer Machine Shop, decided to use his ingenuity and craftsmanship to design and fabricate a tapered steel pole to support electrical wires. Essentially, he cut a piece of steel plate into a trapezoidal shape, made a dozen 30 degree bends and welded the seam to make a tubular steel pole, then welded various connections and attachments, and sold it to a local utility. Soon after, Meyer Industries and the steel utility pole business were born.

Today, tubular steel transmission structures can be designed in a variety of shapes and sizes from small distribution and light-duty poles to 300 ft tall frames (91 meters) that support 345 kV, 500 kV and even 750 kV high-tension electrical lines. A project may consist of a single structure or hundreds, even thousands, of structures.

A basic tubular steel transmission structure is comprised of a tapered tower shaft (with or without base plate), horizontal members called arms that support the wires, splice connections, arm connections, various lifting, climbing and working attachments. Structures over 75 ft (22.8 m) in height are typically comprised of multiple sections that are assembled at the site. The two types of splice connections that connect the sections together are bolted flange plates and slip joints. Bolted flange plates consist of a flat plate on the bottom of the upper section that is bolted to the top plate of the lower section. Slip joints are a telescopic joint connection in which the upper section has a slightly larger diameter than the lower section. The upper section slides over the lower section and hydraulic jacks apply a force to seat the connection.

For large projects, structures are grouped by structure type and loading. Tangent structures are the most common type with lines running parallel from one structure to the next. An Angle structure has the line ahead of the structure running at an angle to the line behind the structure. A Dead-End structure is designed to prevent progressive collapse or a cascading effect on the system. Dead-End structures are strategically placed along the line to mitigate risk to humans and extraordinary damage to the transmission line system.

2 DESIGN CONSIDERATIONS

2.1 *General*

Typically, the design of the transmission pole is performed by an engineer employed by the fabricator and it is designed based on specific loading and performance criteria defined by the transmission line engineer. The transmission line engineer is ordinarily trained as an electrical engineer and usually not as well versed in structural engineering. Also, many of the products, connections and steel members have been researched and developed by the fabricator's engineers and hence, they have more expertise and a better understanding of the design and behavior of these structures. That being said, the owner's engineer typically has a better understanding of the specific site issues and performance concerns. Effective coordination between these engineers is essential to the success of a project.

The primary concerns in the design are life safety and meeting the required performance criteria. However, manufacturing limitations, shipping limitations, site conditions, material availability, costs, ease of construction, and aesthetics may all have some influence on the final design of the members and connections.

2.2 Loading/Standards

For any structure design, the obvious first step is to establish the required loading and design criteria. In the US, the standard used by most transmission line engineers is the NESC 2007 standard (IEEE, 2007), National Electrical Safety Code, IEEE 2007 (commonly referred to as "NESC"). More specifically, section 25 is used to determine general loading requirements for wind and ice loading. In addition, structures are typically designed for line tensions, constructions loads, wire tensions, wire weights, broken wire conditions (to prevent progressive collapse or cascading effects of the system) and second order effects. Furthermore, the transmission line engineer is responsible to specify any special loading or performance criteria and this is typically communicated on the project drawings or specifications. It is interesting to note that seismic loading requirements are clearly omitted in the NESC. In fact, the NESC specifically mentions that as long as the requirements of the code are met, then the structure will have sufficient capability to resist earthquake ground motions. Another omission is any reference to wind induced vibration on conductors or the structure. Under very specific, localized site conditions with open terrain, steady state, lamellar wind flow, the conductors and structures may be susceptible to wind induced vibration. While the structure is typically dampened by the weight of the conductor, special dampeners are usually installed on the conductor to mitigate the risk of conductor galloping.

While NECS is the standard generally used to develop the loading requirements, the design requirements and equations for the design of tubular steel structures come from the ASCE 48-11 standard (ASCE, 2012), Design of Steel Transmission Pole Structures, ASCE 2012 (commonly referred to as "ASCE 48"). The ASCE 48 is more than just a design standard as it also provides requirements and guidance on fabrication, assembly and testing of the steel transmission pole structures. In essence, ASCE 48 is a design manual and a code of standard practice that is widely specified throughout the US and Canada for transmission line projects.

2.3 Structure analysis and design

The ASCE 48 specifies that nonlinear elastic analysis be used to design the structures and consideration for second order effects. In the US and Canada, transmission structures are designed to nominal yield strength of the material without additional safety factors. The grades of steel used have yield strengths of 65 ksi (448 MPa), 60 ksi (414 MPa) or 50 ksi (345 MPa). While some engineers have explored the use of higher strength steels, difficulties in fabrication and welding have prevented their widespread use.

Since the geometry of the tubular members is unique (8, 12 and 16 sided formed tubes), the connection details are unique. ASCE 48 does not specify connection design methodologies but instead provides minimum fabrication or performance requirements for the connection. Therefore, finite element analysis along with full scale testing is often employed by the fabricator's engineer to develop proprietary designs for various connections. These proprietary designs are usually well suited to the fabricator's specific manufacturing capabilities in order to ensure safe and efficient designs.

The typical limit states for member design are tension, compression, shear, torsion, and buckling. However, since these structures are primarily single cantilevered poles or simple frames, in nearly every case, local buckling due to the combined axial and bending compressive stress will control the design. Therefore, this paper will focus on what is commonly referred to as "beam-column" members and specifically width to thickness (w/t) ratios. Based on research conducted by the Electrical Power Research Institute (EPRI), three different sets of equations to determine buckling capacity were developed. Equations (1) and (2) are used for 4, 6, and 8 sided members, Equations (3) and (4) for 12 sided members and Equations (5) and (6) for 16 sided members.

$$Fa = Fy \quad \text{when} \quad w/t \le 260\Omega/\sqrt{Fy} \tag{1}$$

$$Fa = 1.42Fy[(1-0.00114/\Omega)(w/t \times \sqrt{Fy})]$$
$$\text{when} \quad 260\Omega/\sqrt{Fy} < w/t \le 351\Omega/\sqrt{Fy} \tag{2}$$

$$Fa = Fy \quad \text{when} \quad w/t \le 260\Omega/\sqrt{Fy} \tag{3}$$

$$Fa = 145Fy[(1-0.00129/\Omega)(w/t \times \sqrt{Fy})$$
$$\text{when} \quad 240\Omega/\sqrt{Fy} < w/t \le 374\Omega/\sqrt{Fy} \tag{4}$$

$$Fa = Fy \quad \text{when} \quad w/t \le 215\Omega/\sqrt{Fy} \tag{5}$$

$$Fa = 1.42Fy[(1-0.00137/\Omega)(w/t \times \sqrt{Fy})]$$
$$\text{when} \quad 215\Omega/\sqrt{Fy} < w/t \le 412\Omega/\sqrt{Fy} \tag{6}$$

Although rarely used in the transmission structure industry, ASCE 48 also has equations for elliptical tubular members and round members. For guyed structures, ASCE 48 specifies that the tension force in the guy shall not exceed 65% of the rated breaking strength of the guy cable.

2.4 Base plate design

As mentioned previously, there is no industry required design methodology for base plate connections. Therefore, it is essential that design methods are thoroughly researched and tested prior to be placed into service. A sound approach might include finite element analysis to verify bending plane assumptions along with full scale testing to validate both the assumptions and FEA models. In order to be more efficient and reduce lifting and shipping weights, base plates are typically cut in a ring shape with an inside circle of steel removed. This is also necessary for venting and drainage on galvanized structures. Design methods need to account for the removed inside portion of steel as the material inside the tower wall provides added stiffness for the base plate. Maximum bending stress cannot be the only consideration in the design of the base plate. Testing has demonstrated that small, localized deformations can redistribute stresses into the tower wall leading to pre-mature buckling of the tower wall long before yielding of the base plate. With the lack of redundancy inherent in these structures, a sound base plate design is critical to the performance and safety of the structure.

2.5 Connections

In addition to base plate connections, splice connections and arm connections are the other primary connections on a tubular steel transmission structure. As mentioned previously, there are two types of splice connections; bolted flange plates and slip joints.

Slip joint requirements are provided in the ASCE 48 standard. However, the design approach is not based on limit states rather it is based on the geometry of the section. Essentially, the requirement is that the minimum nominal overlap dimension of the top section relative to the bottom section be equal to 1.5 times the maximum inside diameter of the upper section. ASCE 48 requires that the fabricator account for tolerances to ensure the minimum overlap.

Bolted flange plates look and behave similar to base plates and similarly, there are no industry standard design methodologies for these connections. A similar approach to base plate design is typically developed and tested. Arm connections are also based on proprietary research and some designs even have patented components. The ASCE 48 standard provides some information on shear, rupture, bolt spacing and edge distance but development of a complete design methodology is left to the fabricator's engineering and research and development resources.

2.6 Miscellaneous details

In addition to connections for assembling and supporting the structure, there are numerous details and other attachments that include:

- Grounding and Signage attachments
- Details for Climbing Hardware
- Fall Protection Supports
- Lifting and Handling details

The miscellaneous details are typically a combination of standard details developed by the fabricator and requirements specified by the transmission line engineer.

2.7 Research and testing

Full scale testing of components or complete structures is occasionally requested by the owner or transmission line engineer in order to validate a design or provide additional assurance for critical designs. The ASCE 48 standard provides minimum requirements necessary to perform these tests. ASCE 48 also allows different design values than prescribed in the standard if the values are substantiated by testing. Therefore, due to this and a lack of industry funding for research and testing, the bulk of the industry research and testing is proprietary and has been developed by fabricators and their engineering and research and development departments.

While some in the industry might see the lack of standard design methodologies as an obstacle to competition, a more objective view is that this leads to more innovation and improved competition. It challenges engineers to be creative and develop a safe and cost effective design approach rather than making simple computations based on standard formulas from a code. In this challenge lies the opportunity to truly develop the best approach for each project.

For new designs on traditional lattice tower structures, typically the complete structure must be proof tested in order to validate the design. The process of creating a new lattice tower design and testing it can take months. However, because of previous testing and standard design equations, an entire project of tubular steel structures can be designed in a matter of days. This time savings is another great advantage of tubular steel transmission structures.

3 DESIGN CONSIDERATIONS

3.1 General

There are many variables that can affect the design of the foundation including but not limited to the following:

- Soil Conditions
- Site Access
- Base Reactions/Loads
- Construction Expertise and Equipment
- Cost
- Environmental Sensitivity

The foundation design is heavily reliant on the geotechnical survey at the site. Soil borings are usually taken to provide the foundation engineer with the information necessary to assess the soil conditions. The more soil borings, the more complete the information and this will result in a more efficient design.

The two most common foundation types used for tubular steel transmission structures are direct embed and concrete piers.

3.2 Direct embed

This is the simplest method of anchoring a structure to the ground and is widely used for small distribution poles to 150 ft (46 meters) tall transmission structures. The base reactions are resisted primarily by soil bearing and friction. A hole is excavated to a depth determined by the foundation engineer, the structure is set and properly aligned, the hole is then backfilled and compacted around the structure. This foundation type is cost effective, allows for some adjustment while setting the structure and does not require highly skilled labor. It is limited to lower base reactions and loads, highly dependent on soil conditions, it requires the removal or disposal of spoils and is not as well suited for very loose or wet soil conditions.

3.3 Concrete pier with anchor bolts

Designing a concrete pier with anchor bolts for a tubular steel transmission structure requires close coordination with fabricator's engineer designing the steel structure. Typically, the fabricator's engineer will determine the required anchor bolt diameter and length. The foundation engineer designs the concrete pier and determines the diameter, depth and re-bar configuration for the pier. A smaller steel structure diameter may result in a higher structure cost. However, the result would be a smaller concrete pier diameter. There are many parameters that can affect the structure and concrete foundation costs. Therefore, close coordination between the foundation engineer and fabricator's engineer can result in a lower total installed cost.

3.4 Other foundation types

At times, site access and environmental conditions require a more unique foundation solution. For example, marshes, wetlands and swamps are not conducive for excavation or concrete placement. In these conditions, driven caissons are often a very good alternative to the direct embed or concrete piers. The caissons can be driven with a traditional hammer or a vibratory hammer. They do not require the removal of spoils and can even be driven with the use of a helicopter which eliminates the need for bringing heavy equipment to the site and causes minimal disturbance of a sensitive environment.

Another foundation type that reduces the site impact is the screw pile or helical pier. Essentially, a set of helical piers are screwed into the soil in a pattern and a steel transition or adapter piece is bolted to the helical piers. Trinity Meyer Utility Structures has developed and tested several different steel transition configurations to accommodate a variety of loading conditions.

In mountainous or rocky terrain, micro-piles are another option that can be used to attach a tubular steel structure. Sets of micro-piles are driven or screwed into the ground and a steel transition piece is attached to micro-piles allowing placement of the tubular steel structure.

The shape and reduced footprint of the tubular steel transmissions structures provides flexibility for the use of various foundations options.

4 FABRICATION

4.1 General

The fabrication of these tubular steel transmission structures is entirely unlike any other structural steel fabrication for other industries. There is a significant amount of bending, forming and welding that goes into the making of each member. This type of fabrication is not for the uninitiated and it takes a great deal of skill, craftsmanship and experience to fabricate these types of structures that are expected to remain in service for 50 years or more and are constantly exposed to all kinds of environments.

4.2 Shaft fabrication

To create the tubular steel tower or arm shaft, a trapezoidal shape is cut from a steel plate. The steel trapezoids (or traps) are pressed and bent to form an 8, 12, or 16 sided tubular shape. To complete the tube, the seam is welded the entire length. For large diameter structures, due to bending and forming limitations, half shapes and occasionally, quarter shapes are created and then welded together at the seams. Steel mills supply steel plate material to the fabricator within specified tolerances. There are also thermal effects and tolerances during the plate cutting process. Therefore, by the time the trap reaches the press, they already may need to make adjustments in the forming process. The shape prior to seam welding is critical to the overall tolerances for the member because it is difficult to make adjustments in seam welding to account for the other tolerances. Also, an illformed tubular shape may result in difficulty in welding and poor weld quality. Two common fabrication issues are warping and twist. Fabrication tolerances are set to minimize the amount of warping or twist in a tube section. However, it is also common to make adjustments in the lay out and welding of attachments to accommodate these tolerances in order to mitigate potential assembly issues in the field. Clearly, the fabrication of tubular steel transmission structures is not a simple assembly line process of punching and cutting. It requires skilled labor along with robust and proven fabrication and quality control procedures.

4.3 Welding

There may be nothing more critical to fabrication of any steel structure than welding. As mentioned above, seam welding is one of the initial welding processes performed on the structure. Seam welding is typically an automated welding process using the submerged arc welding process (SAW). In addition to seam welding, base plate, flange plate and most arm shaft connections employ complete joint penetration welds (CJP). Due to press length limitations, at times, it is necessary to weld tube sections together to form one piece. These splice welds called circumferential welds (or C welds) are also CJP welds. Various attachments for climbing, lifting and fall protection are welded to the shaft primarily with fillet welds.

The welding requirements and specifications for tubular steel transmission structures are primarily based on the Structural Welding Code – Steel D1.1 (AWS, 2011) (commonly referred to as AWS D1.1). Due to the various grades and types of steel, most weld procedures have to be qualified. Qualification and testing for welders and weld procedures are performed according to AWS D1.1.

For weld inspection, due to the unique geometry and details, AWS D1.1 is used as a guideline with enhanced methods for the tubular steel transmission structure industry. For visual inspection, the requirements in AWS D1.1 for static loading are generally used. For other methods of weld inspection, the fabricator should have rigid and documented weld inspection procedures. Based on decades of weld inspection experience on tubular steel transmission structures, Trinity Meyer Utility Structures has developed inspection procedures to detect structurally significant defects to ensure reliable welds for the required industry performance criteria. Ultrasonic testing (or UT inspection), has been determined to be the most effective at detecting significant weld flaws. The UT inspection methods and criteria in AWS D1.1 may be used as a guideline but adjustments must be made in order to account for the unique geometry and details associated with tubular steel transmission structures. Magnetic particle inspection (or MT) and dye penetrant inspection (or PT) are only used as necessary to verify rejectable surface flaws identified by visual inspection or UT. Other forms of non-destructive inspection have not proven practical or beneficial in identifying structurally significant weld defects in these types of steel structures.

One of the major steel fabrication challenges is mitigating the risk of cracking during the galvanizing process. For the tubular steel transmission structure industry, the highest risk of post-galvanized steel cracking is in the area of the toe of the base plate and flange plate welds. The base plate and flange plate welds are CJP welds in a "T-Joint" configuration. Due to this inherent, and unavoidable, geometry, these joints are highly restrained and have typically accumulated some residual stress. In absence of galvanizing, the risk of cracking or other performance issues is negligible. However, galvanizing processes introduce the opportunity for hydrogen or liquid metal embrittlement which can lead to cracking at the toe of the weld (toe cracks). In an effort to mitigate the risk

of post-galvanizing toe cracks, the ASCE 48 standard recommends that all base plate and flange plate welds be inspected by UT after galvanizing. The most obvious way to mitigate the risk is to avoid galvanizing. Other finishes, such as painted, metalizing and weathering steel are available. However, it is worth noting that, historically, galvanized tubular steel transmission structures have had very few issues.

4.4 Materials

There are different types and grades of steel used in the design and fabrication of tubular steel transmission structures. However, one common requirement throughout the industry is toughness. Toughness is defined as the steel's ability to resist fracture. Toughness is typically measured by the Charpy V-notch test and denoted in foot-pounds (joules) at a specific temperature. In the US and Canada, tubular steel transmission structures are exposed to a variety of temperature extremes and the industry requirement for steel toughness is 20 ft-lbs (27 J) at $-20°$ F ($-29°$C).

4.5 Finishes and coatings

Tubular steel transmission poles are constantly exposed to the environment and often partially buried in the ground. They require some type of protection in order to prevent corrosion. On most projects, there are few options for locating a transmission line and these structures are placed in some very harsh conditions such as extreme cold in mountains, extreme heat in the desert, coastal environments and wetlands.

As discussed previously in section 3.2, tubular steel transmission structures may be embedded into the ground to act as the foundation support. With the steel surrounded by soil and moisture, corrosion can be a concern. Especially, since the steel is in the ground, the corrosion would not visible. However, at a certain depth below the surface, there is not sufficient oxygen for corrosion to take place. Therefore, corrosion protection for the steel is only necessary for the first two or three feet (0.9 m) below the surface. This type of protection is typically referred to as "below grade coating".

A below grade coating is applied much like paint with a sprayer. Surface preparation and cleanliness are critical to ensure that the coating adheres to the steel. The steel is blasted to clean and create a rough surface profile for adherence of the coating. The chemistry of the coating is critical to the performance and durability. It should be nearly 100% solids with negligible solvents. Solvents can evaporate over time and cause tiny voids in the coating. These tiny voids allow moisture to enter and degrade the coating. In past years, there have been some issues with premature failure and peeling of the below grade coatings. This led Trinity Meyer Utility Structures to developed its own brand of coating called Meyerclad with 100% solids (no solvents), that is easier to apply, has UV resistance and excellent flexibility and durability.

Another option for below grade protection is to attach a ground sleeve to the structure. A ground sleeve is a sacrificial steel collar, typically 3/16" to 1/4" thick and 3 feet (0.9 m) long, that is welded to the steel pole and is located near the ground line. The premise is that, over time, the ground sleeve will slowly corrode and experience a loss of material while preserving the primary steel in the tubular steel structure.

Above the ground, the different types of finishes include galvanizing, metalizing, painting and weathering steel. Galvanizing and weathering steel are the most common finishes. Metalizing is considered an reasonable alternative to galvanizing when the size of the structure exceeds practical galvanizing tank limits. Painting requires regular maintenance and does not have as long of a service life as galvanizing or weathering steel. Therefore, the primary focus of this paper will discuss galvanizing and weathering steel.

Galvanizing provides a cost effective and long lasting protective finish for tubular steel transmission structures. It is readily available and widely used in both tubular steel and steel lattice transmission structures. Over time, the galvanizing finish becomes a dull gray finish that some may consider more aesthetically pleasing. However, there are also some concerns with the galvanizing process. One of the primary and immediate concerns is the effects of galvanizing on the fabricated steel. The galvanizing process uses an acid bath to clean the steel, the steel is rinsed and then dipped into a bath of molten zinc to create the finish. Thermal stresses from the galvanizing process increase the residual stresses that may cause cracking problems with copes, bends and welds. Elements such as tin, lead and bismuth are usually added to the zinc bath in order to promote drainage. While reducing the drainage time, this improved drainage can also increase the probability of zinc finding its way into tiny discontinuities and causing cracks. Therefore, fabricators use enhanced inspection procedures for tubular steel transmission structures with a galvanized finish to mitigate the risks of cracked sections being shipped and put into service.

In addition, welding on an already galvanized structure can be problematic. Welding destroys the galvanized coating on the outside and potentially the inside the pole. While the outside can be repaired with a zinc rich paint, there's little that can be done to the inside of the pole and it can be vulnerable to corrosion. There's also a concern with fumes and welder safety along with potential for the galvanized coating to contaminate the weld. Therefore, options for field repairs or alterations are limited on galvanized structures.

Historically, galvanized structures have an excellent record of performance and service. That being said, with the growing awareness of sustainability and impact to the environment, for some, the galvanizing process may be a cause for concern. In general, galvanizing tanks must be kept heated day and night throughout the year which consumes large amounts of energy. The galvanizing process uses caustic chemicals which evaporate and may affect the air quality

and the disposition of these chemicals must also be considered. There's also the potential for zinc to leach into the soil and ground water. Small amounts of zinc exist naturally in soils, however, larger amounts from run-off or leaching may be toxic to fish and wild life.

While galvanizing offers a very robust protective coating, weathering steel provides a more environmentally-friendly finish and excellent protection. Weathering steel has a proven and effective protective finish even in very harsh environmental conditions. It does not use applied chemicals in order to create the protective coating. Essentially, weathering steel oxidizes or rusts on the surface to seal and protect the steel below the surface. Initially, ASTM A588 was used as the primary grade of weathering steel for tubular steel transmission structures. About 25 years ago, engineers at Meyer Industries (now, Trinity Meyer Utility Structures) collaborated with the ASTM committee to develop a higher strength weathering steel, ASTM A871 Grade 65, specifically for the tubular steel transmission structure industry. Today, ASTM A871 grade 65 steel is considered the industry standard. Weathering steel relies on the influence of weather and wet-dry cycles to form the protective finish. Therefore, one of the most important considerations in the design is to develop details that allow drainage and exposure to the air for moisture to evaporate. Details that allow trapped moisture can lead to corrosion issues and undermine the structure's integrity. In addition, the bottom of the tube must be sealed to prevent the accumulation of water inside the tube section. With proper attention to details, weathering steel structures will have excellent resistance to corrosion for decades and offers a more environmentally friendly and low maintenance option when compared to painting or galvanizing.

5 ENVIROMENTAL CONSIDERATIONS

First and foremost, tubular steel transmission structures are efficient and durable. Unlike buildings or bridges, transmission structures are designed to the full yield strength of the material and they have little or no redundancy. Historically, tubular steel transmission structures have performed well in extreme events such as hurricanes and ice storms. Often, when wood poles and lattice steel towers have collapsed, nearby tubular steel transmission structures are left standing. This durability allows for a long life cycle with little or no maintenance.

Secondly, the steel industry is one of the world leaders in the use of recycled material with a recycling rate of 98%. Wood poles are treated with chemicals and generally have limited recycled use. In addition, the removal trees can add to the problems with deforestation. In general, concrete doesn't enjoy the same reputation for recycling as steel. The inherent strength to weight ratio of steel versus concrete or wood may be one of the reasons that it has earned a reputation of being more environmentally friendly because it requires less material. With a transmission line project comprised of hundreds or thousands of structures, using less material and more recycled material is a more sustainable approach.

Finally, transmission lines are located in cities, suburbs, farms, forests, mountains and wetlands. When compared to lattice steel transmission structures, tubular steel has a much smaller foot print which allows for better use of the land and less disturbance of the environment. This translates into fewer trees removed, more land available for farming and housing or simply more land in its natural state.

6 SUMMARY

While elegant and seemingly simple in their configuration, tubular steel transmission structures require engineering expertise, care and craftsmanship in fabrication along with detailed coordination in order to ensure a successful project. However, when compared to other materials and types of transmission structures, they may provide a more cost effective, sustainable and long lasting approach to transmission line design.

Tubular steel transmission structures are durable in their design, have a long life cycle and often perform well in extreme weather events. Engineering expertise, sound choices in the grades of steel used, types of finishes, mature fabrication and quality control practices, along with skilled labor and welders are essential to their performance.

The US, Canada and other developed countries have recognized the long term and sustainable benefits of using tubular steel transmission structures on a wider array of transmission line projects. The smaller foot print, variety of configurations, finishes and more aesthetically pleasing options make them more versatile than the traditional lattice steel structures. It is truly an option that should be considered across the globe by all transmission line owners and engineers.

REFERENCES

ASCE 2012. *Design of Steel Transmission Pole Structures, ASCE/SEI 48-11*. American Society of Civil Engineers, Reston, VA.

IEEE 2007. *National Electric Safety Code*. Institute of Electrical and Electronics Engineers Inc., New York, NY.

AISC 2011. *Steel Construction Manual 14th Edition*. American Institute of Steel Construction, Chicago, IL.

AWS 2011. *Structural Welding Code – Steel D1.1 2010*. American Welding Society, Miami, FL.

Tubular Structures XV – Batista, Vellasco & Lima (eds)
© 2015 Taylor & Francis Group, London, ISBN 978-1-138-02837-1

The continuous strength method for circular hollow sections

C. Buchanan & L. Gardner
Imperial College London, London, UK

A. Liew
Swiss Federal Institute of Technology (ETH), Zurich, Switzerland

ABSTRACT: Circular hollow sections (CHS) are widely used in a range of structural engineering applications. Their design is covered by all major design codes, which currently use elastic, perfectly-plastic material models and cross-section classification to predict cross-section compressive and flexural resistances. Experimental data for stocky sections show that this can result in overly conservative estimates of cross-section capacity. The continuous strength method (CSM) has been developed to reflect better the observed behaviour of metallic materials, with a continuous relationship between cross-section slenderness and deformation capacity, and a strain hardening material model. In this paper, the CSM is extended to cover the design of structural steel, stainless steel and aluminium CHS, underpinned by and validated against 519 stub column and bending test results. Comparisons with the test results show that the CSM offers more accurate and less scattered predictions of axial and flexural capacities than existing design methods.

1 INTRODUCTION

Circular hollow sections (CHS) have been manufactured and used in structures since the early 1800s as columns, beams, tension members and truss elements (Dutta 2002). They have become increasingly attractive to designers due to their aesthetic appearance and their benefits over open sections such as superior torsional resistance, bi-axial bending resistance, reduced drag and loading in a fluid, ability to be filled with concrete to form a composite section and their reduced maintenance requirements with a smaller external area exposed to corrosive environments (Dutta 2002). CHS are primarily thin-walled structural elements, and therefore local buckling, whether prior or subsequent to material yielding, is a primary consideration in their design.

1.1 Traditional CHS design methods

Current design codes use the concept of cross-section classification to separate circular hollow sections into discrete classes depending upon their susceptibility to local buckling. Four classes of cross-section are considered in EN 1993-1-1 (2005) and BS 5950-1 (2000) for structural steelwork, EN 1993-1-4 (2006) for stainless steel and EN 1999-1-1 (2007) for aluminium. In bending, class 1 cross-sections can reach and maintain their full plastic moment capacity M_{pl} with suitable rotation capacity for plastic design. Class 2 cross-sections are also capable of reaching their full plastic moment capacity but with a limited rotation capacity.

Class 3 cross-sections are unable to reach their plastic moment capacity due to local buckling and their bending capacity is limited to the elastic moment capacity M_{el}. Class 4 cross-sections experience local buckling before reaching their elastic moment capacity, and are typically referred to as slender. In terms of axial resistance, the class 3 limit separates the non-slender cross-sections that are fully effective in compression (i.e. classes 1-3) from those that fail by local buckling before reaching their yield load (i.e. class 4). There is no equivalent to class 2 cross-sections in the AISC 360 (2005) and AS 4100 (1998) structural steel codes. These traditional design methods also limit the maximum stress in the cross-section to the yield strength f_y, neglecting the beneficial strain hardening effects in metallic materials. Experimental results have shown that cross-section classification and limiting the maximum stress to the yield stress can be overly conservative in estimating the resistance of stocky cross-sections (Gardner 2008). It is therefore apparent that there are structural efficiency improvements to be sought over existing design methods for CHS.

1.2 The continuous strength method

The continuous strength method (CSM) has been developed in recent years to reflect better the observed characteristics of metallic structural elements. Cross-section classification is replaced with a continuous relationship between cross-section slenderness and deformation capacity (referred to in Section 2.5 as the base curve), reflecting the continuous nature of cross-section capacity varying with local slenderness. A strain hardening material model is also adopted, representing the behaviour seen in material tests, with an increase in strength above the yield strength under plastic deformation.

The CSM has been developed for structural steel (Gardner 2008, Gardner et al. 2011, Foster 2014), stainless steel (Afshan & Gardner 2013) and aluminium (Su et al. 2014) plated cross-sections, such as I-sections, square hollow sections (SHS) and rectangular hollow sections (RHS). The previous work has shown that the CSM predicts enhanced capacities over existing methods; for example, in the case of stainless steel, average enhancements in compressive and bending resistances of 12% and 19% respectively were found (Afshan & Gardner 2013).

The natural progression is to extend the application of the CSM to circular hollow sections, which is the focus of the present paper, and the development process is described herein.

2 CSM CHS EXTENSION

The extension of the CSM to CHS requires identification of the yield slenderness limit (i.e. the local slenderness limit below which significant benefit from strain hardening can be derived), formulation of the CSM base curve and selection of appropriate material models and resistance expressions.

2.1 Cross-section slenderness

The local cross-section slenderness $\bar{\lambda}_c$ is defined in non-dimensional form by Equation 1,

$$\bar{\lambda}_c = \sqrt{f_y / \sigma_{cr}} \tag{1}$$

where f_y is the material yield strength and σ_{cr} is the elastic critical buckling stress, which for a CHS in compression is calculated using Equation 2,

$$\sigma_{cr} = \frac{E}{\sqrt{3(1 - \upsilon^2)}} \frac{2t}{D} \tag{2}$$

where E is the Young's modulus, υ is the Poisson's ratio, D is the outer diameter of the CHS and t is the thickness. The elastic critical buckling stress in bending is also determined using Equation 2. Donnell (1934) and Timoshenko & Gere (1961) suggest taking the local buckling stress in bending to be 1.4 times that in compression based on experimental results. Other literature sources recommend a factor of 1.3 be adopted, such as Gerard & Becker (1957); however this is based upon an incomplete interpretation of the findings of Flügge (1932). Seide & Weingarten (1961) determined analytically that the maximum critical stress in bending is equal to the critical compressive stress, which was confirmed by Reddy & Calladine (1978) and noted by Rotter et al. (2014). Differences

Table 1. Number of CHS specimens used to extend the CSM.

Material	Yield slenderness limit identification		Base curve definition		CSM and Eurocode comparison	
	C	B	C	B	C	B
Hot-finished structural steel	8	14	5	2	8	14
Very high strength structural steel	20	12	1	1	1	1
Cold-formed structural steel	117	34	44	9	43	12
Stainless steel	95	12	41	1	51	3
Aluminium	119	88	50	43	25	43

also exist between international design codes in their treatment of compression and bending. EN 1993-1-1 (2005) and EN 1999-1-1 (2007) utilise the same class 3 limits for both compression and bending, in contrast to the majority of existing design codes (Gardner et al., 2014). Further investigation is required, although assuming identical critical elastic buckling stresses in compression and bending is conservative, and is adopted here.

2.2 CHS experimental database

A dataset of 519 experimental CHS axial (C) and four-point bending (B) results from the literature has been collated for hot-finished, very high strength and cold-formed structural steel, stainless steel and aluminium. Very high strength structural steel has a typical yield stress f_y around 1300 MPa, compared with less than 500 MPa for traditional structural steel (Zhao, 2000). The number of specimens used in the following subsections is shown in Table 1.

The hot-finished structural steel results are taken from Giakoumelis & Lam (2004), Liew & Xiong (2010), Ochi & Choo (1993) and Starossek & Falah (2009) for axial compression and Gresnight & van Foeken (2001), Rondal et al. (1994), Sedlacek et al. (1995) and Sherman (1976) for bending. The very high strength structural steel results are from Jiao & Zhao (2003), Jiao & Zhao (2004) and Zhao (2000). The compressive cold-formed structural steel test results are from Chen & Ross (1977), Elchalakani et al. (2002b), Jiao & Zhao (2003), Johansson & Grylltotft (2002), Kamba (1996), Ochi & Choo (1993), O'Shea & Bridge (1997), Prion & Birkemoe (1992), Sakino et al. (2004), Schmidt (1989), Teng & Hu (2007), Toi et al. (1986), Toi & Ine (1988), Tutuncu & O'Rourke (2006), Wei et al. (1995), Xiao et al. (2005), Yu & Teng (2010) and Zhao et al. (2010). The cold-formed structural steel bending results have been taken from Elchalakani et al. (2002a), Gresnigt & van Foeken (2001), Guo et al. (2013), Haedir et al. (2009), Jirsa et al. (1972),

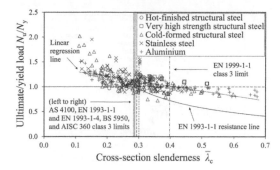

Figure 1. N_u/N_y varying with $\bar{\lambda}_c$.

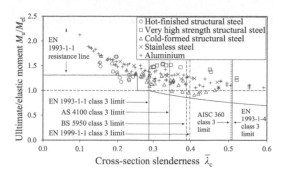

Figure 2. M_u/M_{el} varying with $\bar{\lambda}_c$.

Seica et al. (2006) and Sherman (1976). The stainless steel axial dataset was collated from Bardi & Kyriakides (2006), Burgan et al. (2000), Gardner & Nethercot (2004), Kuwamura (2003), Lam & Gardner (2008), Paquette & Kyriakides (2006), Rasmussen (2000), Rasmussen & Hancock (1993a), Talja (1997), Uy et al. (2011), Young & Hartono (2002) and includes additional finite element modelling carried out by the authors. The stainless steel bending dataset was taken from Burgan et al. (2000), Kiymaz (2005) and Rasmussen & Hancock (1993b). The aluminium dataset was collected from Zhou & Young (2009), Zhu & Young (2006a), Zhu & Young (2006b) and includes currently unpublished experimental and finite element data generated at The University of Hong Kong.

2.3 Yield slenderness limit

The limiting local slenderness that delineates the transition between slender and non-slender cross-sections needs to be defined. Above this limit there is no significant benefit from strain hardening with the cross-section locally buckling below the yield load or elastic moment. This limit is identified by plotting the ultimate capacity of the stub columns normalised by their yield load (N_u/N_y) against cross-section slenderness $\bar{\lambda}_c$, defined by Equation 1, as shown in Figure 1. A linear regression fit can then identify the limiting local slenderness where the ultimate axial load equals the yield load, which from Figure 1 is $\bar{\lambda}_c = 0.41$. The class 3 limits from current design codes are also plotted in Figure 1, and it can be seen that the identified limiting local slenderness is compatible with the aluminium EN 1999-1-1 class 3 limit; however it is above the existing structural steel and stainless steel class 3 limits. There is also substantial scatter in the stub column dataset. Consequently, a lower limit of $\bar{\lambda}_c = 0.3$ is proposed as this represents approximately a lower bound to the assembled dataset and is generally comparable with existing codes.

It is evident from Figure 1 that there are no clear discontinuities in the dataset and that limiting the maximum material stress to the yield stress is overly conservative for stocky cross-sections. For comparison, the ultimate bending moment normalised by the

elastic moment (M_u/M_{el}) is plotted in Figure 2 against the cross-section slenderness $\bar{\lambda}_c$. Similarly, by comparison with the EN 1993-1-1 (2005) resistance line there are no apparent discontinuities in bending capacity that cross-section classification would otherwise suggest and again limiting the material stress to the yield stress can lead to under-prediction of the ultimate cross-section capacity for stocky sections. The previous yield slenderness limit of $\bar{\lambda}_c = 0.3$ can also be applied for bending.

2.4 Normalised deformation capacity (strain ratio)

In the CSM, cross-section classification is replaced with a continuous relationship between local slenderness and deformation capacity. This deformation capacity is called the strain ratio ($\varepsilon_{csm}/\varepsilon_y$) and is defined as the strain at ultimate load normalised by the yield strain. The strain ratio is determined from stub column and four-point bending experiments, as described below.

2.4.1 Axial compression

For stub columns with a local slenderness below the yield slenderness limit ($\bar{\lambda}_c \leq 0.3$) and where the ultimate load exceeds the yield load ($N_u \geq N_y$), the strain ratio is expressed as a function of the strain at ultimate load divided by the yield strain ($\varepsilon_{lb}/\varepsilon_y$) (Eqs. 3 and 4). The strain at ultimate load ε_{lb} can be calculated from the initial specimen length L and the end-shortening δ_u at the ultimate load N_u. A 0.002 strain offset is subtracted for materials with a rounded stress-strain response to be compatible with the material models adopted in Section 2.6, as shown in Equation 3. For specimens that do not exceed the yield load ($N_u < N_y$) or that have slender cross-sections ($\bar{\lambda}_c > 0.3$), the strain ratio is determined as the ratio of the ultimate load to yield load (N_u/N_y), as in Equation 5.

For $\bar{\lambda}_c \leq 0.3$, $N_u \geq N_y$ and a rounded material response:

$$\frac{\varepsilon_{csm}}{\varepsilon_y} = \frac{\varepsilon_{lb} - 0.002}{\varepsilon_y} = \frac{\delta_u/L - 0.002}{\varepsilon_y} \tag{3}$$

For $\bar{\lambda}_c \leq 0.3$, $N_u \geq N_y$ and a sharply defined yield point:

$$\frac{\varepsilon_{csm}}{\varepsilon_y} = \frac{\varepsilon_{lb}}{\varepsilon_y} = \frac{\delta_u/L}{\varepsilon_y} \qquad (4)$$

For $N_u < N_y$ or $\bar{\lambda}_c > 0.3$:

$$\frac{\varepsilon_{csm}}{\varepsilon_y} = \frac{N_u}{N_y} \qquad (5)$$

2.4.2 Four-point bending

The strain ratio for four-point bending is similar in principle to axial compression, and is defined as a function of the maximum strain in the cross-section at the ultimate moment normalised by the yield strain ($\varepsilon_{lb}/\varepsilon_y$) for $\bar{\lambda}_c \leq 0.3$ and where the ultimate moment exceeds the elastic moment ($M_u \geq M_{el}$) (Eqs. 7 and 8). Under uniform bending, the strain can be determined as the product of the curvature κ and the distance from the elastic neutral axis y (Eq. 6); the curvature at the ultimate moment and the elastic moment are termed κ_u and κ_{el} respectively. The 0.002 offset is again subtracted for materials with a rounded stress-strain response as shown in Equation 7. If the cross-section is slender ($\bar{\lambda}_c > 0.3$) or the ultimate moment is less than the elastic moment ($M_u < M_{el}$), the strain ratio is taken as the ultimate moment normalised by the elastic moment (M_u/M_{el}) (Eq. 9).

$$\varepsilon = \kappa y \qquad (6)$$

For $\bar{\lambda}_c \leq 0.3$, $M_u \geq M_{el}$ and a rounded material response:

$$\frac{\varepsilon_{csm}}{\varepsilon_y} = \frac{\varepsilon_{lb} - 0.002}{\varepsilon_y} = \frac{\kappa_u y_{max} - 0.002}{\kappa_{el} y_{max}} \qquad (7)$$

For $\bar{\lambda}_c \leq 0.3$, $M_u \geq M_{el}$ and a sharply defined yield point:

$$\frac{\varepsilon_{csm}}{\varepsilon_y} = \frac{\varepsilon_{lb}}{\varepsilon_y} = \frac{\kappa_u y_{max}}{\kappa_{el} y_{max}} \qquad (8)$$

For $M_u < M_{el}$ or $\bar{\lambda}_c > 0.3$:

$$\frac{\varepsilon_{csm}}{\varepsilon_y} = \frac{M_u}{M_{el}} \qquad (9)$$

2.5 Proposed base curve

A base curve with the form of Equation 10 can be fitted to the experimental strain ratios derived from the axial and bending results from the literature (Table 1), as plotted in Figure 3. Equation 10 is consistent with previous implementations of the CSM, and is similar in form to the relationship between normalised elastic buckling strain $\varepsilon_{cr}/\varepsilon_y$ and local slenderness $\bar{\lambda}_c$

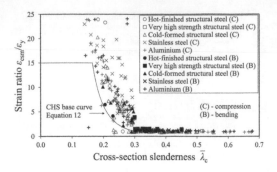

Figure 3.   CHS CSM base curve with experimental data.

(Eq. 11). The chosen base curve generally represents a lower bound to the dataset and passes through the class 3 (yield) slenderness limit previously identified (0.3, 1.0), resulting in Equation 12.

$$\frac{\varepsilon_{csm}}{\varepsilon_y} = \frac{A}{\bar{\lambda}_c^{B}} \qquad (10)$$

$$\frac{\varepsilon_{cr}}{\varepsilon_y} = \frac{1}{\bar{\lambda}_c^{2}} \qquad (11)$$

$$\frac{\varepsilon_{csm}}{\varepsilon_y} = \frac{4.44 \times 10^{-3}}{\bar{\lambda}_c^{4.5}} \text{ but } \frac{\varepsilon_{csm}}{\varepsilon_y} \leq \min\left(15, \frac{C\varepsilon_u}{\varepsilon_y}\right) \qquad (12)$$

Upper limits are placed upon the strain ratio, in effect limiting the extent to which the cross-section can deform. The upper limit of 15 from Equation 12 is the material ductility requirement from EN 1993-1-1 (2005) and is applied to all metallic materials. A second upper limit is applied to cold-formed structural steel ($C = 0.4$), stainless steel ($C = 0.1$) and aluminium ($C = 0.5$) to prevent over-predictions of cross-section resistance due to the chosen simplified material model. ε_u is the strain at the ultimate tensile stress, and is discussed further in Section 2.6.

2.6 Material models

An elastic, linear strain hardening material model (of slope E_{sh}) is adopted in the CSM, replacing the traditional elastic, perfectly-plastic material model.

The CSM limiting stress f_{csm} is defined by Equation 13, which is greater than or equal to the yield strength f_y for strain ratios greater than unity and dependent upon the strain hardening modulus E_{sh} of the metallic material.

$$f_{csm} = f_y + E_{sh}\varepsilon_y \left(\frac{\varepsilon_{csm}}{\varepsilon_y} - 1\right) \qquad (13)$$

2.6.1 Hollow hot-finished structural steel

For hollow hot-finished structural steel the strain hardening modulus E_{sh} proposed by Foster (2014) has been utilised. This is the simplest model adopted as strain hardening is taken as zero (Eq. 14), reducing the model to the traditional linear elastic, perfectly-plastic model. This is due to the extensive yield plateau associated with hot-finished tubes.

$$\frac{E_{sh}}{E} = 0 \qquad (14)$$

2.6.2 Hollow cold-formed structural steel

For cold-formed structural steel, a currently unpublished material model developed at Imperial College London has been adopted, and also applied to very high strength structural steel. The strain hardening modulus E_{sh} is defined by Equations 15 and 16, depending on the ratio $\varepsilon_y/\varepsilon_u$. The material strain ε_u corresponding to the ultimate tensile stress f_u may be predicted using Equation 17.

For $\varepsilon_y/\varepsilon_u < 0.45$:

$$E_{sh} = \frac{f_u - f_y}{0.45\varepsilon_u - \varepsilon_y} \qquad (15)$$

For $\varepsilon_y/\varepsilon_u \geq 0.45$:

$$E_{sh} = 0 \qquad (16)$$

$$\varepsilon_u = 0.6\left(1 - f_y/f_u\right) \qquad (17)$$

2.6.3 Stainless steel

For stainless steel, the material model developed by Afshan & Gardner (2013) is utilised. This model is suitable for austenitic and duplex grades, while work on ferritic grades has been reported by Bock et al. (2015). The strain hardening modulus E_{sh} is predicted by Equation 18 and ε_u is given by Equation 19, which is taken from EN 1993-1-4 (2006).

$$E_{sh} = \frac{f_u - f_y}{0.16\varepsilon_u - \varepsilon_y} \qquad (18)$$

$$\varepsilon_u = 1 - f_y/f_u \qquad (19)$$

2.6.4 Aluminium

For aluminium, the material model proposed by Su et al. (2014) is adopted in this study. The predictive expression for the strain hardening modulus E_{sh} is given by Equation 20. This is similar in form to the previous cold-formed structural steel and stainless steel models. The material ultimate strain ε_u may be predicted from Equation 21. The latter expression is only applicable when the ratio of the ultimate stress f_u to yield stress f_y exceeds 1.01.

$$E_{sh} = \frac{f_u - f_y}{0.5\varepsilon_u - \varepsilon_y} \qquad (20)$$

For $f_u/f_y > 1.01$:

$$\varepsilon_u = 0.13\left(1 - f_y/f_u\right) + 0.06 \qquad (21)$$

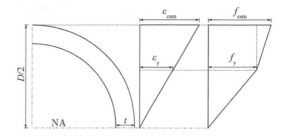

Figure 4. Strain and stress profile for a CHS (a quarter of the cross-section is drawn).

2.7 Cross-section resistance

The cross-section resistance can now be determined utilising the deformation capacity ($\varepsilon_{csm}/\varepsilon_y$) predicted from the base curve, together with the adopted material models. Note that the following is only applicable to cross-sections that are non-slender ($\overline{\lambda}_c \leq 0.3$), although work is being undertaken to extend the CSM beyond the class 3 limit.

2.7.1 Compressive resistance

The CSM axial compressive resistance of the cross-section N_{csm} is calculated as the product of the gross cross-section area A and the limiting material stress, which is the CSM limiting stress f_{csm}, as given by Equation 22. The strength benefit from the CSM arises when the CSM limiting stress f_{csm} exceeds the yield stress f_y. Consequently for hot-finished structural steel there are no strength benefits due to the strain hardening model adopted in Equation 14.

$$N_{csm} = Af_{csm} \qquad (22)$$

2.7.2 Bending resistance

Derivation of the CHS CSM bending resistance expression is outlined in Liew (2014), resulting in Equation 23, which factors the plastic moment capacity M_{pl}. W_{pl} and W_{el} are the plastic and elastic section moduli respectively. As is traditional, it is assumed that plane sections remain plane and normal to the neutral axis in bending, and that the cross-section shape does not significantly distort before the outer-fibre strain ε_{csm} is attained. The expression for the CSM bending resistance M_{csm} is more involved than those currently adopted in design codes, in part due to the stress-profile assumed, as shown in Figure 4.

$$M_{csm} = M_{pl}\left[1 + \frac{E_{sh}}{E}\frac{W_{el}}{W_{pl}}\left(\frac{\varepsilon_{csm}}{\varepsilon_y} - 1\right)\right.$$
$$\left. - \left(1 - \frac{W_{el}}{W_{pl}}\right)\left(\frac{\varepsilon_{csm}}{\varepsilon_y}\right)^{-2}\right] \qquad (23)$$

The variation in bending capacity M_{csm}/M_{pl} with strain ratio $\varepsilon_{csm}/\varepsilon_y$ for various strain hardening ratios is plotted in Figure 5. The bending capacity at a strain ratio of unity is the elastic moment M_{el}. The subsequent

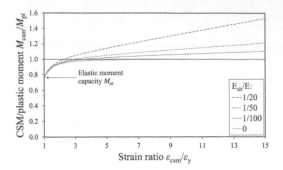

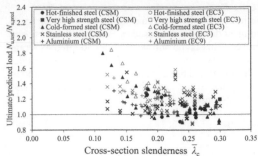

Figure 5. M_{csm}/M_{pl} varying with $\varepsilon_{csm}/\varepsilon_y$.

Figure 6. Compression resistance comparisons with test data.

Table 2. Compression resistance comparisons with test data.

Material	Mean		COV	
	N_u/N_{csm}	N_u/N_{EC}	N_u/N_{csm}	N_u/N_{EC}
Hot-finished structural steel	1.12	1.12	0.15	0.15
Very high strength structural steel	1.19	1.19	–*	–*
Cold-formed structural steel	1.12	1.17	0.17	0.20
Stainless steel	1.15	1.24	0.11	0.13
Aluminium	1.07	1.14	0.12	0.14

*Insufficient experimental data.

Table 3. Bending resistance comparisons with test data.

Material	Mean		COV	
	M_u/M_{csm}	M_u/M_{EC}	M_u/M_{csm}	M_u/M_{EC}
Hot-finished structural steel	1.06	1.05	0.11	0.12
Very high strength structural steel	1.47	1.48	–*	–*
Cold-formed structural steel	1.11	1.18	0.10	0.12
Stainless steel	1.15	1.24	0.01	0.09
Aluminium	1.12	1.26	0.05	0.10

*Insufficient experimental data.

increase in bending capacity for $\varepsilon_{csm}/\varepsilon_y > 1$ is dependent upon the strain hardening ratio, where a larger strain hardening modulus leads to a greater increase in bending capacity with increasing strain ratio $\varepsilon_{csm}/\varepsilon_y$. If the strain hardening modulus is taken as zero, the bending capacity is asymptotic to the plastic moment capacity.

3 COMPARISON WITH TEST DATA AND EXISTING DESIGN METHODS

The CSM predictions for compression and bending resistances have been compared with the measured ultimate values from the collected experiments. This dataset has more experimental results compared with the base curve dataset (Table 1) as occasionally insufficient parameters were reported in the literature to calculate the experimental strain ratios.

The average ultimate test loads N_u and moments M_u normalised by the CSM (N_{csm}, M_{csm}) and Eurocode (N_{EC}, M_{EC}) predictions have been determined for each material type and are summarised in Tables 2 and 3. The coefficients of variation (COV) have also been calculated to quantify the scatter of the predictions. The ultimate experimental loads normalised by their CSM and Eurocode predictions have been plotted for compression ($N_{u,pred}$) and bending ($M_{u,pred}$) in Figures 6 and 7 respectively.

The CSM predicts cross-section resistances that are typically more accurate and consistent compared

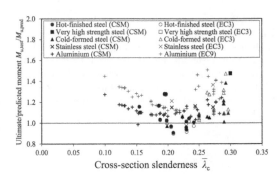

Figure 7. Bending resistance comparisons with test data.

with those from the Eurocodes, particularly for cold-formed structural steel, stainless steel and aluminium. The very high strength structural steel dataset is small and thus it is difficult to make definite conclusions, although the CSM is marginally more accurate in bending. The CSM hot-finished structural steel compressive resistances are equal to those from EN 1993-1-1 (2005), which as discussed previously is due to the material model adopted. It is clear that there is additional capacity to be gained for some hot-finished structural steel sections, which may be achieved through refinements to the material model. In bending, the hot-finished structural steel CSM predictions appear more conservative. This is because the test specimens are primarily within a small specific local slenderness range ($0.20 \leq \bar{\lambda}_c \leq 0.25$) where the predicted CSM capacity is less than the Eurocode capacity, due to the discontinuous relationship between slenderness and cross-section resistance being replaced with a continuous relationship.

4 CONCLUSIONS

The CSM has been extended to cover the design of CHS and has been seen to provide improved cross-section resistance predictions for metallic materials over traditional design methods, with generally more accurate and consistent results. Improved predictions of CHS cross-section resistance will lead to lighter structures with more efficient material use, leading to more sustainable construction.

Further work is currently underway into refining the material models, incorporating a larger CHS dataset and applying the CSM beyond the class 3 limit to a wider range of local slenderness values.

REFERENCES

Afshan, S. & Gardner, L. 2013. The continuous strength method for structural stainless steel design. *Thin-Walled Structures*, 68: 42–49.

American Institute of Steel Construction 2010. *ANSI/AISC 360-10 Specification for Structural Steel Buildings*.

Bardi, F.C. & Kyriakides, S. 2006. Plastic buckling of circular tubes under axial compression—part I: Experiments. *International Journal of Mechanical Sciences*, 48(8): 830–841.

Bock, M., Gardner, L. & Real, E. 2015. Material and local buckling response of ferritic stainless steel sections. *Thin-Walled Structures*, 89: 131–141.

British Standards Institution 2000. *BS 5950-1:2000 Structural use of steelwork in building – Part 1: Code of practice for design – Rolled and welded sections*.

Burgan, B.A., Baddoo, N.R. & Gilsenan, K.A. 2000. Structural design of stainless steel members—comparison between Eurocode 3, Part 1.4 and test results. *Journal of Constructional Steel Research*, 54(1): 51–73.

Chen, W. & Ross, D. 1977. Tests of fabricated tubular columns. *Journal of the Structural Division*, 100(3): 619–634.

Donnell, L. H. 1934. A new theory for the buckling of thin cylinders under axial compression and bending. *Transactions of the American Society of Mechanical Engineers*, 56: 795–806.

Dutta, D. 2002. *Structures with hollow sections*. Weinheim: Wiley VCH.

Elchalakani, M., Zhao, X.L. & Grzebieta, R. 2002a. Bending tests to determine slenderness limits for cold-formed circular hollow sections. *Journal of Constructional Steel Research*, 58(11): 1407–1430.

Elchalakani, M., Zhao, X.L. & Grzebieta, R. 2002b. Tests on concrete filled double-skin (CHS outer and SHS inner) composite short columns under axial compression. *Thin-Walled Structures*, 40(5): 415–441.

European Committee for Standardisation (CEN) 2005. *EN 1993-1-1:2005 Eurocode 3: Design of steel structures – Part 1-1: General rules and rules for buildings*.

European Committee for Standardisation (CEN) 2006. *EN 1993-1-4:2006 Eurocode 3: Design of steel structures – Part 1-4: General rules – supplementary rules for stainless steel*.

European Committee for Standardisation (CEN) 2007. *EN 1999-1-1:2007 Eurocode 9: Design of aluminium structures – Part 1-1: General structural rules*.

Flügge, W. 1932. Die Stabilität der Kreiszylinderschale (in German). *Ingenieur-Archiv*, 3: 463–506.

Foster, A. 2014. *Stability and design of steel beams in the strain-hardening range*. Imperial College London.

Gardner, L. 2008. The Continuous Strength Method. *Proceedings of the Institution of Civil Engineers: Structures and Buildings*, 161(3): 127–133.

Gardner, L., Law, K.H. & Buchanan, C. 2014. Unified slenderness limits for structural steel circular hollow sections. *Romanian Journal of Technical Sciences – Applied Mechanics*, 59(1-2): 153–163.

Gardner, L. & Nethercot, D.A. 2004. Experiments on stainless steel hollow sections—Part 1: Material and cross-sectional behaviour. *Journal of Constructional Steel Research*, 60(9): 1291–1318.

Gardner, L., Wang, F. & Liew, A. 2011. Influence of Strain Hardening on the Behavior and Design of Steel Structures. *International Journal of Structural Stability and Dynamics*, 11(5): 855–875.

Gerard, G. & Becker, H. 1957. *Handbook of Structural Stability*. Washington, D.C.: NACA.

Giakoumelis, G. & Lam, D. 2004. Axial capacity of circular concrete-filled tube columns. *Journal of Constructional Steel Research*, 60(7): 1049–1068.

Gresnigt, A.M. & van Foeken, R.J. 2001. Local buckling of UOE and seamless steel pipes. In J. Chung, T. Matsui & H. Moshagen (eds.), In *Proceedings of the Eleventh (2001) International Offshore and Polar Engineering Conference*. Stavanger.

Guo, L., Yang, S. & Jiao, H. 2013. Behavior of thin-walled circular hollow section tubes subjected to bending. *Thin-Walled Structures*, 73: 281–289.

Haedir, J., Bambach, M.R., Zhao, X.L. & Grzebieta, R.H. 2009. Strength of circular hollow sections (CHS) tubular beams externally reinforced by carbon FRP sheets in pure bending. *Thin-Walled Structures*, 47(10): 1136–1147.

Jiao, H. & Zhao, X.L. 2003. Imperfection, residual stress and yield slenderness limit of very high strength (VHS) circular steel tubes. *Journal of Constructional Steel Research*, 59(2): 233–249.

Jiao, H. & Zhao, X.L. 2004. Section slenderness limits of very high strength circular steel tubes in bending. *Thin-Walled Structures*, 42(9): 1257–1271.

Jirsa, J.O., Lee, F.H., Wilhoit Jr., J.C. & Merwin, J.E. 1972. Ovaling of pipelines under pure bending. *Proceedings of the Offshore Technology Conference*. Dallas.

Johansson, M. & Gylltoft, K. 2002. Mechanical behavior of circular steel-concrete composite stub columns. *Journal of Structural Engineering ASCE*, 128(8): 1073–1081.

Kamba, T. 1996. Stub column test of high-strength CHS steel column with small diameter-to-thickness ratio. In J. Farkas & K. Jarmai (Eds.), *Proceedings of the Seventh International Symposium on Tubular Structures*. Rotterdam: A.A. Balkema.

Kiymaz, G. 2005. Strength and stability criteria for thin-walled stainless steel circular hollow section members under bending. *Thin-Walled Structures*, 43(10): 1534–1549.

Kuwamura, H. 2003. Local buckling of thin-walled stainless steel members. *Steel Structures*, 3: 191–201.

Lam, D. & Gardner, L. 2008. Structural design of stainless steel concrete filled columns. *Journal of Constructional Steel Research*, 64(11): 1275–1282.

Liew, A. 2014. *Design of structural steel elements with the Continuous Strength Method*. Imperial College London.

Liew, J.Y.R. & Xiong, D.X. 2010. Experimental investigation on tubular columns infilled with ultra-high strength concrete. In B. Young (Ed.), *Proceedings of the 13th International Symposium on Tubular Structures*. Hong Kong: Taylor and Francis/Balkema.

O'Shea, M. & Bridge, R. 1997. Local buckling of thin-walled circular steel sections with or without internal restraint. *Journal of Constructional Steel Research*, 41(2): 137–157.

Ochi, K. & Choo, B.S. 1993. Ultimate strength and post-buckling behaviour of CHS columns – A comparison between cold-formed and hot-finished sections. In M.G. Coutie & G. Davies (Eds.), *Proceedings of the Fifth International Symposium on Tubular Structures*. London: E & FN Spon.

Paquette, J.A. & Kyriakides, S. 2006. Plastic buckling of tubes under axial compression and internal pressure. *International Journal of Mechanical Sciences*, 48(8): 855–867.

Prion, H.G.L. & Birkemoe, P.C. 1992. Beam-column behavior of fabricated steel tubular members. *Journal of Structural Engineering ASCE*, 118(5): 1213–1232.

Rasmussen, K. 2000. Recent research on stainless steel tubular structures. *Journal of Constructional Steel Research*, 54(1): 75–88.

Rasmussen, K. & Hancock, G. 1993a. Design of cold-formed stainless steel tubular members. I: columns. *Journal of Structural Engineering ASCE*, 119(8): 2349–2367.

Rasmussen, K. & Hancock, G. 1993b. Design of cold-formed stainless steel tubular members. II: beams. *Journal of Structural Engineering ASCE*, 119(8): 2368–2386.

Reddy, B.D. & Calladine, C.R. 1978. Classical bucking of a thin-walled tube subjected to bending moment and internal pressure. *International Journal of Mechanical Sciences*, 20: 641–650.

Rondal, J., Boeraeve, P., Sedlacek, G. & Langenberg, P. 1995. *Rotation Capacity of Hollow Beam Sections – Research project No. 2P*.

Rotter, J.M., Sadowski, A.J. & Chen, L. 2014. Nonlinear stability of thin elastic cylinders of different length under global bending. *International Journal of Solids and Structures*, 51(15–16), 2826–2839.

Sakino, K., Nakahara, H., Morino, S. & Nishiyama, I. 2004. Behavior of centrally loaded concrete-filled steel-tube short columns. *Journal of Structural Engineering ASCE*, 130(2): 180–188.

Schmidt, H. 1989. Thick-walled tubular members under axial compression. In E. Niemi & P. Makelainen (Eds.), *Proceedings of the 3rd International Symposium on Tubular Structures*. London: Elsevier Applied Science.

Sedlacek, G., Dahl, W., Stranghöner, N., Kalinowski, B., Rondal, J. & Boeraeve, P. 1995. Investigation of the rotation behaviour of hollow section beams. *ECSC Research Project, Final Report, 7210/SA/119*.

Seica, M., Packer, J., Ramirez, P., Bell, S.A.H., & Zhao, X.L. 2006. Rehabilitation of tubular members with carbon reinforced polymers. In J. Packer & S. Willibald (Eds.), *Proceedings of the 11th International Symposium and IIW International Conference on Tubular Structures*. London: Taylor and Francis/Balkema.

Seide, P. & Weingarten, V. I. 1961. On the buckling of circular cylindrical shells under pure bending. *ASME Journal of Applied Mechanics*, 28(1): 112–116.

Sherman, D.R. 1976. Tests of circular steel tubes in bending. *Journal of the Structural Division*, 102(11): 2181–2195.

Standards Australia 1998. *AS 4100-1998 Steel Structures*.

Starossek, U. & Falah, N. 2009. The interaction of steel tube and concrete core in concrete-filled steel tube columns. In Z.Y. Shen, Y.Y. Chen, & X.Z. Zhao (Eds.), *Proceedings of the 12th International Symposium on Tubular Structures*. London: Taylor and Francis/Balkema.

Su, M.N., Young, B. & Gardner, L. 2014. Testing and design of aluminum alloy cross sections in compression. *Journal of Structural Engineering ASCE*, 140(9), Article Number: UNSP 04014047.

Talja, A. 1997. Test report on welded I and CHS beams, columns and beam-columns. *Report to ECSC. VTT Building Technology, Finland*.

Teng, J.G. & Hu, Y.M. 2007. Behaviour of FRP-jacketed circular steel tubes and cylindrical shells under axial compression. *Construction and Building Materials*, 21(4): 827–838.

Timoshenko, S.P. & Gere, J.M. 1961. *Theory of elastic stability* (Second.). McGraw-Hill Book Company.

Toi, Y. & Ine, T. 1988. Basic studies on the crashworthiness of structural elements. Part 5: Axisymmetric crush tests of circular cylinders and finite element analysis (in Japanese). *Journal of The Society of Naval Architects of Japan*, 164: 406–419.

Toi, Y., Yuge, K. & Obata, K. 1986. Basic studies on the crashworthiness of structural elements. Part 2: Non-axisymmetric crush tests of circular cylinders and finite element analysis (in Japanese). *Journal of The Society of Naval Architects of Japan*, 161: 296–305.

Tutuncu, I. & O'Rourke, T. 2006. Compression behavior of nonslender cylindrical steel members with small and large-scale geometric imperfections. *Journal of Structural Engineering ASCE*, 132(8): 1234–1241.

Uy, B., Tao, Z. & Han, L.H. 2011. Behaviour of short and slender concrete-filled stainless steel tubular columns. *Journal of Constructional Steel Research*, 67(3): 360–378.

Wei, S., Mau, S.T., Vipulanandan, C. & Mantrala, S.K. 1995. Performance of a new sandwich tube under axial loading: Experiment. *Journal of Structural Engineering ASCE*, 121(12): 1806–1814.

Xiao, Y., He, W. & Choi, K. 2005. Confined concrete-filled tubular columns. *Journal of Structural Engineering ASCE*, 131(3): 488–497.

Young, B. & Hartono, W. 2002. Compression tests of stainless steel tubular members. *Journal of Structural Engineering ASCE*, 128(6): 754–761.

Yu, T. & Teng, J.G. 2010. Hybrid FRP-concrete-steel double-skin tubular columns with a square outer tube and a circular inner tube: Stub column tests. In B. Young (Ed.), *Proceedings of the 13th International Symposium on Tubular Structures*. London: Taylor and Francis/Balkema.

Zhao, X.L. 2000. Section capacity of very high strength (VHS) circular tubes under compression. *Thin-Walled Structures*, 37(3): 223–240.

Zhao, X.L., Tong, L.W., & Wang, X.Y. 2010. CFDST stub columns subjected to large deformation axial loading. *Engineering Structures*, 32(3): 692–703.

Zhou, F. & Young, B. 2009. Concrete-filled aluminum circular hollow section column tests. *Thin-Walled Structures*, 47(11): 1272–1280.

Zhu, J.H. & Young, B. 2006a. Aluminum alloy circular hollow section beam-columns. *Thin-Walled Structures*, 44(2): 131–140.

Zhu, J.H., & Young, B. 2006b. Experimental investigation of aluminum alloy circular hollow section columns. *Engineering Structures*, 28(2): 207–215.

Tubular Structures XV – Batista, Vellasco & Lima (eds)
© 2015 Taylor & Francis Group, London, ISBN 978-1-138-02837-1

Use of Ramberg-Osgood material laws in the finite element modeling of cold-formed tubes

M. Hayeck, J. Nseir, E. Saloumi & N. Boissonnade

University of Applied Sciences of Western Switzerland, Fribourg, Switzerland

ABSTRACT: The present paper investigates the cross-sectional resistance of cold-formed tubes; in particular, attention is paid to the influence of various material laws on the carrying capacity. Owing to the fabrication process, coupons from cold-formed tubes usually exhibit a rounded stress-strain relationship, whose shape is known to be influenced by many parameters, such as the cross-sectional width-to-thickness ratio and the location of the coupon within the section or the type of steel. The intention is here to bring out to which extent the material law may affect the structural response of such cold-formed tubes, with respect to relevant parameters. The paper first analyzes several tensile test results, and proposes Ramberg-Osgood equations with adjusted parameters evaluated from experimental results. Suitable shell FE models are then used to characterize the influence of the material response roundedness on the cross-section resistance. Eventually, recommendations are proposed for a sound modeling of cold-formed tubes in FE analyses.

1 INTRODUCTION

Cold-formed tubes are mainly manufactured through two different processes: press-braking and roll-forming (Zhao 2005). The press-braking route forms bending sheets by clamping the work piece between the top and bottom tools (Figure 1), and the cold-forming effect is limited to the corners of the section (Akiyama et al. 1992). Besides, roll-forming consists in feeding a long strip of steel through sets of rolls mounted on consecutive stands to form the desired shapes by either direct or indirect forming methods.

The direct forming process consists in roll-forming the strip directly into the desired rectangular shape and then assembling the four plates by welding the corners. The associated cold-forming effect is confined to the area of corners containing the welds whereas the flat faces keep the same properties as the feed material.

The continuous-forming process (Figure 2) consists in roll-forming the strip into a circular shape, welding the edges of the tube, and then reworking it into the desired rectangular or square shape (Sun & Packer 2013). In the latter case, the flat face undergoes high cold-work effects, unlike in the direct roll-forming and press-braking methods.

In order to investigate the structural behavior of cold-formed square and rectangular hollow sections manufactured through the continuous-forming method, experimental results on tensile coupon tests from flat faces and corner areas were collected, both from own tensile tests and from literature; they were related to different sizes of tubular cross sections, different steel grades and different manufacturers in

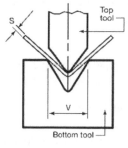

Figure 1. Press-braking manufacturing process.

Figure 2. Roll-forming manufacturing process.

Europe. All tests and measurements were performed in the Laboratories of the University of Applied Sciences of Western Switzerland in Fribourg, the Technical University of Graz, the University of Sydney, and Lappeenranta University of Technology.

The material properties have been characterized through specimens taken from flat sides, corners or welded faces of the cold-formed tubes. Unlike for hot-formed tubes, for which the typical stress-strain curve exhibits a classic behavior with distinct yield plateau and strain hardening effects, the material response of cold-formed tubes shows a pronounced non-linear behavior and thus has no identifiable yield plateau caused by cold-working of the material, and strain hardening immediately follows first yield (see Figure 4 for example). Usually, the 0.2% proof stress is used as a convenient equivalent yield stress; increases in yield and ultimate strengths are also usually observed in the corner regions of cold-formed sections, and the amount of cold working can be shown to be associated with an increase in yield stress and a lower level of ductility at fracture.

With the intention of investigating the influence of all these aspects on the cross-sectional resistance of cold-formed tubes, the present paper is first devoted to the determination of appropriate material coefficients for so-called Ramberg-Osgood equations, either "simple" R.-O. or "double" R.-O. In this respect, the results of many tensile tests from cold-formed tubes reflecting several production sites are collected and analyzed (Section 2).

Then, dedicated finite element models are described, that aim to investigate the influence of a pronounced non-linear material response on the behavior of cold-formed steel hollow sections (Section 3); in particular, analyses on the influence of cold-forming on cross-sectional resistance are reported in paragraph 3.2. Recommendations are finally given for a proper and safe FE modeling and R.-O. coefficients to be used.

2 INFLUENCE OF COLD-FORMING ON MATERIAL SIGMA-EPSILON LAW

2.1 *Experimental results*

Stress-strain properties were obtained by means of tensile coupon tests, for various cold-formed sections (cf. Table 1), from rectangular and square hollow sections. Results of tensile tests carried out at the University of Applied Sciences of Western Switzerland – Fribourg (Nseir 2014) have been combined with other available test data collected from the literature (Kettler 2008, Wilkinson 1999, Pekka 2004).

All coupons specimens originated cold-formed structural hollow sections manufactured by various manufacturers across Europe (Manufacturers A/B/C/D/E/Sarja), and included steel grades S235, S355 and S460 (cf. Table 1). For each of the SHS and RHS specimens, flat coupons were cut from the faces opposite or adjacent to the weld; corner coupons were also tested in order to examine the influence of the high cold-work localized in these regions. At several occasions, the corner coupons were extracted from two corners of the section, "R_{max}" from the corner where the largest corner radius was measured, and "R_{min}" from the corner where the smallest corner radius was measured. Figure 3 illustrates the locations where the various samples were cut, and the adopted dimension labeling system.

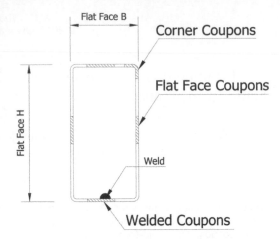

Figure 3. Section notation of typical RHS and SHS.

A summary of all test series containing section sizes, steel grades, position of coupons and manufacturers is given in Table 1. For each of the collected test result, the experimental data was plotted and compared to so-called "Ramberg-Osgood equations" (both simple R.-O. and double R.-O.), where exponent coefficients were deduced from the test in fitting the experimental data (see coefficients n in Table 3).

Typically, the degree of non-linearity depends on the amount of cold-work produced during the fabrication of the section, and the different levels of cold-working generate considerable variability of yield stress around the same section. Figure 4 presents typical stress-strain curves obtained for corners and flat regions of a cold-formed RHS $100 \times 100 \times 8$. It can be seen that the corner portions exhibit a higher yield stress and a lower ultimate-to-yield stress ratio $\sigma_u/\sigma_{0.2}$ than the flat portions, and that the ductility of the corners is very limited compared to that of the flat faces.

2.2 *Simple Ramberg-Osgood formulation*

In the design of aluminum and stainless steel structures, the stress-strain relationship is usually associated to the simple R.-O. formula described by Rasmussen (2003) and given in equation (1):

$$\varepsilon = \frac{\sigma}{E_0} + 0.002 \left(\frac{\sigma}{\sigma_{0.2}} \right)^n \tag{1}$$

where E_0 is the initial Young's modulus, $\sigma_{0.2}$ the equivalent yield stress and n a strain hardening coefficient. Equation 1 was originally proposed by Ramberg and Osgood (1943), then modified by Hill (1944) and has proven suitable for non-linear metals including cold-formed tubes. From the format of Equation 1, it can be noticed that when the value of n increases, the roundness of the stress-strain curve decreases. For high values of n (i.e. $n > 50$), the material response tends to become bi-linear, with a distinct yield plateau such as for hot-formed tubes.

Table 1. Coupon test results used within present study.

Section	Source	Section	Grade	Coupon specimens types	Manufacturer
S1		$200 \times 100 \times 4$	S355	Flat face, corner	
S2	[1]*	$220 \times 120 \times 6$	S355	Flat face, corner	
S3		$200 \times 200 \times 5$	S355	Flat face, corner	
S4		$200 \times 200 \times 6$	S355	Flat face, corner	
S5	[2]*	$180 \times 180 \times 5$	S355	Opposite and adjacent to the weld, corner, weld	
S6		$200 \times 120 \times 4$	S275	Opposite and adjacent to the weld, corner, weld	
S7		$150 \times 50 \times 5$	S450	Opposite to the weld, corner	
S8		$75 \times 50 \times 2$	S450	Adjacent to the weld	
S9	[3]*	$150 \times 50 \times 4$	S350	Opposite to the weld	
S10		$150 \times 50 \times 4$	S450	Web and flange faces	
S11		$100 \times 50 \times 2$	S350	Opposite and adjacent to the weld	
S12		$100 \times 100 \times 3$	S355	Weld, face H, face B, corner R_{max}, corner R_{min}	A
S13		$100 \times 100 \times 5$	S355	Weld, face H, face B, corner R_{max}, corner R_{min}	A
S14		$100 \times 100 \times 6$	S355	Weld, face H, face B, corner R_{max}, corner R_{min}	A
S15		$100 \times 100 \times 8$	S355	Weld, face H, face B, corner R_{max}, corner R_{min}	A
S16		$100 \times 100 \times 10$	S355	Weld, face H, face B, corner R_{max}, corner R_{min}	A
S17		$150 \times 150 \times 8$	S355	Weld, face H, face B, corner R_{max}, corner R_{min}	A
S18		$200 \times 200 \times 10$	S355	Weld, face H, face B, corner R_{max}, corner R_{min}	A
S19		$300 \times 300 \times 6$	S355	Weld, face H, face B, corner R_{max}, corner R_{min}	A
S20		$300 \times 300 \times 12.5$	S355	Weld, face H, face B, corner R_{max}, corner R_{min}	A
S21	[4]*	$200 \times 100 \times 8$	S355	Weld, face H, face B, corner R_{max}, corner R_{min}	A
S22		$200 \times 100 \times 10$	S355	Weld, face H, face B, corner R_{max}, corner R_{min}	A
S23		$50 \times 50 \times 3$	S355	Weld, face H, face B, corner R_{max}, corner R_{min}	B
S24		$50 \times 50 \times 5$	S355	Weld, face H, face B, corner R_{max}, corner R_{min}	B
S25		$100 \times 100 \times 3$	S355	Weld, face H, face B, corner R_{max}, corner R_{min}	B
S26		$100 \times 100 \times 5$	S355	Weld, face H, face B, corner R_{max}, corner R_{min}	B
S27		$100 \times 100 \times 6$	S355	Weld, face H, face B, corner R_{max}, corner R_{min}	B
S28		$100 \times 100 \times 8$	S355	Weld, face H, face B, corner R_{max}, corner R_{min}	B
S29		$100 \times 100 \times 10$	S355	Weld, face H, face B, corner R_{max}, corner R_{min}	B
S30		$150 \times 150 \times 8$	S355	Weld, face H, face B, corner R_{max}, corner R_{min}	B
S31		$200 \times 200 \times 10$	S355	Weld, face H, face B, corner R_{max}, corner R_{min}	B
S32		$200 \times 100 \times 8$	S355	Weld, face H, face B, corner R_{max}, corner R_{min}	B
S33		$200 \times 100 \times 10$	S355	Weld, face H, face B, corner R_{max}, corner R_{min}	B
S34		$100 \times 100 \times 5$	S355	Weld, face H, face B, corner R_{max}, corner R_{min}	C
S35		$200 \times 100 \times 8$	S355	Weld, face H, face B, corner R_{max}, corner R_{min}	C
S36		$160 \times 160 \times 6$	S355	Weld, face H, face B, corner R_{max}, corner R_{min}	C
S37		$100 \times 100 \times 8$	S355	Weld, face H, face B, corner R_{max}, corner R_{min}	D
S38		$100 \times 100 \times 10$	S355	Weld, face H, face B, corner R_{max}, corner R_{min}	D
S39		$150 \times 150 \times 8$	S355	Weld, face H, face B, corner R_{max}, corner R_{min}	D
S40		$200 \times 100 \times 8$	S355	Weld, face H, face B, corner R_{max}, corner R_{min}	D
S41		$50 \times 50 \times 3$	S355	Weld, face H, face B, corner R_{max}, corner R_{min}	E
S42		$50 \times 50 \times 5$	S355	Weld, face H, face B, corner R_{max}, corner R_{min}	E
S43		$100 \times 100 \times 3.2$	S355	Weld, face H, face B, corner R_{max}, corner R_{min}	E
S44		$100 \times 100 \times 6$	S355	Weld, face H, face B, corner R_{max}, corner R_{min}	E
S45		$100 \times 100 \times 10$	S355	Weld, face H, face B, corner R_{max}, corner R_{min}	E
S46		$160 \times 160 \times 6$	S355	Weld, face H, face B, corner R_{max}, corner R_{min}	E
S47		$100 \times 100 \times 3$	S460	Weld, face H, face B, corner R_{max}, corner R_{min}	Sarja
S48		$100 \times 100 \times 5$	S460	Weld, face H, face B, corner R_{max}, corner R_{min}	Sarja
S49		$100 \times 100 \times 6$	S460	Weld, face H, face B, corner R_{max}, corner R_{min}	Sarja
S50		$100 \times 100 \times 8$	S460	Weld, face H, face B, corner R_{max}, corner R_{min}	Sarja

[1]* Results collected from the University of Applied Science of Western Switzerland – Fribourg (Nseir 2014)
[2]* Results collected from Graz Technical University (Kettler 2008)
[3]* Results collected from the University of Sydney (Wilkinson 1999)
[4]* Results collected from Lappeenranta University of Technology (Pekka 2004)

As already explained, all tensile test results on cold-formed tubes collected from different sources have been compared to the curves obtained by the simple R.-O. equation in order to adjust and predict suitable values of parameter n. Figure 5 presents examples of measured stress–strain curves of S3, S16 and S25 specimens, taken from the faces of SHS $200 \times 200 \times 5$, SHS $100 \times 100 \times 10$ and SHS $100 \times 100 \times 3$ profiles respectively, and compared to the simple R.-O. equation. As another example, the key results from

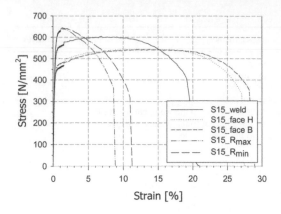

Figure 4. Typical stress-strain curve (S15 specimen).

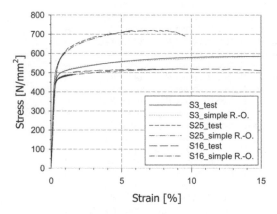

Figure 5. Comparison of test stress-strain curves vs. simple R.-O. equation with fitted n value – S3 specimen.

experimental tensile coupon tests reported by Kettler (2008), with the corresponding material properties, are given in Table 3, where σ_u and ε_u refer to the yield and ultimate strengths and strain of the material respectively, E_0 is the Young's modulus, and $\sigma_{0.2}$ the yield stress taken as the 0.2% proof stress. The parameter n is obtained using Equation 1 for $\varepsilon = \varepsilon_u$ and $\sigma = \sigma_u$.

All available experimental data were analyzed (in total, 235 coupon tests) in order to develop the stress strain equation. The strain hardening coefficients obtained are shown in Figures 6 to 8, where B/t is the width-to-thickness ratio for the weld, flat and corner faces. The simple R.-O. coefficients n were obtained by fitting the experimental curves to the simple R.-O. equation; values of parameters σ_y, σ_u, ε_y, ε_u, and E_0 were determined from the stress-strain curves. For each coupon, the obtained yield stress of the opposite face was found higher than that of the adjacent face to the weld, and the yield stress of the corner was higher than that of the opposite face. Considerable cold-forming work is reported in the corners, and on the flat faces opposite to the weld. For cold-formed sections with low values of B/t ratios, the ductility difference between the flat faces and the corners is

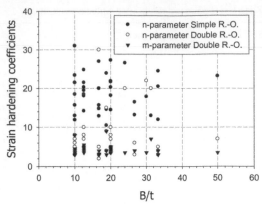

Figure 6. Simple and double R.-O. parameters for welded faces.

Table 2. Obtained n values using simple R.-O.

n values	min	max	average
Flat faces	10.5	33.9	18.6
Corners	8.5	29.7	15.4
Weld faces	10.5	31.0	19.0

minor; however, for sections with high B/t ratios, such differences becomes obvious.

Globally, the values of the simple R.-O. parameter n ranged from 8.5 to 33.9 with an average value of 20 as shown in Figures 6 to 9 and in Table 2.

As Figure 9 clearly shows, no direct correlation could be drawn between the coupon's steel grade and coefficient n, in view of the rather large scatters reported.

2.3 Double Ramberg-Osgood Formulation

Since a simple R.-O. equation was sometimes shown to be inappropriate at high strains, Mirambell and Real (2000) first proposed a derived R.-O. expression for stresses between 0.2% proof stress and the ultimate strength:

$$\varepsilon = \frac{\sigma - \sigma_{0.2}}{E_{0.2}} + \left[\varepsilon_u - \varepsilon_{0.2} - \left(\frac{\sigma_u - \sigma_{0.2}}{E_{0.2}}\right)\right] \cdot \left(\frac{\sigma - \sigma_{0.2}}{\sigma_u - \sigma_{0.2}}\right)^m + \varepsilon_{0.2} \quad (2)$$

Rasmussen (2003) revised the second equation in order to reduce the number of parameters, and proposed the following equation, denoted as "double Ramberg-Osgood" in the following:

$$\varepsilon = \frac{\sigma - \sigma_{0.2}}{E_{0.2}} + \varepsilon_u \left(\frac{\sigma - \sigma_{0.2}}{\sigma_u - \sigma_{0.2}}\right)^m + \varepsilon_{0.2} \text{ for } \sigma > \sigma_{0.2} \quad (3)$$

$$\text{with } E_{0.2} = \frac{E_0}{1 + 0.002 \cdot n/e} \quad (4)$$

Table 3. Examples of fitted values for n coefficient – sections S5 and S6.

Section	Position	E kN/mm^2	$\sigma_{0.2}$ N/mm^2	σ_u N/mm^2	ε_u	n
S5-1	a opposite	184.8	400.6	538.8	0.12	13.74
	b adjacent	179.3	386.3	520	0.13	13.96
	c adjacent	187.7	401.1	543.5	0.11	13.22
	d weld	198.3	421.2	534	0.11	16.94
	e corner	205.3	593.08	649.04	0.01	17.67
	f corner	196.0	595.56	645.66	0.01	18.35
S5-2	a opposite	182.0	399.3	537	0.12	13.65
	b adjacent	186.6	371.1	518.1	0.14	12.68
	c adjacent	188.5	400.2	543.4	0.11	13.05
	d weld	195.1	419.8	530	0.12	17.36
	e corner	182.2	610.29	644.49	0.01	25.79
	f corner	210.1	602.73	659.74	0.01	16.09
S6-1	a opposite	195.1	398.7	494.9	0.13	19.38
	b adjacent	193.8	392.2	502.3	0.13	16.71
	c adjacent	196.7	389.9	496	0.13	17.14
	d weld	201.6	411.5	506.7	0.11	19.06
	e corner	210.7	550.84	619.9	0.02	19.00
	f corner	192.5	569.58	624.51	0.01	19.05
S6-2	a opposite	198.1	411.8	500.7	0.12	20.75
	b adjacent	192.8	386.8	496	0.13	16.67
	c adjacent	199.9	384	498.7	0.13	15.89
	d weld	196.3	407.7	501.6	0.12	19.73
	e corner	205.8	558.67	631.33	0.03	20.86
	f corner	193.4	566.15	632.3	0.02	19.38

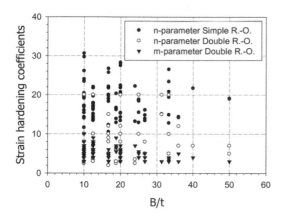

Figure 7. Simple and double R.-O. parameters for flat faces.

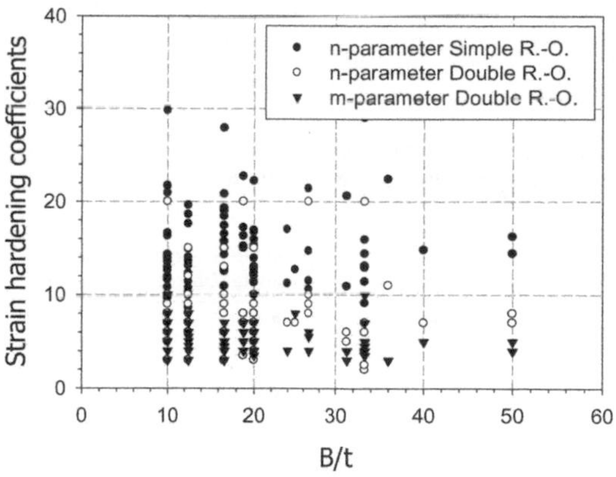

Figure 8. Simple and double R.-O. parameters for corners.

In Equation 4, $E_{0.2}$ is the initial modulus, and $e = \sigma_{0.2}/E_0$ is the non-dimensional proof stress. For stresses up to $\sigma_{0.2}$, the stress strain curve can be determined using Equation 1. Similarly to the procedure followed for the simple R.-O. n values, the experimental curves have been compared to the curves obtained using the double R.-O. equations in order to determine suitable values for the strain hardening parameters. Adjusted n and m values are determined so as to best fit the experimental curve with the predicted one. Figure 10 shows typical measured stress–strain curves of cold-formed material, extracted from the flat face of a SHS 300 × 300 × 6 S19, compared with the double

R.-O. equations for $\sigma \leq \sigma_{0.2}$ and $\sigma > \sigma_{0.2}$. The results of the "best fit procedure" are shown in Figures 6 to 8 with the B/t ratio for the weld, flat and corner coupons, respectively.

Generally, n values were found lower than those obtained from the simple equation, and the m coefficients had the lowest range values. The values of the double R.-O. coefficient n ranged from 2 to 30 with an average of 8.2, and the coefficient m ranged from 3 to 10 with an average of 5.1 as shown in Table 4.

Double R.-O. parameters for the flat faces of steel grades S355 and S460 are presented in Figures 11 to 12. The values of n predicted for the coupons taken

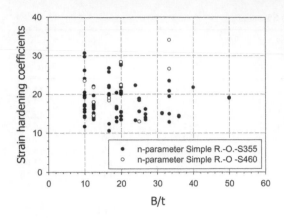

Figure 9. Simple R.-O. *n* parameter for flat faces from steel grades S355 and S460.

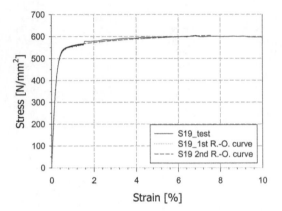

Figure 10. Comparison between experimental and fitted double R.-O. material curve for section S19.

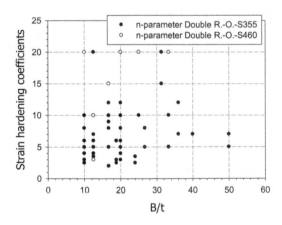

Figure 11. Double R.-O. *n*-parameter $\sigma \leq \sigma_{0.2}$ for flat faces from steel grades S355 and S460.

from the flat face of cold-formed tubes of nominal yield stress 355 N/mm^2 ranged from 2 to 20 with an average of 7.1 and the parameter *m* ranged from 3 to 10 with an average of 5.2; for S460 coupons taken from flat faces, the parameter *n* ranged from 3 to 20

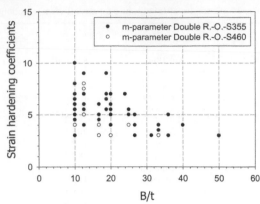

Figure 12. Double R.-O. *m* parameter $\sigma > \sigma_{0.2}$ for flat faces from steel grades S355 and S460.

Table 4. *n and m* values using double R.-O.

n values	min.	max.	average
Flat	2	20	8
Corners	2	20	8
Weld	2	30	8

m values	min	max	average
Flat	3	10	5
Corners	3	10	5.4
Weld	3	9	4

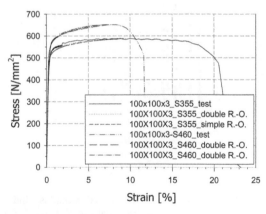

Figure 13. Experimental and analytical stress-strain curves for different steel grades – SHS $100 \times 100 \times 3$.

with an average of 15.7 and the parameter *m* from 3 to 8 with an average of 4.3. Figure 13 finally proposes representative experimental stress-strain curves for coupons extracted from the flat face of the two cold-formed square sections S12 and S47 of dimension $100 \times 100 \times 3$, compared with their associated simple and double R.-O. equations.

3 INFLUENCE OF MATERIAL NON-LINEARITY ON CROSS-SECTIONAL RESISTANCE

3.1 Finite element modeling

In order to investigate how the use of different material parameters may affect the structural response, extensive series of numerical computations have been led with the use of the non-linear FEM software FINELg (ULg and Greisch 2012). Present investigations report so-called "GMNIA" (Geometrically and Materially Non-linear with Imperfections Analysis) results related to the cross-sectional resistance of cold-formed tubes.

Shell modeling has been used to include possible local buckling of the cross section, and the lengths of the specimens considered here have been chosen equal to three times the largest cross-sectional dimension so as to limit the influence of member buckling. The cross-sections were modeled with the use of quadrangular 4-nodes plate-shell finite elements with typical features (Corotational Total Lagrangian formulation, Kirchhoff's theory for bending). Specimens have been modeled with a regular mesh all over the length of the section, with corners modelled with 2 shell elements per corner. The support conditions and the introduction of applied loads have been receiving particular attention, and use of so-called "linear constraints" has been made to ensure a "plane sections remain plane" behavior of the end sections (Nseir 2014). Local geometrical imperfections have been defined as square half-wave patterns in both directions of the flanges and webs, with an amplitude of $a/200$, where a stands for the length of the considered square panel. (see Figure 14). As for material imperfections, membrane and flexural residual stresses with a maximum amplitude of $300 \, \text{N/mm}^2$ were introduced in the numerical model, as suggested by Key (1988).

Two different material laws have been used, one for the base material and one for the corner regions, following either a simple or a double R.-O. material law, in order to account for a higher yield stress in the corner regions.

3.2 Parametric studies – Analysis of results

The present paragraph analyses the influence of various parameters on the resistance of cold-formed tubes. In this respect, GMNIA numerical parametric studies have been conducted, with the following set of parameters:

- Four cross-section shapes: RHS 220 × 120 × 6 and 220 × 120 × 10, and SHS 100 × 5 and 200 × 3. Such dimensions allow to cover plastic to slender responses of the sections, either in compression or in bending;
- Two steel grades: S235 and S460;
- Two different load cases: compression N or major axis bending M_y;

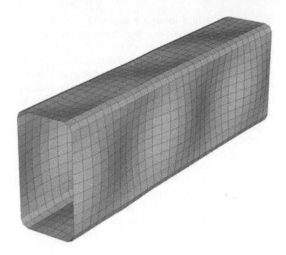

Figure 14. Typical geometrical imperfection pattern adopted.

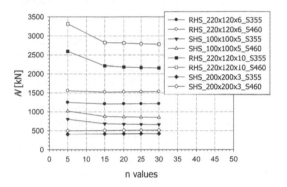

Figure 15. FE results for sections in compression.

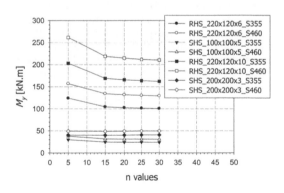

Figure 16. FE results for section under major axis bending.

- Simple and double R.-O. material laws, with different values of R.-O. coefficients n and m, ranging from 3 to 30.

In total, 320 non-linear FE computations have been performed. Figures 15 and 16 present the ultimate loads reached for each value of n used in the parametric study using a simple R.-O material model; they show that, in the case of a simple R.-O. material law, quite close failure loads are achieved, whatever the load case considered; one may however note higher peak loads for the smallest n values, where the influence of strain hardening effects is more pronounced. In all cases, results reached through the use of double R.-O. equations were extremely close to the ones obtained through simple R.-O. equation, and are consequently not reported in the following.

On the basis of these results, coupled with the experimental observations detailed in § 2, a value of the exponent $n = 20$ can then safely be adopted. Besides, when using double R.-O. equations in the parametric study, one may notice that a relatively small dispersion in the results is noted no matter what the strain hardening coefficients are. Smooth and clear tendencies may be observed, similar to the simple R.-O. equation. A pair $n = 8$ and $m = 5$ can safely be recommended for FE modeling.

4 CONCLUSIONS AND RECOMMENDATIONS FOR FE MODELLING

The present paper presented and discussed various tensile tests results on cold-formed tubes collected from literature, with respect to different manufacturing processes; particular attention was paid to the effect of cold-forming on the stress-strain curve. As a consequence of the manufacturing process, the largest influence of cold-forming was found in the corners of the section where the ultimate tensile and yield strengths are higher than in flat faces; the latter also bear the consequences of cold-working as a result of shaping the section from circular to rectangular and longitudinal welding.

Experimental stress-strain curves have been plotted together with simple and double R.-O. analytical ones; for each test, a single value of the parameters n and m are obtained by fitting experimental curve to the R.-O. equations. The values of the simple R.-O. coefficient n was found to range from 8.5 to 33.9 with an average close to 20, while for the double R.-O. equations, n was found to be ranging from 2 to 30 with an average of 8.2, and the coefficient m ranged from 3 to 10 with an average of 5.0.

In a second step, suitable shell FE analyses were performed with different values of strain-hardening coefficients. Results show that for slender sections, the cross section resistance in compression and bending is constant for various values of strain hardening coefficients. However, low values of n lead to higher resistances for stockier sections. A value of $n = 20$ can safely be used in simple R.-O. models, whereas a pair $n = 8$ and $m = 5$ can be recommended for double R.-O. ones.

REFERENCES

Akiyama et al. 1992. Influences of manufacturing processes on the ultimate behavior of box section members. *Tokyo.*

Hill, H.N. 1944. Determination of stress–strain relations from the offset yield strength values. *Technical note no. 927. National Advisory Committee for Aeronautics.*

Sun, M. & Packer, J. 2013. Direct formed and continuous formed rectangular hollow sections. *University of Toronto.*

Kettler, M. 2008. Elastic-Plastic Cross-Sectional Resistance of Semi-Compact H-and Hollow Sections. *TU-Graz, Austria.*

Key, P.W. 1988. The Behaviour of Cold-Formed Square Hollow Section Columns, *University of Sydney, Australia.*

Mirambell, E. & Real, E. 2000. On the calculation of deflections in structural stainless steel beams: an experimental and numerical investigation. *J. Constr. Steel Res. 54:109–33.*

Nseir, J. et al. 2014. Cross-sectional resistance – Experimental tests and validation of FE models. *Hollopoc report, Part B.*

Pekka, S. 2004. Design Rules for Cold-formed Hollow Sections, *Lappeeranta University of Technology.*

Ramberg, W. & Osgood, W.R. 1943. Description of stress–strain curves by three parameters. *Technical note No. 902. National Advisory Committee for Aeronautics.*

Rasmussen, K.J.R. 2003. Full-range stress-strain curves for stainless steel alloys. *J. Constr. Steel Res. 59: 47–61.*

Wilkinson, T. 1999. The Plastic Behavior of Cold-formed Rectangular Hollow Sections. *University of Sydney, Australia.*

Zhao, W. 2005. Behavior and design of cold-formed steel hollow flange section under axial compression. *Queensland University of Technology.*

University of Liège ULg and Greisch Engineering Office, 2012. *Non-linear FEM software FINELg.*

Tubular Structures XV – Batista, Vellasco & Lima (eds)
© 2015 Taylor & Francis Group, London, ISBN 978-1-138-02837-1

Strength and ductility evaluation of steel tubular columns under cyclic multiaxial loading

I.H.P. Mamaghani, B. Dorose & F. Ahmad
Department of Civil Engineering, University of North Dakota, USA

ABSTRACT: This paper deals with the strength and ductility evaluation of steel tubular columns under cyclic multiaxial (constant axial and cyclic bidirectional lateral) loading. Steel columns are very useful in highway bridge pier construction as they offers flexible space requirement and provide speedy construction. The behavior of steel columns under earthquake-induced loads is rather complicated as earthquakes occur in an oblique direction. However, modern seismic design philosophies have been based on the behavior of structures under independent actions of uni-directional loading in orthogonal directions. In this study, the inelastic cyclic behavior of steel columns subjected to constant axial force together with simultaneous bi-directional cyclic lateral loads is investigated using an advanced finite element analyses procedure. Several linear and non-linear idealized loading patterns are employed to check the strength and ductility. The effects of important structural parameters and loading history on the behavior of thin-walled steel tubular columns are examined using the proposed procedure.

1 INTRODUCTION

Thin-walled steel tubular columns used as steel bridge piers have found wide application in highway bridge systems in Japan compared with other countries, where such structures are adopted much less frequently. Steel tubular bridge piers are light and ductile compared to concrete piers. They can be built under severe constructional restrictions, such as in limited spaces in urban areas like New York and Tokyo, where the effective use of limited space is strictly desired. They are also applied to locations where heavy superstructures are unfavorable, such as on soft ground, reclaimed land, and bay areas.

In general, because of these restrictions, steel bridge piers are designed as either single columns of the cantilever type or one- to three-story frames, and they are commonly composed of relatively thin-walled members of closed cross-sections, either box or circular in shape because of their high strength and torsional rigidity (Mamaghani 1996). These make them vulnerable to damage caused by the coupled instability, i.e., the interaction of local and overall buckling, in the event of a major earthquake. For example, Figure 1 shows hollow circular steel bridge piers which suffered severe local buckling damage in the Kobe earthquake. When structural members are composed of thin-walled steel plate elements, the local buckling of the component plates may influence the strength and ductility of those members. As is well known, earthquake waves consist of three-dimensional components. Specifically, the coupling of the two horizontal components is expected

Figure 1. Local buckling of steel bridge piers, Kobe Earthquake, January 1995.

to have an unfavorable effect on the ultimate behavior of columns. Therefore, it is important to examine the ultimate behavior of thin-walled columns under cyclic axial and bidirectional lateral loading.

Present seismic design guidelines for steel columns have been based on numerous numerical and experimental investigations conducted under constant axial load plus uni-directional lateral loads. The superposition of independent action of uni-directional design seismic motion in orthogonal directions or the behavior in the most critical direction is being considered in the present seismic capacity checks. However, it is important to incorporate the bi-axial effects in seismic designs. Several experimental studies have been so far carried out to investigate the effect of bi-directional cyclic loads on the behavior of steel and concrete columns (Watanabe et al. 1999, Takizawa and Aoyama 1976, Saatcioglu and Ozcebe 1989, Bousias et al. 1995, Ohnishi et al. 2003, Zeris and Mahin 1991). Nevertheless, those tests were very costly and the results were inadequate to make firm conclusions. This strongly suggests the importance of having a reliable numerical procedure.

In this study, while keeping the vertical compressive load constant, the behavior of thin-walled steel tubular columns under the cyclic bidirectional lateral loads is examined in comparison to that under the cyclic unidirectional lateral loads shown in Figures 2 and 3. The advanced general purpose finite element program ABAQUS (2014) was employed in the analysis. The

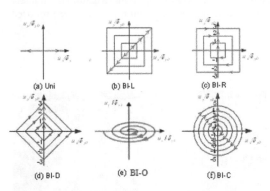

Figure 2. Cyclic loading patterns: (a) Unidirectional (Uni), (b) Bidirectional-linear (BI-L), (c) Bidirectional rectangular (BI-R), (d) Bidirectional- diamond (BI-D), (e) Bidirectional-Oval (BI-O), and (f) Bidirectional-circular (BI-C).

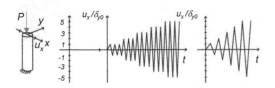

(a) Unidirectional 3 (b) Unidirectional 1

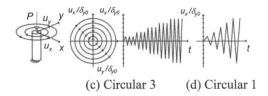

(c) Circular 3   (d) Circular 1

Figure 3. Loading programs.

results obtained from the cyclic bidirectional loading experiment are used to substantiate the validity of geometrically and materially nonlinear finite element analysis.

2 THIN-WALLED STEEL TUBULAR COLUMNS

Steel tubular columns in highway bridge systems are commonly composed of relatively thin-walled members of closed cross-sections, either box or circular in shape because of their high strength and torsional rigidity (Figure 1). Such structures differ considerably from columns in buildings. The former are characterized by: failure attributed to local buckling in the thin-walled members; irregular distribution of the story mass and stiffness; strong beams and weak columns; low rise (1–3 stories); and a need for the evaluation of the residual displacement. These make the columns vulnerable to damage caused by interaction of local and overall buckling in the event of a severe earthquake.

The most important parameters considered in the practical design and ductility evaluation of thin-walled steel hollow box sections are the width-to-thickness ratio parameter of the flange plate R_f for box section, radius-to-thickness ratio parameter of the circular section R_t, and the slenderness ratio parameter of the column λ (Mamaghani 2008). While R_f and R_t influence local buckling of the section, λ controls the global stability. They are given by:

$$R_f = \frac{b}{t}\frac{1}{n\pi}\sqrt{3(1-v^2)\frac{\sigma_y}{E}} \quad \text{(for box section)} \tag{1}$$

$$R_t = \frac{r}{t}\sqrt{3(1-v^2)\frac{\sigma_y}{E}} \quad \text{(for circular section)} \tag{2}$$

$$\lambda = \frac{2h}{r_g}\frac{1}{\pi}\sqrt{\frac{\sigma_y}{E}} \tag{3}$$

in which b = flange width; t = plate thickness; σ_y = yield stress; E = Young's modulus; v = Poisson's ratio; n = number of subpanels divided by longtudinal stiffeners in each plate panel (n = 1 for unstiffened sections); r = radius of the circular section; h = column height; r_g = radius of gyration of the cross section.

The elastic strength and deformation capacity of the column are expressed by the yield strength H_{y0}, and the yield deformation (neglecting shear deformations) δ_{y0}, respectively, corresponding to zero axial load. They are given by:

$$H_{y0} = \frac{M_y}{h} \tag{4}$$

$$\delta_{y0} = \frac{H_{y0}h^3}{3EI} \tag{5}$$

where M_y = yield moment and I = moment of inertia of the cross section. Under the combined action of buckling caused by constant axial and monotonically increasing lateral loads, the yield strength is reduced from H_{y0} to a value denoted by H_y. The corresponding yield deformation is denoted by δ_y. The value H_y is the minimum of yield, local buckling, and instability loads evaluated by the following equations (Usami 1996):

$$\frac{P}{P_u} + \frac{0.85 H_y h}{M_y(1 - P/P_E)} = 1 \tag{6}$$

$$\frac{P}{P_u} + \frac{H_y h}{M_y} = 1 \tag{7}$$

in which $P =$ the axial load; $P_y =$ the yield load; $P_u =$ the ultimate axial load; and $P_E =$ the Euler load.

3 ULTIMATE STRENGTH AND DUCTILITY EVALUATION METHOD

Failure of a tubular steel column is considered when a displacement corresponding to the maximum strength (δ_m) or 95% strength after the peak (δ_{95}) is reached. Therefore, the ductility of steel columns is defined by the index δ_m/δ_y, where δ_m is the displacement at the maximum strength (load) H_m and δ_y is the yield displacement. Another ductility parameter is the ratio δ_{95}/δ_y, where δ_{95} is the displacement at 95% of H_m ($= H_{95}$) beyond the peak. These parameters are schematically depicted on the load-displacement curve in Figure 4. The parameter δ_{95}/δ_y is considered to be more suitable since it includes the effects of cyclic characteristics and takes advantage of the strength of steel under large plastic deformations. Also considering that the strength of thin-walled steel tubular columns deteriorates significantly after the peak due to local buckling and that a crack may occur near the column base after a large drop of strength owing to the low cycle fatigue of columns with small slenderness ratios, the adoption of 95% of the peak is considered to be adequate.

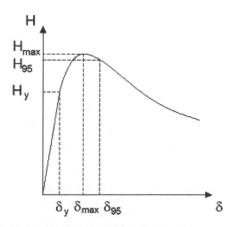

Figure 4. Definition of strength and ductility factors.

4 NUMERICAL ANALYSIS

A finite element analysis procedure is very effective in determining the seismic resisting capacity of structures. The reliability of such an analysis mainly depends on the modeling technique and factors including the type of elements, boundary condition, and material model. In this section, a numerical procedure is explained in view of the geometrical details of the column, the element mesh, the loading procedure including the loading patterns, and the material model.

4.1 Numerical model

The cantilever steel columns with box and circular cross-section subjected to a constant axial force and cyclic lateral loadings are accounted for in the present analysis. The test specimens available in the literature are numerically analyzed following an elastoplastic large displacement finite element analysis procedure. For such thin-walled steel columns, local buckling always occurs near the base of the columns. Therefore, the beam-column element is employed for the upper part of the column; while the shell element that can consider the effect of local buckling is employed for the lower part of the column (Figure 5b). The interface between the shell elements and the beam-column element is modeled using rigid beams (Figure 5b). The column is stiffened by both longitudinal stiffeners and diaphragms (Figures 5a and 5d). The longitudinal stiffeners and each subpanel between longitudinal stiffeners are modeled by using a four-node doubly curved shell element (S4R) available in the general purpose finite element program ABAQUS (2008). The diaphragm is also modeled using the same type of shell element. Shell elements are used only up to the height of the third diaphragm (Figure 5b). The length between the base and the first diaphragm is divided into 18 segments, while the subsequent same lengths are divided into 9 segments along the column length. In the width and depth directions, 24 elements are used. Each subpanel consists of eight columns of shell elements. Five columns of shell elements are assigned in longitudinal stiffeners. For shell elements, five layers are assumed across the thickness, and the spread of the plasticity is considered both through the thickness and along the element plane. The portion of the column beyond the third diaphragm is modeled using a beam-column element (B31). The sectional dimension of this element is chosen in such a way that the moment of inertia and the cross-sectional area of the element section are identical to those of the actual specimen. Ten beam-column elements are adopted to model the upper part of the specimen. The above stated mesh divisions are determined by trial and error. It is found that such mesh divisions can give an accurate result (Mamaghani, 1996). The residual stresses due to welding and the initial deflections of the flange and web plates are not considered in the analysis because their effect is insignificant on the cyclic behavior (Mamaghani 1996, Banno et al. 1998).

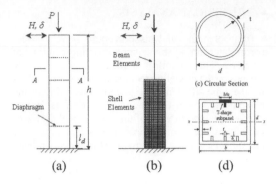

Figure 5. Finite element modeling of steel tubular columns: (a) Steel tubular column (b) Numerical model, (c) circular cross-section, (d) rectangular cross-section.

Table 1. Dimensions of numerical models.

Specimens	h	$b=d$	b_s	t	R_f	λ	γ/γ^*
Uni, BI-L27, BI-l45	2420	450	53	5.8	0.61	0.39	2.4
C35-35	5551	1043	179	6.0	0.35	0.35	3.0
C35-50	8160	1043	105	6.0	0.35	0.50	3.0
C46-35	7559	1364	113	6.0	0.46	0.35	3.0

Unit: millimeters (mm)
Uni, BI-L27, and BI-L45 are test specimens.
C35-35, C35-50, and C46-35 are numerical specimens.
Uni = Unidirectional, BI = Bidirectional

In the following, first analysis of a specimen tested by Ohnishi et al. (2003) under linear loading paths will be presented. Then the results of parametric studies using non-linear loading paths will be presented and discussed. The finite element model of the steel column used in the analysis is shown in Figure 5b. The dimensions of the column used in the analysis are listed in Table 1. The analyzed test specimens (*Uni, BI-L27,* and *BI-L45*) have height of $h = 2420$ mm; cross section size of $b = d = 450$ mm; thickness of $t = 5.8$ mm; and stiffener width of $b_s = 53$ mm. The cross sectional area A and the second moment of inertia I of the section are 1.28×10^4 mm² and 3.92×10^8 mm⁴, respectively.

The structural parameters that play important roles in the earthquake resisting performance of stiffened steel columns are the width-to-thickness ratio R_t and R_f, slenderness ratio λ, and the stiffness rigidity ratio γ/γ^* (Chen and Duan 2000). The values of R_f, λ, and γ/γ^* of the test column are 0.61, 0.39, and 2.4, respectively. In the parametric study, three numerical specimens, namely C35-35 ($R_f = 0.35$, $\lambda = 0.35$), C35-50 ($R_f = 0.35$, $\lambda = 0.50$), and C46-35 ($R_f = 0.46$, $\lambda = 0.35$), were considered. The value of γ/γ^* of all three specimens was 3.0 (Table 1).

Modern seismic design specifications allow steel structures to deform up to a certain displacement level in the inelastic range, which involves both material

Table 2. Material properties.

Specimens	σ_y (MPa)	E (GPa)	ν
Uni, BI-L27, BI-L45	412	206	0.28
C35-35, C35-50, C46-35	315	200	0.30

and geometrical non-linearity. In non-linear analysis, the accuracy of the material model has a large effect on the reliability of the predictions. The modified two-surface plasticity model (2SM) developed by Mamaghani et al. (1995), which has been proved to be very accurate in simulating cyclic behavior of steel (Mamaghani et al. 1996, 1997, Mamaghani 1996, 2005, 2006, Shen et al. 1995), is introduced into the commercial computer program ABAQUS (2008) used in the analysis. The material properties of steel such as yield stress σ_y, Young's modulus E, and Poisson's ratio ν are listed in Table 2.

5 NUMERICAL RESULTS

5.1 Comparison with test

The analyses were carried out using three loading patterns described in Figures 2a and 2b to check the effect of bi-axial cyclic bending. In loading pattern Uni, incremental cyclic lateral displacements were applied along the X-direction only. For comparison purposes, the same loading history as used in the test was employed in the analysis. Figure 6a shows the comparison of test and the numerical results of the loading type Uni. The envelope curves are plotted in Figure 6b. The comparison of numerical and test results for loading patterns BI-L27 and BI-L45 are shown in Figures 7 and 8, respectively. These results indicate that the numerical and test results match very well in all cases; hence, the proposed procedure can be considered to be accurate enough for reliable predictions.

The resultant lateral displacement δ_l and lateral load H_l for the cases BI-L27 and BI-L45 are calculated using the following two equations.

$$H_l = H_x \cos\theta + H_y \sin\theta \qquad (8)$$

$$\delta_l = \delta_x \cos\theta + \delta_y \sin\theta \qquad (9)$$

where θ is the angle between the loading direction and the major axis of the section (i.e., X-axis). The values of strength and ductility indices H_m/H_y, δ_m/δ_y, and δ_{95}/δ_y are calculated in terms of the resultant load H_l and the resultant displacements δ_l using the test and numerical results and are given in Table 3. It is seen here that the $H_{l,m}/H_y$, of uni-directional loading case (Uni) is higher than those of the BI-L27 and BI-L45 cases. On the other hand, values of $\delta_{l,m}/\delta_y$ and $\delta_{l,95}/\delta_y$ of the unidirectional loading case are much lower than the other two cases. This means that bi-directional

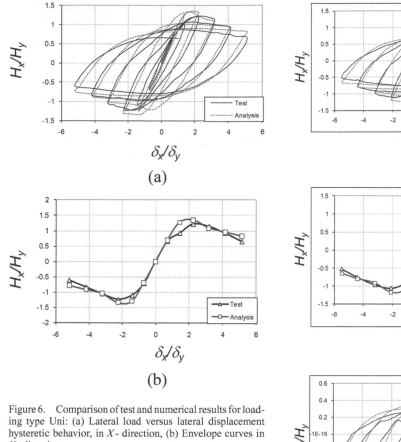

(a)

(b)

Figure 6. Comparison of test and numerical results for loading type Uni: (a) Lateral load versus lateral displacement hysteretic behavior, in X- direction, (b) Envelope curves in X- direction.

loading will result in lower strength while the ductility will be increased. The results clearly prove that the bi-directional loads considerably affect the seismic behavior of columns. As a result, investigating the effects of parameters such as the width-to-thickness ratio (R_t and R_f) and slenderness ratio λ on the strength and ductility performance of columns has significant practical importance when they are subjected to multi-directional cyclic loads.

5.2 Parametric Study

The columns for parametric study were specifically designed in order to check the effects of parameters R_t, R_f, and λ and were analyzed using three types of non-linear loading patterns as shown in Figure 2. The load-displacement envelope curves of each specimen are shown in Figure 9 for the X-direction. The corresponding strength and ductility indices obtained in the X-direction are given in Table 4.

It is revealed from these results that the strength and ductility of columns having the same width-to-thickness ratio and slenderness ratio are different when they are subjected to different loading patterns. Also, as expected, the results were different for different

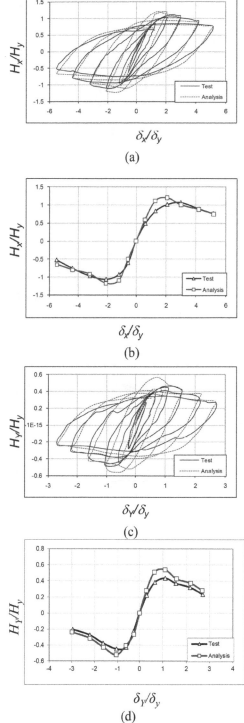

(a)

(b)

(c)

(d)

Figure 7. Comparison of test and numerical results for loading type BI-L27: (a) Lateral load versus lateral displacement hysteretic behavior in Xdirection, (b) Envelope curves in X- direction, (c) Lateral load versus lateral displacement hysteretic behavior in Y-direction, (d) Envelope curves in Y-direction.

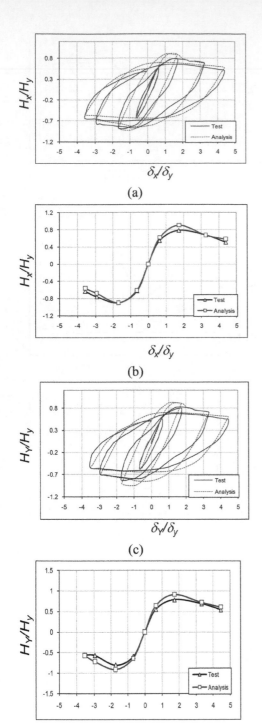

(a)

(b)

(c)

(d)

Figure 8. Comparison of test and numerical results for loading type BI-L45: (a) Lateral load versus lateral displacement hysteretic behavior in X-direction, (b) Envelope curves in X-direction, (c) Lateral load versus lateral displacement hysteretic behavior in Y-direction, (d) Envelope curves in Y-direction.

Table 3. Comparison of test and numerical results.

Loading Patterns		$H_{l,m}/$ H_y	$\partial_{l,m}/$ ∂_y	$\partial_{l,95}/$ ∂_y
Uni	Test	1.23	2.23	2.87
	Analysis	1.35	1.89	2.50
BI-L27	Test	1.15	3.35	3.76
	Analysis	1.32	2.35	2.70
BI-L45	Test	1.11	2.46	3.37
	Analysis	1.29	1.89	2.95

Uni = Unidirectional, BI = Bidirectional

H/Hy

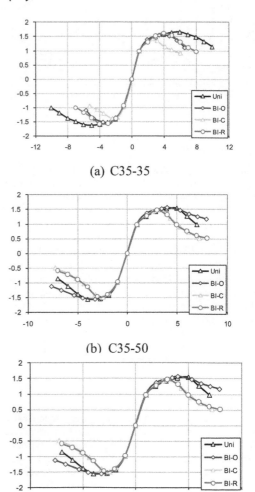

(a) C35-35

(b) C35-50

(c) C46-35

$\delta/\delta y$

Figure 9. Comparison of envelope curves in X-direction for various loading patterns.

Table 4. Comparison of strength and ductility performance for various loading patterns.

Specimen	Loading	$H_{l,m}/H_y$	$\partial_{l,m}/\partial_y$	$\partial_{l,95}/\partial_y$
C35-35	Uni	1.7	6.0	6.4
	BI-O	1.6	2.7	4.3
	BI-C	1.4	2.4	2.9
	BI-R	1.6	4.0	4.4
C35-50	Uni	1.6	4.0	4.5
	BI-O	1.6	3.2	4.8
	BI-C	1.5	2.4	3.1
	BI-R	1.5	3.0	3.2
C46-35	Uni	1.5	3.0	3.4
	BI-O	1.5	2.7	3.1
	BI-C	1.3	1.8	2.0
	BI-R	1.4	2.0	2.4

width-to-thickness and slenderness ratios for a particular loading pattern. The minimum strength and ductility were found to occur under circular (BI-C) loading pattern and the maximum were under unidirectional loading. The results in Figure 9 indicate that the BI-C loading path has caused the highest postpeak strength degradation. It is understood from the results in Table 4 that the values of H_m/H_y, of C35-35 (1.42) and C35-50 (1.48) under loading type BI-C do not differ much. The corresponding values under loading type BI-O (1.55 and 1.57) are very close. This means that the effect of λ on the strength is not significant when circular loading is concerned. On the other hand, under the loading type BI-R, the values of H_m/H_y of C35-35 (1.61) and C35-50 (1.47) differ by about 8%. Thus, it seems that the effect of λ on the strength varies with the loading type. Moreover, similar comparisons revealed that the effects of parameters R_t and R_f on the strength are different with different loading types. Similar to the strength, the effects of the parameters on the ductility also significantly vary with the type of loading.

6 CONCLUSIONS

A finite element modeling procedure for analyzing steel columns subjected to constant axial loads and bi-directional cyclic loads is presented in this paper. The numerical procedure was verified by analyzing previous test specimens. Several columns were designed in view of identifying the effects of structural parameters, such as the width-to-thickness ratio and slenderness ratio, on the behavior when columns undergo different bi-directional loading paths. The obtained results from this study confirm the importance of considering the behavior of steel columns under multidirectional loading. The multidirectional tests and finite element analysis results showed that the behavior of a tubular column under multidirectional loading becomes complex and exhibits a circular trajectory once local buckling occurs. The local buckling bulge in the multidirectional loading case tends to develop monotonically due to the circular trajectory. As a result, the

residual deformation becomes larger. On the contrary, the unidirectional loading test and analysis are likely to underestimate the damage and the residual displacements caused by an earthquake. It is concluded that the effects of multidirectional loading should be considered in ductility evaluation and seismic resistance design of steel structures.

REFERENCES

ABAQUS/STANDARD user's manual, Version 6.11, 2014. HKS Inc., Pawtucket, R.I.

Banno, S., I. H.P. Mamaghani, T. Usami, E. Mizuno, 1998. Cyclic elastoplastic large deflection analysis of thin steel plates. *Journal of Engineering Mechanics, ASCE*, 124(4), 363–370.

Bousias, S., Verzeletti, G., Fardis, M.N., Gutierrez, E., 1995. Load-path effects in column biaxial bending with axial force. *Journal of Structural Mechanics*, pp. 596–605.

Chen, W. F., Duan, L., 2000. *Bridge Engineering Handbook*, CRC Press, Boca Raton, FL.

Mamaghani, I.H.P 2008. Seismic design and ductility evaluation of thin-walled steel bridge piers of box sections, *Transportation Research Record: Journal of the Transportation Research Board*, Volume 2050, pp. 137–142.

Mamaghani, I.H.P., 1996. Cyclic elastoplastic behavior of steel structures: theory and experiment, *Ph.D. Thesis*, Nagoya University, Nagoya, Japan.

Mamaghani, I.H.P. 2006. Inelastic Cyclic Analysis and Stability Evaluation of Steel Braces, *Structural Stability Research Council*, February 8–11, San Antonion, Texas, pp. 281–300.

Mamaghani, I.H.P. 2005. Seismic performance evaluation of thin-walled steel tubular columns, *Structural Stability Research Council*, Montreal, Quebec, Canada. pp. 489–506.

Mamaghani, I. H. P., C. Shen, E. Mizuno, and T. Usami, 1995. Cyclic behavior of structural steels. I: experiments, *J. Engrg. Mech.*, ASCE, 121(11), 1158–1164.

Mamaghani, I. H. P., T. Usami, and E. Mizuno, 1996a. Inelastic large deflection analysis of steel structural members under cyclic loading, *Engineering Structures*, UK, Elsevier Science, 18(9), 659–668.

Mamaghani, I. H. P., T. Usami, and E. Mizuno, 1996b. Cyclic elastoplastic large displacement behaviour of steel compression members, *J. Structural Engineering*, JSCE, Vol. 42A, 135–145.

Mamaghani, I. H. P., T. Usami, and E. Mizuno, 1997. Hysteretic behavior of compact steel box beam-columns. *Journal of Structural Engineering*, JSCE, Japan, Vol. 43A, 187–194.

Ohnishi, A., Susantha, K.A.S., Mizuno, T., Okazaki, S., Takahara, H., Aoki, T., Usami, T., 2003. Experimental study on strength and ductility of steel bridge piers subjected to 2-directional seismic loading. *Proceedings of the Annuals of JSCE*, No. 58, pp. 7–10.

Saatcioglu, M., Ozcebe, G., 1989. Response of reinforced concrete columns to simulated seismic loading." *ACI Structural Journal*, pp. 3–12.

Shen, C., Mamaghani, I.H.P., Mizuno, E. and Usami, T. 1995. Cyclic behavior of structural steels. II: theory. *Journal of Engineering Mechanics*, ASCE, USA, Vol.121, No.11, 1165–1172.

Takizawa, H., Aoyama, H., 1976. Biaxial effects in modeling earthquake response of R/C structures." *Earthquake Engineering and Structural Dynamics*, Vol. 4, pp. 523–552.

Watanabe, E, Sugiura, K, Oyawa, W.O., 1999. Effects of multi-directional displacement paths on the cyclic behavior of rectangular hollow steel columns." *Journal of Structural Mechanics and Earthquake Engineering*, Japan Society of Civil Engineers, JSCE, Vol.17, No 1, pp. 79–94.

Zeris, C.A., Mahin, S.A., 1991. Behavior of reinforced concrete structures subjected to biaxial excitation, *Journal of Structural Engineering*, ASCE, Vol. 117(9), pp. 2657–2673.

Tubular Structures XV – Batista, Vellasco & Lima (eds)
© *2015 Taylor & Francis Group, London, ISBN 978-1-138-02837-1*

Experimental study and associated analysis of inner-stiffened cold-formed SHS steel columns

A.Z. Zhu & H.P. Zhu
School of Civil Engineering & Mechanics, Huazhong University of Science & Technology, Wuhan, P.R. China

Y. Lu
Institute for Infrastructure & Environment, School of Engineering, The University of Edinburgh, UK

ABSTRACT: Axial compression tests were conducted to investigate the compressive behaviour of cold-formed steel stub columns with relatively thick walls. In total, 10 square hollow section (SHS) specimens with wall thicknesses varying from 6 mm to 10 mm and a nominal tubular length of 600mm were tested. Four inner-stiffener arrangements were considered to study the effects of the stiffeners on the behavior of the tubular specimens. A series of tensile coupons were cut from the cold-formed SHSs at different characteristic areas to obtain a full picture of the enhanced material properties due to the cold-forming process. The sectional stresses obtained from the SHS column experiments were compared with the predicted stresses using methods specified in relevant design codes, AISI and Chinese GB codes in particular, taking into account the enhanced material strength of the cold-formed steel. It is found that the AISI method is slightly conservative in predicting the sectional stress of the specimens, whereas the GB method slightly overestimates the stress capacity. In general, both methods produce an acceptable prediction of the actual sectional stress, and they both tend to result in a better prediction for specimens with inner stiffeners.

1 INTRODUCTION

Thin-walled cold-formed structural steel sections with wall thickness ranging from 0.4 to 6 mm are commonly used in construction due to their advantages of superior strength-to-self-weight ratio, ease of construction, and cost-effectiveness. Traditional thin-walled sections with a variety of open and closed section configurations have been studied extensively by Ellobody et al. (2005), Silvestre et al. (2013), Young (2008), Derrick et al. (2011), Nguyen et al. (2012), Gardner (2010), Afshan et al. (2013), and Zhu et. al (2011), etc.

Cold-formed sections are manufactured at ambient temperature and hence undergo plastic deformations causing strain hardening of the material. Due to the varying level of plastic deformation around the cold-formed sections, non-uniformity occurs, with the corner regions being the most affected. The effects of cold-working on the strength enhancements of the cold-formed corners, the flat plates and the section as a whole have been studied in most of the above mentioned studies. Methods for taking into account the corner strength enhancements for the cross-section design using an increased average yield strength have been incorporated into design codes such as the AISI specifications (2007), the AS/NZS standard (2005), and the Chinese GB code (2002).

To provide continuous support to the thin walls so as to enhance the buckling stress of the thin-walled cold-formed sections, effects of stiffeners on cold-formed

channel and angle sections have also been studied by Ellobody et al. (2005), Derrick et al. (2011), and Macdonald et al. (2008).

With the development of cold forming technology, nowadays carbon steel plates of up to 25mm in thickness can be fabricated successfully into structural shapes (Guo et al., 2007 and Yu, 2000). With thicker walls, the overall sectional dimensions of the cold-formed members can be larger, and large square and rectangular hollow sections (SHS and RHS) as 500 × 500 × 16 mm and 400 × 600 × 16 mm have been studied (Tao, 2005). In line with this trend, experimental and numerical investigations on cold-formed steel columns with wall thicknesses ranging from 6 to 16 mm have been studied by researchers in recent years. Guo et al. (2007) investigated the effect of the cold-formed process on the strengths of steel coupons with thicknesses of 8, 10, and 12 mm. Mean strength enhancement of 4% and 44% were respectively obtained in the cold-formed flat and corner coupons as compared to the unformed coupons. Hu et al. (2011) conducted experimental investigations on both the material properties and sectional strengths of the cold-formed steel SHS and RHS tubes with a plate thickness of 9.2mm. Mean strength enhancement of 10% and 47% were respectively obtained for the yield strength of the cold-formed flat and corner coupons. In studies conducted by Afshan et al. (2013) and Rossi et al. (2013), wall thicknesses of 5, 6 and 8 mm were investigated, whereas Tong et al. (2012) conducted

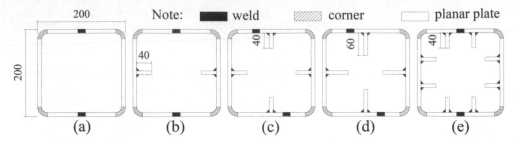

Figure 1. Unstiffened and stiffened sections considered in this study (mm).

Table 1. Test strength of the tensile coupons.

Specimen	A_P (mm^2)	σ_u (MPa)	$\sigma_{0.2}$ (MPa)
PP-6	221.51	537.37	437.88
WP-6	212.57	582.00	535.54
CP-6	94.62	614.00	539.76
S-6	186.73	531.63	410.79
PP-10	341.61	445.67	381.68
WP-10	359.20	526.93	505.00
CP-10	318.34	495.47	433.55
SP-8	244.30	551.07	453.59

a test study on the material properties and residual stresses of cold-formed carbon steel with 10 mm- and 16mm-thick walls.

Despite the above mentioned studies on cold-formed sections with thicker walls, there is still a lack of information about the calculation method for such sections taking the enhanced sectional stresses into consideration. The effectiveness of using longitudinal inner stiffeners to enhance the buckling stress of the SHS and RHS steel sections in the case of thicker walls is also not well understood.

This paper presents an experimental study with associated calculation analysis of the cold-formed SHS carbon steel stub columns with 6 mm- and 10 mm-thick walls. A new type of longitudinal inner stiffener was used to improve the mechanical behavior of the cold-formed SHS steel tubes. Coupon specimens for flat plates, corners, welded plates and stiffeners were tested under axial tension to obtain the enhanced strengths of the cold-formed steel. The effects of different stiffener arrangements, including the number and the width of the stiffeners, on the rigidity, ductility, failure modes and the average sectional stress are investigated. Adequate use of the enhanced yield and ultimate strengths in predicting the sectional stress is examined by comparing the predicted values with the test results.

2 EXPERIMENTAL PROGRAM

2.1 General

A total of 10 cold-formed steel stub column specimens, including 2 unstiffened and 8 longitudinally

inner-stiffened SHS columns were tested under axial compression. Figure 1 shows the unstiffened section and four stiffened sections with different number or width of the inner stiffeners. The thickness of the inner stiffeners in specimens with 6 mm- and 10 mm-thick walls were 6 mm and 8 mm, respectively. The width of the stiffeners was 40 mm, except for section "d" which had 60 mm-wide stiffeners.

All column specimens had the same nominal cross-sectional size of 200 × 200 mm, and the same column length of 600 mm. Steel with nominal yield strength of 345 MPa was used to fabricate the column specimens with 6 mm-thick walls, while steel with a strength of 235 MPa was used to fabricate the specimens with 10 mm-thick walls. Coupon tensile tests were conducted on the flat coupons, the welded coupons, the corner coupons and the stiffener coupons in order to obtain the profile of the enhanced strengths among different parts of the cold-formed SHS sections.

2.2 Test of tensile coupons

Plates cut from flat parts, the welded parts, the corner parts, and the inner stiffeners of the SHS tubes were all processed into short standard tensile coupons according to the procedures described by Tang & Ye (1999). The axial tension test was conducted using a Universal Testing Machine. In total, 24 steel tensile coupons in 8 groups were prepared, with each group consisting of 3 identical samples. The average test strengths of the coupons in different groups are reported in Table 1. The specimen labels in Table 1 are defined according to the coupon configuration and wall thickness. Taking specimen "PP-6" as an example, (1) "PP" indicates coupons cut from the flat plates; "WP", "CP", and "S" indicate coupons from the welded, corner, and inner stiffener plates of the cold-formed section, respectively, and (2) the number "6" that follows indicates the thickness of the coupons, "6" means 6 mm; "8" and "10" means 8 mm and 10 mm, respectively.

The measured stress-strain curves of the tensile coupons with thicknesses of 6mm and 10 mm are shown in Figures 2 and 3, respectively. None of the curves exhibits any significant yielding stage due apparently to the welding or cold-forming fabrication process. The curves from the flat and corner coupons show a marked stress strengthening (hardening) stage,

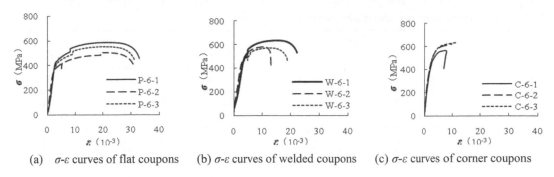

(a) σ-ε curves of flat coupons (b) σ-ε curves of welded coupons (c) σ-ε curves of corner coupons

Figure 2. σ−ε curves of steel coupons with thickness of 6 mm.

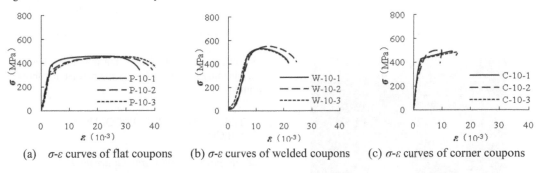

(a) σ-ε curves of flat coupons (b) σ-ε curves of welded coupons (c) σ-ε curves of corner coupons

Figure 3. σ−ε curves of steel coupons with thickness of 10 mm.

(a) Flat coupons (b) Welded coupons (c) Corner coupon

Figure 4. Typical failure mode of coupon specimens.

which represents another significant result of the cold-forming process. The ductility of the corner coupons is markedly smaller than the flat coupons due to the much higher level of plastic deformation that occurred in the corner regions during the cold-forming process.

For a comparison of the yield strength, the 0.2% tensile proof strength $\sigma_{0.2}$ is adopted as the yield stress. Table 1 summarizes the average yield stress $\sigma_{0.2}$ (from three identical coupons) and average ultimate stress σ_u. The average cross-sectional area A_p calculated from the measured dimensions of the coupons is also listed. The final rupture positions of all the tension coupons occurred within the gauge lengths. Thus all the coupons were successfully tested. Typical failure modes of these coupons are shown in Figure 4.

2.3 Test of column specimens

The cold-formed SHS column specimens were classified into two groups according to the two nominal tubular thicknesses of 6 mm and 10 mm. All tubular sections were formed by the cold-forming process, whereby the sheet material was first formed into C-sections, followed by subsequent stiffening with longitudinal stiffeners for the stiffened sections. Two C-sections with or without stiffeners were then welded into the final tubular form. Two end-plates were subsequently welded at the two ends of the closed SHS tubes. The overall size of the end-plates was 240×240 mm, and the thickness was 10 mm for the 6 mm-thick tube specimens and 16 mm for the 10 mm-thick tube specimens. Butt welds and fillet welds were used, respectively, to build up the tubular walls and the inner stiffeners. All the fabrication processes were carried out in a roll forming steel plant.

The measured total sectional area of the SHS columns A and the ratio of the measured sectional area of the inner stiffeners A_s to the area A are summarized in Table 2. The specimen labels in Table 2 are defined according to the section configuration and wall thickness, for example "Pa6", where "Pa" indicates section "a" (or "b" through "e"), as shown in Figure 1; the number "6" indicates the thickness of the SHS tubes, "6" = 6 mm, otherwise "10" = 10 mm.

A compression testing machine with a design capacity of 5000 kN was used to apply the axial compressive

Table 2. Test and calculation strengths of the cold-formed SHS columns.

Specimens	A (mm^2)	A_s/A	N_t (kN)	ΔN_t (%)	f_t (MPa)	Δf_t (%)	f_{ya} (MPa)	$f_{ya,GB}$ (MPa)	f_t/f_{ya}	$f_t/f_{ya,GB}$	$f_t/f_{ya,GB2}$
Pa6	4568.3	0	1830	–	400.59	–	456.80	449.95	0.88	0.86	0.94
Pb6	5104.0	0.10	2070	13.11	405.57	1.24	451.41	444.88	0.90	0.88	0.96
Pc6	5613.7	0.17	2560	35.27	456.03	13.67	445.97	440.97	1.02	0.99	1.08
Pd6	6186.6	0.23	2840	39.45	459.05	12.82	444.17	437.33	1.03	1.00	1.09
Pe6	6647.7	0.30	3050	42.96	458.81	12.68	442.06	434.86	1.04	1.01	1.10
Pa10	7309.4	0	3060	–	418.64	–	400.44	443.48	1.04	0.98	1.02
Pb10	8020.8	0.08	3370	10.13	420.16	0.36	400.64	435.60	1.05	0.98	1.02
Pc10	8631.6	0.15	3570	15.13	413.60	−1.20	398.86	436.04	1.04	0.97	1.00
Pd10	9470.4	0.21	3830	21.57	404.42	−3.44	399.51	423.55	1.01	0.95	0.98
Pe10	9905.6	0.26	4320	32.90	436.12	4.32	402.32	420.24	1.09	1.02	1.06
Mean									1.01	0.96	1.03
SD									0.06	0.05	0.05

Figure 5. Test setup of cold-formed SHS column specimen.

load. The axial compression test was designed to determine the axial bearing capacities and failure patterns of the cold-formed SHS columns. A typical test setup of the column specimens is shown in Figure 5.

Two displacement meters (LVDTs) were used to measure the displacements of the top plate (relative to the bottom of the columns), and four longitudinal strain gauges were installed at the mid-height to measure the axial strain of the steel tubes. To ensure that the load was applied evenly across the cross sections, a thin layer of fine sands was carefully applied at the top and bottom of the specimens. Meanwhile, preliminary tests within the elastic range were conducted during which the position of the specimen was adjusted to ensure that the measured strains from the four strain gauges were eventually distributed. The adjustment was considered complete when the difference between each individual strain and the average strain was no greater than 5%. Considering the estimated capacities ($F_u = \sigma_{0.2} \cdot A$) for each specimen and the convenience of controlling the loading process, a load interval of 200 kN was applied before the total load reached 50% of the estimated capacity. After that a load interval

of 100 kN was applied until the maximum load. At each loading step the load was maintained for about 3 minutes.

3 TEST RESULT OF THE COLUMN SPECIMENS

3.1 Rigidity, ductility, and failure mode

Figures 6 and 7 show the average sectional stress σ versus average strain ε curves of column specimens with 6 mm- and 10 mm-thick tubular walls, respectively. The stress σ was obtained by dividing the total test load by the total area A (including the main section and the stiffeners) shown in Table 2. The average strain ε was calculated according to the measured strains of the four gauges attached at the mid-height of the specimens. It is noted that in some curves the final loading step(s) was missing, and this was due to either or both of the following reasons, (1) the strain gauges attached at the mid-height of the specimens failed before the axial load reached the maximum bearing capacity; (2) the specimen failed with buckling of the steel tubes when the peak load was reached, featuring a rapid decline in the load carrying capacity accompanied by large axial deformation. It was difficult to record the final strains in such cases. It is also noted that specimen Pb10 had failure of two gauges, the corresponding σ versus ε curve was not obtained.

As can be seen from Figures 6 and 7, all $\sigma - \varepsilon$ curves show good linear property during the initial loading stage. The rigidity (slope of the σ versus ε curve) at this stage is almost the same among the specimens with different sections. The slopes of the curves tend to exhibit more notable differences as the load increases, and this is particularly true for specimens with 6 mm-thick tubes; the slopes during the second half loading stage generally increase with the sectional area of the inner stiffeners. For specimens with 10 mm-thick tubes, however, only specimen Pe10, which had densely arranged inner stiffeners (see Figure 1), experienced a marked increase in the second-stage slope as compared with specimens of the same wall-thickness. These results indicate that the rigidity of the specimens

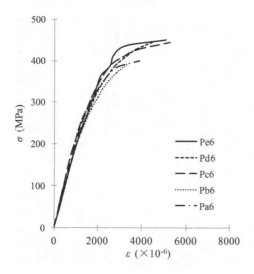

Figure 6. $\sigma-\varepsilon$ curves of specimens with 6 mm-thick walls.

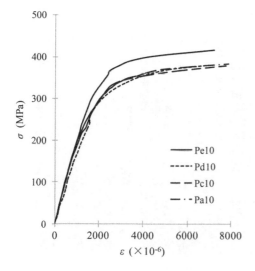

Figure 7. $\sigma-\varepsilon$ curves of specimens with 10 mm-thick walls.

with 10 mm-tubes was not affected by the inner stiffeners unless a considerable amount of stiffeners was used (in this case 8 stiffeners).

Although the whole loading process was not obtained for some specimens, overall the specimens showed good deformability when the axial load approached the sectional capacity. For specimens with more stiffeners or wider stiffeners, the deformability tends to be better. The axial strain at the maximum load reached about 6000 $\mu\varepsilon$ for specimens with 6 mm-thick tubular walls and about 8000 $\mu\varepsilon$ for specimens with 10 mm-thick tubular walls, and this demonstrates that the column specimens with thicker tubular wall have better deformability. There are two possible reasons. One is that the specimens with thicker tubular wall were made from the lower strength (235 MPa) steel which has better deformability in the first place.

The other is that the thicker tubes have smaller width-to-thickness ratios and thus local buckling generally occurs later than the thinner tubes.

Figures 8 and 9 show typical failure modes for specimens with 6 mm- and 10 mm-thick tubular walls, respectively. It is interesting to note that buckling occurred in either outward or inward but there was no clear connection to which sides of the tubular section (whether with butt weld or not) the buckling occurred. Moreover, as shown in Figures 8(b) and 9(b), the buckling deformation of the welded walls without inner stiffeners was more severe than that of the stiffened walls. It also appears that the number of stiffeners affected the buckling deformation more than the width of the stiffeners.

3.2 Sectional capacity and strength

The test capacities N_t and average sectional stress f_t of the cold-formed SHS column specimens are summarized in Table 2. The average strength f_t is calculated as $f_t = N_t/A$. From Table 2, the following observations can be made:

(1) The test capacities N_t of specimens with both 6 mm- and 10 mm-thick tubular walls increase significantly with the ratio A_s/A.

(2) Compared with the unstiffened specimens Pa6, the test capacity of the specimens Pb6, Pc6, and Pe6 increased by 13.1%, 35.3%, and 43%, respectively, as the number of the inner stiffener increased. It is noteworthy that the increase N_t of specimen Pd6 (by 39.5%) is only marginally greater than that of the specimen Pc6, although the stiffener width increased by 50%. This suggests that the overall strength may only be enhanced moderately by using much wider stiffeners. Similar conclusions can be drawn from the results of the specimens with 10 mm-thick tubular walls.

(3) Compared with the unstiffened specimen Pa6, the most significant increase of the stress f_t, i.e. 13.7%, was achieved when four inner stiffeners were used in Specimen Pc6. The insignificant increase of specimens Pb6 (1.24%) indicates that the stiffening effect of using just two inner stiffeners may be negligible. For the 10 mm-thick specimens, the most significant increase of the stress f_t as compared the unstiffened specimen Pa10 occurred in Specimen Pe10 but it only accounted for 4.32%. Therefore, the enhancement effect of the inner stiffeners in the 10 mm-thick specimens was generally not as significant as in the 6 mm-thick specimens.

(4) For the 10 mm-thick specimens, the stress f_t of unstiffened specimen Pa10 is 418.6 MPa, which is about 10% higher than the tensile yield stress of the flat plates shown in Table 1 (381.7 MPa). This is attributable to the contribution of the enhanced corner and welded parts of the tubular sections from the cold-forming process. For the 6 mm-thick reference specimen Pa6, however, the stress f_t is 400.6 MPa, which is about 10% lower than

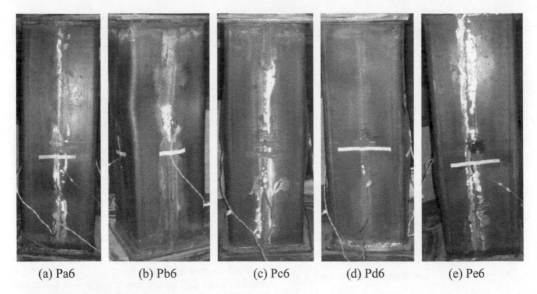

| (a) Pa6 | (b) Pb6 | (c) Pc6 | (d) Pd6 | (e) Pe6 |

Figure 8. Typical failure mode for specimens with 6 mm-thick SHS tubes.

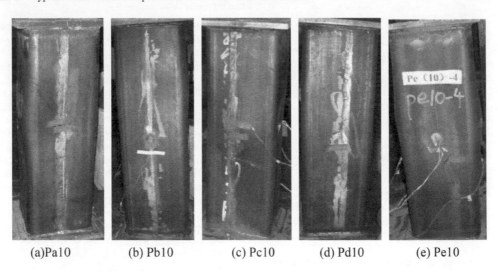

| (a)Pa10 | (b) Pb10 | (c) Pc10 | (d) Pd10 | (e) Pe10 |

Figure 9. Typical failure mode for specimens with 10mm-thick SHS tubes.

the yield stress of the flat plates (437.9 MPa). This may be explained by the fact that, despite the enhanced strength in the corner and welded parts, the overall strength was influenced more by the much larger sectional width to thickness ratio of the tubes.

4 COMPARATIVE CALCULATIONS

4.1 *AISI and similar specification method*

According to Clause B2.1 of the North American Specification for Design of Cold-formed Steel Structural Members (AISI, 2007), as well as Clause 2.2.1 of the Standards Australia/Standards New Zealand for Cold-formed Steel Structures (AS/NZS, 2005), the effective widths of the unstiffened cold-formed SHS tubular

specimens with both 6 mm- and 10 mm-thick walls are equal to the flat plate width. The total sectional area can be used as the effective sectional area when predicting the whole sectional capacity of the specimen.

As a simplification for the calculation of the specimens, especially the stiffened SHS specimens which are not covered by the above specifications and standards, it has been suggested (Guo et al. 2007) that a design yield stress f_{ya} of the full section be analyzed as a replacement of the nominal buckling stress or critical stress f_n. Herein we calculate the stress f_{ya} considering the effects of the cold-forming process by the following formula according to Clause A7.2 of the AISI specification:

$$f_{ya} = Cf_{yc} + Sf_{ys} + (1 - C - S)f_{yf} \qquad (1)$$

where:

- C = ratio of total bend/corner cross-sectional area to total cross-sectional area of the full section,
- S = ratio of total stiffener cross-sectional area to total cross-sectional area of the full section,
- f_{yc} = tensile yield stress of bends/corners and given by $f_{yc} = B_c f_{yv}/(r_i/t)^m$, where
- $B_c = 3.69 f_{uv}/f_{yv} - 0.819(f_{uv}/f_{yv})^2 - 1.79$
- f_{uv} = tensile ultimate stress of unformed steel,
- f_{yv} = tensile yield stress of unformed steel,
- r_i = inside bend radius,
- t = plate thickness,
- $m = 0.192 f_{uv}/f_{yv} - 0.068$,
- f_{ys} = yield stress of the stiffener coupons,
- f_{yf} = yield stress of the flat coupons.

In the calculations, the stresses f_{uv} and f_{yv} are replaced by stresses σ_u and $\sigma_{0.2}$ of flat plates shown in Table 1, respectively. Hence the stress enhancement and the effect of cold-forming process on the ultimate-to-yield stress ratio of the cold-formed steel were both considered. The ratio r_i/t is assumed as the nominal value of 1.5 for all specimens. C and S are determined by the corresponding measured sectional areas. S equals to zero for specimens with unstiffened sections.

4.2 GB-2002 method

According to Appendix C in the Chinese code (GB 2002), the average design yield stress f_{ya} considering the stress enhancement caused by the cold-forming process can be calculated by:

$$f_{ya,GB} = \left[1 + \frac{\eta(12\gamma - 10)t}{l} \sum_{i=1}^{n} \frac{\theta_i}{2\pi}\right] f_{yv} \qquad (2)$$

where the subscript "GB" is used to denote the GB code, and

- η = cold-formed coefficient for SHS and RHS stub columns. For SHS columns, $\eta = 1.0$,
- γ = ratio of ultimate stress to yield stress of steel ($\gamma = 1.58$ for steel with nominal yield stress of 235 MPa, $\gamma = 1.48$ for steel with nominal yield stress of 345 MPa),
- n = total number of the corners in the cross section,
- θ_i = angle of the number i corner (radians),
- l = centre-line length of the cross section,
- t = plate thickness,
- f_{yv} = tensile yield stress of unformed steel.

For the columns considered in this study, each column has four right angles, therefore the sum of the angle ratios $\theta_i/2\pi$ equals to 1.0. The average measured yield stress $\sigma_{0.2}$ of the flat plates shown in Table 1 is used for the stress f_{yv}. The calculation of the centre-line length l is simplified as A/t for the specimens with unstiffened sections according to the recommendation in the GB code. For those specimens with stiffened sections, it is calculated as the sum of the center-line length of the tubes and the nominal center-line length of the inner stiffeners. The former part is calculated using the same simplified method as the specimens with unstiffened sections while the latter length equals the sum widths of all the inner stiffeners. Meanwhile, the estimated yield stresses of the unformed plates with thicknesses of 6 mm and 10 mm are determined according to previous studies about the cold-forming effect by Hu et al. (2011) and Guo et al. (2007). The average stress enhancements of 9% and 4% for the cold-formed flat plates with 6 mm- and 10 mm-thick walls, respectively, are considered when predicting the estimated stress (f_{yv}) of the unformed steel. The corresponding calculated sectional stress is denoted as $f_{ya,GB2}$.

4.3 Analysis and comparison

The calculated stresses $f_{ya}, f_{ya,GB}$ and the ratios of the actually measured stress to the calculated stress, i.e., f_t/f_{ya} and $f_t/f_{ya,GB}$, are presented in Table 2. As a comparative analysis, the ratio $f_t/f_{ya,GB2}$ is also included. The results and comparisons are discussed as follows:

(1) Except for the specimens Pa6 and Pb6, the calculated stress f_{ya} of other specimens are smaller than their respective test stress values, with the ratios f_t/f_{ya} ranging from 1.01 to 1.09. Thus the AISI method appears to overestimate the sectional strength for the unstiffened and partially stiffened 6-mm thick columns (SHS-a-6 and SHS-b-6) by more than 10%, but slightly underestimate in the cases of well-stiffened 6-mm columns and all the 10-mm thick columns. The mean value and the standard deviation (SD) of all the ratios are respectively 1.01 and 0.06. Such results suggest that the AISI specification method appears to predict reasonably well the sectional strength albeit slightly on the conservative side.

(2) The calculated stress $f_{ya,GB}$ is generally higher than the test stress. The mean value and standard deviation (SD) of all the ratios $f_t/f_{ya,GB}$ are 0.96 and 0.05. Thus the GB code method slightly overestimates the sectional stress when the enhanced yield stress of the cold-formed flat plate, instead of the stress of the original/unformed steel, is used. It should be noted that Eq.(2) is a recommended method which takes the stress enhancement resulting from the cold-formed process into account, as described in the Appendix C of the GB code. Hence the stress enhancement has effectively been considered "twice" when the stress f_{yv} is replaced by the enhanced stress, which must lead to an overestimation in the calculated stress $f_{ya,GB}$. However, if the enhancements of 9% and 4% were respectively considered to predict the stresses of the specimens with 6 mm- and 10 mm-thick walls, the GB code method is just slightly conservative with a total mean value of ratios $f_t/f_{ya,GB2}$ equal to 1.03.

(3) As the ratio A_s/A increases in the stiffened columns, the ratios f_t/f_{ya} of specimens with 6 mm-thick tubular walls tend to increase, whereas the ratios of specimens with 10 mm-thick wall exhibit a decreasing trend except for the specimen Pe10. From the mean ratios for specimens with the

6 mm- and 10 mm-thick tubular walls (1.01 and 1.08, respectively), the method tends to yield better results for specimens with thinner tubes or more generally tubes with a larger width-to-thickness ratio.

(4) The ratios $f_t/f_{ya,GB}$, on the other hand, tend to be closer to 1.0 for thinner tubes as the ratio A_s/A increases. The mean ratios of specimens with the 6 mm- and 10 mm-thick tubular walls are 0.95 and 0.98, respectively. For the ratio $f_t/f_{ya,GB2}$, the mean values are 1.03 and 1.02. These results suggest that the GB method can be used to calculate the sectional stress of specimens with both thicker and thinner tubes with an acceptable accuracy, although it is currently recommended only for cold-formed thin-walled steel columns with a thickness $t \leq 6$ mm in the code.

5 CONCLUSIONS AND RECOMMENDATIONS

An experimental program has been conducted in which 10 cold-formed steel stub columns with unstiffened and inner-stiffened SHS sections have been tested under axial compression. The enhanced stresses of the cold-formed steel and the welded plates were obtained via conducting coupon tensile tests. The effects of the inner stiffeners on the rigidity, failure modes, deformability and the average sectional stress of the specimens were evaluated. The results show that, while the number and width of the inner stiffeners do not appear to affect the rigidity, ductility and failure mode significantly, the presence of one or more stiffeners on each wall does enhance both the sectional capacity and buckling stress if the width of the stiffeners is sufficient – based on the present study the width-thickness-ratio would be about 33. If the width-to-thickness ratio is smaller, e.g. 20 based the present experiments, however, only the cross-sectional capacity could be enhanced.

The AISI specification method (2007) and the GB code method (2002) were used to calculate the sectional stress of the column specimens, with the enhanced yield stress of the cold-formed steel flat plates being considered. The calculated stresses were compared with the corresponding experimental results. The AISI specification method appears to slightly underestimate the sectional stress of the specimens, whereas the GB code method tends to slightly overestimate the sectional stress and this is deemed to be attributable to "double-counting" the yield stress enhancement of the cold-formed steel plate. Both the AISI specification and GB code methods tend to predict the sectional stresses for specimens with stiffened sections more accurately.

ACKNOWLEDGEMENT

The authors gratefully acknowledge the support of the National Natural Science Foundation of China (No. 51108204), the Science Foundation of the Ministry of Education of China (No. 20100142120072) and the Fundamental Research Funds for the Central Universities (No. 2014QN210).

REFERENCES

Afshan S., Rossi B., Gardner L. 2013. Strength enhancements in cold-formed structural sections-Part I: Material testing. Journal of constructional steel research, 83: 177–188.

American Iron and Steel Institute (AISI). 2007. North American specification for the design of cold-formed steel structural members (NAS), AISI, Washington, DC.

Chinese Standard 2002. Technical code of cold-formed thin-wall steel structures. In: GB 50018. Beijing (China).

Ellobody E. and Young B. 2005. Behavior of cold-formed steel plain angle columns. Journal of structural engineering, 131: 457–466.

Gardner L., Saari N., Wang F. 2010. Comparative experimental study of hot-rolled and cold-formed rectangular hollow sections. Thin-Walled Structures, 48: 495–507.

Guo Y.J., Zhu A.Z., et al. 2007. Experimental study on compressive strengths of thick-walled cold-formed sections. Journal constructional steel research, 63: 718–23.

Hu S. D., Ye B., Li L. X. 2011. Materials properties of thick-wall cold-rolled welded tube with a rectangular or square hollow section. Construction and Building Materials, 25: 2683–2689.

Macdonald M., Heiyantuduwa M.A., Rhodes J. 2008. Recent developments in the design of cold-formed steel members and structures. Thin-Walled Structures, 46: 1047–1053.

Nguyen V.B., Wang C.J., Mynors D.J. et al. 2012. Compression tests of cold-formed plain and dimpled steel columns. Journal of constructional steel research, 69: 20–29.

Rossi B., S Afshan., Gardner L. 2013. Strength enhancements in cold-formed structural sections -Part II: Predictive models. Journal of constructional steel research, 83: 189–196.

Standards Australia/Standards New Zealand, AS/NZS, 2005. Cold-formed steel structures, 4600: Sydney, Australia.

Silvestre N., Dinis P. B., Camotim D. 2013. Developments on the design of cold-formed steel angles. Journal of structural Engineering, 139: 680–694.

Tang Y.Q., Ye W.M. 1999. A handbook of testing technique in civil engineering. Shanghai, China: Tongji University Press (in Chinese).

Tao Z., Wang Z. B., Han L. H. 2005. Behavior of rectangular cold-formed steel tubular columns filled with concrete. Engineering Mechanics, 23: 147–155 (in Chinese).

Tong L.W., Hou G., Chen Y.Y., et al. 2012. Experimental investigation on longitudinal residual stresses for cold-formed thick-walled square hollow sections. Journal of constructional steel research, 73: 105–116.

Yap D., and Hancock G.J. 2011. Experimental study of high-strength cold-formed stiffened-web C-sections in compression. Journal of structural Engineering, 137: 162–172.

Young B. 2008. Research on cold-formed steel columns. Thin-walled structures, 46: 731–740.

Yu W.W. 2000. Cold-formed steel design (3rd Edition). United States of America.

Zhu A.Z., Luo L.L., Zhu H.P. 2012. Experimental study on concrete-filled cold-formed steel tubular stub columns. Tubular Structures XIV-Proceedings of the 14th International symposium on tubular structures, 73–80, (London).

Zhu J.H., and Young B. 2011. Cold-formed-steel oval hollow sections under axial compression. Journal of structural engineering, 137: 719–727.

Tubular Structures XV – Batista, Vellasco & Lima (eds)
© 2015 Taylor & Francis Group, London, ISBN 978-1-138-02837-1

On the first order and buckling behaviour of thin-walled regular polygonal tubes

R. Gonçalves
CEris, ICIST and Faculdade de Ciências e Tecnologia, Universidade Nova de Lisboa, Portugal

D. Camotim
CEris, ICIST and Instituto Superior Técnico, Universidade de Lisboa, Portugal

ABSTRACT: This paper presents the results of an ongoing investigation concerning the structural behaviour of thin-walled single-cell regular convex polygonal section (RCPS) tubes, such as those widely employed in the construction industry. It is well-known that these tubes are highly susceptible to local (plate-type) deformations but, conversely, it is also true that their distortional behaviour has not been fully understood yet. Taking advantage of the unique modal decomposition features of Generalised Beam Theory (GBT) and its computationally efficient specialization for RCPS recently developed by the authors (Gonçalves & Camotim 2013a), the first-order and buckling (bifurcation) behaviours of RCPS tubes are characterised in this paper. The results presented show that the structural response of the RCPS tubes exhibits several peculiarities, which stem essentially from the cross-section rotational symmetry.

1 INTRODUCTION

Thin-walled steel tubes with single-cell regular convex polygonal cross-section (RCPS) are widely employed in the construction industry, namely for supporting lighting and telecommunication equipment, as well as road signs and overhead power lines. These tubes are characterised by high wall with-to-thickness ratio, which makes them particularly susceptible to cross-section deformation. Although the local (plate-type) buckling behaviour of hollow sections (including RCPS) has been investigated in the past and is by now well understood, until recently no studies were available concerning the distortional behaviour of RCPS, which involves cross-section in-plane and out-of-plane (warping) deformations. This situation is in sharp contrast with that of open section thin-walled members, namely cold-formed steel members with lipped channel, zed, hat or rack sections, whose distortional behaviour has been thoroughly investigated in the past and may be deemed quite well understood nowadays.

In the last year, an ongoing investigation concerning the structural behaviour of RCPS tubes has shed new light on the subject (Gonçalves & Camotim 2013a,b,c). In particular, it was shown that these tubes exhibit several peculiar features, namely that (i) cross-section distortion plays a significant role and (ii) duplicate solutions occur in a wide range of cases. These results were obtained through a Generalized Beam Theory (GBT – see, for instance, Schardt 1989) specialisation for RCPS (Gonçalves & Camotim 2013a), which led to the identification of an orthogonal set of cross-section deformation modes and also made it possible to derive analytical and/or semi-analytical formulae that provide in-depth information concerning the overall structural behaviour of RCPS tubes.

The objective of this paper is to summarize the results of the above ongoing investigation concerning the structural behaviour of thin-walled RCPS tubes. The outline of the paper is as follows. Section 2 focuses on the GBT cross-section deformation modes for RCPS. Then, Section 3 is devoted to the first-order behaviour and Section 4 addresses the linear buckling behaviour under several applied loadings, namely compression, bending and torsion. Finally, the paper closes with Section 5, which includes the main conclusions of the work.

The results presented in this paper are generally expressed and presented in terms of the non-dimensional geometrical parameters

$$\beta_1 = \frac{L}{r}, \qquad \beta_2 = \frac{r}{t}, \qquad (1)$$

where L is the tube length and the parameters r and t are defined in Figure 1, which also shows additional

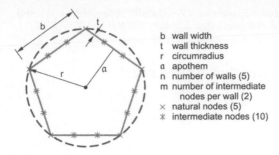

b   wall width
t   wall thickness
r   circumradius
a   apothem
n   number of walls (5)
m   number of intermediate
    nodes per wall (2)
×   natural nodes (5)
*   intermediate nodes (10)

Figure 1.   Parameters for RCPS.

parameters that are relevant for the analysis. Besides presenting critical stresses, buckling factors k are also provided, which satisfy the relation

$$\sigma_{cr} = \frac{k\pi^2 E}{12(1-\nu^2)}\left(\frac{t}{b}\right)^2, \qquad (2)$$

where E (Young's modulus) and ν (Poisson's ratio) are the material parameters.

2   GBT DEFORMATION MODES FOR RCPS

2.1   General aspects

The GBT cross-section deformation modes are obtained from the so-called "cross-section analysis", which involves (i) defining an initial set of kinematically admissible functions (the initial basis functions) and then (ii) finding a set of orthogonal deformation modes (the final basis) by uncoupling the differential equilibrium equation system. The deformation mode set includes the classic prismatic beam theory modes (axial extension, bending about principal axes and torsion about the shear centre), as well as additional ones, namely the so-called "local", "distortional", "shear" and "transverse extension" modes.

RCPS constitute a rather peculiar problem, as they exhibit rotational symmetry of order n. As shown by Gonçalves & Camotim (2013a), (i) it is possible to obtain the deformation modes from numerical procedures that are much more efficient than those employed in the conventional GBT and, within each deformation mode subset, (ii) the ensuing equilibrium equations become fully uncoupled (with the exception of the so-called local w deformation mode subset).

According to the notation introduced by Gonçalves et al. (2010), the deformation modes may be subdivided into the following two families (see Figure 1): (i) "natural modes", generated by imposing displacements at the natural nodes (n wall mid-line intersections for RCPS), and (ii) "local modes", generated by imposing displacements at the intermediate nodes (in the case of the RCPS, m evenly spaced nodes in each wall). Accordingly, the number of natural modes is univocally defined by the cross-section geometry and the

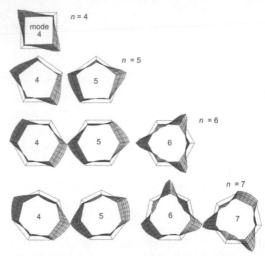

Figure 2.   Shapes of the natural Vlasov distortional warping modes for RCPS with $n = 4$ to 7.

local modes are dependent on the cross-section discretisation. These two deformation mode families are examined, separately, in the next subsections.

2.2   Natural deformation modes

The natural deformation mode set is subdivided into three subsets: (i) "Vlasov warping modes", which involve warping displacements of the natural nodes, null membrane shear strains (Vlasov's hypothesis) and null membrane transverse extensions, (ii) "shear modes", which complement the former by allowing for membrane shear strains, and finally (iii) "transverse extension modes".

The Vlasov mode set is constituted by a total of n modes, where (i) the first ones correspond to axial extension and bending about orthogonal axes (all centroidal axes are principal), and (ii) the remaining modes, termed "distortional", involve cross-section warping and in-plane displacements. The bending and distortional modes appear in pairs, with one exception: the last mode of even-sided polygons (even n). Each pair is associated with a specific warping function, exhibiting a given number of zeros around the cross-section, and a higher-order pair is associated with a higher number of zeros – for instance, the last mode for even n involves alternating warping displacements at consecutive natural nodes). It can be shown that the stiffness properties associated with each deformation mode pair are invariant with respect to a rotation in that particular subspace, a conclusion that constitutes an important generalisation of the bending behaviour of RCPS. For illustrative purposes, Figure 2 shows the shapes of the distortional modes for the cases of $n = 4$ to 7.

The natural (membrane) shear modes can be easily obtained from the Vlasov modes, by (i) removing the warping displacements, while retaining the in-plane

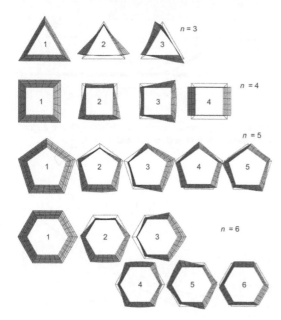

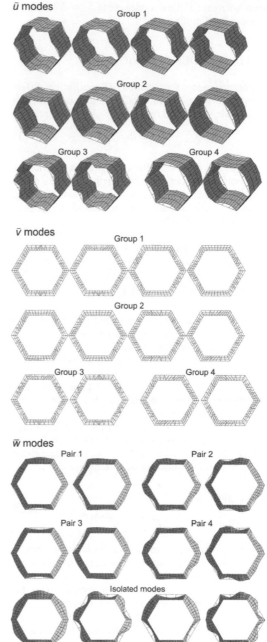

Figure 3. Shapes of the natural transverse extension modes for RCPS with $n = 3$ to 6.

ones, and (ii) replacing the first mode (axial extension) by the torsional mode, which corresponds to a trivial unit rotation about the centroidal axis.

The natural transverse (membrane) extension modes are obtained by eliminating all warping displacements. The first mode is associated with constant transverse extension around the cross-section and, for even n, the last mode involves alternating extensions at consecutive walls. As in the previous deformation mode sets, the intermediate modes appear in pairs and correspond to an increasing variation of transverse extensions around the cross-section. Figure 3 shows the configurations of these modes for cross-sections having $n = 3$ to 6.

2.3 Local deformation modes

The initial basis for the local modes is obtained through the imposition of unit displacements at each intermediate node, leading to a total of nm modes for each displacement component u (warping), v (cross-section in-plane, wall mid-line direction) and w (in-plane, perpendicular to the mid-line).

The local u and v modes appear in groups, whose elements (modes) share the same stiffness properties. In particular, (i) for even n one obtains m groups with 2 modes and m groups with $n-2$ modes, whereas (ii) for odd n one retrieves m single modes and additional m groups with $n - 1$ modes each.

The local w modes constitute the only deformation mode subset that does not become completely orthogonal, i.e., while some GBT matrices become fully diagonal, others become only block-diagonal. For even n, the RCPS cross-section analysis leads to $2m$ isolated modes plus several mode pairs. For odd n,

Figure 4. Shapes of the local modes for $n = 6$ and $m = 2$.

only m isolated modes are obtained (plus several mode pairs).

Figure 4 shows the shapes of all the local deformation modes for a hexagonal tube ($n = 6$) with two intermediate nodes in each wall ($m = 2$). Taking into account the previous observations, it should be noted that (i) the u and v modes comprise 2 groups with 2 modes each and 2 groups with $n - 2 = 4$ modes each, whereas (ii) the w modes contain $2m = 4$ isolated modes plus 4 mode pairs.

3 FIRST-ORDER BEHAVIOUR

The homogeneous form of the GBT equilibrium differential equation system for RCPS is uncoupled within each mode subset and reads

$$C_{ij}\,\phi_{j,xxxx} - D_{ij}\,\phi_{j,xx} + B_{ij}\,\phi_j = 0, \tag{3}$$

where C_{ij}, D_{ij} and B_{ij} are the GBT stiffness matrices, x is a coordinate along the tube length and ϕ_j are mode amplitude functions – the problem unknowns. For long tubes, the analytical solution for individual modes is given by (Schardt 1989)

$$\phi_k = e^{-\alpha x}\left(A_1 \sin \beta x + A_2 \cos \beta x\right),$$
$$\alpha = \sqrt{\sqrt{\frac{B_{kk}}{4C_{kk}}} + \frac{D_{kk}}{4C_{kk}}}, \quad \beta = \sqrt{\sqrt{\frac{B_{kk}}{4C_{kk}}} - \frac{D_{kk}}{4C_{kk}}}, \tag{4}$$

where the exponential decay α and the frequency β (deemed real) provide a measure of the influence length of the deformation mode.

For illustrative purposes, Figure 5 plots the exponential function $e^{-\alpha x}\sin \beta x$ for the distortional modes, with $\beta_2 = 100$ and $\nu = 0.3$. The figure shows that the influence length decreases as the mode number increases and also as n increases.

It is also interesting to examine the benchmark case of simply supported beams of length L acted on by sinusoidal lateral loads. In this case, the single-mode solution (valid for uncoupled systems) reads

$$\phi_k = \bar{\phi}_k \sin \frac{\pi x}{L}, \quad \bar{\phi}_k = \frac{\bar{q}_k}{\frac{\pi^4}{L^4}C_{kk} + \frac{\pi^2}{L^2}D_{kk} + B_{kk}}, \tag{5}$$

where $\bar{q}_k$ are modal loads. This solution shows that, as L increases, the amplitude increases and tends asymptotically to a fixed value that depends on B_{kk} alone. If $B_{kk} = 0$, as in the bending and torsional modes, the amplitude grows unboundedly with the span L.

Figure 6 concerns results for tubes with $n = 8$ subjected to two unit amplitude sinusoidal vertical loads applied at two opposite cross-section walls, forming a couple. The graph plots the evolution of the vertical displacement of the loaded walls with L, at mid-span, as well as the contributions of the individual mode pairs. These GBT analyses were carried out with only the natural distortional Vlasov modes plus the torsional mode, totalling 6 modes. For comparison purposes, shell finite element (SFE) results are also provided. The bottom part of the figure displays GBT deformed configurations, depicting half of the tube only, with its axis directed towards the observer (mid-span is near and the support is far). A virtually perfect match between the SFE and GBT results is observed, showing that the selected modes adequately characterise

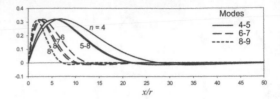

Figure 5. Exponential solution ($e^{-\alpha x}\sin \beta x$) for the distortional modes, assuming $\beta_2 = 100$ and $\nu = 0.3$.

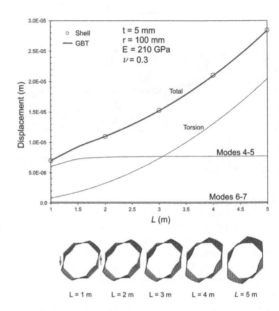

Figure 6. Simply supported beams with $n = 8$, subjected to torsional sinusoidal loads.

the tube behaviour (alt least for this particular case). Moreover, only the torsional mode and the 4-5 distortional mode pair have significant participations, with the latter being more important for the shorter spans. As predicted, as L increases, the contribution of the torsional mode grows unboundedly, whereas those of the distortional modes are limited. Finally, note that the deformed configurations make it possible to visualise the cross-section distortion.

4 BUCKLING BEHAVIOUR

4.1 Axial force

The local and distortional buckling behaviour of uniformly compressed RCPS tubes is investigated, focusing on the standard benchmark case of simply supported members, for which GBT semi-analytical solutions may be sought (sinusoidal mode amplitude functions constitute exact solutions).

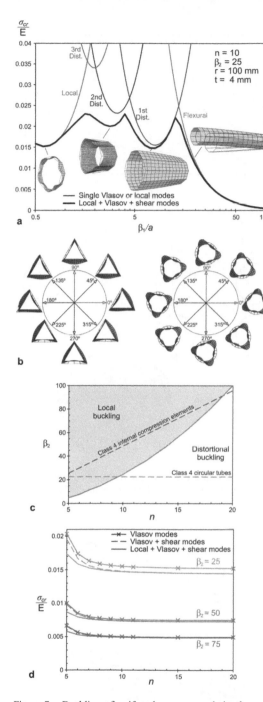

a

b

c

d

Figure 7. Buckling of uniformly compressed simply supported tubes ($\nu = 0.3$): (a) illustrative example, (b) local and distortional mode pair spaces for $n = 3$ and $n = 20$, respectively, (c) parameter ranges associated with local or distortional critical buckling and (d) influence of the local and shear modes on the lowest distortional buckling stress.

Figure 7a shows a typical signature curve for RCPS tubes. Essentially, depending on the β_1/a value (a is the number of longitudinal half-waves), either local (low β_1/a), distortional (intermediate β_1/a) or flexural

(high β_1/a) buckling is critical. Remarkably, as in the case of flexural buckling, both local and distortional buckling modes are generally duplicate. In particular, it can be shown that (i) the critical local mode is always duplicate for odd n and (ii) the critical distortional mode is always duplicate for $n > 4$. Each duplicate solution defines a two-dimensional buckling mode space and, naturally, each mode in that particular space constitutes a possible buckling mode. For illustrative purposes, Figure 7b shows several possible solutions concerning the (i) first local buckling mode for $n = 3$ and (ii) second distortional mode for $n = 20$.

The parameter ranges associated with either local or distortional critical buckling are shown in Figure 7c. These results were calculated for $\nu = 0.3$ and as assuming that no mode interaction occurs. It is concluded that local buckling is critical for high β_2 values and that the transition value increases with n. Practical information for steel tubes may be extracted from the dashed lines, which correspond to the Eurocode 3 limits between class 3 and 4 cross-sections made of S460 steel (which leads to the lowest of such limits) – for a given n value, a β_2 value above a given dashed line indicates, according to that criterion, a class 4 cross-section. It may then be concluded that, according to the internal compression element criterion, a class 4 cross-section with critical distortional buckling can only be obtained for $n > 19$ and $\beta_2 > 90$. However, the circular tube criterion yields a lower limit, which seems to indicate that RCPS with moderate-to-high n values may be class 4 for n and β_2 values below those given by the previous criterion (because the distortional buckling stresses approach the circular tube values as $n \to \infty$).

Finally, the influence of the shear and local deformation modes on distortional buckling may be visualised in Figure 7d. This figure shows the variation of the lowest distortional buckling stress with n, for $\nu = 0.3$ and three β_2 values. These findings evidence that the shear modes have a significant influence, particularly for low β_2 values (up to 5.4% for $\beta_2 = 25$), and that the local modes are only relevant for low n and β_2 values.

4.2 Bending moment

Thin-walled RCPS tubes under bending invariably buckle in local-type modes, even if the wall junctions may undergo non-null displacements as the b/t ratio decreases. Nevertheless, the longitudinal half-wavelength is relatively small and, thus, for high β_1 ratios, the influence of non-uniform bending and end boundary conditions has minor relevance. This means that, once more, one may resort to the analysis of simply supported members subjected to uniform bending.

The assessment of the local buckling behaviour requires estimating the effect of the bending axis inclination, particularly for low n values, when it strongly influences the stress distribution around the cross-section. However, as shown in Gonçalves & Camotim (2013c), the lowest buckling stress is always obtained when the bending axis is parallel to the compressed wall (causing a uniform compressive stress in that wall).

Figure 8 displays buckling factors k for a wide range of RCPS parameter values, as well as the classic solutions for circular tubes (e.g., Timoshenko & Gere 1961). These results show that, in general, the buckling modes exhibit a smooth transition between a plate-like mode (no wall junction displacements, occurring for high β_2) and a circular tube-like mode (wall junction displacements, occurring for low β_2). In particular, the horizontal plateau is associated with the plate-type mode, which is mostly relevant for low n values. Conversely, the circular tube-like mode is mostly relevant for high n values and, remarkably, can be quite accurately predicted by the circular tube solutions. For illustrative purposes, Figure 8 also shows a 3D view concerning the particular case of $n = 40$ and $a = 10$ – it is clearly shown that the RCPS tube buckling mode is virtually identical to that of a circular tube subjected to bending.

4.3 Torsion

RCPS tubes under uniform torsion are subjected to a pre-buckling pure shear stress state. Unlike in the previous cases, sinusoidal amplitude functions are not exact solutions for simply supported members and the buckling loads decrease with L, until a plateau is reached for relatively long members. Both these features render the analyses far more demanding from a computational point of view.

A standard GBT beam finite element was employed to obtain results (see Gonçalves & Camotim 2012). It should be noted that, with the GBT specialization for RCPS, mode coupling only occurs in a few matrix blocks and, thus, significant computational savings may be achieved (with respect to SFE models).

Local (plate-type) buckling is critical for high b/t ratios, but distortional buckling may become critical as the ratio decreases.

As first noted by Wittrick & Curzon (1968), the local buckling solution is similar to that of a simply supported rectangular plate under pure shear, although the nodal lines form a continuous spiral along the tube length. Figure 9 concerns $n = 3$ to 10 and $L/b = 10$ (in this case the buckling stresses are close to the minima). These results were obtained with $m = 5$ and 10–15 finite elements. Although the figure displays only one buckling mode for each n, duplicate modes are obtained in all cases. For the particular cases of $n = 3$ and 4, the solutions of Wittrick & Curzon (1968) for $L = \infty$ are $k = 5.534$ ($n = 3$) and $k = 5.340$ ($n = 4$), falling 2% below those obtained with GBT for $L/b = 10$.

The buckling behaviour in a pure distortional mode may be assessed by considering only each individual natural distortional (Vlasov) deformation mode pair. Duplicate solutions are obtained, both corresponding

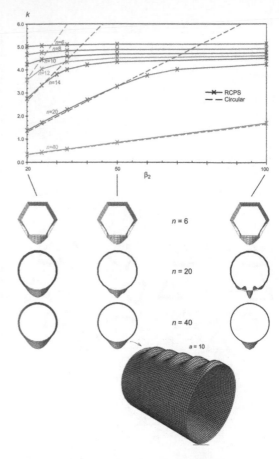

Figure 8. Local buckling of simply supported RCPS tubes under bending: buckling coefficients and mode shapes.

Figure 9. Local buckling of RCPS tubes under uniform torsion for $L/b = 10$ and $n = 3$ to 10.

to a rotation of the modes along the tube axis (a helix). For illustrative purposes, Figure 10 shows results for $n = 10$, $\beta_2 = 100$ and $\nu = 0.3$. The tubes are subjected to a relative torsional twist between their ends, applied

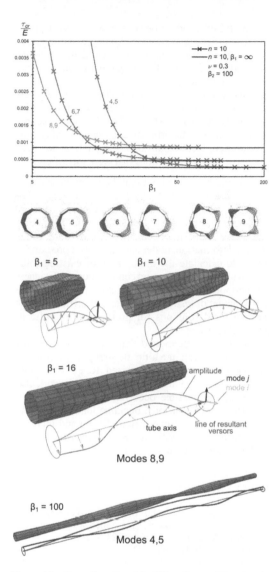

Figure 10. Pure distortional buckling for $n = 10$ and uniform torsion: critical stresses and buckling mode shapes.

through rigid diaphragms. A discretisation with 10–20 finite elements was found to be necessary. For each distortional mode pair, the figure shows (i) the variation of the buckling stresses with β_1 and (ii) the corresponding values for $\beta_1 = \infty$. As before, the buckling stresses decrease with increasing β_1 and approach the solution, asymptotically, for $\beta_1 = \infty$. Moreover, the critical deformation mode pair varies with β_1, with the first one (4, 5) being critical for $\beta_1 = \infty$ (as for circular tubes). This figure also displays buckling mode shapes (only one of the duplicate solutions is shown) for the (8, 9) distortional pair (several β_1 values) and for the (4, 5) distortional pair (only for $\beta_1 = 100$, in which case this pair corresponds to the critical mode), as well as a visualisation of (i) the corresponding "helix" in that

deformation mode space and (ii) the mode amplitude along the axis.

For infinitely long members, it may be assumed that the mode amplitude and helix pitch are constant, making it possible to develop analytical formulas for the lowest load (see Gonçalves & Camotim 2013c). For instance, for a given distortional mode pair, the critical buckling load is associated with a helix pitch θ and shear stress given respectively by

$$\theta_{cr} = \sqrt{\frac{\sqrt{D^2 + 12CB} - D}{6C}}, \qquad (6)$$

$$\tau_{cr} = \frac{\sqrt{6}}{18|X_\tau^w|} \frac{D\sqrt{D^2 + 12CB} - D^2 + 12CB}{\sqrt{C\sqrt{D^2 + 12CB} - CD}}, \qquad (7)$$

where C, D, B and X_τ^w are the GBT matrix components associated with the particular distortional deformation mode pair considered.

Concerning the influence of shear deformation on distortional buckling, calculations for $25 \leq \beta_2 \leq 100$ and $\nu = 0.3$ showed that it is minute (below 1%) and, therefore, may be discarded. However, interaction between local and distortional buckling is not negligible, as shown in Figure 11. To obtain critical stresses close to the lowest value, which occurs for $\beta_1 = \infty$, calculations were performed for $\beta_1 = 100$ and, as before, a relative torsional twist was applied at the member ends, through rigid diaphragms. The cross-section was discretised with two intermediate nodes per wall ($m = 2$) and 10–20 finite elements were found to be necessary. These results show that the buckling stresses decrease with increasing n and approach the circular tube solutions. Moreover, since b vanishes as $n \to \infty$, the k values tend to zero. Although not shown, duplicate buckling modes are obtained in all cases. It is also concluded that the local mode influence is most relevant for $n = 5$ and, to a much lesser extent, for $n = 6$. In the first case, the buckling stresses and k values vary by about 22% for all β_2 values. For illustrative purposes, Figure 11c shows the shapes of one of the duplicate buckling modes, for the particular case of $n = 5$ (where the local/distortional interaction is most relevant), with $\beta_2 = 100, 50, 25$ – it should be noted that, since $\beta_1 = 100$, only the central zone of the tube is shown.

5 CONCLUSION

An up-to-date overview of an ongoing research program, aimed at investigating the structural behaviour of thin-walled single-cell regular convex polygonal section (RCPS) tubes, was presented. The paper focused (i) the GBT cross-section deformation modes, (ii) the first-order behaviour and (iii) the buckling behaviour under compression, bending and torsion.

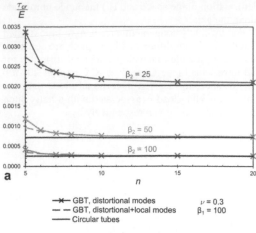

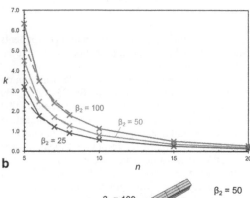

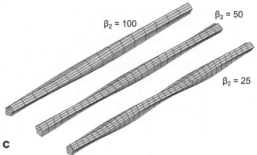

Figure 11. Distortional-local buckling of RCPS tubes under uniform torsion ($\nu = 0.3$): (a) buckling stresses, (b) buckling coefficients and (c) buckling mode shapes for $n = 5$ and $\beta_2 = 100, 50, 25$.

The following results of the work carried out deserve special attention:

(i) The GBT specialisation for RCPS leads to a set of uncoupled and hierarchic deformation modes. Due to the cross-section rotational symmetry, duplicate modes occur and the stiffness properties associated with them are invariant upon a rotation in that 2D mode subspace – a generalisation of the RCPS bending behaviour. Moreover, (i_1) the first-order behaviour may be expressed in terms of deformation mode pairs, rather than individual modes, and (i_2) duplicate buckling solutions occur in several cases.

(ii) Distortional deformation modes play a very relevant role in the structural behaviour (first-order and buckling) of RCPS tubes. For some parameter ranges, distortional buckling is critical. In addition, local-distortional-shear interaction may occur in some situations that were identified.

It should be pointed out that the novel behavioural features reported in this paper could only be unveiled by taking advantage of the unique modal characteristics of the GBT specialization for RCPS. Indeed, they make it possible to identify the relevant cross-section deformation modes and, moreover, provide the means to calculate the multiplicity of existing solutions with both accuracy and computational efficiency. Furthermore, this GBT specialization also enables the development of analytical and semi-analytical formulae that yield solutions for problems of significant practical relevance.

Finally, it is mentioned that the authors are currently investigating the ultimate strength and collapse behaviour of axially compressed RCPS tubes that buckle in local and distortional modes. This work also addresses the possible occurrence of nearly coincident local and distortional buckling loads (i.e., prone to local-distortional interaction effects) and includes imperfection sensitivity studies. The results obtained will be reported in the near future.

REFERENCES

Gonçalves, R. & Camotim D. 2011. GBT-based finite elements for elastoplastic thin-walled metal members. *Thin-Walled Structures* 49(10): 1237–1245.

Gonçalves, R. & Camotim, D. 2013a. On the behaviour of thin-walled steel regular polygonal tubular members. *Thin-Walled Structures* 62:191–205.

Gonçalves, R., Camotim, D. 2013b. Elastic buckling of uniformly compressed thin-walled regular polygonal tubes. *Thin-Walled Structures* 71:35–45.

Gonçalves, R., Camotim, D. 2013c, Buckling behaviour of thin-walled regular polygonal tubes subjected to bending or torsion. *Thin-Walled Structures* 73:185–97.

Gonçalves, R., Ritto-Corrêa, M. & Camotim, D. 2010. A new approach to the calculation of cross-section deformation modes in the framework of Generalized Beam Theory. *Computational Mechanics* 46(5):759–81.

Schardt, R. 1989. *Verallgemeinerte Technische Biegetheorie*. Berlin: Springer-Verlag.

Timoshenko, S., Gere, J. 1961. *Theory of Elastic Stability*. New York: McGraw-Hill.

Wittrick, W. & Curzon, P. 1968. Local buckling of long polygonal tubes in combined compression and torsion. *International Journal of Mechanical Sciences* 10(10): 849–57.

Tubular Structures XV – Batista, Vellasco & Lima (eds)
© 2015 Taylor & Francis Group, London, ISBN 978-1-138-02837-1

CFRP strengthened square hollow section subject to pure torsion

J. Sharrock, C. Wu & X.L. Zhao
Department of Civil Engineering, Monash University, Australia

ABSTRACT: Carbon fiber reinforced polymer (CFRP) has great potential for strengthening aging steel structures, compared to traditional methods of attaching steel plates which are often bulky, heavy and difficult to fix. CFRP provides advanced structural properties with light weight and high tensile characteristics, making it an attractive option to perform the aforementioned role. This paper presents a comprehensive experimental program to investigate the torsional behaviour of strengthened steel square hollow sections (SHS) and strengthened aluminium SHS with various thicknesses. From previous research, a wrapping angle of 45° is the most efficient in terms of increasing the maximum torsional capacity, and was adopted for the tests. It was found that sections with larger depth to thickness ratio (D/t ratio) had a greater percentage increase in maximum torsional capacity. Moreover, it was observed that sections with lower elastic moduli compared with steel will show more improvement in maximum torsional capacity for a given thickness.

1 INTRODUCTION

Carbon fiber reinforced polymer (CFRP) has become increasingly popular for strengthening aging steel structures, compared to traditional methods of attaching steel plates which are often bulky, heavy and difficult to fix (Holloway & Teng 2008). With light weight and high tensile strength characteristics, CFRP provides advanced structural properties and, moreover, a more corrosion resistant solution when compared to steel.

Previous research has found that CFRP bonded onto the steel surface can increase the axial (Zhao 2013) and flexural (Haedir & Zhao 2012) capacity of tubular steel sections. However, there is a lack of understanding of the torsional behaviour of CFRP strengthened tubular sections, with only two studies conducted in this area. Wang et al (2013) uses a theoretical approach to show that the introduction of CFRP externally bonded can increase the torsional capacity of tubular steel sections. Chahkand et al (2013) experimentally tested four CFRP strengthened tubular specimens with different fiber orientation, which all showed increase in torsional capacity. This paper aims to further fill the knowledge gap by undertaking a more comprehensive experimental study of CFRP strengthened tubular sections. By adopting the most efficient wrapping configuration as found in previous research, the influence of the section depth to thickness ratio (D/t) on the torsional capacity was investigated and compared to theoretical studies. Moreover, testing was carried out on both steel and aluminium sections so that torsional behaviour can be compared between the two materials. More improvement in torsional capacity was expected as the modulus ratio of CFRP to aluminum

$(E_{CFRP}/E_{Aluminum})$ is relatively higher than CFRP to steel (E_{CFRP}/E_{Steel}). Failure modes were identified.

2 EXPERIMENTAL STUDIES

Existing experimental results for SHS are very limited, with only one section type tested and a total of four strengthened specimens tested. To fill this knowledge gap, a more complete experimental program has been proposed, testing three different steel SHS in order to validate theoretical investigation into D/t ratio. In addition to this, two different structural grade aluminum SHS were tested and compared with steel. A total of 20 SHS specimens were prepared and tested. Results obtained from the tests were compared to existing experimental data and theoretical studies.

2.1 Strengthening configuration

To get the optimum strengthening effect and due to CFRP wrap being an anisotropic material, the orientation of the fibers is critical. From previous work on CFRP strengthened hollow sections, a wrapping angle of 45 degrees (Figure 1.) is the most efficient in terms of increasing torsional stiffness (Wang et al. 2013). In addition to this, Chahkand et al. (2013) tested four different wrapping configurations and found that the best orientation angle for fibers is in the same direction as the principal tensile stress. Due to normal stresses being negligible in the longitudinal and transverse directions, the ultimate torque will be greatest using a wrapping angle of 45 degrees. When the section is subject to pure torsion, shear stresses can be developed at 45 degrees respective to the longitudinal axis of the section. When the CFRP wrap is engaged, it will be in a state of pure tension which is utilising its superior properties to full potential.

For the present study, the wrapping configuration was held constant to investigate the effect of CFRP on different types of tubular sections. Four layers of CFRP were applied to the samples, with each layer being orientated at a 45 degree spiral wrapping direction so the CFRP works in tension.

2.2 Material properties and dimensions

The type of steel used was cold formed SHS, purchased from Russell Steel in Melbourne and supplied from One Steel Tube Mills Australia. The grade of the steel was C450PLUS and had nominal Elastic Modulus of 200 GPa. The aluminium SHS was structural grade with metallic finish and provided by Capral Aluminium in Brisbane. Tensile coupon tests (two for each section type) were carried out according to Australian Standard AS1391 to determine the elastic modulus and yield strength of each section. The average measured material properties are presented in Table 1.

In order to compare the D/t ratio, all sections had nominal dimensions of 100×100 mm and varied thicknesses of 2, 3 and 6 mm for steel, and 3.2 and 6 mm for aluminium. Width and thicknesses of all sections were measured using a Vernier gauge and the average of the measured dimensions are presented in Table 1.

The type of CFRP wrap used was MBRACE CF 230/4900, which is a unidirectional low modulus carbon fiber wrap, with high tensile strength. A two component epoxy, Araldite 420 A/B was used to bond the carbon fiber to the steel and aluminium, which was also used by Wu et al. (2012). The nominal properties for the CFRP warp and the epoxy are given in Table 1.

2.3 Specimen preparation

Before surface preparation could occur, each section was cut to a length of 460 mm and end plates were welded to each specimen so they could be affixed to the torsion testing rig. Welding end plates to each specimen was undertaken in the Civil Engineering laboratory for steel and Mechanical Engineering laboratory for aluminium at Monash University Australia. The welding was sufficient to carry the torsional load required for the steel specimens, however, the aluminum welding was found to not be sufficient after a sample test (Figure 2(a)). To ensure that welding failure did not occur during tests a further fillet weld (Figure 2(b)) was applied by Crossline Engineering in Melbourne to the inner of the end plate and was found to be sufficient after sample test.

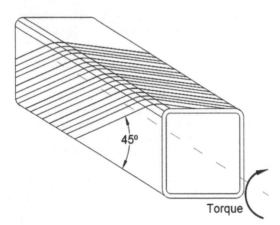

Figure 1.   Strengthening configuration.

Figure 2.   (a) Welding failure, (b) Additional fillet weld.

Table 1.   Material properties and dimensions.

| Material | Section Properties | | | | | | | Material Properties | |
| | Nominal dimensions (mm) | Average measured dimensions (mm) | | | Corner Radii (mm) | | Length (mm) | Elastic Modulus (GPa) | Yield Strength (MPa) |
		Width	Depth	Thickness	Internal	External			
Steel	100x100x6	100.55	100.55	5.95	8.92	14.87	460	201	444
	100x100x3	100.57	100.57	2.93	2.93	5.85	460	201	434
	100x100x2	99.82	99.82	2.04	2.04	4.09	460	202	383
Aluminium	100x100x6	99.70	99.70	5.61	8.42	14.03	460	62	214
	100x100x3.2	99.85	99.85	3.06	3.06	6.13	460	55	207
Carbon Fibre Wrap								230	4900*
Epoxy								1.50	29*

*Ultimate tensile strength.

To achieve a good quality bond between the materials and CFRP, surface preparation was undertaken. For the steel specimens, the coating and other impurities were removed by sand blasting using 30–60 sand (0.6 mm to 0.25 mm grain size) at a pressure of 0.8 MPa. For the aluminium different surface preparation methods were considered as for the concern of removing material from the aluminium sections (Zahurul Islam & Young 2011; Al-Mayah et al. 2005; Bambach et al. 2010). It was decided that 0.226 to 0.310 mm glass beads, at a blasting pressure of 0.3 MPa would be sufficient to remove the galvanized coating and prepare the surface for bonding. Blastmasters in Melbourne undertook the surface blasting for both steel and aluminum.

The samples were then positioned in the carbon fiber wrapping machine, which rotates as layers are applied (Figure 3(a)). The surface was further cleaned using isopropanol to remove any impurities before applying the adhesive. The carbon fiber wrap was cut at an angle of 45 degrees. Parts A and B of Araldite 420 were mixed according to the weight ratio specified by the manufacturer. A thin coat to the surface of the specimens was applied and then the CFRP was wrapped around the surface at the inclined angle of 45 degrees (Figure 3(b)). To ensure that there are no air bubbles and that there is good connection between the epoxy and the CFRP, a roller was used to guarantee good bonding contact (Figure 3(c)). This process was repeated four times for each section to bond the four layers of wrap. The specimens were then kept at room temperature for two weeks to allow for sufficient curing before testing. The final depth and width of the sections were measured to determine the combined thickness of epoxy and carbon fiber, (Table 2).

2.4 Experiment set up and method

All tests were performed with the torsion testing apparatus in the Civil Engineering laboratory at Monash University. Figure 3(d) shows the final position of a tested specimen and Figure 4. shows a typical test set-up prior to commencement of testing. The apparatus consists of one fixed end grip and one rotating end grip which is powered by two 25 ton jacks. The apparatus has the capability of rotating up to 40 degrees. Flow of oil is controlled manually with the aid of a bypass valve to slow the flow of oil and allow for more precise data collection. The load in each of the jacks is measured by pressure transducers attached to each load cell. To measure the rotation angle two string pots were used, which could each independently calculate the rotation angle of the specimen (Figure 4).

2.5 Experimental results and discussion

The experimental results are listed in Table 2 and detailed discussions are presented in following sections.

All specimens have the same outer dimensions of 100×100 mm and same wrapping scheme of four spiral wrapping layers (inclined at 45 degrees so that the CFRP works in tension). S represents steel material and A represents aluminium material. C represents a control specimen with no wrapping and W indicates a CFRP wrapped section. The first number denotes the section thickness. As each section will be tested twice, the second number represents test number one or two. For example SW-3-2 represents the second test for a 3 mm steel section with CFRP wrapping.

2.5.1 Strain gauges

To verify that the specimens are in a state of pure shear when tested in the torsion rig, and hence the wrapping configuration is correct, three strain gauges were attached to each specimen. Two gauges were inclined at 45 degrees, one in compression and the other in tension. A third was inclined parallel to the longitudinal axis of the specimen. Figure 5. shows the torque vs strain for sample SC-3-2.

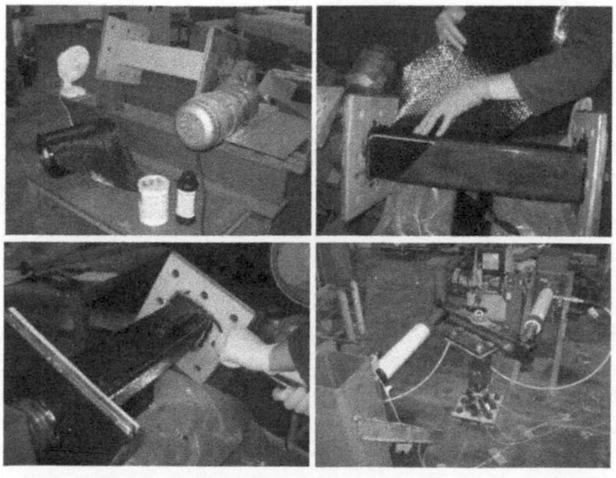

Figure 3. (a) CFRP Wrapping machine, (b) Application of CFRP at 45 degrees, (c) Rolling of fibers for good bond, (d) Torsion testing apparatus after rotation.

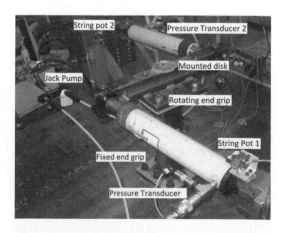

Figure 4. Torsion testing apparatus.

Table 2. Torsion testing apparatus after rotation.

No.	Section Name	CFRP Thickness (mm)	Nominal D/t Ratio	Maximum Torsional Capacity (Nm)	Rotation Angle at maximum (degrees)
1	SC-2-1	0	50	8145	4.27
2	SC-2-2	0	50	8197	5.14
3	SC-3-1	0	33.33	15280	9.22
4	SC-3-2	0	33.33	14725	9.75
5	SC-6-1	0	16.67	30898	16.29
6	SC-6-2	0	16.67	31925	29.30
7	SW-2-1	3.22	50	11837	4.61
8	SW-2-2	3.25	50	11202	4.34
9	SW-3-1	2.99	33.33	17638	7.11
10	SW-3-2	3.26	33.33	16700	5.24
11	SW-6-1	2.97	16.67	36018	11.44
12	SW-6-2	2.86	16.67	33898	11.13
13	AC-3-1	0	31.25	2619	5.12
14	AC-3-2	0	31.25	Fabrication	Error**
15	AC-6-1	0	16.67	10123	24.03
16	AC-6-2	0	16.67	8029	19.81
17	AW-3-1	2.91	31.25	5215	2.90
18	AW-3-2	2.86	31.25	4949	2.38
19	AW-6-1	2.76	16.67	10689*	6.14*
20	AW-6-2	2.90	16.67	6693	15.90

*Noise experienced during data collection.
**6 mm tube section fabricated instead of 3.2 mm section.

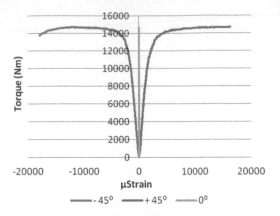

Figure 5. Strain gauge pattern for SC-3-2.

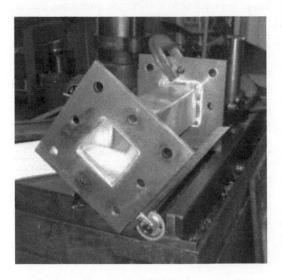

Figure 6. Measurement of residual angle.

The pattern is typical for all control sections and indicates that the direction of the principal stresses is inclined at 45 degrees. This confirms that a wrapping angle of 45 degrees with respect to the longitudinal axis of the section will be most effective in terms of contributing to the maximum torsional capacity.

2.5.2 *Verification of rotation angle*
The test set up allowed for calculation of the rotation angle by two independent methods. Angle 1 was calculated by displacement due to the extension of the jack, which was measured by string pot 1 (Figure 4). The angle can be calculated by trigonometry using the cosine rule. Angle 2 was calculated by the disk mounted to the rotation arm of the torsion rig (Figure 4). As the arm rotates, string pot 2 will extend allowing the rotation angle to be calculated.

In analysing the results, it was found that both angles were slightly different, with angle 2 always slightly higher than angle 1. To confirm which of the two yielded the more precise result, the residual angle (final angle after the specimen has been taken out of the testing apparatus) was measured (Figure 6). This was then compared to the two final calculated angles from the tests for each specimen. It was found that calculated angle 1 was closest to the measured residual angle for a majority of the specimens and hence angle 1 gives the true representation of the rotation angle during the tests.

The difference between the calculated angles can be attributed to local buckling in the specimen and hence tilting of the rotation arm. As only the bottom of the testing apparatus is rigid, when buckling occurs, string pot 2 either shortens or lengthens, giving a rotation much larger or smaller than the actual rotation angle. For the purpose of presenting torsion rotation curves, the rotation angle plotted will be measurement angle 1.

2.5.3 *Torque rotation curves*
Torque vs rotation curves for all specimens are presented in Figure 7 for each thickness of tubular steel and aluminium sections. It can be seen from Figure 7 that a significant increase in the maximum torsional capacity was obtained for thin walled sections that were strengthened by CFRP. Furthermore, CFRP strengthened sections are much stiffer compared to control sections. This means that more torque is required for a CFRP strengthened section to undergo the same rotation as a control section.

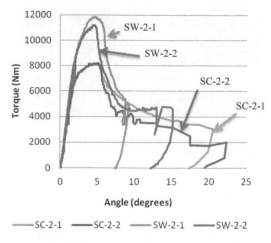

Figure 7a. Torque rotation for 2 mm steel sections.

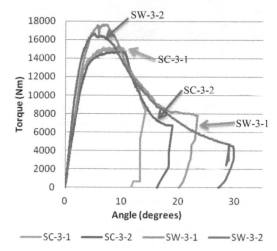

Figure 7b. Torque rotation for 3 mm steel sections.

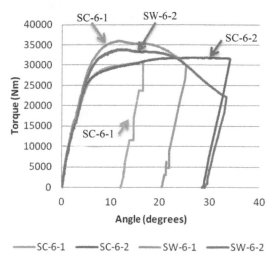

Figure 7c. Torque rotation for 6 mm steel sections.

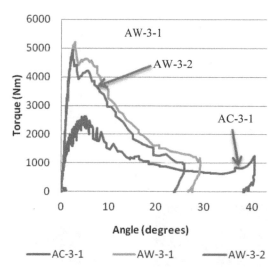

Figure 7d. Torque rotation for 3.2 mm aluminium sections.

2.5.4 Depth to thickness relationships

As each of the specimens has the same outer diameter, it allows for a comparison of the D/t ratio. According to classical theory, increasing thickness should increase the torsional capacity of unwrapped sections (Murray 1984). Therefore, it is expected that the introduction CFRP bonded to the exterior will have more benefit to thin-walled sections (higher D/t ratio). Experimental conformation of the theoretical studies of Wang et al. (2013) will be undertaken in this section.

The maximum torsional capacity of all strengthened steel specimens was larger than their respective control pair. For the thin-walled tubular sections, the improvement was large, especially for the 2 mm steel section which showed 41% improvement on the average maximum torsional capacity. The 3 mm steel section showed an improvement of 14% which was slightly lower than what was expected. For the 6 mm steel section the improvement compared to control

specimens was 11%. Figure 8 provides a visual representation of the percentage improvement for each D/t ratio of steel specimens.

Although there are three tests, the shape of the curve trends upwards with increasing D/t ratio. This verifies that a higher D/t ratio will make the CFRP contribution more effective and will maximize the increase in maximum torsional capacity. This verifies the work of Wang et al. (2013), and proves that CFRP strengthening is most effective in thin-walled SHS.

2.5.5 Failure modes

For steel control sections, non-uniform buckling was experienced at the middle of the sections for both 2 mm and 3 mm sections. This non uniform buckling causes tilting in the section as shown below in Figure 9(a).

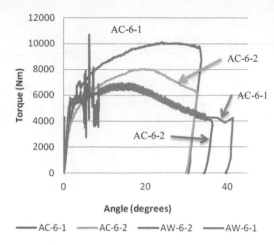

Figure 7e. Torque rotation for 6 mm aluminium sections.

Figure 10. Localized fiber separation at ends of specimens (a) SC-2-1 (b) SW-2-2.

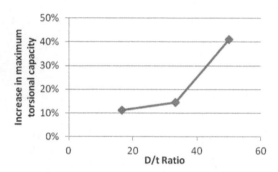

Figure 8. Relation of D/t ratio and percentage improvement.

Figure 11. Buckling in heat affected zone AC-3-1.

Figure 9. Buckling failures (a) SC-2-2 (b) SW-3-1.

Figures 7(a) and (b) show a dramatic drop in the torsional capacity when local buckling is experienced for these two control sections. For the 6mm control steel section, no local buckling was experienced, and the section continued to yield up to a rotation angle of 30 degrees as shown in Figure 7(c).

When introducing CFRP wrap on all steel sections, the local buckling was shifted from the middle of the section to the end of the section in all cases, as shown

by Figure 9(b). The CFRP maintained a good bond with the steel throughout the section. However, there was minor fiber separation (Figure 10) at the ends of specimens. The section is weakest where fiber separation occurs, and hence yield lines form at this point and buckling occurs. Figures 7(a) and (b) show that there is a small increase in the buckled strength of the 2 mm and 3 mm steel sections compared to the control. The strengthened 6 mm steel section experienced local buckling unlike its respective control specimen. Figure 7(c) shows that a CFRP section reaches a larger maximum torsional capacity compared to the control section. However, with the onset of local buckling, the strengthened section is weaker at a larger rotation angle compared to control sections. This phenomenon is due to the 6 mm control section undergoing strain hardening as the torsional load is increased. The 6 mm strengthened section has not undergone the same degree of strain hardening. Therefore, at its maximum torque where the CFRP fibers break, the torsional load is above what the steel section can carry, and so yield lines are formed and local buckling occurs.

As expected, the maximum torsional capacity of aluminium was much lower compared to steel sections. For both thicknesses of the control specimens, buckling failure at the ends occurred in all cases (Figure 11). Knowing that the initial welding failed in the aluminium sections (Section 2.3), each specimen required deeper penetration welding at the ends of the sections. As local buckling has occurred in this area, the properties of the material have decreased in strength, which may be due to the heat-affected zone of the welding. Figure 7(e) also shows that the degree

Figure 12. Deboning failure over additional fillet weld (a) AW-6-1, (b) AW-3-1.

of this heat-affected zone varies with both 6 mm aluminium control sections exhibiting a large difference in maximum torsional capacity.

In preparing the aluminium specimens, CFRP wrapping was applied over the additional welding. It was required as this experiment is looking at the capacity of the CFRP strengthened section not the capacity of the aluminium welding.

During testing in all cases, de-bonding of the aluminium section in this location was experienced as shown in Figure 12. The torsional load of this failure was approximately 5000 Nm, which caused the initial bond to break and then peel back from the aluminium section.

For the strengthened 3.2 mm aluminium section, as the torsional capacity of the bond was greater than the torsional capacity for the control section, improvement of 94% in the maximum torsional capacity was seen for this section. Figure 7(d) shows that at the point where the bond breaks, the torsional load decreases dramatically for both wrapped specimens. Moreover, the buckling capacity of the 3.2 mm aluminium section is improved when wrapped with CFRP.

For the strengthened 6 mm section, the bond between the CFRP and aluminium failed at a torque lower than the capacity of the control specimens. For ductile sections with large thicknesses, the engagement of the CFRP should be delayed to ensure that fibers do not break before buckling occurs. Figure 7(e) shows that there is noise in the data recordings which could indicate how fibers are engaging and then breaking suddenly. For the initial linear portion of Figure 7(e), the CFRP section is much stiffer compared to the control specimen. When the bond begins to break, there is a drop in the torsional load, and buckling occurs as if no CFRP is attached in that area. However, the control specimens are observed to reach a much higher torsional capacity than the wrapped sections that have failed. Again, the heat affected zone may account for this difference. Also, given that the surface has been blasted with glass bead, residual stresses within the aluminium section may have been released, hence making the section weaker.

Comparing the 3.2 mm aluminium specimen to the 3 mm steel specimen, there is a larger increase in the maximum torsional capacity. This shows that there is potential for more increase in torsional load for aluminium section compared to steel sections. If the bond between the aluminium and the CFRP can be improved, there is potential for more increase.

3 CONCLUSION

This paper presents details of an experimental program studying the influence of CFRP externally bonded onto the surface of different types of SHS. In addition to this, it presents the development of a theoretical model that can represent the torque vs rotation curve for a SHS. Based on the results, the following conclusions can be made.

The introduction of CFRP does increase the maximum torsional capacity. However, the amount of increase is related to the thickness of the original section. Thin-walled sections (high D/t ratio) will show a larger improvement compared to thicker-walled tubular sections (low D/t ratio) in terms of maximum torsional capacity.

It is also clear that when strengthening sections with lower Young's moduli, such as aluminium, there is more improvement in the maximum torsional capacity when compared to steel for a given thickness. However, this is relying on no de-bonding between aluminium and CFRP.

Moreover, the theoretical model developed, is in good agreement with the experimental results. This shows that the equivalent thickness method is a good way to represent the torsional behaviour of CFRP externally bonded to tubular sections.

Further research in this area should consider a comprehensive investigation into the influence of reverse wrapping layers of CFRP. The limited past test results demonstrate that there may be delay in the torsional buckling when a reverse wrapping layer is introduced.

For future theoretical analysis, measurement during experimental tests of the out of plane deformation is crucial in developing a true relationship between rotation angle and out of plane deformation.

ACKNOWLEDGMENT

Thanks are given to the Civil Engineering Laboratory at Monash University for fabricating test specimens.

REFERENCES

Al-Mayah, A., Soudki, K., Plumtree, A. (2005). 'Effect of Sandblasting on Interfacial Contact Behavior of Carbon-Fiber-Reinforced Polymer-Metal Couples', *Journal of Composites for Construction*, 9, pp. 289–295.

Bambach, M.R., Zhao, X.L., Jama, H. (2010). 'Energy absorbing characteristics of aluminium beams strengthened with CFRP subjected to transverse blast load', *International Journal of Impact Engineering*, 37, pp. 37–49.

Capral Aluminium Product Data Sheet. (2013). *Industrial solutions National Catalogue*, Brisbane, Australia.

Chahkand, N.A., Jumaat, M.Z., Sulong, N.H.R., Zhao, X.L., Mohammadizadeh, M.R. (2013). 'Experimental and theoretical investigation on torsional behaviour of CFRP strengthened square hollow steel section', *Thin-Walled Structures*, 68, pp. 135–140.

Elchalakani, M., Fernado, D. (2012). 'Plastic mechanism analysis of unstiffened steel I-section beams strengthened with CFRP under 3-point bending', *Thin-Walled Structures*, 53, pp.58–71.

Haedir, J. (2011). *Flexural and Axial Behaviour of Carbon Fibre Reinforced Polymer (CFRP) Strengthened Steel Circular Hollow Sections*, PhD Thesis, Monash University, Melbourne, Australia.

Haedir, J., Zhao, X.L. (2012). 'Design of CFRP-strengthened steel CHS tubular beams', *Journal of Constructional Steel Research*, 72(5), pp. 203–218.

Holloway, L.C., Teng G.J. (2008). *Strengthening and rehabilitation of civil infrastructures using fibre-reinforced polymer (FRP) composites*, Woodhead Publishing/CRC Press, Cambridge, UK.

Mahendran, M., Murray, N.W. (1990). 'Ultimate Load Behaviour of Box-Columns under Combined Loading of Axial Compression and Torsion', *Thin-Walled Structures*, 9, pp. 91–120.

MBRACE Fibre Product data sheet. (2014). *MBRACE CF 230/4900 (High Tensile CF)*, BASF The Chemical Company, Melbourne Australia.

Metallic Materials-Tensile testing at Ambient Temperature (AS1391 – 2007, Standards Association of Australia).

Murray, N.M. (1984). *Introduction to the Theory of Thin-Walled Structures*, Oxford University Press, New York, USA.

One Steel. (2010). *Design Capacity Tables for Structural Steel Hollow Sections, AS/NZS 1163*, One Steel Australian Tube Mills.

Wang, X., Wu, C., Zhao, X.L. (2013). 'Theoretical analysis of CFRP strengthened thin-walled steel square hollow sections (SHS) under torsion', *Fourth Asia-Pacific Conference on FRP in structures, 11–13 December, Melbourne, Australia*.

Wu, C., Zhao, X.L., Duan, W.H., Phipat, P. (2012). 'Improved end bearing capacities of sharp corner aluminium tubular sections with CFRP strengthening', *International Journal of Structural Stability and Dynamics*, 12(1), pp. 109–130.

Zahurul Islam, S.M., Young, B. (2011). 'FRP strengthened aluminium tubular sections subjected to web crippling', *Thin-Walled Structures*, 49, pp. 1392–1403.

Zhao, X.L. (2013). 'Strengthening of compression members', *FRP-strengthened metallic structures*, CRC Press, Boca Raton, FL, pp. 105–158.

Author index